Faber & Kell's Heating and Air-Conditioning of Buildings

Faber & Kell's Heating and Air-Conditioning of Buildings

Tenth edition

D R Oughton MSc FREng CEng HonFCIBSE

S L Hodkinson BTech CEng MCIBSE MEI

AMSTERDAM • BOSTON • HEIDELBERG • LONDON • NEW YORK • OXFORD
PARIS • SAN DIEGO • SAN FRANCISCO • SYDNEY • TOKYO
Butterworth-Heinemann is an imprint of Elsevier

Butterworth-Heinemann is an imprint of Elsevier
Linacre House, Jordan Hill, Oxford OX2 8DP, UK
30 Corporate Drive, Suite 400, Burlington, MA 01803, USA

First published 1936
Second edition 1943
Reprinted 1945, 1948, 1951
Third edition 1957
Reprinted 1958, 1961
Fourth edition 1966
Fifth edition 1971
Reprinted 1974
Sixth edition 1979
Reprinted (with amendments) 1988
Seventh edition 1989
Eighth edition 1995
Reprinted 1995
Paperback edition 1997
Reprinted 1999, 2000
Ninth edition 2002
Tenth edition 2008

British Library Cataloguing in Publication Data
A catalogue record for this book is available from the British Library

Library of Congress Cataloging-in-Publication Data
A catalog record for this book is available from the Library of Congress

ISBN: 978-0-7506-8365-4

Typeset by Charon Tec Ltd (A Macmillan Company), Chennai, India
www.charontec.com
Printed and bound in Great Britain
08 09 10 10 9 8 7 6 5 4 3 2 1

Contents

Preface

This 10th edition of Faber & Kell's Heating and Air Conditioning of Buildings is a milestone in the development of this standard work, some 70 years after it was first published in 1936. In terms of its scope and technical content, this edition has undergone a major revision, reflecting the increasing responsibilities of building services engineers in recent years. The changes have been driven by the critical role of engineers in the quest to reduce global carbon emissions from buildings and by their contribution to the broader sustainability agenda related to building design.

Published in 2006, the new Part L of the Building Regulations Conservation of Fuel and Power has had a significant impact on the design, installation and commissioning of building services systems. In addition to setting performance standards for the building envelope, requiring ever closer collaboration between architects and building services engineers, these new regulations for the first time have set challenging minimum standards for the performance of heating and air-conditioning systems. The Energy Performance of Buildings Directive, published by the EC in 2003, as well as defining energy performance standards for buildings, also includes requirements for energy certification and inspection standards that are progressively being introduced. Such is the scope of this legislation that the majority of chapters in this edition incorporate reference to their requirements.

In recognition of these changes, eight new chapters have been introduced into this edition, including topics covering sustainability, legislation, renewable energy sources, building energy and environmental modelling, commissioning and handover management and the building in operation. The wider environmental agenda has led to revisions to the chapters on refrigeration and combined heat and power. The important interfaces with other disciplines is reflected by new chapters on electric motor drives and starting methods and noise control, and a comprehensive revision to the chapter on controls and building management systems. Safety in design is another new chapter acknowledging that good design includes the proper consideration of health and safety aspects in both the construction and operation of building services systems.

In recognition of the increasing number of specialisms within the building services discipline and the continuing developments in technology that influence modern design techniques, the authorship of the book has been extended to include seventeen experts from Faber Maunsell. This wider contribution, and the contract with the publisher being with the company, will hopefully secure the future of the book. All of the authors have kindly agreed for the royalties from this edition to be given directly to the Chartered Institution of Building Services Engineers Benevolent Fund, to help support their important work.

The authors once again wish to acknowledge the permission of the Chartered Institution of Building Services Engineers to use data from their wide range of publications. To these invaluable works, extensive reference is made to facts and current practice. In some cases presentation of these data may differ from that of the source, with simplifications to match both the style of this book and the inexactitudes of building construction.

Our thanks are due to many colleagues at Faber Maunsell and friends in the industry who have willingly provided contributions and advice, and in particular to the help and support of Jenny Taylor, who has dealt with input from many authors in delivering comprehensible material to the publishers.

Preface to first edition

During 1935 we contributed a series of articles on this subject in *The Architects' Journal*. In view of the considerable interest which these aroused, it has been thought desirable to reproduce them in book form. The materilal has been amplified, and many new sections added.

We are indebted to Mr G. Nelson Haden and to Mr R. E. W. Butt for many helpful suggestions, also to Mr F. G. Russell for reading through the text. Our thanks are also due to Mr J. R. Harrison, for much work in the preparation of the book, and to Mr H. J. Sharpe for reading the proofs.

Oscar Faber
J. R. Kell
1936

Authors

(Left to Right)
Richard Pearce, Richard Brailsford, Chris Dodds, Peter Mann, John Lloyd, Peter Concannon, Mike Campbell, Doug Oughton, Jenny Taylor (co-ordinator), Steve Hodkinson, Rob Manning, Ted Paszynski, Miles Attenborough, Hash Maitra, Ant Wilson, Ray Wilde, Barrie Huggins, Steve Irving.

The authorship of the book originates from members of the firm Oscar Faber. The firm joined the AECOM Technology Corporation, a global multi-disciplinary consultancy, in 2001 to form Faber Maunsell.

The authors for this tenth edition are the following experts from Faber Maunsell.

Doug Oughton after working 6 years with a London-based contractor joined Faber Maunsell in 1967 as an Assistant Engineer. He was made an Associate in 1975; he became a Partner in 1981, a Director in 1983 and a Consultant in 2004. He was awarded an M.Sc. (Arch) by the University of Bristol in 1979 after presentation of a research paper dealing with energy consumption and energy targets for air-conditioned office buildings. He was the CIBSE President for the year 2002–03.

Doug is a Fellow of the Royal Academy of Engineering, a Chartered Engineer and an Honorary Fellow of the Chartered Institution of Building Services Engineers.

Doug is responsible for the authorship of Chapters 1, Fundamentals and 31, The building in operation.

Steve Hodkinson joined Faber Maunsell as an Assistant Engineer in 1976 following 4 years with a national contractor. He was awarded an honours degree in Environmental Engineering by Loughborough University in 1975. He became a Director in 1986 responsible for a wide variety of major projects both in the UK and overseas. He took responsibility for Building Services Engineering in 1995 and for the Building Engineering Division in 2000. Steve is a Chartered Engineer, a Member of the Chartered Institution of Building Services Engineer and a Member of the Energy Institute.

Steve is responsible for the authorship of Chapters 4, The building in winter; 19, Air-conditioning; 20, Air distribution and 23, Calculations for air-conditioning design.

Miles Attenborough is a Director in Faber Maunsell's Sustainable Development Group and was previously a Director of ECD Energy and Environment Ltd before they became part of Faber Maunsell in 2001. He was awarded an honours degree in Civil Engineering from Nottingham University in 1987 before taking a Masters Degree in Energy and Buildings at Cranfield University in 1988. Miles has spent his career developing best practice guidance on the sustainable design and management of buildings and communities, working with clients and their design teams advising on sustainability issues on a wide range of building projects and with Government departments on policy development, research and demonstration.

Miles is responsible for the authorship of Chapter 2, Sustainability.

Richard Brailsford joined Faber Maunsell as an Assistant Engineer in 1983 following 4 years with a national contractor, both in the UK and Overseas. He was awarded an honours degree in Building Engineering (Services) by Liverpool University in 1978. He was awarded an MBA by the Open University in 1995, and he became a Regional Director in 2000, responsible for a group working on projects in both the private and public sectors, including under PFI, and a Director in 2008. For a number of years, Richard also led an internal group looking at the use of advanced design methods within the firm. Richard is a Chartered Engineer, a Member of the Chartered Institution of Building Services Engineers, the Institute of Mechanical Engineers and a Member of the Energy Institute.

Richard is responsible for the authorship of Chapter 28, Controls and building management systems.

Mike Campbell joined Faber Maunsell as an Engineer in 1988 having previously completed a mechanical engineering apprenticeship with a large manufacturing company followed by a number of years experience with a small installation contractor and a heating plant manufacturer. Mike became an Associate Director in 1999 and has been responsible for a wide variety of projects including manufacturing/testing facilities, laboratories and office buildings. Since 2002 Mike has been responsible for the maintenance and development of Faber Maunsell's standard specification and in-house design guidance system for building services. Mike has also contributed to the development of some of BSRIA's recent guidance including 'Design Checks for HVAC'. Mike was awarded a Masters Degree in Building Services by Glasgow University in 1991 and he is a Chartered Engineer and a Member of the Chartered Institution of Building Services Engineers.

Mike is responsible for the authorship of Chapters 7, Heating methods; 8, Electric storage heating; 9, Indirect heating systems and 10, Heat emitting equipment.

Peter Concannon joined Faber Maunsell applied research group as a graduate in 1984 having been awarded an honours degree in Chemical Engineering by Leeds University in the same year. He worked in a wide range of research projects into passive building solutions and low energy technologies such as CHP. In 2002 Peter transferred to the firm's Sustainable Development Group working on renewable and low energy solutions to energy provision in buildings. In 2006 he was appointed as an Associate Director.

Peter is responsible for the authorship of Chapters 14, Renewable technologies and 15, Combined heat and power.

Chris Dodds joined Faber Maunsell in 1973 as an Assistant Public Health Engineer having served an apprenticeship with a plumbing and heating contractor in his native Teesside. Chris is now an Associate Director of the firm and has technical design responsibility for a group of public health engineers. He studied at the Cleveland Technical College and the South Bank Polytechnic (now South Bank University) and is a Member of the Chartered Institution of Water and Environmental Management.

Chris is responsible for the authorship of Chapter 25, Hot water supply systems.

Barrie Huggins joined Faber Maunsell as a Graduate Engineer in 1973 following a 5-year indentured apprenticeship with mechanical services contractors, and obtaining a first class honours degree from the University of Bath. Barrie is an Associate Director of the firm. He is a Chartered Engineer and a Member of the Chartered Institution of Building Services Engineers.

Barrie is responsible for the authorship of Chapters 13, Boilers and burners; 16, Fuels, storage and handling and 17, Combustion emissions and chimneys.

Steve Irving joined Faber Maunsell in 1978 following 6 years with a Research Association investigating industrial fluid flow and heat transfer problems. He has been active for nearly 30 years in researching issues relating to the energy and environmental performance of buildings. For the past 10 years he has been heavily involved in supporting government in successive reviews of the energy efficiency regulations and the UK implementation of the Energy Performance of Buildings Directive. Steve is a Fellow of the Royal Academy of Engineering, a Chartered Engineer, a Fellow of the Chartered Institution of Building Services Engineers, a Fellow of the Society of Façade Engineers, a Member of the Energy Institute and a Member of the American Society of Heating, Refrigeration and Air-conditioning Engineers.

Steve is responsible for the joint authorship with Hash Maitra of Chapter 3, Legislation.

John Lloyd received an Honours degree in Building Services Engineering from the University of Liverpool in 1987 and joined Faber Maunsell where he then read for a second degree in Acoustics at the University of the South Bank. John is now a Director of the firm responsible for a team of engineers providing independent building and architectural acoustics consultancy services throughout the UK. He is a Chartered Engineer, a Member of the Chartered Institution of Building Services Engineers and a Member of the Institute of Acoustics.

John is responsible for the authorship of Chapter 26, Noise control.

Hash Maitra joined Faber Maunsell in 2007 having spent 13 years at the HSE as one of HM Specialist Inspectors of health and safety. Within the firm he is responsible for CDM compliance. He was involved in the revision of the CDM Regulations and headed up HSE's efforts on designer-related issues. Prior to joining HSE, he had extensive experience in the construction industry, in which he has worked since 1979, on a variety of projects both for contractors and consulting engineers. Hash graduated from the University of Newcastle upon Tyne with an honours degree in Civil Engineering and is a Chartered Civil Engineer.

Hash is responsible for the joint authorship with Steve Irving of Chapter 3, Legislation, and for the authorship of Chapter 30, Safety in design.

Peter Mann joined Faber Maunsell as an Associate Director in 1999 following 18 years in consultancy and a 5-year management apprenticeship with a national contractor. He was awarded an honours degree in Electrical and Electronic Engineering by London South Bank University in 1997. He became a Technical Director in 2007 responsible for major projects in London and setting electrical technical design standards, nationally within the firm.

Peter is a Chartered Engineer, a Member of the Institution of Engineering and Technology and is an Associate Member of the Chartered Institution of Building Services Engineers.

Peter is responsible for the authorship of Chapter 27, Motor drives, starting methods and control.

Rob Manning spent the early years of his career in building services with Faber Maunsell between 1972 and 1982. He was awarded an honours degree in Environmental Engineering in 1976. Subsequent experience included building services contracting in Africa and design consultancy directorship before returning to the firm in 2003. His current role as Professional Excellence Director includes development of engineering skills and personnel, promoting information exchange and innovation, developing automation of design processes, design review and safety in design. Rob is a Chartered Engineer and is a Member of the CIBSE Board 2007 and a Member of the CIBSE Policy and Consultation Committee 2005–2007.

Rob is responsible for the authorship of Chapters 11, Pumps and auxiliary equipment and 12, Piping design for indirect systems.

Ted Paszynski joined Faber Maunsell as a Graduate Engineer in 1982 after graduating from Bath University having been awarded an honours degree in Building Environmental Engineering.

Ted is a Director of the firm responsible for major projects throughout the UK and is the company's Sector Head responsible for Sports, Entertainment and Public Assembly buildings. He is also the Faber Maunsell refrigeration specialist. Ted is a Chartered Engineer and a Member of the Chartered Institution of Building Services Engineers.

Ted is responsible for the authorship of Chapter 24, Refrigeration and heat rejection.

Richard Pearce rejoined Faber Maunsell in 1995 having previously worked with BDP and the firm in UK and South Africa. Richard is an Associate Director and Chartered Engineer and is a Member of the Chartered Institution of Building Services Engineers, Member of the Institute of Refrigeration and Fellow of the Institute of Healthcare Engineering and Estate Management. Richard has considerable experience of engineering services in new and refurbished healthcare facilities and is a coauthor of a number of Health Building Notes.

Richard is responsible for the authorship of Chapters 21, Ductwork design and 22, Fans and air treatment.

Ray Wilde joined Faber Maunsell as a Mechanical Services Design Engineer in 1973 following 10 years in the Norwich and London offices of a mechanical services contractor. He attended Norwich Technical College and the National College for Heating, Ventilating, Refrigeration and Fan Engineering (now London South Bank University). Ray became a Director in 1987, is a Chartered Engineer, a Fellow of the Chartered Institution of Building Services Engineers and is a Member of the Institute of Refrigeration.

Ray is responsible for the authorship of Chapters 18, Ventilation and 29, Commissioning and handover.

Ant Wilson joined Faber Maunsell as a Graduate Engineer in 1979 following 4 years at Bath University and was awarded an honours degree in Building Environmental Engineering. He worked within the computer division and the applied research groups based in St. Albans. He is now a Director with responsibility for the firm's Sustainable Development Group within the Building Division.

Ant is a Chartered Engineer, a Fellow of the Energy Institute, a Fellow of the Society of Façade Engineering, a Member of the Chartered Institution of Building Services Engineers and the Society of Light and Lighting.

Ant is responsible for the authorship of Chapters 5, The Building in summer and 6, Building energy and environmental modelling.

Chapter 1

Fundamentals

Fulfilment of the need for satisfactory environmental conditions within a building, whether these be required for human comfort, material storage or in support of some process, is a task which has faced mankind throughout history. With the passage of time, a variety of forms of protection against the elements has been provided by structures suited to individual circumstances, the techniques being related to the severity of the local climatic conditions, to the materials which were available and to the skills of the builders. In this sense, the characteristics of those structures provided a form of inherent yet coarse control over the internal environment: finer control had to wait upon the progressive development, over the last two centuries, of systems able to moderate the impact of the external climate still further. These days the form of the building envelope is determined largely by aesthetics, the Building Regulations in respect of energy conservation and cost.

The human body produces heat, the quantity depending upon the level of physical activity, and for survival this must be in balance with a corresponding heat loss. When the rate of heat generation is greater than the rate of loss, then the body temperature will rise. Similarly, when the rate of heat loss exceeds that of production, then the body temperature will fall. If the level of imbalance is severe, *heat stress* at one extreme and *hypothermia* at the other will result and either may prove fatal.

The processes of heat loss from the body are:

- Conduction to contact surfaces and to clothing
- Convection from exposed skin and clothing surfaces
- Radiation exchange with exposed surfaces to the surroundings
- Exhalation of breath
- Evaporation by sweating.

Involuntary control of these processes is by constriction or dilation of blood vessels, variation in the rate of breathing and variation in the level of sweating, voluntary and involuntary. The individual may assist by removing or adding insulating layers in the form of clothing.

Keeping warm or keeping cool are primitive instincts which have been progressively refined as more sophisticated means have become available to satisfy them. For example, once facilities for the exclusion of the extremes of climate became commonplace for buildings in temperate zones, fashion introduced lighter clothing; this led to greater thermal sensitivity and, in consequence, to less tolerance of temperature variation. By coincidence, however, in the same time span, architectural styles changed also and the substantial buildings of the past, which could moderate the effects of solar heat and winter chill, were succeeded by lightweight structures with substantial areas of glazing having little or no such thermal capacity thereby placing greater reliance upon heating, ventilation and air-conditioning to provide an acceptable indoor climate.

Before any part of this subject is pursued in further detail, however, it would seem appropriate to discuss some of the quantities and units that are relevant, with particular reference to the meaning and use applied to them in the present context.

Units and quantities

The notes included here are not intended to be a comprehensive glossary of terms included in the *Systeme International d'Unites* but, rather, an *aide memoire* covering those which are specific to the subject matter of this book but are not necessarily in everyday use. The four basic units are the kilogram (kg) for mass, the metre (m) for length, the second (s) for time and the kelvin (K) for thermodynamic temperature, all with their multiples and sub-multiples. From these, the following secondary units are derived:

Force: The unit here is the newton (N), which is the force necessary to accelerate a mass of one kilogram to a velocity of one metre per second in one second, i.e. $1\,N = 1\,kg\,m/s^2$. When a mass of 1 kg is subjected to acceleration due to gravity, the force then exerted is 9.81 N.

Heat: The unit of energy, including heat energy, is the joule (J), which is equal to a force of one newton acting through one metre, i.e. 1 J = 1 Nm.

Heat flow: The rate of heat flow is represented by the watt (W), which is equal to one joule produced or expended in one second, i.e. 1 W = 1 J/s = 1 Nm/s.

Pressure: The standard unit is the newton per square metre (N/m^2), also known, more conveniently, as the pascal (Pa). The bar continues to be used in some circumstances and 1 bar = 100 k Pa.

Specific heat capacity: This is the quantity of heat required to raise the temperature of one kilogram of a substance through one kelvin, the units being kJ/kg K. Where heat flow in unit time is involved, the unit becomes kJ/s kg K = kW/kg K.

Specific density: The unit for this quantity is the kilogram per cubic metre (kg/m^3).

Volume: The cubic metre is the preferred unit but the litre ($1\,dm^3$) is in general use since it is much more convenient in terms of a comprehensible size. To avoid printing confusion between the figure 1 and the letter l, this book will spell out the word *litre* in full.

Properties of materials

Table 1.1 lists a variety of materials and provides details of some relevant physical properties. These serve as a source of reference against the following notes:

Latent heat: When the temperature of water at atmospheric pressure, is raised from freezing to boiling point, i.e. through 100°C, the heat added is 420 kJ/kg (4.2 × 100). To convert this hot water to steam, however, still at atmospheric pressure and still at 100°C, will require the addition of a significantly greater quality of heat, i.e. 2257 kJ/kg.[1] This value is the *latent heat of evaporation* of water and its magnitude, 5.4 times that needed to raise water through 100°C shows its importance. A similar phenomenon occurs when water at 0°C becomes ice at the same temperature; this change of state releasing 330 kJ/kg, the *latent heat of fusion* of water.

Specific heat capacity (gases): Whereas, for practical purposes, solids and liquids each have a single specific heat capacity which does not vary significantly, gases have one value for a condition of constant pressure and another for a condition of constant volume. Both values will vary with temperature but, for a given gas, the ratio between them will remain the same. For example, with dry air at 20°C, the value at constant pressure is 1.012 kJ/kg K and the value at constant volume is 0.722 kJ/kg K. At 100°C, the value at constant pressure is 1.017 kJ/kg K.

Thermal expansion (solids and liquids): With very few exceptions, materials expand when heated to an extent which varies directly with their dimensions and with the temperature difference. Metals expand more than most building materials, with the result that, for example, care must be taken to accommodate the differential movement between a long straight pipe and a supporting building structure. For a temperature change from 10°C to 80°C (70 K), a 10 m length of steel pipe (coefficient of linear expansion 11.3×10^{-6}/K) will increase in length by 7.9 mm (70 × 10 × 0.0113).

Coefficients for superficial (area) and cubic (volume) expansion are taken, respectively, as twice and three times those listed for linear movement.

Table 1.1 Properties of materials

Material	*Density* (kg/m^3)	*Thermal conductivity* (W/mK)	*Specific heat capacity* (kJ/kg K)
Masonry materials			
Sandstone	2300	1.8	1.0
Brick exposed	1750	0.77	1.0
Brick (protected)	1750	0.56	1.0
No-fines concrete	2000	1.33	1.0
Concrete block (dense) (exposed)	2300	1.87	1.0
Concrete block (dense) (protected)	2300	1.75	1.0
Precast concrete (dense) (exposed)	2100	1.56	1.0
Precast concrete (dense) (protected)	2100	1.46	1.0
Cast concrete (dense)	2000	1.33	1.0
Cast concrete	1800	1.13	1.0
Screed	1200	0.46	1.0
Ballast (chips or paving slab)	1800	1.10	1.0
Surface materials/finishes			
External render (lime, sand)	1600	0.80	1.0
External render (cement, sand)	1800	1.00	1.0
Plaster (dense)	1300	0.57	1.0
Plaster (lightweight)	600	0.18	1.0
Plasterboard (standard)	700	0.21	1.0
Plasterboard (fire-resisting)	900	0.25	1.0
Miscellaneous materials			
Plywood sheathing	500	0.13	1.6
Timber studding	500	0.13	1.6
Timber battens	500	0.13	1.6
Timber decking	500	0.13	1.6
Timber flooring	500	0.13	1.6
Timber flooring (hardwood)	700	0.18	1.6
Chipboard	600	0.14	1.7
Vinyl floor covering	1390	0.17	0.9
Waterproof roof covering	110	0.23	1.0
Wood blocks	600	0.14	1.7
Floor joists	500	0.13	1.6
Cement-bonded particle board	1200	0.23	1.5
Carpet/underlay	200	0.6	1.3
Steel	7800	50	0.45
Stainless steel	7900	17	0.46
Soil	1500	1.5	1.8
Water at normal pressure			
At 4°C	999.9	0.576	4.206
At 20°C	998.2	0.603	4.183
At 100°C	958.3	0.681	4.219
Ice	918	2.24	2.04
Air at normal pressure			
Dry and at 20°C	1.205	0.026	1.012
Dry and at 100°C	0.88	0.03	1.017

Note
For insulating materials see Table 4.3.

Thermal expansion (gases): A perfect gas will conform to Boyle's and Charles's laws which state that the pressure (P), the volume (V), and the thermodynamic temperature (T) are related such that PV/T is a constant. The coefficient of cubic expansion is therefore temperature dependent and does not have a single value. Most gases conform very closely to the properties of a perfect gas when at a temperature remote from that at which they liquify.

Vapour pressure: Dalton's law of the partial pressures states that if a mixture of gases occupies a given volume at a given temperature, then the total pressure exerted by the mixture will be the sum of the pressures exerted by the components.

Criteria which affect human comfort – definitions

The ancients taught that humans had seven senses,[2] but it is no more than coincidence that the principal influences which affect human comfort are also seven in number:

(1) Temperature
(2) Conduction, convection and radiation
(3) Air movement
(4) Activity and clothing
(5) Air purity
(6) Humidity
(7) Ionisation

Temperature

The direction of heat flow from one substance to another is determined by the temperature of the first relative to that of the second. Thus, in that sense, temperature is akin to a pressure potential and is a relative term: the temperature of boiling water is higher than that of water drawn straight from a cold tap and the temperature of the latter is higher than that of ice. Ice, however, may be said to be hot, or at a high temperature, relative to liquid air at −190°C. Following the adoption of the *Systeme International*, temperature is measured in units of °C or K. The *Celsius* scale has a convenient false zero (0°C), which corresponds to the temperature at which water freezes, and has equal intervals above this to the temperature at which water boils under atmospheric pressure (100°C). The thermodynamic or *absolute* scale, established from the study of pressure effects upon gases, uses the same intervals as the Celsius scale but with a true zero corresponding to the minimum possible temperature obtainable. The intervals here are *kelvin* (K) and, with this scale, water freezes at 273 K.

In order to avoid confusion in terminology, it was the accepted convention in Imperial units that temperature *level* (or potential) should be expressed in terms of °F, whereas temperature *difference* (or interval) was in terms of deg. F. Similarly, under strict SI rules, temperature level is expressed in °C and difference in K and, although this usage is not obligatory, it is an aid to clarity and will be adopted throughout this book. Table 1.2 shows the relationship between the various scales and some notable conversions. Reference will be made later in this chapter to a variety of different temperature notations: *dry bulb*, *wet bulb*, *globe* and *radiant.* These relate to methods of measurement for particular purposes and all use the Celsius or absolute scales noted above.

Conduction, convection and radiation

These three terms have already been mentioned as being associated with heat transfer and it is necessary, for the discussions which follow, that there should be a clear perception of the difference between them.

Table 1.2 Temperature scales

Degrees	*Absolute zero*				*Freezing point*									*Boiling point*	
Centigrade (°C)	−273	−200	−100	−18	0	10	16	20	28	40	60	71	82	100	200
Fahrenheit (°F)	−460	−328	−200	0	32	50	61	68	82	104	140	160	180	212	392
Kelvin (K)	0	–	–	–	273	–	–	–	–	–	–	–	–	373	

Conduction

This may be described as heat transfer from one particle to another by contact. For example, if two blocks of metal, one hot and one cold, were to be placed in contact, then heat would be conducted from the one to the other until both reached an intermediate temperature. If both blocks were of the same metal, then this temperature could be calculated by the simple process of relating the mass and temperature of one to those of the other: but if the materials were not the same, it would be necessary to take account of the different *specific heat capacities* as noted earlier.

Conductivity is a measure of the quantity of heat that will be transferred through unit area and thickness in unit time for an unit temperature difference ($J\,m/s\,m^2K = W/m\,K$). Table 1.1 lists values of this property for various materials and it will be noted that metals have a high conductivity, whereas, at the other end of the scale, materials known as *insulators* have a low conductivity. The conductivity of many materials varies widely with temperature and thus only those values that fall within the range to which they apply should be used. Thermal conductivity, which is the property discussed here, should not be confused with electrical conductivity which is a quite separate quantity.

As far as building materials are concerned, those having higher densities are usually hard and are not particularly good insulators. Porous materials are bad conductors when dry and good conductors when wet, a fact which is sometimes overlooked when a newly constructed building is occupied before the structure has been able to dry out properly, which may take some months in the winter or spring.

Convection

Convection is a process in which heat transfer involves the movement of a fluid medium to convey energy, the particles in the fluid having acquired heat by conduction from a hot surface. An illustration commonly used is that of an ordinary (so-called) radiator which warms the air immediately in contact with it: this expands as it is heated, becomes lighter than the rest of the air in the room and rises to form an upward current from the radiator. A second example is water heated by contact with the hot surfaces around the furnace of a boiler, leading then to expansion and movement upwards as in the preceding instance with air. A medium capable of movement is thus a prerequisite for conduction, which cannot, in consequence, occur in a vacuum where no such medium exists.

Radiation

This is a phenomenon perhaps more familiar in terms of light and was, in Newton's time, explained as being the result of bombardment by infinitesimal particles released from the source of heat. At a later date, radiation, whether of light or heat, was supposed to be a wave action in a *subtile* medium (invented by mathematicians for the purpose) known as the *ether*. In the present context, it is enough to state that radiation is the transfer of energy by an electromagnetic process at wavelengths which correspond to, but extend marginally beyond, the infrared range (10^{-6} to 10^{-4} m). Radiation is completely independent of any intermediate medium and will occur just as readily across a vacuum as across an air space: intensity varies with the square of the distance between the point of origin and the receiving surface. The amount of radiation emitted by surfaces depends upon their texture and colour, matt black surfaces having an *emissivity* rated as unity in an arbitrary scale. Two values only need to be considered in the present context: 0.95 which represents most dull metals or the materials used in building construction and 0.05 which applies to highly polished materials such as aluminium foil. Surfaces which radiate heat well are also found to be good adsorbers; thus, a black felted or black asphalt roof is often seen to be covered with hoar frost on a cold night, due to radiation to space when nearby surfaces having other finishes are unaffected.

Air movement

It is necessary, when considering air movement, to make a clear distinction between the total quantity in circulation and the proportion of it which is admitted from outside a building. It was traditional practice to refer to the

Table 1.3 Typical metabolic rate and heat generation for an adult human

Activity	*Metabolic rate* (met)	*Heat generation* (W)
Sleeping	0.7	75
Sitting	1.0	105
Standing	1.2	125
Typical office work	1.2	125
Walking (1.3 m/s)	2.6	270
Heavy machine work	3.0	315

latter component *as fresh air* but, since pollution in one form or another is a feature of the urban atmosphere, the term *outside air* is now preferred. In many instances the outside air volume may have entered a building by infiltration, in which case it is usually referred to in terms of *air changes* or *room volumes* per hour but, when handled by some form of mechanical equipment, this is rated in either m^3/s or, more usually, in litre/s.

It is not always well understood that air movement within a room is a positive rather than a negative effect. The source will be the position and velocity of admission since the location of an opening for removal has virtually no effect upon distribution. Air movement is measured in terms of air velocity (m/s) and must be selected within the limits of draughts at one extreme, and of stagnation at the other.

Activity and clothing

The interaction between human comfort, activity and clothing has been mentioned earlier. In most instances, the variables derive from the nature of the enclosure and the purpose for which it is to be occupied and thus allow for a group classification with like situations.

Human activity is graded according to the level of physical exertion which is entailed and to the body area, male or female, ranging from a heat output of about 75 W when sleeping, through to in excess of 900 W for some sports activity. Table 1.3 gives some typical rates of heat generation, calculated using a body surface area of $1.8\,m^2$, representing the average for an adult. As to clothing, this is graded according to insulation value, the unit adopted being the *clo.* Unity on this scale represents a traditional business suit, with a value of $0.155\,m^2\,K/W$: zero in the scale is a minimum swim suit and light summer wear has a value of about 0.5 clo. Typically, for people dressed in a traditional suit, a rise in activity rate equivalent to 0.1 met, a measure of metabolic rate, corresponds to a reduction of 0.6 K in the comfort temperature to maintain a similar comfort level.

Air purity

Pollution can derive from sources outside a building or as a result of contaminants generated within it. In the former case, if air enters the building by simple infiltration, then dusts and fume particles, probably mainly carbonaceous in urban areas, nitrous oxides (NO) and sulphurous oxides (SO) will enter with it. The National Air Quality Strategy has set standards to be achieved by local authorities for outdoor air pollution. As a consequence, there is considerable measurement and modelling data available, particularly for urban areas, which can be used to establish concentrations expected local to a site. Table 1.4 provides a summary of the air quality standards objectives for selected pollutants.

Mechanical plant, used for ventilation supply is usually provided with air filtration equipment but this, as a result of indifferent or absent maintenance effort, may well be ineffective. Careful siting and design of ventilation air intakes is required to minimise contamination from external sources.[3] Pollution generated within a commercial building may be from any combination of the sources listed in Table 1.5. Airborne bacteria use dust particles as a form of transport and, if housekeeping is neglected, may present a health hazard. Contaminants are identified by size, the micrometre (μm) being the common unit used for air filter rating.

Table 1.4 National Air Quality Strategy objectives for UK

Pollutant	*Objective*		*Measurement period (mean)*	*Date*
	Value	*Excedance limit*		
Carbon monoxide	10 mg/m³	Daily maximum	8 hour	December 2003
Nitrogen dioxide[a]	200 μ/m³	18 times/annum	1 hour	December 2005
	40 μ/m³		Annual	December 2005
Sulphur dioxide	266 μ/m³	35 times/annum	15 minute	December 2005
	350 μ/m³	24 times/annum	1 hour	December 2004
	125 μ/m³	3 times/annum	24 hour	December 2004
Particulates[b]	50 μ/m³	35 times/annum	24 hour	December 2004
	40 μ/m³		Annual	December 2004

Notes
a Provisional targets.
b Imperial national standards have been established for 2010 (incorporated into legislation for Scotland).

Table 1.5 Indoor contaminants

Basement car parks	*Furnishings*	*Office machines*[a]
Carbon monoxide	Artificial fibres	Ammonia
	Formaldehyde	Formaldehyde
Cleaning agents	Carpet dusts	Ozone
Bleaches		Paper dust
Deodorants	*Occupants*	Methanol
Disinfectants	Carbon dioxide	Nitropyrene
Solvents	Water vapour	Trichloroethylene
	Clothing fibres	Tetrachloroethylene
Cooling towers	Biological aerosols	
Lneumophila	Odours	*Building products*
	Footwear dirt	Material emissions
Activities	Tobacco smoke	Glues
Occupational process		
Combustion products	*Moulds and micro-organisms*	
Cooking fumes	Bacteria	
	Viruses	
	Actinomycetes	
	Moulds	
	House dust mite	

Note
a Photo-printing, copying, duplicating and correcting fluids, etc.

For new construction sites, any evidence of ground source contamination from landfill sites or occurring naturally, such as radon, should be determined and precautions taken to avoid risk to the future occupants. In recent years a considerable amount of research has been carried out into the combined effects of different pollutants.[4,5] Generally it has been found that cooler drier air is perceived to be less contaminated with the sensation of freshness reducing with increasing humidity and temperature.

Humidity

The humidity of air is a measure of the water vapour which it contains and *absolute humidity* is expressed in terms of the mass of water (or water vapour) per unit mass of dry air (kg/kg) and not per unit mass of the

Table 1.6 Moisture content of saturated air at various temperatures (kg/kg of dry air)

Temperature (°C)	*Moisture* (kg/kg)	*Temperature* (°C)	*Moisture* (kg/kg)	*Temperature* (°C)	*Moisture* (kg/kg)
0	0.0038	18	0.0129	36	0.0389
2	0.0044	20	0.0148	38	0.0437
4	0.0050	22	0.0167	40	0.0491
6	0.0058	24	0.0190	42	0.0551
8	0.0067	26	0.0214	44	0.0617
10	0.0077	28	0.0242	46	0.0692
12	0.0088	30	0.0273	48	0.0775
14	0.0100	32	0.0308	50	0.0868
16	0.0114	34	0.0346	52	0.0972

Note
At 20°C, the moisture contents at saturation for altitudes of −1000 m (113.9 kPa) and +2000 m (79.5 kPa) are 0.0131 kg/kg and 0.0189 kg/kg, respectively.

mixture. The greatest mass of moisture which, at atmospheric pressure, can exist in a given quantity of air is dependent upon temperature as may be seen from Table 1.6, the effect of pressure variation, for two extreme values, being given in a footnote. This condition is known as a state of *saturation* and if air, virtually saturated at an elevated temperature, be cooled, then a temperature is soon reached where the excess moisture becomes visible in the form of a mist, or as *dew* or rain.

In most practical situations, both externally and within a building, the air will not be saturated and a water vapour content which may be measured relative to that situation will exist. Two terms are used to quantify the moisture content held, the familiar *relative humidity* and the less well-known *percentage saturation.* In fact, the precise definition of the former is far removed from popular usage and the latter, which is a simple ratio between the moisture content at a given condition and that at saturation, is now more generally accepted for use in calculations.

Humidity levels affect the rate of evaporation from the skin and the mucosal surface. Low humidities may cause eye irritation and dry nose and throat. High humidities reduce the rate of evaporation from the skin and support growth of organisms, fungi and house mites. Levels between 40 per cent and 70 per cent relative humidity are considered to be the limits of acceptability for general applications with control between 45 per cent and 60 per cent for optimum comfort.

Ionisation

Ionisation of gases in the air creates groups of atoms or molecules which have lost or gained electrons and have acquired a positive or negative charge in consequence. In clean, unpolluted air, ions exist in the proportion of 1200 positive to 1000 negative per cm^3 of air but, in a city centre, the quantities and proportion change to 500 positive to 300 negative per cm^3. In a building ventilated through sheet metal ductwork, and furnished with synthetic carpets and plastics furniture, the quantities and proportion may reduce still further to perhaps 150 positive to 50 negative ions per cm^3.

It has been suggested that low concentrations of negative ions (or high concentrations of positive ions) lead to malaise and lethergy, and high concentrations of negative ions produce sensations of freshness, however evidence in support of this theory is inconclusive.

Thermal indices

Numerous attempts have been made to devise a scale against which comfort may be measured and the following deserve attention.

Equivalent temperature

This scale combines the effects of air temperature, radiation and air movement, all as measured by a laboratory instrument named the *Eupatheoscope*, developed at the Building Research Establishment, during the 1920s. The scale takes no account of variations in humidity, and is no longer in use.

Effective temperature

Devised and developed in the USA for particular application to air-conditioning design, this scale combines the effects of air temperature, humidity and air movement but has no point of reference to radiation.

Corrected effective temperature

This is a later version of the scale noted above, modified to include the effect of radiation.

Fanger's comfort criteria

Work in Denmark has produced a series of comfort charts and tables based on analysis of the results of subjective tests which, in this case, took account of two further variables, namely the metabolic rate for various activities and the clothing worn. The volume of data provided is extremely large and enables predictions to be made with a high level of accuracy: for everyday use, however, the calculations which are required seem disproportionate, having regard to the application of the end result.

Subjective temperature

This is an approach not dissimilar to that proposed by Fanger but simpler and more realistic to apply since account is taken of fewer variables.

Dry resultant temperature

Since the publication of the CIBSE Guide A in 2006, dry resultant temperature has been superseded by *operative temperature* as the preferred temperature index for moderate thermal environments. In concept, they are identical and present no change of substance; the new term now aligns UK practice with other international standards.

Operative temperature

As for dry resultant temperature, this thermal index takes account of the effects of temperature, radiation and air movement.

The expression for evaluation of operative temperature is:

$$t_{op} = [t_{ri} + t_{ai}(10v)^{0.5}]/[1 + (10v)^{0.5}]$$

where

t_{op} = operative temperature (°C)
t_{ri} = mean radiant temperature (°C)
t_{ai} = room air temperature (°C)
v = velocity (m/s).

It will be noted that, when the air velocity is 0.1 m/s, this expression may be simplified greatly for use in the general run of calculations as:

$$t_{op} = (0.5t_{ri} + 0.5t_{ai})$$

Table 1.7 Wind-chill equivalent temperatures (°C)[a]

	Air temperature (°C)							
Wind speed (m/s)	*−10.0*	*−5.0*	*0*	*2.0*	*4.0*	*6.0*	*8.0*	*10.0*
2	−11.0	−6.0	−1.0	1.0	3.0	5.0	7.5	9.5
4	−14.5	−9.0	−3.5	−1.5	1.0	3.0	5.0	7.5
6	−17.5	−11.5	−6.0	−3.5	−1.0	1.0	3.5	5.5
8	−20.0	−14.0	−8.0	−5.5	−3.0	−0.5	2.0	3.0
10	−22.0	−16.0	−9.5	−7.0	−5.0	−2.0	−0.5	3.0
12	−24.0	−17.5	−11.0	−8.5	−6.0	−3.5	−1.0	2.0
15	−26.5	−19.5	−13.0	−10.0	−7.5	−5.0	−2.0	0.5
20	−29.5	−22.5	−15.5	−12.5	−10.0	−7.0	−4.0	−1.5

Note
a Dixon, J.C. and Prior, M.J., Wind-chill indices – a review. *The Meteorological Magazine*, 1987, 116, 1.

Excluded from the list are two scales which are not comfort indices but which relate to the calculation of heat losses and gains. *Environmental temperature*, a concept shown by the Building Research Establishment to provide a simplification of the relationship between air and mean radiant temperatures within a space. The second is *sol–air temperatures*, a scale which increments outside air temperature to take account of solar radiation.

Wind-chill indices

Although the expression *wind-chill* will be familiar to most readers, as a result of occasional use by TV weather forecasters, it is not generally appreciated that the concept of wind-enhanced cooling pre-dates the First World War. A paper by Dixon and Prior provides a full history, including a digest of both the empirical and theoretical evaluations and suggests that analyses of wind-chill indices by wind direction may be useful in deciding upon orientation and layout of new buildings, shelter belts, etc. Wind-chill equivalent temperatures for wind speeds of 2–20 m/s (5–45 mph) and air temperatures between −10°C and +10°C are listed in Table 1.7.

Methods of measurement

The descriptions which follow relate to very basic instruments which in many respects have been superseded by electronic or other devices. They do, nevertheless, serve to illustrate requirements and to warn of pitfalls.

Temperature

The readings taken from an ordinary mercury-in-glass thermometer provide, in general, temperature values relating to the gas, liquid or solid in which the instrument is immersed. When applied to the air volume within a building, however, the situation is much more complex. Here, the scale reading will portray a situation of equilibrium, taking account of not only air temperature but also of a variety of heat exchanges between surrounding surfaces. These latter may be partly by conduction, partly by convection and partly by radiation.

If such a thermometer were to be immersed in a hot liquid, conduction and convection would account for practically the whole effect, but if the same instrument were used to provide the temperature of room air, then the effect of radiation might predominate. Thus, in summer, the sun temperature might be 37°C, while the shade temperature reached only 27°C: the *air* temperature in each case might well be the same but in the former case the thermometer would be exposed to the sun, while in the latter case it would not.

In practice, the nature of the source of radiant heat affects questions of measurement. The so-called *diathermic* property of glass, in this case the bulb of the thermometer, permits the inward passage of high temperature radiation but is impervious to an outward passage at lower temperature. A radiation shield, which may be no more than a piece of aluminium foil shading the thermometer bulb, will negate this influence and permit air temperature alone to be read.

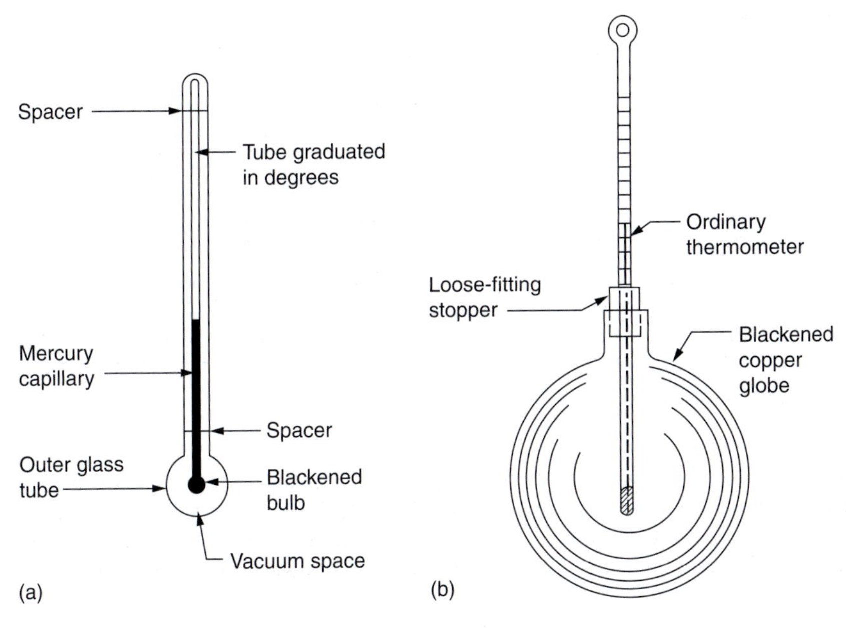

Figure 1.1 (a) Solar and (b) globe thermometers

Measurement of radiant effects thus depends upon a whole variety of circumstances. A *solar* thermometer, for measurement of high temperature radiation, consists of a glass sphere and tube within which a vacuum has been drawn, containing a simple mercury-in-glass instrument having a blackened bulb, as Fig. 1.1(a). Similarly, a *globe* thermometer, used for measurement of low temperature radiation, consists of a hollow copper sphere or cylinder about 150 mm in diameter, blackened on the outside and having a simple mercury-in-glass instrument projecting into it such that the bulb is in the centre, as Fig. 1.1(b). The mean radiant temperature t_{ri} at a single point in a room may, with difficulty, be measured with a globe thermometer or may be calculated from the arithmetic mean of the surrounding areas, each multiplied by the relevant surface temperature.

Humidity

The instrument most commonly used to measure the moisture content of air in a room is a *psychrometer* which makes use of what is called *wet bulb temperature.* This is measured by means of a simple mercury-in-glass thermometer which has its bulb kept wet by means of a water-soaked wick surrounding it. As the water evaporates, it will draw heat from the mercury with the result that a lower temperature will be shown. The rate of evaporation from the wetted bulb depends upon the humidity of the air, i.e. very dry surrounding air will cause a more rapid evaporation – and a lower temperature in consequence – than air which is more moist, although the temperature shown by an ordinary dry bulb thermometer would be the same in each case. The difference between dry and wet bulb temperatures may thus be used as a measure of humidity, individual values being known as the *wet bulb depression.*

The rate of evaporation, and hence the extent of the depression, depends also upon the manner in which the wetted bulb is exposed to the air. Records of external dry and wet bulb temperatures are kept by meteorological authorities worldwide and for this purpose measurement may be made using instruments placed in the open air within a louvred box called a *Stevenson screen.* The louvres are arranged so as to allow a natural circulation of air around the thermometers but, of course, they cannot exclude the effect of radiant heat completely, with the result that what are known as *screen* wet bulb temperatures are always about 0.5 K higher than those read from the alternative instruments used by engineers.

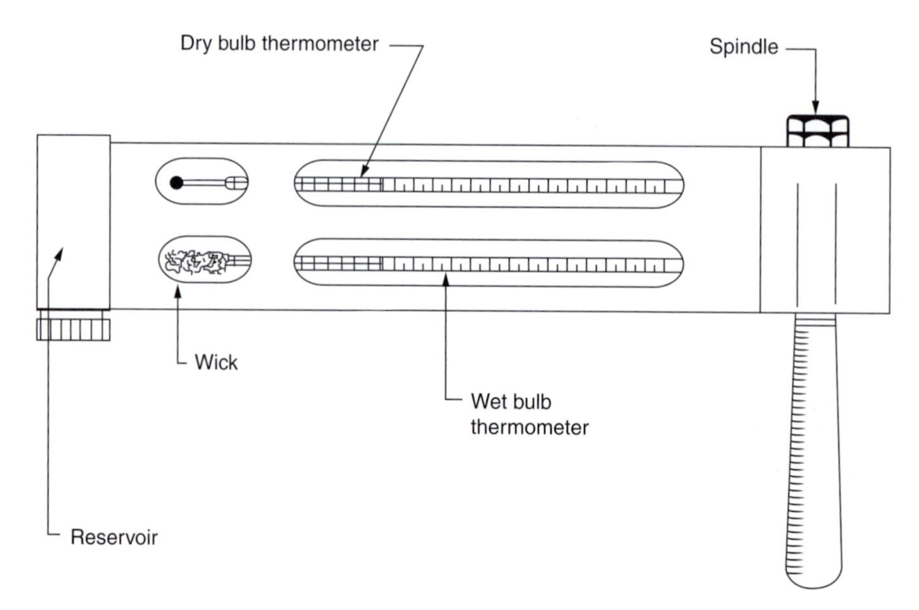

Figure 1.2 A sling psychrometer

The more common of these is the *sling psychrometer* which consists of two thermometers mounted side by side in a frame which is fitted with a handle such that it may be whirled by hand, as shown in Fig. 1.2. One thermometer is fitted with a wick which is fed with water from a small reservoir in the frame and both the water, preferably distilled, and the wick must be clean. Readings are taken after the instrument has been whirled vigorously to simulate an air speed above about 3 m/s. An alternative instrument, the *Assmann* type, consists of a miniature air duct within which the two thermometers are mounted, air being drawn over them by a small fan, clockwork or electrically driven, at a known velocity.

Air velocity

Requirements for reading air velocity fall into two very distinctly different categories. One relates to measurements made for the purpose of determining air volume in a duct or at delivery to, or extract from, a space, and the other to the identification of movement of air within a room. In the former case, velocities are not likely to be less than 4 m/s and (unless things have gone badly astray) the direction of flow should be easily identifiable. For this purpose the traditional field instruments have been *vane anemometers* of one type or another as will be noted in more detail in Chapter 29.

Within a room, however, circumstances are different and it is generally accepted that a tolerable level of air velocity at the back of the neck is not much in excess of 0.1 m/s when the operative temperature is 20°C, although, in extreme summer conditions, with natural ventilation, a transient increase to 0.3 m/s may be acceptable. Where air speeds exceed 0.15 m/s the operative temperature will need to be increased to compensate for the cooling effect of the air movement, as shown in Fig. 1.3. Fluctuations in air speed may also cause discomfort and recourse to analysis using the *draught* rating formula, given in the CIBSE Guide A, may be used to establish whether conditions are likely to be acceptable. For measurement at these low levels, use must be made of either an instrument known as the *Kata thermometer* or of one of the more sophisticated anemometers such as a *hot-wire* type.

The Kata thermometer is best suited to laboratory work and is tedious to use elsewhere. The hot-wire instrument is much more convenient to use than the Kata but neither provides any absolute indication of the *direction* in which the air is moving. It is thus necessary to rely upon a cold smoke, produced chemically, to supply a visual appreciation but this is not wholly satisfactory either, since the smoke tends to diffuse rather too rapidly.

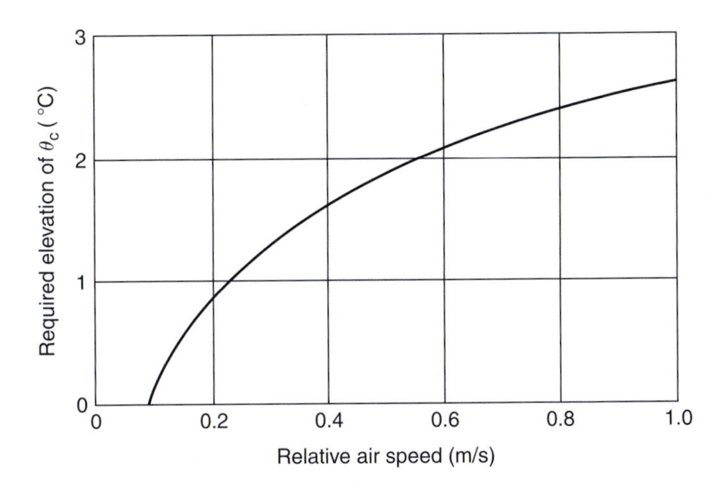

Figure 1.3 Correction to operative temperature to take account of air movement

Application

From preceding paragraphs it will be clear that, while it is possible to define the various factors which relate to comfort and to list the methods used to measure them, application of this knowledge to system design cannot proceed without the more detailed examination of the various aspects, as outlined in the chapters which follow. This brief resume of the fundamentals may thus be best summarised by reference to Table 1.8 which provides, proposed values of operative temperature (t_{op}) for most common applications and outdoor air supply rates; for more comprehensive tables refer to CIBSE Guide A (2006).

The temperatures proposed are for the basis of design and relate to normal clothing for the listed enclosures. Summer temperatures apply to air-conditioned or comfort cooled applications; higher temperatures may be acceptable where such conditioning is not provided. For naturally ventilated spaces, typically, an operative temperature of 25°C may be acceptable if this is not exceeded for more than about 5 per cent of the occupied period, The recommended limiting overheating criteria for daytime occupation is normally taken as an operative temperature of 28°C exceeded for not more than 1 per cent of the occupied period, whilst for bedrooms in dwellings 25°C exceeded for 1 per cent of the occupied hours is the benchmark. In a year-round controlled environment, there is no merit for normal comfort applications in controlling temperatures within close tolerances. Both in terms of comfort and energy use it is desirable to allow the internal temperature to vary from around 20°C to 22°C in winter to 22°C to 24°C in summer.

Research on thermal comfort by Brager and de Dear[6] has produced an empirically based result, from occupancy surveys, establishing a variation in comfort expectation for different external climates, referred to as the *adaptive method.* Figure 1.4 from CIBSE TM36, based upon ASHRAE Standard 55, 2004, shows the relationship between the indoor comfort operative temperature and external temperature, giving the bands of 80 per cent and 90 per cent occupant satisfaction. An alternative approach using computer thermal modelling is covered at the end of the chapter.

Outside air supply rates in spaces are determined by the Building Regulations, 2006, Part F, requiring a minimum rate of 10 litre/s per person for most non-domestic applications. Where smoking is not permitted normally 10 to 12 litre/s per occupant would be provided, except where other factors apply. For example, in factories the rate of ventilation may be determined by the process, and in bathrooms, toilets and changing rooms rates would be determined by odour or humidity control. For dwelling spaces the Building Regulations should be consulted. Where smoking is permitted the CIBSE Guide recommends an outdoor air supply rate of 45 litre/s per person. Ventilation and indoor air quality classifications are given in Table 1.9.

Table 1.8 Proposed comfort criteria

Type of enclosure	*Operative temperature* t_{op} (°C)		*Minimum outdoor air supply rate litre/s per person unless indicated otherwise*
	Winter	*Summer*	
Auditoria	22–23	24–25	10
Banks	19–21	21–23	10
Bars, lounges	20–22	22–24	10
Churches	19–21	22–24	10
Conference, board rooms	22–23	23–25	10
Dwellings			
Bathrooms	20–22	23–25	15
Bedrooms	17–19	23–25	0.4–1.0 ac/h[d]
Kitchens	17–19	21–23	60
Living rooms	22–23	23–25	0.4–1.0 ac/h[d]
Toilets	19–21	21–23	>5 ac/h[d]
Exhibition spaces	19–21	21–23	10
Factories			
Heavy[a]	11–14	Spot cooling	As required for industrial process
Light[a]	16–19	Spot cooling	As required for industrial process
Sedentary	19–21	21–23	As required for industrial process
General areas, corridors, etc.	19–21	21–23	10
Hospital			
Wards/treatment	22–24	23–25	10
Operating theatres	17–19	17–19	0.65–1.0 m^3/s
Hotel			
Bathrooms	20–22	23–25	12
Bedrooms	19–21	21–23	10
Kitchens: Commercial	15–18	18–21	As required for installed equipment
Museums: Galleries[b]	19–21	21–23	10
Offices			
Executive	21–23	22–24	10
General	21–23	22–24	10
Open plan	21–23	22–24	10
Swimming pool			
Changing	23–24	24–25	10 ac/h[d]
Halls[c]	23–26	23–26	Relate to wetted area
Sports centres			
Changing	22–24	24–25	10 ac/h[d]
Halls[c]	13–16	14–16	10
Restaurants	21–23	24–25	10
Shopping centres: malls	12–19	21–25	10
Shops: stores, supermarkets	19–21	21–23	10
Teaching spaces	19–21	21–23	10
TV studios	19–21	21–23	10

Notes
a Subject to legislation.
b Conditions for exhibits may override comfort criteria.
c Audience spaces may require different comfort conditions.
d ac/h = air changes/hour.

The use of operative temperature as a design criterion takes account of radiation in the general sense but not of particular local effects which may arise from *asymmetrical exposure* of the body to:

- Cold radiation in winter to a single-glazed window
- Excessive insolation (short-wave radiation) from unshaded glazing
- Exposure to some internal source of high radiant intensity

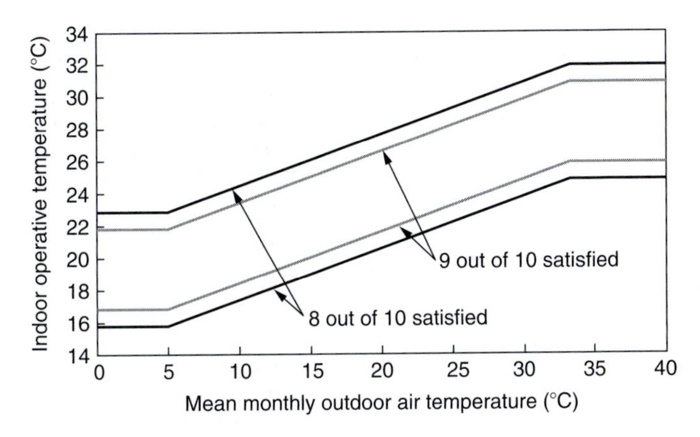

Figure 1.4 Adaptive comfort model

Table 1.9 Ventilation and indoor air classification

Classification	*Indoor air quality standard*	*Ventilation range* (litre/s per person)	*Default value* (litre/s per person)
IDA1	High	>15	20
IDA2	Medium	10–15	12.5
IDA3	Moderate	6–10	8
IDA4	Low	<6	5

Source
BS EN 13779 ventilation for buildings. Performance requirements for ventilation and air conditioning systems.

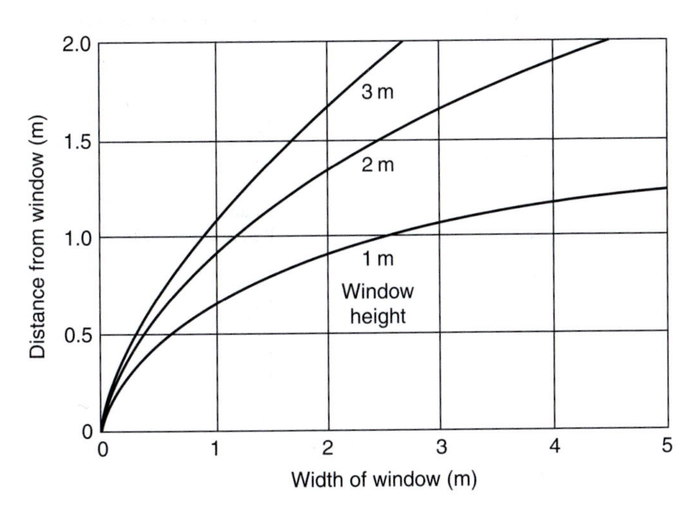

Figure 1.5 Minimum comfortable distances from the centre of a single-glazed window for outside air temperature of -1°C

With regard to cold radiation, an occupant may expect to experience discomfort when seated close to a single-glazed window exposed to cold outside air temperatures; Fig. 1.5, from *Guide Section A1 (1999)*, indicates the discomfort zone when the outside air temperature is close to freezing.

In the case of the discomfort arising from the second situation, there are so many variables that a similar simplification is not possible; in an extreme case, however, a rise in radiant temperature (t_{ri}) of 10–15K might be anticipated and the provision of some form of shading to the glazing, preferably external, is the correct solution. The detail of calculation and measurement of radiant temperature asymmetry is given in BS EN ISO 7726.

As to humidity, too high a level will reduce the ability of the body to lose heat by evaporation, with resultant lassitude; too low a level will produce a sensation of coolness on exposed flesh, a parched throat and dry eyes. Where temperatures are high, these extremes are of great importance but in the maritime climate of the British Isles, variations between 40 per cent and 60 per cent saturation are usually acceptable, although 50 per cent saturation is the usual design target for human comfort. Industrial processes, museums and galleries and associated storage often require a better standard of humidity control to within as little as ±5 per cent and even smaller tolerances for some applications.

There are two quite extraneous aspects which relate to humidity in any building, the first being the likely incidence of condensation on single-glazed windows when their surface temperature falls, in very cold weather, to below the dew point temperature of the building air content. The second is the problem of a build up in static electricity, leading to electrostatic shock when occupants touch earthed building components. This is a function of humidities below about 40 per cent coupled with the material and backing of the floor coverings.

Practical guidelines to comfort

The following are practical guidelines to good practice in providing a comfortable environment for normally clothed people in a sedentary occupation:

- Air temperature within range 21°C in winter to 23°C in summer
- In summer mean radiant temperature below air temperature
- In winter mean radiant temperature above air temperature
- Relative humidity normally within 45–60 per cent
- Air movement not greater than 0.15 m/s
- Carbon dioxide concentration below 0.1 per cent
- Temperature gradient between feet and head not greater than 1.5 per cent
- Floor surface temperature between 17°C and 26°C
- Radiant asymmetry not greater than 10°C horizontal and 5°C vertical.

These conditions may be exceeded for short periods, say up to a maximum of 5 per cent and preferably for not more than 2 per cent of time on average, e.g. in naturally ventilated or passively cooled environments, to enable energy saving measures or other economies. As will be appreciated, the manner in which air temperature, radiation, humidity and air movement interrelate with one another is very complex and nothing short of an exhaustive investigation of each, will suffice to establish the precise situation existing within an enclosure. Even so, when the investigation is complete and all results have been demonstrated to be within a hair's breadth of best practice, the individualities of human sensation are such that it is unlikely that many more than 90 per cent of the occupants of a building will be entirely satisfied with their environment!

Comfort model

Recognising that the various factors affecting thermal comfort of the occupants in a space will vary through the year, for example, due to changes in surface temperatures, changes in clothing, solar conditions, or high summertime temperatures in non-air-conditioned buildings, computer *thermal modelling* (refer to Chapter 6) may be used to establish the percentage of time different levels of operative temperatures are achieved. BS EN ISO 7730, Ergonomics of the Thermal Environment, provides the basis for the determination of thermal comfort from a calculation of the *Predicted Mean Vote* (PMV), which is the average sensation of comfort experienced by a large group of people, and the *Predicted Percentage* (of occupants) *Dissatisfied* (PPD) at different temperatures. The Standard offers design criteria, as given in Table 1.10, related to the activity level indicated and for clothing of 0.5 clo during summer and 1.0 for winter conditions. When using PPD percentages it should be recognised that 5 to 10 per cent of people will be dissatisfied with any given internal environment. In addition,

Table 1.10 Example design temperature criteria

			Operative temperature (°C)	
Application	*Metabolic rate*	*PPD* (%)	*Summer*	*Winter*
Office				
Conference room	70 W/m^2	<6	24.5 ± 1.0	22.0 ± 1.0
Auditorium		<10	24.5 ± 1.5	22.0 ± 2.0
Restaurant		<15	24.5 ± 2.5	22.0 ± 3.0
Classroom				
Retail Store	93 W/m^2	<6	23.0 ± 1.0	19.0 ± 1.5
		<10	23.0 ± 2.0	19.0 ± 3.0
		<15	23.0 ± 3.0	19.0 ± 4.0

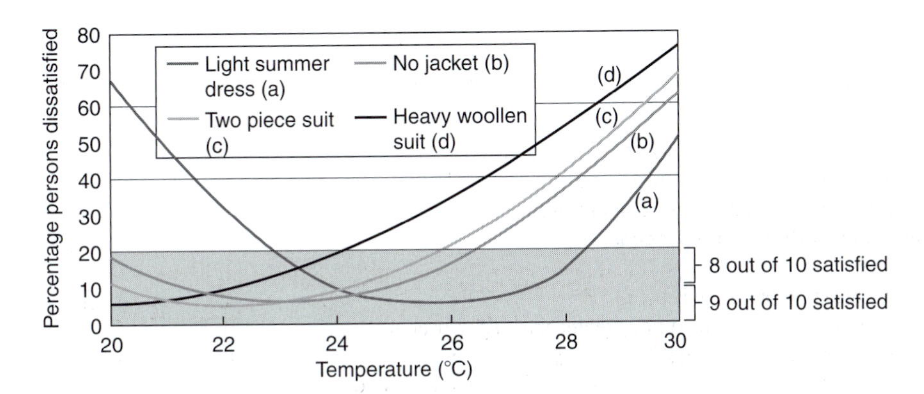

Figure 1.6 Effect of clothing on comfort temperature

where people are able to adjust their clothing they will remain comfortable over a wider range of temperatures. Where this is possible the internal temperature may be controlled over a wider band to effect energy savings; this is illustrated in Fig. 1.6, from CIBSE TM 36 which shows the effect of clothing on comfort.

Determination of an acceptable internal climate has to take in many variables through the year and is a complex area to establish optimal design objectives. Where thermal modelling is based upon weather tapes to represent external conditions through the year it should be recognised that the data is derived from historical records and as a consequence makes no allowance for the effects of global warming and are therefore not necessarily representative of future conditions. This chapter has covered only general principles and further reading is recommended.

Notes

1. If water is evaporated at 20°C, the latent heat of evaporation is 2450 kJ/kg.
2. Animation, feeling, hearing, seeing, smelling, speaking, tasting.
3. Minimising pollution at air intakes, CIBSE Technical Memorandum TM21: 1999.
4. Fanger PO-various important papers published since 1990.
5. CEN Report 1752, Ventilation for Buildings-Design Criteria for the indoor environment, December 1998.
6. Brager G.S. and de Dear R. Climate comfort and natural ventilation: a new adaptive comfort standard for ASHRAE Standard 55.

Further reading

The Air Quality Strategy for England, Scotland, Wales and Northern Ireland, 2000; DETR.
Air Pollution in the UK, 2005; National Environmental Technology Centre, Harwell, UK.
CIBSE Knowledge Series, Comfort.
CIBSE Technical Memorandum, TM36, 2005, Climate Change and the Indoor Environment: Impacts and Adaptation.
CIBSE Guide Book A, Environmental Design, 2006.

Chapter 2

Sustainability

Aim of this chapter

Since the ninth edition of Faber & Kell was published in 2002 there has been a huge increase in awareness of sustainable development (or sustainability) and of the fact that our profligate use of resources is having a detrimental impact on the world's natural systems. It is also now widely accepted that man's activities are responsible for climate change and that climate change is one of the greatest challenges facing humanity. This heightened awareness has lead to a wide range of policy initiatives, internationally, and at national and local levels to help move us towards a more sustainable future.

This chapter aims to provide a summary of the current drivers for sustainable design and construction and the challenges and opportunities these present to those involved in the design of heating and air-conditioning systems. It outlines some of the legislative drivers and design issues and cross references to the following chapters where further guidance is provided.

Sustainability is a broad and fast moving agenda with new policies and guidance emerging all the time, in preparing this introduction to the area we have focused on the drivers most likely to effect those involved in the design of heating and air-conditioning systems and on the policy position at the time of writing. The drivers for sustainability will continue to grow and it will be increasingly important for engineers to keep track of emerging policy and legislation in this area.

What we mean by sustainable development

Sustainable development (or sustainability) is in essence about enabling all people throughout the world to satisfy their basic needs and enjoy a better quality of life without compromising the quality of life for future generations.

As Jonathon Porrit has put it 'it is essentially a different model of progress, balancing the social and economic needs of the human species with the non-negotiable imperative of living within planet Earth's natural limits'. It requires a shift from an ethos of exploitation to one of stewardship and global responsibility.

Awareness of the term sustainable development was heightened following the publication in 1987 of the 'Report of the World Commission on Environment and Development: Our Common Future' popularly known as the 'Brundtland Report' after its author Gro Harlem Brundtland. This report helped establish the widely quoted definition:

> 'Sustainable development is development that meets the needs of the present without compromising the ability of future generations to meet their own needs.'

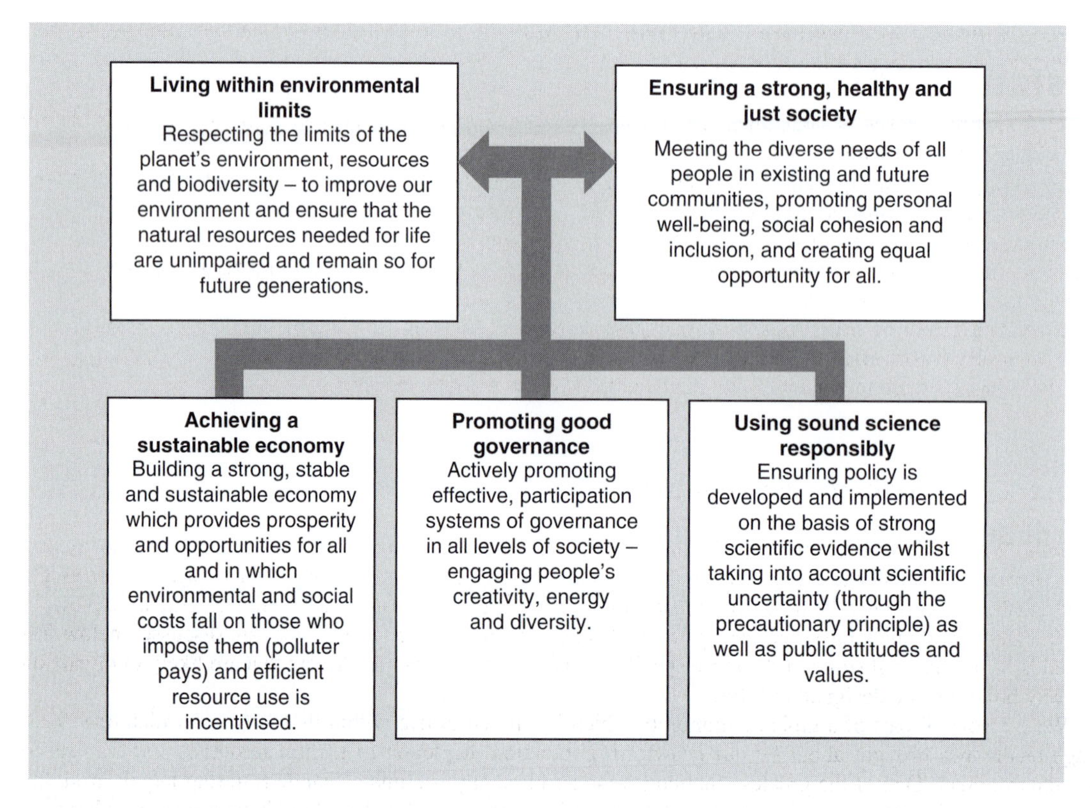

Figure 2.1 UK sustainable development strategy guiding principles (see colour plate section at the end of the book)

The UK Government's current sustainable development strategy 'Securing the Future' sets out five Guiding Principles for sustainable development:

(1) Living within environmental limits
(2) Ensuring a strong, healthy and just society
(3) Achieving a sustainable economy
(4) Promoting good governance
(5) Using sound science responsibly.

These principles are expanded further in Fig. 2.1.

The UK sustainable development strategy can be viewed at: http://www.sustainable-development.gov.uk.

In addition to the overall UK sustainable development strategy there are specific sustainability strategies for particular activities including a new sustainable construction strategy.

Sustainable construction

The way we design, construct and operate buildings has a huge influence on consumption of resources, impacts on the environment and on the well-being of those who occupy them. It also influences the well-being of those involved in extracting natural resources, manufacturing them into building products and assembling them on site.

The Department for Environment, Food and Rural Affairs (DEFRA) has developed the following characteristics to help define a sustainable building:

- A building that leaves as small an environmental footprint as possible, is economic to run over its whole life cycle, and fits well with the needs of the local community.
- A building that is energy and carbon efficient, designed to minimise energy consumption, with effective insulation and the most efficient heating or cooling systems and appliances.
- A building built with good access to public transport in mind.
- A building built with a minimum of waste in its construction and which looks to maximise reuse of on-site materials such as waste soil.
- A building designed and constructed to enable its occupants to use less water, through, for example, the installation of more efficient fittings and appliances.
- A building designed to make recycling and composting easy for the occupants.

Drivers for sustainable development

Over the last decade a wealth of new policies, legislation and fiscal incentives have been introduced to help move the construction industry to more sustainable outcomes. At the same time businesses have begun to accept the need to take responsibility for their own impacts in response to growing consumer and shareholder awareness.

The key global challenges that have prompted action on sustainable development are discussed below followed by a summary of some of the more specific legislative and corporate drivers that are likely to impact on the way buildings are designed in future.

We are now all part of a global community which has failed to live within its limits and which now faces huge challenges brought about by our growth in population, depletion of natural resources and damage to our environment. One of the greatest challenges facing humanity is climate change resulting largely from the burning of fossil fuels to provide energy and the resulting emissions of carbon dioxide into the atmosphere.

Climate change

Climate change results from emissions of 'greenhouse' gases which are helping to trap heat within the atmosphere resulting in a rise in global temperatures. The key contributor to climate change is carbon dioxide but other greenhouse gases include: methane; nitrous oxide; ozone; halocarbons including CFCs (chlorofluorocarbons), HCFCs (hydrochlorofluorocarbons), HFCs (hydrofluorocarbons); halons; perfluorocarbons (PFCs) and sulphur hexafluoride (SF6). A number of these gases are emitted as a result of energy use in buildings and through their use as refrigerants in air-conditioning systems, in fire suppression systems and as insulators in high-voltage switch gear.

Sir Nicholas Stern's report to the UK Government reviewing the Economics of Climate Change 'The Stern Review' concluded:

> 'The scientific evidence is now overwhelming: climate change presents very serious global risks, and it demands an urgent global response.'

The report concluded that even if the annual flows of emissions into the atmosphere do not increase beyond today's rate, the stock of greenhouse gases in the atmosphere will double reaching 550 ppm CO_2 equivalent (CO_2e).[1] At this level there is at least a 77 per cent chance – and perhaps up to a 99 per cent chance, depending on the climate model used – of a global average temperature rise exceeding 2°C. CO_2e is used to measure the impact of a broad range of greenhouse gases.

At present the annual flow of emissions is accelerating, as fast growing economies invest in high carbon infrastructure and as demand for energy and transport increases around the world. The Stern Review highlighted that if these emissions are left unchecked atmospheric concentrations of 550 ppm CO_2e could be reached as early as 2035. The report recommended that the best response to climate change is to take urgent action now to cut greenhouse gas emissions rather than attempting to deal with its consequences later. The cost of keeping CO_2 emissions below 550 ppm CO_2e will be around 1 per cent of global GDP. Under a business as usual scenario with rising emissions resulting in large-scale climate change an average loss in global GDP of 5–10 per cent could result, with poor countries suffering costs in excess of 10 per cent of GDP. This report has paved the way for a stronger and more urgent legislative response on climate change illustrated by the UK Government's proposal to ensure that all new housing has zero net carbon emissions by 2016.

There is no international consensus on the required cut in CO_2 emissions necessary to avoid the worst impacts of climate change. The European Union's view is that surface temperature increase should not be allowed to exceed 2°C and that stabilisation well below 550 ppm CO_2e is likely to be required.[2] The Royal Commission on Environment and Pollution (RCEP) concluded that to achieve stabilisation of carbon emissions at 550 ppm CO_2e, the UK would need to cut its emissions by 60 per cent by 2050 and 80 per cent by 2100.[3] But more recent conclusions from the G8 Science Conference in Exeter in 2005 concluded that a stabilisation of 450 ppm is necessary to have a reasonable chance of remaining below a 2°C rise in global mean temperature, which implies CO_2 reductions greater than those recommended by the RCEP.

The Stern Review considered the impacts of climate change on economic activity, on human life and on the environment. It concluded that on current trends, average global temperatures will rise by 2–3°C within the next 50 years or so, but that if emissions continue to grow the earth will be committed to several degrees more warming (Fig. 2.2).

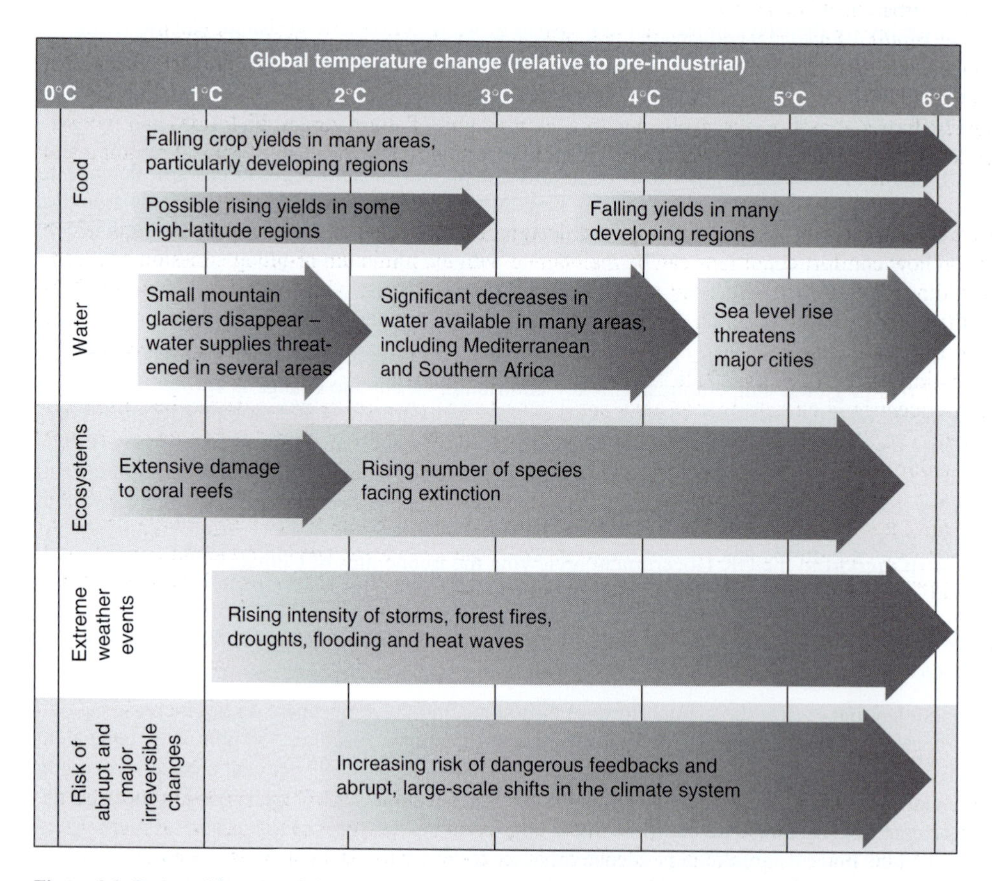

Figure 2.2 Projected impacts of climate change (see colour plate section at the end of the book)

Source
Stern Review, Crown copyright

This warming is expected to have severe global impacts:

- Melting glaciers will initially increase flood risk and then strongly reduce water supplies, eventually threatening one-sixth of the world's population.
- Declining crop yields, especially in Africa, could leave hundreds of millions without the ability to produce or purchase sufficient food.
- In higher latitudes, cold-related deaths will decrease. But climate change will increase worldwide deaths from malnutrition and heat stress.
- Rising sea levels will result in tens to hundreds of millions more people flooded each year with warming of 3°C or 4°C. There will be serious risks and increasing pressures for coastal protection.
- Ecosystems will be particularly vulnerable to climate change, with around 15–40 per cent of species potentially facing extinction after only 2°C of warming. And ocean acidification, a direct result of rising carbon dioxide levels, will have major effects on marine ecosystems, with possible adverse consequences on fish stocks.

The UK Climate Impacts Programme (UKCIP) (www.ukcip.org.uk) has published a series of scenarios for the likely consequences of climate change for the UK. Over the course of the next century the UK is likely to experience the following key changes:

- *Increased temperatures*: The effect of a warmer climate on the comfort of buildings throughout the south-east and east of England will be noticeable, while urban centres will be even hotter than surrounding areas due to the added 'urban heat island' effect.
- *Increased winter rainfall*: Further increasing the risk of flooding associated with rising sea levels.
- *Decreased summer rainfall*: Summers may become drier throughout the UK. This may impact on regional and local water availability and biodiversity.
- *Rising sea levels*: Extreme high water levels, due to a combination of storm surges, high tides and increasing mean sea level, are predicted to become more frequent resulting in a higher risk of coastal flooding and erosion.

These impacts have important consequences for the design of heating and air-conditioning systems which must now focus on how comfort conditions can be maintained with the minimum resulting emission of greenhouse gases. Building designs will also need to consider the future impacts of climate change and be flexible to adapt to these, for example enabling comfortable conditions to be maintained as summer temperatures increase. This does not automatically imply the greater use of mechanical cooling which may serve only to increase CO_2 emissions and exacerbate effects such as urban heat islands.

While the consequences of climate change are alarming, and some are already unavoidable, the problem is not insurmountable. The Stern Review notes 'there is still time to avoid the worst impacts of climate change if strong collective action starts now'.

There are still some who question the science behind climate change, but even in the unlikely event that the world's scientists have got it wrong, there are other good reasons why we should be reducing energy demands in our buildings and reducing our dependence on fossil fuels. These include the need to maintain a secure and reliable energy supply, the opportunity to reduce operational costs for business and to provide homes that are affordable to heat.

A broad range of legislative drivers have been introduced to help cut energy demands and reduce carbon emissions.

Security of fuel supplies

In common with many other countries the UK is no longer a net exporter of oil and in future will be an importer of both oil and gas. By 2020, we could be importing around 60 per cent of the gas we use and up to two-thirds of the oil.[4] This places a greater dependence on other countries to meet our energy needs.

In order to reduce this dependence we need both to reduce our energy demands, to use the energy we have more efficiently and to increase the proportion of energy that is delivered from alternative energy sources including renewables.

Operational costs

While the cost of energy use is a relatively small part of a company's overall costs, compared to items such as salaries and rent, it represents a large part of the costs that a company can influence. Energy costs represent approximately 20 per cent of the service charge paid by occupants of multi-tenanted buildings and thus reducing energy use is likely to have both an economic benefit as well as reducing environmental impacts.

Resource issues

Since the start of the industrial revolution the world's population has increased from around 1 billion to 5.8 billion. By 2025 it is expected to increase to 8.5 billion. At the same time those in developed countries consume greatly more resources than those in developing nations, a situation which is not equitable but also unsustainable as populations increase and as developing nations aspire to the consumer lifestyle enjoyed in the West.

There are also pressing resource issues at a national level. In the UK average household water demand has increased by 55 per cent since 1980. In order to meet increasing demand, new sources of water supply have to be created, but this is costly and potentially damaging to the environment. This problem will be made worse by predicted reductions in summertime rainfall resulting from climate change. This is an issue that can be addressed through the design of heating and cooling systems installed in buildings, for example by exploring the use of recycled rain or grey water for use in cooling systems.

Total household waste per head of population in England and Wales rose by 20 per cent from 1983/1984 to 1997/1998 and in 2003 stood at around 500 kg/person/year.[5,6] Construction and demolition produced waste of 91 million tonnes in England and Wales in 2003 and of this 13 million tonnes consisted of material delivered to sites but never used.[7]

European Union Legislation has been introduced with the aim of increasing the amount of waste that is recovered for useful purposes and to reduce that which is landfilled. There are also increasing fiscal incentives in the UK to cut the amount of waste being landfilled.

The legislative and policy framework

The urgent need to address climate change and make better use of natural resources has resulted in a series of policy responses at both an European Union, national and local level. Some of the key ones are summarised below and those most directly affecting the design of heating and air-conditioning systems are discussed in more detail in Chapter 3 on Legislation.

European Union policy drivers

Under the Kyoto Protocol, the European Union committed to reducing average CO_2 emissions by an average of 8 per cent below 1990 levels by 2008–2012. The UK Government accepted a higher Kyoto target of 12.5 per cent savings. In March 2007 the 27 European Union member states reached an agreement to cut European CO_2 emissions 20 per cent below 1990 levels by 2020. The detailed contributions that each member state will make to this target are the subject of further negotiations.

Energy Performance in Buildings Directive

The Energy Performance in Buildings Directive (EPBD)[8] is a significant driver for improving the energy performance of buildings in the UK. It was published on 4 January 2003 in the Official Journal of the European

Union, becoming European Law the following day. The main requirements of the EPBD and its implementation in the UK are discussed further in Chapter 3, however, key outcomes from its implementation include:

- The introduction of a national methodology for calculating the energy performance of domestic and other buildings, implemented through the 2006 update to Part L of Building Regulations.
- The introduction of Energy Performance Certificates (EPCs) for all buildings at the point of construction or when they are sold or rented out.
- The requirement for Display Energy Certificates (DECs) for public buildings of more than 1000 m^2.
- Regular inspections/assessments of cooling installations.

F-Gas Directive and Regulations

The F-Gas Regulations will come into force with effect from July 2007. The objective is to contain, prevent and thereby reduce emissions of F-Gases covered by the Kyoto Protocol, i.e. fluorinated refrigerant gases. The main focus is on containment and recovery of F-gases, which will impact on the commercial refrigeration, air-conditioning and heat pump sectors, and also the fire protection sector. Initial guidance has been produced by the government and is available on the Department of Trade and Industry (DTI) website,[9] and further information is provided in Chapters 3 and 24.

The Waste Electrical and Electronic Equipment Directive and Regulations

Electrical and electronic waste is one of the fastest growing waste streams. The components used to build electrical and electronic equipment (EEE) contain various hazardous substances. If sent to landfill as untreated waste rather than properly disposed of, these can damage the environment and human health.

In response, European Union member states agreed the waste electrical and electronic equipment (WEEE) Directive in February 2003. Its purpose is to reduce the volume of WEEE generated and ending up in landfill by encouraging better design and production and by setting criteria for its collection for reuse, recycling and safe disposal. It is implemented in the UK by the WEEE Regulations,[10] which entered into force on 2 January 2007. The DTI[11] is the responsible government department and has published 'Non-Statutory Guidance'[12] on implementation.

The DTI guidance states that consumers who use EEE and discard it as waste 'have no legal obligations under the WEEE Regulations', while encouraging them to use the new options that will become available to ensure environmentally sound disposal. The guidance also states that users of non-household EEE 'may, in some circumstances, have a[n] obligation to finance the treatment of EEE … discard[ed] as waste, and its recycling, recovery and environmentally sound disposal' and encourages them to take potential obligations into account when procuring new equipment.

The Directive applies to 10 categories of products, which are listed in Schedule 1 of the UK WEEE Regulations. Those most relevant to engineers are:

- Large household appliances including: large refrigerators and freezers; electric appliances for heating rooms, electric fans; air-conditioner appliances, exhaust ventilation and conditioning equipment.
- Monitoring and control instruments including: heating regulators; thermostats; and other monitoring and control instruments used in industrial installations.

UK policy drivers

Energy White Paper 2003: our energy future – creating a low carbon economy (DTI 2003)

In February 2003 the DTIs published its Energy White Paper, this aimed to set out the policy agenda needed to address climate change, the decline of our indigenous energy supplies and the need to update our energy infrastructure.

The White Paper included 130 commitments which included:

- To be on a path to cut the UK's CO_2 emissions by some 60 per cent by about 2050, with real progress by 2020.
- A target for 10 per cent of electricity to be from renewable sources nationally by 2010 and twice this by 2020.
- A target of at least 10 000 MWe of good quality combined heat and power (CHP) capacity by 2010.

These targets were supported by a broad range of legislative and fiscal incentives.
The full White Paper can be viewed at: http://www.dti.gov.uk/energy/policy-strategy/index.html.

The Sustainable and Secure Buildings Act

The Sustainable and Secure Buildings Act (SSBA) amends the Building Act 1984, extending the range of issues on which the government can make new building regulations and the circumstances where these can be applied. As well as 'furthering the conservation of fuel and power', regulations can now be made for the purposes of:

- preventing waste, undue consumption, misuse or contamination of water;
- furthering the protection or enhancement of the environment;
- facilitating sustainable development;
- furthering the prevention or detection of crime.

Building Regulations previously applied only to new construction and defined works in existing buildings. They may now also be made to apply to 'the demolition of buildings' and 'services, fittings and equipment provided in or in connection with buildings', and may be used to impose continuing requirements on owners and occupiers of buildings.

A key implication of the Act is that it enables Regulations to be applied to existing buildings and also extends the range of issues that can be covered by Building Regulations. In December 2006 Communities and Local Government (CLG) issued a consultation paper on proposals for introducing water consumption targets into Building Regulations (http://www.communities.gov.uk/).

Draft Climate Change Bill (DEFRA 2007)

In March 2007 the Government launched its draft Climate Change Bill. The Climate Change Bill aims to create a strong new legal framework to underpin the UK's contribution to tackling climate change. Legislation will set targets in statute for a 60 per cent reduction in carbon dioxide emissions through domestic and international action by 2050, with a 26–32 per cent reduction by 2020 against a baseline of 1990. This would be based on a new system of 5 yearly 'carbon budgets' set at least 15 years ahead, and with progress reported annually to Parliament. This Bill aims to provide clarity on how UK emissions will be reduced.

If adopted the Bill would also create a new expert Committee on Climate Change to advise the Government on the best pathway to achieving a 60 per cent reduction by 2050.

The draft Bill can be downloaded at: http://www.defra.gov.uk.

A review of sustainable construction (DTI 2006)

In October 2006 DTI published a review of what has been achieved in relation to sustainable design and construction since the last review in 2001. This covers issues such as available skills as well as key environmental issues including energy, waste and water. The report includes targets and visions for sustainable construction up to 2015 and beyond. The review can be downloaded at: http://www.dti.gov.uk.

Building Regulations Part L 2006 (see Chapter 3)

Building Regulations require that reasonable provision should be made for the conservation of fuel and power in buildings. Updates to the Approved Documents for Part L published in April 2006 introduced a number of new approaches to demonstrating compliance with the Regulations. A key change was to replace elemental

U value methods with a calculation procedure for calculating the overall regulated carbon emissions for the proposed building and comparing these against a target emission rate (TER).

The updated regulations were designed to reduce regulated carbon dioxide emissions in new buildings by 20–28 per cent compared to buildings complying with the 2002 regulations. These changes were in part required to implement the EPBD but also to provide greater flexibility in how the required carbon savings can be achieved. This flexibility allows the cost effectiveness of efficiency improvements to be compared against alternative approaches such as low carbon or renewable energy supplies. This is only one of a number of changes introduced in 2006 and further details are provided in Chapter 3.

One of the implications of the changes to the Regulations is the desirability of conducting calculations early in the design process to ensure buildings are meeting the required standards before they are submitted for planning. This will also allow optimum design proposals to be developed that satisfy both Building Regulations and any additional planning requirements.

It should be recognised that Building Regulations are a minimum legal requirement and hence represent the worst standards to which a building can legally be built. Where possible building engineers should seek design solutions which improve significantly upon these minimum standards.

Building a greener future – towards zero carbon buildings

In December 2006 CLG launched draft proposals[13] for future changes to Building Regulations aimed at all new homes achieving zero net carbon emissions for all their energy uses by 2016. The consultation draft includes a number of interim targets as shown in Table 2.1.

The target reductions proposed relate to similar targets established in the Code for Sustainable Homes (CfSH). The implications of the interim targets are that new homes will require a greater proportion of their energy demands to be met from low carbon or renewable sources such as CHP, solar water heating or biomass, and to meet the zero carbon target there will need to be much greater emphasis on reducing electrical demand to a point where they can be met from photovoltaics (PV), wind (or biomass CHP if technically feasible). See Chapter 14 on Renewables.

Proposals for updating Part L 2006 of Building Regulations for buildings other than dwellings have yet to be announced but might be expected to include similarly ambitious targets.

Table 2.1 Proposed updates to Part L of Building Regulations for New Homes[13]

Date to be updated	*2010*	*2013*	*2016*
Carbon improvement as compared to Part L1A (Building Regulations 2006)	25% reduction in regulated emissions[1]	44% reduction in regulated emissions[1]	Zero carbon for regulated and unregulated emissions[2]

Notes
1. Regulated emissions include those resulting from energy use for space heating, hot water, ventilation and fixed lighting, they exclude appliance energy use.
2. Unregulated emissions include emissions for non-fixed lighting and appliances.

Energy Performance Certificates

By October 2008, all buildings that are constructed, sold or rented out will need to have an EPC, which will give an energy rating from A to G and recommendations on how to reduce carbon emissions in the building to which it applies. The rating will be based on the same calculation procedures used in Part L of Building Regulations.

For public buildings of more than 1000 m^2 there is the requirement for DECs that can be seen by visitors to the building.

The aim of the EPC is to allow a purchaser or tenant to make an informed choice about the energy performance of the building they are purchasing or letting and to provide information on the potential improvements they could make to the building. The certificate will also include recommendations on how they can operate the building more efficiently. The potential exists for landlords and developers to use the A–G ratings to establish minimum performance standards for their new or existing property portfolios. This in turn may lead to buildings with a lower-energy demand and carbon emissions, commanding higher values in the market place. The EPC could potentially allow other fiscal incentives linked to energy performance to be introduced in the future, for example the Chancellor has announced that homes with zero net carbon emissions will be exempt from stamp duty.

Planning policy

In the last few years the planning system has become one of the main drivers for low carbon approaches to building design and in particular the promotion of renewables.

Planning policy in England is established through national Planning Policy Statements (PPSs) and Planning Policy Guides (PPGs). PPS 1 sets out the overarching planning policies which then feed into planning policy at a regional level, expressed through regional spatial strategies, and in turn down to policies at a local level expressed through local development documents. Paragraph 3 of PPS 1 states that sustainable development is the core principle underpinning planning.

PPS 1 is supported by a range of further PPSs. Many of these cover aspects relevant to sustainable design and construction. The PPS most relevant to heating and cooling systems in buildings is PPS 22 on Renewable Energy which states that regional spatial strategies and local development documents should contain policies designed to promote and encourage, rather than restrict, the development of renewable energy resources. This in turn has encouraged policies at a regional and local level that have set requirements for a proportion of a development's energy demands to be met from renewables. The first such policy was introduced by the London Borough of Merton which required non-residential schemes of more than $1000\,m^2$ to provide at least 10 per cent of their energy demands from renewables.

Similar policies are now being included in many of the regional spatial strategies and in the core strategies of local development documents. In London proposed alterations to the London Plan (The Spatial Strategy for London) aim to increase the percentage renewable contribution from 10 per cent to 20 per cent and will include separate policies aimed at promoting the use of CHP and combined cooling heat and power (CCHP).

These policies increasingly require an energy statement to be submitted as part of the planning application to demonstrate how energy demands have been reduced, to demonstrate that low carbon energy sources such as CHP have been considered for appropriate sites, and setting out the proposed approach for providing the required proportion of energy from renewable sources.

The implications of these policies for building services engineers are that much earlier input is now required to predict energy demands and assess the opportunities for incorporating both CHP, CCHP and renewable technologies such as solar water heating, ground source heating and cooling, biomass boilers, PV and wind generation.

Further guidance on supplying heating and cooling from renewable resources is set out in Chapter 14 on Renewables and advice on the application of CHP and CCHP is set out in Chapter 15. Advice on early building modelling is provided in Chapter 6.

Draft supplement to PPS 1 – Planning and Climate Change

The CLG recently launched a draft supplement to PPS 1 on Planning and Climate Change.[14] This aims to further strengthen existing policy on mitigation of and adaptation to climate change. One of the policy objectives established in the draft is that local authorities should be seeking a significant proportion of a development's energy demands to be met from renewable and low carbon energy sources. The draft supplement also encourages further use of decentralised energy supply from low carbon and renewable energy sources.

Fiscal incentives

Climate Change Levy

The Climate Change Levy came into effect on 1 April 2001 and applies to energy used in the non-domestic sector (industry, commerce and the public sector). The aim of the levy is to encourage these sectors to improve energy efficiency and reduce emissions of greenhouse gases. The levy is administered by HM Revenue and Customs and further information can be obtained from its website http://www.hmrc.gov.uk/.

Energy generated from CHP systems meeting certain quality standards is exempt from the Climate Change Levy.

Enhanced Capital Allowances

Enhanced Capital Allowances (ECAs)[15] are just one example of the use by government of fiscal incentives to achieve environmental objectives. Capital allowances are a mechanism whereby businesses can reduce their tax liability by the value of certain purchases and investments. Allowances on plant and machinery are generally given at 25 per cent a year on a reducing balance basis. However, with ECAs, businesses can write off 100 per cent of the cost of qualifying equipment against their taxable profits within the first year of investment.

Energy efficient/low carbon and water efficient technologies are currently eligible for ECAs once they have been added to approved 'Technology Lists' administered by the DEFRA and HM Revenue and Customs. The Energy Technology List[16] includes products related to:

- Air to air energy recovery
- Automatic monitoring and targeting
- Boilers
- CHP
- Compact heat exchangers
- Compressed air equipment
- HVAC zone controls
- Heat pumps for space heating
- Lighting
- Motors
- Pipework insulation
- Refrigeration equipment
- Solar thermal systems
- Thermal screens
- Warm air and radiant heaters
- Variable speed drives

The technologies that currently appear on the Water Technology List[17] are as follows:

- Water meters
- Flow controllers
- Leakage detection equipment
- Low flush toilets
- Efficient taps

Engineers can use the energy and water technology lists as a checklist of products that currently meet specified performance requirements relating to energy (and related carbon emissions) and water use and efficiency.

Corporate Social Responsibility

The UK Government sees Corporate Social Responsibility (CSR) – now sometimes referred to under the heading Corporate Responsibility – as business' contribution to sustainable development, i.e. how a business takes account of its economic, social and environmental impacts in the way it operates.[18] The view is that CSR is relevant to all companies, large and small.

Research suggests that benefits of CSR to business include enhanced reputation, attracting investment, and customer loyalty. It can be addressed through a diverse range of measures, from addressing the company's direct environmental impacts – like energy use/carbon emissions and water use – and developing staff more effectively, through to getting involved with local communities and charitable organisations.

One of the implications of CSR is that those companies taking a proactive approach to their corporate governance may increasingly seek to occupy premises that have been designed to reduce carbon emissions and to address other impacts on the environment.

Environmental assessment methods

A range of assessment methods have been established to help measure the environmental performance of buildings and to help establish targets of performance in development briefs and through planning. Some of those most widely used in the UK are discussed below.

Building Research Establishment Environmental Assessment Method

BREEAM (Building Research Establishment Environmental Assessment Method) was launched in 1990 and provides a standard system of assessing the overall environmental impacts of buildings and establishing targets of performance. It is increasingly being used to establish targets both within planning policy and within the development briefs prepared by RDAs and other government clients.

Carried out at the design stage, the scheme sets out a series of environmental best practice criteria against which developments can be compared. These are grouped into the following categories:

- Energy
- Transport
- Pollution
- Materials
- Water
- Land use and ecology
- Health and well-being
- Management

Where the criteria have been met, 'credits' are awarded. Depending on the number of credits obtained and their relative weighting a rating of Pass, Good, Very Good or Excellent is awarded.

There are currently BREEAM schemes for the following building types:

- New industrial and warehouse units
- New and existing offices
- Retail units

The Building Research Establishment (BRE) will also prepare Bespoke BREEAM assessments to deal with specific developments. Assessments are carried out by assessors licensed by BRE. Post-construction reviews are available to ensure that design intent has been followed through into construction.

A large number of the credits relate to issues that are directly within the influence of building services engineers and which can be addressed as part of the design of heating and cooling systems. These range for example from NO_x emissions from boilers to the global warming potential (GWP) of the refrigerants used in cooling systems.

The BREEAM schemes are regularly updated and the latest information can be obtained from BRE's website at: http://www.breeam.org/.

EcoHomes

EcoHomes is effectively the BREEAM scheme for housing and it covers a similar set of issues to those listed above. From April 2008 its application to new housing in England will be replaced by the CfSH. It will continue to be used for the refurbishment of housing and conversion of existing buildings into housing and for new homes in Scotland.

Further details can be obtained on EcoHomes at: http://www.bre.co.uk/service.jsp?id=397.

Table 2.2 CSFH assessment categories and minimum standards

Category	*Application of minimum standards*
• Energy/CO_2 • Water	Minimum standards at each level of the Code
• Materials • Surface water run-off • Waste	Minimum standards at Code entry level
• Pollution • Health and well-being • Management • Ecology	No minimum standards

The Code for Sustainable Homes

The CfSH was published in December 2006 and sets 'a new national standard for sustainable design and construction of new homes'. Unlike the EcoHomes standard, the CfSH is directly backed by the Government. Those building houses with public money will 'be expected to comply' with prescribed levels of the code, and while adoption by private house-builders is currently voluntary, its use will be strongly encouraged.

Housing schemes will be assessed post-completion by an independent assessor and labelled 1-star to 6-star depending on the CfSH 'level' achieved. The CfSH sets minimum standards 'above the level of mandatory building regulations' in five out of nine categories, as shown in Table 2.2.

The CfSH was launched as part of a package of measures intended to contribute towards a government policy that all new housing development in the UK should be zero carbon by 2016. Full technical guidance on how to comply with the Code was published in April 2007 and is available on the Planning Portal website.[19]

A key difference between EcoHomes and the CFSH is that the CFSH will apply to each individual home rather than a development as a whole.

Sustainable strategies for heating and air-conditioning

The role of the engineer

Those involved in the design of heating and cooling systems have a key role to play in addressing the issues set out above. Sustainable development requires a much more holistic and integrated approach to design. It is no longer acceptable for a building to be designed as a glass box and for the engineer's role to be to provide sufficient cooling and heating equipment to make it habitable. Building services engineers need to be involved much earlier in the design process for example in helping to optimise the façade design to provide an appropriate balance of daylight and solar gain while avoiding excessive cooling demands and heat loss in winter.

The increased focus on energy and renewables in the planning system requires that a robust energy strategy is developed at the earliest stages of design and before a planning application is made. This will require assessment of energy demands and increased feasibility work to assess energy efficiency improvement and low carbon and renewable energy supply options. This will for example include the need to undertake building modelling to predict energy demands and assess potential carbon savings as well as outline feasibility work to identify space and access requirements for items of plant such as biomass fuel stores or thermal stores associated with CHP systems.

While payback has often been a key determinate in assessing energy efficiency options, these measures will increasingly be driven through legislation with increasing focus from developers on the cost per tonne of carbon saved.

Energy is not the only issue to be considered and designs will need to address the broad range of sustainability issues including reducing water demand and reducing waste. Engineers need to play a proactive role in working with client's and project teams in identifying opportunities for sustainable design and construction. This may require challenging the client's brief and ensuring the required skills and resources are available to conduct the necessary feasibility and design work.

Summary of opportunities

Key opportunities to be considered in designing heating and ventilation systems sustainably will include the following:

Reducing energy demand and CO_2 emissions

- Reducing energy demand for all energy uses to a minimum.
- Supplying remaining demands with efficient equipment and appliances.
- Utilising renewable and low carbon energy supplies to meet demand.

Adapting to climate change

- Reducing unnecessary heat gains.
- Ensuring building designs take into account future changes to our climate and remain fit for purpose.
- Ensuring building designs do not exacerbate the expected impacts of climate change by for example increasing risks of flooding or unduly adding to the heat island effect.

Reducing water demand

- Reducing water demand for all end uses within the building.
- Meeting this demand with efficient systems and appliances.
- Utilising recycled grey and rainwater where appropriate to reduce mains water demand.

Selecting sustainable materials

- Using life cycle assessment systems to select materials with reduced impacts over their life cycle or which avoid known impacts on the environment.

Reducing waste

- Reducing waste through appropriate design and construction management.
- Reusing existing components and designing for future reuse.
- Utilising products and components with a high recycled materials content.

Reducing pollution emissions

- Reducing risk of pollution to land air and water.
- Limiting nuisance from noise to building occupants and neighbours.

Enhancing ecology

- Identifying opportunities for promoting and enhancing biodiversity.

This list is not exhaustive but identifies some of the key issues relevant to heating and air-conditioning.

Possible approaches to dealing with these issues are described briefly below and then expanded upon in the following chapters.

Energy use and carbon emissions

Carbon emission conversion factors

In determining strategies for reducing carbon emissions from buildings it is important to understand that different fuel sources will result in different carbon emissions.

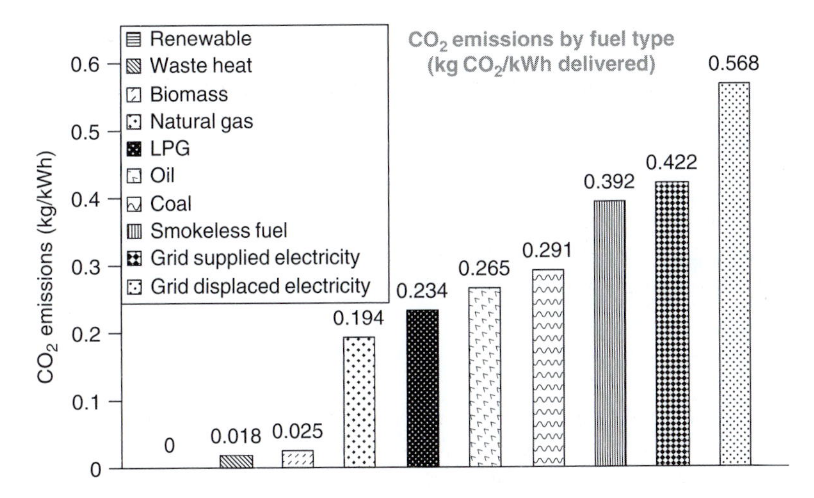

Figure 2.3 CO_2 emission conversion factors by fuel type (see colour plate section at the end of the book)

Figure 2.3 illustrates the typical carbon emissions per unit of energy delivered to a building for a range of fuel sources. These are based on conversion factors presented and used in Part L2A of Building Regulations 2006. It can be seen that emissions associated with grid supplied electricity are more than twice those per unit of gas delivered. A heating system based on direct electric heating might typically result in twice the carbon emissions of a system based on gas, if it is supplied with electricity from the grid. This is due to the large waste of heat that occurs at central power stations when converting fossil fuels to electricity and transmission losses that occur in supplying the power generated to buildings. The figure shown for electricity is based on the average generation mix in the UK which includes a significant proportion of coal and oil generation. This mix may be different in other countries, for example in Norway where the majority of electricity is generated from hydro electricity, the emissions for electricity will be close to zero.

Renewable fuel sources will always result in greatly reduced carbon emissions. It can be seen that a heating system served by biomass in the form of wood chips or wood pellets will have close to zero CO_2 emissions as wood is regarded as a renewable resource, the small emissions shown result from processing and transporting the fuel to site.

The emission figures can be used to convert predicted energy demands for each end use of energy into CO_2 emissions. It can be seen that grid displaced electricity has a higher emission factor than grid supplied electricity. This grid displaced electricity factor is used in determining the carbon savings that result from on-site generation of electricity from renewables such as wind and PV or from CHP schemes. A higher factor is used in these cases as it is generally regarded that local generation will displace the need for peak generation to be deployed at central power stations. Peak generation from the grid typically utilises coal-, gas- and oil-fired plants which have emission rates higher than those for the average generation mix.

CO_2 emissions can be expressed either as carbon emissions (kg C) or CO_2 emissions (kg CO_2). The molecular weight of carbon is 12 and the molecular weight of CO_2 is 44 (12 + 16 + 16). To convert from a kg C to kg CO_2 multiply by 12/44 or 0.2727. To convert from kg CO_2 to kg C multiply by 44/12 or 3.67.

Carbon emissions from homes

Figure 2.4 illustrates the typical CO_2 emissions for a three bed semi-detached house built to comply with the 2006 Building Regulation standards and heated using an efficient condensing gas boiler. It can be seen that energy used for lights and appliances is responsible for around 45 per cent of the CO_2 emissions (the proportions will vary slightly depending on housing mix and density, but lights and appliances are typically the largest

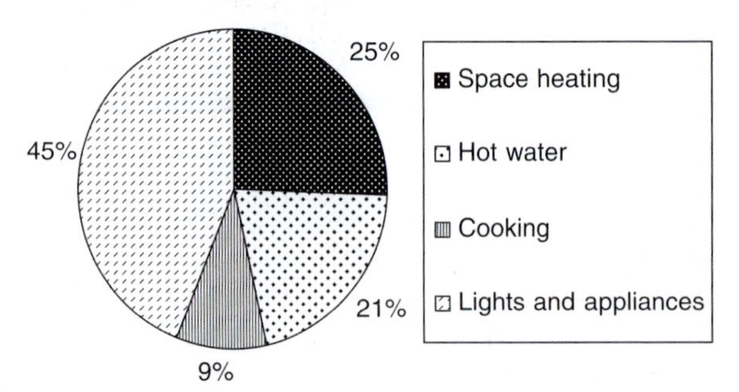

Figure 2.4 Figures based on an example of a typical three bed semi-detached house (see colour plate section at the end of the book)

sources of emissions). CO_2 emissions from hot water heating are very close to those for space heating and in terrace homes or apartments may be greater.

Carbon emissions from offices

The energy demand and CO_2 emissions from non-residential developments vary significantly between different building types – office, retail, industrial, etc. – because of their specific requirements. The Carbon Trust produce information on typical energy use and carbon emissions for common categories of buildings.

Energy Consumption Guide 19 (ECON 19)[20] provides benchmark figures for typical and best practice office buildings. These are shown for both naturally ventilated and air-conditioned buildings. An extract from ECON 19 is shown in Fig. 2.5 and illustrates that a typical purpose designed naturally ventilated office building (type 2) will have carbon emissions that are half those of a typical air-conditioned office building (type 3). This is largely due to the additional energy requirement for fans and pumps and cooling associated with the air-conditioning, but also results from the higher-energy use for lighting in deeper plan air-conditioned buildings.

This diagram illustrates the need to reduce cooling demands to a minimum and where possible to meet them using natural ventilation or mixed mode approaches. This will not always be possible for example where external sources of noise or pollution, or high occupancy densities and internal gains require the need for mechanical cooling.

Reducing energy demands

A systematic approach should be taken to reducing each end use of energy within the building.

Heating demands can be reduced by:

- Using built forms that limit surface area and orientating buildings to utilise solar gain.
- Employing improved standards of insulation (potentially to the point where a conventional heating system can be eliminated).
- Ensuring an airtight fabric and employing heat recovery systems to reduce ventilation loses.
- Avoiding excessive areas of glazing, and using high-performance windows with reduced U values.
- Zoning heating controls and set-point temperatures to only heat rooms to a temperature appropriate for their use.

Cooling demands can be reduced by:

- Specifying low-energy lighting with effective controls that prevent unnecessary use.
- Utilising low-energy appliances controlled to switch off when not in use, i.e. flat screen monitors with power saving switched on.

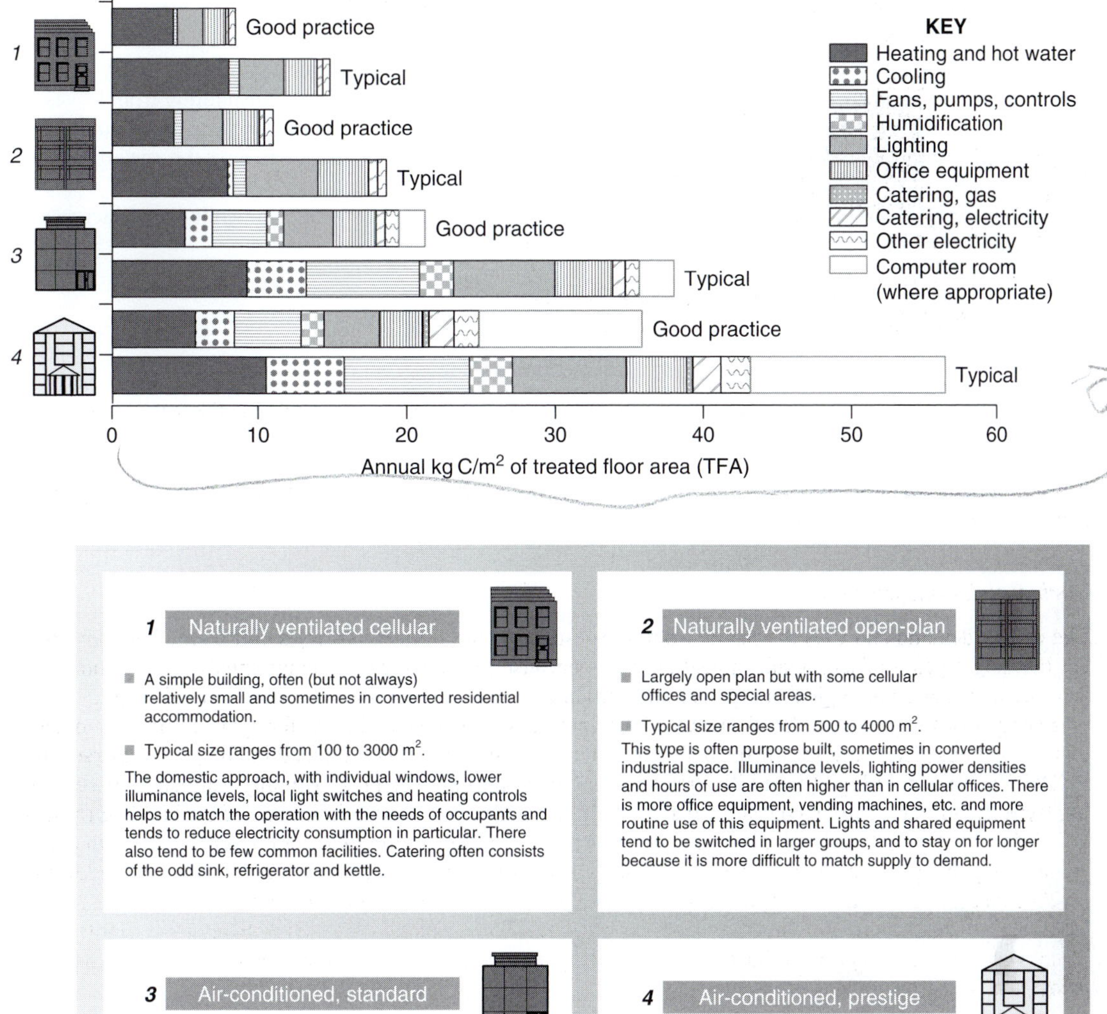

1 Naturally ventilated cellular

- A simple building, often (but not always) relatively small and sometimes in converted residential accommodation.
- Typical size ranges from 100 to 3000 m^2.

The domestic approach, with individual windows, lower illuminance levels, local light switches and heating controls helps to match the operation with the needs of occupants and tends to reduce electricity consumption in particular. There also tend to be few common facilities. Catering often consists of the odd sink, refrigerator and kettle.

2 Naturally ventilated open-plan

- Largely open plan but with some cellular offices and special areas.
- Typical size ranges from 500 to 4000 m^2.

This type is often purpose built, sometimes in converted industrial space. Illuminance levels, lighting power densities and hours of use are often higher than in cellular offices. There is more office equipment, vending machines, etc. and more routine use of this equipment. Lights and shared equipment tend to be switched in larger groups, and to stay on for longer because it is more difficult to match supply to demand.

3 Air-conditioned, standard

- Largely purpose built and often speculatively developed.
- Typical size ranges from 2000 to 8000 m^2.

This type is similar in occupancy and planning to building type 2, but usually with a deeper floor plan, and tinted or shaded windows which reduce daylight still further. These building can often be more intensively used. The benchmarks are based on variable air volume (VAV) air-conditioning with air-cooled water chillers; other systems often have similar overall consumption but a different composition of end use. (See good Practice Guide (GPG) 71 'selecting air conditioning systems. A guide for building clients and their advisors'.)

4 Air-conditioned, prestige

- A national or regional head, office, or technical or administrative centre.
- Typical size ranges form 4000 to 20 000 m^2.

This type is purpose built or refurbished to high standards. Plant running hours are often longer to suit the diverse occupancy. These buildings include catering kitchens (serving hot lunches for about half the staff); air-conditioned rooms for mainframe computers and communications equipment; and sometimes extensive storage, parking and leisure facilities. These facilities may be found in offices of other types, and, if so, should be allowed for.

Figure 2.5 Breakdown in annual carbon emissions for different office types (*Source*: ECON 19[20]) (see colour plate section at the end of the book)

- Avoiding excessive areas of glazing and providing external shading or fixed overhangs to control solar gains.
- Selecting glazing with high light transmission properties but low solar heat gain factor.
- Providing vegetation to provide shading and increase the local cooling effect of transpiration from leaves.
- Zoning building uses to allow high heat gain uses to be served separately from other parts of the building.
- Providing thermal mass to reduce peak radiant surface temperatures and to reduce the mean resultant temperature.
- Avoiding tight control of temperatures within a narrow range and allowing temperatures to increase in summer.

Efficient system designs

The detailed system design will be important with attention paid to:

- Avoiding oversized plant
- Reducing pressure drops and resulting fan and pump power
- Selecting efficient fans and pumps
- Making use of free heating and cooling sources
- Recovering and using waste sources of heat and cooling
- Selecting efficient heating and cooling plant

Detailed guidance on the energy efficient design of service systems is provided in CIBSE Guide F and in the following chapters.

Low and zero carbon energy sources

Having reduced heating and cooling demands to a minimum options should be explored for meeting these demands using renewable (see Chapter 14) or low carbon energy sources such as CHP, CCHP (see Chapter 15).

Options to consider may include:

- Ground source heating and cooling
- Biomass heating (or cooling via absorption chillers)
- Solar water heating
- Local sources of waste heat
- PV and wind for providing power for ventilation and cooling systems

Adapting to climate change

The predicted impacts of climate change include:

- General increases in temperatures, including milder winters and rising summer temperatures
- Enhanced urban heat island effect
- Wetter winters and more intense rainfall
- Drier summers
- Higher daily mean winter wind speeds

Table 2.3 is taken from CIBSE Guide L on Sustainability[21] and summarises some of the issues that will need to be considered in future building design.

Thermal discomfort is likely to become a major issue in the future particularly in built up urban areas which typically experience elevated external temperatures due to the urban heat island effect. There will be more situations where either building occupants will have to accept and adapt to higher internal temperatures, or where some form of mechanical cooling will have to be provided.

Table 2.3 Impact of climate change on future building design

Key predicted climate change effect	*Description*	*Implications for buildings*	*Example measures*
General increase in temperatures	Current modal temperature in London is 10°C: predicted to be 12°C by 2080. High summer temperatures will become frequent and very cold winters will become rare.	Increased outgassing of pollutants from structure and furnishings affecting indoor air quality. Decaying waste more likely to smell and may cause problems with infestation.	Avoid high internal temperatures and VOCs in finishes, construction materials, carpets and furnishings; select labelled low emission products. Allocate adequate space and ensure secure, sealed storage of segregated wastes in regularly cleaned area designed to avoid overheating.
Milder winters	Mean winter temperatures predicted to increase by 2.5°C in the South east and 2°C in the north. Decrease in heating degree days (c.f. 1980s) of 35–40% for London, 30% for Edinburgh by 2080.	Reduced energy use in winter.	Consideration should be given to the potential to reduce system capacity and thereby reduce energy use during winter months, see Section 5.1.3.
Rising summer temperatures	Increase in mean summer temperature.[27] (a) South-east England: 2.5–3°C by 2050; 4.5–5°C by 2080 (b) North England and Scotland: 1.5–2°C by 2050; 2.5–3.5°C by 2080. Occurrence of temperature >28°C increases from 1 to 2 days/year (1989) to 20 days/year by 2080 in London. Increased occurrence of hot spells (number of days when previous 3–5 days have >3 hours at 25°C). Increase in cooling degree-days (c.f. 1989): e.g. +150−200 for London. +20−25 for Edinburgh by 2080.	Higher summertime temperatures will increase overheating risk in buildings. Due to rising external temperatures the traditional mechanism for cooling buildings through ventilation with external air cannot be relied upon. Careful design required to reduce dependence on mechanical cooling and to maintain indoor comfort.	Incorporation of intelligent ventilation systems, preferably automated are set up to limit ventilation during warmer parts of the day and recharging ‘coolth’ reservoirs in high mass buildings when external air temperatures permit.
Enhanced urban heat island effect	In central London the urban heat island effect can currently lead to summer night time temperatures 5–6°C warmer than temperatures in rural areas outside London. This effect is expected to intensify due to climate change, leading to a greater temperature difference between the heat island and surrounding rural areas, and more hours of night time heat island effect per year.[28]	The increase in night time urban temperature due to the urban heat island effect reduces the ability of buildings within the urban heat island to use night time cooling as a strategy. Increased temperatures, particularly at night for any building within urban conurbations, reduce the ability to dissipate heat at night making night time ‘free cooling’ less practicable in the future.	Planting of trees, vegetation and green space for shade and natural cooling through evapotranspiration (estimated to result in 1–5°C reduction in peak summer temperatures.[29] Green roofs (may reduce surface temperature by 20–40°C compared to a conventional dark flat roof).
Wetter winters. More intense rainfall	Winters will become wetter by up to 10% across the country by the 2050s and up to 20–30% across UK by the 2080s. In combination with higher wind speeds, occurrence of driving rain will increase in winter months.	Inability of drainage system to cope with storm surges, damage to some building materials. Increased risk of flash flooding.	Undertake Flood Risk Assessment; design for flood resilience and to reduce flood risk. Use SUDS techniques. Locate ‘resistant’ uses (e.g. car parking) in high-risk areas and raise ground floors. Use of building materials which are resilient to driving rain.

(Continued)

Table 2.3 (Continued)

Key predicted climate change effect	*Description*	*Implications for buildings*	*Example measures*
Drier summers	Decrease in summer rainfall of 40% across most of the UK by the 2080s. Decrease in soil moisture content especially in summer: 30% decrease predicted for most of England and 10–20% for the rest of England, Wales and most of Scotland.	Increased pressure on water resources and increased occurrence of hosepipe bans. Increased pressure for water storage capacity. Possible disruption due to increased subsidence due to drier clay soils.	Apply water efficiency principles in new and existing buildings. Plan for building operation in drought conditions. Avoid siting 'water-hungry' developments in areas prone to drought.
Higher daily mean winter wind speeds	Possible increase of 7% by the 2080s with the greatest increase in the south-east.	Increased risk of wind damage. Higher infiltration and winter heat loss. Pylons carrying electricity and telecommunications may be vulnerable to higher wind speeds. Increased level of power shortages and outages.	Strengthening of tall buildings, increased airtightness and incorporation of cladding materials able to cope with higher wind speeds. Incorporation of local/on-site generation of power and renewables to reduce dependence on the grid.

Source
CIBSE Guide L.[21]

Conventional mechanical cooling results in: additional carbon emissions from energy use, further contributing to climate change and heat rejected from cooling systems is one of the contributors to urban heat islands further exacerbating the problem.

Approaches are needed that utilise renewable or low carbon sources of cooling.

Reducing water demand

In the UK as a whole, water consumption has risen by 70 per cent over the last 30 years. In order to meet increasing demand, new sources of water supply have to be created, but this is costly and potentially damaging to the environment. An alternative and preferable approach is to reduce the demand for water. In offices, an average 43 per cent of water is used for WC flushing, 20 per cent for urinal flushing, 27 per cent for washing and 9 per cent in canteens.[22] Industrial buildings have a similar profile, with additional water used for manufacturing processes. In addition many buildings include external planting which requires irrigation.

A range of simple water demand reduction measures are available for reducing water demand in buildings, some of which will also result in reduced energy demand for water heating, these typically include:

- Low flush WCs
- Low flow showers
- Spray taps on basins
- Waterless urinals
- Low water use appliances such as washing machines and dishwashers.

Water use can also be reduced through the use of leak detection systems and by recycling rain or grey water for reuse in buildings.

There are examples of water conservation strategies being integrated with the cooling systems in buildings. One example is water being extracted from boreholes to provide a free source of cooling for air-conditioning systems and then being utilised for flushing WCs, another is the collection and reuse of recycled rainwater for providing the make-up water supply to wet cooling towers. In applications such as these there are detailed water quality issues that need to be considered as part of the design.

There will be increasing pressure to identify new approaches to avoiding unnecessary water use in buildings.

Pollution emissions

Use of refrigerants in air-conditioning

Under the Kyoto agreement the UK Government is required to reduce emissions of six greenhouse gases including CO_2, methane, nitrous oxide, HFCs, PFCs and sulphur hexafluoride (SF_6) by 12.5 per cent by 2008 to 2012. The large emission of CO_2 from energy use in buildings has already been discussed, but buildings also use HFCs as blowing agents in insulation materials, as gases in fire suppression systems and as refrigerants. SF_6 is used in high-voltage switch gear, circuit breakers and transformers. It is likely that in the coming years greater controls will be introduced to limit emissions of these gases including the F-Gas Regulations which will control the use and handling of fluorinated gases including HFCs, PFCs and SF_6.

HFCs have been widely used as replacements to CFCs and HCFCs for use as refrigerants in air-conditioning systems, which were phased out due to their impact on the ozone layer. To further reduce the GWP and maintain zero ozone depletion potential (ODP), alternative refrigerants are available as HFC replacements as discussed further in Chapter 24.

Refrigerant leak detection should be considered for air-conditioning systems to help reduce refrigerant emissions contributing to ozone depletion or global warming. Emissions can also be reduced by designing systems to reduce the refrigerant charge.

NO_x emissions

NO_x is a naturally occurring gas, produced by biological and atmospheric reactions. It is also released during the combustion process, and the burning of fuels such as coal, oil and natural gas in boilers as described in Chapter 17. While predominantly caused by sulphur dioxide (SO_2) emissions, acid rain results from the conversion in the atmosphere of NO_x to nitric acid (HNO_3), which dissolves in rain and damages ecosystems and buildings. NO_x is also involved in tropospheric (i.e. low-level) ozone chemistry, which results in poor air quality and, in combination with sunlight and hot weather, can result in the formation of photochemical smog. NO_x, ozone and other by-products of the process such as peroxyacetylnitrate (PAN) are known to cause eye irritation, chest pains, coughs, headaches and increased frequency of asthmatic attacks.[23]

While motor vehicles are the main source of NO_x emissions, NO_x emissions can be reduced by specifying low NO_x boilers. The location of air intakes into buildings can also be designed to reduce the occupants' exposure to outside sources of air pollution.

Leakage from fuel tanks

Leakage from fuel tanks has been a significant source of ground contamination, potentially polluting water courses and in the long term requiring significant cost for the remediation of any land affected.

The risk of ground contamination can be reduced by the appropriate design of fuel storage vessels. Further guidance on this issue is provided in Chapter 16.

Selecting sustainable materials

There is often a choice of materials and products available for a given function, for example water can be carried in copper, steel, clay, or a variety of plastic pipes. Each of these will have a range of different impacts on the environment over its life cycle. They will also have different financial costs. Deciding which is the 'best' option for a given use is extremely complex as it involves balancing many different issues, for which there is often no standard means of comparison. For example one product may require less energy to produce and fit on site, but it may produce a potentially carcinogenic by-product as part of its manufacture.

While it can be difficult to reach an overall view on a product's impacts or benefits the issues to be considered will generally include the following:

- *Embodied energy*: This is the primary energy needed to win the raw materials, process them into the product, deliver it to site and install it in the building.

- *Embodied emissions*: These are the pollution emissions that occur in manufacture and can include CO_2, NO_x, SO_x and volatile organic compounds.
- *Toxicity*: This considers potentially toxic compounds that may be released into the environment as part of the manufacturing process, in use or on disposal (e.g. by combustion).
- *Resources*: This looks at the overall use of resources in manufacture and the use of any potentially scarce resources.
- *Waste*: This is the quantity of waste material that is created and which must be disposed of.
- *Recycling*: This looks at the potential to reuse, or recycle the product at the end of its life or whether the product is making use of recycled or reclaimed materials as part of its manufacture.

While guidance has been produced to allow these issues to be compared and balanced for architectural components[24–26] there is currently little definitive guidance available for services components.

In the absence of such guidance materials choice often comes down to decisions being made on single issues. Examples of areas where alternative products can be selected include:

- Selection of insulations materials with zero ODP and GWPs of less than 5.
- Specification of low smoke zero halogen (LSOH) cabling where PVC sheathing is replaced with either polypropylene or polyethylene to reduce the release of smoke and hydrogen chloride in the event of a fire.

Reducing waste

In 1998 a task force led by Sir John Egan published their report Rethinking Construction. It was commissioned by the Deputy Prime Minister to assess the efficiency of the UK construction industry. One of the key findings of the report was that the construction industry should learn from other manufacturing industries and look to utilise Modern Methods of construction that would reduce costs, time, accidents and defects. Since the report was published there has been growing interest in the prefabrication of building components off site in factory conditions, this reduces assembly times on site and potentially reduces the amount of waste in the construction process as waste produced can be recycled or eliminated as part of production process.

There are a range of building services components which are already being prefabricated off site including:

- Whole bathroom pods
- Prefabricated plant rooms
- Prefabricated wiring trays
- Prefabricated pipework.

These can be utilised as part of an approach to reduce waste in construction.

Other options for reducing waste include setting up a site Waste Management Plan to identify the likely waste to arise on site and to look for opportunities to segregate and either reuse or recycle it. This can include returning off cuts to suppliers for reprocessing.

Waste can be reduced by basing design dimensions around standard product sizes to avoid cutting and trimming on site, this may also have a labour saving potential.

Waste can also be reduced as part of any demolition works by ensuring that existing services components are salvaged for reuse or recycling.

Notes

1. CO_2e is used to measure the impact of a broad range of greenhouse gases. The emissions of a greenhouse gas, by weight, are multiplied by its 'GWP' to determine the equivalent emission of CO_2 that would have the same impact on global warming as the gas being considered.
2. Council of the European Union, December 2004.
3. Energy – the changing climate, 22nd Report of the Royal Commission on Environmental Pollution, June 2000.

4. Building on Progress, Energy and Environment, Cabinet Office, 5 June 2007, www.cabinetoffice.gov.uk/policy_review/.
5. Quality of Life Counts, DETR December 1999.
6. Quality of Life Counts 2004 Update, DEFRA March 2004.
7. Review of Sustainable Construction 2006, DTI October 2006.
8. The published text of the Directive can be found at http://europa.eu.int/eur-lex/pri/en/oj/dat/2003/l_001/l_00120030104en00650071.pdf.
9. http://www.dti.gov/uk/innovation/sustainability/fgases/page28889.html.
10. Statutory Instrument 2006 No. 3289. The Waste Electrical and Electronic Equipment Regulations 2006, http://www.opsi.gov.uk/si/si2006/20063289.htm.
11. DTI, http://www.dti.gov.uk. WEEE Regulations page, http://www.dti.gov.uk/innovation/sustainability/weee/page30269.html.
12. WEEE Regulations – Government Guidance Notes Guidance February 2007URN 07/781, http://www.dti.gov.uk/files/file38209.pdf.
13. Building a Greener Future towards Zero Carbon Development, Consultation, CLG December 2006.
14. Planning Policy Statement: Planning and Climate Change, Consultation, CLG December 2006.
15. ECAs, http://www.eca.gov.uk/.
16. Energy Technology List, http://www.eca.gov.uk/etl.
17. Water Technology List, http://www.eca-water.gov.uk/.
18. The UK Government Gateway to Corporate Social Responsibility, http://www.csr.gov.uk.
19. http://www.planningportal.gov.uk/england/professionals/en/1115314116927.html.
20. Energy Consumption Guide 19, Energy Use in Offices, December 2000. The Carbon Trust, http://www.carbontrust.co.uk.
21. Sustainability, CIBSE Guide L, CIBSE, May 2007, ISBN: 978-1-903287-82-8.
22. BREEAM 98 – An Environmental Assessment Method for Office Buildings, BR 350, BRE, 1998.
23. *Chemistry of Atmospheres*, Wayne, Second Edition, Clarendon Press, 1991.
24. The Green Guide to Specification. An Evaluation Profiling System for Building Materials and Components, BRE, September 1998. Available from CRC Ltd Tel: 020 7505 6622.
25. *Sustainable Building Handbook*, Boonstra et al., James and James, 1996, ISBN: 1-873936-38-9.
26. *Green Building Handbook*, Woolley et al., Volumes 1 and 2, E&FN Spon, ISBN: 0 419 226907, 0419 253807.
27. Climate change scenarios for the United Kingdom: The UKCIP02 Scientific Report (Norwich: Tyndall Centre for Climate Change Research, University of East Anglia) (2002) (available from http://www.ukcip.org.uk/scenarios/ukcip02/documentation/ukcip02_scientific_report.asp) (accessed April 2007).
28. Wilby, R.L. Past and projected trends in London's urban heat island. *Weather*, 2003, 58(7), 251–260.
29. London's urban heat island: A summary of decision makers (London: Greater London Authority) (2006) (available from http://www.london.gov.uk/mayor/environment/climate-change/uhi.jsp) (accessed April 2007).

Chapter 3

New legislation

Introduction

Conforming to relevant legislation is a very important activity that must underpin the design, construction and operation of buildings. Understanding the applicable legislation should therefore be an essential part of the building services engineer's training and continual professional development. It is also important to understand that the applicable legislation varies from country to country, and even in the UK, different standards and different compliance procedures exist in the different administrative areas (England and Wales, Scotland and Northern Ireland). This chapter summarises the main legislative requirements that should be considered, and provides pointers to information that the building professional will find useful in meeting the legal requirements. The chapter concentrates on recent developments in legislation that will have a major impact on the work of the building services engineer.

The Building Regulations

The main legal requirements for England and Wales are set out in the Building Act 1984, which enables regulations to be made for the health, safety and welfare of people in and around buildings, and for the conservation of fuel and power.[1] The Regulations themselves contain definitions and set out the administrative procedures that must be followed. Generally, the requirements are set out in a schedule, which summarises the performance goals that should be achieved in the building as-constructed. These performance goals are usually expressed in functional terms such as '*reasonable provision shall be made*'. Supporting the regulations are Approved Documents, which contain detailed technical guidance on what in normal circumstances would be regarded as '*reasonable provision*'. In the particular case of the conservation of fuel and power, the Regulations contain additional requirements (Regulations 4A and 17) over and above those set out in Schedule 1. Some of these requirements are absolute, with no opportunity for making special cases – achieving the new build energy performance standard and the requirements for pressure testing and commissioning are examples.

In the context of the legal obligation to '*make reasonable provision*', this is most easily done by following the guidance in the Approved Documents. If the design team want to follow an innovative approach that is not covered by the Approved Document, they are at liberty to do so, provided they can satisfy the building control body that, in the particular case, they have made reasonable provision, thereby satisfying the legal requirement.

The schedule of requirements contained in the Building Regulations covers many aspects of the design, construction and use of buildings. The various Parts of the schedule are listed below, and it will be apparent that many of these are relevant to the building services engineer. In such cases, pointers are provided to the relevant chapters of this publication so that the reader can follow up the details of interest.

(a) Part A: Structure
(b) Part B: Fire safety
(c) Part C: Site preparation and resistance to contaminants and moisture

(d) Part D: Toxic substances
(e) Part E: Resistance to the passage of sound
(f) Part F: Ventilation – see Chapters 18, 21 and 22
(g) Part G: Hygiene
(h) Part H: Drainage and waste disposal
(i) Part J: Heat producing appliances – see Chapters 13, 16 and 17
(j) Part K: Protection from falling collision and impact
(k) Part L: Conservation of fuel and power – nearly all chapters are relevant
(l) Part M: Access to and use of buildings
(m) Part N: Glazing
(n) Part P: Electrical safety

It should be remembered that the Building Regulations and the associated Approved Documents set out minimum performance standards. In some aspects of performance, notably energy, there is considerable pressure on clients and their agents[2] to go beyond these minimum performance standards. Such issues should be reviewed at the earliest stages of developing the brief – it is much easier and more cost effective to design in improved standards from the outset than trying to add them at a later date.

Part L conservation of fuel and power

Although all parts of the regulations are of equal importance in terms of achieving a safe and efficient building, the conservation of fuel and power has become of particular significance as a result of climate change. This is resulting in a period when energy performance standards of buildings are changing very quickly, requiring that designers and builders adopt new techniques and new approaches.

Part L was amended in 2002 and again in 2006. In a recent consultation paper,[3] the government proposes further amendments in 2010, 2013 and 2016. Figure 3.1 shows how the government propose to improve the standards for dwellings relative to the performance level established in 2002. The exact scale and timing of the improvements will be decided as a result of the responses to the consultation, but the proposals show the level of ambition. Within a decade, the government aims for new dwellings to be zero net carbon over the year, i.e. generating sufficient energy from renewable sources to meet the needs of the heating, hot water, ventilation and lighting and the appliances in the dwelling (fridges, washing machines, etc.). That is why, in 2016, the energy demand goes negative, as the building integrated systems meet not only the heating and lighting needs, but also the appliance energy demands.

The 2006 Part L was a radical departure from previous editions in that it changed the basis on which compliance with the performance standards was demonstrated. The 2006 amendment of Part L was the opportunity to implement aspects of the European Directive on the Energy Performance of Buildings.[4] A key article of

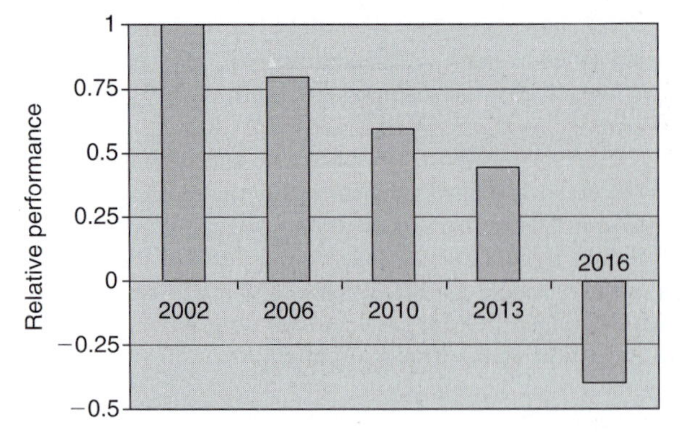

Figure 3.1 Relative levels of energy efficiency as required by recent and prospective future implementations of Part L (see colour plate section at the end of the book)

this Directive required that energy performance standards be set on the basis of defined energy targets determined using a whole building energy calculation model (see Chapter 6 for a detailed discussion of energy calculation methods). This means that compliance now has to be demonstrated by showing that the building as constructed will consume no more energy than that allowed for in the energy target. For national policy reasons, the UK has expressed the energy target in terms of CO_2 emissions.

This approach has many advantages, in that it allows maximum design flexibility in achieving the target. Given the aggressive programme of improvements that are proposed, it is important that the compliance process does not stifle innovation. Some may prefer to concentrate on improving the envelope performance by including more insulation in the walls, floor and roof, and specifying better performance windows. Others may wish to concentrate on the building services plant (e.g. more efficient boilers and chillers, improved lighting or more energy efficient mechanical ventilation), or on renewable energy systems.

What constitutes reasonable target emissions will depend on the size, shape and use of the building. Since the size and shape of the building is often dictated by site constraints and the patterns of occupation will vary according to the intended use, a bespoke target is calculated for each building. The target emission rate is therefore set using the same calculation procedure as is used to estimate the performance of the actual building. A notional building of the same size and shape as the actual building is defined, but the envelope and HVAC and lighting systems are to specific elemental standards. The notional building is also operated according to the same mix of activities as the actual building. The actual design then has to beat the performance of the notional building by a defined improvement factor.

The improvement factor varies between dwellings and non-domestic buildings. In the latter case, the overall improvement factor is made up of two parts:

(1) An improvement in energy efficiency
(2) A contribution to reducing CO_2 emissions by the provision of low and zero carbon (LZC) technologies (renewable energy systems, Combined Heat and Power (CHP), heat pumps, etc.). It is not mandatory to provide any given level of LZC contribution, but the equivalent overall level of improvement must be achieved. This can be done by an increased level of energy efficiency with no LZC systems at all, or indeed with a greater contribution of LZC technology than the benchmark provision of 10 per cent.

The overall compliance process is set out in Fig. 3.2.

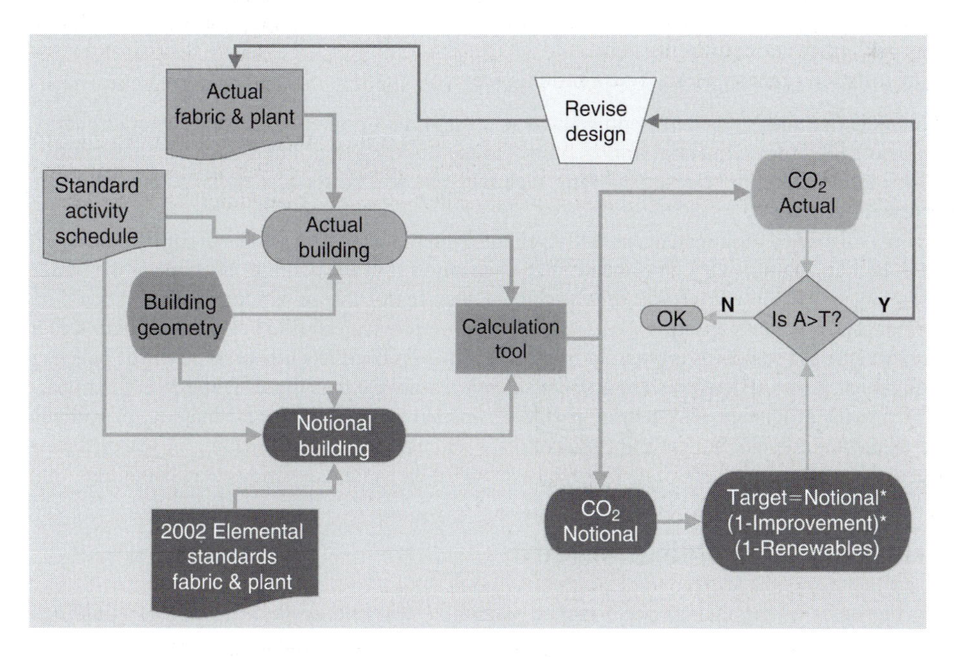

Figure 3.2 Process flow chart for demonstrating compliance with Part L (see colour plate section at the end of the book)

As well as meeting the carbon target, the building has to meet a number of other standards:

(a) Poorest acceptable standards for the various elements of the envelope and the HVAC and lighting systems. This is to make sure that the carbon target is not achieved through a relatively inefficient basic building that is compensated by a large renewable energy system. The reason for this is that renewable energy systems may be removed or fall into disrepair, such that the total energy demand from what is a poor passive building design then has to be met from conventional sources.
(b) Control of overheating: if any zone of the building does not have any mechanical cooling system, it should be designed such that the space will not overheat. This is to minimise the likelihood of a subsequent air-conditioning retrofit. The guidance in ADL2A suggests designers might wish to take into account climate change forecasts in these considerations.
(c) As well as assessing the basic specifications, compliance with the 2006 Part L requires a number of checks to be made on the quality of the completed building. These include:
    (i) An air pressure test; this is often a good proxy for many aspects of construction quality, and it is therefore required that a pressure test be carried out on all new non-domestic buildings and a sample on each new housing development.
    (ii) Confirmation that the building services have been properly commissioned, including air leakage testing of ductwork for ventilation systems with a capacity of greater than 1 m^3/s.
(d) Provision of information: this is intended to make sure that the occupier of the building has the necessary information so that the building can be operated in an efficient way through its life.

Existing buildings

Although a lot of attention is always given to the energy performance of new buildings, there is also increasing attention being given to improving the performance of the existing stock. The most important new element of the 2006 Part L is the requirement to make general improvements once building work of a certain type is proposed, always provided that the marginal costs of so doing are not excessive. The types of building work that trigger this requirement are:

(a) Constructing an extension.
(b) The initial provision of any fixed building service (e.g. the installation of air-conditioning in part of a building that was previously naturally ventilated).
(c) An increase in the installed capacity of any fixed building service.

The logic behind this is that these types of improvement to a building are likely to increase overall energy demand. By making a so-called ‘consequential improvement’ to the building (e.g. improving the solar control to the area where air-conditioning is to be installed), any such increase will be reduced as far as is economically and practically possible.

In addition, the energy efficiency requirements in the Building Regulations also set standards for the refurbishment or replacement of building components. This means that when any part of the building envelope or the building services systems is subject to substantial repair or replacement, then reasonable minimum standards should be achieved. What is reasonable is context specific – e.g. it is easier to improve an unfilled cavity wall than it is a solid wall, and so context specific standards are set out in the relevant Approved Document. In terms of building services plant and equipment, better efficiencies are now considered reasonable than those set out in 2002. In particular, attention is given to the efficiency of systems at part load, since in many buildings the majority of systems spend little time at design condition. For boilers and chillers, this is addressed through the setting of seasonal efficiency standards, where the seasonal efficiency is a weighted average of the system at various load conditions.

The energy performance of buildings directive

As well as defining the basis of the energy performance standards, the Energy Performance of Buildings Directive (EPBD)[4] also covers energy certification and plant inspection requirements. The requirement for certification is to be introduced progressively from 2007, and the paragraphs below outline the main principles that are likely to apply.

Energy certification

Article 7 of the EPBD requires Energy Performance Certificates to be made available in two sets of circumstance:

(a) Whenever a building is constructed, or offered to the market for sale or rent. The purpose of this is that a prospective purchaser can be made aware of the energy standards of the building and take this into account when deciding which building to procure. In order to facilitate the comparison of one building with another, it is necessary to assess performance in a standardised way, using common occupancy patterns and weather data. This is best done by calculation, and the UK has adopted the term Asset Rating for such a performance measure.
(b) On prominent display in buildings '*over 1000 m² occupied by public authorities and institutions providing public services to a large number of persons and therefore frequently visited by these persons*'.[5] Here, the purpose is to use the public sector as a demonstration model for the energy reporting system and offer the public an opportunity to press operators of public buildings to pay attention to energy efficiency. The implementation of this requirement in England and Wales is to be via the Operational Rating, which is based on measured energy performance, which is an indicator of the quality of the building fabric and systems, the actual pattern of use and the effectiveness of the energy management regime.

In all cases, the certificate conveying the rating has to be accompanied by:

(a) Reference values and benchmarks, so that the relative performance of the particular building can be assessed in relation to its peers.
(b) Recommendations for cost effective improvement. These are provided for information only – there is no requirement to implement any of the recommendations.

Asset Ratings

The Asset Rating is a measure of the intrinsic performance potential of the building, and rates the standard of the building fabric, building services equipment and the controls for a standardised pattern of use and climate. The Asset Rating is calculated using accredited software tools in the same way as is used to demonstrate compliance with Part L. This means that for newly constructed buildings the Asset Rating is generated as part of the compliance process. The Asset Rating for an existing building is determined in exactly the same way as for a new building, i.e. via a calculation using accredited calculation tools. The only difference is that some of the input data may not be known with such certainty as for new build. In such a situation, the unknown data will have to be determined using a government approved 'inference engine'. This will infer the performance characteristics based on information about year of construction, type, etc. Government is considering allowing where the building owner has maintained good building records (e.g. via the building logbook as described in Chapter 31) or has invested in a building survey, that any quality assured data from such sources could be used in preference to the results of the inference engine. This means that the energy performance of one building can be compared with that of another on a meaningful and consistent basis.

It is proposed that the reference building against which the actual building is compared for the purpose of generating the rating always has a fixed level of servicing. This is slightly different to the Part L compliance calculation, where the notional building has the same level of servicing as the actual building. The reasons for this difference are that:

(a) Building Regulations are not the vehicle to decide whether a building should be air-conditioned or not. Their purpose is to ensure that if air-conditioning is installed, it is reasonably efficient.
(b) Certification is to inform the market about relative energy performance. Therefore the rating should reflect the fact that an air-conditioned building is likely to be more energy intensive than a naturally ventilated building providing the same levels of accommodation.

The reference values and benchmarks reported with the Asset Rating are likely to be the current Part L standard and an estimate of the stock average for the particular building type. The certificate must contain certain administrative information as part of the quality assurance system. This includes the contact details of the assessors and the accreditation scheme under which they operate. An example of what the certificate might look like is shown in Fig. 3.3.

Operational Ratings

The Operational Rating (OR) is a measure of the in-use performance of the building and is based on actual metered energy consumption, normalised to account for the effects of building size, pattern of use, weather, etc. The in-use performance will be influenced by the quality of the building (as separately measured by the Asset Rating), but also by the way the building is maintained and operated.

The purpose of the OR is to indicate how well the building is being operated and in that sense the most important benchmark is previous performance, i.e. is energy efficiency improving, stable or getting worse? Benchmark data on the performance of other similar buildings will also indicate how well the building is performing relative to its peers. The aim of the public display is to use public pressure as a vehicle for driving improvement.

There are some important differences to the Asset Rating and the calculations used for demonstrating compliance with the Building Regulations in that all end uses of energy are included in the OR, whereas the Asset Rating only measures the energy used by the fixed building services. This is a pragmatic decision because in most buildings (those not built in compliance with the Building Regulations in force from April 2002) the limited availability of meters precludes robust separation of energy demands into its various end uses. For the same reason, the existing benchmark data includes all end uses.

It is important to understand that a building may be a high-energy user for reasons other than poor energy efficiency. A building may be used very intensively, e.g. as a 24/7 call centre, when clearly energy demand will be much more than in an office that operates conventional hours.

Plant inspections

Articles 8 and 9 of the EPBD cover the regular inspection of boiler and air-conditioning systems, respectively. Article 8 allows an alternative approach based on advice campaigns. This is the approach that will be adopted in the UK, at least initially. The purpose of the air-conditioning inspections is to establish:

(a) Whether the size of the installation is appropriate to the requirements of the building. (Oversized plant with poor control is often a significant cause of inefficient operation.)
(b) Whether improvement or replacement of all or part of the existing system would be beneficial. If replacement would be sensible, the inspection should evaluate alternative solutions to meeting the demand, i.e. not presume like-for-like.

Guidance on the inspection of air conditioning systems has been published[6] and this should have an impact on improving the performance of existing buildings. Again, there is no mandatory requirement to implement any of the recommendations that are made, but clearly if any system or component is replaced then it will need to comply with the minimum efficiency standards set through the Building Regulations – and the recommendation might be that significant improvement over this minimum level would be cost-effective in the particular case.

Independent experts

Article 10 of the EPBD requires that certification of buildings (Article 7) and plant inspections (Articles 8 and 9) be '*carried out in an independent manner by qualified and/or accredited experts*'. The government has indicated that for England and Wales, third party accreditation is necessary to achieve sufficient independence. Systems are in place for independently accrediting energy assessors working on dwellings; the systems for non-domestic buildings assessment are still under development.

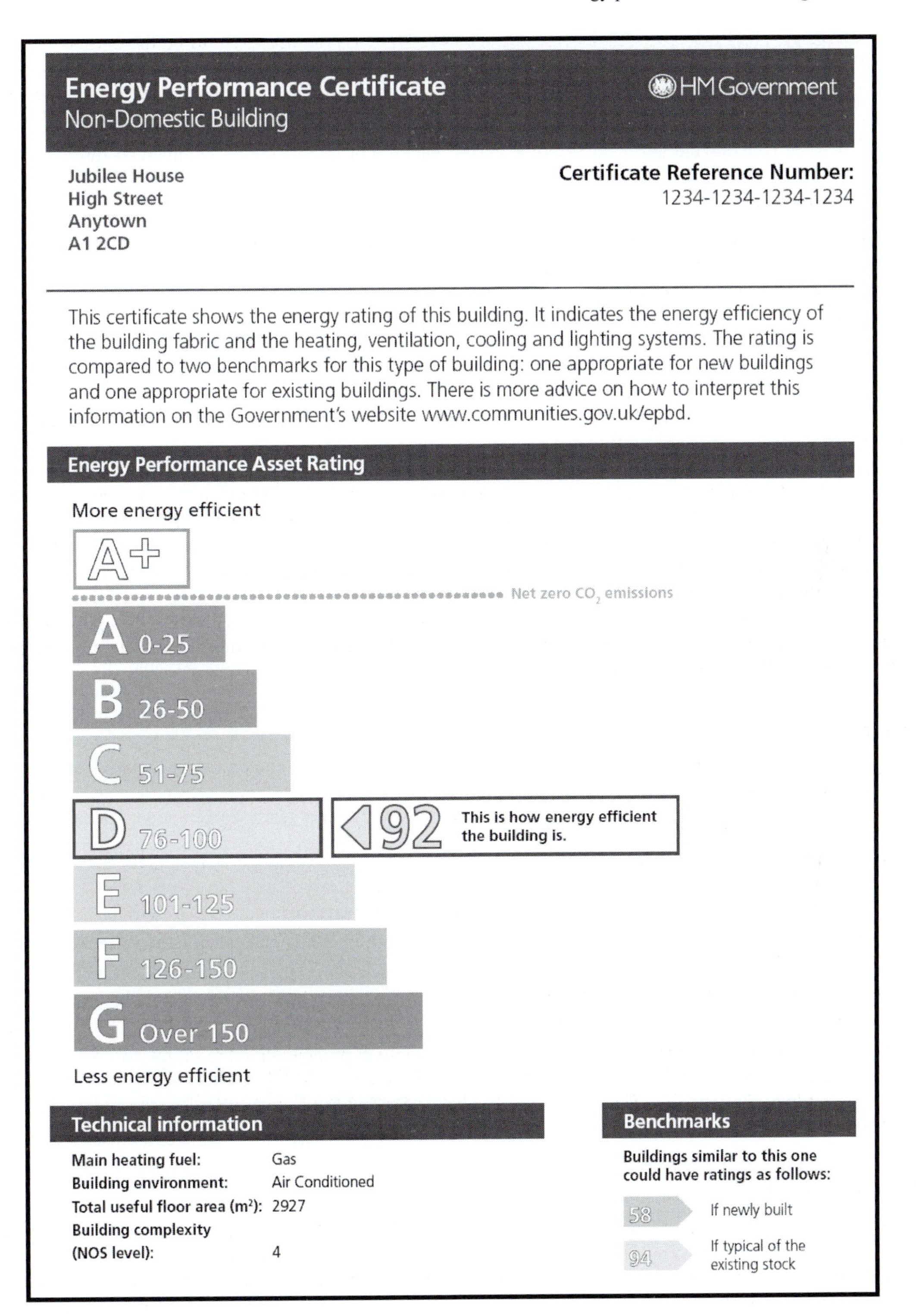

Figure 3.3 An illustrative energy performance certificate (see colour plate section at the end of the book)

The F-Gas Regulations

The F-Gas Regulations will come into force with effect from July 2007. The objective is to contain, prevent and thereby reduce emissions of F-Gases covered by the Kyoto Protocol, i.e. fluorinated refrigerant gases. The main focus is on containment and recovery of F-Gases, which will impact on the commercial refrigeration, air-conditioning and heat pump sectors and also the fire quench protection sector. Operators of relevant systems will have a range of obligations including:

(a) prompt leakage repair,
(b) leakage checking and record keeping,
(c) ensuring appropriately qualified personnel are used.

Initial guidance has been produced by the government and is available on the Department of Trade and Industry (DTI) website.[7]

Health and safety legislation

The Health and Safety at Work etc. Act 1974, is the primary piece of legislation covering occupational health and safety in the UK. It applies to all occupations and to everyone who is 'at work' and requires employers to provide for the health and safety of employees and others, in very general terms. The Act also empowers enforcing authorities to make industry-specific regulations, whenever they deem it necessary to do so. These are secondary types of legislation and cover a wide range of subjects individually: asbestos, dangerous substances, lifting, etc., adding detail to the generalities of the Act.

UK health and safety law is based on the (proven) premise that the most efficacious way of controlling risks is to get rid of the source of the risk, i.e., eliminate the hazard. When this is not possible, the dangerous is replaced by the less dangerous and 'collective (passive) measures must be given priority over individual measures'. The requirement for giving priority to collective measures recognises the reality: that the human interface is where errors occur, either through carelessness, lack of appreciation that there is a risk or straight forward risk taking. Therefore, the safest option is to remove humans from the equation, by providing measures that do not require input by the workers for the measures to be effective.

However, in recognition of the fact that health and safety at any cost is not a viable option, UK health and safety law is moderated by two further principles.

The moderating principles

Almost uniquely, UK health and safety law allows people planning for health and safety to exercise judgement, by allowing for any provisions for health and safety to be 'reasonably practicable' and 'suitable and sufficient'.

There are no set definitions for these terms, which, in another peculiarity of UK law, remain undefined until a judge chooses to interpret them. Consequently, they have come to mean the following:

(a) 'Reasonably practicable': the cost of the provision is sensible in proportion to the safety gain (reasonable) and it must be physically possible (practicable).
(b) 'Suitable and sufficient', is a lot easier to define: The H&S provision is appropriate for the situation (suitable) and there is enough of it (sufficient).

All of the underpinning principles are reasonably easy to apply to a tangible provision. For example, when people work at height it is easy to see that:

(a) Edge protection generally provides a huge safety gain at a reasonable cost, i.e., it is generally reasonably practicable to supply it,
(b) If it complies with the relevant standards and is provided at every edge, it will also be suitable and sufficient, and
(c) It satisfies the principle of collective measure being given priority.

While it is easy to see how designers can contribute to the 'collective measures' principle, it is difficult to translate the meaning of the other two underpinning principles to the design process, i.e. when is a design reasonably practicable or suitable and sufficient?

Enforcing health and safety law in the UK

In the UK, the Health and Safety Executive (HSE) enforces health and safety legislation. The purpose of this enforcement is three-fold. It is to:

(a) Ensure duty holders take action to deal with risks,
(b) Promote and achieve compliance with the law, and
(c) Take action, when it is appropriate to do so, against duty holders who do not discharge their duties.

It is worth remembering that a breach of health and safety law is a criminal offence.

Health and safety in construction

It is recognised that construction is a high-risk industry. People in construction are exposed to many hazards across hundreds of thousands of sites. The nature of the risk, combined with the risk-taking type of person that works in construction means that, inevitably, things go wrong.

Figure 3.4 shows that except for the odd year, we can expect almost 80 people to die each year while working on construction sites. It is worth noting that these figures are a marked improvement on the situation that existed 20 years ago, when in excess of 200 people could expect to die on a construction site.

The situation is as bad for major injuries, as shown in Fig. 3.5

In recognition of the high-risk nature of the industry, construction, uniquely, had sets of regulations devoted to it in an attempt to improve the situation. Up until the mid-nineties, construction was directly regulated by three sets of secondary regulations, i.e., the regulations had the word 'construction' in the name, and by three sets indirectly, i.e., the regulations were not dedicated to construction but dealt with operations and hazards that were common in the industry: lifting, confined spaces and hazardous substances. By the mid-1990s, many of the provisions in the three construction-dedicated regulations were rationalised into one set: The Construction (Health, Safety and Welfare) Regulations 1996 and in 1994, the Construction (Design and Management) (CDM) Regulations were introduced. Of all the secondary legislation that is current, the most important statutory instrument for designers to be aware of is the CDM Regulations.

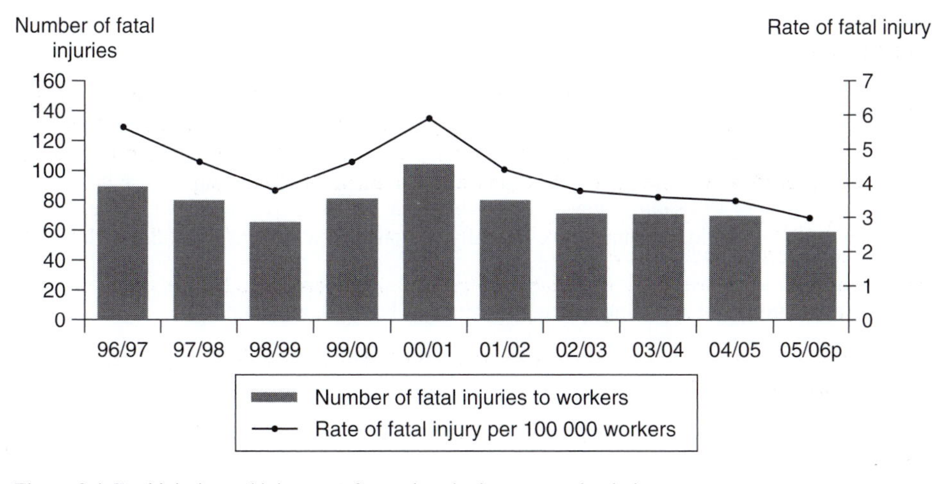

Figure 3.4 Fatal injuries and injury rate for workers in the construction industry

Source
HSE statistics

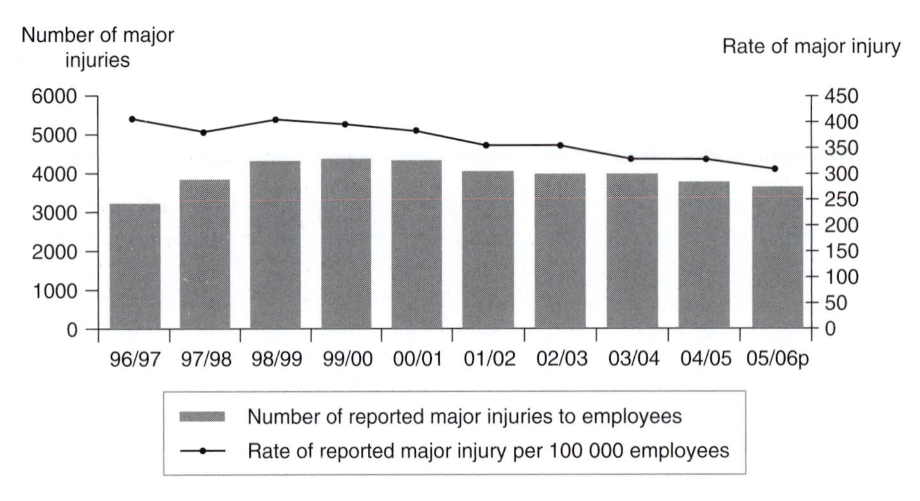

Figure 3.5 Major injuries suffered by workers in the construction industry

Source
HSE statistics

The Construction (Design and Management) Regulations 1994

In 1992, The Temporary or Mobile Construction Sites Directive (Council Directive 92/57/EC) was published. The preamble to the articles of this directive is instructive, highlighting that: 'Whereas unsatisfactory architectural and/ or planning organisational options or poor planning of the works at the project preparation stage have played a role in more than half of the occupational accidents occurring on construction sites in the Community.' In the preamble, it seems that 'architectural' is construed as design, i.e., unsatisfactory designs played a prominent role in the accident statistics, which was the rationale for including articles dealing with design in the Directive.

The CDM Regulations 1994 were the result of the UKs implementation of the main articles of the Temporary or Mobile Construction Sites Directive (92/57EC). Some key aims of the regulations are:

(a) To improve planning and management of construction work,
(b) To ensure that anybody involved in construction is competent in their chosen field,
(c) To instil processes that identify hazards as early as possible so that they can be eliminated or provision made to allow them to be properly managed, and
(d) To allow all involved in construction projects to concentrate their efforts where they can do most to provide for the health and safety of the workers and others.

Two of the four key aims can certainly be applied to designers. Under key aim (b), designers must be competent but the main impact on designers arises out of key aim (c), which places a duty to think about the health and safety of the people who will have to build, maintain and eventually demolish their designs, by and in order:

(a) Eliminating hazards,
(b) Where hazards cannot be eliminated, reducing the risks from these hazards, and then
(c) Providing adequate information about the residual risks.

Designers also have a number of other duties under these regulations but they are mainly administrative.

The Construction (Design and Management) Regulations 2007

In 2007, the CDM Regulations were revised. The key aim of the Regulations has not changed: to focus attention on planning and management of construction projects from the concept stage through to completion, to ensure that the health and safety of people concerned in building, maintaining, using and then demolishing a structure is given due consideration.

Nevertheless, it was felt that the 1994 regulations did not (fully) achieve this aim. Therefore, to improve compliance, the Regulations have created new and more onerous duties for duty holders considered to be able to influence how projects are run. Although the changes for designers are minimal, there are a number of things that designers need to be aware of. The four most important being their duty:

(1) Not to accept an appointment unless they are competent to do so and to ensure that any appointment they accept is adequately resourced.
(2) On notifiable projects,[8] not to continue with work beyond initial design[9] until a Co-ordinator is appointed.
(3) To ensure that any structure that they design complies with the Workplace (Health, Safety and Welfare) Regulations 1992.
(4) To ensure that they pay due regard to the Principles of Prevention set out in Appendix 7 of the Approved Code of Practice (ACoP) that supports the 2007 Regulations.

Who are the designers?

In the CDM Regulations, the term designer is very broadly defined. However, the ACoP stipulates that building services designers are designers. In order to discharge this duty properly, designers will need to have an appreciation of what constitutes a hazard and what constitutes a risk.

Hazard and risk

Unfortunately, hazard and risk are used (incorrectly) interchangeably, which confuses many of the issues that designers have to deal with. They mean different things and a good understanding of each term will help designers in understanding their duties. The accepted definition of each is:

(a) *Hazard*: is the potential to cause harm.
(b) *Risk*: is the likelihood that the hazard will (be realised) and cause harm.

For example, working at height (maintaining ductwork in a ceiling space) is a hazardous activity, because it exposes the worker to the hazard of falling. However, the risks associated with working at height will be very small if suitable and sufficient fall protection is provided. This is an example where elimination of the hazard is extremely difficult, as is the provision of collective protective measures. UK health and safety law allows, under the principle of reasonably practicable, lower levels of safety.

Similarly, working on an existing high-pressure pipe is a hazardous activity, because it exposes the worker to the hazard of sudden (unexpected) release of pressure, which could harm the worker in a number of ways. The risks associated with the work will be controlled if the worker follows exactly the method statements drawn up by competent people. However, the provision of a means of isolating and then depressurising the system would eliminate the hazard and would probably be reasonably practicable. Therefore, designers should choose this option or be able to justify, with reasonably strong arguments and evidence, why this was not done. More information on designing to discharge statutory duties under CDM is given in Chapter 30.

Competence

The revised Regulations place an emphasis on the need to be competent: it is illegal to accept a commission when you are not competent to do so. For the purposes of the 2007 Regulations, competence relates to being able to deliver the duties imposed by these Regulations, i.e., design structures that are safe as possible to build, maintain, use and eventually demolish.

This means, that to be competent within the meaning of the 2007 Regulations, a designer will have to be able to:

(1) Understand any potential hazards related to the work (or equipment) under consideration; meaning that you will need to know:
   (a) the hazards that may be encountered on construction sites;
   (b) under what circumstances they can arise;
   (c) under what circumstances they become:
      (i) not obvious (to a competent contractor),
      (ii) unusual, or
      (iii) difficult to manage under the state-of-the-art.
(2) Detect any technical defects or omissions in that work (or equipment) and be able to specify a remedial action to mitigate those implications; meaning that you will need to be aware of:
   (a) how something might be built;
   (b) how it might subsequently be maintained;
   (c) what work equipment is on the market to build and maintain it;
   (d) the limitations of any work equipment;
   (e) how construction processes may release hazards.

In the ACoP[10] supporting the 2007 regulations, the HSE has set out how it will judge competence. Readers are advised to study the relevant parts of the ACoP.

Other legislation

Summaries of other legislation relevant to building services in respect of sustainability and the operation of plant and systems is given in Chapters 2 and 31, respectively.

Notes

1. Different frameworks apply in Scotland and North Ireland so the reader should refer to the procedures and guidance material appropriate to the administration applicable to the particular project.
2. In some cases, the pressures are binding, e.g. publicly financed new dwellings have to meet level 3 in the Code for Sustainable Homes.
3. Building a greener future; towards zero carbon development, Department for Communities and Local Government (DCLG) consultation paper, December 2006.
4. *Official Journal of the European Communities*, January 4 2003.
5. The Government has signalled the intention of consulting on widening this requirement to apply to smaller public sector buildings and to commercial organisations such as shops and hotels that provide a service to large numbers of members of the public.
6. Inspection of air conditioning systems, CIBSE TM44, 2007.
7. http://www.dti.gov.uk/innovation/sustainability/fgases/page28889.html
8. Any project where the construction phase is likely to involve more than 30 days or 500 person days of construction work.
9. Initial design is defined in the Guidance for designers, available free from CITB-Construction Skills.
10. An ACoP has special legal status. It gives practical advice on how to comply with the law. If you follow the advice you will be doing enough to comply with the law in respect of those specific matters on which the Code gives advice.

Further reading

Further reading on the developments with Building Regulations and the implementation of the EPBD can be found by referring to the '*Planning, building and the environment*' section of the CLG website, www.communities.gov.uk

Chapter 4

The building in winter

As a general principle when approaching the question of space heating, it is desirable that the building and the heating system should be considered as a single entity. The form and construction of the building will have an important effect not only upon the method to be adopted to provide heating service, but also upon subsequent recurrent energy use and carbon dioxide emissions. The amount of heat required to maintain a given internal temperature may be greatly reduced by thermal insulation and by any steps taken to reduce an unwanted intake of outside air. Large areas of glass impose very considerable loads upon any heating system and run counter both to the provision of comfort conditions and to any prospect of energy efficient operation.

The mass of the building structure, light or heavy, has a direct bearing upon the choice of the most appropriate form of system since, in the former case, changes in external temperature will be reflected very quickly within the building and a system having a response rate to match will be the one that is most suitable. On the other hand, a building of traditional heavy construction may well be best served by a system which produces a slow steady output. Tall, multi-storey blocks of offices and dwellings introduce problems related to exposure to wind and solar radiation as well as those related to *chimney* or *stack effects* within the building itself. The form and design of the heating system must take these aspects into account.

Extraneous influences

Within the space to be heated, the energy which is consumed by electrical lighting and by a variety of other items of current-consuming equipment will be released, as will heat from the occupants. The total of these internal emissions will contribute in some measure towards maintenance of the desired space temperature. In addition, even in winter, heat from the sun may sometimes be enough to cause problems of excessive temperature in those rooms exposed to radiation, while at the same time others, in shade, are not so favoured. These fortuitous effects cannot be overlooked, although it is not possible to rely upon them to make a consistent contribution.

It is a matter of importance to consider how far these heat gains should be taken into account: if they are ignored, then the heating system may be so oversized that it will be unwieldy and will rarely, if ever, run at full capacity. On the other hand, if certain reasonable assumptions are made as to the proportion of the total gains which will coincide and, due possibly to a change in building use, these do not apply, then the heating system may well be undersized. In instances where, despite the last comment, some allowance is made, particular thought must be given to the needs of an intermittently heated building where *a pre-occupancy boost* may be required during the time when no internal or solar gains are available.

Past practice in design, for other than off-peak electrical heating systems, has been to ignore the effect of such gains entirely in so far as calculations for the heat necessary to maintain a given internal temperature are concerned. In calculations for energy consumption and running costs per annum, however, the importance of these gains has been brought out by studies of actual fuel use in buildings. The case for adequate thermostatic controls is self evident.

Climate change and the conservation of energy

The past situation, when a seemingly limitless supply of fuels of one sort or another was available came to an end many years ago. A significant change occurred following the 'energy crisis' of the mid-1970s, which has only accelerated in more recent years as the link between carbon dioxide emissions and climate change has been established. Economic forces have led to a startling rise in the cost of supplies, with the result that many of the old standards of comparison no longer apply. The more recent volatility of fuel prices, largely brought about by the price of oil and gas in the international market and the privatisation of the energy market in the UK, may be considered normal in the inevitable upward spiral of energy costs. Government also has a role to play in energy conservation and the introduction of the climate change levy in April 2001 by imposing taxes on an organisation's energy bill and the launch of a Climate Change Bill in March 2007 which sets out a series of targets for reducing carbon dioxide emissions between 26 and 32 per cent by 2020 are examples of this. It is also worth recognising that the security of our future energy supplies will be greatly assisted by conservation and the use of renewable sources.

As a result of this situation, energy conservation in the sense of fuel saving is now a doctrine of political importance as well as one of realism and this whole area of increasing legislation is covered in detail in Chapter 3. In the context of this book, it has been estimated that between 40 and 50 per cent of the national annual consumption of primary energy is used in services to buildings. By the introduction of sensible economy measures, without real detriment to environmental standards or the quality of life, it is possible to make significant savings in the context of heating and air-conditioning systems. But the most dramatic attack upon energy consumption must come about as a result of reconsideration of the building structure, readjustment of capital cost allocations and improved maintenance of buildings and equipment.

It follows that the first step in embarking upon the assessment of heat requirements for a building should be to ensure that they have been reduced to an economic minimum. This will involve collaboration between the architect and the building services engineer in consideration of the building orientation, selection of materials, addition of thermal barriers and reduction in window areas. Ideally, this collaboration should start at an even earlier stage when the basic plan form of the building is being considered, bearing in mind that the major component of the total thermal load is through the perimeter surfaces.[1] In this sense, the most economical shape for maximum volume with minimum surface area is a sphere and although this is hardly a practicable shape for a building, the nearer to it the better. A tall shallow slab building is obviously one of the worst in this respect.

Heat losses

The conventional basis for design of any heating system is the estimation of heat loss and, for the purpose of calculation, it is assumed that a *steady state* exists between inside and outside temperatures although, in fact, such a condition rarely obtains. In the past, air temperature difference has been the sole criterion although the mean radiant temperature within the enclosure, if considered, may well call for a higher or a lower air temperature for equal comfort. The method of calculation recommended in the current edition of the *Guide Sections A5* is in terms of operative temperature within the space to be heated and this, as was explained in Chapter 1, takes account of the mean radiant temperature. This refinement will be discussed subsequently but, in this preliminary introduction, the air temperature difference will be used wherever appropriate.

Each room of a building is taken in turn and an estimate is made of the amount of heat necessary to maintain a given steady temperature within the space, assuming a steady lower air temperature outside. The calculation falls into two parts: one relating to conduction through the various surrounding structural surfaces, walls, floor and ceiling; and the other to the heat necessary to warm to room temperature that outside air which, by accident or design, has infiltrated into the space.

Adjacent rooms maintained at the same temperature will have no heat transfer through the partitions or other surfaces between them and these may thus be ignored. Furthermore, if certain surfaces, such as the ceiling or floor, are used for heat output, then these will also not be taken into account so far as heat loss from the room is concerned; they will, however, have inherent losses upward or downward to unheated areas and those will have to be allowed for separately.

The conduction element is calculable from known properties of the building materials, but the infiltration element presents problems, in that what is called the *air-change rate* or, alternatively in energy terms, the *ventilation allowance*, is not easy to assess other than by experience. This air-change rate is no more than a natural ventilation effect, arising from a number of extraneous circumstances but without which a space would quickly cease to be habitable, and although this element must be dealt with empirically, the ground rules are reasonably well established.

It might be thought that, with so many assumptions and 'rule-of-thumb' estimates, heat loss calculations are very little better than guesswork. In practice, however, they have proved to be a reliable basis for an overall assessment, partly due to the fact that all areas in the building are treated in a like manner and are thus consistent in response. In addition, the building structure is itself a moderator, as a result of its *thermal inertia*, which remains as a significant factor, even in a lightly constructed building which still has floor slabs, partitions, furniture, etc., to absorb and emit heat and thus smooth out any violent fluctuations.

Conduction losses

The conduction of heat through any material depends upon the conductivity of the material itself and upon the temperature difference between the two surfaces. Ignoring for the moment any heat transfer by radiation within a space, it is the air-to-air transfer of heat through building materials which is relevant. On either side of a slab of building material, it may be supposed that there is a film or relatively dead layer of air which retards the flow of heat. This is illustrated, in Fig. 4.1, for a thin material such as a single sheet of glass. The air within the room, being relatively still, offers a higher resistance to heat flow than that outside where wind effects, etc., have to be considered.

The effect of these boundary layers is defined in terms of what is described as *surface resistance*, and representative values are as set out in Table 4.1. The notation here is that given in the *Guide Section A3* and, as noted later, the value of the outside surface resistance (R_{so}) varies with the degree of exposure such that a sheltered surface has a higher resistance to heat flow than one which is exposed to severe wind and other effects. In the extreme case of a very tall building, it might well be supposed that wind forces were such that the boundary layer is totally dispersed and that the value of the surface resistance is zero, i.e. the temperature of the external surface is, effectively, that of the outside air. The figures listed for internal surface resistances (R_{si}) vary, it will be noted, only with respect to the disposition of the area concerned and the direction of heat flow.

The values noted in Table 4.1 for external and internal surface resistances relate to normal building materials, most of which have high emissivity values. An exception may arise in the case of the resistance offered by an air gap in a cavity construction (R_a) where two sets of values are given, one for the situation where the

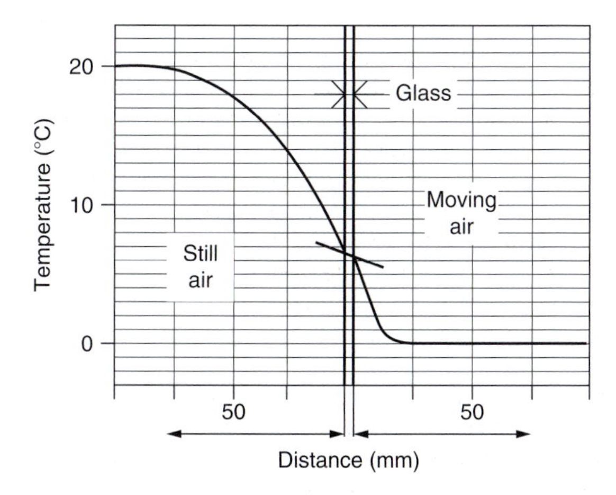

Figure 4.1 Heat transmission gradient through glass

Table 4.1 Thermal/surface resistances (normal design values for *Guide Section A3*)

Building element	*Thermal/surface resistance* (m^2 k/W^{-1})	
	High emissivity	*Low emissivity*
Outside (R_{so})		
Walls	0.04	–
Roofs	0.04	–
Floors	0.04	–
Inside (R_{si})		
Walls (heat flow horizontal)	0.13	–
Roofs (heat flow to above)	0.10	–
Ceilings (heat flow to above)	0.10	–
Floors (heat flow to below)	0.17	–
Unventilated air gap (R_a)		
Heat flow horizontal	0.18	0.44
Heat flow up ward	0.16	0.34
Heat flow down ward	0.19	0.50
Ventilated air gap (R_a)		
Cavity wall (horizontal)	0.18	0.44
Void under pitched roof	0.20	0.30
Void within flat roof	0.14	0.40
Void behind hung tiled wall	0.12	0.30

facing surfaces within the air gap have high emissivity and the other to meet the case where a bright metallic surface, such as aluminium foil, is inserted to provide low emissivity.

Conductivities

Values for the thermal properties of a selection of common building materials were given in Table 1.1, and others are listed in the *Guide Section A3*. The following definitions may be helpful.

Conductivity

The standardised value, in watts for one metre thickness per square metre and kelvin:

$$k = \text{Wm/m}^2\ \text{K} = \text{W/m K}$$

Conductance

The value for a stated thickness (k/L):

$$C = \text{W/m}^2\ \text{K}$$

Resistivity

The reciprocal of the conductivity ($1/k$)

$$r = \text{m K/W}$$

Resistance

The reciprocal of the conductance (L/k):

$$R = \text{m}^2\ \text{K/W}$$

For example, the thermal conductivity (k) of expanded polystyrene is 0.035 W/m K and the thermal conductance (C) of a slab of this material, 25 mm thick, is thus 0.035/0.025 = 1.4 W/m^2 K. Similarly, the thermal resistivity (r) of the material is 1/0.035 = 28.6 m K/W and the thermal resistance (R) of the slab is 1/1.4 (or 28.6 × 0.025) = 0.71 m^2 K/W.

Moisture in masonry materials

The conductivities of all masonry materials bear the same general relationship to their dry densities and follow the same proportional pattern with increasing moisture content. Corrected practical values for moisture content by volume were selected for the *Guide Section C3* in consequence, as follows:

- Brickwork, protected from rain 1 per cent
- Concrete, protected from rain 3 per cent
- Both materials, exposed to rain 5 per cent.

For the inner skin material of a cavity wall and for internal partitions, etc., the protected values are used. In circumstances where materials are exposed to either driving rain or condensation, the 5 per cent content noted above is no longer valid and it would seem that conductivities increase by about 4 per cent for each 1 per cent increase in moisture content.

U values

The rate of heat transmission, in watts per square metre and kelvin (W/m^2 K) is, for the purpose of heat loss calculation, termed the thermal transmittance coefficient (U). Where the individual *leaves* that make up the building envelope are homogeneous this is the reciprocal of the sum of all individual resistances, thus:

$$U = 1/(R_{si} + R_{so} + r_1L_1 + r_2L_2 + r_3L_3 + R_a)$$

where

R_{si} = inside surface resistance (m^2 K/W)
R_{so} = outside surface resistance (m^2 K/W)
R_a = air space resistance (m^2 K/W)
r_1, r_2, etc. = resistivities (m K/W)
L_1, L_2, etc. = thicknesses (m)

In many situations the *leaves* are constructed from more than one building element, for example a timber framed insulated wall. In these situations thermal bridging takes place and the U value is calculated by the combined method using the mean of the upper and lower limits of the thermal resistance of the bridged *leaf* as follows:

Lower Limit

$$R_L = R_{si} + R_{so} + r_1L_1 + \frac{1}{\left(\frac{P_2}{r_2L_2} + \frac{P_3}{r_3L_3}\right)} + R_a$$

Upper Limit

$$R_U = 1\bigg/\left(\frac{P_2}{R_{si} + R_{so} + r_1L_1 + r_2L_2 + R_a}\right) + \left(\frac{P_3}{R_{si} + R_{so} + r_1L_1 + r_3L_3 + R_a}\right)$$

$$U = 1/0.5(R_L + R_U)$$

where P_2, P_3, etc. = proportions of the total surface area occupied by each element in thermally bridged part of envelope.

Pre-calculated values for transmittance coefficients, applicable to a range of typical forms of construction, are provided in Table 4.2. The values listed are for what is called normal exposure but the much more

Table 4.2 Typical transmittance coefficients (U values) for 'normal exposure' (including allowance for moisture content, as appropriate)

Construction (dimensions in mm)	*U value* (W/m^2 K)
External walls	
Solid walls, no insulation	
600 stone, bare	1.72
600 stone, 12 plasterboard on battens	1.35
105 brick, bare	3.27
105 brick, 13 dense plaster	3.0
220 brick, bare	2.26
220 brick, 13 dense plaster	2.14
335 brick, bare	1.73
335 brick, 12 plasterboard on battens	1.53
Dense concrete walls	
19 render, 200 concrete, 13 dense plaster	2.73
19 render, 200 concrete, 25 polyurethane, 12 plasterboard on battens	0.71
Pre-cast dense concrete walls	
80 concrete panel, 25 EPS, 100 concrete, 13 dense plaster	0.95
80 concrete panel, 50 EPS, 100 concrete, 12 plasterboard	0.53
Brick/brick cavity walls	
105 brick, 50 air space, 105 brick, 13 dense plaster	1.47
105 brick, 50 blown fibre, 105 brick, 13 dense plaster	0.57
Brick/dense concrete cavity walls	
105 brick, 50 air space, 100 concrete block, 13 dense plaster	1.75
105 brick, 50 EPS, 100 concrete block, 13 dense plaster	0.55
Brick/lightweight aggregate concrete block	
105 brick, 50 air space, 100 black, 13 dense plaster	0.96
105 brick, 50 urea formaldehyde (UF) foam, 100 block, 13 dense plaster	0.47
Brick/autoclaved aerated concrete block	
105 brick, 50 air space, 100 block, 13 lightweight plaster	1.07
105 brick, 25 air space, 25 EPS, 150 block, 13 lightweight plaster	0.47
Timber frame walls	
105 brick, 50 air space, 19 plywood sheathing, 95 studding, 12 plasterboard	1.13
105 brick, 50 air space, 19 plywood sheathing, 95 studding with mineral fibre between studs, 12 plasterboard	0.29
Party walls and partitions (internal)	
Brick	
13 dense plaster, 220 brick, 13 dense plaster	1.57
13 lightweight plaster, 105 brick, 13 lightweight plaster	1.89
Dense concrete block	
13 dense plaster, 215 block, 13 dense plaster	2.36
Lightweight concrete block	
12 plasterboard on battens, 100 block, 75 air space, 100 block, 12 plasterboard on battens	0.62
Flat roofs	
Cast concrete	
Waterproof covering, 75 screed, 150 concrete, 13 dense plaster	2.05
Waterproof covering, 100 polyurethane, vapour barrier, 75 screed, 150 cast concrete, 13 dense plaster	0.22
Timber	
Waterproof covering, 19 timber decking, ventilated air space, vapour barrier, 12 plasterboard	1.87
Waterproof covering, 35 polyurethane, vapour barrier, 19 timber decking, unventilated air space, 12 plasterboard	0.52
Pitched roofs	
Domestic and commercial	
Tiles on battens, roofing felt and rafters, bare plasterboard ceiling below joists	2.6
Tiles on battens, roofing felt and rafters, 50 glass fibre mat between joists, plasterboard ceiling below	0.6
Tiles on battens, roofing felt and rafters, 100 glass fibre mat between joists, plaster ceiling below	0.35
Industrial	
Corrugated double-skin decking with 25 glass fibre mat	1.1

comprehensive list of examples given in the *Guide Section A3* includes alternative data for other conditions, defined as follows:

Sheltered Up to third floor of buildings in city centres.
Normal Fourth to eighth floors of buildings in city centres; up to fifth floor of suburban and rural buildings.
Severe Ninth floor and above in city centres; sixth floor and above in suburban and rural districts; most buildings on coastal or hill sites.

In general terms, the values listed in Table 4.2 will be increased by up to 20 per cent for severe exposure but, in the particular case of single glazing, the increase is much greater at about 45 per cent.[2]

Calculation of a U value

The wide range of constructions currently encountered, built up from the many composite elements available, often requires that transmittance coefficients (U values) are calculated from first principles. For example, consider a curtain wall construction comprising:

Outside	100 mm thick pre-cast concrete panel
Air gap	25 mm wide
Insulation	75 mm glass fibre slab
Inside	100 mm aerated lightweight concrete block[3]
Plaster	20 mm lightweight

The resistances, from the physical data provided in Tables 1.1, 4.1 and 4.3, may be summated as follows:

Outside surface, R_{so}		$= 0.04$
Concrete panel	$(1/1.56) \times (100/1000)$	$= 0.054$
Air space, R_a		$= 0.18$
Insulation	$(1/0.04) \times (75/1000)$	$= 1.875$
Concrete block	$(1/0.23) \times (100/1000)$	$= 0.435$
Plaster	$(1/0.18) \times (20/1000)$	$= 0.111$
Inside surface, R_{si}		$= 0.13$

$$\sum R = 2.835 \text{ m}^2 \text{ K/W}$$

and thus

$$U = 1/2.835 = 0.35 \text{ W/m}^2 \text{ K}$$

Surface temperatures

It is often necessary, for a variety of reasons, to establish either a surface or an interface temperature for some form of composite construction. The temperature gradient across the structure may be plotted, as shown in Fig. 4.2, using the resistances of the various elements, with the simple case of an unplastered and uninsulated 210 brick wall being illustrated in Fig 4.2(a). In this case, the overall resistance would be made up as follows:

Outside surface, R_s		$= 0.04$
Brickwork	$(1/0.84) \times (210/1000)$	$= 0.25$
Inside surface, R_{si}		$= 0.13$

$$\sum R = 0.42 \text{ m}^2 \text{ K/W}$$

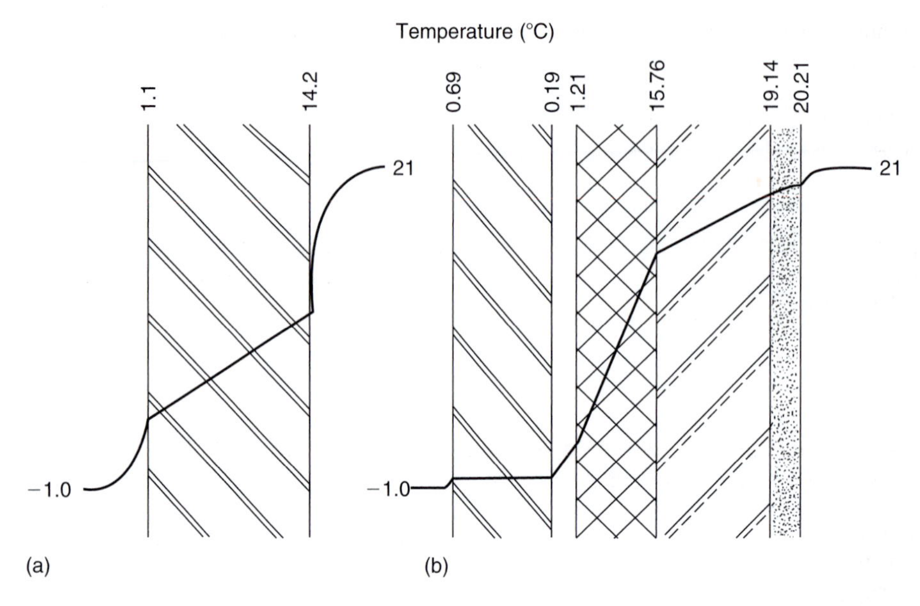

Figure 4.2 Surface and interface temperatures

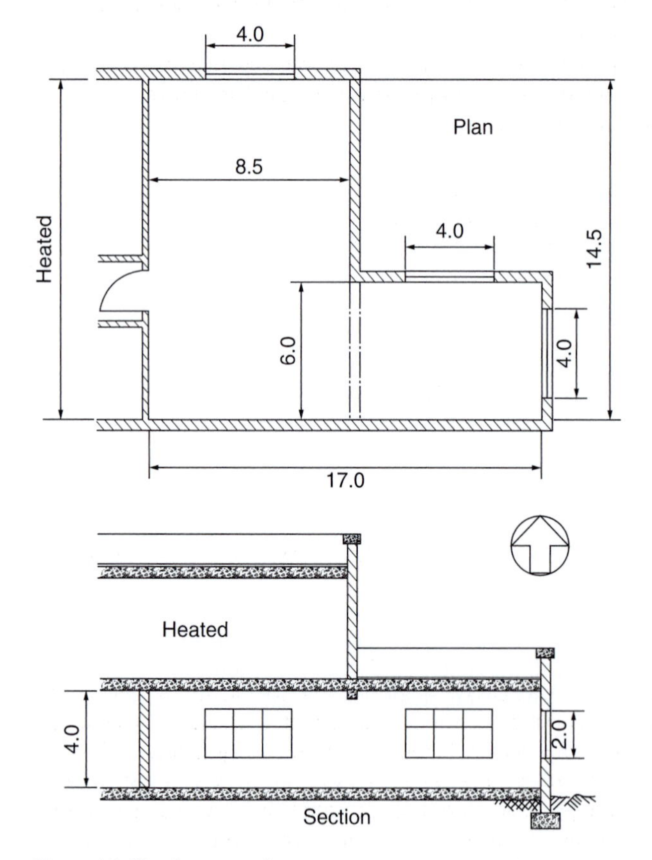

Figure 4.3 Heat loss example

The temperature gradient, including the components at the surfaces, will be *pro rata* to the resistances and, for inside and outside air temperatures of 21°C and −1°C, respectively, would be:

Outside air to outside surface	$= 22 \times (0.04/0.42) = 2.1$°C
Across the brickwork	$= 22 \times (0.25/0.42) = 13.1$°C
Inside surface to inside air	$= 22 \times (0.13/0.42) = 6.8$°C

The surface temperature of the brickwork inside the enclosure would thus be $21 - 6.8 = 14.2$°C.

The more complex case of the curtain wall construction in the previous example is dealt with in the same way, the surface and the interface temperatures being as shown in Fig. 4.2(b).

Application

Having established the transmittance coefficients for the various structural elements enclosing a space, it is then possible to evaluate the conduction component of the heat requirement. This process is best illustrated by an example.

A simple building having a roof which is part exposed to the outside air and part to a heated room above it is shown in Fig. 4.3. It is to be heated to provide an internal air temperature of 21°C when the outside air temperature is −1°C. The construction is as follows:

Walls	100 mm concrete panel, 25 mm air gap, 75 mm insulation, 100 lightweight block, 20 mm lightweight plaster (as previous example).
Roof	19 mm asphalt on 100 mm insulation on 75 mm screed and 150 mm concrete, unplastered.
Floor	75 mm screed on 75 mm EPS on 200 mm solid *in situ* concrete on sand and gravel.
Windows	Double glazing in PVC frames with low e glass.

Surface				*Area (A)* (m^2)	*U* (W/m^2 K)		*AU* (W/K)
Floor	=	14.5 × 8.5	=	123			
		6.0 × 8.5	=	51	174 × 0.25	=	43.5
Roof	=	6.0 × 8.5			51 × 0.25	=	12.75
Windows	=	12.0 × 2.0			24 × 2.2	=	52.80
Walls	=	48.5 × 4.0	=	194			
		Less windows	=	24	170 × 0.35	=	59.50
					419		168.55

Note
perimeter to floor area ratio = 0.279 m^{-1}.

Thus, the conduction loss, air to air, is $169 \times 22 = 3.72$ kW and $\Sigma(AU)/\Sigma A = 169/419 = 0.4$ W/m^2 K.

Building insulation

Consideration of thermal insulation has now come to be regarded as an essential routine during the design process for a new building and a matter for earnest consideration in refurbishment or change of use of any building constructed more than a decade ago. Insulating materials tend to be porous by nature and hence are structurally weak: they are most commonly used as infill material or as an inner skin protected from the weather. When used with flat roofs, insulants are prone to absorb condensation and hence a *vapour seal* or some means of venting is required: recent building methods use a form of construction known as the inverted or 'upside down' roof in which the insulating membrane is placed on top of the structure below the

Table 4.3 Thermal insulating materials for buildings

Material	*Density* (kg/m^3)	*Thermal conductivity* (W/m K)	*Specific heat capacity* (kJ/kg K)
Lightweight aggregate concrete block	600	0.20	1.0
Autoclaved aerated concrete block	700	0.20	1.0
Autoclaved aerated concrete block	500	0.15	1.0
Mineral wool (quilt)	12	0.042	1.03
Mineral wool (batts)	25	0.038	1.03
Expanded polystyrene (EPS)	15	0.040	1.45
Extruded polystyrene	40	0.035	1.4
Polyurethane foam	30	0.025	1.4
Urea formaldehyde (UF) foam	10	0.040	1.4
Blown fibre	12	0.040	1.03

water-proofing layer. Nevertheless, it is frequently possible, by the selection of the right materials and techniques, to achieve a high degree of insulation for little or no overall cost.

Table 4.3 provides details of a range of common types of insulating material.

Envelope insulation standards

The following table defines the limiting insulation standards applicable to non-domestic buildings (similar standards are applicable to dwellings) to comply with the Building Regulations 2006 Part L.

Exposed element	*U value* (W/m^2 K)
Flat roof	0.25
Walls	0.35
Floors and ground floors	0.25
Windows, roof windows and doors (area weighted average for the whole building), glazing in metal frames	2.2
Windows, roof windows and doors (area weighted average for the whole building), glazing in wood or PVC frames	2.2
Roof lights	2.2
Vehicle access and similar large doors	1.5

Minimum standards of airtightness

As insulation standards improve, the energy loss through air infiltration becomes increasingly significant. Air infiltrating through the wall/roof assemblies not only results in a significant direct energy loss, but also reduces building performance in other ways.

- It can reduce the temperature on the warm side of the insulating layer, thereby reducing the effectiveness of the insulation.
- It can also bring substantial amounts of moisture into the assembly (at a rate much greater than that by diffusion). This can cause wetting of the insulation and a consequent degradation of performance. Over the long term, it can also result in structural damage.
- It can create problems of draughts in cold weather.
- It can upset the balance of mechanical ventilation systems, especially in windy or cold weather.

Measurements of air leakage have shown that UK buildings are generally much leakier than their Scandinavian or North American counterparts. In order to stimulate improvement of this critical aspect of energy performance, a maximum air permeability limit of 10 m^3/h m^2 at an applied pressure difference of 50 Pa has been set. Air permeability is the air leakage averaged over the whole envelope area (including ground floors), and is a different measure to the oft-quoted air leakage index, which excludes floors in contact with the ground in the area normalisation. As with avoidance of thermal bridging, achieving a satisfactory air leakage performance requires careful design of the air barrier and of the sealing between elements.

The Department for Communities and Local Government (DCLG) has developed a series of *Accredited Construction Details* to help the construction industry achieve the performance standards required to demonstrate compliance with the energy efficiency requirements (Part L) of the Building Regulations. These *Accredited Construction Details* replace the earlier developed so called *Robust Details*.

An important addition to the regulations is the requirement to pressure test the completed assembly to ensure the required airtightness standard has been achieved. This should be carried out according to the procedures of the Air Tightness and Measurement Association (ATTMA) *Measuring Air Permeability of Building Envelopes 2006*. Should the building fail to meet the standard, identification and sealing of leakage sites will be required until the standard is achieved. The initial airtightness standard of 10 $m^3/h\ m^2$ at 50 Pa is a relatively modest one, with best practice standards being lower by a factor of two or three. It is therefore likely that standards will be made significantly more demanding at subsequent revisions of Part L, especially in non-domestic buildings.

Atypical construction features

The thermal characteristics of the majority of materials and the various composite construction elements built up from them are not very complex and their thermal transmittance, for steady state energy flow, may be calculated by application of simple arithmetic as has been shown earlier on p. 59. Brief notes on some of the exceptions follow here. It should also be recognised that Part L of the current Building Regulations require solutions to be calculated using computer models.

Non-homogenous constructions

These present unique problems arising from discontinuities at corners, at junctions and at heat bridges generally. A rule-of-thumb approach to their solution, which errs on the side of safety, suggests that where insulating blocks are jointed with sand/cement mortar, and where insulation in a timber framed wall is bridged by studs and noggins, the conductivity of the block in the first case and that of the insulation fill in the second, both as quoted in Table 4.3, should be increased notionally by about 15 per cent.

As for hollow blocks, the *Guide Section A3* provides an analytical method of calculation to take account of the resistance of the enclosed air gap (or gaps) but this does not make allowance for mortar infill due to careless workmanship on site. Observation suggests that it is wise to add 50 per cent to any quoted conductivity to allow for the cumulative effect of jointing and of mortar dropped into those cavities.

It is appropriate again to mention that the DCLG has developed a series of *Accredited Construction Details* to help the construction industry achieve the performance standards required to demonstrate compliance with the energy efficiency requirements (Part L) of the Building Regulations. The details focus on the issues of insulation continuity (minimising cold bridging) and achieving airtightness particularly the junctions of sections of the building envelope.

The details contain checklists which should be used by the Designer, Constructor and Building Control Body to demonstrate compliance.

The details have been grouped by generic construction type, as follows:

- Steel frame details
- Timber frame details
- Masonry cavity wall insulation details
- Masonry internal wall insulation details
- Masonry external wall insulation details.

Ground floors, solid and suspended

In cases where a building covers other than a small area, the heat loss from the floor will occur largely around the perimeter since the temperature of the earth at the centre will, over a period of time, approach that within

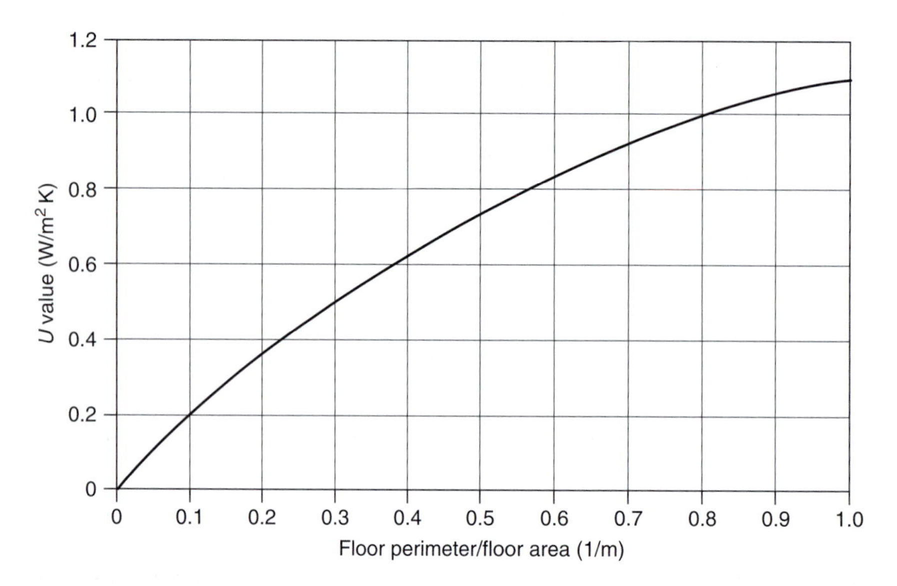

Figure 4.4 *U* values for floors

the room. This situation applies whether the floor is a slab on the ground, a suspended concrete construction or suspended joists and boarding. Whereas previous practice was to select a *U* value according to the floor shape and dimensions, it has been shown by Anderson[4] that a more flexible alternative using the perimeter/area ratio of the floor has much to commend it (see Fig. 4.4).

The boundary between the space to be heated and any unheated areas such as garages is included in the perimeter measurement but the unheated area itself is, of course, excluded. Where the perimeter/area ratio is less than about 0.3 m^{-1} then insulation may not be strictly necessary: however, the provision of a horizontal strip about 1 m wide around the exposed perimeter is good practice (Fig. 4.8). The data plotted in Fig. 4.4 have been adjusted to relate to the air temperature difference ($t_{ai}-t_{ao}$) used for the remaining surfaces.

Glass, glazing and windows

It will be appreciated from Fig. 4.1 that, since the thermal resistance of a sheet of ordinary glass is negligible, the *U* value for single glazing is calculated very simply from the two relevant surface resistances, i.e. $1/(R_{so}+R_{si})$. Similarly, in the case of double and treble glazing, the *U* value may be calculated from those same surface resistances plus, as appropriate, one or two air gap resistances. Thus, for a normal exposure; a 12 mm wide sealed gap between panes and high emissivity surfaces, the notional values for glazing are:

$$\text{Single} = 1/(0.06 + 0.12) = 5.56\ \text{W/m}^2\ \text{K}$$
$$\text{Double} = 1/(0.18 + 0.18) = 2.78\ \text{W/m}^2\ \text{K}$$
$$\text{Treble} = 1/(0.36 + 0.18) = 1.85\ \text{W/m}^2\ \text{K}$$

These values, it will be noted, are not very different from those in the first column of Table 4.4. Both of the remaining examples in this same column relate to double glazing where one of the inner surfaces which faces the gap has been provided with a transparent low emissivity coating ($e = 0.1$). In the first case, a normal air gap is provided but, in the second, the cavity between the panes has been filled with the inert gas *argon* instead of air. The low emissivity coating and the argon fill each improve the thermal performance of the glazing considerably.

The final four columns of Table 4.4 show practical values of the transmittance coefficient for real windows and differentiate between different materials and methods of mounting. The variation from the notional values is due not only to the considerable proportion of the total area available taken up by the frame (often 20 per cent), but also to bridging effects to the structure.

Table 4.4 Thermal transmittance (U value, W/m^2 K) for glazing and windows with frames

	Glazing		*Windows (glazing in frames)*			
				Metal		
Type	*Vertical*	*Horizontal*	*Wood*	*Bare*	*Thermal barrier*	*uPVC*
Single	5.6	7.1	4.8	5.7	5.4	4.8
Double	2.9	3.6	2.7	3.4	3.1	2.7
Triple	1.9	2.6	2.1	2.7	2.3	2.1
Double (low emissivity coating)	2.0	2 2	2.0	2.5	2.1	2.0
Double (low emissivity coating and argon fill)	1.7		1.9	2.4	2.0	1.9

Note
Values quoted are for double-glazing sealed at works; frame area 30% for wood and uPVC, 10% for metal; low emissivity coating; e = 0.2 normal exposure.

It seems appropriate here to draw a clear distinction between double glazing and double windows. The principal difference is that double glazing mounts two sheets of glass, spaced and hermetically sealed in the works of the supplier in a single composite frame, whereas double windows normally consist of a secondary system of glazing, fitted quite independently, on the room side of an existing outer facade window. In consequence, since ready access to the space between the panes of double windows is necessary for good housekeeping, it follows that the air space between them will not be sealed, the resistance R_a will be reduced by about a quarter and the U values will be increased by about 0.5 W/m^2 K.

Windows which are double glazed, if of good quality, may be expected to retain their thermal characteristics whereas double windows, opened regularly for cleaning, are subject to misuse and thus a falling performance. As a contra argument, in the refurbishment of existing buildings, the capital cost of replacing windows *and* frames might be such that adding an additional pane is the only practical alternative.

Further, in terms of noise transfer from outside, double windows have an acoustic advantage arising from the wider air gap, provided that the reveals between the panes are treated suitably. A perfectionist might choose to make use of an expensive compromise and fit double glazing at the outer facade, to ensure the optimum thermal advantage, backing this up with acoustically treated reveals and a moveable inner pane!

Wall cavity fill

The thermal resistance of an unventilated air gap, as stated earlier and demonstrated in Table 4.1, is 0.18 m^2 K/W for building surfaces having a high emissivity, i.e. the majority. However, if that cavity were to be filled with an insulating material having a resistance of, say, $(1/0.035) \times (50/1000) = 1.43$ m^2 K/W, the U value of the structure would be greatly improved. Taking a conventional cavity construction, from the earlier example, having an outer skin of 105 mm brick and an inner skin of 100 mm lightweight concrete block, finished with 20 mm lightweight plaster, the thermal transmittance with an empty cavity would be 0.46 W/m^2 K, whereas with insulation therein it would be 0.35 W/m^2 K, i.e. a 30 per cent increase in performance.

Features to be noted in respect of filling wall cavities, in this and other ways, Figs 4.5 and 4.6, are:

- Cavities in the walls of new buildings may, of course, be filled in a variety of ways during the construction period using materials such as expanded polystyrene sheet, glass or mineral fibre slabs and the like.
- Cavities in the walls of existing buildings, where these are sound and not overly exposed to damp, may be filled by the injection of materials such as either urea formaldehyde foam or beads of mineral wool or polystyrene, treated with a water repellant, blown into the cavity by air jet in a manner not dissimilar to that used with foam.
- Insulation will reduce the initial cost of any heating system.

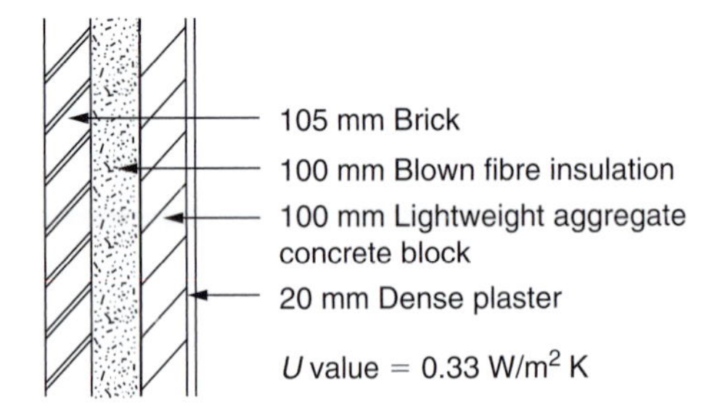

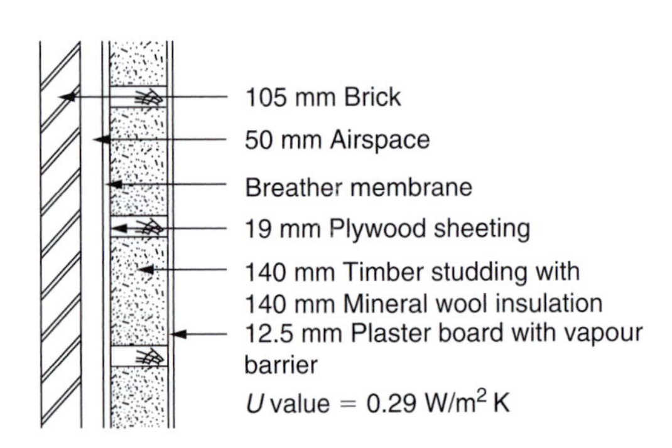

Figure 4.5 Methods of applying insulation to walls

- Cavity fill will reduce noise transmission from dwelling to dwelling as, for instance, at the point of intersection with a party wall.
- When a steady state condition is reached, the inner wall surface temperature will be at a higher temperature than would be the case with an unfilled cavity, thus reducing the risk of condensation.
- If heating is intermittent, the savings theoretically possible may not be achieved since all the heat which has been absorbed in the inner skin during the 'on' period may have been dissipated during the 'off' period.

Safeguards are necessary in adopting this method of treatment, owing to the fact that rain penetration through the outer skin will seek out any discontinuities in the injected or inbuilt material and will thus allow moisture penetration to the inner skin and plaster. It has always been a cardinal rule that mortar dropping onto wall ties and the like must be avoided in the building of cavity walls. It is, in consequence, recommended that application of this form of insulation be entrusted only to firms approved by the British Board of Agreement.

It is, furthermore, essential that the material used be water resistant and resistant to rotting, mould growth and attack by vermin. The principal danger is that of penetration, by driving rain, of the outer skin of the wall construction. Once a cavity has been filled, there is apparently no known method for clearing it completely!

Flat and sloping roofs

As has been noted previously, attack upon the energy loss through factory roofs was the target of early legislation. Methods of insulation more appropriate to the present day are shown in Fig. 4.6.

Floors

It has been noted earlier that the heat loss from a solid floor on earth will occur largely around the perimeter. It follows, therefore, that provision of insulation overall may not be necessary and that a horizontal strip about 1 m wide laid at the exposed edge of the floor, as Fig. 4.7, will be adequate.

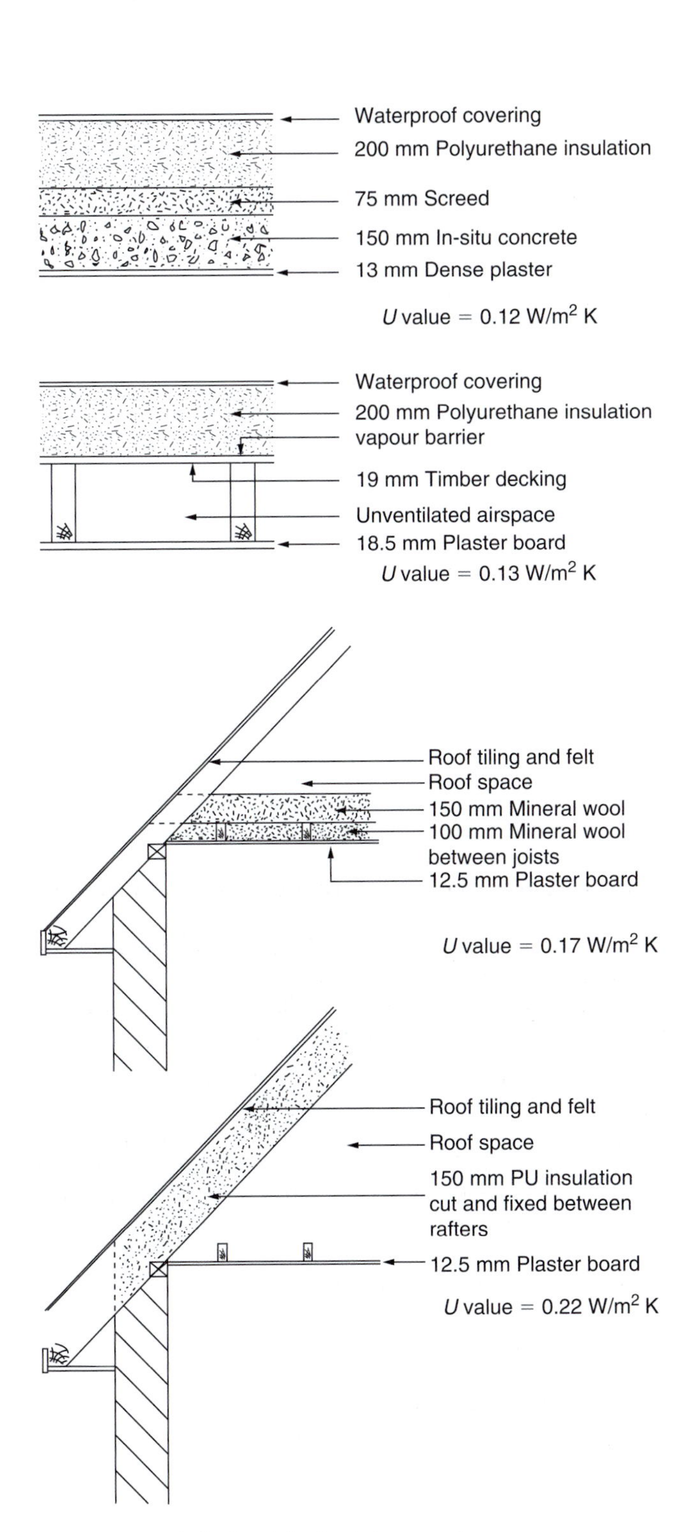

Figure 4.6 Methods of applying insulation to roofs

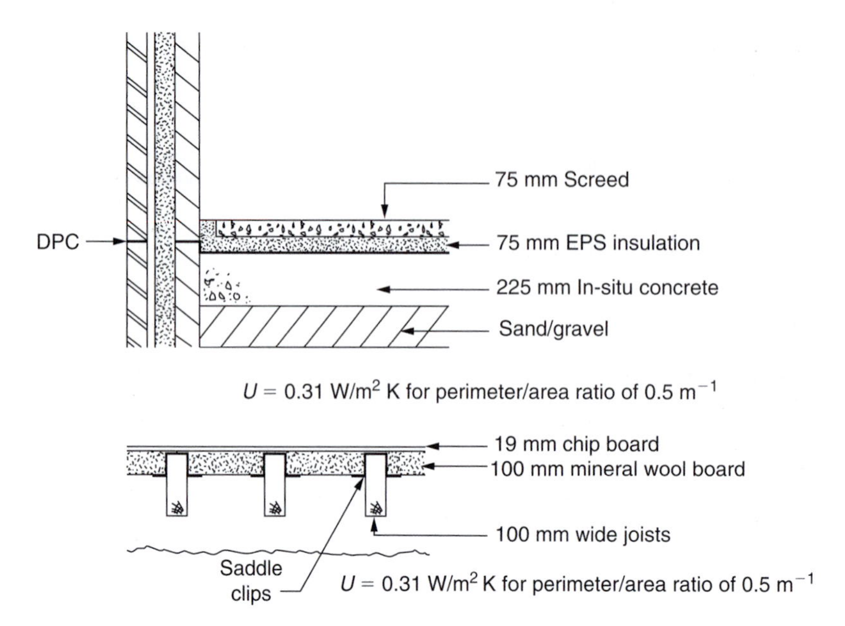

Figure 4.7 Methods of applying insulation to floors

For a timber floor suspended above an enclosed but ventilated air space, insulation is best applied immediately below the actual flooring. One method is to lay rigid mineral fibre or urethane boards, push-fitted between the joists and secured by means of saddle clips or battens as Fig. 4.7. Some means of access to any piping or wiring below the insulation must not be forgotten.

Application

It is now appropriate to reconsider the earlier example of conduction heat loss in the light of possible improvement to the thermal characteristics of the building elements in order to further reduce heat loss and to comply with future editions of Part L of the Building Regulations. For instance, the floor would be covered internally by panels of 75 mm EPS and 75 mm screed and the roof might incorporate 200 mm polystyrene insulation instead of 100 mm. The walls might be improved 100 mm insulation and a 150 mm lightweight concrete block. The internal finish would remain as 20 mm lightweight plaster. As to the windows, those might be replaced with double glazing in a PVC frame having argon injected in the cavity between the panes of glass. Thus, assuming the same temperature difference as before (22 K), the calculation would be revised as follows:

Surface		*Area* (*A*) (m^2)	*U* (W/m^2 K)	*AU* (W/K)
Floor	= 14.5 × 8.5	= 123		
	= 6.0 × 8.5	= 51	174 × 0.25	= 43.5
Roof	= 6.0 × 8.5	=	51 × 0.12	= 6.12
Windows	= 12.0 × 2.0	=	24 × 1.8	= 43.2
Walls	= 48.5 × 4.0	= 194		
	Less windows	= 24	170 × 0.275	= 46.75
			419	= 140

Thus, the conduction loss, air to air, is 140 × 22 = 3.08 kW and 140/419 = 0.33 W/m^2 K. This represents a 17 per cent increase in the thermal performance of the building envelope against the earlier example.

Table 4.5 Transmission coefficients (*U* values) and resistances for a range of typical *uninsulated* constructions for 'normal' exposure (for use in conjunction with Table 4.6)

Construction (dimensions in rara)	*U* (W/m^2 K)	*1/U* (m^2 K/W)
1. 19 render, 220 no fines concrete block : ■ : 12 plasterboard on battens	2.05	0.488
2. 19 render, 220 dense concrete block : ■ : 12 plasterboard on battens	2.40	0.416
3. 80 dense concrete panel : ■ : 100 dense concrete block, 13 dense plaster	2.68	0.373
4. 80 dense concrete panel : ■ : 100 dense concrete block, 13 dense plaster	3.09	0.324
5. 220 solid brick : ■ : 13 dense plaster	2.14	0.468
6. 220 solid brick : ■ : plasterboard on battens	1.93	0.517
7. 105 brick : ■ : 105 brick, 13 lightweight plaster	1.82	0.549
8. 105 brick : ■ : 105 brick, 13 dense plaster	2.0	0.499
9. 105 brick : ■ : 100 dense concrete block, 13 dense plaster	2.55	0.392
10. 105 brick : ■ : 100 lightweight concrete, 13 dense plaster	1.17	0.857
11. 105 brick : ■ : 100 autoclaved aerated concrete, 13 lightweight plaster	1.07	0.936
12. 105 brick, air gap, plywood membrane, 95 stud : ■ : 12 plasterboard on stud	1.44	0.696
13. 105 brick, air gap, plywood membrane, 140 stud : ■ : 12 plasterboard on stud	1.44	0.696
14. Waterproof covering, 75 screed : ■ : 150 concrete roof, 13 dense plaster	2.05	0.488
15. Tiles on battens, roofing felt and rafters : ■ : bare plaster ceiling below joists	2.6	0.385

Note
If a better performance be required, subject to investigation of the risk of condensation, insulation may be placed as shown by the symbol ■.

Table 4.6 Resistances of insulation materials of standard thickness (m^2 K/W)

	Thickness (mm)											
Material	*20*	*25*	*30*	*35*	*40*	*45*	*50*	*55*	*60*	*65*	*70*	*75*
A.1 Expanded polystyrene (board)	0.54	0.68	0.81	0.95	1.08	1.22	1.35	1.49	1.62	1.76	1.89	2.02
A.2 Extruded polystyrene (board)	0.8	1.0	1.2	1.4	1.6	–	2.0	–	2.4	–	–	–
A.3 Glass fibre (rigid slab or batt)	–	0.81	0.97	–	1.29	–	1.61	–	1.94	–	–	2.42
A.4 Glass fibre (flexible and mat)	–	0.63	–	–	1.0	–	1.25	–	1.5	1.63	–	1.88
A.5 Mineral fibre (rigid slab or batt)	–	0.76	0.91	–	1.21	–	1.52	–	1.82	–	–	2.27
A.6 Mineral fibre (flexible and mat)	–	–	–	–	–	–	1.34	–	1.62	–	–	2.02
A.7 Phenolic foam (board)	1.11	1.39	1.67	1.94	2.22	2.5	2.78	3.06	3.33	3.61	3.88	4.17
A.8 Polyurethane foam, CFC blown (board)	1.05	1.32	1.58	1.84	2.11	–	2.63	–	3.16	–	–	–
A.9 Polyurethane foam, PHA blown (board)	0.95	1.19	1.43	1.67	1.9	–	2.38	–	2.86	–	–	–

	Thickness (mm)									
Material	*50*	*55*	*60*	*65*	*70*	*75*	*80*	*100*	*150*	*200*
B.1 Blown mineral fibre (for existing cavity)	1.25	1.38	1.5	1.63	1.75	1.88	–	–	–	–
B.2 Urea formaldehyde foam (for existing cavity)	1.61	1.77	1.94	2.1	2.25	2.42	–	–	–	–
B.3 Glass fibre mat (in roll)	1	–	1.5	–	–	–	20	2.5	3.75	5.0
B.4 Mineral fibre mat (in roll)	–	–	1.62	–	–	–	2.16	2.7	4.05	5.41
B.5 Vermiculite granules (loose fill)	–	–	–	–	–	–	1.23	1.54	2.31	3.08

Selection of the means to be used in upgrading an existing structure or to produce an acceptable standard in a proposed building may follow the use of calculation routines which have been described earlier (p. 59). It may however be found convenient to make use of Tables 4.5 and 4.6 when selecting the type and thickness of insulation to be used. The first of these lists *U* values and the corresponding resistances (1/*U*) for a limited selection of typical *but wholly uninsulated* constructions. It would, of course, be out of the question to provide a fully comprehensive version of such a table in view of the wide and ever-changing availability of composite constructions.

The complementary table (Table 4.6) provides values of resistances for a variety of insulation materials of standard thickness, as manufactured. The method of use of the two tables is best illustrated by a simple example.

The walls of a new building are to be constructed as Item 10 in Table 4.5. The resistance is 0.857 m^2 K/W but Building Regulations require a U value of 0.35 W/m^2 K or a resistance ($1/U$) of 2.857 m^2 K/W.

Resistance deficit = (2.857 − 0.857)	= 2.000
Deduct for an air gap (Table 4.1)	= 0.18
	1.82 m^2 K/W

From Table 4.6, select insulation as, say:

50 mm of extruded polystyrene	= 2.0 m^2 K/W
or 40 mm of polyurethane (board)	= 2.11 m^2 K/W
or 40 mm of phenolic foam (board)	= 2.22 m^2 K/W

Condensation

If the temperature at the internal surface of any element of the building structure falls below the dew point temperature of the air within the space, condensation of the water vapour in the air will take place on that surface. The problem is likely to occur as a result of dense occupancy and of domestic moisture-producing activities such as cooking, bathing, and clothes or dish washing. In commercial or industrial premises, further hazards exist as a result of steam- or vapour-producing activities and of the need for a humid atmosphere to suit certain processes.

In the case of windows, the problem may be ameliorated by either the introduction of double or triple glazing or by provision of a warm air convective current to 'sweep' the glass area. Where the building element is solid, however, such as a wall, a ceiling or a floor, the surface may be absorbent to a greater or lesser degree and thus condensation will be less visible although it will still exist. A concrete floor, suspended above a space open to the outside air, for example an office floor above an open car park, may be subject to condensation on the floor surface, even when insulation of apparently adequate quality and thickness has been incorporated in the structure.

Similarly, in multi-storey blocks of flats, conduction from the edge of an exposed balcony has been known to cause problems where the balcony and the slab forming the ceiling of the flat below have been constructed as a single unit. This type of structural bridging, epitomised by concrete or metal mullions and ribs formed across what should have been a weather barrier, without allowance for discontinuity in the thermal sense, has led to many cases of condensation on surfaces within the building.

The present drive for energy conservation and the consequent introduction of higher standards of structural insulation have led to problems arising from condensation actually within the structure. Perimeters used for the more traditional forms of building construction were homogeneous and to a large degree impermeable and had a wide margin of safety inherent in their character. More modern building structures with the required better level of insulation are, in effect, laminates of diverse internal and external finishes with layers of insulation, etc. sandwiched between. Inevitably, some of the outer layers outside the insulation remain colder and the risk of intermediate or *interstitial* condensation then arises. A brief treatment of this aspect of the subject follows.

Interstitial condensation

The majority of the materials used in building construction and many insulating materials will allow the movement of water vapour through them by diffusion. If a higher vapour pressure exists on one side of the material than on the other, then movement of moisture will take place, subject only to the vapour resistance offered. Table 4.7 provides values of vapour resistivity for a limited range of building and insulating materials.

These values may be thought of as properties parallel to the values of thermal resistivity ($1/k$) listed with them for reference and, like them, to be multiplied by the material thickness to provide individual resistances. Table 4.8 provides approximate values for the resistance of films. The individual vapour resistances may be added to provide a total for a building structure and, although the total should include for surface and air gap resistances, these are so relatively small that they may be ignored.

Table 4.7 Vapour resistivities of some common materials

Material	*Density* (kg/m^3)	*Thermal resistivity* (m K/W)	*Vapour resistivity* [N s/(kg m × 10^9)]
Common brick	1700	1.19	35
Dense concrete	2100	0.71	200
Lightweight concrete	600	4.55[a]	45
Dense plaster	1300	2.00	50
Glass fibre slab	25	28.6	10
EPS slab	25	28.6	100

Note
a Derived from 0.19 (Table 1.1) × 15% for sand cement joints = 0.22 W/m K and thus 1/0.22 = 4.55 m K/W.

Table 4.8 Vapour resistance of some common films

Material	*Thickness* (mm)	*Vapour resistance* (N s/kg × 10^9)
Polythene	0.05	125
Gloss paint	–	8
Varnish	0.05	5
Aluminium foil	–	>4000

As to the rate of vapour transfer, by mass, this may be computed for either an element or a whole structure from:

$$m = \Delta p_v / G$$

where

m = rate of vapour transfer per unit area (kg/m^2 s)
Δp_v = vapour pressure difference across material or structure (Pa)
G = vapour resistance of material or structure (N s/kg)

Application

The application of these details in evaluation of a problem is best illustrated by an example and the curtain wall structure used earlier in Fig. 4.2 is repeated here for convenience as Fig. 4.8. All the temperature data given there are retained and expanded only to include values of percentage saturation, 100 per cent externally and 40 per cent within the room which were not relevant to the thermal calculation. For the necessary listings of vapour pressures, reference would be made in practice to the psychrometric tables in the *Guide Section Cl* or some other source of tabulated data.

In exactly the same manner as that used to calculate the individual surface temperatures using thermal resistances, (p. 61), the vapour pressure gradient through the structure may be found using the vapour resistances from *Guide Section A3*. The overall resistance will be made up thus:

Inside air to surface resistance			–
20 mm plaster	0.02 × 50	=	1.00
150 mm lightweight block	0.150 × 45	=	6.75
100 mm glass fibre slab	0.1 × 100	=	10.0
25 mm air gap			–
100 mm concrete panel	0.040 × 200	=	8.00
Outside air to surface resistance			–
		=	25.75 N s/kg × 10^9

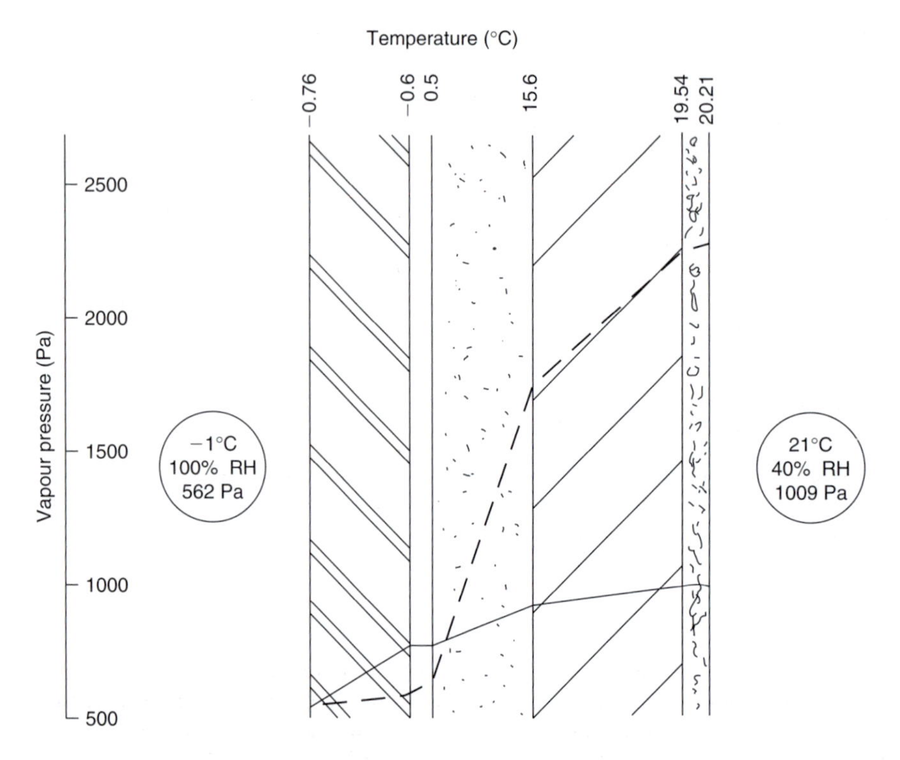

Figure 4.8 Condensation: vapour pressure gradients

The total change in vapour pressure across the structure will be, read or interpolated from *Guide Section Cl*, the vapour pressure at the inside surface condition (20.21°C and 41 per cent RH) minus the vapour pressure at the outside surface condition (−0.76°C and 98 per cent RH) (1005 − 565) = 440 Pa and this must be allocated in line with the resistances of each component of the structure as follows:

Room side surface of plaster	1005 Pa
Interface, plaster to concrete block	993 Pa
Interface, concrete block to glass fibre	914 Pa
Interface, glass fibre to air gap	797 Pa
Interface, air gap to concrete panel	797 Pa
Outside surface of concrete panel	565 Pa

These values may now be plotted, as shown by the heavy full line in Fig. 4.8, to illustrate the vapour pressure gradient through the structure.

To represent the dew points at the various surfaces and interfaces, 100 per cent saturation vapour pressure values may be again read or interpolated from *Guide Section Cl* for the various temperatures noted at the top of Fig. 4.8 and a gradient of saturation vapour pressure may then be plotted, as shown by a heavy broken line. For ease these saturated vapour pressure values as interpolated are as follows:

Room side surface of plaster	2320 Pa
Interface, plaster to concrete block	2200 Pa
Interface, concrete block to glass fibre	1745 Pa
Interface, glass fibre to air gap	650 Pa
Interface, air gap to concrete panel	590 Pa
Outside surface of concrete panel	580 Pa

As will be noted, the two lines cross over one another at vapour pressure values of 790 and 570 Pa, suggesting there is a risk that condensation will occur in the outer *leaf* and close to the interface of the insulation and air gap. Since, of course, it is not possible for the vapour pressure level in the structure to exceed that of the saturation vapour pressure, the former will adjust to the gradient shown by the chain dotted line and indicate that excess moisture will condense out at the interface indicated.

As will be appreciated, the external conditions assumed for this example were the heating design conditions used earlier in this chapter, whilst they serve to illustrate one method used to trace and evaluate interstitial condensation they could be considered severe.

Design conditions are generally selected to suit the purpose of the analysis, which is usually to determine the long term build up of condensation within the thickness of the construction. *BS* 5250 suggests the following that relate to a 60-day period for the UK in winter.

- Outdoors: 5°C, 95 per cent RH
- Indoors (dry-moist occupancy): 15°C, 65 per cent RH
- Indoors (moist-wet occupancy): 15°C, 85 per cent RH.

However it could be considered that these conditions are rather less searching and *Guide Section A7* recommends month-by-month analysis. For further guidance, the reader is referred to two excellent publications by the Building Research Establishment[5] and *Guide Section A7*.

Air infiltration

The subject matter of the last few paragraphs has related to conduction heat loss through the building fabric, a matter which is capable of examination on a rational basis. As was explained much earlier in this chapter, however, the matter of air infiltration from outside the building must be considered also. Leakage through windows and doors, an upward draught through an unsealed staircase, and leakage through the structure itself, particularly in a factory-type sheeted building, will each have an influence.

The importance of air infiltration is that it may well account for as much as half or more of the total heat loss and yet it remains the least amenable to logical and systematic prediction. With improvement to the thermal properties of the building structure through added insulation, air infiltration has increasingly become the dominant component in heat loss. Consequently, temperature guarantees become more difficult to sustain or challenge since any performance test is as much related to the potential for faults in the building as to those in the heating system. This has been recognised in the legislation with the introduction of *accredited construction details* mentioned earlier and the requirement for airtightness testing under Part L of the Building Regulations.

The heat needed to warm infiltration air is calculated using the specific heat capacity of air (at constant pressure) and the specific mass, both at 20°C. Thus, from Table 1.1, the quantity required to raise unit volume through one kelvin is $1.012 \times 1.205 = 1.219$ kJ/m^3 K.

There are two different approaches for making an assessment of air infiltration. One is empirical and is based upon the number of times the air volume within a space will be changed in one hour,[6] this being referred to as the *air-change rate*. The second and more specific approach is confined mainly to heavily glazed commercial buildings and, as will be explained later, relates areas of openings such as assumed lengths of cracks around windows and doors, etc., to rates of air flow. It is also appropriate to mention here the increasing use of computational fluid dynamics (CFD) to evaluate air flow around and through buildings. This modelling tool can be used in complex city centre locations to set up wind pressure coefficients for each facade of the building, it does however require specialist skills to perform this activity.

Air change and ventilation allowance

For application to the air-change concept and in order to work in units consistent with those used for conduction heat loss through the building fabric, the term *ventilation allowance* is now used, this being related to the air-change rate (N):

$$N(1.219 \times 1000)/3600 = (0.339\,N)\,\text{J}/\text{sm}^3\,\text{K} \sim (N/3)\,\text{W/m}^3\,\text{K}$$

Table 4.9 Natural air infiltration for heat losses (air change rate)

	Peak air charges at 4 m/s wind speed		
	Air permeability m^3/m^2, hour @ 50 Pa		
Building type	*Leaky* *20*	*Part L 2006* *10*	*Tight* *5*
Naturally ventilated office up to 4000 m^2 4 storeys	0.75	0.40	0.20
Air conditioned offices up to 8000 m^2 8 storeys	0.85	0.45	0.25
Air conditioned HQ up to 20 000 m^2 12 storeys	0.80	0.40	0.20
Factories/warehouses 5000 m^2	0.55	0.3	0.15
Schools 2 storeys	0.65	0.35	0.20
Hospitals 8 storeys	0.65	0.35	0.20
Hotels 5 storeys	0.8	0.40	0.20
Dwellings/apartments 5 storeys	1.95	1.00	0.50

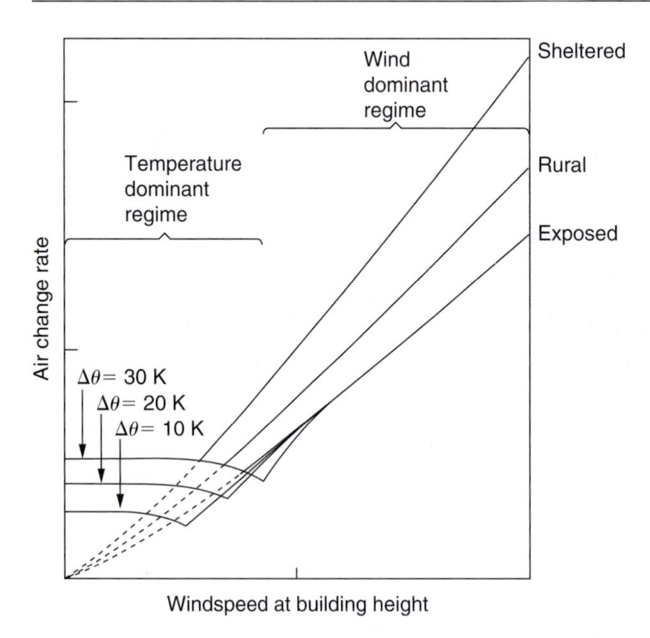

Figure 4.9 Typical infiltration characteristics

Table 4.9 lists the commonly accepted rates of air change for various building types. It should be noted that the values listed in this table are confined to the task of assessing natural air infiltration for heat loss calculations. They do not necessarily represent desirable ventilation rates for the comfort of occupants.

Infiltration through window cracks

Turning now to the more specific method of calculation referred to previously, it should be understood that this is not an alternative to the air-change method, which can be applied generally to buildings other than those for which it has been developed. The two influences considered in this method are that due to wind pressure and that due to temperature induced stack effect in a tall buildings: each of these effects will be dealt with in rather more detail in Chapter 18.[7] From *Guide Section A4* it may be deduced that there is a clearly defined temperature dominant regime at low wind speeds and a wind dominant regime at higher wind speeds, Fig. 4.9 shows graphically these typical infiltration characteristics.

Table 4.10 Values for coefficient k_s and exponent *a*

Terrain	k_s	*a*
Open flat country	0.68	0.17
Country with scattered windbreaks	0.52	0.20
Urban	0.35	0.25
City	0.21	0.33

Air flow through cracks may be evaluated using the following general expression:

$$Q = C(\Delta P)^n$$

where

Q = air volume flow rate per metre run of window-opening joint (litre/s)
C = window air flow coefficient (litre/m s)
n = flow exponent, representing type of opening
ΔP = pressure difference across the window (Pa)

The air flow coefficient depends upon the character of the window, being 0.1 where weather stripping has been applied and 0.2 where it has not. The exponent has been evaluated as 0.5 for large openings and 0.66 for cracks around windows and doors.

In strict terms, solutions from this equation are, however, applicable only to a building with an open plan form, air entering on one side having free access to a similar escape route on the other. In instances where the building has many internal partitions which will impede the cross-flow, then the *building infiltration rate* overall may be only 40 per cent of the calculated value. A typical figure for the generality of buildings might be 70 per cent of that produced by the equation.

The pressure difference ΔP is a function of the prevailing wind speed which will vary according to the terrain surrounding the building and the height of the windows above ground. Wind speed data published by the Meterological Office relate to a height of 10 m above ground in open country but may be corrected for other situations by use of the expression:

$$V = V_m \, k_s z^a$$

where

V = mean wind speed at height z (m/s)
V_m = mean wind speed at 10 m in open country (m/s)
z = height above ground (m)
k_s = a coefficient representing the terrain (Table 4.10)
a = an exponent representing height (Table 4.10)

It is sensible, in establishing infiltration values, to adopt a datum wind speed at the higher end of the scale and, for the greater part of the British Isles, an hourly mean speed of 8 m/s is exceeded for only 10 per cent of the time.

The pattern of wind flow over an exposed building takes a form such as that shown in Fig. 4.10, but this, of course, is a generalisation since the effect of surrounding buildings and obstructions may well disrupt the pattern in an unpredictable manner, as may the aerodynamics of the building shape. However, it would appear that the sum of the positive pressure on the windward side and the negative pressure on the leeward side approximates to unity and that the pressure difference (ΔP) is thus numerically equal to the velocity pressure of the mean wind speed calculated as above, i.e. $p_v = 0.6\ V^2$ Pa.

The method may be summarised as shown in Table 4.11 where the individual values, for a limited range of building heights, are represented as the heat requirements per metre run of window-opening joint, for an air

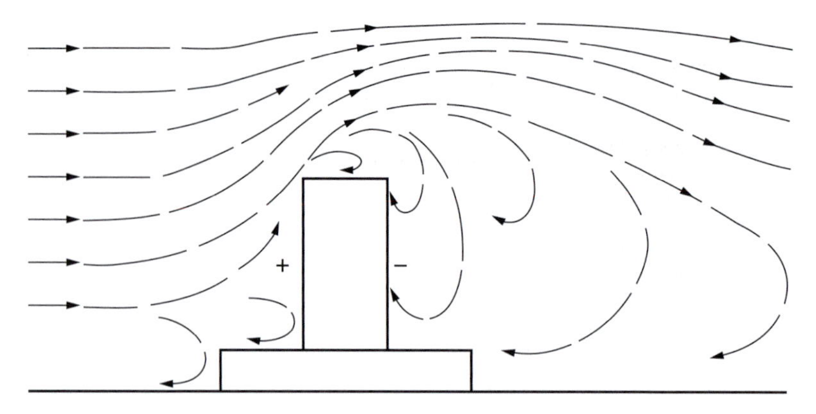

Figure 4.10 Wind currents about a tall building

Table 4.11 Heat requirements for air infiltration through cracks around windows

	Heat requirement per unit length of opening window joint (W/mK)							
	Without weather stripping				*With weather stripping*			
	County				*Country*			
Height of building (m)	*Open*	*With wind breaks*	*Urban*	*City*	*Open*	*With wind breaks*	*Urban*	*City*
6	2.43	1.84	1.22	0.75	1.22	0.92	0.61	0.38
8	2.60	1.98	1.35	0.81	1.30	0.99	0.67	0.40
10	2.71	2.10	1.49	0.94	1.36	1.05	0.72	0.47
15	3.00	2.33	1.66	1.12	1.49	1.17	0.83	0.56
20	3.19	2.52	1.82	1.27	1.59	1.26	0.91	0.64
25	3.35	2.67	1.96	1.40	1.68	1.34	0.98	0.70
30	3.49	2.80	2.08	1.52	1.75	1.40	1.04	0.76
40	3.73	3.03	2.29	1.72	1.86	1.51	1.14	0.86
50	3.92	3.21	2.46	1.90	1.96	1.61	1.23	0.95

Note

Wind speed 8 m/s, at notional height of 10 m in open country, is exceeded for 10% of time only. Other wind speeds derived from this datum.

temperature difference inside to outside of one kelvin. The table lists basic figures which may require adjustment to take account of:

- The reductions which may be appropriate in a building which is liberally compartmented by partitions, as mentioned above. It must be emphasised that these reductions apply to a whole building and *not* to individual rooms.
- The need in a corner room to make an addition of 50 per cent to the tabulated figure to take account of cross-flow.

A further adjustment, which should properly be applied also to results derived from air-change calculations, makes an allowance for stack effect in tall buildings. In winter, the lower floors will have above average infiltration and although the upper floors may have less than the average. The 2006 *Guide Section A4* proposes that on severely exposed sites a 50 per cent increase in air-change rate should be allowed and that air-change rates in rooms in tall buildings may be significantly higher. The design of tall buildings should include barriers against vertical air movement through stairwells and shafts to minimise stack effect.

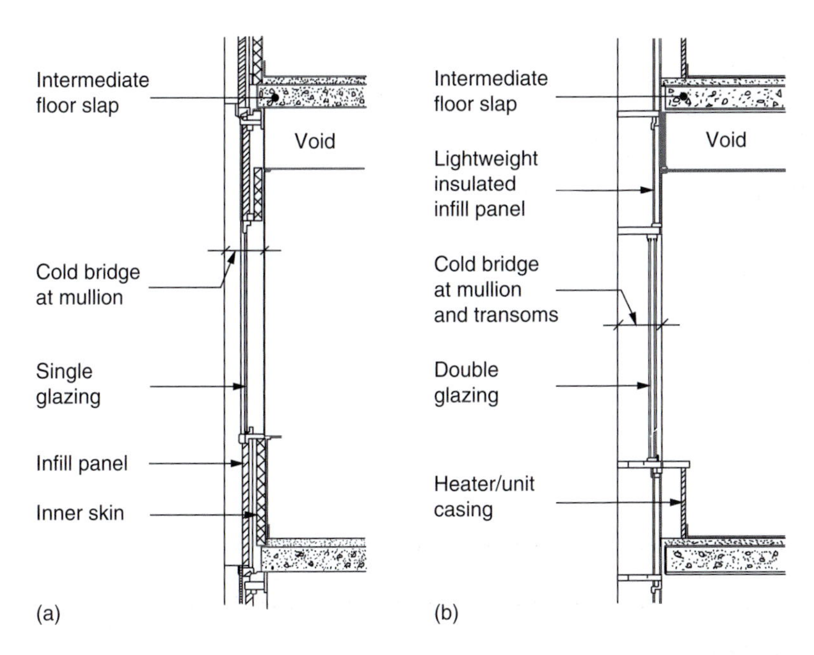

Figure 4. 11 Examples of curtain walling

This last adjustment results from research work[8] published in 1961 which concluded specifically that:

- Swinging door entrances – for a particular set of conditions – infiltrate about 25 m^3 through a single opening, per person entering or leaving a building, and 15 m^3 per person for a vestibule-type entrance.
- Revolving doors under similar conditions infiltrate about 2 m^3 per person, and motor driven doors about half that amount.
- Adequate heating in winter is required in entrances and vestibules at ground floor level by forced warm air heaters or air curtains, augmented by floor panel heating.

The *2006 Guide Section A4* proposes an equation for calculating the pressure difference related to vertical height differences and hence the air flow through openings.

Reference must also be made here to significant problems which have been encountered with buildings clad in curtain walling. As may be seen from Fig. 4.11 if particular care has not been taken to seal the structural joint where the cladding passes the edge beam of the floor slab, leakage can occur and stack effect will cause air to move upwards, floor by floor in cascade. A further example of air flow from an unexpected source can arise where wind pressure is applied to small openings in curtain wall cladding, such as drillings in members provided to drain away condensation, with uncontrolled infiltration to voids above suspended ceilings and consequent excess heat loss. The criteria for tightness of the building envelope has been established and the conclusion reached that site workmanship is the major influence. Part L of the Building Regulations 2006 requires a maximum leakage rate of 10 $m^3/h\ m^2$ at 50 Pa when tested in accordance with ATTMA procedures.

Application

Taking, once again, the simple space shown in Fig. 4.3 and an air-to-air temperature difference of 22 K as before, the infiltration loss on an air-change basis would be:

Room volume

$14.5 \times 8.5 \times 4 = 493$

$6.0 \times 8.5 \times 4 = \underline{204} \quad 697\ m^3$

from Table 4.9, allowing for a leaky construction, take 0.75 air changes. Thus,

$$697 \times 0.75 / 3 \times 22 = 3834 \text{ W}$$

and, in terms of floor area for unit temperature difference,

$$(3834 / 22) / 419 = 0.42 \text{ W/m}^2 \text{ K}$$

If the construction of the building envelope was improved to meet to current Part L standards of air permeability of 10 m^3/m^2 h then the air-change rate might be reduced to 0.4. Thus,

$$697 \times 0.4 / 3 \times 22 = 2045 \text{ W}$$

and, as before,

$$(2045 / 22) / 419 = 0.22 \text{ W/m}^2 \text{ K}$$

It is of interest to compare the magnitude of these two figures with the total conduction losses for the initial and improved versions of this building, as calculated previously, i.e. 3720 W and 3080 W, respectively. If it be assumed that the leaky building is associated with the initial solution and *vice versa*, an overall saving of nearly 50 per cent is revealed. Furthermore, the infiltration losses are shown to be 40 and 50 per cent of the two respective totals, not insignificant proportions! The benefits of airtightness in terms of heating load can also be readily recognised.

As an example of application of the use of Table 4.11, take a typical private office, 5 × 4 × 2.5 m high on the third floor of a 10-storey city building (say 25 m high). The single outer wall has glazing on a 1 m module over the full width with a height of 1.5 m. The crack length of 14 m relates to two window modules, not weather stripped. The air-to-air temperature difference is 22 K.

From the table:

Basic heat requirement = 1.4 W/m K
Allowance for stack effect = add 4 per cent
Thus, $14 \times 1.4 \times 1.04 \times 22 = 448$ W

This requirement may be compared with that which would have resulted from use of an air-change basis:

Room volume of $5 \times 4 \times 2.5 = 50 \text{ m}^3$

from Table 4.9, take 1 air-change or a ventilation allowance of 0.33 W/m K and add 4 per cent for stack effect. Thus,

$$50 \times 0.33 \times 1.04 \times 22 = 378 \text{ W}$$

It will be appreciated that a comparison such as this has no particular significance since addition or reduction in the number of openable windows would not alter the room volume, nor would an increase or decrease in room depth affect the window crack length. In instances where some doubt may exist as to which method of assessment is the more valid, both should be evaluated and the higher result used.

Temperature difference

As will have been noted in the various examples so far provided, the total heat required to maintain a space at the chosen condition is calculated by multiplication of the conduction and air infiltration losses, both in W/K, by a 'temperature' difference between inside and outside. In each of these introductory examples, it was emphasised that air temperature had been used as a simplification.

Inside temperature

In recent years, following the work by Loudon at the Building Research Establishment to which previous reference has been made, the *Guide Section A5* adopted the concept of *environmental temperature* to represent the heat exchange between the surfaces surrounding a space and the space itself. Evaluation of this criterion is dependent upon the configuration of the surfaces and upon the convective and radiant heat transfer coefficients. For the conditions prevailing in the British Isles, environmental temperature may be taken as:

$$t_{ei} = 0.67\,t_{ri} + 0.33\,t_{ai}$$

where

t_{ei} = inside environmental temperature (°C)
t_{ri} = mean inside radiant temperature (°C)[9]
t_{ai} = inside air temperature (°C)

It is now a well-established convention that inside environmental temperature is used for the calculation of conduction loss and that inside air temperature is used for the calculation of infiltration loss. However, as was noted in Chapter 1, *operative* temperature is that which best represents human comfort and a means to facilitate evaluation of the relationship between the various criteria is thus required. This matter will be discussed later in the present chapter.

Outside temperature

To represent conditions external to a building, outside environmental temperature is the appropriate standard. This is more usually known as the *sol–air* temperature which is a notional scale derived from the combined effect of air temperature and solar radiation, to produce the same rate of heat flow as that which would arise if these influences were considered separately. For winter conditions where overcast skies may be assumed to prevail, the outside environmental temperature will be equal to the outside air temperature.

With regard to the outside temperature adopted in the British Isles as a design datum, this will vary depending upon the precise location, the thermal mass of the building and the overload capacity of the heating plant. It may be of interest to follow the development of selection for suitable outside air temperatures for design purposes.

In 1950, a committee was set up by various institutions having an interest in this subject and the report *Basic Design Temperatures for Space Heating*[10] examined the frequency of cold spells and the days per annum when a specified *inside* temperature might not be met by designs based upon an external temperature of −1°C. One argument brought out in the report was that a heavily constructed building has sufficient thermal inertia to tide over a period of a few days of exceptionally cold weather, whereas a lightly constructed building has no such inherent ability. Conclusions drawn from the report were refined by Jamieson[11] and further emphasis given to the importance of thermal time lag related to the overload capacity of the heating system.

Arising from an analysis of background data to international norms, Billington[12] proposed that the British Isles be divided into three zones for the purpose of determining standards for structural insulation and for selection of external design temperatures in place of the earlier single datum of −1°C. A later study[13] analysed data for eight selected locations in order to establish the number of occasions per annum when the mean temperature falls below given levels. The results of this study were used in the *Guide Section A2* as a basis for recommended winter design conditions. The 2006 *Guide Section A2* introduces selection based on variable risk of this occurrence with the choice being agreed between designer and client. *Guide Section A2* does not provide a single design temperature for a particular location. Instead, it offers a range based on frequency of occurrence from which the designer can select, having taken into account the consequences for the occupants or contents of the building. Table 4.12 is derived from this data and assumes that for buildings with low thermal inertia, on average, only 1 day in each heating season had a lower mean temperature and for buildings with a high thermal inertia one 2-day period in each heating season had a lower mean temperature. In order to take account of the 'heat island' conditions in streets and around buildings in towns, adoption of a design basis one or two degrees higher than the levels listed in that table would seem to be reasonable for systems serving buildings in city centres.

Table 4.12 Winter external design temperature (derived from *Guide Section A2*)

	External design temperature, t_{ai} (°C)	
	Low thermal inertia	*High thermal inertia*
Location	*With overload capacity*	*With overload capacity*
Belfast	−2.0	−1.0
Birmingham	−5.0	−3.0
Cardiff	−3.5	−1.5
Edinburgh	−4.0	−2.5
Glasgow	−4.5	−3.0
London	−3.0	−2.0
Manchester	−3.0	−2.0
Plymouth	−1.5	0.0

Temperature ratios

The relationship between the various temperatures noted under the two immediately preceding headings is a function of the thermal characteristics of the method of heating adopted. The inside air and mean radiant temperatures depend, for a given environmental temperature:

- For *convective heating*, upon the conduction loss alone
- For *radiant heating*, upon the infiltration loss alone.

As may be appreciated, there are few methods and even fewer types of heat emitting equipment which fall wholly in one or other of these categories. If, for example, we were to take the same building as used in previous examples with a normal infiltration rate then the inside air temperature would be approximately 1.5 K below the desired operative temperature when heated by radiators and approximately 2 K above the desired operative temperature when heated by a warm air system. It is therefore necessary that a means be provided which will facilitate the performance of routine calculations and whilst there is considerable international debate on the preferred method for these load calculations *Guide Section A5* offers data and guidance. The approach introduced in the *Guide Section A5* offers representative temperature ratios, F_1 and F_2 to group the variables, as follows:

$$F_1 = (t_{ei} - t_{ao})/(t_{res} - t_{ao})$$
$$F_2 = (t_{ai} - t_{ao})/(t_{res} - t_{ao})$$

Values of these ratios, as listed in Table 4.13 which is a summary of data included in the *Guide Section A9, 1986*, are entered into the composite equation:

$$Q_t = [(F_1 Q_u) + (F_2 Q_v)](t_{res} - t_{ao})$$

where

Q_t = the total heat requirement, being the sum of conduction and infiltration losses (W)
Q_u = the conduction loss, per kelvin (W/K)
Q_v = the infiltration loss, per kelvin (W/K)

Application

Take the same building which has been used in preceding examples, as insulated and with an infiltration rate of 1½ air changes per hour. The internal operative temperature is to be 21°C with an external air temperature of −1°C. Heating is to be by single-panel radiators.

Table 4.13 Values for temperature ratios F_1 and F_2

System type and ventilation index		*Loaded average U value ΣAU/ΣA*							
		0.6		*1.0*		*1.5*		*2.0*	
Type of heat emitter (per cent convective)	*Ventilation index 0.33 NV/ΣA*	F_1	F_2	F_1	F_2	F_1	F_2	F_1	F_2
Forced warm air									
(100)	All	0.97	1.10	0.95	1.16	0.92	1.23	0.90	1.30
Natural convectors	0.1	0.97	1.08	0.96	1.13	0.93	1.20	0.91	1.26
Convector radiators	0.2	0.97	1.08	0.96	1.13	0.94	1.19	0.92	1.25
(90)	0.3	0.97	1.07	0.96	1.12	0.94	1.19	0.92	1.25
	0.6	0.98	1.07	0.96	1.12	0.94	1.18	0.92	1.24
	1.0	0.98	1.06	0.96	1.11	0.94	1.17	0.92	1.23
Multi-column radiators	0.1	0.98	1.06	0.96	1.11	0.95	1.16	0.93	1.21
Block storage heaters	0.2	0.98	1.06	0.97	1.10	0.95	1.15	0.93	1.21
(80)	0.3	0.98	1.05	0.97	1.09	0.95	1.14	0.94	1.20
	0.6	0.99	1.04	0.97	1.08	0.96	1.13	0.94	1.18
	1.0	0.99	1.02	0.98	1.06	0.96	1.11	0.95	1.16
Two-column radiators	0.1	0.98	1.05	0.97	1.08	0.96	1.12	0.95	1.16
Multi-panel radiators	0.2	0.99	1.04	0.98	1.07	0.96	1.12	0.95	1.16
(70)	0.3	0.99	1.03	0.98	1.06	0.97	1.11	0.95	1.15
	0.6	1.00	1.01	0.99	1.04	0.97	1.08	0.96	1.13
	1.0	1.01	0.98	0.99	1.02	0.98	1.06	0.97	1.10
Single panel radiators	0.1	1.00	1.01	0.99	1.03	0.98	1.05	0.98	1.07
Embedded floor panels	0.2	1.00	1.00	0.99	1.02	0.99	1.04	0.98	1.06
(50)	0.3	1.01	0.99	1.00	1.00	1.00	1.02	0.99	1.04
	0.6	1.02	0.95	1.01	0.97	1.00	0.99	1.00	1.01
	1.0	1.03	0.91	1.02	0.93	1.02	0.95	1.01	0.96
Ceiling panels	0.1	1.01	0.98	1.01	0.98	1.01	0.98	1.01	0.98
Wall panels	0.2	1.01	0.97	1.01	0.97	1.01	0.97	1.01	0.97
(33)	0.3	1.02	0.94	1.02	0.94	1.02	0.94	1.02	0.94
	0.6	1.03	0.91	1.03	0.91	1.03	0.91	1.03	0.91
	1.0	1.05	0.86	1.05	0.86	1.05	0.86	1.05	0.86
High temperature	0.1	1.02	0.94	1.03	0.92	1.04	0.89	1.05	0.86
radiant heaters	0.2	1.03	0.92	1.03	0.90	1.04	0.87	1.05	0.84
(10)	0.3	1.04	0.90	1.04	0.88	1.05	0.85	1.06	0.82
	0.6	1.05	0.85	1.06	0.83	1.07	0.80	1.08	0.77
	1.0	1.07	0.79	1.08	0.76	1.09	0.74	1.10	0.71

From earlier calculations (p. 70)

$$\Sigma(AU)/\Sigma A = 140/419 = 0.33\ \mathrm{W/m^2\ K}$$

and (p. 80)

$$(0.33\,NV)/\Sigma A = 93/419 = 0.22\ \mathrm{W/m^2\ K}$$

By reference to Table 4.13, and extrapolating,

$$F_1 = 0.98 \text{ and } F_2 = 1.03$$

thus

$$Q_u = 0.98 \times 140 \times 22 = 3018\ \mathrm{W}$$
$$Q_v = 1.03 \times 93 \times 22 = 1903\ \mathrm{W}$$

and hence

$$Q_u = 3018 + 1903 = 4.9\ \mathrm{kW}$$

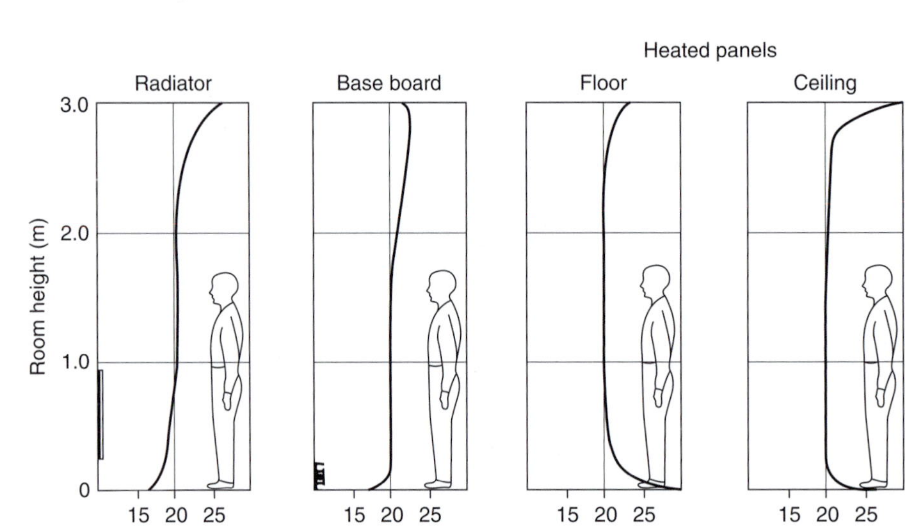

Figure 4.12 Vertical temperature gradients assuming heated rooms above and below

The various internal temperatures may now be established:

$$t_{ei} = (0.98 \times 22) - 1 = 20.56°C$$
$$t_{ai} = (1.03 \times 22) - 1 = 21.66°C$$

The last value above has particular significance since, for the building used as an example when heated by panel radiators, an inside air temperature of 21.66°C is necessary to maintain an operative temperature of 21°C. Had a warm air system been planned, reference to Table 4.13 shows that F_2 would be 1.04, with the result that an inside air temperature of $(1.04 \times 22) - 1 = 21.9$°C would be required to maintain the same operative temperature of 21°C. For this example, because the temperature difference between air and environmental temparature is small, the operative temperature will be very similar to the environmental temperature.

Miscellaneous allowances

The methods outlined for the calculation of heat requirements, as set out in the routines described earlier in this chapter, do not take account of certain elusive factors which may affect the steady state load for a single space or for a whole building. These, in most part, cannot be evaluated other than empirically.

Allowance for height

It appears reasonable to make allowance for the height of a heated space, bearing in mind that warm air rises towards the ceiling, creating a temperature gradient actually within the space. Thus, when heat is provided to maintain a chosen temperature in the lower 1½ −2 m of height, it follows that a higher temperature must exist near the ceiling or roof. In consequence, the conduction loss there will be greater, inevitably, through the surfaces of ceiling, roof, upper parts of walls and windows, etc.

This effect will be greatest with a convection-type system, i.e. one which relies upon the movement of warmed air to heat the space, as is the case with forced and natural convectors and most forms of conventional radiators. Where the radiant component of system output is greater, as is the case with metal radiant panels, heated ceilings and floors, etc., a much more uniform temperature exists over the height of the space and, in the case of floor heating, there is virtually no temperature gradient whatsoever.

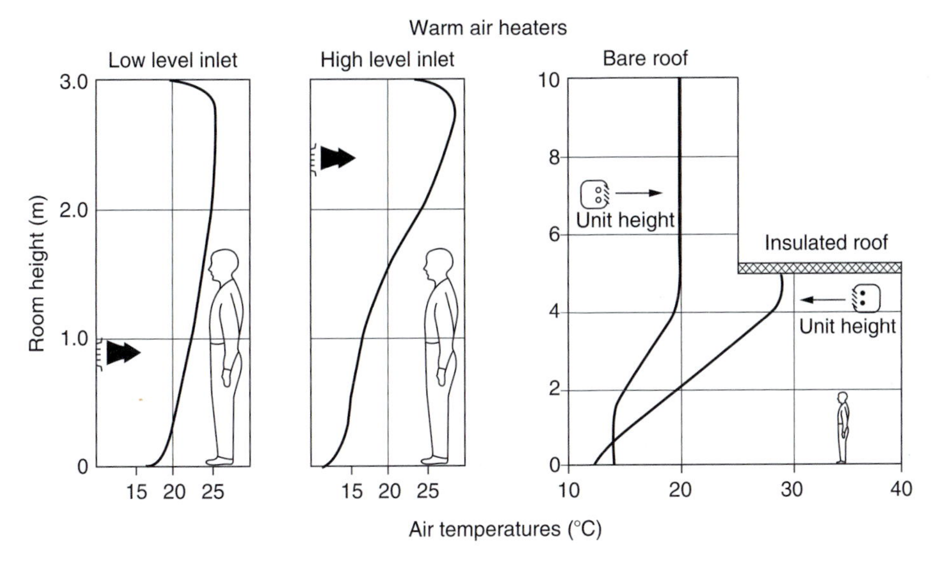

Figure 4.13 Vertical temperature gradients with warm air heating

Table 4.14 Height factors

Type of system	*Type and distribution of heaters*	*Percentage addition for height of heated space*		
		5 m	*5–10 m*	*over 10 m*
Mainly radiant	Warm floor	Nil	Nil	Nil
	Warm ceiling	Nil	0–5	Not applicable to this application
	Ditto, downward radiation from high level	Nil	Nil	0–5
Mainly convective	Natural warm air convection	Nil	0–5	Not applicable to this application
	Forced warm air cross-flow from low level	0–5	5–15	15–30
	Ditto, downward from high level	0–5	5–10	10–20
	Medium and high temperature cross-radiation from intermediate level	Nil	0–5	5–10

This situation is illustrated in Figs 4.12 and 4.13, the latter showing how a warm air system having a discharge at low level produces a temperature gradient which is much less than one discharging at high level due to better mixing with room air. Where radiators and natural convectors are to be used, much can be achieved by correct disposition of these items below large cooling surfaces such as windows. It might, in fact, be said that a system of this type which produces too large a temperature gradient has been badly designed. To make allowance for the effect of room height, the *Guide Section A5* proposes the use of multiplying *height factors*, as listed in Table 4.14.

Whole building air infiltration

The air-change and infiltration data provided in Tables 4.9 and 4.11 relate to allowances appropriate to individual spaces. Where such spaces exist in a building more than one room 'thick', it follows that outside air entering rooms on the windward side is warmed there before passing through rooms on the leeward side and leaving the building. It may be argued, therefore, that although the listed figures should be applied to all individual rooms, the wind direction being unknown, a deduction could be made, in determining the *total* infiltration for the whole building, for this use of 'second-hand' air changes. The *Guide Section A4* includes the concept of a multi-zone model but acknowledges that such models are very complex when balancing the inlet and outlet flows in all of the zones.

In a different category, however, is the parallel proposal that, when a building is unoccupied, the allowance for air infiltration might be reduced. In circumstances where an office building will be occupied for working hours only during a 5-day week, plus an allowance for cleaning, etc., and furthermore where security is such that it may be assumed that both outer and inner doors will be closed out of those hours, then a reduction of air infiltration rates, by some proportion such as half, seems reasonable.

Whole building airtightness can be measured using a fan system temporarily connected to an opening in the building facade. *Guide Section A4* recommends a pressure of 50 Pa being low enough not to cause damage and high enough to overcome the influence of moderate wind speeds. Where problems exist the airtightness test may be accompanied with smoke tests or thermography analysis as a means to identify air leakages routes. Further detailed methodology can be found in ATTMA, Measuring air permeability of building envelopes 2006.

Internal heat gains

Passing reference has already been made to fortuitous heat gains and their potential to make a contribution towards keeping a building warm in winter and thus to reduce the capacity of any heating system which may be provided. Reports[14] by the Building Research Establishment provide a variety of interesting facts regarding the probable annual total of such input, as summarised in Table 4.15. The total of these items approximates to something over a third of the theoretical annual heat requirement of a dwelling so occupied.

In summer, of course, such gains are a penalty to be countered rather than a contribution to be welcomed and they will consequently be dealt with in more detail in the following chapter. For application to a commercial building, the level of reliance which may be placed upon availability may be used to separate winter heat gains into categories:

Reliable: Industrial or office machines operating permanently; electric lighting running permanently; occupied for 24 hours at a constant level.

Relatively reliable: Industrial and office machines that will operate continuously during working hours; electric lighting related to occupancy; permanent staff working to an established attendance routine.

Unreliable: Solar radiation and heat gains from industrial or office machines operated intermittently; electric task lighting; random occupation.

In the case of electric off-peak storage systems using room heaters, it has been suggested that full account should be taken of the 24 hour mean contribution from most of these sources on a 'design day', as related to the thermal capacity of the structure, and this proposition will be discussed further in Chapter 8. However, for conventional heating systems, total dependence upon other than the items listed in the first of these categories would be unwise. As to the second category, where a contribution may be absent over cold weekends and winter holidays, no allowance can be made if comfort temperature in the space is to be recovered in a reasonable time period prior to occupation. This is not to deny that use may be made of all such heat sources, but rather to suggest that a well-designed system will have sufficient capacity to cater for the full design load while retaining the ability to respond to any fortuitous contribution when this may be available.

Table 4.15 Approximate heat input per annum to a dwelling from incidental sources

Source	*Heat input* (GJ/annum)
Two adults, one child (body heat)	5.9
Radiation from sun (15 m^2 glazing)	13.0
Cooking	
Gas	4.3
Electric	3.4
Electrical appliances (lighting, TV, washing machine, etc.)	3.4–8.2
Losses from water-heating pipes and appliances, etc.	10–30

Temperature control

Although this subject is dealt with in some detail in Chapter 28, it is apposite here to emphasise that the ideal system is one where separate automatic control of temperature is available in each room or space. With such a system, heat emitting elements having the capacity to bring the room up to temperature may be provided and then, as heat from occupants, lighting, machines and the sun begins to have effect, the output of the system is reduced to prevent overheating and save unnecessary fuel consumption. The availability today of inexpensive thermostatic radiator control valves has made room-by-room response an economic possibility.

Continuous versus intermittent operation

A building occupied for 24 hours each day, such as a hospital, a police station or a three-shift factory, will require that a continuous supply of heat is available. Most buildings, however, are occupied for a limited number of hours each day and, moreover, not for all days of the week: offices, schools and churches are examples of very different patterns of use. It is, in consequence, a matter of considerable importance as far as energy consumption is concerned to establish whether a pattern of intermittent operation will provide satisfactory comfort conditions for the whole of the occupied period. The manner in which proportions of total energy consumption vary with different operational regimes is shown in Table 4.16.

The *thermal response* or *time lag* of a building *vis-a-vis* a heat supply is perhaps best visualised initially by considering two extreme cases, the first being a lightly constructed building, insulated but overglazed, which will have minimal thermal capacity and hence a relatively short time lag. At the other extreme, a medieval or similar building with thick masonry walls, negligible window areas and massive construction throughout will respond very slowly to a change in external conditions, have a considerable thermal capacity and thus have an extended time lag taking as much as a week or more of heat input to recover from breakdown.

The first building would cool quickly overnight and, without heating, the internal temperature next morning would approximate to that outside. Furthermore, during a clear cloudless night, there might be a risk of condensation as the temperature of parts of the structure fell below the air temperature. Continuous heating would obviously be uneconomic in this case but, equally, a fully intermittent regime would only be adequate in the depth of winter if a lengthy pre-occupancy boost period were provided. The optimum solution might be a compromise with reduced temperature nighttime heating, and a minimal pre-occupancy boost.

Conversely, the air temperature within the heavy building would fall only marginally during a winter night without a heat supply but there would be a considerable time delay the following morning before the daytime temperature was fully recovered. Intermittent heating should, however, provide a reasonably satisfactory

Table 4.16 Per cent annual fuel consumption for various operational regimes

	Internal temperatures maintained					
	20°C for 24 hours		*20°C for 10 hours 15°C for 14 hours*		*20°C for 10 hours 10°C for 14 hours*	
Period	*Day*	*Night*	*Day*	*Night*	*Day*	*Night*
January	51	49	60	40	72	28
February	50	50	60	40	73	27
March	49	51	60	40	76	24
April	48	52	61	39	86	14
May	44	56	65	35	99	1
September	43	57	78	22	100	0
October	47	53	66	34	100	0
November	50	50	62	38	80	20
December	51	49	60	40	73	27
Average over season	48	52	64	36	84	16

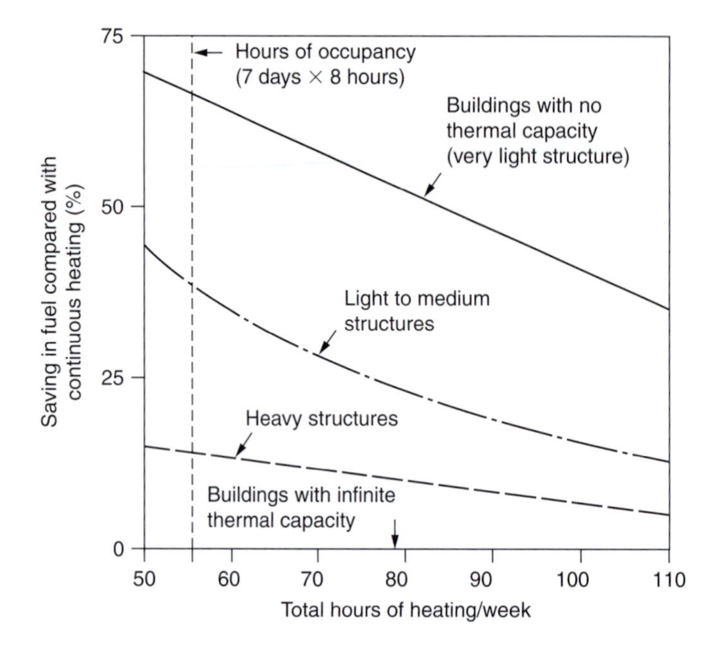

Figure 4.14 Comparative fuel savings for different weekly heating periods, indicating the effect of thermal capacity

performance, particularly if there were an insulating lining added to the walls. Continuous heating, at a marginally lower but constant level over 24 hours, would also meet requirements. Economically, there might be little to choose between the alternative methods in this case.

Between these two extremes lie all the buildings likely to be met in practice, be they of light, medium or heavy construction. Unfortunately, the choice between continuous or intermittent operation does not depend wholly upon the thermal response rate of the building and other factors must be taken into account. These include the proportion of the 24 hours (and perhaps the week also) during which the building is occupied, which may vary as a result of shift-work or flexible working hours in offices: further, the type and density of occupancy and, of course, the characteristics of the heating system will have an influence. Some of these effects have been the subject of a report by BSRIA[15] following extensive examination of a number of buildings of various types. Figure 4.14 provides a summary which illustrates the order of saving in fuel consumption which may be achieved as a result of intermittent operation.

Preheat capacity

When a heating system is operated intermittently, the temperature within the building will fall during the period of system shut-down and, in order to restore this to the proper level by the time the occupants arrive, a preheating period is necessary. Since the outside temperature varies during the winter nights, so will the time required for preheat increase or decrease and control equipment which provides means for dealing with this variation is described in Chapter 28.

Some excess capacity in the heating system is desirable in order to reduce the length of the preheating period and, with the improved standards of insulation and lower rates of air infiltration common in modern buildings, larger plant margins are required than was the case hitherto. During the preheating period:

- The conduction loss is reduced by the use of curtains or shutters and closed windows and doors minimise the loss due to air infiltration.
- The effective output from the system is increased above the design capacity as a result of a higher output from heat emitting equipment due to the lower temperature of the inside air.
- The excess plant capacity allows operation at an elevated temperature.

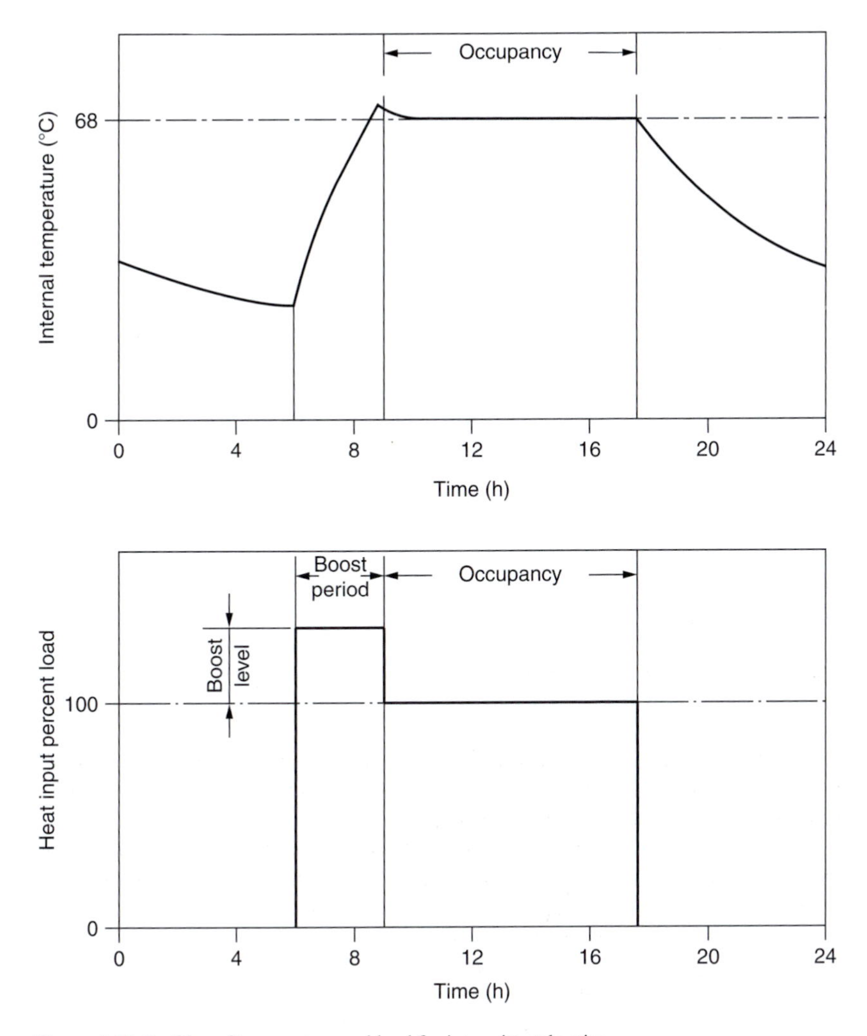

Figure 4.15 Profiles of temperature and load for intermittent heating

The combined effect of these items is illustrated in Fig. 4.15, which represents a normal heating and cooling curve for a building heated intermittently. The effect of the changes noted in the first two groups can be quite significant, as illustrated by an example related, once again, to the simple space shown in Fig. 4.3.

If shutters were used overnight at the windows and the air change reduced by half, then the heat loss might be reduced from 5150 W to about 3840 W. Taking an overnight inside temperature of, say, 10°C at the start of preheating, an average during that period might be 14°C; further, if the mean water temperature in the radiators were raised by as little as 5°C, then the dual effect would be to increase their output by about 20 per cent. This is, of course, a considerable simplification of a complex problem but it serves to show that the overall result might be that a preheat capacity of more than twice normal system output could be available.

Preheat and plant size

It will be obvious from what has been said that the thermal characteristics of both building and heating system, the time lag of the former and the response rate of the latter, have a profound effect upon the preheat time required to restore conditions of comfort in a space after a period of shut-down. The *1986 Guide Section A9* presented data which allowed preheat time to be established relative to values of the other variables: a rounded

Table 4.17 Variations in preheat times

	Plant size ratio[a]			
	Heavyweight structure		*Lightweight structure*	
Preheat (h)	*Quick response system*	*Slow response system*	*Quick response system*	*Slow response system*
1	3.0	–	2.1	–
2	2.5	–	1.6	2.3
3	2.1	2.6	1.5	1.7
4	1.9	2.3	1.4	1.5
5	1.7	1.9	1.2	1.4
6	1.5	1.6	1.2	1.2

Note
a Ratio of maximum plant output to design load at 20°C inside/outside. (Maximum plant output includes inherent capacity of system.)

up summary is presented in Table 4.17. The *2006 Guide Section A5* provides a more-in-depth exploration of this topic and includes correction factors for plant sizing for intermittent use.

It has been shown that there is little advantage to be gained by providing an oversized conventional system having a capacity much greater than that necessary to meet the calculated heat loss, with intent to reduce the preheat period. This is due to the fact that a system, made up of boiler plant, pipework and heat emitters, also has a time lag which increases with size. Margins on conventional heat emitters should be confined therefore to a nominal 10 per cent at most in order to make allowance for the seemingly inevitable poor standard of building maintenance.

As to the influence of system type upon energy consumption with intermittent operation, it has been shown[16] that systems having a small thermal capacity and a rapid response rate, such as those using warm air, are able to achieve greater economies than those subject to time lag of any significance. This is an obvious conclusion and points to the generalisation that with buildings of light construction, greatest economy in energy consumption is achieved with intermittent operation of such systems, but only when the control arrangements provided are sufficiently responsive and are thus able to prevent over-run. Buildings of more solid construction are able to achieve reasonable economies with a steady supply of heat and a less sophisticated level of control. It follows, therefore, that while warm air, convector and radiator systems are well suited to intermittent operation, embedded panels in ceilings and floors are not so suitable.

Preheat for direct and storage heating

In the case of direct heating appliances having a fixed rate of input, some additional capacity is needed to provide for a rapid preheat and this may, to some extent, be available inherently since mass-produced equipment of this type is manufactured in multiples of 2 or 3 kW only, and selection on a generous basis should thus suffice. Heat output from all storage-type systems using room heaters is continuous, even though heat input is intermittent, and thus facilities for controlled preheat in the present sense need not be considered.

Steady state and dynamic response

This chapter would not be complete without mention of the alternative and more thorough methods of establishing not only the room-by-room heat requirements of a building but also preheat times and other operational routines best suited to energy conservation.

Steady state calculations for winter heating, while adequate to produce results satisfactory for all but the most unorthodox forms of building construction, fail to provide data enabling full advantage to be taken of all the sophisticated control systems now available. The essence of that failure lies with the fundamental assumption that heat exchange is a function of values of temperatures inside and outside a building taken coincidentally in time. Except in the case of token weather barriers such as glass or thin sheeting, structural elements have both mass and thermal resistance which together introduce delay – the time lag which has been mentioned previously.

Computer based thermal modelling techniques enable transient heat flow to be analysed and the dynamic response of the building structure to be assessed, on an hour-by-hour basis, a task which would be infinitely tedious, if possible at all, using manual methods. In practice, it is more usual to undertake these calculations using computer models and, indeed, it is a requirement to comply with Part L of the Building Regulations, to undertake these calculations for heating and cooling and then convert these loads into carbon dioxide emissions. Refer to Chapter 6 for further details.

For such dynamic calculations, knowledge of other properties of the elements making up that structure are required, in addition to the *U* value. The more important of these are:

Admittance (*Y value*): This is a measure of the ability of a surface to smooth out temperature variations in a space and represents the rate of energy entry *into* a structure rather than that of passage *through* it. The units are $W/m^2 K$.

Decrement factor (*f*): This represents the ability of a structural element to moderate the magnitude of a temperature change or swing at one face before this penetrates to the other. It is largely a function of the thickness of the element and is dimensionless.

Surface factor (*F*): This relates the admission and absorption of energy to the thermal capacity of a structural element. It is dimensionless.

For values of these factors and further details of these more probing calculations, reference should be made to the *Guide Section A3*.

As mentioned earlier there is considerable international debate on the preferred methods for these load calculations and *Guide Section A5* offers guidance on steady state and dynamic calculation techniques. In practice for most buildings hand calculations should use the simple steady state method. For computer based calculations the simple dynamic model is also usually appropriate. *Guide Section A5* provides an in-depth analysis of the background to these techniques with worked examples where further amplification can be obtained.

Notes

1. Page, J.K., *Energy Requirements for Buildings*. Public Works Congress 1972. Jones, W.P., *Designing Air-Conditioned Buildings to Minimise Energy Use*. RIBA/IHVE Conference 1974.
2. These increased values are of particular relevance only with respect to heat losses in UK winters. Winds in summer may be assumed to have less influence and values for normal exposure are thus appropriate.
3. 0.20 (from Table 4.3) + 15 per cent for sand/cement joints = 0.23 W/m K.
4. Anderson, B.R., *The U-Value of Ground Floors: Application to Building Regulations*. BRE Information Paper IP90, April 1990.
5. Modelling and controlling interstitial condensation in buildings BRE Information Paper IP2/05 (2005).
6. Although the hour is an unacceptable time interval for calculations made in strict SI units, the use of rates (which are at best no more than informed guesses) in multiples of 0.0003 air changes per second would not endear the concept to any practitioner!
7. Air Infiltration and Ventilation Centre, *An Application Guide – Air Infiltration Calculation Techniques*, June 1986.
8. Min, T.C, Engineering concept and design of controlling ventilation and traffic through entrances in tall commercial buildings. *JIHVE*, 1961.
9. In this context, the various room surface areas times their surface temperatures and the product divided by the sum of the areas.
10. Post-War Building Study No. 33. *Choice of Basic Design Temperatures*.

11. Jamieson, H.C, Meteorological data and design temperatures. *JIHVE*, 1954, 22, 465.
12. Billington, N.S., Thermal insulation of buildings. *JIHVE*, 1974, 42, 63.
13. Petherbridge, P. and Oughton, D.R., Weather and solar data. *BSER&T*, 1983, 4, 4.
14. BRE *Domestic Energy Model, Background, Philosophy and Description*. BRE Digest 94: 1985.
15. Billington, N.S., Colthorpe, K.J. and Shorter, D.N., *Intermittent Heating.* HVRA (now BSRIA) Report 26, 1964.
16. Dick, J.B., Experimental and field studies in school heating. *JIHVE*, 1955, 23, 88.

Chapter 5

The building in summer

The effect of global warming and climate change have been well documented over recent years and are seem as key drivers for lowering energy use in buildings as covered previously in Chapter 2. The summer temperatures in the British Isles are increasing and the demand for cooling is increasing due to many buildings overheating. Figure 5.1, which shows the external dry bulb temperatures from London and Manchester Test Reference Year (TRY) weather files shows that temperatures in the 'mid-seasons' of spring and autumn overlap with those of reputed summer and winter. This indicates a temperate maritime climate with London temperatures higher than those in Manchester.

It is impossible to separate those characteristics, desirable in a building envelope during winter, from those which are beneficial in high summer. Good thermal insulation, small window areas, and measures which ensure a reasonable rate of air infiltration are clearly beneficial in either season. It would be inappropriate in consequence to repeat here the comments made in the previous chapter. The effect of infiltration by outside air is less in summer, not only as a result of the reduced temperature difference, outside to inside, but also because wind speeds are usually less than in winter.

Extraneous influences

The main difference between the two seasons is of course that heat gains from people, electric lighting and equipment or machinery combine with heat from the sun to create a penalty during the summer months, as far as human comfort is concerned. These heat gains may lead to high temperatures within a building and, in situations where the occupants are exposed to direct sunshine, there is the added effect of an unpleasantly high radiant intensity from surrounding glazed areas. The control of solar effects is critical to limitation of heat gains in summer and, in consequence, much of this chapter is devoted to that subject.

Not all of the gains noted above related wholly to *sensible heat*, i.e. that which leads to an increase in air temperature. Some part of the output may be in release of *latent heat* which leads to an increase in the amount of moisture in the air and, if the temperature remains constant, to an increase in the relative humidity. Sources of latent heat gains are building occupants, moist air infiltrated from outside and vapour from a variety of activities such as cooking and bathing, to name the simplest.

Conservation of energy

For the first time, the 2002 edition of Building Regulations Part L required designers to give attention to the performance of the building in summer. The requirements affect both the design of the envelope and the engineering systems used to service the building. Previous editions of Part L had prescribed maximum areas of glazing, but this had been to limit excessive conduction losses rather than to control solar gains. The 2006 edition introduces a requirement to limit solar overheating in all types of buildings including dwellings. There are various ways of demonstrating compliance dependant on the type of building, ranging from the procedure in Appendix P of the Standard Assessment Procedure SAP 2005, limiting values of peak daily averaged solar cooling load, to full dynamic calculation of overheating risk. The more complex the method used, the more the design can take advantage of the benefits offered by solar control glass, shading devices, thermal mass and night ventilation, etc.

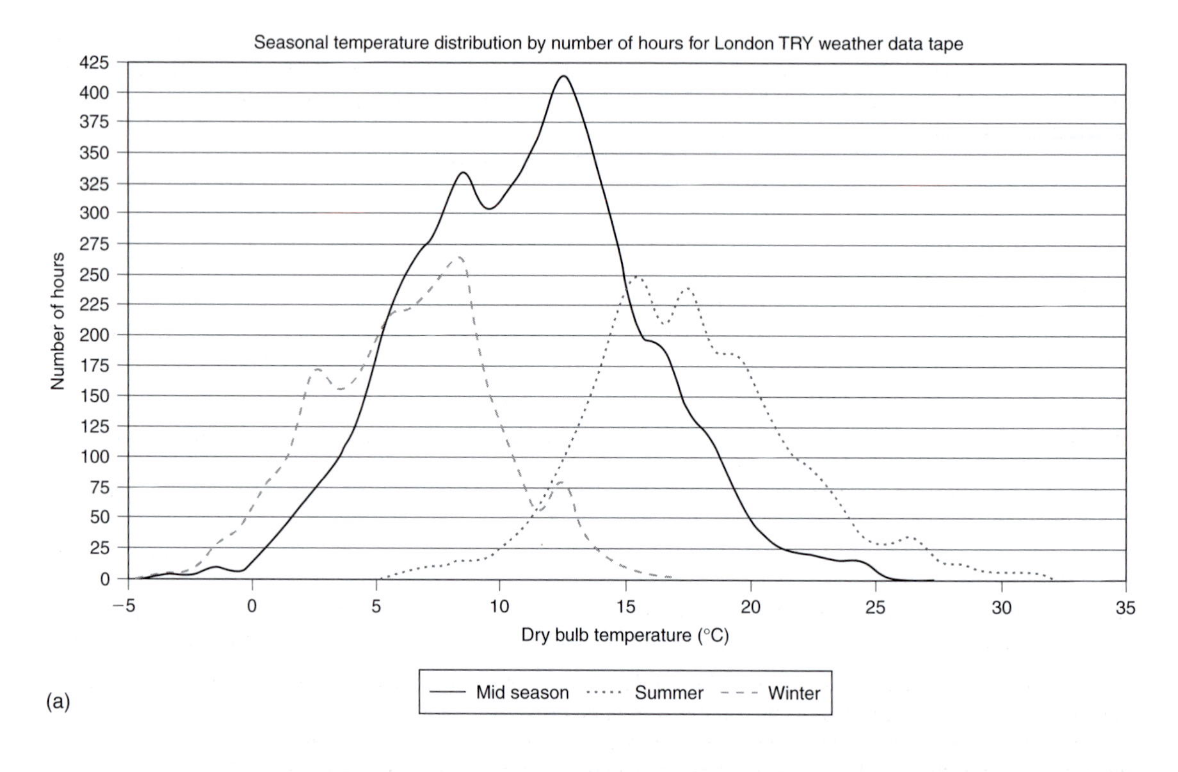

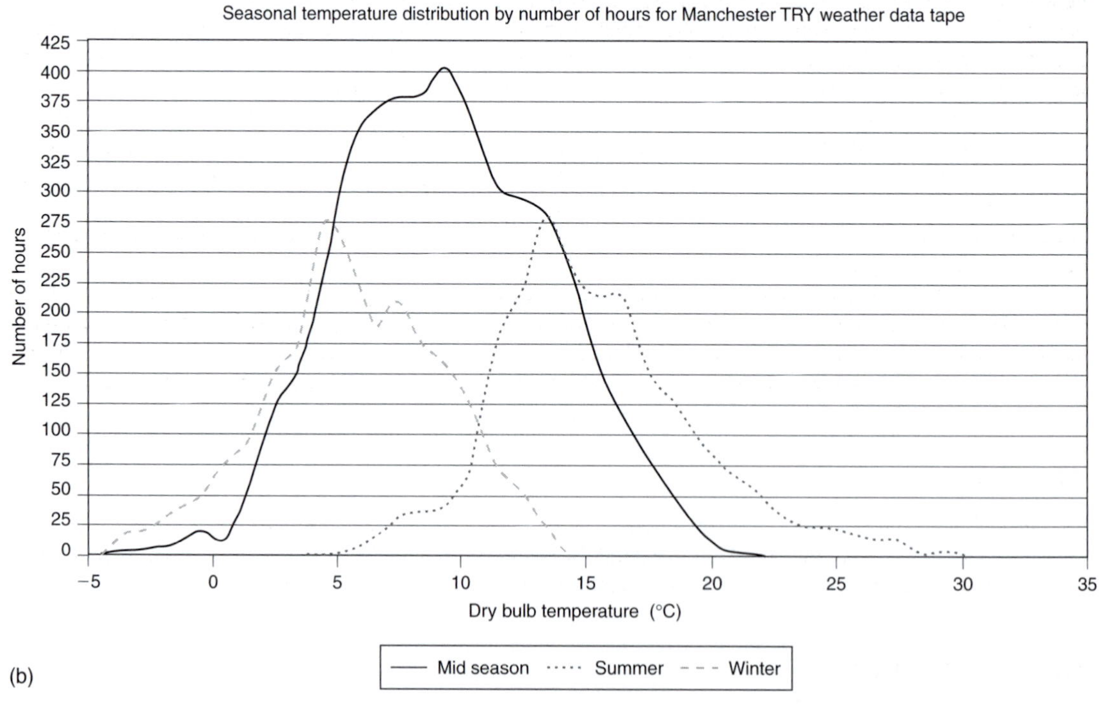

Figure 5.1 Temperature profiles for (a) London and (b) Manchester

It should be noted that the requirement to control solar overheating does not impose any restriction on the use of air-conditioning. What it does do is to make the building capable of being kept reasonably comfortable in the absence of air-conditioning (provided internal gains are modest). The regulations do, however, impose requirements on air-conditioning and mechanical ventilation systems, in that if they are installed, they should be 'reasonably efficient'. The implications of this are covered in Chapters 13, 18, 19 and 24.

Primary influences

The level and pattern of energy use in a building, particularly in summer, has its beginning when an architect accepts a brief from a client, on behalf of a full design team, and the initial composite sketch plans are developed. They are affected throughout the later detailed design and construction process (and until the day of demolition) by a host of influences including:

- *Building exposure*: orientation, shape, modules, mass, thermal insulation, glazing, solar shading, plant room position, space for service distribution.
- *Plant and system design* to match the characteristics of the building and to meet the needs (known and unknown) of the ultimate occupants.
- *Commissioning and testing* of the completed plant and the adjustment to ensure that it operates as designed in all respects.
- *Operational routine* as adapted to match the building use in occupational pattern, working hours and the like.
- *Level of maintenance* provided to both building and plant, energy audits, preservation of records and updating.

Some of the characteristics which lead towards conservation of energy have been dealt with previously in Chapter 4 and will be explored further in this present chapter.

The *thermal balance point* of a building is defined as that outside dry bulb temperature at which, if all heat generated within the building from lighting, occupancy and machines, etc., were to be distributed usefully, energy from a subsidiary source would be unnecessary. A low balance point, then proposed as a standard of excellence, was later demonstrated to be quite the reverse since it implied a requirement for mechanical cooling at any higher outside temperature. The term has since fallen into disrepute.

The last three of the influences listed above are not altogether in the hands of the designers of the building. Commissioning and testing of plant is a matter of increasing importance, as components become more sophisticated, more 'packaged' and thus less susceptible to any level of repair. It seems that little recognition is given to the fact that, although systems in any one building may be made up of standard production line components, the combination is a prototype in every case. The Building Regulations Part L makes it a requirement that the fixed building services are commissioned.

As to operational routines and maintenance, these are often given a low priority by building owners, however as Building Logbooks are a requirement of Building Regulations, they at least have the information at hand to know what is required to keep the building operating in an energy efficient manner. The appointment of a specialist maintenance contractor may be recommended in instances where the building owner has no 'in-house' technicians in this field.

Solar heat gains

The most significant heat gain for the majority of modern buildings, although not always the greatest in terms of magnitude, is that from the sun. To understand the manner in which the sun has an influence upon a building, it is first necessary to have some appreciation of the principles of solar geometry.

Sun-path diagrams

Data providing values for the *altitude* (height as an angle above the horizon) and *azimuth* (compass bearing measured clockwise from the north) of the sun appear as part of the CIBSE reference weather file.

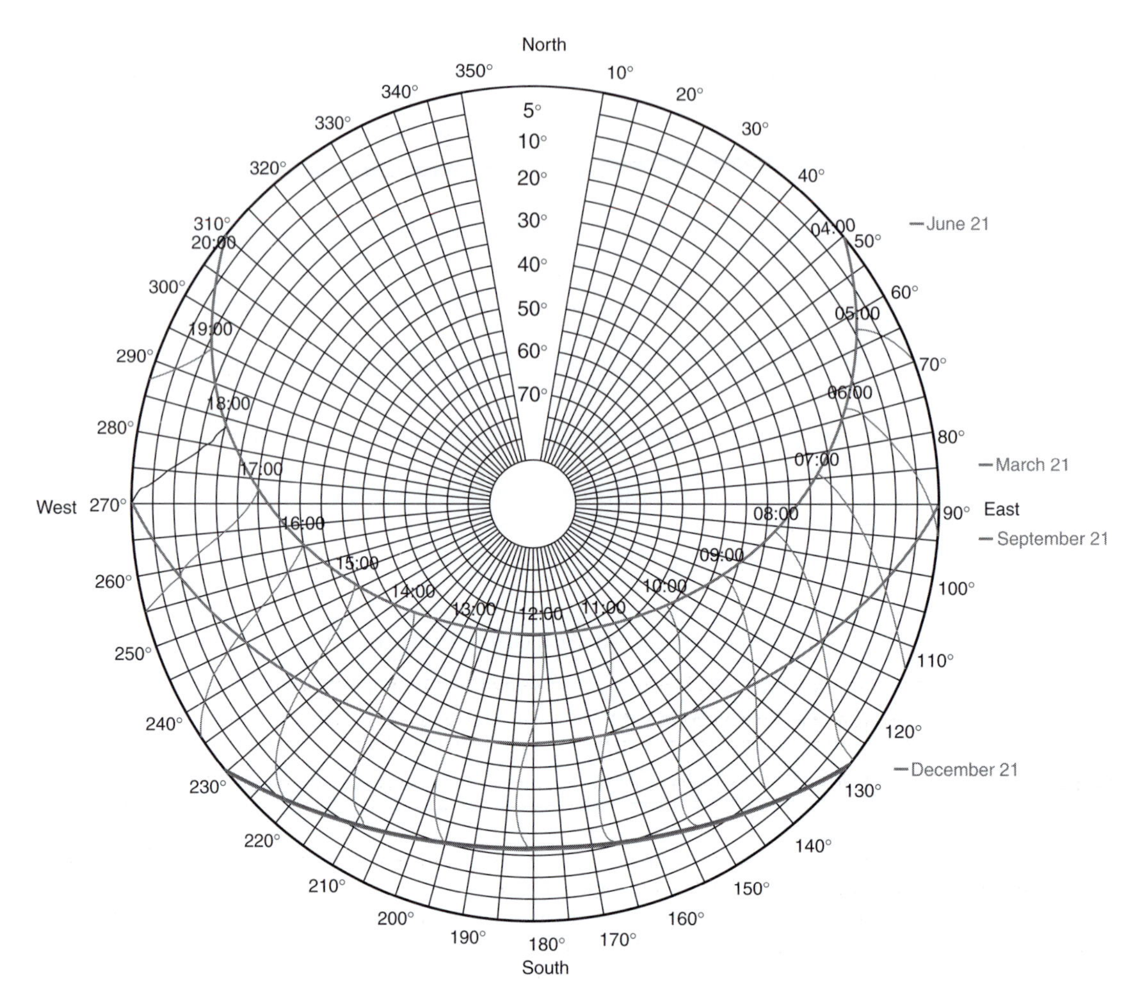

Figure 5.2 Sun-path diagram for London Heathrow (see colour plate section at the end of the book)

This information is more easily understood by inspection of a diagram such as that shown in Fig. 5.2, for London Heathrow at a latitude of 51.48° north. In this example, the position of the sun has been computer plotted from sunrise to sunset on three discrete dates, against a background of altitude and azimuth. Similar sun-path diagrams for other latitudes may be constructed, however computer software to generate these diagrams and shadow animations (movies) are now widely available.

When a miniature plan of a building, correctly orientated, is placed at the centre of a diagram such as this, it is then possible to visualise the likely peak incidence of solar radiation on the various faces throughout the daylight hours of a chosen month. It will be clear that easterly faces will be subjected to solar influence first, followed by the south-east and south, by which time the east face will be going into shadow again and so on.

To make a preliminary assessment of the maximum solar heat gain for the building as a whole, a schedule of the various rooms and spaces may be constructed and, using values of solar intensity for various sun altitudes and azimuths, the coincident load each hour on each face can be summated. In addition, the peak load for each face or any other combination may be established. While this exercise is fairly straightforward in the case of a square or rectangular building it becomes more complex in instances where the plan has re-entrant angles or internal courtyards where other parts of the building cast moving shadows on adjacent faces. While a model can be helpful, extensive analyses of this nature are tedious if attempted manually and use of a suitable computer

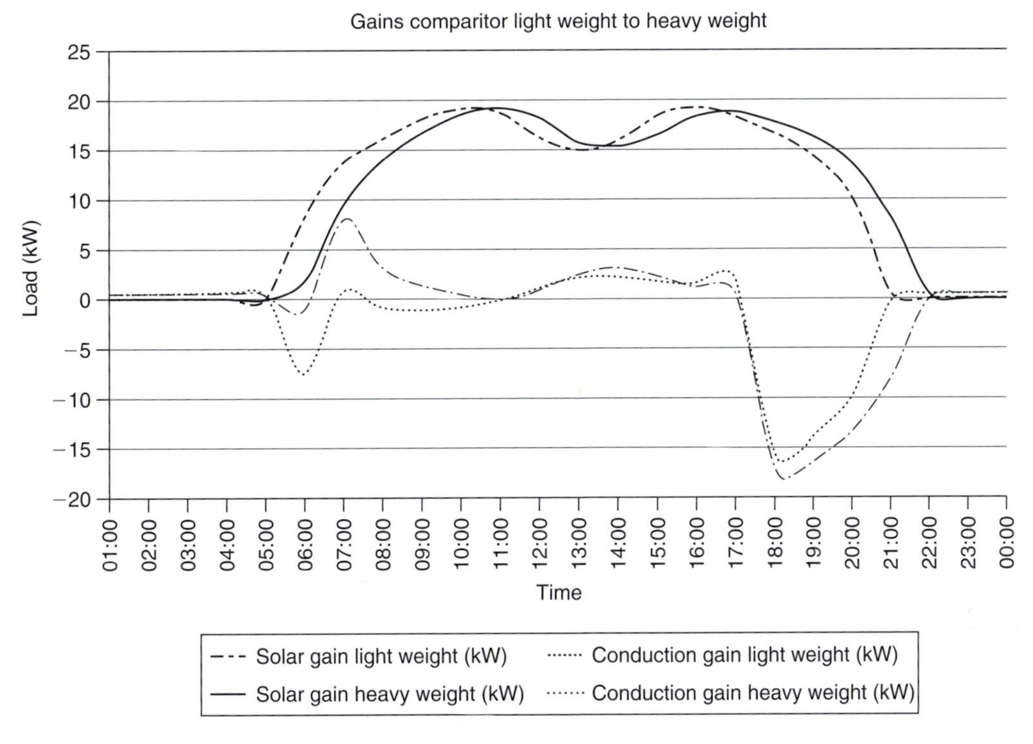

Figure 5.3 Computer plot of building solar gains and fabric conduction for a heavyweight and lightweight building

program is the obvious approach. Figure 5.3 shows a computer generated plot of the solar heat gains and fabric conduction loads for a heavyweight and lightweight building.

A second use for a sun-path diagram is in the design of shading devices taking the form of external sunbreaks. Their shape and angle of incidence may be set up on the plan and separate diagrams made for the various aspects; both vertical and horizontal devices may be so examined. This subject will be discussed in more detail later.

Sun-path diagrams provide a valuable aid in assisting visualisation of the sun position in relation to a building but, in terms of calculation, have been superseded by the development and presentation of data which allow heat gain to be read directly for orientation and time of day without the intervention of a search for incident angles and basic intensities. In turn, computer use, which has opened up facilities for exploring coincident loads and more rigorous examination of time lag, has superseded the refined tabular data. In consequence, the reference here has been abridged and is included to indicate the process rather than the detail.

Incidence of solar radiation

The total of solar radiation to reach the surface of the earth has two components: *direct* and *diffuse*. Of the former, some 1 per cent is ultraviolet, 40 per cent is visible light and the remainder is infrared. Diffuse, sometimes known as *sky* or *scattered* radiation, results from absorption by, and reflectance from, vapours and dusts, etc. in the atmosphere. It is at a maximum with cloud cover and a minimum with a clear sky.

For a latitude of 50° in the British Isles, the accepted value for the intensity of direct solar radiation on a horizontal surface at sea level (or up to about 300 m above) will be about 800 W/m^2 at noon in June under a clear sky. Diffuse radiation received on the same horizontal surface, again at noon in June, will be about 100 W/m^2 and 300 W/m^2 for clear and cloudy skies, respectively. Intensities on surfaces other than the horizontal will be proportional to the *angle of incidence*, which is the angle between the direction of the sun's rays and the perpendicular to the surface.

As far as vertical surfaces (walls and windows) are concerned, some part of the diffuse radiation is received from the sky and the remainder by reflection from adjacent ground surfaces. The intensity of radiation received on the ground will be that for any horizontal surface, as already noted, but the amount reflected will depend upon the nature and colour of the surface. Correction factors related to concrete, grass, snow, tarmac and water can be used but since these were too precise, the choice has been reduced in number to represent light and dark surfaces only. For the case of a city pavement, a convenient manual way to take account *of ground reflection* is to add a margin of 10 per cent to those solar heat gains attributed to glazed areas.

The manner, in which solar radiation varies throughout the day, for north, east, south and west faces at the June summer solstice, is illustrated in Fig. 5.4(a). It will be noted that since the sun altitude is high at midday (63°), the peak solar intensity on the south face is lower than that on the east and west faces which occur at times when the altitude is less. Figure 5.4(b) is a similar plot but shows the situation prevailing at the spring equinox: the intensity on the south face at a noon altitude of 40° now exceeds the peaks on the east and west faces. A measure of the relative intensities of radiation is given by the Design 97.5 percentile 24-hour mean values as stated below the two parts of the diagram taken from CIBSE Guide A2 based on Bracknell data between 1981 and 1992. The numeric values relate to (a) June 21 and (b) March 29. Table 5.1 shows details of Table 2:30 in Guide A2 Design 97.5 percentile of beam and diffuse irradiance on vertical and horizontal surfaces for the London area.

Solar gain through opaque surfaces

When direct and diffuse radiation fall upon the opaque surfaces of a building, a proportion is reflected back into space but the greater part will warm the surface of the material. Of this latter component, some is lost by re-radiation and some by convection to the surrounding air but the remainder is absorbed into the material to a degree which depends upon the nature and colour of the surface. For example, a black non-metallic surface may retain as much as 90 per cent of that remainder, whereas a highly polished metal surface will take in as little as 10 per cent. The majority of building materials absorb some proportion within the range 50–80 per cent and, of this, the greater part will be transmitted through the material by conduction at a rate which will depend upon the U value (W/m^2 K).

In instances where the building element has a negligible thermal capacity, as would be the case with thin unlined cement sheeting, then heat will be transferred at once but where there is mass of any real significance, then a time lag will occur. When the material has both sufficient thickness and substantial mass, this time delay will be prolonged to the extent that solar radiation which falls upon it during the early part of the day may not penetrate to the inside face until the outer surface is in shadow. The curious situation may then occur where the heat stored in the structure will be flowing in both directions, to outside and into the building. In some cases, even on east to south facades which receive solar exposure earlier in the day, the heat absorbed at the time of peak gain may not be released from the inner surfaces until after rooms there have ceased to be occupied.

As will be appreciated therefore, the temperature difference between outside air (t_{ao}) and inside air (t_{ai}), does not represent the situation properly since the outside surface temperature of the material will have been raised by the heat absorbed, as noted above. A convenient but approximate method of dealing with this situation is to make use of the concept *of sol-air temperature* (t_{so}), a hypothetical scale which takes into account not only the outside air temperature but also increments to it which represent the increase due to the effects of solar radiation. Tables of sol-air temperature, hour-by-hour during the critical months of insolation, are given in the *Guide Section A2* for a number of orientations and with respect to both 'light' and 'dark' coloured building surfaces.

The subject of dynamic response, including the influence of Admittance (Y value), time lag (ϕ) and decrement factors (f) was discussed briefly at the end of Chapter 4. In summer, when heat gain is a penalty, some attempt must be made in calculations performed manually to take these factors into consideration. The application of sol-air temperature tables to the problem takes account for such temperatures. Most computer simulation will calculate the sol-air temperature from the dry bulb temperature, total solar radiation and fabric type. Table 5.2 Shows sol-air temperatures for June 21 for light and dark surfaces taken from CIBSE Guide A2, Table 2.34.

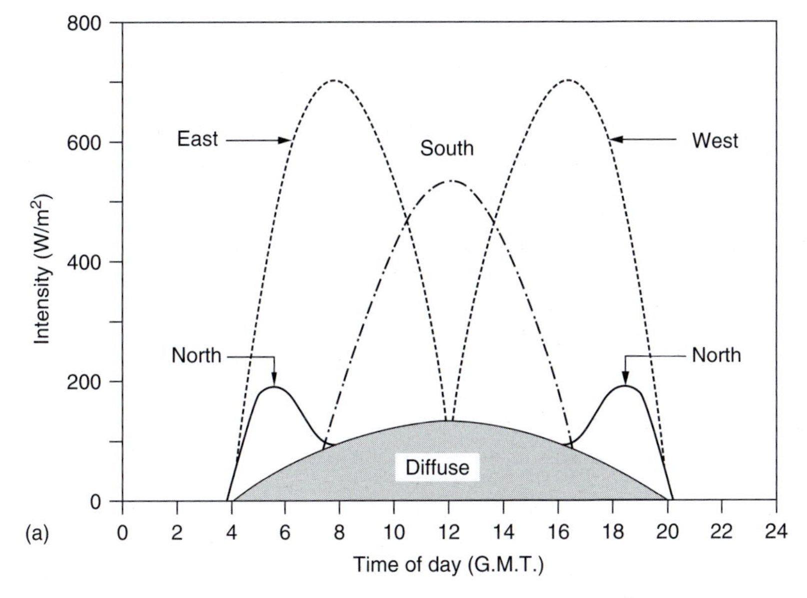

South facing 24 hour mean = 166 W/m^2

East facing 24 hour mean = 203 W/m^2

West facing 24 hour mean = 204 W/m^2

North facing 24 hour mean = 98 W/m^2

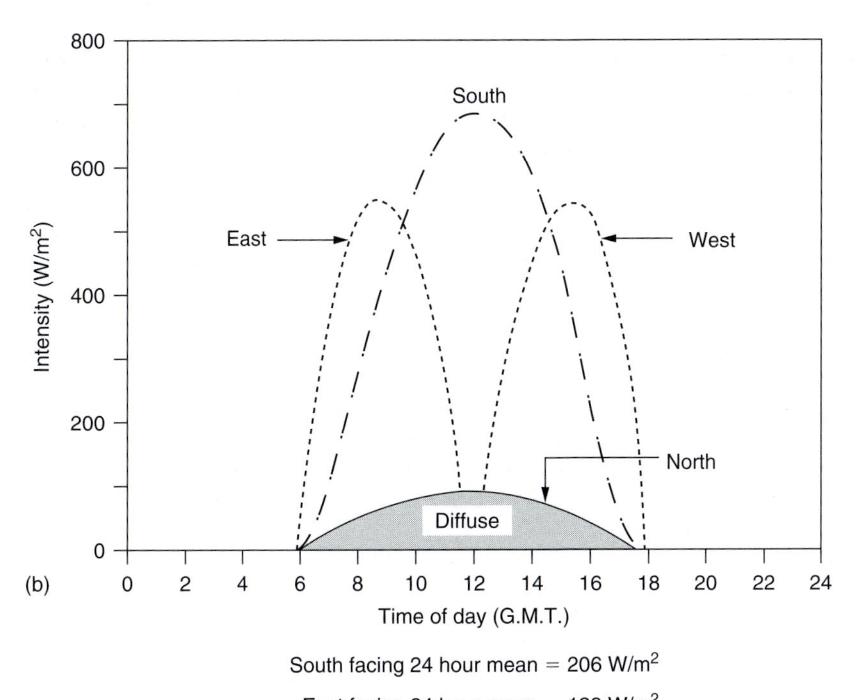

South facing 24 hour mean = 206 W/m^2

East facing 24 hour mean = 129 W/m^2

West facing 24 hour mean = 131 W/m^2

North facing 24 hour mean = 41 W/m^2

Figure 5.4 Total solar heat gains on a vertical surface: (a) 21 June; (b) 22 March

Table 5.1 Irradiance data for London area

Date and times of sunrise/ sunset	Orientation	Type	*Daily mean irradiance* (W/m²) *and mean hourly irradiance* (W/m²) *for stated solar time*																	
			Mean	*0330*	*0430*	*0530*	*0630*	*0730*	*0830*	*0930*	*1030*	*1130*	*1230*	*1330*	*1430*	*1530*	*1630*	*1730*	*1830*	*1930*	*2030*
April 28	Normal to beam		343	–	*74*	235	440	555	674	707	744	765	761	740	724	655	568	408	218	*69*	–
Sunrise: 04:46	N	Beam	7	–	*27*	59	24	0	0	0	0	0	0	0	0	0	0	22	55	*26*	–
Sunset: 19:14		Diffuse	56	–	*25*	43	53	87	101	117	130	139	141	131	120	100	88	52	41	*25*	–
	NE	Beam	47	–	*68*	202	316	298	220	71	0	0	0	0	0	0	0	0	0	*0*	–
		Diffuse	64	–	*33*	86	123	142	142	119	145	139	141	131	120	100	77	47	22	*8*	–
	E	Beam	103	–	*69*	226	423	497	519	417	276	96	0	0	0	0	0	0	0	*0*	–
		Diffuse	74	–	*34*	93	146	183	196	185	171	143	155	131	120	100	77	47	22	*8*	–
	SE	Beam	128	–	*29*	118	282	405	513	519	483	392	253	92	0	0	0	0	0	*0*	–
		Diffuse	77	–	*25*	61	115	164	195	202	205	196	179	136	138	100	77	47	22	*8*	–
	S	Beam	122	–	*0*	0	0	76	207	317	407	458	455	404	324	202	78	0	0	*0*	–
		Diffuse	74	–	*8*	24	57	90	139	168	192	207	212	194	174	138	91	56	23	*8*	–
	SW	Beam	127	–	*0*	0	0	0	0	0	92	255	390	480	531	499	415	262	109	*27*	–
		Diffuse	77	–	*8*	23	48	77	101	134	135	175	201	207	210	192	165	110	58	*25*	–
	W	Beam	101	–	*0*	0	0	0	0	0	0	0	97	275	427	504	509	392	210	*64*	–
		Diffuse	74	–	*8*	23	48	77	101	117	130	152	146	173	192	193	184	139	86	*34*	–
	NW	Beam	45	–	*0*	0	0	0	0	0	0	0	0	0	73	214	305	293	187	*63*	–
		Diffuse	64	–	*8*	23	48	77	101	117	130	139	141	146	122	140	143	117	80	*33*	–
	Horizontal	Beam	199	–	*1*	26	119	235	377	475	558	605	602	555	486	367	240	110	25	*1*	–
	Horizontal	Diffuse	73	–	*18*	44	80	117	138	151	163	170	177	166	157	138	116	80	43	*18*	–
	Horizontal	Global	272	–	*19*	70	199	352	515	626	721	775	779	721	643	505	356	190	68	*19*	–
May 29	Normal to beam		386	–	*145*	350	482	580	685	753	787	794	793	787	758	693	623	511	371	*154*	–
Sunrise: 04:01	N	Beam	22	–	*73*	114	66	0	0	0	0	0	0	0	0	0	0	70	120	*78*	–
Sunset: 19:59		Diffuse	68	–	*49*	72	77	108	116	126	137	143	142	134	125	113	104	73	68	*43*	–
	NE	Beam	63	–	*140*	308	360	333	255	117	0	0	0	0	0	0	0	0	0	*0*	–
		Diffuse	74	–	*64*	121	144	159	154	136	159	143	142	134	125	108	87	60	36	*13*	–
	E	Beam	112	–	*125*	323	444	498	505	426	280	96	0	0	0	0	0	0	0	*0*	–
		Diffuse	81	–	*59*	124	162	191	197	188	170	145	153	134	125	108	87	60	36	*14*	–

	SE	Beam	117	–	*36*	148	268	372	460	486	443	343	207	47	0	0	0	0	0	*0*	–
		Diffuse	80	–	*44*	81	125	167	189	197	194	184	165	126	134	108	87	60	36	*14*	–
	S	Beam	98	–	*0*	0	0	28	145	261	347	389	388	347	262	147	30	0	0	*0*	–
		Diffuse	72	–	*16*	39	72	89	133	163	180	191	190	177	159	128	85	68	97	*14*	–
	SW	Beam	120	–	*0*	0	0	0	0	0	47	207	343	443	489	465	399	284	157	*38*	–
		Diffuse	79	–	*16*	39	64	90	111	137	128	165	184	190	193	183	160	118	76	*39*	–
	W	Beam	116	–	*0*	0	0	0	0	0	0	0	96	280	429	511	535	471	342	*132*	–
		Diffuse	80	–	*16*	39	64	90	111	126	137	153	145	168	184	191	184	153	117	*53*	–
	NW	Beam	66	–	*0*	0	0	0	0	0	0	0	0	0	118	258	357	382	327	*149*	–
		Diffuse	74	–	*17*	39	64	90	111	126	137	143	142	157	134	149	153	136	114	*56*	–
	Horizontal	Beam	243	–	*10*	74	175	296	439	564	649	685	685	649	567	444	318	186	79	*10*	–
	Horizontal	Diffuse	77	–	*33*	70	101	133	146	153	156	162	161	153	148	140	121	91	63	*29*	–
	Horizontal	Global	320	–	43	144	276	429	585	717	805	847	846	802	715	584	439	277	142	39	–
Jun 21	Normal to beam		414	*64*	191	387	554	683	747	797	826	837	840	809	771	705	639	631	390	192	*65*
Sunrise: 03:49	N	Beam	27	*40*	100	132	85	0	0	0	0	0	0	0	0	0	0	82	133	100	*40*
Sunset: 20:11		Diffuse	71	*25*	55	76	85	111	117	125	134	139	140	135	127	121	114	85	77	58	*25*
	NE	Beam	74	*64*	185	342	417	397	285	134	0	0	0	0	0	0	0	0	0	0	*0*
		Diffuse	77	*32*	72	124	157	158	150	134	158	139	140	135	127	113	94	69	42	21	*9*
	E	Beam	124	*50*	162	352	504	579	544	445	290	100	0	0	0	0	0	0	0	0	*0*
		Diffuse	83	*27*	66	126	175	189	189	180	163	140	149	135	127	113	94	69	42	20	*8*
	SE	Beam	122	*7*	44	155	296	422	484	496	447	345	204	36	0	0	0	0	0	0	*0*
		Diffuse	79	*21*	50	82	133	163	180	187	183	173	157	123	134	113	94	69	42	20	*8*
	S	Beam	94	*0*	0	0	0	18	141	256	342	387	389	335	247	133	17	0	0	0	*0*
		Diffuse	72	*8*	18	41	77	83	125	154	170	179	180	172	158	130	87	77	42	20	*8*
	SW	Beam	118	*0*	0	0	0	0	0	0	37	203	346	438	479	457	395	284	157	44	*7*
		Diffuse	80	*8*	18	41	69	91	110	132	122	156	175	185	193	188	169	133	83	52	*21*
	W	Beam	120	*0*	0	0	0	0	0	0	0	0	101	284	431	513	542	483	355	163	*51*
		Diffuse	83	*8*	18	41	69	91	110	125	134	148	140	165	185	197	196	175	128	69	*27*
	NW	Beam	71	*0*	0	0	0	0	0	0	0	0	0	0	130	269	371	399	345	186	*65*
		Diffuse	77	*9*	19	41	69	91	110	125	134	139	140	159	138	156	164	157	126	75	*32*
	Horizontal	Beam	264	*1*	17	91	214	362	493	610	694	735	738	679	590	465	339	205	92	18	*1*
	Horizontal	Diffuse	76	*16*	37	71	104	120	132	140	141	144	146	146	149	146	131	107	72	38	*16*
	Horizontal	Global	341	*17*	54	162	318	482	625	750	835	879	884	825	739	611	470	312	164	56	*17*

Table 5.2 Air and sol-air temperatures for London area

Hour ending	Air temperature	*Sol-air temperature (°C) for stated orientation and surface colour*																	
		Horizontal		*North*		*North-east*		*East*		*South-east*		*South*		*South-west*		*West*		*North-west*	
		Dark	*Light*	*Dark*	*Light*	*Dark*	*Light*	*Dark*	*Light*	*Dark*	*Light*	*Dark*	*Light*	*Dark*	*Light*	*Dart*	*Light*	*Dark*	*Light*
01	13.9	10.5	10.5	11.4	11.4	11.4	11.4	11.4	11.4	11.4	11.4	11.4	11.4	11.4	11.4	11.4	11.4	11.4	11.4
02	13.1	9.4	9.4	10.4	10.4	10.4	10.4	10.4	10.4	10.4	10.4	10.4	10.4	10.4	10.4	10.4	10.4	10.4	10.4
03	12.4	8.8	8.8	9.6	9.6	9.6	9.6	9.6	9.6	9.6	9.6	9.6	9.6	9.6	9.6	9.6	9.6	9.6	9.6
04	12.0	8.5	8.5	9.3	9.3	9.3	9.3	9.3	9.3	9.3	9.3	9.3	9.3	9.3	9.3	9.3	9.3	9.3	9.3
05	12.2	10.3	9.6	11.0	10.3	11.2	10.4	11.1	10.4	11.0	10.3	10.6	10.1	10.6	10.1	10.6	10.1	10.6	10.1
06	13.3	17.8	14.3	24.9	19.0	40.7	28.1	41.4	28.5	26.7	20.0	14.1	12.9	14.1	12.9	14.1	12.9	14.1	12.9
07	15.0	25.7	19.4	24.3	19.7	46.9	32.7	52.5	36.0	38.8	28.0	18.0	16.1	18.0	16.1	18.0	16.1	18.0	16.1
08	16.8	33.4	24.6	21.5	19.2	46.5	33.5	57.3	39.8	48.0	34.4	22.7	19.9	21.5	19.2	21.5	19.2	21.5	19.2
09	18.5	39.8	29.0	24.7	22.1	41.6	31.7	56.3	40.2	53.0	38.3	33.1	26.8	24.7	22.1	24.7	22.1	24.7	22.1
10	20.1	44.9	32.6	27.4	24.6	34.8	28.7	51.8	38.5	54.5	40.1	41.5	32.6	27.4	24.6	27.4	24.6	27.4	24.6
11	21.5	47.8	35.0	29.3	26.5	29.3	26.5	44.3	34.9	52.3	39.5	47.0	36.5	31.0	27.4	29.3	26.5	29.3	26.5
12	22.7	49.3	36.3	30.6	27.8	30.6	27.8	35.4	30.4	47.6	37.4	49.7	38.6	40.6	33.4	30.6	27.8	30.6	27.8
13	23.6	49.3	36.8	31.3	28.5	31.3	28.5	31.3	28.5	41.0	34.0	49.9	39.0	47.9	37.9	35.9	31.1	31.3	28.5
14	24.4	48.1	36.4	31.5	28.8	31.5	28.8	31.5	28.8	33.0	29.7	47.6	37.9	52.4	40.7	45.1	36.5	31.5	28.8
15	24.5	44.8	34.7	30.8	28.3	30.8	28.3	30.8	28.3	30.8	28.3	42.5	34.9	53.3	41.1	51.1	39.8	36.9	31.7
16	24.4	40.5	32.2	29.4	27.2	29.4	27.2	29.4	27.2	29.4	27.2	35.8	30.8	51.3	39.6	53.9	41.1	42.4	34.6
17	24.2	36.1	29.6	27.8	26.0	27.8	26.0	27.8	26.0	27.8	26.0	28.7	26.5	47.4	37.1	54.4	41.1	46.3	36.4
18	23.5	30.8	26.4	30.1	26.7	25.6	24.2	25.6	24.2	25.6	24.2	25.6	24.2	40.6	32.7	50.7	38.4	46.5	36.0
19	22.5	25.2	22.8	30.4	26.2	22.8	22.0	22.8	22.0	22.8	22.0	22.8	22.0	31.6	26.9	42.2	32.9	41.7	32.6
20	20.9	19.5	19.0	19.9	19.4	19.6	19.3	19.6	19.3	19.6	19.3	19.6	19.3	19.9	19.4	19.9	19.4	19.9	19.5
21	19.3	16.6	16.6	17.1	17.1	17.1	17.1	17.1	17.1	17.1	17.1	17.1	17.1	17.1	17.1	17.1	17.1	17.1	17.1
22	17.9	15.1	15.1	15.7	15.7	15.7	15.7	15.7	15.7	15.7	15.7	15.7	15.7	15.7	15.7	15.7	15.7	15.7	15.7
23	16.8	13.7	13.7	14.4	14.4	14.4	14.4	14.4	14.4	14.4	14.4	14.4	14.4	14.4	14.4	14.4	14.4	14.4	14.4
24	15.7	12.5	12.5	13.2	13.2	13.2	13.2	13.2	13.2	13.2	13.2	13.2	13.2	13.2	13.2	13.2	13.2	13.2	13.2
Mean:	18.7	27.4	22.2	21.9	20.1	25.1	21.9	27.9	23.5	27.6	23.3	25.4	22.1	26.4	22.6	26.3	22.5	23.9	21.2

A wall facing south-east is constructed of 105 mm brick, 50 mm Expanded polystyrene (EPS) insulation, 100 mm dense concrete block and 13 mm dense plaster internally. The peak sol-air temperature of 54.5°C occurs at 10 a.m. on 21st June in London (from Table 5.2) and due to the time lag of the wall will be a peak conduction gain into the building 9.4 hours later. Traditionally engineers use their judgment to select the time of peak cooling and work back from that month and hour to see what the gains are. This can lead to under sizing of systems and now most designers would calculate the gains using computer software for a typical day in each of the summer months from May to September at each hour of the day to then know the time and month of the peak gain. Each room in a building may have a different month and hour of peak gain to size the system for that space and the building as a whole will have its own peak system gain for sizing central plant.

It is appropriate to note that, when considered in relation to the customary internal design temperature range 22–24°C for summer, that there are numerous occasions when the heat gain from opaque surfaces may be negative. It is common practice to ignore any such small 'credits' in calculations for plant capacity when calculating by hand, in the same way that small adventitious heat gains are ignored in winter calculations.

Dynamic building energy modelling procedures can pick up on these more detailed interactions and the effects on sizing building services equipment. See Chapter 6 for more information on computer modelling methods and CIBSE Applications Manual AM11: 1998 on 'Building energy and environmental modelling'.

In the special case of lightweight curtain walling, the effect of solar radiation is to raise the outer surface temperature very rapidly when the finish is dark. It can be demonstrated that in the British Isles this temperature may be well over 50°C in the early afternoon on a south-west (SW) facade and, since the time lag of such a form of construction is an hour or less, with negligible decrement, it is inevitable that complaints of thermal discomfort due to radiation will arise.

Flat roofs are subjected to solar radiation during the whole of the daylight hours and a relatively massive construction with a light coloured finish is advantageous. Roof insulation, as may be seen from Table 5.3, does little to increase the time lag but does of course reduce the *U* value. In the case of pitched roofs, the heat gain will depend upon the angle of the slope and may be more severe on one face than another, depending upon orientation. It is worth while making a check on the orientation of so-called 'north-light' roofs since it is not unknown for them to belie their name!

Table 5.3 Representative design factors for solar gain through opaque materials (dimensions in mm)

Construction (dimensions in mm)	*U value* (W/m^2K)	*Time lag* (ϕ h)	*Decrement* (f)
Walls			
220 brick, 13 dense plaster	2.09	7.4	0.42
19 render, 200 dense concrete block, 25 polyurethane insulation between battens, 12.5 plasterboard	0.90	7.5	0.26
105 brick, 50 blown wool insulation, 105 brick, 13 dense plaster	0.59	10.1	0.24
105 brick, 50 air space,19 plywood, 95 studding, 12.5 plasterboard	1.14	4.9	0.67
105 brick, 50 EPS, 50 air space, 100 dense concrete block, 13 dense plaster	1.77	8.1	0.34
105 brick, 50 EPS insulation, 100 dense concrete block, 13 dense plaster	0.64	9.4	0.24
105 brick, 100 blown fibre insulation, 100 lightweight aggregate concrete block, 13 dense plaster	0.33	9.2	0.39
105 brick, 50 air space, 19 plywood sheathing, 140 studding, 140 mineral wool insulation between studs, 12.5 plasterboard	0.29	6.5	0.57
Roofs			
Waterproof roof covering, 200 polyurethane insulation, vapour control layer, 75 screed, 150 cast concrete, 13 dense plaster	0.12	12.2	0.13
Waterproof roof covering, 35 polyurethane insulation, vapour control layer, 19 timber decking, unventilated airspace, 12.5 plasterboard	0.53	1.9	0.93
12.5 plasterboard, 100 mineral wool quilt between ceiling joists, 150 mineral wool quilt over joists, roof space, tiling	0.17	1.3	0.98
12.5 plasterboard, 150 PU insulation between rafters, ventilated airspace, roofing felt, 25 ventilated airspace, clay tiles	0.22	1.2	0.98

Solar gain through glazing

Windows are by far the most significant route by which solar heat enters a building, not least because that entry is without time lag. This is not to say, however, that the effect is instantaneous since this may well depend upon the nature and mass of the internal structure, furniture, carpets and other contents.

A number of computer programs have been produced which predict summertime temperatures. Figure 5.5 shows the relationship between inside and outside temperatures for a heavyweight and lightweight building. The more sophisticated of such programs enable many parallel effects to be examined in detail. The method involves the concept of environmental temperature, which has been referred to previously, thus taking account of both the mean radiant temperature and the air temperature within a space. Advantage may be taken of the same basic approach, including use of the admittance value of the various components of the building structure, to evaluate the extent to which those components are able to smooth out diurnal temperature swings resultant upon intermittent solar heat gains.

Glass has certain unique characteristics as far as heat transfer is concerned in that it has different transmissivities at different wavelengths: it is, moreover, virtually opaque to radiation from any source having a surface temperature less than about 250°C. When a glazed surface is insolated, some part of the total incident radiation will be reflected away, some will be transmitted through the material, and some will be absorbed by it. This simple pattern of heat transfer is confused by re-radiation outwards and by other emissions in both directions, with the result that the final picture is quite complex.

It is of course much more effective to prevent solar radiation reaching glazed areas, in whole or in part, by making use of blinds, etc., fitted *externally*, than to permit exposure, allow penetration into the building, and then to search for means to curb the nuisance. Table 5.4 lists a number of combinations of glazing and shading devices and ranks these in order of efficacy. It will be noted that substitution of double for single glazing offers a much less dramatic improvement in this context than it does in reduction of U value. The computer-based

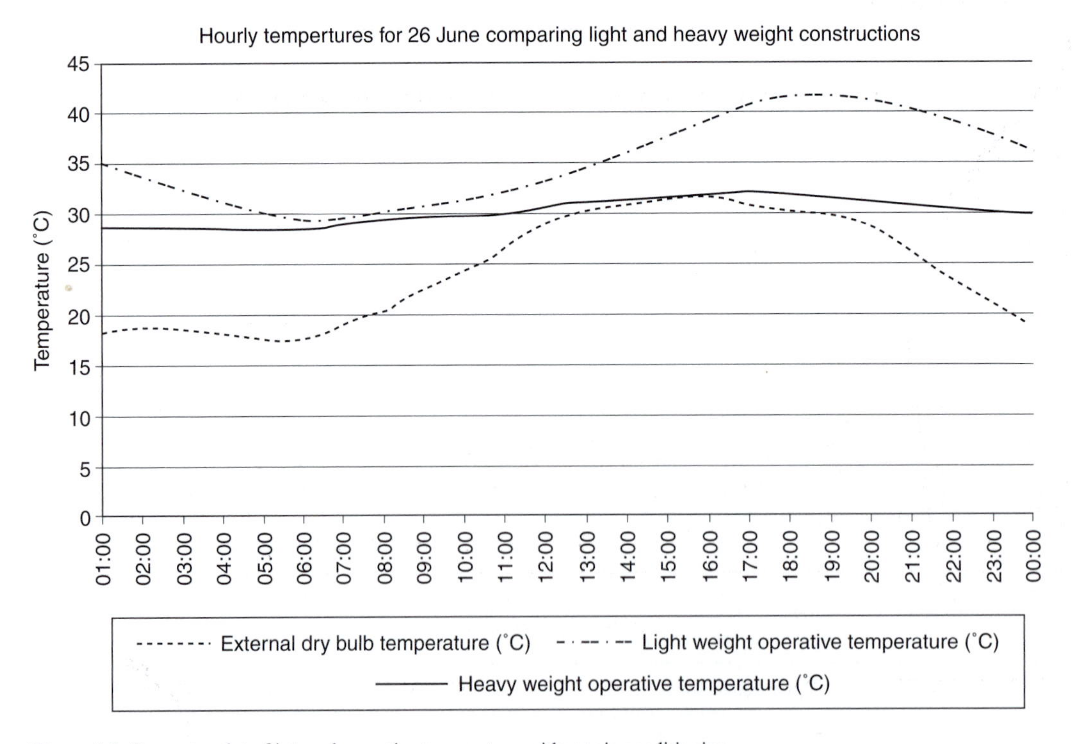

Figure 5.5 Computer plot of internal operative temperature without air-conditioning

plots of Fig. 5.6 provide parallel information and illustrate the effect upon peak air temperature within a south facing room which results from variation in types of protection and rates of air change. The G value of the glazing system is the same as the solar heat gain coefficient (SHGC) and is the total solar energy that comes into the building from directly transmitted and re-emitted absorbed radiation. Figure 5.7 shows these components for a single glazed unit. The following six types of glazing have been used in the example in Fig. 5.6.

Table 5.4 Relative efficiency of shading devices in reducing solar heat gain (plain single glazing (0))

	Relative exclusion efficiency (%)	
Type of solar protection	*Single glazing*	*Double glazing*
External dark green miniature louvred blind	83	87
External canvas roller blind	82	85
External white louvred sunbreaker, 45° blades	82	85
External dark green open weave plastic blind	71	78
Heat reflecting glass, gold	66	67
Clear glass with solar control film, gold	66	67
Densely heat absorbing glass	–	67
Mid-pane white venetian blind	–	63
Internal cream holland blind	57	61
Lightly heat absorbing glass	–	50
Densely heat absorbing glass	49	–
Internal white cotton curtain	46	47
Internal white venetian blind	39	39
Lightly heat absorbing glass	33	–
Internal dark green open weave plastic blind	18	26
Clear glass in double window	–	16

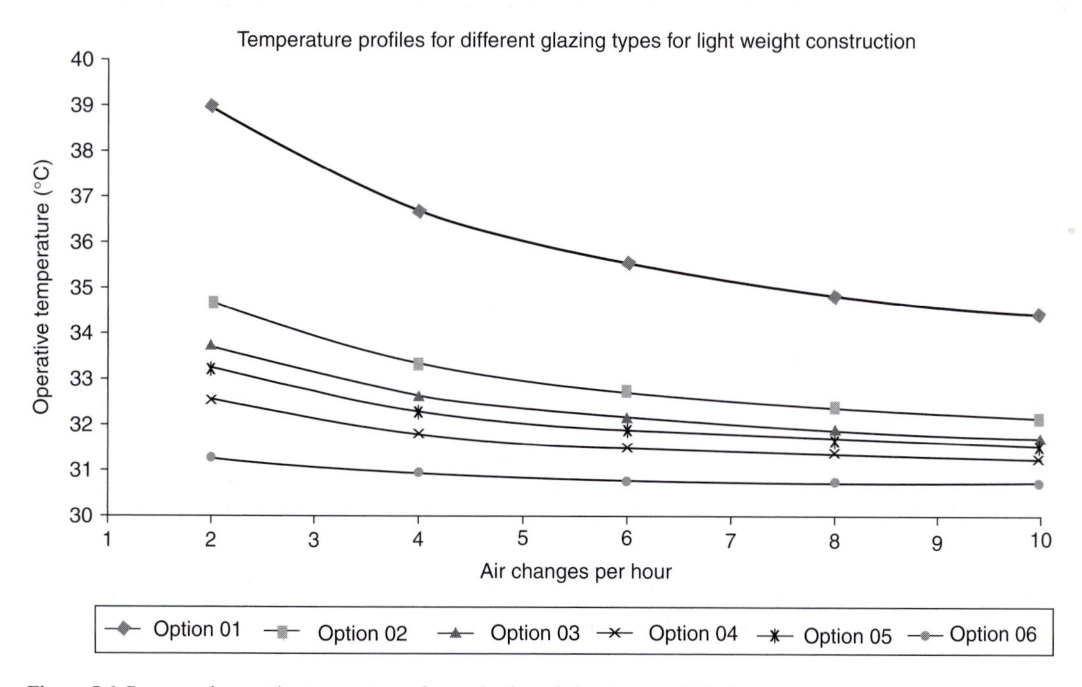

Figure 5.6 Room peak operative temperatures for a selection of glass types and air change rates

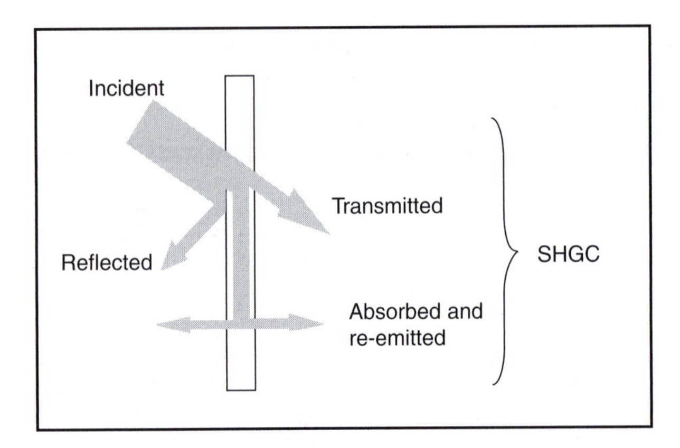

Figure 5.7 Components of SHGC

(1) Option 1 Single clear glass *G* value of 0.82
(2) Option 2 Double glazed with clear glass with *G* value of 0.43
(3) Option 3 Double glazed with absorbing glass with *G* value of 0.36
(4) Option 4 Double glazed with reflective glass with *G* value of 0.26
(5) Option 5 High performance structural triple glazed with *G* value of 0.32
(6) Option 6 Triple glazed with interstitial blind with *G* value of 0.16

Special solar control glasses reduce the total rate of heat transfer by increasing either the absorbed or the reflected component: a variety of different types is available:

Absorbing glasses: There are two main types here: body-tinted and metallic ion coated. The former type (BTG) is coloured throughout its thickness and, in consequence, the colour density and solar control properties increase with glass thickness: colourings available are commonly grey, bronze and green. The alternative variety has a layer of ions, which absorb incident radiation, injected into the surface of clear plate glass. Since the layer is below the surface and therefore protected, the glass may be used as a single pane.

Reflective glasses: These have a metallic oxide or similar coating on one surface and are thus normally used in sealed double glazing units. They are manufactured in a standard colour range of silver, grey, bronze or blue but other colours may be obtained.

Solar control films: Solar control films are available in a wide range of colours, and are applied to the internal side of the glass. They produce an effect similar to that of reflective glass and can provide an economic solution as a retrofit measure to existing windows. A recent addition to the materials available is a film which acts like a photochromic glass, darkening as the incident sunlight increases its intensity.

Smart windows: These so-called materials are in the course of development for commercial use. Typically, the glass has four coatings: a transparent conductor, an electrochromic layer, an ion storage layer and a second transparent conductor. By the application of a voltage to the glass, the transmissivity property of the coating is changed and may be adjusted to control the amount of solar heat entering the space.

Low emissivity coatings: These are, strictly speaking, more appropriate to retention of solar heat within a space than to the control of its entry. There are principally two types known as hard and soft coatings. The hard coatings typically have emissivities from 0.2 to 0.15 and are not as efficient as the soft coatings with emissivities from 0.1 to below 0.026. Most soft coatings also include solar control capability.

In order to take account of the surfaces and other features of a space within a building structure which absorb, to a greater or lesser degree, the solar heat admitted through glazing, the *Guide Section A5* includes a group of tables which list *cooling loads,* as distinct from *heat gains.* These apply to *lightweight* buildings, defined as having demountable partitions, suspended ceilings, and floors which are suspended or, if solid, has either a carpet or a wood block finish: a *heavyweight* building, conversely, has an overall solid construction with no soft finishes. Each such table applies to a single latitude and to a plant operational period of 10 hours: correction factors are

Table 5.5 Solar cooling loads for fast-response building with single clear glazing: SE England, unshaded

		Solar cooling load at stated sun time (W/m²)										
Date	*Orientation*	*0730*	*0830*	*0930*	*1030*	*1130*	*1230*	*1330*	*1430*	*1530*	*1630*	*1730*
March 29	N	38	50	64	74	83	86	86	81	72	60	47
	NE	160	181	135	83	97	98	98	93	84	72	58
	E	252	355	400	361	261	147	133	120	111	99	84
	SE	231	353	456	498	483	426	319	196	142	121	105
	S	109	176	283	384	457	505	497	454	377	270	161
	SW	104	118	135	162	224	347	443	501	514	467	358
	W	87	101	115	126	135	150	175	293	389	420	370
	NW	55	69	83	93	102	105	106	104	97	154	193
April 28	N	70	87	95	103	110	115	116	110	103	93	86
	NE	322	308	236	143	146	143	143	137	131	120	106
	E	439	501	510	430	314	192	179	167	160	149	136
	SE	331	433	511	515	487	413	301	186	169	149	135
	S	115	157	255	346	418	457	455	409	339	236	142
	SW	133	149	162	184	212	330	437	503	535	505	434
	W	138	154	167	176	184	199	226	351	465	518	513
	NW	104	120	133	142	149	154	155	159	167	258	325
May 29	N	118	121	126	131	137	140	139	135	130	123	120
	NE	382	362	290	190	174	166	165	161	155	146	134
	E	475	519	518	450	329	208	194	184	179	169	157
	SE	321	408	471	485	448	367	258	167	166	152	140
	S	110	129	201	289	355	388	384	345	274	182	116
	SW	142	156	168	183	193	290	393	465	497	477	420
	W	170	184	196	204	210	223	247	317	484	544	554
	NW	142	156	168	176	182	185	185	197	221	321	394
June 21	N	144	139	142	147	151	154	154	151	147	144	143
	NE	440	413	320	211	190	180	181	178	173	165	154
	E	533	579	550	469	343	220	207	199	195	186	176
	SE	345	437	482	487	446	363	253	168	171	159	148
	S	111	123	191	276	342	376	374	331	261	174	116
	SW	147	159	170	182	190	282	389	460	492	476	424
	W	185	197	207	215	220	232	257	382	497	561	578
	NW	160	172	182	190	195	198	199	214	244	347	424

provided to cover some other circumstances. These data are represented here for four representative months by Table 5.5, which relates to unprotected clear glass. Correction factors for other glass and building combinations are provided by Table 5.6 respectively.

In the case of the principal Table 5.5 all the values listed assume that calculations for internal design conditions have been based upon operative temperature. If air temperature were to be used, the correction factors at the ends of Table 5.6 must be applied. Diffuse radiation from earth and sky is included in the data provided but, where the ground floor of a building abutts to a wide city pavement subject to solar radiation, an addition of 10 per cent to the listed figure, as mentioned previously, may be justifiable.

There are two recent documents from CIBSE that give a more in-depth understanding of thermal performance of glazing systems and how to design systems to reduce solar gains. CIBSE TM35:2004 'Environmental performance toolkit for glazed façades' and CIBSE TM37:2006 'Design for improved solar shading control'.

Structural shading

In a deep plan building, it is possible that windows are provided more to encourage occupants to retain visual contact with outside rather than in competition with some subsidiary form of illumination. In consequence,

Table 5.6 Correction factors (applying to Table 5.5 only)

Glazing configuration (inside to outside)	*Correction factor for stated building response* Fast	Slow	*Glazing configuration (inside to outside)*	*Correction factor for stated building response* Fast	Slow
Clear	1.00	0.89	Clear/clear/absorbing	0.46	0.39
Absorbing	0.74	0.64			
Reflecting	0.67	0.59	Low-E/clear	0.81	0.69
			Low-E/reflecting	0.52	0.44
Clear/clear	0.83	0.71	Low-E/absorbing	0.55	0.47
Clear/reflecting	0.53	0.46			
Clear/absorbing	0.56	0.47	Low-E/clear/clear	0.70	0.59
			Low-E/clear/reflecting	0.44	0.38
Clear/clear/clear	0.71	0.60	Low-E/clear/absorbing	0.46	0.39
Clear/clear/reflecting	0.44	0.38			
Additional factor for air temperature control	0.86	0.83			

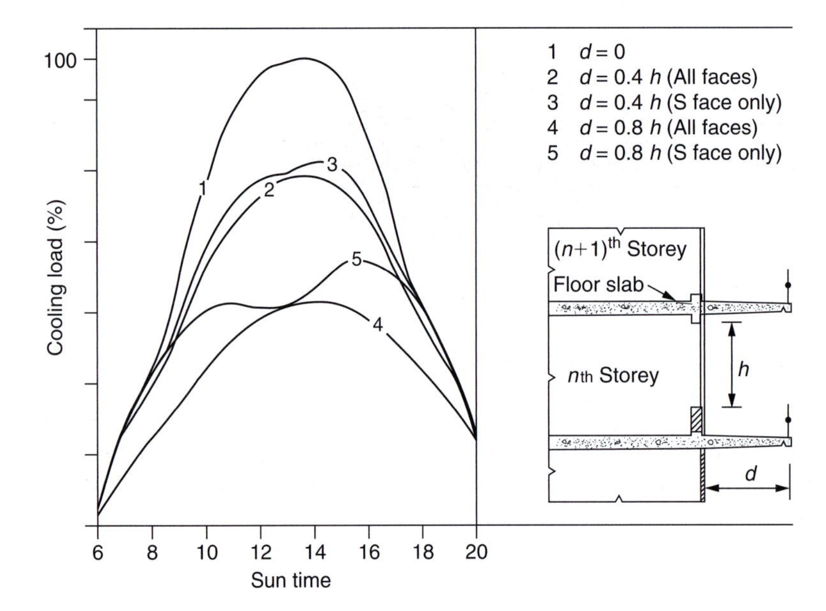

Figure 5.8 Effect of horizontal projections upon cooling load

the use of internal or external blinds may not be acceptable, despite the penalty paid in heat gain. It is therefore necessary to consider the extent to which vertical fins or horizontal projections may be of value. It is self-evident that, in the latitude of the British Isles, vertical fins are useful only on the east and west facades but that their influence there upon the peak cooling load to individual spaces may not be large. In terms of the maximum cooling load in a whole building, however, such fins may contribute to a reduction.

Horizontal projections formed by outward extensions to floor slabs above windows may make a contribution if formed on the southerly facades, south east through to south west, as shown in Fig. 5.8. The effect of such shading, for an air-conditioned commercial building, may be expressed also in terms of annual energy expended in cooling, as predicted for a south facing room using a variety of shading methods and dimensions, and listed in Table 5.6.

Table 5.7 Effect of structural shading upon annual energy consumption for cooling

Method of shading *Type*	*Depth* (mm)	*Annual index*
No shading		1.0
Window recess	150	0.94
	500	0.82
	1000	0.68
Vertical fins	150	0.97
	500	0.94
	1000	0.88
Horizontal overhang	500	0.87
	1000	0.76
	2000	0.57

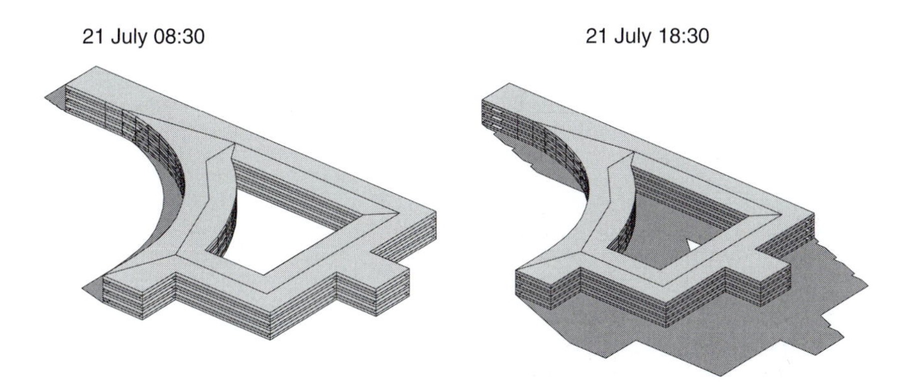

Figure 5.9 Solar shading diagrams from computer simulation

The diurnal cycle of the sun's motion subjects the various faces of a building to a pattern of shading which changes continuously and, as has been noted previously, if a building has wings forming an L or a T or some other more confused shape, the shading pattern may become complicated (Table 5.7). In consequence of demands for detailed data, computer modelling programs have been produced which can not only produce graphical representations of the shading as shown in Fig. 5.9 but can also generate animations of shadows as movies for daily, monthly or a years shading.

Building shape

Deep plan constructions, generally single-storey, have been a commonplace approach to industrial building for many years. In most cases room heights are generous, daylight is available from roof glazing and occupancy levels per unit of floor level are low. Simple heating and ventilating systems will thus, in the majority of cases, provide adequate service.

Exceptions to this general statement, now tending to represent the rule, arise of course where an industrial process such as the assembly of microcomponents, etc. requires close control of temperature, humidity and airborne contaminants. Since requirements of this nature are usually accompanied by demands for a high and constant level of illumination, the logical outcome is the provision of an enclosed environment with full air-conditioning.

In contrast to a deep plan arrangement designed for office use and a high-rise building of equal floor area overall and assuming reasonable proportions, length to width, the deep plan would have four storeys and the tower block twenty. In terms of exposure, the tall building would present more surface to wind, rain, frost and sun at the perimeter but the need for electric lighting in the core of the deep plan building would most probably lead to a requirement there for cooling throughout the year.

Once the size and character of an office building are such that use of natural ventilation, a simple heating system and natural lighting during daylight hours are inappropriate, the ground rules have changed and there is no absolute ideal for building shape. This is not to say that, for a given site and a given accommodation brief, a detailed analysis cannot produce positive advice as to the building form which will allow optimum system design and minimum recurrent cost to be examined.

Building orientation

For the particular case of the rectangular office block, having a width/depth ratio of 3:1 as Fig. 5.10, a calculation may be made to compare heat gains on this building type dependant upon the primary orientation of the glass. Figure 5.11 shows the results at 10° intervals from north to south facing for two zone on June 21 and September 21.

It is evident that, in mid-summer, the orientation of case 1 produces a lower heat gain than that of case 2 but that in spring and autumn the difference will be much less. In reality, with a office block of this plan, the narrow ends are solid and windowless: and the longer walls have 1500 mm high continuous glazing on each facade. This rudimentary example serves to confirm the rule of preference for an east-west axis, now well established.

Conduction

This subject has been explored in detail in Chapter 4 and the mechanism of heat transfer here is similar, being a function of the thermal transmittance coefficient (U) and the temperature difference, inside to outside. A factor may be applied in this instance to allow for the effect of the radiant component of heat transfer which

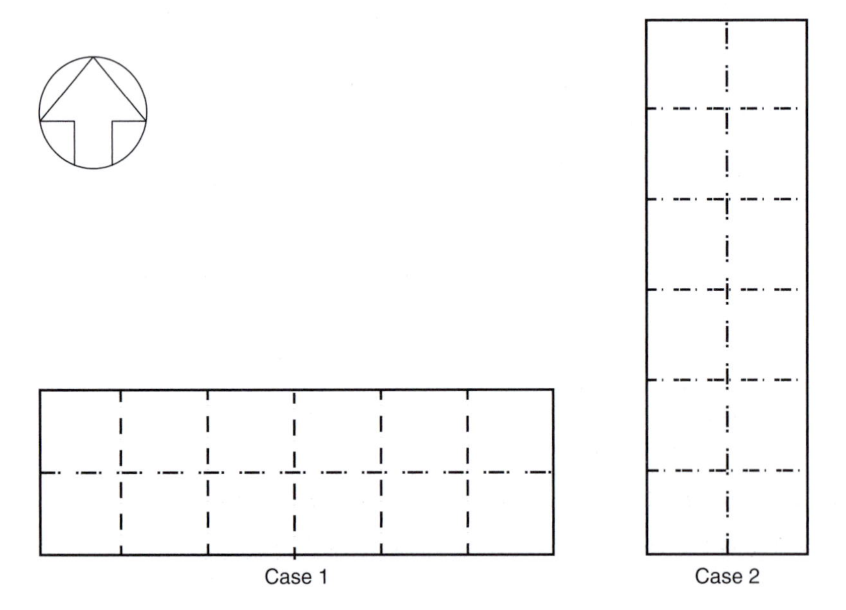

Figure 5.10 Plan of tall slab building

arises from the adoption of operative temperature as the criterion of comfort. For those particular instances where the facade of a building has an extremely high proportion of glazing, correction is necessary and reference to the *Guide Section A5* is advised.

Application

The manner in which these components of total heat gain relate to a building is best illustrated by application to the simple example shown in Fig. 5.12: this represents a multi-purpose hall, in use for concerts and other activities, having the following constructional details, etc.:

- The windows are clear low-e double glazed in metal frames and a U value of 1.95 W/m^{2}K and a G value of 0.64.
- The walls are 105 mm brick, 50 mm urea formaldehyde (UF) foam, 100 mm lightweight concrete block finished with 13 mm gypsum plaster and a U value of 0.35 W/m^{2}K a decrement factor of 0.3 and a decrement factor time lag of 9 hours.
- The roof is flat consisting of light-coloured paving slabs, waterproof membrane, 50 mm extruded polystyrene, 150 mm cast concrete, air cavity and 13 mm ceiling tile and a U value of 0.25 W/m^{2}K a decrement factor of 0.38 and a decrement factor time lag of 7 hours.
- From the orientation shown, the building presents two faces to solar gain, north-east and south-west.
- Inside temperature 22°C and outside air temperature for London of 32°C. (Temperature in London taken from Table 5.10 as 30°C and 2°C added for urban heat island effect, no account of climate change taken into account.)

Hand calculations would assume a peak solar gain to be taken at say 3:30 p.m. sun time. Table 5.5 gives a peak solar cooling load of 492 W/m^2 for SW facing and 173 W/m^2 for north-east (NE) facing.

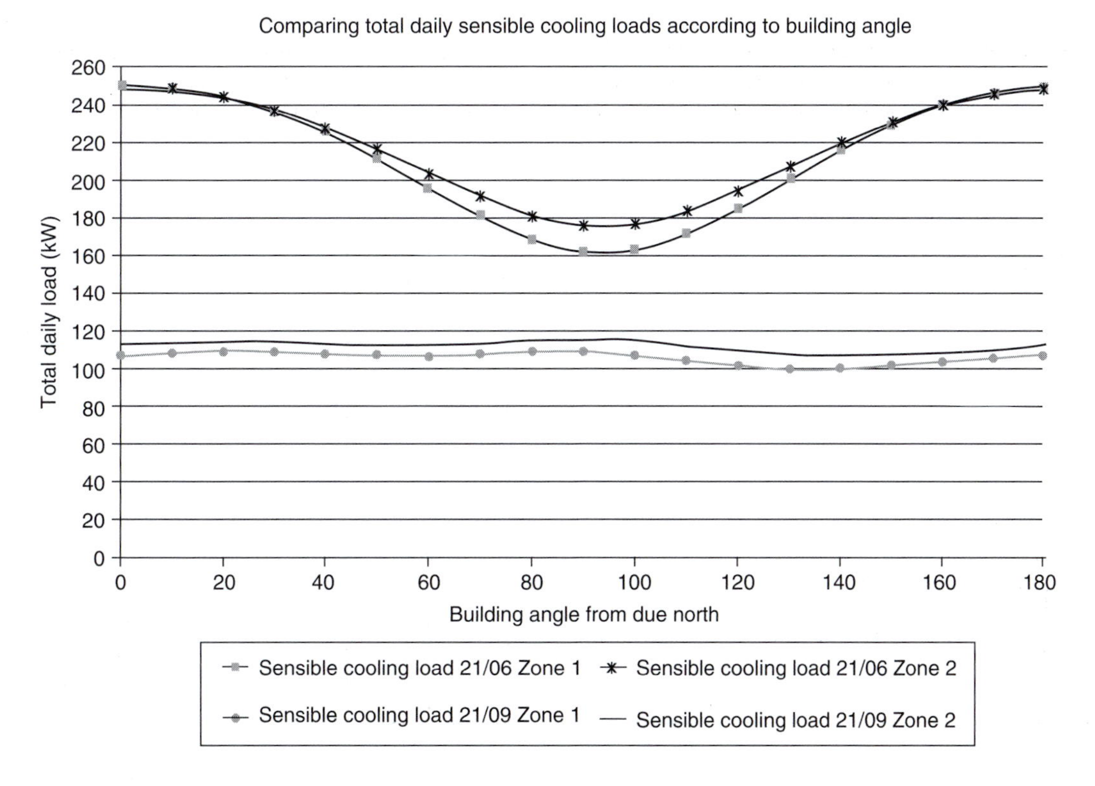

Figure 5.11 Total loads on a building with glass on the longest faces for varying orientations

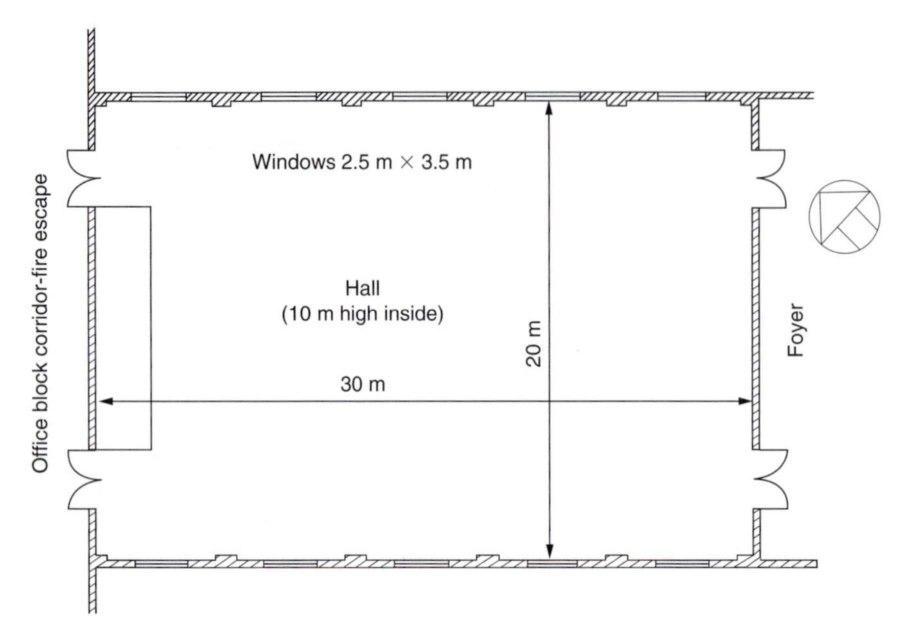

Figure 5.12 Heat gain example

The area of glass facing SW and NE is 5(2.5 × 3.5) = 44 m^2

From Table 5.6 a slow response factor of 0.68 for this heavyweight building and an air node correction factor of 0.86. Adding 10 per cent extra gain to the SW glazing to allow for the pavement.

SW facing glazing = (44 × 492 × 0.68 × 0.86 × 1.1) = 13 925 W
NE facing glazing = (44 × 173 × 0.68 × 0.86) = 4451 W
Total solar gain (13 925 + 4451) = 18 376 W

A detailed solar analysis of the peak solar gain shows a peak solar gain of 21 500 W at 4 p.m. in September. It is always recommended to carry out a detailed assessment of all the gains within buildings using accredited software. It must always be remembered that wrong data input leads to wrong results and care should always be taken to check input data and output results thoroughly.

Conduction gains through glazed elements are due to temperature difference across the element, the surface area and the *U* value.

88 m^2 × 1.95 W/m^{2}K × 10 K = 1716 W

Each fabric element of wall and roof has a different time lag and decrement factor and therefore the sum of all the permutations to get the peak fabric conduction load is time consuming. From the computer analysis the peak conduction gain is 6181 W at 11 p.m. in July. From the analysis for the four months May to September the peak heat gains to the multi-purpose hall occurs at 5 p.m. in August and has a load of 18 200 W.

The total heat gain of 18.2 kW is carried forward to Chapter 22 for calculation of air-conditioning design.

Air infiltration and ventilation

In addition to the heat which enters the building by direct transmission of solar radiation and by conduction, allowance must be made for that which enters with outside air by infiltration. This is not dissimilar to the interchange which takes place in winter except that heat is entering rather than leaving a building.

Table 5.8 Heat emission by occupants in buildings

Activity	*Rate of heat emission for mixture of males and females* (W)		
	Sensible	*Latent*	*Total*
Seated very light work in office	70	45	115
Moderate office work	75	55	130
Walking, standing in a bank slowly	75	70	145
Light bench work in factory	80	140	220
Moderate dancing	90	160	250
Heavy work in factory	185	255	440

Air infiltration

Air tightness testing of buildings is now a legal requirement of all buildings. This will set the air change rate from outside. This subject has been dealt with in some detail in the preceding chapter but, consequent upon the probability of lower wind speeds and less stack induced effects occurring during summer months, lower air change rates may be experienced than in winter conditions.

Ventilation

This matter has been dealt with in Chapters 1 and 4 and it is necessary only to emphasise the difference between *ventilation air* and, as far as system design may be concerned, what is known as *conditioned air*. The former is that quantity of outside air, normally quoted in terms of volume, necessary to dilute contamination from all sources to an acceptable level. The conditioned air quantity, however, normally quoted in terms of mass, is related to its capacity under certain limiting conditions to absorb whatever heat or moisture may be surplus within the space or, conversely, to supply heat or moisture when there is a deficit within the space. The conditioned air will, of necessity, be the means of introducing ventilation but since the whole quantity has been treated at the plant, unlike air which has entered by infiltration, no loss or gain will be involved *within* the space.

Miscellaneous heat gains

Activities which take place within a space produce heat and this, a possible although unreliable bonus in winter, is a certain penalty during summer. CIBSE Guide A6 provides more details on the types of internal heat gains experienced in a variety of building types.

Heat from occupants

The heat emitted by the human body depends upon the temperature, humidity and level of air movement in the occupied space and upon the level of activity. Table 5.8 gives the rate of heat emission from a mixture of male and female occupants within different building and activity types. The sensible heat increases by about 20 per cent with air movements up to 0.5 m/s but latent heat is affected only slightly by air movement for temperatures less than about 18°C, above which temperature the point at which the curve begins to rise is displaced, i.e. at 0.5 m/s, the rise begins at about 26°C.

It is the sensible heat alone which affects temperature within a space, that provided by one person seated at rest being sufficient to heat about 4 litre/s of air through 18 K.

Heat from lighting

All the electrical energy consumed in providing lighting is dissipated as heat, the rate depending upon the type of lamp and the design of the luminaire. For a given level of illuminance, tungsten lamps will emit 4 to 8 times

Table 5.9 Approximate heat gain from lighting (including control gear, if any)

Level of Illumination (lux)	*Approximate electrical loading* (W/m²)						
	Tungsten filament		*Triphosphor fluorescent*		*Metal Halide*		*HP discharge*
	GLS	*CFL*	*Office*	*Industrial*	*Office*	*Industrial*	*Sodium* (SON)
300	45	8	7	6	11	7	6
500	80	14	11	10	18	12	11
750	–	21	17	14	27	17	16

as much heat as efficient fluorescent lamps and the unit emission from the latter varies quite considerably with type, size and colour. Table 5.9 provides outline data for the electrical loading related to floor area but this type of information should be used for preliminary design only, pending details of the final lighting layout.

Extract ventilation from and around recessed luminaries has the effect of increasing luminous flux and improving efficacy (lumens/W) on the older fluorescent lamps but can reduce the efficacy on newer smaller diameter fluorescent lamps. In absence of specific details from a manufacturer, it may be assumed that the output *to the space* of 50 per cent for a conventional luminaire is reduced to about 20 per cent for a type having air-handling facilities. The balance of the output in the two cases, 50 per cent and 80 per cent, respectively, passes to the ceiling void and, although no longer a heat gain *to the space*, remains to be dealt with by the central cooling plant in circumstances where some or all of the air extracted is recirculated.

Heat from office machinery and computers

The rate of heat dissipated from these sources may be high, typically 15 W/m^2 and up to 25 W/m^2 in a commercial office accommodating a large number of items of electronic equipment. Equipment loads in offices are generally specified in W/m^2 without prior knowledge of who will occupy or at what density and care should be taken not to over specify with the resultant inefficiencies this creates. In areas devoted wholly to data processing or financial dealing facilities, a peaks of over 100 W/m^2 may occur. Output is mainly convective and reference should be made to manufacturers' data to establish accurate information. For data centres this can rise to over 2 kW/m^2.

The level of heat gain from office equipment is likely to continue to change in the foreseeable future under sometimes opposing influences. It is predicted that there will be a further increase in the use and that the machines will continue to be more powerful. It is likely, however, that there will be significant improvements in the efficiency of machines thereby reducing energy consumption, flat screen technology being an example of this. The short-term effect is that power requirements may increase from the current typical maximum of about 280 to around 350 W per person, but with a suggested fall in the longer term to 150 W per person.

Heat gains in excess of 600 W/m^2 may be expected in computer rooms and those areas housing support equipment. Such spaces will certainly need to be air-conditioned in order to maintain the electronic equipment within the limits of temperature and humidity set down by the makers and to provide comfort conditions for operators. It is interesting to note signs of return to a practice adopted for early computers whereby certain processing units incorporated a cooling coil and a closed circuit air circulation within the casing. Manufacturers' literature in such cases requires only that a supply of chilled water, specified as to quantity and temperature, be provided.

Heat from machines and process equipment

Electrical energy provided to motor drives is for the most part converted wholly into heat at the time it is consumed, the exceptions being where work is done to produce potential energy, as in the case of water pumped to an elevated tank or goods raised in a hoist. In consequence, the energy input to machines constitutes a heat gain and may be evaluated by reference to the power absorbed and, where appropriate, any diversity of use.

When both the motor and the driven machine are in the space, then the total power input must be considered, but if either are mounted elsewhere, then reference to manufacturers' data will provide guidance as to the proportion of the total which must be allowed for.

No useful guidance can be given as to the amount of heat which may be liberated within a space by equipment such as steam presses, hot plates, drying ovens and gas or electric furnaces and reference must be made to manufacturers' literature. Similarly, for laboratories, etc., standard authorities should be consulted, a wide ranging summary being included in the *Guide Section A6*.

Diversity

As has been mentioned in the two previous paragraphs, it may well be important that the matter of diversity in use be considered. In a workroom, for instance, there may be a variety of machines having 60 or so motors which together have a brochure rating of 300 kW but, at the same time, the total instantaneous loading may be only two-thirds of that total. While it would clearly be wrong to ignore this seeming diversity in use, the detail will often require careful analysis as to both timing and the physical positioning of the machines. At the risk of stating the obvious, a diversity of 66 per cent does not mean that each motor is running at 66 per cent full load but more likely that of each 60 motors fitted, a random maximum of 40 are in use simultaneously. Those at one end of the space might, perhaps, all be idle at one time or, conversely, but one of each pair of adjacent machines might be running. Each case needs to be considered in detail.

Temperature control

The case made in Chapter 4 for temperature control room-by-room in winter, applies equally to buildings in summer. In the case of air-conditioning systems, the control philosophy requires very careful consideration since the plant configuration to deal with heating, cooling, humidification and dehumidification involves much more complex equipment than does a straightforward heating system. Furthermore, it is probable that some spaces within an air-conditioned building may require heating, coincidentally with a demand from others for cooling, as a result of different orientation, time of day and the level of miscellaneous internal gains. This situation will prevail in particular during the mid-seasons between summer and winter and it points to a very positive conclusion that system controls must be considered in detail right from the earliest stages of design.

Temperature difference

The comments made in Chapter 4, with regard to the use of operative temperatures, are equally valid for summer conditions.

Inside temperature

The only comment necessary under this heading is that it is usual for summer design purposes to make a distinction between continuous and transient occupancy, 22°C dry bulb and 24°C dry bulb being the commonly accepted levels, respectively, with 40–60 per cent saturation in each case.

Outside temperature

As was the case for winter conditions, the design temperature datum adopted to represent outside conditions in the British Isles will vary according to the precise location. CIBSE Guide A2, Table 2.15 provides percentage frequencies for which the hourly dry bulb temperatures are exceeded for the period June to September for the years 1983 to 2002, from which a selection can be made for the particular application; Tables 2.7–2.14 give for a similar period the frequency of the combinations of dry bulb and wet bulb temperature. Data from these tables have been used to establish suggested summer external design values for an excedance of 0.3 per cent (equivalent to 9 hours) over the period, as given in Table 5.10 For a higher excedance the design dry bulb

Table 5.10 Summer external design temperatures

	0.3% excedancea (9 hours)		*Possible increases to design dry bulb temperature*	
Location	*Dry bulb* (°C)	*Wet bulb* (°C)	*Climate change*[b] (°C)	*Heat island*[c] (°C)
Belfast	25	19	2	–
Birmingham	29	19	2	1
Cardiff	27	19	2	–
Edinburgh	25	18	1.5	–
Glasgow	26	19	1.5	1
London	30	21	2.5	2
Manchester	28	19	2	1
Plymouth	26	19	2	–

Notes
a Temperature exceeded within period June to September.
b Climate change allowance to 2020.
c Likely maximum increase.

temperature will decrease, e.g. at 1.0 per cent (equivalent to 29 hours) the dry bulb temperature will be 2°C below the 0.3 per cent values given in the table. It will be noted that, in this instance, both dry and wet bulb temperatures will be required for the design of air-conditioning systems.

The design temperatures given in the table are based upon historical meteorological records and a view should be taken on an appropriate allowance for the effects of climate change. CIBSE A2 offers guidance in this respect for scenarios up to 2080, however for the design of air-conditioning an allowance up to 2020 is prudent since the expected economic life of system components is likely to be around 15–20 years. For more details on the design of buildings to take account of climate change see CIBSE TM36:2005 'Climate change and the indoor environment: impacts and adaptation'.

When considering applications for buildings in town centres, it is good practice to use dry bulb temperatures 1° or 2° higher than those listed in order to make allowance for the *heat island* effect created by building density, traffic, etc. Nighttime temperatures in London can be as high as 6° higher. Figure 5.13 shows the heat island intensity distribution on a relatively calm night. More details on the heat island effect in London can be found in the Mayor of London document 'London's Urban Heat Island' October 2006.

For a more accurate assessment of the most appropriate summer design conditions *Guide Section A6* provides data to enable the designer to plot the percentage frequencies of combinations of hourly dry bulb and wet bulb temperatures on a psychrometric chart. This enables the frequency with which the specific enthalpy exceeds given values to be determined, from which summer design conditions may be established.

Intermittent operation

In Chapter 4, when considering intermittent heating, the matter of excess plant and system capacity over and above that calculated for steady state heat losses in order to provide for morning preheating, was discussed in some detail. In the case of plant provided for cooling a building, the situation is not the same since the peak load (for the whole building) will normally occur somewhat later in the day, part way through the occupancy period after a steady rise in heat gain and requirements for cooling. In consequence, it is unlikely in most instances that any useful purpose would be served were the capacity of the plant to be increased above that required to meet the selected design condition.

It is nevertheless possible to go some way towards overcoming a deficit in plant size and to reduce running cost during warm daytime periods, by setting out deliberately to cool a building overnight when unoccupied. This may be achieved to some extent by circulating the relatively cold night air through the occupied areas

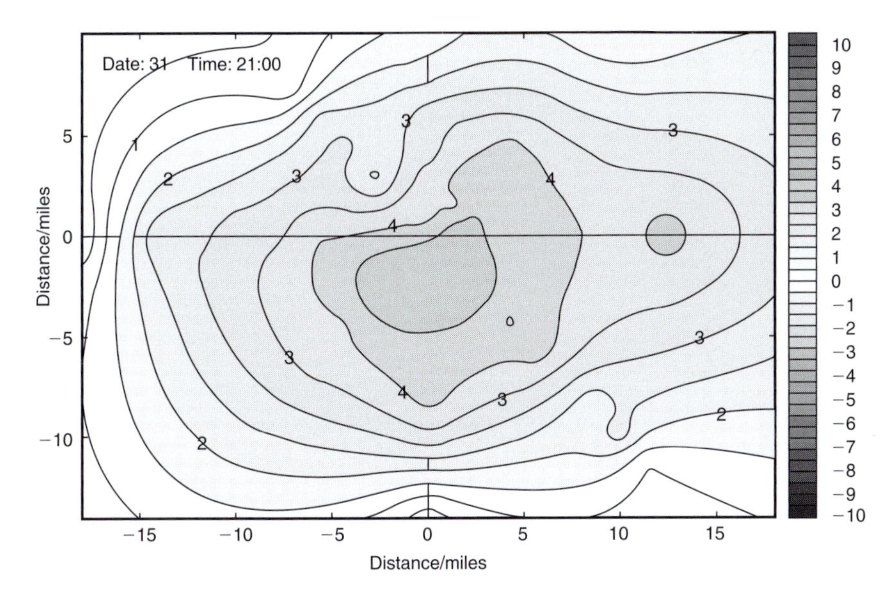

Figure 5.13 Heat Island intensity distribution in London on calm night (CIBSE Guide A Fig. 2.29)

with intent to cool the internal structure and thus create a reservoir against the demands imposed during the following day. This is not a routine to be recommended without a close study of potential problems such as the incidence of condensation, etc.

As to the calculation of summer loads generally, it must be borne in mind that the time at which the peak cooling load occurs in an individual space will vary with orientation. In consequence, the maximum simultaneous demand imposed upon a central plant is not normally the sum of the peak loads occurring in the individual spaces or zones of a building but almost certainly some lower figure. Calculation of the individual and simultaneous demands is further confused by the fact that the maximum cooling demand for a whole building is often displaced in time from what is the apparent peak of solar gain. This situation comes about as a result of an intricate pattern of interplay between fading instantaneous loads on one face and growing loads of similar type on another, all against a background of time delayed loads through opaque surfaces, to say nothing of lighting, occupancy and intermittent loads from machines. It is virtually impossible to make the necessary calculations using manual methods but computer simulation programs exist which take the mixture in their stride.

Chapter 6

Building energy and environmental modelling

This chapter covers different aspects of modelling buildings from simple design calculations through to advanced dynamic thermal modelling procedures. Computers are widely used within the design of heating and air-conditioning systems, and there are essentially four levels of analysis:

(1) Design calculations
(2) Compliance checking
(3) Dynamic thermal modelling
(4) Computational fluid dynamics

The simplest is usually the design of systems following the CIBSE procedures. These come into the 'how many, how much and how big' category and typically would be used for sizing heat emitter output within a room and the calculation of the air flow requirement and supply temperature to provide cooling to an internal space within a building. These procedures could be steady state or simple hourly calculations. The heating and cooling loads for each space can be summed to provide boiler and chiller sizing information.

The next level would be compliance checking calculations, such as those referred to within Building Regulations Part L or for planning purposes to check emissions of carbon dioxide from a building. These could be monthly average outside air temperature methods or more detailed simulation of the building fabric and the associated heating, cooling and ventilation systems.

Dynamic thermal modelling would normally use hourly weather data and simulates the building's performance for up to a 1-year period. The energy used for heating and cooling can be predicted or the occurrences of overheating investigated. The simulation can also determine how quickly buildings will heat up or cool down. In the UK the CIBSE standard weather tapes, either in the Test Reference Year (TRY) or Design Summer Year (DSY) formats should be used. Details on the types of weather tapes are explained later.

The most complex calculations are based on computational fluid dynamics (CFD) which takes account of the very detailed heat transfer mechanisms and fluid flows. These computer programs are used to examine the air flows within buildings and can also analyse air flows around buildings. Topics such as occupant comfort can be assessed by simulating the air flow velocities, radiant and air temperatures. CFD can also check for contaminants and the programs are often used to look at the location of intake and exhaust paths from a building to minimise cross-contamination of air.

Each of the above categories will be covered in more detail below referring back to the requirements from earlier chapters. There are other types of computer software for drawing, visualisation and calculation of mechanical and electrical not covered in this chapter.

Design calculations

The calculation of heating and cooling loads for individual rooms in a building or the complete building can be complex as discussed in the previous chapters. There are various types of data that are needed for these calculations to be performed.

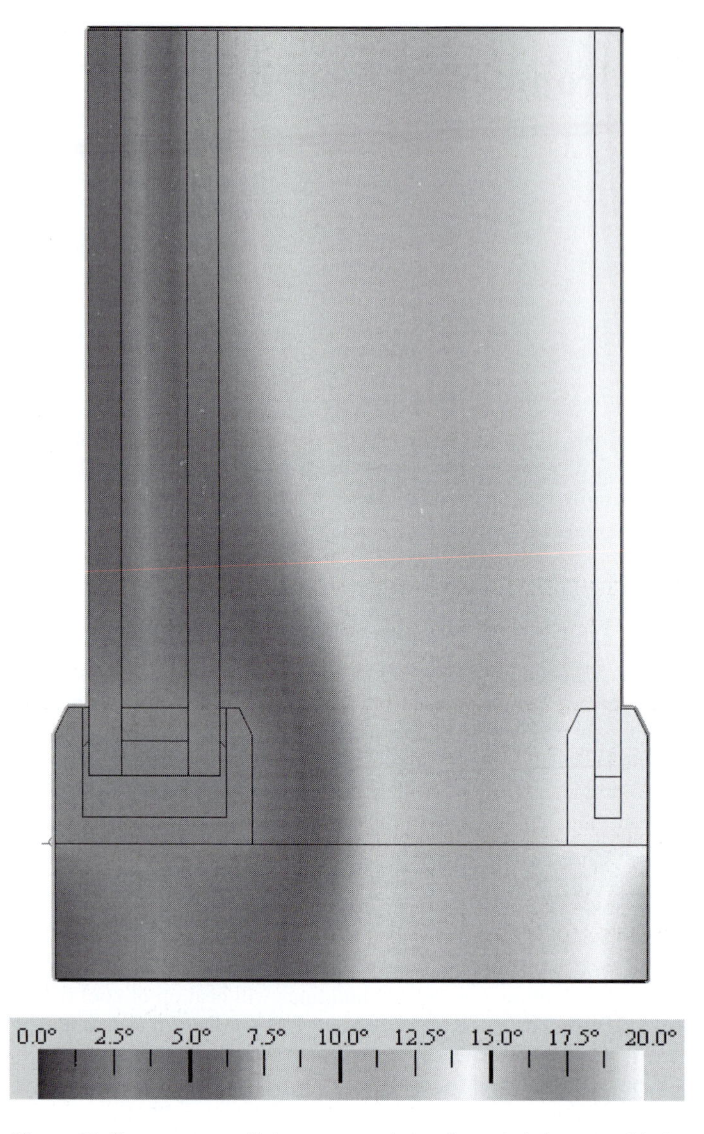

Figure 6.1 Temperature prediction across a window frame and glass assembly (see colour plate section at the end of the book)

Fabric details

The thermal transmittance of a building fabric is the basis for heat loss calculations. Information on detailed calculation methods for various thermal elements such as the wall, floor, roof and windows is provided in the BRE Guide BR443 'Conventions for *U* value calculations'.

This document is referred to within the Approved Documents for Building Regulation Part L and identifies which British, European and International (BS EN ISO) calculation procedures should be followed to assess the *U* value of different building elements. BRE have software for the calculation of *U* values based upon these standards. This is also the starting point for most compliance checking calculations. The basis of the calculations is a summation of the individual thermal resistances of each layer of construction and the internal and external surface resistances as illustrated in Chapter 4. For complex building elements with multiple conduction paths these calculations are no longer simple because of the changes made to comply with the

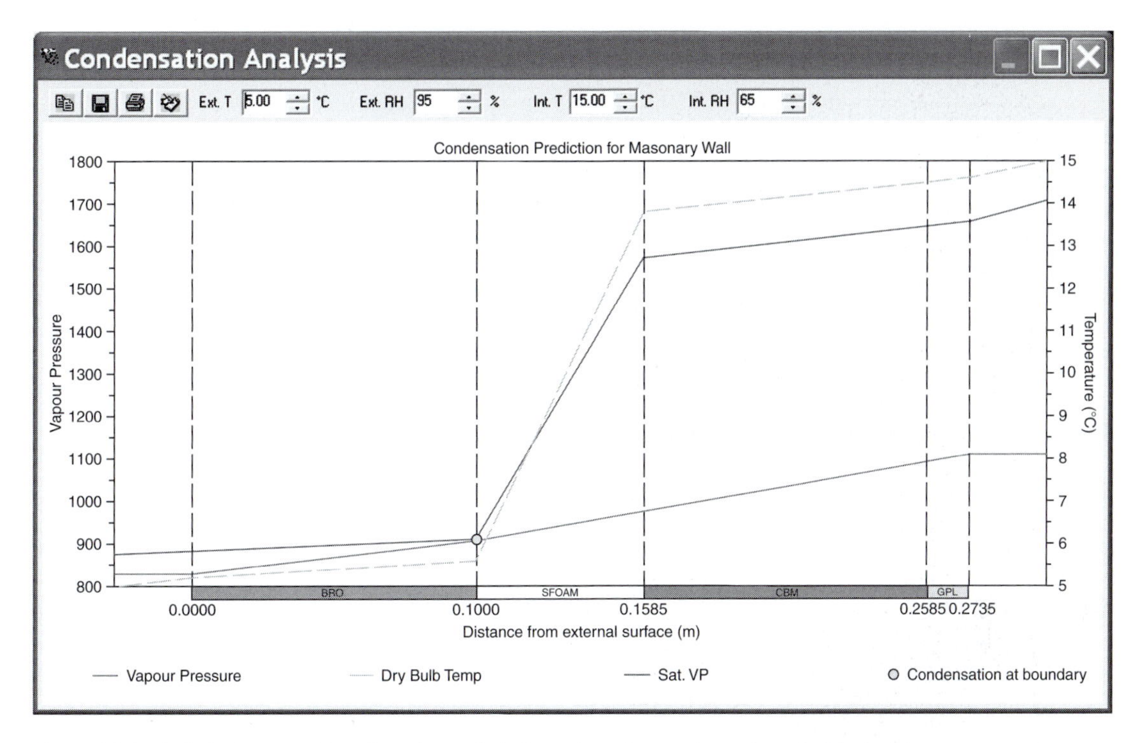

Figure 6.2 Typical plot from a wall condensation analysis (see colour plate section at the end of the book)

European and ISO standards and require computer-based calculation methods. For windows the centre pane *U* value, the window frame *U* value and edge effects of the spacers need to be taken into account, and it should be noted that there is often a large difference on high-performance glazing systems between the centre pane *U* value and that of the complete window unit. Figure 6.1 shows a detailed temperature prediction across a window frame and glass assembly.

Using ISO standard values to set outside conditions for calculating the *U* values of a given fabric element may be inappropriate for a detailed design calculation. When calculating the *U* values for windows, we often use simple values which are required for compliance checking procedures, but this value may not be appropriate for advanced designed conditions in hostile climates.

U value calculation software often contains modules that can check for interstitial condensation within the thermal fabric elements. Standard inside and outside conditions should be used to check for the quantity of moisture on a surface. Figure 6.2 shows a typical plot from a wall condensation analysis.

Heat loss calculations

A steady state heat loss calculation is based on a temperature difference between inside and outside. There are two major heat loss components, the fabric losses and air infiltration/ventilation losses. Traditionally in the British Isles the external temperature is based upon a single dry bulb temperature for the closest CIBSE weather location and the internal operative temperature for the required comfort condition. The calculations ignore solar gains and casual gains within the spaces and often safety margins can be added. Chapter 4 covers these procedures in more detail.

With well-insulated and reasonably airtight buildings, the added heating benefits of internal gains and passive solar gains could be utilised within dynamic thermal modelling software to optimise the sizing of heating systems.

The use of intermittently heated buildings means that great care should be taken when selecting emitter and boiler sizes to make sure enough additional capacity is available to bring the conditions within the building up to temperature. Simple additional margins which have often become 'rules of thumb' may no longer be appropriate with much improved thermal performance of the building fabric and the resulting lower heating demands.

Heat gain/cooling load calculations

Cooling loads add another level of complexity because the calculation of solar gains as outlined in Chapter 5. In addition to instantaneous gains the effect of thermal capacity and the time lag of the building envelope need to be taken into account. The simplest calculations are based upon offsetting the peak heat gains from conduction, ventilation, casual gains and solar gains. These calculations typically analyse a sample day in each of the summer months to predict the hottest conditions within a building or the cooling load required to achieve a given comfort or air temperature within a space. Dynamic thermal modelling is used to take account of the various time lags in the nature of each of the gains to a space.

Compliance calculations

The Building Regulations Part L for conserving fuel and power within buildings requires the calculation of carbon dioxide emissions from fixed building services within the building. The compliance calculations work out the heating and cooling loads and then convert these into carbon dioxide emissions based upon the various CO_2 emission factors provided within the Approved Documents. The Building Regulations will need to assess the energy consumed by any boilers and associated pumps, ventilation fans, air-conditioning equipment and lighting.

The Building Regulations Part L covers new dwellings within Approved Document L1A and new buildings other than dwellings within Approved Document L2A. The Standard Assessment Procedure (SAP 2005),

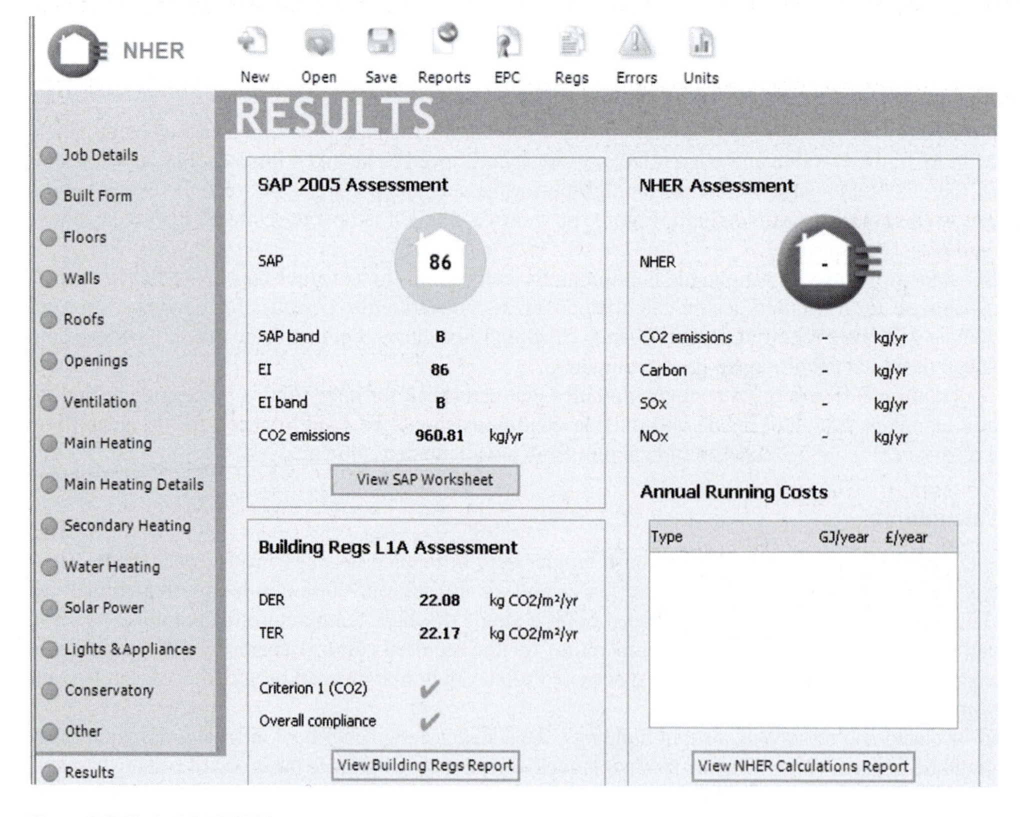

Figure 6.3 Typical SAP 2005 output

developed by BRE, is the procedure used for all dwellings and this uses the same external weather data irrespective of the location within England and Wales. This procedure not only covers the carbon dioxide emissions from the fixed building services within the dwellings, but also contains in Appendix P, the methodology which determines whether the dwelling overheats in summer based upon the SAP 2005 overheating criteria.

There are various computer programs that have been accredited to carry out the calculations contained within SAP 2005 for carbon dioxide emissions, minimum acceptable standards for fabric performance and the summer overheating requirements. It is recommended that only trained individuals that are approved to use the software packages carry out the calculation for submission to Building Control Bodies. Figure 6.3 shows the NHER SAP 2005 software output.

For non-dwelling applications, the Simplified Building Energy Model, SBEM developed by BRE, uses various European committee for standardisation (Comite Europeen de Normalisation, CEN) standards to calculate the heating and cooling requirements of spaces based upon monthly average temperature profiles. A major criterion with compliance calculations is for consistency in results; this is different from design calculations in that standardised occupancy schedules need to be followed for the type of building being analysed. Figure 6.4 shows the iSBEM interface for non-dwelling compliance checking.

The use of dynamic thermal modelling can also be a route for compliance with Part L of the Building Regulations, as long as the standardised occupancy schedules and temperature profiles for buildings are used. For checking carbon dioxide emissions it is important that the appropriate, i.e. closes, site or a location with similar CIBSE TRY weather data is used. For checking on overheating requirements for offices it is important that the CIBSE DSY weather tapes are used to predict the occurrences of high internal temperatures do not exceed the required limits.

All accredited dynamic simulation models (DSM) need to comply with the test requirements of CIBSE TM33 'Tests for Software Verification and Accreditation' and be able to produce the standard Building Regulations UK Part L, BRUKL, output schedules so that the Building Control can check in a consistent way that a building

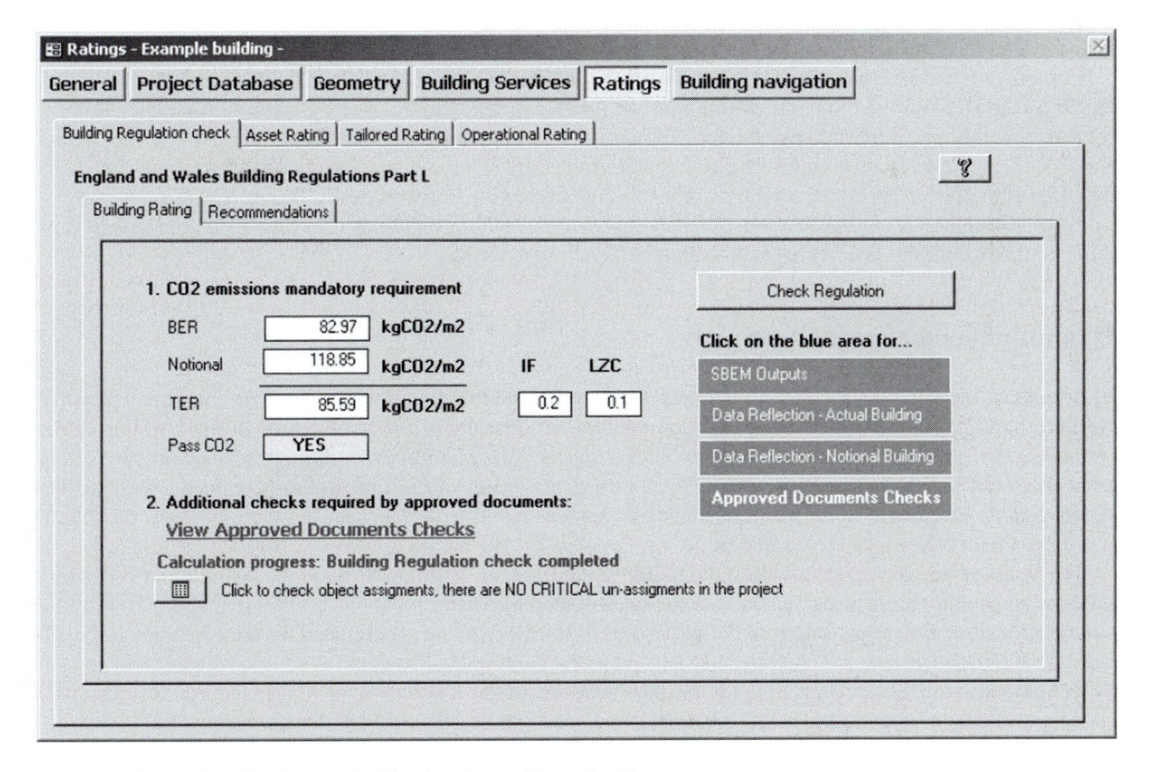

Figure 6.4 iSBEM interface for non-dwelling Part L compliance checking output

Criterion 1 – Predicted CO_2 emission from proposed building does not exceed the target

1.1 Calculated CO_2 emission rate from notional building
50.35 kg CO_2/m^2.annum

1.2 Improvement factor
0.19

1.3 LZC benchmark
0.1

1.4 Target CO_2 emission rate (TER)
36.57 kg CO_2/m^2.annum

1.5 Building CO_2 emission rate (BER)
22.88 kg CO_2/m^2.annum

1.6 Are emissions from building less than or equal to the target?
(BER $\leq$ TER)

1.7 Are as built details the same as used in BER calculation?
Supporting documents in separate submission (developer)

Figure 6.5 BRUKL output schedule from SBEM

meets the requirements of Part L for carbon dioxide emissions and the minimum acceptable standards for fabric elements and fixed building services equipment. Figure 6.5 shows part of the BRUKL output schedule.

Compliance checks need to be carried out using government approved software, a list of which is maintained by the Department of Communities and Local Government and be used by competent people. There is the government funded SBEM calculation and the iSBEM using a basic tabular/numeric input interface. Third party developed interfaces into the SBEM calculation method often use graphical interfaces taking data from drawings. Finally we have approved dynamic thermal simulation software tools. These can use a variety of calculation techniques to analyse more detailed energy flows within buildings. They are also able to deal with more complex building geometries and heating and cooling systems.

Dynamic thermal modelling

This looks at the complex interaction between the external environment and the internal conditions required within a building, taking into account the complexities of heat gains and losses through the building fabric elements. The building model is often input by drawing in the outline of the building and the software calculated areas and volumes of internal spaces. The thermal properties of the various fabric elements then need to be assigned to the rooms along with the activity types associated with the room. Figure 6.6 shows the graphical input for an office located in London.

The building services system within the building will interact with the demand for heating and cooling to produce an overall energy usage which can then be equated to a carbon dioxide emission. Computer software can either include the rooms within buildings as part of the system that is simulated, or use a heating and cooling demand loads file based on output from a heating and cooling load program.

The standard CIBSE weather files for various locations in the UK only have hourly records for quantities such as outside air temperature, solar radiation, wind speed and cloud cover. There are currently 14 locations for the CIBSE UK hourly weather data in TRY and DSY formats and these are listed below:

Belfast, Birmingham, Cardiff, Edinburgh, Glasgow, Leeds, London, Manchester, Newcastle, Norwich, Nottingham, Plymouth, Southampton and Swindon.

Figure 6.6 Graphical input to thermal modelling software

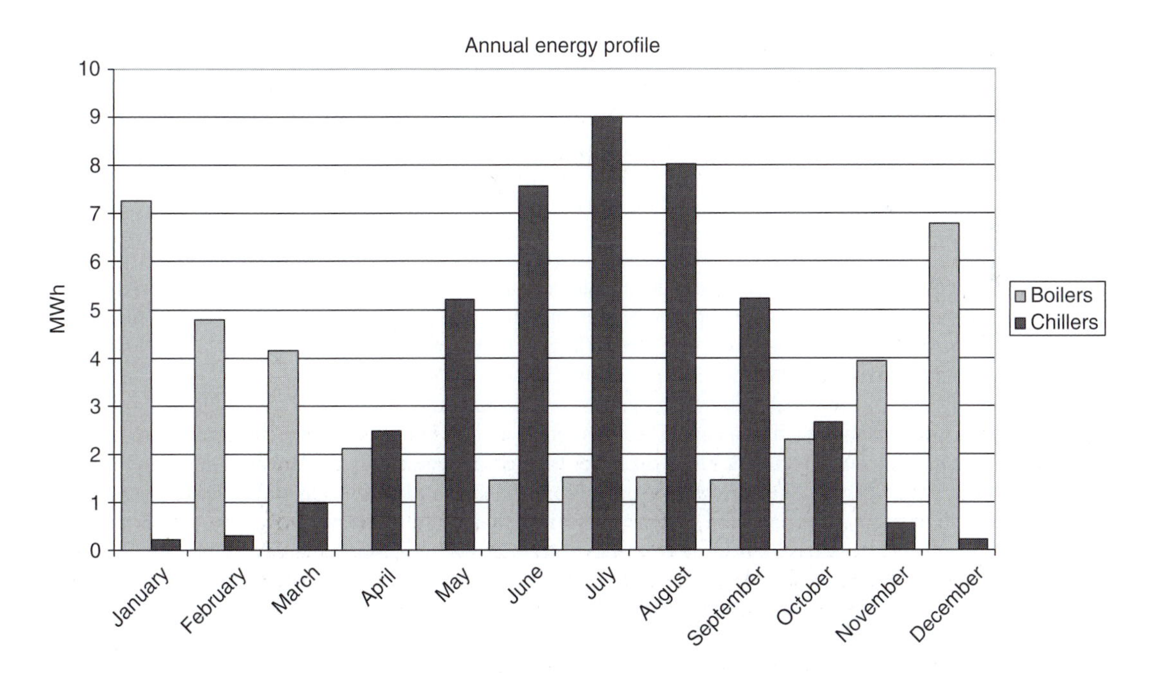

Figure 6.7 Energy prediction from dynamic thermal simulation using TRY weather data

There are two types of weather file, the TRY files which is used for modelling annual energy use and is a year's weather data that is made up from the months of real weather data that are closest to the 20-year average conditions for the month. Figure 6.7 shows the heating and cooling energy requirements-based simulation using the London TRY weather data. The DSY is the weather data that is the third hottest of the 20-year period

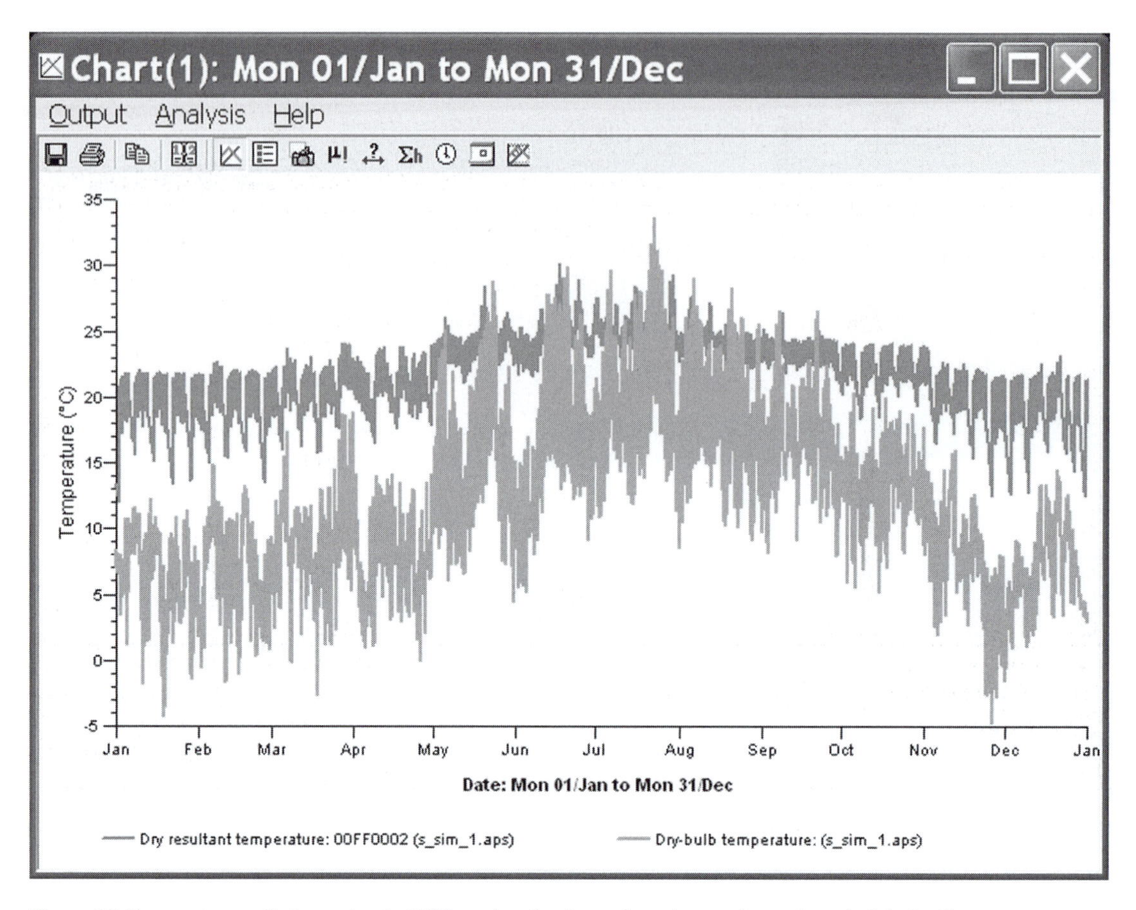

Figure 6.8 Temperature predictions using the DSY weather data (see colour plate section at the end of the book)

and is used to check that buildings do not overheat in the summer. Figure 6.8 shows the inside operative temperature and outside air temperature for a whole year based upon London DSY weather data. The current 20-year period used in Guide A2 is from 1983 to 2002 and more details on the weather files can be found document or the CIBSE Guide J 'Weather, solar and illuminance data'.

However, many dynamic thermal simulation models will evaluate the response of a building at shorter time intervals often down to 1-minute time-steps. The interaction of daylighting within a space and the need for electric lighting can often be considered. This will lower the internal gains from electric lighting into the space based on daylight availability. Bulk air flow models that predict the air change rates within spaces based upon building construction types, external wind pressure coefficients and the weather data allows variations in the air flow in and out of spaces based upon both stack- (thermal) and wind-driven infiltration. Detailed solar shading models are also incorporated so that self-shading and shading by other adjacent buildings can be taken into account and therefore give a more accurate estimate of the solar gains in spaces. Dependent on the sophistication of the thermal model various types of analyses can be carried out.

A full year's weather data can be used to check the peak heating and cooling demands. Once building services systems have been selected and sized then they can be added to the modelling process to predict the energy consumption of the building under a typical year's operation. Some of the models link the HVCA system modelling to the room models whilst others just use an hourly heating and cooling loads file from a building model and run the system simulations based upon the predicted loads. Figure 6.9 shows the use of a graphical interface to enter HVCA systems.

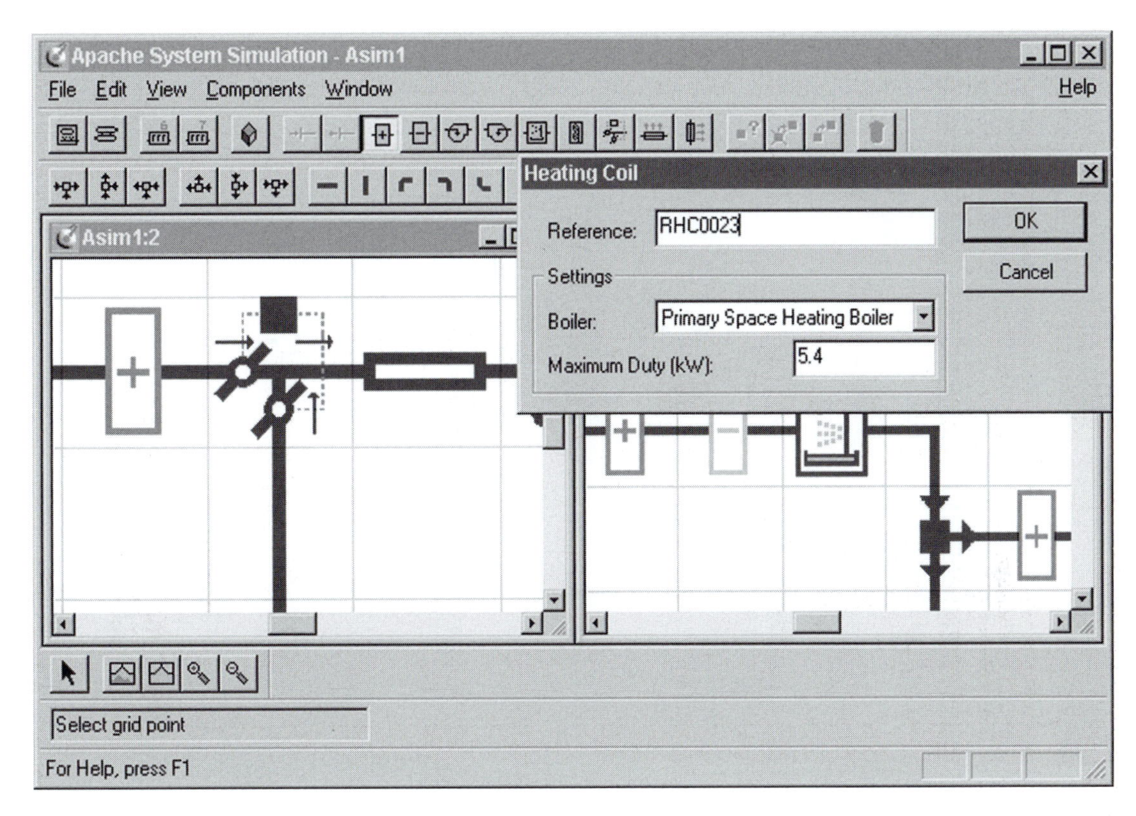

Figure 6.9 IES VE graphical interface to enter HVCA systems

For a greater understanding of these software tools refer to CIBSE Applications Manual AM11: 1998 'Building energy and environmental modelling'.

Computational fluid dynamics

This is the most detailed analysis which is a numerical technique which solves three-dimensional Navier Stokes equations for predicting air movement and when combined with energy conservation and radiation exchanges it can predict mean radiant temperatures, air velocities and air temperatures in spaces. Models using CFD can be performed simplistically as two-dimensional steady state slices through a building or as very complex three-dimensional transient models.

The accuracy of the models depends on many input variables including turbulence model, boundary conditions, computational parameters and the grid definition. The resolution of the geometry, the grid/mesh and the convergence criteria can all affect how accurate a CFD models results will be. Advice from specialist consultants in CFD is recommended until engineers have acquired the recommended experience.

Within buildings the distribution of heating and cooling can be analysed to check that there are no hot or cold spots. Therefore, fairly fine grids are used which may produce thousands of cells to allow for the detailed interaction between the activities within each cell. The detailed modelling of electrical equipment, people and building services equipment will set up often conflicting convection currents and CFD has the ability to analyse the flow of air within the space and the temperature distribution. Within a simple steady state calculation, a room is assumed to have the same air temperature as that in the geometrical centre of the space. However, within CFD the floor could easily be 4°C or 5°C cooler than the ceiling. For a greater understanding of CFD

Figure 6.10 CFD output showing air flow patterns around the building (see colour plate section at the end of the book)

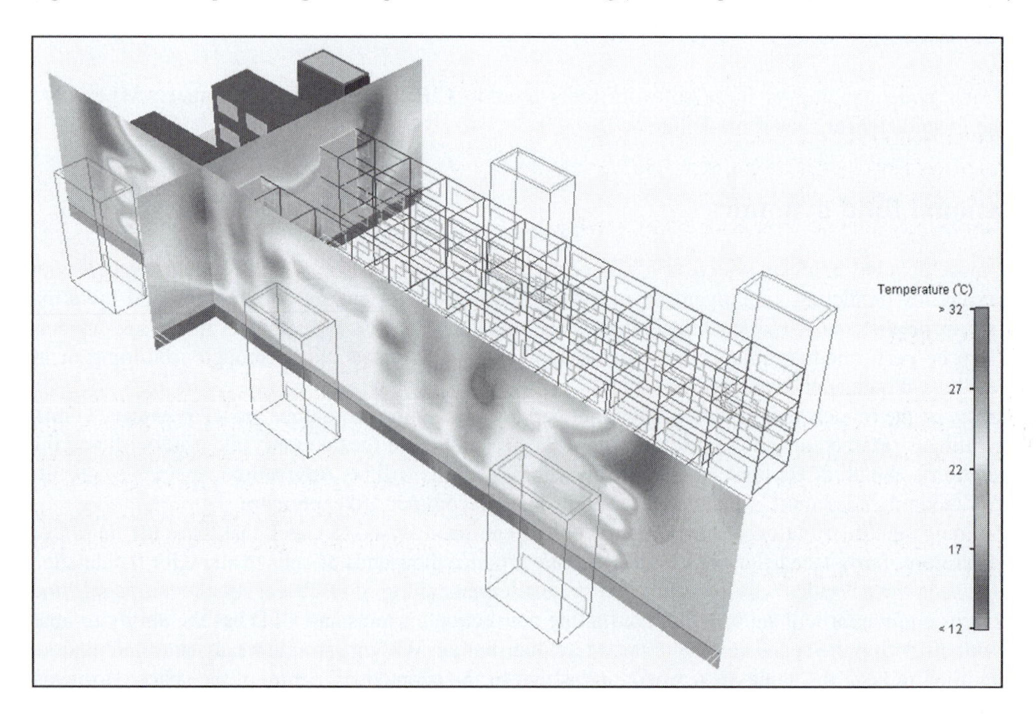

Figure 6.11 CFD output showing air temperature predictions in a data centre (see colour plate section at the end of the book)

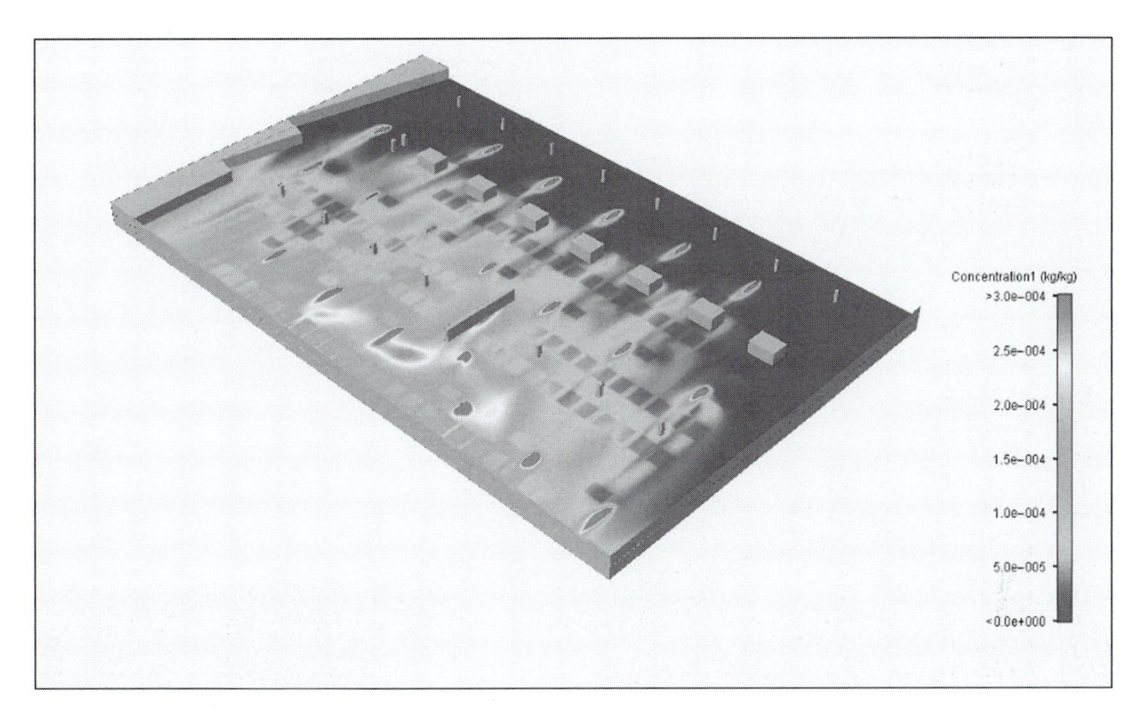

Figure 6.12 CFD output showing gas concentration levels in an underground car park (see colour plate section at the end of the book)

refer to '*An Introduction to Computational Fluid Dynamics: The Finite Volume Method*' by H.K. Versteeg and W. Malalasekra.

Typical applications of CFD analysis would include:

- Detailed thermal performance of glazing systems and downdraughts
- Checks on cross-contamination at air inlets
- Air flow patterns around the building, see Fig. 6.10
- Pedestrians discomfort by generating high wind speeds at pavement level
- Computer data centres, see Fig. 6.11
- Checking air quality in underground car parks, see Fig. 6.12

The way forward

The use of computer software is now a central part of designing and analysing buildings. The thermal performance of a building under both winter and summer conditions has a complex interaction of heating and cooling effects and it will require the use of an appropriate software solution. It must be stressed that approved software must be used to carry out certain compliance checking procedures. The use of competent people who have been fully training in using the approved software is essential. As computer power continues to increase the speed at which calculations can be performed has fallen. This means that more detailed computer models are being developed taking onboard the findings of the research community.

Chapter 7

Heating methods

Chapter 4 (The building in winter) looked in detail at the calculation methodology for determining heating loads and the assessment of plant capacity for the building. This chapter recaps on some of the topics covered in Chapter 4, provides an introduction to heating methods and takes a general overview of the purposes of a heating system and how to achieve comfortable environmental conditions in a building. It also looks at the process of selecting a suitable method of heating; alternative distribution systems; alternative fuel/energy sources and the various types of heat emitter that are available to heat buildings. More details on each of these topics are contained in subsequent chapters.

Legislation

Chapter 3 provides an overview of the key legislative requirements a designer of heating and air-conditioning systems should be familiar with. The following paragraphs provide additional guidance on some of the more important legal requirements that need to be addressed when designing a heating system.

In the UK there are two main types of legislation that are of primary interest when designing a heating system. These are those that control the health and safety/comfort of the occupants (by setting minimum standards for *temperatures in buildings*) and those that aim to *minimise energy use/reduce CO_2 emissions from buildings* and promote the use of low or zero carbon (LZC) technologies otherwise known as renewable energy as described in detail in Chapter 14.

In addition there are also regulations that govern the detail design and installation of the heating systems, these are not discussed in any detail here but include:

- Building Regulations Part J (combustion appliances and fuel storage systems).
- Electricity Supply Regulations.
- Electrical Equipment Regulations.
- Disability Discrimination Act.
- Dangerous Substances Act.
- Gas Safety Regulations.
- Boiler Efficiency Regulations.
- Clean Air Act.
- Pressure Systems and Pressure Equipment Regulations.

Temperatures in buildings

The primary legislation is the Health and Safety at Work etc. Act 1974 (HSW Act). This sets out the general duties that you have towards your employees and members of the public, and those duties which employees have to themselves and to each other. Although it does not mention temperature specifically, it does state that you should ensure, so far as is reasonably practicable, the health, safety and welfare at work of your employees. This includes providing a working environment that is both safe and without risk to health.

The Workplace (Health, Safety and Welfare) Regulations 1992 states '*During working hours, the temperature in all workplaces inside buildings shall be reasonable*'. The Approved Code of Practice (ACOP) to these regulations defines a '*reasonable temperature*' as that which secures the thermal comfort of people at work, without the need for special clothing. This is further defined as being met by '*maintaining a "reasonable" temperature of at least 16°C (or at least 13°C if the work involves physical effort)*'.

These minimum temperatures are for areas such as factories or sports halls where there is significant physical activity and both CIBSE Guide A and BS EN ISO 7730 provide more detailed guidance and recommendations for thermal comfort standards in different circumstances.

Further guidance can be found in the HSE publication, '*Thermal Comfort in the Workplace*', published in 1999, this document notes that 'thermal comfort is difficult to define' and that 'the best that you can realistically hope to achieve is a thermal environment which satisfies the majority of people in the workplace'.

In addition to the above, in October 1980, the then Department of Energy introduced an order called '*The Fuel and Electricity (Heating) (Control) (Amendment) Order 1980*' to prohibit the use of fuel or energy to heat premises above 19°C in order to save energy. There are exceptions to this under certain circumstances and they do not apply to living accommodation. Although not specified, it is generally accepted that this refers to air temperature rather than operative temperature. This is a widely flouted law because many people are uncomfortable at this temperature.

There are also separate regulations for schools. The Education (School Premises) Regulations 1999 state that '*Every school room or other space shall have appropriate heating systems capable of maintaining the following temperatures where the external air temperature is −1°C*':

- Areas with normal level of physical activity 18°C (e.g. classrooms)
- Areas with below normal level of physical activity 21°C (e.g. sick rooms)
- Areas with above normal level of physical activity 15°C (e.g. gymnasia).

Minimising energy usage/reduction of carbon dioxide output from buildings

As discussed in Chapter 3, the 2006 edition of Building Regulations Approved Documents ADL1A&B (dwellings) and ADL2A&B (buildings other than dwellings) cover the conservation of fuel and power in buildings. In addition to ensuring the CO_2 emissions from a building are no worse than a target figure; the installed heating system must meet minimum standards. These minimum standards for space heating systems are defined in two government published '*2nd tier documents*' which are referred to in the Approved Documents.

The '*Non-domestic Heating, Cooling and Ventilation Compliance Guide*' outlines the minimum provisions for efficiency of plant and supplementary measures which will improve efficiency of plant that generates heat, the minimum controls requirements to ensure systems are not generating heat unnecessarily or excessively and minimum requirements for insulation of pipes and ducts serving heating systems. At the time of writing, the minimum boiler seasonal gross efficiencies for boilers in new buildings are: 84 per cent for a single boiler system and for multiple boiler systems any individual boiler should be 80 per cent but the overall boiler system should be 84 per cent.

The '*Domestic Heating Compliance Guide*' is applicable to dwellings and gives minimum provisions for boiler efficiency as expressed by its SEDBUK[1] value. The guide also gives minimum provisions for efficiency for other type of heat source, such as warm air heaters, gas fires, heat pumps, micro-CHP, etc. and gives the minimum requirements for controls and installation requirements of systems including underfloor systems.

In addition to Building Regulations changes, in the drive to reduce carbon emissions, most local authorities now have a policy on the use of renewables or LZC technologies for new buildings. This is driven by '*Planning Policy PPS 22*' on renewable energy which requires planning authorities and developers to consider opportunities for including renewable energy in all new developments. A typical policy will require that between 10 per cent and 20 per cent reduction in the annual CO_2 produced by a building is achieved by the use of renewable energy technologies or low carbon technologies. For heating systems these technologies include:

- Combined heat and power (CHP) plants.
- Biomass boilers that burn either wood chips or pellets that are made by extruding raw sawdust through a die.

- Ground source heat pumps, these are categorised as either *open loop systems* where water is abstracted from the ground (typically from an aquifer) or *closed loop systems*, where a continuous loop of pipework is either buried in the ground or cast into the building foundations.

All the above technologies are described in more detail in the Chapters 13–16.

Heating systems generally

Most space heating systems use a remote heat source such as a boiler, CHP plant, district heating or possibly a heat pump, to supply high-grade heat via a transport medium, usually water but sometimes steam or air, through a distribution network to a series of heat exchangers known as emitters (e.g. radiators or fan convectors). Systems utilising a transport medium are known as *indirect systems*. Alternatively, *direct fired systems* or direct electric systems can be used. These either rely on radiant effects from (usually) overhead tubes or panels, or warm air blower systems. It should be noted that some types of direct gas-fired heaters actually circulate greatly diluted combustion products into the space.

The fundamental components of any heating system are:

- A means of generating heat, i.e. *the heat generator and energy source*.
- A means of distributing the heat around the building or buildings, i.e. *the distribution medium*.
- A means of delivering the heat into the space to be heated, i.e. *the heat emitter*.
- The method by which heat is delivered to the space. This can be either by *convection* or *radiation* (systems are generally considered to be radiant when more than 50 per cent of their output is radiant). Table 7.1 gives an indication of the radiative and convective outputs of various emitter types.

The above gives many possible permutations and options to be considered, some of which are listed in Table 7.1 and explored in detail in other chapters. Examples include the simple use of electric panel heating, using electricity both as the heat source and distribution medium, to a conventional gas boiler system distributing low-temperature water to a convector system. A more complex system would be one serving various buildings by using oil or gas as the energy/heat source to generate high-temperature hot water for the main distribution, which is then reduced in temperature and pressure to low-temperature water via heat exchangers, to serve radiator systems in multiple buildings.

Table 7.1 Components of a heating system

Heat source	Gas-, oil- or biomass-fired boiler or direct fired air heater or direct gas fired radiant panels electricity/off-peak electricity for storage systems CHP units Ground, air or water source heat pumps
Distribution medium	Water – low, medium or high temperature Air Steam Electricity
Emitter types and output characteristics	*Radiators* (between 50% and 80% convective output) *Natural convectors* (90% convective) *Under floor heating* (60% radiant output) *Fan convectors* (100% convective output) *Low-temperature radiant ceiling panels* (50% convective output) *Panel heaters* (50% convective output) *High-temperature radiant panels* (90% radiant output) *Storage heaters* (50% convective output) *Unit heaters* (100% convective output)

In modern comfort cooled office buildings, emitters are usually installed to provide both heating and cooling. These include fan coil units and more recently active chilled beams, as discussed in Chapter 10. Care needs to be taken when using chilled beams for heating by limiting the water flow temperature to between 40°C and 50°C to avoid stratification of warm air occurring at ceiling level.

Factors affecting choice for different building types

Home heating

Choice here may be influenced by personal preference, by the routine of daily occupation, by sales pressure from various fuel interests, by close regard for economy or, where a public authority is concerned, by what is permitted by various local or national regulations.

In an existing building, the choice of system may be limited by the facilities available, i.e. absence of a suitable flue, difficulties in arranging sensible and visually acceptable pipe routes, etc., and the availability, in rural areas, of a public supply of gas or electricity.

Direct heating systems using solid fuel fires and stoves involve labour, dust and dirt and, with modern habits of families being out all day, are often inconvenient. Gas and electric fires are then preferred, but tend to be expensive in running cost. Electric storage radiators offer lower running costs, but provide limited control where occupation is intermittent.

Indirect systems for home heating can give whole-house comfort as well as hot water probably more consistently than any direct system. The capital cost may be higher but the running cost less, subject to the vagaries of the various fuel tariffs. The choice of fuel for a domestic indirect system, however, may not always be a matter of cost even supposing that relative prices were to remain stable. While mains gas is the most common choice, in some rural areas this is not available. In such cases bottled gas, oil or solid fuel may be used.

In new houses gas-fired wet central heating serving radiators have become commonplace. However, some may prefer floor heating in order to avoid the loss of floor or wall space. In an older house, skirting heating might be thought to be less obtrusive than radiators provided that furniture can be suitably disposed: radiators might be cheaper.

The pattern of home occupancy is an important factor as far as running costs are concerned but control systems suitable for domestic use have been developed to a high level of sophistication. Intending purchasers should not be misled by claims made by any one particular fuel or other interest since what can be done by one can probably be done equally well by all.

Thus, as may be seen, the choice for a home heating system is very much a matter of personal circumstances and personal taste. It is impossible to generalise, particularly as to comparative costs, in view of all the variables.

Flats – multi-storey

Systems of heating in common use and from which a choice would no doubt be made are:

- Electric storage radiators off-peak
- Central heating by hot water radiators from a boiler per flat
- Hot water/warm air from a boiler plant per block
- Group heating by hot water to a number of blocks
- Less popular these days, warm air, fuelled by gas, oil or electricity.

The housing authority or estate developer in making a choice will be influenced by the capital cost, fuel cost, maintenance costs and labour to run. There is also the question of how the heat is to be charged – whether by the public utility reading its own meters (the consumer paying direct), or whether the landlord will be responsible for reading meters and collecting the money, or, again, whether the cost of heat is included in the rent or in a service charge, although this can lead to excessive use of energy with resultant increased carbon emissions.

All these matters have to be considered as well as the type of tenant and what kind of expenditure can be afforded before a recommendation can be made.

Commercial and public buildings: schools, universities, halls of residence, swimming pools, hospitals, hotels, etc.

This general class of substantial buildings will in most cases be in the hands of a consulting engineer, or public authority engineer, who can be expected to advise which system and fuel should be used. Matters to be reported on should cover, among other things, energy conservation, life cycle costing, spatial requirements for plant, maintenance demands, amenity control, acoustic treatment and avoidance of pollution.

In schools, radiators are usually the most suitable choice, however in some schools where extensive use of the floor is made underfloor heating is preferred, but would not be appropriate where the floor is likely to be covered for example with insulating mats or where doors are often opened for outside activities due to its slow response. The slow response of underfloor heating systems can also cause unacceptable conditions in classrooms where there is a sudden increase in heat gain, i.e. lighting, plus a class of 30 pupils will add approximately 2.5–3 kW of heat gain to a class room.

Office buildings

Offices being usually in blocks are most economically heated by an indirect central system in some form. Choice is then confined to the kind of emitting surface, radiators, convectors, ceiling heating, etc., and to the kind of fuel.

An office block with large expanses of glass and probably having construction which could be characterised as lightweight will be subject to rapid swings of temperature; hence a system which is quickly responsive to change of output is needed. Thus, underfloor heating systems are not to be recommended although they may be suitable in buildings of heavy construction, including heritage stock.

New offices typically fall into two categories:

(1) Fairly deep plan high rise buildings, in city centres which are built speculatively by developers to accepted standards such as the British Council for Offices (BCO) where the norm is to provide an internal environment where the temperature is only allowed to range between 20°C and 24°C and cannot have openable windows because of noise and traffic pollution.
(2) Narrow plan or atrium-type buildings located on out of town business parks which lend themselves to a natural ventilation or mixed mode strategy, where mechanical cooling is not provided and the temperature in summer is allowed to exceed 25°C for a percentage of the occupied hours.

The first type of office is provided with either heating plus some form of mechanical cooling, or fully air-conditioned with both temperature and humidity control. These systems are developed in a later chapter.

The second category of office is usually heated by a perimeter heating system, either radiators or natural convectors, each with its own means of control that are served by a low-temperature hot water system where the flow temperature is reduced as the outside air temperature increases.

Industrial

Here the choice may be influenced by particular economic considerations related to the process or manufacture concerned, which might be short or long term, and which system will produce adequate conditions with minimum upkeep, fuel consumption and labour.

If steam were to be required for process work, the choice may fall in the direction of using steam for heating also, directly via unit heaters or some form of radiant surface or, alternatively, indirectly using a calorifier to produce hot water. If no steam is available, hot water at medium pressure or high pressure is to be preferred. For heat emission, unit heaters may again be selected on account of low cost, but they involve maintenance which does not apply with radiant panel or strip heating.

Direct oil- or gas-fired air heaters meet certain cases where space limitations or other circumstances preclude the provision of a boiler plant or where extensions cannot be dealt with from existing boilers. Furthermore, such units can be moved to suit any change in building plan or floor layout. Radiant tube systems, heated either by combustion products from gas burners or by hot air, should also be considered having regard to the high efficiency of heating effect and operating economy inherent to these arrangements.

Purposes of a heating system and how to maintain a 'comfortable environment'

Heating systems are generally required to maintain comfortable conditions for people working or living in a building. In some buildings that are not normally occupied by people, heating may not be required to maintain comfort. However it may be necessary to control the temperature to protect the building fabric or its contents from frost or condensation.

Rather than heat our bodies, a heating system regulates the way in which our bodies lose heat due to radiation, convection and evaporation. The type of heating system installed can therefore affect the way our bodies reject heat and the effect this may have on occupant comfort.

The rate at which we lose heat will vary due to the temperature of our surroundings. This temperature is made up of two components, the air temperature and the radiant temperature. Whilst the term air temperature is commonly understood, radiant temperature may need further explanation. If we consider the situation outside, on a cold winter's day when the air temperature is low, an instant feeling of warmth is felt when the sun emerges. Clearly, the air temperature could not have risen in that instant, but the energy radiating from the sun has increased the radiant temperature and thus increased the comfort level. The combination of both the air temperature and mean radiant temperature gives rise to a parameter termed the dry resultant temperature or comfort temperature or more recently defined as '*operative temperature*' (as used in BS EN ISO 7730 for thermal comfort and American Society of Heating, Refrigeration and Air Conditioning Engineers (ASHRAE) standards as discussed in Chapter 1).

Convective heating systems work by heating the air directly. As the heated air circulates within the space, it warms the surrounding walls, floor, ceiling and other surfaces. Although these surfaces are warmed by the air, they generally remain at a lower temperature than the air except in summer conditions or when solar gains heat glazed surfaces. Therefore, during the heating season, the mean radiant temperature and the operative temperature in the room are usually lower than the air temperature. The extent to which this may affect comfort depends upon the building construction.

In winter, inside poorly insulated buildings, the surface temperature of the walls and windows will be significantly lower than that of the inside air. Therefore, to achieve a comfortable operative temperature, the air temperature must be maintained at a higher level.

Radiant heating systems heat the air indirectly. The heat is transmitted from the heat source in the form of electromagnetic rays (mainly infrared), to surrounding cooler objects such as walls, floors and people. As the electromagnetic rays pass through the surrounding air almost no heat is absorbed. Instead, the air is heated by contact with the surrounding surfaces, which have been heated by the radiant source. The mean radiant temperature in the room and hence the operative temperature are therefore higher than the air temperature.

As radiant heating operates at a lower air temperature than convective systems, less heat is lost when air escapes from the building (e.g. through an open door).

In such circumstances, where the infiltration rate is high, radiant heating systems may use less energy than convective systems. Radiant heating also has the advantage of being directional. For example, a floor can be heated by an overhead radiant system, directed at the floor. Compared with a convective system, the temperature at roof level will be lower, thereby reducing the heat loss through the roof. The ability to direct the heat also allows 'spot heating' within a building, where a small area may be heated in isolation. It should be noted that radiant heat transfer also takes place from people to colder objects. This will often give rise to complaints from occupants sitting close to a large window which has a poor U value whereby the internal surface of the glass is significantly lower than the air temperature. Radiant systems are useful in buildings with high air change rates or large volumes that do not require uniform heating throughout, e.g. factories, and intermittently heated buildings with high ceilings. Care must be taken to avoid mounting radiant emitters too low to avoid possible discomfort due to a hot head.

The control of radiant heating can be problematical because ideally control should rely on the measurement of operative temperature which requires the use of a black bulb thermometer located centrally in a zone to avoid the influence of a cold wall for example. If the radiant heating is providing 'total heating' rather than just 'spot heating' the air temperature sensors can be used, however they tend to underestimate operative temperature during warm up and cause waste of energy.

Temperature variations in the space

The ideal for comfort is to have 'warm feet and cool head', i.e. the temperature should be warmer at foot level than at head level. In practice the opposite is often the case as warm air rises, leading to stratification in a space.

Table 7.2 Maximum radiant asymmetry for no more that 5% people dissatisfied

Scenario	*Radiant asymmetry*
	(Difference between radiant temperatures on opposite sides of the human body)
Cool wall (full height glass)	Less than 10 K
Cool ceiling (chilled ceiling)	Less than 14 K
Warm wall (large radiant panel)	Less than 23 K
Warm ceiling (radiant panels)	Less than 5 K

If this is too great then it can feel uncomfortable, with cold feet and a feeling of stuffiness at head level. To avoid discomfort it is recommended that the air temperature rise between ankles and head should not exceed 3°C.

Air and radiant temperature differences

If the radiant temperature is above the air temperature, it will tend to give a feeling of freshness. This can occur with heating systems that have more of a radiant component such as radiant panels or underfloor heating with sunshine entering an air cooled space in summer. If the air temperature is above the radiant temperature it can tend to feel stuffy. This can occur with heating systems that are more convective, such as warm air heating. In order to avoid discomfort the two temperatures should not be too far apart, with ideally the radiant temperature slightly above the air temperature.

Localised radiation

Excessive radiation, particularly if it is on one side of the body only, can cause discomfort, for example if sat next to a cold window surface or next to a roaring fire in winter. Considering the case of the roaring fire in a cold room, one side of the body is excessively hot and the other is cold, and, although the average temperature may well be theoretically acceptable, in practice the imbalance causes discomfort. The same imbalance can be caused to a lesser extent, by heated or cooled surfaces in a room such as overhead radiant heaters, overhead lighting, solar radiation through glass, cold window surfaces, etc. In order to avoid discomfort, large imbalances in radiant temperatures, known as *radiant temperature asymmetry* (defined as the difference between the plane radiant temperatures on opposite sides of the human body) should be avoided. Table 7.2 has been derived from CIBSE Guide A Section 1 and gives the values for radiant asymmetry to achieve no more than 5 per cent people dissatisfied. For example someone adjacent to a fully glazed façade with internal surface temperature of 21°C should not feel uncomfortable due to radiant asymmetry, if the glass surface temperature is approximately 11–12°C.

Warm or cold floors

Localised discomfort can be caused if the floor surface temperatures are too cold or too hot, e.g. if there is underfloor heating. To avoid discomfort it is recommended that floor surface temperatures should be in the range 19–29°C. BS EN 1264 suggests a maximum floor temperature of 29°C for underfloor heating systems.

Heating system selection

Heating system choice depends on many factors. These generally fall into two categories: System installation and system performance/use.

Key design decisions will include the choice of heat generating technology and energy source, the method of delivering the heat to the building and whether the heat generating plant is centralised or decentralised. The heat generating technology selected could either use renewable or fossil energy sources or a combination of both to achieve a reduction in CO_2 emissions and meet Government Legislation on the use of LZC technologies as described in detail in Chapter 14.

System installation factors include:

- Space available – both for plant and for distribution.
- Potential plant room locations related to the spaces to be served.
- Zoning requirements.
- Flexibility – any requirements for future change of use or changes to the fit-out arrangement.
- Ease of installation – access, maintenance, replacement, etc.

Performance and use factors include:

- Installation and operating/life cycle cost.
- Minimising CO_2 emissions and maximising the use of renewable technologies.
- Correct choice of emitter (radiant or convective/noise level) to provide comfort conditions and correct speed of response to changes in use of a space/hours of occupation.
- Provision of localised occupant controls and simplicity of use.

To determine the most appropriate system to meet the client's requirements, an assessment of options against some or all of these factors can be helpful.

Tools to assist the choice of system include the following:

- Flow charts such as the example in Fig. 7.1(reproduced from CIBSE Guide B1).
- Ranking and weighting matrixes to assess the suitability of each system using some of the key usage factors related to system choice. Here each factor is weighted and then each system is allocated a score, typically out of 5 (e.g. for ease of maintenance a radiator system would be allocated 5, whereas a fan convector system may only score 2 or 3 as there would be significantly more maintenance required. The score is then multiplied by the weighting factor. The systems can then be ranked by score, the highest representing the most suitable system.
- Other system selection tools such as the guide to choosing building services produced by BSRIA (Ref. BG9/2004). This is a simple excel program to help clients and other 'non-experts' to decide the suitability of particular building services for different spaces in buildings (it should be noted that the tool covers ventilation, air-conditioning and lighting as well as heating). The tool has look-up tables to match building services to functional space and automatically present a list of systems that are a 'good fit', a 'possibility' or a 'poor

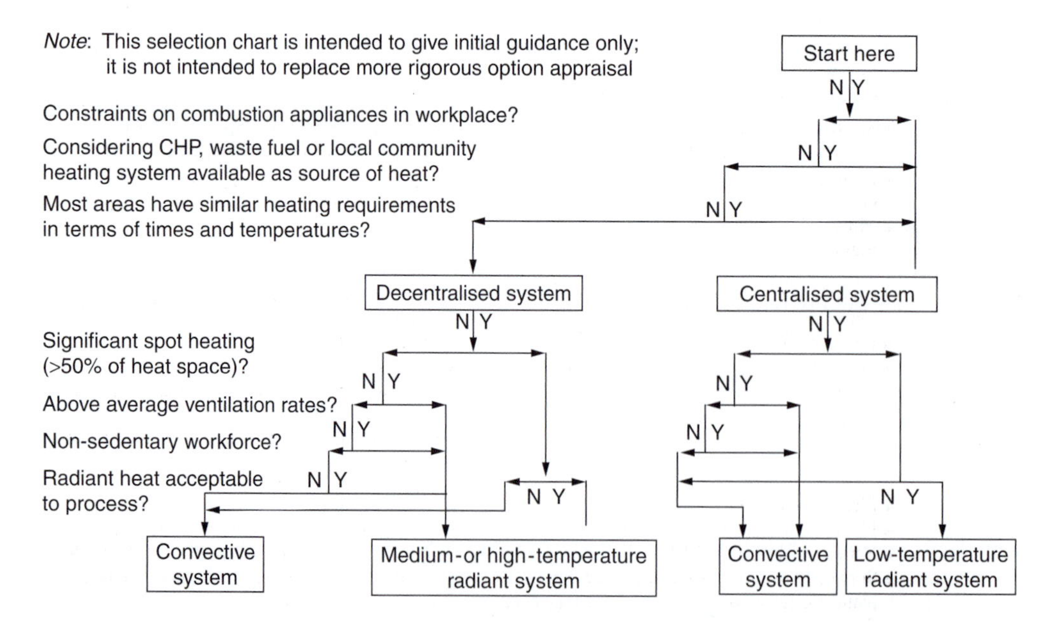

Figure 7.1 Centralised versus non-centralised systems (courtesy of Figure 2.1 in CIBSE Guide B1)

fit' for each type of functional space. The tool is also a decision table that can be filled-in by the client and can form part of the client's functional brief to be used by the building services engineer to develop an options report at the initial design stage.

Centralised versus non-centralised systems

Table 7.3 provides some attributes of both types of system (based on data from CIBSE Guide F Section 10). The case for decentralised heating often hinges on the price and availability of fuels, the space available for plant and distribution pipework, flueing arrangements and the losses in a centralised system and the size of the loads involved.

Table 7.3 Centralised versus non-centralised heating systems

	Centralised	*Non-centralised*
Cost	• Capital cost per unit output falls with increased capacity of central plant, but capital cost of distribution systems can be high • Renewable technologies such as ground source heat pumps and biomass boilers can be integrated	• Low overall capital cost, savings made on minimising the use of air and water distribution systems but possibly lower life cycle costs • Renewable technologies such as ground source heat pumps can be installed locally for heating discrete areas (e.g. underfloor heating of communal open spaces, atria, etc.)
Space requirements	• Space requirements of central plant and distribution systems are significant, particularly ductwork for air systems	• Flueing arrangements can be more difficult
System efficiency	• Central plant tends to be better engineered, operating at higher system efficiencies (where load factors are high) and more durable • As the load factor falls the total efficiency falls as distribution losses become more significant • Greater losses from distribution system • Care needed on plant selection to avoid plant over sizing at low loads and operation at low efficiencies	• Energy performance in buildings with diverse patterns of use is usually better • Distribution systems/losses minimised
System resilience	• Without multiple boilers complete building or site heating will be lost	• Possibility of not providing standby plant as failure will only result in heating lost to one zone
System operation	• Convenient for some institutions to have centralised plant • More difficult to sub-meter energy use for multiple tenants	• May require more control systems • Zoning of the systems can be matched more easily to occupancy patterns
System maintenance and operational life	• Central plant tends to be better engineered, more durable • Less plant to maintain • Purpose built plant room required with adequate space for maintenance	• Can be readily altered and extended • Equipment tends to be less robust with shorter operational life • Plant failure only affects the area served • Possibly maintenance less specialised but may need to be more frequent • Plant space can be small leading to reluctance to carry out adequate maintenance
Fuel choice	• Flexibility in the choice of fuel • Plant is available with dual fuel burners • Better utilisation of CHP, etc. • Some systems will naturally require central plant, e.g. heavy oil and biomass burning plant • Storage is required for biomass fuel	• Fuel needs to be supplied throughout the site or building (all voids/risers containing gas pipework will require ventilating). Refer to the Institute of Gas Engineers guides for details

In a multi-tenanted building with diverse periods of occupancy, the use of decentralised services could match the type of operation better than a centralised arrangement. However, for a building having a large energy demand with a high load factor, the use of centralised plant working at high efficiencies and possibly using dual fuel facilities may provide a better solution.

Choice of fuel/energy source

Details on fuels and combustion emissions can be found in Chapters 16 and 17. The following provides a brief summary of the energy sources available that will enable an educated choice of a suitable fuel.

The flow diagram (Fig. 7.2) is a good starting point.

Environmental considerations

For detailed guidance on the environmental issues reference should be made to Chapter 2 on Sustainability. It is worth noting here, however, that energy use in buildings is responsible for 46 per cent of UK carbon dioxide (CO_2) emissions and most of this CO_2 results from the burning of fossil fuels, either directly in buildings or in the generation of electricity. The burning of fossil fuels also results in emissions of NO_x and SO_x that contribute to acid deposition and poor air quality.

The most common fuel used in the UK for space heating is natural gas, although other fuels are used including fuel oil, LPG and solid fuels. Electricity is also used on some occasions as the primary heat source, although due to relatively high unit cost and the higher pollution emissions during generation, it should not generally be considered as the first choice for space heating.

Figure 2.3 shows the CO_2 emissions by fuel type where it can be seen that electricity from the grid has a figure of 0.422 kg CO_2/kWh compared with 0.194 kg CO_2/kWh for natural gas.

Natural gas

In general, natural gas will be the first choice of heating fuel for most installations in the UK. It is the most widely used fuel and therefore the existing knowledge base and expertise encourages further use and development of gas technologies. Natural gas results in reduced emissions of SO_2 compared with other fossil fuels and greatly reduced emissions of NO_x compared to electricity from conventional power stations.

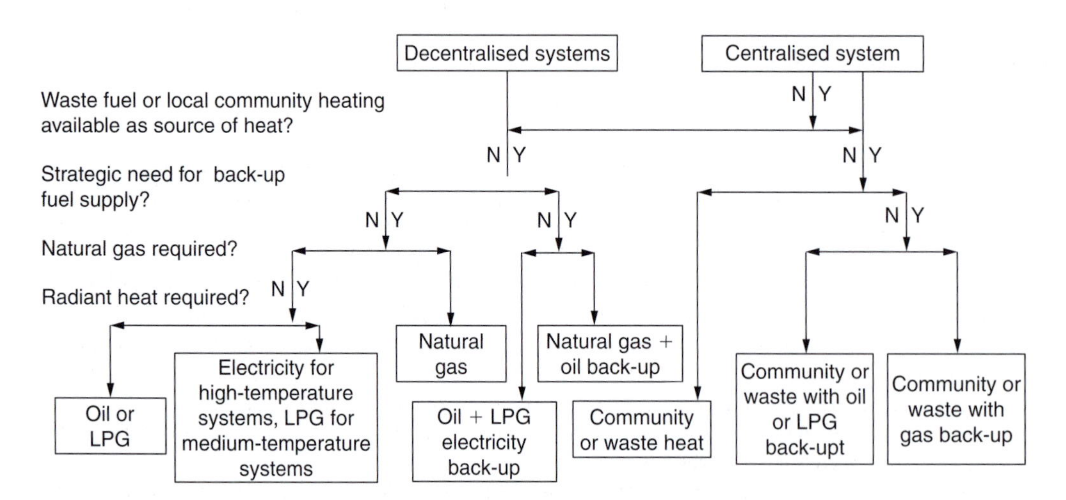

Figure 7.2 Fuel selection chart (reproduced from Figure 2.2 in CIBSE Guide B1)

However, natural gas is a non-renewable fossil fuel and its use will contribute to global warming. For this reason systems should still be designed to operate as efficiently as possible.

Where no gas supply is available consider the use of electric ground source heat pumps or air to air heat pumps such as variable refrigerant volume (VRV) systems in preference to direct electric heating.

Biomass

Energy from biomass is produced by burning organic matter; sources include trees, crops and animal waste. Biomass includes woody biomass which comprises forest residues such as tree thinnings, and short rotation crops such as willow coppice. A common form are wood pellets which are compacted high-density wood with low moisture content having a higher calorific value. Biomass is carbon based so it generates CO_2 when burnt, however the carbon released during combustion is equivalent to the amount that was absorbed during growth so the technology virtually carbon neutral. It does have a CO_2 emission factor of 0.025 and this is due to carbon emissions associated with the treatment required to turn the raw wood into fuel and transportation of the fuel. Harvested crops can be replaced relatively quickly with new plantings ready for harvesting after about 4 years.

Grid electricity

Electricity supplied from the grid is generated from a combination of sources including: coal, gas, oil, nuclear, hydro and other renewable sources such as wind and wave power. The conversion factor for electricity is the average emission rate for electricity from the grid taking all these generation types into account. When electricity is generated from fossil fuels as much as 60 per cent of the primary energy is lost as waste heat, the conversion efficiencies for fossil fuel generation being around 40 per cent and there are further transmission losses that occur delivering electricity to the point of use. These losses explain why CO_2 emissions for each unit of electricity are more than double those of gas.

Due to the higher pollution emissions electricity from the grid should not currently be used for direct resistance heating in buildings, where alternative fuels are available. It should also be avoided for use in storage heating, which while operating at reduced cost will still result in significantly greater CO_2 emissions than gas heating.

Residential storage heating can be used to take advantage of low-cost off-peak electricity and to eliminate standing charges for gas in low-energy dwellings. Unfortunately, due to the way in which many homes are occupied, heating may occur when homes are unoccupied.

Where heating is provided by electric heat pumps the coefficient of performance of the heat pump may offset the higher emissions per unit of electricity delivered and can result in lower CO_2 emissions than gas systems.

The average CO_2 emission rate for electricity has been falling in recent years mainly due to old coal-fired power stations being replaced with more efficient combined cycle gas generation. Greater adoption of renewable energy generation would reduce the figure further. Where electricity is generated on site from a renewable source such as wind, hydro electricity or photovoltaics the emissions will effectively be zero.

Combined heat and power

Refer to Chapter 15 for details on CHP systems, the following notes only provide a simple overview.

Where electricity is generated from fossil fuels in CHP plant the waste heat can be utilised for space and water heating, process loads or other purposes in buildings. The CHP plant can also be situated locally reducing transmission losses. Where waste heat from CHP plant can be utilised effectively, the overall CO_2 emissions for the combined electricity and heat will be significantly reduced. Overall system efficiencies of 60–80 per cent are typical for CHP plant compared to 30–50 per cent for centralised power generation.

CHP will only be viable economically and environmentally where demand for waste heat occurs at the same time as demand for electricity. One way to overcome problems of this nature is to specify a micro-CHP system sized to serve the balanced proportion of the loads with mains electricity meeting the unbalanced demand. As a general rule of thumb CHP systems may be economically viable where there is a demand for heat for more than approximately 4500–5000 hours a year.

Energy supplied from 'good quality' CHP systems is exempt from the Climate Change Levy and new installations may qualify for an Enhanced Capital Allowance and Exemption of Power generating plant and machinery

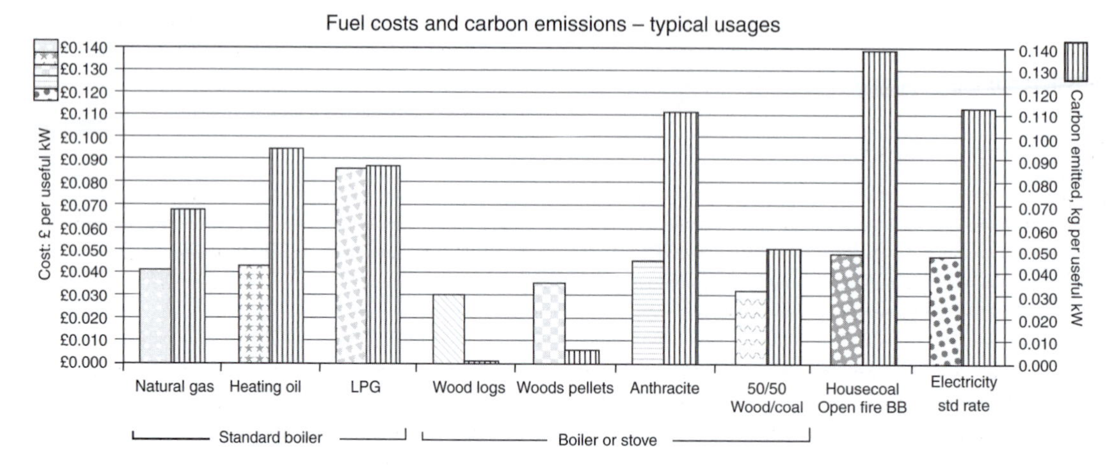

Figure 7.3 Comparison of relative cost and carbon emitted per useful kW output (courtesy of Solid Fuel Technology Institute – October 2006) (see colour plate section at the end of the book)

from business rates. The Government currently operates a CHP Quality Assurance (CHPQA) programme where self-assessment of CHP can be undertaken. Further guidance can be obtained from www.chpqa.com.

Cost considerations

In the UK, the Department of Trade and Industry (DTI) publishes '*Quarterly Energy Prices*', *which is available from their website* www.dti.gov.uk/energy.

At the time of printing the cost of energy is extremely volatile and there are many forecasts of the future trends. Therefore the current cost of energy should not be the sole basis on which the choice of a fuel is made. The carbon emitted per kW of useful heat and the future availability of fuels is a more robust measure by which to make the choice. Figure 7.3 compares the relative cost per useful kW output and carbon emitted per kW useful output.

Heating system types

Heating systems fall into two main groups:

(1) *Direct systems*, in which the energy purchased is consumed as required within the space to be heated, e.g. direct fire gas or oil heaters or direct electric heating. Details and diagrams of the alternative types of heat emitters for direct systems are included in this chapter.
(2) *Indirect systems*, in which the energy purchased is consumed by a boiler, CHP unit or heat pump, and is transferred to emitters in the spaces to be heated via a distribution system. Details of the different types of indirect heating systems can be found in Chapter 9 and details of indirect heat emitters are provided in Chapter 10. However the following notes and tables provide an introduction and overview of the alternative heat sources, heat distribution media and emitters that are available.

Direct systems

The use of direct fired space heating, like any other form of space heating, should be very much dependent on the application. Whilst open fires may be visually attractive, they are, for the most part, extremely inefficient. Efficiencies of over 30 per cent are rare. However, for the central feature for a large dining hall in a hotel for instance, efficiency may well take second place to the visual appeal of the fire itself. Conversely, it is far more likely that the first choice for space heating for a large warehouse is likely to be either suspended gas-fired radiant heating panels or fan assisted gas-fired warm air heating. Particularly in large open spaces with high ceilings, such as warehouses, it is important to include de-stratification fans as part of the design – warm air naturally rises into the roof space, but the personnel are on the floor.

Condensation – a warning

Where a liquid or a gaseous fuel is used to provide an energy supply to a direct system of any description, it is important to appreciate that all equipment does not necessarily provide for the discharge of the products of combustion to outside the building. Where no flue is incorporated for this purpose, there will be a significant emission of water vapour to the heated space, dependent upon the type of fuel and this will in all probability lead to condensation on windows and cold walls.

The *direct* systems may be divided into four primary categories, according to the type of fuel or energy source: solid, liquid, gaseous or electrical. Secondary sub-divisions relate to the characteristics of the terminal heat emitting elements, i.e. whether these *are primarily radiant* or *primarily convective*.

The text and diagrams which follow concentrate upon particulars of the various types of equipment used with direct systems.

Electrical off-peak storage systems

These systems are covered in Chapter 8.

Direct systems – solid fuel

Open fires

These, man's primitive source of heat, were most economic in mediaeval times when the hearth was open to the heated space and gases escaped through a hole in the roof after cooling. With this arrangement little of the energy content of the fuel was lost but there were compensating disadvantages. Later, flues were invented whereby economy was sacrificed in the interests of cleanliness and easier respiration.

The modern version of the open fire is usually arranged for continuous burning, with means for controlling the air inlet and with a restricted throat to the flue, sometimes adjustable. It is designed to burn smokeless fuel and the efficiency may be of the order of 30–40 per cent as compared with 15 per cent or less for the pre-1939 pattern of grate. Some forms of appliance provide airways for a convection output thus increasing efficiency slightly.

Closed stoves

The large coal stoves so common in Europe and those in this country designed to burn anthracite are more efficient, labour saving and draught reducing, than the open fire and may be adjusted to keep alight overnight. Efficiencies of as high as 50 per cent are claimed. Hopper-fed 'closable' anthracite stoves exist in many forms, free standing and brick-set. Examples of the wide variety of wood-burners which have been imported from Scandinavia, however pleasing aesthetically, are rarely either efficient or convenient in view of the need for frequent replenishment of the fuel charge. Logs, even when notionally dry, have a heat content only half that of other more compact solid fuels. Particular care is needed furthermore, but infrequently taken, to ensure that the condensation and tar products deposited in the chimney do not lead to inconvenient soot falls and fire hazards respectively.

Direct systems – liquid fuel (primarily convective)

Industrial warm air systems

The type of unit illustrated in Fig. 7.4 is used for industrial space heating and has the advantage of being simple in both construction and operation. It takes the form of heat transfer passages through which gases from an oil-fired combustion chamber are directed, and over which air is circulated by a fan or fans. Efficiencies of over 90 per cent are claimed for a range of outputs from 30 to about 400 kW. The fuel supply to a number of such units may be piped from a central store.

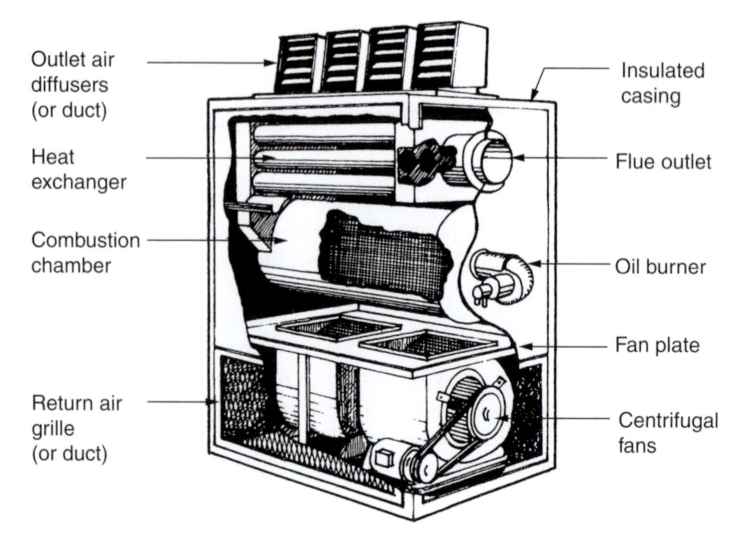

Figure 7.4 Oil-fired industrial warm air unit

The diagram shows a unit with a recirculated-air inlet and cabinet-mounted outlet diffusers but an outside air supply and discharge ductwork can be fitted. It could be argued, however, that the addition of ductwork is not compatible with one particular advantage of such heaters, which is the relative ease with which they may be moved to suit changes in floor layout, etc. Outlet flues are required in all cases.

Direct systems – gaseous fuels (primarily radiant)

Luminous fires

This type of equipment has been improved greatly over the recent decades and, for connection to the public supply, it is now unusual to find a design which does not provide for direct connection to a flue. In addition to the radiant elements, additional heating surface is usually incorporated to provide further output by convection, as shown in Fig. 7.5. Response to control is rapid but the intense radiant effect of the luminous elements may lead to discomfort in some circumstances. A total output of up to 3 kW is common but some patterns have a rating of 5 kW.

Also within this category fall those portable heaters having cabinets designed to house the butane cylinder which provides the fuel source. Such heaters, again generally rated to produce about 3 kW, cannot, by definition, be flued.

Infrared heaters

Designed primarily for industrial applications in higher buildings (but often misapplied to others) heaters of this type are pipe connected to either the public supply or to either butane or propane cylinders. For permanent installations when wall mounted or suspended from a roof, they may be rated at up to 30 kW but a range between 3 and 15 kW is more common.

Construction of one type, Fig. 7.6(a), takes the form of a heavy duty rectangular reflector within which refractory elements and a burner array are fitted behind a safety guard. Another type, Fig. 7.6(b), is cylindrical in plan with a shallow conical reflector over. Portable versions, as in Fig. 7.6(c), are mounted on a telescopic stand which provides means to secure a propane cylinder as a fuel source.

Flue connections cannot usually be provided for heaters of this pattern and responsible manufacturers quote minimum rates of outside air ventilation necessary either per unit or per kW rating. In the case of portable units, these are usually sited in temporary positions within open buildings, stockyards and construction sites, etc., and the absence of a flue is not important.

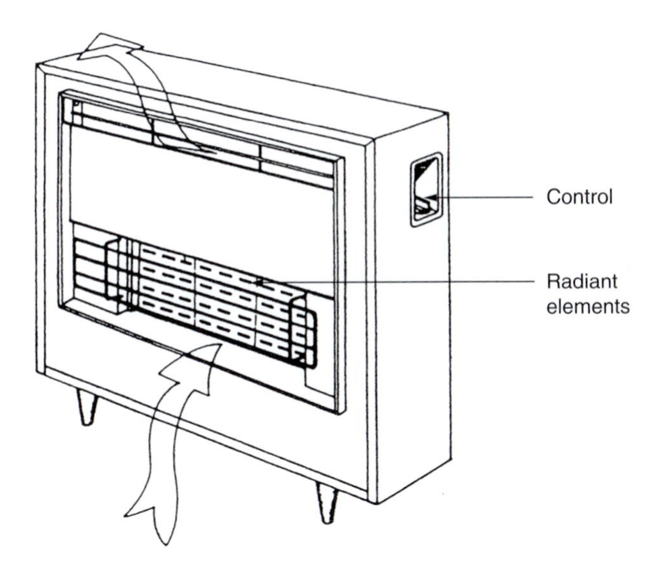

Figure 7.5 Luminous gas fire

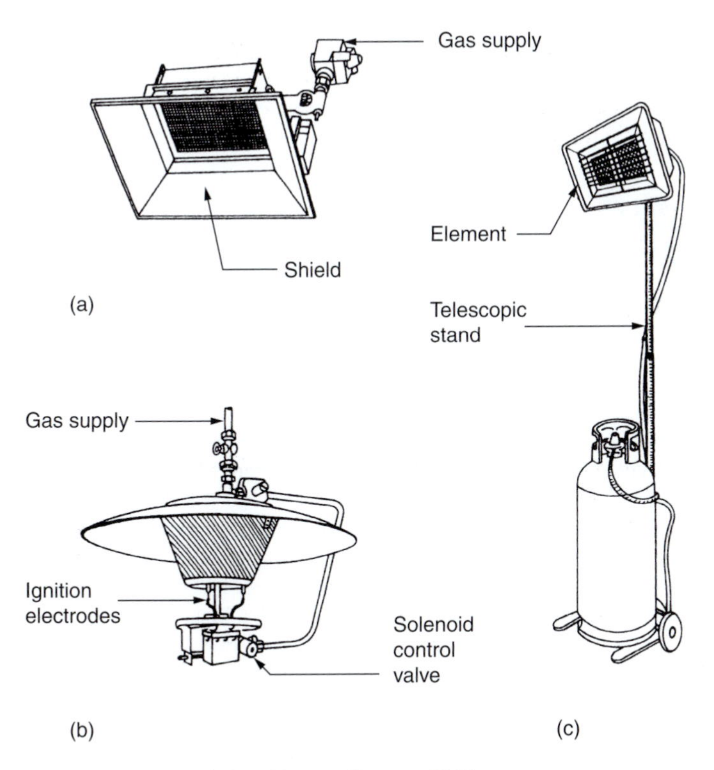

Figure 7.6 Gas-fired infrared heaters (Spaceray/EMC)

Radiant tubes

For such systems, the principal elements are a burner; a steel tube of about 60–70 mm diameter; a reflector plate and a vacuum pump or extractor fan discharging to outside the building. The products of combustion are drawn through the tube which reaches a temperature, below luminosity, of about 540°C. Air for combustion is

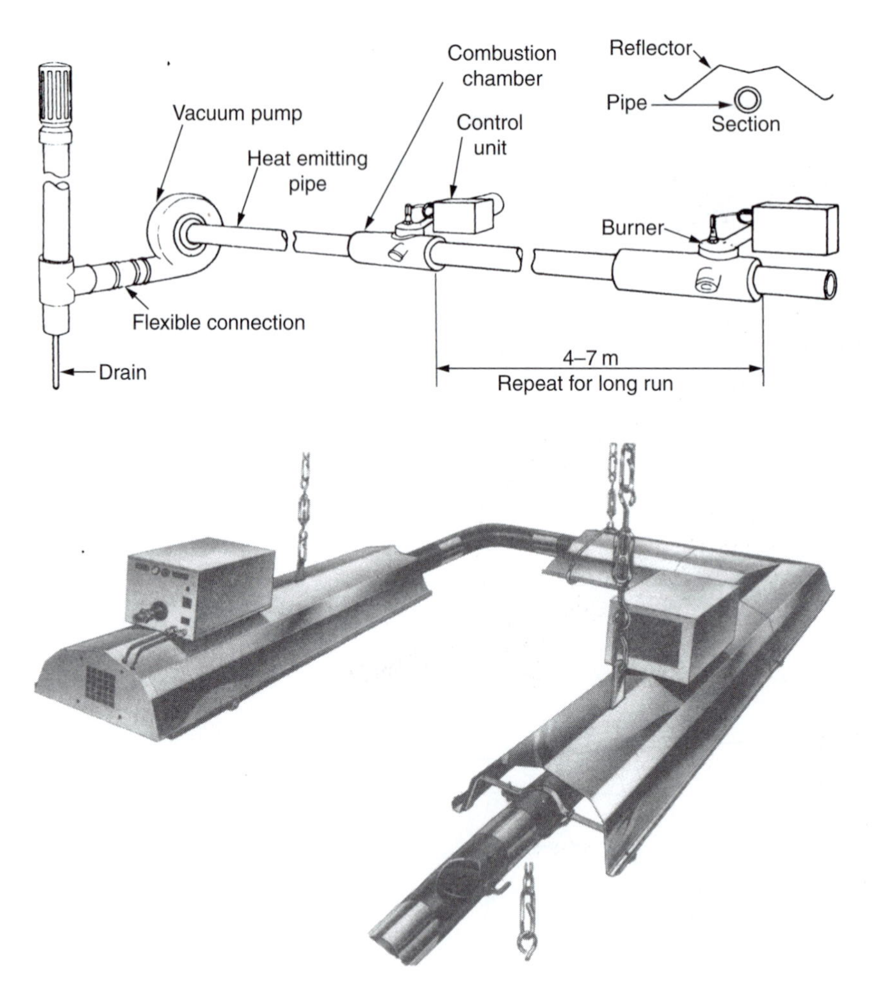

Figure 7.7 Gas-fired radiant tube (Ambirad)

drawn from the heated space, except in circumstances where danger exists as a result of an industrial process; in such applications a ducted air supply is provided. Efficiencies of between 75 per cent and 90 per cent are claimed by the manufacturers.

In one make, the units are in lengths of between 5 and 7 m with a nominal rating of 12 kW at the smaller end of the range and, where circumstances require, lengths are fitted in cascade with outlet gases from the first joining the second to be exhausted finally by a single vacuum pump: this arrangement is shown in Fig. 7.7. Other types have a different configuration in that each element, rated at about 15 kW for a 5 m length, in the form of a 'U' tube with the suction fan adjacent to the burner box, as shown in Fig. 7.8, similar arrangement with a single straight tube is available up to 15 m length with 50 kW rating. The cascade arrangement is thus not practicable, the units requiring a flue for each.

Direct systems – gaseous fuels (primarily convective)

Natural convectors

As in the case of gas-fired luminous fires, the better types of convector are designed to provide means for dispersal of the products of combustion via a conventional outlet or a balanced flue arrangement, as in Fig. 7.9. This type of emitter is very rarely used these days but the diagram is retained here as there will be some buildings

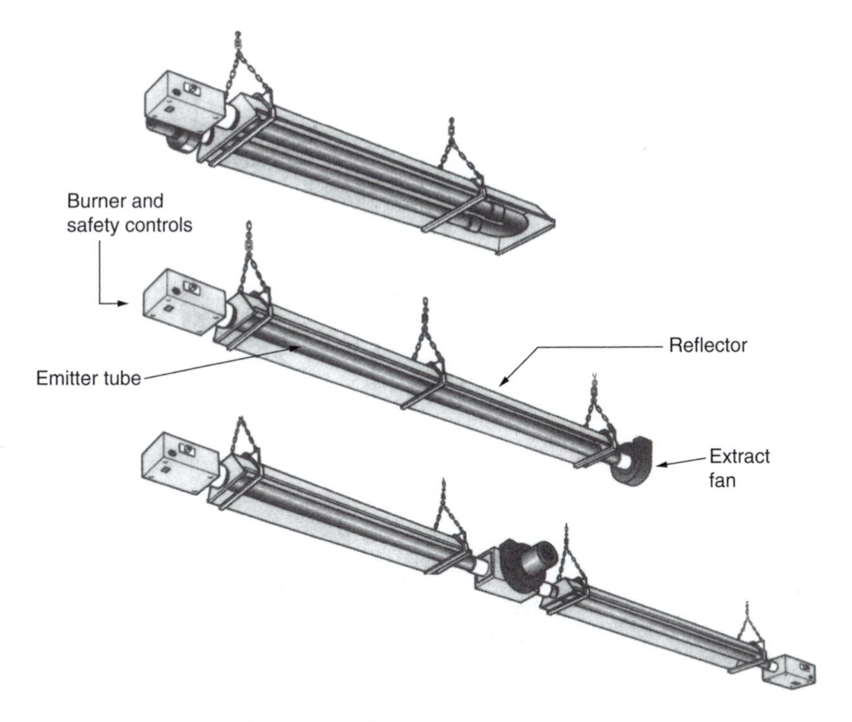

Figure 7.8 Gas-fired radiant tube (Gas-Rad)

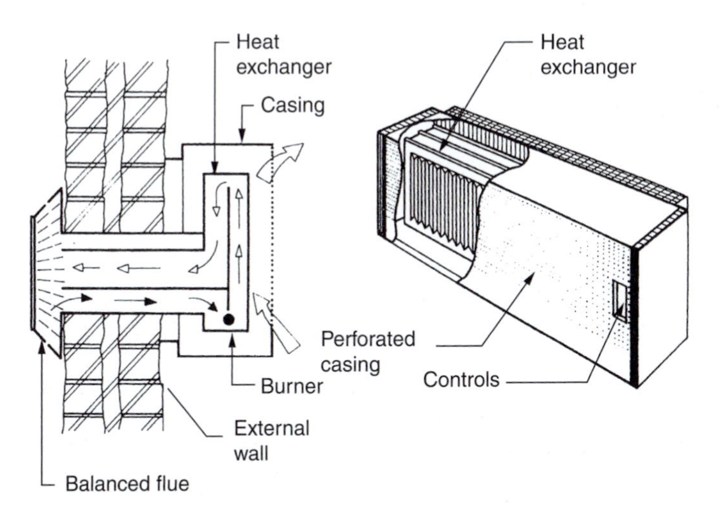

Figure 7.9 Typical gas-fired natural convector

where this type of emitter is still in use. Where direct fired gas heaters are installed in domestic premises to supplement a wet central heating system they are usually the decorative or 'flame effect' type.

Forced convectors

A wide variety of units may be grouped under this heading, ranging from the robust portable blast heaters with axial flow fans used in warehouses and on construction sites, to permanently fixed unit heaters for suspension

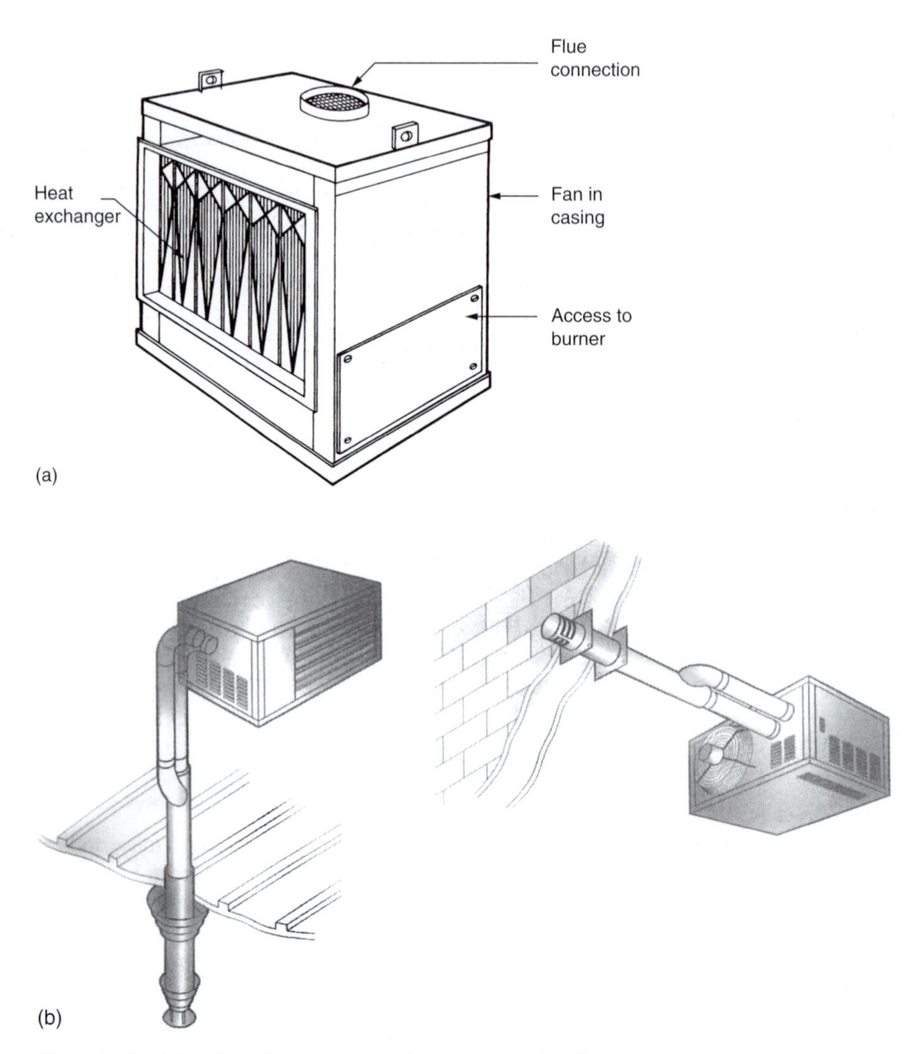

Figure 7.10 (a) Gas-fired forced convector (Reznor). (b) Typical flue arrangements

in industrial buildings. The former are often connected to an equally portable fuel supply in propane cylinders and are, of course, flueless.

The more permanent unit heaters should be connected to a properly arranged piped fuel system which may originate from either the public supply or an LPG tank source. Most manufacturers make provision for units of this type to have flued outlets to outside the building; typical arrangements are shown in Fig. 7.10. Ratings vary with type but most cover the range 10–100 kW.

Domestic warm air systems

A pattern of air heating unit developed for residential use in the 1960s/1970s is shown in Fig. 7.11. It consisted of a combustion chamber formed in some non-corrodible material, a fan for air delivery over that chamber and a casing into which air is drawn from the spaces served and to which it is returned after heating. These systems are no longer used in new buildings as condensing boiler with radiator systems are significantly more efficient.

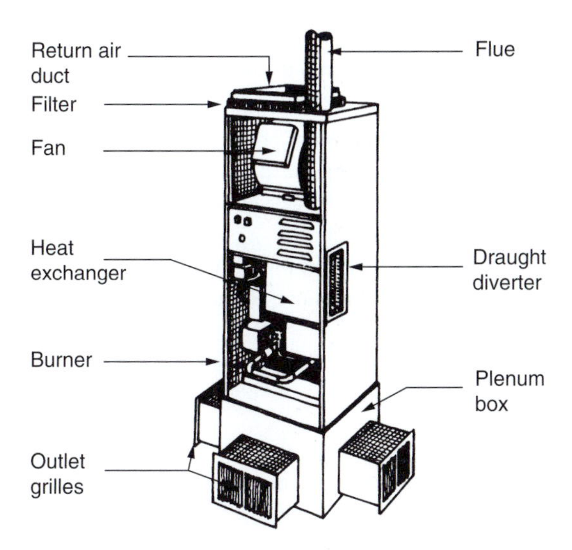

Figure 7.11 Gas-fired domestic warm air unit

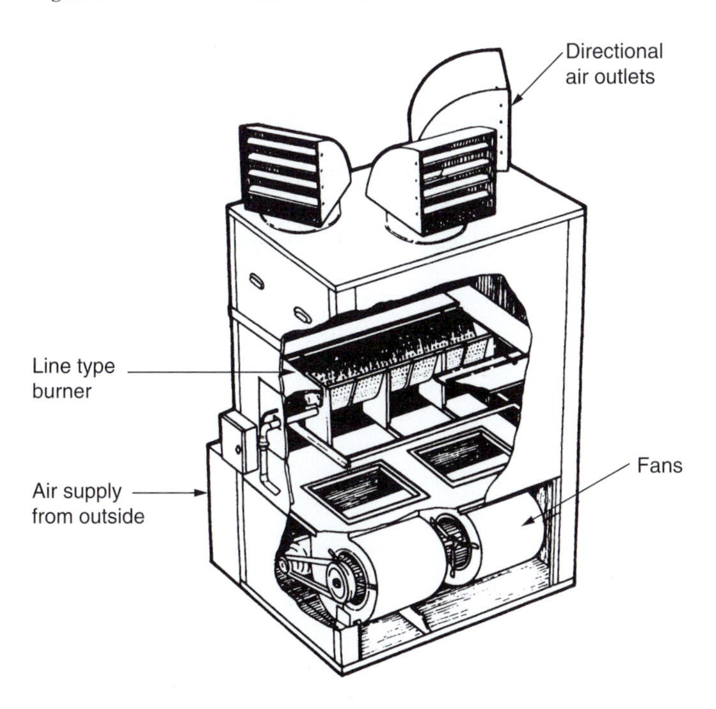

Figure 7.12 Gas-fired industrial warm air unit (Babcock Wanson)

Industrial warm air systems

For industrial application, the type of warm air unit illustrated in Fig. 7.12, but with a burner suitable for gas rather than oil, is applicable. The absence of any necessity for energy storage by way of fuel is an advantage and, as in the case of the oil-fired version, some degree of mobility is retained.

While it is generally to be preferred that such equipment should be provided with a connection such that flue gases are discharged to the outside air, the advent of natural gas and the increasing use of bottled gas have led to the development of 'direct fired' air heaters.

Equipment of this type passes the air for circulation through or directly over an open combustion chamber with no separation or flue, the products of combustion passing out with the air into the heated space. The practicability of this arrangement rests upon two features:

(1) The fuels in question, with the consequent lack of any serious concentration of pollutants in the products of combustion (with the exception of water vapour).
(2) The method of application of the heater whereby the air drawn in for circulation is always *from outside the building* and not recirculated from the heated space.

In these circumstances, the purity of the discharge will be within the accepted threshold values laid down by the Health and Safety Executive. Unit ratings range up to about 650 kW and, since all the energy in the fuel passes to the air circulated, efficiency may be well over 90 per cent.

Direct systems – electrical (primarily radiant)

Luminous fires

The earlier type of luminous fire which consisted of an exposed coiled-wire element, mounted in some manner to a refractory block, is seldom seen today. Elements are now usually silica sheathed and are commonly mounted in front of a polished aluminium reflector having parabolic form. A great variety of types and designs is available with capacities ranging from 500 W to 3 kW. The effect of such heaters is localised and their use is generally confined to domestic premises, hotel lounges and the like.

Infrared heaters

Wall or ceiling models of these are suitable for kitchens and bathrooms in a domestic context and more robust patterns may be used in commercial or industrial premises.

The elements used are similar to those fitted to luminous fires but, for a given rating, are commonly longer, as Fig. 7.13, and arranged to operate at about 900°C. They are sometimes misapplied to churches where, at that temperature, the usual mounting position in the eaves is too high to provide effective radiant cover. Ratings are up to 3 kW per unit.

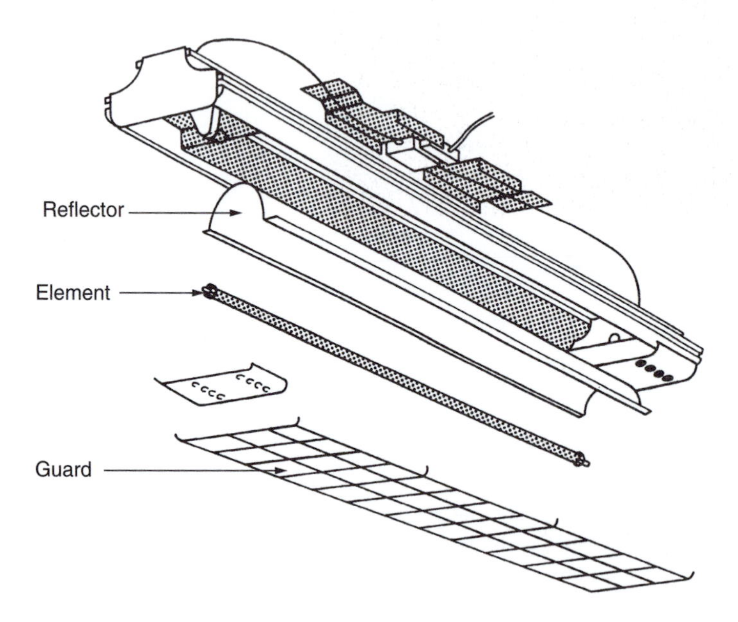

Figure 7.13 Typical infrared electric radiant heater

Quartz lamp heaters
For application to large spaces either where the requirement is intermittent or where only localised areas require spot heating, quartz lamp heaters operate at a temperature of over 2000°C. The elements, each of which is rated at about 1.5 kW, consist of a tungsten wire coil sealed within a quartz tube containing gas and a suitable halide. As illustrated in Fig. 7.14, a rather unlovely casing contains a number of elements (normally a maximum of six), each of which is mounted in front of a polished parabolic reflector.

High-temperature panels
Consisting of either a vitreous enamelled metal plate or a ceramic tile behind which a resistance element is mounted within a casing, panels of this type operate at a temperature of about 250°C and have ratings in the

Figure 7.14 Electrical quartz lamp heater (Dimplex)

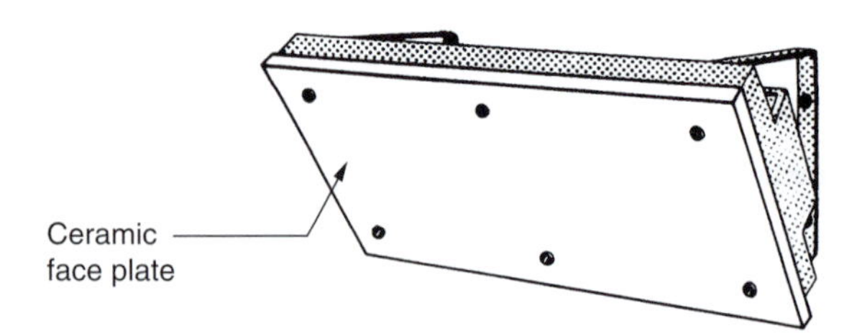

Figure 7.15 Typical electrical high-temperature panel

range 750 W to 2 kW. Although rather more sightly and easier to clean than infrared or quartz lamp units, see Fig. 7.15, application of these panels is usually confined to washrooms and the like in industrial premises.

Low-temperature panels

While equipment of this pattern should, strictly speaking, be listed alongside oil filled radiators and other convective heaters, it is convenient to deal with them here.

Panels of this type are in some cases faced with a metal plate having an enamel finish, other makes use resin impregnated hardboard or a plastic laminate. The heating element may be a conventional form of conductor within an insulating envelope or a polymer coated mesh of synthetic fibre sealed between sheets of polyester film. Many such panels are purpose made for special applications, one being church heating where a panel is fixed to the rear of each pew seat to provide localised warmth to members of the congregation without heating the church fully. Operating temperatures of 70–90°C are usual with ratings of 100–500 W.

Ceiling heating

It is necessary to make a clear distinction here between a purpose designed ceiling heating system and the fortuitous heat input from the underside of an intermediate floor slab which supports a floor heating system (possibly an off-peak storage type) serving the rooms above. It is the former alone which falls wholly within the definition of a *direct system.*

The heating elements used for ceiling heating, which consist of waterproofed conductor strips protected by insulating plastic membranes, are installed immediately above and in contact with the ceiling finish, as shown in Fig. 7.16. A layer of insulating material as thick as possible (preferably 200 mm) is laid over the membrane. The heat emitting surface is, of course, the ceiling itself and some paints and other finishes are unsuitable. Electrical loadings up to a maximum of about 200 W/m^2 of the treated area may be made available. Ceiling heating takes up no floor or wall space and the attributes of an evenly distributed heat output may well be appropriate as a 'top-up' service, complementary to various forms of storage system.

Floor heating

Although the more common form of electrical floor heating is that which is designed to use energy provided during off-peak hours, (as described in Chapter 8) an alternative direct method may be used, superimposed upon a solid ground floor. For this application, a variant of the type of element applied to the manufacture of low-temperature panels is used, having a rating of up to about 150 W/m^2, laid close to the finished floor surface. Typically, the structural floor is covered with about 50 mm of insulating material and the heating elements follow prior to a final layer of chipboard with carpet tiles or some similar finish. Such an arrangement is particularly suitable for buildings which are used only intermittently and for relatively short periods such as churches, etc.

Direct systems – electrical (primarily convective)

Natural convectors

A considerable variety of patterns and sizes of such heaters is available, most now comprising a metal sheathed element or elements fitted within a rectangular casing. The enclosure is arranged to promote air flow over the elements via an opening at the base and a louvered outlet at the top. Convectors may, alternatively, be floor standing and portable or wall mounted: in the latter case, mounting brackets may provide means for easy access to the rear face for cleaning. Ratings range from 500 W to 3 kW, the larger sizes having multiple

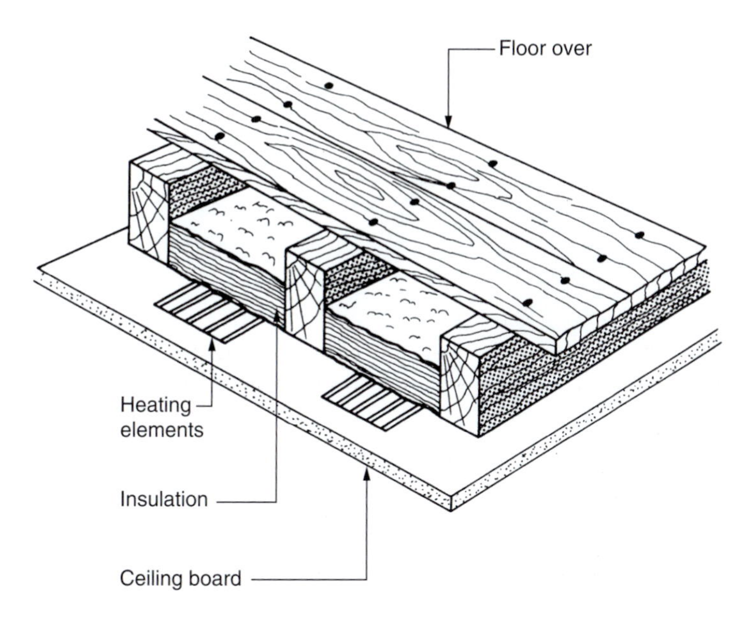

Figure 7.16 Typical electrical ceiling heating

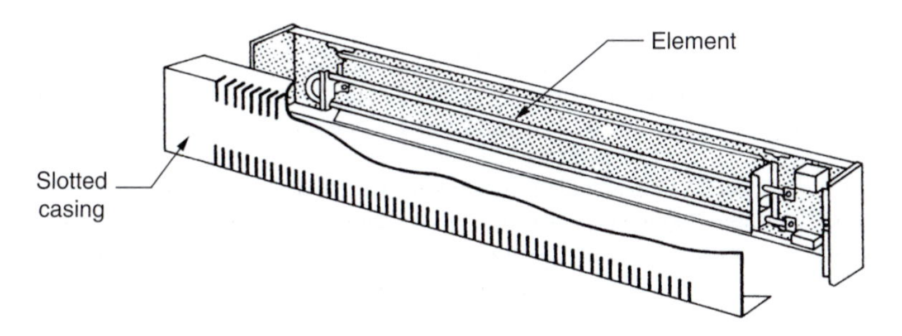

Figure 7.17 Electrical skirting heater (Dimplex)

elements for control purposes. The temperature of the casing remains relatively low in most cases and output by radiation is minimal in consequence. Convectors of this and other types are particularly responsive to thermostatic control.

Skirting heaters

These are a variant of the convector type heaters noted above having a height of about 150 mm with a slotted casing, as in Fig. 7.17, to promote air flow over the element. Most are arranged for either floor or wall mounting and have ratings of between 550 and 750 W/m length. They may be fitted in small rooms or mounted below tall windows.

Oil filled radiators

Similar in appearance to the pattern of steel panel radiators used in hot water heating systems, these have a light oil filling within which is an immersion heater element. The oil begins to circulate when heated and thus acts to transfer heat to the outer surfaces. With a maximum surface temperature of about 90°C, output ratings vary with size up to a normal maximum of 1.5 kW. Wall-mounted or portable models on feet are available: the former are useful in circulation spaces.

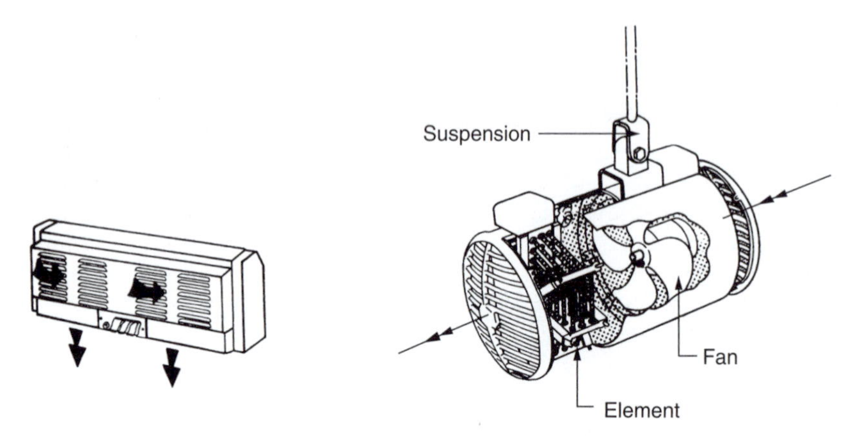

Figure 7.18 Electrical forced convectors (GEC-Expelair)

Tubular heaters
As the name implies, these are steel or aluminium tubes commonly round or oval in cross-section, with no heat transfer filling other than air. The heating element extends from end to end to provide an even surface temperature of about 80°C. A single tube at 50 mm diameter has an output of about 180 W/m length and tubes may be mounted in banks, one above the other, for a greater output per unit length.

Tubulars have been used in churches, placed under the pews, in which position they keep the lower air warm with no attempt to heat the whole enclosure. Fitted below clerestory windows in any tall building, they prevent downdraughts.

Forced convectors
This general heading covers a variety of items from the compact portable domestic fan heater rated at up to 3 kW at one end of the scale to large industrial units rated at 30 kW or more at the other. Between these extremes are commercial-type heaters, fitted with axial flow fans, having ratings of about 3–6 kW and a range of cased tangential-fan units rated at up to 18 kW to meet the particular requirement of a source for warm air curtains at building exits and entrances. Figure 7.18 illustrates two such items.

Indirect systems

Details of the different types of indirect heating systems can be found in Chapter 9 and details of indirect heat emitters are provided in Chapter 10. However the following notes and tables provide an introduction and overview of the alternative heat sources, heat distribution media and emitters that are available.

Heat source type and selection

Details of the three main types of heat source, i.e. boilers, CHP units and heat pumps are provided in other chapters, however the following notes provide some key features of each that should be considered at the start of a design.

Boilers
The many types of boiler are discussed in detail in Chapter 13, however the following points are worth noting here.

The main choice a designer is faced with when choosing boilers for a building is whether they should be condensing or non-condensing. To comply with the current Building Regulations (Part L 2006) for a dwelling, a boiler must be condensing but for buildings other than dwellings, boiler seasonal efficiency only needs to be 84 per cent, therefore a condensing boiler is not required. Designers must remember that boilers will only condense when the return water temperature falls below 55°C. Therefore applying a condensing boiler to a constant temperature system operating at 82°C flow and 71°C return will result in the boiler performing no better than a standard high-efficiency boiler.

One way of ensuring condensing for a high proportion of the year and avoiding having to use larger emitters (that would be required with significantly reduced water temperatures) is to directly weather compensate the boiler flow temperature. Depending on the flow and return temperatures used for the design condition of (possibly −4°C outside air temperature), this will result in boilers condensing whenever the outside air temperature is in the region of 5–6°C which, if the building is intermittently heated, represents a significant proportion of the heating season. Separate condensing-type gas fire hot water generators coupled to thermal storage cylinders would need to be used in this scenario.

Below are some example figures for system seasonal efficiencies – not to be confused with the thermal efficiency of the boiler itself. (*Source*: CIBSE Guide F 2004.)

Condensing boilers:

- Underfloor or warm water system 90 per cent or greater
- Radiators, boilers with direct weather compensation variable temperature 87 per cent
- Standard fixed temperature emitters (82°C flow, 71°C return) 80–84 per cent.

Non-condensing boilers:

- Modern high-efficiency non-condensing boilers matched well to demand 80–82 per cent
- Typical good existing boiler 70 per cent
- Typical existing oversized boiler (atmospheric type) 45–70 per cent.

Combined heat and power units

These systems are discussed in detail in Chapter 15, however the following points are worth noting here.

CHP can provide the single biggest CO_2 reduction that can be achieved in a building where CHP is appropriate. Electricity is generated using an engine or turbine and heat is recovered from the exhaust gases and engine cooling system. They should always operate as the lead boiler to supply a constant base heating load to maximise savings and typically, a CHP plant will produce 1.7 times the electrical output as heat.

The overall efficiency of a CHP plant is in the region of 80 per cent compared to 40 per cent for an average power station. Each 1 kW of electrical capacity provided by CHP has the potential to reduce annual CO_2 emissions by approximately 0.6 tonnes compared to gas-fired boilers and grid-derived electricity.

CHP is most economically viable in buildings which have a 24 hour, 7 day/week demand for heat such as hospitals and hotels or leisure centres with swimming pools. Typically a CHP plant needs to run for at least 4500 hours/year (i.e. at least 12.3 hours/day on average, and a minimum of 14 hours/day is preferable) to achieve an acceptable payback period of 3–5 years.

Heat pumps

Heat pumps are available in a number of different forms and exploit different sources of low-grade heat. Air source heat pumps as used in variable refrigerant flow (VRF) systems may be used to extract heat either from outside air or from ventilation exhaust air. When outside air is used as a heat source, the coefficient of performance tends to decline as the air temperature drops.

Ground or water source heat pumps extract heat from the ground or bodies of water, either at ambient temperature or with temperature raised by the outflow of waste heat.

They have the advantage over air source heat pumps that their heat source has much greater specific heat capacity and the temperature of the source is fairly stable. At about 1.2 m below the surface the annual mean temperature is in the range 9–12°C.

Ground source heat pumps have a seasonal COP of around 3–4 in a typical UK climate and raise the temperature of the heat stored in the ground to approximately 40–50°C; they therefore can be used in conjunction with condensing boilers. A further advantage of ground source heat pumps is that they can operate in reverse and provide cooling in summer and the ground acts as a heat sink to absorb the heat removed from the building. The environmental advantages/disadvantages of heat pumps hinge on their coefficient of performance and the potential CO_2 emission of the fuel used to power them. Gas-fired heat pumps with a relatively low COP may therefore produce lower CO_2 emissions per unit of useful heat output than electrically driven units. For electricity drawn from the UK grid, a seasonal COP of around 2 is required to achieve lower emissions than would be obtained from a gas condensing boiler. Where ground source heat pumps are installed in place of conventional direct electric heating, carbon savings of 50–70 per cent are achievable.

Solar heating

Although solar water heating systems have been used to provide domestic hot water for a number of years, there has been little progress in the use of solar energy for space heating in the UK. Some experimental systems are under development aimed at providing space heating as well as domestic hot water, although these usually require a top-up source of energy.

Heat distribution system selection

Indirect heating systems are detailed in Chapter 9 and piping design for indirect heating systems is detailed in Chapter 12. The following provides a basic introduction to the alternative distribution mediums available. Air and water are the most common choices but steam is still used in many existing buildings and some new buildings where steam is required in large quantities for process loads. Refrigerants are also used in heat pump and VRF systems where both heating and cooling is required. The choice of distribution medium must account for the requirements of radiant and convective output (if air is used there is no opportunity for radiant output).

Table 7.4 gives the principle characteristics together with the advantages and disadvantages of each system.

Air distribution systems

Historically, the culture of the Romans developed a type of heating system which was primarily radiant, an art which was lost for some 1600 years. This system was centred about a furnace room below ground from which hot flue gases were conveyed, via ducts under the floor and flues in the walls, to emerge at various points about the building. Sometimes proper ducts were formed but in others the whole space under the floor (the *hypocaust*) was used. The wall flues took the form of hollow tiles, very similar to their modern counterparts. A similar but rather more sophisticated system was applied to the twentieth-century Anglican Cathedral in Liverpool using heated air, fan circulated through a pattern of structural ducts immediately below the floor finish.

Air has a low specific heat capacity and a low density, which means that large air flow rates are required to meet the heat demands of a space. Fan power is proportional to the cube of the volume of air moved. The result is high running costs and carbon emissions for air distribution system. Ductwork and associated insulation is also expensive and demands significant space requirements, which may be prohibitive where cooling is not required.

Air is usually introduced to a space at high level which can lead to problems of effective distribution of heat throughout. Warm air rises and without good design of the air diffusers, the result can be stratification of heat. This is especially significant in high ceiling and large volume spaces where it is often necessary to install de-stratification fans to blow the warm air down into the occupied zone.

Air systems are able to heat up a space rapidly, which can actually have significant energy advantages in intermittent buildings (short preheat times reduce central plant energy consumption). However a problem in more heavyweight buildings is that the air is heated rapidly but the surfaces/mass take time to warm up, resulting in a low mean radiant temperature. To address the comfort needs of the occupants it may be necessary to have a higher air temperature to offset the cool surfaces, but this in itself can lead to complaints of stuffiness. (By way of example consider the rapid heating of a car interior from cold on a winter's day; to feel comfortable initially, the air temperature needs to be high, but then has to be reduced as the surfaces warm up.)

Air heating is not recommended for large volume spaces (such as factories and warehouses) as it is expensive to heat such a large turnover of air. In these cases it is better to introduce ventilation air (suitably tempered) into the occupied zone and provide local radiant heating to deal with occupant comfort. Such radiant heating may be from high-temperature hot water or steam-fed panels, or direct fired gas heaters. These should be suitably located and guarded for safety. In this way the indoor temperature is reduced, thus reducing the heat loss from the space overall.

Table 7.4 Comparison of alternative heat distribution systems (CIBSE KS 08 – How to design a heating system, 2006)

Medium	*Principal characteristics*	*Advantages*	*Disadvantages*
Air	Low specific heat capacity, low density and small temperature difference permissible between supply and return, compared to water, therefore larger volume needed to transfer given heat quantity	No heat emitters needed No intermediate medium or heat exchanger is needed Fast preheat time (air temperature is raised quickly)	Large volume of air required – large ducts required more distribution space Fans can require high-energy consumption Cannot provide radiant heating Spaces can feel 'stuffy' as air temperature is higher than mean radiant temperature on preheat
Water	High specific heat capacity, high density and large temperature difference permissible between supply and return, compared to air, therefore smaller volume needed to transfer given heat quantity Water temperature choices: LTHW, MTHW, HTHW – see below	Small volume of water required – pipes require little distribution space	Requires heat emitters to transfer heat to occupied space
LTHW	Majority of installation are: low-temperature hot water systems, they operate at low pressures that can be generated by an open or sealed expansion vessel	Generally recognised as simple to install and safe in operation	Output is limited by system temperatures restricted to a maximum of 90°C and 10 bar gauge working pressure. Unless condensing boilers are installed the minimum return water temperature should be 66°C
MTHW	Medium-temperature hot water permits system temperatures up to 120°C and 10 bar gauge and a greater drop in water temperature around the system	Higher temperatures and temperature drops give smaller pipework, which may be an advantage on larger systems	Pressurisation necessitates additional plant and controls, and additional safety requirements
HTHW	High-temperature hot water (up to 10 bar absolute) and temperatures above 120°C with even greater temperature drops in the system	Higher temperatures and temperature drops give even smaller pipework	Safety requires that all pipework must be welded, and to the standards applicable to steam pipework. This is unlikely to be a cost-effective choice except for the transportation of heat over long distances
Steam	Exploits the latent heat of condensation to provide very high transfer capacity. Operates at high pressures. Principally used in hospitals and buildings with large kitchens or processes requiring steam	Can also be used for humidification	High maintenance, insurance and water treatment requirements. Condensate return system required with additional maintenance requirements

Water distribution systems

The design of water systems is discussed in detail in Chapters 9, 11 and 12, but the following key points are worth noting here.

Water has a specific heat more than 4 times that of air (4.18 kJ/kgK), and is over 800 times more dense. The result is the heat carrying capacity almost 3500 times greater than air per unit volume, resulting in small pipes and pump power requirements much smaller than fan loads to deliver the same heat to a space. Water systems can be operated at different temperatures which are summarised in Table 9.1.

When choosing the flow and return temperature difference, care must be taken to ensure water flow rates are not too small to prevent possible problems with laminar flow in the pipe work/terminals that can result in a significant reduction in the heat transfer. Low flow rates can also cause problems with the selection of control and balancing valves with the result that flow rates can sometimes not be measured! Often a minimum pipe size/commissioning valve of 15 mm dia is used where the velocity can be very low, when in reality the final run out to the emitter should be in 12 mm dia pipe to ensure the velocity is high enough to prevent laminar flow. With the reduction in space heating loads in new buildings due to vastly improved insulation levels, this problem is becoming more prevalent. Case studies have shown that temperature differences between flow and return of 15–20 K can cause the problems described.

A special case is underfloor heating systems where a maximum surface temperature of 29°C is required for comfort conditions. To achieve this flow and return, temperatures of 50/40°C or 45/35°C are required (depending on floor construction). These temperatures make underfloor heating systems particularly suited to condensing boilers or ground source heat pumps which both operate at optimum efficiency at these temperatures.

Steam systems

These systems are covered in some detail in Chapters 9, 11 and 12 and therefore not discussed here.

Emitter types and selection

All heat emitters commonly in use in the UK for heating for occupant comfort are either of the convective, combined radiative and convective or radiative type, and are summarised in the notes below and in Table 7.5. Further details on each type of heat emitter are provided in Chapter 10.

Different types of emitter have different output characteristics in terms of the split between convective and radiant heat. Where there is a high ventilation rate, the use of a radiant heating system results in lower-energy consumption. Fully radiant systems heat occupants directly without heating the air to full comfort temperatures. Where convection is the predominant form of heat in tall spaces, care must be taken to avoid vertical temperature stratification. Increased temperatures at high level also increase heat losses through the roof in atria or warehouse/factory buildings.

Convective emitters

Typical convective-type emitters include radiators, natural convectors, fan convectors, unit heaters and gas convector heaters.

Radiators: The name is a little misleading in that the majority of heat is transmitted through convection rather than radiation. These are basically flat or ribbed panels constructed from cast iron, steel or aluminium, and are wall mounted, often below windows to counteract any cold downdraughts. Radiators rely on a convection current of air to draw the heat out of the emitter and carry it into the space, with heat emitted in the approximate proportions of 30 per cent radiative and 70 per cent convective.

Natural convectors: Typical wall-mounted natural convectors have a radiant/convective split of 20:80 and the floor trench types are 100 per cent convective. They tend to produce a more pronounced vertical temperature gradient that can result in inefficient energy use. They come in various forms including wall-mounted panel-type units, and low-level types fitted at skirting level. A natural convector typically consists of a copper or steel pipe covered with an arrangement of fins along its length. This is fitted towards the bottom of the casing. A convection current induced by the warm air above the element is displaced by the cooler air entering below.

Fan convectors: These are similar to natural convectors, but with a fan unit fitted below the finned element, and sometimes a filter as well. These typically have higher outputs than the natural type, and the fan units have speed control. The minimum flow temperature for circuits serving fan convectors should not be below 50–55°C to prevent cold draughts.

Unit heaters: These are fan driven and can be fed from low-temperature hot water, medium-temperature hot water, steam or gas. They are typically used in large open areas where high heat output is required. The units can be mounted at high level or floor standing. They often have louvers fitted on the discharge side of the units to permit directional control of the air path.

Radiative emitters

Typical radiative-type emitters include radiant panels, radiant strip/tube heaters, gas radiant tube heaters and underfloor heating. These are typically used in large volume buildings that have high air change rates, such as

Table 7.5 Features of alternative heat emitters

System	*Design points and applications*	*Advantages*	*Disadvantages*
Radiators	Output up to 70% convective Check for limit on surface temperature Suitable for most applications – special 'low surface temperature' type required for hospitals, care homes and nurseries	Good temperature control Balance of radiant and convective output gives good thermal comfort Low maintenance Cheap to install and replace	Average response to control Occupy wall/floor space, reducing flexibility of use of space Harbour dust, convection currents circulate High surface temperature can restrict use in buildings from a safety viewpoint
Natural convectors	Can be fed by low-temperature hot water or be direct electric type Suitable for all applications	Quicker response to control Skirting convectors or trench type can be unobtrusive and free up floor space Electric type have low capital cost and very good flexibility	Can occupy more floor wall space Can get higher-temperature stratification in space as very high % convective output Need cleaning can harbour dust Electric type expensive to run and produce high carbon emissions
Underfloor heating	Check required output can be achieved with acceptable floor surface temperatures Suitable for: Most domestic applications Buildings or areas with low heat loss, buildings or areas that are continually or frequently used, buildings or areas with high ceilings and large circulation spaces Not suitable for: Buildings or areas that are used very intermittently or infrequently Buildings or areas that have high heat losses, or high sudden losses Applications where large amounts of equipment or fittings will be fixed into the floor, e.g. racking or shelving Buildings where future partitioning may occur	Unobtrusive Good space temperature distribution/minimum stratification between floor and ceiling Increased flexibility for furniture, more lettable floor space Reduced running costs (can be used with condensing boiler and ground source heat pumps) 'Healthy' (do not need cleaning/cannot collect dust)	Heat output limited Slow response to control Output sensitive to floor coverings (e.g. rugs or mats in school classrooms) Future modifications to structure very disruptive to heating system Any leakage from pipework very disruptive/difficult to rectify
Fan convectors/ fan coil units	Can also be used to deliver ventilation air Main application is to office buildings	Quick thermal response Can ducted in ceiling void or be installed in floor voids reducing occupied floor space	Can be noisy/costly to attenuate Higher maintenance, filters to clean Occupies more floor space More controls required
Warm air heaters/ unit heaters	Can be direct fired units Only suitable for industrial or large warehouse type retail units	Quick thermal response Cheap to install for factory/ warehouse heating	Noisy Can get considerable temperature stratification in space
Low-temperature radiant panels	Ceiling panels need relatively low temperatures to avoid discomfort Main application is for hospital and similar	Unobtrusive Low maintenance	Slow response to control
High-temperature radiant heaters	Can be direct fired units Check that irradiance levels are acceptable for comfort	Quicker thermal response Can be used in spaces with high air change rates	Need to be mounted at high level to avoid local high-intensity radiation and discomfort

factories, warehouses and garages. Radiant systems: are more efficient because they only heat the occupants and building fabric and do not generally raise the temperature of the internal air to full comfort levels; generally provide a rapid response, requiring less heat-up time at the beginning of the day. Significant energy savings compared with convective systems are therefore possible.

Radiant panels: These are similar in appearance to smooth-faced radiators but are designed to produce a higher portion of their output by radiation. They are typically mounted at high level, distributing the heat downwards, and are used in areas such as workshops and sports halls where it is not necessary to have a high air temperature. They can also be used with low-temperature hot water in environments such as hospitals where floor-mounted radiators/convectors provide potential health and safety problems (high surface temperature and difficult to clean).

Radiant strip/tube heaters: These are served by steam or medium/high-temperature hot water and consist of heating pipes fixed to a metal panel. The heat in the pipe warms the panel via conduction, which then radiates down to the space. These are also mounted at high level, and have similar uses and applications to radiant panels.

Underfloor heating: The modern underfloor heating system consists of plastic pipe laid in circuits in a floor screed or below a timber floor system, through which low-temperature hot water is passed. The hot water is circulated at a lower temperature than for other forms of heating and provides even heat distribution across the whole installed area. With this type of arrangement, heat is typically emitted in the proportions of 40 per cent convective and 60 per cent radiative.

Note

1. SEDBUK is an acronym for 'Seasonal Efficiency of a Domestic Boiler in the UK'. The system was developed under the UK Government's energy efficiency best practice programme with the co-operation of boiler manufacturers and provides a basis for fair comparison of different models of boilers. The SEDBUK rating is the average annual efficiency achieved in typical domestic situations, making sensible assumptions about climate, control, pattern of usage and other similar factors. The rating is calculated from laboratory tests together with other important factors such as boiler type, fuel used, ignition type, UK climate, boiler water content and typical domestic usage patterns.

Chapter 8

Electrical storage heating

Introduction

This chapter deals with direct acting electrical resistance heating supplied overnight and different methods of storing the heat for use during the day.

It is worth noting that there are a number of British Standards relating to the design and use of electric storage heating systems. These include old standards (that are still current) such as CP 1018: 1971 that covers the design and installation of direct electric floor warming systems and BS 6351-1: 1983 on electric surface heating and modern standards including BS EN 14337:2005 that covers the design and installation of direct electric room heating systems, BS EN 60531; 1999 and BS EN 60379 on methods measuring the performance of household thermal storage room heaters and electric storage water heaters, respectively.

Electric heating systems generally have the following benefits:

- They are inexpensive to install.
- Can reduce space requirements.
- Require little or no maintenance.
- Are highly efficient.
- Provide very quick response to controls and are therefore suitable for intermittently heated areas (particularly when fan assisted).
- Distribution losses associated with central systems are eliminated.

However they have two big disadvantages that will usually outweigh the benefits:

(1) High CO_2 emissions (from electricity generation at power station).
(2) Potentially high running costs.

To overcome the high running costs it is the policy of energy suppliers to offer differential tariffs to both domestic consumers and commercial customers whereby current consumed during the 7 hours after midnight is sold for approximately half the price of that consumed during the remaining 17 hours (known as 'off-peak' rate). The availability of such a reduction to consumers can transform the economics of electrical heating, provided that suitable methods of storing the heat for use during the occupied period are utilised.

In addition energy supply companies will also make additional charges to commercial customers when a user reaches their agreed 'Maximum Demand', which is the highest rate of flow of electricity in any 30-minute period. Using electric storage heating will reduce the building's maximum demand figure resulting in further savings.

Even when the system is designed to ensure that a significant proportion of the energy required is provided at off-peak rate, the use of electric heating of any building should only be considered 'best practice' when all the following conditions have been met:

- The building is very well insulated (probably exceeding the minimum requirements of the Building Regulations).
- The total heating and hot water requirement is small such that when a whole-life cost analysis is undertaken the estimated net present value of the heating and hot water service is lower for electricity than other fuels over the same period.
- The hot water storage cylinder is large enough to supply most of the hot water demand between off-peak periods and is insulated to higher standards than normal.

The main disadvantage of electric storage heating is the limited charging capacity and the difficulty of controlling output, often leading to expensive daytime re-charging. The use of phase change materials (PCMs) can significantly increase the charging capacity and using modern storage heaters with sophisticated controls can overcome the problem of poor output control. These issues are explained in more detail later in this chapter.

It is much more difficult to overcome the high CO_2 emissions. Unless 'grid displaced' electricity is used in the building, i.e. electricity generated by renewable technologies or low carbon technologies such as solar photovoltaic systems, wind turbines and combined heat and power systems, the grid generated electricity used is at low efficiencies, typically 35–45 per cent depending on the mix of generating equipment in use. This results in a carbon emission factor of 0.422 kg CO_2/kWh compared to a carbon emission factor for natural gas of 0.194 kg CO_2/kWh as defined in the 2006 edition of Part L of the Building Regulations.

A number of examples of annual CO_2 emissions for a range of different dwellings are provided in The Carbon Trust *Good Practice Guide 345 'Domestic heating by electricity*'. As an example comparing a typical detached house either heated by electricity or a condensing gas-fired boiler indicates that the electrically heated dwelling will emit approximately 90 per cent more CO_2 per year, if using grid supplied electricity.

Therefore if grid generated electricity is used to heat buildings, high levels of insulation and good control of the electric heating will be essential to ensure compliance with Building Regulations. It should be noted here that the use of electric heating is not precluded in the 2006 edition of Part L. The principle criterion is to demonstrate that the building annual CO_2 emission rate (BER) is no more than a target emission rate (TER) using approved calculation methodologies. The TER is based on an improvement over the 2002 standards (this is 23 per cent for a naturally ventilated building and 28 per cent for a mechanically ventilated or air-conditioned building).

When grid displaced electricity is consumed in a building the associated CO_2 emissions are deducted from the total CO_2 emissions for the building before determining the building emission rate (BER).

In addition there are minimum requirements regarding controls of electric space heating equipment that must be complied with when designing systems and specifying plant. These requirements are detailed in the 2nd tier documents referred to in the Approved Documents (i.e. Non-domestic Heating, Cooling and Ventilation Compliance Guide and Domestic Heating Compliance Guide for dwellings) and for electric storage heaters include:

- Automatic control of the input charge (i.e. the ability to detect the internal temperature and adjust the charging of the heater accordingly).
- Manual controls for adjusting the rate of heat release from the appliance.

The future for electric heating

UK natural-gas supplies are dwindling, and it is expected that by 2020 the level of dependence on imported gas will be as high as 80 per cent. With gas prices continuing to rise, and the increasing generation of electricity from either renewable sources or by nuclear the future for electric heating (if combined with a means of storage – to benefit from cheaper nighttime tariffs) could be very positive, particularly taking into account the following on-going developments:

- There will be an increase in use of renewable generation technologies/low and zero carbon technologies (both building integrated and grid generated) to meet both Building Regulations and Planning requirements.

- Building insulation levels and thermal mass of buildings will increase (reducing heating loads).
- Building air leakage rates will reduce (further reducing heating loads).
- The use of PCMs in the building construction is likely to increase (to enhance the storage capacity of the building).
- Possible improvements to off-peak tariffs for electricity.

Types of electric storage heating

Electric storage heating relies on using cheap off-peak electricity to store heat in either an insulated container of water or bricks or in the building fabric itself. This stored heat can then be gradually released into the building during the day.

Dry storage heating systems tend to be less responsive than central heating systems with a boiler and radiators but these systems are subject to distribution losses and significant additional maintenance costs.

There are two types of storage system. *Uncontrolled* storage systems i.e. heat dissipated continuously throughout 24 hours such as underfloor heating and *controlled* storage systems where heat dissipation takes place only during hours of use. In both cases the storage and charge must be equal to the 24-hour heat demand. The input rating (Q) of the plant for both is calculated using the equation below.

$$Q\,(\text{Watts}) = \frac{24\text{ hour heat requirement (Joules (J))} + 24\text{ hour heat losses (J)}}{3600 \times \text{recharge time (hours)}}$$

If the storage unit is also the space heating emitter the losses can be considered as useful heat.

For uncontrolled storage systems no intermittent heating factors should be applied and the heat requirement is equal to the steady state heat loss calculated at the estimated mean internal temperature (i.e. the design internal temperature) at the mean daily outside air temperature which is assumed to equal the design outside temperature. With these types of system the highest room temperature will be at the end of the charging period and the lowest just prior to recharging. This temperature swing should be limited to 3–4°C.

For controlled storage systems (e.g. controllable room storage heaters or electrically heated water storage systems) the design rate of emission should be calculated as for conventional intermittent systems with the total storage capacity based on the mean internal temperature taking into account preheat and thermal response factor of the building.

Methods of storage

The methods commonly used for storage of heat may, for convenience be grouped as being either room stores or central stores, thus include the following.

Room stores

Storage heaters (radiators) where heat is retained in a solid material contained within an insulated casing and emitted continuously with limited control of output.

Storage fan heaters where heat is retained similarly to the above but where the addition of a fan, switched by a room thermostat, provides that a significant proportion of the heat from the store will be emitted only when required.

Underfloor heating or heated walls where heat is retained in the building structure or structure plus embedded PCM and emitted continuously with no control of output.

Examples of room stores are given later in this chapter.

Central stores

These days this type of unit are seldom used in new commercial buildings due to the equipment losses, distribution losses, additional energy required to deliver the heat and associated high CO_2 emissions and in the case of warm air units space required for the distribution ductwork.

There is still a limited market for domestic scale central stores. These usually comprise a packaged unit that can provide hot water for radiator space heating and instantaneous heating of mains cold water. Further details are given later in this chapter. Looking to the future, some manufacturers are investigating the integration of encapsulated PCMs that melt in the temperature range 80–90°C into the storage cylinder. This can result in a reduction of the cylinder volume to about 25 per cent of the storage volume required for an all-water system. This results in a significant reduction in thermal losses and a smaller footprint for the installation.

Types of central stores

Warm air units which are, in effect, large-scale storage fan heaters arranged as a central heat source for indirect warm air systems.

Dry core boilers in which heat is retained as for storage fan heaters but arranged so that output is used to supply an air/water heat exchanger from which, an indirect system of piping and radiators, is installed.

Wet core boilers in which heat is retained in a hot water vessel, at atmospheric pressure, arranged so that the contents are used for circulation through an indirect system of piping, radiators, etc.

Thermal storage cylinders in which heat is retained in water, at elevated temperature and pressure, to serve indirect systems of all types in large commercial buildings. Electrode boilers, which are used to charge such cylinders, are described in Chapter 10.

Examples of central stores are given later in this chapter.

Capacity of the heat store (room stores)

The amount of energy which must be stored in order to provide a satisfactory service will depend upon which particular system is used. In all instances, however, since heat will be discharged from the store over the whole 24-hour cycle, albeit perhaps at a low level during the period when output is not required, adequate capacity must be provided to include for the static heat output during the charging period. Some proportion of this output will, where the store is within the space to be heated, be absorbed in the surrounding building structure and thus not be wholly wasted.

Where heat output is uncontrolled, as in the case of all underfloor heating and many *input-controlled* storage radiators, the nighttime loss may be between 15 per cent and 20 per cent of the total input to storage. Where equipment such as some more modern storage radiators and most storage fan heaters is *output-controlled,* the situation is better in that nighttime loss is a smaller proportion, about 10–12 per cent of the total input. It must be added, nevertheless, that less than half of the energy which then remains in the store will be available for controlled output, the balance being a daytime static discharge. Equipment for *central stores* which are designed to serve *indirect* systems are not subject to the same dimensional constraints as applied to room-sited units and thus can be better insulated and static losses may thus be of the order of 5 per cent or less of total input.

The foregoing comments are intended to emphasise that conventional methods of calculation for heat requirements must be modified to take account of the abnormal characteristics of storage systems.

The simplistic approach to establishing the rating of a thermal storage unit described in the introduction to this chapter does not, however, take account of a number of the other considerations which were discussed in Chapter 4 (the building in winter), many of which are particularly relevant to a method of heating which is subjected to an intermittent energy supply associated with an output which, although at a level which varies, is continuous. The thermal response of the building structure which is related to its mass; the level of insulation provided; the area of glazing; heat gains during the hours of occupancy from lighting and occupants are all relevant as is, where applicable, the degree of output control available. The complexities of these aspects are compounded in instances where output is primarily radiant and thus is interactive with other surfaces 'seen' by equipment or system. Calculations based upon consideration of all these variables would be out of the question for individual applications since; in any event, meticulous accuracy would be misplaced when applied to selection from a finite range of production equipment.

There are a number of alternative routines to establish the adjustments necessary to calculations of heat loss to account for the abnormal characteristics of storage systems. One such method that is promoted by The *Electric Heating and Ventilation Association* in their guide the *Design of Electric Space Heating Systems DOM 8 (2006)* is outlined below. It should be noted that this is for dwellings, but could easily be applied to commercial buildings.

The principle of sizing storage heating entails calculating the 24-hour heating requirement of a room on a design day, taking into account:

- the heat loss of the structure,
- the air infiltration loss,
- the effectiveness of heat gains to the room from lighting, equipment and people,
- the heating period,
- the thermal mass of the structure,
- the approximate percentage of direct acting heating to supplement the storage heating required for both economic considerations and accurate control.

The calculation of heat losses together with guidance on limiting U values required to comply with current Building Regulations has been dealt with Chapter 4.

In order to evaluate the effectiveness of heat gains, the variation in infiltration rates in various structures and the required heating period, a series of empirical data have been derived from field measurements. These are known as 'z' factors and typical factors are presented in Table 8.1. The z factors will vary from room to room as the heat gains vary depending upon the use. As the level of insulation increases the heat loss of the room will reduce and the effect of heat gains will become greater, thus reducing the z factor.

Storage-based heating systems in dwellings are designed typically to provide between 80 per cent and 100 per cent of the room heating requirements for the season. Hence some supplementary direct heating is usually also installed to give supplementary boost when required, e.g. on cold evenings or fast warm up.

Factors known as the 'y' factors, that describe the proportion of heat met by storage on a design day are given in Table 8.2.

Table 8.1 Values of z for storage heating systems

	Dwelling fabric heat loss		
Room type	*50* W/m^2	*35* W/m^2	*25* W/m^2
Living room	0.7	0.5	0.42
Dining room/study	0.75	0.6	0.6
Kitchen-diner	0.5	0.4	0.34
Hall/landing	0.45	0.36	0.3
Bedroom	0.8	0.64	0.54

Table 8.2 Values of y

	Proportion met by storage on a design day (y)	
Seasonal day energy (SD) %	*Living and dining rooms*	*Hall and bedrooms*
0	1.00	1.00
5	0.87	0.78
10	0.81	0.68
15	0.77	0.61
20	0.73	0.55

The y factors are the relationship between design day stored energy (off peak) and the seasonal day energy (peak). The y factors have been derived empirically from data collected over a number of years from field trials.

In a typical dwelling

For living areas the proportion met by storage on a design day per cent $(y) = 1 - 0.06\sqrt{SD}$
For halls and bedrooms the proportion met by storage on a design day per cent $(y) = 1 - 0.10\sqrt{SD}$
where SD is the seasonal day energy (off peak) percentage required.

This means that on very cold days a higher proportion of direct acting heating will be needed to maintain comfort temperatures. It should also be remembered that some of the direct acting energy would be provided during the off-peak periods as well, thus contributing to the overall seasonal off-peak energy usage.

Storage systems are not recommended for dwellings with chimneys, unless the chimneys have an effective restrictor fitted, because of the substantial amount of heat that can be lost to the atmosphere.

The simple formula for calculating the 24-hour heat requirement (kWh) from the storage heater using off-peak electricity for each room is therefore:

24-hour heat requirement (kWh) $= 24Qzy + c$,

where

- Q is the design day heat loss in kW;
- z is the heat gain and infiltration factor;
- y is the seasonal on-peak energy factor (used when supplementary 'on-peak' heaters will also be installed i.e. – if all off-peak $y = 1$);
- c is the allowance for chimneys (kWh) (this could be between 1 and 8 depending on whether there is an outside air vent into the room with a chimney).

The unit rating in kW would be the 24-hour heat requirement divided by the charge period (normally 7 hours).

An additional allowance is required to account for the warm air drift from rooms with storage heating to the rooms without storage heating. This allowance is to be added to the hall storage heating requirement in homes with cellular rooms, and to the storage heating requirement in the room from which the stairs rise in an open plan home. In two storey homes the warm air drift is 50 per cent of the heat requirement of rooms opening onto the hall and landing. In single storey homes the allowance is 25 per cent. If the storage component is provided by underfloor heating, then an allowance must be included for the additional small amount of heat that is lost downwards through the floor.

Underfloor heating

It is not intended to cover in detail the design and installation of underfloor heating systems in this section. There are numerous publications which provide good detail design guidance but it should be noted that they all focus on water-based systems, which is covered in Chapter 9. These include BSRIA Application Guide AG 12/2001 and parts 2 and 4 of BS EN 1264 describe the method of determination of the thermal output and the installation of underfloor heating systems. At the time of writing this edition there does not seem to be a definitive reference work on electric underfloor storage heating and therefore designers generally rely on the knowledge of manufacturers. The following paragraphs provide an overview of the main considerations when designing this type of system.

For floor heating systems the floor surface temperature should not exceed 29°C for normal occupied spaces (a higher temperature of 33°C is used in bathrooms). This will obviously compromise the storage capacity of the floor; however the space heat requirements in the afternoon, towards the end of usage period will be significantly less than those at the start of the discharge period due to the outside air temperature being above design and internal gains from people, lights and equipment so additional 'top–up' heat from on-peak electricity may not be required. It is worth noting that the heat output from any underfloor heating system is limited to approximately 100 W/m^2 as the surface film heat transfer coefficient is limited to approximately 10–11 W/m^{2}K. (e.g. 19°C in space and 29°C floor temperature).

As stated earlier the successful application of underfloor heating depends to a large extent upon the *swing* in space temperature during the discharge/occupied period. This swing, which is a function of the thermal properties of the building and in particular those of the heated floor, should not exceed about 3–4 K. This means that if it is desired to maintain an average of 19–20°C, then the temperature will be up to approximately 21°C in the morning which will be acceptable but could fall to 18°C in the afternoon/evening which without the influence of internal gains would result in complaints. In domestic premises, some form of top-up direct electric heating is therefore usually required for evening use.

One of the main disadvantages with storage underfloor heating is that it responds slowly to change in demand and requires careful control to provide satisfactory and economic heating performance. It is therefore most suited to buildings/spaces where there are no sudden changes in internal gains.

In dwellings or other buildings with large glazing areas it is often necessary to use what is known as 'rim or perimeter zone' heating. Here a secondary cable is installed in a room to increase the heating close to the windows. The 'rim zone' cable is controlled by separate sensors in the floor and room. The floor sensor is used to limit the amount of heat provided to the perimeter zone.

Recent advances in the development of the use of PCMs in the floor construction can significantly increase the storage capacity of the floor and possibly eliminate the need for 'top–up' heating in the evening. The materials can be integrated into relatively lightweight buildings to increase the 'thermal mass'. This is demonstrated in Fig. 8.1 which shows the latent heat capacity of a proprietary material that has been recently developed for use with underfloor heating systems. The main disadvantage of using PCMs is that they are produced by only a few manufacturers and therefore their cost at the present time is high. If the correct PCM is selected it can also reduce summertime overheating and minimise fluctuations in room temperature by absorbing heat during the day and solidifying at night (if used in conjunction with a nighttime 'free cooling' strategy.

Typical applications of PCMs are indicated in Fig. 8.2 and include:

- The material encapsulated into a strip which is laid on top of the insulation and cast into the screed along with the electric heating cable.
- Granules laid on top of the heating cable and thermal insulation.

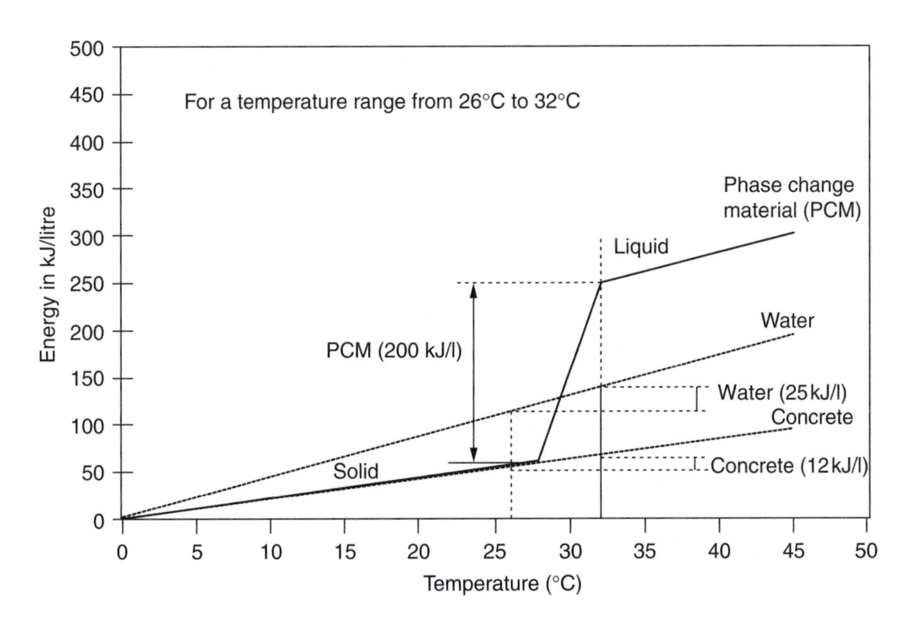

Figure 8.1 Latent heat of typical material that changes phase between 26°C and 32°C

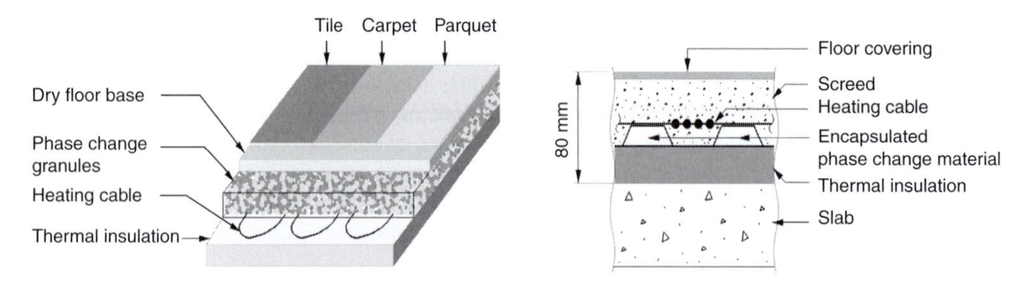

Figure 8.2 Examples of use of PCM with electric underfloor heating

Table 8.3 Maximum upward heat emission from floors charged over 7 hours

Position of floor (see Fig. 8.3)	*Maximum upward emission* (W/m^2)			
	Bare floor		*With carpet*	
	50 mm screed	*100 mm screed*	*50 mm screed*	*100 mm screed*
Ground floor A	17	28	20	31
Ground floor B	31	55	42	68
Intermediate floor	23	39	30	48

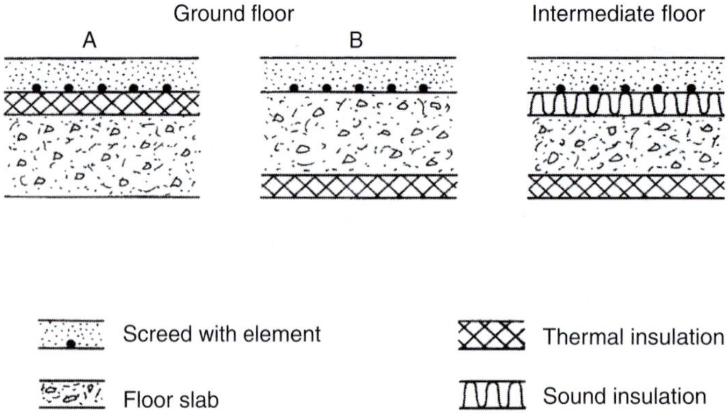

Ground floor
A
B
Intermediate floor

Screed with element
Thermal insulation
Floor slab
Sound insulation

Figure 8.3 Alternative positions for heating elements in floors

Finally, it is necessary to consider the construction of the floor which is to be heated and the floor finish which is to be applied. The thickness and position of any insulation are both critical to performance and affect directly the maximum upward emission which is tolerable. Table 8.3 lists these maximum values for the alternative positions for heating elements shown in Fig. 8.3. Table 8.4 gives nominal correction factors to be applied to the required heat input to account for different finishes (ref BSRIA application Guide AG12/2001). It can be clearly seen from these factors that applying carpet significantly reduces the output of an underfloor heating system; this reduction can be up to 50 per cent.

Once the upward emission has been determined, the loading of the heating element may be established by adding an allowance for the edge loss around the perimeter and downward loss to earth for a ground floor. The edge loss is typically taken as 5 W/m run of perimeter and *BS EN 1264 Part 2* states that downward and edge losses shall not exceed 10 percent of the total calculated losses.

Table 8.4 Correction factors for different floor finishes (ref BSRIA application guide AG12/2001)

Thickness (mm)	*Natural stone*	*Synthetic floor*	*Berber carpet*	*Velvet carpet*	*Parquet*
3		1.09	1.17	1.10	
4		1.12	1.22	1.13	
5	0.95	1.15	1.27	1.17	
6	0.96	1.18	1.32	1.20	
7	0.97	1.21	1.37	1.24	
8	0.98	1.24	1.42	1.27	
9	0.99	1.27	1.47	1.31	
10	1.00	1.30	1.52	1.34	1.30
15	1.03			1.50	
20	1.06				

The design process is completed by calculations, first to ensure that the temperature at the plane of the heating element is not excessive bearing in mind the properties of the chosen element and second to determine the diurnal swing in room temperature in order to establish that this does not exceed 3.5 K, as discussed previously.

In conclusion, it is worth re-emphasising that since the operational characteristics of underfloor heating are relatively critical, it is necessary to pursue the full routine of design calculations although these are somewhat tedious, particularly for multi-storey buildings. The notes here, are therefore no more than a summary of the necessary steps.

Capacity of central stores

As discussed earlier the use of electrically heated central stores is negligible in modern building due to the high cost of electrical energy (even with off-peak tariffs) and resulting high CO_2 emissions. The following notes describe solutions applied in the past, however these cover important design concepts and are included to give the reader an appreciation of the technologies that may, in the future, become more popular as there is a swing towards generation of electricity by low or zero carbon technologies. There will also be advances in the use of PCMs that, when integrated into stores will result in reducing the physical size of central plant and standing losses.

The principal differences between the capacity required of a room store and that of a central store are that it is necessary in the latter case to take account of distribution losses: this penalty may be offset in some instances however, by allowing for any diversity in the primary demand for output, room by room. A further penalty, of more significance in most cases, is that static losses by both night and day may be truly lost if the store is sited outside the spaces to be heated.

Warm air units

The required capacity for such units is determined in the same way as that for storage fan heaters except in circumstances where a ventilation requirement exists and is met by adapting a unit to introduce an outside air supply. In those circumstances, the design heat requirement would be increased during the hours of occupation and the unit output rating would have to be adjusted to suit. Some units have elements rated in excess of the output required to charge the store: the additional capacity may then be used to provide preheat to the spaces served, still using the low-cost supply.

Dry core boilers

The particular characteristics of stores of this type, in conjunction with those of the connected indirect system, lead to a departure from the practice previously noted in that the capacity of the store is normally related to a charging period of about 5½ hours. This period is determined not by the time when a reduced-price supply is available, which is presumed to remain at 7 hours, but in consequence of the capability of the unit to produce a direct output to the connected system. This output is used during the last 1½ hours of the low-cost supply period to provide preheating to the space served without drawing upon the capacity of the store.

Wet core boilers

As for dry core units, the characteristics of both store and system lead to special consideration of their application. The routine of calculation to determine the required rating is similar to that set out above, except that the charging period is commonly reduced to 4½ hours in this case and rather more allowance is made for cold weather and evening requirements for heat to be dealt with by direct methods. Each of these approaches reduces the volume necessary for storage.

Thermal storage cylinders

In this case, where purpose-designed cylinders are used to supply indirect systems in a large building, the method of calculation is a little different. A careful assessment of the true 24-hour load must be made, system by system and hour-by-hour. To the total thus established, it is necessary to add an allowance to take account of the 24-hour static heat losses from the vessels.

Determination of the net storage capacity required is based upon the assumption that *stratification* will exist within the vessels, temperature 'layers' remaining relatively undisturbed as discharge takes place. Hence, reference is made to the temperature at which the store is to be maintained and the volume-weighted average temperature of that returning thereto from the indirect system.

Equipment for room stores

When first introduced storage radiators were clumsy and took up a disproportionate amount of room floor space.

Two particular changes in practice were adopted by most manufacturers of *dry core* storage equipment; the first relating to the material from which the pre-cast storage blocks were made, either a high-density refractory or an iron oxide (*Feolite.*) The second is the use of opacified microsporous silica panels (*Microtherm*) for insulation: this material has a thermal conductivity of only 0.030 W/m K at 800°C, and is thus about 3 times as effective as any material previously used (Fig. 8.4).

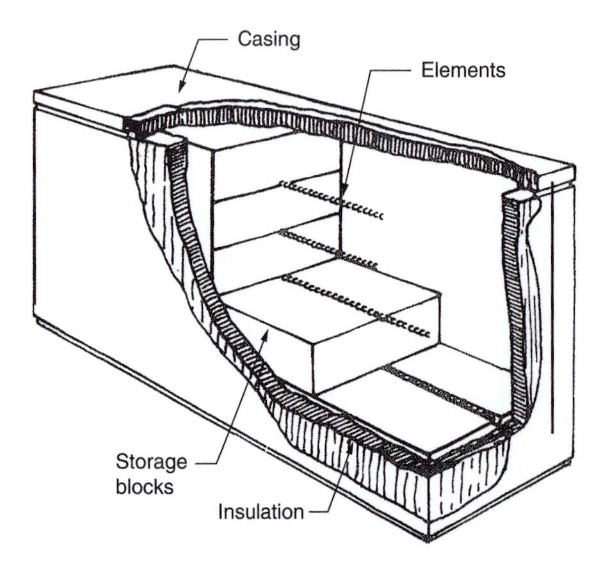

Figure 8.4 Block storage radiator (1979 type)

Storage radiators

Units of this type enable energy to be stored in the space to be heated and in that respect could be said to be 100 per cent efficient. Lack of control of heat output in most patterns, however, erodes this advantage. Such a radiator, as shown in Fig. 8.5, comprises a number of sheathed elements enclosed within blocks of either refractory or Feolite, to form the heated core. This core is surrounded by insulation material which may be fitted in contact with the exterior casing or, in some more advanced patterns, held away from it to provide an airway.

Output is both radiant and convective in almost equal proportions and Fig. 8.6 illustrates the 24-hour pattern of output, the *half-life* indicated being the point in time when output has fallen to 50 per cent of the maximum.

Ratings vary with different makes but are usually 1.7, 2.55 and 3.4 kW, the seemingly odd figures being related to rounded 7-hour charge acceptances of 12, 18 and 24 kWh. Those types provided with airways and a damper may provide control of up to about 20 per cent of the convective output.

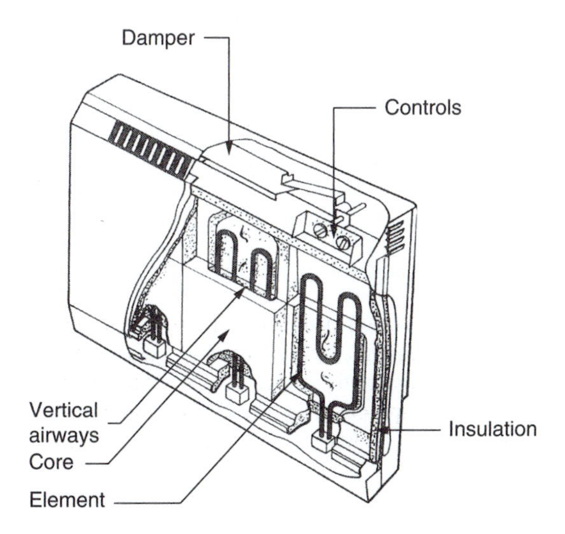

Figure 8.5 Storage radiator (Dimplex)

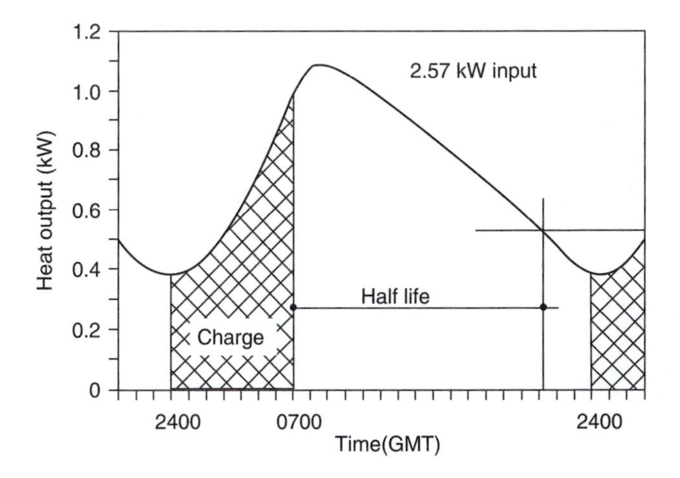

Figure 8.6 Typical 24-hour output pattern from a storage radiator

Storage fan heaters

In most respects these are similar in construction to storage radiators, the differences relating to increased insulation, the design of air passages in the core and the addition of a small fan. In order to provide a uniform air-outlet temperature, irrespective of the store remaining, the casing may incorporate a chamber in which heated air from the core is mixed with room air via a thermostatically controlled damper. Output is about 75 per cent convective.

Heat output ratings are in the range of 2–7 kW, from heaters of the modern type illustrated in Fig. 8.7. The units are well insulated and designed to minimise the heat output when the fan is not running.

It is of interest to note that, in recognition of the fact that the capacity of all types of room storage heaters from storage alone is usually inadequate to cater for the heating requirements of a peak winter's evening, most manufacturers provide a facility for evening 'top-up'. This may take the form of either an additional element or of some means whereby the normal controls to one or more of the core elements may be overridden. The purpose in each case is to provide instant heat output which is quite independent of the off-peak charged supply.

Control of room storage heaters

Storage heaters are normally provided with a *charge controller*. In its simplest form, this comprises a manual control which is set during the previous evening to regulate the energy input overnight. Since, however, it is the next day's weather which needs to be sensed, the results produced can be somewhat inexact unless some means of weather sensing or weather prediction control is incorporated into the facility.

Modern units can be provided with one of the following means of charge control:

- Automatic control where the rate of fall during the night is detected and this brings on the charge accordingly. This is usually incorporated into each heater.
- Weather sensing control which reacts to measured indicators of the next day's requirements such as the present indoor and outdoor temperature. This can be part of each heater or can be applied to control a group of heaters.
- Weather prediction control. This is often offered by the electricity supplier to alter the charging period for all the storage heaters to suit the weather forecast.

Heat output control of storage radiators is possible only when they are of a type which incorporates means to admit and regulate air flow over the core. The level of regulation available is significantly improved in the case of storage fan heaters, the power-assisted component of output then being controlled very easily via a room thermostat and the fan motor. A facility for multi-speed fan operation usually also exists in addition to provide 'low', 'normal' and 'boost' outputs.

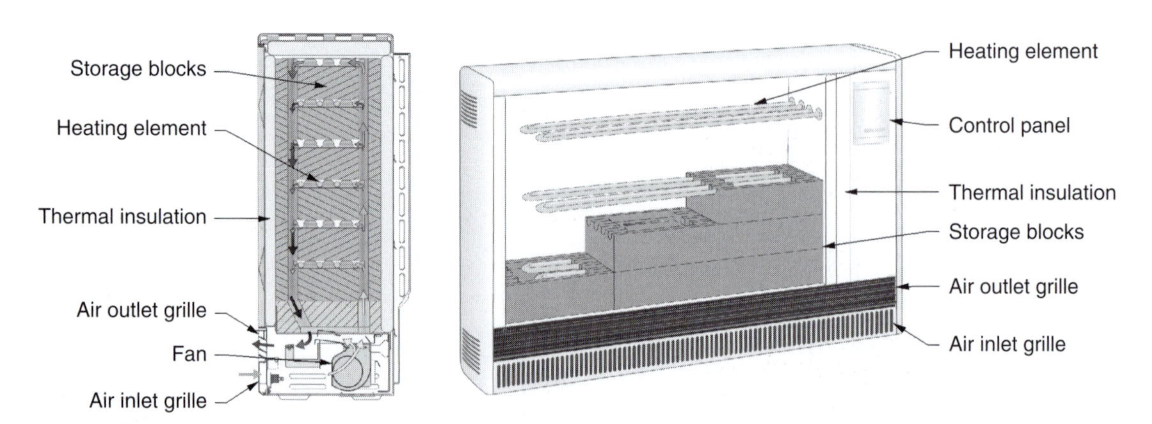

Figure 8.7 Modern storage fan heater (Stiebel Eltron)

Control of underfloor heating, subsequent to the charging period, is not possible but, perhaps fortuitously, some inherent adjustment exists. For example, taking a mean floor surface temperature of 25°C, and an operative temperature in the heated space of, say 18°C, the difference is then 7 K. If the room temperature were to rise by 2°C, then the notional temperature difference would be only 5 K and output from the floor would have fallen (theoretically) by approaching 30 per cent. It is a difficult task nevertheless to explain this situation in an office which is overheated by unseasonable solar radiation in January and yet retains a 'hot' floor!

Equipment for central store

Warm air units

As previously noted these units are very rarely specified for either domestic or commercial applications. They were originally manufacturerd by GEC. Spares and a very small volume of new units are currently manufactured by Hydburn Engineering Services. The original trade GEC trade name of 'Electricaire' is still used however. Such units are no more than large versions of storage fan heaters, arranged to serve as a heat source either to an extended space or to multiple rooms via an indirect system of air-duct distribution. Figure 8.8 illustrates one model where the heat store is mounted above the fan, mixing box and outlet plenum chamber.

Typical equipment of this type will have a range of input ratings between 6 and 100 kW. This is not inconsiderable floor loading imposed by the larger sizes should be noted as should the fact that a three-phase power supply may sometimes be required.

Dry core boilers

This type of heat store is not dissimilar in principle from that provided in a warm air unit having a similar rating, although greatly refined in detail as to configuration.

The Feolite blocks which make up the storage core are disposed about a central *hot draught tube* and have vertical passages formed in them to accommodate the heating elements: they are contained within a casing which is lined with two layers of 'Microtherm' insulation. The core assembly stands upon a substantial layer of insulating blocks which isolate it thermally from the plenum chamber forming the base of the unit.

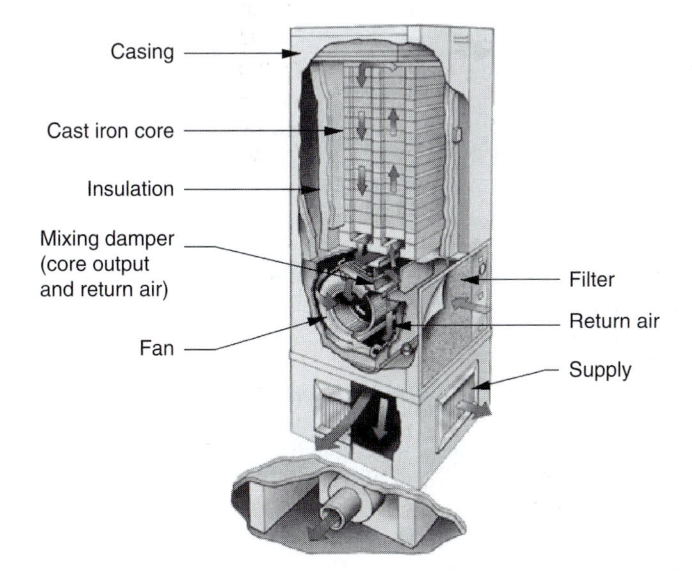

Figure 8.8 Warm air 'Electricaire' unit (Hyndurn Engineering Services)

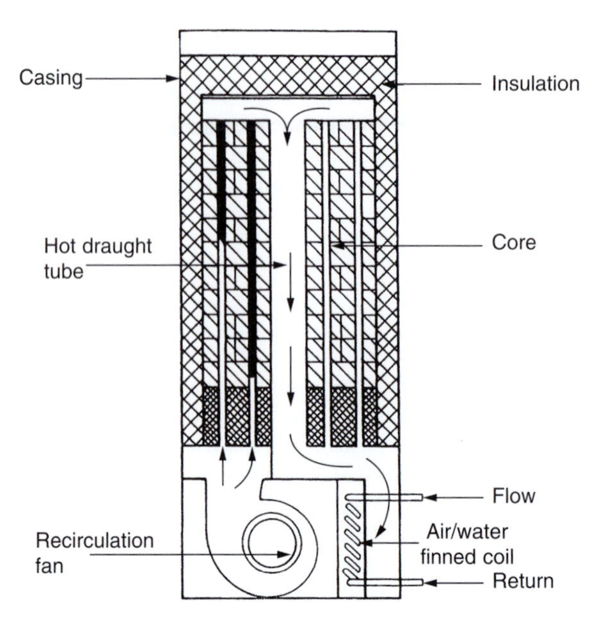

Figure 8.9 Old style dry core boiler (GEC Engineering circa 1970s)

The plenum chamber is a box construction containing a fan, an air to water heat exchanger, a pump and the associated piping connections. A further insulated enclosure surrounds the complete unit and mounts the control equipment, as shown in Fig. 8.9. The elements are arranged so that separate switching of one or more is possible in order to provide a facility for use as a daytime boost on occasions when the store is exhausted.

It should be noted that the characteristics of this equipment are such that water temperatures are not very different from the traditional can be produced and that, in consequence, it may be used to replace a boiler connected to an existing hot water heating system.

Wet core boilers

In the 1970's wet core boilers were developed along the lines of the units shown in Fig. 8.10.

The units comprised packaged equipment typically comprising an insulated cylinder fitted with two banks of immersion heater elements, one near the bottom to produce the overnight charge and another near the top for use as a daytime boost on occasions when the store is exhausted. Water is stored at a temperature which is varied to suit the static head available from a roof-mounted expansion cistern, 95°C being a not uncommon level.

The cylinder was piped to a simple three-way mixing valve set to produce a water-outlet temperature of about 75°C. A pump circulates water round the heating system, the volume being restricted to ensure that water returns to the cylinder at not more than 40°C in order to take maximum advantage of the stored volume.

The modern domestic electric storage boiler typically is a packaged unit comprising an insulated open vented storage cylinder with external electric heating elements as shown in Fig. 8.11. The unit can provide both domestic hot water and space heating. A typical unit can provide up to of 35 litre/minute domestic hot water, raised through 35°C directly from the mains via an external plate heat exchanger, with up to 90 per cent of its energy usage taken at off-peak rates.

In the future it is likely that PCMs may be incorporated into storage cylinders. This would result in either much greater storage capacity for the same volume or significantly reduced physical size and thermal losses to achieve the same storage capacity. A typical PCM with a melting point in the range 86–90°C will have a latent heat of fusion of approximately 150 kJ/kg. It can be shown that the physical size of the required heat storage tank which incorporates a 50/50 mix of PCM and water, could be in the order of 20 per cent of the size of a water storage tank.

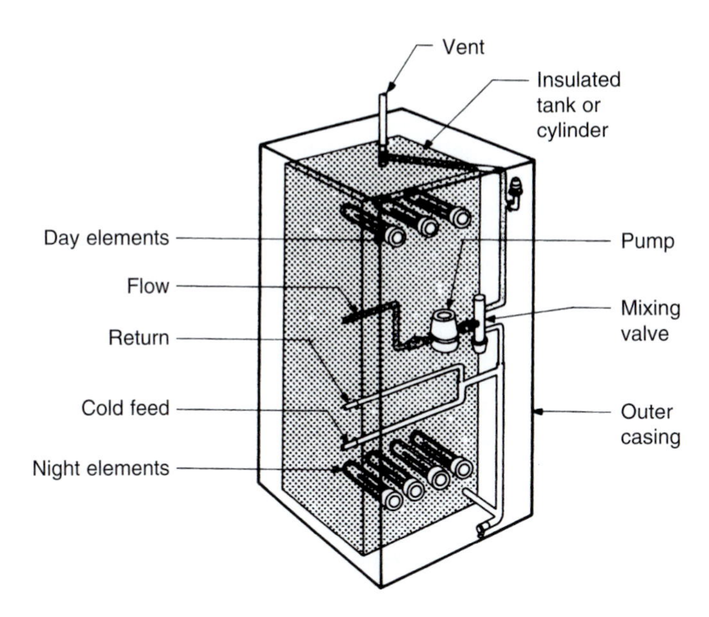

Figure 8.10 Old wet core boiler (circa 1970s)

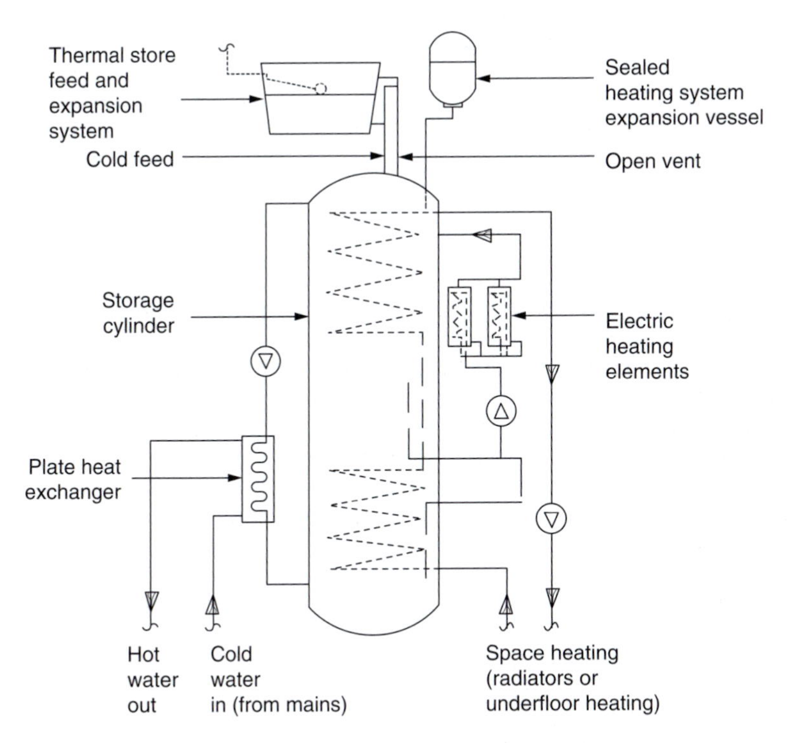

Figure 8.11 Modern electric storage boiler (Gledhill ElectraMate A-Class)

Thermal storage cylinders

These, of course, represent the traditional method of storing heat which has been made available at a cheaper overnight rate. Such vessels were used in systems designed and installed 60 or more years ago and stored energy at relatively high temperatures and pressures. This classifies the systems as 'Pressure systems' in accordance with current regulations and as a result they require additional maintenance, have significantly greater health and safety issues to be overcome and require regular insurance inspections. These issues when combined with the high running costs, thermal losses from the storage vessels and high CO_2 emissions for using grid generated electricity, have resulted in the systems not being employed in modern buildings. There will be existing buildings where this type of system is still in use and the following notes and diagrams are therefore included to demonstrate the principles.

The principal differences between a store of this type and those which are outlined under the previous heading are first: the pressure and, thus, the temperature at which the hot water is stored; second the manner in which electrical energy is transferred to the water and third, a corollary, the overall size of the system. Figure 8.12 illustrates a cylinder arrangement suitable for a smaller installation, where banks of immersion heaters are used, and Fig. 8.13 shows a complete installation incorporating cylinders in association with an *electrode boiler*.

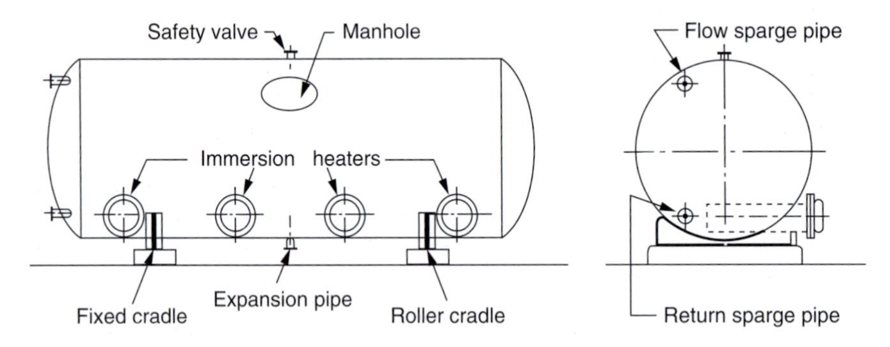

Figure 8.12 Immersion heater-type storage vessels

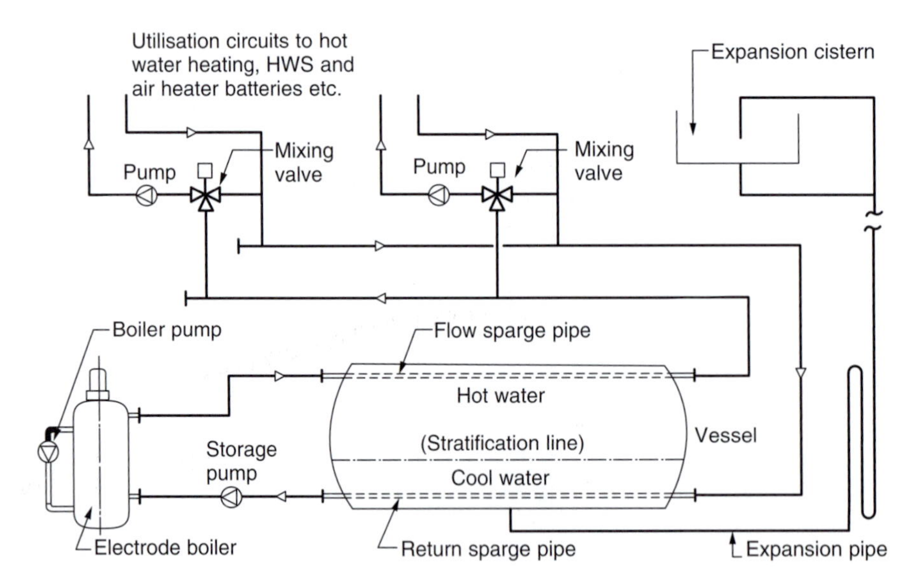

Figure 8.13 Diagram of a large water storage system

Since, for a given heat output, the volume of storage required depends principally upon its temperature, it follows that the design criteria are interactive, i.e. the pressure which can be imposed upon the store and the space which can be made available for the vessels. Early systems made use of the height of the building, and the siting there of an expansion cistern to produce the necessary pressure, but more recent practice has been to use some external means (see Chapter 12) for this purpose. The temperature at which water is stored is normally about 10–17 K less than that at which steam would be produced.

Storage cylinders are almost always mounted horizontally, the size and shape being determined by site conditions. Diameters up to 4 m are common, with lengths of up to 10 m. Loss of heat is reduced as far as possible by insulation, up to 100 mm of glass fibre against the vessel being covered in some instances by 50 mm of dense material prior to a final casing with polished sheet metal. The cradles upon which vessels are mounted, fixed at one end and roller type at the other, stand upon a base which is insulated with hard material in order to minimise heat loss downwards by conduction.

Since effective use of storage depends upon minimising disturbance to potential stratification in the water stored, it is good practice to provide vessels with *sparge pipes* at both outlet and return connections. As shown in Fig. 8.13, such pipes should extend over the whole length of the vessel, except for gaps to permit differential expansion. A rule of thumb suggests that the area of the holes in each sparge should equate to 10 times the diameter thereof and be provided over the whole length in rows arranged to face the critical direction.

The quantity of water stored in cylinders of the size considered is such that, when heated to storage temperature, the increase in volume is considerable. For instance, when water is heated from 50°C to 120°C the volume is increased by about 4.5 per cent, and when heated to 150°C the increase is 8 per cent. It is thus necessary to make provision for this increase by allowing a *dead water space* below the return sparge pipe, equivalent to about one-eighth of the vessel diameter. It is to the bottom of this space that the feed and expansion pipe is connected.

Where pressure is applied by an elevated cistern, as in Fig. 8.14, the syphon loop prevents circulation *within* the expansion pipe and thus, in theory, the dead water space stagnates and only cold water is allowed to pass to the high-level cistern. In practice, some mixing will have taken place with the 'live' content of the vessel and there will also have been heat transfer by conduction.

It is of interest to note that although most applications for thermal storage cylinders on any substantial scale relate to purpose-designed installation, an alternative version is illustrated in Fig. 8.15. This consists of a cylinder which is arranged to retain a space above the level of water fill in order to accommodate the volume of expansion and to provide a compressible pressure cushion. Unique, but inherent in the concept, is the introduction of a heat exchanger as separation between the water store and the indirect system. The mixing valve

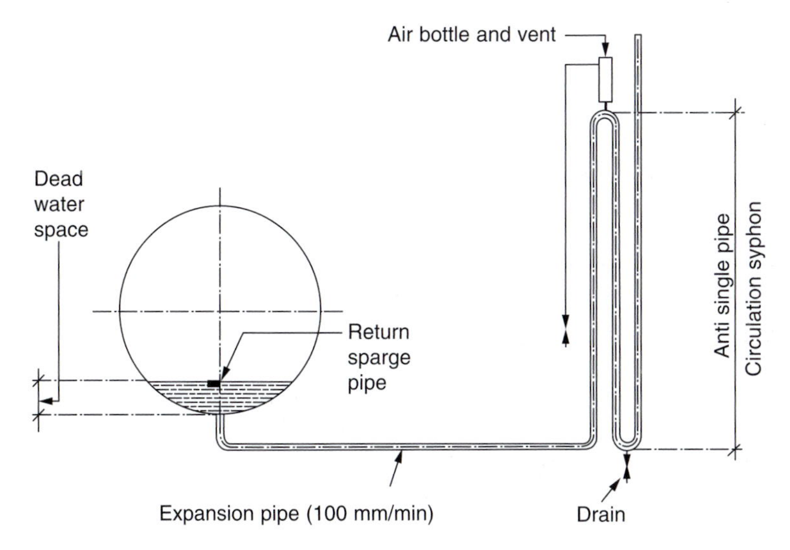

Figure 8.14 Expansion connection to storage vessel

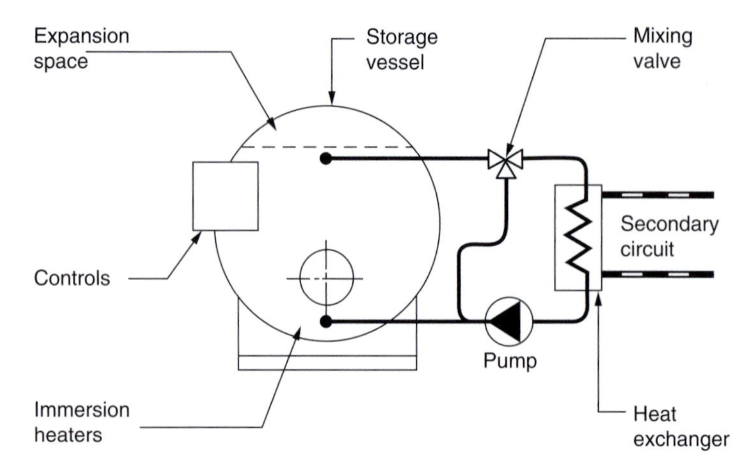

Figure 8.15 Packaged storage vessel and system (Heatrae Industrial)

shown, the variable speed circulating pump and all controls, etc. are part of the design, typically for a range of cylinder sizes from 2000 to 50 000 litre, with loadings of 36–720 kW. The standard operating criteria are 350 kPa and a charged temperature of 138°C.

Control of central stores

As with room stores, some form of *charge controller* for input is provided for central stores. In view of the increased size and capacity in this case, these are rather more complex and sophisticated and will in most instances incorporate some method for automatic sensing of outside temperature during the charging period, in anticipation that this will provide a closer approximation. It is not possible to generalise as to output control from central stores, since this will almost always be related to the connected indirect systems.

Chapter 9

Indirect heating systems

Introduction

As described in Chapter 7 on Heating methods, water (liquid), steam (vapour) and air (gas) are the three mediums most used in indirect heating systems for HVAC systems.

Water is the medium used in over 95 per cent of new and existing buildings and is categorised by the flow temperature (i.e. low, medium or high; see Table 9.1). Each of these categories is discussed in more detail later in this chapter.

The use of steam as a distribution medium is largely limited to process and humidification applications, and is very seldom used directly in heat emitters for space heating due to the significant additional maintenance requirements, water treatment requirements, insurance requirements, and safety issues associated with the distribution media (i.e. high temperatures and pressure – typically 150°C and 4 bar gauge). Where a building such as a laboratory has a high steam requirement for humidification and process heating for autoclaves and washing machines, the steam supply is often used to generate low-temperature hot water via plate heat exchangers installed in the plant room. The low-temperature hot water is then pumped to heat emitters in the occupied spaces. This limits the steam and condensate return distribution pipework to the plant room and process areas, hence localising the maintenance requirements of the steam system.

Where high or very high temperatures are required for process application such as food processing and laundries, other 'thermal fluids' are often used as an alternative to steam. These fluids have the advantage that once installed, their maintenance requirements are very small when compared with a steam system. They are oils and therefore do not require any water treatment or complicated condensate return systems and comprise a simple pumped flow and return system operating at low pressure. The fluids used include synthetic and mineral oils and can operate at temperatures up to 300°C. As this is fundamentally process heating, it is not covered in any further detail in this book, however, should the reader require further information one of the specialist manufacturers such as Babcock Wanson or Fulton should be contacted.

The use of 'all air' systems for heating a building is generally not used today unless the heating function is part of an air-conditioning system that serves, for example, large conference rooms that are internal (i.e. do not have external walls to loose heat to outside). The 'norm' is to provide a ventilation system (incorporating heat recovery facilities) that delivers the quantity of outside 'fresh air' for the occupants at room temperature and install independent heat emitters in the building. This approach has many advantages including: minimising the size of the central plant, minimising the size of the distribution ductwork and minimising losses from distribution ductwork. Also, the ventilation plant would not be required to operate until the building was occupied.

Water systems – characteristics

Water as a heating medium offers many advantages: among the various heat transfer liquids, it has the highest specific heat capacity and the highest thermal conductivity. The properties of water at a number of temperatures are given in Table 9.2.

Table 9.1 Design temperatures for water heating systems (as defined by HVCA TR20 in the UK)

System	*Design temperature*	
	Flow from boiler or water heater (°C)	*Differential at water heater or boiler* (K)
Low-temperature hot water	Up to 90	5–20
Medium-temperature hot water	90–120	25–35
High-temperature hot water	>120	45–65

Table 9.2 Selected properties of water

		Re mass		*Re volume*	
Temperature (°C)	*Absolute vapour pressure* (kPa)	*Specific mass* (kg/m^3)	*Heat capacity* (kJ/litre K)	*Specific volume* (litre/kg)	*Heat capacity* (kJ/litre K)
10	1.23	999.7	4.193	1.000	4.191
40	7.38	992.2	4.179	1.008	4.146
60	19.92	983.2	4.185	1.017	4.115
80	47.36	971.8	4.198	1.029	4.080
100	101.33	958.3	4.219	1.044	4.043
150	475.97	916.9	4.322	1.091	3.961

It is the temperature to which the water is raised in the boiler or other energy conversion plant which dictates the type of system which may be selected. Table 9.1 lists the three temperature ranges used in the UK, as defined in the Heating and Ventilating Contractors Association (HVCA) Guide TR20 on the installation and testing of pipework systems. It is worth noting here that the ASHRAE (American Society of Heating, Refrigeration and Air Conditioning Engineers) use different temperatures to define water heating systems (up to 120°C for low-temperature, 120–175°C for medium-temperature and above 175°C for high-temperature systems).

The majority of new buildings use low-temperature hot water because these systems are simple to install and safe in operation as they operate well below 100°C. Descriptions of medium-temperature hot water and high-temperature hot water systems are included in this chapter, but these systems are generally not used in modern buildings and are therefore included here for historical purposes. They are generally only worth considering if there is a requirement for the transportation of heat over long distances where a significant reduction in pumping power can be realised and much smaller pipework used due to the smaller mass flow rate resulting from the larger drops in water temperatures (e.g. 30–40°C compared to 10–20°C for low-temperature hot water). They are significantly more costly to install and systems where the temperature is above 110°C, fall into the same category as steam systems in the Pressure Systems Safety Regulations 2000, therefore requiring regular inspections by a competent person in accordance with a 'Written Scheme of Examination'.

Low-temperature hot water

In previous editions of this book there was a reference to low-temperature "warm water" that covered water supply temperatures between 40°C and 70°C. In this edition the reference to this category has been omitted and a single definition of low-temperature hot water used, but covering the range 40–90°C.

Historically, the industry norm was to use a flow temperature of 82°C and a return temperature of 71°C. (These temperatures correspond to Fahrenheit temperatures of 180 and 160 to give a nominal 20°F rise across the boiler.) The flow temperature provides a reasonably wide margin with respect to boiling point at atmospheric pressure and also leaves some tolerance within that margin for elevation to perhaps 90°C under subnormal weather conditions, or for fast pre-heat of the building.

Water between 40°C and 70°C was historically used for circulation through embedded panel heating coils or underfloor heating systems where it was necessary that the flow temperature of the water supplied should not be more than about 53°C for ceilings and 43°C for floors. Again historically, conventional (non-condensing) boilers were generally used and these needed to operate at flow temperatures in the region of 80–85°C and have a return temperature of between 60°C and 65°C to operate efficiently and prevent corrosion due to condensation of flue gases at the back end of the boiler.

The low flow temperatures were therefore established by two methods:

(1) Employing an injection-type circuit where the higher-temperature flow from a conventional boiler (operating with a flow temperature of between 80°C and 85°C) is mixed with cooler water returning from the system via a three way valve.
(2) By hydraulically separating the higher-temperature primary circuit from the lower-temperature flow to the heat emitters via a heat exchanger.

Today, however, all new domestic properties must use a condensing boiler to comply with the 2006 edition of Part L of the Building Regulations and although it is not a requirement of these regulations to install condensing boilers in non-domestic buildings, a high proportion of new buildings are being designed with heating systems using condensing boilers. Here the return water temperature needs to be less than 55°C to operate in condensing mode and between 30°C and 40°C to operate most efficiently. Most modern condensing boilers are designed to operate with a temperature rise through the heat exchanger of 20°C, resulting in a flow temperature ranging from 50°C to 70°C, depending on the application, to achieve condensing at all times. (Further information on the different types of boilers is given in Chapter 13.) It can therefore be seen that the over-complication of injection circuits or heat exchangers are generally not required when condensing boilers are used, and in many applications the boiler flow temperature is simply weather compensated to maximise condensing (i.e. 70°C flow at −3°C outside to approximately 40°C at 15°C outside).

Over the next few years there will also be a dramatic increase in the use of ground source heat pumps (that are covered in detail in Chapter 14). They operate most efficiently when producing water at a flow temperature of between 35°C and 40°C but can operate up to a flow temperature of 50–55°C (but at significantly reduced COPs). They can therefore be used in conjunction with condensing boilers to deliver low-temperature hot water for some applications.

Most buildings do not use underfloor heating or embedded ceiling panels, but instead use conventional radiators or convectors, or alternatively are heated by 'air-conditioning units' such as fan coil units or chilled beams. The effect of using water flow temperatures significantly lower than the conventional 80–85°C is that the surface area of the heat emitter has to increase as the water temperature falls. This can result in large radiators as indicated in Table 9.3 (e.g. flow/return temperatures of 65/45°C will result in a mean water temperature of 55°C, if the room is heated to 20°C, then the temperature differential between the air and water will be 35°C – with the result that a typical radiator would need to be approximately twice the surface area).

The above problems can be addressed by installing emitters sized on a higher flow temperature and running the boilers at this temperature only to preheat the building from cold on a winter's morning. When the building is up to design temperature and occupied, it will benefit from heat gains from the people, lighting and

Table 9.3 Variation in size of heat emitting equipment with water temperature

	Comparative length of heat emitting unit for following temperature differentials (K) water to air at 20°C				
Type of heat emitting unit	*55*	*45*	*35*	*25*	*15*
Primarily radiant	1	1.29	1.79	2.83	5.46
Primarily convective	1	1.35	1.95	3.31	7.14

computers. Therefore in most cases it should be possible to automatically adjust the flow temperature from the boilers as soon as the building is occupied, based on both the outside and internal air temperatures. This will ensure the boilers operate in condensing mode for as much of the year as possible and any heat from ground source heat pumps can be utilised effectively. It will also ensure that the emitter physical sizes and associated costs are minimised.

The water in a heating system expands when the temperature is raised. There are two ways of accounting for this increase in volume (refer to Chapter 11 for further details).

Installation of a feed and expansion system

This method is now mainly only seen in domestic installations and comprises a water tank installed above the highest point in the system (refer to Chapter 11 for details). The purpose of the feed and expansion cistern is to receive the additional volume when the system is hot and return it when the system cools down. In following this cycle, the water content remains relatively unchanged and the scaling or corrosion which might arise from the admission and rejection of fresh water each time is largely avoided.

Installation of a packaged pressurisation unit with expansion vessel to form a 'sealed system'

This is the preferred method in almost all commercial buildings and comprises two parts: a water break tank, pump and control system to automatically fill the system and maintain pressure at a 'cold fill' set point, plus an expansion vessel which comprises a metal container divided in two by a rubber diaphragm. One side is connected to the pipework of the heating system, the other, the dry side, contains air or nitrogen under pressure with a valve for checking pressures and adding air. When the heating system is empty or at the low end of the normal range of working pressure, the diaphragm will be pushed against the water inlet. As the water expands so the diaphragm moves, compressing the air and gives rise to an increased pressure that will be seen on the pressure gauge.

The volume of expansion at a range of temperatures may be found from Table 9.2 since the mass remains constant. Thus, between 10°C and 100°C, for example, it is (999.7 − 958.3)/958.3 = 4.32 per cent or one twenty-third of the original volume. Manufacturers' literature must be consulted for the water content of equipment, since this varies greatly with pattern: Table 9.4 provides that information for pipework.

Medium- and high-temperature hot water

Detail design guidance on both medium- and high-temperature hot water systems is given in Chapter 14 of *ASHRAE HVAC Systems and Equipment 2004*. Some of the following notes describe solutions applied in the past, some of which are still in use but mainly in process applications. They are included in this edition for historical completeness.

Medium-temperature hot water

It could be argued that systems of this type originated from those which enjoyed a vogue in about 1920, when fitted with a device known as a *heat generator*. This was an item of static equipment, containing a mercury

Table 9.4 Water content of steel (BS EN 10255: medium weight) and copper (BS EN 1057 R 250/BS 2871 Table X: pipes)

Nominal pipe size (mm)	*Water content* (litre/m)		*Nominal pipe size* (mm)	*Water content* (litre/m)	
	Steel	*Copper*		*Steel*	*Copper*
15	0.205	0.145	40	1.376	1.234
20	0.367	0.321	50	2.205	2.095
25	0.586	0.540	65	3.700	3.245
32	1.016	0.837	80	5.115	4.210
			100	8.680	8.680
			150	18.95	

column and reservoir, which was interposed between the feed and expansion cistern and the boiler. The effect was to multiply the pressure available by density difference and thus permit the working temperature to be elevated accordingly.

As now understood, however, systems of this type were advocated as being a compromise between low- and high-temperature practice. As will be noted later, high-temperature systems at that time used steam boiler techniques to apply pressure; *split-casing* circulating pumps with water cooled glands and either high-quality welding or flanged joints for pipework.

In the 1950s it was suggested that medium-temperature systems could be constructed using what were, effectively, the best of low-temperature equipment and methods. The use of screwed joints was proposed and adopted for valves and other fittings, but for most pipework joints, welding of an adequate commercial standard was becoming the more economical approach. These days, in order to comply with current legislation including the Pressure Systems Safety Regulations 2000, great care must be taken to ensure all parts of the system and jointing methods are suitable for the temperatures and pressures. A further concurrent development was the introduction of the factory made *gas pressurisation sets* which will be referred to in Chapter 11 on pumps and pipeline equipment.

Most medium-temperature systems are equipped with one or other of the packaged pressurising sets which are described in more detail in Chapter 11 on pumps and pipeline equipment. However, it is possible in some instances to provide adequate pressure from a cistern elevated above the highest point of the system. A central boiler plant providing primary water to heat exchangers at basement or ground level in each of a group of buildings might be so arranged. As in the case of pressure so applied to thermal storage vessels, Fig. 11.12, it is necessary to take steps to ensure that heated water from the system does not circulate *within* the expansion pipe and rise to the cistern.

In view of the elevated surface temperature that will result if medium-temperature hot water is used, it is normal to install plate heat exchangers and then pump low-temperature hot water to local heat emitters such as radiators. This also minimises the extent of pipework distribution that will need regular inspections to comply with the Pressure Systems Safety Regulations. Emitters that are connected directly to a medium-temperature hot water system are generally limited to fan convectors/unit heaters of various patterns, radiant strip heaters and radiant panels fitted at higher levels where they are out of reach, e.g. in a factory or other large space where the use of radiant heating would be suitable.

High-temperature hot water

The father of all high-temperature systems was Perkins, who filed a patent in 1831. In this system, the wrought iron piping was about 22 mm bore (32 mm outside diameter, jointed with right- and left-hand threads) and each circuit formed one continuous coil, part of it inside the boiler with the remainder forming the heating surface in the building, Fig. 9.1: one such coil was measured as being over 180 m in length! Partial allowance for expansion of the water content was provided by means of closed vessels. Operating temperatures of 180°C

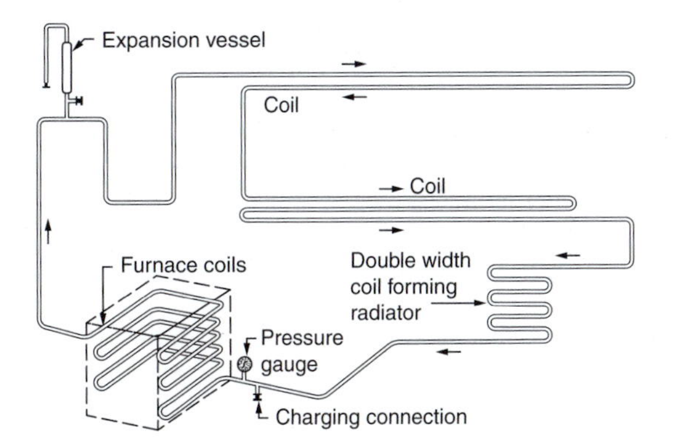

Figure 9.1 Diagram of Perkins' high-temperature hot water system (1831)

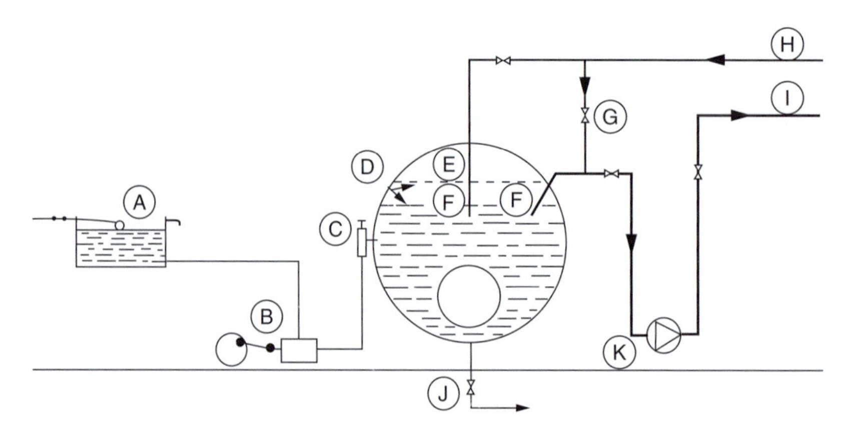

Figure 9.2 Steam pressurisation (shell boiler): A, feed cistern; B, feed-pump; C, feed check valve; D, water line (upper and lower); E, steam space; F, dip-pipes; G, cooling water bypass; H, return; I, flow; J, blowdown; K, system circulating pump

were common and, firing being by hand with no real control, frequently reached 280°C (7 MPa!) near the boiler. Some examples of the system may remain in use, some converted to oil firing, in churches and chapels built towards the end of the nineteenth century.

The basic concept of water circulation at elevated temperatures and at pressures above atmospheric boiling point was revived during the years between the wars and applied, in particular, to large industrial premises. The essential differences between this application and the Perkins system were that circulation was by pump, through conventional pipe circuits arranged in parallel, and that the working pressure and temperature were controlled at the boiler plant to more specific levels. Prior to the reintroduction of this system, the type of building to which it was particularly suited had been supplied by steam plant and arguments among advocates of the alternative methods, as to the superiority of each, continued for 20 years. The flexibility of the water system, both in the potential available to vary the output temperature to suit weather conditions and in the physical freedom to route pipework largely irrespective of site levels, were among the considerations which led to its increasing popularity. In terms of operation and maintenance, virtually all equipment needing attention is, for the water system, concentrated in plant rooms.

In order to achieve the elevated temperatures associated with these systems, the practice adopted initially was as shown in Fig. 9.2, and this arrangement still has its advocates. Steam is generated in a more-or-less conventional steam boiler and the water to be circulated is drawn from below the steam/water separation line, to be returned there also after passing through the system. The connections are made in the form of *dip-pipes* from the top of the boiler in order to avoid draining the boiler, with consequent danger, should a serious leak occur in the system.

The temperatures of both the steam and water are the same, at saturation level, and were pressure to fall in the water pipework without any parallel loss of temperature, the water would flash into steam and create an unstable condition. In order to prevent this happening, a small proportion of the water returning from the system is injected into the flow outlet as it leaves the boiler, so reducing the temperature there to below the point of boiling. Although Fig. 9.2 is no more than a diagram, it does illustrate a desirable arrangement whereby flow pipework is routed to below the steam/water interface in order to increase static pressure before connection to the circulating pump. Where the size of the installation requires that more than one boiler be installed, their respective water levels are kept uniform by use of balance pipes, one such joining the water spaces and another the steam spaces.

For much larger installations, high-velocity water tube boilers with connections to a *steam drum* to take up expansion are generally used: where a number of boilers of this type are required, a common drum or drums are used, as shown in Fig. 9.3 The water of expansion which has to be dealt with as a result of diurnal temperature fluctuations may well be contained by variations in level within the drum but when the contents of a large system are heated up from cold, discharge either to a cistern or to drain is necessary.

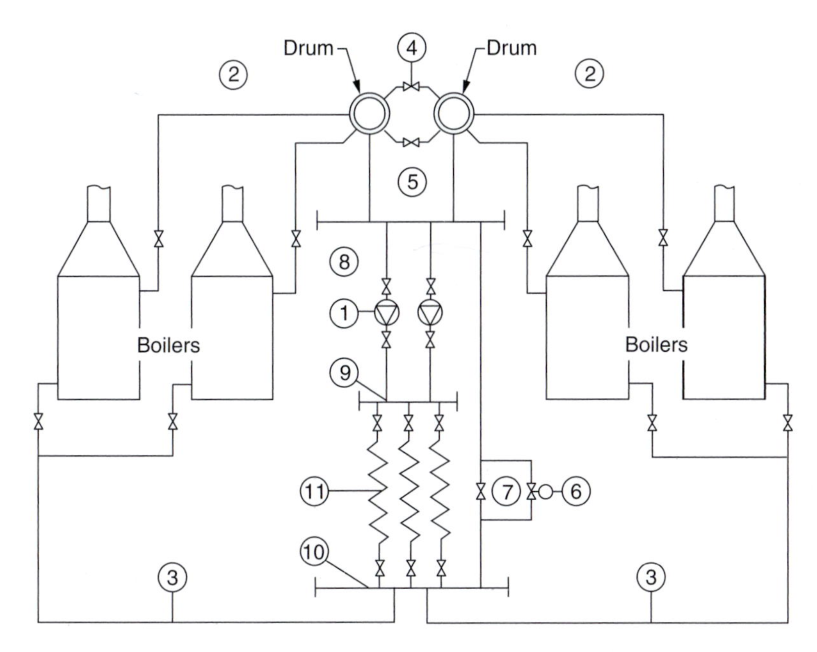

Figure 9.3 Steam pressurisation (water tube boilers): 1, system circulating pumps; 2, flows, boilers to drums; 3, returns, boilers to drums; 4, steam balance pipe; 5, water balance pipe; 6, mixing valve; 7, manual bypass; 8, pump suction header; 9, flow header; 10, return header; 11, external circuits

Make-up water for installations of this type, to replace the inevitable losses at valve and pump glands, is provided via a conventional boiler feed-pump which draws its supply from a cistern containing treated water. The boiler water level is monitored and the make-up pump is usually controlled by hand.

The principal disadvantage of steam pressurisation is that the system is inherently unstable since the steam pressure applied is so closely associated with the temperature of the water circulated. Any minor variation in the applied load will be relatively critical to performance: experienced operatives and skilled maintenance personnel are necessary if satisfactory service is to be assured. As a result, most existing high-temperature installations use either inert gas pressurisation or pump pressurisation.

Nitrogen is the most commonly used inert gas, compressed air is not recommended because the oxygen in air will contribute to corrosion in the system. The expansion vessel (pressurisation tank) replaces the steam expansion drum and is usually installed vertically to minimise the area of contact between gas and water. When the water expands and rises in the tank, the inert gas is relieved either to atmosphere or recovered in a low-pressure gas receiver. This low-pressure gas is then passed through a compressor back to a high-pressure receiver. A method of sizing the expansion vessel is given in the ASHRAE Guide described earlier.

Simplistically, pump pressurisation comprises a feed-pump and a regulating valve. The pump runs continuously introducing water into the system from a make-up tank of treated water and the pressure regulating valve, bleeds continuously back to the make-up tank. This type of system is limited to small process heating systems. For larger high-temperature hot water systems, pump pressurisation is combined with a fixed quantity gas compression tank that acts as a buffer. When the pressure rises in the buffer tank, a control valve opens to relieve water from the balance line into the make-up storage tank. When the pressure falls, the feed-pump is started to pump water from the make-up tank into the system.

As in the case of medium-temperature systems, it is the surface temperature of the heat emitting equipment which restricts selection of type. However, the higher temperatures available, permit the use of most forms of radiant surface and these will show to advantage where they can be fitted at levels beyond reach. For a building or group of buildings where heating from an available high-temperature source is unsuitable, one or other of the arrangements shown in Fig. 9.4 may be used. The second of these, in Fig. 9.4(b), it should be noted, operates at the same *pressure* as the high-temperature system.

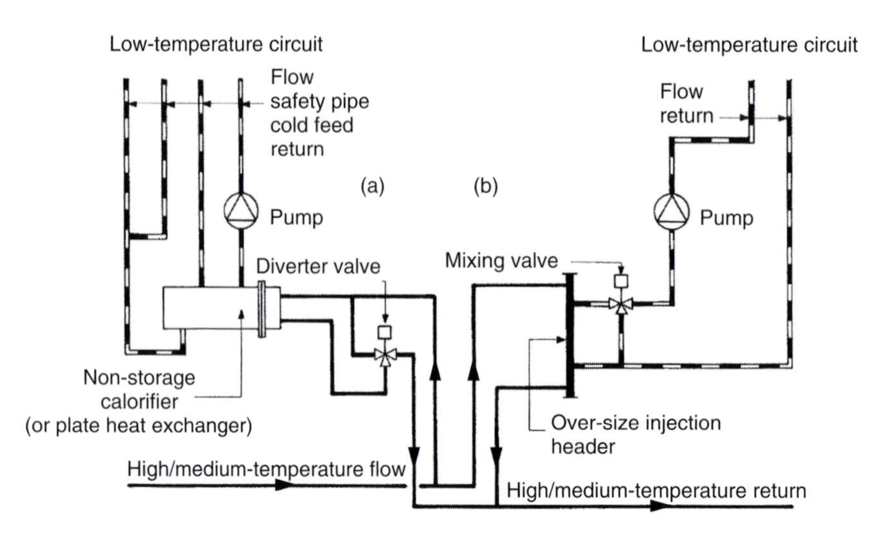

Figure 9.4 Low-temperature circuits served from a high-temperature system (isolating and regulating valves are omitted for clarity)

Low-temperature hot water systems – piping arrangements

Some of the following notes describe solutions applied in the past that are not used in modern buildings. However, these cover important design concepts and are included in this edition as there are likely to be many existing buildings with such systems installed which may remain in use from many years to come.

Although the materials used and the techniques of construction span a wide range from the smallest domestic premises to the largest industrial building, piping arrangements for systems using water as a heat distributing medium differ only in detail. A line diagram of any system will reveal that the various circuits and sub-circuits form a network of parallel paths for water flow. If the arrangement of these is kept simple and they are arranged in a logical manner, a design has a sound basis from which it may be developed further.

The diagrams included under the subsequent headings do not show circulating pumps, feed and expansion pipes, valves, or connections to pressurising equipment, etc. These important aspects will be dealt with separately.

The single-pipe circuit

These types of circuit are theoretically the simplest possible and here hot water from the boiler is fed to each radiator in turn with the cooler water from each radiator being fed back to the same pipe. Natural convection causes the hot water to rise into the radiator displacing cooler water back into the pipe. As a result the temperature of the water is gradually reduced as it enters each successive radiator. This makes the control of the distribution of heat difficult. In order to compensate for this decay in temperature, the size of successive radiators is usually increased and care is taken to select a suitable temperature drop across both of them and the piping circuit. For example, consider a system having 10 radiators, each of which is required to provide the same output. The temperature drop across the circuit might be chosen to be 10 K and that across each individual radiator, to be 15 K. The water temperature in the single pipe would fall by an average of 1 K after each radiator and thus the mean temperature of the first and the last would be 72.5°C and 63.5°C, respectively.

One could argue that this type of system is the easiest and cheapest to install, but control of individual radiators is virtually ineffective as the first radiator gets hotter than the second, etc. The emitters must be sized to offer negligible resistance and therefore have large waterways making them bulky and costly. These systems were popular 30–40 years ago and many of them are still in operation. They have been superseded by two-pipe circuits as these are generally simple to design and install, requiring smaller pipework and all commercially available heat emitters are now designed for integration into two-pipe systems.

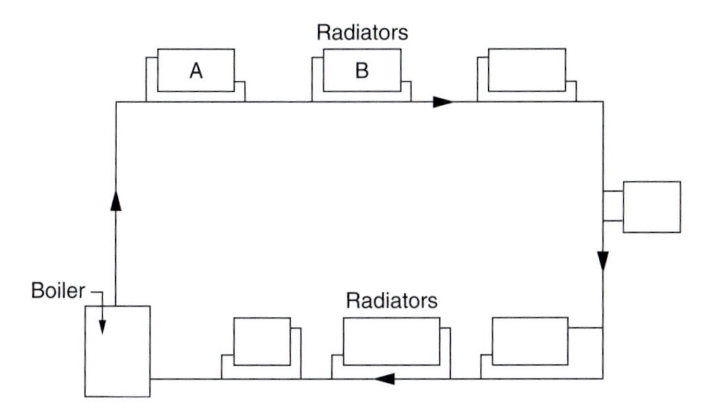

Figure 9.5 A simple, single-pipe hot water circuit

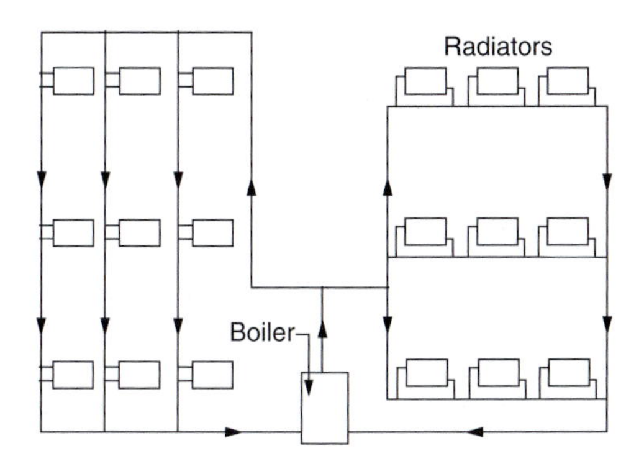

Figure 9.6 Multiple, single-pipe hot water circuits in parallel

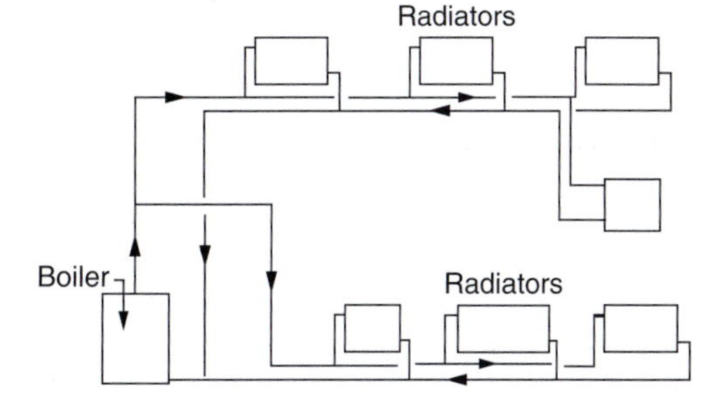

Figure 9.7 A simple, two-pipe hot water circuit

The calculations for one-pipe systems are tricky and tedious and it could be argued that one of these reasons they went out of favour is because of the failure in the past by engineers and plumbers to use simple arithmetic calculations for heating systems designs.

Examples of these systems are shown in Figs 9.5 and 9.6.

The two-pipe circuit

This arrangement has, in most respects, come to supersede the single-pipe configuration and has an inherent logic as far as parallel circuits are concerned, as shown in Fig. 9.7. Flow and return mains originate from the boiler plant and each main or sub-circuit consists of branches from them. Each branch conveys an appropriate quantity of water to and from whatever heat emitting terminals are connected to it. In an ideal world, all circulating pipework would be insulated perfectly, and the water inlet temperature at each terminal would be exactly the same as that leaving the boiler. Similarly, the temperature of water returned to the boiler would be exactly the same as that leaving each terminal. In practice, heat output from the pipework – heat loss in this context – will reduce the temperature at the various inlets to a level which will vary, roughly, according to how distant each is from the boiler. Likewise, the water returned to the boiler will be at a lower temperature than that at which it leaves the outlets of the various terminals.

In terms of hydraulic balance, systems arranged as shown suffer from a number of difficulties. It will be obvious from the diagram that the most distant heat emitter is disadvantaged, by comparison with that nearest to the boiler, in this respect. The problem will be reduced if either the heat emitters or the final sub-circuit pipework to them offer a high resistance to water flow, relative to that of the remainder of the system.

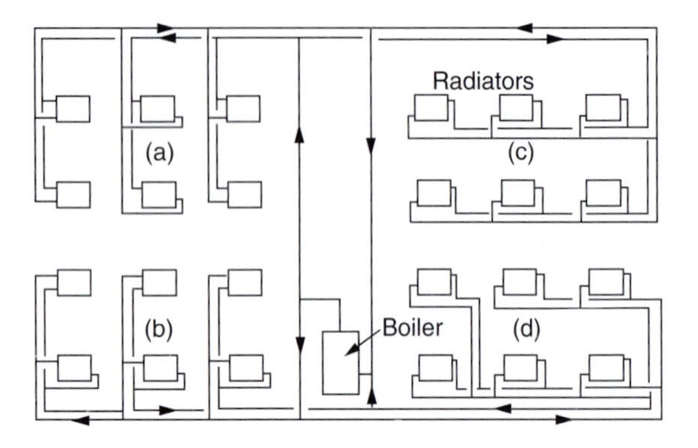

Figure 9.8 Multiple, two-pipe hot water circuits: (a) drop feed, (b) riser feed, (c) drop feed to horizontals and (d) riser feed to horizontals

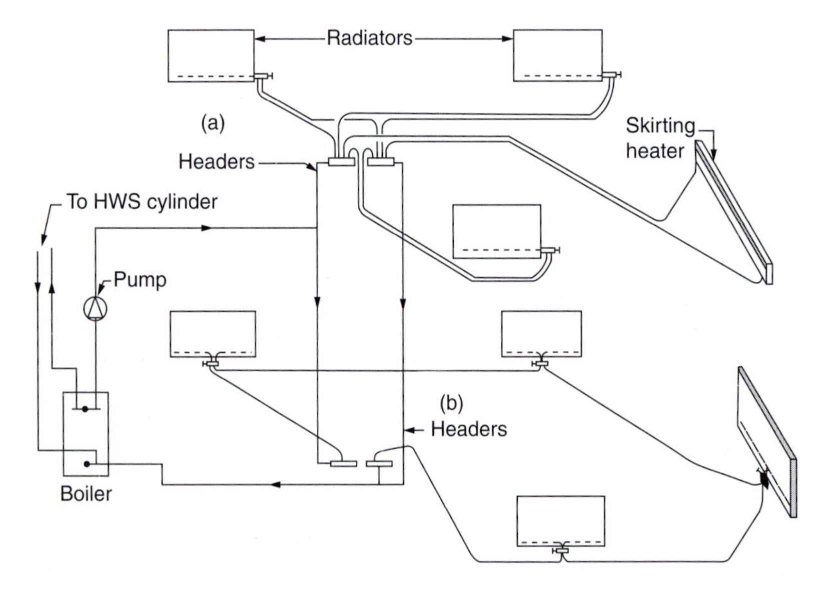

Figure 9.9 Micro-bore circuits: (a) two pipe and (b) single pipe (generally obsolete)

In terms of physical arrangements, two-pipe circuits are extremely flexible and may be set out to suit the building facilities available, as shown in Fig. 9.8. They may be applied to all types of hot water systems whatever the water temperature and to supply all types of heat emitting equipment.

The *micro-bore system* was developed in the 1980s mainly for domestic heating installations to further exploit the use of even smaller copper pipes, normally 8 or 10 mm. The two-pipe design concept, as shown in Fig. 9.9(a), differs somewhat from those described earlier under this heading in that it is strictly radial, i.e. the flow and return connections to each heat emitter originate quite separately, each from a central manifold. From each manifold, the micro-bore pipework is connected to a series of radiators. The pipework between the manifold and each radiator is usually no longer than 5 m to minimise pressure losses in the system. Special twin entry radiator valves and fittings were used such that both the supply and return micro-bore pipes are connected to the same end of each radiator. If these systems are fitted with thermostatic radiator valves (TRVs), then normally they are omitted from three radiators to avoid the need to fit a pressure relief/automatic bypass valve to protect the boiler and pump should the TRVs close.

The only advantage of the micro-bore system is that the smaller pipes loose less heat and it would seem the real reason for their rise in popularity was to reduce installation time and hence save cost. The main disadvantage

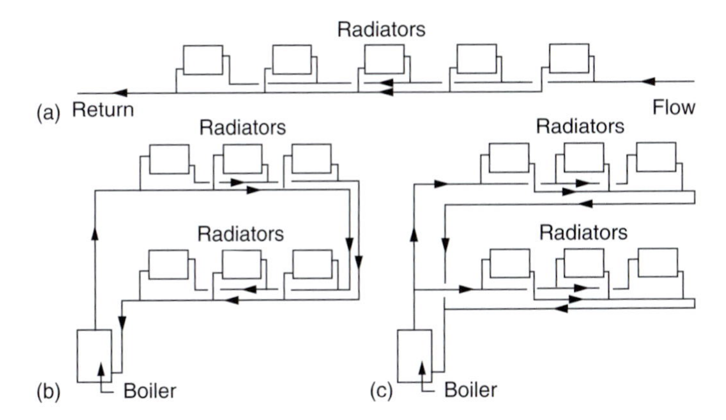

Figure 9.10 Reversed return circuits: (a) basic principle, (b) logical layout and (c) contrived layout

and the reason for their decline is that the small bore pipes can become easily blocked due to internal sediment and in hard water areas lime scale can build up if adequate water treatment is not provided and the water quality checked regularly. The blocked pipes can cause premature pump and even boiler failure.

The reversed return circuit

It is almost always possible to arrange circuits in a manner which will ameliorate the hydraulic problems mentioned above for two-pipe circuits. This is achieved, in principle, by laying out the system so that the distance water has to travel, from the boiler to each individual terminal item and then back again to the boiler, is the same, as illustrated in Fig. 9.10(a). Whether this is a practical proposition in terms of capital cost depends upon the building configuration and the ingenuity of the designer. The alternative layouts in Fig. 9.10(b) and (c) show that in some cases it may be possible to produce a *reversed return* at minimal additional cost, whereas in others there are complications.

For a given heat output, the water quantity in circulation around a system is a function of the temperature difference, flow to return. A low-temperature hot water system is likely to have over 4 times more water flowing in each circuit and sub-circuit, than would be the case for a high-temperature hot water system having the same output. This comparison serves to show that the use of reversed return arrangements, although always desirable, is much easier to justify on technical grounds for medium- or high-temperature systems.

Hybrid circuits

The descriptions given of the three principal circuit configurations should not be read as suggesting that they are mutually exclusive. Provided that a clear design logic is followed, no harm is done by mixtures if they go to make up a total arrangement which suits the use and style of a building. Examples that have been used in buildings in the past are described below:

Ladder circuits: For high-rise buildings, a reversed return two-pipe system as in Fig. 9.11(a) serves single-pipe circuits at each floor. In contrast, a simple two-pipe system serves reversed return circuits at each floor, as in Fig. 9.11(b).

Series circuits: Again for high-rise developments, a system much used in Europe has been applied to some buildings in the British Isles. It consists, as shown in Fig. 9.12, of a medium-temperature hot water circuit, which on a single-pipe basis, serves cased convectors (having 'hairpin' elements) in series: the return water therefrom passes to a two-pipe system serving radiators. By this means, the potential of the medium-temperature water is exploited to the full.

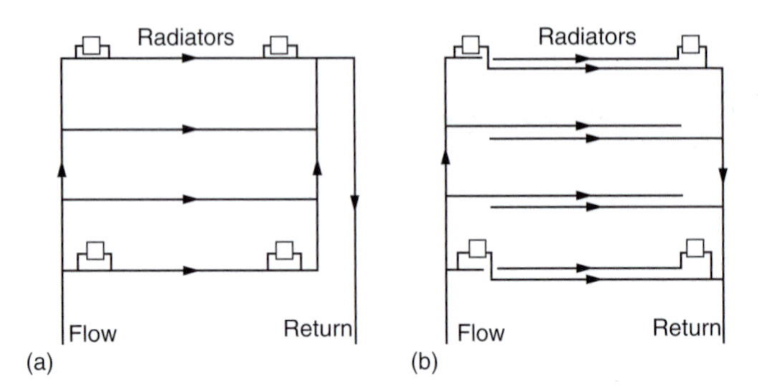

Figure 9.11 Hybrid reversed return circuits with: (a) simple sub-circuits and (b) simple supply mains

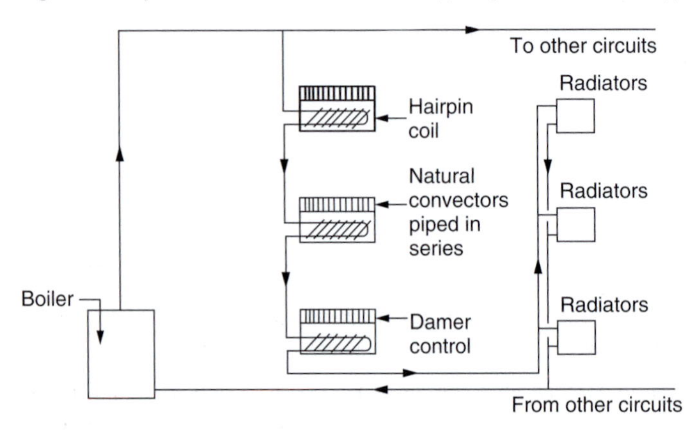

Figure 9.12 A hybrid series circuit

Underfloor heating systems

As described in Chapter 8, there are numerous publications which provide detail design guidance on these systems including BSRIA Application Guide AG 12/2001; and Parts 2 and 4 of BS EN 1264 describe the method of determination of the thermal output and the installation of underfloor heating systems using hot water. The following paragraphs provide an overview of the main considerations when designing this type of system.

Details on the floor construction, the position of insulation and the benefits of integrating phase change materials into the floor construction associated with underfloor heating are covered in Chapter 8 (Electrical storage heating).

Underfloor heating systems operate by means of embedded loops of pipe connected via a manifold as in Fig. 9.13, to the flow and return sides of the heat source.

Each loop or circuit can be controlled and/or isolated on both the flow and return.

Systems are normally designed to operate at low water temperatures of between 40°C and 50°C and a temperature drop of between 5 and 10 K across the system. They can therefore be used in conjunction with condensing boilers and ground source heat pumps.

Most systems today use non-ferrous plastic pipe or composite pipes instead of ferrous or copper material. By laying modern polymer plastic pipe in continuous coils without joints, it is possible to avoid many of the problems associated with ferrous or copper material pipe systems in the past. Modern plastics do not corrode or attract scale and are in many cases are capable of outliving the useful life of the building.

The most common materials in use today are:

PEX: cross-linked polyethylene
PP: co-polymer of polypropylene

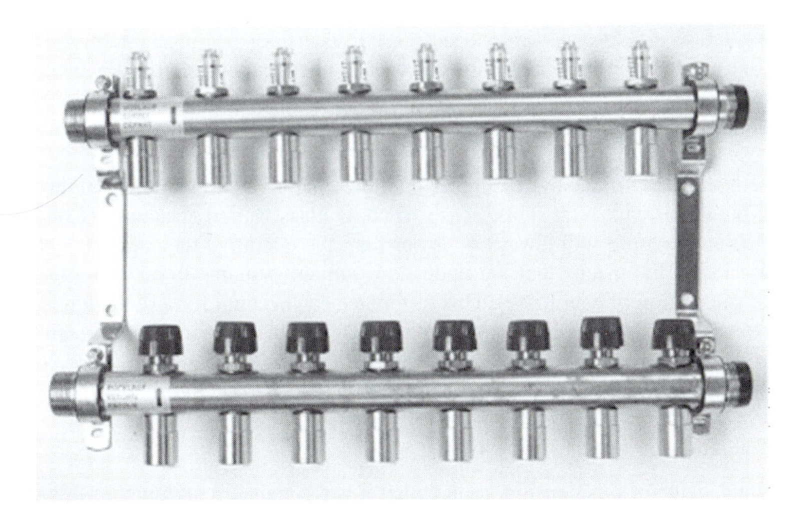

Figure 9.13 Typical pipe manifold for underfloor heating system

PB: polybutylene

PEX/aluminium/PEX pipe, sometimes known as 'multi-layer' or 'sandwich' pipe: The inner and outer polyethylene pipes prevent scaling and corrosion, the pipe is 100 per cent oxygen-tight (as discussed below) and the butt-welded aluminium ensures that this pipe is the only full strength alternative to copper. It gives the installer the advantages of both metal and plastic, but with the disadvantages of neither.

All pipes should ideally incorporate a diffusion barrier, which can be either integral or applied to the outside of the pipe as a coating. The purpose of the barrier is to reduce the amount of oxygen that can migrate through the pipe wall. Pipes without a diffusion barrier will pass much higher rates of oxygen thereby providing a highly oxygenated water circulation around the heating system. If the heating system is not fully protected by a corrosion inhibitor, then rapid corrosion of any steel components within the heating system can occur.

Better insulation standards in our buildings have meant that most floors are now insulated as standard. This means that for most buildings the installation of the underfloor heating will be no more difficult than any other form of heating.

The human foot is a highly effective thermostat for the whole body. Research shows that a temperature of 21°C on the floor surface will give an ideal sensation of comfort. International Standards indicate that the comfortable range is between 19°C and 29°C and dependent upon the required air temperature, floor heating systems should stay within this band. BS EN 1264–3 also states that the floor surface temperature should not be more than 9°C above the room temperature for most living areas or sedentary environments such as offices, and temperatures above this are generally perceived as 'uncomfortable' to the human foot at normal room temperatures of 18–22°C and should be avoided. Temperatures up to 35°C are however acceptable for specific areas such as pool surrounds, changing rooms, bathrooms and the first metre of space adjacent to walls with high fabric losses. This may occur where there is extensive floor to ceiling glazing where it is necessary to offset as much as possible the cold downdraught that occurs.

Modern well-insulated buildings no longer need high surface temperatures in order to provide sufficient output to meet the heat losses of a typical building. Average floor emissions of between 50 and 75 W/m^2 will often be more than sufficient. Low-temperature floor systems provide inherent, passive self-regulation. Floor temperatures generally need only 3–5 K higher than room air. Any rise in air temperature due to solar gain or increased occupancy means the air will begin to approach the floor temperature. As this temperature difference decreases the heat emission from the floor reduces. This process is rapid and precise. Heat emission from the floor will begin to decrease as soon as air temperature rises. Given an air temperature of 20°C and a floor temperature of 23°C, heat emission from the floor will decrease by about a third for each degree of temperature that the air temperature rises. A three degree rise in air temperature will thus be sufficient to neutralise the system.

Theoretically, this inherent self-regulation makes it possible to design an underfloor heating system with no other form of room temperature control. However, to account for variations in heat gains from people/solar, and variations in wind direction/velocity in practice, the usual approach to achieve maximum comfort is to provide local zone control, comprising air temperature sensor and motorised control valve to each zone. This gives the end user individual control over each room or area.

With convective systems, the air temperature will always be higher than the operative temperature and this difference can be as much as 5 K in buildings with high fabric losses such as glass walled structures. Conversely, in floor radiant systems the air temperature will always be lower than the operative temperature and is not so affected by the rate of fabric loss. This means that the underfloor radiant system's lower operative temperature can safely be used for the calculation of heat losses. This difference can account for a reduction in heat loss of 5–10 per cent for most types of buildings. It is also not necessary to allow any margins for height factors in buildings when considering the use of radiant floor heating systems.

The temperature drop across the pipe loops should be kept low, i.e. approximately 5–10 K in order to maintain even floor temperatures.

There are three types of loop configuration for underfloor heating:

(1) *Serpentine coil layout*: The main advantage of this configuration is that it is adaptable to all kinds of floor structures and can be easily modified for different energy requirements by altering the pitch of the pipes.
(2) *Double coil layout*: The characteristic of this configuration is that supply and return pipes in the layout run in parallel. This provides an even mean temperature, but will result in a higher variation of temperature within small areas. It is suitable for heating larger areas with higher heat demands, e.g. churches.
(3) *Spiral coil layout*: Here supply and return pipes are run in parallel, but in the form of a spiral. This approach is suitable for buildings with a higher heat demand, but is less suitable for installation in association with wooden floor structures.

A number of different approaches may be applied to the control of water temperature in underfloor heating systems. One of the simplest is to maintain a constant supply water temperature from the boiler to the system by means of a three port control valve and mixing circuit. This type of circuit works by mixing some of the return water from the underfloor heating with the hot water from the boiler to maintain a fixed supply water temperature.

Whilst this is satisfactory for small buildings such as domestic housing, heat demand for buildings is generally a function of outdoor temperature, therefore it is normal practice to introduce a weather compensation control strategy, reducing the flow temperature as the outside air temperature falls.

One of the main advantages using outdoor temperature compensation is the shorter reaction time, particularly for systems installed in concrete floors. The lower the water temperature is in the system, the smaller the heat sink effect of the floor and the quicker the response, a disadvantage being the possible rapid changes in the outside air temperatures.

An example of a low-temperature hot water heating system with a mix of underfloor heating, requiring water at a flow temperature of approximately 50°C and other circuits that require water at higher temperature (e.g. for hot water generation) is shown in Fig. 9.14.

Steam systems – characteristics

As described in the introduction to this chapter, steam as a medium for heating radiators and the like is a thing of the past.

Generation of steam

When heat is applied to water in a closed vessel such as a boiler which is only partially filled, the temperature will rise to boiling point (100°C at atmospheric pressure): application of further heat will cause a change of state and water will become steam. The quantity of heat involved in this latter process – *the latent heat of evaporation* – is considerable, at 2257 kJ/kg, when this is compared with the 420 kJ/kg required to raise water from 0°C to 100°C. Since the vessel is closed, the steam has no means of escape and the addition of yet further heat will cause the pressure to rise and with it the temperature of the steam and of any water remaining.

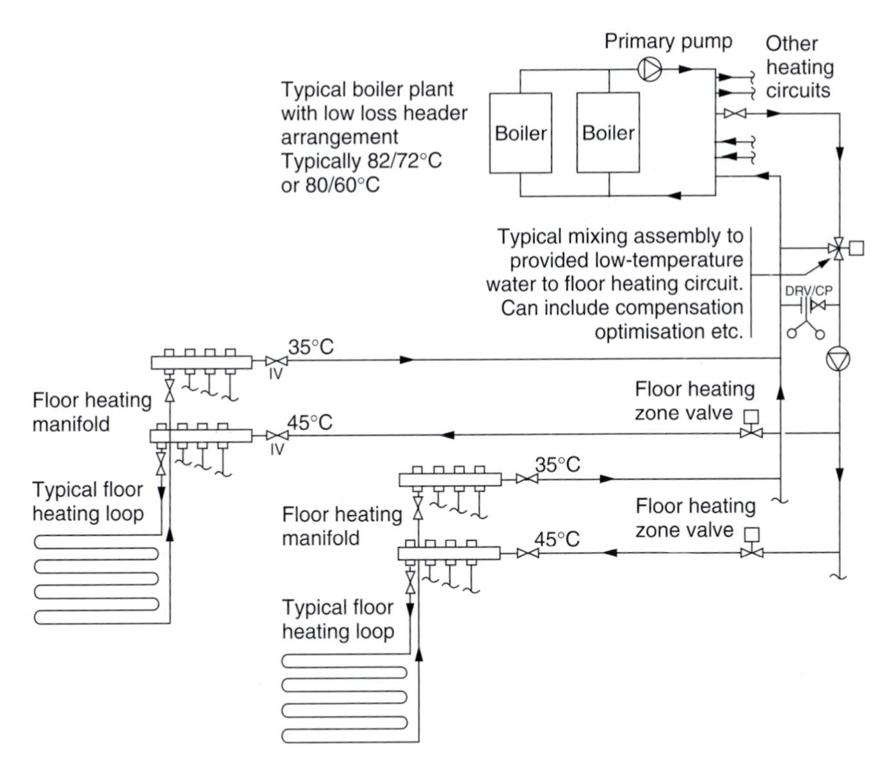

Figure 9.14 Typical underfloor heating circuits connected to a low-temperature hot water system

Steam in contact with water is described as being *saturated*: if it carries some water droplets in suspension, then it is referred to as being *wet*. Saturated steam removed from the vessel in which it has been generated and subjected to further heat is said to be *superheated*. Superheated steam is mostly used in power generation applications and on large sites supplied by large-scale combined heat and power (CHP) plant and is therefore not considered further here.

Utilisation involves the process of condensation, in which the latent heat is removed from the steam, by the use of heat exchangers of one sort or another, with the result that the steam reverts to water at the same pressure and temperature. This hot water or *condensate* must be removed from the heat exchangers as soon as it is formed, in order to prevent them from becoming waterlogged and thus useless. On release to atmospheric pressure, the condensate will fall in temperature since that part of the heat within it which is above atmospheric boiling point will act to re-evaporate a proportion of the remaining liquid. The vapour so produced is known as *flash steam* and this may be used in a variety of ways to economise in the use of what is known, conversely, as *live steam*.

Condensate is, for practical purposes, distilled water and will attract uncondensable and unattractive gases such as chlorine, carbon dioxide and oxygen. The resultant admixture will most probably be acidic and thus liable to cause corrosion in pipelines and elsewhere in the system. For this reason heavyweight copper pipework is usually used for condensate return to a boiler.

Automatically controlled steam boilers

In the past all steam raising plant was traditionally monitored at all times by a trained boiler technician. These days, however, most modern steam raising boiler plants are fitted with automatic controls, are usually unmanned and at best only inspected daily by general facilities technicians. Because of this the Heath and Safety Executive have produced Guidance note PM5 titled 'Automatically controlled steam and hot water

Table 9.5 Extract from steam tables

Pressure bar gauge	*Saturation temperature* (°C)	*Enthalpy (kJ/kg)*			*Volume of dry saturated steam* (m^3/kg)
		Water h_f	*Evaporation* h_{fg}	*Steam* h_g	
0	100	419	2257	2676	1.673
1	120	506	2201	2707	0.881
2	134	562	2163	2725	0.603
3	144	605	2133	2738	0.461
4	152	641	2108	2749	0.374
5	159	671	2086	2757	0.315
6	165	697	2066	2763	0.272
7	170	721	2048	2769	0.240

boilers'. This draws attention to the causes of damage and explosions, some of which have resulted in fatal injuries and makes recommendations designed to prevent such occurrences.

The minimum recommended requirements for automatic controls for boilers not continuously supervised are as follows:

(a) Automatic water level controls *so* arranged, that they positively control the boiler feed-pumps or regulate the water supply to the boilers and effectively maintain the level of water in the boiler between certain predetermined limits.
(b) Automatic firing controls so arranged, that they effectively control the supply of fuel to the burners on oil or gas-fired boilers, or air to solid fuel-fired boilers' and shut off the supply in the event of a critical alarm.
(c) Independent overriding controls that must be hand reset to cut off the fuel supply to the boilers and cause an audible alarm to sound when the water level in the boiler falls to a predetermined low water level. This alarm must be located remote from the boiler house in a location that is manned by a person competent to respond to alarms 24/7, such as a security office. This person should press the shut-off button and call the trained boiler engineer.

Properties of steam

Comprehensive tables listing the properties of steam are published in CIBSE Guide C or they can be obtained from steam equipment manufacturers such as Spirax Sarco. Table 9.5 provides an extract from the tables and Fig. 9.15 provides a graphical representation of the tables:

Pressure: It should be noted that the pressure in the CIBSE steam tables is stated in absolute terms. Gauge pressure is used in some tables as in the example above. This is the pressure above atmospheric pressure of 101.325 kPa or approximately 1 bar (e.g. 1 bar or 100 kPa absolute pressure = 0 bar gauge pressure.)
Temperature: This is stated in relation to the rounded values for pressure. A water temperature of 100°C occurs at the standard atmospheric pressure of 101.325 kPa.
Enthalpy: These three columns give values for heat content or specific enthalpy, in terms of kJ/kg, for the water (hf), for the latent heat (evaporation) in the steam (hfg) and for the sum of the two (total heat of saturated vapour). The decline in latent heat content with pressure increase should be noted.
Volume: The last column gives values for the specific volume of the steam.

System pressure

It will be clear from examination of the steam table that high pressures are not necessarily desirable for heat transfer, as there is negligible difference in the total heat at 1 or 7 bar gauge. For industrial purposes, there may be a need for high temperatures (above 170°C) requiring pressures above 7 bar gauge.

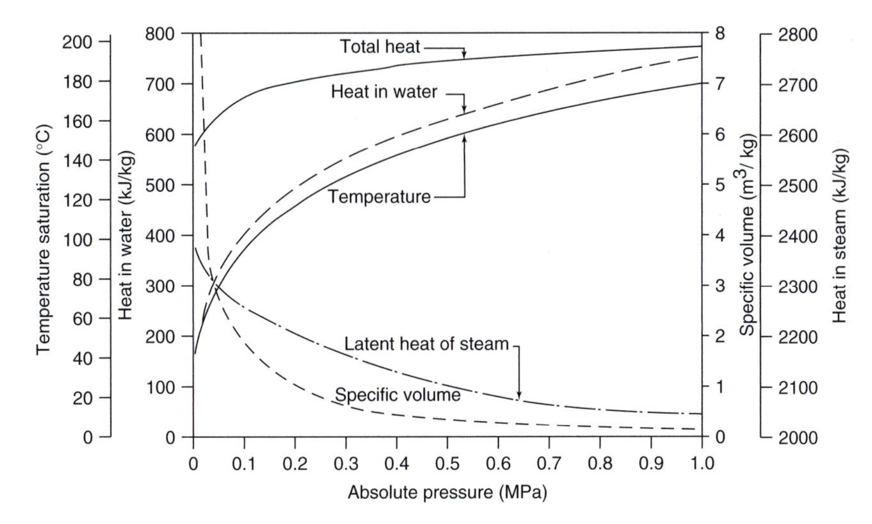

Figure 9.15 Properties of steam: graphical representation

Most steam boilers however are designed to work at relatively high pressures and should not be run at lower pressures, since wet steam is likely to be produced. By doing this the thermal storage capacity of the boiler is increased, helping it cope more efficiently with fluctuating loads. It is also more economical to produce and distribute steam at a higher pressure and reduce pressure upstream of any items of plant designed to operate at a lower pressure. This type of arrangement has the added advantage that relatively smaller distribution mains can be used due to the relatively small volume occupied by steam at high pressure.

The effect of air, oxygen and other gases in a steam system

The presence of air has a devastating effect on steam systems and processes.

The presence of air and other incondensable gases is a problem in any steam system, whatever the pressure. They exist as a result of aeration of condensate, decomposition of calcium bicarbonate to form calcium carbonate and CO_2 and the inevitable inward leakage through valve glands, etc., which follows any seasonal plant shutdown. Within any heat exchanger, a stagnant air film at the steam-side surface will act as an insulant and prevent condensation and the desired surrender of latent heat. At low pressures this effect is particularly persistent.

Air is present within steam pipes and steam equipment at start-up. Even if the system were filled with pure steam when used, the condensing steam would cause a vacuum and draw air into the pipes at shut-down. Even the best physical and chemical treatments will still allow some untreated incondensable gas to leave the boiler with the steam. Air is more widespread in steam systems than believed and is the cause of both limitation of output and equipment corrosion.

Air can also enter the system in solution in the feedwater. At 80°C, water can dissolve about 0.6 per cent of its volume, of air. The solubility of oxygen is roughly twice that of nitrogen, so the air which dissolves in water contains nearly one part of oxygen to two of nitrogen, rather than the one part to four parts in atmospheric air. Carbon dioxide has a higher solubility, roughly 30 times greater than oxygen. Boiler feedwater, and condensate exposed to the atmosphere, can readily absorb these gases. When the water is heated in the boiler, the gases are released with the steam and carried into the distribution system. Unless boiler make-up water is fully demineralised and degassed, it will often contain soluble sodium carbonate from the chemical exchange of water treatment processes. The sodium carbonate can also be released in the boiler and again carbon dioxide is formed.

Air removal

The most efficient means of air venting is with an automatic device. Air mixed with steam lowers the mix temperature. This enables a thermostatic device (based on either the balanced pressure or bimetallic principle) to vent air from the steam system.

Requirements for boiler feedwater treatment and deaeration

Oxygen is the main cause of corrosion in steam boiler feedwater tanks, feedlines, feed-pumps and boilers. If carbon dioxide is also present then the pH will be low, the water will tend to be acidic and the rate of corrosion will be increased. Typically the corrosion is of the pitting type where, although the metal loss may not be great, deep penetration and perforation can occur in a short period.

Elimination of the dissolved oxygen may be achieved by chemical or physical methods, but more usually by a combination of both.

The essential requirements to reduce corrosion are to maintain the feedwater at a pH of not less than 8.5–9, the lowest level at which carbon dioxide is absent, and to remove all traces of oxygen. It should also be noted that the return of condensate from the distribution system will have a significant impact on boiler feedwater treatment – condensate is hot and already chemically treated, consequently as more condensate is returned, less feedwater treatment is required.

Water exposed to air can become saturated with oxygen, and the concentration will vary with temperature: the higher the temperature, the lower the oxygen content. Therefore the first step in feedwater treatment is to heat the water as high as practically possible to drive off the oxygen. Typically a boiler feedtank should be operated at 85–90°C. This leaves an oxygen content of around 2 mg/litre (ppm) which is usually removed by the addition of an oxygen scavenging chemical (sodium sulphite, hydrazine or tannin) to prevent corrosion. Operation at higher temperatures than this at atmospheric pressure can be difficult due to the close proximity of saturation temperature and the probability of cavitation in the feed-pump, unless the feedtank is installed at a very high level above the boiler feed-pump.

This is the normal treatment for industrial boiler plant in the UK. However, to reduce the quantity of water treatment chemicals and with higher pressure and larger boilers, the alternative is to use a pressurised deaerator which can reduce the oxygen levels to 3 ppm in water. This comprises a domed 'head section' fitted to a pressurised deaerator vessel that replaces the traditional heated feedwater tank or 'hot well'.

Operating principles of a pressure deaerator

If a liquid is at its saturation temperature, the solubility of a gas in it is zero, although the liquid must be strongly agitated or boiled to ensure it is completely deaerated.

This is achieved in the head section of a deaerator by breaking the water into as many small drops as possible, and surrounding these drops with an atmosphere of steam. This gives a high surface area to mass ratio and allows rapid heat transfer from the steam to the water, which quickly attains steam saturation temperature. This releases the dissolved gases, which are then carried with the excess steam to be vented to atmosphere. (This mixture of gases and steam is at a lower than saturation temperature and the vent will operate thermostatically.) The deaerated water then falls to the storage section of the vessel.

A blanket of steam is maintained above the stored water to ensure that gases are not re-absorbed.

Flash steam and recovery

'Flash steam' is released from hot condensate when its pressure is reduced. Even water at an ambient room temperature of 20°C would boil if its pressure were lowered far enough. It may be worth noting that water at 170°C will boil at any pressure below 6.9 bar gauge. The steam released by the flashing process is as useful as steam released from a steam boiler.

As an example, when steam is taken from a boiler and the boiler pressure drops, some of the water content of the boiler will flash off to supplement the 'live' steam produced by the heat from the boiler fuel. Because both types of steam are produced in the boiler, it is impossible to differentiate between them. Only when flashing takes place at relatively low pressure, such as at the discharge side of steam traps, is the term 'flash steam' widely used. Unfortunately, this usage has led to the erroneous conclusion that flash steam is in some way less valuable than the so-called live steam.

In any steam system seeking to maximise efficiency, flash steam will be separated from the condensate, and used to supplement any low-pressure heating application. Every kilogram of flash steam used in this way

Table 9.6 Availability of flash steam

Initial absolute pressure (kPa)	*Per cent condensate available as flash steam at following reduced absolute pressures* (kPa)				
	100	*150*	*200*	*250*	*300*
1200	16.9	14.8	13.3	12.1	11.0
1100	16.1	14.1	12.5	11.3	10.2
1000	15.3	13.3	11.7	10.4	9.3
900	14.4	12.4	10.8	9.5	8.4
800	13.5	11.4	9.8	8.5	7.4
700	12.5	10.3	8.8	7.5	6.3
600	11.2	9.1	7.5	6.2	5.0
500	9.9	7.7	6.1	4.8	3.7
400	8.3	6.2	4.5	3.2	2.0
300	6.4	4.2	2.5	1.2	–

is a kilogram of steam that does not need to be supplied by the boiler. It is also a kilogram of steam not vented to atmosphere, from where it would otherwise be lost.

Flash steam, if not recovered, represents a loss in efficiency and the magnitude of this may be seen from Table 9.6.

Steam boiler blowdown

Steam boilers are blown down at regular intervals to remove some of the suspended and dissolved solids which result from raising steam. These solids cause foaming and overheating of the heat transfer surfaces of the boiler and it is necessary to control their concentration by 'blowing them down' into a tank for future disposal to drain. Blowdown can be either intermittent or continuous depending on the boiler design. It should be noted that in the UK, for example, water above 43°C cannot be returned to the public sewer by law, because it is detrimental to the environment and may damage earthenware pipes.

When a continuous blowdown system is installed, it is usual to incorporate heat recovery to preheat the feedwater. It may then also be possible to discharge the cool blowdown water directly to drain rather than discharge it into a blowdown tank. Intermittent blowdown is essential for boilers operating with feedwater treatment, as it is the most effective method of removing sludge by being applied in short sharp bursts.

The sudden or rapid discharge of high-temperature hot water is hazardous because a considerable amount of energy is released. Therefore all blowdown tanks must be designed as pressure vessels and are usually designed to withstand at least 25 per cent of the maximum allowable pressure of the boiler to which it is attached and have adequate structural strength to sustain shock loading associated with intermittent blowdown.

The Health and Safety Executive produce a Guidance note PM60 covering steam boiler blowdown systems that gives advice on the design and installation of safe discharge of blowdown from steam boilers.

Pressure reduction

The main reason for reducing steam pressure is fundamental. Every item of steam using equipment has a maximum allowable working pressure (MAWP). If this is lower than the steam supply pressure, a pressure reducing valve must be employed to limit the supply pressure to the MAWP. In the event that the pressure reducing valve should fail, a safety valve must also be incorporated into the system.

It is worth noting that in the process of expansion of steam through a pressure reducing valve, the total heat would remain constant (as at the higher pressure). This is higher than the total heat of saturated steam for the lower pressure and results in a small amount of superheating of the steam above the anticipated temperature of saturated steam at the lower pressure.

As discussed earlier, steam boilers are designed to work at relatively high pressures and should not be run at lower pressures as they produce wet steam. Steam is therefore usually generated and distributed at high pressure (which also results in smaller distribution pipework with reduced distribution losses) and pressure reduction facilities are installed local to items of plant that are designed to operate at a lower pressure.

Since the temperature of saturated steam is closely related to its pressure, control of pressure can be a simple but effective method of providing accurate temperature control. This fact is used as good effect on applications such as sterilisers and dryers where the control of surface temperature is difficult to achieve using temperature sensors.

Plant operating at low steam pressure can tend to reduce the amount of steam produced by the boiler due to the higher enthalpy of evaporation in lower-pressure steam but will reduce the loss of flash steam produced from open vents on condensate collecting tanks.

Most pressure reducing valves currently available fall into two main groups:

(1) *Direct acting valves*: With this type of valve the downstream pressure will increase as the load falls and will be highest when the valve is closed. This change in pressure relative to a change in load means that the downstream pressure will only equal the set pressure at one load, i.e. it suffers from proportional offset as the steam flow changes. This means the valve is not suitable for critical applications where the pressure cannot increase and a pilot-operated valve should be used. They are however perfectly adequate for a substantial range of simple applications where accurate control is not essential and where steam flow is fairly small and reasonably constant.
(2) *Pilot-operated valves*: Where accurate control of pressure or a large flow capacity is required, a pilot-operated pressure reducing valve should be used. This works by balancing the downstream pressure via a pressure sensing pipe against a pressure adjustment control spring. This moves a pilot valve to modulate a control pressure. Any variations in load or pressure will immediately be sensed on the pilot diaphragm, which will act to adjust the position of the main valve accordingly, ensuring a constant downstream pressure.

Further information on steam pressure reduction can be found in Chapter 11.

Steam systems – piping arrangements

A much quoted sentence from the well-loved first edition of *The Efficient Use of Steam by Oliver Lyle* postulates that 'A steam pipe should carry steam by the shortest route in the smallest pipe with the least heat loss and the smallest pressure drop that circumstances will allow.' As succinct guidance in relation to the design of distribution pipework, these few words gather together all the salient criteria.

Historical arrangements

The following solutions are not used in modern buildings. However, these cover important design concepts and are included in this edition for historical reasons.

Disposal of condensate and elimination of air were the twin problems which dominated the design of steam pipework in early years, where system pressures adopted were in the range of 120–170 kPa absolute. Single-pipe arrangements were not uncommon, Fig. 9.16(a), and these were featured in early North American textbooks. Horizontal pipework, where this occurred, was laid to a generous 'pitch' to encourage condensate return and vertical risers were oversized to allow steam and condensate to flow in opposite directions in the same pipe. Air vents were fitted at the end of each horizontal pipe run and at mid-height on each radiator. Water hammer was endemic.

Two-pipe gravity systems were classified as having 'wet' or 'dry' returns, dependent upon whether the condensate pipework was above or below the water level in the boiler, as in Fig. 9.16(b). As in the case of the single-pipe system, radiators were either on, at temperatures between 105°C and 115°C, or off and cold; there was no intermediate level. Water make-up to both single- and two-pipe systems was made from a cistern, fitted at a height sufficient to overcome the system pressure, to a 'boiler feeder' which consisted of a ball valve within a closed casing connected top and bottom to the steam and water contents of the boiler.

The subsequent *vacuum* systems where, as the name suggests, use was made of a vacuum pump to assist in removal of air and to return condensate to the boiler, were a substantial advance and permitted system size to be increased considerably. The later introduction of *sub-atmospheric* systems was perhaps the ultimate development since these permitted steam pressures to be both lowered and controlled such that temperatures could

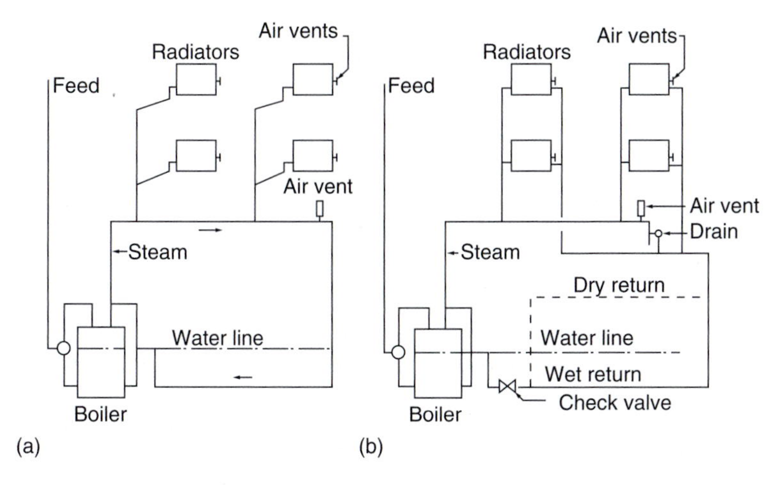

Figure 9.16 Low-pressure steam systems: (a) single-pipe gravity return and (b) two-pipe gravity return

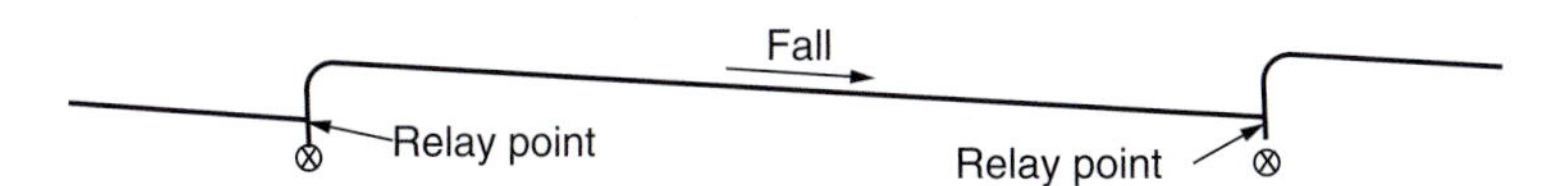

Figure 9.17 Steam main relay points

be varied between about 55°C and 105°C at source, according to weather conditions. Such precision was not unaccompanied by disadvantages, one of the more significant being the necessity to use specialised components such as glandless valves in an attempt to produce a piping system which remained absolutely 'airtight', both in service and after shut-down.

Current practice

Facilities for the removal of condensate so that the steam supply remains dry are among the more important features and supply pipework must be provided with *relay points*, as in Fig. 9.17, to assist in this respect (see Fig. 11.23 also). The relay points should be installed at intervals of 30–50 m and the pipework should be installed with a fall of not less than 1:100. If the steam mains are not drained of condensate effectively, the result will be what is known as 'water hammer'. This is when a slug of condensate collides with pipe fittings or plant at high velocity (if the steam pipework is sized correctly the velocity will be approximately 25–30 m/s). Water hammer will cause a banging noise, and perhaps movement of the pipe. In severe cases, water hammer may fracture pipeline equipment with almost explosive effect, with consequent loss of live steam at the fracture, leading to an extremely hazardous situation.

Pipework follows a two-pipe pattern very much as might be anticipated, incorporating specialised equipment as described in Chapter 11.

Condensate will, in most normal circumstances, be collected and returned to a tank or *hot well* near to the boiler. Where distance or site levels make direct return impossible, intermediate collection and pumping units may be required. At the hot well, to make up for any losses which may have occurred, a treated water supply will be provided and this may alternatively be directly 'on line' or via an intermediate store. The optimum size of a hot well will depend upon a number of factors such as the order of fluctuation in the steam demand, the corresponding – but not necessarily equivalent – rate of condensate return and the operational routine of the water treatment plant. It is good practice, nevertheless, to provide a minimum size related to the evaporation capacity of the boiler plant over 2 hours.

The temperature of the returned condensate is likely to be approximately 90°C, but the hot well should nevertheless be provided with a steam coil to maintain that level. To avoid cavitation at the suction connections of the pumping equipment provided to return condensate to the boilers, the hot well should be elevated

Table 9.7 Approximate properties of dry air (at atmospheric pressure)

	Re mass		*Re volume*	
Temperature (°C)	*Specific mass* (kg/m^3)	*Specific heat capacity* (kJ/kg K)	*Specific volume* (m^3/kg)	*Specific heat capacity* (kJ/m^{3}K)
10	1.247	1.011	0.802	1.257
20	1.205	1.012	0.830	1.219
30	1.165	1.013	0.858	1.180
40	1.128	1.013	0.887	1.143
50	1.093	1.014	0.915	1.108
60	1.060	1.015	0.943	1.076
70	1.029	1.015	0.972	1.045
80	1.0	1.016	1.0	1.016
90	0.973	1.016	1.028	0.988
100	0.946	1.017	1.057	0.963
110	0.922	1.018	1.085	0.938
120	0.898	1.018	1.113	0.915
130	0.876	1.019	1.142	0.893
140	0.855	1.020	1.170	0.838
150	0.835	1.020	1.198	0.851

where this is practicable, to about 5 m above that level. Where packaged boilers are used, each is likely to be provided, at works, with a small cylindrical 'condense receiver' in association with the pre-wired feed-pump and associated controls. The size of such receivers is usually quite inadequate to provide the necessary storage capacity but their volume may be deducted from the capacity of the hot well as proposed above.

Air systems – characteristics

Passing reference was made in Chapter 7 to the use of hot air as a heat transfer medium within floor ducts but such an application is a rarity and no purpose would be achieved by pursuing it further here. Table 9.7 lists the principal characteristics of dry air in context with its use for space heating and it is of interest to compare these with the equivalent values for hot water as listed in Table 9.2. The specific heat capacity of water in terms of mass is 4 times that of air at the same temperature and in terms of volume, the difference is immensely greater.

Ventilation systems are dealt with later (Chapter 18) and it is necessary to draw quite a clear distinction between the function which they fulfil and that of air systems, that are used to both heat and ventilate a building that historically were called *plenum* systems, which fall under this present heading. The essence of the difference lies in the fact that a minimum outside air ventilation system, properly designed, will do no more than introduce a supply of air into the space served at a temperature which is either at or just below that required in the space. Heating is provided by a separate system, usually low-temperature hot water serving radiators, convectors or other heat emitters located throughout the building.

The plenum system, conversely, introduces a supply of air at a temperature well in excess of that required in the space: ventilation, in effect, is a by-product. In cooling to the space temperature, the air provided surrenders the heat it carried to windows, walls, floor and roof, etc., in just the same way that heated air convected from a hot water 'radiator' cools in circulation within a room. Where a building is heated by an air *plenum* system, the central plant will usually comprise an air handling unit complete with a mixing box. To preheat the building from cold, all of the air will be recirculated. At occupancy the dampers in the mixing box will open to a predetermined position to mix the required proportion of outside air with recirculated air.

The traditional plenum system

This is best discussed using, as a simple illustration, the building shown in Fig. 4.3. As may be seen the component of the calculated heat loss associated with the building fabric is all that is required, in the case of a fully convective system, to determine the value of the two necessary temperature ratios. Thus, from earlier calculations (pp. 70 and 83):

$$\sum AU/\sum A = 140/419 = 0.33\ \text{W/m}^2\ \text{K}$$

By reference to Table 4.13 and extrapolating

$$F_1 = 0.98 \text{ and } F_2 = 1.04$$

To produce a positive pressure differential and to enable a satisfactory air distribution, a supply air quantity from outside, equivalent to say three air changes per hour of supply air would be necessary, (i.e. $697\ \text{m}^3 \times 3 = 2091\ \text{m}^3$/hour, or 580 litre/s. Thus:

$$Q_u = 0.98 \times 140 \times 22 = 3018\ \text{W}$$
$$Q_v = 1.03 \times 2091 \times 0.33 \times 22 = 15636\ \text{W}$$

and hence

$$Q_{u+v} = 18654\ \text{W}\ (\simeq 18.7\ \text{kW})$$

The various internal temperatures may now be established:

$$t_{ei} = (0.98 \times 22) - 1 = 20.6°\text{C}$$
$$t_{ai} = (1.04 \times 22) - 1 = 21.9°\text{C}$$

The significance of this last value, for the building used as an example when heated by a plenum system, is that it shows that an inside air temperature of approximately 22°C is necessary to maintain an operative temperature of approximately 21°C. This figure of 22°C would, taking a volumetric specific heat of air from Table 9.9, now be used to calculate the required air temperature reaching the room. (For this example because the temperature difference between air and environmental temperatures is small, the operative temperature will be similar to the environmental temperature.) Thus:

$$t = (22 + 1) + [(3018)/(580 \times 1.18)] = 23 + 4.4 = 27.4°\text{C}$$

The heat load noted above is some 13.8 kW more than the total calculated on p. 83 for a panel radiator system, a substantial part of the difference coming about as a result of the necessarily increased rate of air change. The supply air will, of course, cool to only about 23°C in the space served and then be exhausted to outside at that temperature. Thus, about a third of the heat input is wasted unless either some form of heat reclaim equipment (pp. 538–545) is provided, or a part of the supply air quantity is returned from the space via an extract system, for recirculation. The greater the proportion recirculated, however, the less the differential pressure and the greater the probability that strong winds and thermal forces will upset the theoretical pressure balance.

For commercial and institutional buildings having a multitude of compartments, the plenum system was misapplied since it could not be expected to provide satisfactory service in a situation where the requirements varied constantly from room to room. In terms of comfort, moreover, the elevated temperature of the air supply often as high as 50°C led to complaints that the atmosphere was oppressive. For low-cost installations in factories and buildings where the volume per occupant was relatively large and where floor space was valuable, application was more successful but was largely superseded when unit heaters became popular.

A relatively recent development that uses a low-velocity air system to heat, ventilate and also cool modern, well-insulated buildings, usually with small glazed areas and infiltration losses, is the system that uses hollow core concrete slabs with mechanical ventilation. This system is typically referred to by the trade name 'TermoDeck'.

This is a fan-assisted, heating, cooling and ventilation system that uses the high thermal mass of structural, hollow core floor slabs through which warmed or cooled fresh air is distributed. The supply air is heated in a central air handling unit that also contains heat recovery facilities. It is then distributed via insulated ductwork which connects to the hollow cores of the concrete structural slabs. The air passes through the concrete slabs at low velocities, allowing prolonged contact between the air and the slabs, enabling them to act as passive heat exchange elements. The air is then distributed via diffusers in the ceiling or floor. There is no requirement for wet heating or air-conditioning and the temperature difference, between the slab and the air that exits the slab, is not more than 1°C or 2°C.

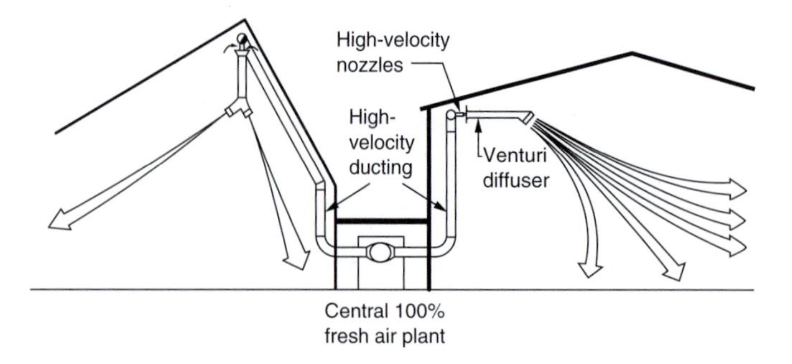

Figure 9.18 High-velocity plenum system (Casaire)

As long as the building is well insulated, airtight and has heat recovery, the utilisation of the thermal mass of the building will produce a very efficient environmental control solution that damps down fluctuations in temperature, resulting in lowering a building's requirements for heating and cooling.

The high-velocity system

The majority of the disadvantages associated with the plenum system can be overcome by using direct firing of heaters within central air plant, coupled with a high-velocity air distribution system to specialist *terminal diffusers*. The main application of this type of system is to large industrial spaces, aircraft hangars, workshops, warehouses, etc. There was significant take up of this type of system in the 1970s/1980s but in recent years there has been limited application of the technology, with these types of spaces either being heated by local unit heaters (served by low-temperature hot water or gas fired) or by gas-fired radiant systems.

The central plant consists of a forced convection air heater within which the products of combustion mix with the heated air, which is provided for circulation to the space served. By this means, an outlet temperature of about 140°C may be achieved with a combustion efficiency which approaches 100 per cent. This concept, which applies only when fuels containing minimum potential for pollution are used, is an extension of the principle illustrated in Fig. 7.12. The basis of design used for the system considered here restricts the concentration of carbon dioxide, by volume, to 2800 ppm in the air circulated, a quantity not much more than half the maximum permitted by the Health and Safety Executive. A similar plant arrangement may be used with the products of combustion flued to outside the building, thereby reducing the carbon dioxide and moisture levels within space, but with a resultant reduction in thermal efficiency.

The heated air, which represents only about half an air change within the space served, is distributed from the central plant through insulated sheet-metal ducts at a velocity of about 35 m/s, to an array of terminal diffusers. Each diffuser is equipped with one or more nozzles which *induce* a supply of air from the space into circulation via a venturi arrangement. The quantity of air induced, which may be drawn from an area within the building where temperature gradient has created a potential wastage, is about 4 times that of the supply from the central plant and the total is thus enough to allow control over the final distribution. The output discharge temperature is of the order of 50°C at a velocity of about 5 m/s. Fig. 9.18 provides a diagram of the system arrangement.

Pipework heat emission

Whatever the heating medium may be; water, steam or air, the piping distribution arrangements have certain features in common. They will expand when heated, as described later in Chapter 11 and they will give out heat throughout their length, whether this is desired or not. The magnitude and effect of this second matter must be evaluated and dealt with by application of thermal insulation where necessary.

Pipework heat emission

Exposed pipework, as a form of heating surface, is rarely used in current practice other than for the provision of supplementary heating for anti-condensation measures to large areas of glazing such as large roof lights

Table 9.8 Theoretical heat emission from pipes with different emissivities[a]

Water or steam temperature in pipe (°C)	*Heat emission from stated nominal pipe sizes, horizontal mounted, freely exposed in surrounding at 20°C* (W/m)							
	15 mm diameter		*25 mm diameter*		*50 mm diameter*		*100 mm diameter*	
	Steel	*Copper*	*Steel*	*Copper*	*Steel*	*Copper*	*Steel*	*Copper*
50	29	17	44	29	72	50	126	90
60	42	25	62	42	102	72	179	128
80	69	41	102	69	169	119	297	212
100	100	60	149	99	246	171	431	307
120	135	80	200	133	331	230	582	412

a Steel pipe emissivity = 0.95; copper pipe emissivity = 0.5.

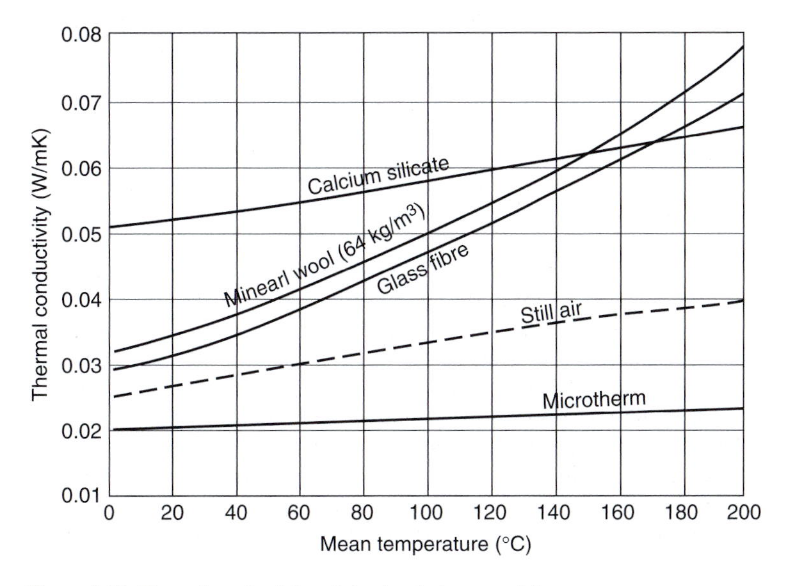

Figure 9.19 Thermal conductivity of pipe insulating materials

or atrium roof glazing. However in any system of distribution, main and branch piping running through the spaces to be heated may offer a contribution. The theoretical emission from bare horizontal pipework may be calculated from expressions given in the CIBSE Guide C (2007). Section 3 of this guide also provides various tables of heat emissions for different types of pipes freely exposed in surroundings at 20°C. The figures in Table 9.8 have been extracted from these CIBSE tables to give an indication of the heat emissions per m length of pipe and demonstrate the effect of the surface emissivity on the heat emission. (From radiation heat transfer theory, emissivity is the ratio of the total emissive power of a body to the total emissive power of a black body at the same temperature. A polished aluminium surface has a very low emissivity, typically 0.05, whereas an oxidised steel pipe or a painted surface either black or of zinc has a high emissivity of 0.95.)

Emission from pipes fixed vertically, varies from the listed values and is about 25 per cent less for small and 5 per cent less for large pipes.

Pipework thermal insulation

In order to reduce the unwanted heat output from distribution mains running in trenches or ducts and through basements or any other spaces not requiring heat, thermal insulation is applied. The effect of such action depends upon the conductivity of the insulation material (shown in Fig. 9.19), its thickness and the surface finish.

Table 9.9 Maximum permissible heat losses (W/m) from pipework to comply with Part L of the Building Regulations 2006 (reproduced from the TIMSA Guide)

	Maximum permissible heat losses for different pipe sizes			
Outside pipe diameter (mm)	*Hot water*[1]	*Low-temperature Heating*[2] $\leqslant 95°C$	*Medium-temp Heating*[3] *96–120°C*	*High-temp heating*[4] *121–150°C*
17.2	6.60	8.90	13.34	17.92
21.3	7.13	9.28	13.56	18.32
26.9	7.83	10.06	13.83	18.70
33.7	8.62	11.07	14.39	19.02
42.4	9.72	12.30	15.66	19.25
48.3	10.21	12.94	16.67	20.17
60.3	11.57	14.45	18.25	21.96
76.1	13.09	16.35	20.42	24.21
88.9	14.58	17.91	22.09	25.99
114.3	17.20	20.77	25.31	29.32
139.7	19.65	23.71	28.23	32.47
168.3	22.31	26.89	31.61	36.04
219.1	27.52	32.54	37.66	42.16
273.0 and above	32.40	38.83	43.72	48.48

Notes
1 Horizontal pipe at 60°C in still air at 15°C
2 Horizontal pipe at 75°C in still air at 15°C
3 Horizontal pipe at 100°C in still air at 15°C
4 Horizontal pipe at 125°C in still air at 15°C

A bright metal cladding to insulation (e.g. aluminium with a low emissivity of approximately 0.05) will reduce heat loss from that resulting from a dull painted finish by up to 10 per cent.

In order to comply with the current UK Building Regulations Part L, 2006 the following provisions must be met:

- Pipework serving space heating (and hot water) should be insulated in all areas outside the heated building envelope.
- All pipes should be insulated in voids and in spaces which will normally be heated if there is a possibility that those spaces might be maintained at temperatures different to those maintained in other zones.
- The control of the heating water temperature should be maximised and heat loss from uninsulated pipes should only be permitted where the heat is 'useful' to the space.

The Thermal Insulation Manufacturers and Suppliers Association (TIMSA) have produced a guidance document for achieving compliance with Part L of the Building Regulations that is cited as a 'second tier' document in the Approved Documents. This document provides a series of maximum permissible heat losses and insulation thicknesses that can demonstrate a payback no longer than 7 years, where simple payback is given as the marginal cost of the energy efficiency measure divided by the value of the fuel savings.

The maximum permissible heat losses for different pipe sizes and temperatures and the minimum thicknesses of insulation for a range of thermal conductivities are shown in Tables 9.9 and 9.10. (These have both been reproduced from the TIMSA Guide.)

The insulation materials used most often on heating pipework are mineral wool, phenolic foam, and nitrile rubber. These are all supplied in preformed cylindrical shapes and generally include a vapour barrier enabling them to be used on cold pipes in addition to hot pipes.

All modern insulating materials are manufactured from materials that are chlorofluorocarbon (CFC), hydrochlorofluorocarbon (HCFC) and hydroflurocarbon (HFC) free to give the products zero ozone depletion potential and are normally manufactured to achieve a global warming potential (GWP) of less than 5. This ensures the materials have a very low environmental impact.

Mineral wool is the most popular choice as this does not have any temperature limitations. If used on external pipework the insulation should be protected by a vapour sealing barrier. If this is not provided the insulation

Table 9.10 Calculated minimum thicknesses of insulation comply with Part L of Building Regulations 2006 (reproduced from the TIMSA Guide). Thicknesses calculated according to BS EN ISO 12241 using standardized assumptions

Indicative thickness of insulation for non-domestic low-temperature heating service areas to control heat loss

Outside diameter of pipe on which insulation thickness has been based (mm)	*Water temperature of 75°C: ambient temperature 15°C* — *Thermal conductivity at insulation mean temperature* (W/mK) *(low emissivity facing: 0.05)*							*Maximum pemissible heat loss* W/m
	0.025	*0.030*	*0.035*	*0.040*	*0.045*	*0.050*	*0.055*	
	Thickness of insulation (mm)							
17.2	12	17	22	30	39	51	66	8.90
21.3	14	20	28	35	46	59	75	9.28
26.9	16	22	29	38	49	62	78	10.06
33.7	18	24	31	40	51	64	79	11.07
42.4	20	26	33	42	52	65	79	12.30
48.3	21	27	35	44	55	67	82	12.94
60.3	23	29	37	46	56	68	82	14.45
76.1	24	31	39	48	58	70	83	16.35
88.9	25	32	40	49	59	70	82	17.91
114.3	27	34	42	51	61	71	83	20.77
139.7	28	35	43	52	61	71	82	23.71
168.3	29	37	44	53	62	72	82	26.89
219.1	30	38	45	54	62	72	82	32.54
273.0 and above	31	38	46	54	62	71	80	38.83

can become wet, resulting in significant increase in the conductivity. If the water were to freeze, the insulation could become permanently damaged. For this reason mineral wool which is an open pore, fibrous material is often avoided on external applications. Mineral wool is chemically inert; this means that unlike some foamed plastic insulants which can be inherently acidic, when a leak occurs there is no problem of the water picking up any chemicals with potential corrosive effects on steel. The main disadvantage of mineral wool is that it has a relatively high conductivity (approximately 0.037 W/mK compared to 0.21 W/mK for phenolic foam). This results in significantly thicker sections to achieve the same heat loss, therefore when space for services distribution is at a premium, designers often choose phenolic foam.

Most foam-type materials including phenolic foam have an upper limit of 120°C, which means they can only really be used on low-temperature hot water systems.

Closed cell nitrile rubber is also limited to an upper operating temperature of just over 100°C. Because of its flexible nature it is generally used on refrigeration pipework rather than heating pipework. It has a low thermal conductivity and is resistant to water, unlike mineral wool or phenolic foam.

Both mineral wool and phenolic foam are supplied in preformed sections and provided with a factory applied foil facing which meets the requirements of Class O, as defined within the Building Regulations (i.e. a material with limited combustibility) with joints sealed using a suitable foil tape.

In areas where this finish could be easily damaged, i.e. plant rooms, riser cupboards and external plant areas, the insulation is usually protected by one of three methods:

(1) Aluminium or other metal cladding
(2) Synthetic rubber-type cladding (polyisobutylene or PIB)
(3) Laminated foil/film protection.

The recent trend is to use the laminated film protection particularly for external applications, this is mainly because it significantly cheaper than aluminium cladding but it also can achieve a complete vapour barrier, is more robust than PIB cladding, and will not split due to expansion and contraction. It also can be easily repaired without the use of solvents and when applied to ductwork it cannot 'balloon' from air leakage through the ductwork.

Table 9.11 Advantages and disadvantages of push-fit and press-fit copper/steel systems

Advantages	*Disadvantages*
Heat free jointing, therefore no requirements for hot working permits	Potential that electrical continuity is not maintained once the joints are made; depending on system used earth continuity straps may be required. Checks must be made with the manufacturers.
Advantage can be made of relatively unskilled labour	A quality assurance system is required on site to ensure all joints are made in accordance with the manufacturers recommendations, to ensure 'O' rings are not damaged, pipes ends are prepared correctly, pipes are inserted into fittings correctly, pipes are supported correctly and press-fit joints are made correctly.
System is intrinsically clean after installation, therefore no or minimum requirements for flushing to remove flux residue	Failure of a poor joint can pass a pressure test but failure can be catastrophic, rather than a leak that would be observed during testing. There have been examples of system failure, weeks or months after a successful pressure test.
Significantly reduced installation time	Pressure and temperature of push fit systems generally limited to 90°C and 6 bar gauge. Press fit systems limited to 110°C and 16 bar gauge. The systems use 'O' rings which will have a limited life expectancy depending on operating temperature/pressure.
An overall cost saving can usually be demonstrated	The system manufacturer's proprietary tools must be used for 'pres-fit' joints.
	Special care must be taken in the design and installation of pipe supports and facilities for thermal expansion to ensure joints are not misaligned which could cause failure of 'O' rings.

Internal pipework distribution systems and jointing methods

Historically the majority of heating pipework was installed using one of the two traditional materials/jointing methods below:

(1) *Copper*, with either capillary soldered or compression joints was used for almost all domestic installations.
(2) *Steel*, with a mixture of screwed and welded/flanged joints was used for commercial buildings. Typically pipework was screwed up to 50 mm diameter and welded/flanged for 65 mm diameter and above.

Over the past 5–10 years there has been a significant shift away from these traditional techniques towards the use of both alternative materials and jointing methods. The main drivers for this change is the overall cost of these systems and the lack of skilled labour available for the traditional methods. The materials are generally more expensive than traditional copper or steel but significant savings can be achieved on the installation costs. Independent case studies have been carried out by BSRIA as part of their series of 'Innovative M&E Data Sheets' Ref. ACT 5/2002, where it has been shown that up to a 25 per cent saving can be achieved by using one of these techniques in lieu of the traditional method. The materials/techniques include:

Copper and thin walled, carbon and stainless steel tube with either 'pres-fit' or 'push-fit' joints. Push-fit fittings have been around for about 15 years and there are a number of these systems available from different manufacturers, many of which have been available for over 10 years. Press-fit fittings are a newer technology in the UK and they are compressed onto the appropriate tube using a special tool. The advantages of press-fit over push-fit systems are that they are available in larger sizes (up to 108 mm diameter) and suitable for higher working temperatures.

Plastic and multi-layer pipe systems (usually aluminium pipe between layers of either HDPE – high-density polyethylene or PE-X – cross-linked polyethylene) with a variety of jointing methods including solvent, push-fit, compression fusion and electrofusion. The HVCA produce a guide to the use of plastic pipework (Ref. TR/11) which compares the various materials and jointing methods currently available.

Tables 9.11 and 9.12 provide a summary of the main advantages and disadvantages of the push- and press-fit systems and of the plastic systems compared to traditionally joined copper/steel systems.

Table 9.12 Advantages and disadvantages of plastic pipe systems

Advantages	*Disadvantages*
Plastic pipes usually supplied in coils, therefore individual pipe runs can be cut to length on site to minimize joints	Plastic pipe can be subject to oxygen diffusion, where oxygen molecules can penetrate through the tubing wall and if there are ferrous materials in the system (e.g. radiators) general corrosion will occur. Therefore all pipes used must have an oxygen diffusion barrier either integral or applied as a coating externally.
No painting of pipes required	Maximum temperature will vary with plastic type but limited to 80–90°C. To achieve a life expectancy of over 25–30 years the temperature should generally be limited to 70°C and 6 bar gauge.
Heat free jointing, therefore no requirements for hot working permits and lower skill level may be required	Materials cost higher than steel or copper pipe.
System is intrinsically clean after installation, and internal surfaces will not corrode or, an therefore no or minimum requirements for flushing to remove flux residue	Additional supports required as plastic pipes sag under their own weight.
Reduced installation time	Additional allowance required for thermal expansion as coefficient of linear expansion of plastic is much higher than copper or steel.
Lightweight, making off site prefabrication site, maneuvering into position and installation easier.	Less robust than steel, therefore probably not suitable for plant rooms and risers.

External distribution systems

It is not unusual that, in the design of a heating system for a large site, the matter of external pipe distribution arrangements has to be considered. There are three possible solutions to this problem:

(1) Arrange that piping be run overground, with particular care as to topography and any building features which may assist.
(2) Arrange that piping be routed through basements or, if none exist, through purpose-built walk-in subways.
(3) Provide for excavation and either formation of underground ducts or direct underground burial of piping enclosures.

The last of these approaches will most probably be thought of as being the only practical solution in terms of first cost. The margin of advantage, however, will be less than first imagined if the whole of the site preparation and subsequent building construction methods are of the necessary quality. If they are not of that quality, excavation for and replacement of the pipework is not only inevitable in the long term but likely to be required within 5 years of installation.

Overground arrangements

On industrial and semi-industrial sites, such as dispersed hospitals, etc., where appearance may not be a first-order priority, pipework may be run externally and be supported either at waist level on road verges or from gantries where vehicular traffic must have passage. The thermal insulation must be provided with a finish which is weatherproof and reasonably resistant to vandalism: supports, etc., must be arranged so that the finish is not punctured. Use of the roofing felt and wire netting combination so often seen is inadequate since it is not fully weatherproof with the result that the insulation material absorbs water and corrosion follows. Sheet metal cladding or a proprietary finish, such as laminated foil film, is to be preferred.

In some city centre and housing developments, there has been a tendency to separate pedestrian and vehicular traffic by use of bridges and overhead walkways. These offer the prospect of other routes for external piping, as shown in Fig. 9.20 where protection to insulation by way of cladding may not be required.

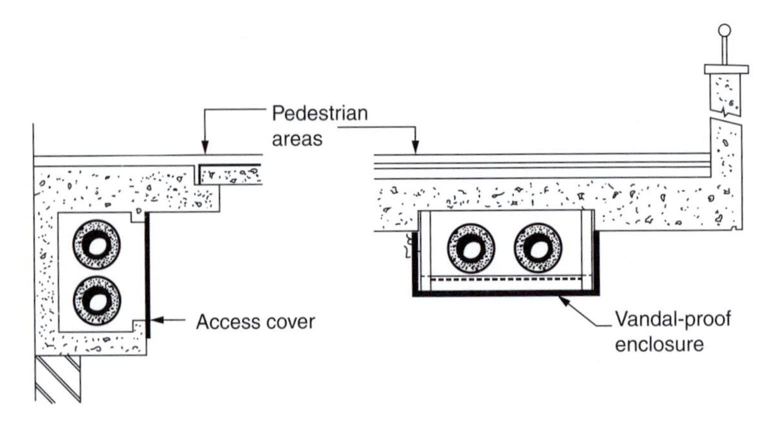

Figure 9.20 External pipework protected by overhead walkways

Basements and subways

In an era when the availability of skilled plant operators is limited, it is more than ever desirable that full consideration should be given to facilities for the day-to-day routine of preventative maintenance. The existence of a basement corridor through which pipework may be routed is a bonus in this respect, in particular where no suspended ceiling is fitted below the service route. Between buildings, a tunnel or subway of sensible size is needed: by the time that an excavation has been made and a waterproof construction formed, it is barely relevant in terms of real cost whether the cross-section is 1.5 m by 1.5 m or a metre bigger either way.

In terms of pipework installation, what little saving might have been made in constructing the small tunnel, rather than the larger subway, will have been swallowed up as a result of the sheer difficulty of working on hands and knees. As to subsequent maintenance, leaking valve glands, dripping joints, distorted flanges, corroded supports and damaged insulation would all go unnoticed were they to occur in the small tunnel.

Underground ducts and buried distribution systems

Most modern heating pipework distribution systems from CHP systems serving district heating schemes use buried pre-insulated steel or plastic pipes that have an outer casing of polyethylene. There are however many older buildings including hospitals where some or all of the alternative techniques described in the following paragraphs and sketches are still in use.

The principal hazard to which underground enclosures are exposed is ground water since, when an excavation is made preparatory to the construction work proper, the natural water table is disturbed and any seepage tends to follow the boundary between the undisturbed ground and the later backfill. Most of the pipeline enclosures used in the past have been site formed, as in Fig. 9.21(a)–(e), and are thus vulnerable as far as site settlement is concerned. Ventilation ducts within buildings, constructed in a manner not unlike that of examples shown in the diagram, are rarely more than 50 per cent airtight when new: it seems unlikely that a better result can be obtained under the more adverse conditions prevailing externally. Finally, many such ducts have been constructed within the constraints of an inadequate cost allocation, which has led to less than perfect detailing and workmanship. A brief description of the enclosures shown may, however, be useful, although none can be recommended:

(a) Insulated pipes laid within a duct having brick or concrete walls built as a preformed or *in situ* channel and supporting a preformed concrete cover.
(b) Insulated pipes laid within a duct having a preformed half-cylindrical cover set on an *in situ* concrete base.
(c) Construction as (a), but with a loose granular insulating fill poured round the bare pipes.
(d) Construction as (a), but with a fill of aerated concrete poured round the bare pipes.

Figure 9.21 Underground pipe ducts (past practice – now superceded by buried pre-insulated pipes with outer plastic pipe)

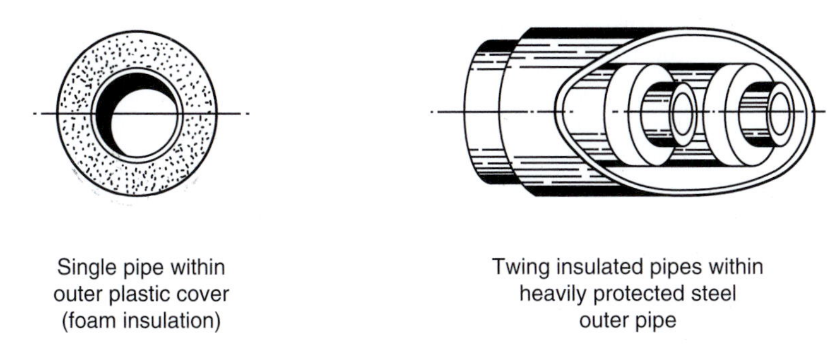

Figure 9.22 Pipe-in-pipe systems for underground heat distribution

(e) Raw excavation, sometimes lined with tar paper, where bare pipes are laid for insulation by a hydrophobic powder which cures when it is heated to form a protective skin on the pipe.

Instances of failure of each of these methods of construction have resulted in these methods not being used for new distribution systems. Instead the use of one of the factory formed *pipe-in-pipe* systems (Fig. 9.22) is now accepted as the industry standard. These systems comprise a steel service pipe with polyurethane thermal insulation and an outer casing of polyethylene. BS EN 13941: 2003 specifies the rules for design, calculation

Table 9.13 Heat loss from pipe-in-pipe system buried 600 mm below ground level

Size of service pipes (mm)	*Insulation thickness on pipes* (mm)	*Diameter of enclosing conduit* (mm)	*Heat loss (W/m run of enclosing conduit for following mean water temperatures)*		
			75°C	*100°C*	*125°C*
Two × 20	25	250	39	53	63
Two × 25	25	250	42	58	73
Two × 32	25	300	50	65	83
Two × 40	25	300	54	71	89
Two × 50	25	300	58	79	99
Two × 65	25	350	69	92	117
Two × 80	37	350	62	84	108
Two × 100	37	400	73	98	123
Two × 125	37	450	83	110	140
Two × 150	37	600	96	130	165

and installation for these pre-insulated bonded pipe systems which are generally used for district heating systems. The pipe assemblies themselves are manufactured in accordance with EN 253, for continuous operation with hot water up to 120°C and occasionally with peak temperatures up to 140°C and maximum internal pressure 25 bar.

Table 9.13 lists heat loss data for one such system. The Achilles heel of site application of these is the point at which they enter a valve pit or other site formed structure, where the watertight continuity of the outer pipe is broken.

The use of pre-insulated plastic pipe systems for underground district heating networks of low-temperature hot water up to 90°C and 6 bar pressure is becoming more popular as an alternative to the steel systems described above. These proprietary systems are available up to pipe sizes of 150 mm. The systems include pre-insulated proprietary fittings for tee pieces, elbows, branches, wall entries, etc. The internal carrier pipe is usually manufactured from polybutylene and the jointing method is electrofusion welding/butt fusion welding. The insulation is either polyethylene or polyurethane foam with the outer casing made from corrugated black polyethylene. The pipelines are either available on coils or in straight 6 or 12 m lengths.

Most manufacturers will guarantee their systems for at least 50 years when operating at 80°C for a minimum of 12 hours, 365 days/year.

Chapter 10

Heat emitting equipment

The heat emitting equipment described in this chapter is generally for use with the indirect low-temperature hot water systems described in Chapter 9. Heat emitters for direct systems (electric and gas and oil fired) are described in Chapter 7 and electric storage heat emitters are described in Chapter 8.

Some of the emitter types described are no longer manufactured or utilised in modern buildings. They are however included to give the reader an appreciation of the technologies that were used in the past and may still be found in older buildings.

With the exception of certain rather specialist applications involving pipework embedded in the building structure, there are very few items of heat emitting equipment which could not be used in conjunction with the whole range of water and steam distribution media. That is not to say however that such use would be equally effective in all cases, nor would it always be acceptable from the point of view of avoidance of burns. Exposed heating surfaces must not be accessible to touch if the temperature exceeds 80°C and in many circumstances including healthcare facilities, the surface temperature of space heating devices and any exposed pipework within 2 m of the floor should not exceed 43°C when the system is running at the maximum design output. This can either be achieved by using purpose built low surface temperature heat emitters/radiators and boxing out any low-level pipework, or by using water with a flow temperature below 45°C.[1]

Principal criteria

Normal design temperatures

For water systems, the maximum design flow temperature from a boiler for 'Low' and 'Medium'- temperature hot water systems are given in Table 9.1 in Chapter 9. The recommended operating temperatures at the standard UK design external temperature (see Chapter 4) are given in Table 10.1.

Table 10.1 Design temperatures for hot water systems with various emitters

	Temperature (°C) at water heater	
System	*Flow*	*Return*
Low-temperature hot water		
Underfloor heating	50	40–45
Radiators	70–80	50–60
Radiant panels and strips	80	70
Low surface temperature radiators	50–60	40
Natural and forced convectors	80	60
Fan coil units	70	50
Chilled beams	40–50	40–30
Industrial unit heaters	80	60
Medium-temperature hot water		
Radiant panels and strips	120	85
Forced convectors and unit heaters	120	90

Table 10.2 Values of coefficients and exponents for natural convection

	Theoretical		*Simplified*	
Surface and aspect	*C*	*n*	*C*	*n*
Horizontal facing up	1.7	1.33	2.5	1.25
Vertical	1.4	1.33	1.9	1.25
Horizontal facing down	0.64	1.25	1.3	1.25

Fundamentals of heat transfer

The various types of heat emitting equipment which are the subject of this chapter are those fitted within or immediately adjacent to the space served. Output will be both by convection, radiation and by a combination of radiation and convection, the proportion of each being determined by the form which the equipment takes. There are two empirical relationships applying to emission from plane and cylindrical surfaces which are used to put values to such output, these being:

Convection $h_c = C(T_s - T_a)^n$

Radiation $h_r = 5.67e(T_1 - T_2)$

where

h_c = heat output by convection (W/m^2)
h_r = heat output by radiation (W/m^2)
C = a coefficient (Table 10.2)
e = emissivity of the heated surface
N = an exponent (Table 10.2)
T_s = absolute temperature of the heated surface (K)
T_a = absolute temperature of room air (K)
$T_1 = (T_s/100)^4$
$T_2 = (T_a/100)^4$

The coefficient and the exponent in the convection equation are varied, as shown in Table 10.2, in order to reflect the direction of heat emission, upwards or downwards, and the attitude of the surface, horizontal or vertical. These circumstances affect the nature of the air current pattern over the surface, which arises with buoyancy change, to the extent that movement may be either smooth or turbulent. In most practical cases of natural convection where no fan or other external force is involved, flow will be in a transition stage no longer smooth, but not yet fully turbulent.

Use of the convection equation may be simplified in consequence, without significant error for this transition stage, by using a single value of 1.25 for the exponent and the alternative values, as listed also in Table 10.2, for the coefficient. As to the radiation equation, the value for emissivity may be taken as being between 0.8 and 0.95 for metal which is either tarnished or painted and also for most building surfaces.

Under the headings which follow, the various types of heating equipment are dealt with in terms of the predominant component of the output, radiant or convective. The suitability of each type, for use with the various distribution media is discussed.

Combined radiant and convective heating

Pipework

As discussed in Chapter 9, exposed pipes are rarely used as heat emitters, however it can be seen from Table 9.8 that a 100 diameter steel pipe with a high emissivity surface of 0.95 will provide approximately 300 W/m with a water temperature of 80°C, with this output increasing to approximately 600 W/m if high-temperature hot water or steam is used. Tabulated figures for heat emissions from pipe sizes up to 400 diameter and a range of temperatures can be found in Section 3 of the CIBSE Guide C: 2007.

Radiators

The name is a little misleading in that the majority of heat is transmitted through convection rather than radiation, typically in the region of 30 per cent radiative and 70 per cent convective. Table 10.3 gives the proportion of radiant and convective emissions from the two most common types of radiator (panel and column).

Modern radiators are tested and rated in accordance with BS EN 442 part 2 which dictates that the outputs should be given at water flow and return temperatures of 75/65°C and a room temperature of 20°C (i.e. temperature difference or Δt of 50 K between air and mean water temperature). All manufacturers' information should state the output at these conditions. Table 10.4 gives a series of correction factors by which manufacturers' outputs at 50 K should be multiplied to give the actual radiator output at the alternative temperature difference. Alternatively the correction factor can be calculated by using the equation.

$$(\Delta t/50)^{1.3}$$

The nominal output of a range of radiator types is given in Table 10.5. It should be noted that there is an increasing usage of chrome finish tubular/column style architectural radiators where a significant proportion of the output is radiative. Manufacturers will usually quote outputs from these types of radiator with a white

Table 10.3 Emission from radiators in an isothermal enclosure: proportion radiant and convective

	Proportion of output (%)	
Radiator type	*Radiation*	*Convection*
Pressed steel single panel without convector fins	50	50
Pressed steel double panel each with convector fins	30	70
2 column	30	70
4 column	19	81
6 column	17	83

Table 10.4 Temperature corrections $(\Delta t/50)^{1.3}$

Temperature difference between mean water temperature and room temperature (K)	*Correction factor*
20	0.304
25	0.406
30	0.515
35	0.629
40	0.748
45	0.872
50	1.0
52	1.052
53	1.079
54	1.105
55	1.132
60	1.276

Table 10.5 Typical emissions from different radiator types for a temperature difference of 50 K between air and mean water. (W/m^2 of elevation)

Radiator type	*Nominal heat emission W/m^2 of elevation*
Pressed steel single panel without convector fins	1000
Pressed steel single panel with convector fins	1650
Pressed steel double panel with single convector fins	2400
Pressed steel double panel each with convector fins	3000
Cast iron sectional column	2100
Architectural style vertical steel panel (typically 2 m high)	1000–1400
Architectural style tubular or vertical column (typically 1.8 m high)	1000–1900
This type is also used as bathroom and kitchen radiators and special versions are available as heated towel rails	

Figure 10.1 Bundy Radiator 1877

painted finish, it is worth noting that the output of these types of radiators could be reduced by approximately 20 per cent due to the low emissivity finish.

Cast iron column type radiators

The first cast iron sectional radiators were produced in the USA in 1877 an example of which is shown in Fig. 10.1 and James Keith took out the first English patent in 1882.

Figure 10.2 shows two types of cast iron column type radiators. This type of emitter can be seen in numerous older buildings and are still used in the refurbishment of period buildings. They are however generally no longer used in most modern buildings and only manufactured in small quantities by a few manufacturers for replacement of units in period homes, etc. Numerous manufacturers however manufacture column type radiators in steel in a number of styles.

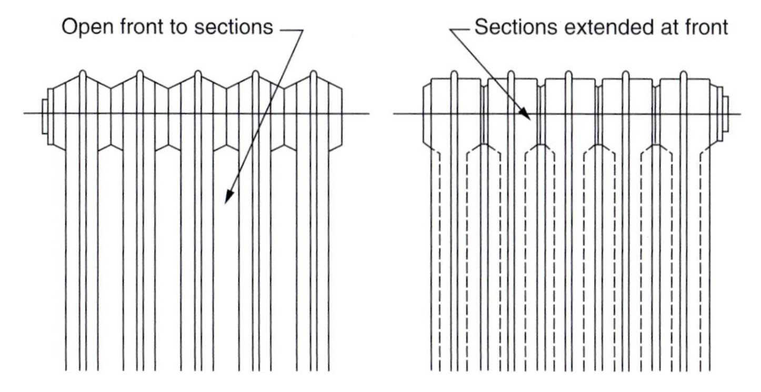

Figure 10.2 Types of cast iron radiator

Steel radiators

The majority of modern radiators are fabricated from steel. Steel is more susceptible to corrosion than cast iron and thus it is essential that corrosion inhibitors are used in the heating system.

Steel radiators are generally manufactured by two distinct techniques:

- Lightweight steel sheet pressings welded together with extended surface (or fins) welded to the rear to increase the heat transfer surface area and ensure they emit most of their heat by convection. These are manufactured in four types: Single panel, single panel with convector, double panel with single convector, double panel each with convector fins.
- Welded tubular or panel type construction, with or without extended surfaces to increase the convective component of their output. These come in a multitude of shapes, sizes and finishes by many manufacturers, the selection of which will often be dictated by aesthetics not engineering principles.

Some examples of the variations of steel radiator that are available are shown in Figs 10.3 and 10.4.

The particular merits of steel radiators result from their small mass and their comparatively narrow waterways: they are light to handle on a building site and respond quickly to temperature control.

In technical terms, siting a radiator under a window is preferred because:

- Heat is provided where most needed, at the point of maximum heat loss.
- Cold downdraughts from glazing are eliminated.
- Cold 'negative' radiation is countered at source and direction by hot 'positive' radiation.
- The temperature gradient in a room with radiators fitted below windows is less than if they were sited elsewhere.
- Marking of a wall surface by dust carried in rising convection currents is avoided.

It is important to remember that wherever possible the length of a radiator should be matched to that of the window beneath which it is fitted. A narrow high output radiator beneath a wide window will produce an equally narrow 'fountain' of warm air moving upwards, with a cascade of cooler air falling at either end of the window (Fig. 10.5). This was a significant problem on older buildings with single glazing with high '*U*' values but is much less pronounced on new buildings with doubled glazed windows with very low '*U*' values.

Painting of radiators was a sore subject at one time as a result of the industrial and commercial practice of using metallic paints, aluminium or bronze. These reduced the emissivity of the surfaces, and in consequence the total output, by a significant extent. It was later found that a coat of clear varnish over the metallic paint resolved the problem.

The concept of energy saving by provision of a reflective surface behind a radiator has been widely publicised and extravagent claims of economy made. Two methods have been advocated, first by the provision of a metallic-foil covering to the hidden wall surface and second, by attachment of polished metal strips to the rear

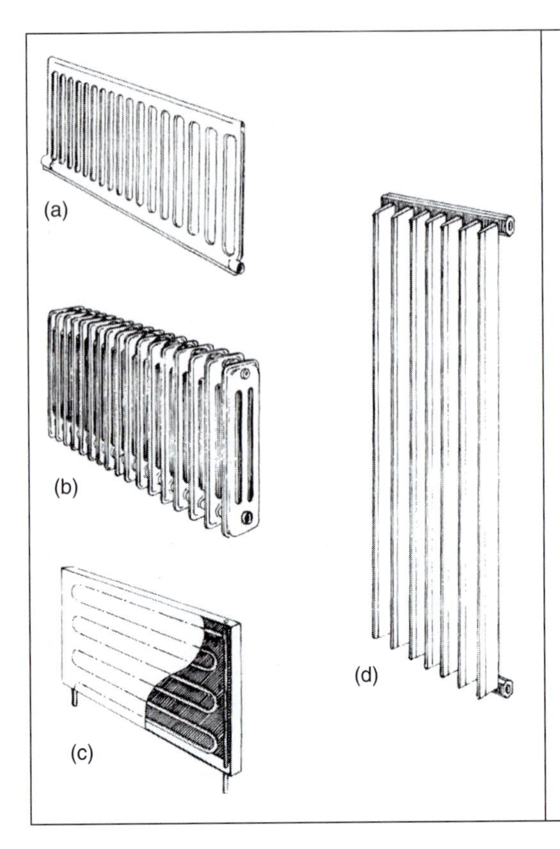

(a) Pressed steel single panel without convector fins. These can either have a seamed top or rolled/ round top.

(b) Column type (these reproduce the appearance of the classic cast iron radiators).

(c) Panel type, with a pipe coil attached to a flat panel. This acts effectively as a radiant panel with negligible convective output.

(d) Tubular type with top and bottom headers and pipe arrays connecting them. This type is often specified by architects.

Figure 10.3 Steel radiators (some radiant component)

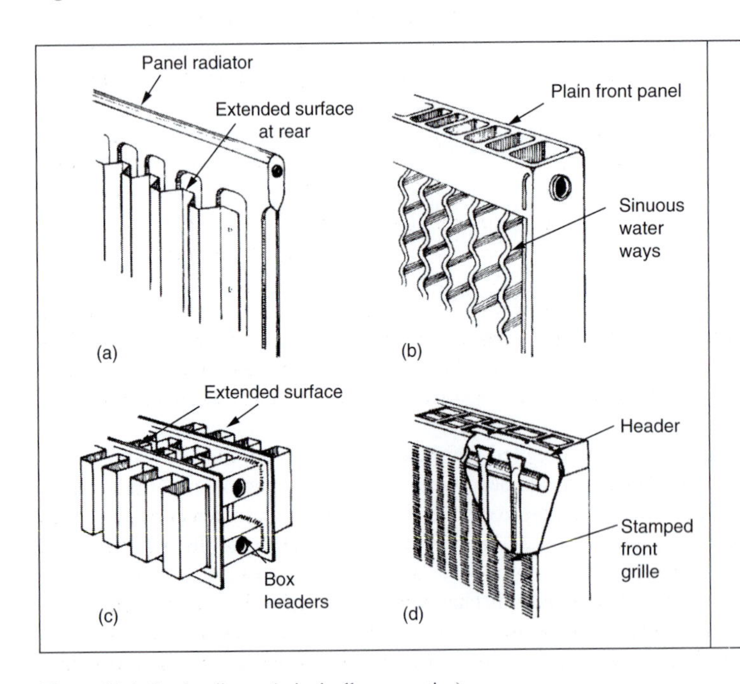

(a) Pressed steel single panel with convector fins (extended surface at rear). This type is also available as a double panel, with or without the extended surface.

(b) A plain fronted panel type with additional sinuous waterways at the rear. This type is more costly that the pressed steel type and provide a flat front.

(c) This is effectively a convector rather than a radiator.

(d) This again is effectively a convector and will operate as a low surface temperature heat emitter suitable for healthcare facilities and children's nurseries. It comprises a tubular element attached to a front plate complete with top and bottom louvres.

Figure 10.4 Steel radiators (principally convective)

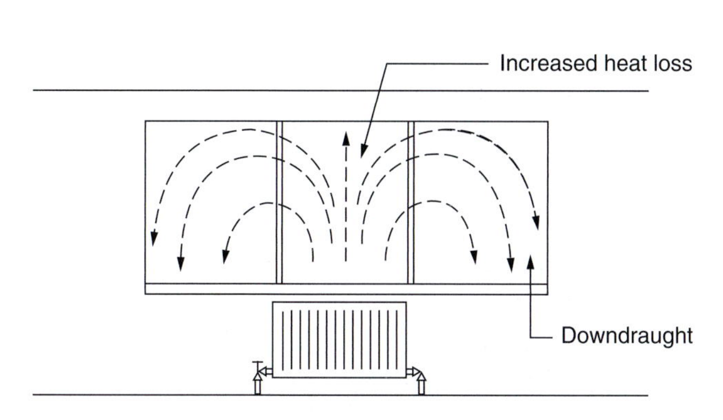

Figure 10.5 Mismatch of radiator to window

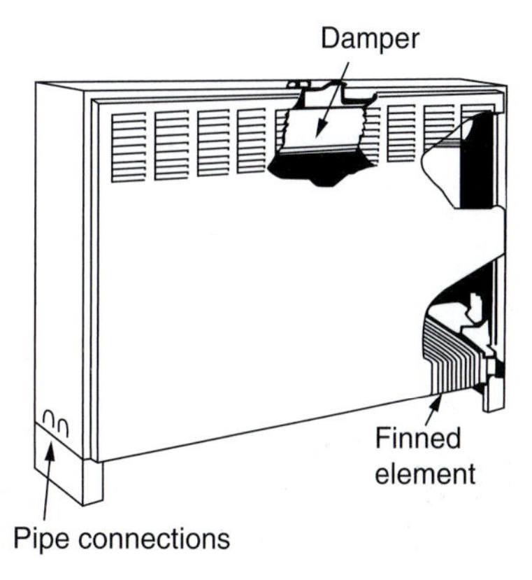

Figure 10.6 Cabinet type natural convector

of the radiator itself. The efficacy of either approach depends to a large extent upon how well the wall construction behind the radiator has been insulated (i.e. the *U* value). In the case of a solid 220 mm brick wall, the energy saving for a typical domestic living room might be of the order of 3 per cent but for a 260 mm cavity wall with an effective insulant between the leaves, it would be less than 1 per cent. There would of course be no saving at all if the radiator were fitted on a partition wall dividing two heated rooms.

Convective heating

Cased natural convectors

Although a wide variety of equipment falls within this general category, it is used here to identify the cabinet type only, other patterns being dealt with separately. The principal components are a finned tubular element mounted near the bottom of a sheet metal casing such that a *chimney effect* is created and a rising column of warm air flows from the top, inducing an inlet of room air at or near the base. Figure 10.6 shows the general arrangement of an old style unit.

The heating element will normally span between two headers which accept the external pipe connections and, depending upon the required duty and the particular manufacturer, take a variety of forms. The relationships are complex since the tubes can vary as to number, size, and shape (round or oval) and the fins as to area, spacing, thickness, material and method of bonding to the tubes (solder or mechanical expansion). The casing height has an important effect upon output as it provides the *chimney effect* (the larger the distance between the top of the convector and the outlet grille the greater the unit output).

Control of output may be either by adjustment of the temperature of the heating medium, or locally by use of a damper fitted within the casing. Ideally the damper should be mounted just above the element, but it is more usually fitted at a convenient hand level as shown in Fig. 10.6. Emission with the damper closed is by radiation from the casing, which then has an increased surface temperature and amounts to about 20 per cent of the normal output.

This old style of convector has, in recent years, been superseded by more compact low water content convectors (that are often marketed as radiators). These can be fitted with fans to boost the output of the unit to preheat the space in the morning making them a hybrid between a natural convector and a fan convector, (see Fig. 10.14). This type of emitter can also have special fresh air intake units integrated into the casing of the convector that are ducted to outside. The fresh air intake unit has a variable speed fan that can modulate its speed to maintain acceptable CO_2 levels in an occupied space. This outside air intake fan can also run overnight to provide night cooling in summer. Space temperature control is achieved by using weather compensated variable temperature hot water, combined with local thermostatic radiator valves to provide local final trimming control in conditions of high solar gain.

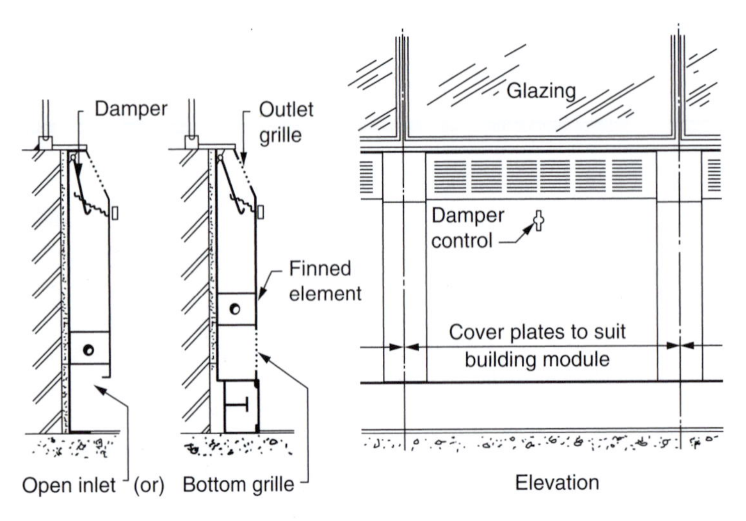

Figure 10.7 Continuous type natural convector

Continuous natural convectors

This application of the convector was used extensively in 1960s/1970s style office buildings where there were severe cost constraints. Where a continuous glazing strip was formed above still level, the finned heating element is continuous from end to end of an elevation, interrupted only by axial type expansion bellows. The metal casing is continuous also, as in Fig. 10.7. The louvered outlets at the top are in sections, arranged to suit the modules, and each is provided with a damper, no other form of control being suitable. The air inlet may be at an open base, as shown, or through a low-level grille. In circumstances where the sill height permits, the base of the unit may be raised and the casing integrated with electrical distribution trunking. Where partitions occur, a soundproof barrier is provided within the casing.

There is, or should be, a limit to the length of any single run since output will decay from module to module as the water temperature falls. A much better arrangement is to use the fundamental concept but adapt it such that only a limited number of modules (say two structural bays) is served in series from pipe mains running within the casing. The appearance remains the same but there is less variation in water temperature from the first to the last module.

Skirting (baseboard) convectors

A form of heating that was also popular in the 1960s/1970s was skirting heating. This approach provides an unobtrusive method of heating for some parts of domestic premises and for halls and corridors, etc. in commercial or institutional buildings. The best known pattern is as shown in Fig. 10.8.

Trench heating (natural convectors and fan assisted types)

This type of heating was historically installed to heat churches and similar buildings, where cast iron pipes were run in shallow structural trenches half full of dust and covered by decorative cast iron grilles is now available. Their use is primarily to counteract downdraughts when recessed in the floor immediately adjacent to room height windows. The components usually include a metal channel section designed to accommodate a finned heating element, which may be fitted to a single pipe or a hairpin loop, and a transverse bar type grille. The grille is substantial, lying flush with the floor finish and may be either in sections or a full length 'roll-up' type.

A number of modern versions of both natural and fan assisted trench heaters are described below.

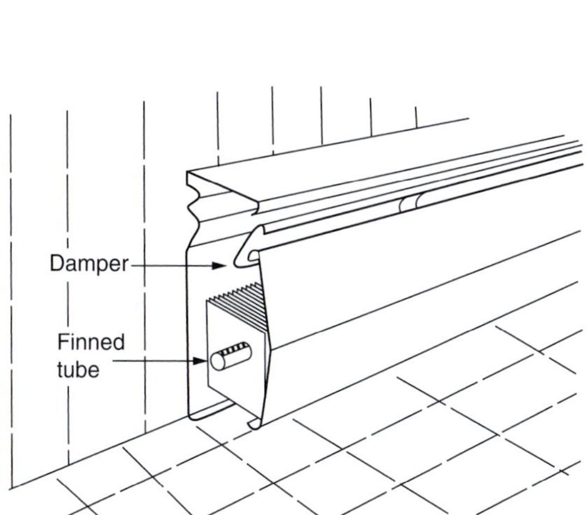

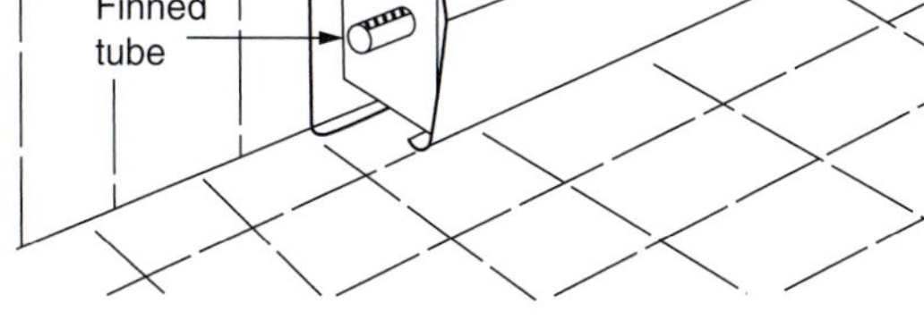

Figure 10.8 Skirting heater

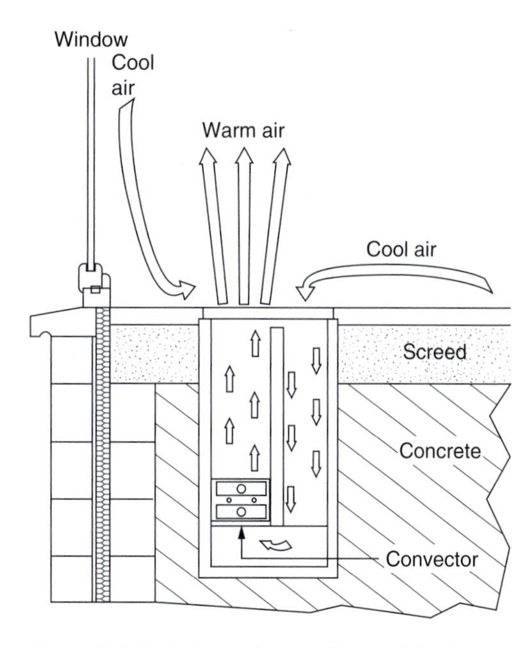

Figure 10.9 Typical natural convection trench heating using chimney effect (Jaga)

Natural convection

The principle of operation of this type of convector is shown in Fig. 10.9, which comprises a finned heating element mounted in a channel casing. The output of this type of unit is generally limited by the depth of floor void available. The significant heat outputs which are often required to preheat a building with a fully glazed façade can only be achieved by utilising the 'chimney' effect, which results in relatively deep units that are only suitable for installation into a ground floor slab.

Fan assisted trench heating

To increase the output of trench heating and make it suitable for a façade with full height glazing and installation within a typical raised access floor or a screeded floor, there has been a significant increase in recent years in the use of fan assisted trench heating. Figures 10.10–10.12 show two alternative types of fan assisted trench heaters, one of which can provide both heating and cooling.

Figures 10.10 and 10.11 show a unit constructed from a galvanised steel duct with a depth of only 112 mm and up to 5 m long. The unit is fitted with a factory wired variable speed fan with voltage regulation speed control that pressurises a primary air duct under the convector and air from this duct is blown through a series of nozzles. This process induces further 'secondary' air from the room to increase the air volume across the heating coil and hence increase the output.

The typical output for a 400 mm wide version of this type is approximately 850 W/linear metre of duct (using low-temperature hot water at 82/71°C). This output is with the fan at 50 per cent of its maximum duty creating a room noise level of 35–38 dBA. This output increases to approximately 1300 W/m with the fan at maximum speed. This increased output could be used to preheat a building from cold before occupancy and the fan speed limited to 50 per cent to offset the design fabric heat loss during the occupied period.

Figure 10.12 shows a new type of fan assisted trench heater where multiple cross-flow fans and motors with external rotor motors are used along the length of the convector. These are manufactured with duct depths of either 132 or 150 mm and up to lengths of 2.75 m long and are therefore suitable for either installing

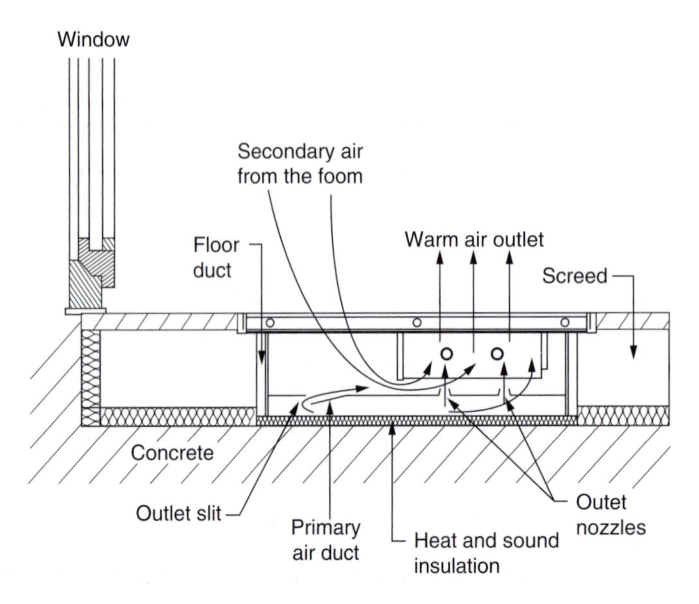

Figure 10.10 Fan assisted trench heating (Kampmann type GK – Functional view)

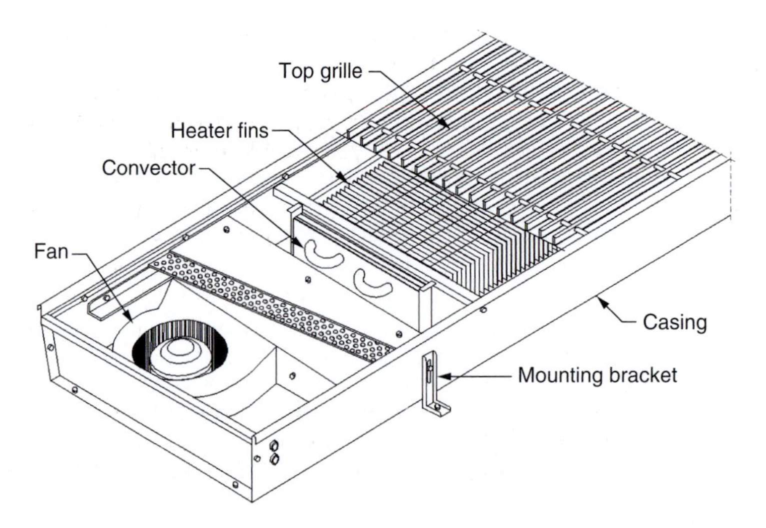

Figure 10.11 Fan assisted trench heating (Kampmann type GK – Isometric view)

in a raised floor or in the screed depth. (A 2.75 long unit has six fans and three motors and output is usually by speed control.)

This type of unit is designed to provide either heating or cooling. In cooling mode it can operate 'dry', i.e. without condensation forming, using chilled water at 16/18°C or can give greater output by using chilled water at 6/12°C, however at the lower temperature a condensate pump is required to be installed in the duct, which is a welded construction and doubles up as a condensate tray. The unit can be configured to either discharge air on the room side or towards the window if there is likely to be a problem with drafts from cold downdraughts from full height glazing. A feature of this type of unit is that a slot is incorporated between the convector at

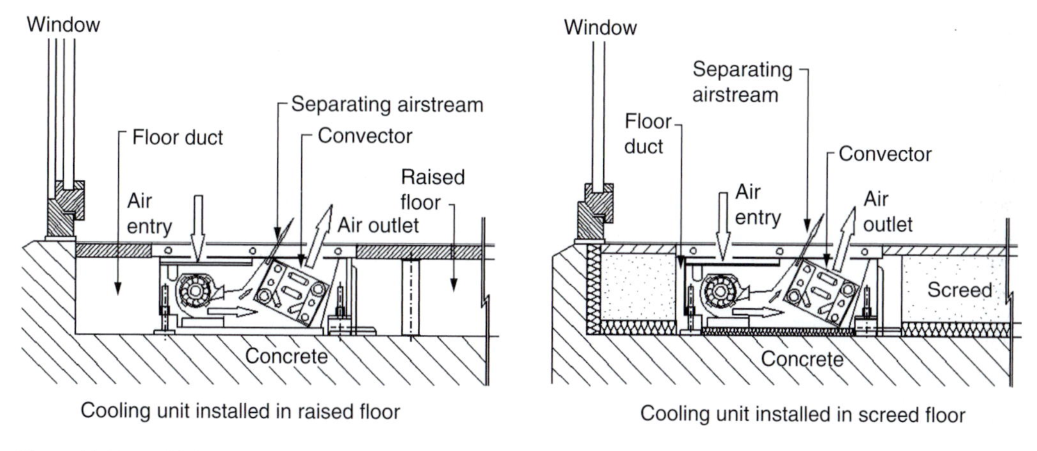

Figure 10.12 Multiple cross-flow fan type fan assisted trench heating (Kampmann HK)

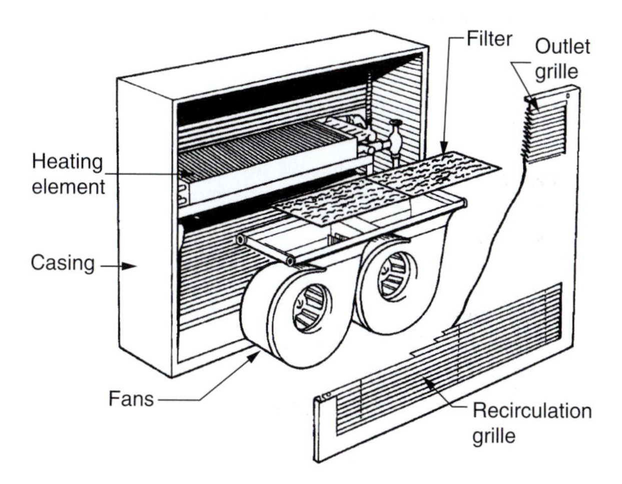

Figure 10.13 Forced (fan) convector

the fan discharge casing to create a jet that acts as a higher velocity separating airstream to prevent the warm discharged air from mixing/recirculating with the cooler entering air.

The typical heat output from this type of unit depends on whether the convector is piped up as a two-pipe unit (where all the heat transfer area of the coil provides either heating or cooling) or a four-pipe configuration (where the coil is split into heating and cooling sections) For a typical 340 mm wide, 2705 mm long, 150 mm deep 4-pipe unit, unit using low-temperature hot water at 82/71°C and running at medium speed the heat output is approximately 2000 W/linear metre of duct. The sensible cooling output is approximately 400 W/m with 16/18 chilled water increasing to approximately 800 W/m with 6/12 chilled water.

Forced convectors

Units of this type are best described by reference to an illustration such as Fig. 10.13 which, for a commercial size of convector, shows the various component parts in separation. The finned tube element will probably differ from the pattern used in a natural convector in that the headers, to which pipe connections are made, are likely to be one above the other at the same end of the casing, the tubes being in hairpin form to make up a ‘two deep’ arrangement. In the earliest models made, propellor type fans were used. These were superseded

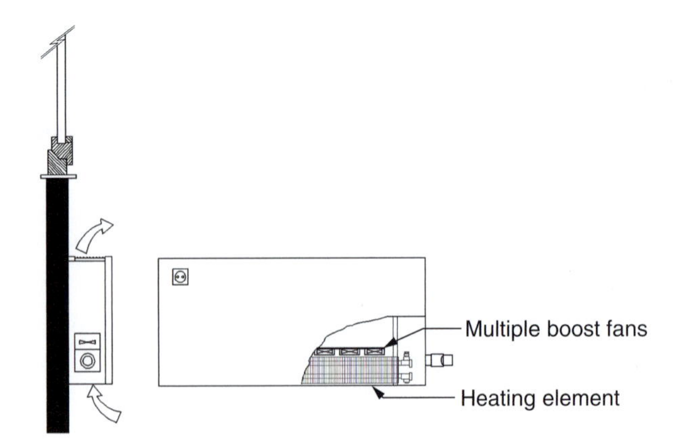

Figure 10.14 Natural convector with boost fans (Jaga Strada DBE)

by small quiet running centrifugal units having long-life motors mounted within the suction eye, which in turn were replaced in some models by tangential fans.

The heating element and fans are contained within a casing which, where free-standing, is similar to that provided for a natural convector. For the type of unit illustrated, provision is made at the fan discharge for an air filter. This is often a washable type, consisting of no more than a thin mat of synthetic fibre stretched on a wire grid, and able to prevent carpet fluff and other airborne dust from fouling the heat exchanger. Such an arrangement does not keep the fan clean and a better alternative is to provide a simple modular disposable type of filter immediately behind the recirculation grille.

Fan motors are usually provided with facilities for two or three running speeds and the lower or middle speed rating is that used for selection, operation then being acceptably quiet. The higher or boost speeds are available to provide an overheat facility after periods of shut-down when an increase in noise level may be tolerated for a short time. In some makes, the casing is provided with an acoustical lining in an endeavour to reduce the noise level. Use is often made of fan speed control by thermostat, timer or other device.

The size range available in the commercial pattern of unit is considerable and application is correspondingly wide. Medium- and high-temperature hot water and steam was used in the past but most modern units are now only manufactured for use with low-temperature hot water. In a block of flats, each separate dwelling may be provided with a unit fed from a central boiler plant. In a school, one heater may be provided for each classroom and several for service to a gymnasium or assembly hall. In lecture theatres and large spaces, which require mechanical ventilation when occupied but are not always in use, a fan convector may be used as a form of background heating.

A modern development of the fan convector as shown in Fig. 10.14, is the hybrid unit described earlier under "cased natural convectors" which comprises a low water content finned convector with a series of fans mounted on top of the convector assembled into a stylish casing. These units act as natural convectors in normal operation but can provide a 'boost' effect for morning warm-up or for extreme outside conditions.

Using active chilled beams to provide heating and cooling

An active chilled beam is essentially an air driven, ceiling-mounted induction unit. A dehumidified outside air supply is connected to the beam and this primary air is blown through a series of nozzles along the length of the beam at a relatively high pressure. This induces air from the room to flow across the cooling or heating coil which is then mixed with the primary air and diffused into the space from linear slots on either both sides, or only one side of the chilled beam. This principle is shown in Fig. 10.15.

A typical method of providing heating and cooling from an active chilled beam is by dividing the number of passes on the coil between the heating and cooling elements. For example, for a coil with eight passes across the length of the beam, six would be used in cooling mode and the remaining two would be used for heating.

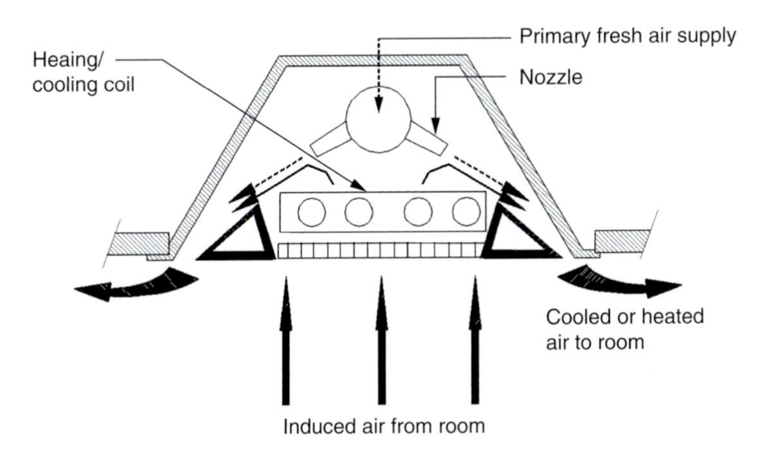

Figure 10.15 Typical active chilled beam operation

Most modern highly insulated buildings with windows with very good *U* values are suited to the application of heating from active chilled beams. However buildings with a high proportion of glazing or particularly leaky façades that are inherent in older buildings will require higher heating duties and therefore chilled beam will not be suitable.

Typically low-temperature hot water is used at flow temperature of around 40–50°C and a return temperature of 30–40°C (dependent on manufacturer and the duty required). These temperatures are significantly less than that of the industry accepted norms for low-temperature hot water (typically 82–71°C or 80–60°C), to avoid the air discharge from the beam resulting in stratification of the air. However these low temperatures are well suited to the use of condensing boilers or ground source heat pumps and thus this method of heating can result in a high seasonal efficiency for the heating system.

For comfortable conditions it is important that an occupant in a space does not experience a temperature gradient of more than 3°C. If a temperature difference of more than 3°C is experienced then occupants may complain of draughts. In cases where the glazing is significantly colder than the room temperature (i.e. single glazing) or where full height glazing is installed, downdraughts may be experienced and should be addressed by supplementing with low-level perimeter heating such as trench heating described previously.

To operate correctly in heating mode, the chilled beams will need the ventilation (primary air) to be operating. If no other means of heating is provided to preheat the building before occupancy, it is essential to include a recirculation facility in the central air handling plant in addition to any heat recovery devices that are installed to minimise energy use to heat the outside air.

Care will also need to be taken with the control of the primary supply air temperature and this is often scheduled against outside air temperature. The air being supplied at 14–16°C when the outside air temperature is above about 12°C (when there should either be a demand for cooling or the building's internal heat gains are adequate to maintain comfortable conditions without supplementary heat input), to about 20–21°C when the outside air temperature falls to approximately 1°C when almost full heating is required from the beams.

The height of a space is also a determining factor in the suitability of using chilled beams for heating. Most manufacturers use a ceiling height of approximately 2700 mm, this is important because a height much above this may lead to poor distribution of the conditioned air and bigger temperature gradients, resulting in stratification.

The heat output from a chilled beam is typically in the range 150–250 W/active metre of beam depending on the beam width and temperature of water supply. This relates to a capacity of 40–60 W/m^2 of floor area for beams typically spaced at 3 m centres.

Unit heaters

The use of unit heaters served by hot water or steam for industrial premises, has in recent years given way to the use of either radiant panels that are either served by low- or medium- temperature hot water, are gas-fired

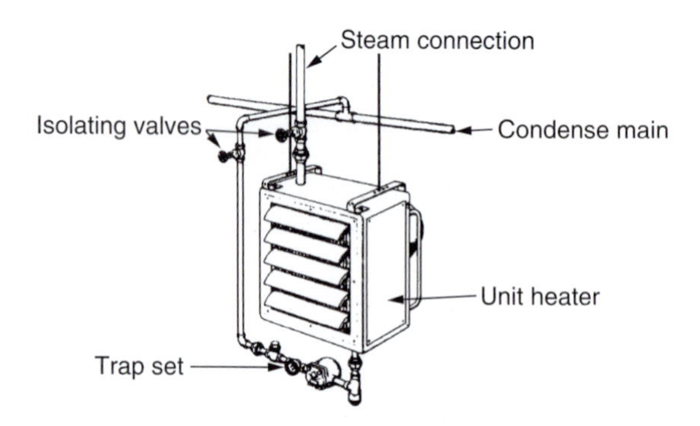

Figure 10.16 Suspended type unit heater, recirculating

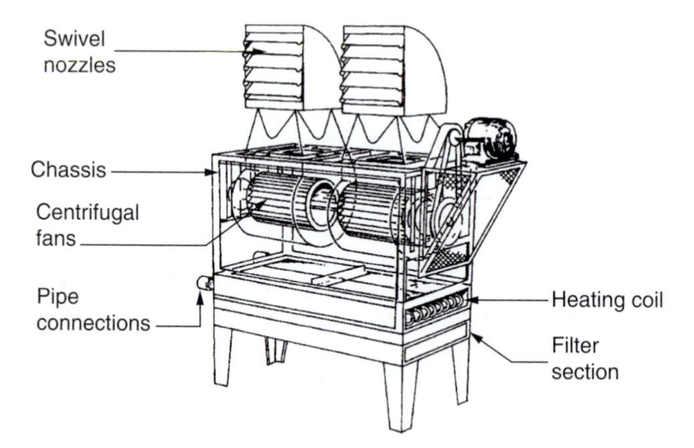

Figure 10.17 Projector type unit heater, recirculating

or are unit heaters of the direct gas-fired type (as described in Chapter 7). This method will reduce the overall cost by omitting the requirement for a central low-temperature hot water boiler and distribution system and is more energy-efficient.

In older factories as illustrated in Fig. 10.16, may still be in use using steam as the heat distribution medium.

Downward and horizontal discharge units are also used, the latter being floor mounted as shown in Fig. 10.17. It is important, when applying the latter, to ensure that the discharge air pattern is unimpeded by partitions or other large objects.

Since the capacity range of unit heaters of the various types is from 10 to 30 kW, it would serve no purpose to include a list here, particularly since dimensions are small and most are suspended rather than floor mounted. Selection must be from makers' lists, taking account of air volume, discharge temperature, mounting height and the *throw* which is produced. The principal criteria determining choice, are that a relatively lower air discharge temperature (40–50°C) and a larger air volume are to be preferred over the converse, since the buoyancy of air at higher temperatures will affect the throw available from a given mounting height, as well as encouraging an adverse temperature gradient.

Two alternative types of unit heater that were common in the 1970s to provide an outside air supply to industrial premises are shown in Fig. 10.18. Item (a) being a purpose made unit, again protection against wind pressure effects and heating element freezing are desirable – the filter is omitted for clarity. In instances where the design incorporates some units having an outside air supply and others which recirculate, it is desirable

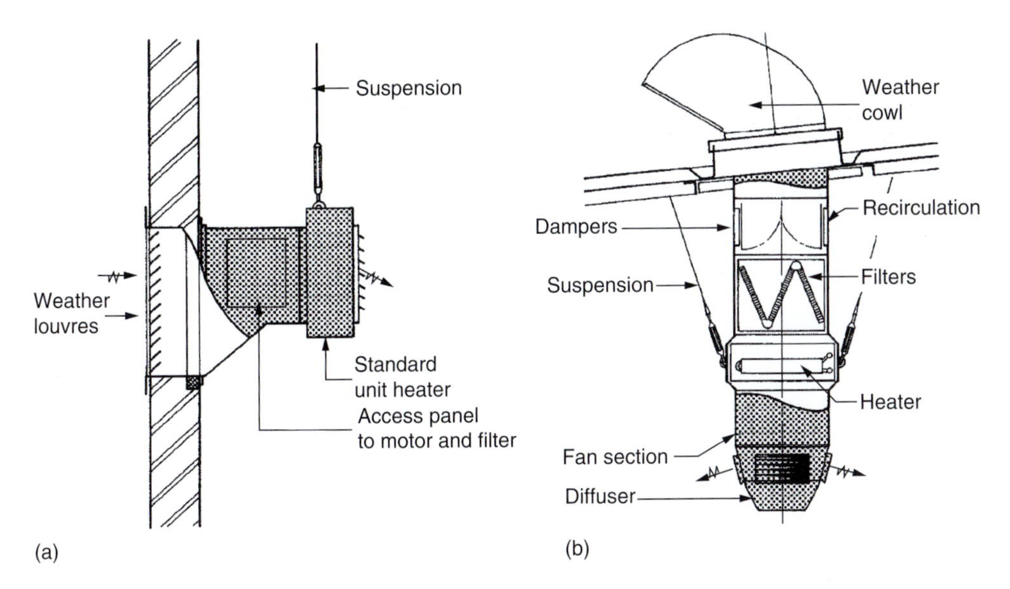

Figure 10.18 Unit heaters with outside air inlets

that output ratings for the two different duties are so selected that all will provide the same discharge temperature. The recirculation opening and damper on the duct in example (b) are provided to allow for a boost discharge after a period of shut-down and also to provide a measure of frost protection. It should be noted that if the fan motor to such a unit were to fail or be shut off by thermostatic or timer control, a reversed air flow will be induced by the heating element and unless the damper is in the recirculating position, air flow will then be discharged to outside. For this and other reasons, it is as well to provide for motorised operation of the damper, interlinked with the supply to the unit motor.

Modern industrial type heating and ventilating units

There is generally a requirement for heat recovery from the exhaust air from modern industrial type buildings, particularly for manufacturing areas that may have high fresh air requirements for replacement of exhaust from process areas or high occupancy.

A number of manufacturers produce units of the type shown in Fig. 10.19. These types of units comprise an above roof unit containing fans (which can be fitted with speed control), dampers and plate heat exchanger and a below roof unit comprising filters, low-temperature hot water heating coil and air diffuser section. The units can operate in a number modes including:

- Ventilation with heat recovery (as indicated in Fig. 10.19).
- Recirculation to preheat the building before occupancy.
- Exhaust only to rapidly clear a zone of smoke or other pollutants.
- Night summer cooling – similar to ventilation mode but where the exhaust air bypasses the plate heat exchanger.

Convectors generally

When control of natural convectors and similar equipment is to be by adjustment of the temperature of the heating medium, it should be noted that their output falls away more quickly than that from radiant equipment. It has been found in practice that the fundamental equations quoted at the beginning of this chapter must be

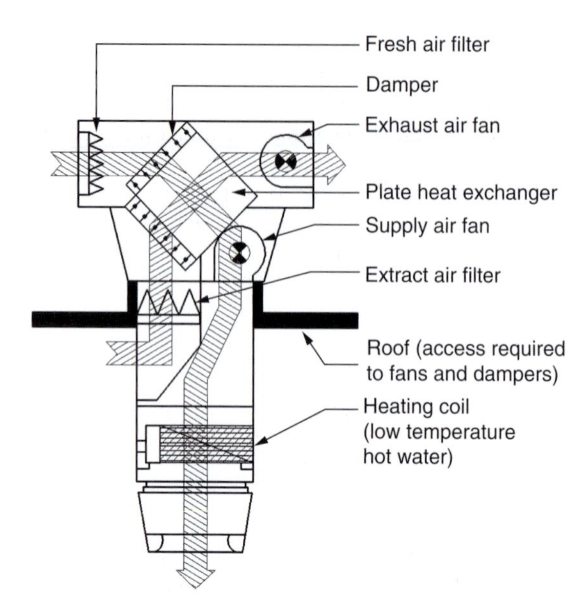

Figure 10.19 Modern industrial unit heater/ventilation system with heat recovery (Hoval Roofvent LHW)

adapted slightly to reflect the performance of actual equipment, output being proportional to the temperature difference, heating medium to air, as follows:

Radiant equipment Proportional to $(t_m - t_a)^{1.3}$

Natural convectors, etc. Proportional to $(t_m - t_a)^{1.5}$

For example if the temperature difference between mean water temperature and air temperature reduced from 60 to 20 K (i.e. water temperature of 40°C and room temperature of 20°C) the output of a radiant emitter would fall to approximately 25 per cent but the output of a convective emitter would fall to about 20 per cent of the output at 60 K temperature differential.

It is for this reason inadvisable to supply equipment of the two categories from the same piping circuit, but to provide for them separately, using different rates of temperature adjustment. Historically is was good practice to supply equipment of this type from a constant temperature circuit, control then being by either 'on/off switching of the fan motor, or speed change as mentioned previously thereby avoiding the forced circulation of 'cooler' air creating possible discomfort from draughts. However in modern buildings the prime objective is to minimise the use of energy and heat losses from distribution pipework. This practice is therefore adopted much less these days, with much greater use being made of weather compensated variable temperature circuits, operating in conjunction with condensing boilers to serve convectors and coils in air handling plant.

In earlier paragraphs covering convective heating, no mention has been made of builders work enclosures for natural or forced convectors, skirting heaters, etc. These of course, may be used in any situation where appearance is of importance and maltreated sheet metal is not acceptable. It is nevertheless important that the form and the dimensions of the enclosure and the dimensions of inlet and outlet grilles, be discussed with the heater manufacturer. Enclosures having grilles set into a flat top perform badly when covered with a variety of books and papers: a top angled at 45° is an encouraging first step in the movement of letters towards files (and files towards cabinets).

There are many further aspects in application of convective heating equipment which are unique to a given manufacturer and reference must in those circumstances be made to the technical data published. One single point, however, which seems to receive inadequate attention, is that the heating elements of most natural and forced convectors are particularly susceptible to reduced output as a result of accumulation of air in the waterways. As will be appreciated, there is little free space above the tubes in the element headers to accept air, in contrast with that available in a radiator. With those waters which tend to encourage the initial formation of gas and air mixtures, some engineers fit air bottles above the element header to allieviate the problem.

Any form of convector, natural or forced, requires some level of maintenance to keep the finned heating element free from dust, etc. even if this is no more than good housekeeping. It is for this reason that particular attention to convenience of access is not only desirable but absolutely essential.

Radiant heating

Hot air rises, 'So why put heating elements in the ceiling?' is a question often asked. As described in earlier chapters, a radiant ceiling heating system directly heats the room surfaces and people by radiation, so the heat cannot build up at ceiling level as can occur with a convective system. Typically 60 per cent of the output from a radiant ceiling system is by radiation, the remainder by convection.

People also perceive they may have a *hot head* with a radiant heating system. Most radiant systems that are installed in spaces such as sports halls, classrooms, hospital wards, corridors and offices use low-temperature hot water at temperatures between 30°C and 80°C. Therefore the perceived problem of a hot head is not usually realised as long as the panels are mounted high enough.

The radiant temperature asymmetry or RTA that is defined in BS EN ISO 7730: 2005 on thermal comfort, as the difference between the plane radiant temperatures on opposite sides of the human body should be less than 5 K for a warm ceiling. The BS also states that the percentage of people dissatisfied should be no greater than 5 per cent with a RTA of 5 K. RTA is measured at a plane 0.6 m above the floor with sitting activities or at 1.1 m above the floor with standing activities and depends on the installation height, mean surface temperature of radiant panel, the size (width and length) of the radiant panel and the temperature of the other surfaces in the room. It is worth noting that the panel mean surface temperature will be no more than 2–5 K lower than the mean water temperature.

The hotter the surfaces, the greater will be the required heater's installation height. Therefore when selecting and spacing radiant panels in a ceiling, the manufacturers should always be consulted, as it will usually be necessary to 'optimise' the above parameters to achieve a satisfactory solution and ensure the RTA of 5 K is not exceeded. The graph in Fig. 10.20 has been reproduced from information published by a manufacturer and gives an indication of the lowest installation height for a series of radiant panel widths at different mean heating panel surface temperatures with a radiant asymmetry of 5 K. Typically the smaller the surface area of the panels, the lower they can be installed without exceeding the limiting radiant asymmetry.

The use of radiant heating in high, narrow areas generally gives poor results. This is not due to the distance to the floor but is due to the fact that a large part of the heat radiation will be taken up by the walls.

Other questions that are often asked regarding the use of radiant heating are:

- Will it be cold under the tables giving rise to cold feet?
- Will there be downdraughts at the windows?
- Does radiant heating save energy?

Generally it will not be cold under tables, the radiated heat is absorbed by the walls and floor but some of the radiation is reflected to heat all the room's surfaces including the floor under desks. This means the difference between air temperature and radiant temperature is very small. It is worth noting that the radiant heating is indirectly heating the floor, resulting in the floor surface being 2–3°C warmer that the air just above the floor.

In modern well-insulated buildings with double or even triple glazing with windows forming 30–40 per cent of the façade it is not always necessary to provide heat under the window. A radiant ceiling panel close to the façade will heat the window's surface directly, as heat radiation is distributed to the room's surfaces in proportion to its surface temperature, more heat will go to the cold window surface, the frame and window sill. Provided sedentary occupants are not seated directly adjacent to windows, there should not be problems with cold

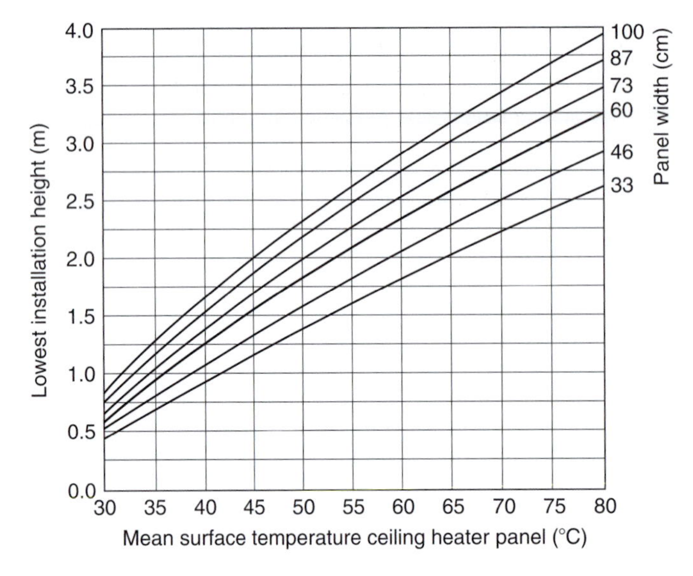

Figure 10.20 Lowest installation heights for radiant panels up to 3.6 m long such that RTA does not exceed 5°C (courtesy of Frenger Systems)

downdraughts or cool wall radiant asymmetry. If full height glazing is installed however, it is usual to provide some form of trench heating, particularly in offices, as occupants are usually very close to the windows.

A radiant heating system can usually demonstrate an energy saving because of the ability to maintain a lower air temperature in the occupied zone (typically 1–2°C lower), without the operative temperature (measure of comfort) being lowered. With high-level radiant heating, the temperature difference between ceiling and floor temperature is also lower, resulting in further energy savings by reducing the heat losses through the roof, particularly in high-ceiling areas, warehouses and factory type spaces. For most buildings where the ceiling height is between 2–3 m, savings can be in the order of 2–7 per cent. In buildings with higher ceilings, particularly if the building is old, has a high infiltration rate due to poor fabric or has large doors or openings, the energy savings can be much greater (studies have shown that savings of up to 30 per cent are possible).

Modern radiant heating panels

Offices, healthcare and similar applications

All modern radiant heating panels should be manufactured, rated and tested to BS EN 14037:2003 and the two most common forms of radiant panels that are commonly used in applications with sedentary occupants with ceiling heights of approximately 3 m are shown in Figs 10.21 and 10.22. These emitters are usually supplied with low-temperature hot water and have a design life expectancy of 25 years.

Figure 10.21 shows a typical custom made, smooth faced panel that is either manufactured from aluminium or steel. Copper pipes are rigidly fixed into extruded aluminium pipe seats, which are fixed to the rear of the aluminium panels to give the copper pipes an enhanced heat transfer surface. The panels are backed by a foil-backed layer of insulation. This type of panel can be free hanging, surface mounted or recessed/integrated into a suspended ceiling. The panels can also be wall mounted for applications such as sports halls and other large open plan areas.

A specific application of this type of panel is for prison heating where flush continuous panels are installed either at low level, as cornice type or at ceiling level. These units are designed for use with low-temperature hot water and include anti-vandal construction and an absence of ligature points. The units are purpose designed to meet the exacting requirements of prison 'Safe Cell' environments.

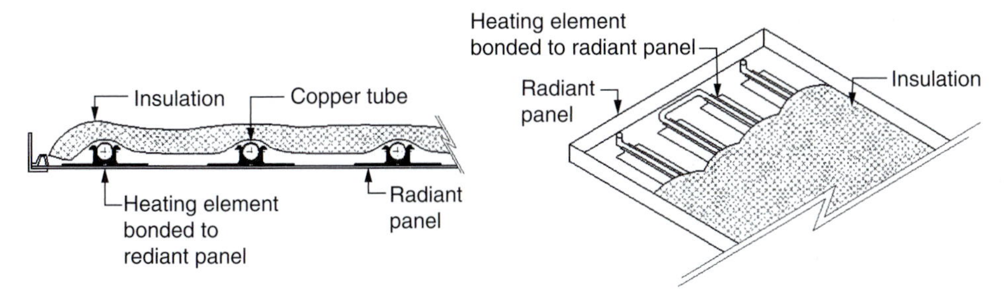

Figure 10.21 Radiant panel with copper tube fixed in aluminium pipe seats (Frenger Warm)

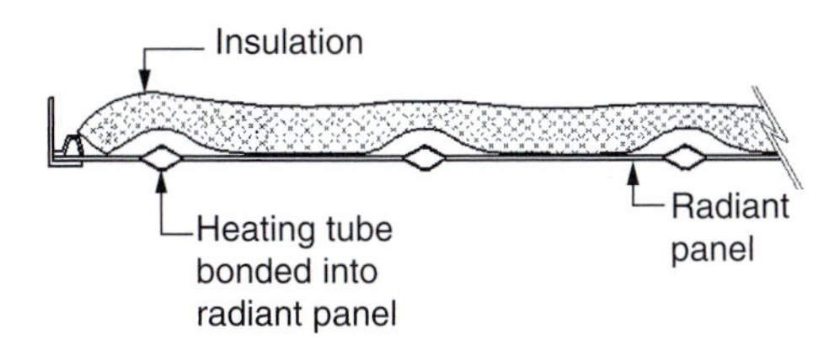

Figure 10.22 Radiant panel with copper heating tubes metallurgically bonded into aluminium plate (Frenger Atrium)

Figure 10.22 shows a high performance type of radiant panel that can deliver in excess of 500 W/m^2 (based on low-temperature hot water flow and return temperatures of 80/70°C giving a temperature difference between the air and the mean surface temperature of the panel of 55 K). This type of panel is fabricated from aluminium and encapsulates the water carrying copper pipes by metallurgically bonding the aluminium plate to the copper tubes. This gives 100 per cent metal to metal contact between the waterways and the heat emitting surface. As for the smooth faced panels this type of panel is backed by a layer of insulation and can be integrated into a suspended ceiling or be free hanging.

Industrial type building and sports centres

This type of heat emitter comprises of a robust steel panel fixed to a grid of 12 mm steel pipework, which are usually spaced at 150 or 100 mm centres to create the radiating surface to create a strip type radiant panel. The strips are backed with mineral wool insulation with foil finish. The panels are manufactured in standard widths of 300, 600, 900 and 1200 mm. Standard factory assembled strips are made up to 6 m long comprising two end panels, with headers and an intermediate panel as indicated in Fig. 10.23. Much longer strips (up to 100 m) can be fabricated on site by joining together two end panels and multiple intermediate panels. The jointing method can be either by welding or by using proprietary 'press-fittings' as described in Chapter 9.

The strips are usually suspended by chains from the building roof or structural soffit, such that the thermal expansion can be accommodated with a fall from the supply connection to the return connection to facilitate air venting and draining.

Low-temperature hot water at 80°C is usually used but this type of radiant heating can be used with medium temperature or high-temperature hot water or even steam, which are often available on industrial sites. The output of the radiant strip will depend on the temperature of the heating medium but Table 10.6 gives an indication of the nominal output for typical strips (with pipes at 100 mm centres) of various widths using either low- or medium-temperature hot water.

This type of heating is an ideal solution for the industrial heating of larger buildings such as aircraft hangars and factories and sports halls.

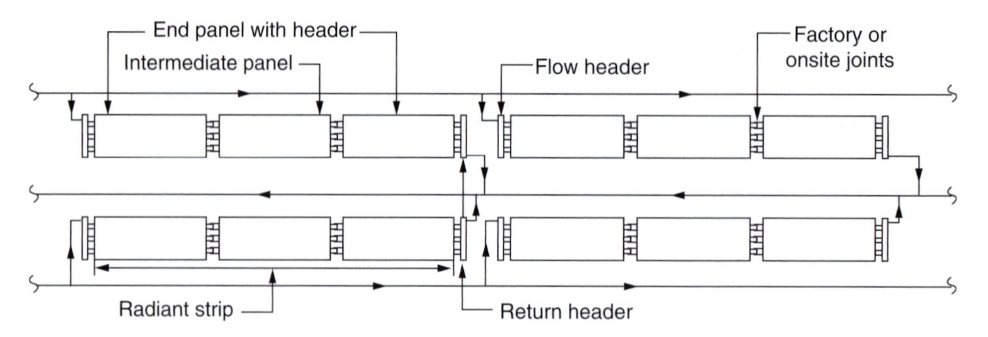

Figure 10.23 Typical installation of radiant strip heating in a factory

Table 10.6 Typical output of radiant strip heating (Frenger EcoStrip)

	Heat output (Watts per linear metre)	
	Low-temperature hot water (80/60°C)	*Medium-temperature hot water* (120/90°C)
Width of radiant strip (mm)	*Nominal Delta T between air temperature and mean surface temperature of 50 K*	*Nominal Delta T between air temperature and mean surface temperature of 85 K*
300	174	329
600	308	584
900	448	850
1200	571	1080

Other older style radiant heating systems

The following paragraphs describe solutions applied in the past, however these cover important design concepts and are included in the edition to give the reader an historical appreciation of the technologies used in older buildings, many of which may still be in operation today.

Metal radiant ceilings

The equipment to which reference has been made in previous paragraphs provides either a *linear* or a *point source* of radiation. Another form of radiant heating used in the past comprised a metal plate suspended ceiling. This in contrast, originates from a relatively wide area and may thus operate at rather lower face temperatures. Such a ceiling combines acoustical treatment with a radiant heat output and, in one proprietary make, consists of thin aluminium pans about 600 mm square having regular perforations. The pans are clipped to a piping grid at their junctions, and support a blanket layer of insulation material spread above them, as in Fig. 10.24.

It was usual to arrange that the whole surface of the ceiling is so treated; those parts of the coil which are necessary for heating being supplied with a hot water circulation and the remainder used only as a means of suspension. This aspect is illustrated in Fig. 10.25, which is a typical application layout serving a number of rooms. The total heat emission from a ceiling of this type, for a mean water temperature of 70°C and a room temperature of 20°C, would be about 160 W/m^2 downwards and 15 W/m^2 upwards to a heated room over. Two-thirds of the downward emission would be radiant.

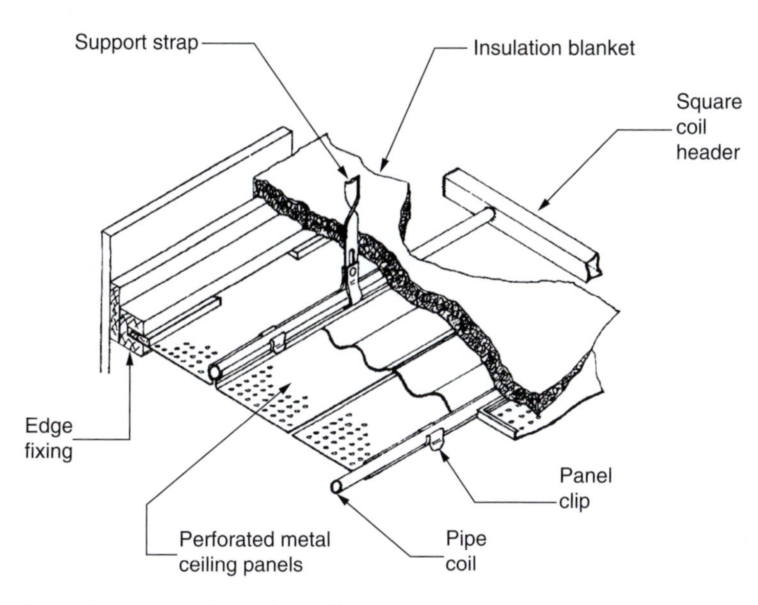

Figure 10.24 Metal plate radiant ceiling

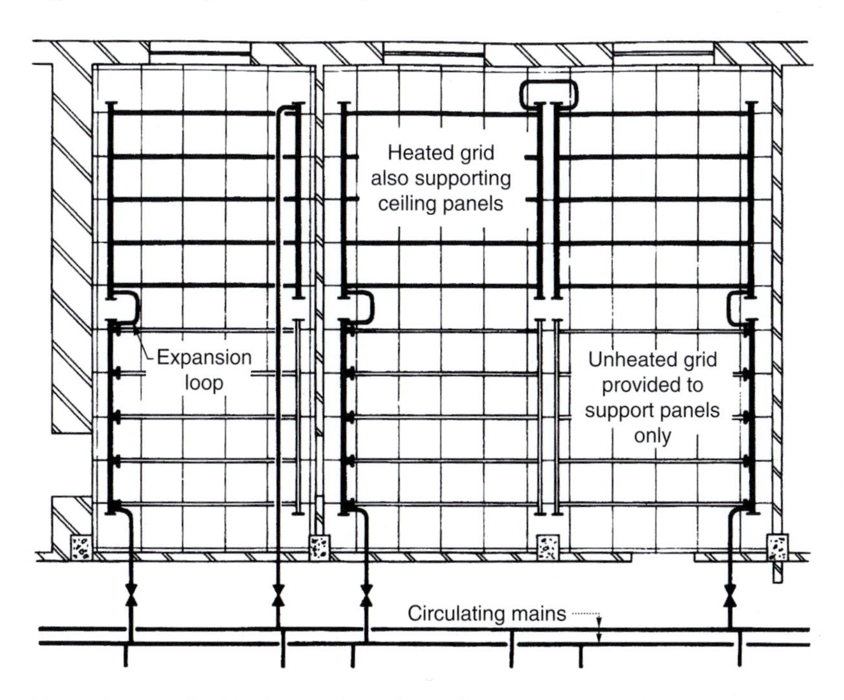

Figure 10.25 Application of metal plate radiant ceiling

Fibrous plaster radiant ceilings

Here a medium temperature hot water pipe coil is supported above and independently of the fibrous plaster panels which provide the acoustical treatment and radiant surface. Above the pipe coil an insulating blanket is supported on wire mesh to contain the heat output. The extent and temperature of the pipe coil determine the level of heat output which, in round terms, is similar to that of the metal type described previously.

Embedded ceiling panels

The practice of heating rooms by means of embedded ceiling panels was common for the older type of building, pre about 1950, where a long time lag was inherent in the heavy construction and thus matched that of such a system. Two techniques were used: Either sinuous coils usually of 15 mm bore steel or copper pipe were laid at 150 mm centres on the shuttering for a concrete slab prior to the forming and laying of bottom reinforcement or panel coils, which were incorporated within suspended ceilings, formed and plastered *in situ*. This type of system is obsolete in modern buildings with extensive glazing and little thermal capacity, which require systems able to provide a more rapid response to variations in outside conditions.

Underfloor heating

Modern underfloor heating systems that utilise cross-linked polyethylene (PEX) or polybutylene (PB) plastic pipework, are described in some detail in Chapter 9 and electric 'storage type' underfloor heating systems, which use PCMs described in Chapter 8.

Note

1. NHS Estates Health Guidance note 'Safe hot water and surface temperature' 1998.

Chapter 11

Pumps and other auxiliary equipment

The principal component parts which must be brought together to make up an indirect heating or cooling system are the energy source, the distribution pipework and the energy emitting equipment. There are, however, a number of important auxiliary items which each make a contribution to the functioning of the whole. The subject matter of the present chapter is devoted to consideration of these enabling items, some understanding of which is fundamental to appreciation of system operation.

Pumps

There are two basic categories of pump used in connection with distribution systems, *positive displacement* and *centrifugal.* Of the former, the rotary gear type is used exclusively (in the present context) for liquid fuel handling, as described in Chapter 16. Direct acting positive displacement pumps were applied to early heating systems, but they are now rarely used, even for those boiler feed duties for which they were once popular as a result of their robust reliability. The wide use of the centrifugal type for other applications is such that no further introduction is required.

Reciprocating pumps

As the name implies, pumps of this type produce a discharge as a result of the axial reciprocating movement of a plunger within a cylinder, displacing fluid from suction to delivery. Pumps may be arranged vertically or horizontally and be either single acting with drive from a piston rod and crank (often electrically driven) or double acting with a direct drive through a shaft which is common to the plunger and to a steam piston and cylinder. A twin cylinder balanced action is usual where a pump is single acting. Pumps of this type operate at low speeds and are particularly suited to providing output against high pressures in circumstances where an inherent minor intermittency in delivery is not of consequence.

Rotary gear pumps

These take the form of two interlinked and contra-rotating gears, set with close clearances within a single casing. On the suction side, as the gears disengage, fluid fills the spaces between the teeth and is conveyed round the periphery of the casing. It is then discharged as the gears re-engage, to provide a practically constant level of delivery against any chosen pressure.

Centrifugal pumps (general)

Some form of centrifugal pump, be it single- or multi-stage, is now used to meet most duties including that of boiler feed against pressure. In principle, such pumps consist of an impeller, having backward curved

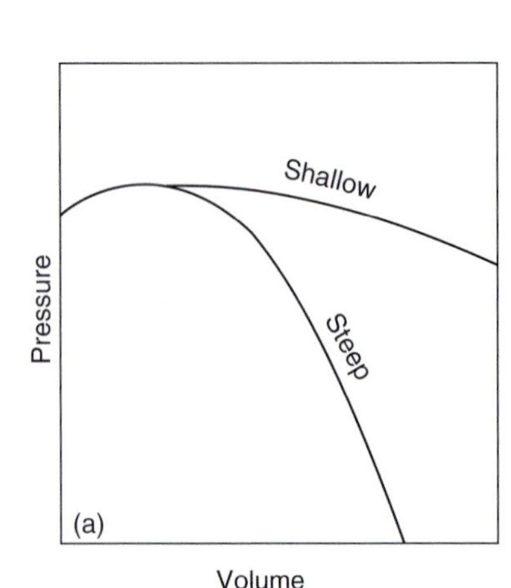

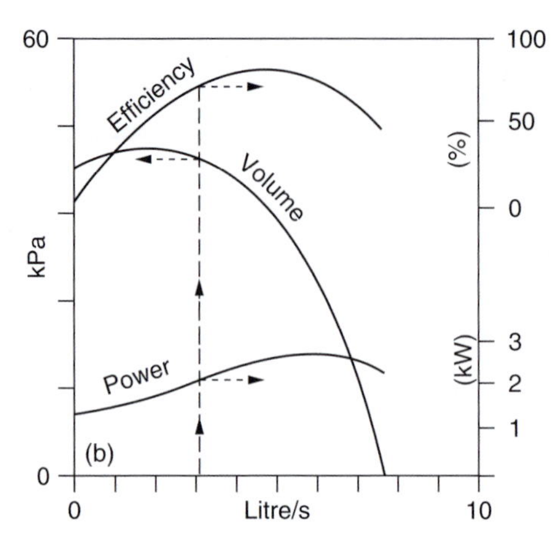

Figure 11.1 Centrifugal pumps. Typical characteristic curves

blades, which rotates within a scroll casing. This casing, more properly called a *volute,* has a profile which is so formed that the flow, discharged at high velocity by the impeller to the circumference, passes smoothly through an increasing area. As a result, when the point of discharge is reached, most of the velocity energy has been converted into pressure. Multi-stage units consist of a grouping of such volutes, arranged in series such that the discharge from the first passes to the inlet of the second and so on.

A number of simple relationships shows how, for a given diameter of impeller, the performance of a centrifugal pump will change as the speed is varied, as follows:

- Volume varies directly with speed.
- Pressure varies with the square of the speed.
- Power absorbed varies with the cube of the speed.

To these may be added a further series of similar relationships which show that, for a given speed, the performance of a pump will alter as the diameter of the impeller is changed. To expand further upon these and other similar matters is outside the scope of this book.

The performance of a typical centrifugal pump may be represented by *characteristic curves* as illustrated in Fig. 11.1 Part (a) of this diagram shows alternative relationships between the volume flow and the pressure developed. It will be seen that, dependent upon the pump design, the shape of the curve may be either steep or shallow to suit the particular application. Part (b) of the diagram repeats the steep curve and adds further characteristics which illustrates how *efficiency* varies over the operating range and, consequently, the *power absorbed.* It will be noted that, in this instance, near-to-peak efficiency is retained over a relatively wide range of volumes.

The manner in which the volume/pressure characteristic may be adjusted either by varying the speed or fitting an impeller of a different diameter is shown in Fig. 11.2(a). Centrifugal pumps may be arranged to operate either in series or in parallel and the outputs which result from such arrangements are shown in Fig. 11.2(b), taking the same volume/pressure characteristic as was used in part (b) of the preceding diagram, but drawn to a smaller scale. It will be seen that two *identical* pumps which are operating in parallel will deliver twice the original volume for any given pressure and, similarly, that two *identical* pumps which are operating in series will produce twice the original pressure for any given volume. Note that taking account of the system curve the pump/system combination delivers less then twice the flow rate. In instances where two *dissimilar* pumps are arranged to operate in either parallel or series, the result may be unsatisfactory and a detailed examination of the characteristic performance of each must be undertaken.

Net positive suction head and cavitation

It is important to consider the pressure and temperature conditions at the inlet to a centrifugal pump. *Net positive suction head* is the term used to describe the absolute pressure of the fluid at the inlet to the pump minus the vapour pressure of the fluid. It is known as net positive suction head available (NPSHa).

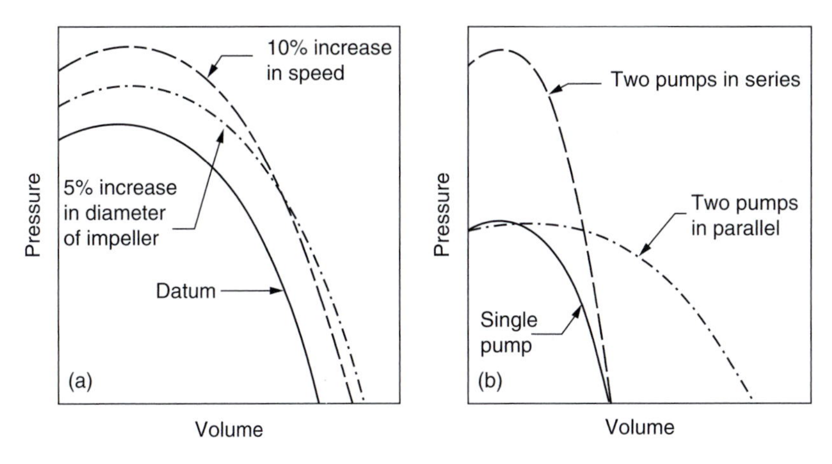

Figure 11.2 Centrifugal pumps. Duty change and alternative applications

Pump manufacturers use a similar terminology to describe the amount of pressure in excess of the vapour pressure required to prevent air bubbles coming out of solution at the impeller inlet. This is called the net positive suction head required (NPSHr). NPSHr is a feature of each particular pump and varies with speed, impellor diameter, inlet type and flow rate. NPSHr is established by the manufacturer and is included on the pump performance curves.

If the absolute pressure at the pump inlet becomes close to the vapour pressure of the fluid at its operating temperature then air can come out of solution and form bubbles. As these bubbles pass to the discharge of the impellor the increased pressure causes them to implode violently. This effect is known as *cavitation* and there are two major impacts: (1) dependent upon the amount of vapour generated the impellor alternately impacts upon water and air which leads to considerable vibration (2) the velocity of the implosion is so rapid that it has the force to wear grooves in both the impellor and the casing. These grooves can be as much as 10 mm wide and deep.

NPSHr is particularly relevant when hot liquids are being pumped (steam condensate return, steam boiler feeds), when liquid is being drawn from open tanks (cooling tower trays) and when a suction lift is involved (drawing cooling water from a lake). It is important to avoid high pressure losses that would reduce the NPSHa below the NPSHr, e.g. blocked strainers can lead to cavitation.

Each application has its own NPSHa and this can be calculated, with reference to the detail in Fig. 11.3 as follows:

NPSHa = Absolute pressure at the fluid surface + static pressure of liquid above the centerline of the pump (note that becomes negative for liquid surface below the pump centerline) – pressure losses in the suction piping – absolute vapour pressure at the pumping temperature. All units must be the same e.g. Pascals.

If NPSHa is greater than NPSHr cavitation should not occur.
If NPSHa is lower than NPSHr cavitation is likely to occur.

*To increase the NPSHa try to:

Increase the suction pipe size to reduce friction losses.
Reduce the friction loss from pipeline components e.g. strainers, bends, valves.
Raise the liquid level surface in any open tank.
Lower the centerline of the pump.
Increase the pressure on the pump inlet e.g. with a pressurisation vessel connection on a closed circuit.

Note that cavitation can occur wherever the pressure of the fluid drops below the vapour pressure of that fluid. This reduction in pressure can occur as liquid velocity increases through a valve and the bubbles can then implode downstream. The implosion has the same effect of eroding the valve and the immediate downstream pipework.

Cavitation can manifest itself as severe pump vibration and cracking sounds like a handful of pebbles clattering around in the pump impeller.

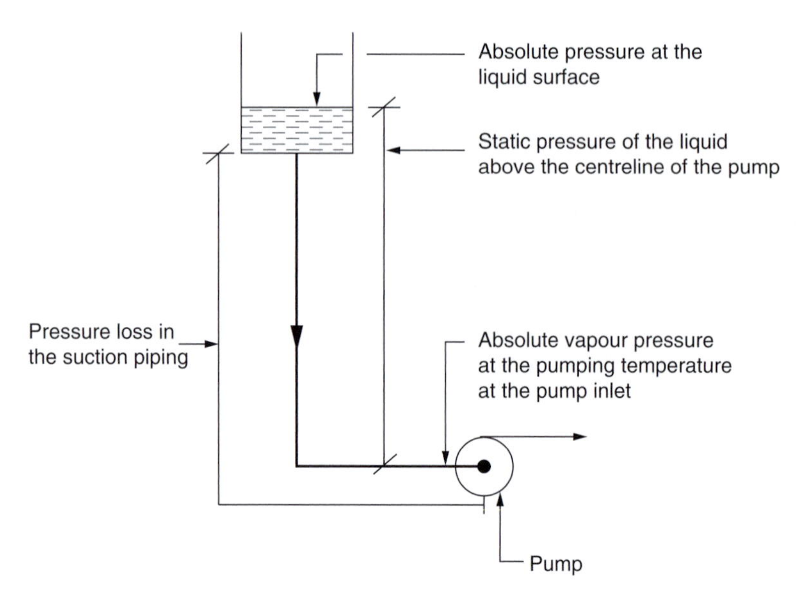

Figure 11.3 Design features affecting HPSHa

Water hammer

Water hammer can occur when pressure pulsations are set up by any moving liquid being stopped quickly e.g. by closing a valve too quickly. It is independent of the system working pressure. It is prevalent with quarter turn valves and ball valves. Water hammer manifests itself by a sound similar to a pipe being struck by a hammer, hence the name.

To avoid water hammer:

- Close susceptible valves slowly.
- Avoid quick closing automatic valves.
- Reduce the liquid flow velocity.

Centrifugal pumps for water systems

Passing from the general to the specific, there are a number of types or styles of centrifugal pump available for application to water heating and cooling systems, and these fall into groups which may be listed as follows:

- Form of volute – split casing or end suction.
- Drive arrangement – close-coupled, direct or belt.
- Shaft axis – horizontal or vertical.

Dealing with these in turn, the split casing type, as in Fig. 11.4(a), is used only for the largest installations where access to the impeller or casing is necessary for cleaning purposes. End suction is the most common arrangement as shown in Fig. 11.4(b).

A close-coupled drive, where the pump is attached to a flange on the motor casing and the impeller is mounted on an extension to the motor shaft, is a compact arrangement well suited to industrial use or pressure development. Direct drive, where motor and pump are mounted on a common baseplate, with drives coaxial one to the other but joined only by a flexible coupling, is a convenient arrangement and will provide long service without problems provided that the pump and motor shafts are aligned at manufacturers' works. Adjustments to the duty of the pump are achieved by speed control. Drives which employ generously proportioned vee belts and pulleys, as in Fig. 11.5, are preferred by some designers. Here, the pump shaft runs at a speed which is quite independent of that of the motor and waterborne noise is reduced by the separation of the two components. Adjustments to the duty of the pump may be achieved by a simple change of pulleys and belts.

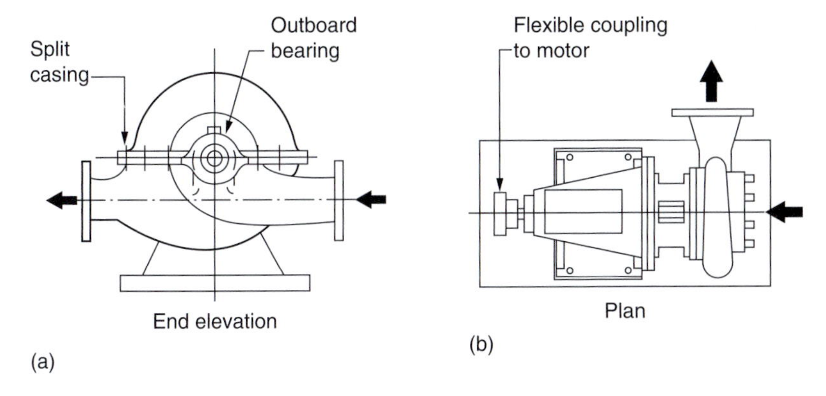

Figure 11.4 Centrifugal pumps. Forms of volute

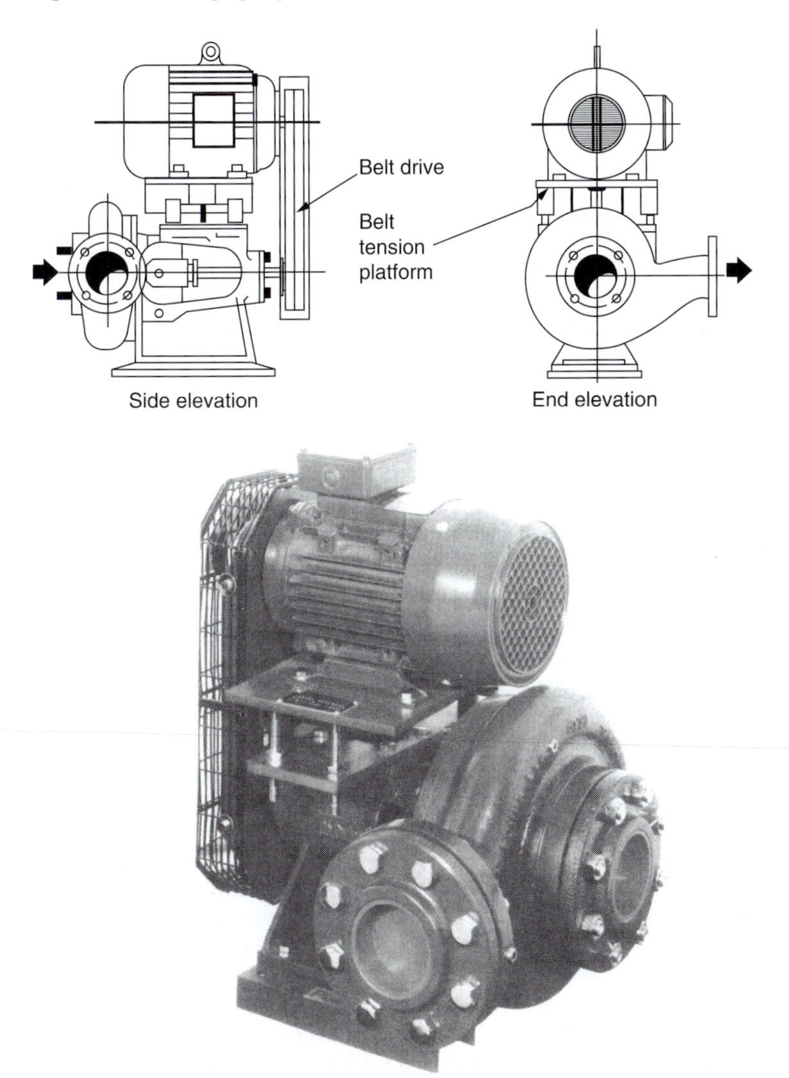

Figure 11.5 Centrifugal pumps. Belt driven (Pullen)

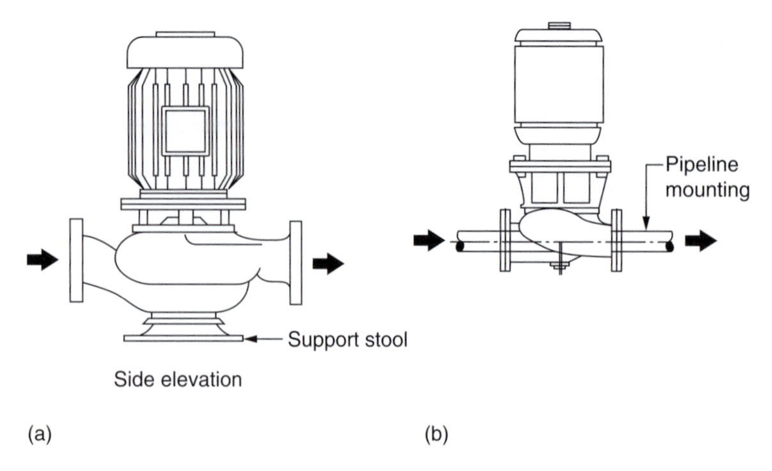

Figure 11.6 Centrifugal pumps. Vertical spindle types

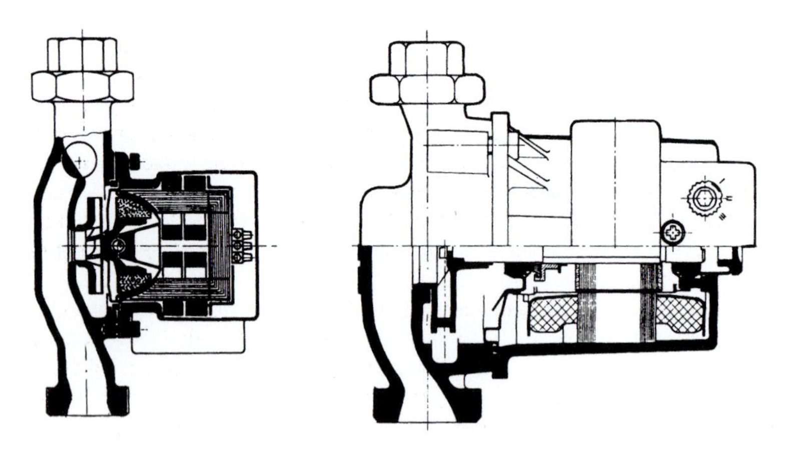

Figure 11.7 Centrifugal pumps. Submerged (canned) rotor type (Brefco)

The axis of the driving shaft between motor and pump is most commonly arranged to be horizontal and supported by intermediate bearings: there is in consequence much to commend this arrangement. As an alternative, which may be more economic of space than the horizontal direct drive arrangement, pumps having vertical shafts (sometimes known in this context as *spindles*) are available, as in Fig. 11.6. Some, as part (a) of the diagram, are provided with stools or feet for floor mounting but smaller sizes, as part (b), are sold as being suitable for pipeline installation. It is wise to establish that the latter have thrust bearings and shaft seals which have been designed to suit a vertical configuration.

The types and styles noted above do not include that which is probably best known: the glandless submerged (canned) rotor pattern as fitted in most domestic systems. This is effectively an end suction type, and has the advantage of being without a gland or seal since the driving shaft does not pass through the casing. Figure 11.7 provides a cross-section through two pumps of this pattern. The penalty paid for this considerable advantage is that the clearances between the stator and the rotor are necessarily small and, in consequence, any foreign matter in the water circulated may lead to seizure. Most such pumps incorporate devices whereby output may be varied: by an electrical adjustment to motor speed; by use of a hydraulic *spoiler bypass* or by some combination of these two methods. Most manufacturers now advise that pumps of this type should in no circumstances be mounted with the shaft in a vertical plane.

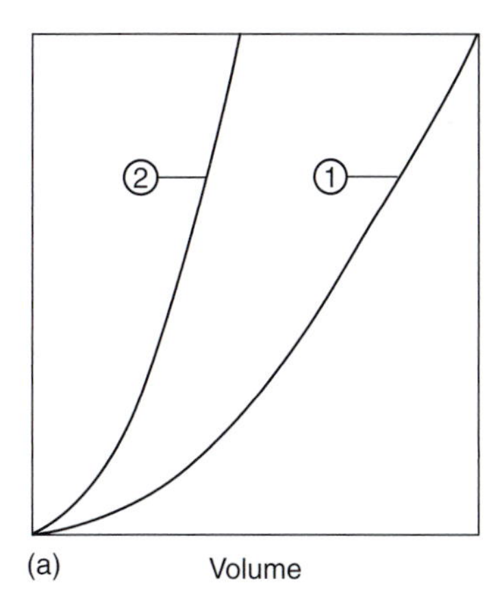

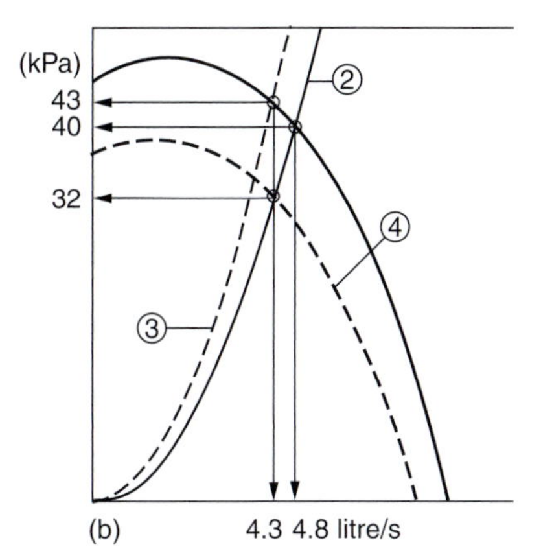

Figure 11.8 Centrifugal pumps

Centrifugal pumps for condensate

A separation of condensate pumps from those handling water circulation may seem artificial but there are several important differences in the present context. First is the matter of pressure differential in that condensate is most usually pumped from suction at a near-to-atmospheric condition to discharge at a much higher potential. Second is the difficulty of quantifying the temperature/pressure relationship at the pump suction connection where condensate may be less 'stable' than is the case for water circulated in a closed system subjected overall to pressure above ebullition point. Third is the question of performance, since practice is to select a condensate pump having a capacity several times the calculated mean demand, in the realisation that performance must meet peaks well in excess of that level. Lastly, since condensate pumps are in most cases installed adjacent to plant items producing a much higher noise level, this will affect choice.

In consequence of these circumstances, the centrifugal pump most commonly used to handle condensate is, in comparison with those used for circulation purposes, designed to produce an output against a significant pressure differential, with good access for cleaning and without much regard for noise production: close-coupled drive is the type most usually adopted and the volute is often split at right-angles to the shaft as an aid to cleaning. Reference will be made later in this chapter to factory made condensate receiver sets incorporating such pumps.

System characteristics

The operation of a centrifugal pump cannot be considered in isolation from the distribution system with which it is associated, and which system is made up of a pipework reticulation and various single items such as boilers and terminal equipment. As water flows through such a system, surface friction and other parallel effects will cause a loss of pressure which, for practical purposes, may be considered as varying in proportion to the square of the water velocity and hence of the quantity flowing. Taking any given system therefore, once the pressure loss has been determined for one flow quantity (see Chapter 12), other values may be produced as the basis for plotting a curve which represents the pressure reaction of that system to varying rates of water throughput.

Two characteristic curves produced in this manner are plotted in Fig. 11.8(a), that marked '1' being for a *system* having a relatively low resistance in comparison with that marked '2'. The use of such curves is illustrated in Fig. 11.8(b) where that having the higher resistance is superimposed upon the pump volume/pressure characteristic reproduced from Fig. 11.1(b). The point at which the system curve intersects with the pump curve represents the duty at which that particular combination would operate, i.e. 4.8 litre/s against a pressure of 40 kPa. As an example of the type of adjustment to such an intersection which may be required in practice, let us assume that a flow rate of 4.3 litre/s is critical to the operation of the system. There are two obvious ways in which this may be produced, as shown in the diagram:

- By adding resistance to the system (part closing a valve or some other similar action) such that a new *system* characteristic is developed, as '3', where 4.3 litre/s will be delivered against a pressure of 43 kPa.

or

- By reducing the speed of the pump by 10 per cent (changing a pulley or an impeller) such that a new *pump* characteristic is developed, as '4', where 4.3 litre/s will be delivered against a pressure of 32 kPa.

The second alternative would be preferred in this case since less energy will be expended.

Pump application

The duties required of centrifugal pumps may be related to the heating or cooling capacity of a system, these then being translated to litre/s. Mass flow of water is thus, in this respect, the energy output in kW divided by the specific heat capacity of water and by the temperature differential across the system, flow to return. For all practical purposes, therefore, over the temperature range 10–180°C which is encountered in practice, the flow rate in litre/s = 0.238 x kW/K.

To limit a circulating pump to this exact duty would mean that the water flow through the many parallel circuits and, in effect, each terminal fitting, must be precisely the calculated quantity. This is obviously impracticable bearing in mind that the pipework system will be built up on site from commercially available materials under less than ideal working conditions and may, in any event, deviate slightly from the design. In consequence, it is usual to make the best calculation possible at the design stage and, this having been done, to add a margin to the calculated pump volume requirement. The size of the margin, which will vary between 10 and 20 per cent, depends upon the complexity of the system arrangement: for either a simple single-pipe layout or for a reversed return system, 10 per cent would be adequate. There is a difference of opinion among designers as to the validity of adding some similar margin to the calculated pressure loss: experience suggests that such an addition should be made only after careful study of the pump characteristics.

For larger systems it is usual to provide a duplicate pump for each circuit, in order to provide some insurance against failure. The two pumps are piped to the system in parallel and each is fitted with isolating valves on the suction and delivery connections. Where pumps are provided in duplicate for a high-temperature system, it is good practice to arrange for a small circulation to be maintained through the stand-by pump so that it may be brought into service at the system working temperature, thus avoiding thermal shock.

It used to be thought good practice to provide a non-return valve on the delivery connection from each pump so that change-over could be a simple matter of electrical switching, without making use of the isolating valves. Such arrangements resulted, however, in the pumps and the valves being totally neglected, there being no need for the plant operator to visit them. In consequence, non-return valves may be considered to be an unnecessary and inappropriate provision, except in the case of packaged twin pump sets, as Fig. 11.9, where a single flap is integral to the construction. The particular merit of providing duplicate pumps in this way is

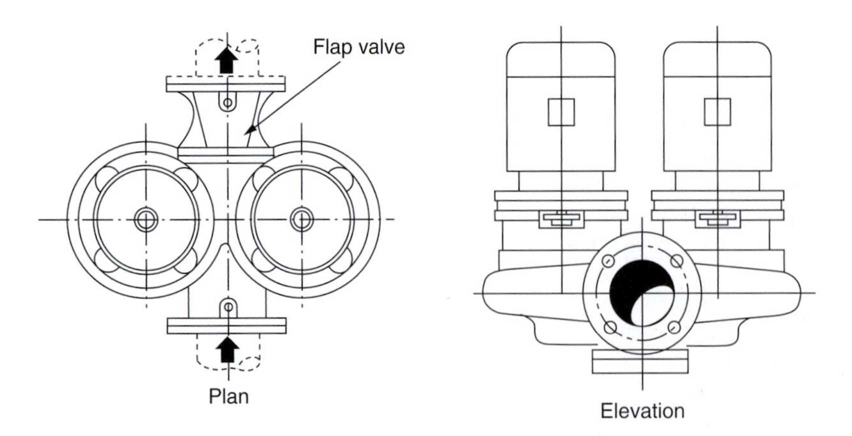

Figure 11.9 Centrifugal pumps. Dual vertical spindle type

that the combination is very compact: some such dual sets are made with pumps of different sizes in order to provide for day/night or winter/summer duties.

Pump speed control

Variable flow pumped circuits, normally associated with two-way modulating valve control (Chapter 28), can give significant savings in electrical energy. Such control these days is achieved by changing the speed of rotation of the pump impeller using inverter control on the power supply to the pump. An inverter functions by rectifying the mains supply to direct current and converting this back to an alternating current with a variable amplitude and frequency. Inverter drives may be fitted to both single and three-phase supplies.

Pump construction

Casings for centrifugal pumps are commonly manufactured in close-grained cast iron although a copper alloy may be used in particular circumstances. Impellers are of cast gunmetal, machined and balanced to close tolerances, and are mounted on stainless steel shafts. A problem which is common to almost all such pumps, no matter how arranged, is that the rotating shaft must pass through some form of gland or seal in the casing to whatever form of drive is to be used. This gland must, of course, allow for free rotation of the shaft with the minimum practicable leakage of the fluid pumped. Mechanical seals are generally preferred to packed glands for most applications but special designs are necessary for higher pressures and for temperatures above atmospheric boiling point. In the latter instance, water cooling arrangements as in Fig. 11.10 are required.

Pump mountings

In addition to the isolating valves mentioned previously, each pump, or pair of pumps where in duplicate, should be provided with facilities for establishing the suction and delivery pressures. These may take the form of a pair of pressure gauges, a single differential gauge which is connected to both suction and delivery or, at the very least, a pair of pressure tapping points. Pressure readings will allow the performance of the pump to be checked against the characteristic curve when the system is commissioned and will provide a means for monitoring any fall off in performance during the life of the plant.

Some designers provide a small bore (say 25 mm) cross-connection between the suction and delivery pipework at the pump and fit this with isolating valves and a strainer. The pump capacity is increased by a small percentage and the arrangement ensures that a proportion of the system water content is being filtered continuously.

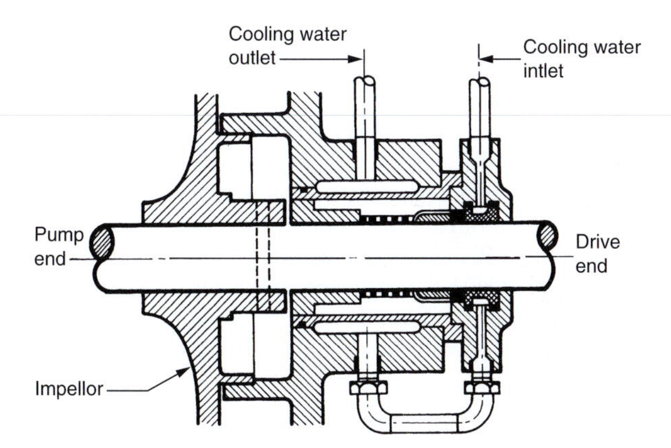

Figure 11.10 Centrifugal pumps. Water cooled bearing for high temperatures

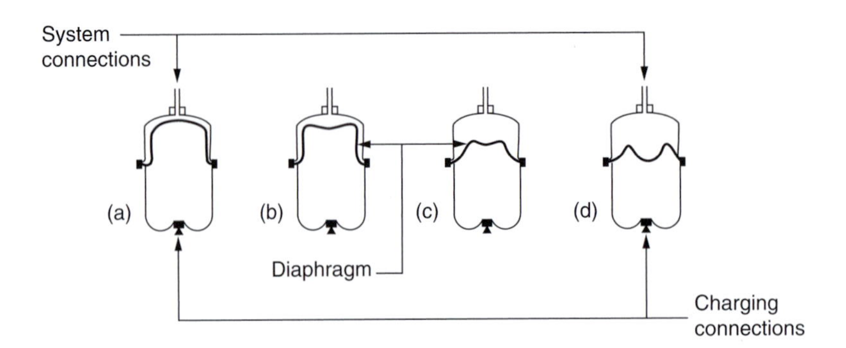

Figure 11.11 Application of a simple expansion vessel

System pressurisation

Low-temperature hot water heating systems, where the operating temperature is held below atmospheric boiling point, derive an adequate working pressure as a result of connection to an open feed and expansion cistern fitted above the highest point. Nevertheless, a cistern so sited suffers, particularly in domestic situations, from being out of sight and out of mind with the result that odd noises may be the first sign that it is empty, following seizure of the ball valve. Furthermore, the very requirement that the cistern be fitted high may well lead to it being positioned where frost is a potential hazard in exceptionally severe weather.

It has been explained previously (Chapter 9), that the early high-temperature hot water systems were pressurised by means of a steam space which was either, in the case of shell boilers, within the shell or, in the case of water tube boilers, within a separate steam drum. The disadvantage of this method, as explained, was the inherent instability resulting from the close association between the pressure applied and the temperature of the water circulated, with the result that skilled operators were required.

At either end of the temperature spectrum, therefore, the adoption of some alternative method of pressurisation needed consideration.[1]

Pressurisation by expansion

This, at the simplest domestic level, involves little more than the addition of an unvented expansion vessel to a heating system which is then charged with water and sealed. The function of the vessel is to take up the increased volume of the water content of the system as it is heated and, by so doing, apply additional pressure. In practice, proprietary type vessels are used which incorporate a flexible rubber diaphragm separating the water content of the system on one side, from a factory applied charge of nitrogen on the other. The size of vessel required is a function of the initial and final pressures and the water capacity of the system: suppliers rate a standard range accordingly.

The way in which such a vessel performs is shown in Fig. 11.11, the sequence being (a) before connection, with the diaphragm held to the vessel wall by the nitrogen charge; (b) connected to the system which has been filled; (c) during heating, as the water expands; (d) at system working temperature, the water content now fully expanded. In addition to the vessel, other fittings necessary are a safety valve fitted to the boiler (of a rather better quality than is normally provided for domestic systems), a fill/non-return valve which will accept a temporary hose connection from a water supply and an automatic air release valve, possibly associated with a centrifugal air separator.

For larger systems operating at low temperature, the principles of operation remain the same. The expansion vessel will increase in size and may even be duplicated. A small filling unit is usually provided for 'topping-up' purposes, consisting of a cistern with a ball valve for water supply and an electrically driven pump all as illustrated in Fig. 11.12. A low pressure switch fitted to or near the boiler will control the operation of the pump to ensure that a minimum water pressure exists and a parallel high pressure switch may be incorporated to stop firing of the boiler in the event of over pressure, at a level below safety valve operation. The various components of a system of this type may be built up into a complete packaged unit.

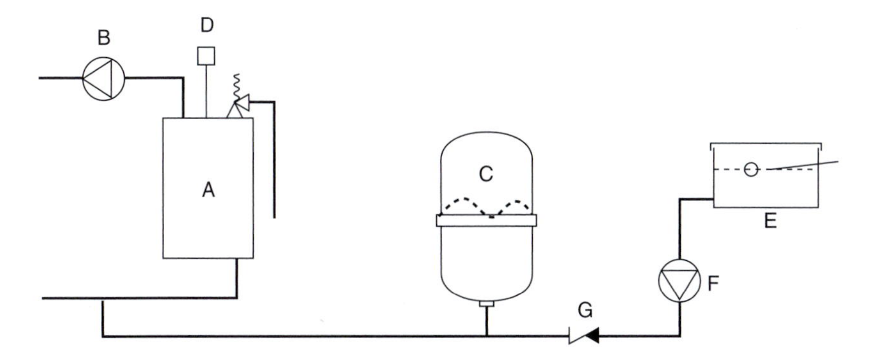

Figure 11.12 Components of a simple packaged expansion unit. A, boiler; B, system circulating pump; C, expansion vessel; D, topping-up control; E, feed cistern; F, pressurising pump; G, non-return valve

Table 11.1 Percentage expansion of water heating up from 4°C

Temperature (°C)	*Expansion* (%)
40	0.79
50	1.21
60	1.71
70	2.27
80	2.90
90	3.63
100	4.34
110	5.20
120	6.00
130	7.00
140	8.00
150	9.10
160	10.2
170	11.4
180	12.8
190	14.2
200	15.7

Sizing of water expansion vessel

The following calculation is an example of how an expansion vessel performance can be established and is a direct extract from CIBSE Guide B1.

It is required to determine the size of the sealed expansion vessel required if the pressure in the system is not to exceed 3.5 bar gauge (i.e. kPa gauge).

Initial data:

- water volume of system = 450 litre
- design flow temperature = 60°C
- design return temperature = 50°C
- height of system = 7.5 m
- design pump pressure rise = 42 kPa
- temperature of plant room during plant operation = 25°C (i.e. 298 K).

Consider first the water. It may be assumed that, at its coolest, the temperature of the water in the system will be 4°C (i.e. 277 K). Similarly, depending on the control system, the greatest expansion could occur at part load, when the entire water system is at the design flow temperature. Table 11.1 gives the expansion between 4°C and 60°C as 1.71 per cent.

$$\Delta V = 0.0171 \times 450 = 7.7\,\text{litre}$$

Ideally the pressure vessel should be connected to a position of low water pressure. This would reduce the required volume of a sealed vessel. However it is more convenient for items of plant to be located in close proximity, and in this example the expansion vessel is being connected to pipework 7.5 m below the highest position of the circuit. It must, however, be positioned on the return side of the pump, not the outlet.

The 'cold fill' pressure at the pump, due to the head of water, will partially compress the air within the expansion vessel, thus necessitating a larger expansion vessel. Thus it is advisable to pre-pressurise the air within the vessel to this pressure so that, once connected, it will still be full of air. Thus the initial air volume will be the same as the vessel volume. No further head of water should be applied as it would serve no useful purpose and would increase the operating pressure of the system, which is undesirable.

Pre-pressurisation required for a head of 7.5 m is given by

$$P_1 = \rho g z$$

where

$$P_1 = 1000 \times 9.81 \times 7.5$$

$$= 73.58 \text{ kPa gauge} = 174 \text{ kPa absolute}$$

Maximum permissible pressure, P_2, at inlet to the pump:

$$P_2 = (350\text{–}42) \text{ kPa}$$

$$= 308 \text{ kPa gauge} = 408 \text{ kPa absolute}$$

Initial volume of air in vessel $= V_1$; final volume of air $= V_2 - (V_1 - 7.7)$ litre.

For the air, the ideal gas equation will apply, using absolute values of temperature and pressure. Note that since no hot water flows through the vessel, there should be no effect upon the temperature of the air cushion within the vessel. However, it could be affected by the plant room temperature.

The ideal gas equation is:

$$\frac{P_2 V_2}{T_2} = \frac{P_1 V_1}{T_1}$$

Therefore:

$$\frac{408(V_1 - 7.7)}{298} = \frac{174 V_1}{277}$$

Hence, the minimum volume of expansion vessel required, $V_1 = 14.23$ litre.

As the calculation was carried out based on the maximum permissible pressure, the next size up must be selected. Since sealed pressure vessels constitute such a small portion of the equipment cost, consideration should always be given to selecting one which is larger than necessary, the advantage being a reduced operating pressure for the system.

The following offer some practical comments:

- The higher pressure is usually set by the maximum allowable system pressure for its pipework and components and account is taken of safety relief valve settings.
- The pressure at all points in the system must exceed the saturation pressure at the operating system temperature.
- Pumps have sufficient NPSHa to avoid cavitation.

Pressurisation by pump

For medium-temperature hot water systems, operating at temperatures up to about 110°C and at pressures of about 400 kPa, an alternative approach to pressurisation has led to the introduction of units of the type shown in Fig. 11.13. These rely principally upon pressure-pump operation in conjunction with a *spill valve* and a

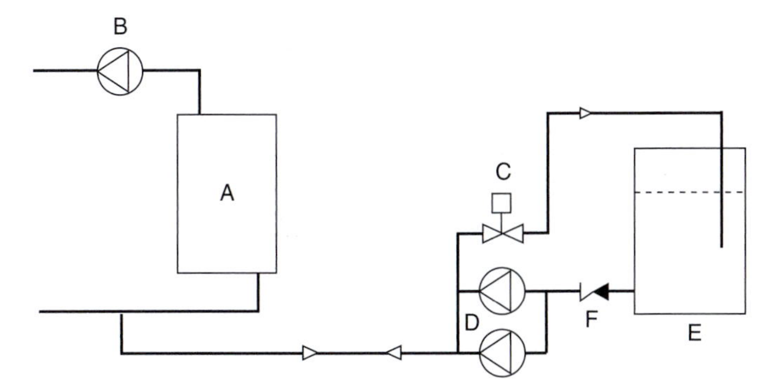

Figure 11.13 Components of pump-pressurising packaged unit. A, boiler; B, system circulating pump; C, water pressure spill valve; D, pressure pumps; E, expansion water spill cistern; F, non-return valve

cistern fitted at low level to take up the water of expansion. Some units of this type are factory assembled and are provided with duplicate pumps and controls.

Starting a system so pressurised from cold, the spill valve allows the water of expansion to escape into the cistern once a preset pressure is reached. While the system remains at the design working temperature, and thus at constant pressure, the spill valve remains closed and the pump is idle. When the system temperature and pressure fall as a result of decreasing demand or boiler cycling, the pump will start and the preset pressure will be restored. Since the *rate* of fall in pressure may vary considerably, the pump does not run continuously on demand but is arranged to cycle 'on/off' by means of timed interruption of the electrical supply. This arrangement provides, with admirable simplicity, the equivalent of a variable pump output. Arrangements such as this are sometimes criticised as being too reliant upon the availability of an electrical supply to maintain the necessary working pressure. Since automatic boiler firing depends also upon an electrical power source, avoidance of an unbalanced condition may be overcome by prudent design and the provision of self-actuating override controls.

Pressurisation by gas

This was, in chronological sequence, probably the first independent method to be introduced. It was conceived as a result of the difficulties experienced in application of steam as a pressurising agent for high-temperature hot water systems, as noted previously. Early versions were bespoke to suit the characteristics of particular installations and were built up on site from standard components to the requirements of the system designer. A range of packaged units, rather more, sophisticated as to detail, soon followed but these retained the original principles of operation.

The logic of the design sequence was that a pressure cylinder would be connected to the pipework arrangements in a chosen position, generally in the main return near the boiler where the water is coolest. This cylinder would be filled in part by water and in part by air or an inert gas, the initial supply of which derived from a small air compressor or from a gas bottle. By these means, an initial pressure could be applied at a level suited to the limits imposed by the system construction but in any event well above the boiling point of the water content. The water of expansion would be discharged from the system through a spill valve to a cistern open to atmosphere and, as the system cooled and contracted in turn, a pressure pump would draw water from the spill cistern and return it to the system.

The basic elements of the arrangement are shown in Fig. 11.14, one of the more important being the pressure controller which regulates the admission of water from the pump or its expulsion through the spill valve. The use of an inert gas such as nitrogen as the cushion, in place of air, offers the advantage that it is less soluble in water: at a later stage in the development of packaged units, the spill cistern also was provided with a cover and was pressurised very lightly with nitrogen for the same reason. In most units, two pumps are provided, to run in parallel when required to meet an unusual demand, and the water of expansion passes through some form of heat exchanger in order to lower its temperature and, if possible, prevent flash.

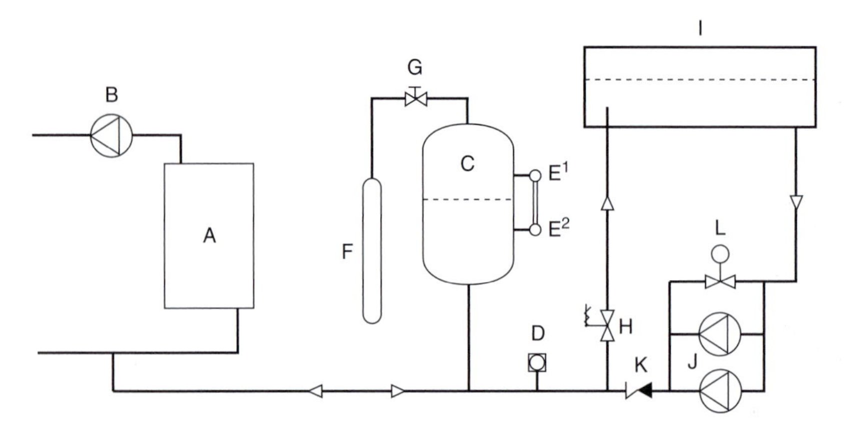

Figure 11.14 Components of gas-pressurising packaged unit. A, boiler; B, system circulating pump; C, pressure cylinder; D, pressure control; E_1 and E_2, high and low level cut-outs; F, gas cylinder (or compressed air supply); G, manual top-up valve; H, water pressure spill valve; I, expansion water spill cistern; J, pressure pumps; K, non-return valve; L, pump pressure relief valve

The calculation of the working pressure, to which components of the system and of the pressurising set are exposed, starts from a decision as to the flow temperature required at the highest point in the system, such as a unit heater or a high-level main pipe. This, when established, is used with values interpolated from Table 9.2 as follows:

Assume that the boiler flow temperature is 160°C and that, at a high point 10 m above the pressure cylinder water level, the temperature in the pipework is 5 K less, i.e. 155°C. Adding 15 K as an anti-flash margin to the water temperature at the high point, the vapour pressure equivalent to 170°C = 792 kPa.

Allowing for:

a 10 m rise to the high point, add (10 × 9.81) the	= 98 kPa
differential on the pressure controller	= 50 kPa
Normal system operating pressure	= 940 kPa

And, hence

Pressure pump starts at 890 kPa
stops at 940 kPa

Pump relief bypass
opens at 990 kPa

Spill valve
starts to open at 955 kPa and
fully open at 965 kPa

Safety valve settings
on cylinder 1015 kPa on
boiler 1040 kPa

It will be noted that an anti-flash margin of 15 K has been added for the purpose of this example: this would be varied to suit the particular circumstances in each case. The pressure exerted by the circulating pump will, in normal working conditions, augment this anti-flash margin but it is almost inevitable that at some time during the life of the system there will be a pump failure of some sort: it is usual therefore not to rely upon this added margin.

The required volume of the cylinder, if left uninsulated, is a function of the polytropic expression where PV^n is a constant. It has been found by experiment that $n = 1.26$ approximately and thus the movement of the water level may be calculated from:

$$a = h(1 - e^{0.794})$$

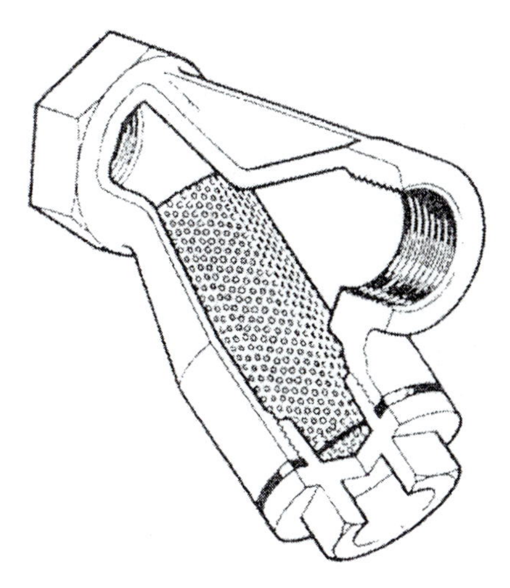

Figure 11.15 Y-type basket strainer

where

a = movement in the water level (m)
h = height of gas space prior to the start of compression (m)
e = P_1/P_2
P_1 = initial absolute pressure in gas space (kPa)
P_2 = final absolute pressure in gas space (kPa)

Control of the pump and spill valve is greatly simplified if the movement of the water line in the pressure cylinder is at a maximum: this movement is independent of the diameter of the cylinder but that dimension must be considered in conjunction with the duty of the pressure pump which should have a realistic running time of at least 2 minutes between start and stop when restoring the water level. The pump duty is a function of the rapid contraction which would take place if the boilers were to be shut down in emergency while the circulation continued.

Application of pressurisation

Although the equipment and packaged units which have been described under the preceding three headings were suggested as being appropriate to low, medium and high-temperature heating systems in that sequence, this was merely to note that they were often so applied. There is of course no reason why packages for pumped or gas pressurisation should not be applied to any category of system provided that they have been designed to suit the pressure and volume conditions obtaining and that their use can be justified in economic terms.

Strainers

Strainers are used in water (and steam) systems to remove dirt and debris most of which will have come from installation works e.g. welding slag, swarf and jointing materials. They are used upstream of major equipment, the performance of which could be impaired by such materials e.g. pumps, heat exchangers and control valves. Current practice on most water heating and cooling applications is to use strainers before pumps, boilers, water to water heating/cooling exchangers, and major water to air heating/cooling coils. Water to multiple small terminal units such as fan coil units is usually cleaned by a strainer on branch pipework where it can be readily maintained and avoiding the blockage and costly maintenance of numerous small strainers on each terminal unit.

The most common type is the Y-type basket strainer. A perforated plate or mesh basket sits inside the strainer as shown in the Fig. 11.15.

Water enters the centre of the basket and passes through the perforated plate or mesh such that debris is caught in the basket. The debris can be cleaned from the basket by removing the end cover and flushing it with water.

The basket is available with differing perforation sizes e.g. 0.8 mm, is typically used for branch pipework to fan coil units and 1.5 mm is typically used for pumps, boilers, chillers.

It is common to call for fine mesh strainers to be incorporated inside the basket to protect small valves and terminal units during flushing and these can very quickly block e.g. a 40 mesh is 400 micron and a 200 mesh is 76 micron. There should be a decision about how long they are retained because they influence system resistance, balancing and maintenance commitment.

Larger systems can incorporate dual basket strainers with two separate baskets. One basket can thereby be removed and cleaned whilst the other remains in operation. Strainers should be capable of isolation for cleaning.

Hydrocyclones, as in Fig. 11.16 are now commonly used to remove debris. Water enters the unit tangentially and the spiral water movement creates a centrifugal force that separates out particles with a relative density greater than water. Solids are deposited at the base of the unit and are cleaned out by opening a valve and using system pressure.

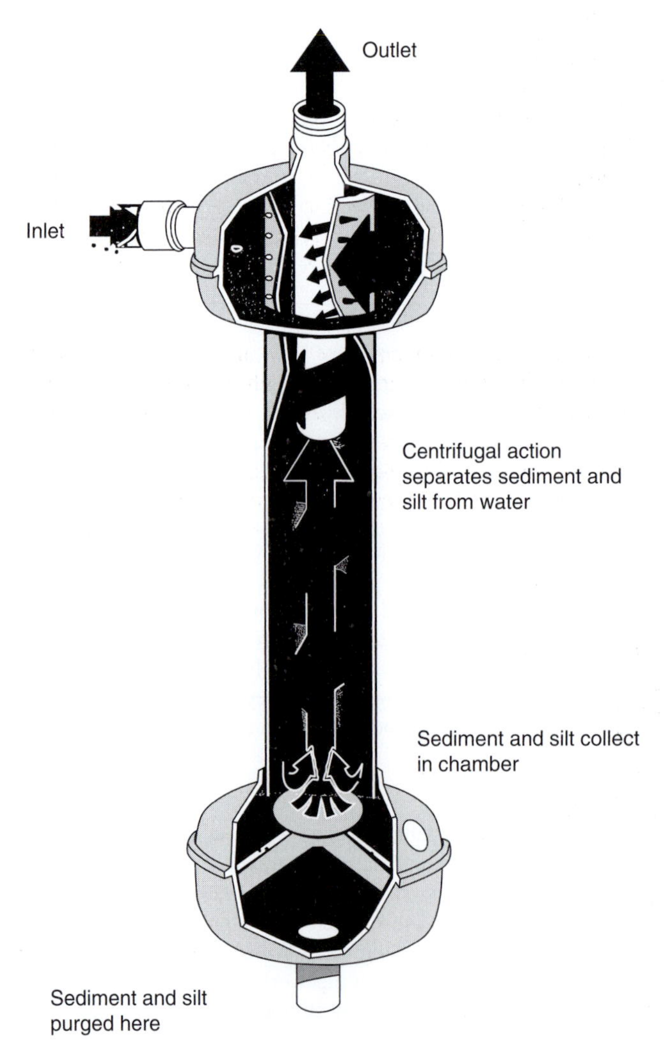

Figure 11.16 Example of typical hydrocyclone

In preparation of the foregoing reference has been made to training material from the Commissioning Specialist Association (CSA).

Non-storage calorifiers

A calorifier is a heat exchanger, i.e. an item of equipment used to transmit heat from one fluid, at a higher temperature, to another at a lower temperature. The *non-storage* pattern, which is the subject of the following paragraphs, is so called to distinguish it from those related to hot-tap supplies which are large by comparison and provide a reservoir of hot water. While non-storage calorifiers are manufactured to suit either steam or water as the higher temperature source, the latter are now less commonly used for duties such as that which was shown in Fig. 9.5(a) the function they fulfilled now being dealt with by injection circuits as Fig. 9.5(b). They are, however, still required for particular applications, such as the heating of chlorinated swimming pool water, where separation between the higher-temperature (primary) and lower-temperature (secondary) contents is necessary. Subsequent references here are directed principally to steam-to-water types but apply with little revision to water-to-water equipment.

Simple non-storage calorifiers

A considerable number of patterns of non-storage units has been produced but all have had four principal components: an *outer shell*; a tubular *heater battery*; one or more *tube plates* and a *steam chest* for pipe connections. The primary heating medium is passed through the tubes with the secondary contents in the surrounding shell, this disposition reducing heat losses with the lower temperature fluid against the outside surface. The various patterns of calorifier have generally evolved from the manner in which the tubes were arranged, in hairpin or 'U' form connected to a single tube plate or as straight runs between tube plates at each end of the shell.

In most cases tubes have been plain but one manufacturer produced an indented type which offered a high rate of heat exchange per unit length and an ability to shed scale with movement. For most applications, however, the pattern shown in Fig. 11.17 is suitable but an alternative vertical form, with the steam chest at the bottom supported clear of the floor, is sometimes used. This vertical form is economic in plan space but height is required to allow the shell to be lifted from the 'U' tube battery.

The outer shell is often constructed in cast iron but may be of welded mild steel. Tubes may be formed in steel or copper, the latter being the more usual material, and are individually expanded into the tube plate which is commonly of brass, the formation in end view being staggered as shown in Fig. 11.18(a). The steam chest, usually of cast iron, mounts the tube plate to the shell and provides facilities for steam and condensate pipe connections as Fig. 11.18(b). Since the tubes and tube plate are a heavy assembly, they are sometimes

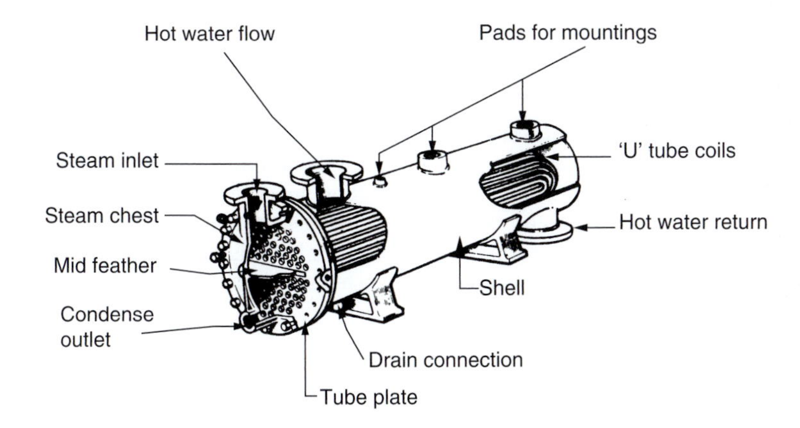

Figure 11.17 Horizontal 'U' tube steam-to-water calorifier

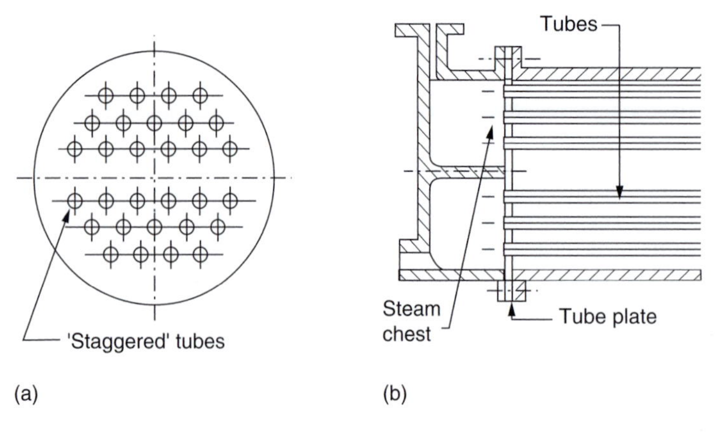

Figure 11.18 Calorifier tube and steam chest arrangements

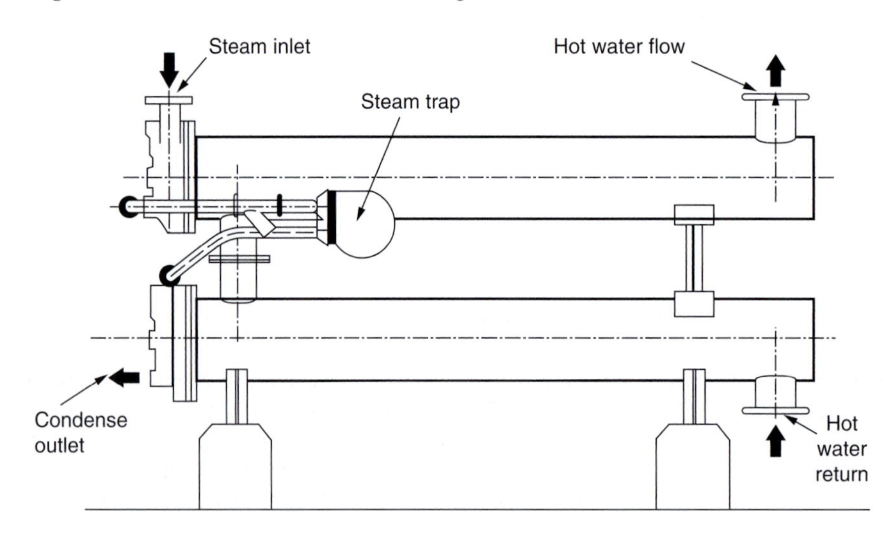

Figure 11.19 Steam-to-water calorifier and condensate cooler assembly

provided with a trolley type runway within the shell to facilitate removal for inspection: the pipe connections should be arranged to make this convenient.

Calorifier-condensate cooler units

Reference has been made previously in Chapter 9 to the recovery of flash steam. In circumstances where non-storage calorifiers are supplied with other than low pressure steam, discharge of the resulting condensate may give rise to problems and, in any event, a potential for energy saving will exist. Equipment suited to flash steam recovery is available in the form of two heat exchanger shells close-coupled, as illustrated in Fig. 11.19. The upper unit is a normal steam-to-water calorifier equipped with a trapped outlet which discharges a mixture of flash steam and condensate to a tubular heater battery in the lower unit. The hot water circuit is directed through the two heat exchangers in series and in counterflow.

Plate heat exchangers

Constructed as a 'sandwich' from a number of corrugated plates which provide passages through which separated primary steam (or water) and secondary water circuits flow, the appropriate number of plates to produce

Figure 11.20 Steam-to-water plate heat exchanger (Spirax Sarco/Alfa Laval)

the required rate of heat transfer are mounted to a frame pulled together by long compression bolts; flanges on one end cover plate providing pipework connections. Figure 11.20 illustrates a typical steam-to-water type.

The extended area for heat transfer and high velocity flows in both primary and secondary circuits enable the equipment to be very compact. Any deposition on the heat exchange surfaces will affect performance and water quality is critical as a consequence. Manufacturers claim the units are easy to dismantle and reassemble for maintenance.

Calorifier ratings and mountings

Since, in most respects, a non-storage calorifier is a 'boiler substitute', calculations to determine the rating and number of units required should follow almost the same routine as that used for boiler selection. The only difference of any significance is that low load efficiency need to be considered only as a function of the standing heat loss from a calorifier which is heated when output is not required. As in the case of a boiler, reference should be made to manufacturers' lists for actual outputs and sizes particularly since the use by some of controlled flow paths and high velocities within the shell produce an enhanced performance.

As to mountings, those required for a non-storage calorifier are precisely the same as the items which would be fitted to a boiler having an equivalent rating, as noted in Chapter 13.

Condensate handling equipment

When steam has given up its latent content in heating equipment of any type, the condensate remains at the temperature of the steam and it would be inexcusably wasteful to discharge it to drain since it might well represent 20 per cent or so of the energy input to the boiler plant. The condensate must, however, be removed as soon as it is formed, in order to prevent water-logging, and returned to the boiler plant for re-evaporation.

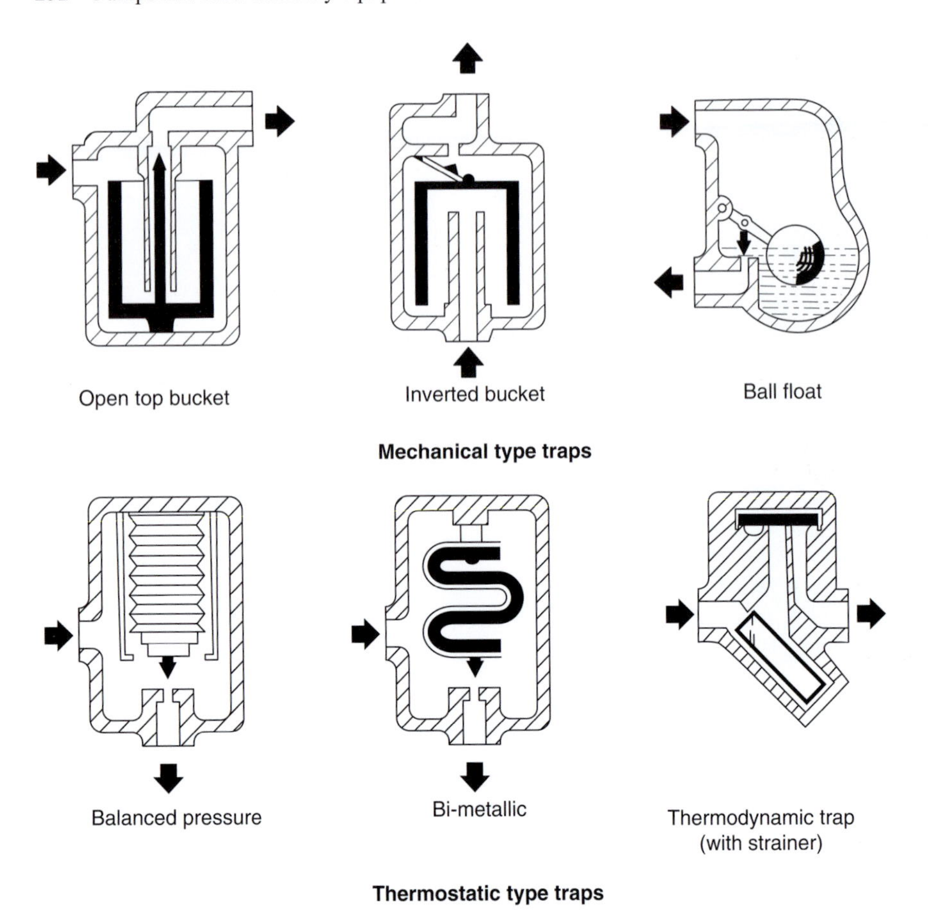

Figure 11.21 Types of steam trap (various manufacturers)

Steam traps

At the exit point of the heating equipment it is necessary to introduce a device which will allow condensate to pass but not steam: this is a function fulfilled by the *steam trap*. There are various types of trap and these fall into three broad categories, identified by the means adopted to distinguish and separate condensate from live steam, as follows:

- *Mechanical, incorporating*
  Open top bucket
  Inverted bucket
  Ball float.
- *Thermostatic, with various elements*
  Balanced pressure
  Liquid expansion
  Bi-metallic.
- *Miscellaneous, including*
  Labyrinth thermodynamic
  Impulse.

Some of these types are illustrated in Fig. 11.21 but specialist texts[2] and manufacturers' literature should be consulted for specific details. It is usual to add auxiliary components to the trap proper in order to make

Table 11.2 Steam traps: general application

Application	*Type of steam trap*
Drain points on steam mains	Mechanical open top, inverted bucket or thermodynamic
Fan convectors (large)	Mechanical ball float or inverted bucket
Natural convectors and radiators	Thermostatic balanced pressure
Oil storage tank coils and outflow heaters	Mechanical open top or inverted bucket
Oil tracing lines	Thermostatic bi-metallic or thermodynamic
Plenum heating or air-conditioning heater batteries	Mechanical ball float or inverted bucket: may be multiple for large units
Storage and non-storage calorifiers	Mechanical ball float or open bucket
Unit beaters (small)	Thermostatic balanced pressure
Unit heaters (large)	Mechanical ball float or inverted bucket

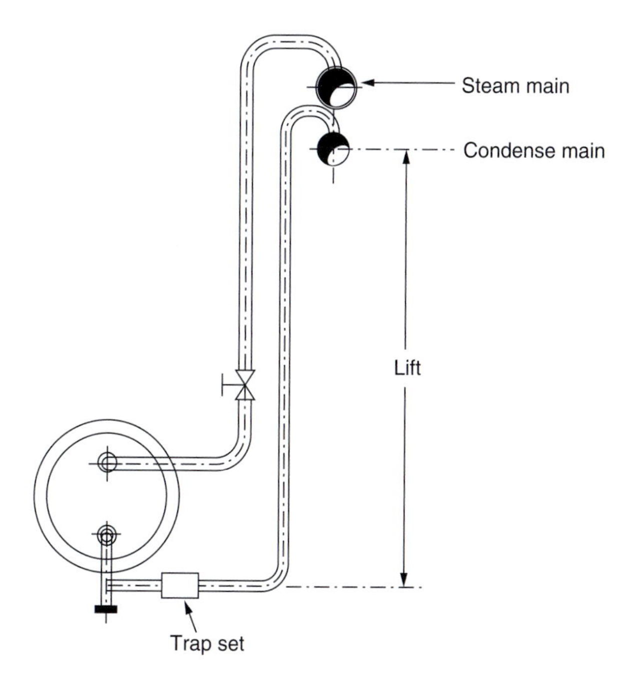

Figure 11.22 Lifting condensate to a higher level

up what is called a *trap set* and these will probably include some or all of: a strainer at entry; a check valve at exit; a sight glass and isolating valves when appropriate. The choice of a trap type suitable for any particular application depends upon a number of factors; load characteristics (constant or fluctuating); inlet and outlet pressures; associated thermostatic or other controls to the steam supply and the relative levels of trap and condensate piping, to name but a few. While it is not possible to generalise for all varieties of application, the trap patterns listed in Table 11.2 are generally suited to equipment within the scope of this book.

In an ideal situation, a trap would discharge into a condensate main run below it and the main would then fall in level back to the boiler plant. Such an arrangement may sometimes be possible in an industrial building but it is very often the case that a trap must discharge into a main above it; this is quite practicable *provided that* the steam pressure at the inlet to the trap is always adequate to overcome the back pressure imposed by the water in the discharge pipe, Fig. 11.22. In simple terms, an available steam pressure of at least 10 kPa is required for each metre height of vertical 'lift'.

In instances where the steam using equipment is fitted with some type of thermostatic or other automatic control which takes the form of a throttling valve, it must be remembered that such a valve acts by causing a *reduction in steam pressure* to the equipment (and thus to the steam trap). In consequence, although the initial

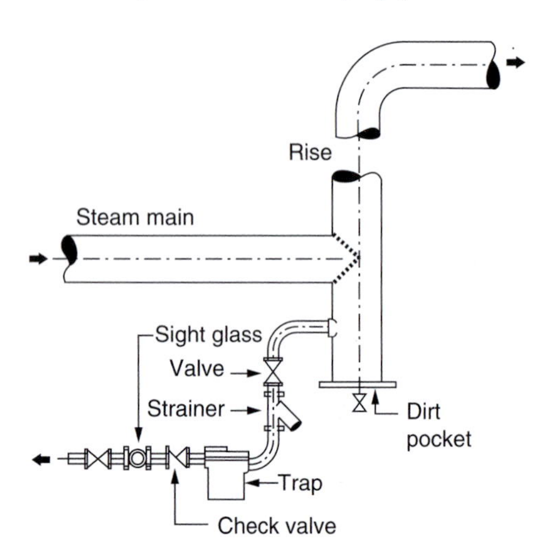

Figure 11.23 Steam main relay (to a higher level) and drain (drip) points

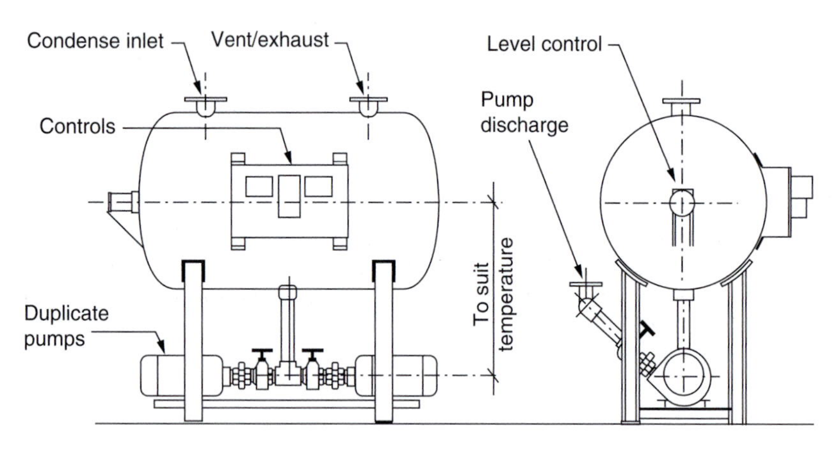

Figure 11.24 Condensate pump and receiver unit

steam supply pressure to the control valve may be adequate to provide the required lift, this will not be the case at low loads. It is *always* good practice, therefore, to avoid a situation where condensate has to be lifted from a trap serving equipment fitted with automatic control of steam input.

Condensate which is formed at individual items of equipment is easy to identify as to source and, hence, not difficult to deal with. It must not be forgotten, however, that steam mains, however well insulated, emit heat and that condensate will thus be formed along their run. Mains should always be arranged to fall in the direction of steam flow and arrangements made to provide drain (or 'drip') points at intervals along the runs, as in Fig. 11.23.

Condensate return pumping units

In instances where either distance or site levels make it impossible for condensate to be returned to the boiler plant by gravity flow, it may be delivered either piecemeal or in quantity to some convenient central point where a built-up pumping unit of the type shown in Fig. 11.24 is sited. Such *condensate receiver sets* are made to a range of manufacturers' ratings and consist of a cylinder fitted with a float operated switch which is mounted above a pump or pumps. The cylinder is sized to accommodate about one and a half to twice the maximum

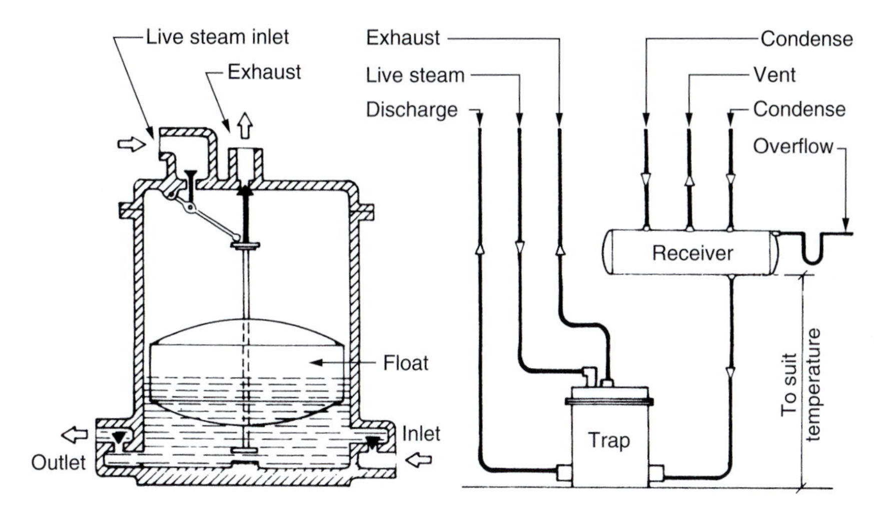

Figure 11.25 Super-lifting or pumping trap (Spirax-Sarco)

quantity of condensate returned per minute and each pump is rated to empty the cylinder in about 2 minutes. In order to avoid any possibility of pressure build up, the cylinder must be provided with an open vent to atmosphere, terminating in what is known as an *exhaust head* from which a drain pipe must be run to a safe position.

Super-lifting steam traps

Where it is difficult to provide for a pipe run from isolated items of equipment to convey condensate back to a condensate receiver set, use may be made of a *super-lifting trap* which is, in effect, a steam-pressure-operated pump, Fig. 11.25. A small vented receiver is required to collect incoming condensate which then feeds the trap, in preparation for intermittent discharge.

Steam pressure reduction

Where, for some industrial process, it may have been necessary to generate steam at a pressure quite unsuitable for heating purposes, some means of reduction will be necessary. For this purpose, pressure-reducing valves are available, as in Fig. 11.26. It has been explained earlier, in Chapter 9, that dry-saturated steam entering a valve of this type will produce a super-heated output. However, if there were some level of wetness in the entering stream, as is probable at the end of a long pipe run, and some losses following pressure reduction then the condition at the reduced pressure may approach that of saturation.

As in the case of steam traps, the provision of a pressure-reducing valve involves installation of a number of auxiliary components as shown in Fig. 11.27, the complete array being known as a *pressure-reducing set.*

Heat meters

Circumstances arise from time to time where it is necessary to consider methods of making charges for heat supply. These may relate to internal audits within a single industrial organisation, to multiple tenancies on an industrial estate served from a single boiler plant or to households supplied from a group heating system. Such charges may be assessed and levied in a number of ways, the four principal methods being by:

- a flat rate proportional charge based on floor area,
- a flat rate charge based upon an agreed estimated usage,

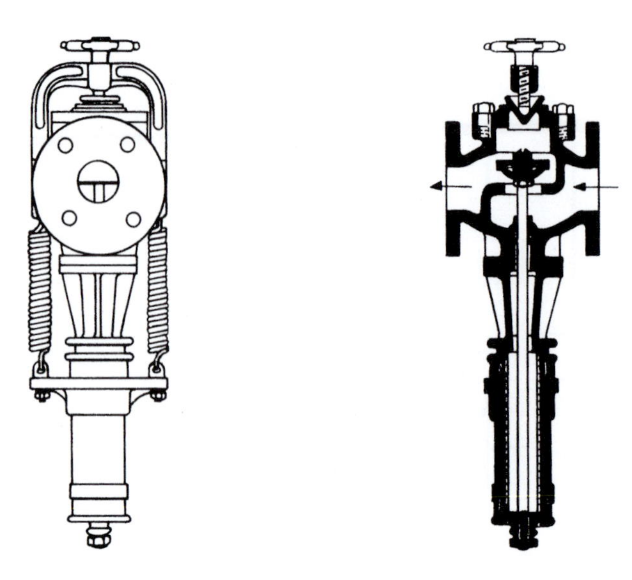

Figure 11.26 Steam pressure-reducing valve

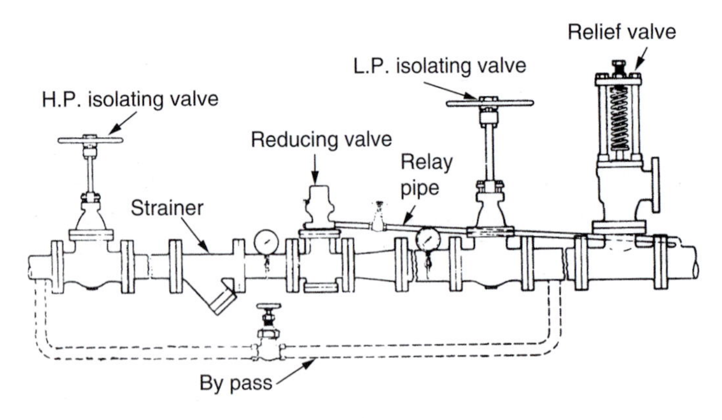

Figure 11.27 Steam pressure-reducing set

- a fixed charge plus a unit rate based upon metering,
- a unit rate based wholly upon metering.

The first two of the alternatives offer the advantage that they may be assessed before the event and thus included in any periodic exercise of forward planning. In an era of rising fuel and labour costs, however, they must, in equity, be subjected to a periodic review and may be thought not to reflect any individual efforts made towards conservation of energy. In consequence, the situation overall must be considered in relation to the practicability of using some form of meter to determine heat consumption.

Metering of steam systems

In a context which is wholly industrial and where steam is the medium for heat distribution, the boiler plant will most probably be equipped with an accurate *inferential type* steam flow meter. Although it would be possible to use similar equipment at the remote points of use, the application of a mechanical meter of some sort to condensate flow would be a more economic solution. Rotary piston, rotary vane and rotary helix patterns are available and have an accuracy of about plus or minus 2 per cent. The rotary vane type has generous clearances and is thus particularly suitable for use with condensate.

Heat meters for hot water systems

Complex hot water meters have been produced. An acceptable level of accuracy in the integration of the two variables, flow quantity and temperature difference, has been the problem requiring an economic solution. Three quite different methods have been adopted by various manufacturers as follows:

Mechanical. Where the drive between components of a conventional water meter is adjusted mechanically by temperature sensitive elements.
Electrical. Where both flow rate and temperature difference are measured and integrated by electronics.
Inferential. Where a measured but very small quantity of return water is heated electrically back to the supply temperature, as sensed by thermostat. The energy then taken is measured by a simple kWh meter.

All such meters, as shown in Fig. 11.28, are quite large, relatively expensive and subject to error if not maintained with care and returned regularly to the manufacturers for recalibration. Their application is, in consequence, best confined to central plant rooms where they may be used to provide records of system output overall. For use at remote points of heat consumption it is necessary to fall back on a range of simpler devices.

More recent heat metering involves the use of ultrasonics and magnetic induction techniques to measure the water flow rate so that there is no intrusion upon the water flow within pipework.

Apportioning meters

For radiator systems, the type of device shown in Fig. 11.29, one of which must be clamped to each radiator, was used in Europe for many years. It consisted of a housing bearing a scale, to the rear of which was a *replaceable* open phial containing a volatile liquid. The amount of liquid evaporated by the heat from the radiator was a broadly approximate measure of usage and this was read from the scale each season or other period. This type of device was subject to vandalism and has now been superseded by a relatively simple and secure electronic counter which integrates time and temperature and provides a numerical LED display. As before, readings are taken and totalled for the whole system and used as a basis to apportion the overall running cost to individual radiators or groups of radiators. The more obvious disadvantages include the necessity to provide access for a meter reader to each radiator, not always achieved easily in domestic premises. It is understood, however, that the manufacturers continue to offer an all-in meter reading and accounting service in given circumstances.

Water flow meters

An alternative method, for radiator systems, uses a simple rotary water meter, as for condensate measurement, associated with a *temperature limiting controller.* This device, when fitted to the return pipe of the circuit being metered, acts to maintain a constant outlet temperature and, by doing so, adjusts water flow in proportion to heat demand.

Hours-run meters

For any part of a system which requires a power supply to an individual fan or circulating pump, it is possible to arrange for a cyclometer type *hours-run meter* to be wired in parallel therewith and sited so that it may be read from outside the premises. The products of the reading taken, and factors representing the heat capacity of the individual units powered, are totalled as for other apportioning meters and used to determine the overall running cost. Allowance has to be made of course for any useful heat output from the metered components, by radiation or natural convection, which may take place when the fan or pump is not running.

Air venting, etc.

Mention has been made in Chapter 9 of the adverse effect of a stagnant air film upon steam-side heat transfer. As far as water systems are concerned, there will be an initial presence of both dissolved and free air and,

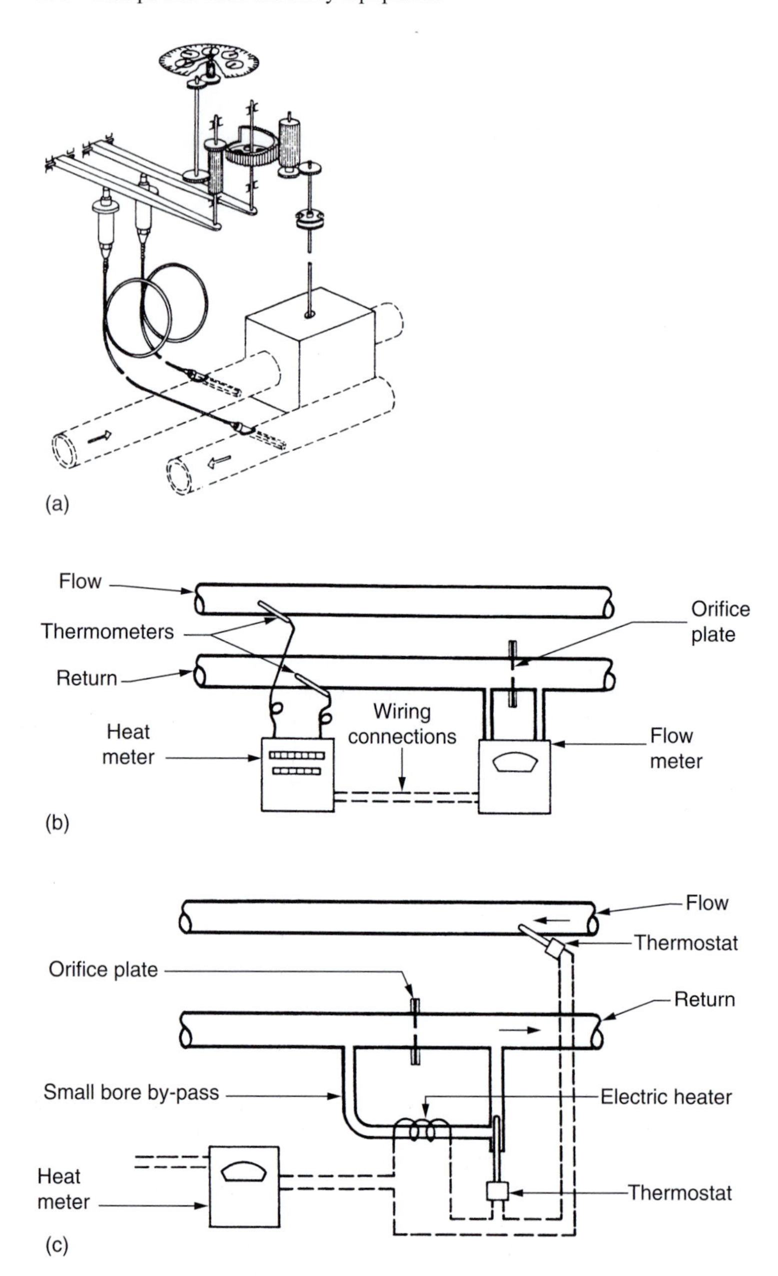

Figure 11.28 Mechanical, electrical and 'shunt' heat meters for hot water

since the solubility of air in water falls with increasing temperature, the free component will increase rather than decrease as the system is brought into service. The water content of an open system, or of one pressurised by equipment incorporating an open spill cistern, will be subjected to a process of continual re-aeration as a result of expansion and contraction. Furthermore, small leakages at pump and valve glands, etc. and evaporation from almost any type of cistern will require that raw water be admitted regularly to make up the deficit.

If the presence of air is not to impede pipe circulation; reduce heat output from emitters; lead to cavitation difficulties in low pressure areas such as pump suction connections and, primarily in domestic systems,

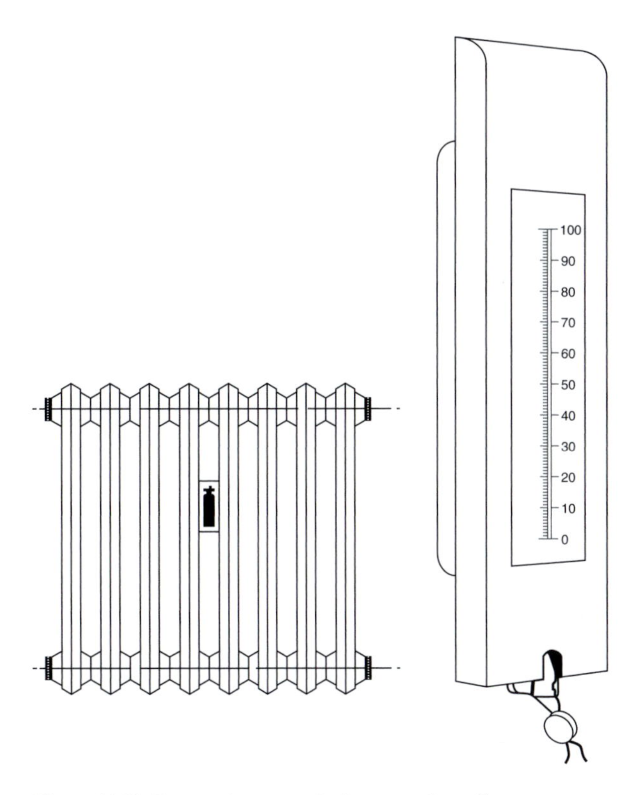

Figure 11.29 Evaporative proportioning meter for radiators

produce 'kettling' noises in boilers; then active measures must be taken towards its elimination. The matter of corrosion arising from air in the water content will be considered separately.

Air venting

The traditional approach to the disposal of air from low-temperature hot water systems was to route a full-bore flow pipe *immediately* from the boiler plant to the highest level in the building and, at that point, to provide an open air vent to above the feed and expansion cistern. This approach, although a counsel of perfection in some respects, is rarely practicable in modern buildings and in any event would be a considerable impediment to design development.

For low-temperature hot water systems, it is current practice to take advantage of any sensible opportunity to dispose of air without using components which require maintenance. A simple open vent taken straight from the boiler plant, as shown in Fig. 13.14 although provided for a different purpose, is favourably placed to carry away air coming out of solution in the boiler. Elsewhere, if high points in the system cannot be provided with air bottles having small bore vents to outside, then robust and simple automatic air valves, as shown in Fig. 11.30, must be fitted. Centrifugal and other types of air separators are sometimes added to domestic systems but there should be no need for such devices in a system which has been designed properly.

For medium- and high-temperature water systems, where air release will – inevitably – be accompanied by some flash steam, full-bore air bottles as shown in Fig. 11.31 must be used with the discharge pipe led down to a position clear of any traffic. The needle valve fitted to the discharge pipe should be provided with a lock-shield type terminal or other means to avoid unauthorised use.

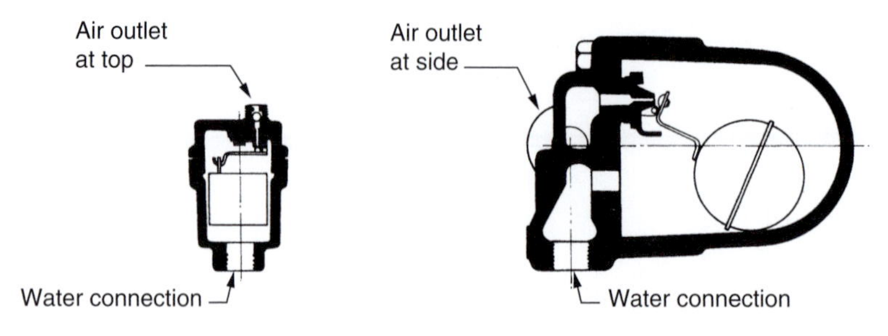

Figure 11.30 Simple automatic air valves (air eliminators)

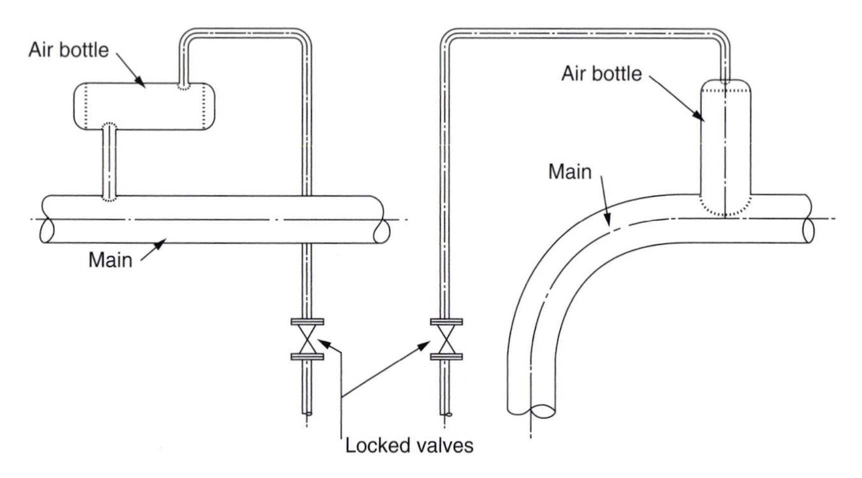

Figure 11.31 Air bottles for high-temperature systems

System draining

Over the life of any pipework system carrying a fluid it will be necessary to replace components, carry out repairs, make alterations, etc. To avoid the need to drain down the whole system to carry out localised work, isolating valves should be installed on both flow and return pipes at strategic positions around the system, e.g. at floor branches off main risers, enabling the section to be worked on to be isolated from the remainder. Within each section capable of being isolated in this way one, or a number of drain valves (cocks), each with a hose connection, should be provided at all low points to facilitate, as near as practical, complete drain-down of each section.

Corrosion

A full-scale treatment of this subject would be quite beyond the scope of this book as would be any attempt at discussion in depth of water treatment and other ameliorative measures. It would, however, be wrong to ignore the subjects altogether in view of their importance: a reader in search of further information might refer either to the *Guide Section B7, 1986* or, better still, consult a professional specialist adviser. The necessity to treat feedwater to steam systems has been well known and well explored for many years and it is not necessary to deal with that aspect of the matter here, other than by reference to Table 11.3 which summarises the additive processes commonly used.

As far as water systems are concerned, a radical change in the availability of equipment and consequently in design practice has taken place during the last 25 or more years. Boilers of solid construction with

Table 11.3 Water treatment additives for low pressure steam and hot water systems

Treatment	*Process/additive*	
Elimination		
dissolved oxygen	Hydrazine	N_2H_4
	Sodium hydroxide	NaOH
	Sodium sulphate	Na_2SO_4
Inhibition		
corrosion	Sodium compounds	
	Benzoate	
	Borate	$Na_2B_4O_7$
	Nitrite	$NaNO_3$
	Phosphate	Na_2HPO
	Silicate	
	Tannins	
Scale	Lignins	
	Organic polymers	
	Phosphates	
	Phosphonates	
Neutralising CO_2	Filming amines	
pH control	Alkalis	
	Neutralising amines	
	Sodium hydroxide	NaOH

Note
This list is for information only; specialist advice should be sought for all water treatment problems.

wide waterways, whether cast iron sectional or welded steel shell-type, have given way generally to higher-efficiency units of smaller size manufactured to designs meeting quite different criteria. Similarly, heat emitting equipment is no longer made in the British Isles in substantially proportioned cast iron but in light gauge pressed-steel sheet or aluminium. Lastly, the various new-type components are now very often piped together with small-bore light-gauge copper, rather than with sensibly thick mild steel pipes. The former are commonly made by a process which leaves a *carbonaceous* film on the internal surfaces which may lead to accelerated deterioration when these are exposed to certain hard waters.

Ideally, a water system should be filled with a stable, non-corrosive, non-scaling water, softened or de-mineralised as may be appropriate. No engineer who has seen a swimming pool filled by water straight from a mains supply, without first being passed through the filtration plant as is good practice, will remain under any illusion as to the condition of water from such a source! During the construction of the heating system, mill scale and foundry sand will have accompanied material deliveries and welding beads, cement, plaster and other site debris will have been introduced: not all of this will have been removed by any flushing process.

The mixture of diverse materials noted above, when brought into contact with mains water, lays the foundation for ultimate disaster if no positive remedial action is taken. Dependent upon the composition of the particular water supply, corrosive or scale forming or both, some level of treatment will be necessary. The symptoms of a problem relate to evidence of:

- corrosion at exposed parts
- deposition of salts at valve glands, etc.
- gassing at any tested air vents (hydrogen)
- sludge formation in the system (magnetite)
- bacteriological growth (anaerobic).

Corrosion, deposition and gassing, although apparently diverse, may well all result from the decomposition of the calcium bicarbonate ($Ca(HCO_3)_2$), common in all raw waters, to produce calcium carbonate ($CaCO_3$) and carbon dioxide (CO_2). As has been noted under the previous heading, the presence of air within a hot water system is inevitable at some time: the oxygen content of that air is a significant source of corrosion problems, in combination with the differing electrical potentials of the various metals involved. The formation *of magnetite sludge* (Fe_3O_4) results from oxygen combination with ferrous components. The deposition of a hard calcium carbonate scale on the internal surfaces of pipes may provide some measure of protection against corrosion but where such a coating breaks down due to thermal movement, attack may then be concentrated upon a small area of the pipe wall thus exposed. The presence of *anaerobic* bacteria within the quiet dark warmth of a pressurising vessel, where they ingest sulphates to produce corrosive sulphides, may generally be disturbed by raising the temperature of the water content to as far above atmospheric boiling temperature as the design of the system will permit.

As far as most conventional domestic and commercial systems are concerned, failing a detailed expert analysis of the situation by an independent professional adviser, the application of one of the commercially available corrosion inhibitors must be recommended. These take the form of chemical additives to the water content of the system at a calculated concentration. In the case of an existing system which may already have been so treated, it is imperative that details of any previous inhibitors be identified since the various specialist suppliers approach the subject in different ways and a disastrous colloidal interaction might result if two separate chemical compounds, however dilute, were to be mixed. Additives having reliable antecedents are available, based upon a selection from the processes listed in Table 11.3 or their equivalents.

Notes

1. Kell, J.R., A survey of methods of pressurization. *JIHVE*, 1958, 261, 1.
2. Northcroft, L.G., *Steam Trapping and Air Venting.* Hutchinson, 1945. *Spirax-Sarco.* Various excellent instructional courses and texts.

Chapter 12

Piping design for indirect heating and cooling systems

Heating systems

Having selected the distribution medium, the general arrangement of the system and the type, position and size of the heat emitters, the next logical stage in the design process is to select, from the range of commercial pipes sizes available, those which will provide the most economical and effective pattern of connection. The principal factors which may influence this choice may be considered, simply as a matter of convenience, as falling under two headings:

- *External constraints*
  Appearance; noise due to water turbulence or equipment; space occupied (i.e. pipe shaft size); force imposed on building components by mass; physical strength of pipe between supports.
- *Technical constraints*
  Velocity to avoid erosion, turbulence, air entrainment; dynamic pressure and pump availability; total pressure with regard to cavitation; profile of hydraulic gradient for water flow.

Superimposed upon these aspects is the matter of achieving a long-term balance between capital and recurrent costs in that a large pipe will be expensive to install initially and will emit heat (probably unwanted) in quantity. The energy required to induce fluid to flow through it, however, will be relatively low. A small pipe will require a greater energy to induce flow but will emit less heat and be cheaper to install.

Water systems – principles

Circulation of water in a piping system may be maintained either by a thermo-syphon effect or by means of a pump. A thermo-syphon or *gravity* circulation is produced by a temperature variation, the water in the hot flow pipe being less dense than that in the cooler return pipe. The consequent difference in mass between a rising hot water column and a falling cold water column stimulates movement throughout the pipework system. As to circulating pumps, the centrifugal types which are most used were discussed in Chapter 11.

Flow of water in pipes

It would be inappropriate to discuss this wide subject in any depth here: a considerable literature in terms of textbooks and technical papers is available and an adequate bibliography is presented in the *Guide Section C4.* For present purposes it is enough to note that the relationship between mass flow rate and loss of pressure is complex and depends upon the velocity of flow; the diameter of the pipe; the roughness of the internal wetted surfaces and certain temperature-dependent characteristics of the fluid flowing such as specific mass and viscosity. The *Guide Section C4* provides a selection of tables showing pressure loss and mass flow of water at several temperatures and for a range of steel and copper pipes.

The mass flow of water required to convey a given quantity of heat from a central heat source to the various heat emitters, in unit time, is a function of the temperature difference (Δt) between the flow (t_f) and the return (t_r) and of the specific heat capacity of water. This latter may, for practical purposes over the range of temperatures considered here, be taken as constant as 4.2 kJ/kg K. Mass flow may thus be calculated from:

$$\text{kg/s} = (\text{kJ/s})/4.2\Delta t$$

In a complex system where heat loads are known only in energy (kW) terms, this is a tedious calculation to work out a great many times for each and every pipework section, as required by the tables included in the *Guide Section C4*, and one which may lead to nostalgic recollections of the Imperial system where units of heat flow had a calorimetric base. However, since 1 kW = 1 kJ/s, it is easy to produce data in more direct terms (kW/K) by simple substitution in the expression set out above. By adoption of this approach, it is possible to retain the kW as the base unit throughout all calculations: heat loss; selection of heat emitting equipment; piping design and, finally, the overall totals for boiler power. It must be remembered always that the kW is no more than a measure of energy flow and use of it in this manner is an aid towards understanding the thread of continuity which persists from first to last.

Figure 12.1 has been prepared on this basis to give a vertical scale of energy flow in kW/K against a base scale of pressure loss in Pa/m run, for water flow at 75°C in steel pipes. Similar data could be provided for medium- and high-temperature hot water systems but the temperature ranges at which these operate are less well established: since they are not, use is made of correction factors and the *Guide Section C4* includes a table listing those appropriate to a water temperature of 150°C. Since the error involved in using Fig. 12.1 without correction is only about 3 per cent, it would be out of place to pursue this added complexity here. Where small bore copper tube is to be used for low-temperature systems, however, it is necessary to use alternative data since both the roughness of the wetted surfaces and the internal pipe diameters vary from those of steel. Reference to Fig. 12.2 provides the alternative information, again presented on the basis of energy flow. For water flow in PVC pipework, refer to basic data in Chapter 25 and to manufacturers' literature.

Flow of water through single resistances

In parallel with the loss of pressure consequent on flow through straight pipes, additional losses arise at bends, branches, valves, etc., and at major components such as boilers, heat emitters and so on. These come about as a result of disruption to the velocity profile which is developed in flow along straight pipes, the turbulence created at the *single resistance* persisting for a considerable distance downstream. The pressure loss characteristics of most types of single resistance have been determined experimentally, as a function of flow velocity, and are represented by the symbol ζ.

A convenient way of making allowance for single resistances is to add to the measured length of straight pipe an *equivalent length (EL)* for each such item. Values for unit equivalent length are marked on the charts in Fig. 12.1 and 12.2 and these, multiplied by the appropriate value of ζ, provide the actual equivalent length to be added. The *Guide Section C4* lists hundreds of values of ζ for various types of pipe fitting, etc., but Table 12.1 provides a short summary adequate for approximate calculations to suit most normal situations.

Velocity

Experimental data regarding acceptable maximum and minimum water velocities in pipes are meagre but there seems to be general agreement that erosion is unlikely to become a problem at less than about 2.5–3 m/s: similarly, it has been suggested that velocities of less than 2–3 m/s in straight pipes are not likely to give rise to a noise level which is unacceptable. As to minimum velocities, these are of importance as far as air entrainment and self-flushing are concerned and a value of 0.5 m/s has been proposed. The ranges of values listed in Table 12.2 represent normal practice for pumped circulations.

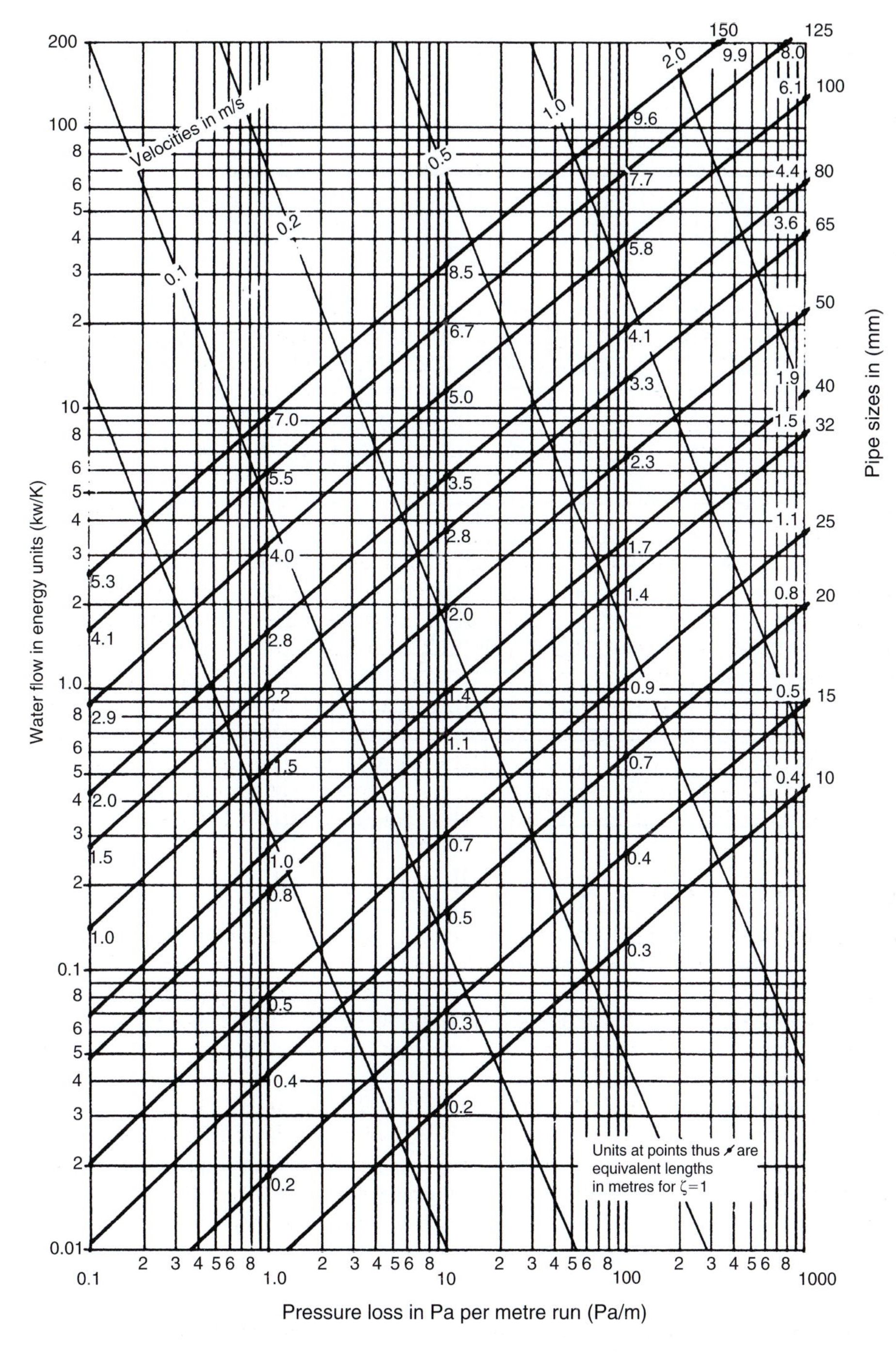

Figure 12.1 Sizing chart in energy units (kW/K) for water flow at 75°C in steel pipes (BS 1387: medium)

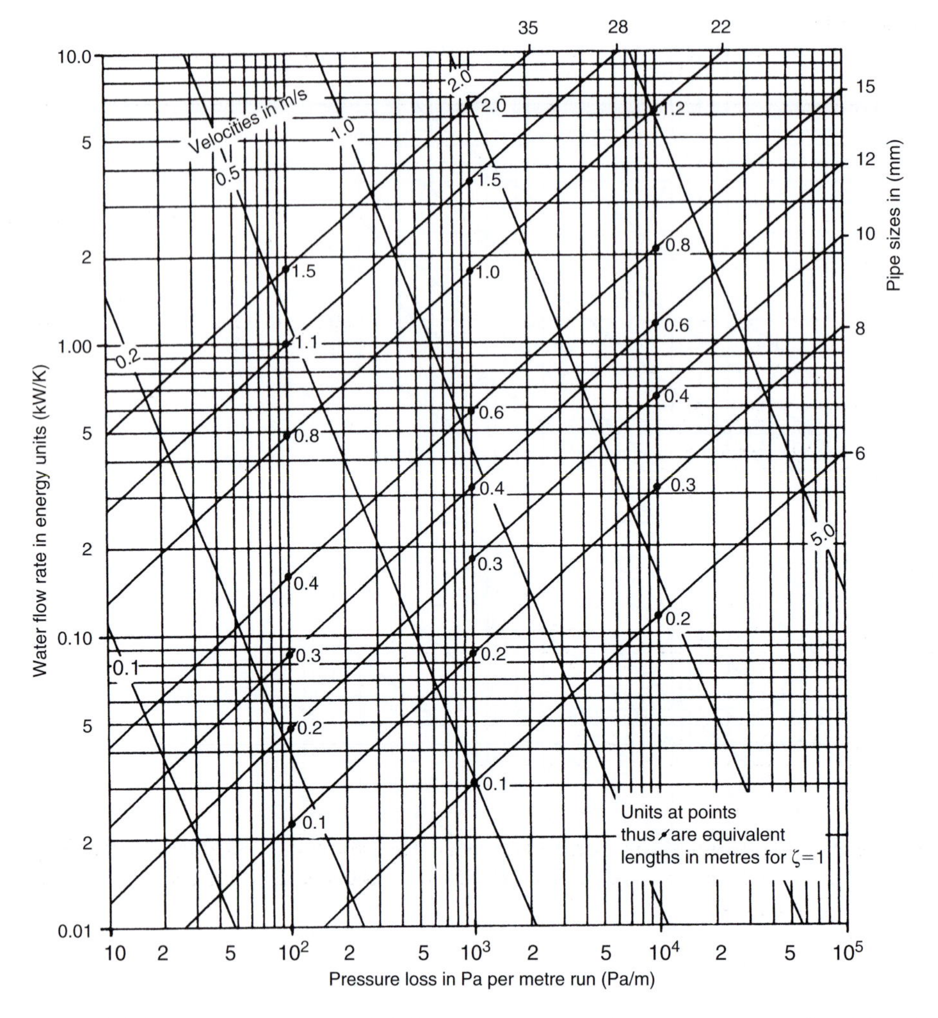

Figure 12.2 Sizing chart in energy units (kW/K) for water flow at 75°C in copper pipes (BS 2871: Table X)

Water systems – applications

In applying theory to practice, it is as well to remember that a system made up from commercial pipework bears little resemblance to a laboratory test rig. Manufacturing tolerances for internal diameters vary with the material, being about ±10 per cent for steel, and other variables such as internal roughness with ageing, thickness of scale deposition, etc. add to the uncertainty. Furthermore, for one reason or another, site work may not proceed exactly to the design: additional bends may be introduced, tee branches may be arranged in a different way and welded joints in awkward positions may not be quite as free of protrusions into the bore as they should be. These comments are not intended to suggest that pipework sizes should either be selected solely by eye or that data included in the *Guide Section C4* is too exact for practical use. They are offered rather to emphasise that approximate calculations are more pertinent in this context than results displayed on a pocket calculator, or from computer output, tuned to provide answers to many decimal places.

Except for the smallest heating systems and for primary circuits to some indirect hot water supply systems, water circulation by gravity has now been supplanted by use of centrifugal pumps. The origins of this change were the sponsorship of domestic small bore systems in the mid-1950s by a solid fuel research association and

Table 12.1 Pressure loss factors for single resistances

Item		*ζ*	*Item*		*ζ*	*Item*		*ζ*
Sharp elbow (90°)	20 mm	1.7	Reducer d_i/d_2	0.2	0.5	Values		
	50 mm	1.0		0.5	0.4	Standard globe	20 mm	8.0
	100 mm	0.7		0.8	0.2		40 mm	4.9
Long elbow/bend (90°)	20 mm	0.9	Expansion d_i/d_2	0.2	0.9		100 mm	4.1
	50 mm	0.4		0.5	0.6	Angle globe	20 mm	2.7
	100 mm	0.2		0.8	0.1		40 mm	2.4
Pipe joints			Entry to vessel		1.0		100 mm	2.2
Metal		Negligible	Exit from vessel		1.0	Gate spherical seal		0.03
Welded plastic		0.4	Column radiator		5.0	Gate plain parallel		0.3
Flanged plastic		0.13	Panel radiator		2.5	Non-return flap	20°	1.7
Through tees	= 0.5 plus reducer/expander						40°	6.6
Branch tees	= 0.2 plus equivalent elbow/bend						60°	30.0
180° bend	= 1.6 × 90°elbow/bend							
45° bend	= 0.7 × 90°elbow/bend							

Table 12.2 Water velocities in pipework

Pipes nominal bore (mm)	*Application group*	*Velocities in normal use* (m/s)
10, 15, 20	Small domestic	0.25–0.75
25, 32, 40	Domestic	0.5–1.0
50, 65, 80	Small commercial	0.75–1.75
100, 125, 150	Commercial	1.25–2.5

the coincident introduction of the small relatively inexpensive submerged rotor pumps noted in the preceding chapter. In these circumstances, it is logical to give precedence here to consideration of pumped systems and to refer to gravity circulation subsequently.

A single-pipe circuit

To consider the simplest of circuits first, Fig. 12.3(a) shows a single-pipe system serving six radiators. The pipe must carry sufficient energy, via the water flowing in it, to cater for the emission from all the radiators plus that of the piping. This total will determine the net requirement imposed upon the boiler.

Assume that the radiators emit 6 × 4 kW = 24 kW
and that pipe emission (taken as 32 mm) = 3 kW
Total load = 27 kW
If the temperature drop, flow to return,
Δt = 10 K, then energy flow = 27/10 = 2.7 kW/K
Assume that the piping measured length = 30 m
and that the equivalent length (for fittings, etc.) = 5 m
Total length = 35 m
Referring now to Fig. 12.1, a choice is available:
40 mm pipe with a pressure loss of 65 Pa/m, thus 65 × 35 = 2.3 kPa
32 mm pipe with a pressure loss of 120 Pa/m, thus 120 × 35 = 4.2 kPa
25 mm pipe with a pressure loss of 550 Pa/m, thus 550 × 35 = 19.3 kPa

Both pressure loss and velocity (0.45 m/s) are low in the 40 mm pipe and pressure loss is a little on the high side in a 25 mm pipe: thus, it would seem that the original assumption of 32 mm was correct. The net water duty required

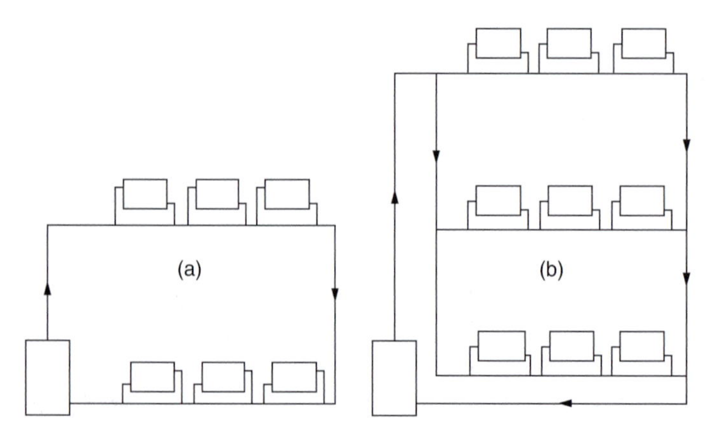

Figure 12.3 Single-pipe circuits: (a) simple and (b) multiple

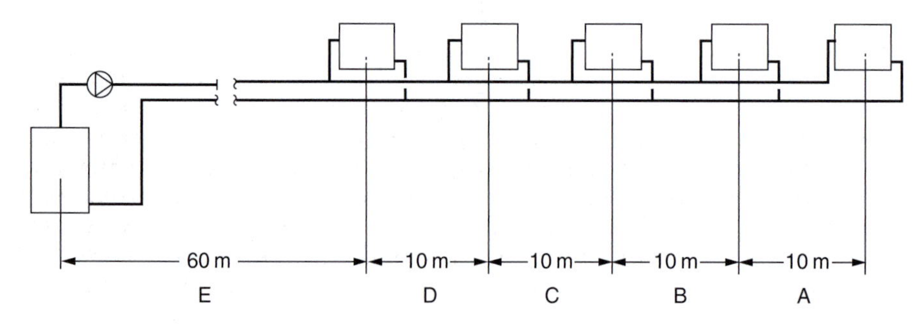

Figure 12.4 Example of a two-pipe circuit

would be (2.7/4.2) = 0.65 litre/s against a pressure of 4.2 kPa and a pump rated at perhaps 0.7 litre/s against 5 kPa would be chosen. The piping connections to individual radiators would be sized separately as will be shown later.

For reasons which will become apparent, parallel single-pipe circuits are dealt with later.

A two-pipe circuit

A simple example of a two-pipe circuit is shown in Fig. 12.4 and this will serve to illustrate how the pipe sizes might be selected in an approximate way, perhaps to produce a preliminary cost.

Assume that each of the heat emitters		
is a 20 kW fan convector, 5×20 kW	=	100 kW
and that pipe emission is say 10%	=	10 kW
Total load	=	100 kW
If the temperature drop, flow to return,		
$\Delta t = 12$ K, then energy flow = 110/12	=	9.1 kW/K
Measured length of piping, from Fig. 12.4	=	200 m
assume that equivalent length	=	20 m
Total length	=	220 m

If we assume that a pump pressure of say 20 kPa would be appropriate, then the unit pressure drop for the longest (*index*) pipe run, from the boiler to the most distant fan convector, would be 20 000/220 = 90 Pa/m.

From Fig. 12.1, using this value, pipe sizes for energy loads between 1 and 10 kW/K may be read off and the following schedule constructed:

Pipe size (mm)	kW/K	kW for 12 K Δt
25	1.1	13.2
32	2.2	26.4
40	3.1	37.2
50	6.3	75.6
65	10.2	122.4

Assuming that the pipe emission is constant at 10 per cent of the net energy load, a second schedule may be constructed:

Pipe section	*Total load* (kW)	*Pipe size* (mm)
A	22	32
B	44	40
C	66	50
D	88	50
E	110	65

It will be noted that the loads on pipe sections B and D, 44 and 88 kW, are slightly above those listed in the last column of the first schedule for 40 and 50 mm pipes respectively. By reference to Fig. 12.1, however, it will be apparent that the unit pressure loss for these sections will be increased only slightly if these sizes are selected: the next pipe size up in each case would be much too large. The short connections to the individual fan convectors would, continuing with the same approach, be 32 mm in each case but, alternatively, might each be considered separately and used to balance the system a little better. For example, since the total length related to the first item served is not much more than half that of the index circuit, 25 mm pipe might be used for the connections.

Parallel single-pipe circuits

By inspection of Fig. 12.3(b), it will be apparent that each single-pipe circuit here may be treated as if the points of connection to pipework common to all circuits were flow and return connections to a boiler. The size of the single pipe may then be selected on that basis, as discussed previously. Thereafter, each single-pipe circuit may be considered as if it were a compact heat emitting unit connected to a two-pipe circuit and sizes for the common pipework then selected accordingly.

The pressure loss which is used to select the pump will be that of the longest or *index* circuit and the volume handled will be the sum of the individual requirements. If a single value of unit pressure loss were to be used then those circuits which are shorter or carry less load than the index run will lose less pressure and it will be necessary to take account of this when sizing the common pipework and, in all probability, also necessary to provide regulating valves to overcome out-of-balance conditions.

As in the case of the example for a simple two-pipe arrangement, the process outlined above is approximate only but would serve also as a basis for making a preliminary cost estimate.

Calculations for final design

Moving now from approximations to the calculation routine required to complete a design, the system shown in Fig. 12.5 will serve as a basis to illustrate the necessary steps which are as follows:

- The *iterative* process of allocating guessed preliminary pipe sizes to a system in order to establish the amount of heat emitted by each section (guessed or estimated as a percentage in the previous examples) is tedious and use may be made of Fig. 12.6 as a shortcut. From the curves included, a near approximation to heat emission from a pipe may be read from one or other of the vertical scales for a chosen temperature drop t,

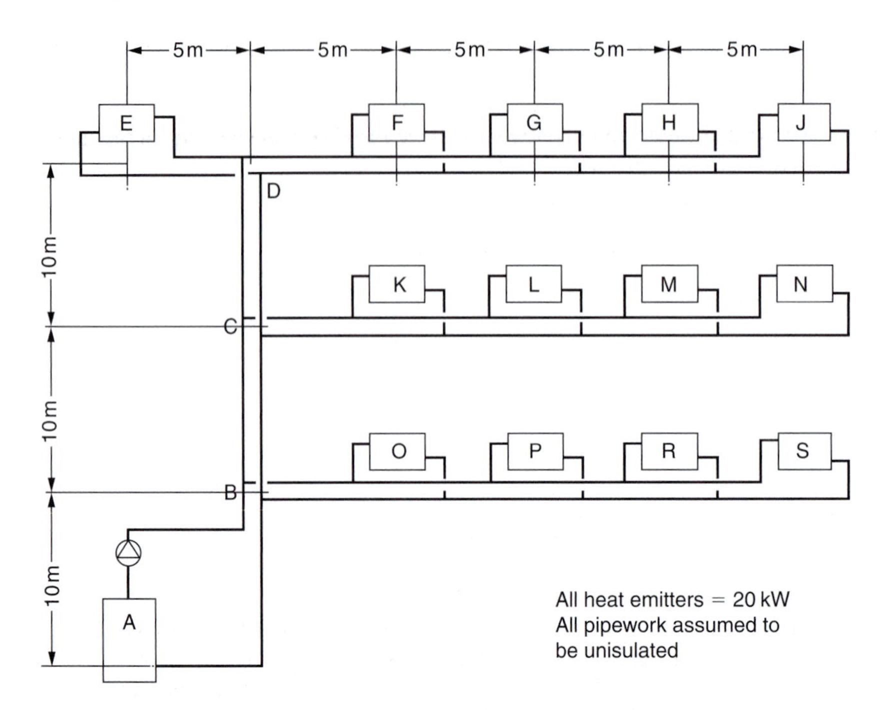

Figure 12.5 Example of two-pipe sizing exercise

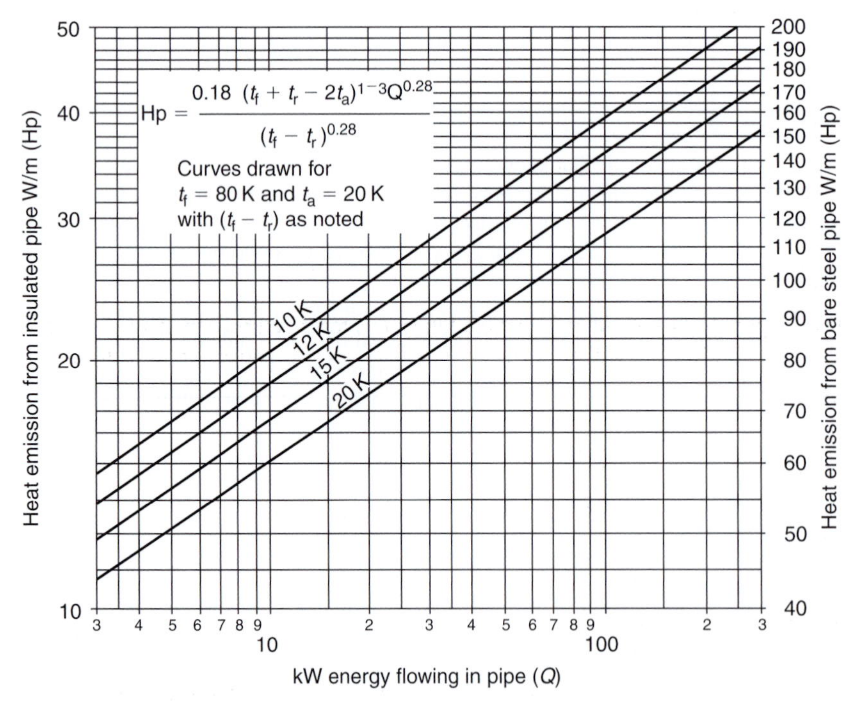

Figure 12.6 Piping heat emission related to energy flow

flow to return, against the energy flow rate shown on the bottom scale. Thus, for the system in Fig. 12.5, a list may be made as follows for a 12 K temperature drop:

- The apportionment of these heat emissions, section by section, must now be carried out systematically on a 'compound interest' basis. A first step would be to consider the three identical main circuits on the right-hand side of the figure, taking the upper identifications as follows:

	Heater			
	F	*G*	*H*	*J*
Rated emission (kW)				
Heat from pipe HJ is allocated to J alone	20.00	20.00	20.00	20.00
				0.92
Heat from pipe GH is allocated to H and J	20.00	20.00	20.00	20.92
			0.55	0.57
Heat from pipe FG is allocated to G, H and J	20.00	20.00	20.55	21.49
		0.40	0.41	0.44
Heat from pipe DF is allocated to all	20.00	20.40	20.96	21.93
	0.32	0.33	0.34	0.35
Final total	20.32	0.33	0.34	0.35

- Proceeding similarly, the heat loss from pipe CD is allocated to heaters E, F, G, H and J, that from pipe BC to heaters E, F, G, H, J, K, L, M and N, and lastly, that from pipe AB is allocated to all the heaters shown. This process produces the total loadings shown below and it will be seen that, although the *average* mains loss is just over 10 per cent, that for heater J is 18.9 per cent while that for heater O is only 3.5 per cent.

Heater	*Gross load* (kW)	*Mains loss* (%)
O	20.71	3.5
P	21.15	5.7
R	21.74	8.7
S	22.73	13.6
K	21.09	5.4
L	21.53	7.1
M	22.12	10.6
N	23.14	15.7
E	22.35	11.7
F	21.69	8.4

The effect of the mains losses upon the heat emitters is that the full *system* temperature drop of 12 K will not be available at the individual flow and return connections to those heaters. Taking the two extreme examples:

*Heater O, temperature drops**

in flow main	=	$(12 \times 0.35)/20.71$	=	0.25 K
across heater	=	$(12 \times 20)/20.71$	=	11.5 K
in return main	=	$(12 \times 0.35)/20.71$	=	0.25 K

*Heater J, temperature drops**

in flow main	=	$(12 \times 1.89)/23.79$	=	0.95 K
across heater	=	$12 \times 20)/23.79$	=	10.1 K
in return main	=	$(12 \times 1.89)/23.79$	=	0.95 K

* These are not strictly correct since the flow pipe is hotter than the return, and will thus have a slightly higher heat loss.

The object of this calculation is to produce energy loadings (and hence water quantities) which will lead to the *mean* temperature at each heat emitter being the same throughout the system. Naturally, this small simple example does not show up the relative importance of mains loss as would be the case in an extensive system, but no doubt it will serve to show the principle.

It will be seen that, for a large installation involving a great many branches of different lengths and sub-circuits of varying size, the apportionment of mains losses, if pursued to an ultimate refinement, can be a very laborious process. Various methods have been devised to simplify this task, such as have appeared in previous editions of this book. By way of rough compromise, if the total mains loss of one circuit from the boiler be calculated and divided by the number of branches on that circuit, even if not of uniform load, some attempt at apportionment can generally be made by sight to achieve a percentage basis which is probably not far from reality.

However arrived at, the mains loss for each section is added to the emitter load of the branch, and these are added progressively back to the boiler, or to the headers if there are several main circuits.

- The loads which each section of main must carry are now available and the size can be judged from a starting basis of say 100 Pa/m unit pressure drop. The length of each section, flow plus return, plus single resistances can be set down and a table prepared, thus, for heater J in the example.
- Next come the other branches and sub-circuits. At each off-take from the index circuit there will be some surplus pressure available: this must be dissipated in the branch connections, otherwise short circuiting will occur.

Each branch taken in turn then becomes a fresh exercise to be retabulated as above, sizes being adjusted to absorb surplus pressure. There is of course a limit to the practicability of such adjustments since the range of commercial pipe sizes is not infinite and, because pressure loss varies with the square of the velocity, 'steps' between pipe sizes are quite large. For example, take an energy flow rate of 3 kW/K and note from Fig. 12.1 that, whereas unit pressure loss in a 32 mm pipe is 140 Pa/m, it *is*, *5 times* greater in a 25 mm pipe.

- Having established the total pressure loss of the system based upon a number of initial assumptions, it is necessary to consider whether the result achieved has invalidated any of these to the extent where it might be necessary to repeat the calculation routine using corrected figures. Furthermore, the question arises as to whether, by selecting a different temperature drop or by varying the pump pressure, a more economical solution might have been produced.

Section of pipe *(1)*	*Load carried* (kW/K) *(2)*	*Total length (L + EL)* (m) *(3)*	*Pipe size* (mm) *(4)*	*Unit pressure loss* (Pa/m) *(5)*	*Section pressure loss (3) × (5)* *(6)*
AB	23.85	41.1	80	140	5760
BC	16.70	23.6	80	70	1652
CD	9.37	21.3	65	58	1235
DF	7.52	11.1	50	120	1332
FG	5.72	11.2	50	70	784
GH	3.88	11.1	40	120	1332
HJ	1.98	14.0	32	60	840
Emitter	1.98	–	–	Catalogue	5000

Total for circuit = 17 935 Pa
Say, 18 kPa

Hydraulic gradients

Although, as noted above, it is rarely possible to select pipe sizes for intermediate circuits such that the pressure loss there is in exact balance with that of the index circuit, it is possible to reduce the disparity by careful manipulation of the *hydraulic gradient*. Figure 12.7 shows, at the top, a string of seven heat emitting elements served by a two-pipe arrangement. Below this are plots of three alternative hydraulic gradients against a base scale of pipe length and a vertical scale of pressure loss. The vertical lines at the bottom of the plot represent pressure loss through the actual heat emitters. It will be noted that the total pressure loss represented by each curve is the same.

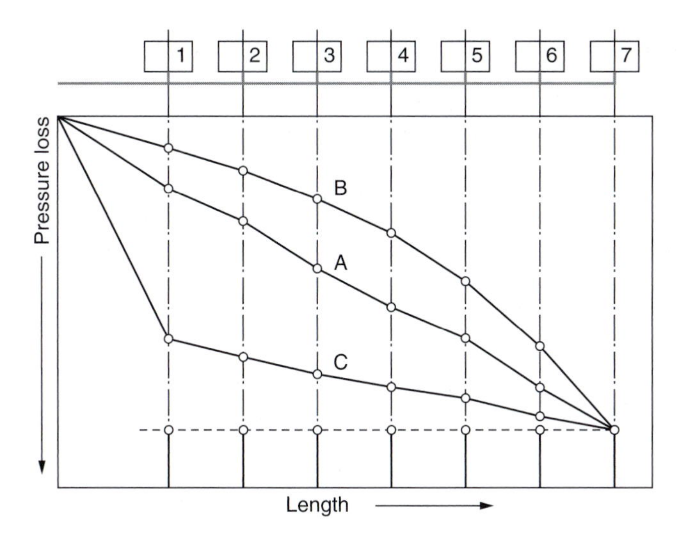

Figure 12.7 Hydraulic gradients

From the point of origin at top left, curve A is the profile of a pipe system chosen on the basis of constant pressure drop: it will be seen that a significant amount of valve regulation will be required to balance all the circuits. The concave-down curve B is a profile where unit pressure drop is low in the early sections of the system and becomes higher towards the index heat emitter: an even greater out-of-balance situation exists here. Curve C in contrast, which is concave-up, is the profile of a system which is all but self-balancing. Self-balancing performance can be achieved by selecting the pipework system with a low-pressure drop and the terminal units with a relatively high-pressure drop such that the terminals are the controlling resistance.

The lesson to be learnt from this simple example is that, for a given *overall* pressure loss through a multi-section pipe run, neither constant unit pressure loss nor constant velocity are suitable design criteria: ideally, pipe sizes should be such that the velocity in the index circuit will *decrease* stage by stage. That said, many of the constraints which have been mentioned earlier in this chapter may militate against consistent adoption of this principle.

Reversed-return systems

It will be obvious that with this arrangement (Fig 9.10) the loads to be carried by the piping are added progressively forwards for the return and backwards for the flow. Branches, however, are dealt with in the same way as for a two-pipe system. Although, in theory, there is no index circuit, the old bugbear of commercial pipe sizes confuses matters and it is sometimes found that an energy load in the *centre* of the string of circuits is disadvantaged. Moreover, a further word of caution is necessary since a condition can arise at an intermediate branch where a greater than average pressure loss has occurred in the flow sections and a less than average in the return sections. As a result, the pressure differential at the branch connections may even be negative! A careful check of pressures is required where loads or lengths of sections of pipe run vary greatly from their average.

Multiple zone systems

Previous reference has been made to those low-temperature hot water systems which are fed from a high- or medium-temperature source (Fig. 9.4). The same principle may be applied in circumstances where it is necessary to provide individual temperature or other control to a number of circuits all fed from a single low-temperature heat source as Fig. 12.8. This application is mentioned here, in the context of pipe size selection, simply because the water quantity circulating in the ring main must be slightly more than the total of that handled by the zone-circuit pumps: a margin of the order of 8–10 per cent has been found adequate. The connections to

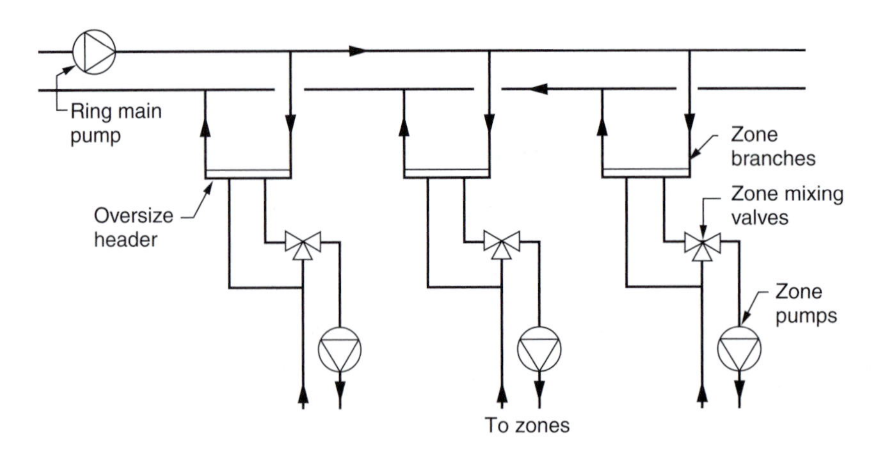

Figure 12.8 Header arrangement from ring main to multiple zones

the zone circuits are often taken from an oversize section of pipework fed from the ring main: some engineers will select a pipe one size greater than the zone connections and others prefer to use a standard size of, say, 100–150 mm in all cases. Neither practice is right (or wrong).

The objective is to ensure that there is a minimal pressure drop between the flow and return connections of the secondary circuit so that there is hydraulic independence such that both ports of the mixing control valve effectively see the same pressure and mix flow rates proportional to valve opening.

Embedded panel systems

The particular circumstances which are unique to this type of system simplify the routine of pipe size selection since:

- The design temperature drop is usually small with the result that the water quantity in circulation is correspondingly large.
- The pressure loss in the heat emitting elements (the pipe coils) is high relative to that in the main pipework.
- The pipe coils are, as a result of care in design, of much the same length and thus carry similar water quantities with similar pressure loss.

In consequence of these characteristics, such systems are very nearly self-balancing. In determining the volume which is to be handled by the circulating pump, note must be taken of both the upward and downward components of emission.

Commissioning procedures (refer to Chapter 29)

These have been established for many years. The CIBSE Commissioning Codes (A – Air distribution, B – Boiler plant, C – Automatic controls, R – Refrigeration systems, W – Water systems) set down the principles to be followed, including those for *proportional balance* of water systems; applicable to heating, chilled and other closed water recirculating systems.

Supplementary information is available in the BSRIA Commissioning Guide publications.

Regulating valves

Reference has been made here, in several earlier paragraphs, to the virtual impossibility of achieving an exact balance between various branch circuits by selection of pipe sizes alone. In order to overcome this problem, branches are provided with regulating valves which may be adjusted to take up over-pressure. Traditionally, gate valves were used for this purpose but they, in common with most other commercially available angle

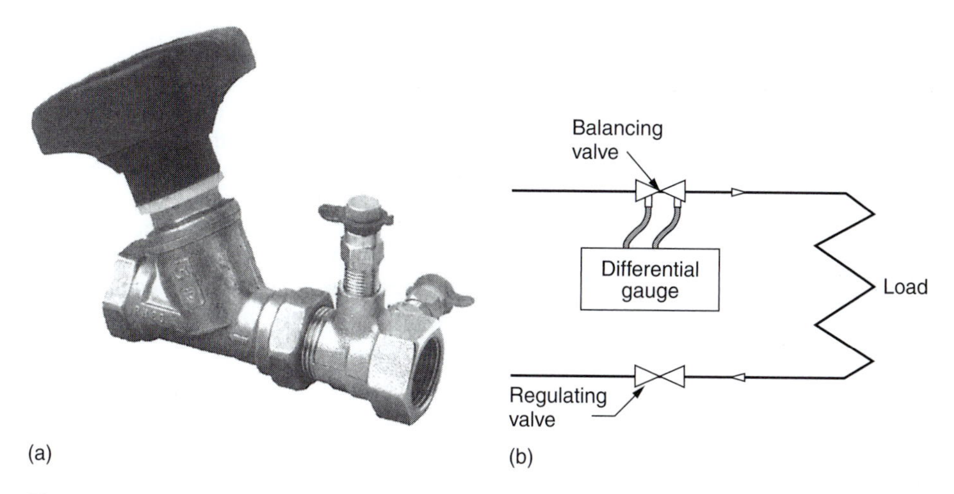

Figure 12.9 Special circuit balancing valve: (a) combined regulating value and flow metering device (Hatterstey) and (b) separate regulating and orifice measuring value arrangement

radiator valves, were virtually useless for this application since the relation between control of water flow and spindle rotation was far too critical. Valves having a good regulation performance are now available and it is accepted practice to provide major branches with balancing valves of the type shown in Fig. 12.9. These have connections to which a sensitive manometer or other test instrument may be fitted to provide data for comparison with curves or tables provided by the manufacturer to show pressure loss and water flow characteristics.

An alternative to manually adjusted regulating valves is the *constant flow regulator*. These are claimed to provide a constant flow rate over a considerable range of operating conditions. Generally, these devices comprise an inner cartridge with precision machined holes through which water passes. The cartridge is positioned against a spring within the body of the regulator and as the pressure differential increases the cartridge moves to compress the spring causing the holes open to flow to become blocked, thereby reducing and regulating the rate of flow. Cartridges are interchangeable to suit different system operating conditions. Such devices are only required at terminals, thereby reducing the total number of regulating devices in the system.

Use of constant flow regulators should avoid the need for system balancing during commissioning, but double handling is required since the devices have to be removed from the system during flushing, otherwise the holes may become blocked with debris. Some concern is often expressed over the inevitable development of dirt within a system in use fouling the holes and affecting the ability of the device to regulate to the required flow rate but design development is reducing this risk for some products.

Since, however, it is not practicable to integrate flow measurement with a flow regulator of this type, independent measuring devices are required to check that flow rates are set up and maintained at the correct levels.

Water systems – generally

'Hydraulic' design

This ill-named concept, which may have originated from study of the self-balancing characteristics of the embedded panel system, is worth noting as an oddity if nothing else. The basis of the routine is that the system pipework of the index circuit alone is selected in the normal way and the pressure loss through it established for the required water quantity. Balance is achieved for other circuits not by adjustment of pipe size but by selection of a water quantity which will produce equal pressure loss. With this approach, the temperature drop across the various circuits will differ but the argument is that this, in terms of emission (i.e. the difference between mean water temperature and air temperature in the space to be heated), will be so small as to be irrelevant. Application of the concept means, of course, that the volume duty of the circulating pump will be greater than necessary, as will be the energy which it consumes. This approach is not recommended for normal usage.

Computerised design

A totally refined solution to routine calculations directed at selection of pipe sizes would be a lengthy task using long-hand methods, having regard to the large number of interdependent variables. Traditional methods have followed principles and limitations which, in the main, have worked in practice. With the advent of the computer, new techniques have been developed to assimilate all the variables and to produce iterative solutions which would be too time consuming to deal with by any other means.

A number of computer programs are now available, some written by engineers, and experience suggests that a project of suitable size may be processed in a fraction of the time which would apply to an equivalent manual exercise. In this context, however, the limitations imposed by commercial pipe sizes and other imponderables apply equally to both computer and manual solutions. In the end, despite the means, circuits must be balanced one with the other by someone's skill in adjustment of regulating valves!

System pressure condition

The point at which the feed and expansion pipe (or the equivalent influence from a pressurisation unit) is connected to a piping system is of some significance as far as boiler and pump operation are concerned. This point of connection is the only position at which any external pressure is applied and is known as the *neutral point* of the system. With respect to the boiler or other energy source, this neutral point could be connected either to the flow or to the return but common sense dictates that it should be at the position where the water of expansion will be at the lower of the two temperatures, i.e. the return. As to position *vis a vis* the circulating pump, the connection should also be on the suction side to reduce the risk of low system pressure which could both draw air out of solution and create a risk of cavitation in the pump.

Since, as will be obvious, the pressure at the neutral point can be no greater or less than that imposed by the external influence, it follows that connection to pump delivery will result in parts of the piping system near to the pump return being under *less* than external pressure whereas connection to pump suction will place the whole system under external pressure *plus* whatever residual pump pressure is available at any given point.

Four situations are shown in Fig. 12.10

- In case A, the vent pipe must be carried above the water level in the cistern, to a height exceeding the pressure exerted there by the pump, if water discharge is to be avoided in normal operation. With a high pump pressure this may not be possible. Avoid it.
- In case B, the vent pipe is not subject to pump pressure and no additional height is necessary but there is some tendency to draw air into the system at any air vent or other open high point. Avoid it.
- In case C, with the pump in the flow pipe, the vent and the feed and expansion pipe are in balance. Since virtually the whole system is under pump pressure, air release will present no problems. This is the preferred solution.
- In case D, the feed and expansion pipe is combined with the vent. This is a solution shown to be very dangerous more than 50 years ago but still sometimes used. Don't do it.

Water systems – gravity circulation

A simple single-pipe system is shown in Fig. 12.11 and, with a gravity circulation, the force creating water movement will be that due to the difference in mass between the positive column P_1 at temperature t_2 and the negative column N_1 at temperature t_1. The force available will be that of unit mass, subject to acceleration due to gravity, which may be taken as 9.81 m/s^2, and thus:

$$CP = 9.81(p_2 - p_1)$$

where

CP = circulating pressure per m height (Pa/m)
p_2 = specific mass of water at temperature t_2 (kg/m^3)
p_1 = specific mass of water at temperature t_1 (kg/m^3)

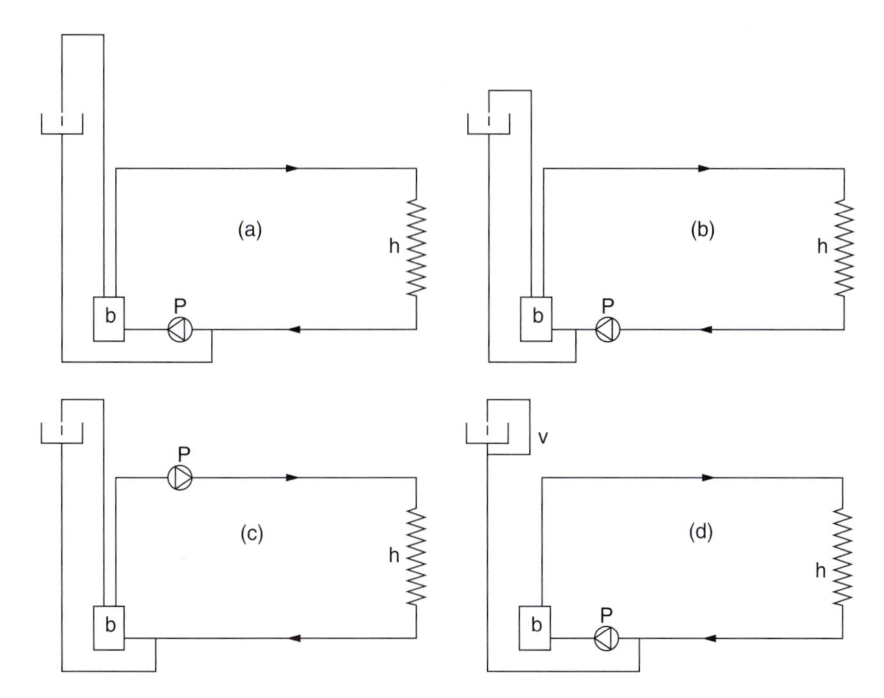

Figure 12.10 Alternative arrangements for cold water feed and expansion pipe

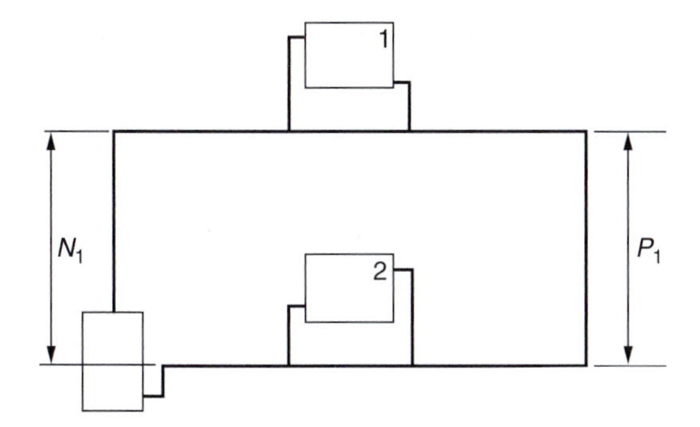

Figure 12.11 Simple example of single-pipe gravity circulation

As a matter of convenience, values of *circulating pressure (CP)* have been pre-calculated and are listed in Table 12.3 for a range of flow temperatures and temperature drops, flow to return.

It will be obvious that the water temperature at any point around the piping circuit may be determined knowing the ratio between the heat emitted between the boiler and the point in question and the heat emission overall. For example, if the output of radiator '1' and the upper horizontal pipework were 25 per cent of the overall emission, and the temperatures at the boiler were 80°C flow and 60°C return ($\Delta t = 20$ K), then the temperature of the water at the top of the drop P would be $80 - (0.25 \times 20) = 75$°C. The circulating pressure available may be determined by use of this simple relationship.

Equilibrium conditions

For a given quantity of heat to be transmitted from the boiler to radiators '1' and '2', water must flow through the pipework, the mass depending upon the temperature drop.

Table 12.3 Circulating pressures for gravity hot water systems

Flow temperature (°C)	*Circulating pressure (Pa) per metre height differences (K) flow, for the following temperature to return*						
	10	*11*	*12*	*14*	*16*	*18*	*20*
60	47.4	56.1	64.6	72.8	80.7	88.4	95.7
65	50.3	59.7	68.8	77.6	86.2	94.6	103
70	53.1	63.1	72.8	82.3	91.5	101	109
75	55.9	66.4	76.7	86.8	96.6	106	116
80	58.5	69.6	80.5	91.1	102	112	122
85	61.1	72.7	84.1	95.3	106	117	128

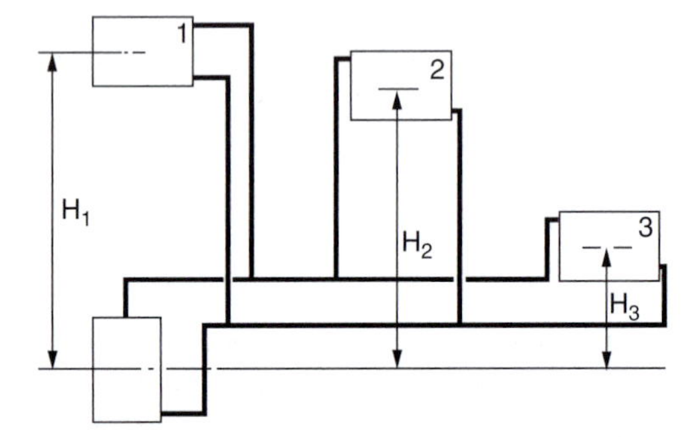

Figure 12.12 Simple example of two-pipe gravity circulation

The higher the temperature drop, t_1 to t_2, the more heat will be carried by each kilogram of water in circulation. On the other hand, the greater the difference between t_1 and t_2, the greater the circulating pressure and, in consequence, the greater the mass flowing. In practice, with constant heat output from the boiler, these effects fall into equilibrium such that the temperatures t_1 and t_2 adjust themselves to produce precisely that circulating pressure necessary to maintain a mass of water in motion which is sufficient, in turn, to release the quantity of energy necessary to create the appropriate temperature difference between them.

Thermal centrelines

For the purpose of calculation, heights are measured from the *thermal centreline* of boilers and radiators, i.e. the midpoint between the flow and return connections. For a system having more than one circuit, for example the two-pipe arrangement illustrated in Fig. 12.12, it is necessary to calculate the circulating pressure for each individual circuit and to divide this by the appropriate total length (measured plus equivalent length) in order to obtain a unit of available pressure (CP/metre length) and thus determine the least favourably placed, or index circuit. This, when identified (e.g. that serving radiator '3' in the diagram), is used to select sizes for all common pipework, those for the remaining circuits then being chosen to absorb the residual of their individual values of CP/length.

Radiator connections

It is necessary, in conclusion, to mention a particular example of gravity circulation still in use to augment pump pressure. This occurs in the case of any radiator fed from a single-pipe system. The gravity circulation through the radiator is determined as though there were a boiler in the centre of the single pipe, Fig. 12.13. The circulating pressure available may thus be calculated from the temperature difference across the radiator, which will be other than the system temperature drop, and this may then be added to the pressure differential between the two 'T' junctions resultant from any pumped circulation through the single pipe.

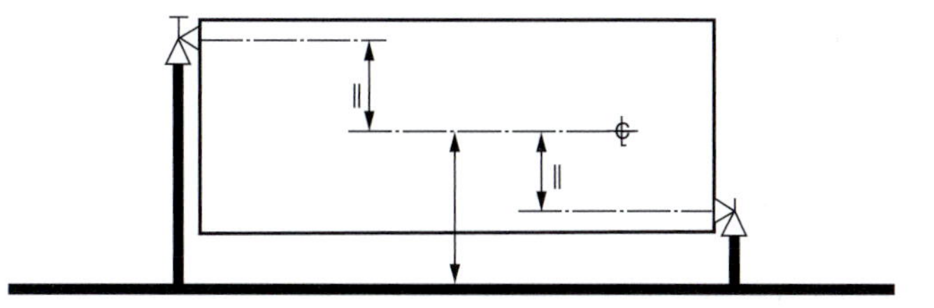

Figure 12.13 Connections to a radiator from a single-pipe circuit

These few notes are intended to do no more than provide a brief insight into the principles adopted to design gravity systems. Those readers who wish to have more knowledge of the genuine 'art' involved may consult earlier editions of this book or other works produced before 1950. The very fact that systems having connected loads of up to 3 MW were designed on this basis may come as a surprise when viewed in the context of pressure availabilities of not much more than 4 or 5 Pa/m in comparison with the 100–500 available today with pumped systems. It is no wonder that pipes of 200 mm diameter were required in boiler houses serving loads of this order and that the mass of calculations necessary was equally formidable!

Steam systems – principles

As will be obvious, no external motive force is required in any normal circumstances to produce steam flow since pressure is inherent in the state. An argument can be developed as to the practical unit to be used for steam mass, as distinct from that which best suits the purist approach. The most recent edition of the *Guide Section C4* has retained kg/s in preference to the far more useful unit of g/s listed in the 1970 edition. The latter equates to an energy supply unit of 2.25 kW and, moreover, permits working in whole numbers rather than with three or four decimal places.

While reasonable care must be taken in the calculations leading to selection of pipe sizes for steam service, it must be remembered that a number of imponderables exist in that heat losses from the pipework, whether insulated or not, will lead to formation of condensate. In consequence, the steam pipe will be carrying a mixture of vapour and liquid, the proportions of the two components varying from time to time dependent upon the external ambient air temperature and fluctuations in the rate of mass steam flow. The variables relating to flow of condensate are no less obscure since air and flash steam will accompany the boiling water in pipes which are often only half full and are under a changing mixture of pressure influences.

Calculation routines

For selection of steam pipe sizes, the following information must be available:

- The mass flow required, in g/s. For heating service this will equate to the energy requirement in watts divided by the latent heat at the pressure of utilisation. For kitchen or other equipment, the manufacturers' data must be consulted.
- Where the steam main pipework is comparatively short, heat loss from it may be ignored, but for extensive runs. Provision must be made for heat losses.
- The initial pressure available at the steam source.
- The minimum pressure required at the point of consumption.
- Lengths of the various pipeline sections, with details of all single resistances.

Selection by steam velocity

Main headers in boiler houses and other plant rooms, off-takes from them and actual connections to principal items of steam consuming plant are most conveniently sized on the basis of velocity. Volume flow may be

Table 12.4 Cross-sectional areas of steel pipes (BS 1387: medium)

Pipe size (mm)	*Area* (m^2)	*Pipe size* (mm)	*Area* (m^2)	*Pipe size* (mm)	*Area* (m^2)
15	0.000175	40	0.001272	100	0.00838
20	0.000326	50	0.002070	125	0.01305
25	0.000518	65	0.003530	150	0.01870
32	0.000927	80	0.004905		

Table 12 .5 Conventional velocities for saturated steam

Position	*Velocity* (m/s)	*Position*	*Velocity* (m/s)
Boiler outlet connections (LP)	5–10	Trunk steam mains	30–50
Boiler outlet connections (HP)	15–20	Unit heater connections	25–30
Boiler headers (HP and LP)	20–30	Calorifier connections	25–30

calculated from the energy load using steam tables data as given in Table 9.5 and, from that, velocity using the internal cross-sectional areas of pipes, in m^2, given in Table 12.4. Velocities which are generally recommended for use in various applications are quoted in Table 12.5.

Selection by pressure drop

Elsewhere in the piping system, selection of size is usually made on the basis of overall pressure drop. If no particular criteria were imposed and depending upon the extent of the distribution system, it is normal to allow for a pressure drop of between 5 and 10 per cent of the initial pressure, subject to a limiting maximum velocity of 50 m/s. Since steam is compressible, both the specific mass and the viscosity change with pressure: as a result, presentation of data relating pressure drop to flow is more complex than is the case with water. Bearing in mind the many extraneous influences mentioned earlier, an approximate approach to the relationship is required and in the context of the subject matter of this book, one will provide adequate accuracy for velocities between 5 and 50 m/s and pressures between about 100 and 1000 kPa.

A method which was developed originally by Rietschel[1] in 1922, and used in early editions of this book, fulfilled this specification but became slightly dated, and in any event referred to a different quality of pipe than that available in the British Isles. For the 1965 edition of the *Guide*, the basis for the expression was re-examined and a slightly modified version was used to produce the data included both there and in later editions including the current *Section C4*. This approach defines the relationship between steam flow and pressure loss as:

$$Z_1 - Z_2 = 3.648(M^{1.889} \times L)/(10^6 \times d^{5.027})$$

where

Z_1 = initial pressure factor = $P_1^{1.929}$
Z_2 = final pressure factor = $P_2^{1.929}$
P = steam pressure, absolute (kPa)
M = mass of steam flowing (kg/s)
L = length of pipe run (m)
d = diameter of pipe (mm)

Solutions may be presented in tabular or graphical form as in Fig. 12.14 where the *difference* between the pressure factors Z_1 and Z_2 is plotted on the base scale against the mass flow of steam on the vertical scale. It should be noted that, to avoid decimal places and use whole numbers, mass flow in the figure is plotted in

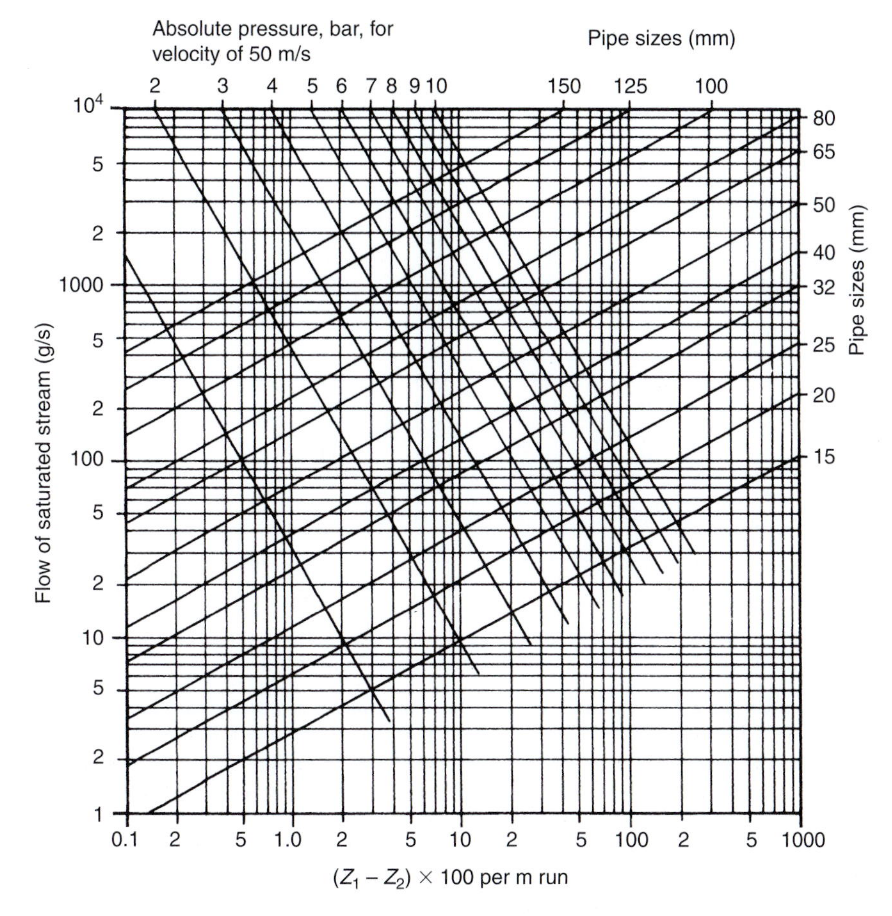

Figure 12.14 Sizing chart for flow of saturated steam in steel pipes (BS 1387: heavy)

terms of g/s and the pressure factors listed in Table 12.6 are actually $100Z$. The values within the figure and table, nevertheless, have been adjusted to be compatible. An example best illustrates the method of use:

Initial steam pressure	=	600 kPa
Pressure loss, say 10%	=	60 kPa
Final steam pressure	=	540 kPa
Energy to be transmitted	=	1000 kW
Latent heat, final (Table 6.6)	=	2099 kJ/kg
Steam flow thus is $10^6/2099$	=	476 g/s
Z for 600 kPa (Table 9.6)	=	3170
Z for 540 kPa (Table 9.6)	=	2587
Length including single resistances	=	55 m

Thus

$$(Z_1-Z_2)/L = (3170-2587)/55 \quad = \quad 10.6$$

Reading this last value on the base scale of Fig. 9.14, against 476 g/s on the vertical scale, it will be seen that a 65 mm pipe will carry about 510 g/s. By reference to Tables 6.6 and 9.4, the velocity of flow will be $0.476 \times 0.349/0.00353 = 47$ m/s and if drainage of the main presents any problem, it would be better to select an 80 mm pipe since the 10 per cent pressure drop assumed is near the top of the preferred range of velocities.

Table 12.6 Values of Z for use with Fig. 12.14 (P = absolute pressure in kPa)

P	Z	P	Z	P	Z	P	Z
100	100	240	541	460	1899	740	4751
110	120	250	586	480	2061	760	5001
120	142	260	632	500	2230	780	5258
130	166	270	679	520	2405	800	5522
140	191	280	729	540	2587	820	5791
150	219	290	780	560	2775	840	6066
160	248	300	832	580	2969	860	6348
170	278	320	943	600	3170	880	6636
180	311	340	1060	620	3377	900	6930
190	345	360	1183	640	3590	920	7230
200	381	380	1313	660	3810	940	7536
210	418	400	1450	680	4036	960	7849
220	458	420	1593	700	4268	980	8167
230	499	440	1743	720	4506	1000	8492

Note
For convenience in using whole numbers, the values given are actually ($100Z$). Figure 12.14 has been adjusted similarly.

Condensate pipe sizes

Pipes carrying condensate fall into three classes as far as selection of size is concerned:

(1) Pump discharges following condensate collection in a receiver which is so vented to atmosphere that the majority of the air and flash steam has been removed. These may be considered as flowing full and may be sized using the water flow data of Figs 12.1 and 12.2, plus a 25 per cent margin on the pressure drop read therefrom.
(2) Gravity flow at low pressure where extraneous means have provided that the majority of the air and flash steam has been discharged. A pressure drop 3 times due to water flow alone may be anticipated.
(3) Gravity flow immediately following trap discharge where air and flash steam will be present in quantity. A drop in pressure 10 times that due to water flow alone may be anticipated.

For these three very different situations, Table 12.7 provides values for condensate flow in g/s based upon a pressure drop of 40 Pa/m run of condensate main. It will be appreciated that these values are not the result of a deeply theoretical analysis: the same could be said, however, of the proven traditional approach which was to use condensate pipes one size smaller than the associated steam pipe! This somewhat intractable problem may be summarised by advocating that a well-designed system of condensate collection will provide for flow by gravity, using consistent grading, to an adequate number of pump and receiver units. In no circumstances should a trap discharge be connected to the pump delivery pipe from such a unit.

Provision for thermal expansion

Whilst the coefficients of linear expansion for steel and copper are low in comparison with those of other metals and only a tenth of those for pipeline plastics, this does not mean that their effects may be ignored. Table 12.8 shows the extent of the growth in length which arises, relative to temperature increase. When carrying out a design for thermal expansion, the temperature at which the piping is installed should always be taken into consideration, and this should be regarded as 0°C. Provision to accommodate these changes must be made in the physical layout of system and, as is obvious, the seriousness of the matter increases for those which operate at higher temperatures or where piping is in straight runs and rigidly fixed between critical points.

In all instances, short rigid connections between the various system components, boilers, pumps, etc. should be avoided since situations may arise where a temperature differential exists between piping and component, positive or negative as the case may be. Dependent upon the pipework configuration, the shape change of such

Table 12.7 Approximate data for condensate flow in steel (BS 1387: heavy) and copper (BS 2871: Table X)

Piping nominal bore (mm)		*Flow of condensate* (g/s) *for listed conditions and pressure loss of 4 mm/m run* (40 Pa/m)					
		Liquid condensate without air or vapour		*Liquid condensate with some air and vapour*		*Mixture of liquid condensate with air and vapour*	
Steel	*Copper*	*Steel*	*Copper*	*Steel*	*Copper*	*Steel*	*Copper*
10	12	15	10	10	10	5	–
15	15	30	20	20	15	10	5
20	22	65	65	40	40	20	20
25	28	120	135	75	85	35	40
32	35	260	240	170	155	80	75
40	42	400	410	260	265	125	125
50	54	765	835	500	540	240	255
65	67	1560	1500	1030	975	500	460
80	76	2420	2130	1600	1380	780	655
100	108	4920	5580	3260	3640	1590	1740

Table 12.8 Expansion of pipework for temperature difference (K)

Temperature difference (K)	*Expansion* (mm/m) *for the following materials*		*Temperature difference* (K)	*Expansion* (mm/m) *for the following materials*	
	Steel	*Copper*		*Steel*	*Copper*
50	0.567	0.846	110	1.247	1.861
60	0.680	1.105	120	2.361	2.030
70	0.794	1.184	130	1.474	2.200
80	0.907	1.354	140	1.588	2.369
90	1.021	1.523	150	1701	2.538
100	1.134	1.692	160	1.814	2.707

a connection may impose a torsional movement upon materials not best suited to the resulting stress, even at temperatures below 100°C. Given a fair length and plenty of bends, a pipework system can often accommodate itself to cope with some movement but an expansion specialist should be consulted.

Expansion joints and loops

Pipework expansion may be provided for in a number of ways:

- By length and changes in direction as commented upon above. An example is provided in Fig. 12.15(a)and it will be noted that certain points are designated as anchors or guides to ensure full control of the expansion. This is known as *natural flexibility*.
- By purpose made expansion loops, set into the pipework of the rectangular shape shown in Fig. 12.15(b). This is known as *induced flexibility*.
- By expansion joints manufactured from *bellows* inserts of spirally wrapped multi-ply stainless steel which, although quite thin in individual section, become extremely robust and provide system security and personnel safety, Fig. 12.15(c).

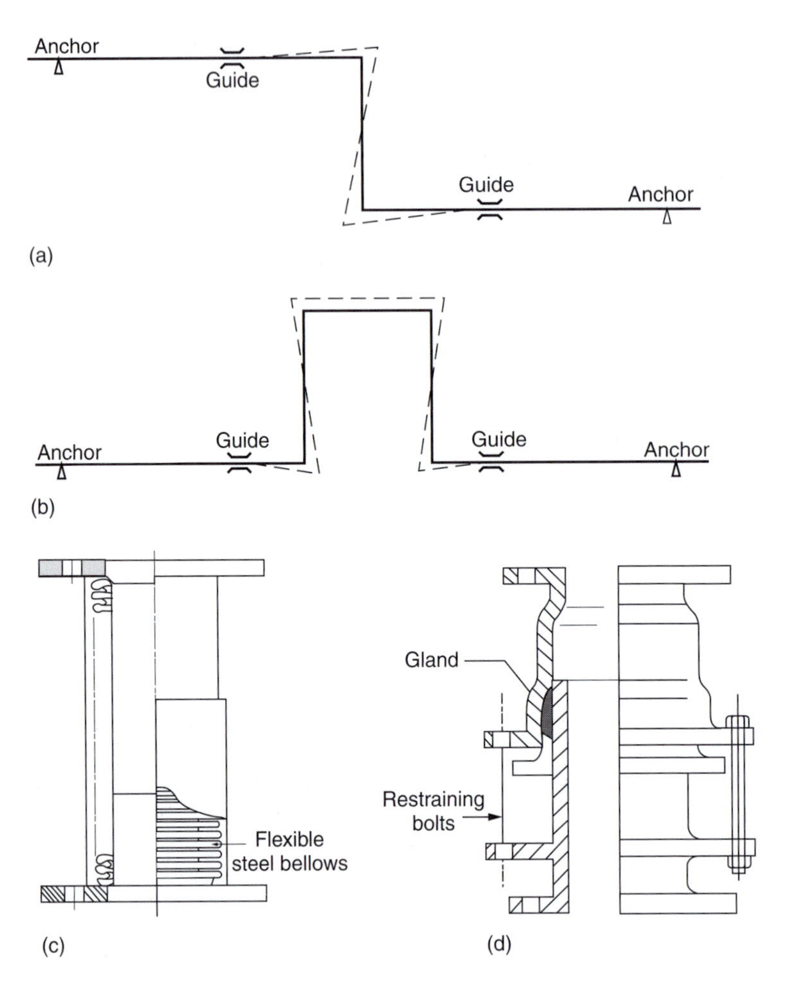

Figure 12.15 Expansion offsets and loops: unrestrained axial bellows and telescopic joints

- By sliding *telescopic* type joints, Fig. 12.15(d), which rely upon a packed gland for integrity and are restrained by bolts to prevent separation of the parts under pressure. Included here, but are no longer in common use since they could accommodate only a completely axial movement and, in any case, was prone to persistent leakage.

A *pipe stress evaluation* of the complete system should always be considered. This will be deemed a requirement from May 2002 to comply with the Pressure Equipments Directive (PED) for systems operating at temperatures in excess of 110°C.

Manufacturers should be consulted to obtain the acceptable forces that their equipment, such as boilers, pumps, etc. are capable of absorbing. Any resultant stresses in branches should be evaluated along with any building structural constraints as these affect transmitted forces. When designing any system the first approach must always be natural flexibility, but application of anchors and guides should be employed to fully control the resultant expansion. Where natural flexibility is not practical induced flexibility (expansion loops), as Figs 12.15(b) and 12.17, would be the preferred option. Anchors and guides must always be applied to expansion loops to ensure correct operation. The lack of space in modern buildings often prohibits the use of loops.

Expansion joints fall into two categories, unrestrained (axial bellows) and restrained (lateral and angular articulated bellows). Axials must always operate in a perfectly straight line without any offset, but anchors and

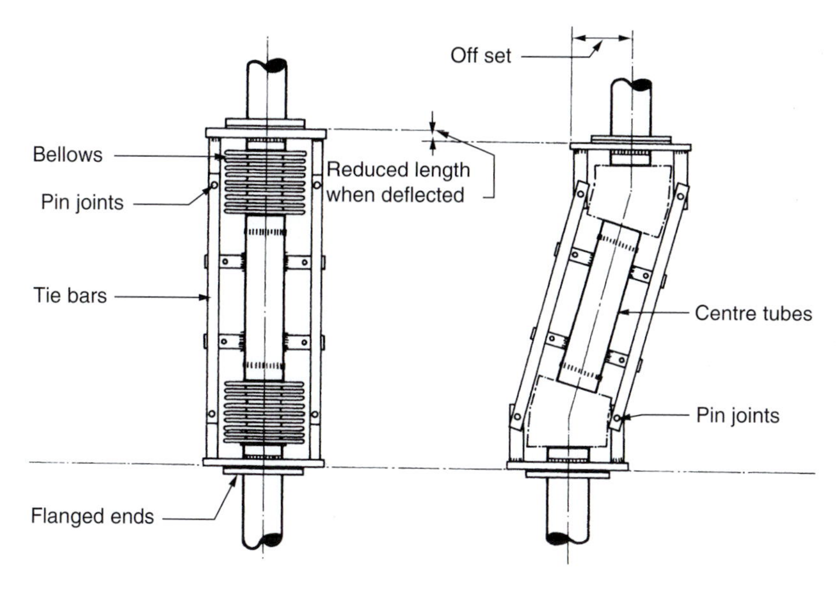

Figure 12.16 Restrained bellows and their application (engineering appliances)

guides must always be installed correctly. The reason for this is that greater forces are exerted upon the anchors and guides due to *pressure thrust* (the force necessary to restrain the effects of pressure × cross-sectional area attempting to 'straighten' out the convolutions that form the expansion joint) and in addition there is a lateral force on guides equivalent to 15 per cent of the axial force. The restrained type of joint, Fig. 12.16, can be introduced into an offset to eliminate the high forces evident in axial installations, since the tie-bars absorb the pressure thrust. The forces exerted on the anchors are greatly reduced; the larger force being due to the friction developed by the guides and supports during expansion and contraction. Normally lateral expansion joints can provide up to ±100 mm movement, and for greater movement angular expansion joints are normally installed.

Finally, the PED requirement will deem more use *of pipe stressing software* for systems where temperatures are in excess of 110°C. This could reduce the number of expansion joints required, but normally a compromise is reached as expansion joints have the inherent benefit of de-stressing a piping installation.

Cold draw

It is usual, in making calculations for dealing with the results of thermal expansion, to take account of what is called *cold draw*. In effect, this represents action taken during installation whereby the loop or offset is shortened to take up a proportion (often half) of the anticipated expansion in service. By this expedient, the force on the anchor points is reduced *pro rata*.

When applying *natural flexibility* cold draw should not be used to reduce the offset as this is difficult to monitor on site and it actually increases the anchor forces due to the increased rigidity of shorter offset pipes.

Offsets and loops

The *Guide Section B16, 1986*, presents a collection of data which enable the dimensions of offsets and rectangular loops to be evaluated and the thrust upon anchors determined. The following five salient points, not in order of merit, require greater emphasis:

- Contrary to previous usage, it is the distance between any *guides*, as distinct from *anchors*, in the pipework which is the critical dimension.

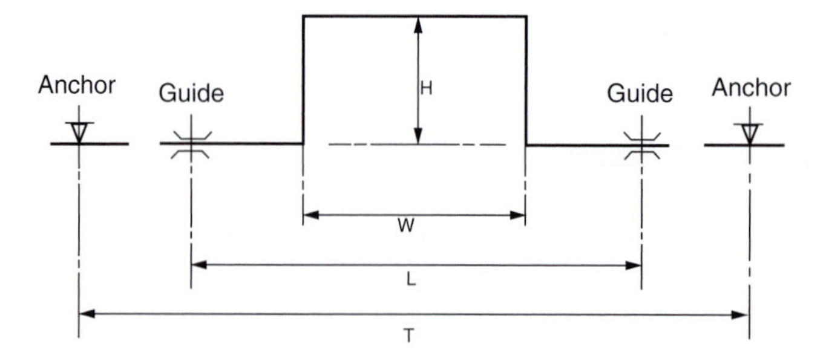

Figure 12.17 Leading dimensions for analysis of an expansion loop

Table 12.9 Fabricated expansion loops (see Fig. 12.17 for symbols)

	Pipework dimensions (m)				*Total thrust on anchors* (kN)		
	Pipe run		*Loop*		*System temperature* (°C)		
Nominal bore (K)	*T*	*L*	*W*	*H*	*80*	*110*	*140*
25	21	4	1.00	0.66	0.34	0.49	0.63
32	24	5	1.25	0.71	0.50	0.71	0.91
40	27	6	1.50	0.75	0.72	1.01	1.29
50	30	8	1.75	0.88	1.02	1.41	1.80
65	34	11	2.00	1.10	1.54	2.10	2.65
80	38	14	2.25	1.16	2.16	2.90	3.65
100	43	19	2.50	1.58	3.15	4.16	5.16
125	48	25	2.75	1.79	4.54	5.87	7.20
150	52	30	3.00	2.14	6.03	7.69	9.36

- A *symmetrical* offset or loop, between guides, produces the least thrust at the anchors.
- Anchor points in long runs of pipework may be subjected to equal thrust from each direction and thus to a balanced force of compression.
- Least thrust is imposed upon anchor points when the height of a rectangular loop is both twice its width and equal to the distance between the associated guides.
- When loops are fabricated, short radius elbows should be used rather than wide bends. The latter offer stiffness, and consequent increased thrust on anchor points, in their tendency to assume an oval cross-section.

The leading dimensions necessary to initiate calculation of the forces and moments acting as a result of thermal expansion are identified in Fig. 12.17 and, from these, results such as those listed in Table 12.9 may be produced. These latter are not, it must be emphasised, recommended values but no more than a collection made to illustrate the order of the numbers which may be expected. In practice, it is usual to use a computer to undertake what is otherwise a laborious and time-consuming exercise.

Cooling systems

The fundamental pipework design principles for chilled water systems are the same as outlined for heating systems.

Practical differences relate to the lower temperatures adopted and the significance of flow and return temperatures related to heat transfer.

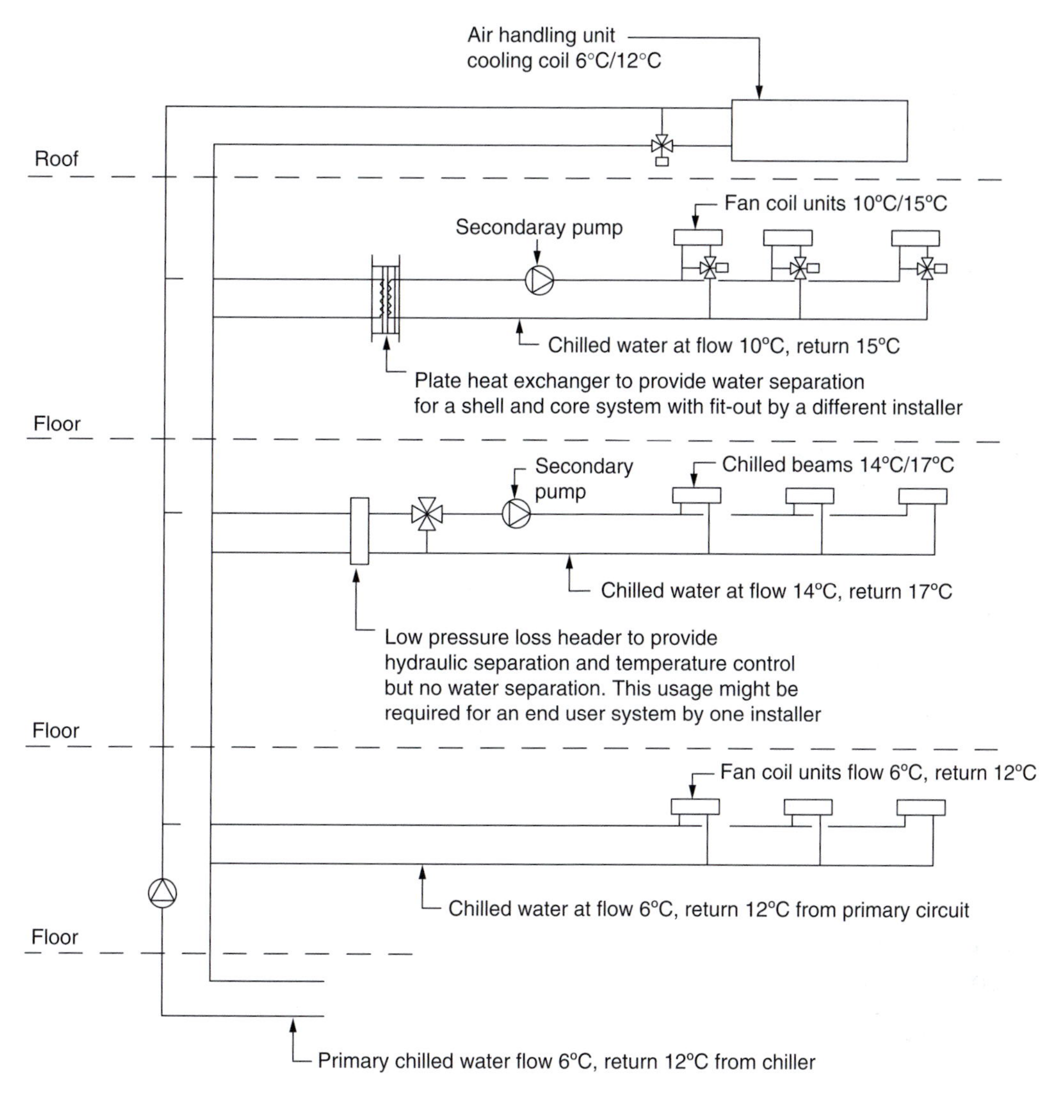

Figure 12.18 Chilled water distribution types and indicative temperatures

Temperatures

Figure 12.18 shows various temperature selections and system configurations. Primary chilled water is typically generated at 6/7°C flow to 12/13°C return giving a sufficiently low apparatus dew point on air handling unit cooling coils to enable the condensing out of moisture content from the cooled air. Water at 7°C flow/13°C return can be used for fan coil units where condensation is acceptable and drain pipework can be provided to remove the condensed water either by gravity or by the more high maintenance, leak prone pumped system.

Designers often seek to avoid condensation on fan coil unit coils either because of psychrometric considerations or because of a biological contamination risk or because the provision of drainage and/or the leakage risk is not acceptable. In these cases water temperatures typically of 10°C flow to 15°C return are used such that condensation does not occur.

When using chilled beams or chilled ceilings it is imperative to avoid condensation on the heat exchange surface which by the nature of the beam/ceiling panel are exposed to the room air. Any condensation forming as drips upon the cooling surfaces will simply rain upon the furnishings and fittings beneath them and will give rise to numerous complaints. For these applications the chilled water temperatures are typically raised to 14°C flow to 17°C return. Other key preventative measures are to ensure that the building is well sealed against moist air leakage from outside, internal moist air sources are isolated and incoming ventilation air is dehumidified usually by water at 6°C flow/12°C return to a moisture content below the dew point of the cooling surfaces in the room.

Pipe sizes

With lower-temperature differences, the flow rates increase and so chilled water pipes are generally larger than heating pipes for the same heat transfer.

To optimise pipe sizes distribution design for large buildings using chilled beams or ceilings, designers often provide a primary circuit with a 6° temperature difference which serves multiple areas and/or floors where a heat exchanger or a flow mixing circuit is then used to provide a secondary pumped circuit with a 3° temperature difference serving a smaller load.

Isolation of water systems

Where a building is equipped with primary equipment and distribution by a developer (shell and core) and then finally completed by a tenant (fit out) it is sometimes considered prudent to provide isolation of the shell/core and fit-out water systems to avoid contractual disputes about either pump performance or water quality and cross-contamination. In these cases it is sometimes chosen to use plate heat exchangers to provide isolation of water systems. For buildings constructed under one contract for end users it is common to use low loss headers to provide hydraulic independence without isolation of water systems.

Significance of water temperatures related to heat carrying capacity

For typical low-temperature heating systems the difference between the mean water temperature (75°C) and the air being heated (20°C) is 55°C.

For a typical chilled water cooling system the difference between the mean water temperature (9°C) and the air being cooled (28°C) is 19°C.

For the above heating temperature differences and a water flow rate of 1 kg/s the impact of a heat loss of 1°C in the distribution system will reduce the heating carrying capacity of the water by 1.7 per cent.

For the above cooling temperature differences and a water flow rate of 1 kg/s the impact of a heat gain of 1°C in the distribution system will reduce the cooling by 5.3 per cent.

Insulation

For both systems the provision of pipework insulation to control heat gain and heat loss is of great importance both to the capacity of the equipment and to prevent overheating in areas which contain heating pipework and to prevent unwanted condensation on chilled water pipework which has not been adequately vapour sealed.

Condenser water systems

Pipework design for open circuit cooling towers

Condenser cooling water pipework connected to an open circuit cooling tower must be routed such that only a minimum length is installed above the water level in the cooling tower tray or pond. The pipework to the cooling tower distribution arrangement must rise at a point close to the tower. Figure 12.19 shows a simplified piping

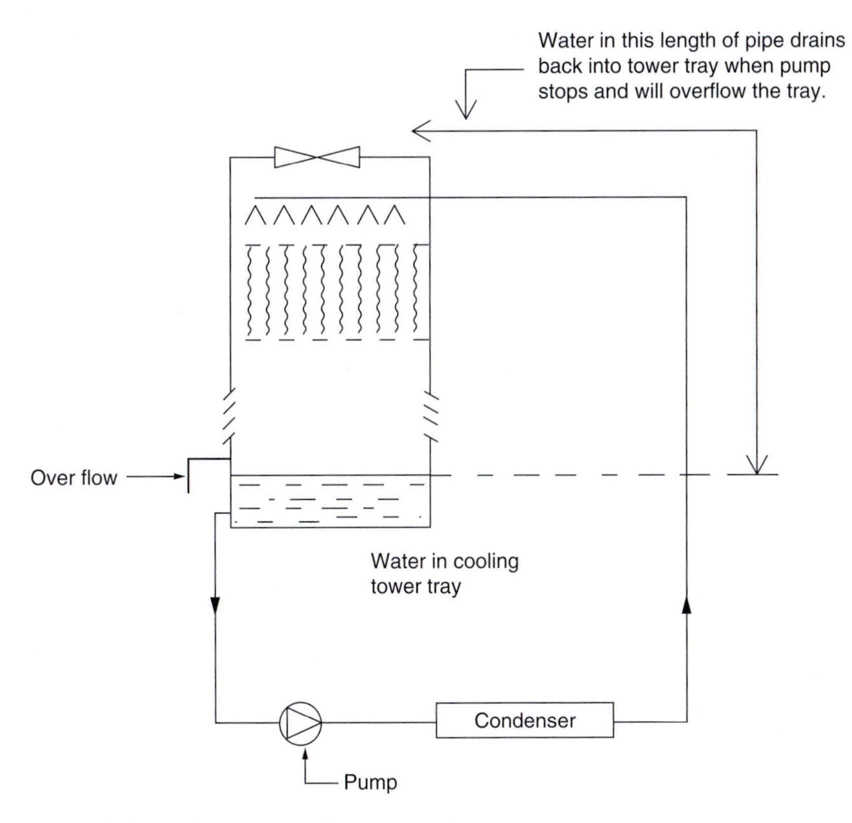

Figure 12.19 Cooling tower operation

arrangement to an open cooling tower and if long lengths of high-level pipework are installed the sequence of events will be as follows:

(1) *Pump operating*: Pipes fully charged and tray water level normal.
(2) *Pump stops*: Water in pipe above water level in tray gravitates to tray, water level in tray rises and is discharged and wasted through the overflow.
(3) *Pump starts*: Water in tray is drawn out but not immediately replenished because high-level pipe is empty, water level in tray falls, air is drawn in suction pipe and circulation stops.

Water treatment

The full text of the following summary of key points can be obtained from BSRIA Guide AG 1/2001.1 Pre-commission Cleaning of Pipework Systems.

Pipework system contaminants

The main contaminants in heating and cooling systems are installation debris, scale, corrosion products and biological fouling.

Installation debris

Welding slag, swarf, mill scale, rust and building debris can usually be removed by water flushing. Surface oxides are often removed by inhibited acid cleaners.

Cutting fluid, soldering flux, jointing compounds and grease usually require chemical degreasing agents to remove.

Scale

Scale is a problem to existing systems and occurs when water is heated or when the pH value of the water is changed. Hard water contains calcium and/or magnesium salts which can form carbonates, sulphates or silicates, and adhere as a hard scale on internal boiler, heat exchanger and pipe surfaces. Soft water is much less at risk of scaling. The impact is to greatly reduce heat transfer, reduce water flow and in the worst cases in boilers can lead to overheating and damage of combustion side surfaces because of overheating. Scale is usually removed by chemical treatment.

Corrosion products

Usually electrolytic action or bacteria induced corrosion.

Electrolytic action is the deposition of ions from one material to another through the water which is the electrolyte. There is a hierarchy of electrolytic corrosion risk among dissimilar metals and it is important to consider the metals present in a system during discussion with a water treatment specialist.

Bacteria can induce corrosion by changing the chemical content of otherwise stable water.

Corrosion rates are determined by temperature, types and concentration of impurities and water flow rate. Chemical factors are the amount of dissolved oxygen, the carbon dioxide content, pH value and dissolved solids.

Passivators, inhibitors and biocides are usually used to control corrosion rate and bacteria.

Biological fouling

Bacteria can be introduced during installation, or by filling a system but not using it, or by a contaminated water source which can multiply causing gassing, sludge formation and biofilm layers on internal surfaces. During the 1990s it was found that a high level of bacteria called Pseudomonas was often associated with the above problems which led to sludge blocking valves/strainers and gassing which affected flow rate measurements. Although Pseudomonas may not have been exclusively responsible its level is now used as an indicator of biological cleanliness and bacteria levels. Bacteria levels are controlled by biocides and biodispersants.

Pipework cleaning process

There is a sequence which can involve all or some of the following stages:

(1) Static flushing
(2) Dynamic flushing
(3) Degreasing
(4) Biocide wash
(5) Removal of surface oxides by inhibited acids
(6) Effluent disposal/final flushing
(7) Neutralisation after inhibited acid cleans
(8) Passivation
(9) Dosing with corrosion inhibitor and biocide

Generally, this is a prerequisite to water flow regulation as described in Chapter 29.

Pipework design to facilitate water treatment

It is important to maintain a positive pressure in all parts of the system relative to atmospheric pressure in order to prevent air/oxygen ingress.

To enable static flushing flow rates the make-up water flow rate to break tanks at the top of the building should be at a pressure of about 1.3 bar and pipes should be sized at a minimum of 25 mm for system volume up to 2000 litre, 40 mm for system volume 2000–10 000 litre and 50 mm for system volumes above 10 000 litre.

Regulating valves should be sized to be no less than 25 per cent open at the design flow rate in order to prevent blockage.

At terminal units thought should be given to pipe sizes to try to maintain minimum flushing velocities.

Flushing bypasses should be installed to enable control valves and regulating valves to be kept clean during initial flushing.

Up to 50 mm diameter full pipe diameter drain points should be provided at all low points and points of low flow velocity where debris can accumulate. Above 50 mm pipe diameter drains should be at 50 per cent pipe diameter. Dead legs should be avoided.

Stagnant water trapped in pressurisation units has led to bacterial growth so consider units with recirculating circuits.

The size of perforations in strainer baskets and the grade of any additional mesh should be selected to suit the application. Mesh should be removed for initial flushing in most cases. Strainers should have pressure test points to enable checking for blockage.

Pressure test points should be located at the side of pipes or components to avoid dirt or air blockage.

Air vents should be provided at high points, at the end of horizontal pipe runs and at the top of self-draining sections to prevent air blockage. Water velocities in air vent locations should be low (less than 0.4 m/s) to allow air to collect at the top of the pipe rather than pass by in the centre stream flow.

De-aerators can be used where dissolved oxygen content is high.

All major plant should be provided with bypassing and isolation valves to enable flushing bypass if required.

Required dynamic flushing velocities should be established and the velocities can be achieved either by the duty pump operating on its own or by standby and duty pumps operating together or by standby and duty pumps running in parallel or series with a temporary flushing pump.

Foul drains must be provided close to flushing out points, 100 mm diameter generally with 200 mm diameter in plant rooms.

For large diameter pipework where flushing velocities cannot be achieved, provide easily demountable pipe sections at dirt collection points to enable manual clean out. Provide strainers on all plant items. Provide dirt pockets along large pipes to trap dirt. Consider the use of chemical dispersants to keep particles in suspension.

Provide manual dosing pots and/or automatic dosing pots to enable on-going chemical dosing.

Consider the use of side stream filtration, hydrocyclones or full-flow duplex filters for sensitive applications.

On-going water treatment

Closed low-pressure hot water heating and chilled water systems

Cleaned systems must be treated with corrosion inhibitors and biocides to limit corrosion, sludge formation and bacterial slime formation. Soft water make-up can be considered to reduce water treatment requirements in hard water areas.

Condenser water cooling systems

Cleaned water circuits must be treated with inhibitors to limit corrosion, dispersants to keep particles in suspension and alternating biocide treatment to minimise the risk of legionella and algae and slime growth. Systems must enable periodic disinfection.

Note

1. Rietschel, H. and Brabbee, K., *Leitfaden der Heiz-und Luftungstechnik*, Julius Springer, 1922, 2, 28.

Chapter 13

Boilers and burners

In subsequent chapters (Chapters 16 and 17), the properties of available fuels or energy sources and the chemistry of combustion will be examined. As an introduction to this subject, however, we now turn to the plant items employed in practice to burn those fuels and to transfer the energy so liberated into a heating medium. Features that are particular to those boilers designed to serve domestic hot water systems only are described in Chapter 25.

Basic considerations

Boiler power

The load imposed upon a boiler plant is derived from knowledge of not only the sum of the heat losses calculated for the individual spaces under basic design-temperature conditions, but also of any interaction or diversity between the components thereof. The total of the heat losses may well exceed the actual peak demand due to the fact that infiltration air entering rooms on one side of the building may leave via rooms on the other side. As a result, an adjustment is necessary to avoid making allowance twice for the same air change. In addition, where heating is continuous, it is known that further diversities exist and use of the following correction factors has been proposed:

Single space	1.0
Single building or zone, controlled centrally	0.9
Single building or zone, controlled per room	0.8
Group of buildings with similar pattern of use	0.8

The appropriate corrections having been made and the peak load thus estimated, an arbitrary allowance or a calculated estimate must be added to cover unwanted heat losses from distribution pipework and fittings. It is then necessary to consider by how much the boiler rating should exceed this corrected total in order to make sure that the design temperatures within the various heated spaces may be reached in a reasonable time. When heating is continuous, of course, the excess capacity will be a minimum. On the other hand, the more intermittent the usage, the greater the excess capacity required.

There is a further case for some margin of capacity where a boiler plant is thermostatically controlled, in order to provide what might be called *acceleration* – the ability of the plant to surmount the load under peak conditions whilst still remaining under control. In addition, it is wise to remember that boiler plants are generally not well maintained and boiler efficiencies can deteriorate. The result is that output over time is likely to decline from the rated level due to irrational attention to combustion equipment and consequent fouling of heat transfer surface by soot or other deposits.

To some extent the various corrections cancel out and in the past these refinements in calculation were often ignored, a blanket margin of 25 per cent being added to the total of heat losses to deal with such adjustments, to make allowance for distribution heat losses and to cover a multitude of other sins. More recent thinking has concluded that much smaller allowances, of the order of 10–15 per cent, are adequate but there is no proper alternative to a painstaking analysis of the various aspects and the addition then of no more than a notional 5 per cent or so to the calculated total. By such means, the designer will know just what allowance he has made for what eventuality.

By way of example to illustrate the associated problems, if the total heat loss for a building, after adjustment for diversity, etc., were 490 kW and a calculated addition of 25 kW were made for heat losses from piping outside the occupied space, then a net boiler capacity of 515 kW would be needed. Adding an arbitrary margin of 10 per cent to that figure would produce a gross requirement of about 570 kW. Boilers, as will be appreciated, are not purpose made to size and a side issue might be that for physical or other reasons one particular make was preferred. The product range then examined would perhaps offer a choice between output ratings of 550 or 600 kW and since these would, respectively, represent margins of either 6.8 or 16.5 per cent, the former would probably be chosen as being the more suitable.

In days when hand stoking with coke was the norm, a generously oversized boiler with a large firebox would allow longer periods of running without attention on one charge of fuel. With automated firing, however, and in particular with fuel oil or gas, it is a positive penalty on running cost to select a single boiler with too great a margin. The consequent extended periods of firing at low load will lead to frequent cycling of the firing mechanism and a reduced annual average combustion efficiency will result, to say nothing of smoke and other nuisances.

Boiler efficiency

Heat *input* to boilers of all types must not be confused with rated *output*, both figures are often quoted by manufacturers. *Input* is measured in a rational manner best suited to the fuel: coal is weighed and the volume of liquid or gaseous fuel is metered. Detailed procedures for boiler efficiency tests are set out in various British and European Standards, the duration usually being a 6-hour test period with pre- and post-control periods. From such a test and taking account of an associated analysis of the fuel, a heat balance may be struck. The complete routine is extensive but the final summary may be brief, for example:

		(Per cent)
Heat content of fuel		100.0
Heat loss in flue gases	11.4	
Loss due to unburnt carbon monoxide	1.9	
Heat losses (from boiler casing)	4.2	17.5
Overall thermal efficiency		82.5

The heat content of a fuel is referred to either by its net or gross calorific value. Gross calorific value includes the latent heat within any water vapour formed as a result of the combustion of hydrogen in the fuel; net calorific value excludes this constituent. In the UK and the USA, boiler manufacturers tend to use the gross value when publishing efficiency figures. The net value is extensively used in continental Europe in quoting boiler efficiency figures. Using the net calorific value a higher percentage efficiency figure is obtained compared with using the gross calorific value. In the case of condensing boilers, efficiency figures calculated using the net calorific value of the fuel could appear to be in excess of 100 per cent.

Heat *output* in the case of hot water boilers is troublesome and costly to measure on site since simultaneous integration of both water quantity and temperature difference, inlet to outlet, is needed over an extended period of steady state conditions. In contrast, for steam boilers, the water pumped in and then evaporated may be measured easily and conveniently over the test period using a simple turbine-type water meter. Thence, from the inlet water temperature and the outlet steam pressure, the total heat added per unit quantity of water may be calculated, which represents the heat output.

Quoted figures for works test efficiencies for most modern boilers range between 85 and 90 per cent and in the particular case of condensing boilers, referred to later, may be somewhat higher. In daily use, however, efficiencies may be expected to average at least 5 and more probably 10 per cent below those recorded under controlled conditions on a test bed.

Criteria for boiler selection

Hot water boilers are normally rated in kW and some manufacturers of steam boilers use this same basis in parallel with the more conventional mass approach of *kg/s from and at 100°C*. This alternative rating for steam boilers denotes the output which would be achieved if steam were being generated at atmospheric pressure from boiling water, i.e. it assumes that latent heat only is added. Such a rating is of little practical use since the feedwater supplied is seldom at 100°C and the pressure is usually above atmospheric level. For any set of conditions other than those listed, the actual performance of a steam boiler may be calculated from:

$$E_a = 2258(E_r)(H_t - H_w)$$

where

E_a = actual evaporation (kg/s)
E_r = rated evaporation (kg/s)
H_t = total heat at required pressure (kJ/kg)
H_w = heat in water at feed temperature (kJ/kg)

Ratings for all boilers fall broadly into the following three ranges:

Small	10–50 kW (mainly domestic)
Medium	50–500 kW
Large	500 kW and above

Boiler margins

In any sizeable installation of over say 300 kW, there is a case for providing more than one boiler. If there are two boilers then one will suffice in mild weather, working near to its full output which is advantageous in terms of efficiency and avoidance of corrosion. The second boiler is then brought into use during cold weather and can act as a standby for a large part of the year against breakdown or outage for maintenance. In the past it was routine to select each of two boilers to have two-thirds of the total required capacity, but more modern practice, although not with universal agreement, suggests that each should meet no more than half of the total requirement. For larger installations, three, four or more boilers may be used, giving greater flexibility still, since several boilers that each have a small margin may then give almost a complete one-boiler standby.

Selection of the size of individual units for a multi-boiler plant, nonetheless, is a matter for compromise and cannot be determined wholly by a statistical approach. On the one hand, limitation of the number of spare parts to be stocked suggests the use of equally sized units, e.g. three boilers at 167 kW to meet a gross total of 500 kW. Conversely, to hold maximum combustion efficiency whilst meeting a load that varies with the external temperature, a case might be made for selecting units of unequal size, e.g. for the same gross total of 500 kW, one boiler at 100 and two at 200 kW, which would provide five steps in output instead of three.

The first of these two loading arrangements would require no more than a simple sequence controller, manually or automatically reset at intervals to change the order of firing (1–2–3, 2–3–1, 3–1–2) to avoid undue use of any one boiler. The second example, however, would need a very much more complex sensing system with automatic, though limited potential (and this beyond the ability of an operative to monitor) for change of sequence. The end result might be that the costs associated with an elaborate sequence arrangement would outweigh any saving resulting from better annual combustion efficiency.

Packaged modules

The modular approach to the provision of boiler power differs from the conventional in that it represents the provision of an array of small boiler–burner units, each of which may have a rating in the range of 40–100 kW dependent upon the maker. The units are assembled with their waterways connected in parallel, as Fig. 13.1, so that, for instance, an array of six 60 kW units would substitute for a single large boiler rated at 360 kW.

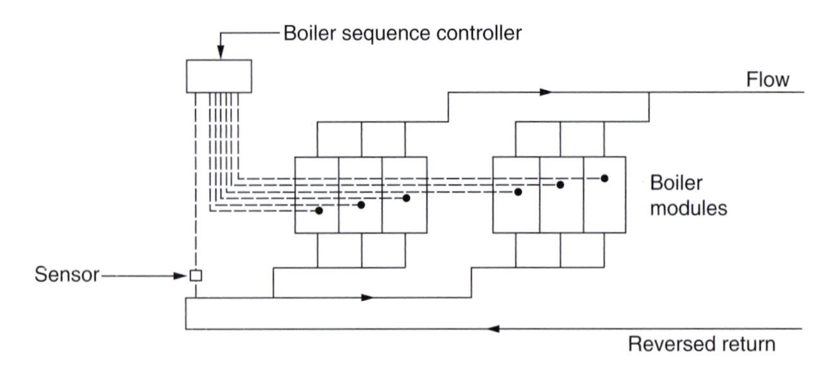

Figure 13.1 Typical pipework arrangements for modular boilers

The principal gain arising from such an arrangement is in flexibility to meet variable loads at high efficiency. For the example quoted, if only one of the units were to be fired then the *turn-down ratio* would be 6:1 at approximately the same efficiency as if all units were in use.

Essential to the modular concept is the in-built system of control which, via a single monitor, will arrange for the optimum number of modules to be fired to meet the imposed thermal load. Facilities are provided in the control arrangements to allow changes in the order in which the modules are fired to produce even running across all units. Care must be taken in the design of the associated water circuits with particular reference to possible interaction between the boiler load controller and any diverting or mixing valves fitted there: some means to maintain constant water flow is required.

The advantages claimed for the concept, which is, as will be noted, no more than a works designed package echoing the *ad hoc* arrangement for load matching 'steps' described under the first heading here, include the low thermal capacity of the boiler modules and the consequent quick response to both firing and control.

Thermal storage

An alternative approach, which is in some respects diametrically opposite to that noted above, takes account of not only the cold weather peak load but also the extent by which this will be reduced on even the most severe winter day by lighting, solar radiation, occupancy and other heat sources. As may have been noted from Chapter 8, such aspects are taken into account as a matter of course when electrical off-peak systems are designed.

In essence, the heat storage method noted here consists of a package made up from a well insulated water vessel interposed between the boiler unit and the distribution system. The boiler charges the store and maintains the water temperature in it, quite irrespective of the concurrent output of the system. By this means, it is claimed that the frequency of the on/off cycling of the firing arrangements is greatly reduced, particularly under the less than full load conditions which persist for most of the year, thus increasing annual efficiency. It is further claimed that as a result of adoption of such an arrangement, the capacity of the boiler plant may be reduced. Figure 13.2 shows the arrangement of such a system.

Certainly, on a domestic scale, this approach has proved effective and further reference will be found in Chapter 25 to the situation where a so-called *combination boiler* is used and both heating and domestic hot water supply systems are served from it.

As a word of warning, however, it is worth remembering, when considering this or any other design hypothesis (however promising in theory) which relies for success upon a reduction in plant potential, that building occupiers are inclined to deplore ingenuity when full capacity is unavailable on a cold winter morning.

Condensing

In Chapter 17, an explanation is given regarding corrosion problems that may arise at boiler heat exchange surfaces when combustion flue gases from sulphur bearing fuels are allowed to cool below about 250°C.

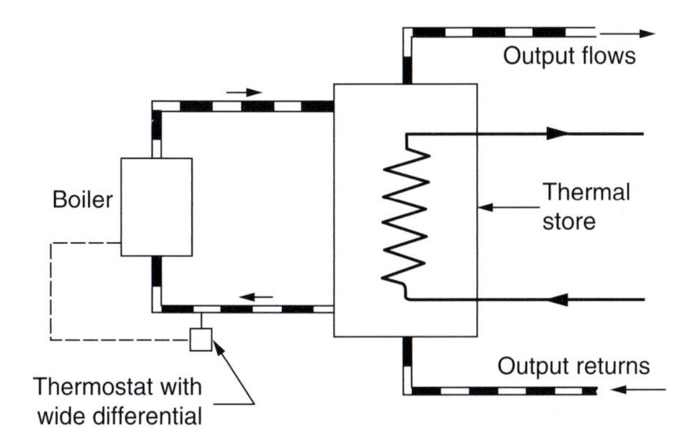

Figure 13.2 Pipework arrangements for heat storage

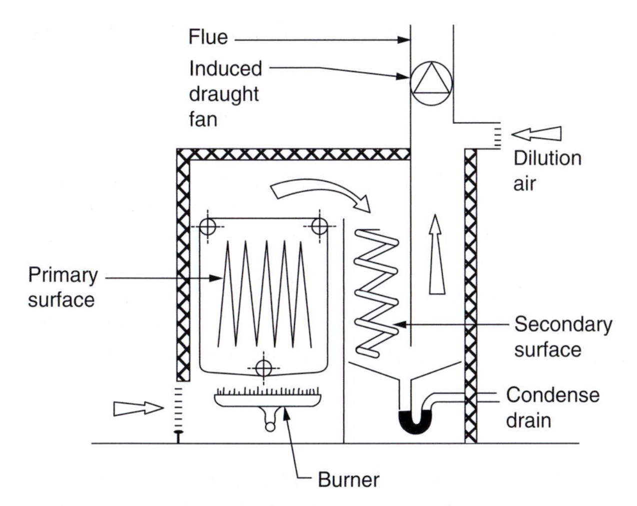

Figure 13.3 Principle of condensing boilers

It has been appreciated, nonetheless, that a very considerable quantity of heat is lost as a result, in particular that within the water vapour which has not been condensed (i.e. the difference between the gross and the net calorific values of the fuel, see Chapter 16). The advent of fuels having negligible or no sulphur content, coupled with the need for ever-increasing boiler efficiencies has seen a dramatic rise in boilers specifically designed to operate in a condensing mode.

First developed in the domestic range of sizes but later introduced for commercial use with hot water systems, the principle underlying the design of this type of boiler is the introduction of a second heat exchanger of stainless steel or a protected material, through which the flue gases pass after leaving the boiler proper. As a result of the additional resistance to gas flow so caused, it is usually necessary that forced or induced draught be provided. The outline arrangement of such a boiler, now made for loads between 12 and about 800 kW, is illustrated in Fig. 13.3.

The full potential of the *condensing boiler* is, of course, available only when the flue gases are reduced in temperature to that of the ingoing combustion air, ideally about 15°C. In practice, the factor that determines performance is the temperature at which water returns to the boiler from the associated system, since the heat exchanger surfaces upon which condensation takes place are normally about 5 K above this. The practical order of efficiencies which may be obtained, with various return water temperatures, is shown in Fig. 13.4.

The condensate from the flue gases is mildly acidic (average 3.5 pH) and a means of neutralising prior to discharge to drain may be necessary. Proposals need to be discussed and agreed with the local authority.

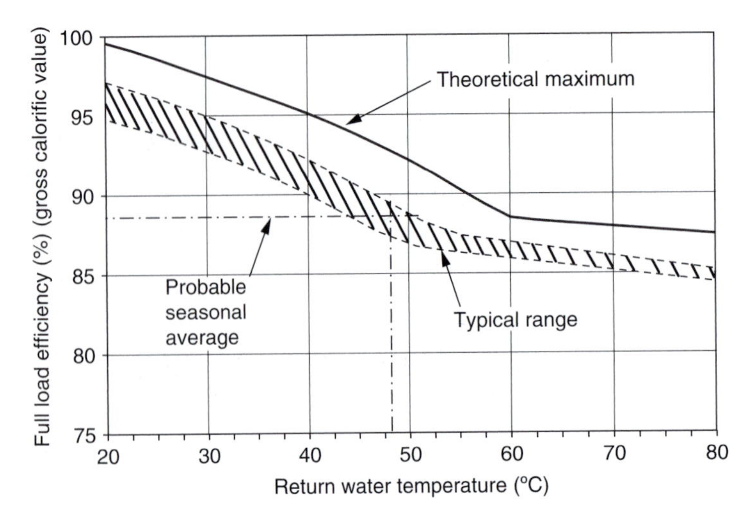

Figure 13.4 Efficiency of condensing boilers

Care should be exercised when considering the use of, and application for, condensing boilers. It is rarely a matter of choosing a condensing boiler simply because it has greater potential efficiency than a conventional boiler.

A constant temperature, hot water circuit, unless specifically designed for low temperature operation will return water at a temperature well above the condensing band. Should the highest efficiencies, with water returning at a low temperature, only be achieved for short periods throughout the year, then the extra costs associated with condensing boilers may not be justified. Nevertheless, the higher efficiencies attainable mean that the viability of using condensing boilers should always be considered in any new or refurbished development.

One technique often adopted where several boilers are used is for the lead boiler, taking most of the load, to be of a condensing pattern, with the remainder conventional. This arrangement has the advantage of reducing the extra capital cost of the boiler plant whilst maximising operating efficiency during part load periods requiring lower temperatures, which occur for the majority of the heating season. Full load is still available during very cold weather. Such an arrangement, however, does not allow for full rotation of boilers for even usage.

Balanced flue

In the sense that a balanced flue relates to the provision of combustion and ventilation air into a sealed boiler compartment, either ducted around, or adjacent to the flue conveying the products of combustion away from the boiler, this subject is dealt with in Chapter 17. However, many domestic/small wall-mounted boilers with outputs in the range 12–80 kW incorporate integral purpose-designed balanced flues, as shown in Fig. 13.5.

Hot water supply

It sometimes happens that it is convenient to provide service to a domestic hot water supply system during the heating season, by use of a combined boiler plant serving a heat exchanger or calorifier. Since demand for hot water is in most circumstances spasmodic, it is usually possible to provide that service by 'robbing' the heating system for short periods. In other words, when selecting the boiler power for such a combined system, it may not be necessary to add the whole or even a part of the heat exchanger rating to that of the heating system. This approach is not valid, of course, where a very large hot water load occurs in a building having a small heating load, e.g. for a large kitchen or laundry, (although the converse might possibly apply in these circumstances!)

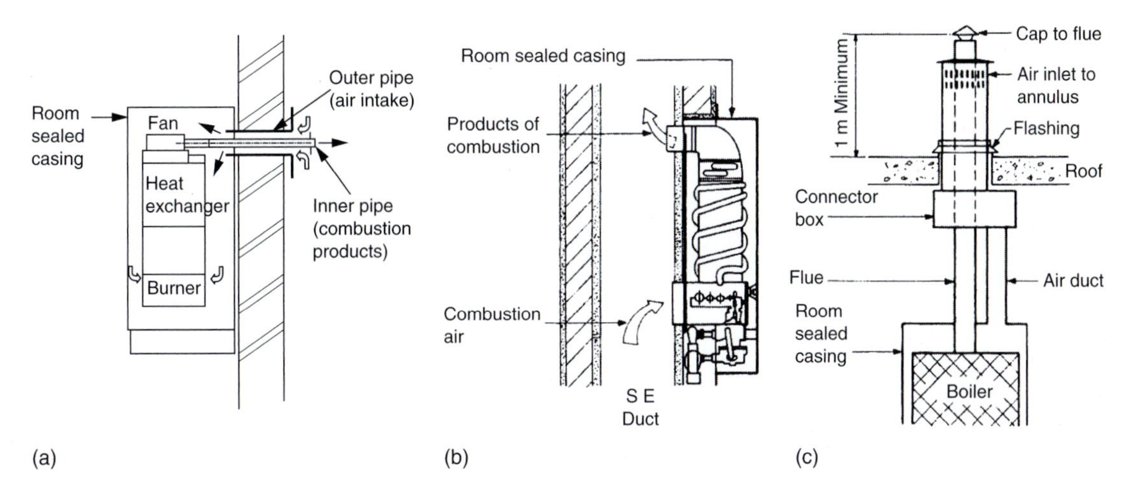

Figure 13.5 Balanced flues for boilers: (a) Fan assisted twin pipe, (b) horizontal and (c) vertical

In summer, when the heating plant is not in use, service to the heat exchanger might still be dealt with in the case of a single boiler if this were not disproportionately large. Similarly, where a multi-boiler plant is envisaged, it could be that the selection overall would offer the opportunity to include one boiler which was small enough to be kept in use during the summer.

Boiler houses – size and location

Various attempts have been made to provide guidance for provision, in a new building, of boiler and other plant rooms of adequate size and suitable location. In general terms, it is easier to suggest that the site should be at the thermal centre of the load, rather than to define the size in either area or length × width.

Size

It is often the case that information in this respect is requested long before adequate details of the building structure are available, with the result that the required boiler capacity cannot be determined to greater accuracy than plus or minus 25 per cent or more. Furthermore, there are so many types of boiler and kinds of arrangement that advice at this stage of building development cannot be given in more than very general terms. Some boilers are long and narrow; some are short and tall; some are sectional whereas others are factory packaged with the result that, in that instance, availability of access for delivery may be a critical factor.

As good a rule of thumb as any is that a boiler house should have an area of never less than 35 m^2 plus a further 35 m^2 per MW of boiler capacity. The aspect ratio should not exceed 2½ to 1 and additions must be made for fuel storage, etc., where appropriate. Height should never be less than 4 m although for smaller boiler installations, say up to 500 kW, 3 m height may be possible. Any lesser dimensions may present an escape hazard and fall foul of a local authority for that reason alone.

Location

Past practice was to house the boiler plant in a basement, under the stairs in a position useless for any other purposes, a relic from the days when coke was delivered by horse and cart. Construction below ground is relatively expensive and accommodation through the building, to roof level, must be found for a flue. In many ways preferable from the point of view of operatives and such maintenance staff as may be available, a purpose-built detached plant house may be sited at ground level, particularly where several buildings are served from a central plant.

Alternatively, where planning permission and practicality permit, there is sometimes a case for siting a boiler house on the roof. With mechanical draught, the flue may be quite short and both ventilation and access are straightforward. Roof top boiler houses are really only sensible when firing is by oil or gas. Oil is pumped up to a daily service tank, as shown in Chapter 16 (Fig. 16.7), but design considerations in all instances must include provision against any noise or vibration which might, via the structure, affect occupants.

Types of boiler

Cast iron sectional

By far the most familiar type of heating boiler for commercial applications is the cast iron sectional pattern which, in one form or another, has been in use for more than a century. Originally designed to burn coke by hand stoking onto a set of fixed firebars, it has been developed and refined for use with solid fuel, automatically fired, and for application of oil or gas firing at a much improved efficiency. Sizes range from the smallest up to 3 MW; however, few over 1.8 MW are available these days.

Generally, the boiler construction consists of individual sections connected together by use of *machined nipples*, usually three in number, all pulled together and held water tight by steel tie-bars externally. Some of the larger sizes have half-sections, with multiple nipples but pulled together as before. Front and back sections differ from those in the centre as they make provision for firing and cleaning at the front and for a flue outlet at the back. Intermediate sections may be added for extensions or as replacements on failure.

Figure 13.6 shows a type of cast iron boiler which has been designed especially for oil or gas firing and in which the flue passages have been arranged for higher gas velocities and hence greater output for a given physical size. This type has a *waterway* bottom instead of being open and hence the structural base, in brick or concrete, on which it stands is subject only to relatively low temperatures, thus avoiding the complexities of floor insulation.

For use with atmospheric gas burners, as described later, a special type of cast iron boiler is also available. The outer surfaces of each casting, facing onto the internal flue-way, may also have integral nodules or studs

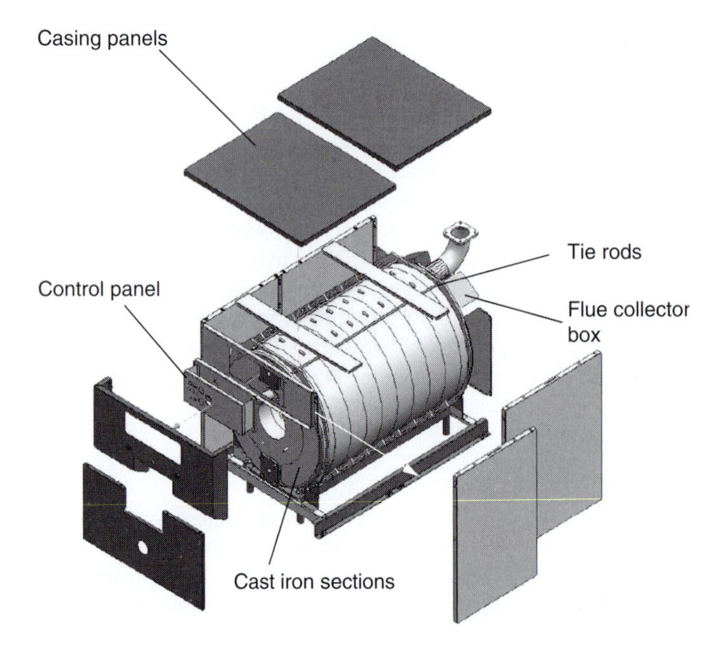

Figure 13.6 Cast iron sectional boiler (Hamworthy)

cast onto them to provide additional surface for heat transfer. Sizes range from 20 kW to 1 MW per boiler unit and combustion efficiency may be about 80 per cent. Other small, modular boilers may have sections built up vertically. Figure 13.7 shows a typical assembly of a modular, cast iron sectional boiler with atmospheric burner, in this case also with a condensing heat exchanger.

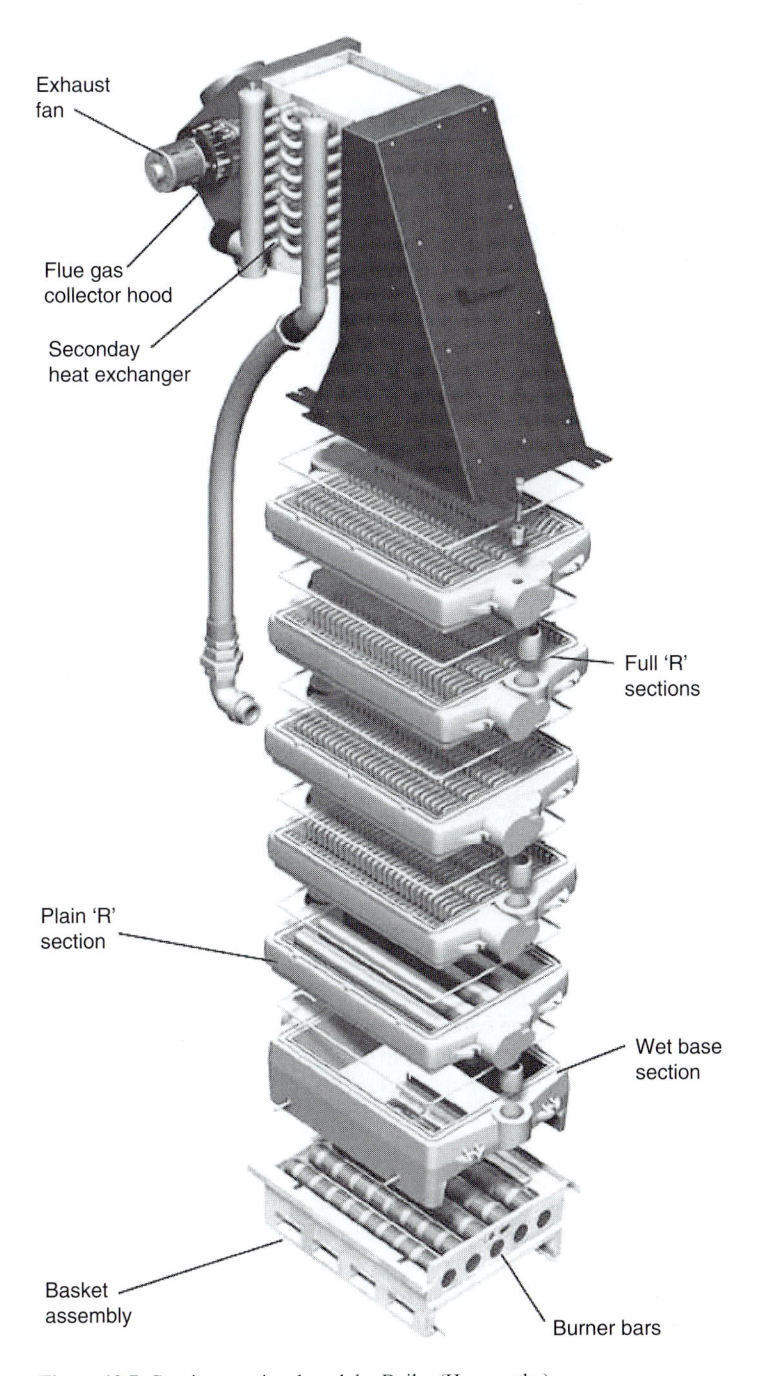

Figure 13.7 Cast iron sectional modular Boiler (Hamworthy)

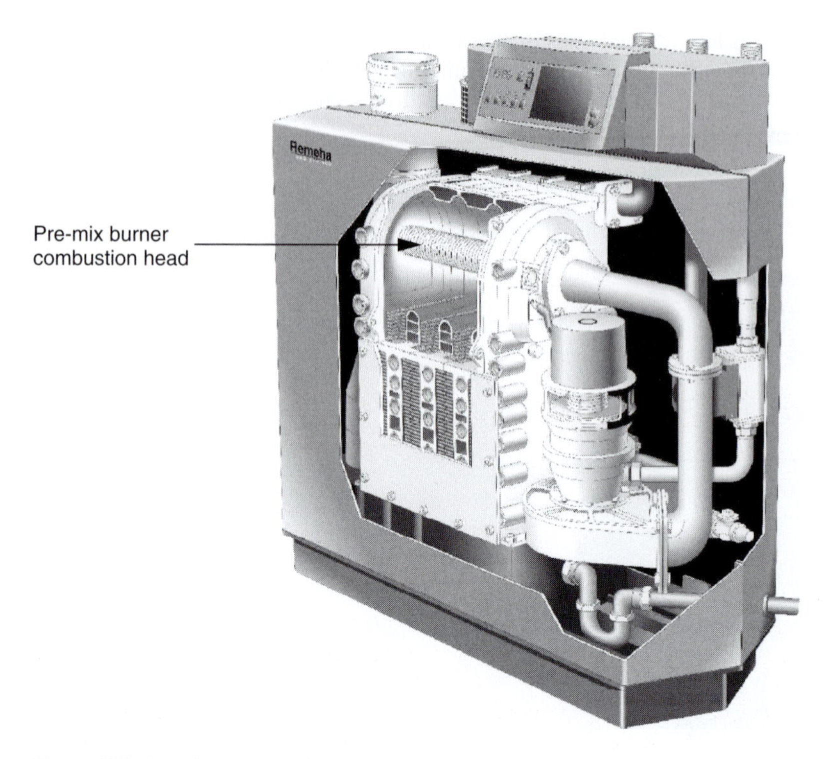

Figure 13.8 Cast aluminum condensing boiler (Broag – Remeha)

The normal range of cast iron boilers is designed for operation under water pressures of up to about 400 kPa (40 m head of water) which covers most building sites except in high rise city centre developments. Certain manufacturers, however, do produce higher rated boilers using a spheroidal graphite grade of cast iron which will withstand water pressures of up to 1 MPa (100 m head of water), although 600 kPa (60 m of water) working pressure is the typical maximum.

To supply low pressure steam, cast iron sectional boilers were very commonly used during the era when this heating medium was popular. The earlier types had a limited steam space made available actually within the sections but later patterns mounted a horizontal steam drum above them. Ratings available were up to 750 kW for working at not much more than 200 kPa absolute pressure (2 bar).

Cast aluminium

Cast sectional aluminium heat exchangers are available offering compact lightweight construction. Figure 13.8 illustrates a high efficiency, low NO_x and condensing unit, available from one manufacturer with an output range 80–160 kW non-condensing and 86–170 kW in condensing mode. This particular boiler is fitted with an integral pre-mix burner. This type of burner is described later in this chapter.

Steel – sectional

The place once held by wrought iron as a favourite material for boiler construction, due to its ductility and ability to resist corrosion, has largely been taken by mild steel as a result of the disappearance of the manual skills in *puddling*. Steel is homogeneous in structure, in contrast to the more laminar nature of wrought iron, and is very liable to attack by sulphurous corrosion products.

For service to hot water systems, mild steel sectional boilers were obtained in a variety of forms to cover a wide range of duties from about 30 kW to 1.5 MW. However, steel sectional boilers have generally gone out of favour.

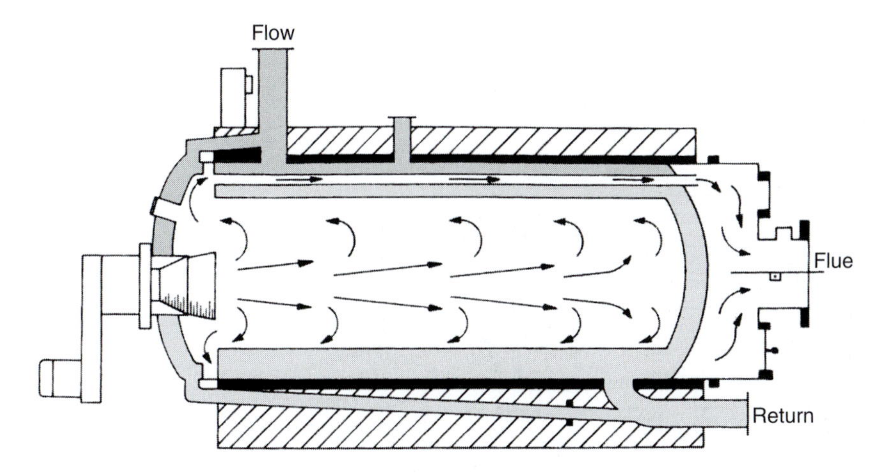

Figure 13.9 Steel reverse flow boiler

Steel shell – reverse flow

For long life, a mild steel water boiler requires that the system which it serves should be designed to operate at a temperature outside the region where severe corrosive attack may be anticipated. Given these conditions, steel has advantages over cast iron in that it is more versatile: the more modern developments of the last 40 years demonstrate this versatility as shown in Fig. 13.9.

In this type, a pressurised combustion chamber is provided in the form of a welded cylinder having a blind rear end. In consequence, the burner discharge is reversed to provide, *in counterflow*, a second pass of flue gases within the chamber. This reversal causes considerable turbulence in the flame zone before the gases enter the third and final pass through secondary surfaces arranged as a circumferential ring of fire tubes around the combustion chamber. At the end of the tubes the gases pass into a rear smoke hood and thence to the discharge flue. This boiler is usually referred to as a three pass reverse flame, but is also sometimes known as a 'thimble' boiler. On larger plant the front access door has a double skin and may be water cooled. A range of ratings from 100 kW to 3.5 MW is available for operation at absolute pressures up to 1 MPa (10 bar).

This type of boiler is now also available in stainless steel construction for condensing applications. Outputs range from 75 to 640 kW.

Steel shell and fire tube type

It is convenient to separate boilers falling into this category from those others which are manufactured from the same material, namely mild steel, and to make brief mention of earlier patterns from which the present range has evolved. Single-flue *Cornish* and twin-flue *Lancashire* boilers, both types brick set, are now rarely seen. Nevertheless, in their day, they had the merits of sturdiness, extreme simplicity and vast thermal storage capacity. All of these were useful attributes when hand fired using coal of indifferent quality to meet fluctuating loads, but none of them seem relevant today when viewed in conjunction with a combustion efficiency, on test, of only about 60 per cent.

For both water and steam service, with ratings between 1 and 10 MW, shell and fire tube boilers of both *economic* and *super-economic* type are available, one being as illustrated in Fig. 13.10. Without any brick setting, either type may have one or two furnace tubes within the pressure shell, from which the flue gases pass to a combustion chamber at the rear of the boiler. In the *two-pass* economic design, the gases return through a secondary array of fire tubes to a smoke hood at the boiler front, from which they are discharged. In the *three-pass* super-economic design, a transfer box takes the place of the front smoke hood and a further array of fire tubes conveys the gases back again to the rear of the boiler for collection prior to dispersal.

The manner in which the combustion chamber is constructed, for either type, as a refractory lined external box at the back of the boiler or as a water-immersed pressure vessel within the main shell, types the design as being either *dry back* or *wet back*.

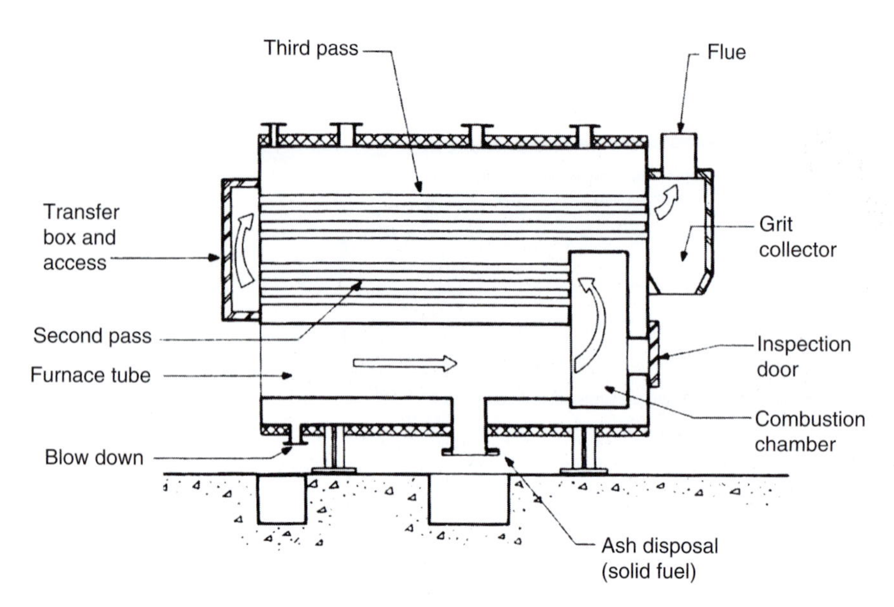

Figure 13.10 Shell boiler of three-pass super-economic type

Apart from the different patterns of mountings fitted, as described later, the principal difference between the steam and hot water variants of this type is that for steam storage whereas the latter are commonly *drowned*, i.e. are completely water filled. Mention has been made in Chapter 9 of a method adopted in some high temperature hot water systems whereby pressurisation by steam was employed: with this technique, the steam storage space was retained even though the boiler served a hot water system.

For applications where a substantial, large capacity boiler is required, this type will provide a combustion efficiency of the order of 80 per cent. A further advantage is that the type is not only capable of burning either solid, liquid or gaseous fuel but also of site conversion once or more during its working life to a fuel other than that for which it was initially equipped.

Packaged

As boiler design grew in sophistication and the associated firing equipment, controls, etc., became increasingly complex the market became established for a range of large, factory-built, boiler–burner units, incorporating all necessary working parts and pre-commissioned at the manufacturers' works. The *packaged boiler* may be no more than an assembly of production line components but with the very important difference that all those components are, as it were, hand picked for compatibility.

The firing equipment will have been chosen to give the optimum flame shape for the combustion chamber; the forced or induced draught arrangements will be matched to provide the most suitable gas flow pattern and the control system will be fitted, wired and proven. Site tasks after delivery, and time for erection and commissioning should be reduced in consequence.

In some respects it is unfortunate that the introduction of fire tube shell boilers in packaged form coincided with a trend towards smaller furnace volumes, reduced diameter fire tubes and a general move to squeeze the last gram of heat transfer from each kilogram of metal. The consequent need for greater care in water treatment and greater skill in maintenance coincided with a period when both care and skill are in short supply. A typical packaged boiler is shown in Fig. 13.11.

Water tube

Boilers of this type, while available in sizes down to as little as 500 kW, are commonly used only for much larger duties of up to about 10 MW. The most typical application has been for service to high temperature hot

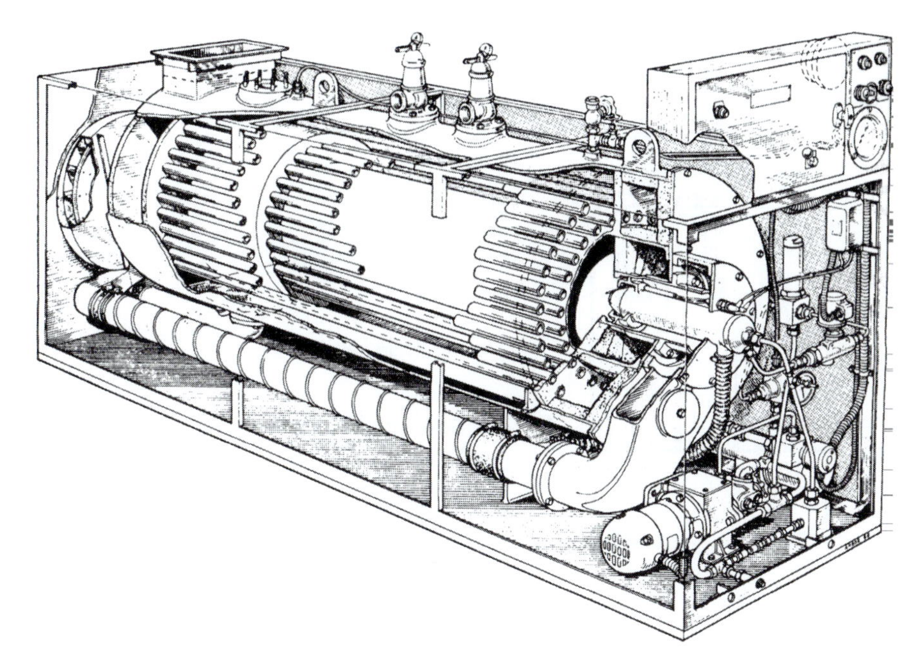

Figure 13.11 Packaged hot water boiler

water plants for very large industrial sites and steam generation in power stations. Since fire tube boilers are now available in larger sizes than was the earlier practice, for absolute pressures up to about 1.5 MPa (15 bar), a tendency now exists to use these for such systems.

Low water content boilers

Small duty, low water content gas boilers have been used for many years in the domestic and modular boiler markets. Recent developments for the commercial sector have resulted in this type of boiler being available with greater output, typically up to 650 kW from a single module. The heat exchangers come in a variety of patterns and configurations to maximise surface area. Common materials used for the heat exchangers are aluminium, copper and stainless steel. Both natural and forced draught types of boiler are available.

Forced draught types often incorporate pre-mix burners as an integral part of the boiler. Here the gas and air is mixed before entry to the burner, thereby maximising efficiency within a small combustion chamber space. Improvements in burner control have resulted in achieving high turn-down ratios, further improving fuel consumption on part load with reduced carbon dioxide and NO_x emissions. When designed for, and used in condensing mode, even greater levels of efficiency are achievable.

Good water circulation is essential to the operation of low water content boilers and a means of flow sensing is usually required, interlocked with the burner controls.

The main attributes of low water content boilers are rapid heat up, high efficiency, compact size, and lightweight. They are particularly useful where plant space is at a premium. The main disadvantage with low water content boilers is the limited life expectancy, which may be less than half that of a robust cast iron sectional or steel boiler.

Electrode

For use with off-peak current and the large-scale thermal storage cylinders described in Chapter 8, a heating source may be provided by banks of immersion heaters, but is more likely to be one or more electrode boilers. These may be connected to either a medium voltage supply (up to 650 V) or to high voltage (3.3, 6.6 or

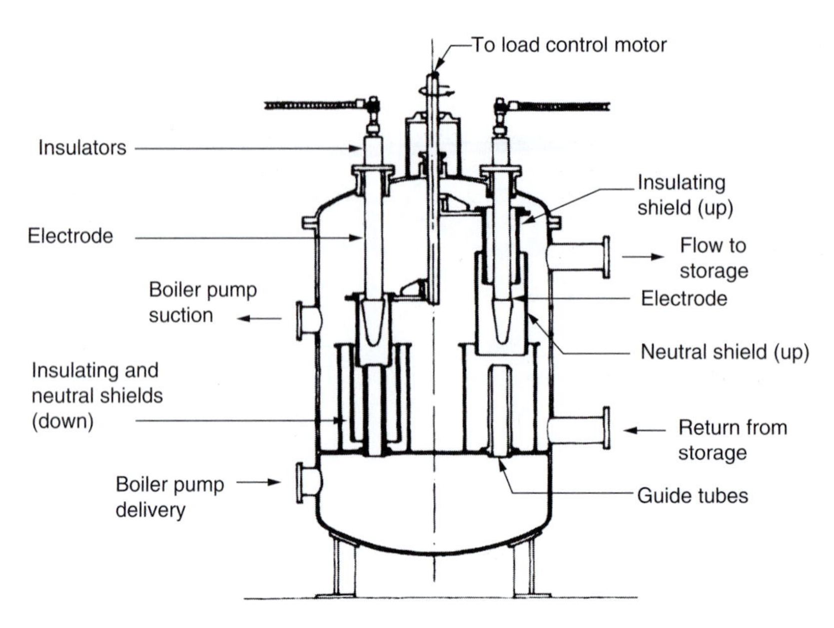

Figure 13.12 High-voltage electrode water heater

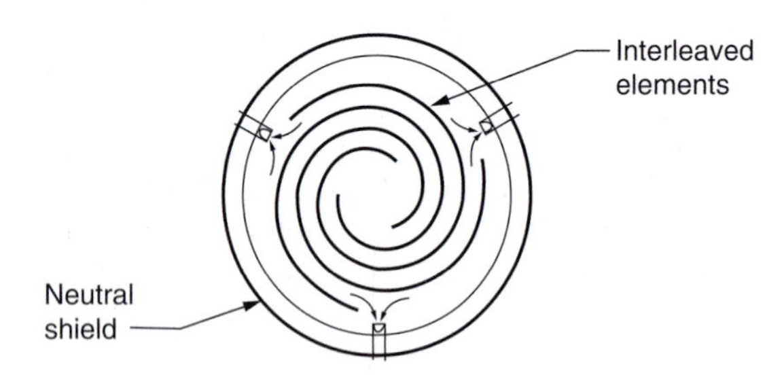

Figure 13.13 Interleaved elements for electrode steam boiler (plan view)

11 kV), the latter being the more common for installations of any significant size. Such boilers, as illustrated in Fig. 13.12, are available in ratings up to about 2.5 MW, the principal difference between the various designs being in the detail of the method adopted for load regulation.

Current is passed from electrode to electrode, the resistance of the water in which they are immersed acting as the heating element. The load is varied by increasing or decreasing the length of the path which the current has to take between and around the non-conducting *neutral shields* which shroud the electrodes. The material from which these shields are made and the actual mechanism used to raise and lower them varies from maker to maker. A brisk water movement is required around the electrodes and most manufacturers mount a centrifugal pump to the boiler shell for this purpose. This pump must not be confused with that which must be provided to circulate water, during the charging period, between the boiler and the storage cylinder or cylinders.

Where electrode boilers are used for steam generation, at ratings between 20 kW and 2.5 MW, it is most probable that they will operate 'on-peak' although some form of thermal storage may still be used to meet short fluctuations in demand. The arrangement of the electrodes in a steam boiler is usually relatively simple since the imposed load depends upon how much of the electrode is immersed. Load regulation is thus achieved by adjustment of the water level within the shell. A typical form for electrodes is as three interleaved *scrolls*, as shown in Fig. 13.13, one per phase within a single-neutral shield. For both steam and water operation, the conductivity of the water may have to be adjusted from time to time by the addition of soda, or other salts if need be, in order to preserve the required electrical resistance.

Boiler fittings and mountings

The much-loved first edition of *The Efficient Use of Fuel* provided a definition which distinguished between the two classes of attachments to boilers. 'Generally speaking', the author wrote, 'the term *mounting* implies that the equipment is mounted on a pad or stool riveted onto the fabric of the boiler as distinct from a *fitting* which may or may not be attached to the boiler, but for which there is neither pad nor stool' This distinction served for an era when boilers were purchased 'raw' and both mountings and fittings were selected and bought quite separately: it seems less appropriate to the complete packages now available.

Hot water boilers

The principal fittings and mountings in this case include:

- Relief valve
- Altitude gauge
- Thermometer
- Drain cock
- Control devices
- Feed and expansion pipe (where fitted)
- Open vent pipe (where fitted).

The first four of these items require no comment other than emphasis that relief valves should be of a type approved by the building owner's insurance company, purpose set for the particular installation and discharging to a safe location. Similarly, the altitude gauge and thermometer should be of a pattern having large legible figures; the former calibrated in kPa, m head of water or other accepted SI unit. It is not possible to generalise as to provision for control devices since these may be either rudimentary or complex.

The feed and expansion pipe and the open vent pipe are, in many respects, a part of the connected system but both are very closely associated with the boiler plant. The former dictates the pressure at which the plant will operate and the latter is a safety feature paralleled only by the relief valve. In the past it was common for boilers to include independent mountings for feed and expansion and open vent pipes. However, few commercial sized boilers these days are provided with such connections directly on the boiler and it is left to the system designer to make adequate provision. This provision may take the form of a 'conventional' feed and expansion pipe from a header tank plus an open vent pipe terminating over the tank, or, perhaps more commonly these days, a sealed pressurisation/make-up unit coupled with an expansion vessel to accommodate the system expansion volume. These systems are discussed further in Chapters 11 and 12.

Regarding the size of feed and expansion and open vent pipes, where appropriate to the chosen system, and the size of relief valves, this information is generally based upon the size of boiler plant and fuel. It is therefore appropriate to include this information here, in Table 13.1.

Table 13.1 Feed and expansion pipes, open vents and relief valves

	Pipe size (mm)		*Relief valve, minimum clear bore* (mm)	
Boiling rating (kW)	*Feed and expansion pipe*	*Vent pipe*	*Solid fuel firing*	*Oil or gas firing*
Up to 74	20	25	20	25
75 to 224	25	32	20	25
225 to 349	32	40	25	32
350 to 399	32	40	32	40
400 to 449	40	50	32	40
450 to 499	40	50	40	50
500 to 749	40	50	50	65
750 to 900	50	65	65	80

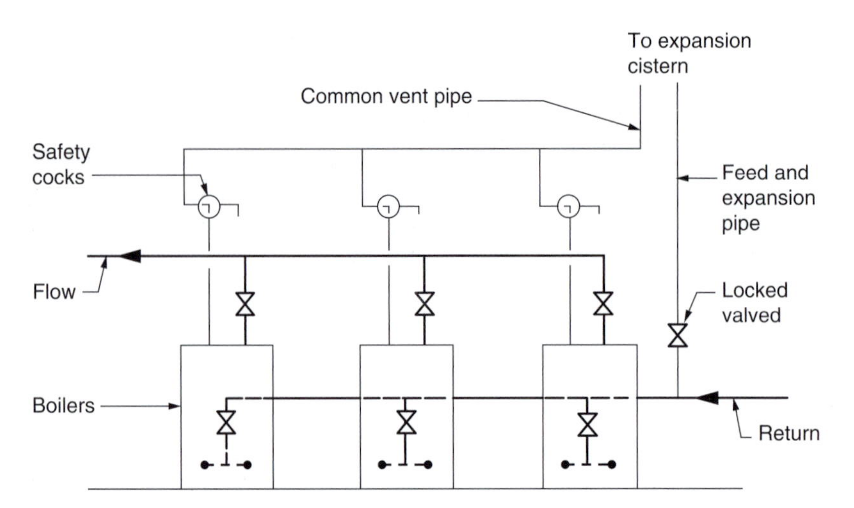

Figure 13.14 Open vent arrangement for multi-boiler plant

Where a plant has only a single boiler, an open vent pipe may be routed from it to a point above the expansion cistern. Where more than one boiler is provided, it is necessary to fit a special type of 'safety cock' (approved by insurance companies) so that individual boilers are, first, open to atmosphere when working and, second, may be isolated from the system for service. Figure 13.14 illustrates this arrangement.

Steam boilers

As far as fittings and mountings for steam boilers are concerned, stringent requirements for these are set out in various of the publications of the Health and Safety Executive. Certain of the mountings are required to be duplicated and the following is no more than a list of the principal items:

- Safety valves
- Pressure gauge
- Water level gauges
- Steam stop valve (crown valve)
- Auxiliary steam stop valve
- Feed and check valve
- Blow-down valve
- Alarm devices (high and low water, etc.)
- Control devices.

Packaged steam boilers are normally provided with an electrically driven boiler feedpump supplied from a higher level, possibly frame-mounted condensate vessel. This latter must not be confused with the larger-scale hot well required in the case of most steam systems to collect the intermittent flow of returning condensate (see Chapter 9).

Furnace linings

The fire-brick construction of a combustion chamber subjected to oil firing, where this is provided, needs to be in a material which will withstand temperatures of between 1400°C and 1600°C without fusing or premature disintegration. Linings must not be carried solid up to the metal of the boiler construction but must allow an air gap of 15 mm or so for brickwork expansion. Many boilers of the packaged type dispense with any brickwork apart from the quarl.

Under-hearth insulation

Where a boiler does not have a waterway below the combustion chamber, it is necessary to provide a layer of insulation above the floor to prevent damage to the structure and, in particular, to any damp-proof membrane. For boilers up to a size of, say, 600 kW, a 150 mm layer of *moler* insulating brick may be enough. For larger boilers some form of open honeycomb construction should be arranged to provide additional cooling through circulation of air. To encourage this, it may be possible to arrange that a forced draught fan draws its air supply via passages formed within the honeycomb, so serving a dual purpose by preheating this supply by conduction. It is desirable that a concrete slab below a boiler should not exceed a temperature of about 70°C.

Boiler firing – solid fuel

Biomass fuel

Environmental considerations, especially the drive for CO_2 emission reductions have seen a dramatic demand recently for burning carbon-neutral biomass fuels, particularly wood chip or pellets. In many ways the fuel handling, combustion and ash removal associated with wood burning is not dissimilar to coal. Boilers and fuel handling systems for biomass fuels deserve special mention and are detailed in Chapter 14.

Fossil fuel

In the last 20 years or so there has been little interest in the use of solid fossil fuels for firing either large or small boiler plants. That which follows in this chapter is included here mainly for reference purposes.

Automatic stokers

Of the overall choice available, there are several specific types of automatic stoker which are particularly appropriate for use with the size range of boiler which falls within the scope of this book. These include those which are most commonly used, namely the various patterns of underfeed stoker and also, as shown in Fig. 13.15, sprinklers, coking types and moving beds.

All modern types of stoker may be arranged for some form of semi- or fully automatic control whether this be variable through speed control of the mechanism or merely 'on/off'. In the latter case a '*kindling*' control is a necessary feature, the stoker operating for a few minutes in each and every hour in order to keep the fire alive. A recent development, for use when starting the fire, is a form of electric ignition.

It must be emphasised that no single type of automatic stoker is suited to all types and sizes of solid fuel. It is fundamental to the selection of the appropriate type of machine that the characteristics of the fuel are carefully considered, in consultation with the local solid fuel agency.

The principal design features of the various generic types of automatic stoker may be summarised as follows.

Underfeed (Fig. 13.15(a))

This has always been the most popular type of stoker for application to boilers serving heating systems since it is simple in construction and application; tolerant of a low level of intelligence for maintenance and is relatively inexpensive. Requirements as to fuel are that coal be weakly caking, low in ash content and of even size at not much over 20 mm (coal Rank 600–900). More details of this type are set out in later paragraphs.

Sprinkler (Fig. 13.15(b))

The original arrangement for this type was conceived early in the nineteenth century to imitate the spreading motion achieved by hand-firing using a shovel. Several basic patterns of equipment have since been developed. One, as illustrated, spreads fuel via a *rotary distributor* mounted at the mouth of an overhead hopper.

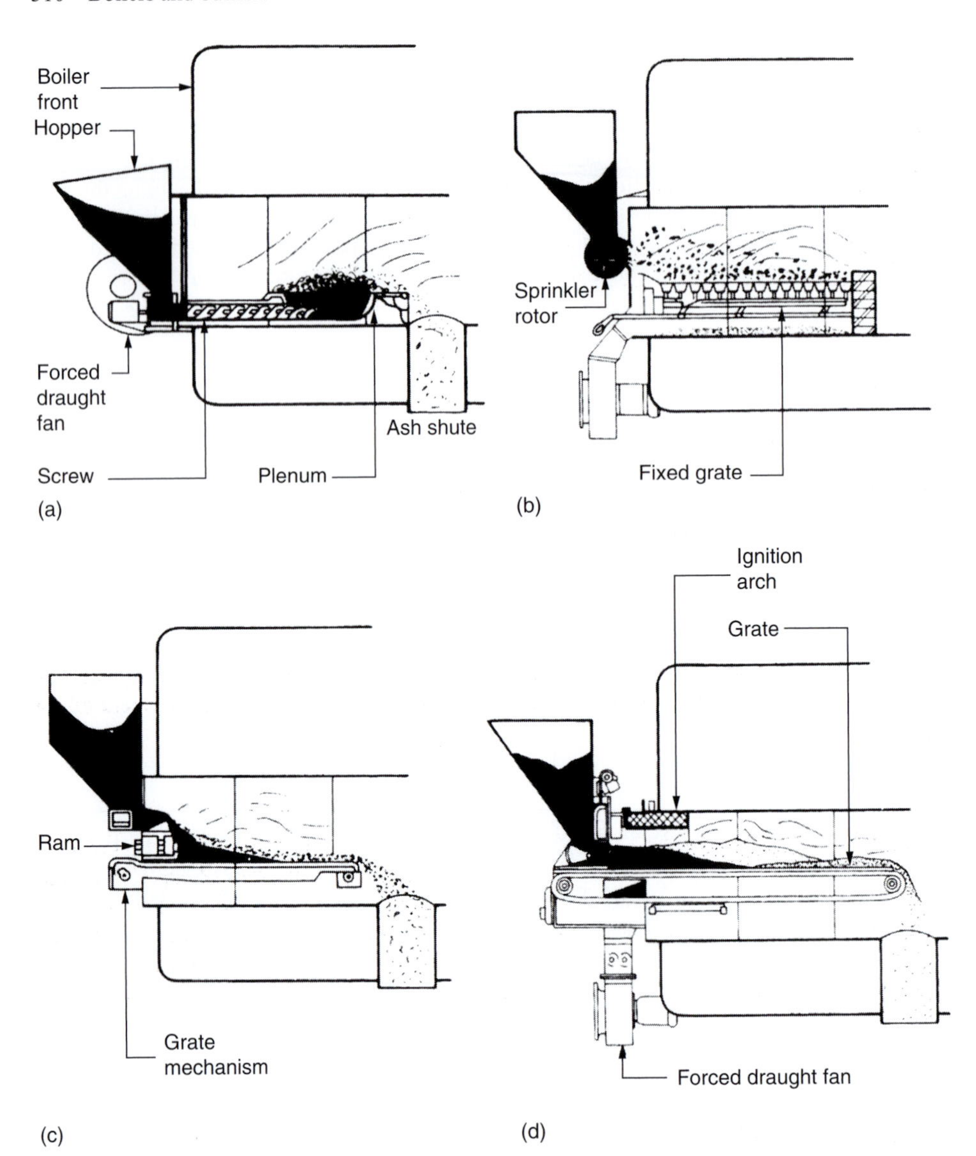

Figure 13.15 Types of automatic stoker: (a) underfeed; (b) sprinkler; (c) coking and (d) moving bed

A second type sprays pneumatically conveyed fuel from a pipe terminal fitted to the furnace front plate. Yet a third, which is integral to the boiler design, also sprays fuel on the fire bed but in this case from a vertical manifold penetrating the pressure shell as shown in Fig. 13.16. Sprinklers will burn most types of small washed coal except anthracite: low ash content is preferred (coal Rank 600–700).

Coking (Fig. 13.15(c))

The operating principle in this equipment rests with the action of a reciprocating ram mechanism which pushes fuel forward from an overhead hopper at the boiler front and deposits it on a coking plate in the hottest area of the furnace. At that point, the volatiles are consumed and the residual coke is then passed towards the boiler rear by the action of a moving grate. Early designs used a deep fuel bed and were, in consequence,

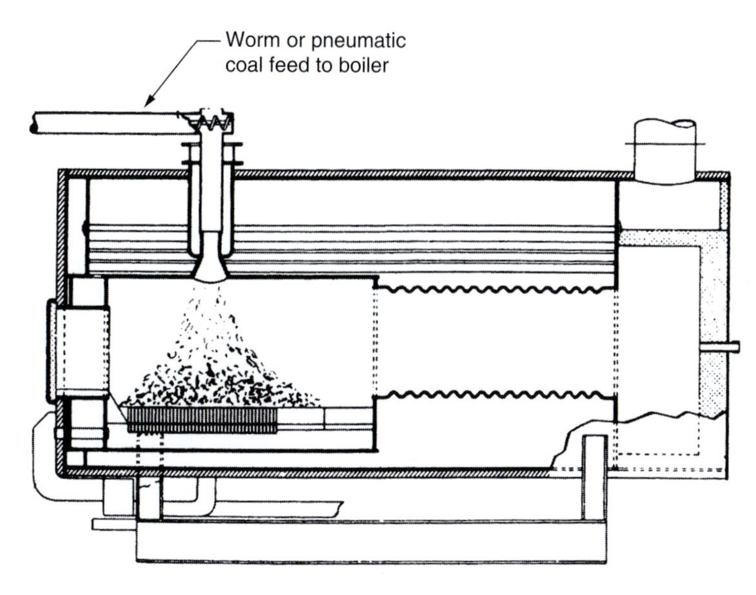

Figure 13.16 Overhead fuel feed (worm feed type)

unsuited to any type of shell boiler: this problem was later overcome when '*low-ram*' patterns appeared. Singles and doubles are the appropriate sizes of a coal having slight caking properties and an ash content of not more than about 8 per cent (coal Rank 700–900).

Moving bed (Fig. 13.15(d))

A chain grate stoker, as the name suggests, has a fire grate made up of a continuous chain of links or bars which run over driving sprockets at the boiler front. Fuel is fed to the grate by gravity from a hopper, the depth of the bed being controlled by a guillotine plate mounted above the furnace entry. The position of an '*ignition arch*' behind the guillotine is arranged to expose it to maximum radiation so that it may raise and maintain the temperature of incoming fuel. Combustion takes place progressively, downwards from the ignited upper layer of fuel towards that in contact with the grate. Wet smalls are the preferred fuel (coal Rank 700–900). Moving bed stokers are also used for the larger types of biomass fuel boilers

Fixed bed

Fixed beds for burning either wood chip or pellets are sloped, stepped or flat. The stoker feeds fuel to the fire either from one side of the combustion chamber or upwards through openings in the centre of the grate.

Hopper or bunker underfeed stokers

In the simplest form of such equipment, fuel is fed from a hopper down to a worm or screw at its base: this is rotated at slow speed through reduction gearing from a motor drive. Beyond the feed point the worm is enclosed within a tube and serves to convey the fuel into a fire-pot which is built into a refractory base inside the boiler, as shown in Fig. 13.17. Safeguards are provided, in the form of *shear-pins* or a *slipping clutch*, to prevent damage to the worm if jammed by a foreign body or by oversize coal.

A second tube delivers air to the fire-pot from a fan driven by the same motor that powers the worm. This air is discharged through a series of slots or openings, disposed around the fire-pot to form *tuyeres* which will not be fouled by fuel or ash. Secondary air is supplied by the same fan and preheated in order to control smoke emission.

Thus, forced draught is provided and a very high combustion rate is possible, so high in fact that no grate is necessary. The fuel is burnt as it passes over the edge of the fire-pot and all the ash is reduced to clinker in the

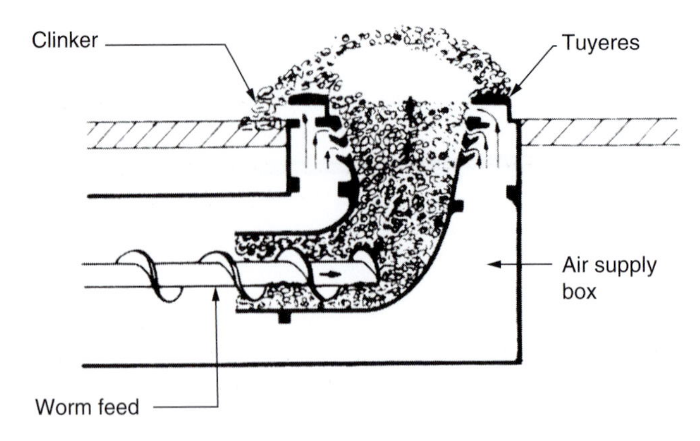

Figure 13.17 Fire-pot and fuel bed for underfeed stoker

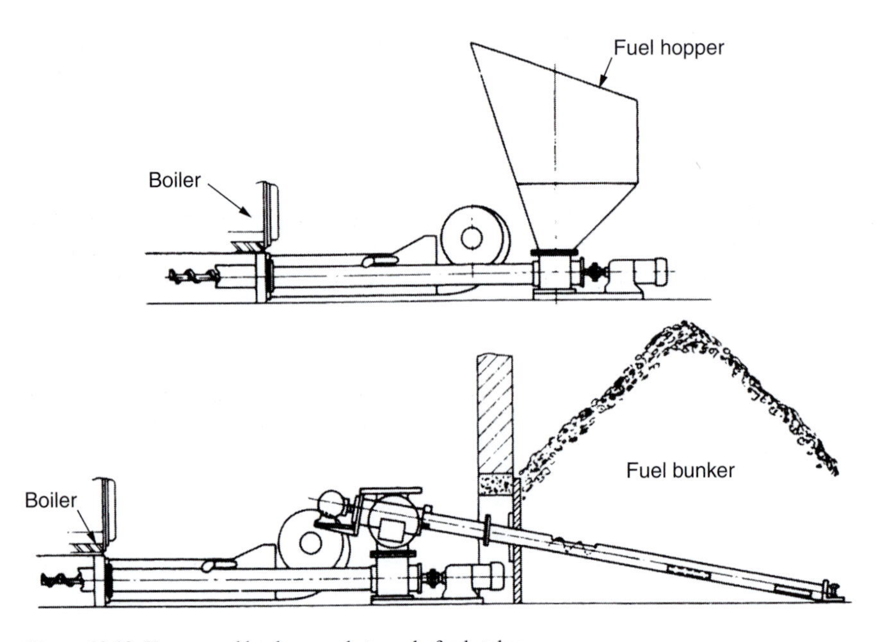

Figure 13.18 Hopper and bunker supply to underfeed stoker

process. Any accumulation of clinker, as an impediment to combustion, is prevented as the fresh coal brought in by the worm pushes it to one side from which position it is removed at intervals.

An alternative method of feeding the fuel is direct from a fuel bunker. In this application, either the firing worm or a subsidiary extends into the main fuel store and the hopper is thus eliminated. Both hopper and bunker feed methods are illustrated in Fig. 13.18.

Ash Removal

All solid fuel systems produce ash and this must be removed regularly. Most ash accumulates in the combustion chamber and a system of ashing screws are often used to move the ash from the grate away to a collection container, for vehicle removal and disposal.

Smaller systems may adopt manual ashing, especially where the boiler is designed such that the ash can be raked and shoveled out by hand without the need for the boiler to be shut-down.

Boiler firing – oil fuel

Achievement of clean and efficient burning with oil fuel rests almost entirely with the matter of atomisation, that is to say the intimate mixing on a molecular scale of the carbon in the fuel with the oxygen in the air supply. All manner of methods have been used over the years in attempts to produce an ideal solution but, for the size and type of boilers which are chosen to serve heating plants, it is necessary to consider only the following.

Vaporisation

This may be compared with the principle used in a blow-lamp or a primus stove and is applied only to very light oils, such as kerosene. On start-up, the oil is preheated, often electrically, to form a vapour which is then ignited: subsequent vaporisation is produced by heat from the flame. Pot-type burners, now probably obsolete, which used this principle were notoriously unstable.

Pressure atomisation

In this arrangement, the oil supply is fed at a controlled rate to a nozzle, the discharge from which meets a stream of air (or sometimes steam) at high, medium or low pressure. The primary air quantity so provided is less than that required for combustion and acts principally to atomise the fuel.

Mechanical atomisation

Oil is fed, at a controlled rate, typically to the inside surface of a conical cup which is rotated at a high speed. As the oil leaves the edge of the cup under centrifugal force, it is atomised by a primary air stream supplied concentrically around the cup and contra-rotating vis-a-vis the oil supply. More details of this type are given later.

Pressure jet

At a relatively high pressure, oil is supplied to a fine nozzle which is so designed as to apply a swirling motion to the spray of droplets discharged. An air supply, primary or total, is provided in a contra-rotating swirl. Additional information regarding this type is given later.

Emulsification

This process may be applied to the principle of either pressure atomisation or pressure jet, the difference being that the oil is pre-mixed with a controlled quantity of primary air before it is delivered to the nozzle in the form of an emulsion.

Combustion air

As noted above, the total air supply to an oil burner falls into two categories: the primary air which is intimately involved in the atomisation process, and the secondary air which makes up the balance necessary to complete the combustion reactions, including any excess air which may be required.

The secondary air may be induced to flow through preset registers at the burner front, either by natural draught from the chimney or as a result of a mechanical *induced draught* fan at the boiler exit, in which latter circumstances the boiler combustion chamber is under suction. Alternatively, and as is now more usual practice for applications of substantial size, the secondary air is supplied by a *forced draught* fan into what then becomes a pressurised combustion chamber.

It will be seen that there is a variety of possible combinations of atomising methods with systems of primary and secondary air supply and hence many possible variations in burner design. For fully automatic operation the choice is generally limited to either pressure jet or mechanical atomisation rotary cup equipment.

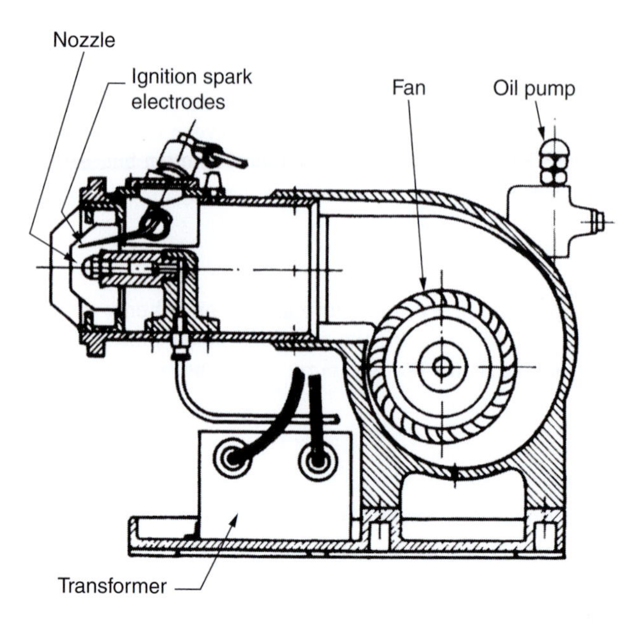

Figure 13.19 Gun-type pressure jet oil burner

Gun-type pressure jet burners

By far the most adaptable type of burner, the gun-type unit, is available in sizes suitable for boiler outputs ranging from 10 kW to 2.5 MW. Burners of this pattern consist, in the simplest form, of a direct electrical drive to a centrifugal fan which produces the whole of the air required for combustion, including any necessary excess. A positive displacement oil pump is coupled to the fan drive and supplies oil at an absolute pressure of up to about 1 MPa (10 bar) to a fine-calibrated jet. Air from the fan is delivered via swirling vanes disposed round the nozzle following the general pattern shown in Fig. 13.19.

The air quantity supplied by the fan is set by slots or dampers at the inlet and the oil quantity delivered is adjusted by control of the pump output pressure and the size of the jet orifice. The flame shape produced by the burner may be varied to suit the geometry of the boiler combustion chamber by changing the angle of the swirling vanes and by selecting the angle of divergence of the jet orifice.

When the fuel used is light oil (class C2 or class D), there is no requirement for preheating. For heavier grades, an electrical heater is incorporated as part of the burner unit to raise the oil temperature to between 60°C and 70°C. Ignition is by spark between two electrodes located near the nozzle tip which have a high tension feed from a transformer. These suffice for smaller size burners and lighter oils but, in other instances, a two-stage process for ignition is used via a gas supply, the electrodes igniting a gas pilot and the pilot igniting the oil.

Axial air flow pressure jet burners

A variant on the gun type, this burner uses an in-line mixed flow axial fan with variable speed drive instead of a centrifugal fan (Fig. 13.20). A very uniform air distribution to the burner head, without the need for straightening splitters or vanes is claimed, together with optimum combustion maintained throughout all levels of firing, resulting in lower overall fuel consumption and low NO_x levels.

Rotary cup burners

The principal components of this type of burner are illustrated in Fig. 13.21, the secondary air *quarl* shown being at the point where the burner is mounted to the front of the boiler combustion chamber. Drive from the motor is not only to the primary air fan and the oil pump but also, via a hollow shaft, to the spinning cup

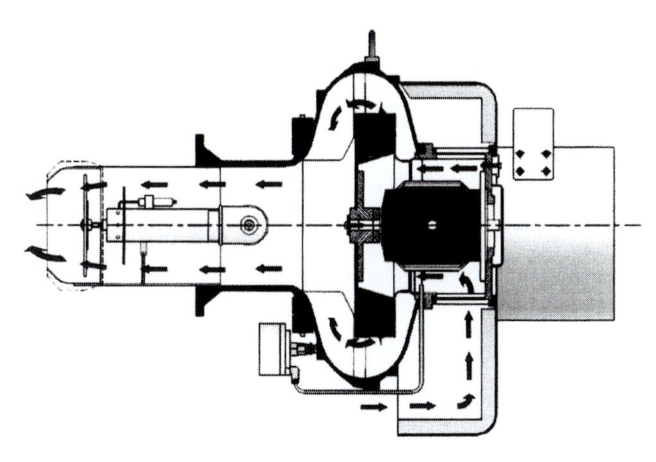

Figure 13.20 Axial-type pressure jet oil burner (Dunphy)

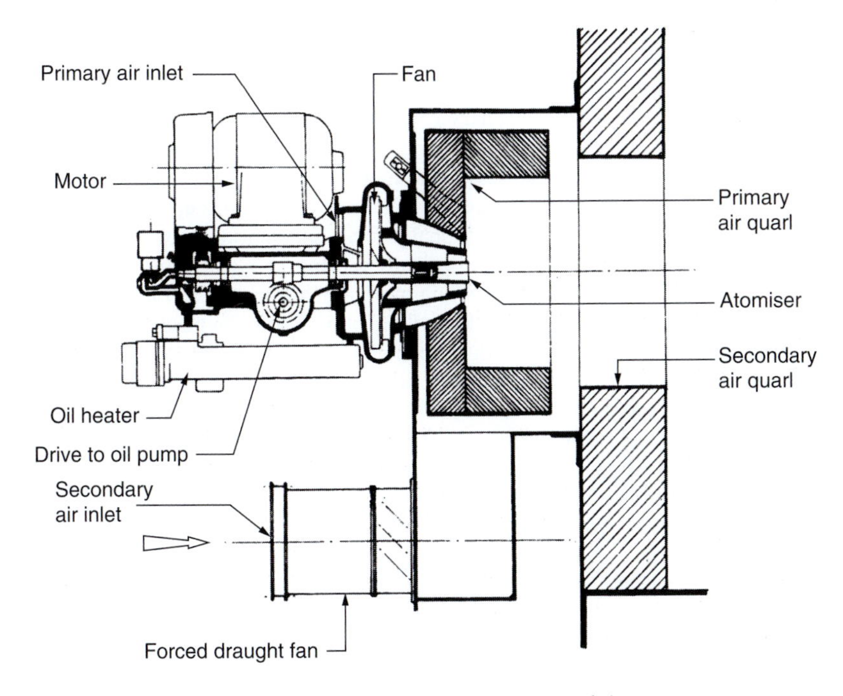

Figure 13.21 Rotary cup oil burner with forced draught (Hamworthy)

atomiser. Oil is delivered through this shaft, which rotates at about 100 rev/s, to the cup. The primary air supply from the fan passes through an annular chamber having internal swirling vanes to be delivered by nozzles circumferential to the cup.

The secondary air required for combustion is introduced at a lower pressure by a further forced draught fan, usually an axial flow type, into the burner box where it can be preheated to some extent by radiation from the quarls. Characterised dampers are provided for both primary and secondary air and are linked to a metering valve in the oil supply.

As noted above for larger gun-type burners, ignition here is two stages, using a gas pilot ignited by electrodes. Temperature control of the fuel supply, which will usually be a heavier grade of oil for the 150 kW to

5 MW range of this type of burner, is provided for at the boiler front but topping up of temperature is all that is necessary since fuel of these grades must be supplied from a heated oil ring main.

Regulations, etc.

Since oil storage and oil firing both impose some elements of fire hazard, various bodies have drawn up regulations relating to these installations. A British Standard *Code of Practice* (BS 5410: Parts 1 and 2) outlines many of the hazards and proposes ways in which they may be reduced. In addition, insurance companies and the fire service departments of some of the larger local authorities have their own regulations and these should be ascertained when any new installation is under consideration. Local authorities have certain responsibilities under the Clean Air Act which may affect the grade of oil used and the dimensions of the chimney, as referred to in Chapter 17.

Burner controls

The notes which follow are not intended to be a full and complete dissertation covering the whole subject of automatic control to oil burners, particularly since many of the primary devices may be applied with equal validity to the control of automatic firing equipment for other fuels. Primary functional control is certainly common to all, but unique requirements regarding safety and function, particular to individual fuels, are subjects which cannot be dealt with satisfactorily here.

Fully automatic types of oil burner have been referred to and it will be obvious that these, by definition, incorporate means for self-igniting and self-extinguishing. Primary control of these functions is normally by either two separate devices, be they thermostats or pressure sensors, or one single dual-service device. One such item will be the basic controller and the other, normally a hand-reset pattern, will be a high-limit safety override.

Whether the firing rate is capacity controlled between the extremes of on and off depends upon the method of operation and, to some extent, the size of the fired boiler. Traditionally small units normally operated in the 'on/off' mode only, whereas the larger units incorporated some means of varying or modulating the output. Apart from the nuisance which would arise resultant upon the cycling of a large burner to meet a small load, 'on/off' operation in those circumstances would create thermal shock to the boiler. With the need to operate at ever higher efficiencies small burners are now also available with automatic output modulation.

Obviously, both fuel and combustion air must be controlled in parallel, which tends to involve complexity if adjustment is envisaged right across the spectrum of capacity. This difficulty is commonly overcome in one of two ways, the simpler being to arrange for two preset levels of operation, low and high, i.e. perhaps a sequence of 40 and 100 per cent capacity. More complex, but still simpler than the ideal, is an 'on/off' operation up to one-third load, with full modulation to match the load thereafter.

Improvements in design and modern, sophisticated controls are such that full modulation is becoming increasingly popular, coupled with the drive for ever-increasing efficiencies and minimum pollution. For oil firing, the minimum efficient operating capacity is about one quarter of the maximum (4:1) and is referred to as the turn-down ratio.

With a self-contained burner in which one motor drives both fan and fuel pump, there is little likelihood of fuel being delivered into the boiler without a parallel supply of air. In types where separate drives are necessary, suitable interlocks are available. Residual problems arise however when a flame is not established due to failure of ignition or, having been established, is later extinguished for some reason such as an intermittent stoppage in the fuel supply. A variety of flame failure devices has been used in the past but it is now almost universal practice to incorporate a photo-electric cell either within the burner housing or at some other site where the flame may be viewed. Arrangements are incorporated in the control sequence such that this cell is out of circuit for a predetermined few seconds on start-up but that, thereafter, the burner will cut out if no flame is sensed. In some instances the control circuits provide a facility which will allow a second attempt at ignition to be made after a suitable time delay but will ensure that if that second attempt fails, the burner is deactivated until reset by hand.

With larger and more sophisticated burners, a commonplace addition to the control system is a purging sequence which arranges for the air supply fan to run for a period before fuel is made available to the burner head and also to run for a similar period after supply has been shut-off. This sequence is, of course, interlocked with the flame failure circuit and will be initiated whether or not a flame is established.

Boiler firing – gaseous fuel

A gaseous fuel discharged at a low velocity from a simple nozzle will burn with a soft lazy flame at the interface between the gas envelope and the surrounding air. Such a flame is unstable with a low heat flux and if the velocity be increased with intent to produce a more useful combustion characteristic, then there will be a tendency for the flame to lift from the nozzle and extinguish. A variety of complex jets has been produced to substitute for the simple nozzle and other methods, principally involving some process of aeration, have been developed to address the problem. Those most commonly used in boiler firing are:

Neat flame burner: As the name suggests, these are naturally aerated burners having shielded jets and their use is generally confined to conversion exercises for boilers rated at not much more than 50 kW.

Atmospheric burners: Since a fuel gas reaches the point of use under pressure, it is logical to make use of this potential energy to induce a proportion of the air required for combustion into a mixing tube. This, the principle used in the Bunsen burner, is adapted also for the atmospheric burner.

Packaged burners: The introduction of a fan powered air supply is the principle distinguishing feature of this general type. There are two main configurations: gun type and pre-mix.

A gun-type gas burner is similar to the type for oil firing, the air supply mixing with the fuel at the burner face, and the flame discharging from control ports into the boiler combustion chamber. (This burner type can also be readily designed for dual-fuel use with a relatively simple changeover arrangement.)

A pre-mix burner mixes air and gas in precise proportions before entering the actual burner, with ignition usually through multiple openings (and very small flames) in a ceramic combustion head.

It should be noted that whilst the general thrust of the comments made above and of those which follow is directed to the burning of natural gas derived from the public supply, this is not to suggest that the equipment described is exclusive to that fuel. On the contrary, much if not all of it may be used equally satisfactorily with LPG when appropriate adjustments have been made to jet sizes and to air supplies.

Atmospheric burners

These represent a type of equipment which follows the traditional pattern associated with the firing of cast iron sectional boilers purpose designed for this fuel. They are extremely quiet in operation and, as multiple units, may be obtained in ratings up to about 1 MW. A gas pressure of about 1.5–2 kPa is required at the burner, down stream from the governor.

Gas is supplied to a calibrated nozzle fitted to a manifold and flows through a venturi to induce a supply of primary air. The resultant mixture is delivered by the manifold to a series of outlet ports where the flames entrain secondary air from the surrounding space at the boiler base.

Combustion gases rise by convection through channels between the heat exchange surfaces, boiler height providing an adequate motive force. As illustrated in Fig. 13.22, it is necessary that the flue terminal of the boiler incorporates a *draught diverter* and gas dilution device to prevent any interaction between that small natural draught and the effect of a connected chimney. An atmospheric burner must not be used in any situation where draught is not so stabilised.

Packaged burners

Known also as forced draught burners, units of this type are built up around an integral fan which is arranged to provide, under pressure, all the air required for combustion. In most models, such fans operate with direct coupled drives at relatively high speed and, as a result are often noisy.

Figure 13.23 illustrates a gun-type burner, where both gas and air under pressure are delivered to nozzles in the combustion head. The design arrangement and the geometry of the burner heads are very important in ensuring that the gas and air are properly mixed to provide efficient combustion, and ensure the correct flame shape and size. The precise detail varies from manufacturer to manufacturer. Turn-down ratios of up to 10:1 are possible with this type of gas burner.

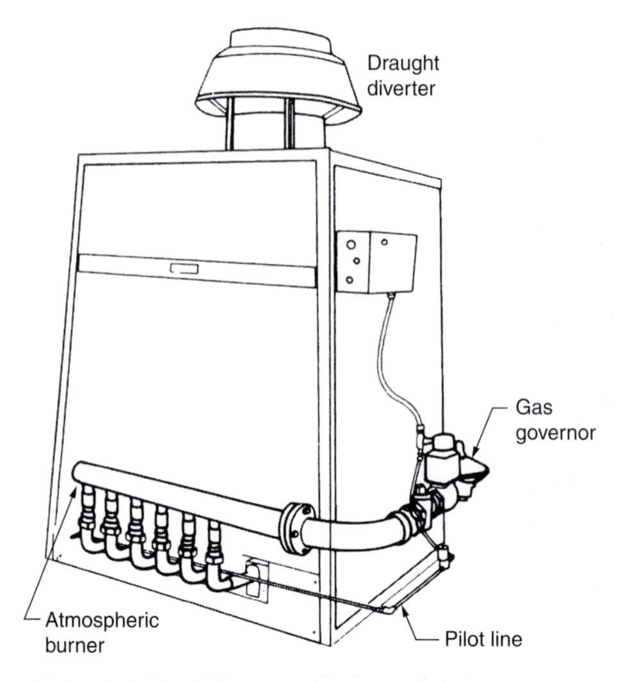

Figure 13.22 Atmospheric gas burner and boiler

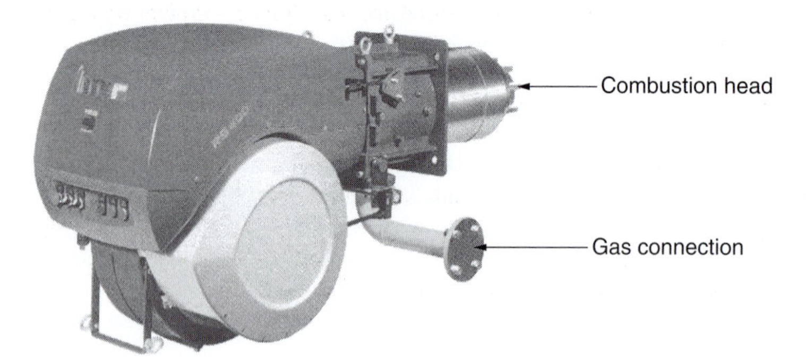

Figure 13.23 Packaged forced draught gas burner (Riello)

Since the boiler combustion space is pressurised by the forced draught air supply, the flue connection may be made without regard to the limitations existing in the case of an atmospheric burner and no draught diverter is required. In order to overcome the noise nuisance mentioned previously, some manufacturers provide insulating shrouds tailored to allow for relatively easy removal and consequent access for burner maintenance.

Pre-mix burners

Pre-mix burners deliver a controlled mixture of gas and air at relatively high pressure to a combustion head located within the boiler chamber. The high temperature combustion head provides a very even temperature distribution. The combustion chamber of a boiler operating with a pre-mix burner is considerably smaller than an equivalent capacity boiler operating with a gun-type burner. The burner combustion head and boiler chamber are often specifically design 'matched' to maximize efficiency within a very compact assembly. As noted earlier in this chapter, a pre-mix burner/boiler, low water capacity unit is very much smaller than a 'conventional' boiler/burner arrangement.

Figure 13.24 shows a typical pre-mix burner, purpose designed to suit a vertical condensing boiler.

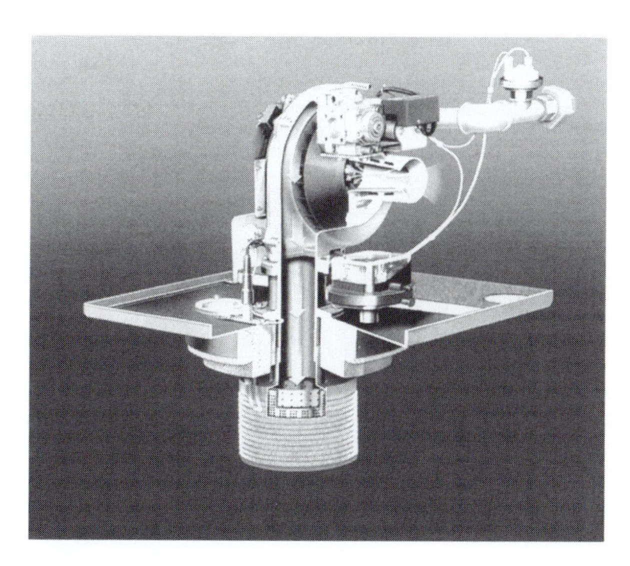

Figure 13.24 Pre-mix burner (Hoval)

Burner controls

In the case of atmospheric burners, the control systems are comparatively simple, usually consisting of an electrical ignition device which operates in conjunction with a gas ladder and a flame failure sensor connected to a safety shut-off valve, which will be activated also by sub-normal gas pressure. Basic and high-limit controllers are either separate fittings or a single dual-function type with the high limit arranged such that hand reset is necessary.

As in the case of oil burner controls, the operational sequence will vary with the size of the burner from simple 'on/off' for the smaller range to more complex systems for larger sizes. Some makes of burner are now provided with a further safety thermostat which is fitted to sense flue downdraught.

For packaged burners, the same principles apply but the details are somewhat more complex to take account of fan controls and, where larger sizes are involved, the sequencing arrangements will probably provide for purging periods during both start-up and shut-down cycles. The method used for detection of flame failure is either by probe or by ultra-violet scanner, the luminosity of a gas flame having been found insufficient to operate the type of light sensitive cell used to view oil combustion.

Dual-fuel burners

Brief mention has been made already of equipment which offers the facility of use with more than one fuel. Combination dual-fuel burners, capable of firing oil or gas are as shown in Fig. 13.25. It is usual for both supplies to be connected and change-over is thus a relatively simple matter of shut-down; isolation of one supply; activation of the alternative and start-up.

Gas boosters

In certain circumstances it may be found necessary to provide pressure boosting equipment to serve burners which require a fuel supply at a pressure higher than that available at the meter. Such a situation may arise in particular where a dual-fuel burner is selected and the pressure at which the combustion air is provided exceeds that at which gas is available. Pressures higher than normal offer advantages in that piping, automatic control valves, etc., may be of smaller size but it would rarely be economic to introduce boosting equipment for such reasons alone.

Booster equipment may take the form of one centrifugal unit per burner or a common unit serving a range of burners. Operation must be fully automatic and cross-linked with the burner control system. The circuit arrangement illustrated in Fig. 13.26 shows a single booster serving two burners.

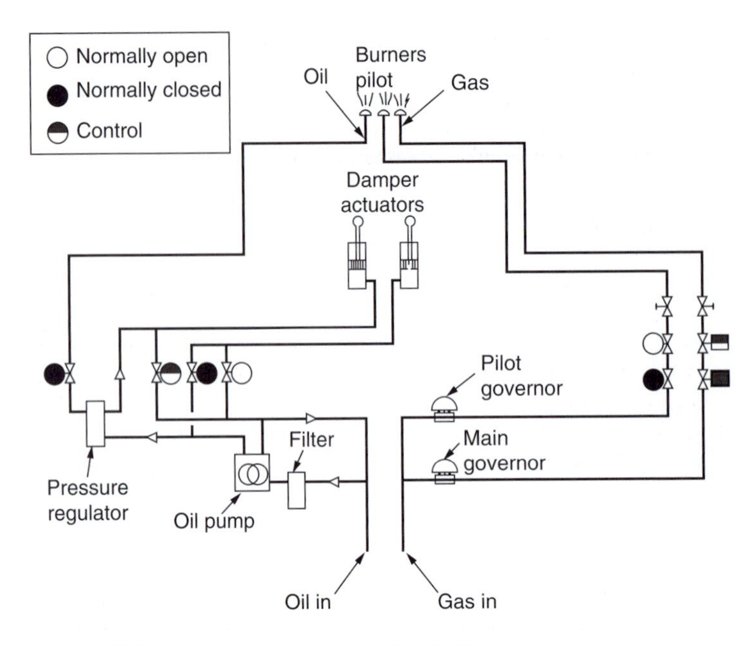

Figure 13.25 Schematic connections for a dual-fuel burner

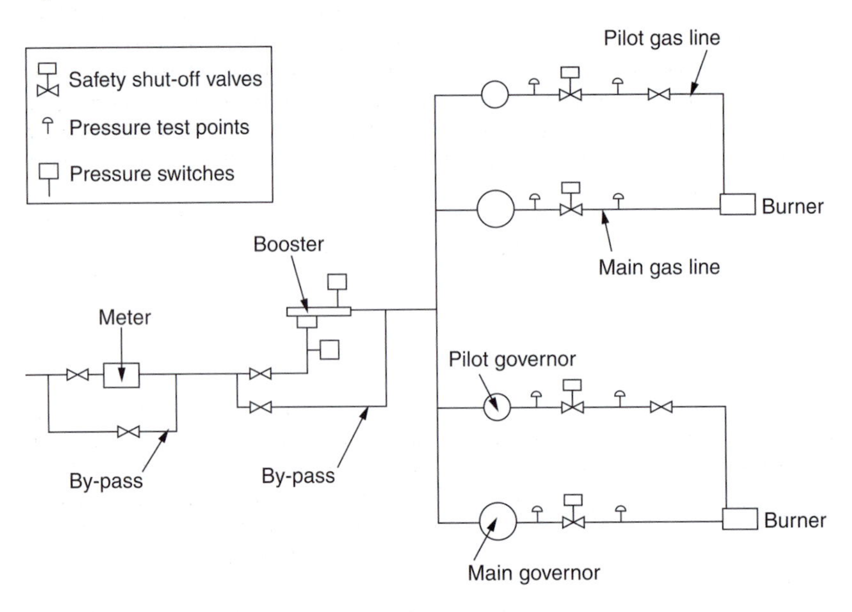

Figure 13.26 Pipework arrangements for gas booster

Safety precautions

In addition to the various devices incorporated in burner control systems, a number of further items must be considered. These apply equally to oil- and gas-fired plants, as follows:

- Automatic fuel shut-off at the entry to the boiler house, either by a solenoid or deadweight operated valve. The valve is usually held open by thermal links located above each boiler, plus emergency push/lock-off buttons at each boiler house exit.

- Boiler dampers should be either removed, locked open with an indicating plate showing the position of the vane, or provided with an interlock with the burner control.
- Both boiler house and fuel tank room should be provided either with piped foam connections for use by the fire brigade or with chemical fire extinguishers of appropriate type.
- Plant operators should be certificated as competent to undertake the tasks expected of them.

Miscellaneous boiler/burner equipment

The need to reduce carbon emissions from the burning of fossil fuels has increased interest in the consumption of a wide variety of waste products for heat production and to the development of specialist boilers and burners suitable for their disposal. Renewable energy sources are covered separately in Chapter 14. The burning of municipal waste is mentioned at the end of Chapter 16. However, the types of equipment necessary to handle and burn such material is too great to be dealt with in sufficient detail here, especially when considering all the attendant ramifications of incomplete incineration, furnace fouling by non-degradable material and atmospheric pollution by exhaust gases. However, two items, from opposite ends of the technological spectrum, are considered here, as being of special interest.

Straw boilers

Straw as a rural waste product is considered a biomass fuel with carbon-neutral emissions. A number of purpose-made boilers have appeared on the market, together with a variety of ancillary equipment. Bales are produced in two sizes, the small rectangular size which measures 350×450×900 and weighs about 18 kg and the large cylindrical sizes some of which measure up to 2 m in diameter and weigh upwards of 250 kg. Moisture content is quoted at 17 per cent and the net (calorific value) CV may be expected to be about 12 MJ/kg.

The original ranges of boilers were designed to burn the small bales and had ratings of between 20 and 100 kW based upon a 4-hour loading cycle: the combustion chamber of a 100 kW unit, for instance, being large enough to accept seven small bales. Later models designed to burn the larger cylindrical bales are rated at between 100 and 250 kW based upon a 6-hour loading cycle. The loading doors for all sizes are, of course, extremely large and seem to be arranged in most cases for water cooling.

To overcome loading difficulties which, with the larger bales, must be considerable, automatic feed systems are available. These use hydraulic power to divide the bales into smaller compressed portions and to provide a piston feed to the boiler via a conveyor tube. This tube is fitted with a water drenching system and a spring-return fire guillotine.

Fluidised bed boilers

The concept of combustion within a fluidised bed is so different from conventional practice that it deserves further explanation of the principles involved. If a bed of sand, or a similar inert material such as crushed refractory, is mounted over a plenum box then, when a critical air velocity is reached through the bed, it will behave very much as if it were boiling, with the bed particles mixing rapidly throughout the depth. If such a bed is heated to a temperature of say 750°C and fuel is then added, combustion will be self-sustaining and the entire bed will become incandescent. In this way low grade fuels can be burned with reasonable overall efficiency. The principles are illustrated in Fig. 13.27.

Heat exchange within a combustion chamber provided with a fluidised bed is extremely complex, including radiant and convective transfer to the surrounding surfaces and direct transfer to any surfaces immersed in the bed, the latter being particularly significant. What is important, however, is that combustion is highly efficient and almost complete with minimal ash residuals.

Instrumentation

The theoretical background to analysis of the products of combustion, as a measure of the 'efficiency' of the combustion process, is dealt with in Chapter 17. There is now available a number of sophisticated instruments

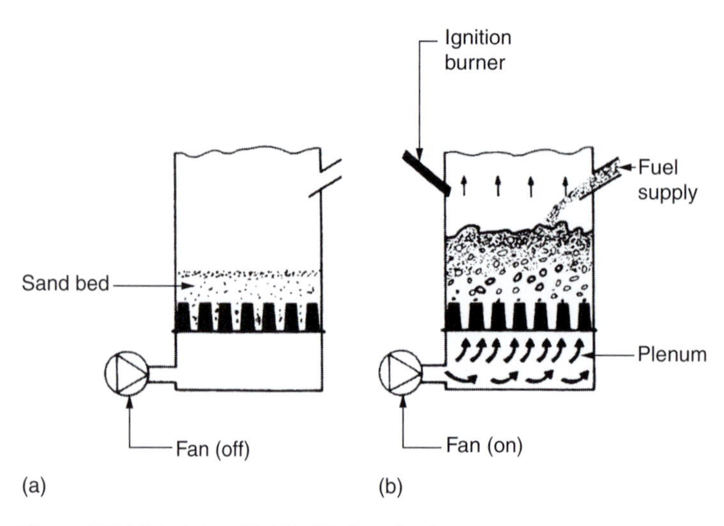

Figure 13.27 Principles of fluidised bed combustion

which enable the necessary measurements to be taken but, for site use, the more robust and simple devices of the past may continue to be of value.

Flue gas analysis

The standard apparatus is the *Orsat* with which a sample of the gas is first dried to remove water vapour and is then subjected to an adsorption process by three liquids in turn. The first is a solution of caustic soda which adsorbs CO_2 and the second an alkaline solution of pyrogallic acid plus caustic soda which adsorbs O_2. The third is a solution of cuprous chloride in hydrochloric acid which reacts with CO. Measuring burettes enable the proportion of gas adsorbed in each case to be assessed with comparative ease and the volumetric content of the whole established.

Flue gas temperature

The temperature of combustion gases at boiler exit are measured using either pyrometers or high temperature thermometers.

Smoke indication

The *Ringlemann* charts are a series of grids of increasing darkness that provide a scale against which smoke emission from a boiler plant may be compared. The charts are enshrined in legislation in that the Clean Air Act prohibits emission of smoke darker than Scale No. 2 except in smokeless zones where no levels on the scale are acceptable. Another scale is the *Bacharach* which uses a filter paper through which a controlled volume of flue gas is passed. Discoloration is matched against standard shades to which are allocated numbers in a scale of 0–9.

Larger sizes of boiler plant are often fitted with smoke density indicating or, in some cases, recording equipment. These devices rely upon an arrangement whereby a light source on one side of the flue relates to a light sensitive cell on the other. In practice, deposition of soot on either the viewing or the transmitting ports may distort the readings.

Chapter 14

Renewable technologies

Introduction

Environmental concerns and the drive to reduce CO_2 emissions from buildings have led to an increased interest in renewable energy as sources of heat, cooling and power for buildings. Planning Policies are increasingly demanding the inclusion of renewable energy systems into major developments thus placing a requirement on the building services designer to have a good knowledge in this area. Further information on Planning and other policy drivers is given in Chapter 2.

This chapter is intended to provide an introduction to building integrated renewable systems; most authorities agree that these include:

- Solar thermal
- Biomass heating
- Ground source heat pumps (GSHP)
- Solar photovoltaics (PV)
- Small scale wind turbines

Two electrical generating technologies have been included in the above list but in a book on HVAC systems these will only be touched on for completeness and the main focus will be on those renewable systems that are directly related to heating and cooling. Other options not included in the above list are:

- Biogas fired heating
- Biogas combined heat and power (CHP)
- Biomass CHP

At the time of writing biogas has a limited use in buildings, but increased interest in a range of technologies could lead to the wider use of biogas fuels in the future. Similarly biomass CHP is a technology under development. Therefore these technologies are not covered in this chapter. More information on CHP is given in Chapter 15.

Solar thermal systems

In an effort to maximise the use of renewable energy, reduce parasitic energy use such as pumping energy and reduce capital cost a very wide variety of systems have been developed. The following discussion is intended to highlight the main types of equipment and systems available.

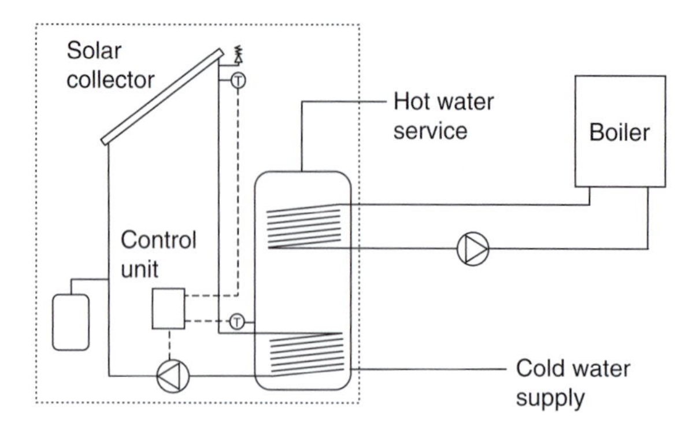

Figure 14.1 Typical solar hot water system

System principles

Solar thermal systems use energy from the sun to heat water, most commonly in the UK for domestic hot water needs or swimming pool heating. The systems use a heat collector, generally mounted on the roof or a south facing façade in which a fluid is heated by the sun. This fluid is pumped to a thermal store, where the heat is released to the store. Figure 14.1 shows the general arrangement of a typical system.

Solar collectors

There are two standard types of collectors used; flat plate collectors and evacuated tube collectors.

Flat plat collectors, as the name implies, have a flat plat absorber across which the heat transfer fluid is passed through channels in the plate or pipes bonded to the back. A header at each end of the plate allow for easy connection to the heat transfer circuit. The front of the plate is coated with a special absorptive layer which maximises the capture of the solar irradiation that falls onto the plate. Typically the plates are supplied as standard modules in frames, insulated on the back and with a glass or plastic screen on the front. These glass or plastic screens reduce heat loss while being specifically chosen to allow as much solar irradiation to pass through as possible. Figure 14.2 illustrates a typical panel. In the UK flat plate collectors might be expected to deliver between 300 and 600 kWh of heat per year per square metre of collector, but actual outputs will vary from location to location.

Evacuated tube collectors use a specially coated metal collector closely coupled to either a pipe through which a heat transfer fluid is passed or a metal rod that is heated and transfers the heat via conduction. The collector plates are long and narrow and enclosed in an evacuated tube. The evacuated tube reduces heat loss through conduction and convection while allowing solar irradiation to fall onto the collector plate. Manufacturers usually supply standard panels with a number of individual tubes connected to a common header. Figure 14.3 illustrates a typical panel. In the UK evacuated tube collectors might be expected to deliver between 400 and 850 kWh of heat per year per metre square of collector, but as with flat plat collectors the actual output will be location dependent.

While evacuated tube collectors generally deliver a higher energy output per unit area than flat plate collectors, they are usually more expensive due to a more complex manufacturing process (to achieve the vacuum).

There are other types of collector that have been developed, usually aimed at meeting a specific need such as reducing cost where energy yield is not a major constraint.

Heat transfer system

The most common system for transferring heat from the collector to the thermal store or building load is via a pumped circuit containing water or a water/antifreeze mix. These provide good control, reduced system size and limited overheating capacity for short periods. However, they are relatively expensive and reduce CO_2 savings

Figure 14.2 Flat plat collector (Filsol Solar Ltd)

Figure 14.3 Evacuated tube collector (Riomay)

due to the pump electrical energy requirements. In some systems the circulating pumps use power derived from solar PV, as these will tend to generate electrical power at the same time as there is a significant amount of heat to deliver to the building. At the same time the system energy delivery becomes totally renewable.

An alternative approach is to use the temperature difference generated by the collector to drive flow around the heat transfer circuit by gravity circulation as described in Chapter 12. The advantage of this type of system is that flow will only occur when sufficient temperature gradient exists, i.e. there is a benefit in circulating water, and the elimination of pumping energy.

In some systems the water to be used in the building is directly circulated through the solar collector.

Thermal store

In most applications there is a need to store the solar energy collected for when it is needed.

In domestic applications of hot water supply, the thermal store is usually the hot water cylinder, where the water to be used for the supply service is heated directly by the solar energy collected. Because solar energy will vary with time of day and day of the year, a secondary conventional heat source is usually provided to ensure that the stored water is heated to an appropriate temperature throughout the year.

Non-domestic applications can use direct heating of the hot water supply, but more often tend to use the thermal store as an indirect preheat for the supply water. This approach has the advantage of ensuring that all the water stored for the supply service is maintained at the required storage temperature while maximising the potential for using the solar energy collected.

One application that does not usually require a thermal store is for swimming pool water heating. Here the pool water is usually of sufficient volume that it can absorb all the heat collected directly.

System design

Basis of design

Unlike conventional heating equipment which is designed to meet a specific instantaneous heat output, solar thermal systems are designed to deliver a given amount of energy (measured in kWh) over a year. This is because, while the maximum output of a given area of collector can be calculated, it is not possible to ensure a specific output at any one time.

Using energy demand rather than peak instantaneous output as the basis of design has led to the use of empirical rules of thumb for designing solar thermal systems. These work particularly well with domestic systems where as a rule of thumb 1 m^2 of panel area is installed per occupant.

Solar collectors

The actual output from the solar collector will depend on location, orientation and angle of tilt, as well as the actual efficiency of the collector itself.

Solar irradiation on the collector varies considerably with location. Figure 14.4 shows the typical range of annual solar radiation in kWh/m^2 falling on a south facing panel tilted at 30°.

For locations such as the Caribbean or India the average annual solar irradiation is likely to be between 1800 and 2250 kWh/m^2 pa, while the Middle East could be as high as 2700 kWh/m^2 pa.

In the UK the maximum annual energy output from solar collectors will be obtained when the collector is oriented due south at an angle of 30–40° to the vertical. As one travels north the optimum angle of tilt will increase, while decreasing to the south. As one departs from this optimum then the total amount of energy collected will reduce. However, it is still possible to obtain outputs close to 95 per cent of the maximum for orientations within 45° of south and tilt angles of 10–50°. Figure 14.5 shows the percentage outputs relative to the optimum angle and tilt for the south of England.

It is recommended that when obtaining manufacturer's information about solar panels that the annual energy outputs are quoted against the standard En12975-2. This will give a standard for comparison purposes, but does not necessarily indicate the energy output for the particular site under development.

The choice of collector system will depend on a number of factors including area and locations available for mounting panels, required energy output, costs and aesthetics. Evacuated tube collectors deliver a higher energy output for a given area and therefore will be the more appropriate choice when space is limited. In many cases they can also be mounted at any angle of tilt, from vertical to horizontal, as the individual tubes can be rotated to the optimum tilt within the panel. Flat plate panels have the advantage of being cheaper per square metre, thus making them the more appropriate choice where a fixed area is to be installed.

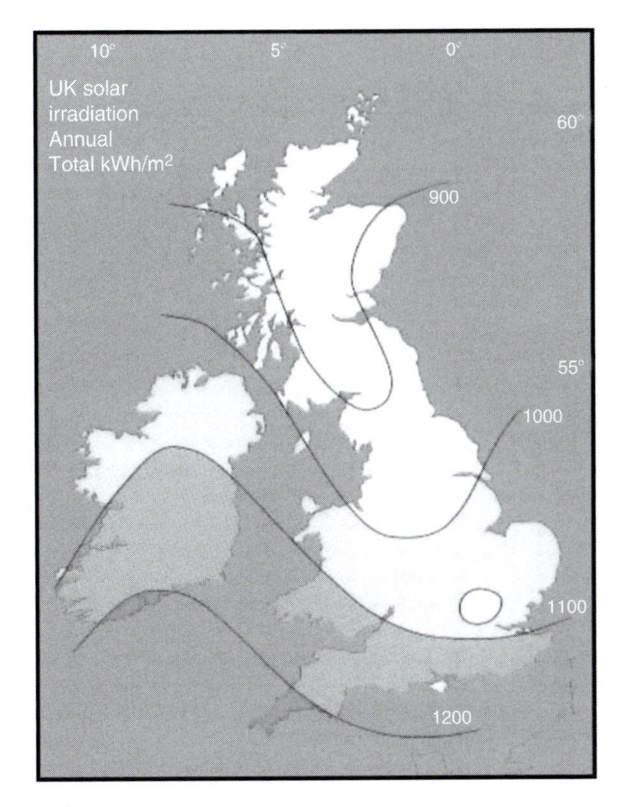

Figure 14.4 UK average solar irradiation (kWh/m^2) facing south at 30° incline (see colour plate section at the end of the book)

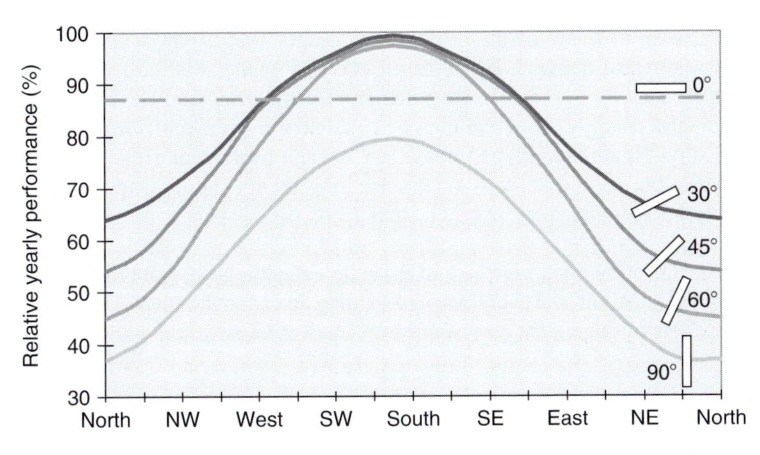

Figure 14.5 Percentage outputs of solar systems for South-East England (see colour plate section at the end of the book)

Where the solar system is to be used as a heat source for the hot water supply system then it is usual to design the solar system to meet a specific proportion of the estimated total heat demand. While in theory it is possible to design a system to meet 100 per cent of the heat demand, previous discussions regarding the availability of solar energy will show that this is not easy to actually achieve. More typically a solar system would be sized to meet between 30 per cent and 70 per cent of the total annual heat demand. The aim is to provide

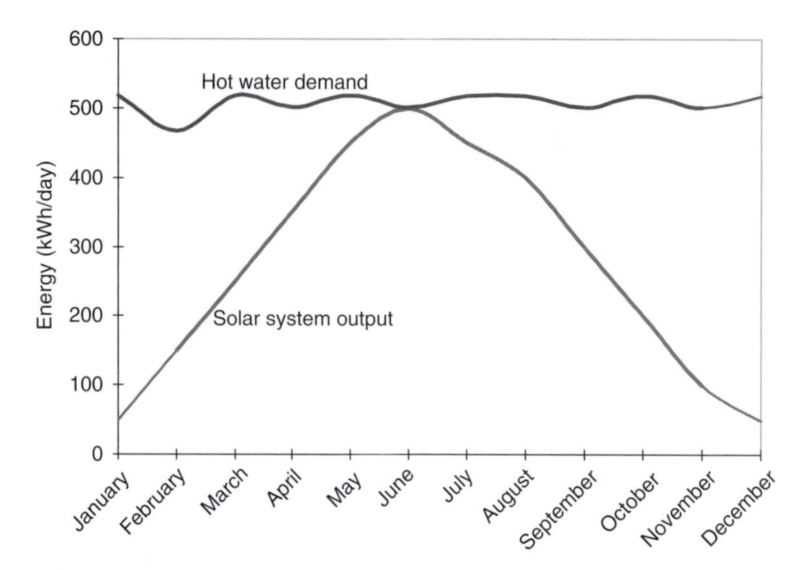

Figure 14.6 Matching solar system output to hot water demand (see colour plate section at the end of the book)

as close to 100 per cent of the heat demand in the summer months and a proportion of the heat demand in the winter, illustrated in Fig. 14.6. This will generally maximise the use of solar energy while not over producing heat in the summer months that must be wasted.

Given the target energy requirement from the solar system, manufacturers' data on annual energy outputs can be used to determine the size of system required.

An alternative approach to sizing a system might be to determine the maximum size of system from solar irradiation over a peak day for the building location, the efficiency of the chosen panel and the area of collector proposed. These quantities will give an estimate of the peak energy output over a day, which can be compared with the estimated daily hot water heat demand. By adjusting the size of the solar collector array a balance can be struck between the solar array output and the heat demand for the peak day. Having identified the size of the solar collector the annual energy output can then be calculated from manufacturers' data. It should be noted that the solar irradiation figures mentioned above are not the peak solar figures quoted in CIBSE Guide A or other design guides, but should be daily total solar irradiation figures. One source of such information may be the CIBSE Test Reference Year weather tapes (refer to Chapter 6 for details) or Meteorological Office records.

For swimming pools it is possible to estimate the maximum size of the solar collector array from the instantaneous maximum solar gain for the chosen location and the estimated heating load for the pool. In practice some over sizing may be possible as the effect of raising the pool water by a small amount would not usually be great enough to discomfort pool users.

Thermal store

Typically the thermal store will be designed to balance heat demand and supply over a day. Where heat demand occurs throughout the day this may mean that storage for only a few hours of collector output is required. Where heat demand occurs at the beginning or end of the day then a full-day's storage may be required.

The optimum storage capacity will enable full use to be made from the energy available from the collectors without being over sized. Over sizing of the thermal store reduces the temperatures that can be achieved at times of low collector output, thus reducing the benefit available from the solar system as a whole.

An industry rule of thumb for store size is 50 litre/m^2 of panel.

If storage for 1 day is to be provided then the maximum storage capacity would be determined by the peak day's solar irradiation and the expected temperature rise in the store. An example calculation is shown below:

Peak day solar irradiance South-East England at 30° incline approximately 5 kWh/m^2
If active solar panel area = 8 m^2 and panels are 50 per cent efficient
Hence Peak day heat gain = $5 \times 8 \times 0.5 = 20$ kWh
For a maximum temperature rise in the store from 20°C to 80°C = 60°C
Size of store = $(20 \times 3600)/(4.2 \times 60) = 285$ litre

where

specific heat capacity of water = 4.2 kJ/kg K
3600 seconds per hour converts from kJ to kWh

In winter, when only 10 per cent of peak solar gain can be expected, the total energy input will be 2 kWh. Hence store temperature rise will be only 6°C.

The heat transfer circuit

The choice of which type of heat transfer circuit to use will depend on cost, available space and whether there is a need to achieve a totally renewable installation. The design of the system will be similar to that for any heating circuit of the chosen type.

Where the system is to remain unused for periods during the year there is a risk of the system overheating. Under these circumstances it is necessary to consider how to safely shut the system down without losing the heat transfer fluid. There are a number of proprietary approaches to this. These are based on providing an expansion or drain-down tank that collects the heat transfer fluid and stores it such that the system can automatically recharge itself on start-up.

Controls

Where the system is a pumped system, an electronic controller is usually included. This controller monitors temperatures at the solar collector and in the thermal store and operates the pump when the temperature difference is such that some benefit will be derived. Controllers can also monitor the thermal store for excess temperature where the circuit is able to shut-down without releasing a pressure relief valve. Controllers can often be used to report on conditions within the system and provide information on the renewable energy being gathered.

Gravity circulation systems require no such controller as the heat transfer circuit will only operate when a positive temperature gradient exists. A non-return valve will be required, however, to ensure that heat is not transferred from the store to the collector.

Integration into buildings

As previously stated the most common use of solar thermal systems in the UK is to generate domestic hot water or provide heating to swimming pools. While solar systems can be installed to meet almost any level of hot water demand, buildings with high hot water demands will be most suitable as the benefits from the solar system will be proportionally higher. The time of the heat demand is also important as the solar system will deliver most of its energy in the summer months. Schools, which do not operate during the summer months, may therefore not be suitable for solar systems unless some extra curricular activity during the summer requires the heat. Applications for which solar thermal technology is well suited will include:

- Hotels
- Swimming pools
- Leisure centres
- Catering facilities
- Domestic dwellings.

Compatibility with other energy systems

The uncontrollability of the availability of solar energy means that there will nearly always be a need to integrate solar thermal technology with other heat generating systems. These other systems will need to be able to provide the full heat demand served by the solar system.

Gas or oil-fired boiler plant can be easily integrated with solar thermal systems as the plant is flexible in operation and relatively cheap.

CHP (see Chapter 15) systems or biomass boilers are less readily integrated with solar thermal as they both operate to their best advantage when meeting base heat loads such as hot water supply. Thus solar water heating will be in competition with CHP or biomass for the base load, leading to a sub-optimal system.

Other renewable energy systems such as GSHPs or solar PV are, however, compatible with solar thermal systems.

Biomass

Introduction

Biomass is nominally considered a carbon neutral fuel, as the carbon dioxide emitted on burning has been (relatively) recently absorbed from the atmosphere by photosynthesis. The circular carbon cycle assumed for biomass fuels is illustrated in Fig. 14.7. In practice there are CO_2 emissions associated with the processing and transportation of the fuel and so a nominal emissions rating is usually associated with the fuel. Building Regulations 2006 Approved Document L2A sets an average factor of 0.025 kg/kWh.

The main focus of this chapter will be the direct combustion of biomass to provide heat.

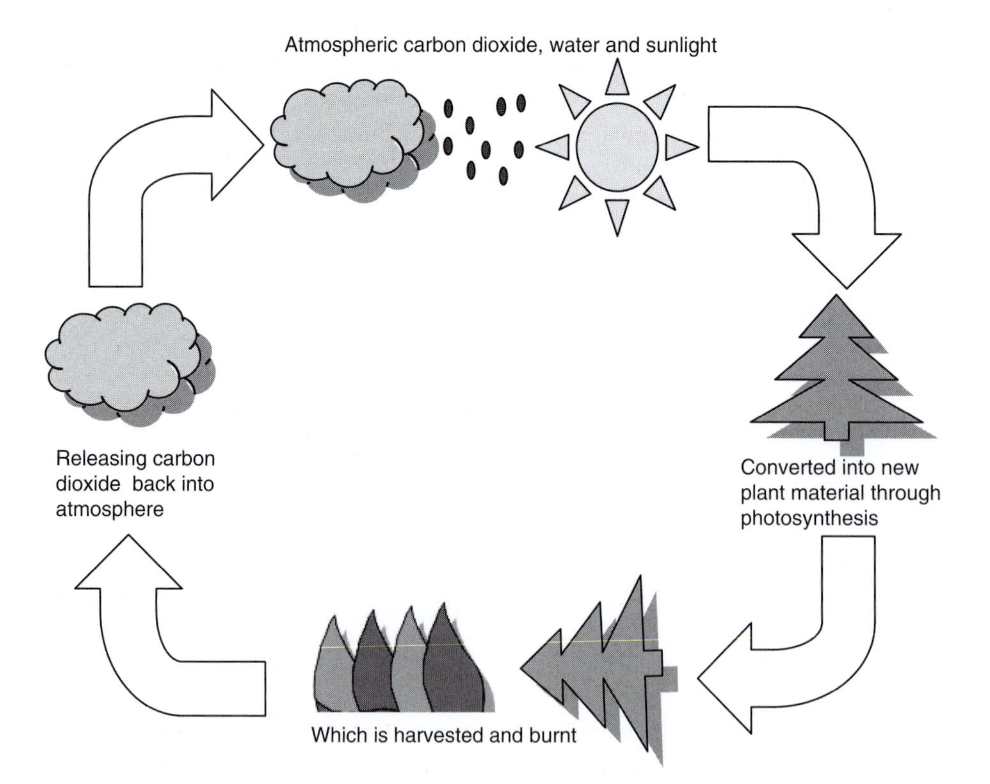

Figure 14.7 Biomass carbon cycle (see colour plate section at the end of the book)

Fuel

Sources of biomass

Biomass can come from a range of sources including wood from forests, urban tree pruning, farmed coppices, factory wood waste, miscanthus grass or straw from cereal crops. This type of biomass can be burnt directly to release energy. Other methods exist to release energy from biomass such as gasification or pyrolysis, however, at the time of writing, these are not widely applicable to individual buildings.

Other organic waste material (vegetable waste, the biodegradable component of municipal waste, chicken litter, etc.) is also classified as biomass and can be used to create energy through anaerobic digestion. This source of fuel is more appropriate to large scale systems.

Wood in the form of chip or manufactured pellets is the most readily available form of biomass in the UK and these fuel sources will be the main focus of this chapter.

The chosen fuel can have a significant impact on delivery and storage requirements. Wood chip is widely available from a range of sources and is therefore relatively cheap. However, energy density is low compared to pellet fuel, which is processed to give a more even quality and a low moisture content. A rough rule of thumb indicates that to deliver the same energy to site the volumetric ratio between oil, wood pellet and wood chip is in the ratio of 1:3:9.

Fuel quality

Fuel quality is very important as it determines combustion efficiency, emissions from the combustion process and ease of fuel handling.

Typically wood fuel with a moisture content of 35 per cent or less is required to achieve high combustion efficiencies. Moving grate boilers can often cope with fuel with moisture contents up to 50 per cent but efficiencies will be reduced.

If a fuel contains significant amounts of bark or pollutants such as plastic then this can lead to harmful emissions from the combustion process. It may be tempting to use wood obtained from waste sources but it is important to ensure that such material is free from substances such as paints or some preservatives that could lead to pollution problems.

Other foreign bodies such as minerals or metal will also cause problems with fuel handling with the risk of jamming or damaging fuel handling equipment.

When identifying a fuel source it is important to ensure that moisture content, consistency and quality are all assessed and that supply contracts specify standards to be met by the supplier.

Biomass installation components

Boilers

Modern biomass boilers have been developed to a very high level of sophistication. Smokeless combustion, to comply with the Clean Air Act, is now achievable with efficiencies of over 90 per cent.

Many day-to-day operation and maintenance tasks are automated to reduce operating costs and boiler downtime. Common features include:

- Automatic fuel feed
- Automatic ignition
- Modulation typically down to 40 per cent rated output but can go as low as 20 per cent for fuels with moisture content of less than 35 per cent
- De-ashing – ash content is very low for wood fuel, 0.5–1.5 per cent
- Automatic boiler tube cleaning – removal of fly ash.

Boilers come in a wide range of sizes from 10 kW up to multi-megawatt. There are three basic types:

- Underfeed stoker – requires fuel with low moisture content (<35 per cent), not generally used for larger multi-megawatt boilers.

- Moving grate, sometimes known as a stepped grate – capable of burning high moisture content fuels (<50 per cent), not used for smaller boilers below 200–300 kW.
- Drop feed used with pellet fuel only.

The boilers all have two major sections, the combustion chamber and the heat exchanger. The combustion chamber is refractory lined and designed to aid complete combustion of the fuel. The heat exchanger is a multi-pass shell and tube heat exchanger. Both components are very heavy (a 1 MW boiler could have a dry weight of 12 Tonnes) and suitable allowances for the weight of the boiler need to be made when considering installation and replacement. The larger boilers can usually be delivered split into the two major components described above. In this way delivery weight can be reduced by approximately half. However, suitable space must be provided on site for recombining the components in their final installation positions.

Biomass combined heat and power

There are a number of technologies being developed to use biomass to generate electrical power as well as heat. While large scale systems in the 2 MW range are well developed, smaller systems suitable for use in buildings are at the development stage; Chapter 15 has further details.

Fuel handling and storage

Most modern boilers can be fed automatically. Small units may have an integral hopper with around a days fuel supply, larger units will have an external fuel supply delivered to the burner by an auger arrangement. For the larger systems fuel stores will typically hold two or more week's fuel supply.

There are a range of ways fuel can be stored from hoppers specifically designed for purpose to bunkers or brick stores. Depending on the fuel store size different techniques are used to deliver the fuel from the store to the boiler. For smaller stores up to 5 m in diameter rotating arms are used. Larger systems, 500 kW boilers and above, use walking floors. Walking floors are mechanically actuated grids that progressively move fuel towards an auger for transfer to the boiler. In all cases an auger system is then used to deliver the fuel to the boiler.

A number of precautions are typically included in the fuel supply system to prevent the accidental spread of fire from the boiler to the fuel store. Many auger systems are designed with a physical barrier within the supply line to limit the spread of flame. Automatic monitoring and drench systems are also included. Further, fuel supplies are usually separated from the boiler room by a fire wall of half hour or more rating.

Flues

The combustion gases will require an external flue. Clean wood fuel is classed as low sulphur fuel. Where the boiler is certified as complying with the Clean Air Act then the chimney height can be calculated using the same procedures as for a gas boiler. If, however, the fuel is of unknown or inconsistent quality or the chosen boiler does not comply with the Clean Air Act then the procedures for solid fuel boilers may need to be followed. It is recommended that the boiler manufacturer's advice is sought regarding the appropriate calculation method to follow for their boiler. Chapter 17 has further information on flues and chimneys. One specific point to make with regard to biomass boilers is the need to keep flues separate from other boiler flue systems.

System design

Boiler sizing

Biomass boilers have a slow response to changes in demand due to their high thermal inertia. They are therefore best used to meet base loads rather than peak demands. The sizing of boiler installed will therefore need an alternative approach to that used for conventional boilers. One approach would be to use the load duration curve to identify a suitable base load.

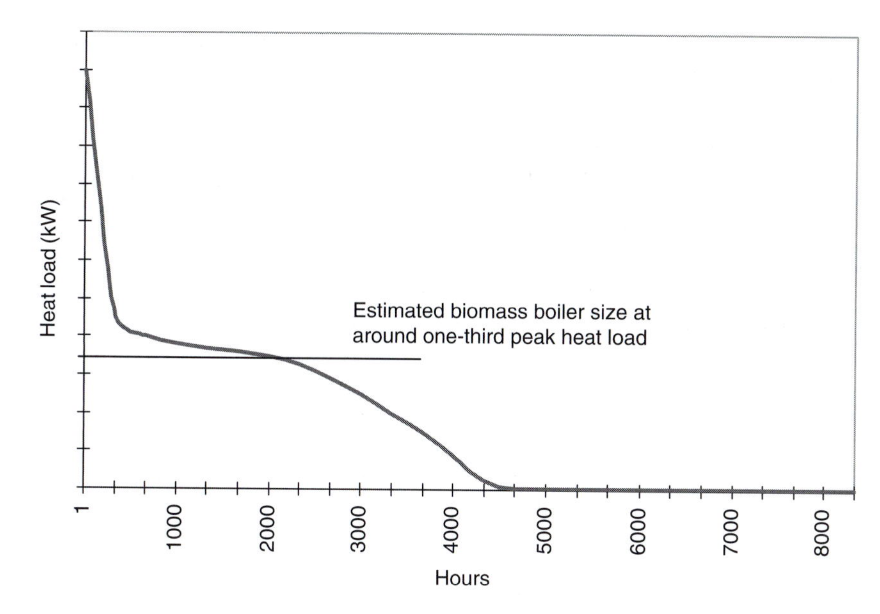

Figure 14.8 Load duration curve for modern well-insulated building

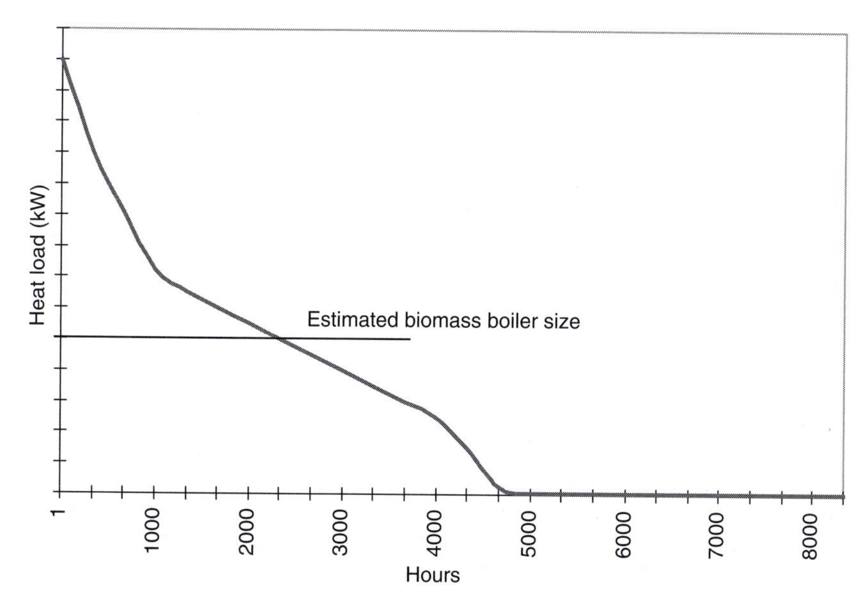

Figure 14.9 Load duration curve for older poorly insulated building

Modern buildings are likely to have low heat demands and hence boiler sizes could be as low as one-third peak demand as illustrated in Fig. 14.8. Older buildings are likely to have higher heat demands and may support larger biomass boilers as illustrated in Fig. 14.9.

The remaining demand can be met by more conventional gas or oil-fired boilers. Not only will these be better matched to the variable operation but they will also be less expensive to install.

It is worth noting that despite the biomass boiler being only a relatively small proportion of the total installed capacity, often 70 per cent or more of the total annual heat load will be met by such a boiler.

Thermal store/buffer vessel

The slow response of biomass boilers to changes in heat load mean that some form of thermal store or buffer vessel is always recommended. Different manufactures will have different rules of thumb for buffer vessel size. Examples are 10 litre storage per kW rated boiler output, or sufficient capacity to raise the maximum boilers output by one-third for 3–4 hours.

Larger thermal stores can be very beneficial where there are significant variations in loads across a day. Such a thermal store will even out the fluctuations between heat loads and biomass boiler output, thus increasing the proportion of the thermal load that can be met by the biomass boiler.

The sizing of the thermal store will depend on the period of time over which variations are to be smoothed, the size of the boiler and its capacity for modulation. In the case of community heating a thermal store could be sized to meet a full day hot water supply demand.

Plant space

Biomass boilers are significantly larger than the equivalent output gas or oil-fired boilers. In addition to the boiler, sufficient space must be provided for the cleaning and maintenance of the heat exchanger tubes. While manual cleaning can be reduced to once or twice a year there is still a need to regularly carry out this operation. Fuel storage and handling must also be born in mind, together with the space requirements for a thermal store. Figure 14.10 illustrates a typical arrangement for a biomass boiler.

Fuel storage will be a function of the fuel to be used, the expected rate of consumption and the desired frequency of delivery. Rules of thumb can be used to estimate the likely fuel storage.

Typical energy content of:

Wood chip (<35 per cent moisture content) 3.5 kWh/kg
Wood pellets 4.8 kWh/kg

Typical volumes of:

Wood chip 0.0045–0.0055 m^3/kg
Wood pellets 0.00165 m^3/kg

Hence if a 200 kW boiler is anticipated to operate for 10 hours per day and the desired delivery rate is once every 2 weeks the fuel storage requirement can be estimated as follows:

Energy demand = 200 kW×10 hours per day × 14 days = 28 000 kWh.
If the average boiler efficiency is 85 per cent then the fuel requirement will be 28 000/0.85=32 900 kWh.
Assume fuel is wood chip at moisture content less than 35 per cent then volume of biomass fuel to be stored = (32 900 kWh/3.5 kWh/kg)×0.005 m^3/kg = 47 m^3.

This could be met by a store 4 m by 4 m by 3 m high. In practice it would probably be better to have the store slightly taller than the minimum required as not all the space is useful due to the limitations of the fuel handling systems.

The delivery weight equates to around 47 m^3/0.005 m^3/kg = 9400 kg or 9.4 Tonnes.

Fuel delivery

Wood chip and pellets can be delivered by a range of truck sizes, but larger loads reduce frequency of delivery and hence transport and handling costs. The simplest delivery mechanism is to be able to tip the fuel into a suitably placed bunker or store. Direct access is therefore an important consideration when planning fuel supply to a site.

Where this is not possible fuel will need to be moved from the delivery bay to the store by a vehicle with a suitable bucket thus increasing fuel handling costs.

Fuel can be blown from truck to store but distances are limited, up to 20 m for pellet fuel. This method has higher costs as the time of delivery is increased while the fuel is blown into the store.

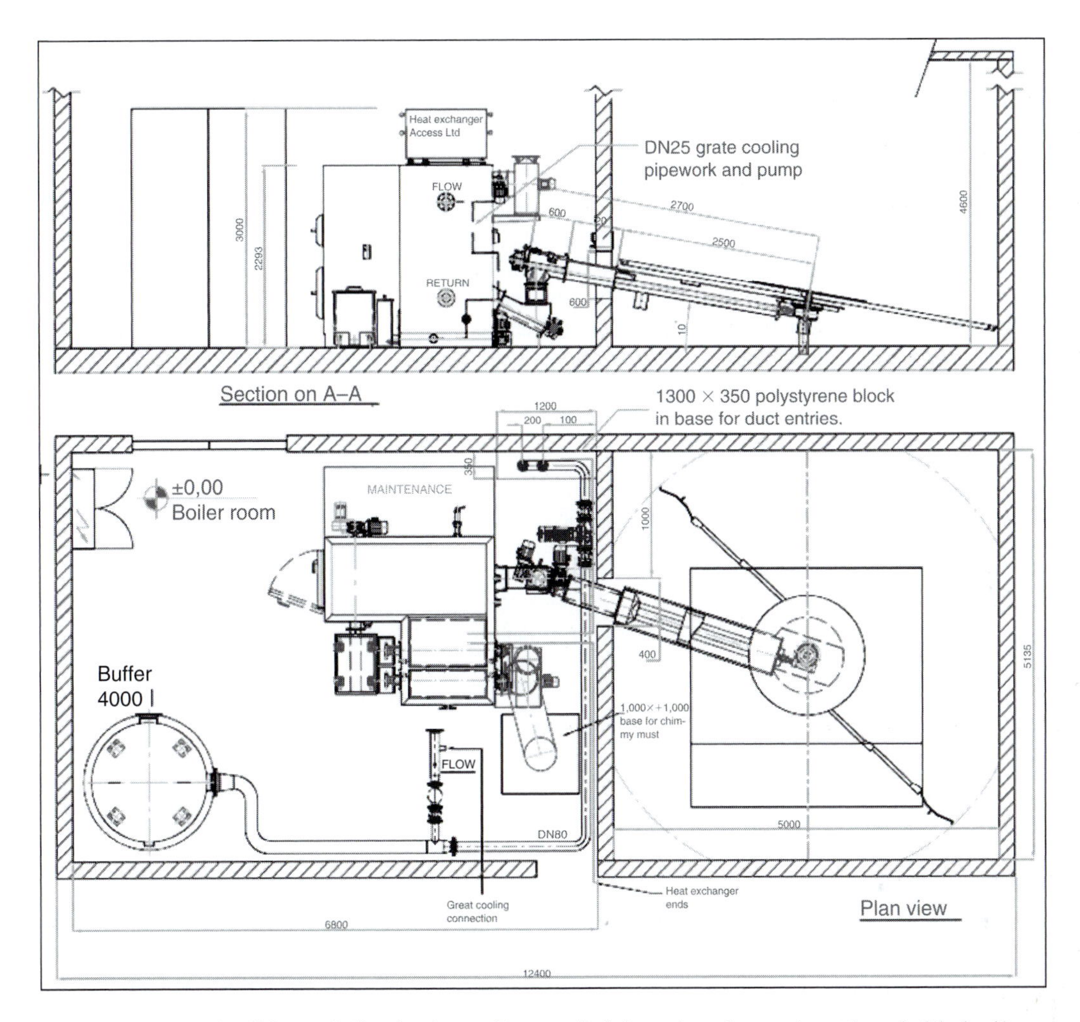

Figure 14.10 Example of biomass boiler plant layout (Econergy Ltd) (see colour plate section at the end of the book)

Typically wood chip will be delivered by either trailer or truck. Trailers are usually used where small deliveries are required and the source of the biomass is a local farmer or arboreal contractor. Larger loads will be delivered by truck and these can be up to 16 Tonnes (approximately 80 m^3).

There are limited sources of pellet fuel and so deliveries tend to be by truck, usually in the range of 10–12 Tonnes but can be up to 22 Tonnes (approximately 36 m^3). For smaller installations pelletted fuel can be supplied in bags suitable for manual handling.

Compatibility with other systems

Biomass boilers can be integrated with gas or oil-fired boiler plant. However, the characteristics of the biomass plant need to be born in mind when designing the installation.

Biomass boilers can deliver hot water or steam at a range of temperatures, but usually provide 80–90°C hot water. Return temperatures are more restricted than gas boiler plant as biomass boilers are not suitable for condensing on return water temperature. Therefore return temperatures usually need to be in the range 60–70°C. Where lower temperature water circuits are required then secondary side mixing or boiler backend protection will need to be included in the design.

In order to maximise the benefit from the biomass boiler in a multi-boiler arrangement it is recommended that the boilers be installed in parallel but with the biomass boiler as the lead. This is not as straight forward as a multiple gas or oil boiler installation as the slow response of the biomass boiler is out of step with the faster acting gas or oil boilers.

Start-up offers and example of the problems of controlling combined gas and biomass boiler installations:

- The building optimum start control calls for the boilers in the morning to bring the building up to operating temperature at the required time.
- The lead biomass boiler begins its heat up process, which takes 1 hour. During this time the boiler uses any heat output to increase the rate of its own heat up process and thus does not deliver heat to the building.
- The building heat demand is not being met so the gas boilers start-up.
- After 1 hour, when the biomass boiler is ready to deliver heat, the building has reached operating temperature and the boiler plant is only required to meet standing losses.
- The gas boiler plant switches off and the biomass boiler turns down to minimum output.

Thus most of the heat needed to bring the building up to temperature in the morning is actually delivered by the gas boilers, even though the biomass boiler is set as the lead boiler.

While biomass boilers can be integrated with CHP systems (see Chapter 15), they are not an ideal combination as both systems are best used to deliver the base heat load. When the biomass boiler is operating it will either reduce or take all of the heat load required to justify the operation of the CHP unit. Calculations will usually show that biomass and CHP systems will reduce CO_2 emissions by about the same amount. It is therefore prudent to carry out an economic assessment to determine which solution is most suitable for the particular application.

Other low temperature systems such as solar hot water and GSHPs can be combined with biomass boilers but the above comments on low temperature circuits and avoiding competing for base load need to be born in mind.

Ground source heating and cooling

Introduction

GSHP systems use the ground as the heat source/sink for a heat pump or chiller. Below 2–3 m the ground remains at a stable temperature throughout the year, typically 12°C, thus offering an advantage over ambient air as a heat source/sink. This more stable temperature leads to higher seasonal efficiencies in both heating and cooling mode than air source heat pumps. Typically GSHP seasonal CoPs will be around 3.5 in heating mode (compared to 2.5 for an air source heat pump) and 4 to 5 in cooling mode. It should be noted that while a ground temperature of 12°C has been identified above, there can be substantial variations dependent on location. London offers an example where ground temperatures can be higher due to human activity above and below ground.

Most Local Authorities and Government organisations consider energy extracted from the ground by a GSHP system as renewable (while air source heat pumps are not). However, the heat pump will require electrical power together with any additional circulating pumps, and this energy consumption must be subtracted from the energy delivered to get the net renewable energy impact of the technology.

GSHP systems are often characterised by the method used to enable access to the ground. Therefore the following discussion will examine each of the major approaches in turn.

Open loop

An open loop system extracts ground water via a borehole, uses this water as a heat source/sink for the heat pump and then returns the water to ground (or rejects to a water course/sewer). Several boreholes can be linked together to increase the capacity of the system. Figure 14.11 illustrates the principle of the open loop system.

This type of system has the potential to meet large energy loads, but the capacity of the system is dependent on the volume of water that can be extracted and the temperature change that is allowed.

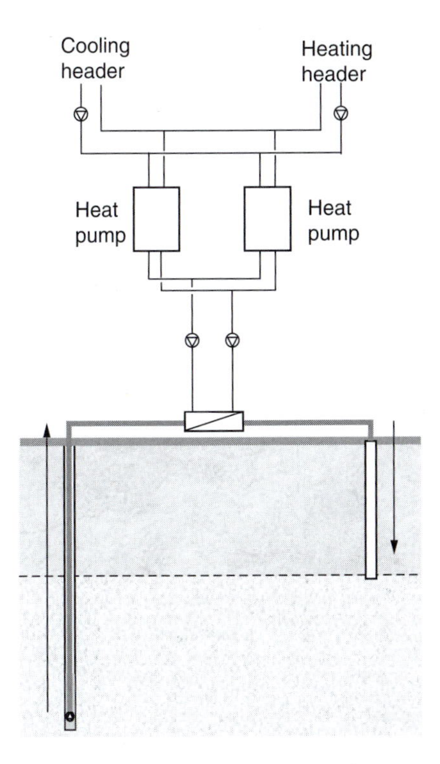

Figure 14.11 Principle of open ground loop system

In the UK, the Environment Agency has control of ground water resources and an extract licence will be required in order to use this type of system. This licence is not guaranteed and can be refused. There is also a limit to the life of the licence, which is becoming shorter, with no guarantee of renewal. If the ground water is to be returned to the ground then a discharge license will also be required. It is generally advisable to contact the Environment Agency at the earliest opportunity when considering open loop GSHP systems to identify whether this type of system is likely to be acceptable to them.

While an indication of potential borehole yield can be obtained from geological records and existing local borehole information, the actual yield will only be known when a test bore is drilled. Drilling of boreholes for water extraction can be very expensive, depending on depth and geology.

Closed loop

Closed loop systems pass a heat transfer fluid (usually water or water antifreeze mix) around a closed loop of high-density polyethylene pipe (HDPE) that is in close contact with the ground. There are a number of ways in which the closed loop is installed, the three main ways being vertical loops, horizontal loops or energy piles.

A much better estimate of energy yields can be made for closed loop systems before the need to carry out ground works. Local geological information, together with any existing borehole data can provide an indication of the probable yields from typical ground conductivity data. This allows a reasonable estimate of the length of loop that is required to meet the desired energy yield. When on-site work begins it may be worth carrying out a conductivity test on the first bore to confirm the expected yield. This may lead to an adjustment in the loop length required, but this is usually limited. For smaller systems such a test may not offer sufficient reduction in loop size to justify the cost of carrying out the test.

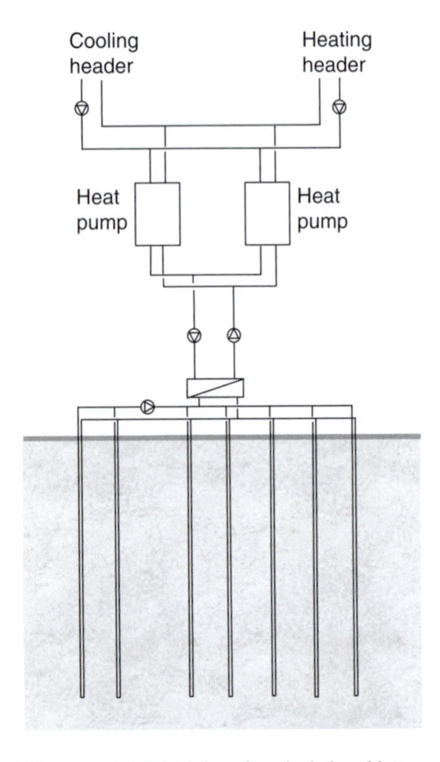

Figure 14.12 Principles of vertical closed loop

At the time of writing the Environment Agency have limited powers to prevent a closed loop systems being installed, unlike the open loop system. However, the Environment Agency is looking at obtaining the right to control close loops in order to avoid significant changes in ground temperature. Such changes could affect ground water conditions and could also jeopardise the viability of existing systems.

Vertical loops

Vertical loops consist of one or more borehole down which a pipe loop is placed. The borehole is then back-filled with sand or a high heat transfer grout. These individual vertical loops are then joined together to form one or more circuits, connected to the heat pump (Fig. 14.12). Bores can be up to 100 m deep depending on geology, the energy output required from the system and the space available. Where land area is limited this type of system may be the only closed loop solution capable of meeting the required energy outputs.

This type of system will often require a specialist to drill to these depths and boreholes must be kept open while the heat transfer loop is installed. Costs are therefore relatively high.

It is possible to install a vertical loop system below a building where there is no land available around the building. In such cases it is important to investigate the ground for obstructions and contaminants and to understand the proposed building foundations before taking the design too far. Where the Environment Agency perceives there may be a risk to ground water from drilling through a contaminated layer in the ground they are likely to raise objections to the installation. Sequencing of on-site works will also be important, especially with the foundation work. If drilling is to be sequential with the foundations then the construction time for the building could be increased by several weeks, increasing costs. In some cases these delay costs can exceed the cost of the actual GSHP system. It is therefore important that the whole design and contractual team understand the work required at the earliest stage.

Figure 14.13 Energy pile installation (see colour plate section at the end of the book)

Energy piles

A specific application of vertical loops is to incorporate the loop into foundation piles. This technique reduces costs by removing the need for a specialist contractor and also reduces the construction time required compared to drilling dedicated piles where the ground loop must be installed under the building.

The GSHP loop is attached to a reinforcing bar or cage and then lowered into the foundation bore, before concrete is poured (Fig. 14.13). This can slow the installation of the foundations slightly but is much quicker than installing a conventional vertical loop.

Outputs may be limited from such systems as foundation piles do not normally go as deep (up to 20–30 m) as the conventional vertical loop bores.

It will also be necessary to ensure that the structural engineer is happy with the proposal and to co-ordinate the design of the loop with the structural design.

Horizontal loops

Horizontal loops are buried 1–3 m below the surface in a stretched coil (often referred to as a slinky). The advantage of this approach is the simplicity of installation. The ground works contractor will be able to carry out the necessary excavation and backfill operations, thus reducing costs by eliminating a specialist drilling contractor (Fig. 14.14). The system does, however, require large land areas outside the building footprint. Car parks (for retail parks) or playing fields (in schools) offer two opportunities for installing horizontal loops systems.

The heat pump

Water-to-water heat pumps which may be used in ground source systems are described in Chapter 24.

Figure 14.14 Horizontal loop installation (see colour plate section at the end of the book)

System design

The balanced system

It is important to recognise that the ground does not represent and infinite energy source. There is a limit to the rate at which energy can be transferred, and then the ground must be allowed time to dissipate the effects of the input/extraction of this energy.

The ideal use of a GSHP system is where the heating and cooling loads match each other. In this way the heat extracted from the ground during the winter months can be put back into the ground during the summer.

Where the ground is to be used for heating or cooling only then the expected yield will be lower due to local heating or cooling and the spacing between the sections of the ground loop will need to be wider apart to avoid interference between the different parts of the loop.

Open loop systems or closed loop systems where there is considerable ground water flow are best suited to single service applications or where an imbalance exists between heating and cooling.

When developing the design of the GSHP system it is important to make good estimates of the heating and cooling loads to be met by the heat pump system and to aim to match these loads where possible. Specialist contractors will usually ask for this information, often on a monthly basis.

Matching building services

The efficiency of the GSHP is dependent upon the temperature differences that the heat pump is required to work against. With the heat source/sink typically being at around 12°C, cooling to conventional chiller water temperatures of 6/12°C is possible at good efficiencies. In heating mode, however, there will be a limit to the temperature at which heat can be delivered to the building. The limiting temperature is around 50–55°C, but in practice lower continuous operating temperatures are required to benefit from high efficiencies. More typically the building heating will need to operate in at 45°C flow/38°C return, or lower if possible.

GSHP therefore best suit low temperature heating applications such as underfloor heating, but can be used where other heating systems can be suitably adjusted to operate at the reduced system temperatures.

GSHP systems can also be used to preheat domestic hot water services. The heat pump system raises the incoming water from say 10°C to around 35–40°C and a conventional heat source can then be used to raise the water to its desired storage temperature of 60°C.

System sizing

When sizing the ground source system it is important to consider not just the peak demands to be met but the actual delivered energy required. It should be remembered that a system sized to meet 30–50 per cent of the peak load may well deliver 80 per cent or more of the annual energy demand.

Limitations in access to the ground may limit the peak loads that the GSHP system can achieve. Alternatively an imbalance in heating and cooling demand, irrespective of the peak demands, may limit the use of a GSHP system.

Open loop peak capacities are determined by the anticipated extraction rate and the allowable temperature rise of the water. If the water is to be returned to the aquifer then the temperature rise is likely to be limited to around 3–5°C. The Environment Agency will usually look to limit ground water temperature to less than 20°C but this does not necessarily mean that return water temperatures will be allowed this high. Extract and return wells need to be well spaced (around 50 m apart) to avoid the risk of recirculation of water from return to extract. Where multiple wells are required to meet the expected demand these must be placed sufficiently far apart to avoid interference.

Closed loop systems typically yield around 50 W/m length of buried pipe. However, this is dependent on the ground type and the existence of any water movement through the ground. Energy piles can have similar yields but can be lower for large diameter piles or piled walls. In order to prevent interference between each part of the ground loop vertical bores should be at least 6 m apart for well balanced systems and 10 m or more apart where systems are either imbalanced or used for heating or cooling only.

Environmental benefits

Unlike other technologies classed as renewable, GSHP require the use of electricity. Environmental benefits, in the form of reductions in CO_2 emissions are therefore not as straight forward to calculate as other types of renewable system. Unless a renewable source of electricity is provided there will be emissions associated with the heat pump compressor and any pumps used for the ground loop. These must be offset against the reduced emissions resulting from the provision of heating and cooling.

An estimate of the reductions in CO_2 achieved can be provided from:

(Emissions factor of boiler fuel × Displaced heat / Boiler efficiency) +
(Emissions factor of electricity × Displaced cooling / Conventional chiller CoP) –
(Emissions factor of electricity × Displaced heat / Heat pump heating CoP) –
(Emissions factor of electricity × Displace cooling / Heat pump cooling CoP) –
(Emissions factor of electricity × Ground loop pump energy)

Taking:

- Standard CO_2 emission factors used in Approved Document L2A (see Chapter 3) for gas of 0.194 kg/kWh and for electricity of 0.422 kg/kWh.
- Conventional boiler seasonal efficiency of 86 per cent.
- Conventional chiller seasonal CoP of 2.5.
- Heat pump CoP in heating mode of 3.0.
- Heat pump CoP in cooling mode of 5.0.
- The heating load met by the heat pump as H kWh pa.
- The cooling load met by the heat pump as C kWh pa.
- The ground loop pump energy as P kWh pa.

The above equation can be simplified to:

$$CO_2 \text{ savings} = (0.194 \times H / 0.86) + (0.422 \times C / 2.5) - (0.422 \times H / 3) - (0.422 \times C / 5) - (0.422 \times P)$$

Integration into building

GSHPs can be readily integrated into most non-domestic buildings. It should be remembered that even if the peak heating and cooling loads cannot be met it may still be possible to derive benefit from GSHP systems for part of the load, or for specific circuits.

Residential applications are mainly for heating only. As such, large communal systems are not very suitable for ground source. While space heating demand may be met from the ground source, an additional heat source will be required for the hot water supply system to achieve the desired storage temperature of 60°C. This may need to be installed in each dwelling and thus offset some of the benefits from having a central system. GSHP are more applicable to individual dwellings, particularly in areas where mains gas is not available for heating and hot water generation.

Compatibility with other systems

Due to the operating temperatures of GSHP systems, they are not readily integrated with systems that need to operate at high temperatures such as biomass boilers. However, combining GSHP with systems capable of operating at low temperatures, such as condensing boilers or solar water systems, can be very effective.

There are few restrictions from the cooling point of view as the GSHP works well at convention cooling temperatures.

Photovoltaics

PV systems convert energy from the sun into electricity through semi-conductor cells. Systems consist of semi-conductor cells connected together and mounted into modules. The PV modules produce direct current (DC), which can either be used directly by dedicated DC equipment or converted to alternating current (AC) via an inverter for connection to the building electrical distribution system. The latter is the more usual approach for building integrated PV systems as it allows the power generated to be used by a number of standard pieces of equipment or exported from the building if there is no demand.

As with the solar thermal technologies, in the UK PV works most effectively when orientated due south and at an incline of 30° to the horizontal. Orientations within 45° of south are acceptable as are tilts between 10° and 50°. It is essential that the system is unshaded, as even a small shadow may significantly reduce output. Figure 14.15 shows how PV output varies depending on panel orientation and pitch, the optimum being taken as 100 per cent.

PVs are available in a number of forms including monocrystalline, polycrystalline, amorphous silicon (thin film) or hybrid panels that are mounted on or integrated into the roof or facades of buildings. Monocrystalline cells are the most efficient (up to 20 per cent), with polycrystalline being the next most (around 12 per cent) and amorphous being the least efficient (around 6 per cent). Costs reflect these efficiencies, with monocrystalline having the highest per square metre cost and the amorphous having the lowest. Amorphous PV has advantages of generating more power at low sun angles and being less affected by overshading of part of a panel then the crystalline systems. The hybrid systems have been developed to take advantage of the characteristics of both monocrystalline and amorphous panels to increase the energy generated over the year.

PV system size is measured in kWp (peak). This is a nominal output rating that a given installation will produce under test conditions. It should not be taken as the continuous output rating of the system as the output will usually be much lower as actual solar radiation levels are not always be as high as the test conditions. The

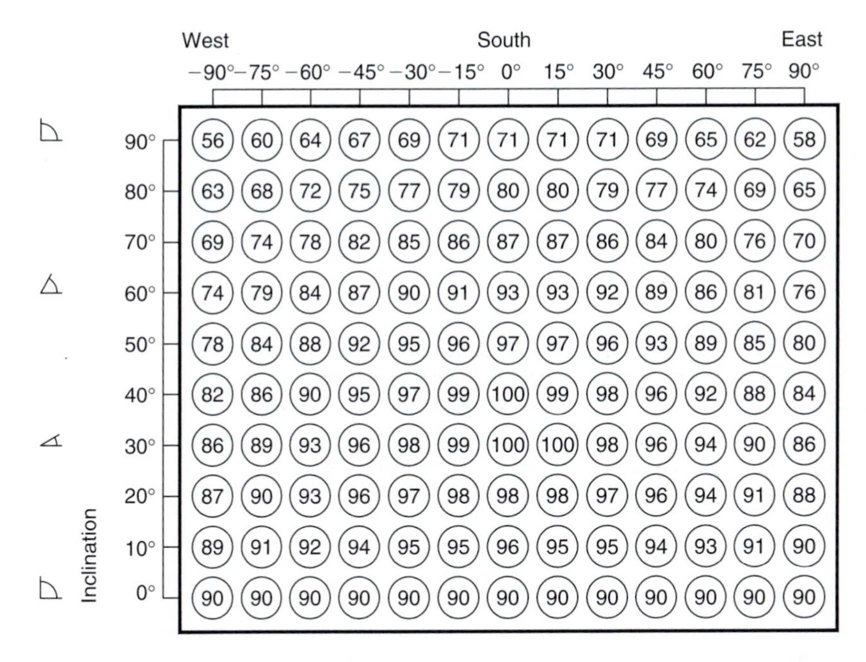

Figure 14.15 Effect of orientation and pitch of PV output (Solar Century)

differences in efficiency mentioned above mean that each of the systems will require a different area to generate a given kWp. However, the cost per kWp will remain nearly the same regardless of the type of PV chosen.

When sizing a PV installation it is usual to aim to generate a specific amount of energy over the year. The choice of technology will then depend on the building design. Where there are large areas where PV can be installed then the amorphous products may offer a good solution. Where space is at a premium then monocrystalline or hybrid products may offer the only way of meeting the target energy production.

A range of products are available from standard panels, through to roof slates and metal cladding and roofing systems as illustrated in Fig. 14.16.

Wind turbines

Wind energy can be one of the most cost effective methods of renewable power generation, producing electricity without carbon dioxide emissions with outputs ranging from 500 W to 2 MW. As with PV systems, turbines will rarely achieve thier nominal output.

Turbine output is related to the wind speed squared and so lower than anticipated wind speeds will have a significant effect on the power generated. Location, height and obstructions to air flow will all affect the wind speed that a turbine sees. Open or exposed areas generally offer higher wind speeds than built-up areas, while wind speed also increases with height.

For large scale turbines it is essential to gather actual wind data for the location and turbine hub height before proceeding with design and purchase of the system. At least 6 months on-site data gathering is recommended, but it is more preferable to have a year's data.

Smaller turbines may not justify the expense and time necessary for carrying out such monitoring. In these cases an estimate of the potential wind available for the site can be obtained from sources such as the computer generated Noabl database available on the internet or wind rose data available from the Meteorological Office and in CIBSE Guide A.

Figure 14.16 Examples of PV (Solar Century)

Turbines can be building mounted or stand alone. In both cases there are a number of common issues that need to be considered before the technology is implemented.

- *Visual impact*: Turbines are highly visible and an early approach to planning authorities is recommended.
- *Noise*: Often believed to be a major problem for planning purposes, noise issues are reducing with better design. Small scale turbines (<6 kW) are not generally a problem as they are usually designed for integration with buildings. Acoustic performance for larger turbines has improved considerably in recent years with gearless generators and better blade design.

Figure 14.17 Examples of building integrated turbines

Building integrated wind turbines

Small turbines of 0.5 up to 6 kW can be mounted on buildings and there are a number of manufactures geared up to this market. Figure 14.17 shows examples of some of these turbines.

The small scale or micro turbines have a diameter of around 2 metres and typically require mounting above roof level to take advantage of the increased wind speeds. Outputs from building mounted turbines can vary over a wide range depending on the local conditions. Turbulence around buildings can have an adverse effect on turbine output and make it difficult to predict the exact output for an urban installation. Where turbines are to be mounted on tall buildings there may be an enhanced output due to the increase wind speeds at height, especially if the building is not surrounded by other tall buildings. At the time of writing there is a monitoring programme in place to gather data on a number of micro turbine installation to confirm the actual energy outputs achieved.

Structural issues must also be addressed. As well as supporting the loads imposed by the turbine, it will also be necessary to ensure that structural born noise or vibration does not become a problem.

Stand-alone turbines

Larger turbines (see Fig. 14.18) will need to be installed on land adjacent to the building(s) they are to serve. It is rare for very large scale turbines to be installed in this way, but there are examples of turbines up to 1 MW being installed specifically to serve a given building or development. The most common examples are for retail distribution warehouses or industrial parks.

Figure 14.18 Examples of stand-alone turbines

In addition to the noise and visual impact mentioned above, stroboscopic effects from large scale turbines need to be considered. This is due to sunlight glinting off the blades as they rotate.

For very large turbines (500 kW+) it may not be possible to build any type of residential building within 400 m of the turbine.

Chapter 15

Combined heat and power

Introduction

Combined heat and power (CHP), also known as total power or co-generation, is the simultaneous generation of heat and electrical power from the same system. The concept has been in existence since the late 1800s and has been applied to district wide energy systems, industry and buildings. Of these applications this chapter will focus on district and building applications.

Heat generated by conventional boiler plant is obtained at a reasonably high thermal efficiency of 80–90 per cent based on the high or gross calorific value (see Chapter 17 for full description). However, the generation of electrical power is achieved at much lower efficiencies, with the national grid delivering power to the end user at an efficiency of 35–40 per cent (depending on the generating technology mix). CHP makes use of the 'waste' heat resulting from electrical generation, thus increasing the overall thermal efficiency of the process typically to 75–85 per cent. Figure 15.1 illustrates the reductions in overall fuel consumption and the resulting reductions in CO_2 emissions.

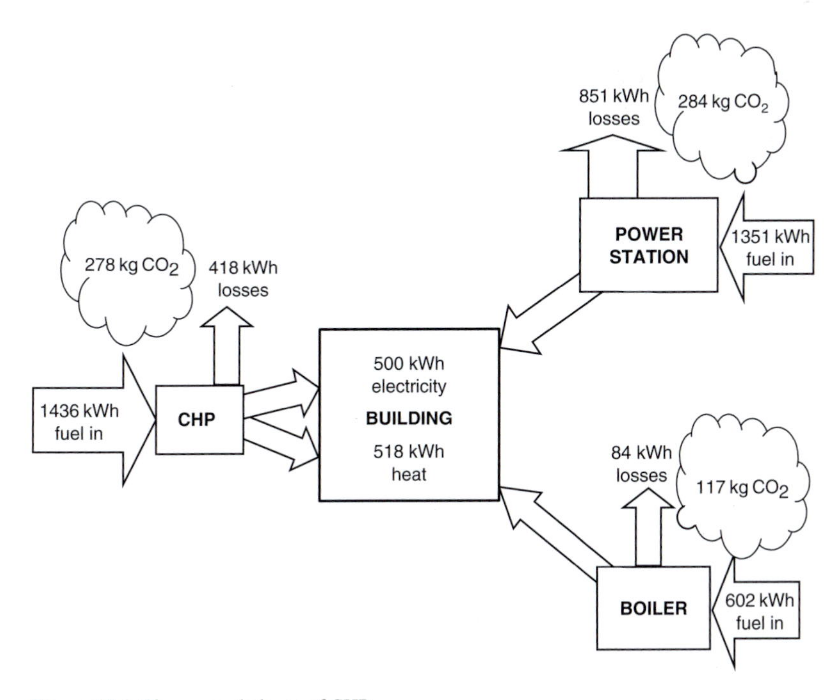

Figure 15.1 The energy balance of CHP

This improved efficiency results in reduced fuel costs and environmental benefits due to reduced CO_2 emissions. However, the improved efficiency comes at a cost in complexity and hence capital and maintenance outlay required to buy and operate the plant.

In the past, CHP plant has mainly been installed to provide a more cost-effective solution to delivering heat and power. Under these circumstances it is necessary to run the plant for long periods throughout the year to generate the cost savings necessary to justify the higher capital outlay. Typically, CHP needs to operate for 4000–4500 hours per year to meet a purely financial case unless there are some special factors such as a low-cost source of fuel.

More recently the improved environmental performance of CHP has been the main focus of interest in the technology. At the time of writing, government has set targets for installed CHP capacity and local planning policies are frequently requiring the assessment of the suitability of CHP for major developments (refer to Chapter 2) with the aim of reducing CO_2 emissions. Where environmental concerns are the main driver then CHP may offer a very cost-effective solution to reducing CO_2 emissions compared to alternative options, even when not meeting a conventional financial test for viability. Under these circumstances, reduced operating hours may be acceptable, provided that the required environmental benefits are achieved, and usually providing there is at least a small positive operating cost benefit.

Basic components

A CHP installation will necessarily comprise four fundamental items of plant:

- A prime mover
- An electrical generator/alternator
- Heat recovery equipment
- Controls

To these essential items must be added a number of ancillaries which will include:

- Exhaust stack
- Acoustic attenuation

Packaged plant

For most building applications, whether single building or groups connected to a district distribution system, packaged CHP plant provides the most common and convenient solution. Suppliers will provide units covering a wide range of outputs from 50 kWe up to 3 MWe as standard packages. In all cases the prime mover will be matched with the generator, heat recovery equipment and controls before delivery to site. In most cases these components, together with the necessary ancillary equipment, will be integrated in the factory and delivered to site pre-tested and ready for connection to the heating and electrical systems. Figure 15.2 illustrates a typical 200 kWe packaged installation. The advantage of these packaged units is the improved reliability achieved and the reduced installation and commissioning time required on site.

Prime mover

The main prime mover used in building applications is the internal combustion engine. Gas turbines compete for larger applications above 2 MWe, while the more recent micro-turbine technology is an alternative that has established itself at the smaller end of the market (100 kWe).

Other technologies are either emerging, such as the fuel cell, or have a limited application to buildings, such as the steam turbine. These technologies are dealt with later in this chapter.

Internal combustion engine

The most common internal combustion engine used in packaged CHP units is the spark ignition engine operated on mains gas, with modern engines being specifically designed to operate on this fuel. Compression ignition engines are also readily available where diesel is the chosen fuel. In both cases the emphasis on

Figure 15.2 200 kWe packaged CHP unit (Ener-G) (see colour plate section at the end of the book)

environmental benefits means that bio-gases and bio-diesel can also be used as fuels where the appropriate energy content can be delivered.

The electrical efficiency of the internal combustion driven CHP units is typically in the range 30–40 per cent with heat to power ratios of 1.1–1.5:1. Heat is usually delivered at 70–85°C but can be raised to 110–120°C for special applications. Where heat is required at elevated temperatures then there is usually a reduction in the overall heat output unless a use can be found for the remaining low-grade heat (see Fig. 15.3).

Gas turbines

Gas turbines driven CHP by contrast deliver lower electrical efficiencies, typically 25–30 per cent, with slightly higher heat to power ratios at 1.5–2.0:1. The heat can also be more readily obtained at higher temperatures if required.

Micro-turbines are also now available on the market but have a much shorter operating history than either large-scale turbines or internal combustion engines. They tend to have similar electrical characteristics to their larger cousins, but deliver their heat at similar temperatures to the internal combustion engine, with heat to power ratios of between 1.5 and 2.5:1.

Generator

The detailed design of generators does not form part of this publication. However, it is important to note the two types of generator that can be used in CHP applications.

Synchronous

Synchronous alternators use battery start and are suitable for standby generators. They are more complex and therefore a more costly option but require no power factor correction.

Asynchronous

Asynchronous units use the mains as the excitation current and are not, therefore, able to operate as standby generators. Although they are simpler than synchronous units and therefore cheaper they require power factor correction.

Heat recovery

There are three sources of heat from internal combustion engines, the engine cooling jacket, oil and exhaust gases. Where units are turbocharged then the intercooler can provide a fourth. Each of these sources will deliver heat at a different temperature and the specification of the temperature of the heat to be delivered to the building services will determine from which heat is collected and from which heat is rejected. Typically heat is collected from the engine water jacket and from the exhaust gases via a pumped water circuit, with other heat being rejected. The heat recovered from the engine is then delivered to the building services, typically at 70–85°C via a heat exchanger to maintain hydraulic separation. Figure 15.3 illustrates a typical heat recovery arrangement.

For special applications requiring higher temperatures heat can be recovered from the exhaust only. However, unless there is a use for the lower-grade heat from the engine jacket then this heat will have to be rejected, thus reducing the overall amount of heat recovered.

It is possible to increase the heat recovered by including a condensing heat exchanger in the exhaust. However, it will be necessary to ensure that the return side from the building services is at a low enough temperature to cause condensing for this approach to be fully realised.

Heat rejection circuits are included in Fig. 15.3. The actual level of heat rejection capacity provided will depend on the intended operation of the plant and to some extent the supplier's preference for protecting the

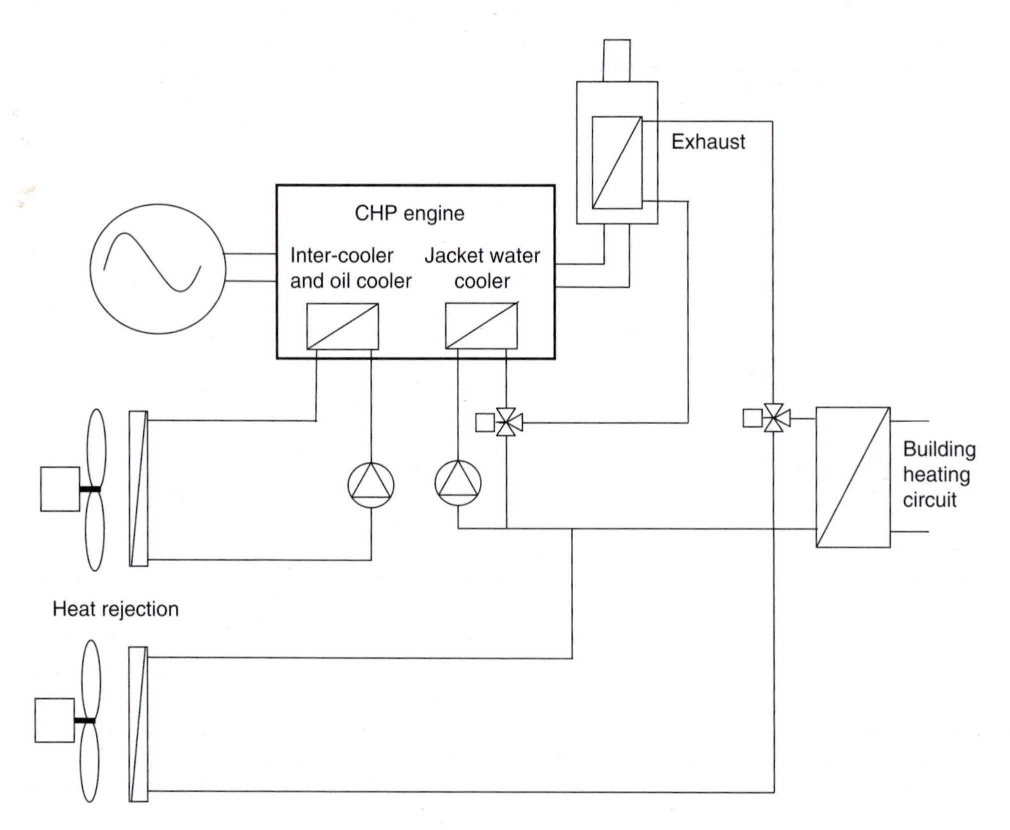

Figure 15.3 Heat recovery arrangement for gas engine CHP unit

CHP plant. Where it is envisaged that the CHP unit may be used to generate electrical power when full use cannot be made of the heat generated then the heat rejection system needs to be able to cope with the anticipated level of heat rejection. Alternatively where the CHP unit will be heat lead with no anticipated heat rejection then there may only be a requirement to provide sufficient heat rejection capacity to protect the unit under abnormal shut-down.

Heat recovery from gas turbines is from exhaust gases, which will be between 450°C and 550°C at full load. It is therefore possible to generate medium-temperature hot water at 110–120°C (refer to Chapter 7) should it be required with little loss of overall heat output.

Controls

The control and instrumentation systems have three main functions:

(1) To integrate the heating and electrical outputs with the building systems; stop/start the engine, modulate output, connect/disconnect the generator and synchronisation.
(2) Monitor performance.
(3) Safety; reverting to predetermined safe conditions on any failure.

Modern systems are usually microprocessor based, designed to communicate with a remote control point, typically in the CHP supplier's office.

Acoustic attenuation

Internal combustion engines and gas turbines generate noise and vibration that need to be attenuated where CHP is to be integrated into buildings. Packaged units typically come in attenuated enclosures, but care is still required when installing and integrating the units into the building. Manufacture's advice should be closely followed and if necessary independent specialist advice sought.

The application of combined heat and power in buildings

Any building or groups of buildings that have extended hours of use and a coincidental demand for heat and power is worth considering as an application for CHP. Experience has shown that swimming pools, hotels, leisure facilities, hospitals, residential homes, universities and communal heating are all good applications for CHP. However, almost any building can be considered under the correct circumstances.

Load profiles

As previously stated, CHP systems need to operate for as long a period throughout the year as possible. The conventional approach to plant sizing based on the peak demand will therefore not be suitable as the plant will be under utilised. In the case of CHP it is necessary to look rather at the minimum loads and the building load profiles through an annual cycle to identify the optimum solution. While well-tested procedures exist to determine maximum heat loads (refer to Chapter 4) and electric loads, determine the minimum loads is not as straightforward.

For an existing building it may be possible to obtain energy consumption data, ideally specifically for those loads to be served by the CHP. Such data may well be provided as half hourly profiles for electricity, but gas is more likely to be provided on a monthly basis. Monthly data can be acceptable for an initial assessment, but final plant sizing will require more detailed information. This may only be provided by carrying out direct monitoring over a representative period to enable full understanding of the load profiles.

New buildings will have no historic data for the designer to draw on. In such cases it may be possible to consider buildings of a similar type and design, or to consult published benchmark data. Both these sources of data need to be treated with caution as such data is rarely directly applicable. It also does not usually offer

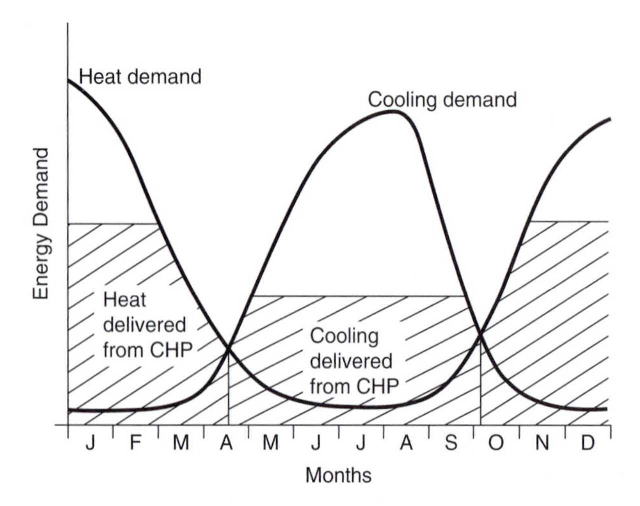

Figure 15.4 Extending CHP operating hours by using heat for cooling

the level of detail required for accurate plant sizing. The increased use of thermal modelling as described in Chapter 6 offers another source of information. It should be noted, however, that models developed to show compliance with Part L of the Building Regulations may not be the most appropriate for designing CHP plant as they are intended for comparison purposes only.

Extending heat demand

The main sources of heat demand in buildings will be from space heating, for domestic hot water (DHW), and in some cases from swimming pools. DHW and swimming pools offer a year-round heat demand and CHP units can usually be sized to meet such demands. Space heating, however, only offers limited scope of CHP unless the building is occupied for very long hours in the winter months. Where there is a summer cooling demand, absorption chillers, fed by heat from the CHP unit, can be used to extend the heat demand of a building by providing a heat demand in the summer when the space heating is not required. This type of system is typically referred to as combined cooling, heat and power (CCHP) or tri-generation. It is therefore advisable to identify the building cooling loads separately so that a decision can be made regarding the inclusion of absorption cooling. CCHP can be particularly effective for office or government buildings. Figure 15.4 illustrates the principle of using CCHP to extend the operating hours of CHP.

Combinations of different types of buildings can also offer good opportunities for CHP when served by a central or district energy system. Different occupancy patterns mean that as the demand in one type of building reduces it increases in another, hence providing long overall operating hours. Figure 15.5 illustrates the principle of combining different building types to extend the operating hours of CHP.

Plant sizing

When determining the size of a CHP unit the aim is to develop a system that will allow the CHP plant to operate at its maximum electrical rating for as many hours as possible, while all available heat is utilised within the building.

While sizing the CHP system to meet the minimum building demands will lead to the longest running hours, it may not lead to the optimum economic or environmental benefits. Gas-fired engines can often be operated at electrical outputs as low as 25 per cent of their maximum continuous rated output. This feature can increase the size of CHP installed by allowing the plant to be run at reduced output for short periods when demands fall below the rated output. However, it is not usually economic to operate at low outputs for long periods. An alternative approach could be to operate the CHP for a reduced number of hours during times of the year with low loads, provided it is possible to operate for long hours during other times of the year.

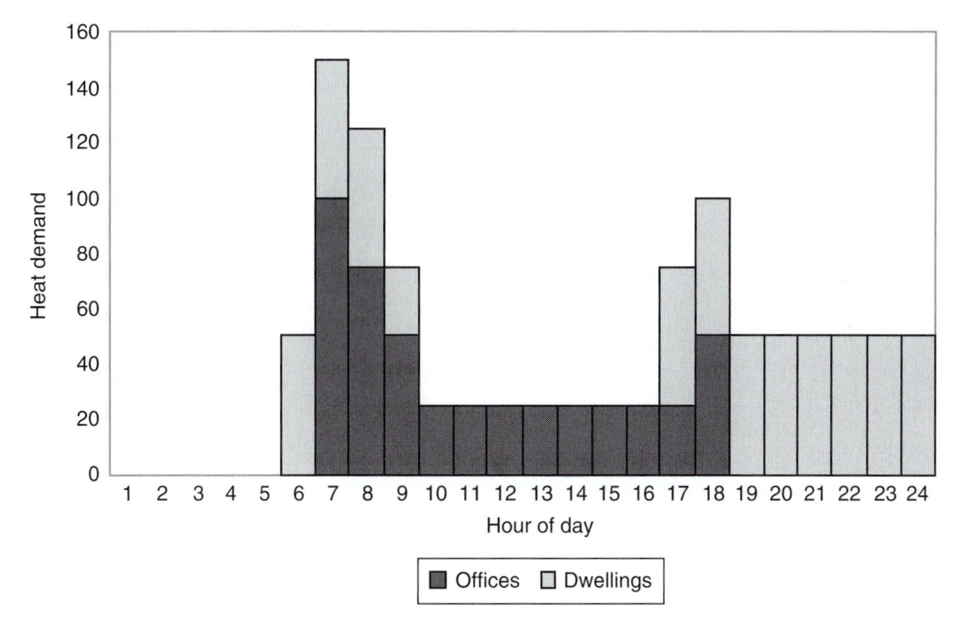

Figure 15.5 Mixed used development extending CHP hours of operation

More than one CHP unit could be installed where, say, one unit meets the summer demand or base load and a second unit is sized such that the combined output meets a higher winter demand. Such arrangements can increase to operating time and maximise the energy and environmental benefits from the CHP. However, there will be a penalty in terms of increased capital cost, maintenance cost and plant space requirement.

Thermal stores

The inclusion of a thermal store can be very beneficial where there are significant variations in loads across a day. Such a thermal store will even out the fluctuations between heat loads and CHP output. Thus, the full load operating hours of the CHP unit and the proportion of the heat demand met by the CHP can be increased. Figure 15.6 illustrates the principles of using a thermal store to increase the usable heat from CHP.

The sizing of the thermal store will depend on the period of time over which variations are to be smoothed and the size of the CHP unit. In the case of community heating a thermal store could be sized to meet a full days DHW demand.

Sizing for heat or power

The heat to power ratio will also determine the design of the CHP. CIBSE suggests in AM12 that where the ratio between the chosen heat and power loads is equal or greater than 1.5 then the CHP unit is designed to meet the electrical demand. This is because at this ratio between heat to power, most CHP units will generate less than the building heat load regardless of the electrical load. Conversely where the heat to power ratio of the building is 1.3 or less then the CHP plant should be sized on the building heat load.

Integration into buildings

The successful operation of the CHP will not only depend on the correct sizing of the unit, but also requires careful integration with the building services. This not only requires integration of the generator output with

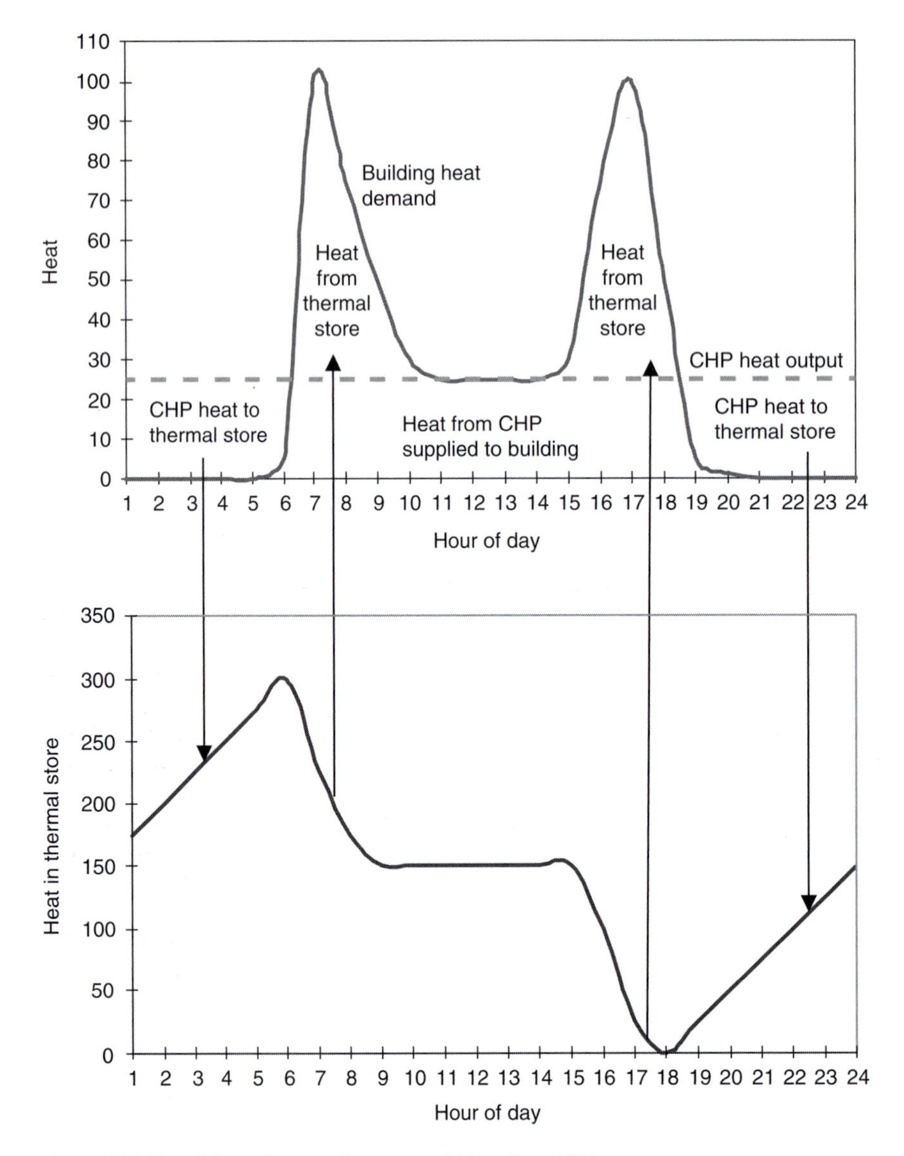

Figure 15.6 Use of thermal store to increase useful heat from CHP

the electrical distribution system but also the integration of the heat recovery circuit(s) with the building heat distribution system and possibly an absorption chiller.

Integration with heating distribution circuits

The heat recovery circuit(s) from the CHP can be connected in either series or parallel to other heat generators such as boilers. CIBSE AM12 suggests that if the building system maximum flow temperature is less than or equal to the CHP maximum delivery temperature then the CHP should be connected in parallel, otherwise the CHP should be connected in series. In either arrangement it is necessary to provide the CHP unit with a separate circuit pump to ensure that the unit receives the required water flow when it is operating.

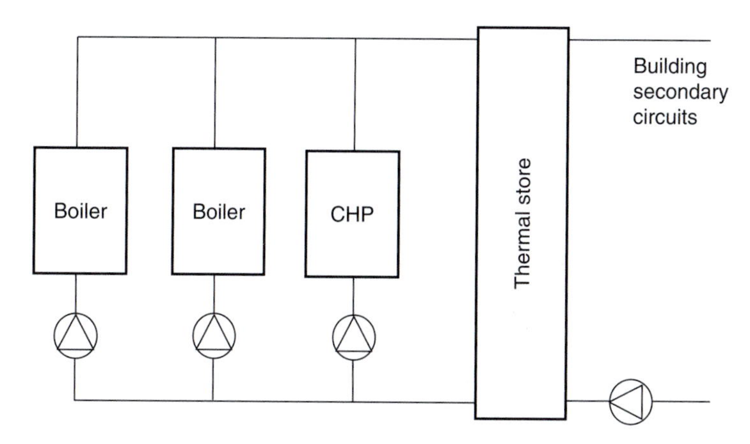

Figure 15.7 CHP unit connected in parallel

Connection in parallel

Figure 15.7 shows a simplified schematic for a CHP unit connected in parallel with two boilers and a thermal store. This arrangement is simple to control, allowing heat to be put into or drawn out of the store as necessary with limited need for heat exchangers and resultant losses.

The choice of building system temperatures in such an arrangement will have a significant impact of the physical size of the thermal store and on the system pumping energy. Where possible, designing the building system to provide low return water temperatures will not only help to reduce the size of the thermal store by increasing the system temperature difference, but will also encourage condensing, thus increasing plant efficiency.

It is important that the CHP is designed to cope with the range of return temperatures it may encounter in such an arrangement. Where the CHP unit has been designed to deliver heat at 80°C based on a return temperature of 70°C then any lowering of the return temperature will lead to a lowering of the resultant flow temperature.

In some cases it may be desirable to make use of condensing, but not possible to deliver all the return water at condensing temperatures due to restrictions on the building heating services. In such cases, where a separate condensing circuit is installed, a condensing heat exchanger in the CHP exhaust should also be connected into the circuit (Fig. 15.8).

Connection in series

Figure 15.9 shows a simplified schematic for a CHP unit connected in series with two boilers. In this type of arrangement it is less likely that a thermal store will be required. However, it would be possible to install a thermal store in parallel to the CHP unit such that the CHP and thermal store are both in series with the conventional boiler plant.

Integration with cooling systems

The single pass absorption chiller typically used in CCHP installations is described in Chapter 24. While Fig. 24.8 illustrates an absorption chiller using steam as the heat input medium, it is possible to operate this type of chiller successfully using water at 80–90°C. This means that CHP plant can be linked to absorption chillers to make use of the available heat during the summer months when there is a cooling requirement.

Many suppliers of CHP plant are now offering to also provide integrated CCHP systems where the CHP provider has pre-matched a CHP unit to an absorption chiller. For such systems based on a gas internal combustion engine, the cooling output will typically be close to but less than the rated electrical output.

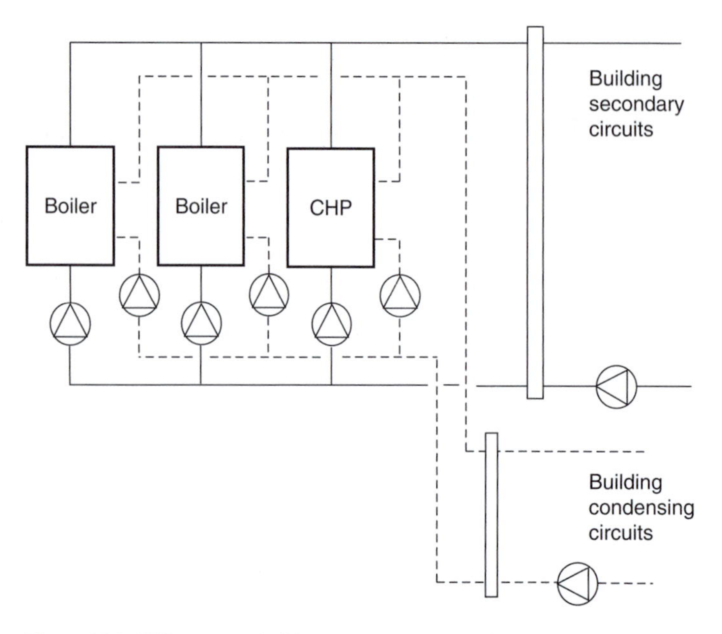

Figure 15.8 CHP connected with separate condensing circuit

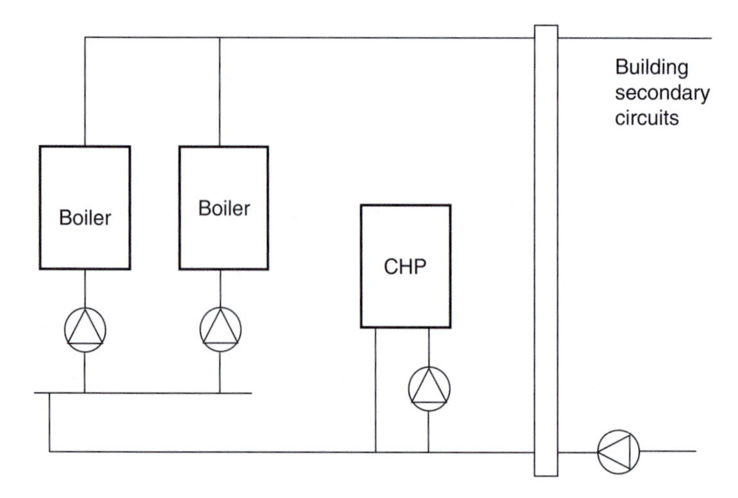

Figure 15.9 CHP unit connected in series

It should be noted that while CCHP will extend the usable hours of the CHP there will usually be higher savings in CO_2 and fuel costs when the heat from the CHP is used directly as heat in the building rather than for driving the absorption chiller. This is due to the differing efficiencies of a boiler and a vapour compression chiller. This difference is becoming even more marked with the advent of high-efficiency chillers with coefficient of performances (CoPs) in the order of 4–6. By way of an example assuming 1000 kWh of heat is available from a CHP unit at a nominal 85°C.

If this heat is used to displace space heating then the resulting boiler gas displaced can be calculated as:

$$1000 / 0.86 = 1163\,\text{kWh}$$

Assuming a gas unit price of 1.8 p/kWh the cost boiler gas costs displaced will be:

$$1163 \times 0.018 = £21$$

Taking the Building Regulations CO_2 emissions factor for gas of 0.194 kg CO_2/kWh the displaced CO_2 emissions will be:

$$1163 \times 0.194 = 225\,\text{kg}\,CO_2$$

Alternatively the CHP heat available can be used to generate cooling via an absorption chiller, displacing electrically powered vapour compression cooling.

Assuming the absorption chiller has a CoP of 0.7 the cooling generated will be:

$$1000 \times 0.7 = 700\,\text{kWh}$$

Taking the CoP of the displaced vapour compression chiller as 2.5, the electricity displaced can be calculated as:

$$700 / 2.5 = 280\,\text{kWh}$$

Assuming an electrical unit price of 6 p/kWh the cost of electricity displaced will be:

$$280 \times 0.06 = £17$$

Taking the Building Regulations CO_2 emissions factor for electricity of 0.422 kg CO_2/kWh the displaced CO_2 emissions will be:

$$280 \times 0.422 = 118\,\text{kg}\,CO_2$$

Integration with electrical systems

CHP units in the size range normally met in buildings will generate electrical power at 400 V three phase. Where the unit is to operate *in parallel with mains* supply then permission will need to be sought from the local electrical supply company for connection of a generator to the network. There will also be a requirement to meet with the necessary safety standards covered under the Electrical Contractors Association publication G59/1.

When determining the connection point it will be necessary to consider whether the system is to operate in *island mode* (i.e. disconnected from the mains supply) should the mains electrical supply fail. If the CHP unit is to act as a standby unit in island mode then care must be taken to ensure that the connected loads will not exceed the rated output of the unit.

Emergent technologies

In recent years a number of new technologies have emerged. At the time of writing most are under development, but while some technologies are at the field trial/demonstration stage, others have market ready products. These emergent technologies may offer an increased range of opportunities for integrating CHP into buildings; offering a wider range of sizes and heat to power ratios. The following paragraphs give a brief description of three such groups of technologies.

Fuel cells

Fuel cells electrochemically combine hydrogen and oxygen to produce electricity, heat and water. Originally developed for use on space missions, fuel cells have generated considerable interest in recent years as a pollution free, reliable and quiet solution to power generation. There are a limited number of commercial units on the market, but those that are available can demonstrate significant operating experience across the world.

Fuel cells are generally split into five different types, depending upon the type of electrolyte used.

Proton exchange membrane

Proton exchange membrane (PEM) fuel cells use a moist polymer membrane and are sometimes referred to as solid polymer fuel cells. The advantages of this type of cell include its high-power density, low weight, relatively low operating temperatures (around 90°C), quick start-up at full power capacity and quick adjustment to variable power demands. These advantages make this technology very attractive to the automotive industry as well as for CHP and distributed power applications. The main disadvantages of this technology include high costs and difficulties with maintenance.

Phosphoric acid fuel cells

Technically well developed, phosphoric acid fuel cells (PAFCs) are the first commercially available type of fuel cell. PAFCs are a candidate for use in a wide range of applications. However, their relatively high operating temperature (around 200°C) may favour CHP and distributed power/uninterruptible power supply applications. Electrical efficiency is typically around 40 per cent.

Alkaline fuel cells

Alkaline fuel cells (AFCs) are relatively simple, operate at low temperatures (around 70°C) and have been extensively used in the space industry. Commercial development of this technology is well underway, oriented towards transport applications. This type of cell can achieve power efficiencies of up to 70 per cent.

Molten carbonate fuel cells

This type of cell uses a molten alkali metal carbonate as the electrolyte and operates at high temperatures (around 650°C). Therefore, they are a candidate for large-scale stationary power and CHP applications. Demonstration projects in the order of 2 MWe (electrical output) have already been carried out. Advantages of operating at such high temperatures include the wide range of fuels that can be used and higher overall system efficiencies compared to PAFC systems.

Solid oxide fuel cells

This type of fuel cell uses solid, non-porous metal oxide electrolytes and operates at high temperatures (around 1000°C). They have similar advantages and uses to molten carbonate cells. Power generation efficiencies could reach 60 per cent for this type of cell.

Commercial systems

At the time of publication the commercial fuel cells systems do not usually use hydrogen and oxygen gases directly. Reformers will use a variety of hydrocarbon fuels to provide hydrogen rich gas to the fuel cell, while oxygen will be taken directly from air. This process will produce some carbon emissions, but emission levels will be very small compared to even the best combustion process.

Current commercial CHP/power units are around several hundred kilowatts electrical output with electrical efficiencies around 40 per cent and heat to power ratios of around 1:1.

Stirling engines

Stirling engines are predominantly being developed for the domestic market. Their more simple mechanics is expected to lead to better and cheaper reliability at these small outputs compared to conventional internal combustion engines. Typical electrical outputs are in the range 0.5–3 kW at an electrical efficiency of between 15 and 25 per cent. Heat to power ratios vary in the range 2.5:1 up to 4.5:1.

This type of CHP unit is anticipated to be used as a direct replacement for a conventional gas-fired domestic boiler. Operation of such units will therefore differ from conventional CHP in that operating hours may be relatively low and electrical production may not match demand. Connection in parallel with the electrical grid will therefore be essential to absorb excess power and to supply any power shortfalls.

Biomass-fired combined heat and power

The use of renewable energy supplies to drive CHP is not new. Gas-fired engines have been adapted to operate using a variety of bio-gases derived from sources such as sewage, land fill, coal mines or anaerobic digestion. More recently biomass fuel CHP systems have also been under development in an attempt to make CHP emissions free on a wider scale and to make the best use of a limited biomass resource.

Large-scale projects using biomass to fire steam turbines are in existence; however, the technologies aimed at building scale CHP systems are at the demonstration phase with very few units installed in real buildings. Three of the technologies under development are briefly described below.

Gasification

This process generates gas from biomass that can be used in a modified gas engine in the same way as other bio-gases are used. Small- to medium-scale units (<1 MW) have been trialled by a number of developers, one of the most well known being the 100 kW unit installed at the BedZed housing development. Gasification under 1000 kWe capacity appears to be problematical with these smaller installations having either very poor availability or being shut-down. Work is continuing, however, into the development of this technology.

Air cycle turbine

This technology uses biomass to heat air that is then expanded through a turbine generating electrical power. The air is then passed into the biomass combustion process while excess heat left in the exhaust gases from the combustion process is collected for use. The heat to power ratio is around 3:1 with an electrical efficiency of around 20 per cent. At the time of publication, a small number of 100 kWe units have been sold, one of which is being used as an operational test bed.

Organic Rankin cycle

This process uses biomass to heat an organic oil, which is then used to drive a turbine. There are commercial units on offer but there are relatively few installations. Electrical efficiencies are in the order of 18 per cent with heat to power ratios in the order 3.5:1.

Chapter 16

Fuels, storage and handling

Except where circumstances require that energy cannot be provided by means other than electricity or gas from a public supply, provision must be made on site to accommodate fuel storage. The correct position for the fuel store can only be decided after taking into account a number of determining factors including aesthetics, reasonable proximity to the boiler plant and convenience of access for deliveries. It is a mistake to agree upon a position, however well suited to the fuel chosen to fire the plant initially, which may then inhibit absolutely any change to an alternative fuel at some time in the future.

Storage capacity

It is usual to provide facilities to store fuel in a sufficient quantity to enable the boiler plant to run at full load for a period of 3 weeks. Except in special cases, such as a hospital unit, the full load figure would not apply to a 24-hour period but to say 12 hours per day. This rule is one of convenience and is related to an arbitrary estimate of duration for exceptionally severe weather, for an industrial dispute or some other cause for cessation of supply.

For plant that is quite critical to an industrial process or to, say, a computer building, and also in rural situations remote from trunk roads, it may be wise to provide storage for a period longer than 3 weeks. In terms of economics, seasonal price differentials may also make it attractive to stock-pile fuel in summer. In the special case of small installations, particularly domestic premises, the fuel supplier may propose a storage capacity suited to his delivery routine and offer a reduced unit cost if this level of storage is available.

Solid fuel

An explanation is provided in the following chapter of the various types of analysis to which solid fuel is subjected. *Approximate* analysis will provide information relevant to the practical aspects of combustion, the most important items being the proportion of *volatile* matter, the *caking qualities* and the properties, etc., of the *ash*. Volatiles, a grouping of tars and gases produced as the fuel decomposes in burning, affect not only the rate of combustion but also the length of flame produced. After the volatiles have been burnt off, the solid residue (less any ash) is known as *the fixed carbon*. The cohesive nature of the fixed carbon is an indication of the caking and swelling properties, which are important since they affect the porosity of a fuel bed to air flow. The ash content has an indirect influence upon the caking properties and upon the formation of clinker.

The principal solid fuel for consideration in this chapter is coal. Characteristics of the various categories (or ranks) of coal which are available, are summarised in Tables 16.1 and 16.2, the details given being from typical analyses. Size nomenclature is given in Table 16.3, the more common grades being singles and smalls. The former is the more consistent as to quality and size mixture and is thus the more conveniently handled by mechanical equipment.

Table 16.1 Properties of coal (proximate analysis)

Rank No.	Volatiles %[a]	Volatiles Type	Caking properties	Air-dried moisture (%)	CV (MJ/kg) Gross	CV (MJ/kg) Net
100	<10	Very low	Non-caking	2	32.4	31.6
201	12	Low	Non-caking[b]	1	33.0	32.2
202	15	Low	Weakly[b]	1	33.2	32.3
204	19	Low	Strongly[b]	1	33.2	32.6
300	25	Medium	Weakly[c]	1	33.2	32.3
301	25	Medium	Strongly[c]	1	32.9	31.8
400	>30	High	Very strongly[c]	2	32.3	31.2
500	>30	High	Strongly[c]	2.5	31.5	30.3
600	>30	High	Medium[c]	4	30.1	29.1
700	>30	High	Weakly[c]	5	29.1	27.9
800	>30	High	Very weakly[c]	8	27.9	26.6
900	>30	High	Non-caking[c]	10	25.4	24.1

Notes
a Average for singles.
b Sub- and semi-bituminous.
c Bituminous.

Table 16.2 Properties of coal (ultimate analysis)

Rank No.	Content (%) Moisture[a]	Ash	Carbon	Hydrogen	Nitrogen	Oxygen	Sulphur
100	3.5	5	84.5	3.0	1.1	1.8	1.1
201	3	5	83.7	3.7	1.3	2.2	1.1
202	3	5	83.4	3.8	1.3	2.4	1.1
204	3	5	83.1	4.1	1.3	2.4	1.1
300	3	5	82.1	4.4	1.4	2.8	1.3
301	3	5	81.0	4.5	1.4	3.8	1.3
400	4	5	78.5	4.7	1.7	4.2	1.9
500	5	5	76.3	4.8	1.6	5.4	1.9
600	6.5	5	73.9	4.7	1.5	6.5	1.9
700	9	5	71.1	4.5	1.5	7.1	1.8
800	11	5	67.8	4.4	1.4	8.5	1.9
900	16	5	63.0	3.9	1.3	9.0	1.8

Note
a As fired.

Table 16.3 Properties of coal (physical data)

Group name	Bulk density[a] (kg/m^3)	Size range (mm) Upper	Size range (mm) Lower	Angle to horizontal (degree) In repose	Hopper sides Steel	Hopper sides Concrete
Trebles		90–64	50–40			
Doubles	580–700	57–45	40–25	40	40	47
Singles		40–25	25–13			
Peas		20–13	13–6			
Grains	700–800	11–6	6–3	55	55	60
Smalls[b]		50–25	None			

Note
a Dry coal loosely packed.
b Smalls are not a graded group.

Solid fuels from sustainable sources are becoming increasingly popular at the present time, under the heading of biomass fuels. These are generally classified as carbon neutral and are discussed later in this chapter and also in Chapter 14.

Delivery

Coal delivery, except to very large industrial users able to take a 400 tonne train load, will normally be by road either by a 5–20 tonne tipper or by a 15–20 tonne special purpose vehicle. The latter will be equipped with either a belt conveyor or with facilities for pneumatic discharge. It should be the aim of delivery methods to avoid degradation and segregation of the fuel, i.e. not to break up the larger pieces and not to separate out the large from the small within the mix.

Delivery by tipper vehicle is more appropriate to larger plants, rated perhaps at above 8 MW although there can be no absolute rule as to minimum rating, and to those which are sited away from occupied buildings. Tipping might be through a road grid into an underground store or into a captive tippler hopper as described later. For smaller plant or one sited near to areas sensitive to noise and dust, totally enclosed pneumatic handling may be preferable.

Storage

Solid fuel may, of course, be stored in the open without cover, and for very large plants in industry this may be the method adopted. In other circumstances however, the site and arrangements selected for storage may very well depend upon the economics of excavation for underground bunkers, the practicability of arranging overground silos, the necessity for mechanised fuel handling and the method of boiler firing to be adopted. An example of the last influence is the use of automatic underfeed stokers, particularly where these are to be of bunker-to-boiler type (see Chapter 13).

It is necessary in designing all types of silo or bunker to take into account the natural angle of repose which the fuel will adopt when heaped and the consequent necessity to slope the sides of the store to encourage movement, both as shown in Table 16.3. When silos or bunkers are constructed on a permanent basis in concrete, either underground or at surface level, it is wise to provide flexibility for a possible future change in use (to house oil tanks for example) by arranging for the structural walls to be vertical and forming the internal sloping faces in falsework.

An overground silo may be built in any of a number of materials and in some cases may be only a semi-permanent structure built up from prefabricated sections. Ferrous sheets of cast iron or mild steel, unless provided with very substantial protection, are not suitable materials for use in the construction of storage for a wet and acidic material such as coal.

Handling

The range of equipment available for conveying coal between a central store and the point of combustion is wide. In the case of a large plant, handling may be dealt with in more than one stage, possibly making use of several different methods. The principal criteria to be met are cleanliness, quietness and amenability to automatic control, all achieved with an acceptable level of efficiency. Traditional methods, using open drag-lines, cables, belts, rollers, buckets and other medieval devices are now acceptable only on remote industrial sites.

A prime requirement for any mechanical handling system is that it should be fully enclosed, a prerequisite that is often incompatible with an adequate level of maintenance. Two methods alone seem to meet with this demand; screw conveyors and what are known as the *en masse* systems. The former operate using the principle of an Archimedean screw, the latter transport the fuel by means of a chain that moves slowly through an enclosed casing. Attached to the sides of the chain are 'flukes', several times wider than the driving links, which take advantage of cohesion within the fuel mass to induce surrounding material to flow as a column.

Alternatives to mechanical means are pneumatic systems of conveying by what are known as either *lean phase* or *dense phase* methods. In the former case, the solids are conveyed at a relatively high velocity (20–30 m/s) in a consistent but dilute fluidised mixture with the transporting air volume exceeding that of the

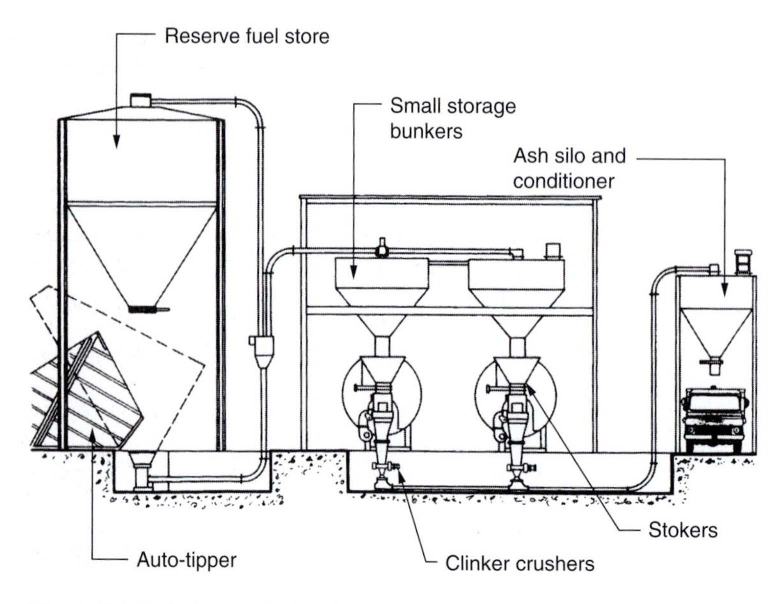

Figure 16.1 Typical composite handling plant for coal and ash

fuel moved by as much as 300:1. Using the dense phase approach, the solids are conveyed in compact 'slugs' at a velocity of only about 3 m/s with an air/fuel ratio of perhaps 25:1. The latter method is to be preferred since it is less demanding of energy and causes less equipment damage by abrasion. An important component in the dense phase system is the shutter or valve which admits the fuel intermittently into the conveying pipeline.

A complete handling system, from fuel intake to stoker hoppers is illustrated in Fig. 16.1. Delivery is to a tipper hopper and thence either direct to service bunkers feeding the stoker hoppers or to an intermediate storage silo. Once fuel has been unloaded into the tippler hopper, it is distributed thereafter within a completely enclosed system under automatic control. The fuel levels in the various bunkers and hoppers are sensed by a variety of probes, either rotating paddle or self-cleaning capacitance type, and the appropriate section of the system is energised. The various components illustrated may be disposed to suit other configurations or methods of delivery, intermediate storage and boiler firing.

Ash and clinker handling

Whilst, strictly speaking, the treatment and handling of ash and clinker falls outside the subject matter of this chapter, it is convenient to deal with it here. It will be noted that the overall arrangement illustrated in Fig. 16.1 includes a pneumatic system of waste disposal, integrated with that for fuel supply. Use of the word 'waste' may perhaps be an error in this context since ash and clinker have a value and may be sold in some circumstances to offset the cost of collection.

Clinker, of course, must be crushed before any sensible action can be taken to deal with its disposal. In some applications, crushing equipment may be incorporated adjacent to each individual boiler but such an arrangement is not always possible. In many respects, a pneumatic lean phase arrangement has advantages as far as handling ash and clinker may be concerned but each boiler plant will have individual characteristics which will dictate the most appropriate system to be adopted. In a fully automated system, final disposal of ash and clinker will be via a cyclone hopper to a suitably enclosed vehicle.

Liquid fuel

Of the two principal types of liquid fuel, petroleum oil and the coal tar series, it is the former only which need concern us here. Crude petroleum oil, a complex mixture of paraffins, aromatics and naphthalenes, is subjected at the refinery to heat treatment and other processes leading to progressive decomposition. The various 'families'

Table 16.4 Properties of fuel oils

Description	Grade/Class							
	A1 (Grade)	*A2*	*C1*	*C2*	*D*	*E*	*F*	*G*
	(Class) Middle distillate	*Middle distillate*	*Kerosene (paraffin)*	*Kerosene*	*Middle distillate (gas oil)*	*Residue (light)*	*Residue (medium)*	*Residue (heavy)*
Density at 15°C (kg/m³)	820–845	820–845	817	803	850	940	970	980
Flash point (closed) (°C)	56	56	43	38	56	66	66	66
Kinematic viscosity (mm²/s)								
at 40°C	2–5.5	2–5.5	–	1.0–2.0	1.5–5.5			
at 100°C	–	–	–	–	–	≤8.2	≤20.0	≤40.0
Maximum pour point (°C)	–	–	–	–	−18	−6	24	30
Calorific value (MJ/kg)								
gross	45.5–46.8	45.5–46.8	45.2	46.4	45.5	42.5	41.8	42.7
net				43.6	42.7	40.1	39.5	40.3
Impurities (% mass)								
sulphur	–	0.2[a]	0.04	0.2	0.2[a]	1.0	1.0	1.0
ash				–	0.01	0.10	0.10	0.15
Temperature (°C)								
storage (minimum)						10	25	35
handling (minimum)						10	30	45
Primary application	Automotive diesel fuel	Agricultural engine fuel	Flue-less heating appliances	Vaporising or atomising domestic heating appliances	Atomising burners for domestic, commercial or industrial applications	Atomising burners for boilers or certain industrial engines which may require pre-treatment or additives		

Note
a From 1 January 2008 the sulphur content of Class A2 and Class D fuels is to be reduced to 0.1% (BS 2869: 2006).

of components are segregated and subsequently re-grouped into commercially usable products, bitumen being the ultimate residual. The resulting fuels are sub-divided into the following classifications: middle distillate fuels, kerosene fuels and residual fuels. In the case of middle distillate fuels, up to 5 per cent by volume of fatty acid methyl ester (FAME) may be added. It will be appreciated, therefore, that the grades of petroleum oil which are now available, as listed in Table 16.4, are not unique compounds but products created by selective blending. Rising or falling demand for a given product in the chain may upset the viability of the whole and lead to a quite significant reappraisal of unit costs overall.

As far as liquid fuel for burners is concerned, kerosene (grade C2) is normally only used for firing small domestic boilers rated at not more than about 50 kW. Of the remainder, the heavier the grade the less the unit volume cost, but the greater and more costly the facilities required for storage, handling and burning. A total economic judgement has to be undertaken, taking all these aspects into account before a selection can be made. For instance, light fuel oil (grade E), was for many years a much used technical/economic compromise, but can no longer be so classified and thus is now little used.

As far as the various characteristics of the oils listed in Table 16.4 are concerned, the following should be noted:

Flash point (closed): This is the minimum temperature at which, when the oil is heated, a flash may be obtained within the apparatus. Whilst the flash point is a limiting factor upon the amount of low boiling-point material which may be incorporated in the oil, it is not very significant from the point of view of combustion.

Kinematic viscosity: This quantity is important in as it affects the ability of oil to flow in delivery and from storage.

Pour point: This generally follows viscosity and is the temperature at which the oil ceases to run freely. For practical purposes, this must be below that of the normal temperature of a storage vessel.

Sulphur content: Very significant as far as the potential for corrosion is concerned and is also a determinant of statutory requirements for chimney height. Recent legislation (Directive 1999/32/EC) has resulted in oils produced with far lower sulphur content levels than has previously been the case.
Ash: Although present in small quantities, this has little significance.

Blended biofuels are becoming increasingly popular with the current trend towards burning fuels that can be classified as 'renewable'. Biofuels are a blended mixture of heating oil and liquid fuel(s) produced from renewable resources. Rapeseed oil, palm oil and used cooking (vegetable) oil have all been successfully blended in various proportions with middle distillate and residual oils to produce a liquid fuel that can be used in boilers and burners with minimum change or adjustment. The present target for utilising 20 per cent renewables can be achieved by such means. This fuel is discussed further in Chapter 14.

Delivery

Except in the case of very large installations, oil delivery is made by road tanker. Thus, site facilities must be provided for tanker parking and the position of the fill terminal must be within an agreed distance from the park. The fill terminal should be a standard 50, 65 or 80 mm male gas thread, the size being determined by the length of the fill pipe and the grade of oil. The thread should be suitable for hose coupling and fitted with a non-ferrous cap as protection. As may be seen from Table 16.4 the heavier grades of oil are delivered already heated and for such usage the fill pipe should be insulated and provided with electrical trace heating.

Where more than one tank exists, filling should preferably be through separate pipes, one per tank, although a single fill pipe and an arrangement of three-way cocks or valves at the tanks may be used.

Fill pipes should, where practicable, be laid to drain into the tank and be fitted with an isolating valve at the terminal end. Although the tanker driver should purge the fill pipe with air on completion of each delivery, the provision of a permanent drip-tray arrangement is a prudent measure.

Storage

Since the storage of a quantity of oil in or near a building provides the elements of a fire hazard, various bodies have drawn up regulations covering the installation of storage tanks and other aspects of oil firing. A number of British Standards has been published and reference to the latest (currently BS 5410, Parts 1, 2 and 3) is advisable, always bearing in mind that these publications provide no more than minimum specifications. Local authorities, fire authorities and insurance companies have their own requirements insofar as oil storage is concerned and these too should be taken into account.

At domestic level, a tank capacity of not less than 3500 litres makes sensible allowance for the minimum delivery of 2500 litres preferred by oil companies. If a facility exists whereby a metered 'milk round' service is offered, then a tank holding as little as 1300 litres may be sufficient to serve a house under normal conditions. However, as abnormal conditions can arise, particularly in the depths of winter, limiting the storage volume is not recommended.

In larger installations, it is frequently necessary to provide two or more storage tanks. This has the advantage that filling and usage may be rotated, thus allowing any sludge or water to settle out and maintenance to be effected. Table 16.5 gives the consumption of boiler plants of various sizes based upon a 3-week operation at full load, 12 hours per day.

Tanks used for oil storage in commercial and industrial situations are most often of welded steel construction and may be either rectangular or cylindrical. A generality might be that vessels mounted in the open air are more commonly cylindrical, horizontal or vertical, and that tanks housed within a building are more commonly rectangular. Construction may be prefabricated completely or carried out *in situ* from steel plate or formed sections. The latter method is obviously necessary for confined situations. Capacities and sizes of cylindrical and rectangular tanks are given in Table 16.6.

Cylindrical vessels for mounting in the open should, where horizontal, be supported on sleeper walls topped by steel cradles. Vertical cylinders should be constructed in a manner that keeps the bottom plate free from overall contact with the supporting base. Rectangular tanks sited in the open should have top plates hipped to reject rainwater and be supported on sleeper walls topped by steel joists. Cradles and joists should

Table 16.5 Approximate requirements for oil storage

Boiler rating (kW)	*Storage for 3 weeks* (litre)	*Boiler rating* (kW)	*Storage for 3 weeks* (litre)
20	700	200	6720
40	1390	300	9870
60	2080	400	12 880
80	2760	500	15 750
100	3430	750	22 310
150	5090	1000	28 000

Assumptions
1. 21 days × 12 hours.
2. Boiler efficiency = 75%.
3. Oil CV = 41 MJ/kg.
4. Error at 1000 kW = 10% and pro rate below.

Table 16.6 Oil tank capacities

Diameter (m)	*Length* (m)	*Gross capacity* (litre)	*Net capacity*[a] (litre)
Cylindrical			
1	2	1570	1300
	2.5	1960	1600
1.5	2	3530	3200
	2.5	4420	4000
	3	5300	4800
2	3	9430	8800
	3.5	11 000	10 200
	4	12 570	11 800
2.5	3.5	17 180	16 200
	4	19 640	18 400
	4.5	22 090	20 800
3	4	28 280	26 900
	5	35 350	33 600
	6	42 420	40 200

Length (m)	*Width* (m)	*Depth* (m)	*Gross capacity* (litre)	*Net capacity*[a] (litre)
Rectangular				
1	1	1	1000	700
1.5	1	1	1500	1100
2	1	1	2000	1500
3	1.5	1	4500	3300
3	1.5	1.5	6750	5600
3	2	1.5	9000	7500
4	2	2	16 000	14 000
4	3	2	24 000	21 000
4	4	2	32 000	28 000
6	4	2	48 000	42 000

Length (m)	*Width* (m)	*Depth* (m)	*Diameter* (m)	*Net capacity* (litre)
Plastic				
1.37	1.06	1.25	–	1250
1.8	–	–	1.1	1350
2.02	1.36	1.36	–	2500
1.35	–	–	1.61	2600

Note
a Allowance 150 mm up to outlet and 100 mm ullage (at top), rounded to nearest 100 litres lower.

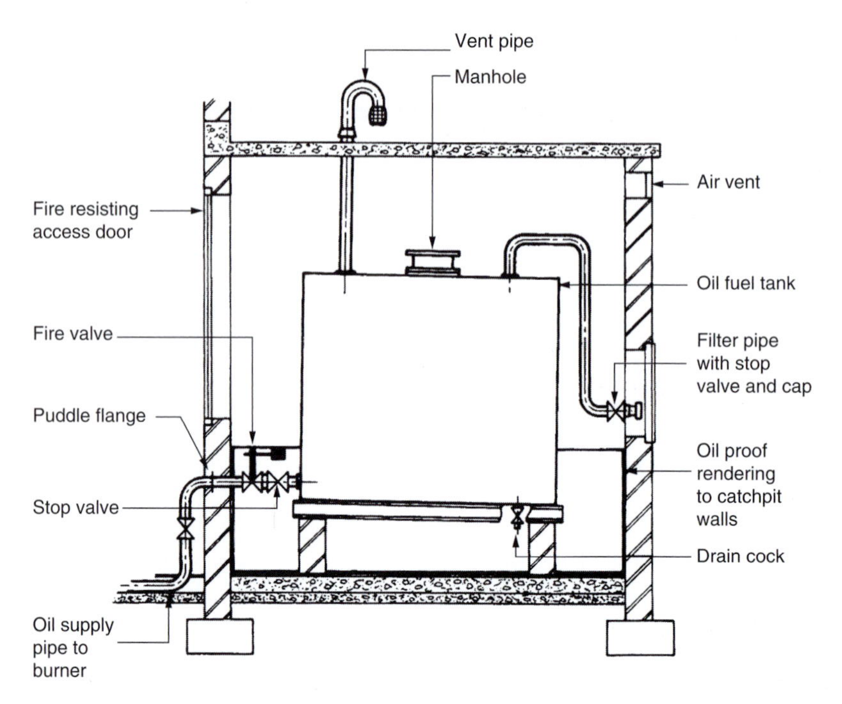

Figure 16.2 Oil tank and connections, above ground

both be bedded on bituminous felt or lead packing. All tanks must be arranged such that water, sludge and other accumulations may be drawn off at a convenient low point.

A domestic range of oil tanks manufactured in medium density polythene is available. The material is rotationally moulded in one piece to provide a stress-free container of uniform wall thickness that is resistant to cracking. The rectangular tanks, with domed top, and the vertical cylindrical pattern must be mounted on a flat surface and not on piers. The capacities and dimensions are listed in Table 16.6.

Bunded tanks

Modern steel tanks are available in bunded form comprising an inner storage tank with an outer weatherproof enclosure. The space between the inner and outer casings is of sufficient volume to contain the full contents of the inner tank. In all other respects bunded tanks are similar to single skin tanks.

Tank rooms

Where located within a building, a tank room to house oil storage must be separated from a boiler house by brick or concrete construction. Unless double skin bunded tanks are used, the floor and lower walls of the tank room must be treated with an oil-proof render to contain the whole capacity of storage, as a precaution against leakage. The access opening must have the sill at an elevated level, with ladders inside and outside, and a fire resisting door. The general arrangement is illustrated in Fig. 16.2. Provision for inlet and outlet ventilation needs to be made and this should be separate from other ventilation systems.

There are three methods used for providing external underground storage tanks: buried, within a chamber and formed underground.

Buried: The cylindrical vessel is literally buried in a prepared pit. Substantial protection of the external steel surface is essential and the bottom of the excavation should be drained. In waterlogged ground, it is generally necessary to anchor the vessel to a heavy block of concrete to overcome buoyancy when the tank is less than full. This method is not very satisfactory, and is little used these days.

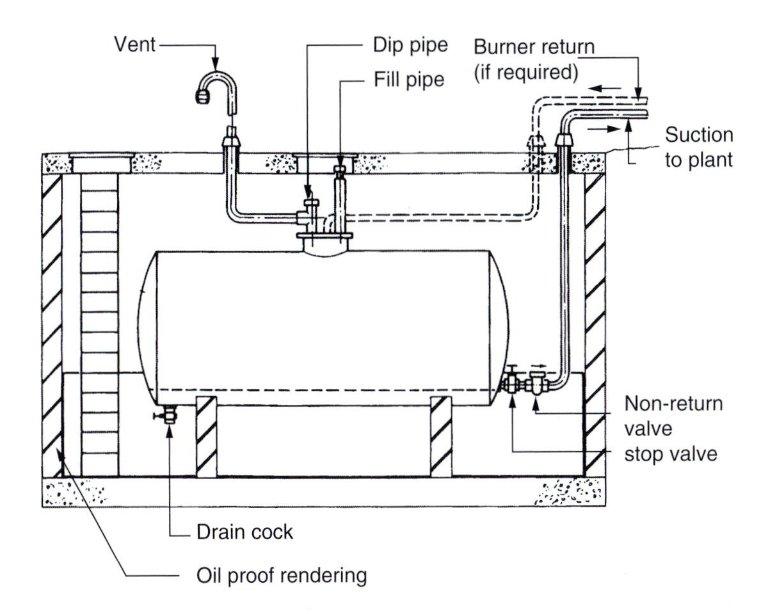

Figure 16.3 Oil tank and connections, below ground

Within a chamber: This method involves the provision of an underground chamber, as illustrated in Fig. 16.3 Construction must be such that there cannot be either ground water leakage inwards or oil leakage outwards. Manholes for access must be substantial and convenient if maintenance is to be adequate. After allowing for a walking space around the tank and for shape factors, the ratio of excavation to oil storage volume is likely to be at least 2:1.

Formed underground: This method is based upon a radically different approach in that the tank shell is itself a concrete structure lined with special tiles. This concept, which is a proprietary design, has much to commend it, not least being the fact that the whole of the available cube internal to the underground structure is used for storage.

Tank fittings

Storage tanks must be provided with a variety of fittings, some as part of the tank and others related to filling. The following is a brief summary of the principal items:

Manhole: Except in the case of small domestic tanks, a manhole with an air-tight cover joint must be fitted. Internal and external ladders should be fitted.

Vent pipe: A vent from each tank, preferably of the same size as the fill pipe, is required. It should be carried up to some point where any fumes will not be troublesome, perhaps to the roof of an adjacent building. A bird guard should be fitted at the terminal. Oil may rise in the vent pipe should the tank be over-filled, and thus impose an excessive pressure on the tank shell. An alarm device is necessary to give warning of this condition and an oil-seal trap or one of the proprietary types of unloading device should be fitted.

Sludge valve: Water, being denser than oil, collects at the bottom of a tank together with sludge and other solid matter. This accumulation must be removed at intervals through a sludge valve which should be easily accessible. In the case of buried or underground tanks, this can prove to be a messy business.

Level indicator: Gauge glass indicators, although positive, are not very durable; dip-sticks are inconvenient; direct mounted dial gauges are subject to damage and what are known as 'cat and mouse' float levels are thought by some to be fallible. There is a variety of hydrostatic remote reading gauges available, as Fig. 16.4 and electrically or (better) pneumatically operated versions are available.

Heaters: In the case of heavy oils, grades F and G, it is necessary to provide means of maintaining a suitable minimum temperature in storage (Table 16.4) either by pipe coils or by electrical heaters. These are commonly of very simple construction and provided with a coarse control of temperature.

Outlet valve: For isolation, the outlet from each tank must be provided with a simple and reliable stop valve.

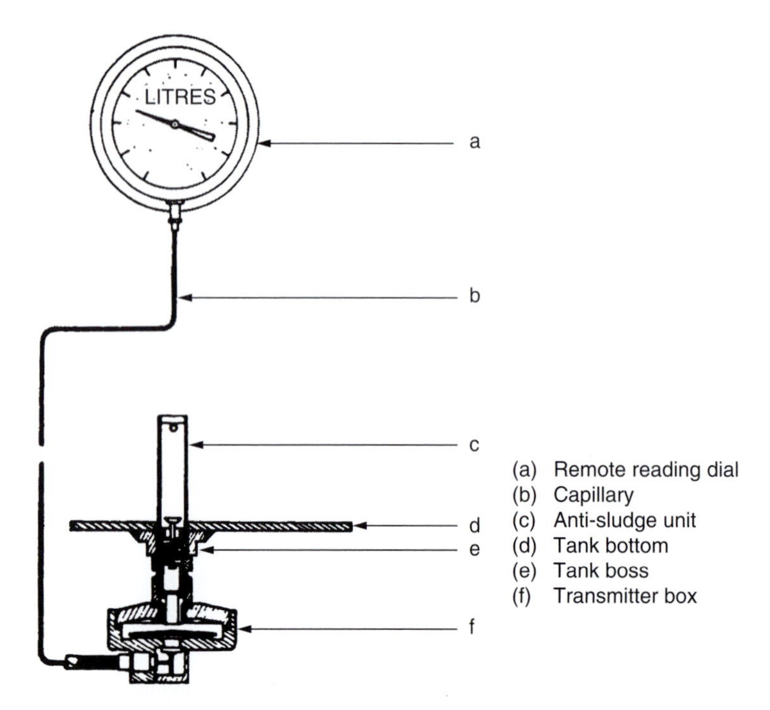

Figure 16.4 Oil tank contents gauge

Handling

The treatment required to handle the flow of oil between storage and the point of utilisation and the associated equipment, varies greatly with the grade of oil used. Necessary in all cases however, is an oil filter, often in a duplex arrangement to continue supply during maintenance, and a fire valve mounted preferably externally, but always as near as possible to the oil pipe entry to the boiler house.

In its crudest form, a fire valve consists of a lever type isolator, heavily weighted, which is kept open by a taut wire stretched across the boiler house with a fusible link of low melting-point alloy over the oil burner and with a hand operated release/test point at the boiler house door. Such devices however, rely for operation upon the flexibility of the wire which, inevitably becomes corroded or deformed where it passes over pulleys and, in any event, is rarely if ever tested for operation. It is current practice to provide a much superior system where the weight-closed valve is kept open by a solenoid operated catch, current to the circuit being maintained through heat sensitive device(s) (thermal fuses) mounted above the burner(s). Alarm bell or other contacts are easily added to such a circuit, which is inherently 'fail safe'.

For heavy oils, grades F and G, it is necessary to provide heating and pumping equipment to take fuel from storage and deliver it to the burners. At the tank, an outflow heater may be provided, as in Fig. 16.5 designed to preheat oil flow at the point of exit rather than in store. The principal item of equipment may be a prefabricated pumping and heating unit as shown in Fig. 16.6 which incorporates not only the necessary positive displacement pumps but also hot water or steam heaters with electrical backup for use when starting the plant from cold.

From the pumping and heating set, the hot oil circulation is extended right up to the burner to avoid any problems of smoke on start up. However, variations in the actual arrangement may be found to be necessary with burners that do not incorporate integral pumping equipment. Pipework carrying oil at the temperatures noted in Table 16.4 should be insulated and provided with *trace heating,* either electrically or by hot water/steam as may be available.

The viscosities and velocities encountered in oil distribution piping are normally such that the flow is streamline or laminar. The consequent loss in pressure may be calculated from a simplified approximate equation:

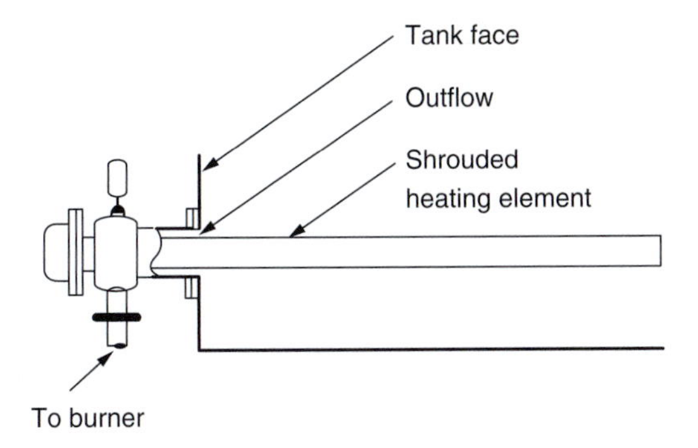

Figure 16.5 Oil tank outflow heater

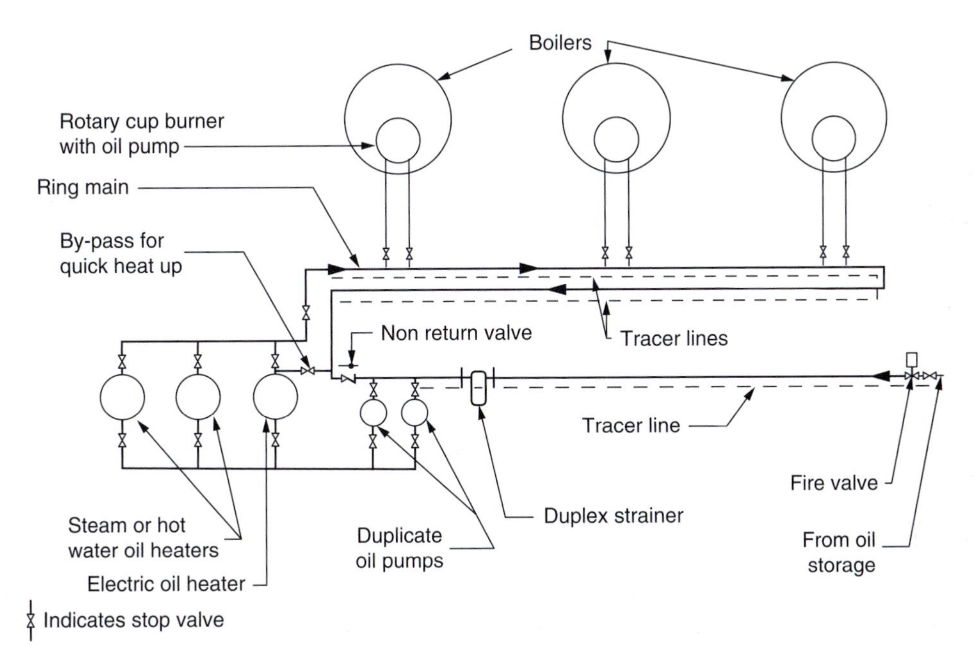

Figure 16.6 Pumping and heating set with oil ring main

$$\Delta p = \theta \cdot Q$$

where

p = pressure drop per m run (Pa/m),
θ = a factor read from Table 16.7,
Q = boiler power, at 75 per cent efficiency (kW).

Subsidiary storage and handling

The subject of roof-top boiler houses has been referred to in Chapter 13, particularly in the context of multi-storey buildings. In such cases, it is usual for the main oil store to be retained at ground level and for a supply to be pumped to a small *daily service tank* in or near the boiler house, as shown in Fig. 16.7. The size of the service tank is dictated by regulations as are the methods of control for the transfer pumping equipment. Pumps are usually provided in duplicate and arranged to be started and stopped automatically by level controls fitted in the service tank.

Table 16.7 Flow of oil in steel pipes (BS 1387: medium)

	Value of factor θ for pipes having stated nominal bore (mm)								
Grade of oil	*10*	*15*	*20*	*25*	*32*	*40*	*50*	*65*	*80*
D	0.56	0.19	0.06	0.02	0.01	–	–	–	–
E	–	7.69	2.43	0.96	0.32	0.17	0.07	0.02	0.01
F	–	–	3.04	1.19	0.40	0.22	0.08	0.03	0.02
G	–	–	4.86	1.91	0.64	0.34	0.13	0.05	0.03

Notes
Factors apply to oil at handling temperature, Table 16.4.

Single resistances
For approximate purposes, equivalent lengths of bends, tees, etc., may be ignored.

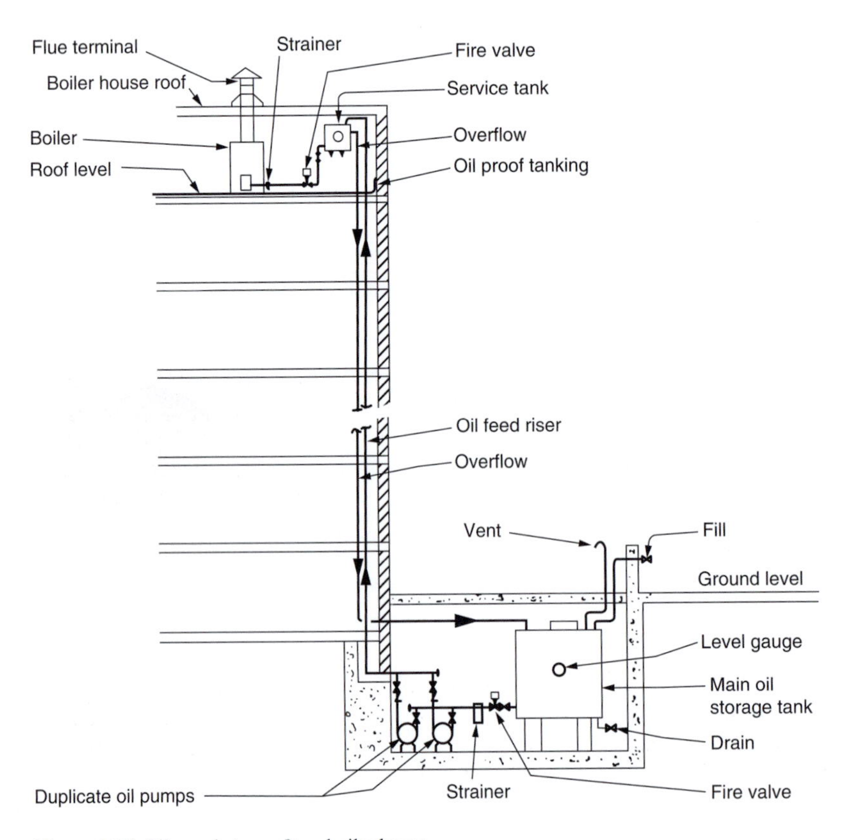

Figure 16.7 Oil supply to roof-top boiler house

Natural gas

Natural gas provides effectively all of the national '*mains*' supply in the British Isles and, in consequence, manufactured gas need be considered only in the context of history and as a basis for comparison. Gas distribution is through a high-pressure pipe network serving the various areas of the country. To cater for the inevitable diurnal fluctuations in demand, the not inconsiderable volumetric capacity of the primary high pressure distribution pipework is supplemented at national level from storage underground and, for local low pressure storage, by the familiar above-ground gasometers.

Table 16.8 Properties of natural and manufactured gas

Item	*Natural gas*	*Manufactured gas*
Specific mass		
kg/m^3 at 15°C	0.78	0.6
re air = 1	0.6	0.48
Calorific value (MJ/m^3)		
gross	41.6	20.9
net	38.7	18.6
Wobbe No.	49.9	27.0
Burning velocity (m/s)	0.35	0.8
Ignition temperature (°C)	650	600
Operating pressure (kPa)	2	1
Toxicity	Nil	Toxic
Constituents (% volume)		
methane	92.6	33.5
ethane	3.6	–
propane	0.8	–
butane	0.2	–
pentane and above	0.1	–
hydrogen	–	47.9
nitrogen	2.6	–
CO_2	0.1	4.9
CO	–	13.7

In order to make provision for exceptional peak demands upon the system, special '*interruptible*' tariffs have been offered to large industrial users, the arrangement being that their supply may be cut, after due warning, for specified periods. In such instances an alternative fuel supply is needed in the form of individual bulk stores, on site, of either liquefied petroleum gas (LPG) or fuel oil. The boiler plants concerned must, of course, be equipped to utilise both natural gas and the alternative fuel. In general terms, however, natural gas offers the advantage of requiring no storage facilities at the point of use.

Table 16.8 lists the properties of both natural and manufactured gas, the former is a methane-rich mixture of a number of hydrocarbons.

Delivery

A statutory requirement exists whereby the natural gas suppliers are obliged to provide and maintain gas to premises sited within 23 m of a distribution main. In principle, the cost of making a connection to that main and of running the first 9 m of service pipe is met by the supplier and the remainder is a charge to the building owner. The service pipe terminates in an isolating valve or cock and is followed by pressure reduction/regulation equipment and a meter.

Pipework

From the meter outlet, the gas supply distribution system becomes the responsibility of the building owner. Earlier versions of the CIBSE Guide C4 provided data in tabular form for flow of fluids in pipes. The latest CIBSE Guide however, provides simplified equations for calculating pressure drops along pipes and ducts, avoiding the need for the re-printing of precalculated tables. However, for simplification, a digest from the original table for the flow of gas in steel pipes is given in Table 16.9. Retaining the kilowatt as a basis for calculation, a unit flow of natural gas (1 litre/s) is equivalent to an energy supply of 39 kW (strictly 38.6). Thus the input requirement, in kW, of any boiler or other appliance may be converted into gas flow by using this equivalence. Alternatively, manufacturers sometimes list input gas rates volumetrically and these may be used.

Table 16.9 Flow of natural gas in steel pipes (BS 1387: medium)

Pa per m run	Litre per second in pipes having stated nominal bore (mm)										
	15	20	25	32	40	50	65	80	100	125	150
0.5	0.07	0.24	0.47	1.02	1.6	3.0	6.1	9.5	20	35	57
1.0	0.14	0.37	0.71	1.54	2.4	4.5	9.1	14.1	29	52	84
1.5	0.21	0.48	0.90	1.96	3.0	5.7	11.5	17.9	37	65	106
2.0	0.25	0.56	1.08	2.31	3.5	6.7	13.6	21.0	43	76	124
2.5	0.29	0.64	1.23	2.64	4.0	7.6	15.4	23.9	49	86	140
3.5	0.34	0.78	1.50	3.20	4.8	9.2	18.6	28.9	59	104	169
5.0	0.43	0.96	1.83	3.92	5.9	11.2	22.7	35.1	72	127	205
7.0	0.51	1.17	2.23	4.75	7.2	13.6	27.4	42.3	86	152	246
10.0	0.64	1.44	2.72	5.81	8.8	16.6	33.4	51.5	105	185	298
15.0	0.81	1.82	3.43	7.30	11.0	20.7	41.8	64.3	130	230	370
20.0	0.95	2.14	4.04	8.57	12.9	24.3	48.9	75.3	152	269	432
25.0	1.09	2.43	4.58	9.71	14.7	27.5	55.2	84.9	172	303	487
	0.5	0.5	1	1	1.5	2	2.5	3.5	5	7	8

Note
Single resistances
For approximate purpose, equivalent lengths may be taken as length in metres as listed in the bottom line of the table for each bend, tee, etc.

In terms of pressure loss, it is usual to limit this to between 75 and 125 Pa from the meter to the point of use. Thus, e.g.:

Input rating to boiler		= 700 kW
Natural gas rate	= 700/39	= 17.95 litre/s
Pressure available, say,		= 100 Pa
Pipe run from intake to boiler		= 30 m
Pressure available/m run	= 100/30	= 3.3 Pa
Thus, from Table 16.9, pipe size		= 65 mm

The pressure available for distribution of gas within a building is bound up with governor settings, requirements for burner intake pressure and with demands for boosting equipment. Such matters need investigation in the light of manufacturers' application data in individual cases and with the gas suppliers.

Liquefied petroleum gas

Many households in Great Britain are not connected to gas from a public '*mains*' supply. LPG is a convenient alternative in circumstances where a supply of natural mains gas is not available. It may be used not only for the familiar portable heaters and other domestic appliances but also for firing static boilers, etc. Whereas natural gas is a mixture of various hydrocarbon compounds, LPG is available as commercially pure butane and propane. Use of butane is normally confined to the smaller applications. Table 16.10 lists the properties of both gases and, for comparison, those of natural gas.

Delivery and storage

LPG is stored on site, as a liquid, in pressure vessels which are fitted by the gas suppliers and remain their property. They are refilled as may be necessary by delivery from a road tanker. It is a general recommendation that vessels for larger installations are provided in duplicate so that inspection and any maintenance may be carried out without interrupting output. Dimensions and other important criteria of LPG storage vessels are given in Table 16.11.

Regulations exist as to the siting of storage vessels and these are set out in a variety of publications, some by HMSO, some by British Standards and others (*perhaps the most useful*) by the distributors' Trade Association. A prime requirement is that vessels must not be installed within buildings, particularly basements; LPG

Table 16.10 Properties of natural and liquefied petroleum gas

		Commercial LPG	
Item	*Natural gas*	*Butane*	*Propane*
Specific mass			
as gas at 15°C (kg/m^3)	0.78	2.45	1.85
as gas at 15°C (*re* air = 1)	0.60	2.0	1.5
as liquid (kg/litre)	–	0.575	0.512
Volume of gas			
per volume of liquid (m^3/m^3)	–	233	274
per mass of liquid (m^3/kg)	–	0.41	0.54
Calorific value (gross)			
re mass (MJ/kg)	53.3	49.2	50
as gas (MJ/m^3)	41.6	121.5	95
as liquid (MJ/litre)	–	28.2	25.5
Ignition temperature	650	510	510
Operating pressure (kPa)	2	3.7	3.7
Toxicity	Nil	Nil	Nil

Table 16.11 Storage of liquefied petroleum gases (LPG)

	Cylinder		*Capacity*			*Size* (mm)		
Gas	*Colour*	*Pressure* (kPa)	kg	*Liquid*[a] (litre)	*Gas* (m^3)	*Height or length*	*Diameter*	*Normal offtake* (litre/s)
Butane	Blue	172	4.5	7.8	1.8	340	240	0.04
			7	12.2	2.8	495	256	0.06
			15	26.1	6.1	580	318	0.08
Propane	Red	690	13	25.4	7.0	580	318	0.16
			19	37.2	10.2	810	318	0.2
			47	92	25.2	1290	375	0.35
	White	690	615	1200	330	2000	1000	1.6
	or		1020	2000	550	3000	1000	2.0
	Green		1735	3400	930	3800	1200	3.7
	White	690	715	1400	380	Sphere	1500	1.8
			2040	4000	1100	3800	1200	4.4

Note
Larger bulk storage tanks of 24 m^3 are available but no standard sizes exist.
a National capacity 87% fill at 15°C.

being heavier than air. According to the capacity of a store, the distance by which it is separated from sources of ignition (such as balanced flue boilers having low-level terminals), site boundaries and buildings is laid down, and varies from 3 m for a small domestic size vessel to 15 m for the largest. Access by tanker must not be more than 30 m distant from the store.

Storage pressure for propane is 690 kPa and this is reduced to about 75 kPa at a first stage regulator which supplies the primary distribution main running up to the outside of a building. A second stage regulator fitted there will have an output pressure set at 3.7 kPa to suit the standard low-pressure input rating for which most utilisation equipment is designed.

Pipework

From the second stage regulator, pipework for LPG should be sized using the same principles as have been set out previously for natural gas but using the data in Table 16.12.

Table 16.12 Flow of LPG in steel and copper pipes

Pa per m run	*Litre per second in pipes having stated nominal size* (mm)								
	Medium steel (BS 1387)				*Copper (BS 2871 (Table X))*				
	8	*15*	*20*	*25*	*6*	*10*	*12*	*22*	*28*
4	0.030	0.23	0.46	0.93	0.006	0.044	0.075	0.39	0.78
5	0.034	0.26	0.53	1.11	0.007	0.049	0.083	0.42	0.89
7	0.039	0.31	0.63	1.32	0.008	0.059	0.103	0.54	1.06
10	0.047	0.38	0.75	1.61	0.010	0.072	0.122	0.66	1.31
15	0.058	0.46	0.92	2.00	0.012	0.091	0.156	0.83	1.64
20	0.068	0.53	1.08	2.33	0.015	0.107	0.183	0.99	1.94
25	0.076	0.60	1.22	2.61	0.017	0.121	0.208	1.11	2.19
30	0.083	0.66	1.33	2.89	0.019	0.135	0.230	1.25	2.47
40	0.097	0.77	1.56	3.33	0.022	0.158	0.272	1.47	2.86
50	0.108	0.86	1.74	3.75	0.025	0.178	0.306	1.67	3.31
70	0.119	0.95	1.93	4.17	0.027	0.201	0.342	1.86	3.64
80	0.139	1.11	2.25	4.86	0.032	0.236	0.403	2.17	4.31
	0.3	0.5	0.7	1.3	0.1	0.3	0.4	0.7	1.0

Notes
Single resistances.
For approximate purposes, equivalent lengths may be taken as lengths in metres as listed in the bottom line of the table for each bend, tee, etc.

A pressure drop of about 250 Pa between the second stage regulator and the point of use is the accepted criterion. Once again, retaining the kilowatt as a basis for calculation, a unit flow of propane (1 litre/s) is equivalent to an energy supply of 95 kW.

Electricity

The latest statistics available (2005[1]) show that in the British Isles, output from the major electricity generating companies was produced by the following energy sources:

Gas	38.5%
Coal	34.3%
Nuclear	20.5%
Oil	1.4%
Other	1.2%
Renewables (hydro)	1.3%
Renewables (other)	2.8%
	100.0%

Imports accounted for 2.7 per cent of the total. Each of these main types has a different capital/running cost relationship and, in consequence, each is best suited to contribute to a particular pattern of loading. Nuclear stations, for instance, are more economical when meeting a base load, and combined cycle gas-turbine stations for dealing with a short-term peak load. Furthermore, stations of the same type will have differing costs of production depending upon age, design, etc.

Growth in generation from renewable sources is increasing, albeit more slowly than many would wish. The impact of targets set out in the Energy White Paper 2003 (noted in Chapter 2) to achieving 10 per cent of the UK's electricity supplies from renewable sources by 2010, plus the increased use and output from Combined Heat and Power (CHP) plant, will result in changes to the above statistics in the next few years.

A typical electricity demand load curve has a morning peak that usually extends from 8 a.m. onwards until about noon, with a lesser peak in the late afternoon. Such changes in demand, insofar as they relate to motive power and lighting, etc., are inevitable but nevertheless create problems for the supply industry. Electricity is unique among energy supplies in that it cannot be stored. However, it can be converted without loss into heat which may then be stored using both old and new techniques. Those most relevant to the present subject are described in Chapter 8.

Tariff structures

Plant operating at night has a higher overall efficiency than when operating during the day. Electricity generation costs are therefore lower at night. If the costs of metering were of no consequence either to the supply industry or to the consumer, then a price structure which took account of both demand pattern and energy use may be the universal and preferable tariff practice.

For the larger consumer, a tariff including separate components for maximum demand and energy use is commonly used and a variety of such arrangements exists to suit different circumstances. In each case, the charge for energy used may be at a single rate or be biased as to time of use. The family of tariffs used in the domestic sector may either take the form of a standing charge which reflects the cost of making the supply available, plus unit rates set at a level appropriate to recovery of both the demand and energy-related costs, or a rate combining these two elements of the charge. The unit rate may, again, be as a single rate for all units no matter when consumed; as a dual rate where units consumed for a specific purpose during the night are charged separately or in simple day/night form where one rate applies during the day and a lower rate is charged for all use during the night. For domestic consumers, the off-peak rate is commonly 50 per cent or less of that charged for daytime use.

Since the various supply companies throughout the country are independent organisations, the details of their tariffs are related to the particular supply and distribution circumstances arising within the geographical area which they serve. In general terms, however, roughly 7 hours of off-peak supply are available each 24 hours, commonly from midnight (24.00 GMT) until 7 a.m. (07.00 GMT) and normally referred to as Economy 7 (see also Chapter 8). During the summer months, when British Summer Time is in force, the availability of the off-peak supply remains related to GMT and is thus out of phase.

Space heating

When a daytime tariff applies, electricity is an expensive source of energy but, in spite of this, it is used extensively as a result of its almost universal availability and the flexibility it offers. It is, however, as a supply taken off-peak that electricity has rather more to offer the user. At any time this supply offers:

- Transmissibility to any point regardless of physical levels or similar limitations.
- Absence, local to the point of use, of dust from fuel and ash, fumes, etc.
- Reduction in on-site labour for plant operation availability at actual point of use.
- Relative ease of control for energy input.

Use of electricity for heating is discouraged under the current Building Regulations where the high carbon emission rate estimated for the overall production of electricity has a significant impact on building compliance calculations.

It is often argued that to produce electricity from a raw fuel involves much wastage since more than three-quarters of the calorific value of that fuel, as supplied to the generating station, may be wasted in cooling towers, rivers and canals or is lost in the distribution system. The contra-argument was that the grade of raw fuel used in generating stations is so low that it would be difficult if not impossible to burn it elsewhere and, whilst this statement is valid as advanced, the value of that low-grade fuel as a chemical feedstock cannot be ignored. The construction or conversion of a generating station designed to burn natural gas alone may be attractive in the sense of capital expenditure. Properly equipped and sited near to a demand for heat supply (see Chapter 15), a case just might be made for such usage. Otherwise, the profligate consumption of such a refined source of energy (and one of limited reserves) is highly questionable.

Biomass fuels

Chapter 14 describes the various types of biomass fuel currently available, and also includes delivery, storage and handling issues. Biomass is essentially organic material from recently living organisms. Both biomass and

Table 16.13 Calorific value for various fuels

Fuel	*Energy density by mass* (GJ/tonne)	*Energy density by mass* (kWh/tonne)	*Bulk density* (kg/m³)	*Energy density by volume* (MJ/m³)	*Energy density by volume* (kWh/m³)
Wood chips	7–15	2–4	175–350	2000–3600	600–1000
Log wood (stacked – air dry: 20% moisture content)	15	4.2	300–550	4500–8300	1300–2300
Wood (solid – oven dry)	18–21	5–5.8	450–800	8100–16 800	2300–4600
Wood pellets	18	5	600–700	10 800–12 600	3000–3500
Miscanthus (bale)	17	4.7	120–160	2000–2700	560–750
Coal (lignite to anthracite)	20–30	5.6–8.3	800–1100	16 000–33 000	4500–9100
Oil	42	11.7	870	36 500	10 200
Natural gas (NTP)	54	15	0.7	39	10.8

Source
Biomass Energy Centre.

fossil fuels release carbon dioxide into the air on burning. However, biomass fuels release carbon dioxide that was captured during recent times. Fossil fuels, on the other hand, captured carbon dioxide millions of years ago and its re-release is effectively increasing atmospheric concentrations.

Delivery and the large storage requirements for biomass fuel need to be considered at an early stage in the design process, particularly for use in built-up areas and city centres.

Wood

Typical calorific values of wood fuels are given in Table 16.13, together with, for comparison, those for coal, oil and gas. Moisture content, as mentioned in Chapter 14, has a very significant impact upon the calorific value of the fuel.

Waste products

These, from a very wide variety of sources in manufacturing industries and from municipal collection, may be burnt at a relatively low efficiency in large-scale specialist furnaces equipped with accessible heat exchangers, and most importantly effective flue gas cleaning methods. The characteristics of such waste material vary so widely that no sensible generalisations can be made.

At the present time only approximately 9 per cent of municipal waste is incinerated in 17 regulated incinerator plants in England and Wales.[2] Strict controls on the incineration of waste in recent years have greatly reduced harmful dioxin and furan emissions. Heat from incinerator plant is often used for local heating and power. However, for practical reasons not all incinerators can make use of the heat produced. Most municipal waste still goes to landfill but EU Directives will reduce this means of waste disposal, possibly leading to an increase in the demand for waste incineration in the future. The capacity of existing incinerators together with the overall number of incineration plants is likely to increase over the course of the next few years.

Organisations such as the Carbon Trust have undertaken considerable research in recent years to assist in developing the technology, identifying markets and generally promoting biomass heating in the UK.

Notes

1. *Digest of UK Energy Statistics* 2006, DTI Publication.
2. *Source*: Environment Agency.

Chapter 17

Combustion, emissions and chimneys

As an extension to the understanding of fuel characteristics and of the methods used for storage, handling and burning, some knowledge of the chemistry of combustion is useful. In addition, this has a bearing upon the selection of and design for the chimneys necessary to disperse the products of combustion. The text which follows is no more than an introduction to the subject, with particular reference to those fuels which have been noted in the preceding chapter as being commonly used for plant of the scale necessary to service individual buildings or relatively compact groups of buildings.

Fuels encountered in practice consist of carbon, hydrogen and oxygen combined to form relatively complex mixtures or compounds; the hydrogen may be uncombined and will be found present in the *free* state in most gaseous fuels. In addition, there are generally small quantities of sulphur, nitrogen and – in the case of most solid fuels – ash. Moisture is also a constituent of solid fuels and takes two forms: the superficial or free moisture resulting from either pit head water screening or subsequent open air storage, and the *inherent* or air dried content.

Combustion processes

The combustion of fuel is an *oxidation* process which is accompanied by the liberation of heat energy. The reactions occur as the carbon, hydrogen and, when present, sulphur combine with the oxygen in the air supplied. These reactions can take place only at a relatively high temperature, known as the *ignition temperature*, which varies between 400°C and 700°C according to the fuel. If an adequate supply of air has been made available, then the carbon will burn completely to form *carbon dioxide* but, when combustion is incomplete due to shortage of air, *carbon monoxide* will be formed. The hydrogen will burn to form water vapour and any sulphur present will burn to produce *sulphur dioxide* which may perhaps later combine with more oxygen to form *sulphur trioxide*.

On average, by mass, air contains 23.21 per cent oxygen and 75.81 per cent nitrogen, plus traces of argon, helium, krypton and other inert gases: without sensible error, therefore, the proportions may be considered as 23.2 per cent oxygen and 76.8 per cent incombustible. By volume, the proportions are 20.9 per cent oxygen and 79.1 per cent incombustible. The various elements previously mentioned combine with oxygen in proportion to their respective molecular weights which are: oxygen ($O_2 = 32$), carbon ($C = 12$), hydrogen ($H_2 = 2$), nitrogen ($N_2 = 28$) and sulphur ($S = 32$). Table 17.1 lists the fundamental combustion equations.

Fuel analysis

The *ultimate analysis* of a fuel gives the percentage by mass of the various elements or compounds contained in a sample. Results from this type of analysis are used in the majority of calculations made in combustion problems. The *proximate analysis* of a fuel (specifically a solid fuel) gives the percentage by mass of certain

Table 17.1 Combustion reactions (elements)

	kg/kg of combustible								
	Requirements		*Products of combustion*						
Reaction	O_2	*Air*	CO_2	*CO*	H_2O	SO_2	N_2		*Heat liberated* (MJ/kg combustible)
$C + O_2 = CO_2$ $12 + 32 = 44$	2.67	11.51	3.67	–	–	–	8.84		33.6
$2C + O_2 = 2CO$ $24 + 32 = 56$	1.33	5.73	–	2.33	–	–	4.40		10.1
$2CO + O_2 = 2CO_2$ $56 + 32 = 88$	0.57	2.46	1.57	–	–	–	1.89		23.5
$2H + O_2 = 2H_2O$ $4 + 32 = 36$	8.0	34.48	–	–	9.0	–	26.48		142.7
$S_2 + 2O_2 = 2SO_2$ $64 + 64 = 128$	1.0	4.31	–	–	–	2.0	3.31		9.2

characteristic groupings of elements which, as a result, indicates the probable physical nature of the combustion performance.

Excess air

The requirements for oxygen, and thus of air, which are listed in Table 17.1 are those calculated from the equations to be precisely the quantities necessary for the various reactions to take place. They represent the ideal situation or what is known as the *stoichiometric* condition. In practice, complete combustion cannot be achieved unless more air is provided than that which is theoretically required.

This situation comes about for very practical reasons related to the parallel difficulties of ensuring that the combustible elements will be mixed intimately with the air and that the consequent reactions will be completed before the products of combustion are discharged. The additional or *excess* air should not, however, be more than is necessary to prevent the discharge of unburnt fuel since the presence of a surplus will result not only in unnecessary energy losses, as noted later, but also in corrosion hazards associated with the formation of sulphur trioxide.

Flue gas temperature

The temperature at which the products of combustion are finally discharged raises difficulties since, if this is allowed to fall to too low a level, condensation will take place. Any water vapour in the gases, resultant from the combustion of hydrogen, will be deposited at what is called the *water dew point*, approximately 60°C. In the case of sulphur bearing fuels there is also the *acid dew point*, at which the water vapour combines with any sulphur trioxide present to form sulphuric acid. This occurs when temperatures fall much below about 130°C.

In order to reduce the prospects of corrosion resultant upon the presence of sulphur trioxide from sulphur bearing fuels, normal practice is to design for discharge temperatures of 250–270°C (some 240 K average above the surrounding atmosphere). Hence, the combustion products hold a considerable amount of sensible heat in addition to the latent heat within the uncondensed water vapour which has been referred to previously. This sensible heat may, in turn, be quantified as being proportional to the mass of the total products *including any excess air* carried over, which serves to re-emphasise the importance of setting and maintaining an optimum quantity of excess.

With non-sulphur bearing fuels acid dew point is not an issue and flue gas temperatures can be allowed to reduce to below the water dew point. A significant improvement in boiler efficiency is achieved from the extraction of heat from the condensed water vapour. Condensing boilers are described in Chapter 13. The lower the return water temperature the more condensate is formed, the more heat is extracted and the greater the boiler efficiency. Depending upon the system, return temperatures are unlikely to be below about 25°C and therefore the minimum flue gas temperature exiting a condensing boiler is likely to be around 30°C.

Flue gas analysis

Carbon being one of the principal components of most practical fuels, knowledge of the proportion of carbon dioxide present in the products, provides a useful indication of completeness of combustion. Dilution of these products with excess air leads to a parallel decrease in the proportion of carbon dioxide present and thus measurement of the CO_2 content provides a relatively simple criterion which indicates not only the degree to which the combustion reactions have been completed, but also the proportion of excess air admitted.

Taking the simplest case of pure carbon as an example, it may be seen from Table 17.1 that, for the ideal state of complete combustion, the CO_2 content of the products by volume is:

$$\begin{aligned} CO_2 &= 100(3.67/44)/[(3.67/44)+(8.84/28)] \\ &= 8.3/(0.083+0.317) = \underline{21}\text{ per cent (max)} \end{aligned}$$

If 50 per cent excess air were admitted, the CO_2 content would then be:

$$\begin{aligned} CO_2 &= 8.3/[0.4+0.5(2.67/32+0.317)] \\ &= 8.3/(0.4+0.2) = \underline{13.8}\text{ per cent} \end{aligned}$$

When analysing flue gases it is necessary to be aware not only of the carbon dioxide content of the products of combustion, but also that of other components such as oxygen, carbon monoxide, sulphur oxides and nitrogen oxides.

Harmful emissions

The process of combustion produces CO_2, CO, H_2O, SO_2 and N_2 in various proportions according to the composition and type of fuel and the conditions under which combustion has taken place.

Carbon dioxide is always formed when burning a carbon-based fuel. Its impact as a 'greenhouse' gas is well known and no further wording on this subject is necessary here.

Carbon monoxide is formed when combustion is incomplete. CO is very poisonous and it is essential that CO in flue gases discharges directly to atmosphere, without leakage into rooms.

Sulphur dioxide in the atmosphere mixes with water vapour to produce sulphuric acid and ultimately acid rain. The amount of SO_2 formed is proportional to the sulphur content in the fuel and has a significant impact in determining the height of the chimney.

Nitrogen is normally very unreactive due to its strong triple bond. However, at high temperature, such as occurs during combustion, the bonds break down and combine with oxygen to form nitrogen monoxide (NO) and nitrogen dioxide (NO_2). NO and NO_2 are collectively known as nitrogen oxides (NO_x). Whilst nitrogen oxides are important ingredients in almost all atmospheric reactions there is a critical balance. Excessive NO_x in the atmosphere can combine with water vapour to form nitric acid (HNO_3) another major contributor to acid rain.

The impact that harmful products of combustion have within the atmosphere is discussed further in Chapter 2.

Calorific value

The calorific value of a fuel is the quantity of heat energy released as a result of the complete combustion of unit mass. As described in Chapter 13, two values are normally quoted, the *gross* or higher and the *net* or lower. For ease of reference, these are defined here again: the *gross calorific value* includes the latent heat within any water vapour formed as a result of the combustion of hydrogen and the *net calorific value* excludes this constituent.

With conventional boilers it is not practicable to recover this latent heat by condensing the water vapour and therefore the net calorific value is perhaps the more useful figure. However, in the case of condensing boilers, latent heat is recovered and the gross calorific value is the more meaningful.

Since, again as may be seen from Table 17.1, the mass of water produced is 36/4 = 9 times the mass of hydrogen burnt, it follows that the latent heat within the uncondensed vapour may be calculated as:

Latent heat *re* 15°C = 2450 kJ/kg

Thus

$$CT_g - CV_n = H(9 \times 2450)/(100 \times 1000)$$
$$= \underline{(0.221H)} \text{ MJ/kg}$$

Since the heat liberated as a result of the oxidation of various elements is known, a calculation may be made using a suitable analysis to derive a theoretical calorific value. For solid fuels in particular, the result of such a calculation will lack accuracy as a result of the presence of a variety of extraneous elements or impurities. Practical evaluations are therefore made using a *calorimeter*.

Chimney loss

The chimney loss, as the name suggests, is a summation of all those losses which have accumulated by the time that the products of combustion reach the chimney. They are directly proportional to the mass gas flow, to the temperature at which discharge from the boiler occurs and to the specific heat capacity of the gases. Furthermore, the total also includes the latent and sensible heat content of water vapour arising from the combustion of hydrogen. The inter-relation of the various components of the total is complex and it is convenient to make use of graphs as Figs 17.1–17.3, which are extracted from the *CIBSE Guide Section C5 – published in April 2007*.

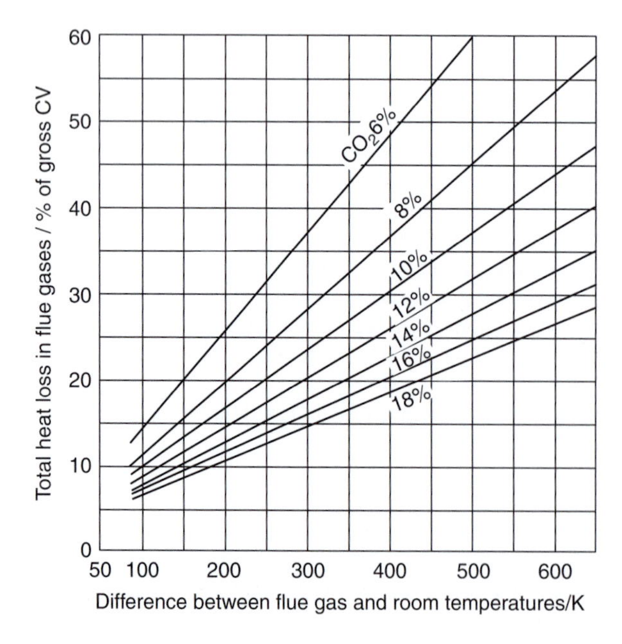

Figure 17.1 Chimney loss for coal firing

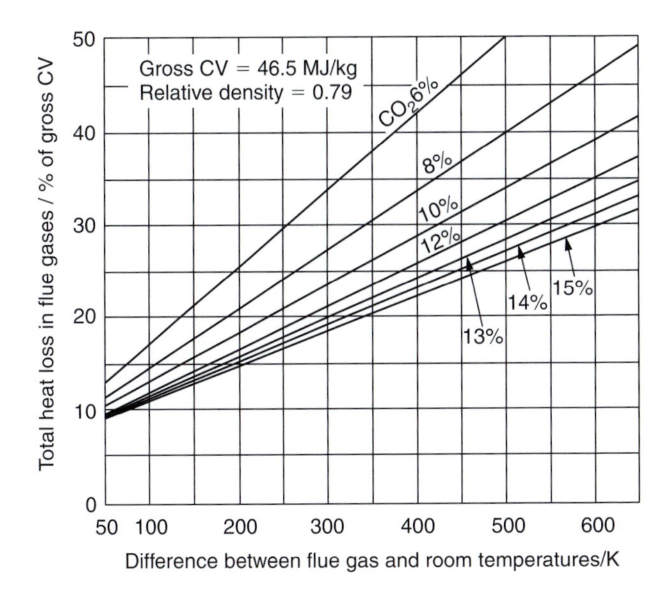

Figure 17.2 Chimney loss for oil firing

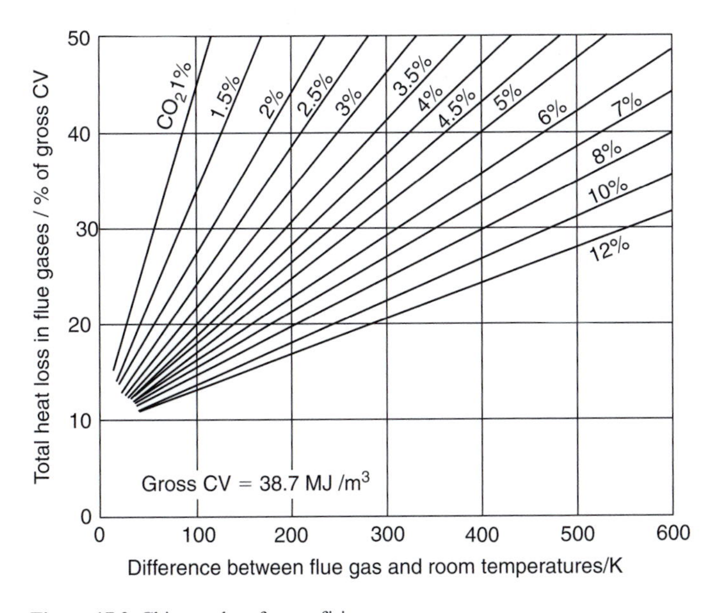

Figure 17.3 Chimney loss for gas firing

Sample calculations

Solid or liquid fuels

A single example common to both these fuels may be taken to introduce the use of Table 17.1 by assuming that a sample has, by mass, the following ultimate analysis:

Carbon 80 per cent, hydrogen 4 per cent, sulphur 2 per cent, ash and moisture 12.5 per cent.

The air required for stoichiometric combustion would be:
By mass:

C to CO_2	$= 11.51 \times 0.80$	=	9.21
$\left.\begin{matrix}H\\O\end{matrix}\right\} H_2O$	$= 34.48 \times (0.04 - 0.015/8)$	=	1.32
S to SO_2	$= 4.31 \times 0.02$	=	0.09
∴ Air required (kg/kg)		=	10.62

By volume (air at 15°C = 0.816 m^3/kg)
∴ Air required (m^3/kg) = 8.66

Note: The oxygen content of 1.5 per cent combines with one-eighth of its weight of hydrogen. An external supply of air is required to burn the remaining hydrogen content.

The CO_2 present in the combustion products, with 75 per cent excess air would be:

Combustion products, by mass:

CO_2 =	3.67×0.80	=		2.94
H_2O =	9.0×0.04	=		0.36
SO_2 =	2.0×0.02	=		0.04
N_2 =	8.84×0.80	=	7.07	
	26.48×0.04	=	1.06	
	3.31×0.02	=	0.07	8.20
(check: [10.62(76.8/100)])		=	8.16	
N_2 in excess air 0.75×8.20		=		6.17
O_2 in excess air $6.17 \times 23.2/76.8$		=		1.86

The water vapour may be ignored since it will have been condensed out before the CO_2 is measured. Hence, by volume:

CO_2 =	100(2.94/44)	=	6.68	
SO_2 =	100(0.04/64)	=	0.06	
N_2 =	100(14.35/28)	=	51.25	
O_2 =	100(1.86/32)	=	5.81	63.80

Thus,

CO_2 = 6.80/63.80 = 10.5 per cent

If the flue gas temperature were to be 300°C with a CO_2 level of 10.5 per cent, the chimney loss in the case of solid fuel firing, as read from Fig. 17.1, would be 23.5 per cent.

Gaseous fuels

In this case, although the principles remain exactly the same as before, it is convenient to approach any calculations by making use of the data for the various hydrocarbon compounds listed in Table 17.2. Hence, we consider a fuel having the following composition, by volume:

Methane 90 per cent, ethane 5 per cent, propane 1 per cent, nitrogen 3.5 per cent, carbon dioxide 0.5 per cent.

Table 17.2 Combustion reactions (gases)

Reaction	Requirements kg/kg O_2	Requirements kg/kg Air	Requirements m^3/m^3 O_2	Requirements m^3/m^3 Air	*Heat liberated* (MJ/m^3 combustible)
Methane					
$CH_4 + 2O_2 = CO_2 + 2H_2O$	4		2		40
$16 + 64 = 44 + 36$		17.24		9.57	
Ethane					
$2C_2H_6 + 7O_2 = 4CO_2 + 6H_2O$	3.73		3.5		69
$60 + 224 = 176 + 108$		16.08		16.75	
Propane					
$2C_3H_8 + 10O_2 = 6CO_2 + 8H_2O$	3.64		5		95
$88 + 320 = 264 + 144$		15.69		23.92	
Butane					
$2C_4H_{10} + 13O_2 = 8CO_2 + 10H_2O$	3.59		6.5		121
$116 + 416 = 352 + 180$		15.47		31.1	
Pentane and above					
C_5 +	3.1		8.3		164
		13.36		39.7	

The air required for stoichiometric combustion would be:

By volume:

Methane	0.9×2	=	1.8
Ethane	0.05×3.5	=	0.175
Propane	0.01×5	=	0.05
∴ Oxygen required (m^3/m^3)		=	2.025

Thus

air required (m^3/m^3) = 2.025/(20.9/100)
= 9.7

The CO_2 present in the combustion products with 25 per cent excess air would be:

Combustion products, by volume:

$0.9\ m^3$ of methane

combines with 0.9×2 = $1.8\ m^3$ of O_2
produces 1.8(79.1/20.9) = $6.8\ m^3$ of N_2
produces 0.9×1 = $0.9\ m^3$ of CO_2

$0.05\ m^3$ of ethane

combines with 0.05×3.5 = $1.175\ m^3$ of O_2
produces 0.175(79.1/20.9) = $0.66\ m^3$ of N_2
produces 0.05×2 = $0.1\ m^3$ of CO_2

$0.01\ m^3$ of propane

combines with 0.01×5 = $0.05\ m^3$ of O_2
produces 0.05(79.1/20.9) = $0.19\ m^3$ of N_2
produces 0.01×3 = $0.03\ m^3$ of CO_2

Thus,

volume of CO_2 in products (m^3)
= 0.005 + (0.9 + 0.1 + 0.03) = 1.035

and volume of N_2 in products (m^3)
= 0.035 + (6.8 + 0.66 + 0.19) = 7.685
and excess air volume (m^3)
= 0.25[7.685/(79.1/100)] = 2.429

Thus,

CO_2 = 1.035(1.035 + 7.685 + 2.429) = 9.3 per cent

Chimneys

The purpose of a chimney is to provide a means whereby the products of combustion may be exhausted in such a manner as either to avoid pollution or, at worst, to provide means of dispersion such that pollution is diluted to an acceptable level. Obviously the nature of the pollutants will vary with the nature of the fuel used. Particulate matter may be carried over in some cases, acid vapours in others and in some the discharge may be no more than a thermal carrier-plume.

Chimney design is a task which may involve a number of disciplines in the solution of problems related to the aesthetic, structural and functional aspects. Very many chimneys are concealed for much of their height within the massing of a tall building, but this approach to concealment may lead to local difficulties as a result of lee-side down-wash or vortex effects. It is not possible to discuss these matters in greater detail here and, in any event, they are often best resolved through model tests in a wind tunnel using techniques that are now well established.

There are certain principles applying to all chimneys, regardless of fuel type, which are now recognised as being fundamental to good design. These are:

- The velocity of the gases through the stack should be as high as possible commensurate with the draught available.
- The temperature of the gases should be maintained, throughout the stack, at near to the entry level by effective insulation.
- The use of draught stabilisers (as distinct from draught diverters) and leakages which admit cold air should be avoided.
- The inner surfaces of the flue construction should be smooth and changes in cross-section, bends, etc. should be designed aerodynamically.
- The use of a single flue-way to serve more than one boiler should be avoided wherever possible. In the rare case where a common flue for a multi-boiler plant is genuinely unavoidable, means should be provided to isolate the outlets of idle boilers and to maintain a constant efflux velocity at the chimney terminal.

Chimneys for solid or liquid fuels

For the purpose of this present text the design process may be simplified by reducing the number of variables which relate to the particular cases of solid fuel and oil. Since the remainder, including the calorific value of the fuel envisaged for design purposes, will almost certainly vary during the life of the chimney, further precision will serve little purpose.

Products of combustion, quantity

Assume excess air provided to be 75 per cent.

Assume flue gas temperature, on average, to be 200–300°C. Assume boiler efficiency to be 75 per cent.

For these conditions, it may be shown that the products of combustion per MJ of boiler output are between 1.1 and 1.3 m^3/MJ. It is convenient to take a mean of 1.2 m^3/MJ. The temperatures assumed cover the range mentioned previously.

Time

Bringing a time scale into consideration, the volume of the products as above will derive from combustion associated with 1 MJ/s, i.e. 1 MW.

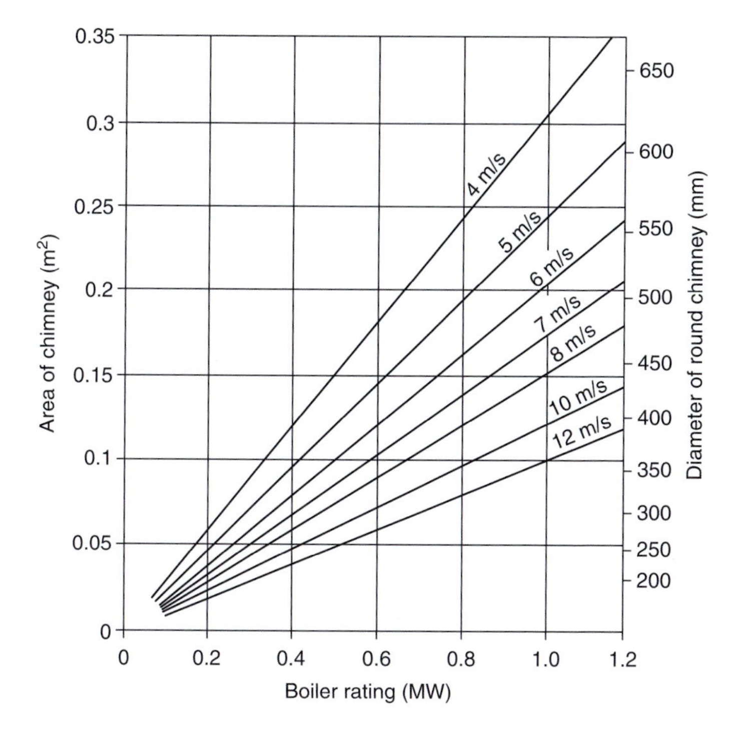

Figure 17.4 Chimney areas for stated velocities

Velocity of flue gases in the chimney

It may be assumed that the range of velocities appropriate to natural and to mechanical draught are, respectively, 4–8 and 10–12 m/s.

Area

Figure 17.4 gives areas and diameters for round chimneys direct, having selected a velocity. The *equivalent diameter* of a square or rectangular chimney, in mm, is 1000 × the square root of the area as read from the figure.

The cross-sectional area of the chimney may then be derived according to boiler duty, fuel and type of draught. Thus, if the boiler duty were 1 MW, oil-fired, with natural draught (5 m/s) and a chimney temperature average of 250°C,

$$1.2/5 = 0.24\ \text{m}^2 = 550\ \text{mm in diameter}$$

Efflux velocity

In order that the plume of flue gas should rise clear of the chimney top and not flow down the outside (*downwash*), the diameter at the top should be reduced so as to maintain as high a velocity as practicable. For small boilers with natural draught, a velocity of 6 m/s is advised. Larger boilers with mechanical draught should achieve 7.5–15 m/s.

The velocity pressure corresponding to these rates may be read from the bottom line of Table 17.3 as appropriate to a flue gas temperature of 250°C.

Draught required

The following may be calculated from the data on duct sizing in Chapter 21, adjusted for temperature.

Table 17.3 Pressure loss per metre height of smooth chimney (multiply values by 4 for brick or rough cement rendering)

Effective Chimney diameter (mm)	*Pressure loss* (Pa per m height or run) *for gas flow at the following velocities* (m/s)								
	4	*5*	*6*	*7*	*8*	*9*	*10*	*11*	*12*
200	0.45	0.71	1.02	1.39	1.81	2.29	2.82	3.41	4.07
250	0.34	0.52	0.76	1.02	1.34	1.70	2.10	2.53	3.04
300	0.27	0.41	0.60	0.81	1.06	1.35	1.66	2.00	2.39
350	0.22	0.35	0.49	0.68	0.88	1.12	1.38	1.67	1.99
400	0.18	0.29	0.41	0.56	0.73	0.93	1.14	1.38	1.65
450	0.16	0.25	0.36	0.49	0.64	0.81	1.00	1.21	1.44
500	0.14	0.22	0.32	0.44	0.57	0.72	0.89	1.07	1.28
550	0.13	0.20	0.28	0.38	0.50	0.64	0.78	0.95	1.12
600	0.11	0.18	0.25	0.35	0.45	0.57	0.71	0.85	1.02
650	0.10	0.16	0.23	0.32	0.41	0.52	0.64	0.78	0.92
700	–	0.14	0.21	0.28	0.37	0.47	0.57	0.69	0.83
750	–	0.13	0.19	0.26	0.34	0.43	0.53	0.64	0.76
Velocity pressure	5.4	8.4	12.1	16.5	21.5	27.3	33.6	40.6	48.4

At boiler exit, consult makers' data but:

Oil-fired boilers vary from	7 to 50 Pa
Solid fuel fired, if burning rate is 5 kg/m^2 grate area	70 Pa

Flue connections, boiler to chimney, depending on number of bends and other losses, average	15 to 30 Pa
Efflux velocity pressure, e.g. for 6 m/s	12.1 Pa

The new total for an oil-fired boiler may then be between 50 and 100 Pa and for a solid fuel boiler 100 to 150 Pa.

Draught produced by chimney

The theoretical draught of a chimney at the two temperatures named varies with the external ambient temperature. Assuming that this is 20°C in summer and 0°C in winter, unit values are:

Winter

per metre height at 300°C	6.7 Pa
at 200°C	5.5 Pa

Summer

per metre height at 300°C	5.8 Pa
at 200°C	4.5 Pa

Figure 17.5 is drawn on this basis and thus a chimney of 30 m height will in summer produce, theoretically, a draught of 155 Pa with flue gases at 250°C.

Draught loss in chimney

The pressure loss per metre height may be taken from Table 17.3. Values for other velocities may be interpolated. Figure 17.6 gives the pressure loss for chimneys of given heights, the loss per unit length having been determined from Table 17.3. The pressure loss arising from the flow of gases in rough brickwork or concrete chimneys will be as much as 3 or 4 times greater than that for relatively smooth sheet steel.

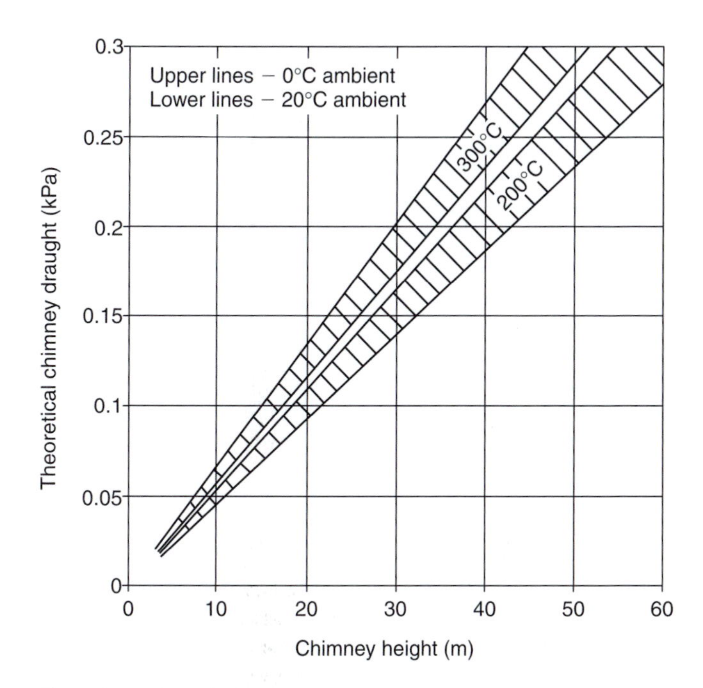

Figure 17.5 Theoretical draught for a chimney of given height

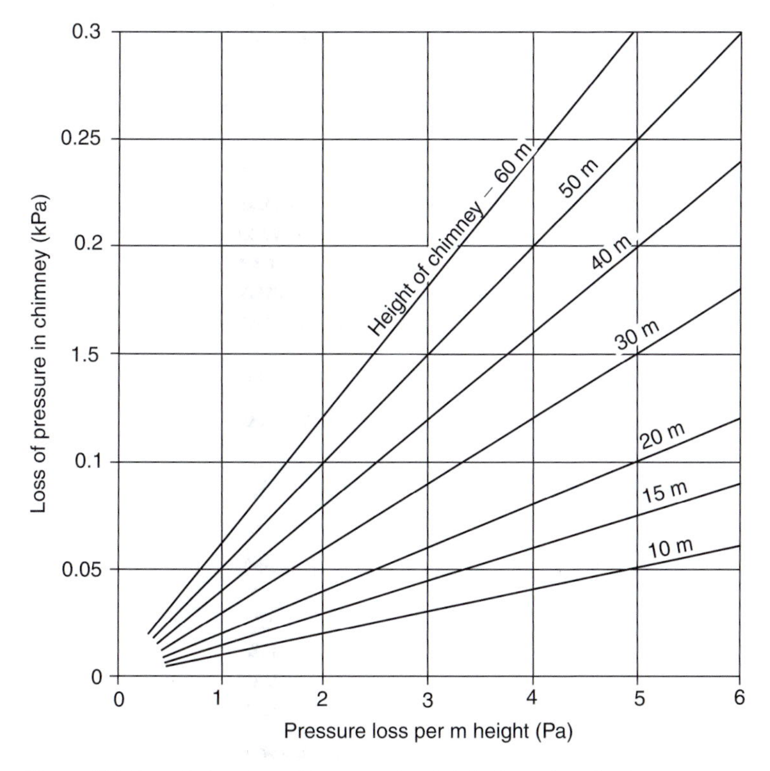

Figure 17.6 Draught loss for a chimney of given height (smooth surface)

Rectangular equivalents

For square or rectangular chimneys, the effective areas are those of the circle or ellipse which may be inscribed within them. The *equivalent diameter* of such flues is therefore the square root of the square or rectangular area. If the flue must be rectangular, it is a general rule that a ratio of sides of 3:1 should not be exceeded.

In conclusion:

- Using Fig. 17.4 select a velocity according to whether draught is natural or mechanical and find chimney area and diameter.
- From Table 17.3 determine pressure loss per metre of height for this diameter.
- Make an assumption as to chimney height and hence from Fig. 17.6 note pressure loss in chimney. Add for loss through boiler and flue connection and for efflux velocity.
- Using Fig. 17.5, the available draught may be found for the same assumed chimney height. If this were equal to or in excess of the sum of the losses, the assumption as to height may stand. If the available draught were insufficient, either the height must be increased or the velocity reduced, or both. If the calculated draught were much in excess of requirements, a smaller chimney or less height might be the solution, or if neither were possible, or desirable, a damper may be used.

Clean Air Act

The chimney height determined by following the routine just described is that which is required for combustion of sulphur bearing fuels. It is however also necessary to consider the height in relation to the mandatory requirements of the Clean Air Act (1993). For this purpose, reference should be made to the third (1981) edition of the document *Memorandum on Chimney Heights* and also to Appendix A2 of the CIBSE Guide B1 published in July 2002.

Memorandum on chimney heights

The purpose of the *Memorandum* is to provide a basis for limiting the pollution due to sulphur dioxide (SO_2) near to the chimney by increasing the height as the rate of emission increases. In addition, the relationship between chimney height and building height is taken into account as well as the type of locality, here classified in précis as follows:

A An undeveloped area where background pollution is low with no other emission nearby.
B A partially developed area with low background pollution and no other emission nearby.
C A built-up residential area with only moderate background pollution and no other emission nearby.
D An urban area of mixed industrial and residential development, with considerable background pollution and with other emission nearby.
E A large city, or an urban area of dense residential and heavy industrial development with severe background pollution.

The concentration of SO_2 is determined from equations quoted in the *Memorandum* which require knowledge of the maximum rate of fuel consumption, the sulphur content of the fuel and the boiler efficiency. For calculation from first principles, the required information is given in Chapter 16 (Tables 16.2 and 16.4) but, as an approximation for the coals most in use and the heavier oils, emission of SO_2 in g/s may be approximated by multiplying the boiler rating in MW by either 1.7 for coal firing or 1.0 for oil firing.

If the calculated SO_2 content is less than 0.38 g/s then the chimney height need be only 3 m higher than that which has been calculated as necessary for combustion. For SO_2 contents greater than this, reference must be made either to the *Memorandum*, which contains a set of nomograms, or to Fig. 17.7, from which an 'uncorrected' chimney height may be read, taking account of the location, from the left-hand scale. Where the fuel has a sulphur content of more than 2 per cent, as used to be the case for most oils, a primary correction must then be made by adding 10 per cent to the height read from the diagram. If, after this increase, the dimension is more than $2V_2$ times the height of the building or any other building in the immediate vicinity, then no further correction is necessary.

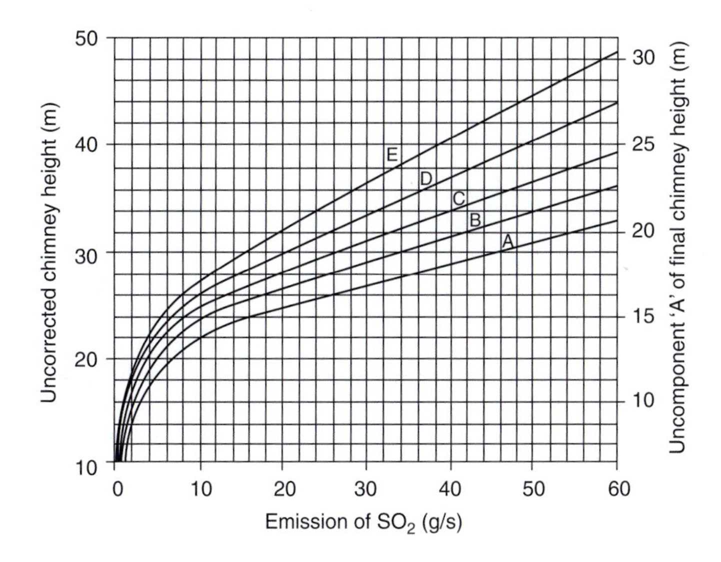

Figure 17.7 Emission of SO_2 and chimney height

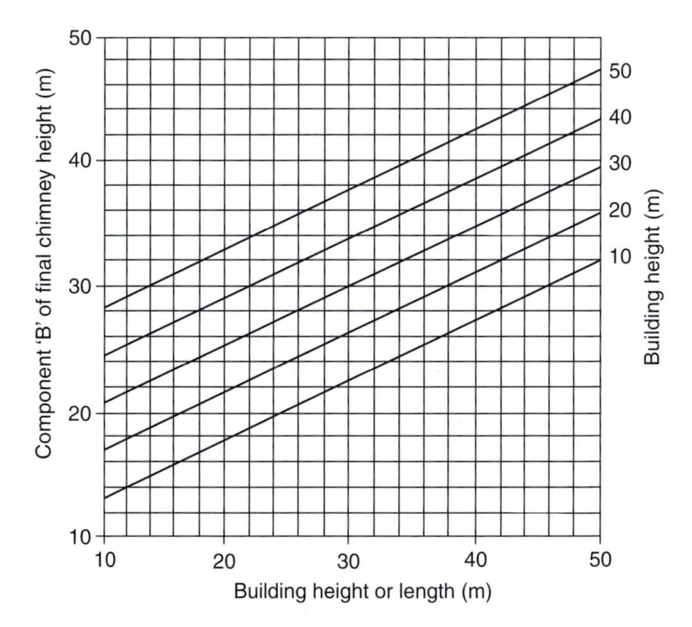

Figure 17.8 Component 'B' for chimney height

In other circumstances, it is necessary to take the dimensions of the building into account. A convenient way of dealing with this is to consider the final chimney height as having two separate components, the first of which (A) is read from the right-hand scale of Fig. 17.7. The second (B) is read from Fig. 17.8 by entering building height or length, whichever is the greater, to the base scale. The sum of the two components, the final chimney height, must be corrected as before by the addition of 10 per cent if the sulphur content of the fuel is greater than 2 per cent.

It should be noted that the use of the shortcuts given here does not wholly invalidate the final result produced since the basic processes laid down in the *Memorandum* are, at best, only pseudo-scientific approximations. Nevertheless, the rule book should always be followed when making the necessary applications to a local authority.

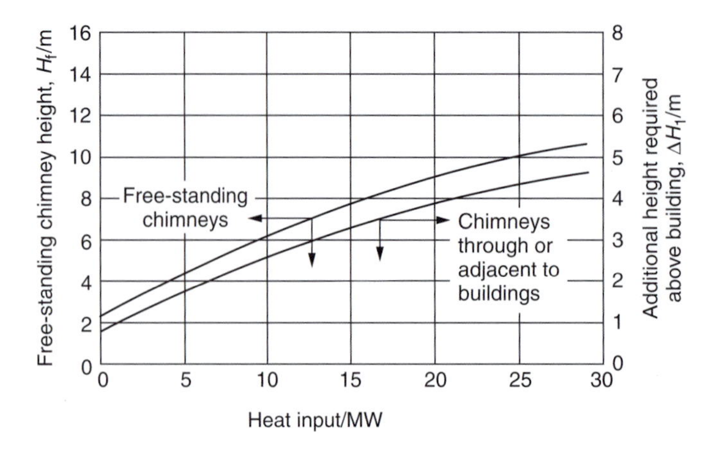

Figure 17.9 Heights for single chimneys for non-sulphur bearing fuels

Mechanical draught

Any solid fuel- or oil-fired boiler may be fitted with an *induced draught* fan for exhausting the products of combustion and discharging them up the chimney. Many boilers are obtainable with such a fan fitted as part of the standard unit. In others, the fan supplying combustion air has sufficient power to expel the products under pressure, i.e. using *forced draught.*

The advantages of mechanical draught are: first, that the boiler can be designed for higher velocities over the heating surface, so giving a higher rating for a given size; second, that the natural draught produced by stack height is no longer of importance, and the chimney may thus be short (subject to the Clean Air Act), or, if the boiler is on the roof, notional only.

Induced draught fans usually take in some diluting air and makers' data should be consulted for the volume to be handled by the chimney, the area of which will then be determined by the velocity decided upon, such as 10–12 m/s. The resistance would be calculated as before. It is usual, where the fan is an integral part of the boiler package, for about 60 Pa surplus pressure to be made available for chimney loss and efflux velocity.

Chimneys for gaseous, non-sulphur bearing fuels

The required chimney height for non-sulphur bearing fuels is that which will provide for adequate dispersion of the products of combustion, such that their concentration at ground level does not exceed a critical value.

Natural draught

The requirements for chimneys for gas firing equipment operating under natural draught differ from those previously discussed. Here air is admitted by a *draught diverter* that acts as a dilutant to the combustion gases. The CO_2 content of the latter, at the boiler discharge, may be 9 per cent with a gas temperature of 240°C, but the normal for the 'secondary' flue after the draught diverter is about 4 per cent with a gas temperature of about 120°C.

The negligible sulphur content of natural gas means that the provisions of the third edition of the *Memorandum on Chimney Heights* apply only to the very largest installations. The required height can be read from Fig. 17.9 for free-standing chimneys and for chimneys through or immediately adjacent to a building. As can be seen, for heat inputs of upto 5 MW, the flue terminal for gas-fired boilers need only be about 2 m above roof level of a building, or 4 m above ground level where the chimney is free standing.

The volume of the dilutant gases from a natural draught boiler is greater than from a forced draught boiler and hence the flue area is also greater. The flue area may be read from Fig. 17.10 where the value of the factor F has been calculated from:

$$F = H_n/(S + 2 + b/2)$$

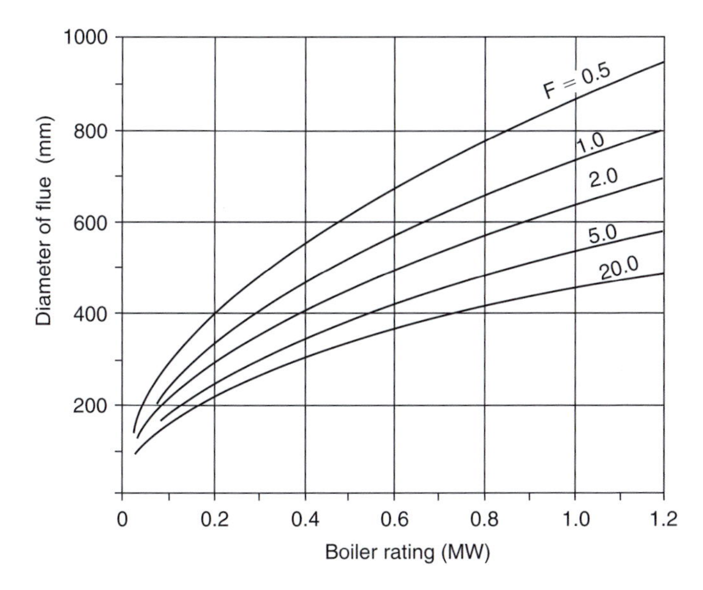

Figure 17.10 Chimney areas

where

H_n = height of flue above boiler outlet (m)
S = two-thirds of any suction required at flue base (Pa)
b = number of bends in the route of the flue

Forced draught

For a forced draught gas burner, the chimney height is again obtained from Fig. 17.9. Chimney sizing does not vary greatly from that described earlier for solid fuel and oil equipment. The data provided by Figs 17.5 and 17.6 are equally valid. Some manufacturers list technical information regarding performance of burners and of draught requirements at the point of connection to the chimney: most do not. These data must be obtained before design of the chimney can be tackled with confidence.

Chimney construction

Domestic chimneys

Building Regulations require that all new masonry constructed domestic chimneys be lined with some impervious materials such as tile. Proprietary flexible flue liners are only permitted in retrofit applications.

Larger installations

The importance of keeping the products of combustion in the chimney warm has been emphasised earlier. With this in mind some form of insulation is a necessity and various forms of construction are illustrated in Fig. 17.11. They are:

A An outer stack in brickwork or concrete enclosing an independent lining of firebrick or moler brick with an air gap between.
B An outer stack in brickwork enclosing a moler brick lining which may be bonded in.
C An outer stack in brickwork lined with fire clay tiles.

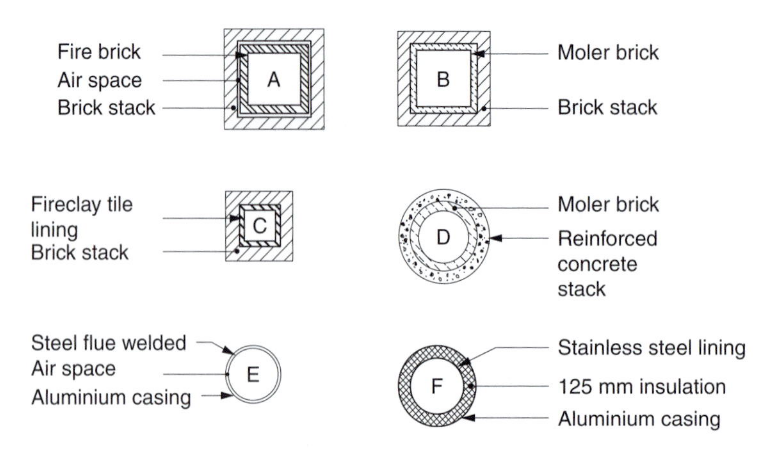

Figure 17.11 Methods for single-flue construction

D An outer stack in reinforced concrete sections lined with insulating moler brick or concrete.
E A welded steel flue within aluminium cladding. The construction may be guyed or self-supporting. The annular space between flue and cladding may be packed with insulation material.
F One of a family of prefabricated, single- or twin-wall sheet metal chimneys manufactured in lengths of about 1 m with socket-and-spigot or twist-lock joints. A wide variety of constructions is available, insulated and un-insulated.

The relative heat losses from these constructions may be calculated as for *U* values – the inner surface resistance being assumed as zero. Cost, permanence and appearance, if free standing, will always be determining factors.

Multiple flues

It will be apparent that, where more than one boiler occurs and there is a common chimney, it would be impossible to maintain the design efflux velocity with anything less than the full number of boilers in use. Furthermore, where mechanical draught is used and the flue is under pressure, back draught might occur to any boilers which are not being fired. Hence, it is now advocated, as has been noted previously, that each boiler should have its own individual flue connection and chimney, and that they should not be combined into one large stack as was the practice in the past. The separate vertical flues may be grouped into one stack, as in Fig. 17.12.

Smoke problems

The corrosive effect of sulphurous gases has been alluded to previously. If the products of combustion, on entering the chimney, are allowed to cool to the region of the acid dew point, a new hazard is set up with oil firing, namely acid smuts.

It appears that the minute unburnt particles of carbon always present in flue gases are liable to form nuclei on which condensation takes place, and they then agglomerate into visible black oily specks. In a chimney they are particularly liable to collect on any roughnesses, or at points of change of velocity. In so doing, on starting up, the sudden shock may cause them to be discharged from the top – often giving rise to complaints from surrounding property.

Acid smut formation is less of a problem with a low sulphur oil such as Class D, even though draught stabilisers are often part of standard boilers on a domestic scale.

For the control of draught on plants burning the heavier oils, in lieu of a draught stabiliser admitting cold air, some form of damper control is necessary which may be automatic, of a type such as Fig. 17.13.

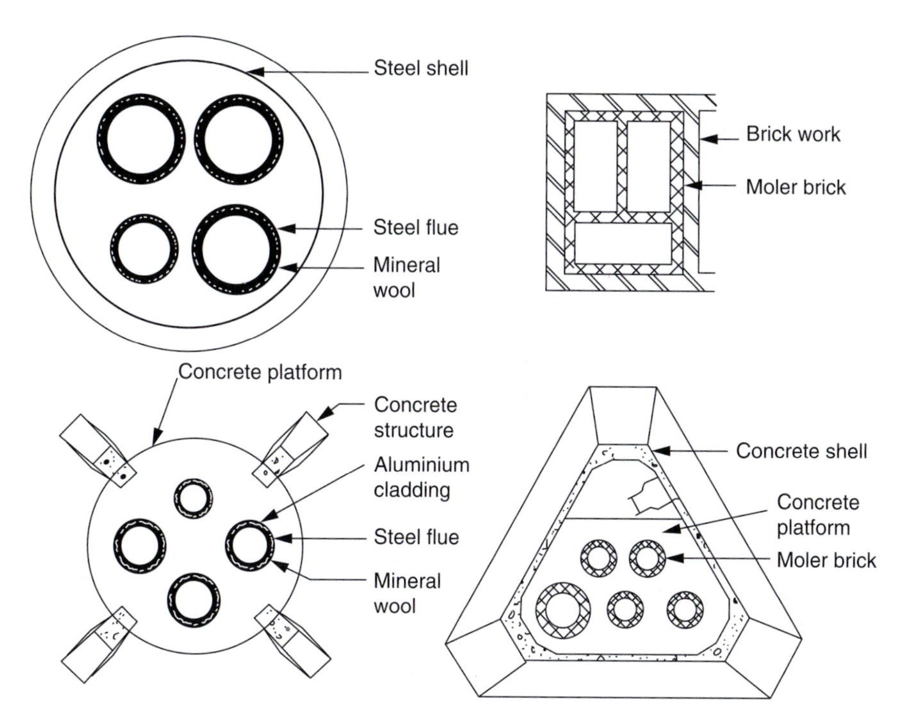

Figure 17.12 Methods for multiple-flue construction

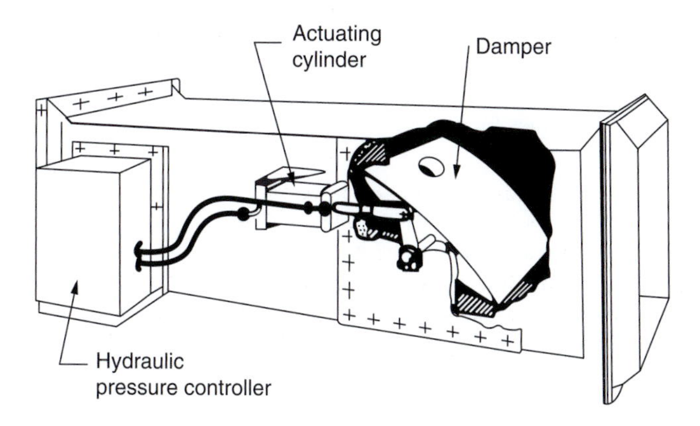

Figure 17.13 Damper for automatic draught control

Treatment of flue gases by means of a proprietary additive to the fuel or in the hot-gas outlet may be considered where smut nuisance is particularly troublesome.

Materials for chimneys and flues (gas-fired boilers)

For each unit of heat released, combustion of natural gas will produce between one and a half and two and a half times as much water vapour as will combustion of oil or coal. Hence, condensation is likely to occur in gas flues and chimneys, particularly on start-up. Masonry constructed chimneys should be lined with an impervious and acid-resisting material such as glazed stoneware or refractory lining. It is desirable to keep the chimney warm to assist draught, either by enclosing it in brickwork or concrete, or by insulation. Connecting flues between such boilers and vertical chimneys are usually of double-wall stainless steel.

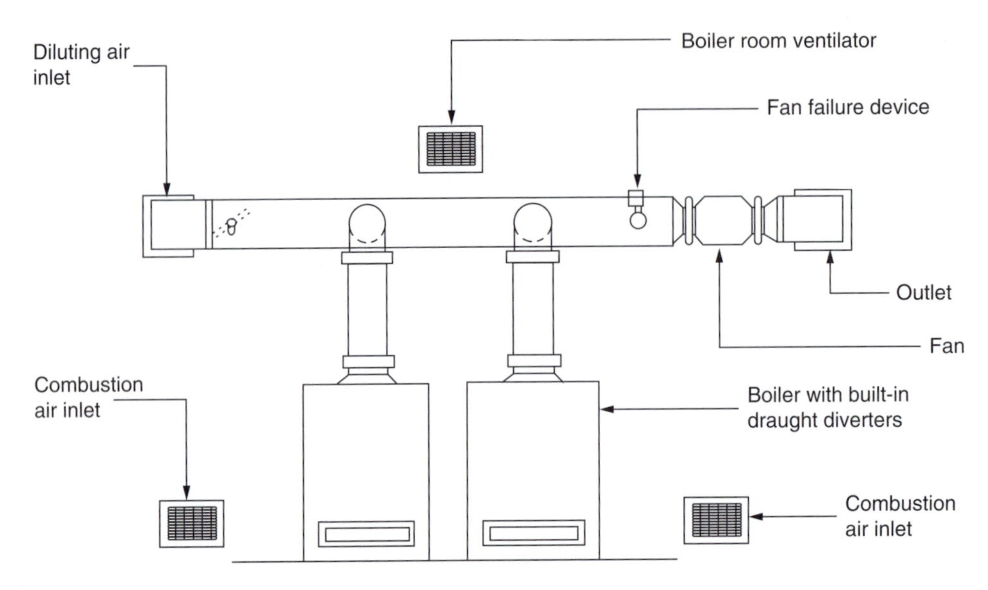

Figure 17.14 Fan diluted draught

Special methods for flue gas disposal (natural gas)

Apart from the conventional chimneys mentioned, a number of special methods for flue gas disposal have been developed, principally by the various research bodies within the gas industry. Some of these have also been adapted for use with equipment burning the lightest grades of fuel oil but such applications apply only on a domestic scale. Reference should be made to BS 5440 Part 1 2000 that describes the requirements, locations, clearances, etc., applicable to the installation of such flues for gas appliances up to 70 kW.

Fan diluted draught

In this system, one form of which is shown in Fig. 17.14, the combustion products, together with a quantity of diluting air, are exhausted to the atmosphere by a fan. By this means it is possible to dispense with a chimney entirely, for a gas-fired boiler plant, by discharging through the wall of the boiler house. The diluting air quantity must be such as to bring the CO_2 content of the mixture down to 1 per cent, which involves a fan to handle approximately 100 m^3/m^3 of natural gas burnt. Included in the system is a fan failure device to shut off the burners in the event of draught failure. The duct sizing with this system should be on the basis of a gas velocity of 3–5 m/s.

The outlet location requires careful consideration. It must be at least 2 m and preferably 3 m above ground level. The outlet must not discharge directly into a courtyard or similar partially enclosed space. The minimum distance to the nearest building is 3.54 m.

Domestic multi-appliance flues

Another flue system is the SE, as shown in Fig. 17.15(a). This is suitable for multi-storey buildings and, as can be seen, air enters at the base and the products of combustion from the various appliances discharge into the same duct. Thus the gas combustion space is virtually sealed off from the occupied space. An alternative, for use where a bottom inlet is not possible, is the 'U' duct shown in Fig. 17.15(b). Fresh air taken in at the roof is conveyed down a shaft running parallel with the rising shaft, which acts as a shared flue as in the SE duct, the products of combustion being exhausted at the roof from a terminal adjacent to the intake, but at a slightly higher level. All gas burning appliances take their combustion air from, and return the products of combustion to, the rising duct.

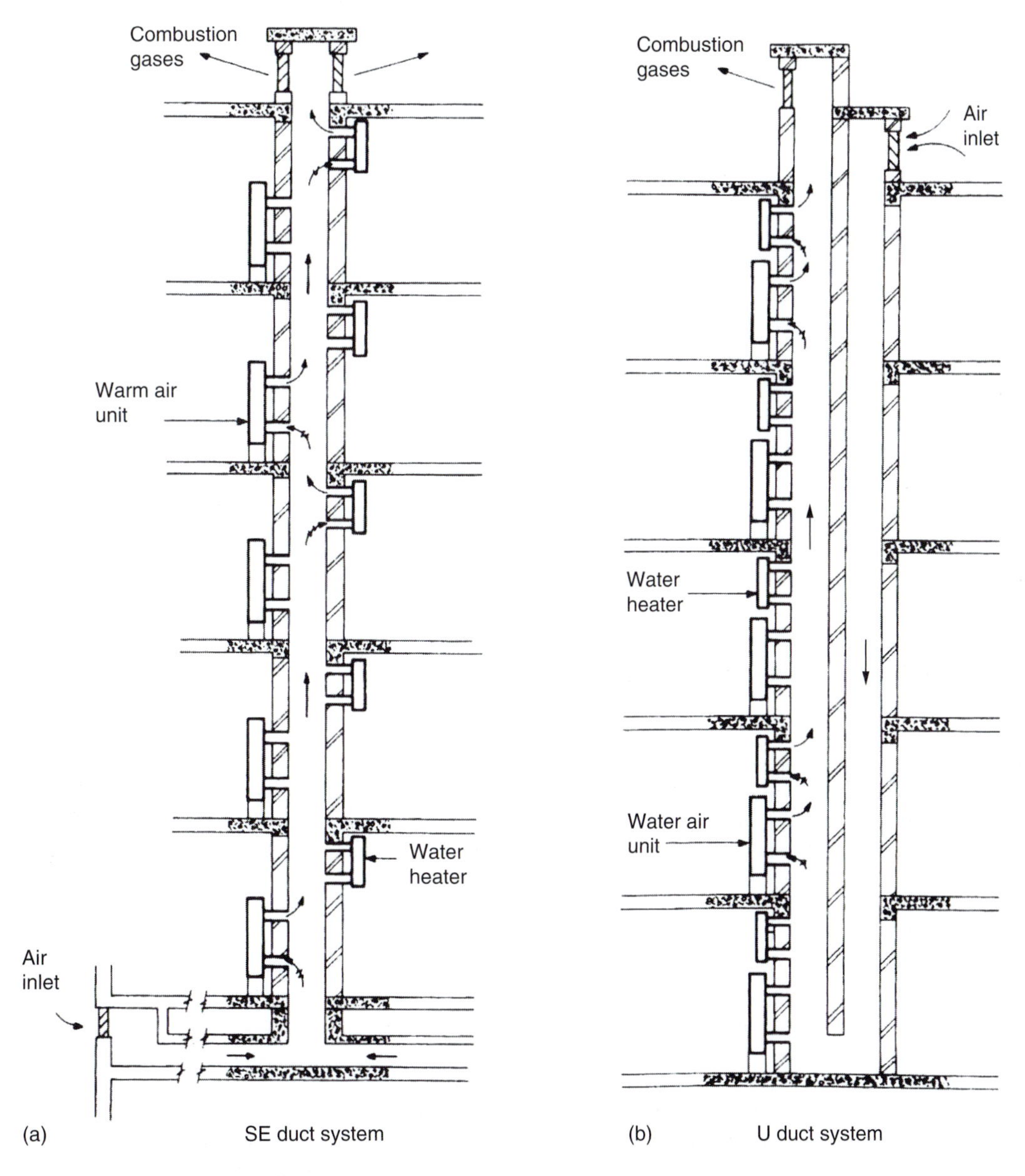

Figure 17.15 SE and 'U' duct flue systems

Balanced flues

The now conventional balanced flue boiler terminal has adjacent openings for drawing in air for combustion and discharging the products of combustion, configured such that wind effects are substantially balanced. Balanced flue boilers are included in Chapter 13.

Balanced flue domestic boilers are usually room sealed, where the combustion system is sealed from the room in which the appliance is situated. It is not practical, however, for larger boilers to be room sealed. In these situations a balanced compartment arrangement can be adopted. Here, the open-flued appliance is located in a separate non-habitable room, sealed from other rooms. The door into the room should be self-closing. Fresh air into the room is taken from a place adjacent to, but just below the flue discharge point. Figure 17.16, sourced

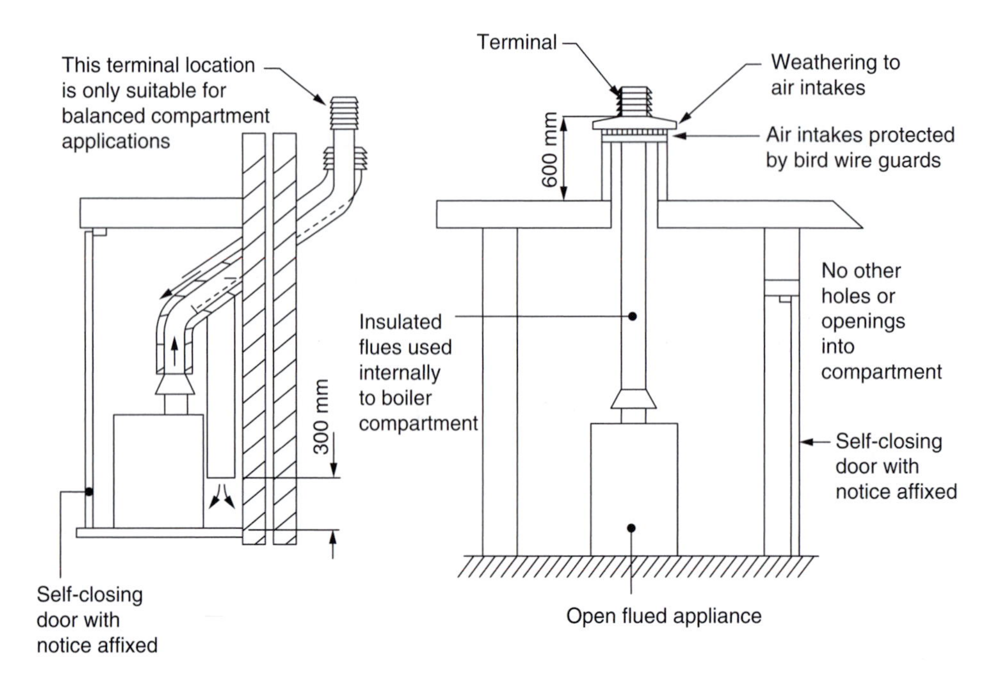

Figure 17.16 Balanced compartment installations (reference BS 5440 Part 1 2000) (a) adjacent termination and (b) roof termination

from BS 5440 Part 1, shows typical arrangements. The sizing of this fresh air duct must be generous, since it has to convey not only the air for combustion, but also that drawn in by the draught diverter where fitted; and the resistance should, in any event, be kept as low as possible. Since limited quantities of ventilation air may be introduced there may be a risk of overheating. It should be noted that the room is to enclose the boiler(s) only. All other plant, pumping, etc. should be contained in a separate, perhaps adjacent plant room.

Terminals

Care must be taken to ensure that flues discharge the products of combustion freely to the atmosphere. In buildings with return walls, areas, etc., peculiar atmospheric pressure conditions may occur and it is advisable to extend the flue to a height of at least 500 mm above the eaves of the roof. The possibilities of downdraught occurring are then much reduced.

Every flue should be fitted with a terminal of approved design to prevent birds nesting, reduce rain penetration, deflect wind and reduce staining due to downdraught.

Chimney heights and terminal locations for domestic and small commercial boiler plant are covered under the Building Regulations and B5 5440 Part 1. There are numerous permutations described in the Standard to suit particular circumstances and it should be noted that these are legal requirements and the final terminal proposal must satisfy the minimum criteria stipulated.

Roof-top boiler house

Discharging combustion gases are considerably simplified where the boilers are located on the roof of a building. Figure 17.17 shows such an arrangement with short outlets from the boilers carried through the roof of the boiler house to atmosphere. Gas boilers on the roof are obviously simpler to deal with than oil – there is no delivery pumping and negligible fire risk. Thus, in many examples of multi-storey buildings, the use of gas in this location for boilers has much to commend it.

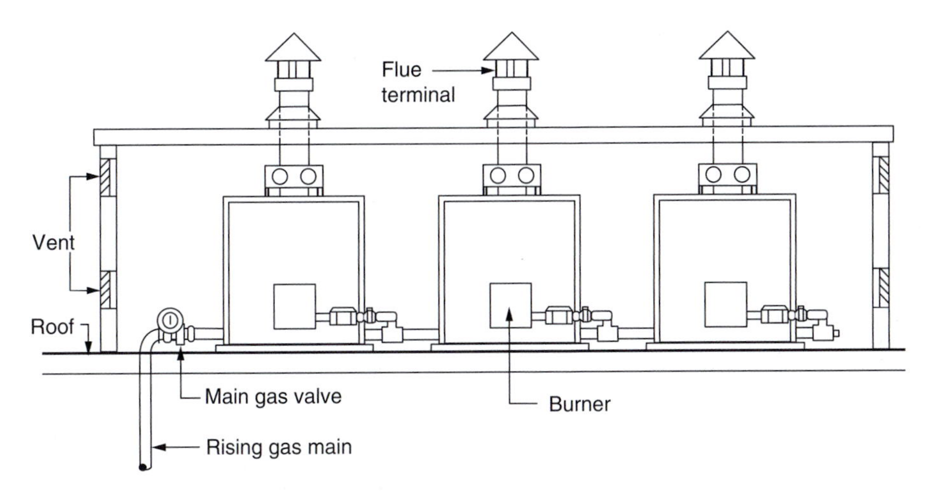

Figure 17.17 Roof-top boiler house

Table 17.4 Air required for combustion of various fuels

Fuel and burner	*Theoretical* m³/kg	*Theoretical* m³/m³	*Excess air* (%)	*Practical* m³/kg	*Practical* m³/m³	*Practical* litre/s per kW[a]
Bituminous coal	8	–	50	12	–	0.53
Fuel oil	12	–	30	16	–	0.51
Natural gas						
Atmospheric burner (primary)	–	10	60	–	16	0.56
Atmospheric burner (total)	–	10	180	–	28	0.98
Packaged burner	–	10	30	–	13	0.46
Propane						
Atmospheric burner (primary)	–	24	60	–	38	0.53
Atmospheric burner (total)	–	24	180	–	67	0.94
Packaged burner	–	24	30	–	31	0.44

Note
a litre/s per kW boiler rating, at 75% efficiency.

Air supply to boiler houses

The manner in which calculations are made in order to determine the quantity of air which is required theoretically for combustion has been set out earlier in this chapter. In the design of a boiler house, provision must be made for this air to enter, together with any excess which may be necessary. Where atmospheric burners are used with gaseous fuels, the associated boiler flue must be equipped with draught diverters, and further air will be necessary to replace that which is used as the dilutant.

Generally, air supply for combustion appliances is covered in Building Regulations Approved Document J. However, this is primarily concerned with domestic installations. The requirements for the provision of air for combustion and cooling of boiler houses are covered by BS 5440 Part 2 for gas-fired boilers not exceeding 70 kW, and BS 6644 2005 for boilers between 70 kW and 1.8 MW.

Table 17.4 sets out, in round figures, the air quantities required for combustion for various fuels and situations. The last column of this table makes use of the respective calorific values of the fuels and expresses the requirements as a function of boiler rating in kW. It will be noted that, with the exception of those associated with atmospheric gas burners, the figures in the last column all fall within the range 0.44–0.56 litre/s/kW.

As far as is possible, air should be supplied to the boiler house by natural induction through door or wall louvres and other openings. A velocity of between 1 and 2 m/s is commonly used in calculating the free area of opening needed. Taking the higher figure of 0.56 litre/s/kW (= 0.56 m^3/s/MW), and a face velocity of 1 m/s, we arrive at a free area for inlet louvres of, say, 0.6 m^2/MW of boiler rating.

In parallel with the need to supply combustion air, it is necessary also to consider the extent to which the boiler house may require ventilation to prevent overheating. A separate lightweight building at ground level would very probably not require any such treatment but, at the other extreme, a basement enclosure might become intolerably hot during some seasons.

The 1986 edition of the CIBSE Guide B (Section 13) provides an estimation of the boiler house heat gain. A calculation to arrive at the necessary peak ventilation rate can then be carried out. As previously mentioned, in the case of atmospheric burners the ventilation air volume also needs to include the additional requirement for the diluting air into the draught diverter(s).

In addition to the inlet louvres, which should be sited at low level, it is necessary to provide for disposal of the surplus ventilating air. As this is warmer than the supply, it will collect near to the roof or ceiling and exhaust louvres should be sited as high as possible in the boiler house. The possibility of noise nuisance through any openings, inlet or outlet, to the boiler house must not be overlooked and special louvres having acoustic linings may be required.

Where the boiler plant is sited within a building, away from the building perimeter, it may be necessary to provide mechanical systems for the inlet and possibly also for the extract ventilation. It is worth emphasising here that whilst mechanical inlet and natural extract ventilation, or mechanical inlet and (reduced volume) mechanical extract ventilation are both permissible, natural air inlet and mechanical extract ventilation are prohibited. Practical difficulties arise with mechanical ventilation systems for this application since boiler loading and consequent combustion air requirements will vary throughout the year, leading to imbalance between inlet and extract unless some form of capacity control is provided. Also as the mechanical ventilation is an essential part of the combustion process the ventilation system operation and air proving will need to be interlocked with the fuel burner controls.

Chapter 18

Ventilation

The act or art of ventilating is, according to the *Oxford Dictionary*, to 'expose to fresh air' and to 'cause air to circulate freely in an enclosed space' (*L. ventilatio*). As now understood, the purpose may be either to supply the untainted air necessary for human existence or to provide for dispersal of smells, noxious gases and dangerous concentrations of fumes and smoke. It has become practice to avoid the use of the definition *fresh air*, with its connotation of purity, and to refer instead to *outside air* leaving open the possibility that pre-treatment may be necessary.

The need for a supply of outside air as a contribution to human comfort, the parallel requirement for the removal of contaminants and some of the limiting considerations have been touched upon in the first chapter of this book. The subject matter which follows is concerned primarily therefore with quantification of both need and requirement followed by consideration of some of the methods used to meet them.

For the first time the 2002 edition of Part L of the *Building Regulations* impose requirements on air-conditioning and mechanical ventilation systems, in that if they are installed, they should be 'reasonably efficient'. Clearly the most efficient approach to ventilation is by natural means. In *BRE Digest 399 Natural Ventilation of Non-domestic Buildings*, natural ventilation is defined as ventilation driven by the natural forces of wind and temperature. It is intentional and, ideally, controlled. It should not be confused with infiltration, which is unintentional and uncontrolled entry of outdoor air through cracks and gaps in the external fabric of the building. This aspect is covered in Chapter 4 of this book.

Natural ventilation may, in certain circumstances, provide tolerable comfort conditions in the climate of the British Isles for office and other buildings provided that:

- There will be, in the vicinity of the site, no excessive atmospheric or noise pollution which will require the building to be sealed.
- The purpose of the building is such that there will be no activities or processes which require close control of the internal temperature or humidity.
- The design of the building is such that excessive solar gains will not occur.
- There will be no excessive heat gains arising from office machinery, etc.

Considerable work has been undertaken into past few years to understand the complexities and variability of natural ventilation solutions mainly as a result of increased concern over the adverse environmental impact of energy use. This has encouraged the design and construction of energy efficient buildings, many of which are suited to natural ventilation because of the holistic approach in the design of the building fabric, the structure and the engineering systems. These buildings can, in certain circumstances, provide year round comfort, with good user control, usually at lower capital and maintenance costs. This integrated approach to design of ventilation systems is incorporated within the 2001 edition of the *Guide Section* B.2. The *CIBSE Applications Manual AM 10: 2005* provides a comprehensive analysis of the subject including, calculation methods and case studies and readers requiring more detail information are advised to refer to this document.

It is appropriate to mention here so-called 'mixed-mode' solutions that combine natural ventilation with mechanical ventilation and/or cooling in the most effective manner to achieve satisfactory year-round conditions

Table 18.1 Ventilation rates required to limit CO_2 concentrations

	Minimum ventilation required (litre/s per person)		
Activity	*0.1% CO_2*	*0.25% CO_2*	*0.5% CO_2*
At rest	5.7	1.8	0.85
Light work	8.6–18.5	2.7–5.9	1.3–2.8
Moderate work	–	5.9–9.1	2.8–4.2
Heavy work	–	9.1–11.8	4.2–5.5
 Very heavy work	–	11.8–14.5	5.5–6.8

whilst minimising energy use. The *CIBSE Application Manual AM 13: 2000* provides guidance on this subject. Mixed-mode or hybrid ventilation is described in more detail later.

Air supply for human emissions

The volume of air necessary to provide for human occupancy may be considered under the following principal headings:

- Provision of oxygen for respiration
- Removal of products of exhalation
- Removal of body odour
- Removal of unwanted heat
- Removal of unwanted moisture
- Removal of contaminants.

Respiration and exhalation

At rest, the normal adult inhales between 0.10 and 0.12 1itre/s of air and of this only about some 5 per cent is absorbed as oxygen by the lungs. The exhaled breath contains between 3 and 4 per cent of carbon dioxide (CO_2) which amounts to about 0.004 litre/s.[1] The accepted level for a maximum concentration of CO_2 within an occupied space is 5000 parts per million, or 0.5 per cent by volume, for an exposure of 8 hours. The outside air requirement, at an equilibrium condition, to restrict the level of concentration to the maximum level permitted, may be found by using the contaminant expression given on page 404. This indicates a very low rate of 0.847 litre/s per person.

Table 18.1 lists ventilation rates for concentrations other than the maximum and for other levels of activity but there are parallel aspects of human occupation of a space which are more important criteria for the need for ventilation.

As a result of these parallel effects, a room may feel fresh and pleasant with CO_2 content higher than that noted above and yet feel stuffy with a much lower content. In consequence, it has in the past been considered that it is an unsatisfactory datum for assessment of ventilation quality. However, the results of research at BSRIA and elsewhere suggest that it might well be used as a means for ventilation control.

Body odour

One of the essentials of good ventilation is the removal of odours arising from human occupation which can become serious in crowded places. A supply of outside air at a rate of at least 5 litre/s per person has been found to be the minimum quantity necessary to be reasonably sure that no trouble will arise from this source but a rate of about 8 litre/s is to be preferred. This quantity should be adjusted upwards in instances where occupation is particularly dense and, for factory canteens and the like, a rate of 10–15 litre/s per person may be appropriate.

Table 18.2 Heat and moisture from occupants

	Heat emission per occupant (W)		
Activity[1]	*Sensible*	*Latent*	*Total*
Sedentary worker	100	40	140

Note
1. See Table 5.8 for a range of activities.

Table 18.3 Contaminants released to an enclosure by the combustion of fuel in a direct heating appliance

	Rate of release of contaminants (g/h per kW)		
Fuel	CO_2	SO_2	*Water vapour*
Kerosene	244	0.155	151
Natural gas	187	–	98
Butane	220	–	113
Propane	216	–	117

Unwanted heat

A sedentary worker, as may be noted from Table 18.2, will emit sensible heat at a rate of about 0.1 kW. If it be assumed that ventilation air is provided to the workplace at the rate of 16 litre/s per person, double the minimum quantity required to combat body odour, then this sensible heat will lead to the air temperature rising by $(0.1 \times 1000)/(16 \times 1.205 \times 1.012) = 5.1$ K, assuming that there is no heat loss from the room.

Unwanted moisture

Similarly, Table 18.2 shows that the same sedentary worker will produce latent heat at a rate of about 0.04 kW. This represents a moisture output of $0.04 \times 3600 \times 1000/2450 = 59$ g of water vapour per hour or, with respect to the ventilation air quantity of 16 litre/s per person noted above, 1.02 g/m^3. Table 18.3, last column, shows the moisture produced by un-flued direct heating appliances.

Contaminants

Much recent research has been directed to assessment of the ventilation rate required to deal with pollution in offices and workrooms arising from tobacco smoke. Smoking within offices is now usually restricted to smoking rooms or banned completely. In the near future this will also apply to all buildings used by the general public in the UK. Current recommendations are that, where smoking is allowed, mechanical ventilation should be 4–5 times the normal rate of 10 litre/s.

Air supply for other reasons

As in the case of emissions from occupants, those from other sources may be grouped as being unwanted heat, unwanted moisture and contaminants, as follows:

Unwanted heat

Incidental sources of heat have been referred to in Chapter 5, up to 40 W/m^2 being quite normal in a commercial office, and where these occur it may be possible to introduce a sufficient quantity of outside air to prevent the internal temperature from rising above some predetermined level. That quantity may be found from the expression:

$$Q = (1000\ H)/pc$$

where

Q = outside air quantity for 1 K rise (litre/s)
H = heat emission (kW)
p = specific mass of air (kg/m^3)
c = specific heat capacity of air (kJ/kg K)

Similarly heat gains from the sun, particularly during the summer months, may in some cases be removed by a simple ventilation system but in many instances the magnitude of such a gain is such that a very large air volume would be needed which may be undesirable from other points of view. Where the building structure is heavy, it may be possible to improve daytime conditions during the summer by use of a high ventilation rate during the night, when the outside air is cooler, and a lower rate during the day.

In general terms, however, it must be remembered always that the removal of heat by simple ventilation is achieved by raising the temperature of the air introduced. In any normal circumstance therefore, with reliance upon ventilation alone, the result may well be that the temperature within the space is higher than that prevailing outside. If such a situation were considered to be unacceptable then it would be necessary to consider the application of air-conditioning, as dealt with in later chapters.

Unwanted moisture

In certain special circumstances, the need to combat condensation may be the criterion for ventilation rate. One such instance is that of a swimming pool hall where, of the requirements to supply air to the occupants – removal of the familiar 'chemical' smell and action to combat condensation – the latter is by far the more important. It has been shown that, with double glazing, the outside air supply rate should not be less than 10 litre/s per m^2 of total pool hall area and a minimum of 12 litre/s per person.[2]

In housing, it has been shown that a mean ventilation rate of one air change per hour (see later comment upon this quantity) is required to prevent condensation on single-glazed windows in a dwelling which is heated uniformly. In domestic kitchens, the ventilation requirement to avoid condensation is about 100 litre/s for electric cooking and half as much again for gas cooking. Packaged units which provide for whole-dwelling supply and discharge ventilation, including energy recovery, appear to provide for an outside air volume of 70 litre/s or about 1.5 air changes per hour. Minimum and recommended rates for dwellings are given in Part F of the Building Regulations 2000 (2006 edition).

Contaminants

Under this heading, a whole range of industrial hazards arises in addition to the particular requirements for medical buildings, laboratories, animal rooms, horticulture, etc., which are too diverse and too specialised to be dealt with here. It is not possible to offer any generalised observations in this respect other than to refer to the equilibrium expression which follows here. This may be used when details of the process and the permissible concentration of the contaminant are known in any set of consistent units: a brief list of common contaminants is given in Table 18.4.

$$Q = P[(1-C_1)/(C_1-C_2)]$$

where

Q = rate of supply of outside air
P = rate of contaminant release
C_1 = permissible concentration in room
C_2 = contaminant concentration in outside air

Air change rates

Where the occupancy is unknown or variable, an arbitrary basis for the ventilation rate must be taken. Table 18.5 may be used as a guide but the cubic content of the space and the length of time for occupation may have an effect upon the rate of air supply necessary. For example, the air volume of a very large hall might be considered as holding a 'store' of air which could be used to reduce the quantity provided over a short period of occupancy.

Criteria for air supply to occupied spaces

Good ventilation cannot be defined in simple terms which can be quantified as hard and fast rules. Reference must be made to conditions which have been found in practice to give reasonably satisfactory results, as outlined

Table 18.4 Limiting values of some common contaminants (Titon)

Contaminant	*Exposure limit* (ppm) *8 hours*	*10 minutes*	*Contaminant*	*Exposure limit* (ppm) *8 hours*	*10 minutes*
Acetone	750	1500	Hydrogen peroxide	1	1.5
Ammonia	25	35	LPG	1000	1250
Bromine	0.1	0.3	Methanol	200	250
Butane	600	750	Nitrous oxide	100	–
Carbon dioxide	5000	15 000	Ozone	0.1	0.3
Carbon monoxide	50	300	Sulphur dioxide	2	5
Carbon tetrachloride	2	–	Trichloroethane	100	150
Chlorine	0.5	1	Turpentine	100	150
Formaldehyde	2	2	White spirit	100	125

Note
These data are presented to illustrate the wide range of limiting values. Reference in practice should be made in all cases to the current edition of the Health and Safety Executive publication *Occupational Exposure Limits*, *EH* 40/– which is published annually.

Table 18.5 Indication of ventilation requirements based on air change rates

Type of enclosure	*Air changes per hour*	*Ventilation Allowance* ($W/m^3 K$)
Assembly halls	3–6	1.0–2.0
Bedrooms	0.4–1.0	0.13–0.33
Boiler rooms and engine rooms	10–15	3.33–5.0
Class rooms (refer to Building Bulletin 101)	3–4	1.0–1.33
Corridors	2–3	0.67–1.0
Entrance halls	3–4	1.0–1.33
Factories, large open type	1–4	0.33–1.33
Factories, densely occupied workrooms	6–8	2.0–2.67
Foundries, with exhaust plant	8–10	2.67–3.33
Foundries, without exhaust plant	10–20	3.33–6.67
Hospital operating rooms	20	6.67
Hospital treatment rooms	10	3.33
Kitchens above ground	20–30	6.67–10.0
Kitchens below ground	40–80	13.33–26.67
Laboratories	10–15	3.33–5.0
Laundries, dye houses and spinning mills	10–20	3.33–6.67
Libraries	3–4	1.0–1.33
Living rooms	0.4–1.0	0.13–0.33
Offices above ground	2–6	0.67–2.0
Offices below ground	10–20	3.33–6.67
Restaurants and canteens	10–15	3.33–5.0
Rolling mills	8–10	2.67–3.33
Stores and warehouses	1–2	0.33–0.67
Workshops with unhealthy fumes	20–30	6.67–10.0

Note
Also refer to relevant CIBSE Guides, Building Bulletins, Building Regulations etc.

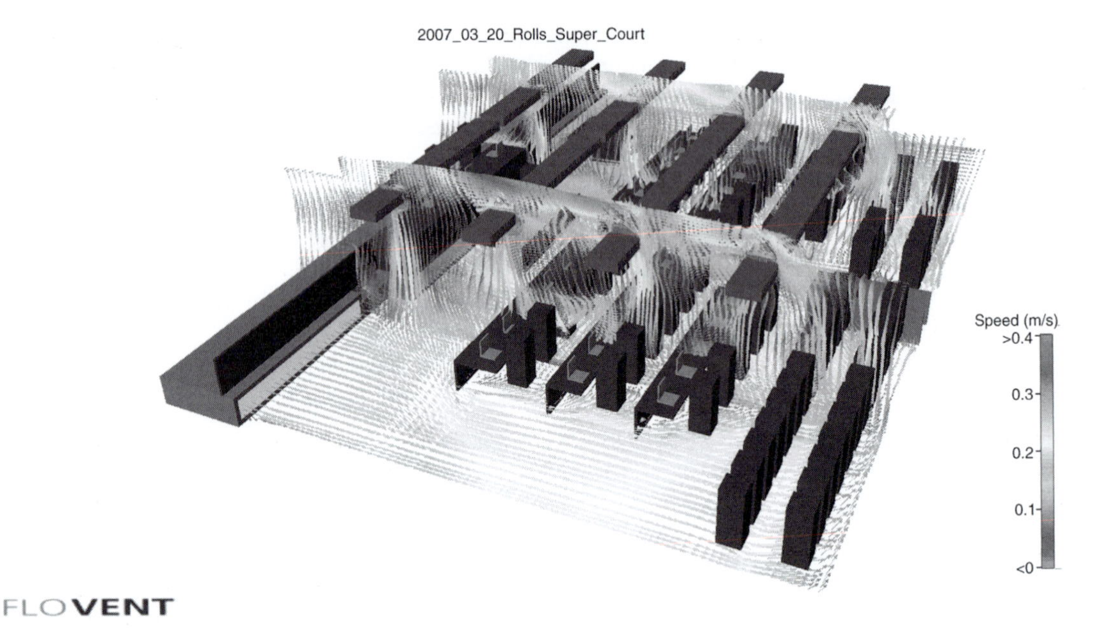

Figure 18.1 Simulation of air flow within a court room (see colour plate section at the end of the book)

in the following few paragraphs. It may be of interest to compare these with the fundamentals of natural ventilation given by Walker[3] in 1850:

- Windows are to admit light and not air; ventilation should be catered for separately
- Both inlets and outlets are necessary
- Incoming air should be warmed to avoid draughts
- Inlets and outlets should be well distributed
- Ventilating openings should be permanent, realising that once closed, they will remain closed.

Nowadays we are able to use modelling software which will simulate the air flow. This type of software is generically known as 'computational fluid dynamics' (CFD) and is available from a number of sources. The output from a CFD analysis is a 3D model which can provide data of air velocities, temperatures and contamination levels. These are widely used to check air flow performance in large spaces such as atria or in unusual applications to predict and eliminate any problems in the air flow design and are further described in Chapter 6. Figure 18.1 shows the simulation of air flow within a large court room.

Distribution and air movement

The admission of outside air into an occupied space is discussed at some length in Chapter 20 on Air Distribution, the cardinal principles being that:

- It is evenly diffused over the whole area served, particularly at breathing level
- It should not strike directly upon the occupants
- It should, nevertheless, provide a feeling of air movement and not allow any areas of stagnation

The method of distribution adopted often determines the volume of air to be circulated. Though a small quantity of outside air may be all that is required for a sparse population in a large room, it may be impossible to diffuse this evenly over the whole area and to avoid pockets of stagnation. It will be necessary in consequence to increase the volume in circulation either by adding further outside air or by arranging to add a proportion of room air as recirculation.

Conversely, in a small crowded room, distribution problems may make it impossible to introduce the required quantity of outside air without causing draughts. In this case, the volume could be reduced which might lead to an unavoidable temperature rise which may not be acceptable: in this situation, the only solution would be to cool the air before it is admitted.

Given a room of reasonable proportions, the limits of effective distribution consistent with the maintenance of comfort appear to range between a minimum of about four and a maximum of up to 20 air changes per hour. Obviously, the higher rates cannot be achieved unless mechanical means are employed to provide the necessary propulsion.

The extraction of vitiated air from a space is not to be relied upon to create a directional effect. Movement upwards, downwards or sideways will, in a general sense, be towards the point of removal but with no positive momentum.

Temperature

If the temperature of the air admitted were too much below that of the room, it would fall to the floor without proper mixing and might cause cold draughts. On the other hand, the temperature must not be too much above that of the room since in that case the air would rise to the ceiling with the result that stagnation could occur in the breathing zone.

In the case of a straightforward ventilation system, which is under discussion here, it is assumed that the heating of the building is dealt with by some separate radiator or other similar plant. This will have been designed to balance the heat loss through the building fabric and to maintain an internal temperature of, say, 20°C during winter weather. In these circumstances, the ventilation air supply should be delivered to the room at about 17°C, or 3 K lower than the desired temperature, so that it has the potential to reach 20°C as it absorbs unwanted heat prior to leaving the room.

Under summer conditions, without provision for mechanical cooling, no control can be exercised over the temperature of the air delivered to the room other than by the small reduction made possible by a process of evaporation, referred to later.

Humidity

Ventilation air admitted from outside will have the same moisture content as that prevailing at the source. In cold weather, as was mentioned in an earlier chapter, this moisture content may be low and although the quantity will not alter in the *absolute* sense when the air is heated to say 17°C, the *relative* humidity will have fallen. This situation may be corrected, where mechanical inlet plant is used, by incorporating humidification equipment in the form of steam injection or some other device for producing water vapour. By such means, the amount of water added may be controlled to give a desirable relative humidity of 40–60 per cent, at room temperature.

In the normal mid-season mild weather of spring and autumn in the British Isles, the humidity prevailing outside will generally be at a level which will permit some temperature adjustment without any necessity to take corrective action. The human body, as was explained earlier, is not too critical of humidity variations at the temperature levels then prevailing. There is, in addition, a reservoir effect in most buildings as a result of the hygroscopic retention of fabrics, timber, paper, etc., which serves to steady changes in humidity taken over a long period.

In summer, when the humidity outside is high, there can be no complete control of internal humidity without full air-conditioning which, by definition, includes means for dehumidification.

Air purity

Most buildings requiring some form of ventilating system are in the centre of towns or cities where atmospheric pollution is at the highest level normally encountered. As a result of the action which followed the implementation of the 1968 Clean Air Act, the soot, tar, ash and sulphur dioxide which had previously been a major problem as a result of indiscriminate use of coal and heavy oil, are no longer the principal pollutant.

Their place has largely been taken by the effluents produced by motor vehicles, in contrast to those complained of by Florence Nightingale in 1860, when she referred to:

> Dirty air coming in from without, soiled by sewer emanations, the evaporation from dirty streets, bits of unburnt fuel, bits of straw and bits of horse dung.

When air is introduced into a building by ventilation, it brings with it a proportion of whatever pollutants exist outside. If inlet is to be by natural means, i.e. through open windows and the like, then there is very little to be done to prevent this incursion. Where the supply of air is to be by mechanical means then a whole variety of types of filters is available. Some of these are discussed later in Chapter 22.

The use of *activated carbon* filters will revive vitiated air by adsorption of many impurities such as body odours, sulphur dioxide, petrol fumes, and other noxious products of civilisation. Filters of this type are, however, high in first cost and have a relatively short service life before they are either replaced or where possible returned to the manufacturers for re-activation.

Methods of ventilation

Ventilation can be either fully 'natural', induced by wind or temperature, fully 'mechanical', where air movement results from power drive applied to a fan or fans or it can be a combination of natural or mechanical. Where a combination of natural and mechanical ventilation is used intermittently this is known as a mixed-mode or hybrid system.

The options for ventilation systems are described under the following sub-headings:

- Natural inlet and natural outlet
- Natural inlet and mechanical extract
- Mechanical extract and natural outlet
- Mechanical inlet and outlet
- Mixed mode or hybrid.

Natural inlet and natural extract

Arrangements of this sort are used in buildings which seem to fall into three categories which, for want of better descriptions, might be called traditional, commercial and industrial. We have retained in this section descriptions of the approach taken to natural ventilation in the past, e.g. solid fuel open fireplaces inducing air flow. This gives the reader information on the development of methods of ventilation and an appreciation of current approach including mixed-mode systems and CFD analysis.

Traditional

This category includes most school and university buildings, hospitals, shops, office buildings constructed between the wars, almost all domestic premises and other rooms and buildings with low levels of occupancy. Conditions internally depend upon clean outside air and upon other external features which permit windows to be opened.

Where an open fireplace exists, the flue provides a route for exhausting air from the room. Further, in instances where a fireplace is in use for burning solid fuel, the flue serves the dual role of not only carrying away products of combustion but also of inducing a flow of room air much greater in quantity than is normally required, as illustrated in Fig. 18.2: an adjustable restriction at the throat of the flue is advised in order to ameliorate this problem. It must be added that a gas fire connected to a flue will have the same effect but at a much reduced level. Various methods of providing natural ventilation to rooms have been devised as shown in Fig. 18.3:

(a) The Dr Arnott ventilator of the 1850s which, built into a chimney breast at high level in the room, acted to remove 'used' air as induced by the flow of products of combustion in the flue.

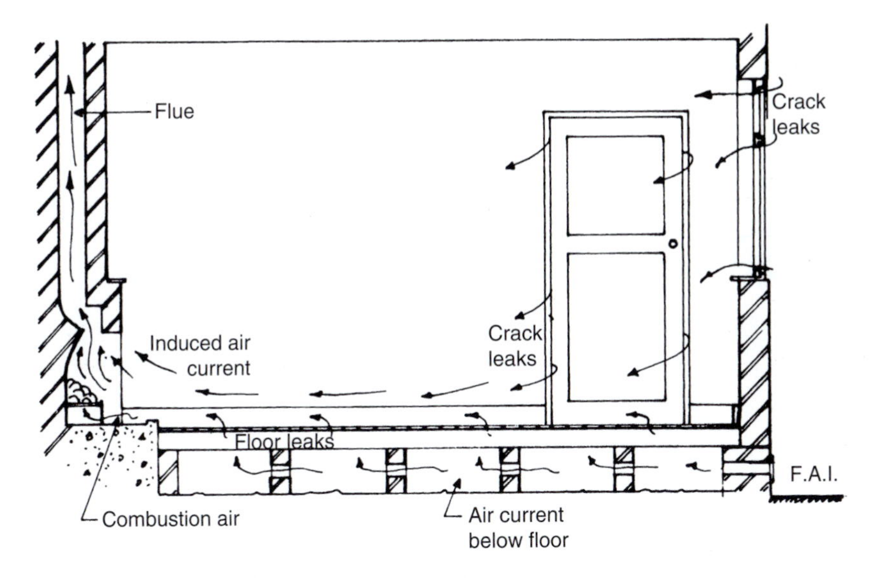

Figure 18.2 Ventilation by open fire

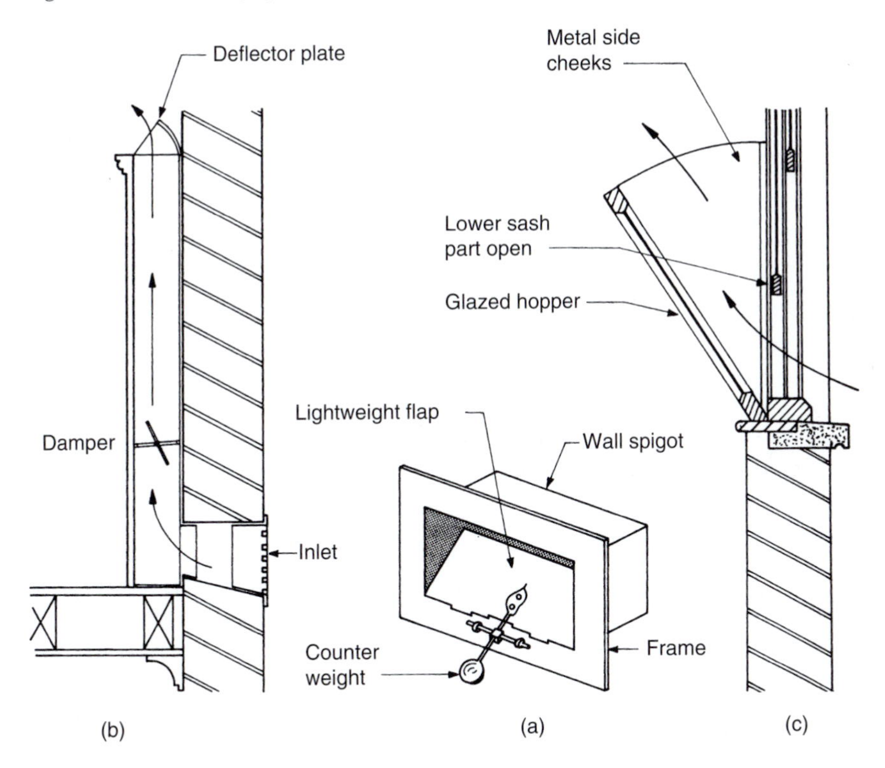

Figure 18.3 Natural ventilators

(b) The Tobin tube of the same era which sometimes had a water tray fitted at the low-level inlet 'to cleanse the entering air from smuts or organic impurities' or a wetted fabric 'sock' suspended from the outlet as a terminal filter.

(c) The hopper window of Edwardian school rooms which, when the lower sash was opened, diverted entering air to a higher level in the room.

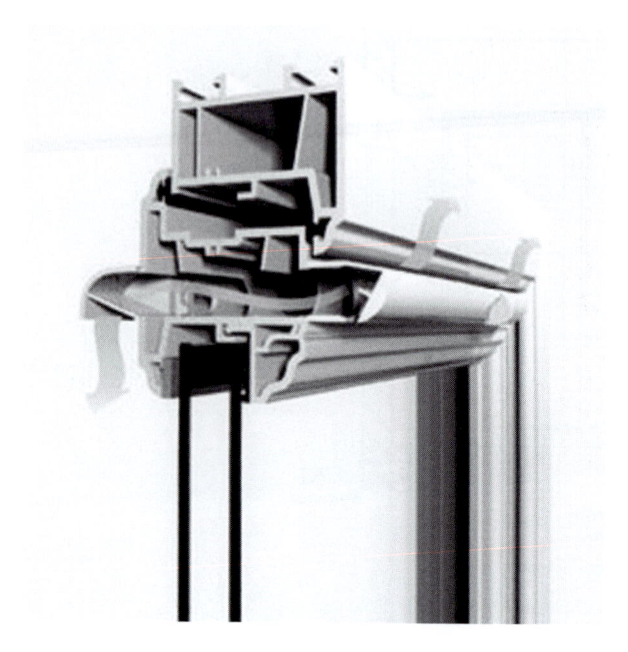

Figure 18.4 Trickle vent (Titon) (see colour plate section at the end of the book)

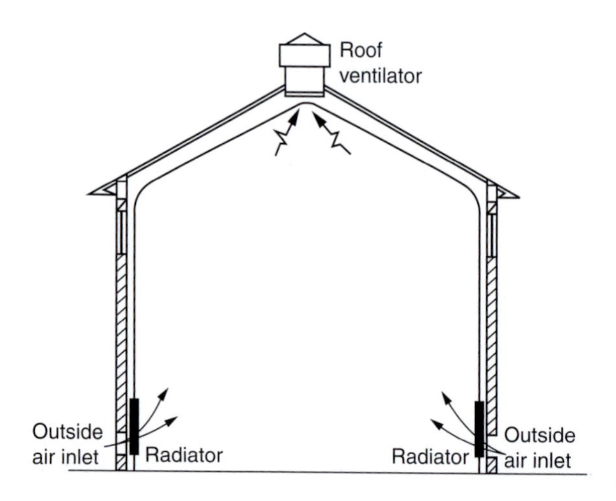

Figure 18.5 Natural inlet and natural extract

Many modern windows of better quality, whether metal or timber, provide some means for 'trickle' or 'night' ventilation. This feature is sometimes incorporated in the window proper or in the fastening mechanism but more usually at the head of the sub-frame, see Fig. 18.4.

Commercial

For larger rooms, assembly halls, workshops, etc., simple high-level outlet ventilators, combined with low-level inlets for outside air, provide a solution which is low in capital cost but unreliable in operation. A simple system is illustrated in Fig. 18.5 where air inlet is through grilles fitted behind radiators, an arrangement much favoured in the past. The grille openings, inside and outside, soon become dirt and insect traps and are often

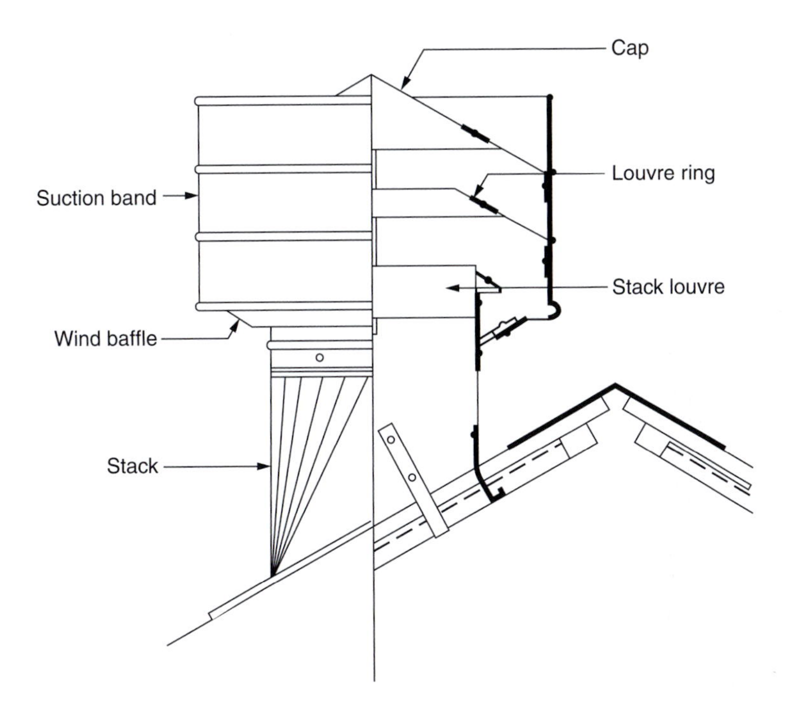

Figure 18.6 Robertson's ventilator

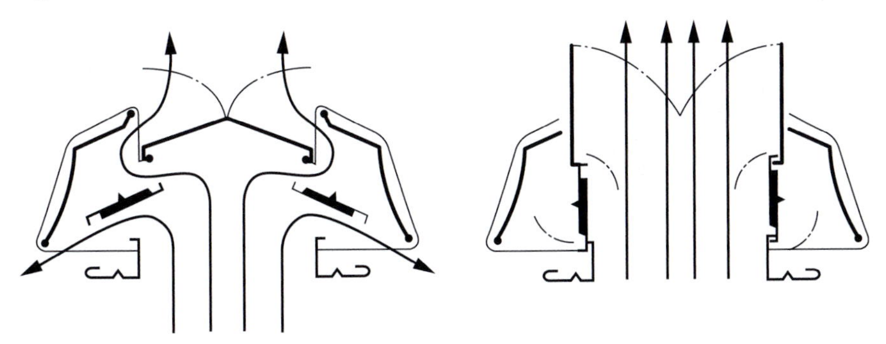

Figure 18.7 Colt type MF ventilator

sealed by occupants in order to avoid winter draughts. In hot weather, when ventilation may be most needed, little appreciable air movement will occur in consequence.

Industrial

This, in effect, is an extension to the previous category but with the clear distinction that the type of building is different as is the equipment used. There are great many heavy industries where natural ventilation has been, and will be applied with success to single-storey shed type constructions. These industries are those in which there is a considerable release of heat or steam, such as foundries, rolling mills, welding shops, dye houses and power stations, etc. The quantity of heat to be removed is significant and a temperature rise of 2 K/m height may often be encountered. The equipment used in this application has been evolved over the years with a proven capability, an example being the well-known Robertson's roof extract unit shown in Fig. 18.6.

An alternative, with a less obstrusive profile, is a roof ventilator panel with labyrinth air path louvres. A further alternative form of roof unit is illustrated in Fig. 18.7, this being a multi-function device, having

Figure 18.8 Louvred wall panel for natural inlet

side dampers which may be closed by remote operation at times when no ventilation is required, and a weather cap, shown in the Fig. 18.7(a) diagram, which may be opened similarly when a completely unobstructed outlet is necessary. Such units are manufactured to provide outlet areas of between about 0.8 and 2.5 m^2. Any natural ventilators in the roof can offer the considerable advantage of providing smoke venting in the event of fire by encouraging rapid clearance of smoke. On an industrial scale, inlets for natural ventilation may well present a problem because of their size and possible proximity to those operatives who may not be affected directly by excessive heat. Wall openings, provided with louvres as Fig. 18.8, may be acceptable and be arranged for automatic opening in parallel with the associated roof outlets. Air admitted by such louvres is unlikely to be heated at the point of entry in winter, since the volume required may well be too large to pass over a conventional heater without mechanical assistance.

Temperature difference effects

In some respects, the motive force arising as a result of temperature difference is similar to that utilised in a gravity hot water heating system. This so-called *stack effect* relates a column of air within a building to one of similar cross section outside and considers the average specific mass of each, together with the acceleration effect of gravity (9.81 m/s^2). The mass varies inversely with absolute temperature, approximately as $353/(t + 273)$, and thus the following expression may be derived:

$$\Delta P = (9.81 \times 353 \times H)\,[1/(t_1 + 273) - 1/(t_2 + 273)]$$

and thence, for the range of temperatures encountered in practice:

$$Q = (268 \times A)[H(t_1 - t_2)]^{0.5}$$

where

ΔP = pressure difference (Pa)
Q = air volume (litre/s)
A = free area of inlets or outlets (m^2)
H = height, centre of inlets to centre of outlets (m)
t_1 = average indoor temperature over height, H (°C)
t_2 = average outdoor temperature over height, H (°C)

Table 18.6 lists values calculated from this expression, including a resistance factor as noted later, for a range of heights and temperature differences.

Table 18.6 Air volumes for natural ventilation due to stack effect

Effective Height (m)	*Air volume* (litre/s per m²) *free opening for temperature difference* (K)								
	2	*4*	*6*	*8*	*10*	*15*	*20*	*25*	*30*
2	240	340	420	480	540	660	760	850	930
4	340	480	590	680	760	930	1070	1200	1310
6	420	590	720	830	930	1140	1310	1470	1610
8	480	680	830	960	1070	1310	1520	1700	1860
10	540	760	930	1070	1200	1470	1700	1900	2080
15	660	930	1140	1310	1470	1800	2080	2320	2550
20	760	1070	1310	1520	1700	2080	2400	2680	2940
25	850	1200	1470	1700	1900	2320	2680	3000	3290
30	930	1310	1610	1860	2080	2550	2940	3290	3600

Notes
1. Areas of openings are assumed to be equal, i.e. upper = lower.
2. Openings are assumed to be 45 per cent effective.

Table 18.7 Air volumes for natural ventilation due to wind

Effective height above ground (m)	*Air volume* (litre/s per m²) *free opening with notional wind speed of* 3 m/s							
	Openings normal to wind				*Openings at angle to wind*			
	Country				*Country*			
	Open	*With wind breaks*	*Urban*	*City*	*Open*	*With wind breaks*	*Urban*	*City*
2	1260	990	690	440	690	540	380	240
4	1420	1130	820	550	720	620	450	300
6	1520	1230	900	630	830	670	490	340
8	1610	1300	970	690	880	710	530	370
10	1660	1360	1040	740	910	740	570	400
15	1780	1470	1140	850	970	800	620	460
20	1870	1560	1220	930	1020	850	670	510
25	1940	1630	1290	1000	1060	890	700	550
30	2000	1690	1350	1060	1090	920	740	580

Notes
1. Areas of openings are assumed to be equal, i.e. inlet = outlet.
2. Wind speed is 3 m/s (exceeded for 80 per cent of time).
3. Wind speed is at notional height of 10 m in open country; other speeds are derived from this datum.

Wind effects

Natural ventilation resulting from wind is unpredictable and any attempt to quantify the effect should be based upon data which give some indication of the frequency of occurrence. Statistics which are reproduced in the *Guide Section A2* relate to the *meteorological wind speed*, i.e. as measured in open country at a height of 10 m. For use in practice, factors are published which enable this speed to be adjusted to suit other heights and alternative topographies. In addition, *isopleths* (contours of hourly mean wind speeds) are available together with multipliers which allow frequency of occurrence to be estimated. In the context of ventilation, the design criterion must be aimed at the low end of the scale and, for a speed which will be exceeded for 80 per cent of the time; a figure of about 3 m/s would appear to be a reasonable choice. Table 18.7 has been prepared, from the simple relationship $Q = A \times V$, using this wind speed for the columns headed 'open country' and adjusted values for other locations. A resistance factor, as noted in the following paragraph, has been applied to produce the volumes listed.

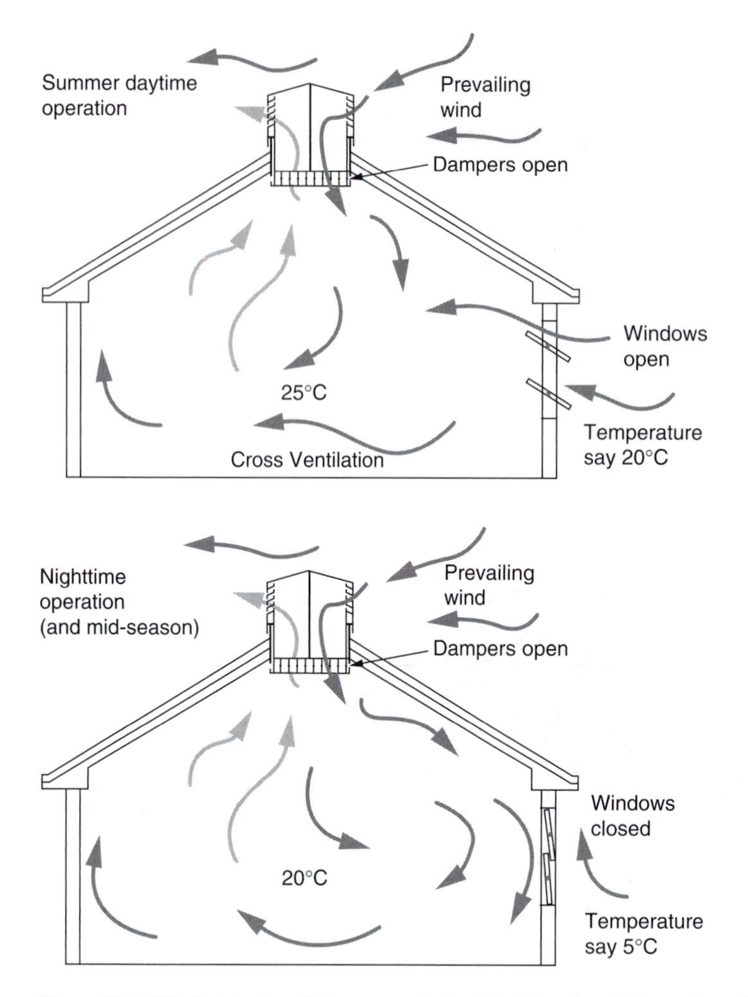

Figure 18.9 'Windcatcher' ventilator as manufactured by Monodraught (see colour plate section at the end of the book)

Resistance factors

Theoretical air flow quantities due to temperature and wind effects are not achieved in practice due to flow resistances and the physical arrangement of inlet and outlet openings. For the flow due to temperature effects, a correction factor of 0.45 was used in preparing Table 18.6 and, for Table 18.7, correction factors of 0.55 and 0.3 were used for the first and second sets of columns, respectively. These factors are probably appropriate to buildings where simple louvred ventilators are used but should not be confused with the comprehensive performance data published by specialist manufacturers of ventilators for the industrial category.

Other considerations

The combined results of temperature difference and wind effects may be estimated by their addition, with respect to sign: that is to say, with the assurance that both are acting to produce air flow in the same direction (which is by no means always the case!). Similarly, the matter of a positive wind effect upon an exposed face of a building and a negative effect upon the lee side must be considered, particularly in circumstances where openings may not be of the same size. The entire subject cannot be dealt with in the abstract since the variables are complex and, to some extent, depend upon the shape of the building. The essential factors are given in the *Guide Section A4*. Figure 18.9 shows a 'Windcatcher' ventilator which uses compartmentalised vertical

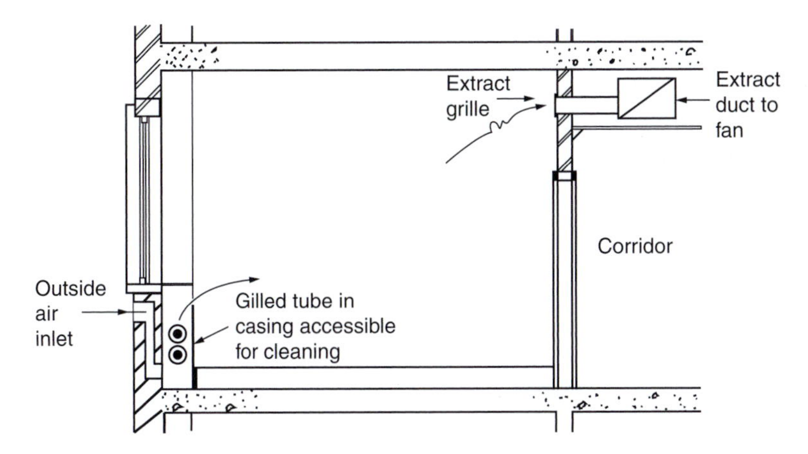

Figure 18.10 Natural inlet and mechanical extract

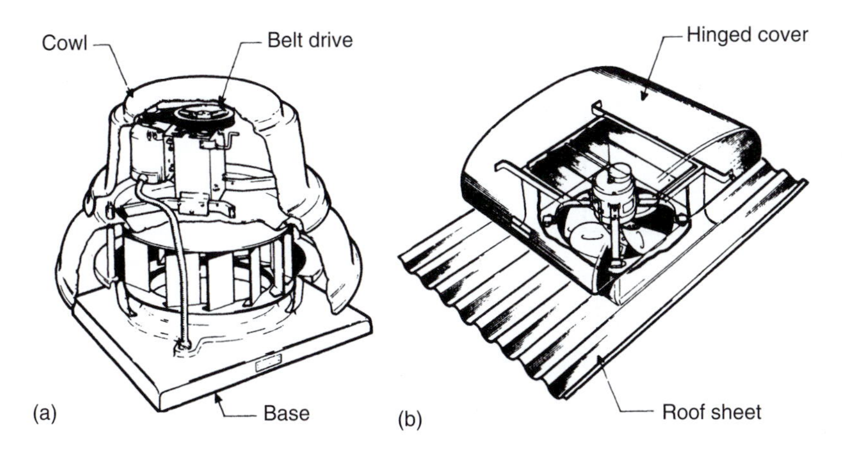

Figure 18.11 Fan-powered extract units: (a) centrifugal and (b) propellor

vents utilising the effects of wind and temperature to naturally ventilate a space or room. The diagram shows the principles for summer operation together with the mid-season and night-time cooling operation.

Natural inlet and mechanical extract

A mechanical extract system will function irrespective of wind and temperature difference and will be positive in action. Since the air to be extracted from the space must be replaced and the means provided for admission of outside air will present some resistance to flow, leakage inward from surrounding spaces is more likely than leakage outward. In consequence, escape of steam, fumes and noxious vapours generated within the ventilated space is less likely than would be the case if reliance were placed upon natural extract alone.

A difficulty arises however in providing a satisfactory means of admitting the bulk of the air required to balance the extract volume, and of heating it in winter. Fresh air inlets behind ranges of gilled tube or some other form of natural convector, as in Fig. 18.10, need regular cleaning and have only a limited application.

For summer use only in a small building, infiltration via the type of window ventilator ('trickle' or 'night') may suffice. In some circumstances, replacement air may be drawn from another part of the building to serve a dual purpose; a good example being extract ventilation from a small kitchen which, by drawing air through serving hatches, serves to dilute the spread of cooking smells to the associated dining area.

For industrial applications, the natural roof ventilator shown earlier might be replaced by some form of fan-powered extract unit as Fig. 18.11. Units of this type are made in a variety of patterns with mounting arrangements

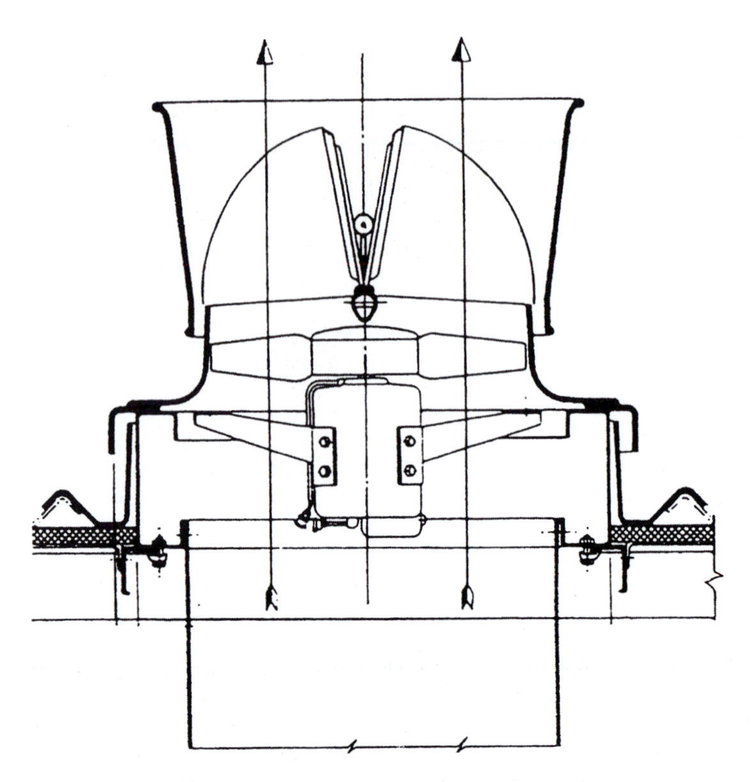

Figure 18.12 Vertical jet-discharge roof unit

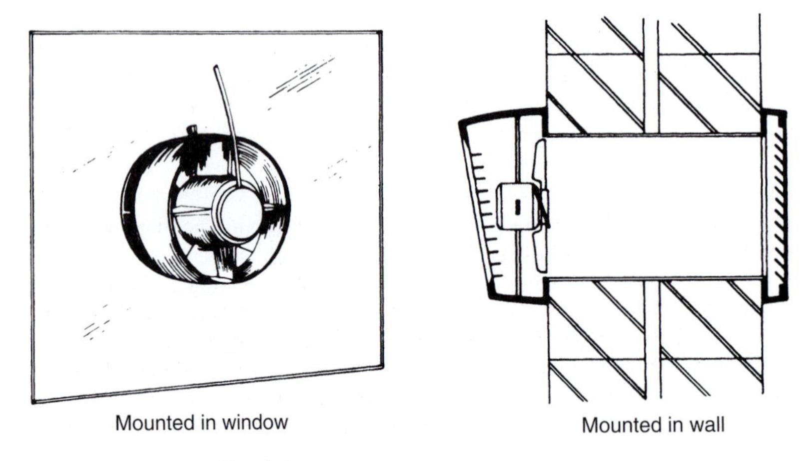

Figure 18.13 Mounted in window

and weathering covers to suit most types of roof construction, pitched and flat. The air extracted by such units is discharged at a relatively low velocity and has a tendency to hug the roof profile, thus, where fumes and other pollutants are carried in that air, it is better practice to use a *vertical jet-discharge* unit as Fig. 18.12. Axial flow fans, which are commonly more noisy, are often fitted to such units, together with hinged vanes or dampers, for rain protection when the fan is idle. These units may be connected to either horizontal or vertical ductwork.

For domestic use and indeed for application to small rooms generally, the familiar window or wall-mounted extract units manufactured in moulded plastic, Fig. 18.13, may well provide a solution to a local ventilation

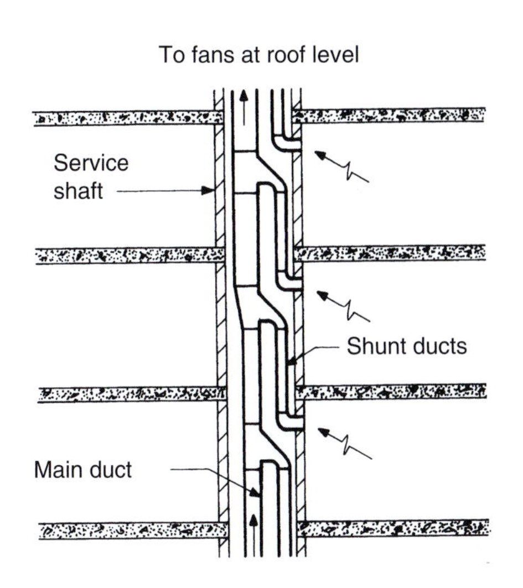

Figure 18.14 Extract duct from toilets in multi-storey flats (fire barriers not shown)

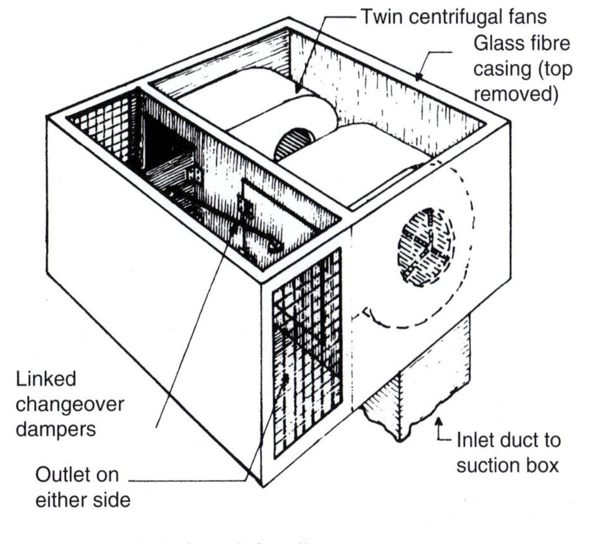

Figure 18.15 Twin fan unit for toilet extract

problem which is adequate for most of the time. Such units may be provided with speed control and, in some instances, are reversible so that they may, alternately, be used for either an inlet or an extract duty.

A specific area where Building Regulation requirements bear upon the provision of mechanical extract with natural inlet relates to internal toilets and bathrooms, particularly in multi-storey housing. The principles of an investigation which was reported over 40 years ago remains valid,[4] the principal conclusion being that a ventilation volume of 5.5 litre/s should be provided for each fixture, i.e. per WC pan or bath, and thus 11 litre/s for a bathroom.

Current regulations require an intermittent extract rate of 15 litre/s or continuous extract rate of 8 litre/s. In order to overcome stack effect and other chance pressure differences due to open doors, etc., a duct system design as Fig. 18.14 was recommended; this has the added advantage of providing an extended travel for voice transfer between dwellings. In the case of office blocks and other large buildings, an extract rate of 10–12 air changes per hour is desirable.

In this context, most authorities require that the exhaust fan must be duplicated, with arrangements for automatic changeover in the event of failure. Conventional fans may be fitted in parallel but it is more convenient to fit a purpose-made twin fan unit of the type shown in Fig. 18.15.

As in other instances, provision must be made for a supply of inlet air from outside and, in a block of flats where toilets and bathrooms are likely to be arranged one above the other, inlets may connect to a rising shaft with, perhaps, some form of heater at the base. Subject to decision by the fire authority, it may be possible to use the shaft provided for rising water pipes for ductwork but not, of course, if space be allocated also to wastes from sanitary fittings. Some authorities require that a lobby be introduced at the entrance to toilets in public buildings and that this is provided with a supply of outside air by an inlet plant serving no other duty.

Mechanical inlet and natural extract

This proposition should not be confused with the plenum heating system described in Chapter 9. It is typified, for application to an office building, by the arrangement shown in Fig. 18.16. As may be seen, an air supply from some central source is ducted through a corridor ceiling void to individual rooms; the vitiated air escapes therefrom, through a low-level register, to the corridor proper and thence to outside. It is now unlikely that a fire authority would permit the construction of such a system since it has the potential to endanger the atmosphere of an escape route. A single room, perhaps an office, might have an outside air supply from some form of reversible fan unit as described earlier but this hardly falls within the context of the present heading.

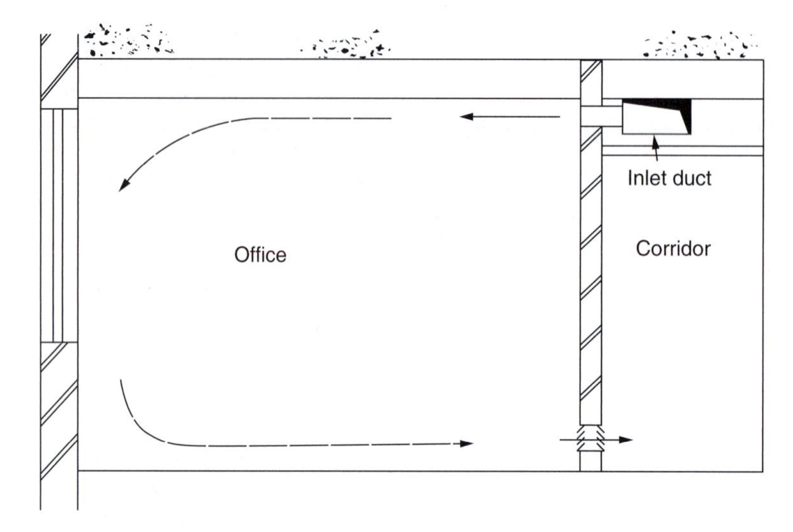

Figure 18.16 Mechanical inlet and natural extract

For an industrial application, such a system might consist of a number of individual unit heaters, each with an outside air inlet as described in Chapter 10 and working in parallel with natural roof ventilators. But this arrangement is barely a credible proposition for present-day requirements.

The ventilation of boiler houses is a special case as a result of the air consumed in combustion. The volume required may be estimated, as was described in Chapter 17, and, where the boiler house is above ground, this quantity may be adequate without any augmentation. Where a boiler house is sited in a basement, or otherwise remote from outside air, the amount of heat emitted by the plant may produce conditions which are unacceptable to maintenance staff and thus be the criterion for determination of the ventilation air quantity. As a notional figure representing emission by a heating boiler plant in winter, a supply of between 5 and 6 times the combustion air quantity is a suitable volume for preliminary design purposes.

It is most desirable that this outside air supply be provided by mechanical means in order to maintain a positive pressure to the boiler plant. The siting of the louvres which admit outside air and any parallel openings for exhaust to atmosphere should be chosen with care since they will act also as a route for the escape of noise. The provision of louvres which have been designed to absorb this noise is a prudent step towards avoidance of what may be a major nuisance and source of complaint.

Mechanical inlet and extract

This final combination is that which must be noted as not only having the widest application but also as providing the greatest challenge since the air distribution, in terms of quantity, pressure and temperature, is wholly in the hands of the designer. It may be applied to all manner of spaces. This provides better control than is offered by the alternatives discussed earlier, but carries a penalty of higher energy consumption. In application, the ratio between the air volume duties of the inlet and extract systems must be selected with care in order to suit the particular application.

For instance, in normal living and working spaces where no noxious fumes are generated, the extract volume should be arranged to be slightly less (by, say, 10–20 per cent) than that provided by the inlet system: any air movement will, in consequence, be outward rather than inward. Conversely, in cases where fumes of any sort might be generated in the working space and should not be distributed, the balance would best be reversed and the inlet volume arranged to be slightly less (by, say, 10–20 per cent) than that handled by the extract system. These two examples serve only to show that each case must be considered on its merits and that there is no single preferred solution.

It will be obvious from what has been said earlier that there is a great number of combinations of inlet and extract arrangements, with and without ductwork, etc., to suit various purposes. It is now commonplace to provide balanced inlet and extract systems for office and other commercial or recreational accommodation but, possibly because of the large air volumes involved, a similar approach is not always applied to industrial buildings. Wherever any form of mechanical extract is used, it is a parallel necessity that replacement air be introduced in a suitable way and at an appropriate temperature. In the most general terms, a satisfactory solution can be produced only by the addition of some form of mechanical inlet.

In this context, the supply arrangement need not necessarily be a complex system and one solution is to provide unit heaters having an outside air inlet, suitably filtered. These may serve also as a means of preheating the building for occupation, when fitted with a damper mechanism which allows recirculation. An alternative method of warming replacement air is the use of direct fired oil or gas heaters.

Mixed-mode or hybrid systems

A mixed-mode or hybrid system is one which combines natural and mechanical ventilation to achieve a comfortable indoor environment and to minimise energy consumption. The system will be controlled and operated to use either natural ventilation, mechanical ventilation or combination of the systems at different times of the day or night and different seasons of the year. The control system will switch between the systems to suit comfort and energy requirements.

The mixed-mode strategy can be based upon natural and mechanical ventilation operating concurrently or the design could be based upon changeover on a seasonal or day/night basis. Concurrent operation could be where a mechanical system provides a base requirement of ventilation with the option for occupants to open the windows as desired.

Changeover systems could be based upon natural ventilation during the day with mechanical ventilation at night to cool the structure in summer. This operation could be changed in winter when mechanical ventilation utilising heat recovery may be the best option for comfort and energy. During mid-season natural ventilation could suffice without any requirement for mechanical backup. If top-up cooling by air-conditioning units is included within the strategy, it is essential that the control system should be linked to any opening of windows so that cooling cannot take place with open windows.

The mixed-mode strategy should be easily understood by the occupants, management and maintenance personnel. The handover documentation should fully explain the principles of operation and should include clear guidance on any limitations or provisions for future fine tuning or additional features.

Kitchens

The removal of cooking odours from a kitchen and prevention of spread to adjacent rooms is most desirable in dwellings but essential as far as hotels, restaurants and institutions are concerned. The ventilation rate must be high if the system is to be a success and air change rates per hour from a minimum of 30 to as many as 100 are not unusual. In order to prevent the spread of odours from the kitchen, the replacement air delivered by a supply plant should generally be 15–20 per cent *less* than the volume extracted, the balance being drawn either from the restaurant or some other intervening area through serving hatches and transfer grilles associated with them. Air velocities through hatches should not be allowed to exceed about 0.2 m/s if complaints are to be avoided. Table 18.8, taken from the 1986 edition of *Guide Section B2*, lists the more common appliances and the notional exhaust air requirement per unit and per m^2 area of appliance. A comprehensive list of equipment extract rates is given in HVCA (Heating and Ventilating Contractors Association) standard specification DW/172.

Ventilation by canopy

An extract ventilation system, using collecting hoods, or more properly canopies, over all the principal items of equipment is the method most commonly adopted, as shown in Fig. 18.17, which illustrates a large institutional

Table 18.8 Nominal exhaust rates for kitchen appliances

	Air extraction rates (litre/s)	
Equipment	*Unit*	*Per* m^2 *net area of appliances*
Roasting and grilling		
Ranges, unit type (approximately 1 m^2)	300	300
Pastry ovens	300	300
Fish fryers	450	600
Grills	250–300	450
Steak grills	450	900
Speciality grills	450	900
Steaming and vapour producing		
Boiling pans (140–180 litre)	300	600
Steamers	300	600
Sterilising sinks	250	600
Bains-marie	200	300
Tea sets	150–250	300

kitchen. Figure 18.18 shows a typical single-sided extract canopy. The design and construction of canopies, commonly finished in polished aluminium or stainless steel, is usually dealt with by specialist manufacturers.

If canopies are to be effective, their size in plan should be such that they overlap the area of the block of appliances which they serve by about 300 to 400 mm on all open sides. The *capture velocity* of the air extracted over this plan area should be not much less than 0.4 m/s. Provision must be made for drainage of the condensate which will form on the inside surfaces, by means of a perimeter channel, and grease filters must be fitted, at the point of air exit into ductwork, from any canopy which collects fumes from a process generating oily vapours. Cleanliness in a kitchen is a prime requirement and all exposed surfaces, including the not inconsiderable internals and vertical outside enclosures of canopies, need regular cleaning *in situ* with detergents (Figs 18.17 and 18.18).

Energy recovery

Bearing in mind the large volume of air extracted from a kitchen through conventional canopies and the consequent need to supply replacement air which must not be much below kitchen temperature which, inevitably, is high, it follows that an appreciable waste of energy will occur. An energy recovery system of conventional plate type may be used and Fig. 18.19 shows how a compact arrangement may be mounted to the roof of a small kitchen to operate in conjunction with a simple canopy arrangement. It is of course necessary in all cases to provide a high level of secondary filtration to the exhaust air. Carry-over of grease deposits which will occur if maintenance effort is neglected will soon lead to a commensurate reduction in the efficiency of recovery. This filtration will mean an increase in fan energy consumption which must be less than the energy recovered on an annual basis to make the investment worthwhile.

Energy-saving canopies

A relatively recent development has gone a long way towards overcoming the problems of energy recovery with the introduction of canopies which use supply air at outside temperature ducted to a venturi slot or slots within the canopy to produce high velocity jets which induce kitchen air and fumes into the canopy prior to exhaust therefrom. With this arrangement, as shown in Fig. 18.20, only about one-third of the total supply air quantity provided to the kitchen needs to be raised or cooled to near kitchen temperature. An added advantage is that the resultant air mixture passing through the grease filters is at a lower temperature than it would be in a conventional canopy and the efficiency of filtration is thus improved.

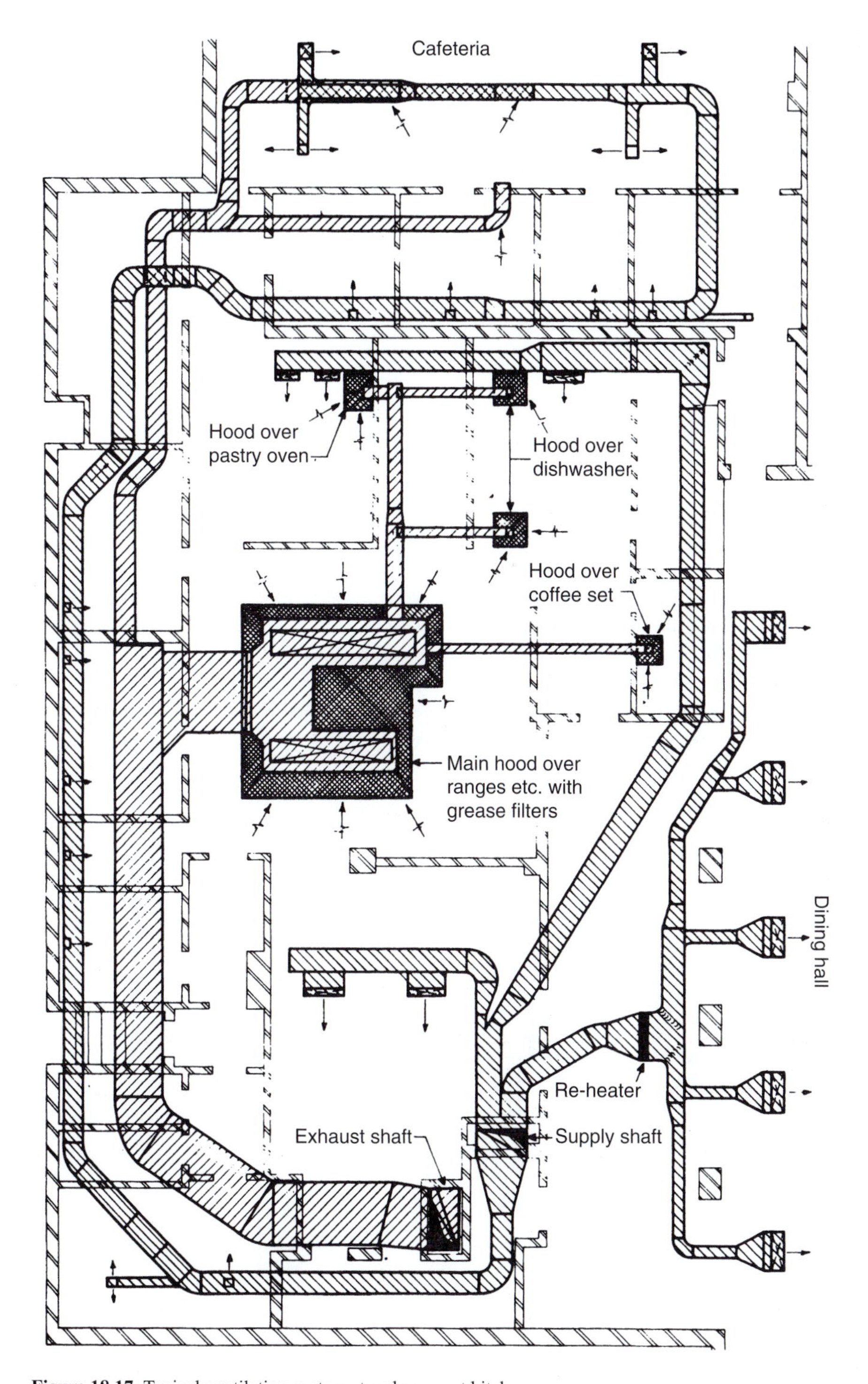

Figure 18.17 Typical ventilation systems to a basement kitchen

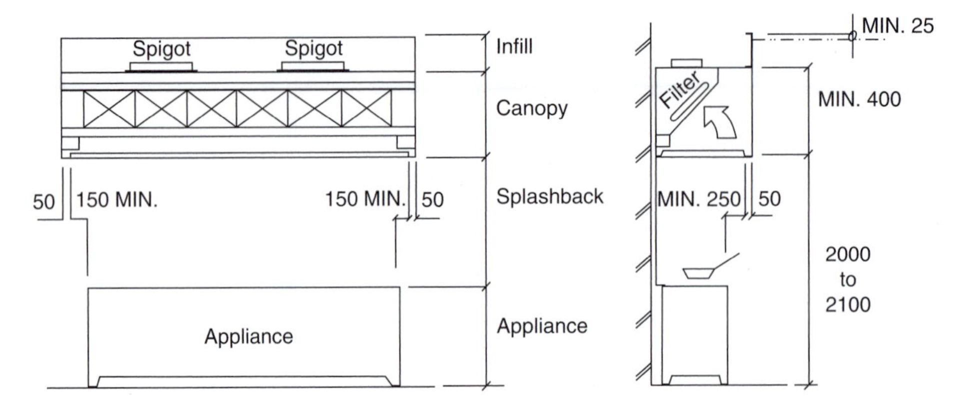

Figure 18.18 Extract only canopy (HVCA DW/172)

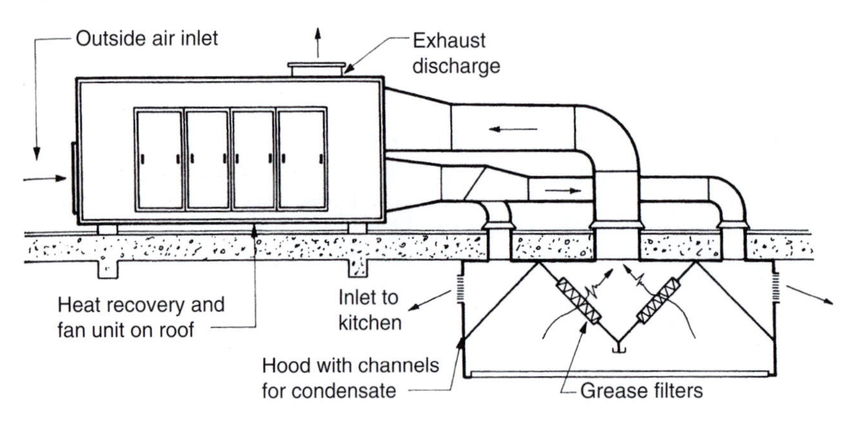

Figure 18.19 Heat recovery from kitchen extract system

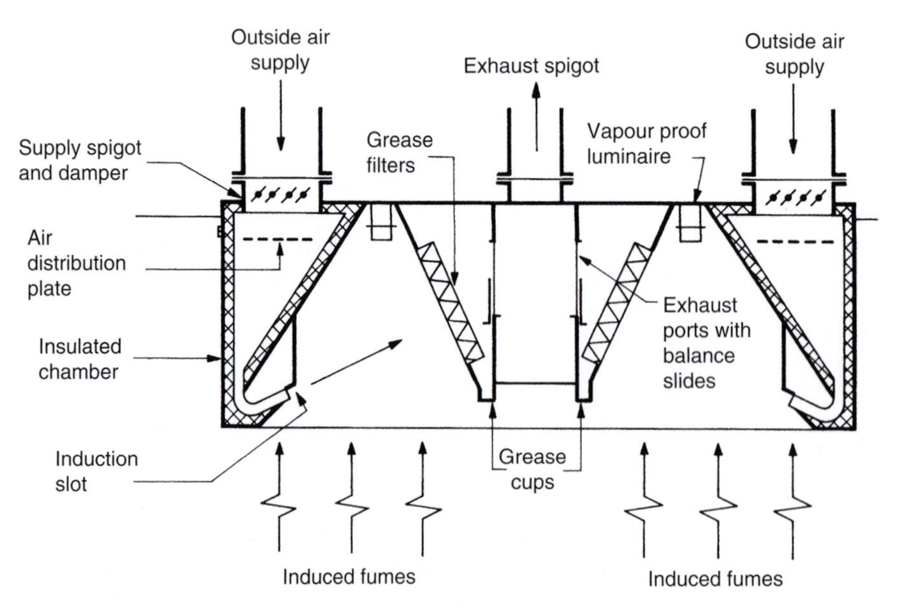

Figure 18.20 Energy-saving canopy (Stott-Benham)

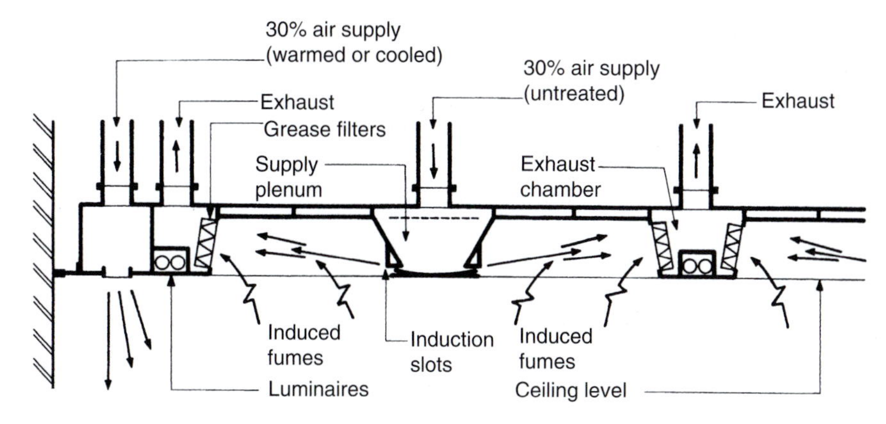

Figure 18.21 Typical energy-saving ceilings

It would be wrong, nevertheless, to pretend that the use of canopies of any design in a kitchen is not without problems, particularly in spaces where the floor to ceiling height is, as so often seems to be the case, less than it should be. The effective form of a canopy catchment, a hollow pyramid or cone, is usually concealed within an enclosure having vertical sides, supposedly in the interests of cleanliness, but the visual effect of such large vertical plane surfaces suspended from the ceiling down almost to head height, is claustrophobic.

Energy-saving ceilings

An alternative form of 'ventilating ceiling' is available which offers the advantage of energy economy in that it makes use of venturi slots within a series of shallow concavities in the ceiling profile to echo the performance of the previously described energy-saving canopies. Figure 18.21 shows this arrangement. The whole of the ceiling is manufactured in stainless steel or anodised aluminium and is insulated to avoid condensation. It should be noted that, as a result of the direction of a large proportion of the air supply to the venturi slots within the ceiling troughs, the air change rate in the occupied area is reduced *pro rata*. This reduced level will lead to a reduction in air velocity through servery hatches and, in the majority of cases, provide a more comfortable working environment for the kitchen staff.

Ductwork connections

Ductwork generally is dealt with in Chapter 21 but that for kitchens, whether supply or exhaust, is a special case. In many instances, with a view to cleanliness, modern kitchens with canopies are provided with a false ceiling to conceal down-standing beams, ducting, pipework and electrical conduits, etc. It is often argued that the resultant void could, if adequate in depth, be used as a plenum for supply air, with ducting reduced to short lengths and bell mouths, or as a suction chamber at negative pressure for exhaust air. Poor site workmanship by building contractors and the natural permeability of most structures and materials should be taken into account and also the inevitability of a long-term cleaning problem.

As far as supply air is concerned, use of such a ceiling void would be practicable only if there were absolutely no possibility of fouling by a noxious leak from, say, a waste pipe or a gas service or, furthermore, that there would be no possibility that a requirement for access, perhaps for maintenance, would ever occur.

It goes without saying that canopies should always be connected directly by sheet metal ducting to the suction side of the exhaust fan. Similarly, in the case of exhaust from a kitchen through a ventilated ceiling, sheet metal suction chambers above the ceiling and sheet metal ducting to the fan suction therefrom is much to be preferred. In the interests of cleanliness and as a precaution against fire, facilities for cleaning, by way of access holes and covers, should be provided at regular intervals along all exhaust ductwork. As a counsel of perfection, similar access on supply ductwork is desirable.

Special applications

It is not possible within the scope of this present chapter to deal with the whole range of demands for ventilation which are imposed by specialist applications. The notes included in the following paragraphs are therefore no more than an introduction to the wide variety of those demands.

Smoke ventilation and pressurisation

Ventilation systems are used to assist in the control and clearance of smoke to improve life safety, reduce damage and to assist in firefighting. Ventilation systems in large buildings will form part of the 'Fire Engineering' design which is a specialist subject and is addressed in detail in CIBSE Guide E. This guide should be used in conjunction with BS 5588 (Fire precautions in the design and construction of buildings), particularly Part 4, Code of Practice for Smoke Control using Pressure Differentials.

The CIBSE Guide addresses the need for ventilation systems to protect the area where the fire starts so that:

- A smoke free escape route or smoke free layer is maintained
- The smoke is diluted within the space to improve visibility to assist means of escape.

The ventilation system must also assist in protecting other areas of the building during a fire. This could include:

- Creating barriers to contain the smoke
- Pressurising escape routes.

Systems designed to create a smoke free layer of 2 m clearance would normally use reservoir sizes of 2000–3000 m^2 which would be created by screens or curtains which would drop to form the reservoir in the event of a fire. Systems used to dilute the smoke within a fire can also be used to remove cold smoke after the fire has been extinguished. This can be achieved by natural cross ventilation if sufficient perimeter openings are available or if mechanical extract is required, a rate of 6 to 8 air changes per hour is appropriate. Where natural ventilation cannot be provided in car parks, a mechanical extract system providing 10 air changes per hour is required.

Where pressurisation is required to maintain escape routes, BS 5588 Part 4 designates stairwells and lifts and other lobbies, with associated corridors in certain circumstances, as being areas in which air pressure should be maintained at levels in excess of that which exists in surrounding accommodation zones. It is proposed that this situation should be achieved by the admission of an outside air supply to the designated areas, via an independent plant or plants, such that a positive pressure of about 50 Pa, with respect to those surrounding areas, be maintained with all doors closed, a reduced pressure of 10 Pa is to be maintained when the main escape door to the outside is open. A minimum air velocity of 0.75 m/s across an open door between the pressurised space and the accommodation area is also required. Detailed requirements exist for different types of buildings, for example where the occupants could be sleeping, and for different fire strategies, where the evacuation of the building is phased. These are classified A to E within the code of practice. With such large air volume flow rates involved in the design of these systems it is essential to provide an air release path from the accommodation area to outside. This can be achieved using external wall vent, which can include automatically openable windows, vertical shafts or mechanical extract.

Obviously, a high degree of reliability is required from smoke control systems and it is essential that standby equipment is provided including the electrical supplies to the fans. Regular testing and strict maintenance programmes are also required.

The calculations necessary to determine both volume and pressure requirements to meet the duty concerned are tedious and must take into account stack wind buoyancy and thermal expansion effects which may oppose the required air movement. Account must also be taken of the likely airtightness of the building fabric. It is, in consequence, appropriate to consider the magnitude of the margins which must be added to any solution produced by the current calculation methods in order to take account of the many variables.

Table 18.9 Air velocities through *open* sashes in fume cupboards

Category of cupboard use	*Velocity* (m/s)
Teaching	0.3 minimum
Research and analytical	0.5–0.6
Highly corrosive or toxic	0.5–0.75
Radioactive[1]	0.5–2.0

Note
Sash openings are normally 750 mm to 1 m, velocities quoted are through *open* sash.
1. Grade of radioactivity determines velocity.

Local fume extract systems

Under this heading fall the familiar fume cupboards used in laboratories which, if they are to operate with any success, must be integrated with any inlet or extract ventilation system serving the same space. In general terms, the air volume which is extracted from a cupboard should be such that a face velocity of between 0.25 and 0.75 m/s is produced with the sash fully open (Table 18.9) and a much higher level when the opening is reduced to the working height of between 25 and 50 mm.

An interesting design problem arises in heavily serviced pharmaceutical laboratories where as many as 25 or more fume cupboards may be required in a single room but are not subjected to any predictable pattern of use. As will be appreciated, the requirement for warmed, or conditioned, air make-up may be considerable, with a consequent heavy use of energy, if extraction from the cupboards continues whether they are in use or not.

The arrangement of the extract system might take the form of a fan connected to each cupboard, discharging either individually to atmosphere or into a common-discharge duct. The former of these alternatives would produce a forest of unsightly terminals and the latter, without suitable precautions, might lead to a situation of potential air flow imbalance in the common collecting duct and consequent recirculation to any inactive cupboard. Either solution would be prodigal of energy and of maintenance effort if assured service were to be demanded of all the fans.

A preferred solution would be to provide extraction using a duplicate set of centrifugal exhaust fans, sited in a convenient plant room, drawing from a common-discharge duct provided at the remote end with a dilution entry for outside air. The capacity of each fan would be somewhat in excess of the total extract requirement of all the fume cupboards and the connection to the common duct from each individual cupboard would be fitted with a motorised damper, arranged for fast response variable control. A closed loop control system would sense the air flow requirement related to the sash position of the fume cupboards in use at any one time and adjust the position of the dampers accordingly to maintain the recommended air velocities. At the same time a constant efflux velocity is achieved at the discharge point to atmosphere. *Guide Section B2* provides some additional basic information for laboratories with fume cupboards.

As to replacement air, this would be provided from a suitable central plant through one or more variable volume units which, via the control system, would modulate the quantity of supply air to match that extracted at any time. With this solution, optimum energy consumption and convenient centralised maintenance would be provided.

Exhaust of industrial fumes

In an industrial context, local extraction of fumes from benches is more effective and more economic than an attempt to deal with them by treatment of the whole volume of a workshop. For a welding bench, a volume of 200–300 litre/s needs to be removed and a convenient way of meeting this requirement is by use of a flexible tube supported from a wall-mounted swivel as shown in Fig. 18.22. The tube may be connected to an individual fan or, where multiple benches exist, to a header duct and a central fan.

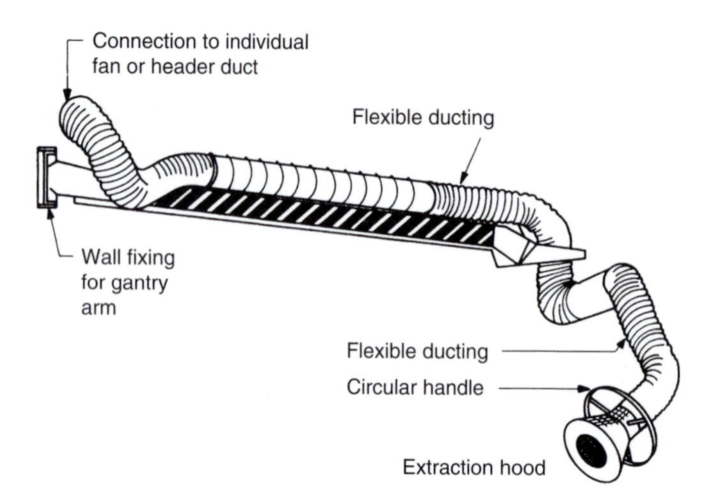

Figure 18.22 Fume extract from a welding bench (Plymovent)

Fabric tube supply systems

The need exists in some industrial buildings for the introduction of large air volumes in circumstances where the process cannot tolerate any significant air movement. A solution which makes use of permeable fabric tubes, in place of the more conventional ductwork plus diffuser assemblies, has been applied with success to tobacco processing and textile spinning workrooms. For a multi-tube arrangement, a sheet metal header duct is connected to the appropriate number of tubes which are inflated and maintained under pressure by a conventional centrifugal fan. Supports take the form of a 'curtain track' above each length which allows a deflated tube to be drawn to the header for disconnection.

Tube sizes range from 150 to 750 mm diameter and lengths of 20 to 50 m are used, dependent upon diameter. The initial velocity of air supply into a tube is between 12 and 15 m/s and a typical pressure drop, for a flow of 500 litre/s through a 300 mm diameter tube, would be about 100 Pa. The efficacy of the system is a function of adequate inflation which, in turn, depends upon a constant air supply and upon a good standard of filtration at the plant. Tubes require laundering at intervals and the support system permits easy removal for this purpose and replacement by a spare.

Plenum systems

In an earlier paragraph of this chapter, it was suggested that a straightforward ventilation system would be designed on the assumption that heating of the building was dealt with by a radiator or other similar plant. The plenum system, which is a method of providing heat using air as the distribution medium, is thus noted here only to show that it has not been forgotten. The reader is referred to the more comprehensive description provided in Chapter 9.

Demand controlled ventilation

It is a generally accepted principle that successful distribution of a supply of ventilating air within a space, from a simple terminal such as a grille, is a function of the design discharge velocity. An increase over this level might lead to draughts, due to impingement of the air stream on walls, etc., and a decrease might lead to 'dumping' of the air near to the terminal. Refer to Chapter 20 for general principles of 'Air distribution'.

As will be explained in later chapters, the development of variable air flow devices has enabled the concept of *demand controlled ventilation* to be introduced. This method, it is claimed, will allow more effective operation by adjusting the air flow rate to the needs of the occupants at any one time and thus lead to improved air quality, better energy efficiency or both.

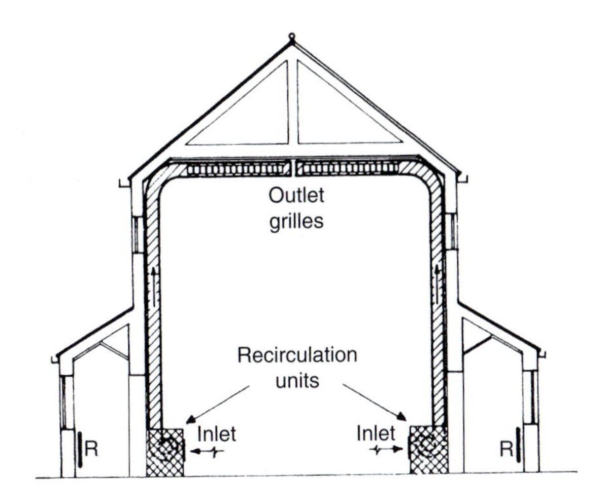

Figure 18.23 Upward air-transfer arrangement in a church

The system relies upon control of the supply air volume, or, in some cases, the rate of extraction, using sensors which detect the concentration level of contaminants and decrease, or increase where possible, the ventilation volume to suit. As far as human occupancy is concerned, the level of carbon dioxide in an enclosure may be sensed with comparative ease and is a convenient measure of density of occupation.

Recirculation units

Although the use of recirculation units and equipment should not be considered as providing ventilation in the accepted sense, there are applications where they can be of value. Reference has been made in earlier chapters to temperature gradient within lofty rooms and two diametrically opposite methods have been used to overcome this. For soaring buildings such as churches, a technique used with success in the USA is shown in Fig. 18.23. Here, with conventional methods of space heating at low level, heating effect and air movement have been greatly improved by displacement downwards of warm air at high level using a supply drawn from the occupied zone. It has been found that the temperature gradient may be reduced to about 2 K with an air circulation rate equivalent to 3 room volumes per hour.

The alternative approach, perhaps more suited to an industrial application, is the suspension of fan units at high level which take in warm air from above them and discharge it to the working level: a unit of the type used is shown in Fig. 18.24, the duty recommended being between 1.5 and 2 room volumes per hour. Lastly under this heading, the large diameter *Punkah* fans which were installed by the thousand in tropical countries before the days of air-conditioning, are being increasingly used in a variety of buildings. They do no more than agitate the air at high level in the room but, in so doing, movement is induced in the lower occupied area which produces evaporation from the skin and consequent cooling. The models used in the past were noisy and ugly but modern versions are not only much quieter but greatly improved in appearance.

Ventilation efficiency

To conclude this chapter, it is appropriate to note the term *ventilation efficiency* which has been used increasingly in recent years, as a result of the sensitivity of those non-smokers who may be required to share spaces with those who smoke. This situation has been the subject of considerable research and many published papers.[5]

The term represents an attempt to quantify the rate at which contaminants are removed from a space and, in particular, from the occupied zone. No simple definition of the term exists but it may be expressed as the ratio between the average concentration of a contaminant within a space and that concentration present at the

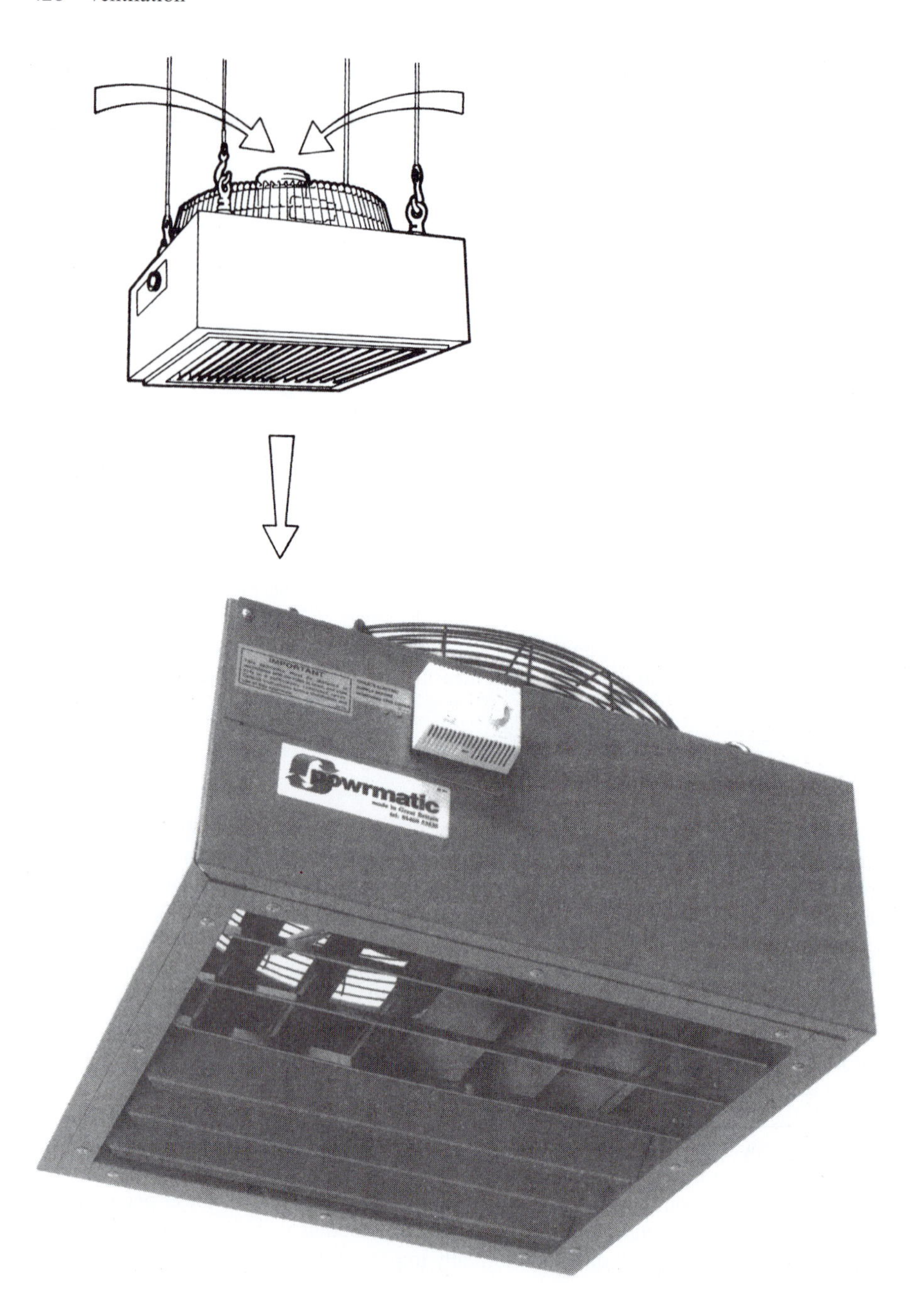

Figure 18.24 Downward air-transfer unit (Powrmatic)

point of extract. This ratio will depend upon many factors including the distribution, properties and period of discharge of the contaminant, the quality and rate of supply of replacement ventilation air and the pattern of air movement within the space.

For contaminant emissions which are distributed, ventilation efficiency improves with the rate of air supply and with those air movement patterns able to carry the contaminant to the extract positions quickly as may be produced by an effective piston displacement flow. Complete mixing of the air supply with room air produces no more than an average ventilation efficiency and any short circuits in room air flow patterns, supply to exhaust, reduce that efficiency further.

Notes

1. This is sometimes quoted as 0.004719 litre/s which is no more than an exact metric conversion of 0.01 ft^3/min which, in turn, is an average between values of 0.513 and 0.684 ft^3/h. Since these last two figures were derived from an estimate of the CO_2 content of the average exhalation from average human lungs, 3–4 per cent of 17 ft^3/h, the authors are inclined to believe that a somewhat rounder figure is to be preferred.
2. CIBSE Guide B2; 3.21.7.
3. Walker, W. *Useful Hints on Ventilation*. Parkes, Manchester.
4. Wise, A.F.E. and Curtis, M., Ventilation of internal bathrooms and water closets in multi-storey flats. *JIHVE*, 1964, 32, 180.
5. Shandret, E. and Sandberg, M., Air exchange and ventilation efficiency – new aids for the ventilation industry, *Norsk VVS (Norway)*, 1985, 7, 527–534. Design and Performance of Mechanical Ventilation Systems – Ventilation Effectiveness. *BSRIA Contract 7190A*, 1987.

Further reading

Auditoria. The *CIBSE* occupancy requirement is that an outside air quantity of 10 litre/s per person be provided by a ventilation system. As a result of variable occupancy rates, carbon dioxide controlled mechanical ventilation systems are permitted.

BSRIA TN19 Airtightness Testing.

Building Regulations, Part F sets out the ventilation to be provided for people in buildings, both dwellings and other buildings.

Building Ventilation – The State of the Art, AIVC – Mat Santamouris and Peter Wouters.

Car parks. *Building Regulations*, etc., require, for natural ventilation, that openings to outside should have an area of 5 per cent of floor area. Between 6 and 10 air changes are required, dependent upon circumstances, if ventilation is to be mechanical.

CIBSE Guide A, Section 1.4 sets out recommendations for ventilation for comfort.

CIBSE Guide B2, ventilation and air-conditioning.

CIBSE Guide E, fire engineering.

Dwellings. *Building Regulations* Part F sets out minimum ventilation rates and provisional dwelling and room types.

Factories. Since almost all work places fall within this one category, specific Orders under the Factories Act, the Offices, Shops and Railway Premises Act and the Health and Safety at Work Act will apply. An Order published by the Health and Safety Executive suggests that a minimum of 6 litre/s of outside air should be provided per person.

Hospitals. Publications by the Department of Health and Social Security cover requirements for Buildings over a very wide field of activities.

Kitchens. HVCA (Heating and Ventilating Contractors Association) DW/171 Standard Specification for kitchen ventilation systems.

Schools. The DfES Building Bulletin 88 Fume Cupboards in Schools. Building Bulletin 101 Ventilation of School Buildings.

Smoke Control – BS 5588 Part 4 Code of Practice for smoke control using pressure differentials.

Theatres. These, with other places of public entertainment such as dance halls, fall under the same category as auditoria.

Toilets. Many local authorities (and *Building Regulations*) require three air changes per hour or 6 litre/s of outside air per WC pan. (For public use, it is good practice to provide double this air quantity.) Duplicate fans and motors are commonly required.

Chapter 19

Air-conditioning

The science of air-conditioning may be defined as that of providing and maintaining a desirable internal atmospheric environment irrespective of external conditions. As a rule 'ventilation' involves the delivery of air which may be warmed, while 'air-conditioning' involves delivery of air which may be warmed or cooled and have the moisture content (humidity) raised or lowered.

National and international concern directed at the global environmental effects arising from release of refrigerants into the atmosphere, as described in Chapter 24, and the energy used in mechanical cooling systems have, together, opened the door to a period of fundamental change in attitudes towards the use of air-conditioning in certain buildings.

For the first time the 2002 edition of Part L of the Building Regulations imposed requirements on air-conditioning and mechanical ventilation systems, in that if they are installed, they should be 'reasonably efficient'. Further, the 2006 edition of Part L requires that the systems are suitably efficient with effective control systems. These requirements are set out in the Non-domestic Heating, Cooling and Ventilation Compliance Guide, May 2006, published by the Department of Communities and Local Government. Also, it should be noted that the requirement to control solar overheating does not impose any restriction on the use of air-conditioning. What it does do is to make the building capable of being kept reasonably comfortable either in the absence of air-conditioning (provided internal gains are modest) or at reduced air-conditioning loads.

The difficulty, as far as the efficiency of the system is concerned, is that the energy consumed by an air-conditioning system is dependent on a number of different aspects of the building design:

- The form and fabric of the envelope, especially in relation to the solar gains to the space.
- The efficiency of the cooling plant (chiller coefficient of performance, CoP, or the energy efficiency rating, EER).
- The method of 'coolth' distribution (air, water or refrigerant).
- The design of the distribution network (high velocity/low velocity), and the efficiency of the prime movers (fans, pumps and auxiliary equipment energies).
- The control system (use of free cooling, etc.).

In order to provide a simple but robust mechanism for assessing the diversity of designs, the estimated annual carbon emissions from an air-conditioned building are calibrated against the estimated annual carbon emissions for a so-called *notional* building. This is achieved by the designer initially selecting systems and equipment that comply with the Non-domestic Heating, Cooling and Ventilation Compliance Guide mentioned above and then running an accredited energy model for the actual building and comparing the results with the *notional* building also produced by the model. This simple model has been calibrated against a dataset of energy consumption in buildings to arrive at a target design figure for carbon emissions. The targets in the 2002 edition of Part L were not particularly demanding where the actual had to be less than the *notional*, however, the 2006 edition requires a 28 per cent reduction when compared against the *notional* model. It is also considered likely that these targets will be made much tighter at subsequent revisions, once industry has become familiar with the requirements.

Because refurbishments of buildings also have to comply with Part L, a set of identical standards are defined for new or refurbished systems in existing buildings. However, there is no requirement to undertake the comparative modelling described for the new buildings. This is in recognition of the fact that there is less design flexibility in such situations. A similar set of design targets is provided for buildings that are mechanically ventilated.

Other important new features of 2002 Part L were the emphasis on effective commissioning and on the provision of adequate energy metering. Consideration of both these issues is required at the design stage, so as to ensure the building user is able to operate the systems reasonably efficiently, and to monitor on-going performance as a means of ensuring the required performance is maintained. To further reinforce this requirement there is an obligation under the 2006 Part L to carry out an assessment *as designed* and a second assessment *as built*.

Whilst statutory regulation may not be appropriate, there is already a quite discernible trend to question whether air-conditioning is a necessary prerequisite for quality commercial premises, for example, Chapter 18 deals with the suitability of natural ventilation solutions.

Some years ago in response to a *Code of Practice on Environmental Issues* by the Engineering Council CIBSE stated the following which still holds good today:

If the requirement for air-conditioning has been fully established, the following principles should be adopted and adhered to:

- *The system should be energy efficient (with due regard being given to the inclusion of cost-effective energy-saving methods such as free cooling) and also controlled to minimise energy use.*
- *Operation and maintenance strategies should be devised and the necessary regimes adopted to deliver economy, efficiency and effectiveness in the working of systems throughout their life cycles.*
- *System design, construction and commissioning should be carried out in accordance with current national and European standards, codes of practice and statutory requirements.*
- *Cooling system refrigerants should be used in accordance with the policy laid down in the CIBSE Guidance Note on CFCs, HCFCs and halons. (See Chapter 24 for current policy.)*

General principles

The desired atmospheric condition for comfort applications usually involves a temperature of 18–21°C in winter and 22–24°C in summer; a relative humidity of about 40–60 per cent and a high degree of air purity. This requires different treatments according to climate, latitude, and season, but in temperate zones such as the British Isles it involves:

In winter: A supply of air which has been cleaned and warmed. As the warming lowers the relative humidity, some form of humidifying plant, such as a steam injector, with preheater and main heater whereby the humidity is under control, may be necessary.

In summer: A supply of air which has been cleaned and cooled. As the cooling is normally accomplished by exposing the air to cold surfaces, the excess moisture is condensed and the air is left nearly saturated at a lower temperature. Inherent in this process therefore is a measure of dehumidification which counters the increase in relative humidity that results from cooling the air. The temperature of air has then to be increased, to give a more agreeable relative humidity, which can be done by warming or by mixing with air which has not been cooled.

Dehumidifying may also be brought about by passing the air over certain substances which absorb moisture. Thus, in laboratories, a vessel is kept dry by keeping a bowl of strong sulphuric acid in it or a dish of calcium chloride, both of which have a strong affinity for moisture. Silica gel, a form of silica in a fine state of division exposing a great absorbing surface, is used also for drying air on this principle, but this process is complicated by the need for regeneration of the medium by heat and subsequent cooling, and is not generally used in comfort air-conditioning applications.

Establishment of need

The application of air-conditioning may be considered necessary to meet a variety of circumstances:

- Where the type of building and usage thereof involves *high heat gains* from sources such as solar effects, electronic equipment, computers, lighting and the occupants.
- In buildings which are *effectively sealed*, for example where double glazing is installed to reduce the nuisance caused by external noise.
- The *core areas* of deep-planned buildings where the accommodation in the core is remote from natural ventilation and windows, and is subject to internal heat gains from equipment, lights and occupants.
- Where there is a *high density of occupation*, such as in theatres, cinemas, restaurants, conference rooms, dealing rooms and the like.
- Where the process to be carried out requires *close control of temperature and humidity*, such as in computer suites, or where stored material, or artefacts, require stable and close control of conditions, such as in museums and paper stores.
- Where work has to be carried out in a confined space, *the task being of a high precision and intensive character*, such as in operating theatres and laboratories.
- Where the *exclusion of airborne dust* and contaminants is essential, such as in microchip assembly and animal houses.

In tropical and sub-tropical countries, air-conditioning is required primarily to reduce the high ambient temperature to one in which working and living conditions are more tolerable. In the temperate maritime climate of the British Isles and in similar parts of the world, long spells of warm weather are the exception rather than the rule, but modern forms of building and modern modes of living and working have produced conditions in which, to produce some tolerable state of comfort, air-conditioning is the best answer. Thus we find buildings of the present day incorporating to a greater or lesser extent, almost as a common rule, some form of air-conditioning. This great variety of applications has produced an almost equally great variety of systems, although all are fundamentally the same in basic intention: that is, to achieve a controlled atmospheric condition in both summer and winter, as referred to earlier, using air as the principal medium of circulation and environmental control.

The installation of complete air-conditioning in a building often eliminates the necessity for heating by direct radiation, and it naturally incorporates the function of ventilation, thus eliminating the need for opening windows or reliance on other means for the introduction of outside air. Indeed, opening windows in an air-conditioned building should be discouraged since, otherwise, the effectiveness of the system to maintain conditions will be reduced and the running costs will be increased.

All air-conditioning systems involve the handling of air as a means for cooling or warming, dehumidifying or humidifying. If the space to be air-conditioned has no occupancy, no supply of outside air is necessary, that inside the room being recirculated continually. In most practical cases, however, ventilation air for occupancy has to be included and in the design for maximum economy of heating and cooling, this quantity is usually kept to a minimum depending on the number of people to be served. Thus, in most instances, it will be found that the total air in circulation in an air-conditioning system greatly exceeds the amount of outside air brought in and exhausted. Where, however, it is a matter of contamination of the air, such as in a hospital operating theatre, or where some chemical process or dust producing plant is involved, 100 per cent outside air may be needed and no recirculation is then possible.

Weather data

With certain designs of plant it may be cost effective to arrange for 100 per cent outside air to be handled, normally during mid-season periods when untreated it can provide useful cooling. Figure 19.1 shows, for various months of the year, the proportion of daytime hours when the London outside temperature, for the CIBSE Test Reference Year, is below 13°C and is thus available for a cooling duty; a second set of data is also shown in the figure for periods below 15°C. Meteorological data for other stations in Britain show similar availability.

When designing an air-conditioning system it is not enough to consider only the winter and summer peak design temperatures. It is important that the system should operate satisfactorily through the range of annual

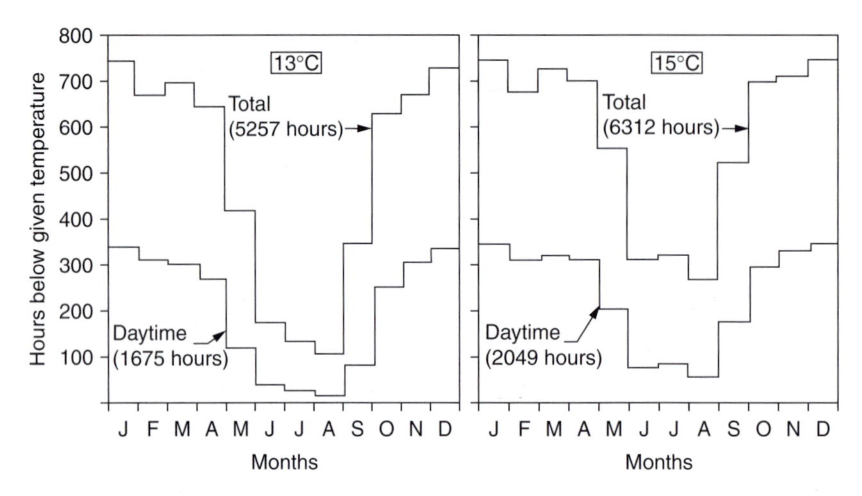

Figure 19.1 Annual hours when 'free cooling' is available

external conditions. To this end it is necessary to have an understanding of the range and frequency of occurrence of coincident wet bulb and dry bulb external conditions. Figure 19.2 shows, on a psychrometric chart, the percentage occurrence between June and September of external conditions lying within specified limits for London Heathrow.

Central plant

The basic elements of air-conditioning systems of whatever form are:

Fans for moving air.
Filters for cleaning air, either fresh or recirculated, or both.
Cooling plant connected to heat exchange surface, such as finned coils for cooling and dehumidifying air.
Heater batteries for warming the air, such as hot water or steam heated coils or electrical resistance elements.
Humidifiers, such as by steam injection.
A control system to regulate automatically the amount of cooling, warming, humidification or dehumidification.

The type of system shown in Fig. 19.3 is suitable for air-conditioning large single spaces, such as theatres, cinemas, restaurants, exhibition halls, or big factory spaces where no sub-division exists. The manner in which the various elements just referred to are incorporated in the plant will be obvious from the caption. It will be noted that in this example the cooling is performed by means of chilled water cooling coils and the humidification is by means of steam injection. A humidifier would be provided only when humidification is required in winter.

In an alternative version of a central plant system, the cooling coil may be connected directly to the refrigerating plant and contain the refrigerating gas. On expansion of the gas in the evaporator, cooling takes place and hence this system is known as a direct expansion (DX) system. It is suitable for small- to medium-size plants. Humidification could be by means of a capillary washer or water spray into the air stream, but a non-storage type such as steam injection is recommended, due to the risk of *Legionnaires' disease* associated with types which incorporate a water pond to facilitate recirculation within the humidifier. For the same reason, an air cooled condenser could be used as a means of rejecting unwanted heat from the refrigeration machine in place of the cooling tower shown.

In Fig. 19.3 it will be noted that there is a separate extract fan shown exhausting from the ceiling of the room. This would apply particularly in cases where smoking takes place, such as in a restaurant, to remove fumes which might otherwise collect in a pocket at high level. Sometimes this exhaust may be designed to remove the quantity equivalent to the outside air intake, in which case the discharge shown to atmosphere from the return air fan would not be necessary.

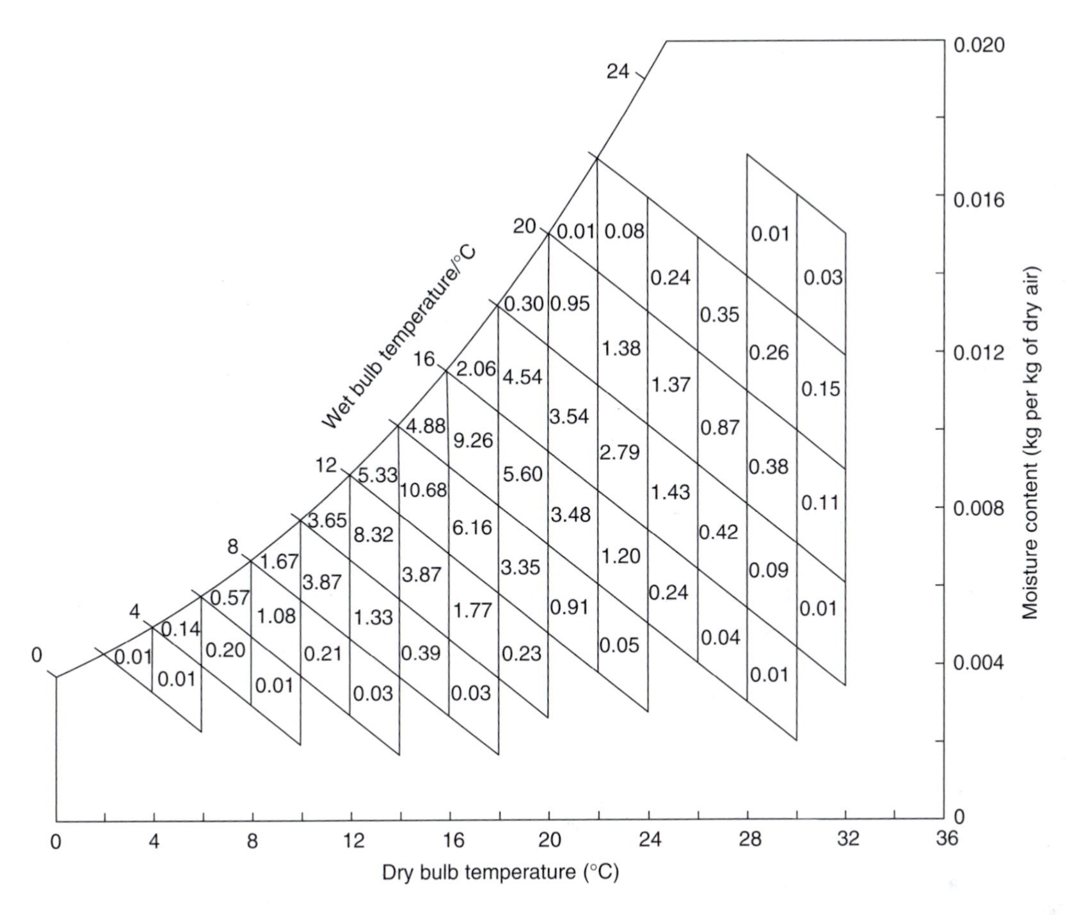

Figure 19.2 Percentage frequencies of wet and dry bulb temperatures plotted on a psychrometric chart: London (Heathrow) (1982–2002) *Guide Section A2*

Motorised dampers are shown in the air intake, discharge and recirculation ducts to allow the proportion of outside and recirculation air flow rates to be varied to effect economy in plant operation. Reduction in energy consumption may also be achieved by transferring heat between the exhaust and intake air streams; this is not illustrated in the diagram but is described in Chapter 22.

Where there is a number of rooms or floors in a building to be served, it is necessary to consider means by which the varying heat gains in the different compartments may be dealt with. Some rooms may have solar gains, and others none; some may be crowded and others empty; and some may contain heat producing equipment. Variations in requirements of this kind are the most common case with which air-conditioning has to deal and for this a simple central system is unsuitable. The ideal, of course, would be a separate system for each room but this is rarely practicable unless the individual spaces are very large or important.

Zoned systems

A building may be divided into a number of zones for air-conditioning purposes. The sub-division could be dictated by spatial constraints, by the requirements for sub-letting or by the hours of use. Each zone may be served by a separate 'central' plant, perhaps located in a common plant room, or alternatively one plant located on each floor might be an appropriate arrangement. There is a number of ways in which a central plant, broadly on the lines of that previously described, may be modified in design to serve a number of groups of rooms or zones. To take the simplest case, if such a plant served one large space of major importance and a subsidiary

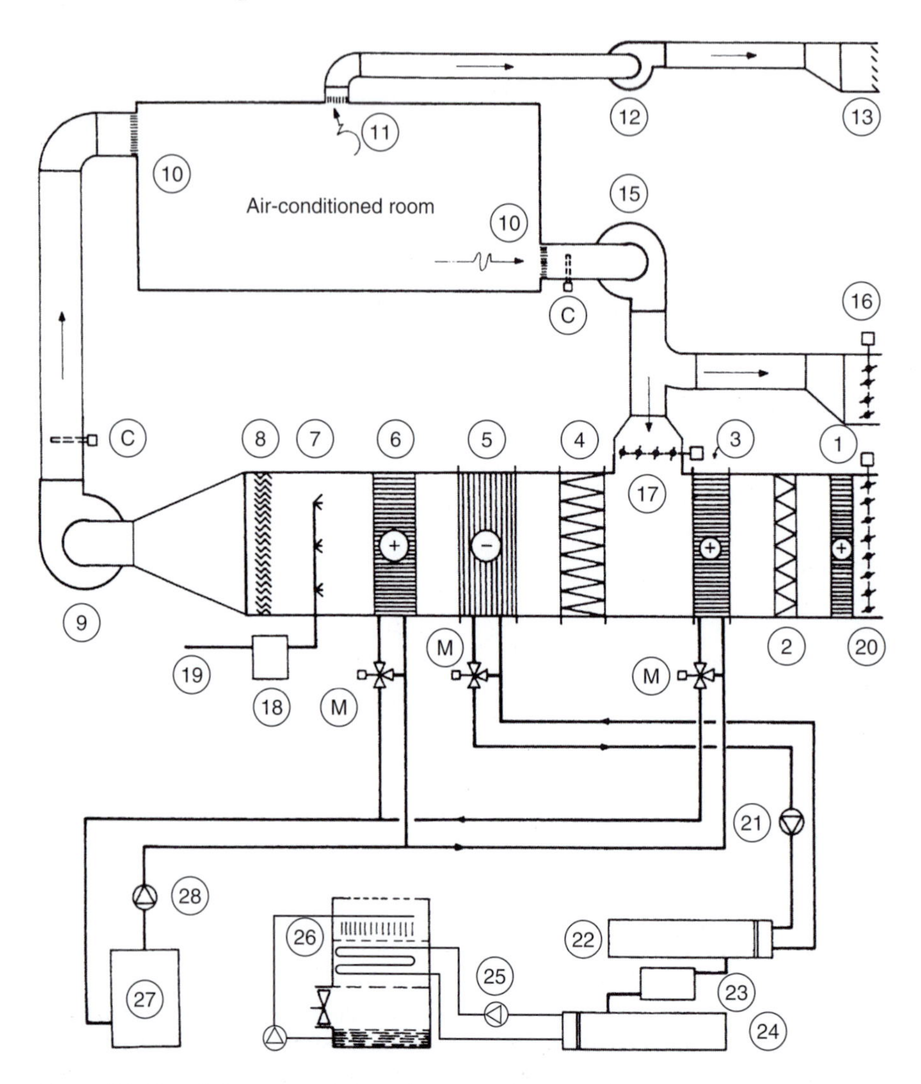

Figure 19.3 Central plant system: 1, frost coil (optional, for fog elimination); 2, pre-filter; 3, preheater; 4, secondary filter; 5, chilled water cooling coil; 6, reheater; 7, direct injection steam humidifier; 8, eliminators (optional); 9, supply fan; 10, conditioned air supply; 11, high-level smoke extract; 12, smoke exhaust fan; 13, smoke discharge to outside via louvre; 14, extract and recirculation; 15, main extract fan; 16, discharge to outside via motorised damper and louvre; 17, recirculation; 18, steam generator; 19, water make-up; 20, outside air intake via motorised damper and louvre; 21, chilled water pump; 22, water chiller evaporator; 23, refrigeration compressor; 24, shell and tube condenser; 25, condenser water pump; 26, closed circuit cooling tower; 27, hot water boiler; 28, hot water pump; 'M' denotes motorised valves; 'C' denotes temperature and humidity control points

room having a different air supply temperature requirement, it would be possible to fit an additional small heater or cooler, or both, on the branch air duct to the subsidiary room. Separate control of temperature would thus be available.

High-rise buildings, say over 12 storeys, served from a central plant may require plant rooms at intermediate levels to reduce the air quantity conveyed in any one duct and thereby reducing the space taken by vertical service shafts. Similar vertical sub-division would also be appropriate for closed and open piped distributions.

An extension of this principle, to serve two or more zones more or less equal in size from a single plant, may be achieved by deleting the reheater of Fig. 19.3; dividing the supply fan outlet into the appropriate number of ducts and fitting a separate reheater to each. Since, however, the output of the central plant cooling

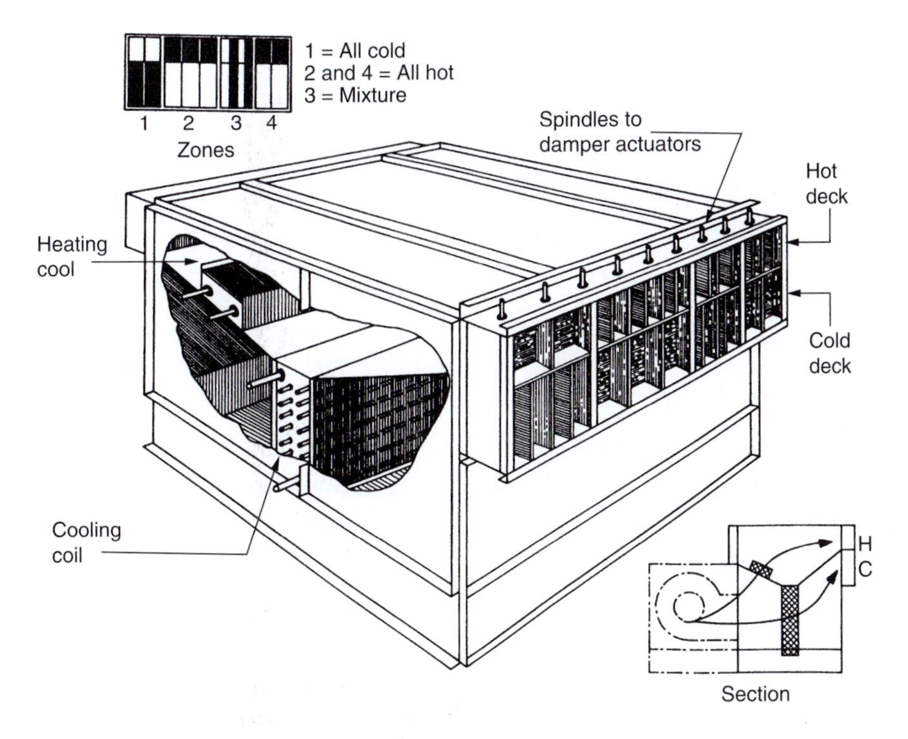

Figure 19.4 Multi-zone plant showing damper arrangement

coil would have to be arranged to meet the demand of whichever zone requires the maximum cooling, extravagant use of reheat would result and lead to uneconomic running costs.

The plant illustrated in Fig. 19.3 is arranged on the 'draw-through' principle, the various components being on the suction side of the supply fan. It is, of course, possible to adapt this sequence to produce a 'blow through' arrangement with the fan moved to a position immediately following the secondary filter. With such a redisposition, what is known as a *multi-zone* arrangement may be produced, as illustrated in Fig. 19.4. Here, the fan discharge is directed through either a cooling coil or a heating coil into one or other of two plenum boxes: a *cold deck* or a *hot deck*. Each building zone is supplied with conditioned air via a separate duct which, at the plant, is connected to both plenum boxes. A system of interlinked dampers, per zone, is arranged such that a constant air supply is delivered to each zone duct which may be all hot, all cold, or any mixture of the two as required to meet local demand. When providing a mixture, such a plant wastes energy and as a result is not in common use today.

Zone units (air handling units)

An alternative approach to the problem of providing local control to a variety of building zones having differing demands involves the provision of a recirculating air handling unit, having a booster reheater or recooler, within each zone. Outside air, in a quantity suitable to provide for occupancy, is conditioned as to temperature dependent upon outside conditions and corrected for moisture content by means of what is, in effect, a conventional central plant. This supply is delivered to the local zone plant where it is heated or cooled as may be required to suit the zone conditions. Figure 19.5 shows two alternative local plant arrangements, incorporating cooling only, suitable in this case to provide for conditions within the apparatus area of a major telephone switching centre. Note how integration of the air-conditioning system within the structure has been arranged.

For a more conventional application to an office building, Fig. 19.6 shows plants arranged in floor service rooms. In case (a), air recirculation from the individual rooms passes through louvres above the doors into

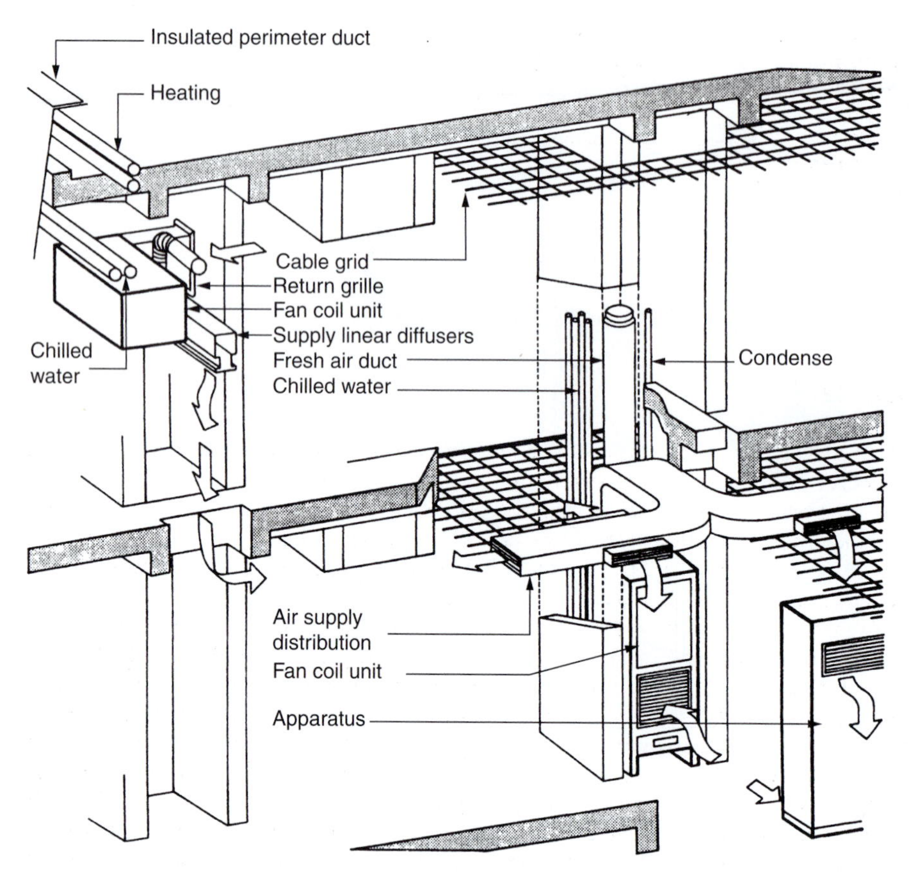

Figure 19.5 Fan coil unit with ducted outside air

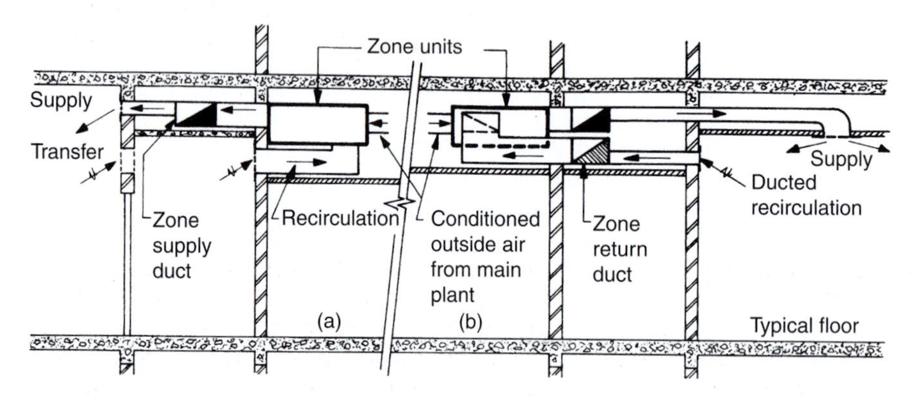

Figure 19.6 Zoned system in multi-storey building

the corridor and thence back to the zone plant. While such an arrangement has the great merit of simplicity, coupled with relatively low cost, current thinking and fire regulations disapprove of the use of a corridor *means of escape* as a return air path on the grounds that fire and/or smoke generated in any one room could be transferred into an escape route and lead to disruption and panic. It is now, in consequence, more usual to provide a quite independent return air collection system, via duct branches from each room, back to the local zone plant, as shown in case (b).

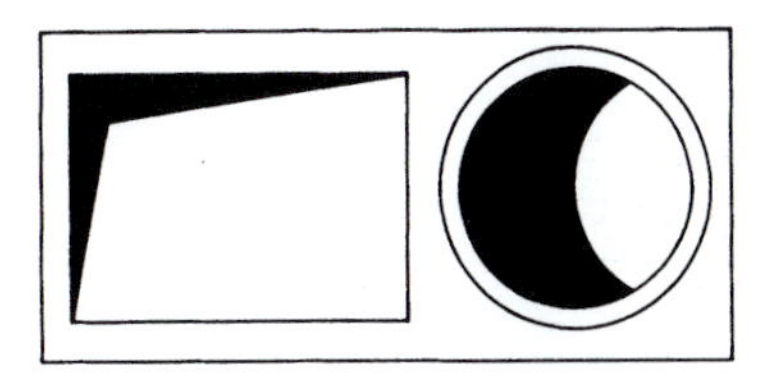

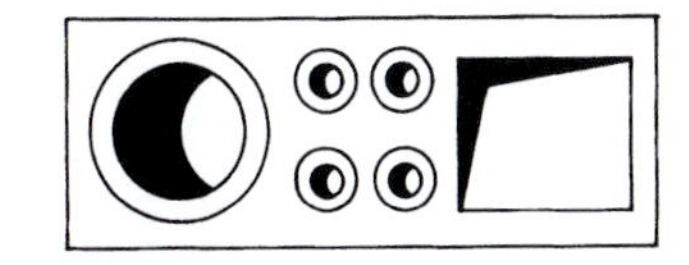

All-air
(supply and extract)

Air–water
(Supply, extract, heating and chilled water)

Figure 19.7 Space requirements for all-air and air–water distribution arrangements

High-velocity systems

Traditionally until the mid-1950s, air-conditioning systems were designed to operate with duct velocities of not much more than about 8–10 m/s and fan pressures of 0.5–1 kPa. With the advent of high-rise buildings and, concurrent with their introduction, demands for improved working environments coupled with *less* space availability for services there was a requirement that tradition be overthrown. This situation led to a radical rethink and to the introduction of a number of new approaches to air-conditioning design using duct velocities and fan pressures twice and more greater than those previously in use.

Whilst the principal characteristic of the new generation of systems relates to the methods adopted for distribution of conditioned air and exploits these to the full, the principles previously described in this chapter remain unchanged. As before, the conditioning medium may be all air or air–water dependent upon a variety of circumstances. Figure 19.7 shows comparative space requirements for the alternative supply media. The low-velocity extract ducts associated with the respective supply air quantities will increase the space requirements further in favour of the air–water systems. Added to which, the increase in space cooling loads arising from greater use of electronic equipment increases yet again the spatial benefits of the air–water system since the additional cooling load is handled by the water circuits; the air ducts being unaffected. There is, in consequence, a practical limit to the level of heat gain that can be dealt with satisfactorily by an all-air system.

To complement high-velocity supply systems, where space for ductwork distribution is limited, consideration may be given to using a high-velocity extract system, but this will impose additional initial cost and subsequent running costs.

All-air systems

Simple systems

These, the most primitive of high-velocity systems, differ from their traditional counterparts by the form of the terminals used to overcome problems arising from noise generated in the air distribution system. To this end, a variety of *single-duct* air volume control devices has been developed to provide for the transition between a high-velocity distribution system and air outlets local to the conditioned spaces. The terminal box, with *octopus* distribution section illustrated in Fig. 19.8, is typical of equipment produced for this purpose. It consists, in principle, of an acoustically lined chamber provided with a sophisticated air volume damper or 'pressure reducing valve': such dampers are in some instances fitted with self-actuated devices, others have powered actuators, arranged so that they may be set to provide constant or variable output volume under conditions of varying input pressure. Such devices are a considerable aid in regulating air flow quantities during the commissioning process.

Simple all-air systems will provide adequate service in circumstances where the load imposed is either constant or will vary in a uniform manner for the area served thus allowing temperature control by means of adjustment to temperature of the supply air at the central plant.

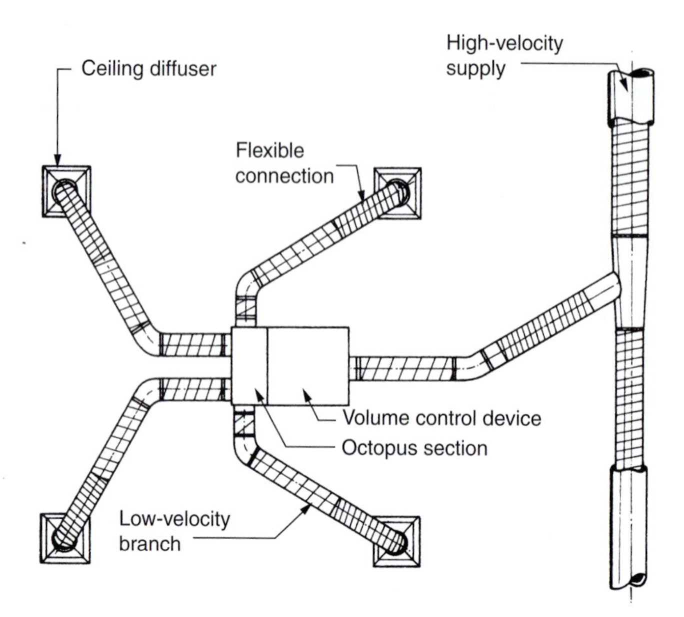

Figure 19.8 Single-duct high-velocity terminal box

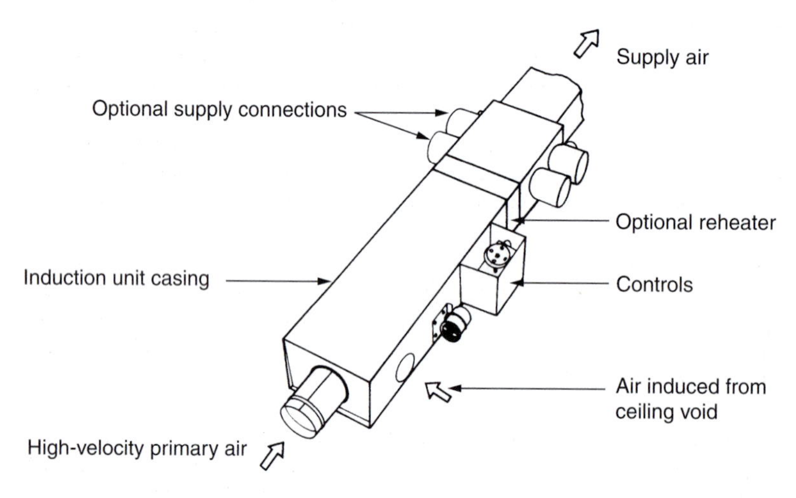

Figure 19.9 Typical all-air induction box

All-air induction systems

For the particular case where air is returned to the central plant via a ceiling void and, further, where lighting fittings (luminaires) are arranged such that the bulk of the heat output (which may be as much as 80 per cent) is transferred to this return air, an all-air induction system may be used.

With such an arrangement, conditioned air is ducted to induction boxes mounted in the ceiling void, as shown in Fig. 19.9. Each box incorporates damper assemblies or other devices which, under the control of a room thermostat, act to permit the conditioned air flow to induce a variable proportion of warm air from the ceiling void into the discharge stream. Reheat and consequent local control is thus achieved such that, with one type of unit, the cooling capacity may be controlled down to about 45 per cent of maximum. The subsequent introduction of either an automatic switching system, which will minimise the period that heat gain from lighting is available, or the availability of more efficient lamps (both being changes which reduce the potential

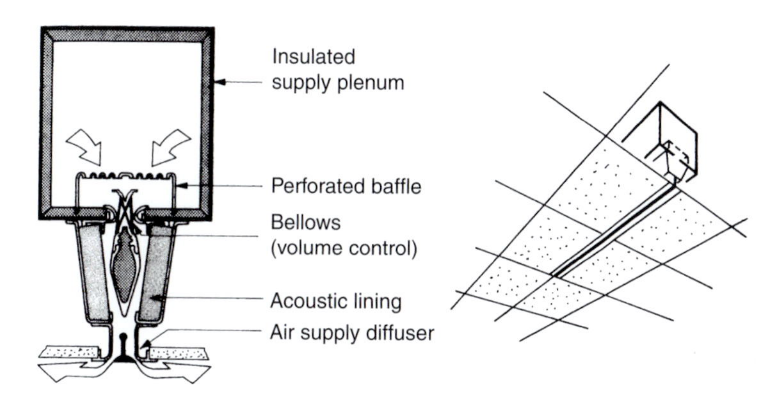

Figure 19.10 Linear diffusers for variable volume (Carrier)

for reheat) would have a detrimental effect upon the operation of such systems. To counter such eventualities, reheater batteries are available as an optional feature and terminal units which vary the quantity of the primary air supply are also available.

All-air variable volume systems

The traditional approach to air-conditioning design placed, as a first principle, insistence upon the concept of maintaining air discharge to the spaces served at constant volume. Load variations were catered for by adjustment to air temperature. This axiom arose, no doubt, from the known sensitivity of building occupants to air movement and, furthermore, from the relative crudity of the air diffusion equipment then available.

With the advent of terminal equipment not only more sophisticated but also with performance characteristics backed by adequate test data, circumstances have changed. The activities of BSRIA and of a variety of manufacturers in this area must be applauded. Hence the availability of potential for abandonment of the traditional approach.

In principle, the variable volume system may be considered as a refinement to the simple all-air system whereby changes in local load conditions are catered for not by adjustment of the temperature of the conditioned air delivered, at constant volume, but by adjustment of the volume, at constant temperature. This effect may be achieved by means of metering under thermostatic control of the air quantity delivered either to individual positions of actual discharge, as shown in Fig. 19.10, or to groups of such positions via a terminal unit of the type illustrated in Fig. 19.11.

In such cases, a true variable volume arrangement is possible since the effect of reduced output at the terminal units or at the discharge positions may be sensed by a central pressure controller, this being arranged to operate devices which reduce the volume output of the central plant correspondingly. Economies in overall operation in energy consumption and in cost will thus result.

If an adequate supply of outside air is to be maintained and problems of distribution within the conditioned space avoided, volume cannot be reduced beyond a certain level. Good practice suggests that minimum delivery should not be arranged to fall below about 40 per cent of the designed quantity. For different reasons, such a limitation may present problems to both internal and perimeter zones. At internal zones, if lights were individually switched in each room, then the load in an unoccupied or sparsely occupied space could be greatly reduced, in consequence of which such rooms would be overcooled. In the case of perimeter zones, where conduction and solar gain form a high proportion of the design load, it may be necessary to introduce some level of reheat to augment capacity control by volume reduction.

A solution for both internal and perimeter zones where the reduced air supply quantity is required to be below that to produce satisfactory air distribution in the space, without causing the cold supply air to *dump* into the occupied area, is to install variable geometry diffusers: a volume flow reduction down to about 25 per cent of the maximum is claimed for this type of outlet. Figure 19.12(a) shows a typical example which

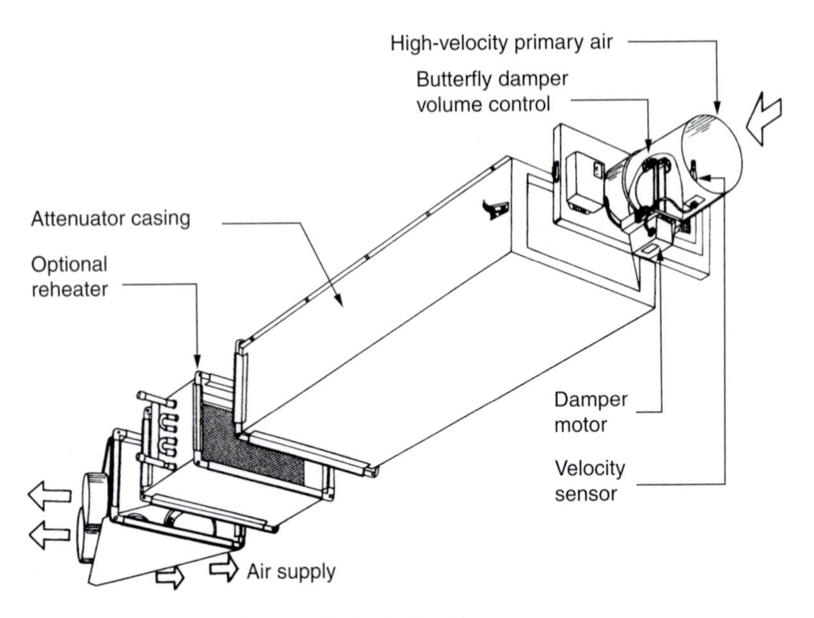

Figure 19.11 Variable volume terminal unit (Wozair)

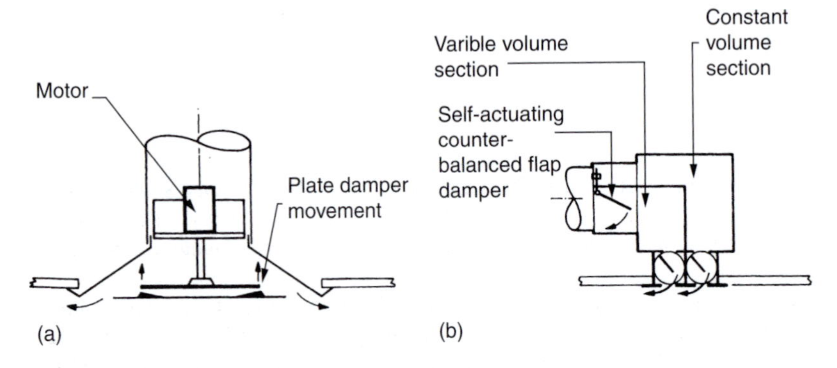

Figure 19.12 Variable geometry supply diffusers: (a) variable orifice (Wozair) and (b) variable bypass (Trox)

operates to maintain the air velocity at discharge to the space by changing the area of opening, thereby ensuring adequate diffusion of the supply air. A similar effect may be obtained using a two-section bypass type air diffuser in which the supply is divided into two passages, one of which is controlled at a constant volume to maintain the velocity of the air stream from the diffuser and hence satisfactory air distribution, as Fig. 19.12(b).

While most variable volume terminal units can be adapted to incorporate reheaters, the required effect may equally well be achieved at perimeter zones via control of a constant volume perimeter heating system or even space heating units such as hot water radiators either of which may, in any event, be required at windows to deal with downdraughts.

The difference in the space heating and cooling loads within internal and perimeter areas may require a two or three zone supply system, with the supply temperature of each zone controlled to suit the particular characteristics of the area served. For single or multi-zone systems, the zone temperatures may be controlled at a constant level, be varied to suit outside conditions (scheduled), or be varied in response to feedback from the controls at each terminal device to provide the optimum supply with a maximum operating economy. Figure 19.13 shows a typical variable air volume system arrangement. To maintain the volume of outside air above the

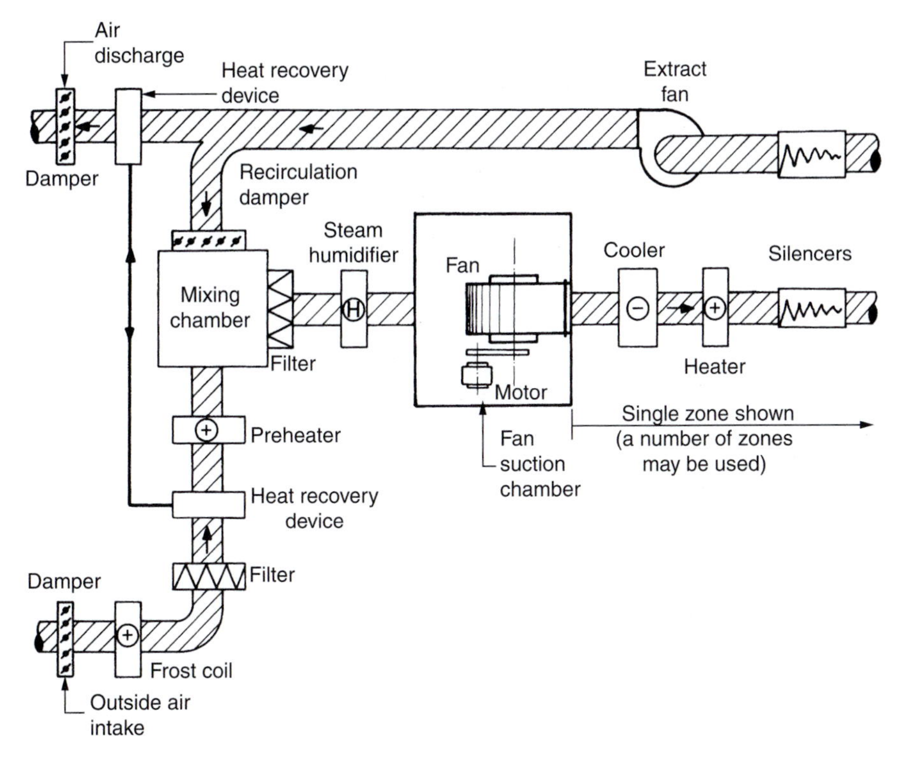

Figure 19.13 Typical plant arrangement for a variable volume system

minimum requirement for occupant ventilation, it may be necessary to add air velocity sensors to the plant controls in the outside air intake duct, to actuate the dampers in the outside air, exhaust air and recirculation ducts. This control must be arranged to override any damper controls provided to achieve economy of operation.

The increased requirement for cooling, currently arising from the proliferation of electronic equipment in modern buildings, has led to the development of variable air volume terminals which incorporate a means to provide additional capacity. This has been achieved by introducing a secondary cooling coil in the terminal. Typically, a recirculation fan draws air from the ceiling void through a filter, and discharges this across the coil where it is cooled before being introduced into the space. Two types of fan-assisted variable air volume terminal devices have been developed: one has the secondary fan *in parallel* with the primary air supply and the alternative is an *in series* configuration. Diagrammatic representations of these units are shown in Fig. 19.14. The parallel flow type is currently more popular on the grounds that the secondary fan runs only when the cooling load so demands, and thus uses less energy than the series flow terminal where the fan runs continuously. However, the variation in supply air quantity with the parallel arrangement could result in poor air distribution, and the intermittency of the fan operation might cause more disturbance than the continuous running in the series flow unit.

One advantage of the series flow arrangement is that, due to the inherent mixing of primary and recirculated air, the former may be delivered to the unit at a much lower temperature, typically down to 8°C. In consequence, primary air volume and duct sizes may be reduced, which may be an important consideration in refurbishment work. Lower supply temperatures, however, reduce the efficiency of the associated refrigeration machines as explained in Chapter 24.

Since the principle of operation with all variable air volume systems is to vary the quantity of primary air supplied by the plant, it follows that the associated extract fan must respond to the changes in supply volume to avoid under- or over-pressurisation of the building.

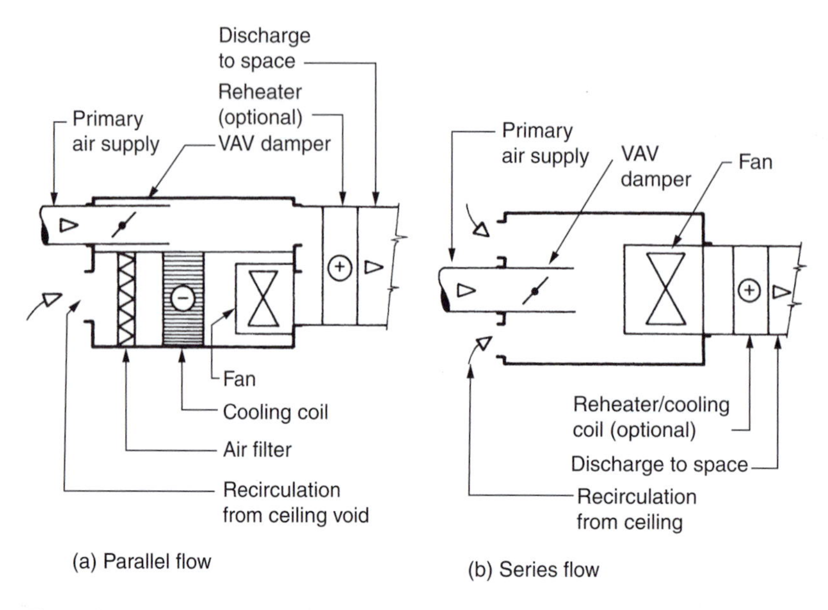

Figure 19.14 Fan-assisted variable volume terminal devices

Variable air volume systems are currently a popular choice because the associated energy consumption is lower than that of other equivalent systems. They are particularly suited to buildings subjected to long periods of cooling load, and are easily adapted to changes in office partition layouts. It is quite practical to incorporate constant volume terminal boxes within a variable volume system in circumstances where, for example, it is desirable to maintain a room at a positive pressure. In this application a terminal reheater, either an electric element type or a coil connected to a heating circuit, would normally be installed downstream of the terminal to provide temperature control of the space.

The variable volume principle allows the ductwork to be sized to handle the air quantity required to offset the maximum simultaneous heat gain that could occur and not the sum of the quantities demanded by the peak loads in each of the individual spaces. Extensive use is made of computer analyses to establish the maximum simultaneous design conditions.

All-air dual-duct systems

The multi-zone system previously described is arranged to mix, at the central plant, supplies of hot and cold air in such proportions as to meet load variations in building zones. An extension of this concept would be that an individual mixed-air duct was provided for each separate room in the building but this, on grounds of space alone, would not be practicable. The same degree of control may however be achieved by use of the dual-duct system where the mixing is transferred from the central plant to either individual rooms or small groups of rooms having similar characteristics with respect to load variation.

Use is made of two ducts, one conveying warm air and one conveying cool air, and each room contains a blender, or mixing box, so arranged with air valves or dampers that all warm, all cool, or some mixture of both is delivered into the room. Figure 19.15 illustrates such a unit, incorporated therein being a means for regulating the total air delivery automatically such that, regardless of variations in pressure in the system, each unit delivers its correct air quantity. Referred to as 'constant volume control', this facility is an essential part of such a system.

Owing to the fact that air alone is employed, the air quantity necessary to carry the cooling and heating load is greater than that used in an air–water system. Air delivery rates with the dual-duct system are frequently of the order of 5 or 6 changes/hour, compared with the 1½–2 required for introduction of outside air. Owing to the considerable quantity of air in circulation throughout the building, dual-duct systems usually incorporate means for recirculation back to the main plant and this involves return air ducts and shafts in some form.

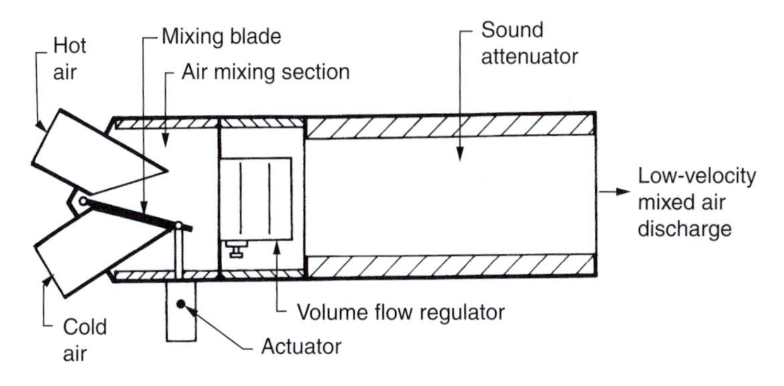

Figure 19.15 Dual-duct mixing box (Trox)

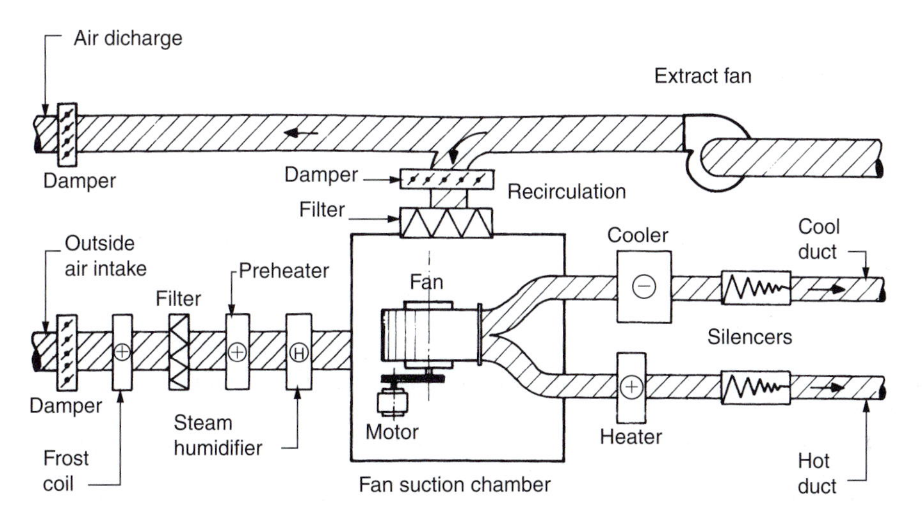

Figure 19.16 Typical plant arrangement for a dual-duct system

An advantage of the dual-duct system is that any room may be warmed or cooled according to need without zoning or any problem of changeover thermostats. Furthermore, core areas of a building, or rooms requiring high rates of ventilation, may equally be served from the same system, no separate plants being necessary.

Figure 19.16 shows the plant arrangement of a system in one form, though there are variations of this using two fans, one for the cool duct, one for the warm duct.

To avoid undue pressure differences in the duct system, due for instance to a greater number of the units taking warm air than cool air, static pressure control may be incorporated so as to relieve the constant pressure devices in the units of too great a difference of pressure, such as might otherwise occur under conditions where the greater proportion of units are taking air from one duct than from both.

Since dual-duct systems supply a constant air volume to the conditioned areas this overcomes the potential problems arising from maintaining adequate quantities of outside air and from providing satisfactory air distribution experienced with some variable volume systems. The disadvantages of dual-duct systems are high-energy use and the need for large shafts and ceiling voids to accommodate the ductwork.

In special circumstances, a dual-duct system may be provided with variable volume terminal equipment to combine the best features of each system. With such an arrangement, when volume has been reduced to the practical minimum, control is achieved by reheat from the hot duct supply. Typically, the dual-duct supply would serve perimeter zones and a single-duct supply of cool air would serve internal areas.

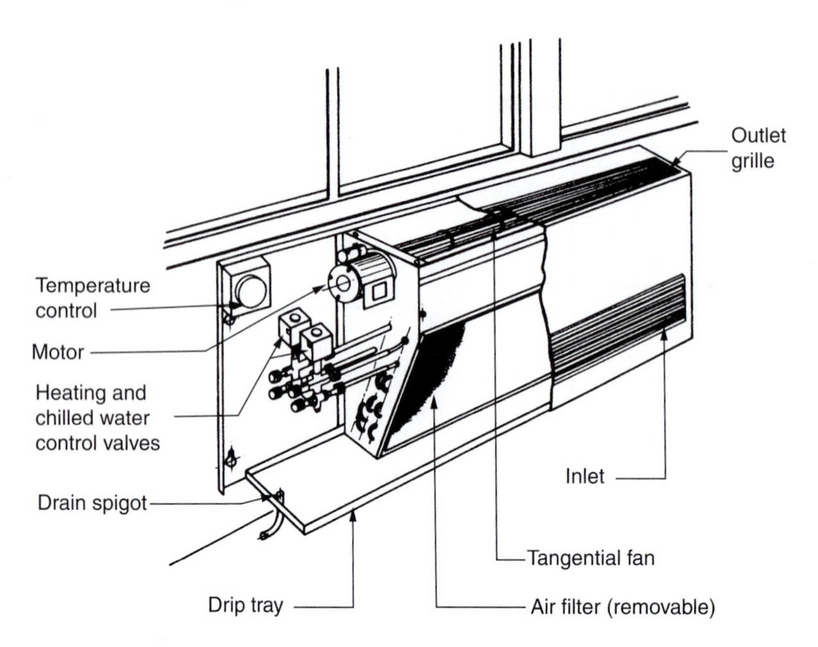

Figure 19.17 Under-window fan coil unit (four pipe)

Air–water systems

Fan coil system

From the point of view of economy in building space, the use of water rather than air as a distribution medium for cooling and heating, from plant rooms to occupied spaces, has much to commend it. *The fan coil* system exploits this saving in space and has the added advantage of offering facilities for relatively simple local temperature control in each individual room. The fan coil terminals each consist of a chassis which mounts a silent running fan, either centrifugal or cross-flow (*tangential flow*), a simple air filter and either a single water-to-air heat exchange coil or a pair of such coils. They are made in a limited range of sizes and may be fitted, one or more to each occupied room, using either the manufacturer's sheet metal casings or some form of concealment in purpose-designed enclosures. A number of different patterns of fan coil unit are available to suit various mounting positions and Fig. 19.17 shows the form which historically was probably most familiar, an under-window cabinet type. For mounting horizontally at high level, perhaps above a false ceiling and ducted locally to outlet and recirculation terminals, the type shown in Fig. 19.18 is available. Perhaps the most common arrangement today is where the fan coil units are located in ceiling voids with octopus-type discharge connections to ceiling diffusers. Air recirculation is usually drawn over the luminaires through the ceiling void where it mixes with the fresh air supply being discharged locally to the fan coil unit intake. The perimeter units will be arranged to provide the control modularity of the building, usually at 3, 4.5 or 6 m centres.

Since each fan coil unit serves principally to recirculate room air, the necessary volume of outside air to meet the ventilation needs of occupants must be provided quite independently. This requirement may be met in a number of ways, depending upon the sophistication of the individual system. At one end of the scale this air, pre-filtered and conditioned, is introduced to the space through a ducted system from a central plant, and the exhaust air is ducted to an energy recovery unit of some sort. The diametrically opposite, and obviously much cheaper method (which cannot be recommended), is for each unit to be placed below a window and outside air admitted to it through an aperture in the wall in a manner similar to that shown in Fig. 18.10. The hazards of dust and noise pollution to say nothing of unit overload due to wind pressure cannot be exaggerated.

For a high proportion of the year, external conditions in the British Isles are such that rooms on certain elevations of a building may require heating while others require cooling. It is thus desirable to provide units that

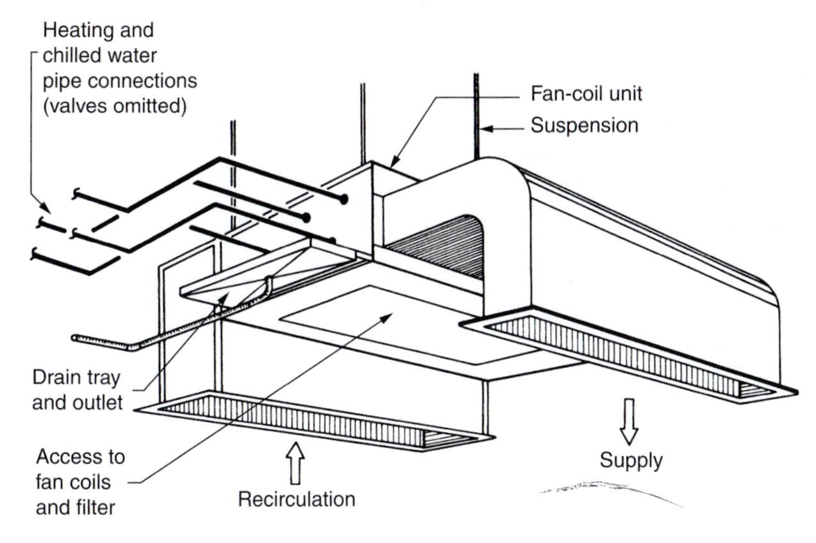

Figure 19.18 Fan coil unit fitted above a suspended ceiling

are able to meet either demand at any time. Such an arrangement is known as *a four-pipe* system, heating flow and return and cooling flow, and return pipes being connected to separate coils. A *two-pipe* system provides heated water to a coil in winter and chilled water to the same coil in summer and problems arise in mid-season when some spaces require heating while others are calling for cooling: clearly both requirements cannot be satisfied simultaneously with this two-pipe arrangement. Such a system is suitable only where the period of climate change between summer and winter is short, with little or no mid-season (some parts of the USA have such a climate), or where the internal loads are such as to require only local cooling.

The chilled water flow to the units may be circulated at an elevated temperature, as described later for air–water induction systems, eliminating the formation of condensate on the cooling coils and thus the need to pipe this from each unit to drain. Very basic control of either individual units or groups may be effected by switching the fan motor(s) on and off. The preferred alternative for the climate of the British Isles, however, is to keep the motor running and to control either the water temperature using automatic valves as Fig. 19.17 or the airside using mixing dampers as Fig. 19.19.

Airside controlled fan coil units are an alternative particularly where the units are to be mounted within ceiling voids in commercial offices. The units being controlled by a damper arrangement avoids the need for a large number of small water control valves with their inherent reliability and maintenance issues. Installation, balancing and commissioning times can also be reduced. However, with both the heating and cooling coils operating at design water flow rates thermal insulation and air leakage between the hot and cold decks needs to be effective. This aspect is sometimes not tested by manufacturers and individual tests may be required. Outside temperature compensation of the heating circuit is an energy efficient way of minimising this effect.

Fan coil systems are inherently flexible and are well suited to refurbishment projects since any central air handling plant and duct distribution system will be relatively small in size. The cabinet type of unit, as Fig. 19.19, has been designed specifically for application to existing buildings, the component parts being stacked vertically in a small plan area, space being available there also to enclose vertical water piping and a supply air duct.

Air–water induction systems

Although there has been virtually no development in the design of this type of unit for some years, it is appropriate, nevertheless, to outline the principles of their operation and application. As the name implies, the principle of induction is employed in this system as a means to provide for an adequate air circulation within a

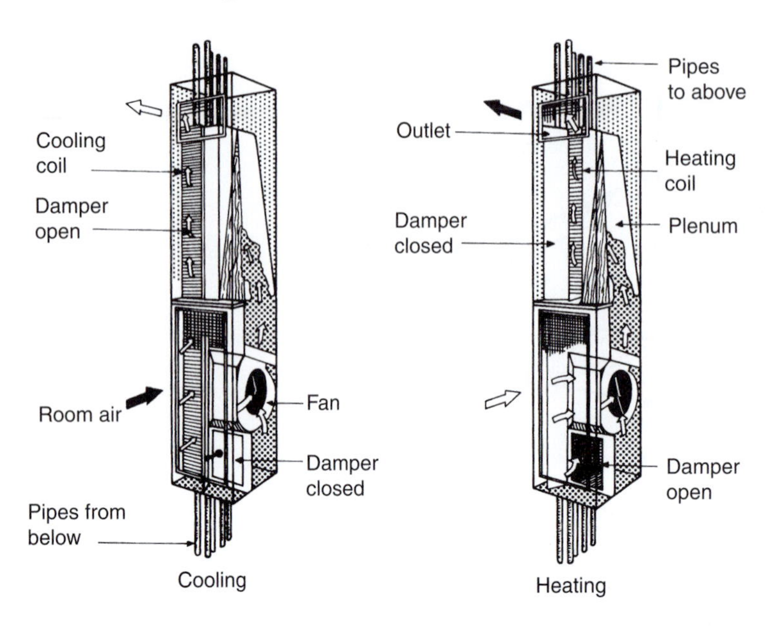

Figure 19.19 Typical vertical-type fan coil unit

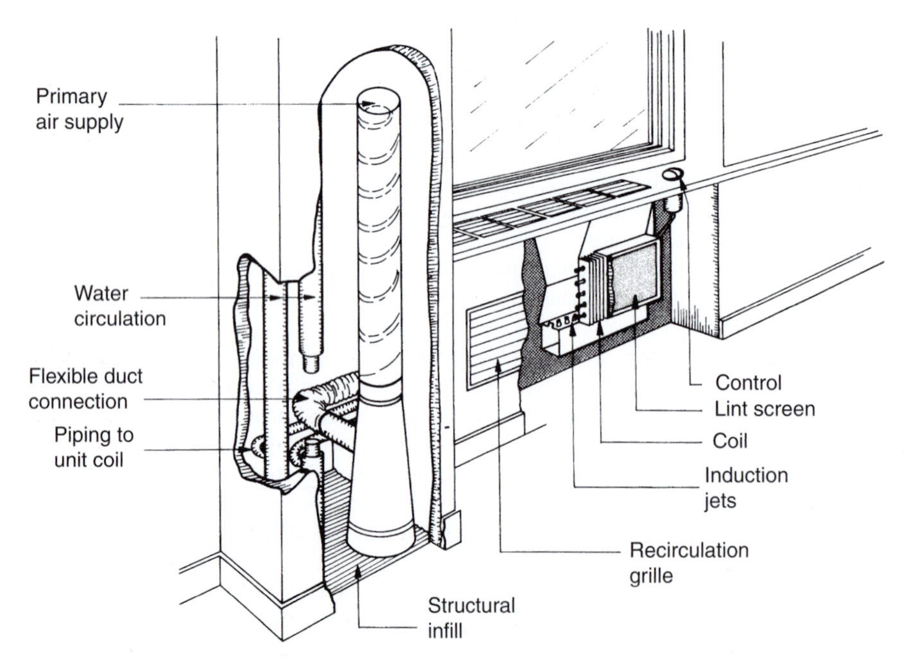

Figure 19.20 Air–water induction unit (two pipe)

conditioned room. Primary air, conditioned in a central plant, is supplied under pressure to terminal units, generally placed below the window with vertical discharge, each of which incorporates a series of jets or nozzles as shown in Fig. 19.20. The air induced from the room flows over the cooling or heating coils and the mixture of primary and induced air is delivered from a grille in the sill. The induction ratio is from three to one to six to one. The primary air supply provides the quantity for ventilation purposes, and the means to humidify in winter and deal with latent loads in summer.

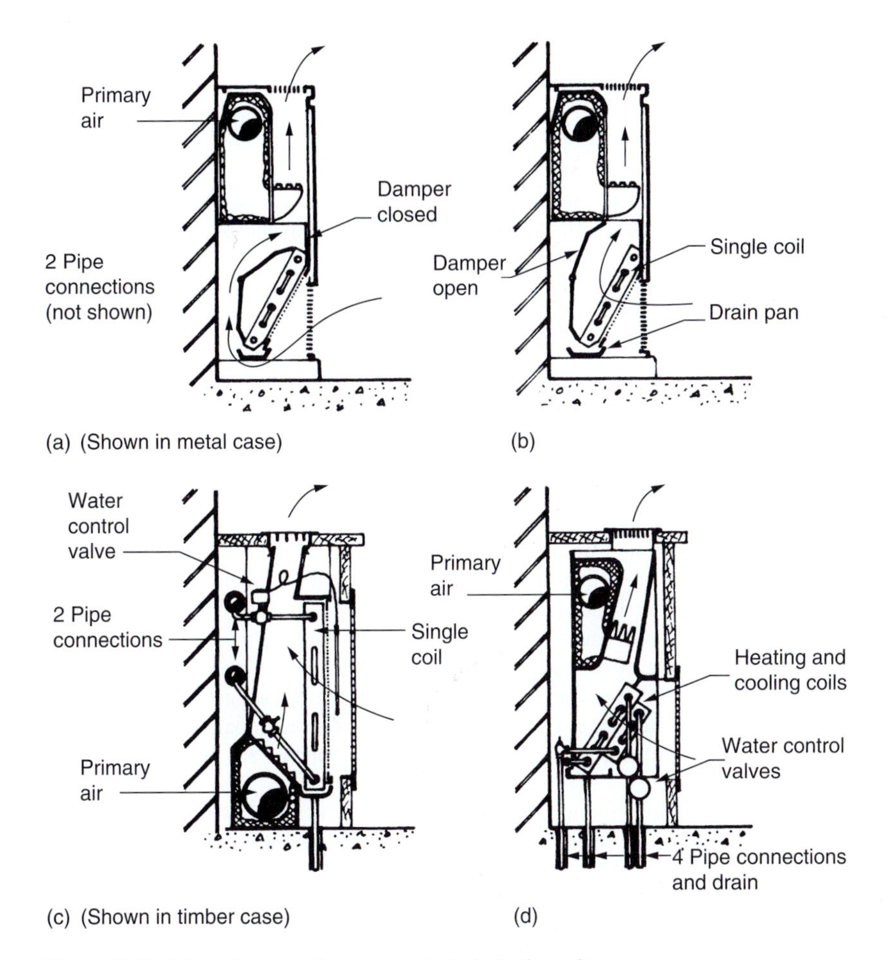

Figure 19.21 Alternative control arrangements for induction units

The coils are fed with circulating water which, in the so-called *changeover* system, is cooled in summer and warmed in winter, an arrangement more suited to sharply defined seasons than the unpredictable long springs and autumns of the British climate. In Fig. 19.21(a) and (b), control is achieved by an arrangement of dampers, such that the return air from the room is drawn either through the coils for heating or cooling, or the coils are bypassed to a greater or lesser extent. In the type shown in Fig. 19.21(c), control is by variation of water flow: increase in water flow is required to lower the air temperature during the cooling cycle, and increase in flow is required to raise the air temperature in the heating cycle. There is thus required some means of changeover of thermostat operation according to whether the winter or summer cycle is required.

An alternative method, using the so-called *non-changeover* system, avoids this problem by always circulating cool water through the coils of the induction units and varying the primary air temperature according to weather only. Thus, throughout the year, the heating or cooling potential of the primary air is adjusted to suit that component of demand imposed upon the system, or zone of the system, by orientation, by outside temperature or by wind effect. Any other variant – solar radiation, heat from lighting or occupancy – will necessarily produce a local heat *gain* and the sensible cooling needed to offset this will be provided by the capacity of the unit coils under local control. In so far as such an arrangement acts as 'terminal recool in winter, it is uneconomic in terms of energy wastage.

A variation of the two-pipe induction system is the *three pipe*, in which both warm and cool water are available at each unit, with a common return, and the control arrangement is so devised as to select from one or the other. Likewise, in the main system the return is diverted either to the cooling plant or to the heating plant, according to the mean temperature condition. Such an arrangement suffers from problems related to hydraulic instability due to the changes in water quantities flowing through the alternative paths.

The preferred system for the British climate is *the four-pipe system*, two heating and two cooling, but it is correspondingly expensive. A unit having two coils, one for heating and one for cooling, is shown in Fig. 19.21(d).

Induction systems might be expected to be noisy, due to the high-velocity air issuing from the jets, but the units have been developed with suitable acoustical treatment such that this disadvantage does not arise in practice.

Morning preheating may be achieved by circulating heating water through the secondary coils allowing the unit to function as a simple natural convector. This avoids the need to run the primary air fan and is therefore energy efficient.

Figure 19.22 is a simplified diagram of an induction system showing the primary conditioning plant, the primary ducting and the water circulation. The heat exchanger shown is for warming the water circulated to the units, and this would be fed from a boiler or other heat source. It will be noted that the chilled water supply to the coils of the induction units is arranged to be in the form of a subsidiary circuit to that serving the main cooling coil of the central plant. Such a system has the advantage of providing a degree *of free cooling* when the outside air supply to the central plant is at low temperature during winter. Furthermore, since the flow to the room units is connected to the return pipe from the cooling coil it has an elevated temperature and, as a result, this circuit arrangement provides an in-built protection against excessive condensation on the unit coils such that local drain piping may therefore be dispensed with in most cases.

The induction system involves the distribution of minimum primary air, often as little as 1½–2 air changes/hour, and has been widely applied to low-cost multi-storey office blocks or hotels where in either case there is a large number of separate rooms to be served on the perimeter of a building. Practice suggests that, provided application of the system is confined to perimeter areas not deeper than 4 m, with relatively low occupancy, satisfactory service will result. Interior zones of such buildings that require cooling year-round are usually dealt with by an all-air system.

Induction systems inherently cause any dust in the atmosphere of the room to be drawn in and over the finned coil surfaces, and, to prevent a build-up of deposit thereon, some form of coarse lint screen, easily removable for cleaning, is usually incorporated.

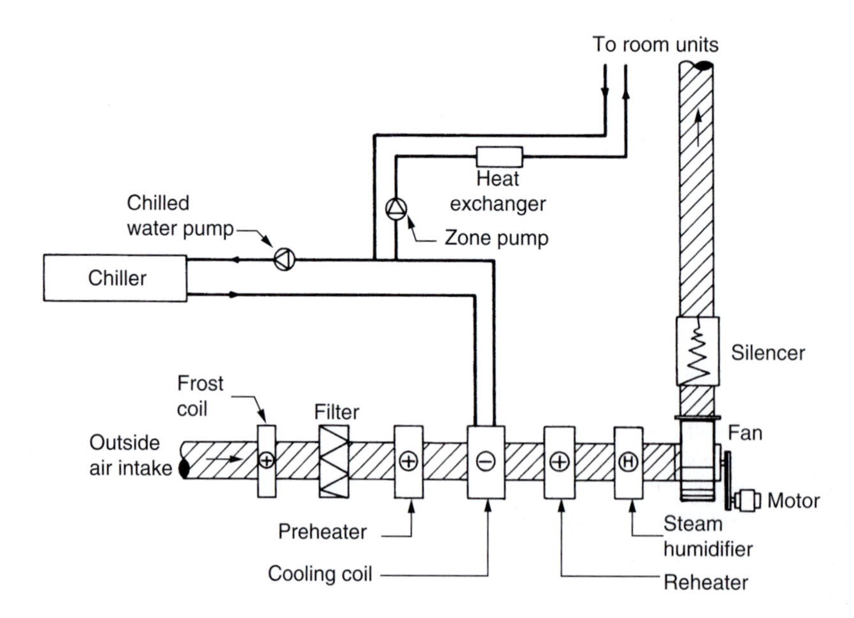

Figure 19.22 Typical plant arrangement for an induction system

Other systems

Upward air flow systems

The most common arrangement for introducing conditioned air into a space is from supply diffusers or grilles positioned at high level. Becoming more popular however is the use of upward air flow systems, where the air is distributed within a false floor, often required in any event for routing electrical power and communications networks, and introduced via floor-mounted outlets, perhaps supplemented by desktop supply terminals. Such systems may function using either the constant or the variable air flow principle.

Typical arrangements of this type of system are shown in Fig. 19.23, from which it may be seen that the desktop outlet can provide, effectively, a micro-climate for the occupant. It follows therefore that the temperature swing in the general area may be allowed to be slightly greater with this type of system than with one relying on conditioning of the general space. It is claimed that such systems have inherent flexibility to provide for changing the location of terminal devices to suit variations in an office layout. Special air terminal devices have been developed for floor and desk distribution, as described in Chapter 20.

A development of the upward air distribution principle is an arrangement where both the supply and the extract positions are at floor level, with the floor void divided into a supply plenum and a return air space. With this arrangement, no false ceiling is necessary since all servicing may be from low level, including up-lighting. This system uses fan-assisted conditioning modules to filter, cool, heat and humidify recirculated air, the units being either free standing in the space served or incorporated into service zones. In most cases, they will be supplied with heating, chilled and mains water piping and with a power supply. A typical arrangement of such a system is illustrated in Fig. 19.24.

The conditioned air is discharged into the supply plenum to which are connected fan-assisted terminal units, which may be either wall mounted or of an underfloor type, the latter being connected to supply grilles integrated into a standard 600 mm square floor panel. Return air is collected through similar panel-mounted grilles into the return air section of the floor void from which it is drawn into the conditioning module. Outside air for ventilation purposes is introduced into the return air section of the floor void and temperature

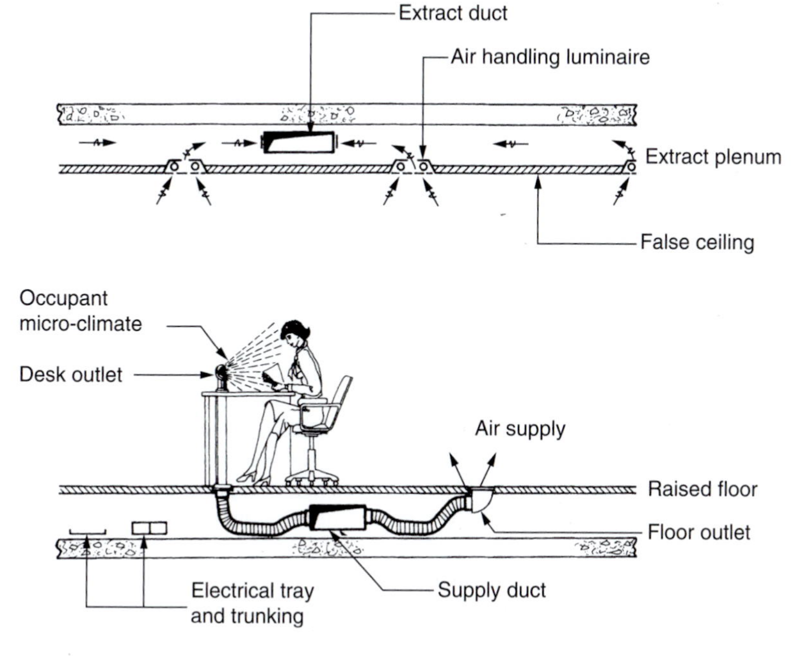

Figure 19.23 Typical upward air flow arrangements

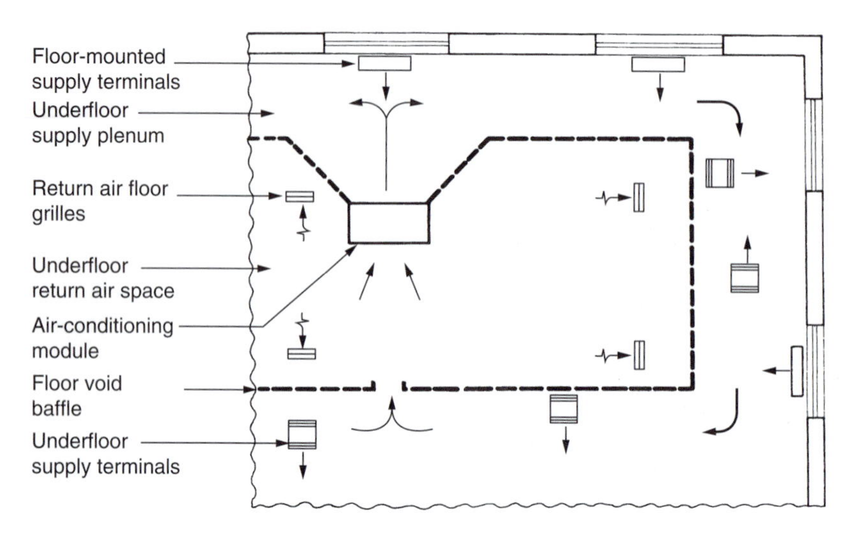

Figure 19.24 Upward air flow system with floor distribution (Liebert Hiross)

control in the space is achieved by varying the ratio of conditioned and recirculated room air introduced; alternatively, electric trim heaters may be used.

A clear underfloor depth of 200 mm will be required and, typically, up to 300 m^2 may be served from a single module. As will be appreciated, airtightness around the perimeter and at baffles between the supply plenums and the return air zones is critical to maintain performance. These systems provide good flexibility for changes in furniture layout and all components, including the conditioning modules, terminal units, false floor panels and void baffles, are available as a proprietary system. Individual desktop controls are available, as a proprietary package, to allow the occupant to set the conditions to suit his or her preferred working temperature: some systems also incorporate a presence sensor as an energy-saving feature to shut off the air supply at the workplace when the occupant is away from the desk.

Displacement ventilation

Indoor pollutants are diluted by 'mixing' with outside air when traditional methods of ventilation are employed and the same principle is adopted for dealing with heat gains and losses. An alternative approach is to introduce the supply at one position, and at low velocity, such that it moves in a single direction through the room using a piston effect to take the pollutants, including thermal effects, with it. In such systems, supply air is introduced close to the floor at a few degrees below the room design temperature allowing the input to flow across the floor forming a 'pond'. Heat sources within the room produce upward convective currents resulting in a gentle upward air flow towards high-level extract positions. For effective operation, the air in the space should not be subjected to continuous disturbance by rapid movement of occupants neither should there be high rates of infiltration nor downdraughts due to poor insulation.

A number of limiting performance and comfort factors impose restrictions upon the use of displacement systems.[1] The warm and often polluted upward air flow spreads out beneath the ceiling and, since the lower boundary of this layer should be kept above the zone of normal occupancy, application is limited to rooms with high ceilings. Comfort factors which limit the temperature difference between head and feet determine that the air supply temperature should be in the range 18–20°C for seated sedentary occupations, and this, in turn, restricts the cooling capacity of the system to 30–40 W/m^2. Air supply terminals should be selected to achieve a uniform air distribution pattern across the floor whilst keeping air velocity low, not exceeding 0.4 m/s close to the point of discharge. Terminals of suitable pattern for this application are illustrated in Chapter 20.

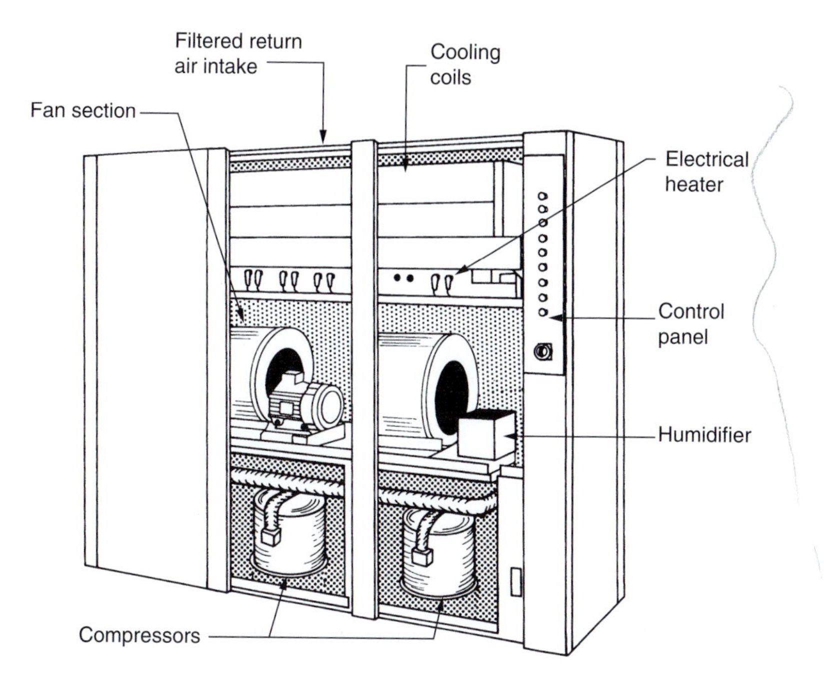

Figure 19.25 Typical down-flow-type room air-conditioning unit

Room air-conditioning units

Developed mainly for computer room applications, such units have been produced in recent years to provide close control of both temperature and humidity over wide ranges of load variation. The units may be self-contained, that is with an integral refrigerant compressor, or be served from a central chilled water source. Self-contained units may have remote condensing units or be connected to a water cooling circuit. Heating may be direct electric or by coils served from a central heating circuit.

Humidification is normally by steam injection from electrically heated units integral within the package. Air filtration is a requirement for such applications. Figure 19.25 shows a typical cabinet-type unit. Close control of conditions in the space together with the need for quick analysis of component failure has led to the general use of microprocessor controls for these units.

For computer room applications, a relatively small quantity of outside air is needed for ventilation purposes and normally no extract system is provided, thereby allowing the supply air to pressurise the space. The air-conditioning units would normally be duplicated, or in larger installations at least one redundant unit would be provided as a standby in case of a unit failure or for use during routine maintenance operations.

Room coolers

This heading covers a separate field in that such units are commonly complete in themselves, containing compressor, air filter, fan and cooling coil. Electric resistance heaters may be incorporated for winter use and, rarely, means for humidification. Fresh air may be introduced if required. Being of unit construction, alternatively described as *packaged*, they are not purpose made to suit any single application and thus may well be economical in first cost. In some cases, a so-called 'split type' of unit may be found where the condenser and compressor are mounted remotely from that part of the equipment which serves the room concerned. Bulk and noise at the point of use are thus much reduced.

Units of small size generally have the condenser of the refrigerator air cooled, but in larger sizes the condenser may be water cooled in which case water piping connections are required. Apart from this, the only

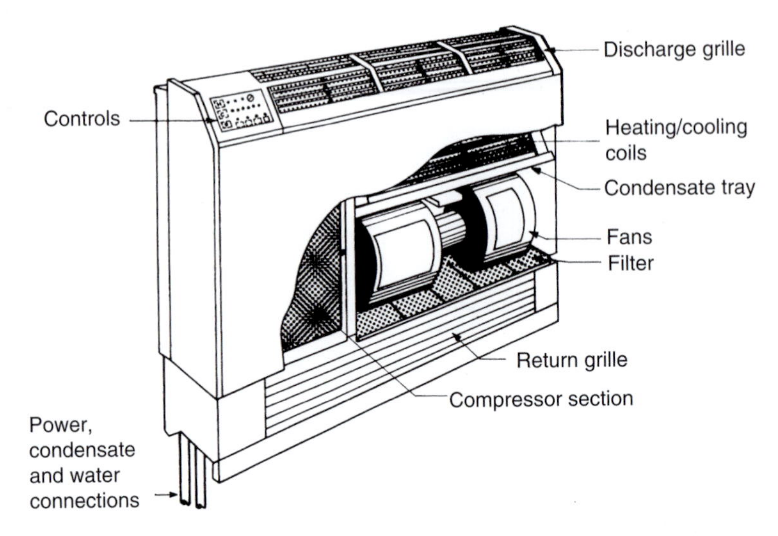

Figure 19.26 Under-window reverse cycle heat pump unit

services needed are an electric supply and a connection to drain to conduct away any moisture condensed out of the atmosphere during dehumidification. Compressors in most units are now hermetic and are therefore relatively quiet in running.

Sizes vary from small units suitable for a single room, sometimes mounted under a window or in a cabinet, similar to that shown in Fig. 19.25 and the range goes up to units of considerable size suitable for industrial application, in which case ducting may be connected for distribution.

Reverse cycle heat pumps

The heat pump principle described in Chapter 24 has been applied to general air-conditioning applications. In a similar fashion to the application of induction units and fan coil systems, heat pumps may be arranged in a modular configuration around the building. Units that operate both as heaters and as coolers are normally installed and these are termed *reverse cycle* heat pumps. Floor, ceiling and under-window types are available; a typical under-window unit is illustrated in Fig. 19.26.

Each unit incorporates a reversible refrigeration machine comprising a hermetic refrigeration compressor, a refrigerant/room air coil, a refrigerant/water heat exchanger, a cycle reversing valve and a refrigerant expansion device. When the space requires heating, the air coil acts as a condenser, drawing heat from a water circuit through the heat exchanger acting as an evaporator, upgraded by the compressor. When the space demands cooling the air coil becomes the evaporator, the heat being rejected to the water circuit via the waterside heat exchanger acting as a condenser. Figure 19.27 illustrates the components and operating cycles of such a system.

Simultaneous heating and cooling can be provided by individual units to suit the thermal loads around the building. There is a running cost benefit arising from such operation due to condenser heat from those units performing as coolers being rejected into the water circuit, thereby reducing the heat input required from central boilers. A diagrammatic arrangement of the system is given in Fig. 19.28, from which it can be seen that a two-pipe closed water circuit is maintained at around constant temperature, typically 27°C, to provide the heat source and heat sink for the heat pumps.

The CoP of the smaller distributed refrigeration compressors is lower than that obtained from central plant; the ratio being of the order of two to one. However, the distribution losses and the electrical power absorbed by the chilled water pumps associated with a central plant have the effect of reducing the effective difference between the energy requirements of the alternative methods of cooling.

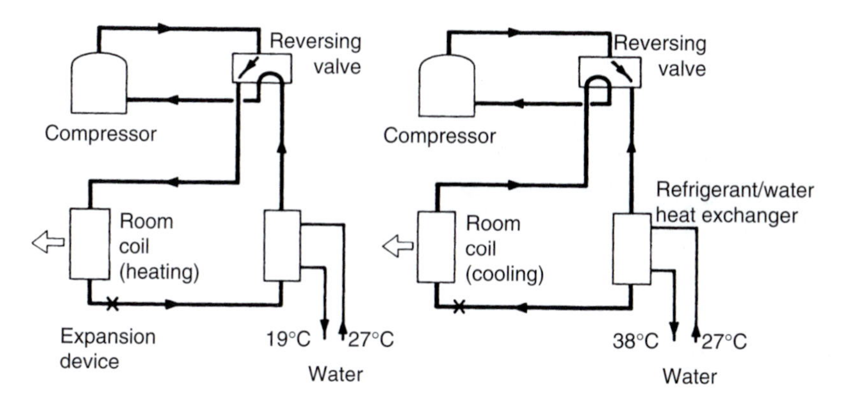

Figure 19.27 Operating cycles for a reverse cycle heat pump system

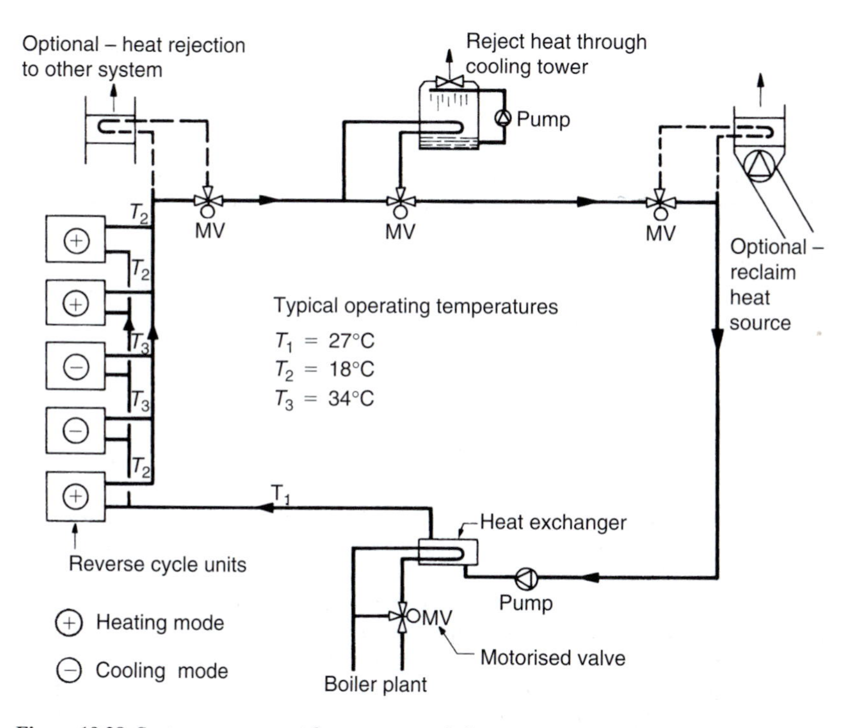

Figure 19.28 System arrangement for a reverse cycle heat pump system

Outside air may be introduced to the space through a central plant either independent from the units, connected to each unit, or alternatively drawn from outside directly into each unit. Air supplied from a central plant is preferred since this provides better control and reduced maintenance.

Temperature control is normally by a thermostat sensing return air to the unit which sequences the compressor and reversing valve. Only coarse temperature control is achieved and control over humidity is poor with such a system. Noise can be a problem with the units located in the space, particularly since the compressors may start and stop fairly frequently.

Larger heat pump units may be used to serve areas such as shops and department stores. Typically, these would operate on the reverse cycle principle, but would use outside air both as the heat source in the heating cycle and as the heat sink for the cooling cycle.

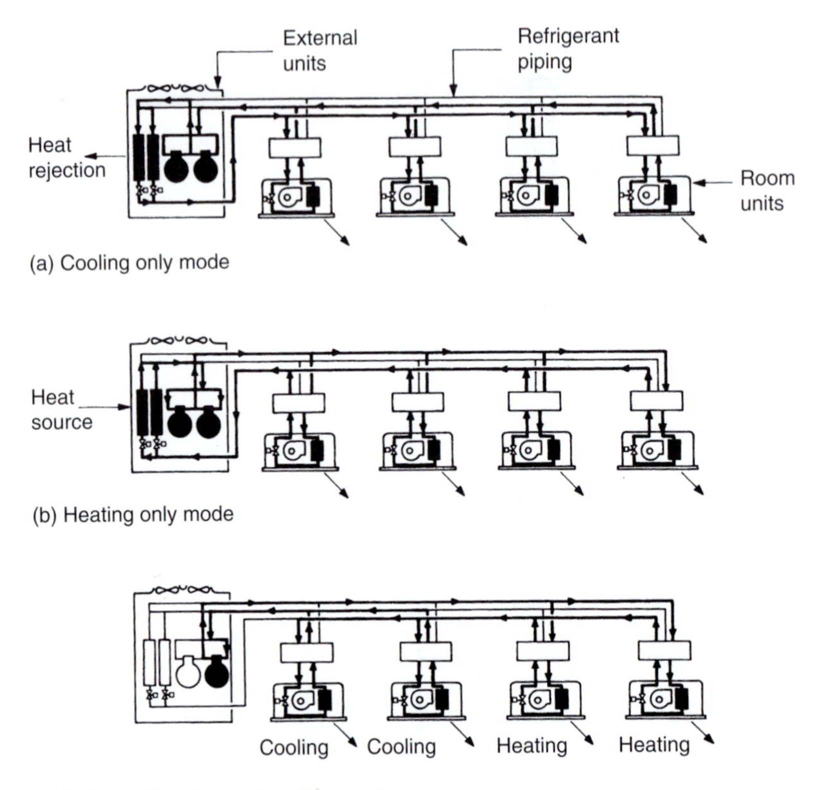

Figure 19.29 Variable refrigerant volume system: modes of operation

Variable refrigerant volume systems

An understanding of the operating principle of a variable refrigerant volume system will be assisted by reference to the description of the split-system heat pump (Fig. 24.26(b)). In the case of the variable refrigerant volume system, the external unit, which is usually roof mounted, comprises twin compressors; heat exchangers and air circulation fans. This external unit has refrigerant pipe connections to remote room terminals each of which incorporates a refrigerant-to-air heat transfer coil, a filter and a fan to recirculate room air (Fig. 19.29).

In the cooling mode of operation, the heat exchangers of the external unit function as a refrigerant condenser producing liquid which is circulated to the remote room terminals. As the refrigerant liquid passes through the coils where heat is absorbed, evaporation takes place and the gas is returned to the compressors. In the heating mode of operation, the heat exchangers of the external unit will function as a refrigerant evaporator, absorbing heat from the outside air and boiling off the liquid prior to compression. Hot gas from the compressors is then circulated to the room terminals, where useful heat is rejected at the coils and the resultant liquid is returned to the external unit.

The system is also able to operate in a dual mode offering both cooling and heating service to satisfy the requirements of the spaces served, including the thermal balance situation where the room heating and cooling demands are in equilibrium, in which case high thermal efficiencies will result. A three-pipe distribution system is needed for this dual function as shown in part (c) of Fig. 19.29.

In all modes, capacity control of the system, as the name implies, is by varying the quantity of refrigerant in circulation by means of speed variation of the compressors, with temperature control in individual rooms achieved by throttling the flow of refrigerant through the coil of the terminal concerned in response to a signal from a thermostat.

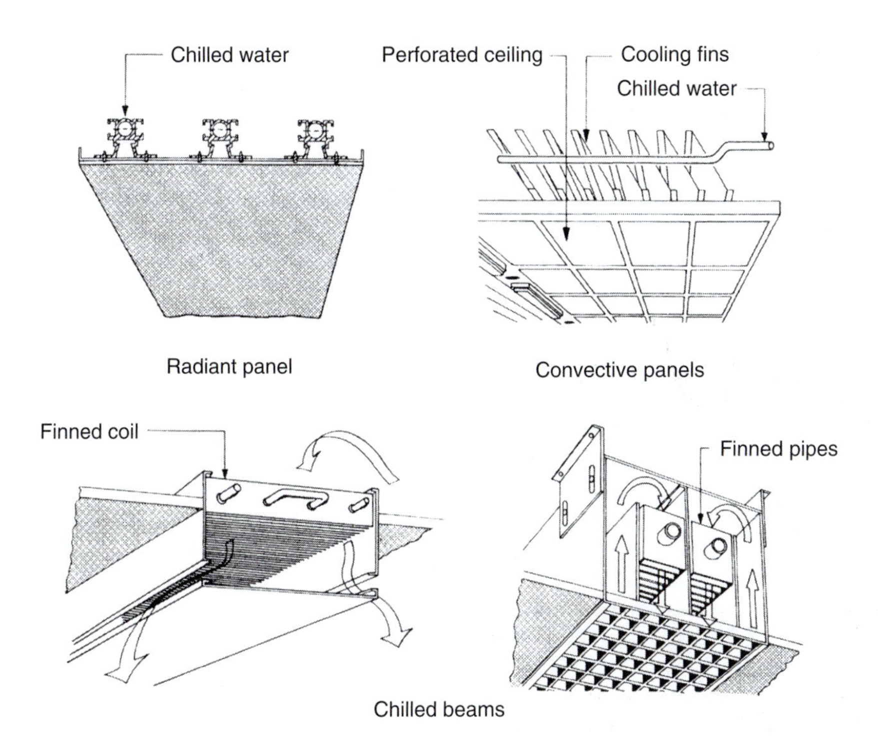

Figure 19.30 Chilled ceiling systems: radiant (Trox); convective (Krantz); chilled beams (Stifab Farex)

The maximum cooling capacity of an external unit is of the order of 30 kW and up to eight room terminals, having typical outputs in the range 2.5–15 kW, may be served from one external unit. There are limitations to the length of pipework between the external unit and the most remote room terminals, normally 100 m, with a maximum height difference of 50 m. The units and terminals (quite often in the form of ceiling-mounted cassette units), their controls and the distribution fittings are available as a proprietary system.

The extent of the internal piping system gives rise to some concern bearing in mind the potential for refrigerant leakage, the resultant concentration in an occupied space and thence into the atmosphere generally.

Chilled ceilings/beams

A fundamentally different approach to the whole matter of temperature control and ventilation, covered by the general term 'air-conditioning', is a system in which surfaces within the ceiling are cooled by chilled water circulation for the removal of heat gains, leaving to the air distribution system the sole purpose of ventilation and humidity control.

An essential feature of systems of this type is that the entering chilled water temperature should be above the room dew point, by at least 1.5 K to allow for control tolerance, in order to avoid any possibility of condensation forming on the cooling surfaces. Typically, chilled ceiling systems have a flow water temperature of 14–15°C and a temperature increase across the exchange device of 2–3 K. The dehumidifying capacity of the air supply is also important for control of the dew point and, in consequence, a design margin of the order of 20 per cent should be provided.

The cooling surfaces may take any of a number of forms which may be classified into one of the following categories, typical examples from which are illustrated in Fig. 19.30:

- Radiant panels
- Convective panels
- Chilled beams.

In the case of both the radiant and convective panels, the cooling surface covers large areas of the ceiling. In the former case, where the radiant component provides up to 40 per cent of the cooling effect, pipe coils may be either fixed, via a thermal conducting plate, to the upper surface of the ceiling panel or be embedded within the panel itself in which case the material may be small bore polypropylene in the form of a pipe mat. Radiant panels may be accommodated in shallow ceiling voids, as little as 65 mm for some proprietary designs.

Convective panels take the form of finned pipe coils which are located *within* the ceiling void, typically 250 mm deep, above a perforated or slotted ceiling which has at least 20 per cent free area to the room space. With such a system warm air from the room rises into the ceiling void where it is cooled by the coil and, being now more dense, the air then falls through the ceiling to provide room cooling.

Chilled beams operate in a similar manner to convective panels but in this case the finned coils are concentrated into a smaller unit which may also incorporate a ventilation air supply. These smaller units may be positioned either above the ceiling, with their underside flush therewith or be partly below the ceiling surface to form a dropped beam effect. A minimum depth of ceiling void of about 300 mm is required for installation in this instance. Some later models are suitable for suspension below the ceiling. Others incorporate the facility to introduce ventilation air via nozzles within the unit to increase, by induction, the air flow across the coil and thus extend the sensible cooling capability; these are known as *active chilled beam*.

One of the advantages of a chilled ceiling system is that it has capacity to offset high cooling loads without producing an unacceptable air movement within the space, a result which is often difficult to achieve with air-based cooling systems. With all types of chilled ceiling, temperature control is via a simple control valve acting in response to a thermostat, with chosen sections being controlled independently to provide zone control. This same control principle may be applied also to areas of perimeter radiant heating which may be integrated with the chilled ceiling panels to provide for winter weather.

Although convective systems have a higher sensible cooling capacity compared with radiant panels, 160 and 120 W/m^2, respectively, it is important to recognise that radiant systems reduce the mean radiant temperature in the space and, in consequence, reduce also the resultant temperature sensed by occupants by as much as 2 K.

Recent solutions incorporating chilled beams also integrate other service outlets and lighting within a common prefabricated casing. Figure 19.31 shows typical examples of these solutions commonly known as multi-service chilled beams.

Alternative methods of cooling

Reference was made in the preamble of this chapter to the need for avoidance, where possible, of methods for dealing with overheating in occupied spaces which rely upon the use of mechanical cooling plant. There are three obvious routes to follow in pursuit of a solution to the problem of overheating:

(1) Examination of the source of the problem with intent to reduce *heat gains* through the structure and from internal sources such as office equipment and lighting.
(2) Investigation of methods to improve the use and effectiveness of *natural ventilation*.
(3) Consideration of the use of *innovative cooling* techniques.[2]

Since the first two of these headings have been touched upon in earlier chapters, the following notes relate to remaining items and are to some extent anecdotal.

Innovative cooling techniques

Strictly speaking, such methods may be categorised as either *active* or *passive*, the former using some minimum amount of energy and the latter none at all. As might be appreciated, the distinction between the two, in practice, is a matter of degree.

Active (or perhaps passive!) cooling may be achieved, but not exclusively, by one or a combination of the following methods:

- *Night ventilation*, coupled with high thermal mass: low-temperature nighttime air is passed through the building to cool the structure which, next day, then acts to offset heat gains. (Unfortunately, in modern

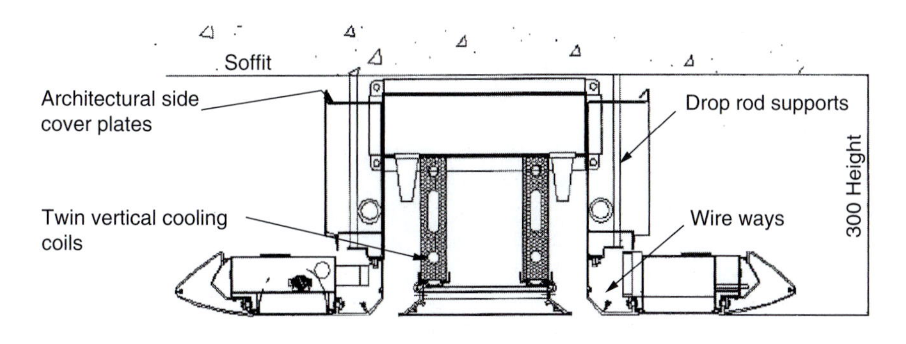

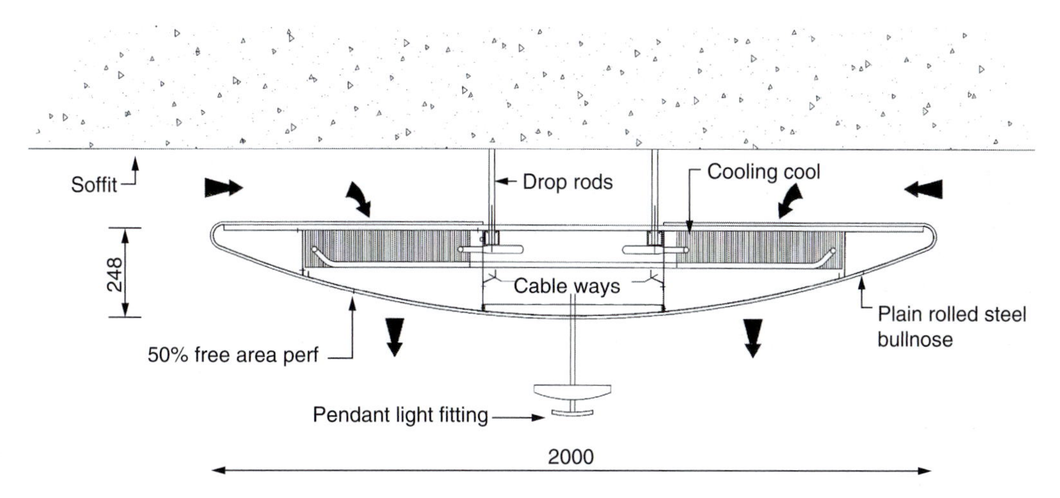

Figure 19.31 Typical examples of multi-service chilled beams solutions (Trox)

office buildings with lightweight partitions, false ceilings and carpeted floors, the mass of the actual structure is degraded thermally to 'lightweight' and thus no longer functions well as a store.)

- *Ground cooling*, using stable year-round ground temperature, normally 8–12°C, as a cooling source for circulated water, air or other transfer media.
- *Evaporative cooling* of air streams: passed through a water spray, the temperature of the air may be reduced by 3–4 K. The quite significant associated increase in humidity may not be tolerated in summer.

To achieve optimum conditions for effective cooling by such means, detailed examination of the thermal performance of the building fabric and any engineering systems *acting together* must be carried out through the whole range of coincident external and internal conditions within which the building will function. Means for analysis on this scale are not only available but commonplace.

Hollow floor system

A proprietary arrangement, using a combination of nighttime ventilation and structural mass, has been developed in Sweden, supply air being passed through the cores in hollow concrete floor planks before being introduced into the space. To be most effective, both the ceiling soffit and the floor surface should remain 'hard' and uncovered in order not to dilute heat transfer by any covering which might act as an insulating layer. The air, supplied at constant volume to meet the ventilation requirement, is further treated when necessary at a conventional central air handling plant providing filtration and heating or cooling.

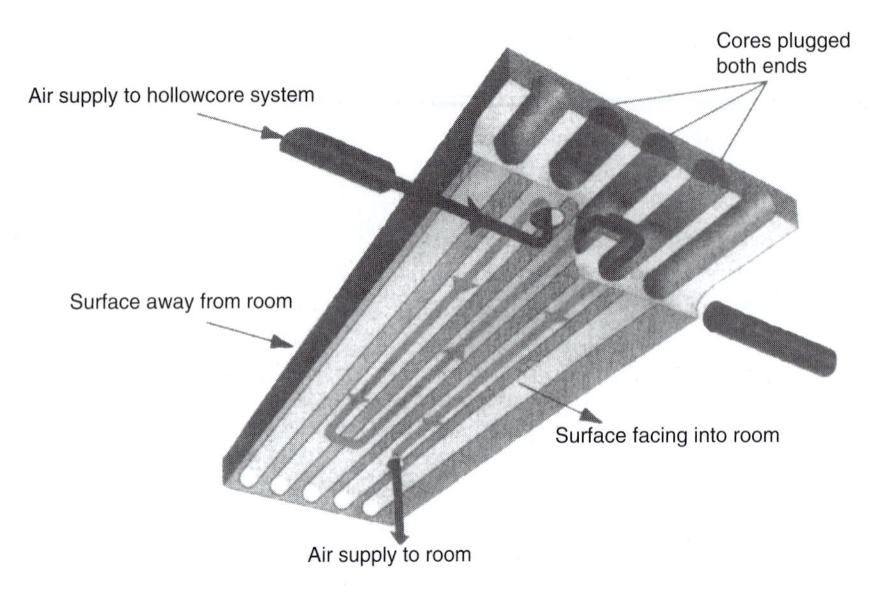

Figure 19.32 Structural slab system with integral airways (TermoDeck)

The planks, each up to 18 m in length and normally about 1.2 m wide, are precast with five smooth faced cores, 180 mm diameter, per plank. These are modified on site, by core drilling, to provide holes for air inlet and outlet to cores 2 and 4 and cross-passages between these and the central core 3. The end holes are plugged by backfilling with concrete to produce the arrangement shown in Fig. 19.32, and are then pressure tested to 400 Pa. In use, the air velocity in the cores will be about 1 m/s.

The planks, when laid, provide airways to cover the whole floor area. However, a consequence of the standard plank dimensions is that the system is capable of providing no more than coarse control which imposes a limitation in respect of handling diverse thermal loads across a whole floor. Furthermore, although the high thermal mass provides temperature stability, it is slow in consequence to respond to load changes in the space served. A damper box facility is available which may be fitted such that the air inlet is short circuited to the outlet core, thus increasing the thermal capacity of the air supply.

In summer, the supply air fan will normally run over 24 hours, continuously, to take advantage of diurnal variations in outside temperature. Cool air at nighttime will reduce the temperature of the building fabric and, during the day, warmer outside air will be cooled by the slab which, at the same time, will provide direct radiant cooling. Additional cooling, available at the central plant, may be used to further cool the supply air if required. Typically, air will leave the plant at about 13°C and enter the space at about 3 K below room temperature, a suitable level for use in a displacement ventilation system. In winter, the supply air will be heated, to a maximum of 40°C, and the slab will then perform as a large low-temperature radiant surface to supplement the heated air supply.

The concept is a simple one requiring relatively low capital expenditure and little maintenance. Exposed surface duct runs and connections to the core airways may be difficult to integrate aesthetically and, since the performance of the system is inextricably involved with the integrity of the floor structure, great care is required in defining contractual responsibilities. Nevertheless, tests at the Building Research Establishment suggest that, although supplementary cooling from a mechanical source will be required for applications in the climate of the British Isles, use of the system may be economic in energy terms.[3]

Embedded coil systems

Pipe coils embedded in a floor slab have been used with some success in a number of installations, using well-established technology and taking advantage of the good energy transport characteristics of water as the heat transfer medium. In the present context, it is the means of cooling the water which is of interest, this being

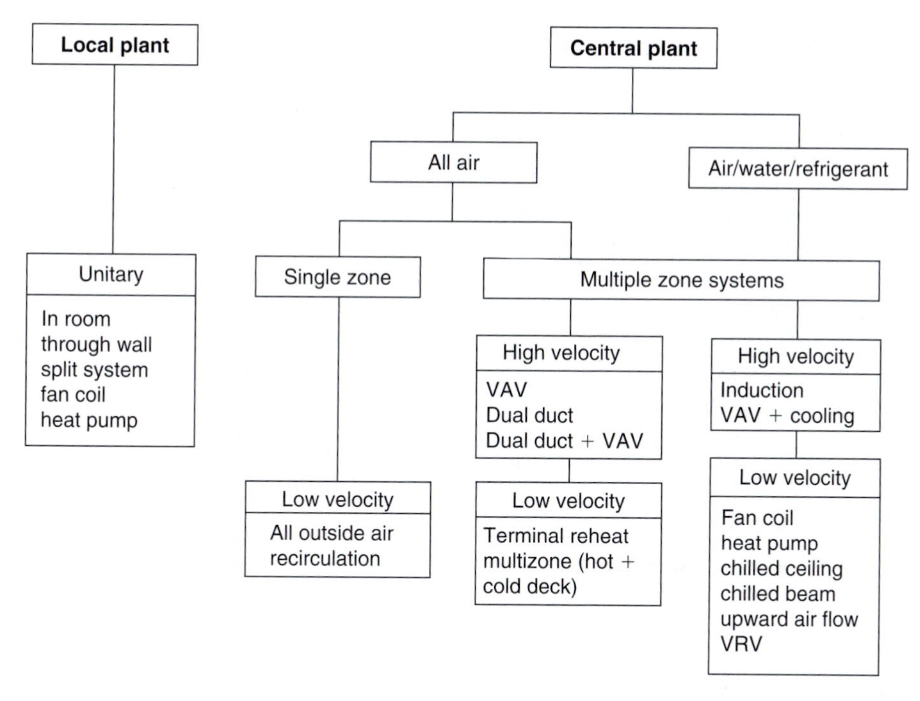

Figure 19.33 Principal characteristics of systems in common use: VAV, variable air volume; VRV, variable refrigerant volume

Table 19.1 Factors, other than thermal, affecting the choice of system

	System type				
	Fan-coil system, four-pipe with independent air supply	*Variable air volume with perimeter heating*	*Passive chilled beam with perimeter heating*	*Displacement ventilation with thermal mass*	*Reverse cycle heat pump, independent supply*
Spatial impact					
Plant space	Average	Poor	Average	Good	Good
Riser shafts	Average	Poor	Average	Good	Average
Floor space encroachment	Poor	Average	Good	Poor	Poor
Ceiling depth	Average	Poor	Average	Good	Average
Quality of performance					
Temperature control	Good	Good	Excellent	Adequate	Satisfactory
Humidity control	Adequate	Satisfactory	Good	Adequate	Poor
Air distribution	Adequate	Satisfactory	Good	Good	Poor
Noise	Adequate	Good	Good	Excellent	Poor
Costs					
Capital	Average	Average/high	Average	Average	Low
Operating energy	Average/high	Average	Low/average	Low	Average
Maintenance	High	Average	Low	Low	High
Flexibility					
To suit partitioning arrangements	Good	Good	Good	Poor	Poor
Increase cooling load	Good	Average	Good	Poor	Poor
Increase ventilation	Average	Good	Excellent	Good	Average

either by a dry air cooler or by a closed circuit water cooler, either of which may be operated at nighttime when the outside air temperature is low. The principles of operation of these cooling devices are described under the heading 'Free cooling' in Chapter 24.

Such a system for an office block in Switzerland serves a building of 8000 m^2 floor area with some 60 km of embedded pipe coils operating in conjunction with roof-mounted dry cooler heat exchangers. The water temperature between 8 a.m. and 10 a.m. is reported to be 19°C as is the supply air temperature which is, at most periods of the year, not cooled mechanically. Air delivery to the space is through air handling luminaires which raise the temperature to 1 or 2 K below room level.

Summary of systems and application

Figure 19.33 sets out the various air-conditioning system types in common use and identifies their principal characteristics.

To give an indication of the relative merits of the more popular systems, Table 19.1 provides a summary of some of the important design parameters with an indication of how well the various systems are able to satisfy these.

Notes

1. Jackman, P.J., *Displacement Ventilation*. BSRIA Technical Memorandum 2/90.
2. International Energy Agency. Energy conservation in buildings and community systems. *Innovative Cooling Systems*. Workshop Report, 1992.
3. Willis, S. and Wilkins, J., Indoor climate control – mass appeal. *CIBSE Journal*, 1993, 19, 25.

Chapter 20

Air distribution

Effective distribution of air within an occupied space is the key to successful operation of a ventilation or air-conditioning system. It is of little use to provide plant and ductwork distribution arrangements ideally suited to meet load demands if the methods used to introduce the air supply do not provide for human comfort or process needs.

Successful air distribution requires that an even supply of air over the whole area be provided without direct impingement on the occupants and without stagnant pockets, at the same time creating sufficient air movement to cause a feeling of freshness.

This definition indicates what is probably the key to the problem of successful distribution: that unduly low velocities of inlet are to be avoided just as much as excessively high ones and that distribution above head level not directly discharging towards the occupants will give the necessary air movement to ensure proper distribution over the whole area without draughts. Low-level floor supplies, however, introduce air directly into the occupied zone and so need special attention.

To produce satisfactory conditions in the comfort zone of a space to be held at normal temperature, the distribution system should produce an air velocity, at a measurement point 1.8 m above the floor and not less than 0.15 m from a wall, of between 0.1 and 0.25 m/s and never less than 0.05 m/s. Where activity is high and spot cooling needed, as in a factory, a velocity of up to 1 m/s might be acceptable. *Laminar flow*, used in cleanrooms, and other special distribution methods are outside the scope of this book.

General principles

There are five general methods of air distribution.

(1) upward
(2) downward
(3) mixed upward and downward
(4) mixed upward and lateral
(5) lateral.

The choice of system will depend on:

- whether simple ventilation or complete air-conditioning is employed
- the size, height and type of building or room
- the position of occupants and/or heat sources
- the location of the central plant, and economy of duct design
- constraints imposed by the building structure and internal layout.

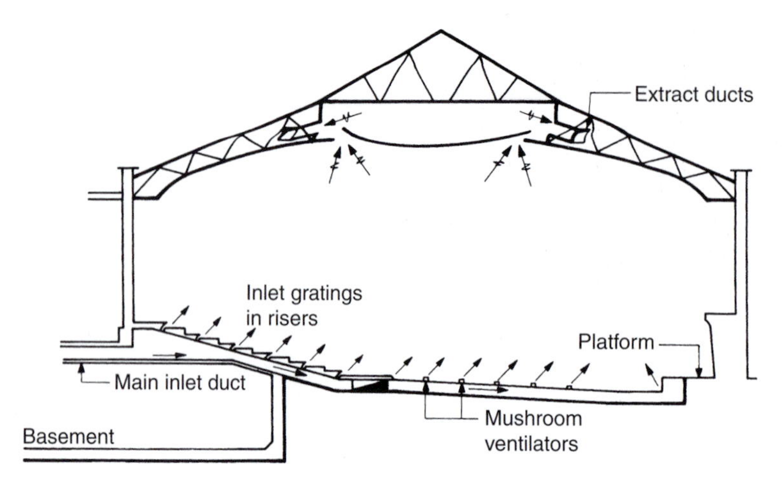

Figure 20.1 Upward air distribution

Upward system

The air is introduced at low level and exhausted at high level as in Fig. 20.1 which shows a section through an auditorium, with mushroom inlets under the seats and riser gratings (see Figs 20.30 and 20.31) in the gallery risers. The air is exhausted around the central laylight in the roof. Such a system could be designed so as to be reversible, i.e. operated as a downward system.

When working upwards, the air generally appears to be somewhat 'dead', due to the very low velocity of inlet (about 0.5 m/s) necessary with floor outlets to prevent draughts in this type of application. When working downwards more turbulence is set up in the air stream, with a greater feeling of freshness.

The upward system is not, however, confined to one with floor inlets. The inlets may equally well be in the side-walls, with extract in the ceiling as before. The limitation of an upward system is that in a large hall it may be difficult to get the air to carry right across without picking up heat *en route* and rising before it reaches the centre.

The upward system is used successfully with simple ventilation systems or where air is being introduced 2 or 3 K below room temperature. When the air is cooled to conventional temperatures of, as in a complete air-conditioning system, it will tend to fall too early, before diffusion, and thus cause cold draughts unless it is introduced through specially designed outlets carefully selected to suit the application. The upward system, however, lends itself to simple extract by propeller fans in the roof in the case of a hall, factory, etc., and is thus generally the cheapest to install.

Another application of the upward system is the swimming pool hall, as shown in Fig. 20.2. Here the supply air is introduced through special plastic discharge spouts situated below a large area of glazing and is exhausted by specially treated roof-exhaust units.

On a smaller scale, upward distribution has been successfully applied to computer rooms and offices. In the machine areas of computer suites, where occupancy is transient, air velocities within the occupied zone are less critical than is the ability of the system to maintain close control over temperature and humidity. The large air quantities required in such applications to counter the high heat gain from equipment give rise to air velocities above the recommended comfort criteria since air supplies are distributed to match the positioning of the space loads.

In office applications air movement is more critical and for floor supply systems it is important that the air is introduced by using a relatively large number of small outlets, the design of these being such as to produce a high induction effect, the supply air mixing quickly with the room air to reduce velocity and temperature differential. A typical arrangement is shown in Fig. 19.23. A floor supply system may be supplemented by desk outlets fed from the same plenum thus providing desk-bound operatives with a degree of control over their

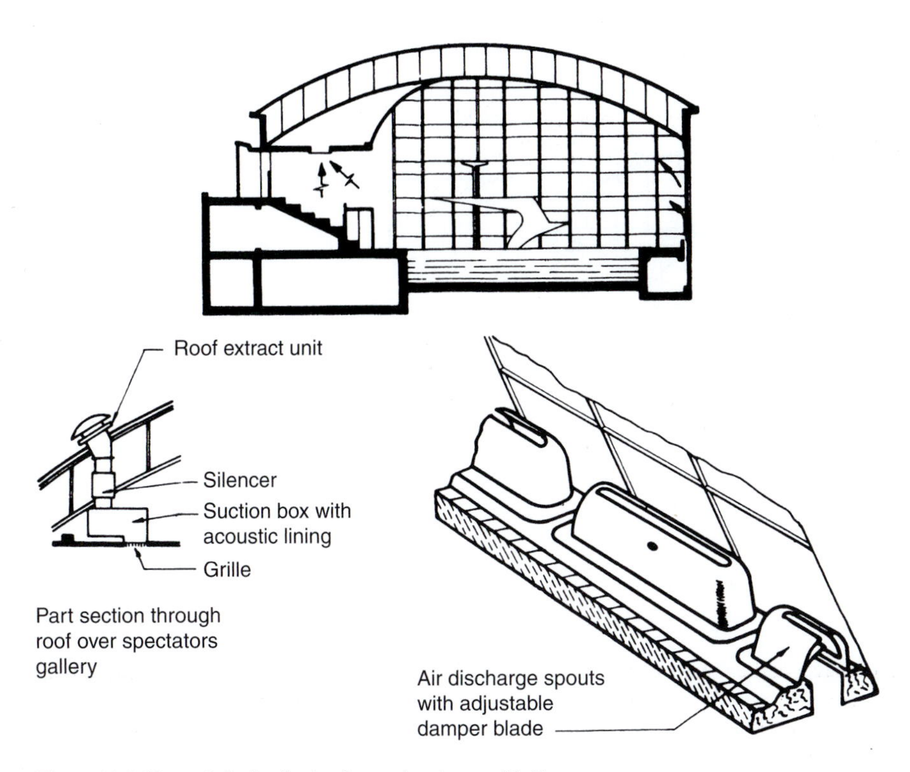

Figure 20.2 Upward air distribution in a swimming pool hall

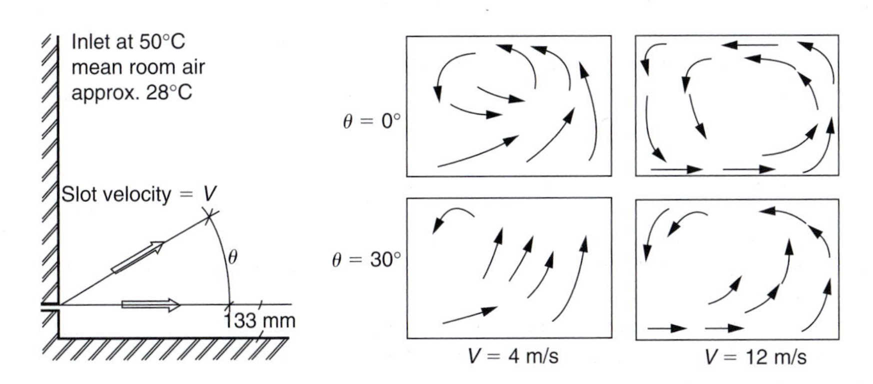

Figure 20.3 Upward air distribution from low level side-wall outlets

micro-climate. Displacement air-conditioning systems, following the pattern described in Chapter 19, rely for their operation on minimal mixing of supply air with room air and, in consequence, low level side-wall or floor outlets having low discharge velocities are selected for this application. With such systems, perimeter heating loads are dealt with conveniently by an independent system.

On a scale appropriate to single rooms, experimental work has been reported which suggests that for winter use it is possible to introduce warm air via a long slot near floor level, as shown in Fig. 20.3, and by this means much reduce the temperature gradient within the room.[1] Discharge velocities of up to nearly 12 m/s were used without reports of discomfort from occupants.

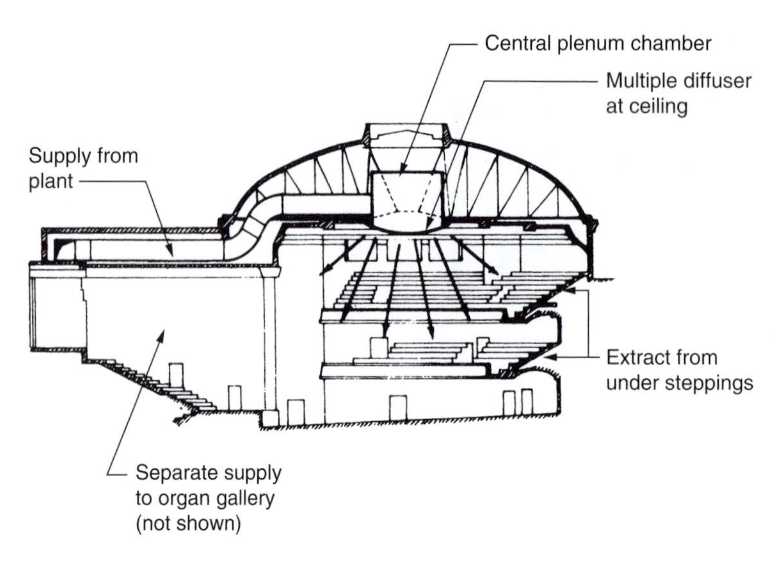

Figure 20.4 Downward air distribution in a concert hall

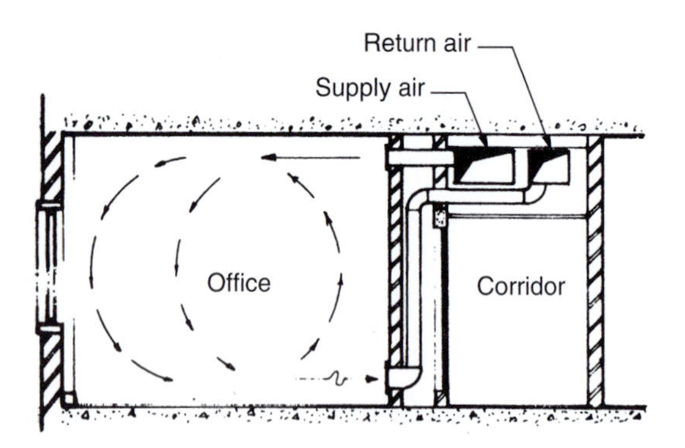

Figure 20.5 Downward air distribution in offices

Downward system

In this type of system the air is introduced at high level and exhausted at low level, as in Fig. 20.4. It is commonly used with full air-conditioning where, due to the air admitted being cooled, it has a tendency to fall. The object of distribution in this case is so to diffuse the inlet that the incoming air mixes with room air before falling. Thus, the inlets shown in the diagram as discharging downwards, in practice deliver in part horizontally at sufficient speed to ensure that the air completely traverses the auditorium. Turbulence is thus caused with the desirable effect already mentioned. On a smaller scale, as applied to an office building, this system appears as in Fig. 20.5

Provided the height of room is not abnormal, the extract opening may be at high level as in Fig. 20.6. Short circuiting is avoided by the velocity of the inlet air carrying over to the far side of the room. Another possible arrangement is a variation of this, namely, 'downward-upward', as in Fig. 20.7(a).

Computer rooms may also be conditioned using a ceiling supply, normally through a perforated ceiling or an equivalent system, with the extract taken out via floor extracts or low level side-wall grilles. The supply arrangement is described then under the heading of *perforated ceilings*.

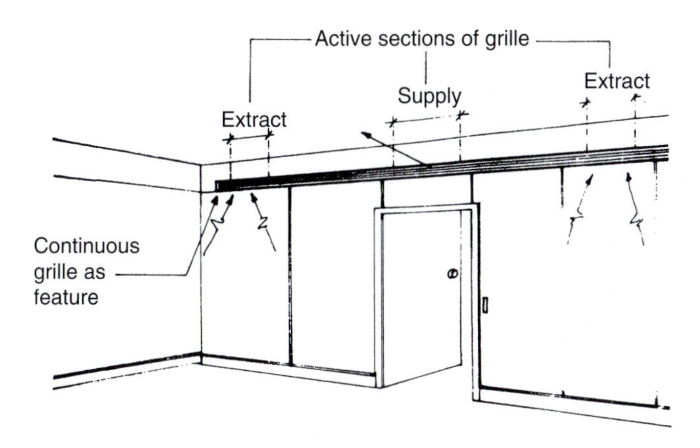

Figure 20.6 High-level inlet and extract

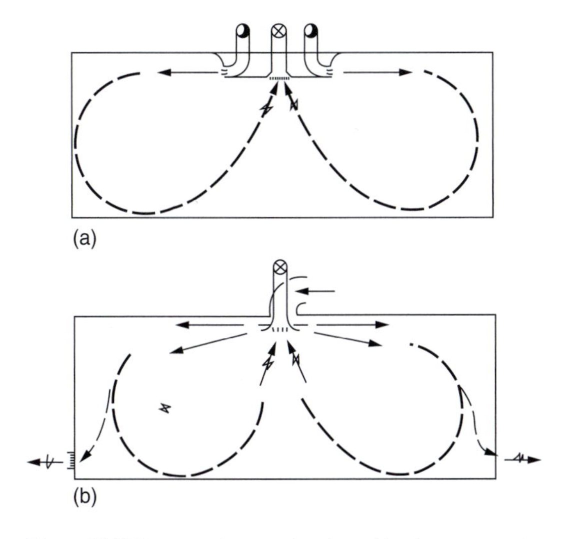

Figure 20.7 Downward–upward and combined arrangements

An application of downward inlet with both downward and upward extract, suitable for rooms of greater height, is shown in Fig. 20.7(b). This is usually adopted where smoking occurs and it is necessary to provide some top extract to remove the smoke. In this case the top exhaust is discharged to atmosphere by a separate fan and the low-level extract constitutes the recirculated air. The low-level extract also serves to ensure that satisfactory air movement is achieved at the occupied level, in cases where the room height is over 4 m. Care must be taken when placing low-level extract grilles close to areas where people are seated: such grilles must be selected for a very low velocity through the free area and be well spaced out so that excessive air movement will not occur.

Mixed upward and downward

Such a system is shown in Fig. 20.8, illustrating a typical swimming pool hall application. It will be clear from the previous descriptions that the principle of the air distribution is, in effect, an upward system providing good mixing. Normally about 25 per cent of the extract air quantity will be at low level also, the remainder being exhausted at high level.

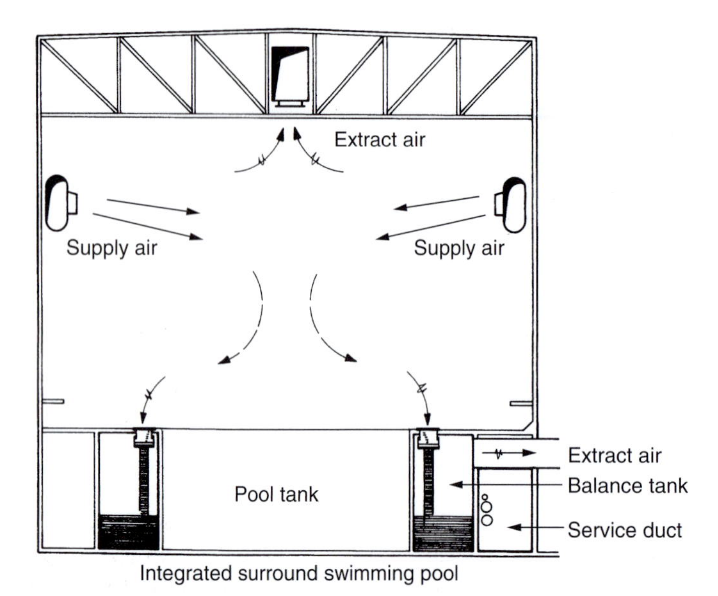

Figure 20.8 Mixed system of air distribution

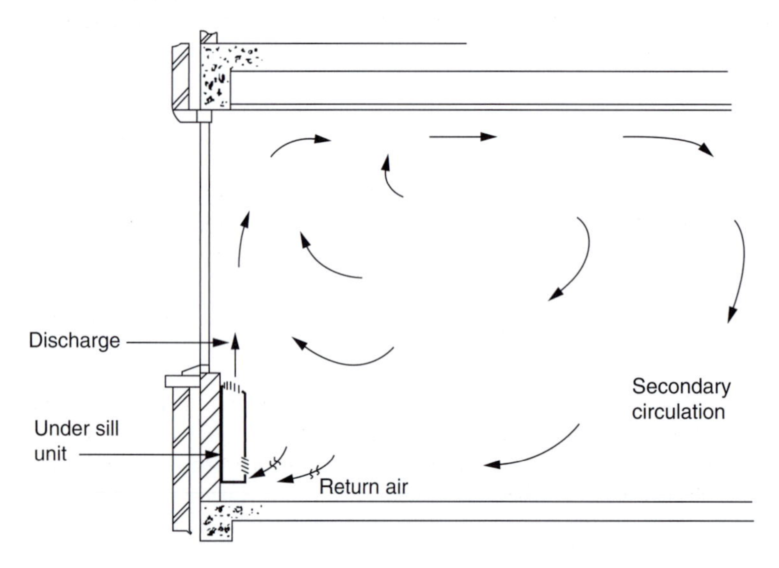

Figure 20.9 Vertical air distribution from sill level

Mixed upward and lateral

Such a method has been used to describe a system where the air is introduced vertically upwards from beneath a window. The flow pattern is vertical, or virtually so, up to ceiling level and then horizontal across part of the ceiling. Secondary room air is induced into the air stream producing a flow pattern as shown in Fig. 20.9. This method of air distribution is in common use with induction unit, fan coil and reverse cycle heat pump systems, where the terminals are located at low-level under windows. Fixed blade linear diffusers are normally used and the flow may be vertical or angled towards or away from the window. Vertical throw is preferred as this avoids the risk of the air, when cool, 'dumping' into the room. However, where there is a recess formed at the junction

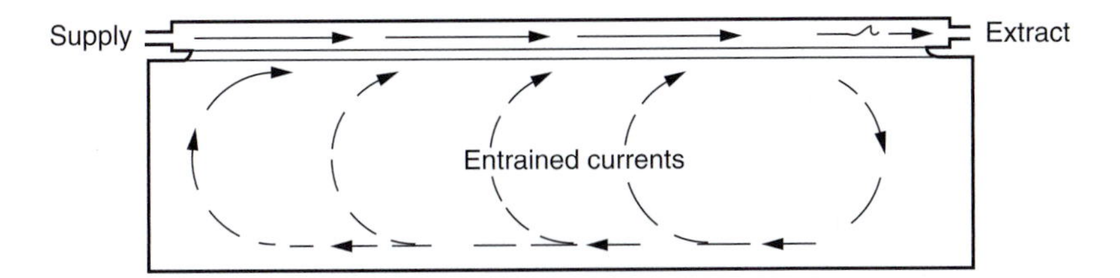

Figure 20.10 Lateral air distribution

of ceiling and window, perhaps where there is a dropped ceiling, the air has to be directed away from the window to avoid interruption of flow at the recess; an angle of 5° or 10° from the vertical is common. With upward flow over glass, care must be taken to ensure that the air stream when cool will not cause the insolated glazing to crack under the conditions of thermal stress then created: the glass manufacturer should be consulted in this respect.

Lateral

This arrangement may sometimes be necessary where dictates of planning preclude more orthodox solutions. Air is introduced near the ceiling on one side of a long low room with a smooth flush ceiling, and is exhausted at the opposite end at the same level taking advantage of the 'Coanda' effect which will be referred to in more detail later (see Fig. 20.10). The inlet is at a high velocity and strong secondary currents are set up in the reverse direction at the lower levels, as shown. It is these secondary currents which are important in many distribution systems, creating the turbulent mixed flow pattern already mentioned. Care should be taken, however, to avoid high velocity reverse flow at floor level creating draughts.

Distribution for air-conditioning

Air-conditioning usually involves the handling of large air quantities, and one of the chief problems of successful design is how to introduce and extract these quantities without giving rise to complaints of draught or of causing noise. A great variety of methods has been adopted and there are innumerable devices available to suit different conditions or architectural tastes.

In installations involving unusual air distribution patterns, or abnormally high or low air change rates, the system should be tested as a mock-up on site or in a laboratory, such as BSRIA, to ensure that satisfactory air movement will be achieved.

The selection of fixed pattern terminal devices for use with variable flow rate systems must be suitable for the full range of supply volumes. Manufacturers' data are often adequate for the general run of cases, but where a detailed study of the physics of air distribution is required, reference should be made to relevant independent laboratory reports such as those produced by BSRIA.[2]

Air diffusion terminology

Definitions and recommended terminology used in this subject are given in an ISO Standard,[3] supplemented in the CIBSE *Guide Section B2*. The following are relevant to the text:

Hovel: A contour of equal air velocity around a terminal device.
Throw: The distance from the terminal to the position where the velocity has decayed to 0.5 m/s, i.e. the 0.5 m/s isovel.

Normally, velocities for air entering the occupied zone would be limited to 0.25 m/s for cooling and 0.15 m/s for heating.

Spread: The width of the 0.5 m/s isovel (manufacturers may quote for 0.25 m/s isovel).
Drop: The vertical distance from the terminal to the lower part of the 0.25 m/s isovel.

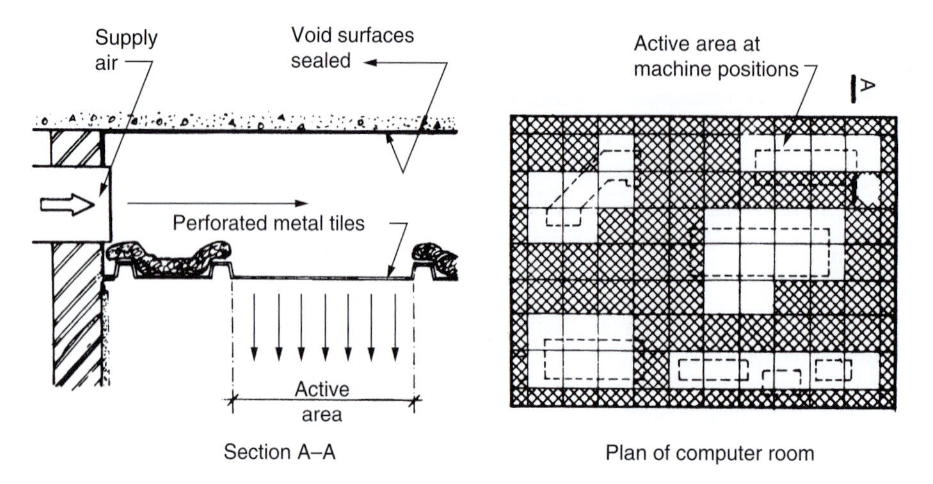

Figure 20.11 Inlet via a perforated ceiling

It is also necessary to understand the *Coanda* effect. This occurs when an air stream is discharged along an unobstructed flat surface. The jet entrains room air from one side only and friction loss between the jet and the surface causes the air stream to 'cling' to the surface until the velocity of discharge has decayed sufficiently as a result of entrainment of room air. A projection into, or a large gap in the surface may well destroy the Coanda effect causing the air stream to become detached prematurely from the surface; throw is reduced by about one-third when the Coanda effect does not occur.

Typical air flow patterns from a selection of terminals and applications are given in the *Guide Section B2.*

Perforated ceilings

Probably the most completely diffused form of inlet is the system which employs a perforated ceiling as shown in Fig. 20.11. In this case air is discharged into the space above the ceiling, which is usually divided up so as to ensure uniformity of distribution, and the air enters the room through the perforations. Where extremely large air quantities are involved the whole ceiling may be used, but in the normal case only selected areas of panels serve as inlets, acoustic or other baffles being positioned behind the 'non-active' panels. The ceiling inlet system destroys all turbulence, which may be a good thing in some cases but not in others. This system has been applied successfully in law courts, department stores and in confined spaces such as radio commentators' boxes where no other distribution arrangements would be possible. An alternative to a perforated ceiling, but which achieves an equivalent effect, is the use of an array of strip diffusers, installed at say 300 mm intervals. An example is shown in Fig. 20.12.

Cone-type ceiling diffusers

The type of ceiling diffuser shown in Fig. 20.13(a) has certain interesting characteristics. Figure 20.13(b) illustrates the kind of flow pattern to be expected from such a unit and it will be noted that entrainment air is drawn up in the centre immediately under the diffuser, being caught up in the nearly horizontal delivery from the unit itself. Due to the fact that air is discharged radially, the air stream velocity falls off rapidly as the distance from the centre increases, and hence this kind of unit may be used with temperature differentials of up to 20 K.

There is a great variety of makes of this type of ceiling diffuser, all under different trade names and some with special features, such as:

- An adjustable arrangement whereby the inner cones may be raised or lowered in relation to the periphery, so causing a variation in the flow pattern to be achieved. Instead of the bulk of the air travelling horizontally,

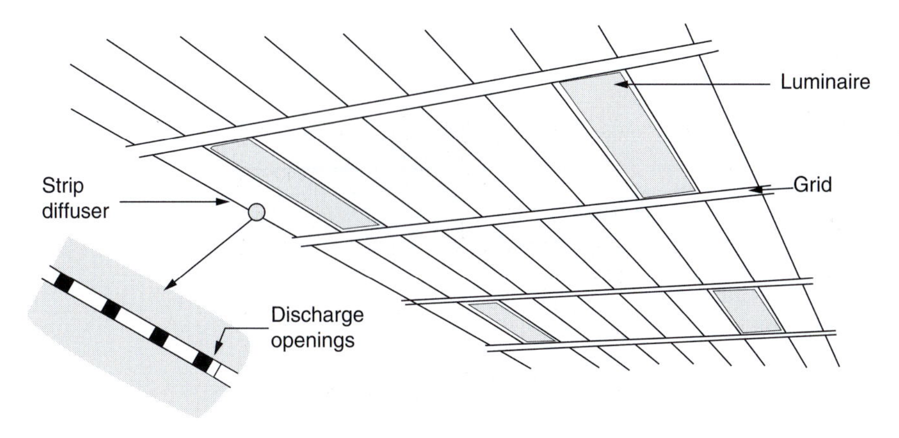

Figure 20.12 Ceiling distribution using multiple strip diffusers

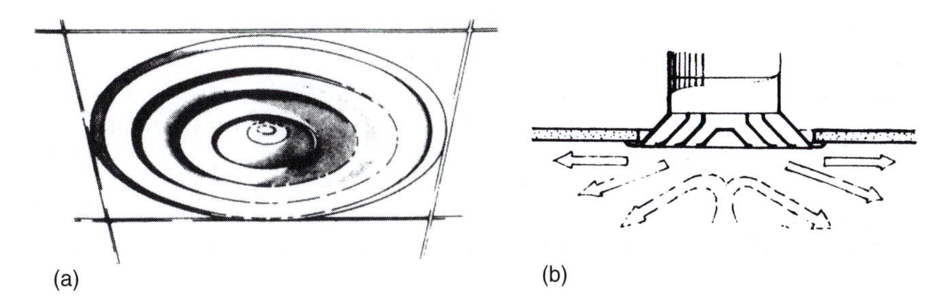

Figure 20.13 Cone-type ceiling diffuser

it is possible by this kind of adjustment to make it discharge vertically downwards or at any intermediate flow pattern desired. This might be of advantage in cases where it is desirable to cause the air to descend quickly, such as in a hot kitchen, rather than that it should be dispersed and lost at high level. Terminals of the pattern may be provided with removable cores for cleaning, which would be suitable for applications such as hospital operating theatres.

- In another type, the unit is square instead of circular, this sometimes being necessary to match ceiling tiles, etc.
- In other examples, the diffuser is flush with the ceiling instead of projecting.

Perforated-face ceiling diffusers

This type, shown in Fig. 20.14, has been developed from the early *pan* devices and has characteristics similar to the cone pattern: it is often used to blend in with perforated acoustic-tile ceilings. The deflecting pan may be removed if a predominantly vertical air distribution is required. Both the cone and perforated plate types may be fitted with internal blanking pieces to limit the spread of air discharge.

Multi-directional ceiling diffusers

Such units provide facilities for positive directional control of air discharge, the blades being individually adjustable as illustrated in Fig. 20.15. They are useful for applications where partition changes may occur or where movement of office machinery may create change in demand for air movement.

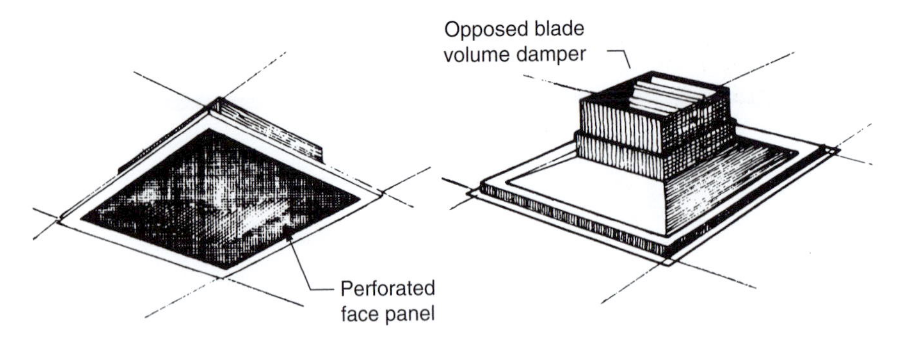

Figure 20.14 Perforated late-type ceiling diffuser

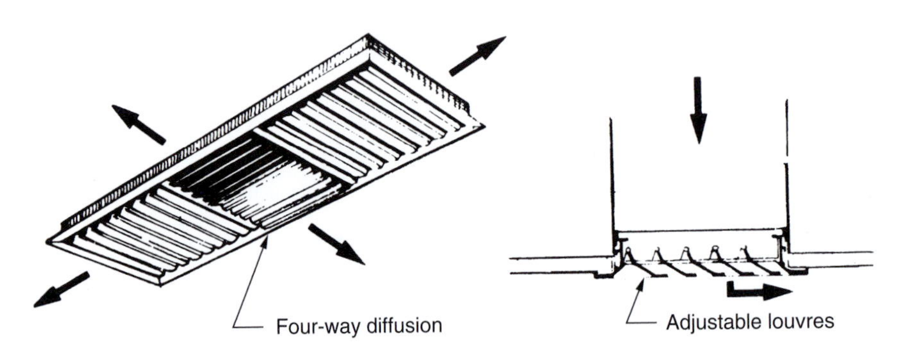

Figure 20.15 Multi-directional type ceiling diffuser

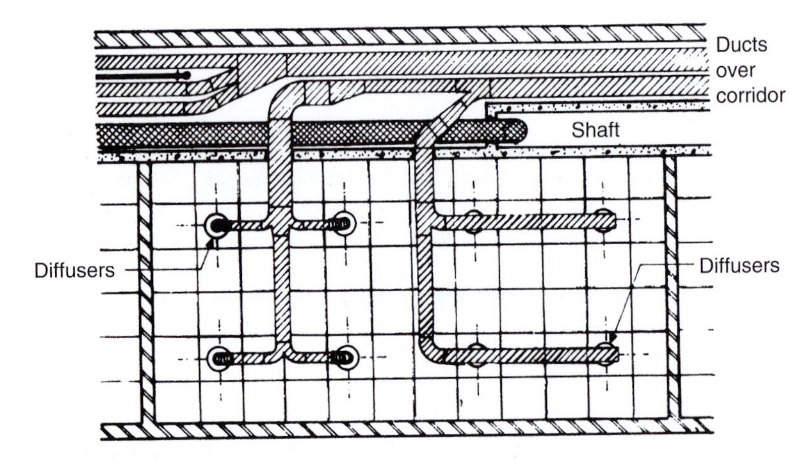

Figure 20.16 Typical layout of ceiling diffusers

A typical layout of ceiling diffusers is shown in Fig. 20.16. It will be noted that use is made of the false ceiling space for the concealment of the connecting ducts, the final connections to the diffusers being of flexible ducting, thereby providing dimensional tolerance between the duct and the terminal position. To ensure that air distribution is effective, the room is divided into approximate squares with one diffuser to each. Alternatively, if it be necessary to use a non-symmetrical spacing, segments of diffusers may be blanked off to avoid, e.g.

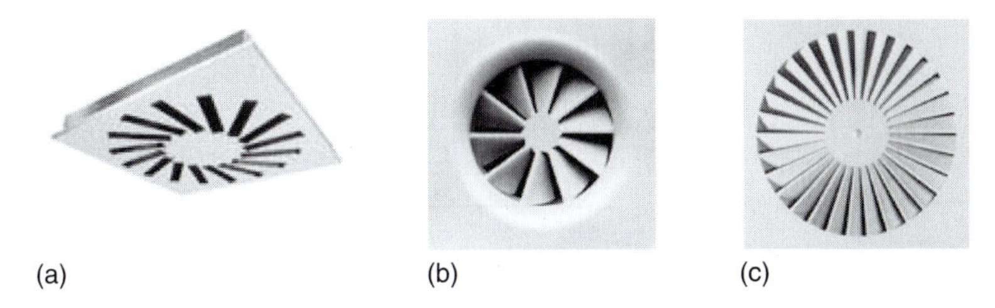

Figure 20.17 Typical designs for swirl ceiling diffusers (Trox)

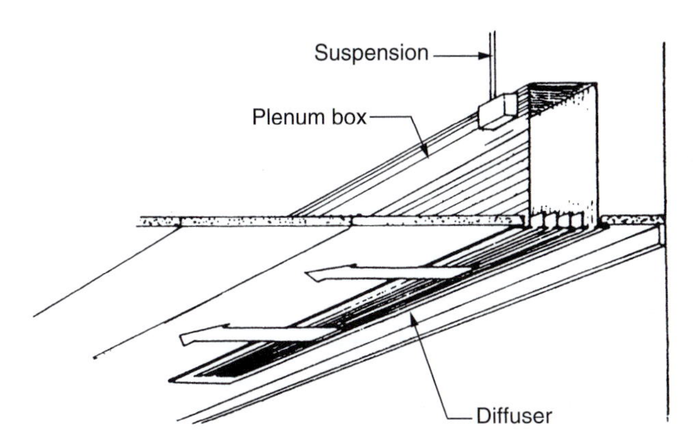

Figure 20.18 Multi-vane type linear diffuser

impingement of high velocity air flow at walls which would otherwise create excessive air movement at the occupied level.

Swirl ceiling diffusers

Originally used more in industrial applications to introduce large volumes of air, swirl diffusers are now used in many comfort air-conditioning solutions and ranges of smaller diameter diffusers have been developed. The design of the diffuser produces a swirling rotational type of discharge of the supply air creating a high level of induction and a rapid reduction of temperature difference. At the same time the noise levels are also low. In variable volume application the diffusers, because of the high induction, will maintain good performance down to about 25 per cent of maximum duty.

The layout and integration of the diffusers with false ceiling arrangements is similar to other ceiling diffusers described previously. Figure 20.17 shows typical arrangements.

Linear diffusers – ceiling

For use in open plan offices and in avoidance of interference with ceiling pattern, the linear diffuser in many different forms is commonly used. In its original form, Fig. 20.18, it was an adaptation of the continuous side-wall grille but took advantage of the Coanda effect.

Development of this type of diffuser takes many forms, as shown in Fig. 20.19. As may be seen, facilities are available for adjustment of the air flow pattern. Individual sections of the continuous length are supplied by means of air plenum boxes which may, in turn, be integrated with luminaires as in Fig. 20.20.

A further development of linear ceiling diffusers is the high induction slot which results in a rapid decay of the supply air velocity and temperature differential. These highly adjustable diffusers are made up from air control blades inserted within the slot diffuser. The direction of air discharge can be adapted to the room

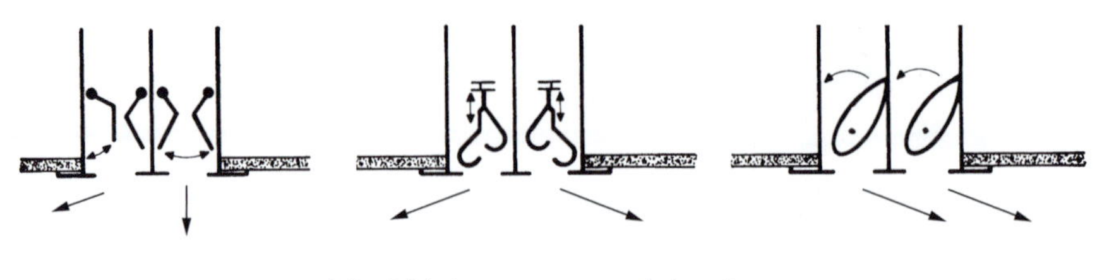

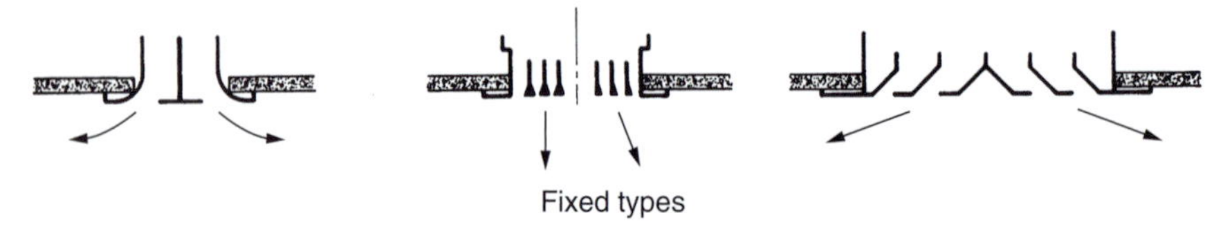

Figure 20.19 Various forms of ceiling-integrated linear diffusers

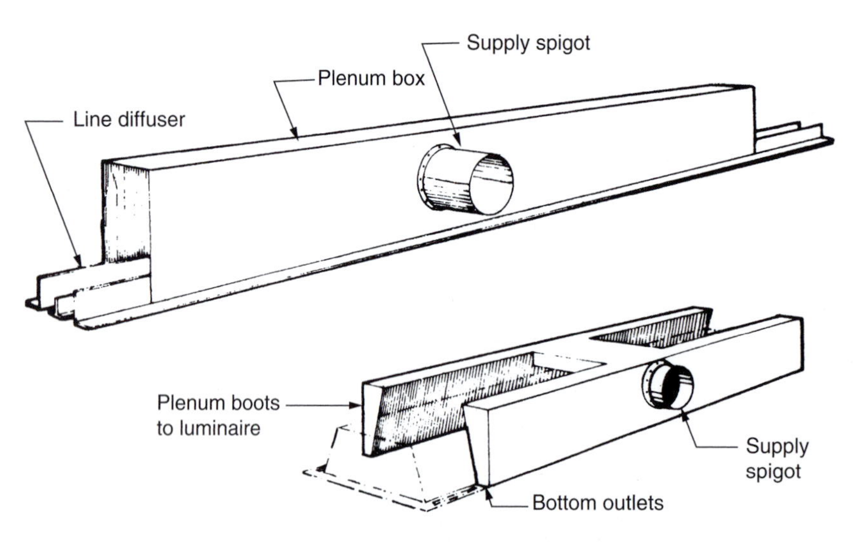

Figure 20.20 Plenum boxes for linear diffusers

conditions as required and are suitable for applications with high-cooling loads. Their stable discharge characteristics also means they are suitable for use in systems with constant or variable volume air flows. Figure 20.21 illustrates an alternating discharge arrangements using high induction slot diffusers.

Side-wall inlets

Where there is no false ceiling or other means of introducing the air through ceiling diffusers, it is necessary to adopt single wall inlets, these usually taking the form of a series of grilles distributed at intervals along the inner partition wall with ducts in the corridor false ceiling behind. Each inlet is equipped with a grille, so designed as to enable the requisite quantity of air to be introduced without draught.

A common and effective form of grille is that known as the *double-deflection* type, as shown in Fig. 20.22. In this form of grille there are two sets of adjustable louvres, one controlling the air delivery in the vertical plane and one in the horizontal plane. The vanes are usually independently adjustable by means of a special tool, and when once set are not altered. A variety of flow patterns can be achieved according to the width of room, aspect ratio and velocity.

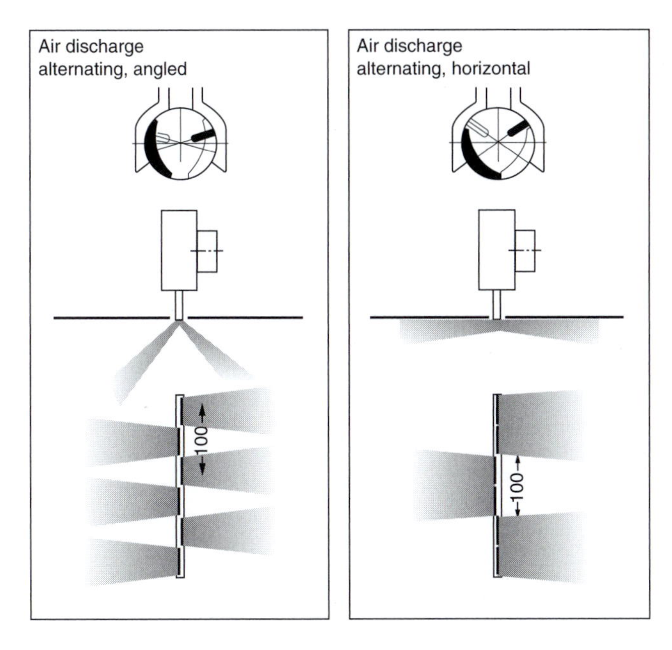

Figure 20.21 Alternating air discharge arrangements for high induction slot diffusers

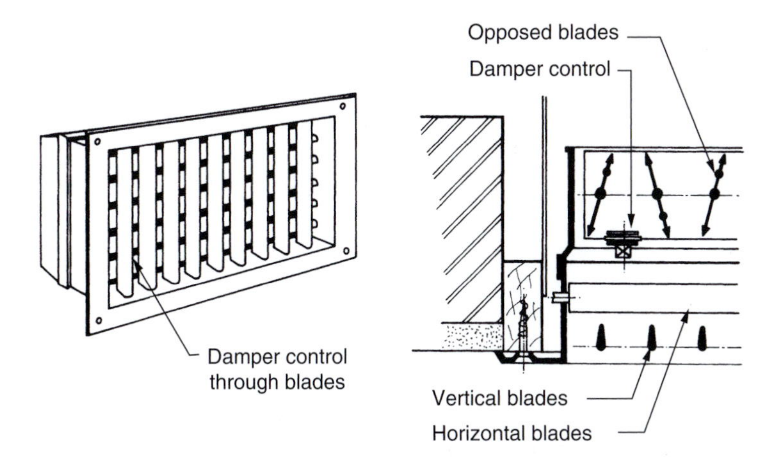

Figure 20.22 Double-deflection type side-wall grilles

For side-wall applications, however, much use is made of the linear diffuser as illustrated, for ceiling application, in Fig. 20.19. By careful selection and making use of the Coanda effect by direction of the air pattern towards the ceiling, good overall air distribution may be achieved from such diffusers.

Single deflection type grilles are also available, i.e. with one set of adjustable blades either horizontal or vertical. Such terminals have limited use for air supply and are more suited to extracts. Fixed single-blade type grilles are also available.

Nozzles

Introduction of air by nozzles was, in the past, usually confined to small compartments such as ship's cabins and to aircraft, where it was desired to obtain the maximum effect under the direct and easy control of the occupant, both in the matter of quantity and direction: terminals of this type are known as *punkah louvres.* Nozzles may however be a useful manner of introducing large quantities of air in some vast arena, where it would be impossible to achieve the throws required by using any other type of terminal. In effect, the more the better. Examples of this kind are the arrangements used to ventilate the Earls Court Exhibition building, and the Sheffield and Belfast Arenas. In this last example, automatic adjustment of the discharge velocity is provided so as to maintain satisfactory air speeds at the occupied levels under both heating and cooling modes of operation.

For the auditorium illustrated in Fig. 20.4, conditioned air was introduced solely from the central ceiling feature making use of 65 jet diffusers of the type shown in Fig. 20.23. Each handled approximately 280 litre/s at an outlet velocity of approximately 3 m/s.[4] Figure 20.24 shows another but similar approach where *drum* type punkah louvres have been used to introduce large air quantities into a swimming pool hall and similar applications have been used for shopping centres.

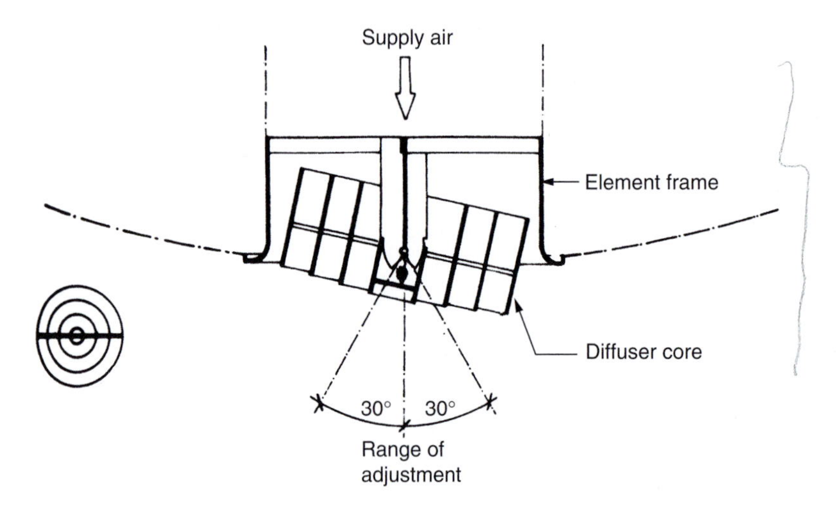

Figure 20.23 Jet-type diffuser

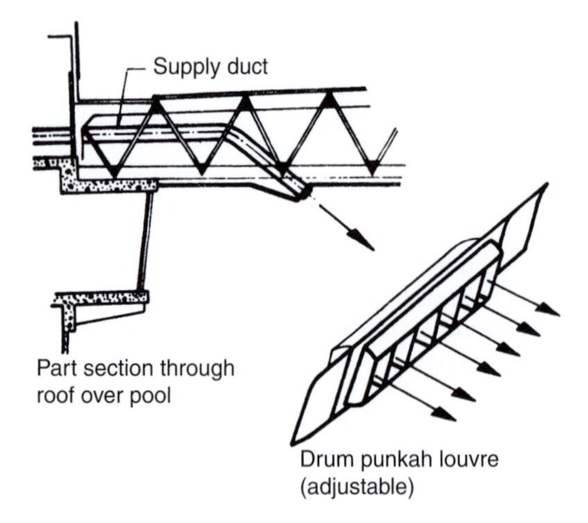

Figure 20.24 Punkah-type louvres in a swimming pool hall

Variable geometry diffusers

Linear and cone-type ceiling diffusers are available with variable outlets which are able to adjust to provide the correct air pattern with varying air flow quantity. Such diffusers are used with variable air volume systems where the variation in air quantity is greater than that for which a fixed device is able to provide a satisfactory distribution within the space. Typical designs for these were illustrated in Fig. 19.12.

Floor outlets

The principle to be applied where floor distribution is to be used is 'little and often'. To avoid unacceptable air movement being experienced by the occupants, no more than a relatively small air quantity can be introduced at each outlet. There are effectively two types of floor outlet used in comfort application: the 'twist' pattern and the straight pattern, both being circular and non-adjustable. These are shown in Fig. 20.25 together with the resulting air distribution patterns. The limiting air quantity handled by this type of outlet is about 12 litre/s, and the supply temperature should not be lower than 5 K below the occupied zone temperature, and less in the case of displacement ventilation. Outlets should be positioned clear of furniture and not closer than 1 m to the nearest work station. For computer room applications, where underfloor distribution is employed, the outlets normally take the form of a 600 mm × 600 mm floor plate either perforated or with a fixed linear grille over the face; both types may be fitted with opposed blade regulating dampers. The volume handled by a 600 mm square plate would be of the order of 200–300 litre/s.

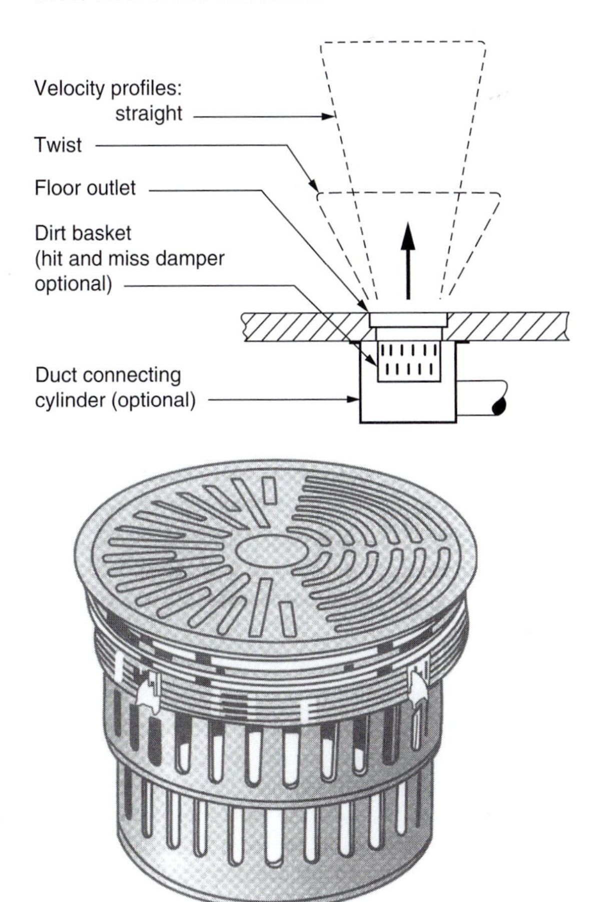

Figure 20.25 Typical floor outlets (Krantz)

Desktop outlets

There are many designs for these devices, some of which are shown in Fig. 20.26, and they are usually used in conjunction with a floor distribution system. The purpose of desktop terminals is to provide occupants with considerable control over their immediate micro-climate. In practice, the air flow quantity supplied in this fashion would be limited to about 14 litre/s. Similar devices may be used in other applications, such as at the rear of theatre seating.

Low level side-wall outlets

Supply air panels are available for discharging usually cooled air, with a low supply to room temperature differential, into a space at low level, from a side-wall or column position (Fig. 20.27). The principle of design is to introduce a supply at low velocity, evenly distributed over the panel face and the effect on the air distribution in the space is that of displacement ventilation. Pressure drop across the panel is 50–100 Pa and the maximum face velocity 0.5 m/s.

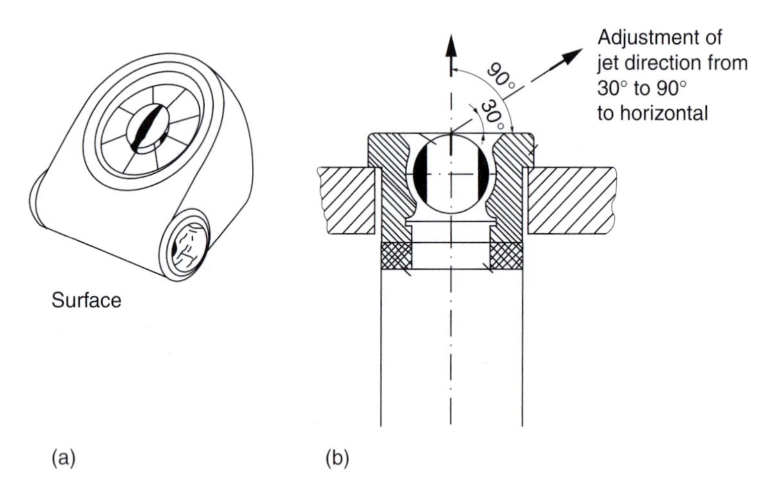

Figure 20.26 Typical desktop outlets (Krantz): (a) Surface; (b) Section through linear

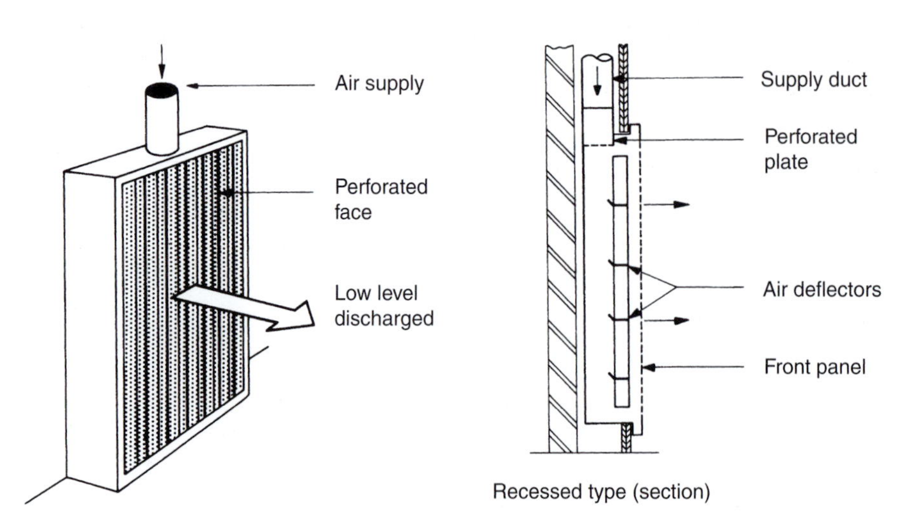

Figure 20.27 Low-level, low velocity side-wall outlets

Volume control

Means for controlling as accurately as possible the volume of air issuing from each grille, diffuser or other inlet, are essential. Probably the commonest form of control is the opposed blade damper. For ceiling diffusers and the like, a variety of multi-louvre dampers is available, or plain butterfly dampers in the ducts may be used. Any damper is a potential source of noise and its location and duty require careful consideration if such is to be avoided. Where there is a risk of a noise problem due to excessive throttling at the terminal, the major part of the regulation must be applied further upstream in the branch ductwork, and where necessary the ductwork will need to be designed to suit.

Selection of supply terminals

The selection of the best number and size of inlet grilles, diffusers and the like, involves a choice to meet a number of variables at the same time:

- Air velocities produced at head or foot level must not exceed those given at the beginning of this chapter.
- If the air entering is cooler than the room, as in an air-conditioning installation, there will be a tendency to fall, best avoided by keeping a reasonable entering velocity with the object of inducing air entrainment from the room; the air mixture then being warmer will have less tendency to fall.
- The velocity of inlet must not be so high that the air will impinge on the wall opposite, thereby causing undue turbulence. The throw should be about three quarters of the distance to the point of impingement with a wall, or opposing air stream.
- The distance apart of inlets, particularly in the case of ceiling diffusers, must be such that the streams from two adjacent units do not collide at such a velocity that a strong downward current results.
- The velocity selected for the grille or diffuser must be such that the sound level produced therefrom is below the design standard for the room.
- Appearance, layout and pattern are all architectural matters, which must also as a rule be taken into account and this sometimes creates a difficulty where the desired spacing or size is not acceptable on these grounds: system performance must take priority.
- The throw must not be directed towards projections that will deflect the air stream from the intended direction.
- The distribution should be arranged to deal with loads at source.

Primary selection should be by use of the design guides published by BSRIA[5] or CIBSE *Guide Section B2* prior to reference to manufacturers' data. These latter are now freely available for all types of equipment and, from a study of them, a variety of alternative solutions to a particular problem may be put down with the object of a final selection being made best suited to meet all the other conditions. The reader is referred to such data for precise information, but a few sample sizes are given in Table 20.1. It will be clear that, with

Table 20.1 Approximate sizes (mm) of air inlets (representative only)

	Air quantity (litre/s)				
Type	*50*	*100*	*200*	*300*	*500*
Perforated ceiling panels (face size)	300 × 150	300 × 300	500 × 500	600 × 600	1200 × 600
Line diffuser (duct size)	750 × 50	1000 × 90	2000 × 90	3000 × 90	2400 × 165
Ceiling diffuser					
(neck diameter)	100	150	200	250	380
(overall diameter)	330	330	450	600	860
Side-wall grille double deflection type (face size)	200 × 150	300 × 200	450 × 250	450 × 400	660 × 450

Note
Above are all selected at approximately the same sound level of 40 dB.

Table 20.2 Typical maximum air change rates based on a cooling differential of 10 K

Device	*Air change per hour*
Side-wall grilles	8
Linear grilles	10
Slot and linear diffusers	15
Rectangular diffusers	15
Perforated diffusers	15
Circular diffusers	20

Table 20.3 Typical maximum cooling temperature differentials

Application	*Temperature difference* (K)
High ceiling large heat gains/under window input	12
Low ceiling air handling luminaires/ under window input	10
Low ceiling downward discharge	5
Floor discharge	5

the large number of variables, air distribution design is perhaps more an art than a science but on it to a large measure, as has previously been emphasised, depends on the success or otherwise of any air handling installation, particularly in the case of air-conditioning. It is worth noting again here that particular attention must be given to the selection of terminals for variable air volume applications to ensure that satisfactory distribution will be achieved through the full range of operating volumes.

The *Guide Section B2* gives useful rule-of-thumb guidance as to both the maximum air change rates that can be achieved using various devices and as to typical maximum cooling temperature differentials (supply air to room) for various applications. This is used here for Tables 20.2 and 20.3, respectively.

Air distribution performance

An air distribution performance index (ADPI), a method of assessing the effectiveness of an installation to meet set comfort criteria, based upon air velocity and air temperature has been established,[6] but is not in common use in British practice.

High velocity supply fittings

The use of high velocity air distribution in ducts has been referred to in connection with the induction, dual-duct and variable air volume systems. High velocity air distribution may also be the most practical and economical method both in cost and space for use with any, otherwise normal, ventilation or air-conditioning system where extensive ductwork is involved. This is particularly the case in multi-storey buildings where duct sizes are much reduced by use of higher velocities. Duct velocities up to 30 m/s may be used, although it is more usual to limit these to the range of 15–20 m/s.

At the terminal end, it is necessary to break down the high velocity to low velocity for introduction into the room, and a silencing or attenuating box in some form is required: this subject has been referred to previously and illustrated in Chapter 19.

Combined lighting and air distribution

Mention has previously been made of the use of special fittings which enable some part of the heat generated by lighting to be dealt with at source before it enters the room. Early types of such fittings used 'boots' mounted to conventional lighting enclosures, as Fig. 20.21, but air handling functions have been developed in an integrated design, as shown in Fig. 20.28. The performance of the lighting apparatus can be improved by such arrangements, where the air extracted is drawn over the tubes thus producing temperature stability, and the heat entering the room may be reduced very considerably, as described in Chapter 5. However, if the return air flow rate is too high over the tubes they can overcool and loose efficacy (lumens per Watt). In these cases it is important to refer to the lighting manufacturers' requirements, if the return air flow rate is too high over the tubes they can overcool and loose efficiency.

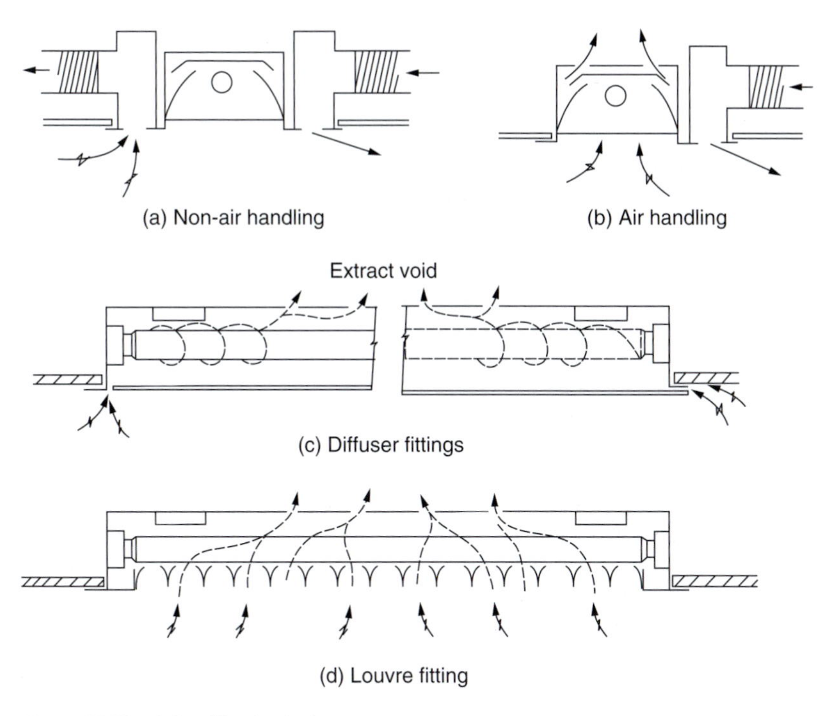

Figure 20.28 Air handling luminaires

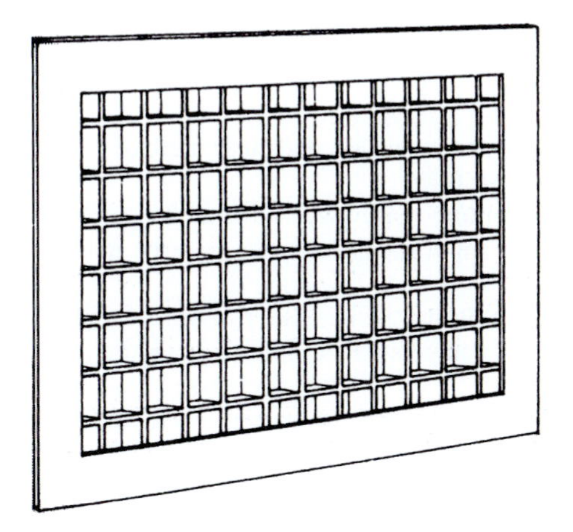

Figure 20.29 Standard 'egg-crate' register

Dust staining

The streaks of dirt that appear sometimes at the point of air discharge from supply terminals are caused either by dust in the room becoming entrained in the air stream, or by dirty supply air. The former is the most usual cause.

Extract or return air grilles

The particular form of grille for extract is unimportant since air approach to a return fitting is no aid to distribution. It may, for instance, be of egg-crate plastic within an aluminium frame as Fig. 20.29, or louvred, or any

Table 20.4 Typical maximum face velocities for extract grilles

Location	*Face velocity* (m/s)
Above occupied zone	4
In occupied zone	3
Door or wall transfer	1.5
Under-cut doors	1.0

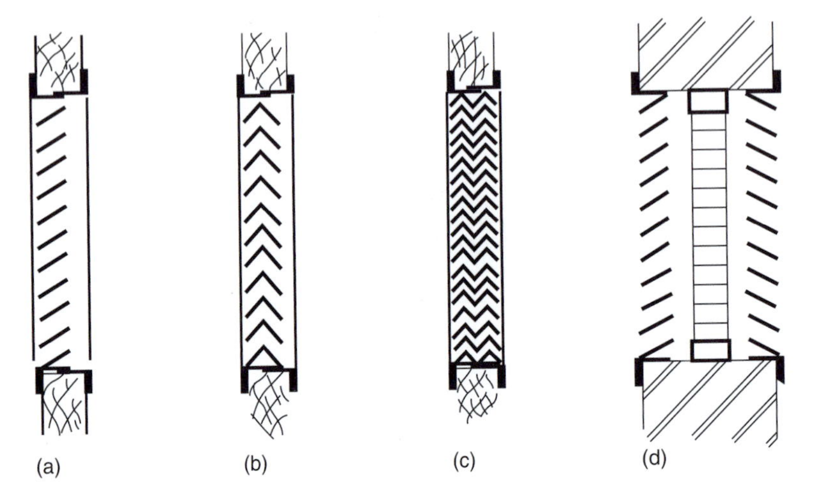

Figure 20.30 Cross-sections through transfer grilles: (a) simple; (b) vision proof; (c) light proof; (d) with fire damper

design giving the required free area. Dust collects on extract gratings on the outside, and close mesh or closely placed slats are undesirable as they quickly block up and impede the air flow.

The sizing of extract and return grilles is normally based on limiting noise levels. Recommended maximum grille face velocities are given in Table 20.4, but for critical applications reference must be made to manufacturers' selection data.

Where extraction takes place naturally from one space to another, it may be necessary to use a light-trap grille to avoid direct vision. Transfer grilles may be fitted with fire dampers where necessary. Figure 20.30 shows alternative fittings suitable for this purpose.

A method of extracting room air which has been used with some success is illustrated in Fig. 20.31. It has been claimed that heat loss and gain to the occupied space is much reduced by passing room return air through the cavity of the double window. A glazing U value equivalent to about 0.6 W/m^2 K is produced and, of course, hot and cold radiation, in summer and winter respectively, is reduced proportionately. Although this arrangement, with the blinds outside the glazing, provides the optimum result in terms of energy saving, costs relating to repairs to and cleaning of the blinds may be high. An alternative, which provides an equally effective but more accessible arrangement, is to position the blinds within the ventilated cavity between the inner and outer panes.

Details of the mushroom ventilator and gallery-riser vent referred to earlier are shown in Figs 20.32 and 20.33. These types are more commonly used as extracts, though they may be used as inlets at low velocities.

Toilet extract

For this purpose, use may be made of the special wall-type mushroom ventilators shown in Fig. 20.34. These combine a facility for air volume regulation with a degree of acoustic attenuation.

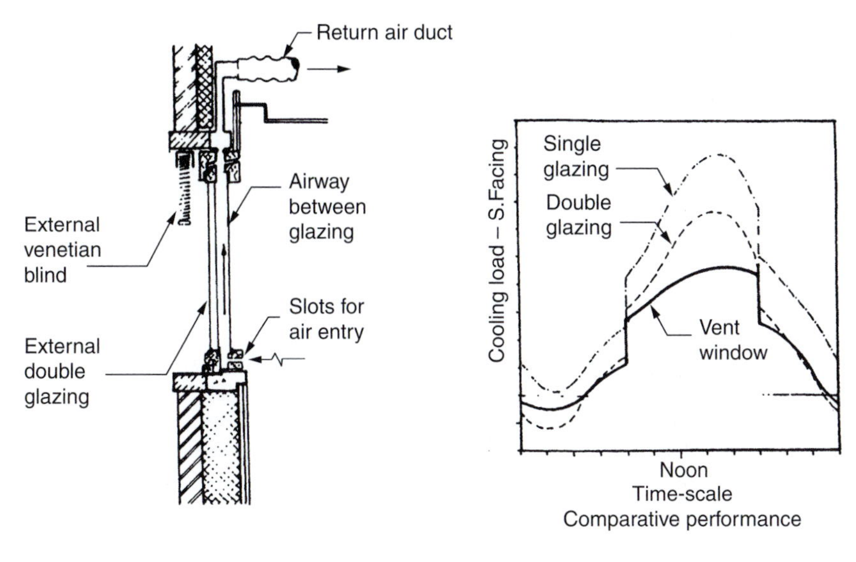

Figure 20.31 Ventilated double window

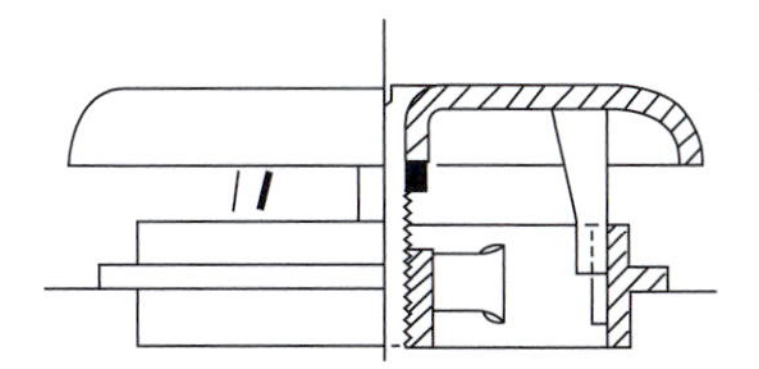

Figure 20.32 Mushroom-type floor level ventilator

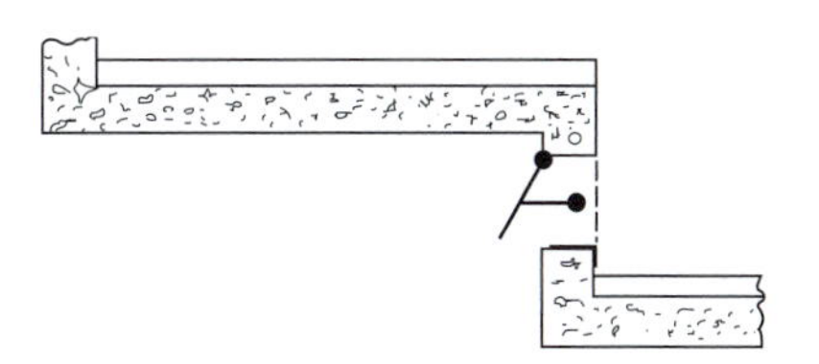

Figure 20.33 Ventilator in gallery steppings

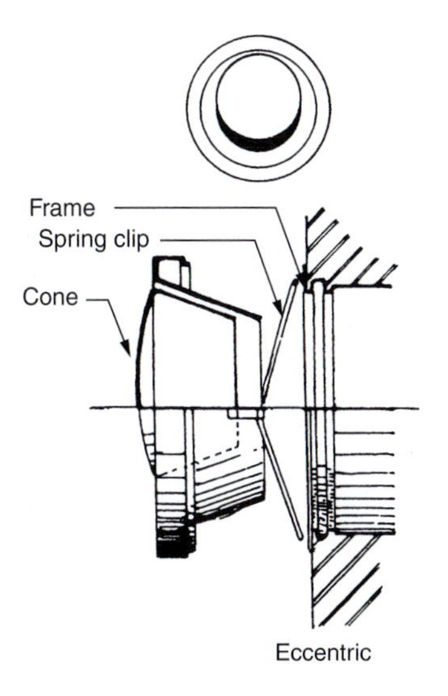

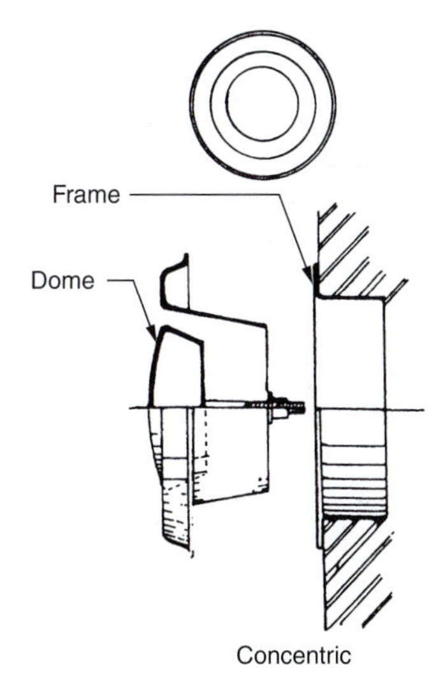

Figure 20.34 Mushroom-type wall ventilators

Special air distribution applications

The following provides the reader with a brief introduction to some special air distribution applications. For further information the reader should refer to *Guide Section B2*.

Cleanrooms

Cleanrooms require a special approach to air distribution. The origins of cleanroom design and management go back more than 100 years and arose as a result of man's desire for the control of infection in hospitals and more recently the need for a clean environment for industrial manufacturing, pharmaceutical preparations and food production of today's modern society.

Cleanrooms have evolved into two major types, which are differentiated by their method of ventilation and containment. These are conventional and unidirectional flow. Conventional cleanrooms are known as turbulently ventilated cleanrooms or (US Federal Standard 209E) non-unidirectional. Unidirectional flow cleanrooms are also known as laminar flow or ultra cleanrooms. Figure 20.35 illustrates the different air flow patterns used in the design of cleanrooms.

Some clean and containment room applications are as follows:

- electronics and opto-electronic manufacturing
- biotechnology

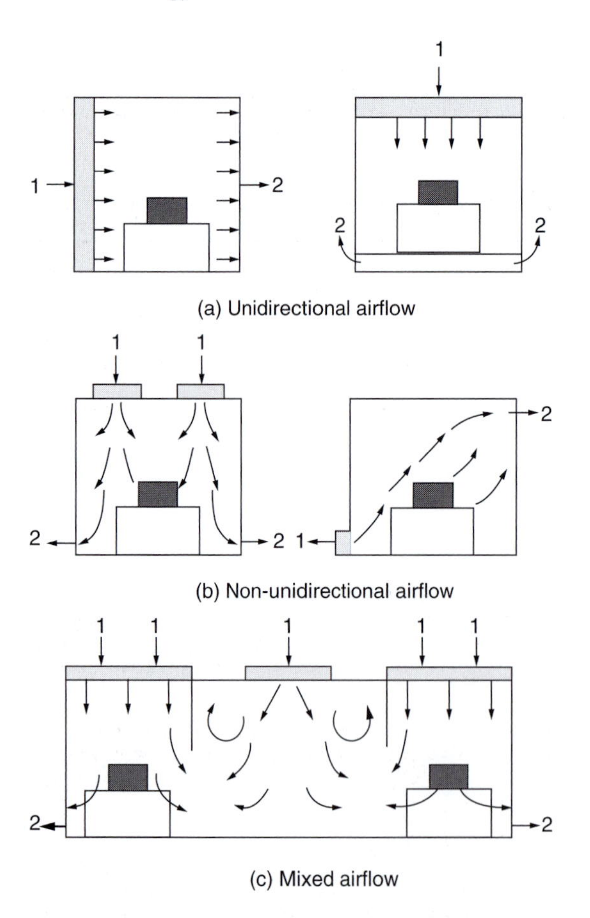

Figure 20.35 Air flow patterns in cleanrooms from European Standard EN ISO 14644-4: 2001

- pharmaceutical manufacture
- food and drink production
- hospitals.

British Standard 5295 defines a cleanroom as:

> A room with control of particulate contamination, constructed and used in such a way as to minimise the introduction, generation and retention of particles inside the room and in which the temperature, humidity and pressure shall be controlled as is necessary.

The American Federal Standard 209E describes air cleanliness classes related to the number of particles in each size category as follows:

	Measured particle size (μm)				
Class	*0.1*	*0.2*	*0.3*	*0.5*	*5.0*
1	35	7.5	3	1	N/A
10	350	75	30	10	N/A
100	N/A	750	300	100	N/A
1000	N/A	N/A	N/A	1000	7
10 000	N/A	N/A	N/A	10 000	70
100 000	N/A	N/A	N/A	100 000	700

There is one additional class recognised by the industry but not yet included in Federal Standard 209E based on 0.1 μm and larger particles.

Typical design objectives for a pharmaceutical cleanroom suite can be summarised as follows:

- exclusion of the environment external to the suite
- removal or dilution of contamination arising from the local manufacturing process
- removal or dilution of contamination arising from personnel working in the area
- containment of hazards arising from the product
- control product to product cross-contamination
- protection of personnel
- control and management of flow of material
- control and management of flow of personnel
- optimum comfort conditions for personnel
- adjustable environmental conditions for products
- accommodation of process plant and equipment to ensure safe and easy use, as well as
- good access for maintenance
- effective monitoring of the conditions of the suite.

Assembly halls and auditoria

Assembly halls and auditoria are generally characterised by large but variable occupancy levels, relatively high floor to ceiling heights, sedentary occupation, usually non-smoking and stringent acoustic requirements. Specific issues that need to be addressed include the following:

- flexible layout to suit possible alternative seating layouts
- acoustic control measures – plant location, vibration, noise breakout, attenuators, duct linings, etc.
- integration of large air handling plant and distribution ductwork
- occupancy patterns and part load operation
- consider viability of heat recovery devices and possible variable speed operation
- zoning of the plant in large auditoria
- treatment and integration of builderswork plenums
- careful selection of air terminal device, integration with seats, control of noise and draughts

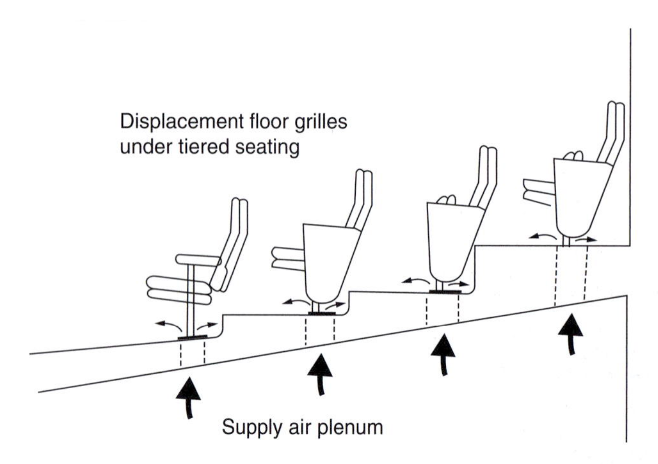

Figure 20.36 Typical arrangement for displacement ventilation outlets located under auditorium seating

- stage ventilation and assessment of lighting gain for cooling
- temperature control at rear of auditorium where there is reduced height
- background and out of occupancy hours heating
- back of house zoning of plant
- local cooling and ventilation of control rooms, translation booths, etc.

Air distribution strategies include:

- Mechanical displacement ventilation: Low-level supply and high-level extract. Low-level supply is often via a plenum beneath the seating, as shown in Fig. 20.36. Air is extracted at high level, returned to the central plant for heat recovery or exhausted to atmosphere. This approach is suitable for raked fixed seating halls and auditoria. Advantages are that only the occupied zone is conditioned, not the entire space, and the potential for 'free cooling' is maximised as supply air temperatures are usually 19–20°C. Air volumes, energy consumption and maintenance costs are usually less than when compared to high-level supply systems.
- Mechanical ventilation: High-level supply and extract. This system is usually selected where a flexible space is required, seating is removable or where it is not feasible or cost prohibitive to provide under seat plenums.
- Natural ventilation: Supply inlets are usually provided by attenuated builderwork ducts at low level and extracted by attenuated outlets at high level, relying on stack effect to ventilate and cool the area. This approach has potentially the lowest running costs but may require a number of provisions to ensure adequate air flow rates and to limit peak summertime temperatures. Particular considerations include providing suitable air paths, inlets and exhaust positions, solar protection, thermal mass exposure and night cooling.

Control strategies include:

- demand controlled ventilation and cooling depending on occupancy or carbon dioxide levels
- space temperature and humidity
- time control
- night time purging of the space and possible precooling of the structure.

Television studios

The general requirement is to provide a comfortable environment within the constraints imposed by the production of television programmes. Specific issues that need to be addressed include:

- high lighting loads in studios
- high occupancies for shows with audiences
- rapid changes in load

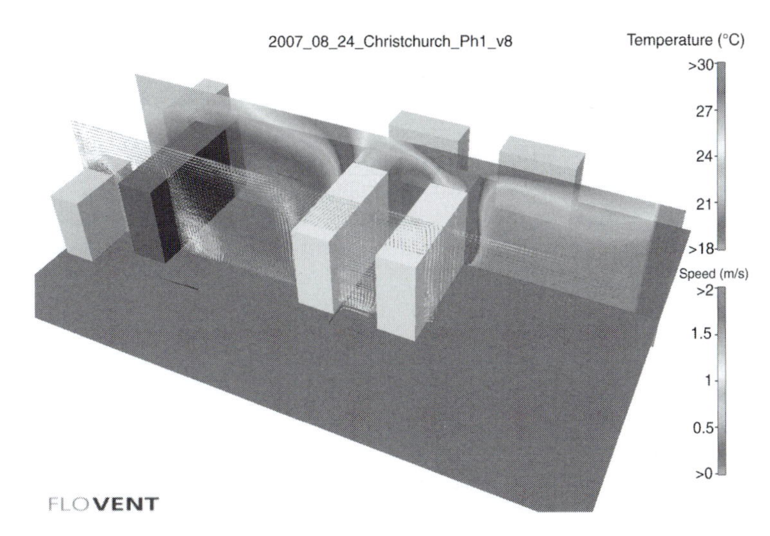

Figure 20.37 CFD analysis of a data centre (see colour plate section at the end of the book)

- variable operation times and periods
- sensitivity to air movement and noise
- high equipment loads in technical areas
- critical areas requiring a high degree of reliability
- multiplicity of studio arrangements
- adaptability to respond to changing requirements.

Air distribution strategies include:

- Systems need to be able to cope with high loads and rapid changes in load. All air systems are preferred because of concerns over water in the space and central plant is preferred to local units because of maintenance access restrictions.
- Whilst variable air volume systems may provide an energy efficient solution for studios, constant volume systems provide an even air flow and at a constant noise level which may be critical.
- Blow through coils with airside damper control is preferred to waterside control to respond to rapid load changes. Steam injection is used for fast response to meet humidity requirements.
- High reliability for critical area is normally provided by redundancy on individual units and/or the number of units provided. Dual power supplies and generator backup are also provided.
- High loads can lead to rapid temperature rises, which could have a knock-on effect on fire systems.
- To separate audience and set areas for control purposes, studios may be zoned into quartiles by multiple damper arrangements.
- Attenuation should be provided to reduce noise ingress from outside and central plant.
- Noise from air balancing dampers can be a particular problem and should be avoided if possible.
- Air velocities inside the studio are critical regard to noise.
- Particular problems can arise with boom microphones when located close to high-level supply diffusers both due to noise from the diffuser and wind generated noise from local excessive air movement.
- False floors are normally provided in studios but not generally for air supply as they are usually filled with extensive cabling.
- Equipment heat gains in technical areas can sometimes be treated by providing dedicated supply and/or extract ducts to equipment cabinets.

An example of the application of computational fluid dynamics (CFD) techniques, referred to in Chapter 6, to prove the performance of design solutions for a datacenter is illustrated in Fig. 20.37.

Notes

1. Howarth, A.T., Sherratt, A.F.C. and Morton, A.S., Air movement in an enclosure with a single heated wall. *JIHVE*, 2072, 40, 211.
2. BSRIA Application Guide AG 1/74, and Laboratory Report LR 83.
3. International Standard, ISO 3258, 'Air distribution and air diffusion-vocabulary'.
4. Clark, I.T., Air conditioning the Usher Hall, Edinburgh. *JIHVE*, 1977, 45, 125.
5. Laboratory Reports 65, 71, 79, 81 and 83. Application Guides AG 1/74 and 2/75, and Technical Notes TN 3/76, 4/86 and 3/90. The principal authors are P.J. Jackman and M.J. Holmes.
6. *ASHRAE Handbook*, Fundamentals, 1993.

Further reading

Cleanrooms

The European Standard EN ISO 14644-4:2001.

Assembly halls and auditoria

BSRIA LB/90 Conference Centres and Lecture Theatres.
BSRIA TM2/90 Displacement Ventilation.
BSRIA LB18/93 Concert Halls and Theatres.
BSRIA LB30/93 Demand Controlled Ventilation.
BSRIA TN12/94.1 Carbon dioxide Controlled Mechanical Ventilation Systems, 1994.

Television studios

3.5.7 Guide to Acoustic practice, 2nd Edition, BBC Engineering Information Department, ISBN 0 563 36079 8.

Chapter 21

Ductwork design

Having decided on the type of air system to be employed, made the calculations of air quantity and temperature, and considered the type and location of the air terminal devices and central plant, it is now necessary to consider in more detail the characteristics of the ductwork system which will convey the air about the building. Most of the discussion which follows applies equally to ventilating or air-conditioning systems. One of the fundamental relationships is that between air speed and pressure, given by:

$$p_v = 0.5pv^2$$

where

p_v = velocity pressure (Pa)
p = specific mass of fluid (kg/m^3)
v = velocity (m/s)

for standard air

$$p = 1.2 \text{ kg/m}^3$$

Thus

$$pv = 0.6v^2$$

Figure 21.1 gives the relationship graphically for air velocities encountered in ventilation work. For other temperatures and pressures, correction is necessary:

$$P_v^2 = pv(P/101.325)[293/(273+t)]$$

where

P = alternative pressure (kPa)
t = alternative temperature (°C)

Other properties of air used in deriving the standard data for air flow in ducts are:

temperature = 20°C
atmospheric pressure = 101.325 kPa
relative humidity = 43 per cent

Ductwork

Ducts for ventilation and air-conditioning are commonly constructed of galvanised sheet steel. Other materials, including *builders' work*, may be used and these are discussed later. The objective in duct design is to

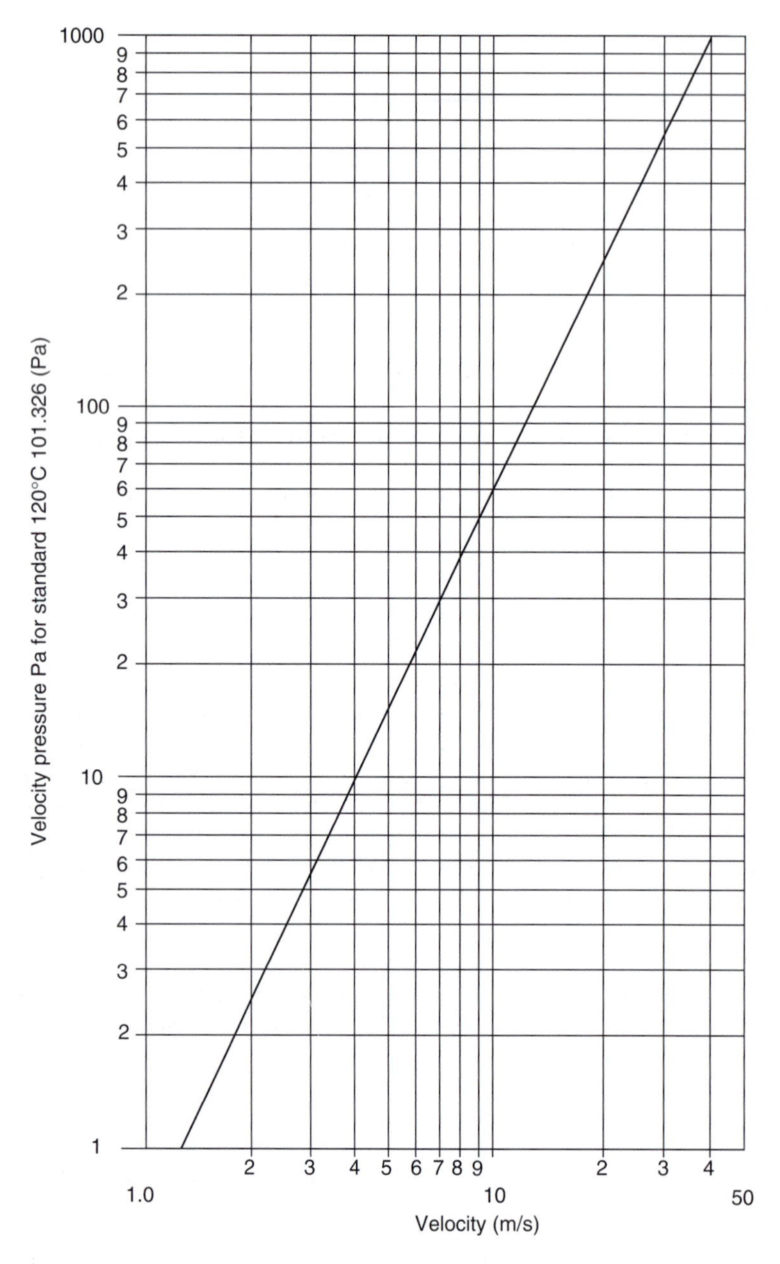

Figure 21.1 Velocity pressure related to air speed for standard air

provide a system in which the air stream follows the line of the duct with no disturbance to the established velocity profile across the section and that excessive turbulence is avoided. The principles to be followed in the design of such ducts are:

- Avoidance of sudden restrictions or enlargements, or any arrangement producing abrupt changes of velocity.
- Bends to be kept to a minimum, but where required should be radiused, not mitred. Deflectors may be fitted to improve the air flow characteristics.
- Where branches occur, they should be taken off at a gradual angle or radiused to avoid abrupt turning.

Table 21.1 Ductwork classifications

Duct pressure class	*Static pressure limit*		*Mean air velocity* (max.) (m/s)	*Air leakage limit*[a] (litre/s m^2 duct surface area)
	Positive (Pa)	*Negative* (Pa)		
Low	500	500	10	Class A – 0.027 $p^{0.65}$
Medium	1000	750	20	Class B – 0.009 $p^{0.65}$
High	2000	750	40	Class C – 0.003 $p^{0.65}$

Note
a p, differential pressure (Pa).

- Sharp edges should be avoided, as these will be the cause of noise which may travel a considerable distance through the duct system.
- Rectangular ducts should be as nearly as possible square: the more they depart from this the more uneconomic they become.
- The ductwork configuration must allow the air flow to be regulated without the need for excessive throttling, which will lead to noise being generated.

These principles apply to all ductwork classifications but are clearly more important the higher air velocity.

Ductwork classification

The current classification for ductwork is given in an HVCA specification[1] according to the operating pressure, air velocity and leakage rate. These are summarised in Table 21.1. The air leakage rates quoted are satisfactory for all general ventilation and air-conditioning applications, but where hazardous or obnoxious substances are handled, the ducting must be completely airtight. Specially designed and constructed ductwork is required for pressures higher than those stated.

The Building Regulations part L states that ductwork leakage testing should be carried out in accordance with the procedures set out in HVCA DW/143 on systems served by fans with a design flow rate of $\geq$1 m^3/s and for those sections of ductwork where

- the pressure class is such that DW/143 recommends testing;
- the building emission rate (BER) calculation assumes a leakage rate that is lower than that defined in DW/144 for its particular pressure class, in which case low-pressure ductwork should be tested to DW/143 provisions for medium pressure ductwork.

Galvanised steel ducts

The gauges of metal and form of construction commonly used are set out in the HVCA specification[1] for all of the classifications.

Ducts of galvanised sheet steel are often priced by weight. Sheet metal is made in metric thickness and the mass of metal will thus merely be thickness times area times 7800 kg/m^3 (for steel) or whatever other specific mass is appropriate. One of the aims when designing ductwork is to achieve as much standardisation and repetition in the configuration and components as is practical, since this will give a more economic installation.

Rectangular ducts are constructed of flat sheets by bending, folding and riveting and are erected in sections with slip joints or bolted angle ring joints. Larger sizes tend to drum and adequate stiffening is necessary. Sometimes a *diamond break* is used to assist in stiffening, as Fig. 21.2, the sheets being pressed to provide, in effect, a very shallow pyramid form; alternative methods include *pleading* or *beading.* Where rectangular ducts are used in high velocity systems, stiffening is particularly important using bracing angles, and internal tie rods for larger sizes. Aspect ratios in excess of 4:1 are to be avoided. Sizes should be selected from the range given in the HVCA specification.

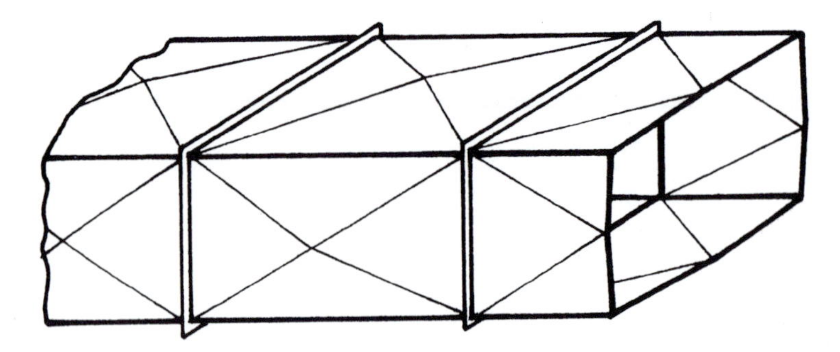

Figure 21.2 Diamond break stiffening for ducting

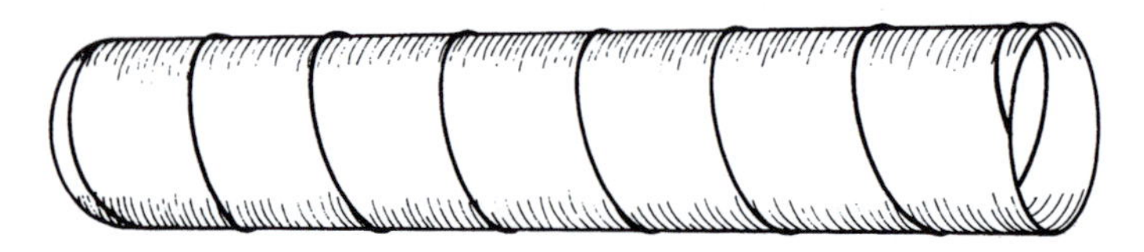

Figure 21.3 Spiral wound circular ducting

Circular ducts may similarly be formed from flat sheet with folded or riveted seams and are inherently free from risk of drumming. Traditionally, the use of good quality circular ductwork was largely confined to industrial applications. Mass produced *snap-lock* duct lengths were used for low-cost domestic warm air heating, reliance being placed upon jointing tape for airtightness.

The advent of circular ducts made from galvanised strip steel with a special locking seam, as in Fig. 21.3, has led to their adoption generally. Sizes are available from 63 mm up to 1600 mm diameter. Ducts of this type are particularly suitable for high velocity systems due to their rigidity, 'deadness' and airtightness. Ranges of standard tees, bends, reducers and other fittings are made and these, together with lengths of straight ducting cut off on site as required, facilitate the task of erection compared with the tailor-made methods previously adopted.

Jointing is by means of a special mastic, with some riveting, and in best practice a *heat shrink* plastic sleeve or a *chemical-reaction* tape is used externally to seal the joint.

Alternative methods of jointing ductwork are available, e.g. where the duct fittings have a groove formed at each end which houses a sealing gasket and locking collar. Connection is made by pushing the duct and fitting together which forces the locking collar into position to provide an 'airtight' joint. Such forms of construction should be restricted in use strictly to manufacturers' recommendations.

Flat oval or spirally wound rectangular ducts are in common use in air-conditioning. These are formed from spirally wound circular ducts to the profile illustrated in Fig. 21.4 and are available in sizes from 75 mm × 320 mm to 500 mm × 1640 mm. A limited range of standard tees, bends and reducers is made. Such ducts have the advantage of being mass produced and can be used in conjunction with circular sections where space is limited. Site jointing, as before, is completed by a heat shrink or chemical reaction sleeve.

Builders' work ducts

The suggestion is sometimes put forward, in early stages of design for a building, that main air ducts and rising shafts should be constructed as a part of the fabric. The following advantages are claimed for such an expedient:

- Cheapness in that the space very often exists as a structural wind brace or in association with one or more lift shafts.

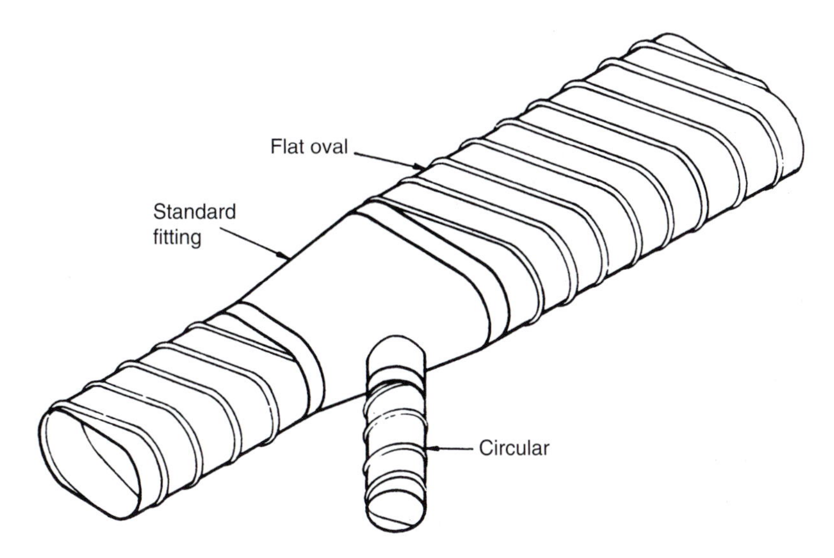

Figure 21.4 Flat-oval ducting and branches

- Permanence.
- Reduction in heat gains and losses, due to the heavy construction as compared with metal.
- Rigidity and hence reduction in noise transmission, also as a result of the heavier construction.
- Accessibility for cleaning.

Several of these claimed advantages may have been valid half a century ago when the techniques of sheet metal duct construction were less well developed and when time lag in response to automatic control was of less significance. They must now be questioned and considered in the light of the proven difficulty of producing any form of builders' work construction which is reasonably airtight.

As a result of structural settlement following completion; later movement as a result of either drying out or temperature changes; and the high permeability of masonry material of all descriptions, builders' work ducts are always suspect. They may be considered reasonably viable only when lined with a compound having some measure of elasticity and which has been applied by a specialist contractor. Following such treatment, air leakage may possibly be reduced but only to a level which is still many times greater than that which would be acceptable for low pressure sheet metal ductwork, as in Table 21.1.

Consequent upon these difficulties, the use of builders' work construction should be considered only as a last resort and then for extract ducts alone, at low velocities and pressure differentials. Where the air extracted may be contaminated, such construction should not be used under positive pressure, i.e. on the delivery side of the fan.

Figure 21.5 shows an example of a duct constructed over a basement corridor and serving as a main distribution trunk from the plant to rising shafts about the building: such a construction cannot be recommended since it will, inevitably, not be airtight.

Ducts of other materials

Other materials used for ducts are:

- *Welded sheet steel* is mainly used in industrial work, or where the airtightness of the joints is of great importance; such ducts may be 'galvanised after made'.
- *Copper, aluminium, stainless steel*: Ducts constructed of these materials are used in special cases where permanence or a high degree of finish is required, or where special corrosion problems arise.
- *uPVC and Polypropylene* which are chiefly used for chemical fume extraction and tend to be expensive due to the cost of moulds, etc., for fittings and special sections. Reference HVCA specification DW/154.which

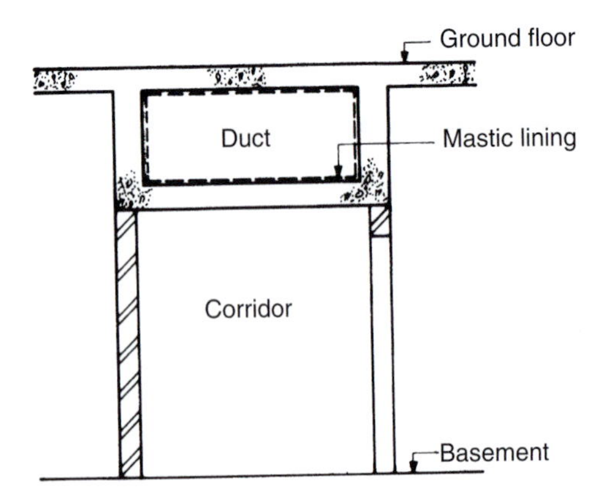

Figure 21.5 Structural air duct over corridor

stipulates that all uPVC materials used shall be low flammability and self–extinguishing. The Laboratory of the Government Chemist has developed a combination of uPVC coated externally with fire-retardant glass fibre reinforced filled polyester resin (GRP) for use with fume extract systems. This is inherently stronger and more durable than uPVC used alone.

- *Glass fibre resin bonded* slabs can be cut and adapted at site, and jointed to form a smooth continuous duct. This technique was favoured in the USA but has not been much used in this country although a HVCA specification exists (DW 191) to codify construction methods. The material is, of course, suitable for the lowest pressures and velocities only and will not, even then, be airtight.
- *Phenolic foam boards* are available in the form of a proprietary duct system. As with the resin-bonded glass fibre alternative its use and application are fairly limited. However as this is light in weight, requires no insulation and is a finished product it can be used to advantage, e.g. for a short-term fit-out project requiring relatively simple ductwork systems.
- *Fabric ducts* can be used as a room terminal device as diffusers in applications where a large volume of air is to be delivered in a space with the minimum of air movement. A typical application would be a low-temperature food preparation area. A variety of woven materials are available including polyethylene and polyester which are formed into a cylindrical tube supported on hoops with fixings so that the fabric tube can be removed for cleaning or replacement.
- *Chlorinated rubber paint or PVC coating* protection to galvanised steel is recommended for use in swimming pool applications.[2]

A useful list of materials, their application and limiting characteristics is given in CIBSE *Guide B3*.

Ductwork components and auxiliaries

To achieve a distribution system that will function efficiently, be capable of regulation, and be easily maintained, it is necessary to incorporate a variety of components to assist towards these objectives. The following gives a brief overview of some of these items.

Ductwork fittings

Figure 21.6 serves to illustrate the basic forms that bends can take. *Radius bends* are to be preferred, and where practical a centreline radius not less than 1.5 times the diameter or duct width should be provided (Fig. 21.6(a)). Splitters (as shown in Fig. 16.6(b)) should be fitted where it is important to maintain the upstream air velocity profile around the bend, e.g. at approach to equipment or measuring points. *Mitre bends,*

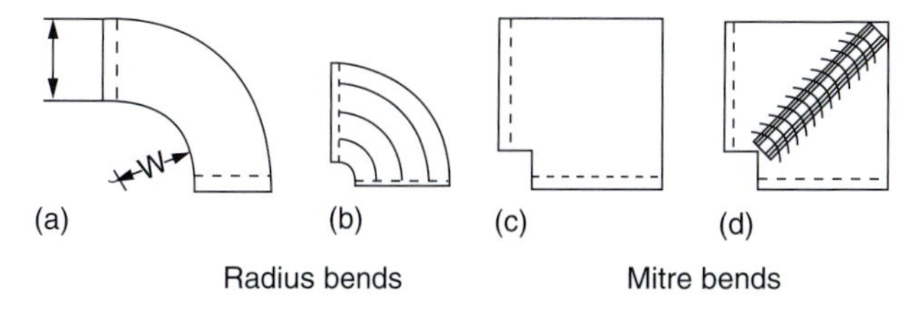

Figure 21.6 Typical ductwork bends

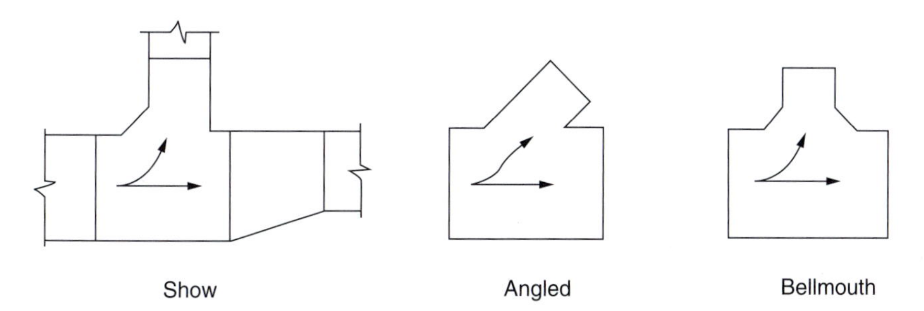

Figure 21.7 Typical ductwork branch connections

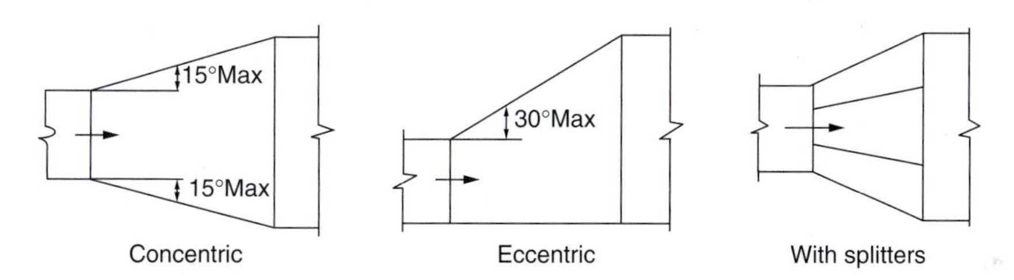

Figure 21.8 Typical ductwork expansion and contraction sections

without turning vanes, (Fig. 21.6(c)), should not be used since pressure drop and noise generation are high. Mitre bends with turning vanes (Fig. 21.6(d)) are quite satisfactory so long as the turning vanes are of the correct design and are constructed to a good standard. Radius bends alone should be used on medium and high pressure/velocity systems.

Branch connections should be constructed such that the air flow is divided, or combined in the case of an extract system, with minimum turbulence. Changes of duct size should be made downstream, or upstream, of the fitting, not at the fitting. Typical branch connections are shown in Fig. 21.7. Branches on medium and high pressure/velocity systems should be taken off at 45°, or be made to a recommended pattern for such applications.

At a *change of section* in a duct, the aim is to maintain the air stream attached to the duct sides. If the air becomes detached, high turbulence occurs with resulting high pressure loss, noise and a non-uniform air velocity profile. At an expansion, the included angle should be limited to 30°, above this internal splitters should be used, as in Fig. 21.8. On duct contractions the angle of taper is less critical, but 40° should be considered a limit for good design.

Fan and plant connections

Catalogue fan performance figures will not be obtained in practice if the inlet air stream is other than uniform and without turbulence. At the fan discharge, all transitions should be gradual, and bends closer than three duct diameters should be arranged to turn in such a manner that the velocity profile from the fan is maintained. Air flow upstream of plant items such as filters, humidifiers, heating and cooling coils must be uniform across the whole cross-section of the equipment, otherwise performance will be reduced to below the rated duty.

There are too many possible combinations of duct configuration to show here, and reference should be made to *CIBSE TM 8*.

Data have been established to enable the reduction in fan performance arising from poor intake or discharge ductwork configurations to be calculated. This is expressed as an additional system resistance to be added to the ductwork and component pressure loss and is termed *system effect.*

$$\text{System effect (Pa)} = \text{system effect factor} \times \text{velocity pressure at the fan intake or discharge (Pa)}$$

Some typical system effect factors are given in Fig. 21.9.

Access openings

Access into ductwork should be provided at every component, at both sides of plant items, and at regular intervals along the run to allow for inspection and cleaning.

Balancing dampers

Permanently set dampers are used in low pressure systems to regulate the air flow quantities at the commissioning stage. These are installed downstream of the fan in the main duct, in every branch duct from the main duct, in every sub-branch serving three or more terminals and at every terminal. The butterfly type is normally used in circular ducts and a multi-bladed type in rectangular sections, with some form of locking devices. The contra-rotating type, commonly called opposed blade dampers, is superior from the regulating point of view, to the type where all vanes rotate in the same direction. A selection of the more popular types of damper is shown in Fig. 21.10. Where tight shut-off is essential, it is necessary for the dampers to be felt-tipped and to close on to a felted frame, and with edge seals.

Dampers should only be installed in medium or high (pressure) velocity systems where absolutely necessary. Where these must be provided, it will probably be necessary to provide attenuators upstream and downstream together with protection to reduce noise breakout from the duct.

Controllable dampers, adjusted either manually or automatically, are dealt with in Chapter 28.

Fire and smoke dampers

In principle, fire dampers must be installed where ducts pass through fire compartment walls, floors or other elements, with the possible exception of low-level penetrations less than 0.013 m^2 in section. Ideally, dampers should be built in to the compartment element. Where this is not practical, the damper should be located close to the element and connected to it by 6 mm thick (minimum) steel plate with flanged joints. Fire/smoke dampers may also be required where ducts penetrate smoke barriers in ceiling voids. The requirements may be relaxed in toilet blocks, where a 'shunt' duct arrangement, as shown in Fig. 18.14, may be an acceptable alternative.

Figure 21.11 illustrates some typical fire damper constructions. The swing blade type, Fig. 21.11(a), consists of a heavy steel casing with the steel damper blade kept open by a fusible link. In the event of a fire the link melts and the damper swings closed. Another pattern, Fig. 21.11(b), takes the form of a number of substantial shutter blades retained out of the air stream by a fusible link as before; this is normally referred to as a 'curtain' type. A different form of fire damper is one where intumescent material is secured in a steel frame and inserted into the air duct, as Fig. 21.11(c). This material has the property of swelling to many times its original volume when heated and thus forming a barrier to air flow. Intumescent collars can be used on circular plastic extract ducts where steel dampers would corrode.

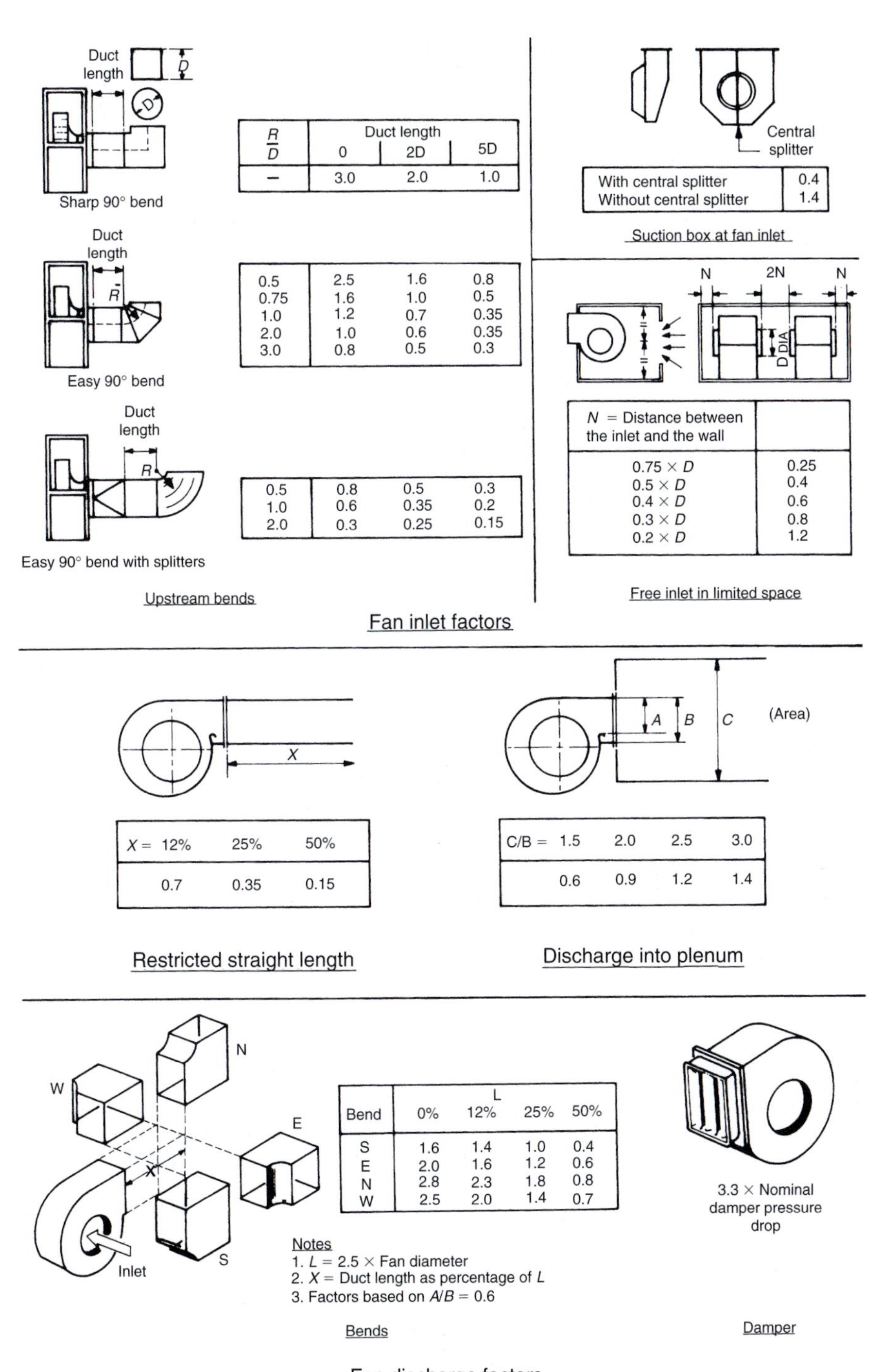

Figure 21.9 System effect factors

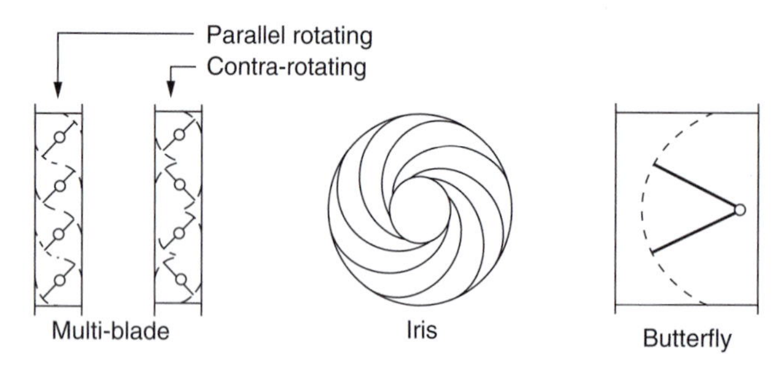

Figure 21.10 Typical balancing dampers for insertion in docting

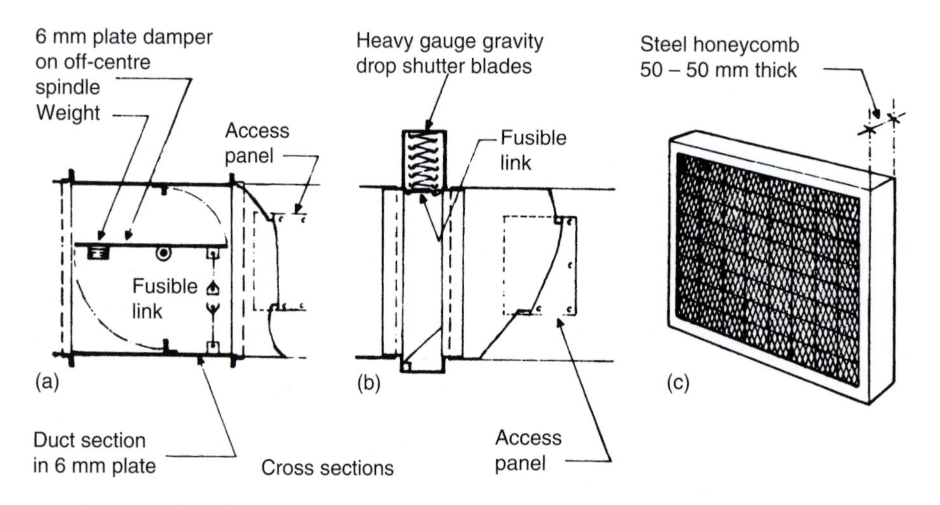

Figure 21.11 Typical construction methods for fire dampers

Combination fire and smoke dampers are designed to function both as a fire damper, operated by a fusible link or thermal actuator, or as a smoke control damper powered by an actuator from a remote signal. Remote indication of the damper position can also be provided.

Smoke release dampers may also be required at air inlets and exhausts. Operation normally would be a fail-safe spring loading, via a remote signal, which would cause the damper to open thereby allowing any smoke in the system to discharge to atmosphere.

Flexible ductwork

This is available in treated fabric on a helix of various materials and finishes, and in spirally wound 'bendable' metal of various materials. Selection is made for the particular application. Flexible ductwork is normally used at the final connection to air terminals and at connections to variable air volume and mixing boxes, induction units and the like. The length of these connections should be limited to 600 mm, and never be more than the equivalent of 6 times the duct diameter.

Whatever the choice of material, it must meet the fire requirements, and the overall air leakage and frictional resistance criteria for the system.

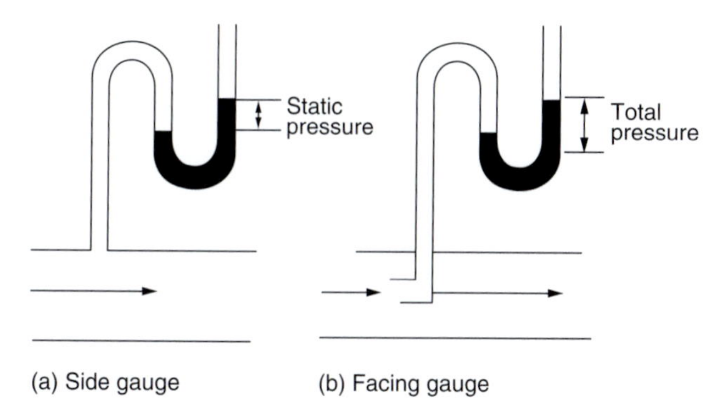

Figure 21.12 Air pressure gauges

Pressure distribution in ducts

Static pressure

When air is moved in a duct or through a filtering, heating, cooling or humidification plant, a resistance to flow is set up.

The air is slightly compressed by the fan on its outlet side, so setting up a *static pressure* in the duct or plant. This pressure is tending to 'burst' the duct, and may be read by means of a U tube partly filled with water, connected at right-angles to the air stream at any point in the duct: this is called a *side gauge* (Fig. 21.12(a)). On the suction side of the fan the static pressure is negative with respect to the surrounding atmosphere, tending to collapse the duct.

As the air proceeds along the duct from the fan, the compression is released gradually until at the end of the duct open to atmosphere, the air is at atmospheric pressure. This falling away of the static pressure proportionately with the length of travel is called the *resistance* of the duct. Similarly all obstructions, such as heaters, filters, dampers, etc., cause a loss of pressure when air is passing through them.

It should be noted that as the static pressure becomes reduced, the air in effect expands such that pressure × volume = a constant (or nearly so, as explained earlier). This expansion therefore signifies an increase in velocity of the air if the size of the duct is unchanged.

Velocity pressure

A fan, in addition to generating static pressure, supplies the force to accelerate the air and give it velocity. This force is termed the *velocity pressure*, and is proportional to the square of the velocity (see Fig. 21.1). It is measured by a 'U' tube connected to a pipe facing the direction of air flow in a duct, etc., which is called a *facing gauge* (see Fig. 21.12(b)). But, obviously, the pressure so measured will in addition include the static pressure which occurs throughout the duct, as mentioned earlier, and the reading so obtained will thus be the *total* pressure. Thus, the velocity pressure alone may be found by deducting the static pressure from the total pressure reading, or by connecting one side of the U tube to the facing gauge and the other to the side gauge, provided that the two gauges are at the same point. A Pitot tube, as illustrated in Fig. 29.4, combines a side gauge with a facing gauge in one standard instrument.

If a fan discharges into an expanding duct (Fig. 21.13), the velocity will obviously decrease as the distance from the fan increases, and at the same time the velocity pressure will be converted into static pressure (not at 100 per cent efficiency, but about 75 per cent if the expansion is sufficiently gradual). If the fan discharges into a large box (Fig. 21.14), from which at some point a duct connects, the fan velocity pressure will be entirely lost in eddies, and at the duct entrance must be recreated by a corresponding reduction in static pressure.

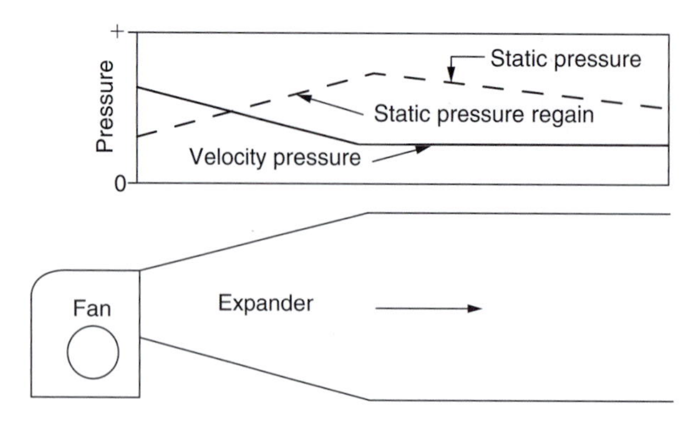

Figure 21.13 Velocity and pressure changes at an expander

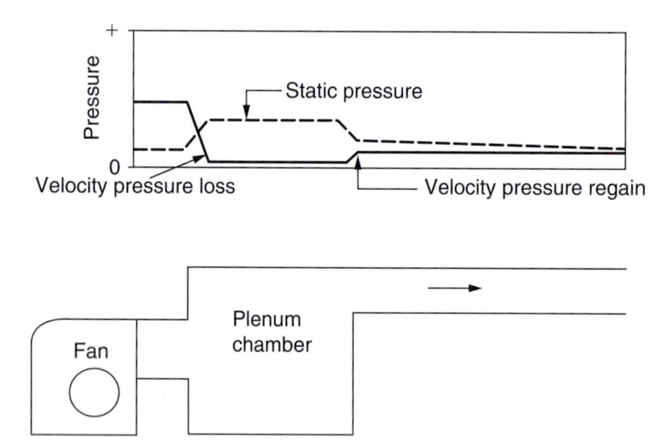

Figure 21.14 Velocity and pressure changes at a plenum box

Fan pressure

In all air flow considerations as affecting resistances of ducts, plant, etc., it is the static pressure alone which is of importance, this being the pressure which changes with such restrictions. It is the static pressure set up by a fan which is, therefore, a criterion of its performance. The velocity pressure, if taken as supplementing the fan duty, may be more misleading than useful, owing to the uncertainty of friction losses which occur at points of varying velocities. The velocity pressure is more generally not recovered, though sufficient must remain at the duct termination to eject the air at the required velocity.

Where, however, by careful design of the fan discharge expander, the velocity pressure is converted to static pressure (probably to the extent of about 75 per cent), this additional pressure may be reckoned as augmenting the static pressure of the fan.

The pressure generated by a fan may be better understood by study of Fig. 21.15:

- The total fan pressure is defined as the algebraic difference between the mean total pressure at the fan outlet and the mean total pressure at the fan inlet.
- The total pressure on the suction side, as will be seen from this figure, is TP_s, i.e. the negative pressure *AO* minus the velocity pressure equivalent to *AB*.
- The total pressure TP_D on the discharge side is similarly the static pressure *OC* plus the velocity pressure *CD*.

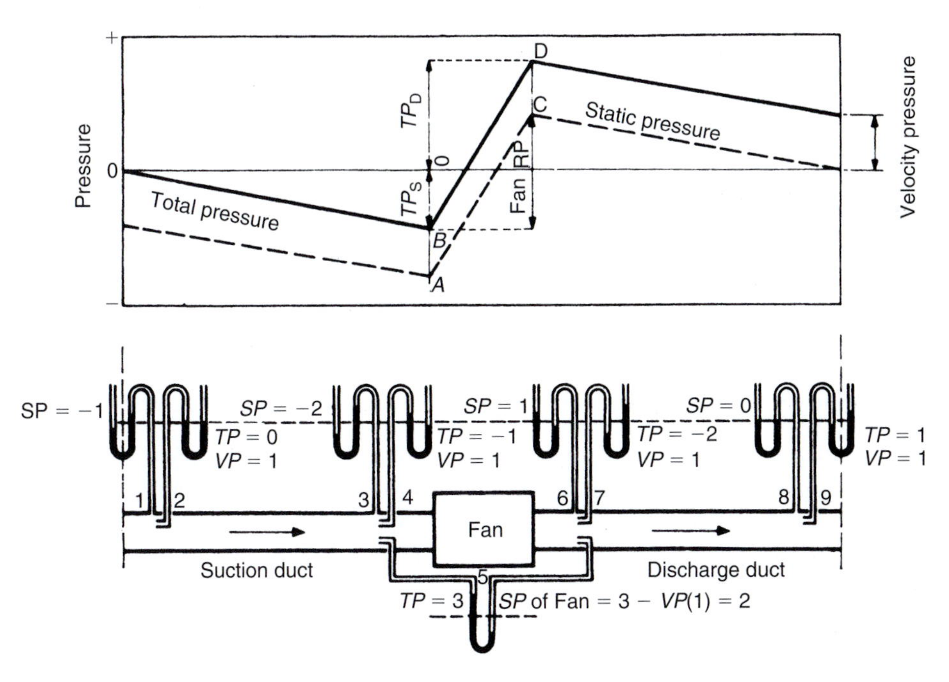

Figure 21.15 Total and static pressure changes at a fan

It has previously been explained that we are concerned only with the resistance pressure set up by the fan and this, as may be seen, is the difference in pressure of points *B* and *C*.

We can arrive at the static pressure by measuring the total pressure of the fan as by the U tube 5, and deducting therefrom the velocity pressure as given by difference between gauges 6 and 7. Such a method is valid only if the velocity in suction and discharge ducts is the same.

The other U tubes indicate the pressures at the various points along suction and discharge ducts, and their meaning will be apparent.

If a fan has suction ducting only, the static pressure produced for overcoming friction of ducting will be represented by *OB* (a negative pressure), since the discharge will be at atmospheric pressure. Similarly, if the fan has only discharge ducting, the static pressure will be represented by *OC*, the suction being at atmospheric pressure.

Flow of air in ducts

The static pressure which a fan will produce is utilised as the motive force required to overcome the sum of all resistances to air flow throughout the system of air ducting to which the fan is connected. This book is not the place for a dissertation upon the theory of air flow in ducts save to say that an air stream moving within a long straight duct develops a constant and smooth velocity profile, slower adjacent to the perimeter, due to a drag effect from the walls and faster at the centre. Any changes in direction; junctions and variations in the cross-section of the duct break up this profile and lead to turbulence and added resistance. Generally speaking, therefore, all resistance to flow results in a drop in static pressure and may be considered as falling into one of three categories:

(1) That due to the frictional resistance arising from air flow through lengths of straight ductwork made of whatever material.
(2) That due to individual ductwork fittings such as changes in direction, contractions, enlargements and branches, etc.
(3) That arising from plant and other built-up items which have characteristics unique to one specialist manufacturer or another.

In many respects, air flow in ducts is analogous to water flow in pipes but an important difference exists in practice in that, whereas pipes are available in a commercial range of diameters, air ducting is more usually purpose made up from flat sheet metal to bespoke sizes and cross-sections. Exceptions to this rule are, of course, spiral-wound circular ducts and the parallel flat-oval sections.

Straight ductwork

To represent resistance, and consequent pressure loss due to air flow in straight circular ducting, the *Guide Section C4* presents a chart, similar to Fig. 21.16. The base scale of the chart is graduated in units of pressure loss (Pa per metre run) and the vertical scale in units of air flow (litre/s). Diameters of circular ducts in mm and consequent velocities in m/s may be read directly. For example, a volume flow of 200 litre/s through a duct of 225 mm diameter, will be at a velocity of 5 m/s and will produce a unit pressure loss of 1.4 Pa/m.

An economic rate of pressure loss takes into account the capital cost of ductwork and balances this against the recurrent cost of energy to develop the fan pressure. Traditionally, a unit pressure loss of between 0.75 and 1.25 Pa/m has been the aim.

More recently the Building Regulations Approved Document L states maximum specific fan powers, W/l/s, which has seriously challenged traditional duct velocities and pressure losses in ducts, fittings and across equipment. Consequently if traditional duct velocities were reduced by 20 per cent, e.g. the corresponding pressure loss would be 64 per cent and the resulting fan power 51 per cent of the original value. Limiting specific fan powers are included in Chapter 22.

The chart, strictly speaking, applies only to ducts manufactured from clean galvanised mild steel sheet: thus, for other materials, pressure loss read from the chart must be corrected, *up or down*, by use of the factors listed in Table 21.2. Since, in the majority of duct installations at low velocity, square or rectangular ducts are used in order to suit building configurations, some method of representing these as *equivalent diameters* to suit the chart is required. Without going into detail, the relationship is complex since both the cross-sectional area and the internal circumference of the alternative shapes must be taken into consideration. A simplified method of calculating a diameter is available, using the expression and values given in Table 21.3 For tabulated data on rectangular and flat-oval ducts, reference should be made to *Guide Section C4.*

Ductwork fittings

The conventional method of dealing with the pressure loss arising from the presence of all manner of duct fittings, is to evaluate the effect of each separately as a fraction of the velocity pressure existing at the position in the system where they occur. A selection of pressure loss factors, ζ, is illustrated in Fig. 21.17 for use in conjunction with velocity pressure values read from Fig. 21.1, thus:

Factor for a simple 90° bend having an aspect ratio, H/W, of 0.5 and a throat radius, R, of 0.5 = 0.29

If the air velocity in the duct is 4 m/s then, from Fig. 21.1, the velocity pressure = 9.5 Pa

Thus,

pressure loss due to bend = 0.29×9.5 = 2.8 Pa

Plant components

For plant components, heater batteries, cooling coils, air filters and the like no generalisation is possible since the resistance to air flow will vary from manufacturer to manufacturer and may well be a function of performance. For a given air flow quantity, pressure loss resulting from such items may well be relatively large. They should be quoted to the designer directly in Pa.

Application

The mean air velocities quoted in Table 21.1 are intended to be broadband classifications for various qualities of ductwork construction: they are *not* system definitions. For use in design, the data listed in Table 21.4

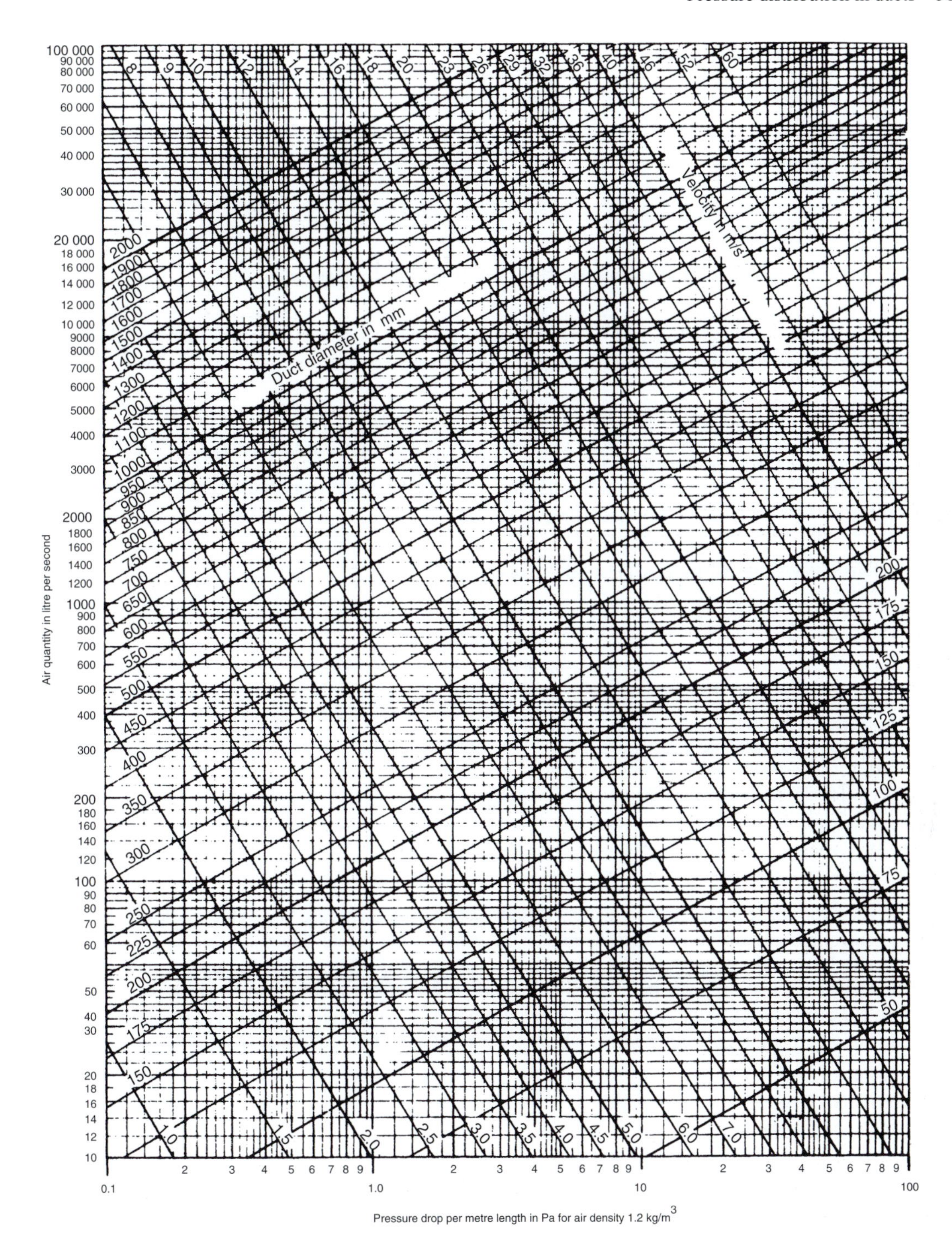

Figure 21.16 Duct sizing chart

Table 21.2 Correction factors for ducts of other materials

uPVC	0.85
Aluminium	0.9
Plastered, smooth cement	1.1
Fair faced brick, concrete	1.5
Rough brickwork	2.5

Table 21.3 Equivalent diameters (d) of rectangular ducts ($a \times b$) for equal volume and pressure drop

a/b	*x*	*a/b*	*x*	*a/b*	*x*	*a/b*	*x*	*a/b*	*x*	*a/b*	*x*
1.0	0.908	1.6	0.710	2.2	0.622	2.8	0.557	3.4	0.510	4.0	0.475
1.1	0.865	1.7	0.701	2.3	0.608	2.9	0.548	3.5	0.504	4.2	0.465
1.2	0.829	1.8	0.683	2.4	0.597	3.0	0.540	3.6	0.498	4.4	0.455
1.3	0.798	1.9	0.665	2.5	0.586	3.1	0.532	3.7	0.491	4.6	0.447
1.4	0.770	2.0	0.650	2.6	0.576	3.2	0.524	3.8	0.486	4.8	0.438
1.5	0.745	2.1	0.635	2.7	0.566	3.3	0.517	3.9	0.480	5.0	0.431

$d = 1.265\,[(ab)^3/(a+b)]^{0.2}$ and thence $b = xd$, where b is the shorter side

Table 21.4 Maximum velocities for low pressure ducting systems

	Velocity (m/s)	
Application	*Main duct*	*Branch duct*
Domestic	3	2
Theatres, auditoria, studios	4	3
Hotel bedrooms, conference halls, operating theatres	5	3
Private offices, libraries, cinemas, hospital wards	6	4
General offices, restaurants, department stores	7.5	5
Cafeteria, supermarkets, machine rooms	9	6
Factories, workshops	10–12	7.5

should be used for conventional low velocity ducting systems. In each case the value quoted is arbitrary but takes general account of the fact that the higher the velocity, the greater the risk of noise disturbance in the space served. For high velocity systems, the situation is rather different since some noise is inherent in the concept in this case: Table 21.5 lists practical values for maximum air velocities in such systems. The routine of carrying out manual duct sizing calculations depends upon a systematic approach and in all cases it is best to make use of a purpose designed *pro-forma* which, for present and future reference, lists at the very least:

- The type of system.
- The material to be used for the ductwork.
- The method of calculation.
- Any allowances included for commissioning, etc.
- Headings as listed below.

Location section	*Volume*		*Equivalent diameter*	*Rectangular size*	*Air velocity*	*Pressure drop/m*	*Duct length*	*Fitting*		*Velocity pressure*	*Pressure drop*
	Design	*Revised*						*Type*	ξ		
(a)	(b)	(c)	(d)	(e)	(f)	(g)	(h)	(j)	(k)	(l)	(m)

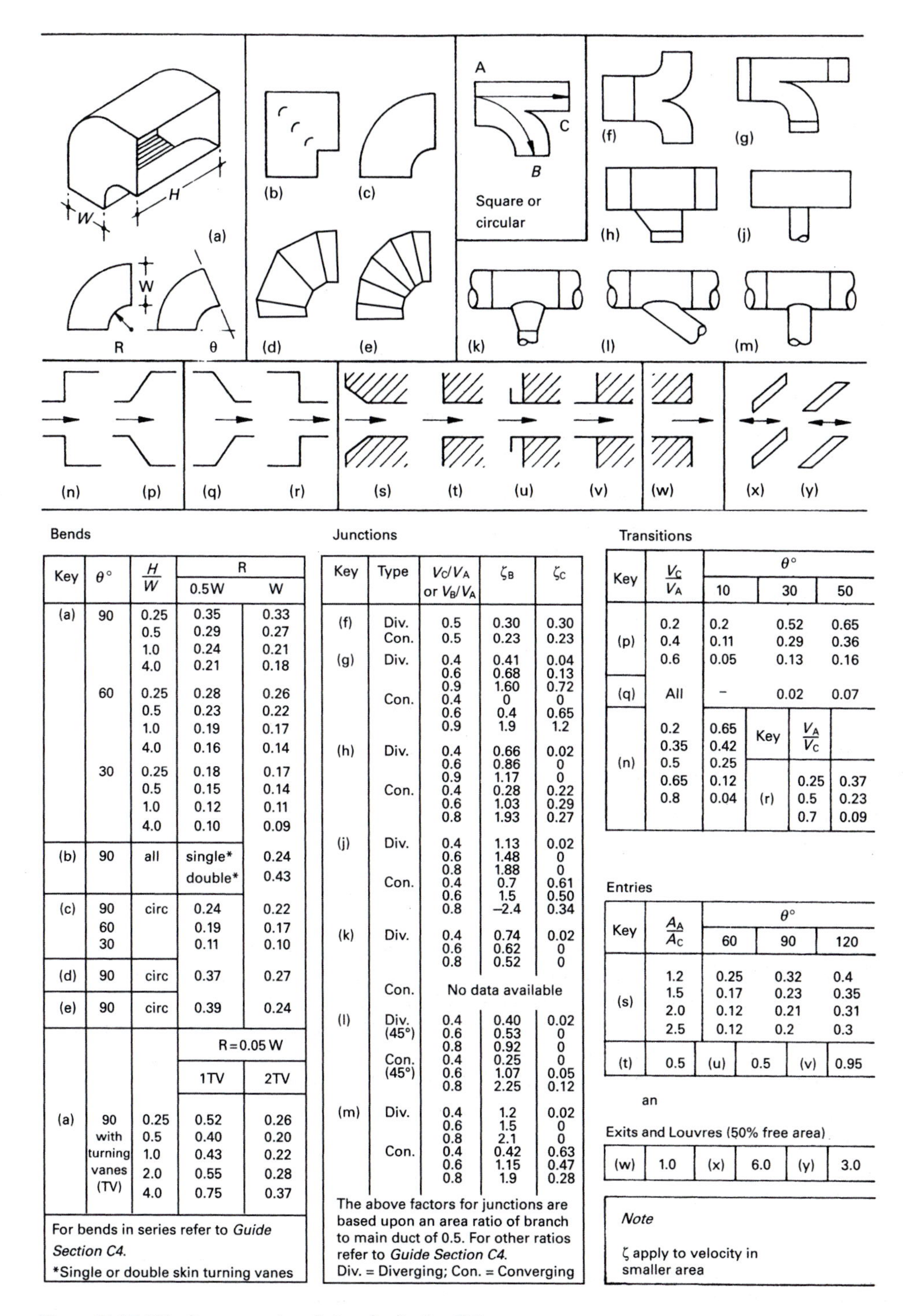

Bends

Key	θ°	H/W	R: 0.5W	R: W
(a)	90	0.25	0.35	0.33
		0.5	0.29	0.27
		1.0	0.24	0.21
		4.0	0.21	0.18
	60	0.25	0.28	0.26
		0.5	0.23	0.22
		1.0	0.19	0.17
		4.0	0.16	0.14
	30	0.25	0.18	0.17
		0.5	0.15	0.14
		1.0	0.12	0.11
		4.0	0.10	0.09
(b)	90	all	single*	0.24
			double*	0.43
(c)	90	circ	0.24	0.22
	60		0.19	0.17
	30		0.11	0.10
(d)	90	circ	0.37	0.27
(e)	90	circ	0.39	0.24
			R = 0.05 W	
			1TV	2TV
(a)	90 with turning vanes (TV)	0.25	0.52	0.26
		0.5	0.40	0.20
		1.0	0.43	0.22
		2.0	0.55	0.28
		4.0	0.75	0.37

For bends in series refer to *Guide Section C4.*
*Single or double skin turning vanes

Junctions

Key	Type	V_C/V_A or V_B/V_A	ζ_B	ζ_C
(f)	Div.	0.5	0.30	0.30
	Con.	0.5	0.23	0.23
(g)	Div.	0.4	0.41	0.04
		0.6	0.68	0.13
		0.9	1.60	0.72
	Con.	0.4	0	0
		0.6	0.4	0.65
		0.9	1.9	1.2
(h)	Div.	0.4	0.66	0.02
		0.6	0.86	0
		0.9	1.17	0
	Con.	0.4	0.28	0.22
		0.6	1.03	0.29
		0.8	1.93	0.27
(j)	Div.	0.4	1.13	0.02
		0.6	1.48	0
		0.8	1.88	0
	Con.	0.4	0.7	0.61
		0.6	1.5	0.50
		0.8	−2.4	0.34
(k)	Div.	0.4	0.74	0.02
		0.6	0.62	0
		0.8	0.52	0
	Con.	No data available		
(l)	Div. (45°)	0.4	0.40	0.02
		0.6	0.53	0
		0.8	0.92	0
	Con. (45°)	0.4	0.25	0
		0.6	1.07	0.05
		0.8	2.25	0.12
(m)	Div.	0.4	1.2	0.02
		0.6	1.5	0
		0.8	2.1	0
	Con.	0.4	0.42	0.63
		0.6	1.15	0.47
		0.8	1.9	0.28

The above factors for junctions are based upon an area ratio of branch to main duct of 0.5. For other ratios refer to *Guide Section C4.*
Div. = Diverging; Con. = Converging

Transitions

Key	V_C/V_A	θ° 10	30	50
(p)	0.2	0.2	0.52	0.65
	0.4	0.11	0.29	0.36
	0.6	0.05	0.13	0.16
(q)	All	–	0.02	0.07
(n)	0.2	0.65		
	0.35	0.42		
	0.5	0.25		
	0.65	0.12		
	0.8	0.04		

Key	V_A/V_C	
(r)	0.25	0.37
	0.5	0.23
	0.7	0.09

Entries

Key	A_A/A_C	θ° 60	90	120
(s)	1.2	0.25	0.32	0.4
	1.5	0.17	0.23	0.35
	2.0	0.12	0.21	0.31
	2.5	0.12	0.2	0.3

(t)	0.5	(u)	0.5	(v)	0.95

an

Exits and Louvres (50% free area)

(w)	1.0	(x)	6.0	(y)	3.0

Note

ζ apply to velocity in smaller area

Figure 21.17 Velocity pressure loss factors for ducting fittings

Table 21.5 Maximum air velocities for medium and high ducting pressure systems

	Velocity (m/s)	
Air quantity (litres/s)	*Medium*	*High*
Less than 100	8	9
100–500	9	11
500–1500	11	15
Over 1500	15	20

Low velocity systems

The most convenient method of sizing ducts in a low velocity system is by using the *equal pressure loss* basis. For this purpose, it is necessary to establish the maximum velocity in the main duct leaving the fan from Table 21.4 and, knowing the air volume to be carried, locate the intersection point of the volume and velocity lines on the chart, Fig. 21.16. From this point, a line followed vertically downwards will indicate the unit pressure loss, Pa/m, on the base scale. The main duct diameter may be read, at the same time, from the chart.

If the unit pressure loss so found were to be considered acceptable, the remainder of the duct sections following on (or those in the most disadvantaged duct run, the *index run*, in a system having many such) would be sized using the same unit rate. However, if that rate were thought to be too high or too low then another velocity might be chosen as the datum using the expression:

$$V = Q / A$$

where

V = the chosen velocity (m/s)
Q = air volume leaving fan (m^3/s)
A = duct area (m^2)

Table 21.4 provides traditional economic limits on velocity and as a guide to minimise the regeneration of noise within the ductwork according to application.

In a complex system, some attempt should be made to balance the pressure loss through other duct runs with that of the index run but velocity and possible noise generation would have to borne in mind if any variations from those given in Table 21.4 were extreme. Finally, when all circular duct sizes have been chosen, reference to Table 21.3 should be made and square or rectangular shapes chosen to suit the space available. In carrying out this last routine, it is as well to check that unnecessary minor reductions in shape or size are not being made to suit relatively small changes in velocity, since to do so is uneconomic.

It is sometimes advantageous, when sizing a long run of ducting with air supply grilles along its full length, to take what is called *static pressure regain* into account. This is done by so reducing the velocity in the duct in stages that the regain in static pressure compensates for the frictional loss in the duct up to the point of reduction. In practice it has been found that this can normally be achieved in low velocity systems by sizing at constant pressure loss in duct runs not exceeding 5 m/s. This method provides an approximately equal pressure at each grille and this assists in balancing one with another.

High velocity systems

In view of the impact of the Building Regulations (Part L) it is now unlikely that systems other than low velocity would be used in order to achieve the required specific fan power and references to high velocity systems are included largely for historic reasons. However systems with specific 'process' requirements may be exempt where it can be demonstrated that the system has to incorporate features such as high velocity or high pressure for the system to be effective. For example an extract system designed to capture and convey solids would require a high 'transport' velocity.

With the use of the higher duct velocities inherent in dual duct, induction unit and variable air volume systems, etc., as in Table 21.5, the use of static pressure regain is much more important, the potential availability being higher. All the static pressure gain calculated, theoretically, at the expanding duct section will not be available since the total pressure there will be reduced due to turbulence and added wall friction. Thus:

$$\text{Static regain} = (p_{vi} - p_{vo}) - \Delta p_t$$

where

p_{vi} = inlet velocity pressure at expander
P_{vo} = outlet velocity pressure at expander
Δp_t = reduction in total pressure

The reduction in total pressure depends upon the design of the expanding duct section and another way of stating the expression above is:

$$\text{Static regain} = f(p_{vi} - p_{vo})$$

where

f = a factor representing the 'efficiency' of the expanding duct section, say, 75 per cent.

It is normal practice to use the availability of static pressure regain in sizing the main ducts only of high velocity systems since the branch ducts tend to be self-balancing as a result of the high pressure drop through the terminal fittings, be they double-duct, variable volume or induction units. The routine adopted in sizing a duct system by this method is first to consider the penultimate section of the main duct and so to select the size that the static pressure regain at the expander is equal to the pressure loss in the final section, repeating this procedure in regression, section by section, back to the plant outlet.

By these means, the entire system will be inherently balanced and the need for dampers in upstream ducts either much reduced or, ideally, avoided altogether. Ductwork for high velocity systems is almost always constructed with spiral-wound circular or flat-oval sections except at plant and final connections to terminals.

Computer design of duct systems

It is now quite common practice for ductwork sizing calculations, other than those for simple distribution networks where an equal pressure drop approach is quite adequate, to be made using computer programs. These are particularly useful when the static regain method is used since this is very tedious to apply manually. An example output is shown in Fig. 21.18.

Fan pressure and fan duty

From the various descriptions of volume flow and pressure loss above it will be obvious that the duct sizing operation is but one part of an exercise to determine the total and the static pressure which must be developed by the connected fan. A summation of the various system losses, those through straight ductwork, through duct fittings and through plant components, with reductions where available to take account of static pressure regain, will go to produce the static pressure against which a fan must operate.

This, of course, is the net or calculated duty and a margin of 10 per cent is usually added to that pressure to allow for minor changes in the duct configuration during the installation process. Similarly, margins should be added to the calculated volume of say up to 5 per cent for leakage (reference Table 21.1) plus another 5 per cent as an allowance for contingencies in commissioning.

Thermal insulation of ducts

To conserve energy and to minimise heat loss or gain the ductwork should be externally insulated. This applies to supply air ducts and to return air ducts passing through untreated spaces upstream of heat recovery

Nodes Near / Far	Sizing limits	Duct Size (mm)	Shape code	Duct length (m)	Diversity	Flowrate (l/s)	Velocity (m/s)	Total *K*-Factor	Pressure gradient (Pa/m)	Pressure loss in duct (Pa)	/ damper (Pa)		Index path
10/110	FX /	1000/800	R	19.20	0.800	10960.	13.67	0.53	1.829	171./	0.		★
110/120	/	600/600	R	4.30	1.000	3250.	9.02	1.23	1.319	67./	0.		
120/121	/	400/400	R	0.20	1.000	1050.	6.56	1.08	1.173	29./	7.		
121/125	/	740/250	F	3.00	1.000	1050.	6.12	0.03	1.060	4./	0.		
125/126	MD/300	605/200	F	2.00	1.000	450.	4.00	1.12	0.621	37./	5.	T	
135/137	FD / 300	955/300	F	2.00	1.000	750.	2.81	1.38	0.191	37./	0.	T	
130/140	VL / 3.0	600/600	R	4.25	1.000	1000.	2.78	4.19	0.144	70./	0.	T	
210/220	/	800/800	R	4.30	1.000	4050.	6.32	1.66	0.475	42./	12.		
220/221	/	600/600	R	0.20	1.000	1500.	4.17	1.39	0.305	15./	14.		
221/225	MW/1000	835/500	F	3.00	1.000	1500.	4.12	0.00	0.277	1./	0.		
335/336	/	735/400	F	2.00	1.000	900.	3.47	1.13	0.252	49./	0.	T	
335/337	/	735/400	F	2.00	1.000	900.	3.47	1.13	0.252	49./	0.	T	
330/340	/	600/600	R	4.25	1.000	1200.	3.33	2.61	0.201	69./	0.	T	★
110/210	SM/	1000/800	R	8.90	0.800	8360.	10.44	0.28	1.087	28./	0.		★
210/310	/	800/800	R	11.10	1.000	6400.	9.99	0.10	1.136	19./	0.		★
430/440	/	400	I	2.10	1.000	850.	6.77	0.01	1.283	3./	0.		
440/450	/	355	I	2.10	1.000	600.	6.06	0.06	1.207	4./	0.		
450/460	/	315	I	3.40	1.000	350.	4.49	1.26	0.796	18./	1.	T	
430/435	VL / 3.0	355	I	2.10	1.000	250.	2.53	5.08	0.237	20./	6.	T	
440/445	VL / 3.0	355	I	2.10	1.000	250.	2.53	5.08	0.237	20./	3.	T	
450/455	VL / 3.0	355	I	2.10	1.000	250.	2.53	4.78	0.237	19./	0.	T	

$$★$$ fan requirements $$★$$

	Operating conditions	Standard conditions
Fan total pressure	319.9 (Pa)	314.8 (Pa)
Fan static pressure	205.9 (Pa)	202.7 (Pa)
Total flowrate	10960.0 (l/s)	10960.0 (l/s)
Air power	3505.9 (W)	3450.1 (W)

The following TM-8 defined margins are added to give design values

10.0 percent on flow for leakage and balancing
10.0 percent on pressure for flow margin
10.0 percent on pressure for calculation uncertainties

★★★★★★★★★★★★★★★★★★★★★★★★★★★★★★★★★★★★★★

Design duty at standard conditions –

12056. l/s at 378. Pa total pressure

★★★★★★★★★★★★★★★★★★★★★★★★★★★★★★★★★★★★★★

Figure 21.18 Sample computer output for duct sizing problem

devices. Supply air ducts conveying cooled air should be provided with thermal insulation and vapour sealed (i.e. sealed against ingress of atmospheric air, otherwise the insulation will quickly become waterlogged and useless).

Air intakes including plenums and intake ducts connected to the air handling plant, and where necessary exhaust air ducts downstream of a heat recovery device, should be insulated and vapour sealed to prevent condensation.

Materials for duct insulation, internal and external to the duct, must be such as not to support fire. When tested to BS 476: Part 6: 1989 and Part 7: 1997,[3] the insulating material must have a fire propagation index of performance not exceeding 12, with not more than 6 from the initial period. The surface flame spread, as defined in Part 7, must meet the Fire Authorities' requirements. Some suitable materials are glass fibre, mineral wool and phenolic foam.

Table 21.6 Insulation thickness to control heat loss

Ambient temperature °C	*15 (internal)*	*−5 (external)*
Heat loss (W/m^2)	16.34	16.34
Thickness of mineral wool insulation (mm)	40	80
Thickness of phenolic foam (mm)	25	45

Table 21.7 Insulation thickness to control heat gain

	Thickness (mm)			
	Ambient temperature 25°C *(internal)*		*Ambient temperature* 30°C *(external)*	
Air temperature in duct °C	*Mineral wool*	*Phenolic foam*	*Mineral wool*	*Phenolic foam*
20	25	25	40	30
18	25	25	55	35
16	35	25	65	40
13	50	35	80	50
10	65	45	95	60
5	95	60	120	80

Table 21.6 provides the minimum insulation thickness to satisfy the maximum permissible heat loss of 16.34 W/m^2 stated in Building Regulations Part L Non-domestic Heating, Ventilation and Cooling Guide. Table 21.7 similarly provides the minimum insulation thickness, provided with a low emissivity facing, to limit the maximum heat gain to 6.42 W/m^2. Where ductwork is exposed to solar gain the insulation thickness should be increased accordingly.

Insulated ducts located outside a building must be provided with a commercial weatherproof finish such as bitumen coated roofing felt, with well-sealed joints, secured with galvanised wire netting, or some better and more permanent protection. Horizontal ducts fitted externally should preferably be circular or, where they must be rectangular, be provided with a sloping cover to prevent water lying on the upper surface.

Ductwork cleanliness

Measures need to be taken to ensure that ductwork achieves an appropriate level of cleanliness before the ventilation system is put into use. These include the prevention of internal contamination during transit, site storage and installation. This is described in HVCA TR/19 Internal Cleanliness of Ventilation Systems which categorises protection, delivery and installation appropriate to application in three levels of protection, with level 3 requiring precommissioning specialist cleaning. It should be recognised that conditions on site may result in contamination of ductwork, plant and duct mounted equipment after their installation and that internal cleaning may be required irrespective of the precautions taken.

All ductwork should be designed with access/cleaning panels such that the whole of the system can be inspected and cleaned if necessary when the installation is completed and subsequently during use. The location, size and type of access would be dependent upon the system design and the type of ductwork cleaning, inspection and testing methods to be adopted which may require advice at the design stage from a specialist cleaning contractor. CIBSE *Guide B3* provides further guidance on provisions for cleaning ductwork systems.

There are a number of techniques used for cleaning including both wet and dry methods which are described in HVCA TR/19 which also includes methods of system testing and surface deposit limits.

The use of internal acoustic linings is not recommended due to their susceptibility to surface damage during cleaning operations. Acoustic treatment for ductwork systems is described in Chapter 26.

Notes

1. Heating and Ventilating Contractors Association, Specification DW/144.
2. Corrosion of Heat Recovery Exchangers Serving Swimming Pool Halls. Electricity Council Research Report, August 1961.
3. BS 476, Fire tests on building materials and structures, parts 6 and 7.

Chapter 22

Fans and air treatment equipment

Air handling plant often consumes large quantities of energy both in the fan power required to overcome the resistance of the plant components and the ductwork distribution and also the energy required to heat, humidify and to cool and dehumidify the air.

The Building Regulations have set stringent limits for *specific fan power* (*SFP*) as set out below. If these requirements are to be met much attention is required in the design of the air handling and distribution system to minimise the static fan pressure and to maximise the efficiency of the fan. The former can be achieved by reducing the velocity of the air through plant components and ductwork compared with hitherto customary values. Overall fan efficiency and the resulting power requirement should be optimised by selection of suitable fan type, appropriate selection against the fan characteristic curve, the selection of efficient motor drive and speed control arrangements and the use of high efficiency motors. Variable speed drives are required on fans rated at more than 1.1 kW to provide flexibility for change and to aid commissioning.

There is also potential for energy savings by avoidance of 'over design' for example the use of comfort cooling in lieu of air-conditioning where this is appropriate and if humidification may not be justified. To meet the carbon emission targets required by Part L, it may be necessary to recover heat or coolth from the extract air to preheat or precool the incoming outside air.

Fans

The fan is the one item of equipment which every mechanical ventilation and air-conditioning system has in common. A fan is simply a device for impelling air through the ducts or channels and other resistances forming part of the distribution system. It takes the form of a series of blades attached to a shaft rotated by a motor or other source of power. The blades are either in the plane of a disc (propeller and axial fans), or in the form of a drum (centrifugal fan), however a recent development, the mixed flow fan, is a suitable alternative for certain specific applications.

There is as yet no other practicable commercial method of moving air for ventilation purposes, but fans in general suffer from various disadvantages such as low efficiency and noise. It is the latter which is probably the most troublesome to designers, and to which much careful attention must be given if acceptable acoustic conditions are to be achieved in the system as a whole: factors involved are air speeds, fan speeds, duct design, materials of construction, acoustical treatment of ducts and provision for absorption of vibration.

Fan types and performance

Fan characteristics

A comparison of the operation of fans of various types is best understood by studying their characteristic curves. For this purpose consider a fan connected to a duct with an adjustable orifice at the end, as in Fig. 22.1. Pressures are measured by water gauges connected to a standard Pitot tube, see Fig. 29.4. The perforated portion gives the static pressure, and the facing tube the total pressure.

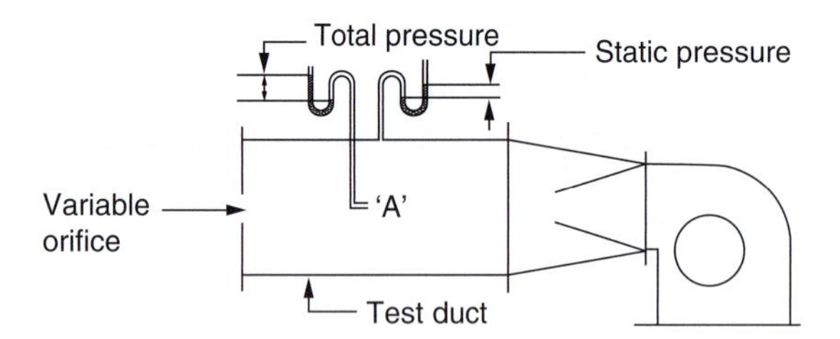

Figure 22.1 Simplified diagram of fan test arrangement

If the fan is running with the orifice shut, no air will be delivered. Static pressure will be at a maximum, and velocity pressure nil. As the orifice is opened the static pressure will fall and the velocity pressure increase until, with the orifice fully open, the static pressure will be negligible and velocity pressure at a maximum. Over this range the power required to drive the fan will have increased from minimum to maximum, and perhaps will fall away as the total pressure falls off. The power required to drive the fan, if 100 per cent efficient, would be:

$$H_t = VP$$

where

H_t = total air power (W)
V = air volume handled (m^3/s)
P = total pressure (Pa)

The mechanical efficiency of a given fan will be the ratio between this air power and the actual mechanical power supplied to the fan shaft (fan shaft power): this will depend on design, type of fan, speed, and proportion of full discharge. If the static pressure is used, the efficiency derived will be *static efficiency*: if the total pressure is used the efficiency will be *total efficiency*.

The standard air[1] for testing fans is taken at a density of 1.2 kg/m^3. Any fan at constant speed will deliver a constant volume at any temperature; as the temperature varies the density will increase or decrease proportionately with the absolute temperature, hence the power input will vary in the same ratio. With increase of temperature the power will be reduced and vice versa. Similarly, decrease of atmospheric pressure (as in the case of a fan working at high altitude) will cause a reduction in power and conversely.

If a fan running at a certain speed be rearranged to run at some higher speed, the system to which it is connected remaining the same, the volume will increase directly as the speed; the total pressure will increase in the ratio of the speeds squared and the power input will increase in the ratio of the speeds cubed.

These relationships are known as the *fan laws* which, for practical purposes, may be expressed for impeller diameter (d), speed of rotation (n) and air density (p) as:

Volume flow is proportional to d^3 and n Fan
Pressure is proportional to d^2, n^2 and p Fan
Power is proportional to d^5, n^3 and p

Characteristic curves

Fans are of three main types, with sub-divisions as follows:

- Centrifugal type:
  - Multivane, forward bladed
  - Multivane, radial bladed
  - Multivane, backward bladed
  - Paddle wheel

  The three types of runner (a), (b) and (c), are shown in Fig. 22.2.

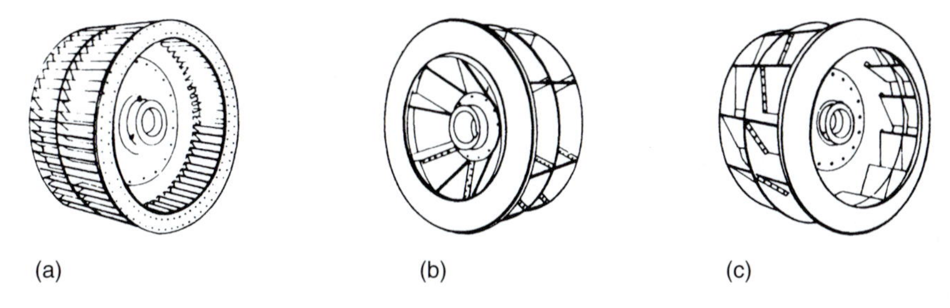

Figure 22.2 Types of runner in centrifugal form: (a) forward curved, (b) radial and (c) backward curved.

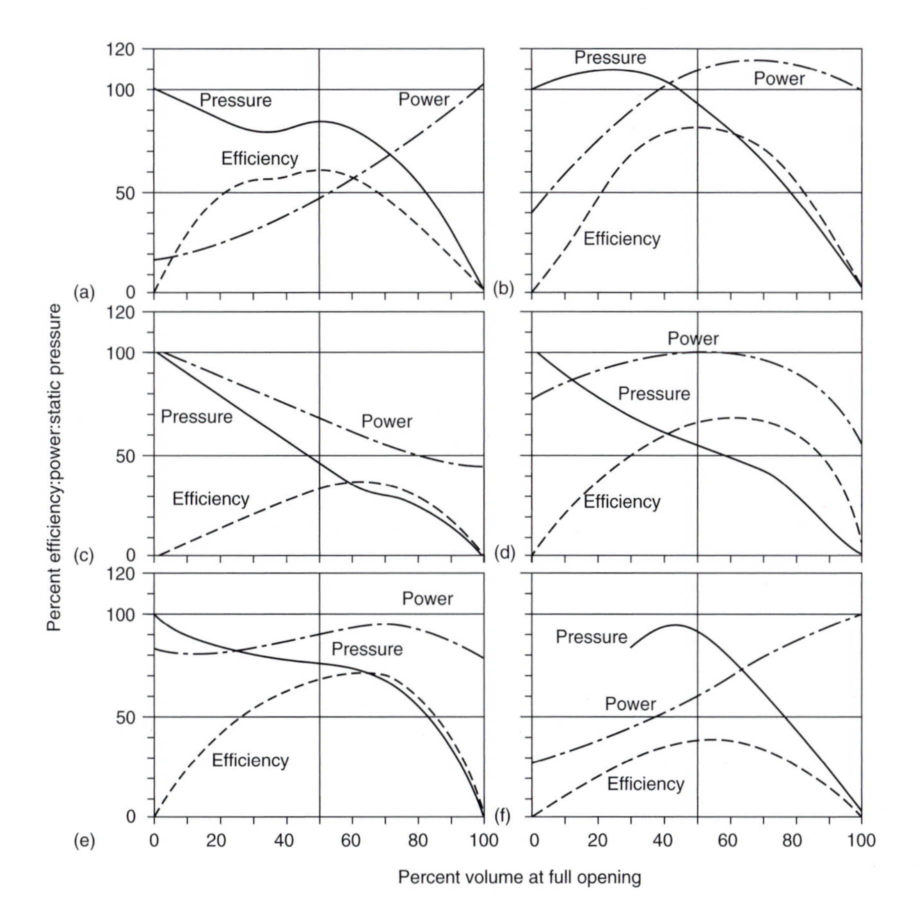

Figure 22.3 Fan characteristics: (a) forward curve, (b) backward curve, (c) propellor, (d) axial flow, (e) mixed flow and (f) tangential

- Propeller type:
  - Ordinary propeller or disc fan
  - Axial flow
- Mixed flow type

Figure 22.3(a)–(f) gives typical curves for the main types. Paddle-wheel fans are not given, as they are now little used in ventilating work on account of their noise, being confined chiefly to dust removal and industrial uses. Radial-bladed fans have characteristics which are similar to those of the backward curved type but without the advantage of power limitation. They are not commonly used in ventilation applications.

The curves for static pressure, power input, and static efficiency are drawn from tests at constant speed, as already described. The base of the curve is percentage of full opening of the orifice. The vertical scale is percentage of pressure, efficiency or power input.

Forward curved

It will be noted that the forward-bladed centrifugal fan most commonly used in ventilation reaches a maximum efficiency at about 50 per cent opening, where at the same time the static pressure is fairly high. Fans are generally selected to work near this point. It will also be observed that the power curve rises continuously. Thus, if in a duct system the pressure loss is less than calculated, the air delivered will be more and the power absorbed more, which may lead to overloading of the motor.

Backward curved

This type of fan runs at a higher speed to achieve the same output as a forward curved. The efficiency reaches a maximum at about the same point, and the power, after reaching a peak, begins to fall. This is called a *self-limiting characteristic*, and means that if the motor installed is large enough to cover this peak it cannot be overloaded. This is often useful in cases where the pressure is variable or indeterminate. The pressure curve is smooth without the dip of the forward curved; for this reason this kind of fan is to be preferred where two fans are working in parallel. The forward curved type under such conditions is apt to hunt from one peak to the next, so that one fan may take more than its share of the load and the other much less. The backward-bladed fan is often made with aerofoil blades, so raising efficiency. It is much used in high velocity systems where high pressures are required.

Ordinary propeller

From the curves it will be observed how the pressure falls away continuously, and the static efficiency reaches but a low figure. Thus, this type is unsuitable where any considerable run of ducting is used. To generate any appreciable pressure its speed becomes unduly high, and hence the fan is noisy. Its main purpose is for free air discharge where its velocity curve would rise towards a maximum at full opening. It should be noted that the fan power is a maximum at closed discharge, and, as the motors supplied with these fans are not usually rated to work at such a condition, the closing or baffling of the discharge or suction may cause overloading.

Axial flow

This type shows a great improvement over the ordinary propeller fan, both as regards efficiency and pressure. The fan–power curve is self-limiting. Hence these fans may safely be used in conjunction with a system of ductwork, being often more convenient than a centrifugal, particularly for exhausting. They run at higher speed than a centrifugal fan to produce a given pressure, and are liable to be more noisy: this may be overcome, to some extent, by the use of an attenuator. Their limited pressure development may be overcome by installing multiple fans to operate in series.

Mixed flow

Such fans are designed for in-line flow or radial-discharge and comprise an impeller having a number of blades, on a conical-shaped hub. The efficiency and pressure developed are generally higher than for the axial flow type, with lower noise levels. Pressures up to 1750 Pa can be achieved with the range of fans currently available. Variable duty can be obtained using either variable speed drives or inlet guide vanes. Either direct drive or belt drive configurations are available. Figure 22.4 shows diagrammatically the operation of a mixed flow fan.

The fan curves indicate a non-overloading power characteristic, making it suitable for use with ducted systems. Efficiency is good at about 80 per cent over a limited range of duty, but the pressure characteristic, being similar to that of a forward curved centrifugal, makes it unsuitable for parallel operation.

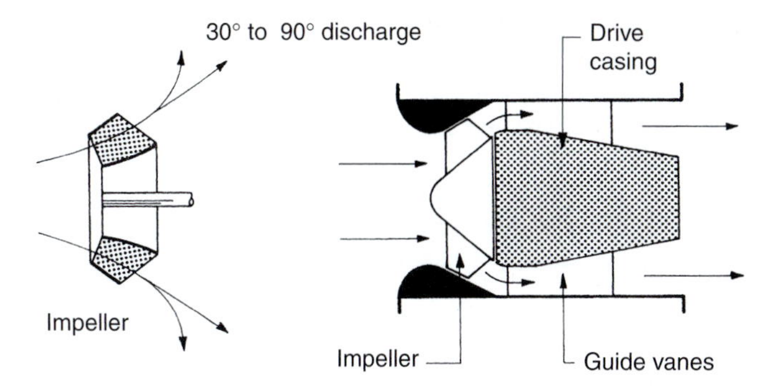

Figure 22.4 Mixed flow fans

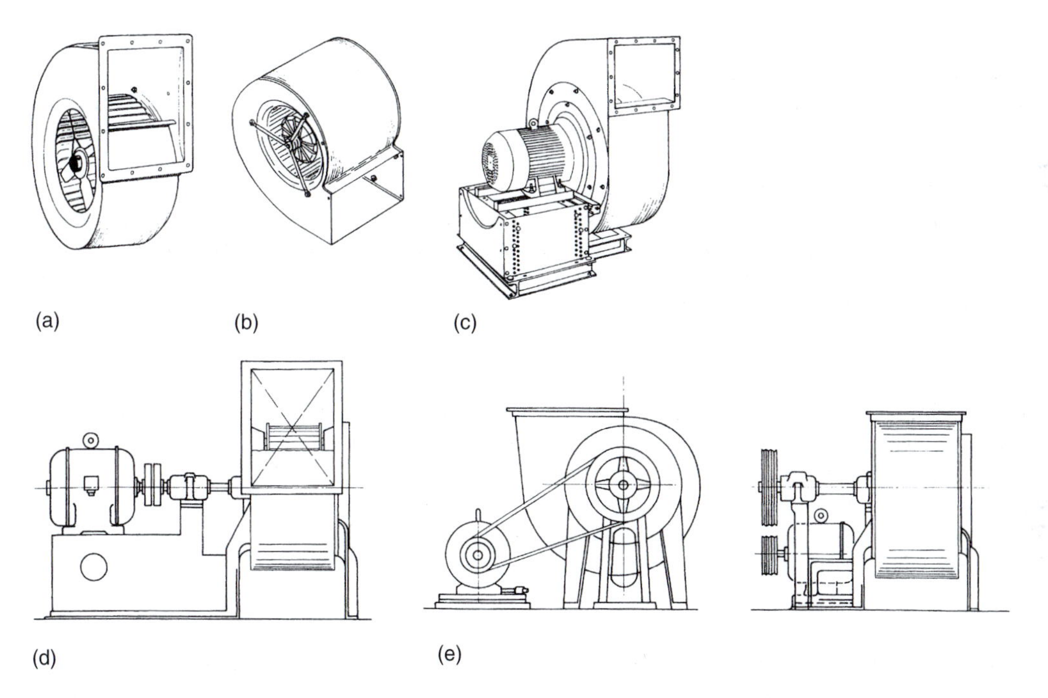

Figure 22.5 Arrangements for centrifugal fans: (a) single inlet, (b) double inlet, (c) close-coupled with motor, (d) flexible coupling in motor drive shaft and (e) belt driven

Centrifugal fan arrangements and drives

Centrifugal fans may be *open* or *cased*. When open they can only be used for exhausting, and the discharge is tangential from the perimeter of the impeller, as might be suitable in a large roof turret.

The usual arrangement is the cased type, and the suction is then either on one side, as in Fig. 22.5(a), with a single inlet, or both sides, as Fig. 22.5(b), with double inlet. The double inlet double-width fan is useful in packaged air handling plant or where large volumes are concerned as it gives double the capacity of the single inlet with the same height of casing.

Fans are almost invariably driven by electric motor except for cases where there may be a requirement for independent drive from a petrol or diesel engine. Figure 22.5(c) shows a typical arrangement with the fan impeller mounted on a shaft extension of the motor. This is a compact arrangement, but generally used for small- or medium-sized fans only.

A motor direct coupled to a fan with a flexible coupling is illustrated in Fig. 22.5(d). The fan shaft runs in its own bearings independently of the motor. This is obviously to be preferred for heavy duty and for large fans. The motor can be removed and replaced without affecting the fan.

The arrangement shown in Fig. 22.5(e), where the motor drives the fan via pulleys and vee belts, has the great advantage that the *motor* speed may be a standard, such as 16 or 24 rev/s, whilst *the fan* speed is that best suited to the duty. A further advantage is that if on testing it is found that the pressure loss of the system is less or more than allowed for, the fan duty may be corrected by merely changing the pulleys.

Certain applications require fans to have an automatic standby capability. The belt drive configuration enables duty and standby motors to be installed to drive shaft of a single fan, although this increases the load on the running motor. The alternative is to use twin fans, with the resultant increase in space and capital cost.

Centrifugal

It will be noted that in the illustrations a variety of different positions of the discharge opening in relation to the suction eye of the fan is given in each case. It is usually possible to obtain a fan with its discharge at any angle, vertical, horizontal top, horizontal bottom, downwards, and intermediately at an angle of 45°.

Inlet guide vanes, or variable speed drive using static inverters are usually adopted to give variable duty.

Axial flow fan arrangements

This type of fan can produce pressures up to a maximum of about 1000 Pa, as a single unit within the normal range of noise generation. To achieve such performance, guide vanes would be fitted to improve the operating efficiency by reducing the *swirl* effect.

These fans can be built in one, two or three stages, to obtain increased pressure, the volume remaining the same. Alternatively they may have two sets of blades made counter-rotating. Both direct drive and belt drive types are available.

Such fans are illustrated in Fig. 22.6, many are manufactured to permit adjustment of the pitch angle of the blade. Since this angle, for a given fan diameter and speed, determines the volume delivered it follows that adjustment facilities permit output to be matched to the duty required with some precision. Figure 22.6(d) and (e) show how pitch may be adjusted.

Axial flow fans of the *bifurcated* type, where the air stream is directed around the motor which is enclosed in a protective casing, are suitable for handling corrosive gases, from fume cupboards and the like, and also high temperature gases/air such as experienced in smoke extract during fire conditions.

Standby arrangements are also possible with axial flow fans. A common way is to install the two fans in series with the increased pressure loss over the idling impeller to be considered. Alternatively, it is also possible to mount two fans in a parallel arrangement with automatic changeover dampers coupled to the fan controls. A standby motor arrangement is also possible where the motors are out of the air stream driving the shaft via pulleys and belts.

The range of fan types, speeds, pressures, and volumes is too great for any indication to be given here of sizes, duties, power requirements, etc., or of the problem of motor types suitable for fan drives.

Where it is necessary, as in the case of a variable volume air-conditioning system, to exercise control over fan output this may be achieved in a number of ways, which may be summarised conveniently under two headings:

Constant fan speed

- Throttling dampers
- Inlet guide vanes
- Variable pitch blade angle (axial flow fans)

Variable fan speed

- Eddy current coupling
- Variable voltage (fixed frequency)
- Variable frequency, variable voltage

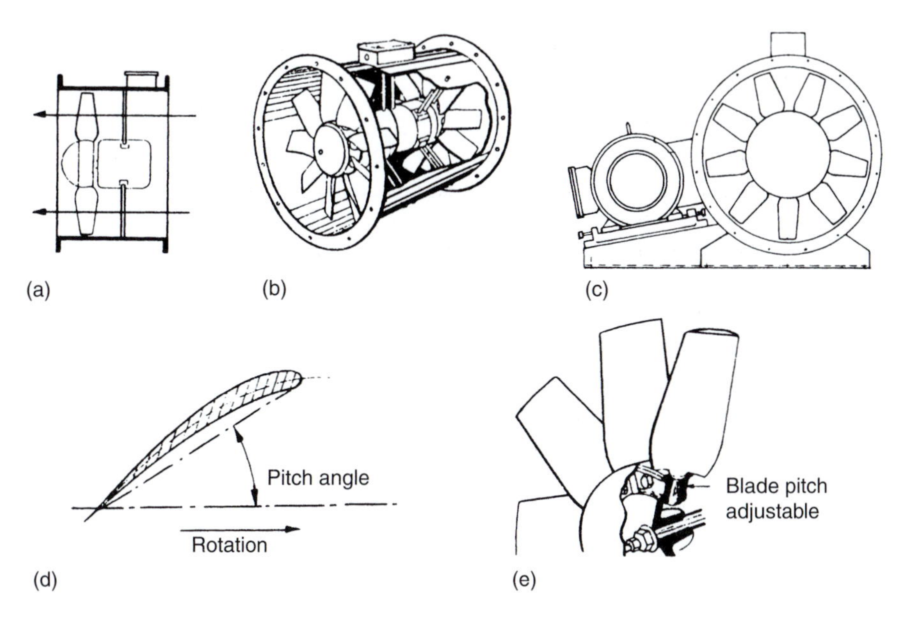

Figure 22.6 Arrangements for axial flow fans: (a) single-stage fan, (b) two-stage fan (contra-rotating), (c) belt-driven fan, (d) blade section and (e) variable pitch

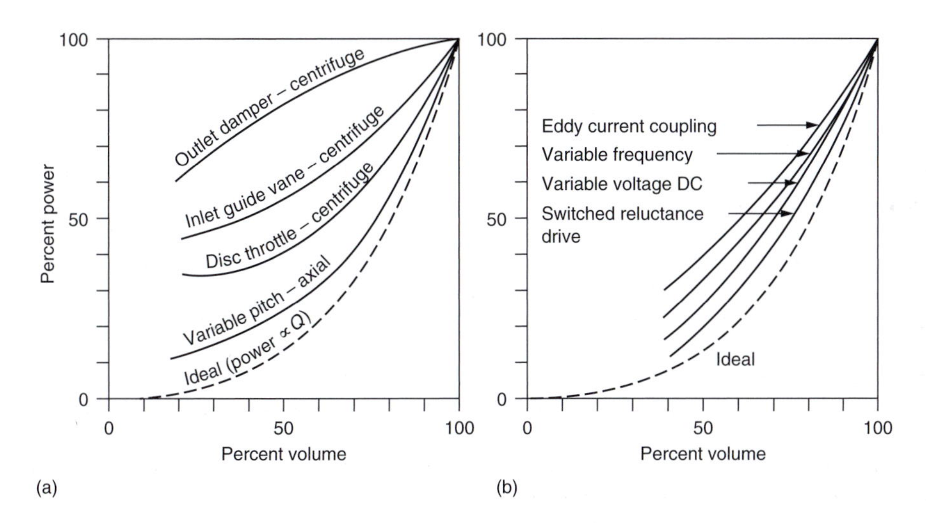

Figure 22.7 Power input to variable volume fans: (a) constant and (b) variable fan speed

- Thyristor converter (AC to variable voltage DC)[2]
- Fluid couplings
- Variable pulley drive
- Slip-ring motor
- Switched reluctance drive

Figure 22.7(a) and (b) shows for constant and variable fan speed drives respectively, the percentage power input to a fan assembly relative to volume flow.

Where *throttling dampers* are introduced into the system, volume reduction will be achieved at the expense of efficiency as illustrated in Fig. 22.8(a) and (c). In (a), the full volume operation of a system is represented

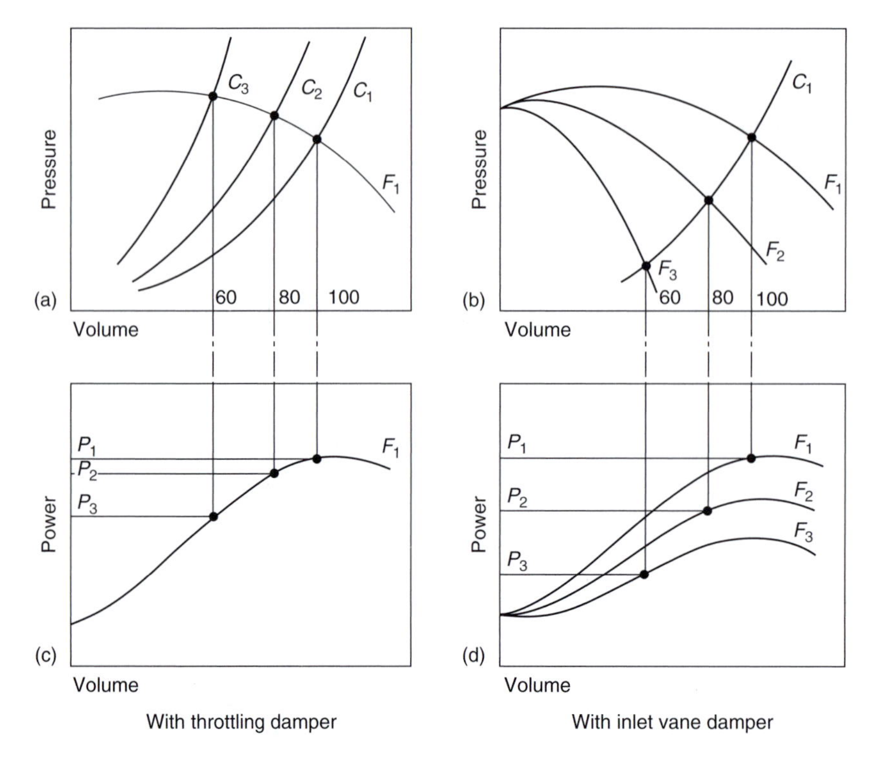

Figure 22.8 Control of fan volume using dampers

by the intersection of the fan characteristic F_1 with the system characteristic C_1. The use of throttling dampers changes the *system* characteristic to C_2 and C_3 at 80 and 60 per cent volume, respectively. The resultant small savings in power absorbed are shown in (c).

The use of radial *inlet vane dampers* fitted to the eye of a centrifugal fan provides a much more effective means of volume control since such devices act, by changing the air flow pattern at the fan inlet, to modify the performance of the fan impeller in much the same way as would result from change in the speed of drive. This effect is illustrated in Fig. 22.8(b) and (d). As before, F_1 represents full volume operation with the system characteristic C_1: operation of the radial inlet vanes changes *the fan* characteristic to curves F_2 and F_3,which represent 80 and 60 per cent volume, respectively, against a constant system characteristic curve. The resultant dramatic savings in power are shown in Fig. 22.8(d). Throttling dampers would only be considered for use in very small systems used intermittently.

Where axial flow fans are used with systems demanding volume reduction, the facility of change in *pitch angle* of the blades may be automated. This permits both volume and power to be reduced as required. Figure 22.9 shows the effect of blade pitch angle on power consumption at various duties. As with inlet guide vanes on centrifugal fans, a significant power reduction results.

The *variable fan speed* method is well suited to variable air volume systems in comfort air-conditioning applications and to maintain constant air volume irrespective of filter condition in systems where the filter represents a large proportion of the system pressure loss. In these systems, the torque varies with the square of the speed and the power with the cube of the speed. Usually a drive is supplied capable of working against a torque, with the reduced power and torque requirements at lower speeds taken into account in the design of the motor. Special attention should be given where a constant pressure has to be maintained to all, or to a major part of, the system distribution since, to achieve the required performance over the range of volume variation, the motor characteristics would need to match the particular system requirements. Fan operating characteristics for centrifugal fans with variable speed control are shown in Fig. 22.10.

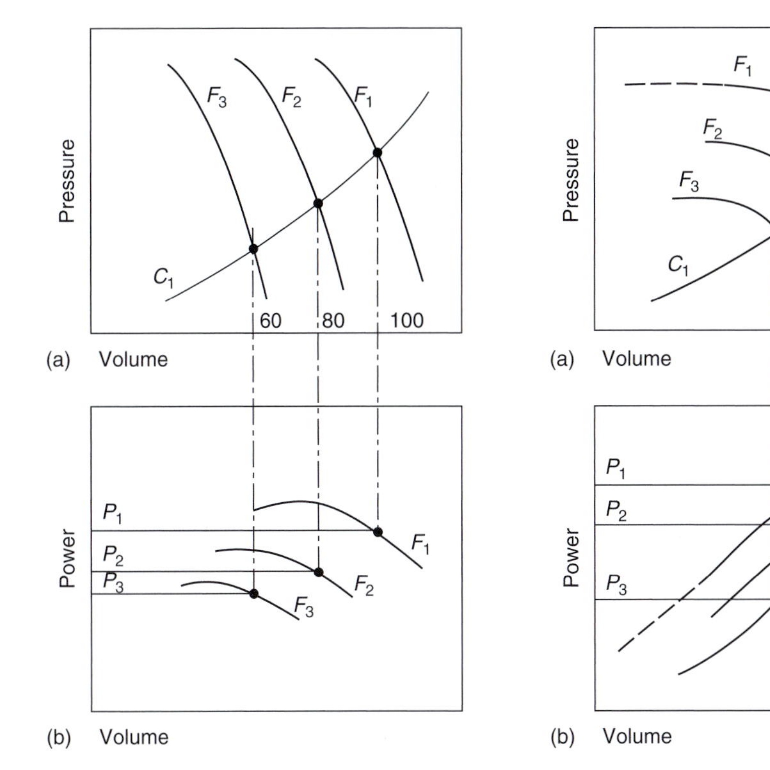

Figure 22.9 Control of fan volume (axial flow) by variable pitch

Figure 22.10 Control of fan volume (centrifugal) by variable speed

Multi-speed dual wound or pole-change motors may be used where the operating requirements are in clearly defined steps, such as winter/summer or day/night operation.

Current practice on all but the smallest systems is to use centrifugal fans with backward curved impellors with inverter speed control. BSEN 3053:2006 states that for air handling units the fans should be provided with backward curved blades for energy reasons and that energy-saving motors are preferred.

Specific fan powers

Part L of the Building Regulations, Non-Domestic Heating, Cooling and Ventilation Compliance Guide, requires that ventilation systems achieve the target SFP figures in new buildings, and in existing buildings where new systems are installed or whenever air handling plant is provided or replaced. Limiting SFP values are given in Table 22.1.

It should be noted that the SFP relate to the combined fan powers of both the supply and extract systems.

External louvres

It might be thought that the purest air supply would be obtainable at the roof of a building, but experience shows that in many instances this is not the case. At roof level, air is certainly free from street dust, but chimneys or contaminated exhausts from kitchens, toilets or fume cupboards of the same or neighbouring buildings may, with certain states of the wind, deliver fumes into the intake. Particular care must be taken to ensure that

Table 22.1 Limiting specific fan powers W/(litre/s)

System type	*New buildings*	*Existing buildings*
Central mechanical ventilation including heating, cooling and heat recovery	2.5	3.0
Central mechanical ventilation with heating and cooling	2.0	2.5
All other central systems	1.8	2.0
Local ventilation units within the local areas, such as window/wall/roof units, serving one room or area	0.5	0.5
Local ventilation units remote from the area such as ceiling void or roof mounted units, serving one room or area[a]	1.2	1.5
Other local units, e.g. fan coil units (rating weighted average – refer to Guide)	0.8	0.8

Note
a Includes fan assisted terminal VAV units where the primary air and cooling is provided by central plant.

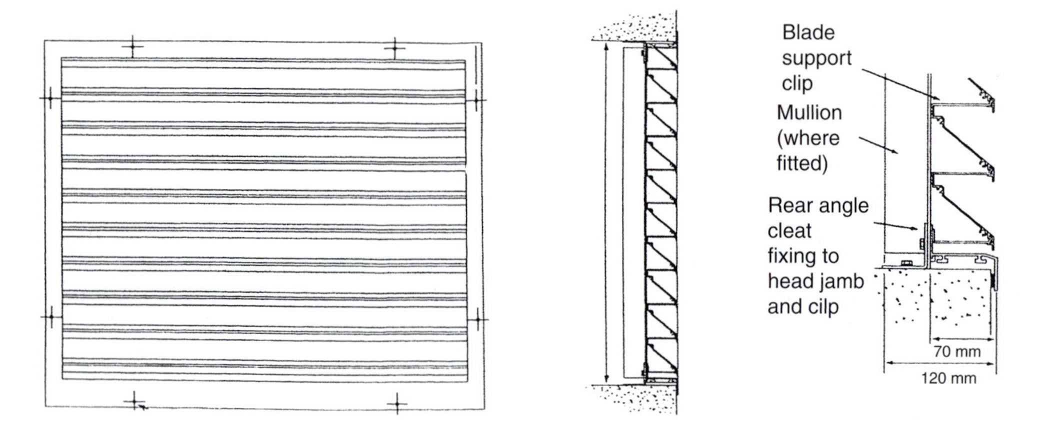

Figure 22.11 Section through typical weather louvre

drift from cooling towers, or other possible sources of *Legionnaires' disease*, cannot be carried over into air intakes.

An intake at low level, such as near a busy street, would be liable to draw in much road dust and exhaust fumes from motor vehicles. If a point half way up the elevation of a building can be found, this is probably the best. It must, however, be clear of windows where fire or smoke might occur, and particularly of lavatory windows.

There is no general solution to the problem of the outside air intake position, as obviously every case requires examination of orientation, possible sources of contamination, risk of terrorist action, position of air-conditioning plant, and so on.

It is equally important to have adequate separation of air intakes and exhaust air outlets which can be best achieved by locating on a different face or plane, or if located on the same face of the building separated with sufficient distance typically 10 m to minimise the risk of recirculation.

The selection of the weather louvre is obviously important but quite often left to the architect to be provided as an integral part of the cladding. It is important that an appropriate type is provided to prevent the ingress of rain irrespective of the operation of the system and applicable to both supply and exhaust louvres. There are commercially available weather louvres which consist of a number of profiled blades or series of blades which are designed to trap rain or snow and provided with internal drainage channels. Various levels of ingress protection can be provided according to requirements, by appropriate selection. Figure 22.11 shows a typical weather louvre.

Where use is to be made of a band of architectural louvres offering little weather protection a section or area of louvre could be activated for ventilation and the remainder blanked or plated off behind. The activated

Table 22.2 Typical mass of solids in the atmosphere

Locality	*Total mass* (mg/m^3)
Rural and suburban	0.05–0.5
Metropolitan	0.1–1.0
Industrial	0.2–5.0
Factories or workrooms	0.5–10.0

Table 22.3 Typical analysis of atmospheric dust

	Amount of solid (%)	
Range of particle size, diameter (μm)	*Number of particles*	*Total mass of particles*
30–10	0.005	28
10–5	0.175	52
5–3	0.25	11
3–1	1.1	6
1–0.5	6.97	2
Below 0.5	91.5	1

areas should then be provided with a plenum between the louvre and the duct connection to the ventilation plant and the plenum designed with a water tight drained section below the level of the louvre.

To achieve the appropriate level of weather protection and to minimise resistance the velocity across the face of the louvre would not normally exceed 2.2 m/s. The free area of the louvre will depend on the particular arrangement but typically will be 40–50 per cent of the total area. BS EN 13030 provides a method for measuring the water rejection performance of louvres subjected to rain and wind pressures. For most application class B having an effectiveness of 0.99 to 0.95 will suffice.

Air filtration

Air contaminants

Atmospheric air is contaminated by a variety of particles, such as soot, ash, pollens, mould spores, fibrous materials, dust, grit and disintegrated rubber from roads, metallic dust and bacteria. The heavier particles may be such that under calm conditions they will settle out of their own volition. These are termed ‘temporary’. The smokes, fumes and lighter particulate matter remain in suspension and are termed ‘permanent’. Non-particulate contaminants, such as vapours and gases, also exist in the air. Sulphur dioxide is the most damaging, affecting building fabric, vegetation and artifacts; carbon monoxide and other oxides of nitrogen are also present but normally in small concentrations.

The unit of measurement for dust particles is the *micron* (1 mm = 1000 μm). The human hair has a diameter of about 100 μm, and the smallest particle visible to the naked eye is about 15 μm. The smallest range of particles we need consider here is of the order of 0.01–0.1 μm, which is represented by smokes of various kinds, such as tobacco smoke. The upper range of particle size we need consider is about 15 μm.

Pollution in all its forms, and especially atmospheric pollution, continues to be a subject of increasing public concern, though apart from smoke, fumes and soot it appears doubtful whether much of the other airborne dusts and dirt are susceptible of reduction. Tests are regularly carried out and records kept of suspended pollution material and of sulphur dioxide concentrations.[3]

Tables 22.2 and 22.3, extracted from the *Guide Section B2*, give typical values for solids in the air for different localities and typical analyses of dust contamination with respect to particle size. These values will vary between locations and with the season; winter conditions normally producing the highest values.

Necessity for air cleaning

If, in a mechanically ventilated or air-conditioned building, air is blown in without some means for filtration, deposits of dust will be found to occur throughout the rooms and the system of ducts will in itself become coated with solid matter. Heater batteries, cooling coils and fans will also become fouled so that in time the efficiency of the system as a whole will fall off at an increasing rate. Except in certain industrial applications, ventilation and air-conditioning systems therefore invariably include some means for filtration of the air.

The removal of the larger particles is, of course, a simple matter, since any mesh of fine enough aperture will arrest such particles. A plain mesh is however liable to become clogged very quickly, and hence is of little use for the purpose. The finer material and the smokes are, however, much more difficult to arrest and yet it is these which are largely responsible for the staining of decorations, the soiling of shirts and garments and, to some extent, also, the bearing of harmful bacteria. Apart from outside air, recirculated air carries fluff from carpets, blankets and clothes, dust from paper and brought in on shoes and, in an industrial application, any dust resulting from the process.

The greater the degree of filtration, as a rule, the higher the cost of the equipment and the greater the space occupied. The selection of the best filter for a particular application therefore depends on whether great value is placed on a high degree of cleanliness or not. It is perhaps worthy of note that the staining on ceilings close to the point of air supply to a space, seen in many installations after a period of use, is more likely to be due to contamination generated from within the space than to dirt in the outside air.

Tests for filters

The filter efficiency is a measure of its ability to remove dust from the air, expressed in terms of the contaminant concentrations upstream and downstream of the filter, thus:

$$\eta = 100\left(\frac{C_1 - C_2}{C_1}\right)$$

where

η = filter efficiency (%)
C_1 = upstream concentration
C_2 = downstream concentration

In view of the wide range of filters and their capabilities it is not possible to define their effectiveness in the same way. Consequently their relative efficiency is expressed differently for course, fine and high efficiency air filters.

In the case of *coarse filters* which are often used as pre-filters their efficiency is measured by 'arrestance' or weight and the current applicable standard is BSEN 779 – arrestance.

For *fine filters* most commonly used as the main or final filter in most general applications efficiency is measured against atmospheric dust spot efficiency and the related standard is BSEN 779 – atmospheric dust spot efficiency.

High efficiency filters are normally only used in specialist applications and the related standard is BSEN 1822 – high efficiency test which determines the 'most penetrating particle size' at which the filter media has the lowest efficiency.

Weight or gravimetric test

With this method, a carefully metered quantity of air containing a known quantity of synthetic dust is drawn through a filter paper from the unfiltered intake, and a similar quantity of air is drawn through another filter paper downstream from the filter. These are weighted on an accurate balance and a comparison of the two weights gives the gravimetric efficiency. The heavier particles, as explained, are the most easily collected and these constitute the greater part of the weight, hence even a poor filter will give a high efficiency of perhaps 90 per cent by the gravimetric method. In consequence, the weight test has effectively been superseded.

Dust spot test[4]

Using this method, sample quantities of air are drawn, as before, from upstream and downstream of the filter under test through filter papers and the resultant stains are viewed optically. The light penetration is measured by a photo-sensitive cell and a comparison of the relative intensities then provides dust spot test efficiency. This, known previously as the *blackness* test, is a much more stringent criterion and many filters which may provide a 90 per cent result by the gravimetric method are able to produce no more than 50 per cent following a dust spot test.

Arrestance tests

These establish the ability of the air cleaning medium to remove injected dusts from an air stream and may be carried out as a part of the dust spot test procedures. Arrestance is expressed as a percentage, calculated as for efficiency, but using mass in place of concentration values. Both the dust spot and arrestance tests may be used for on-site testing.

Where high filter efficiencies are vital to the installation the quality of the seal between the filter and supporting frame is as important as the efficiency of the filter media. On-site testing of such high efficiency particulate air filter (HEPA) installations is in consequence necessary before the plant is put into service and at frequent intervals during use.

Filter Categories

In selecting a filter it is necessary to know by what method the maker's guarantee of test efficiency has been determined, and what type of dust was used, since it is unlikely that a determination was made under the particular conditions of atmospheric pollution obtaining at the site in question.

The *dust holding capacity* of a filter, that is the mass of dust a filter can retain between its 'clean' and 'dirty' condition, is also an important feature to consider when selecting a filter because this will affect the frequency of maintenance or replacement.

Air filters fall into five main categories as follows:

Viscous impingement: Usually of some form of corrugated metal plates or metal coils or turnings or the like, in each case covered with a viscous oily liquid to arrest the particles on impingement. Other materials used instead of metal are glass fibres, similarly coated with a sticky fluid.

Fabric: The material used in this type of filter is some form of textile, normally glass or synthetic fibres in a random matting bonded together. The dirt particles are in this case arrested partly by being trapped in the interstices of the material, and partly by being caught on the fibres.

Electrostatic: In this system of filtration, dust particles entering the filter are subjected to an electrostatic ionising charge and, on subsequently passing through parallel plates which are alternately charged and earthed, the particles are repelled by the charge plates and adhere to the earthed plates. The electrical charge is supplied from a power-pack containing the necessary transformers and rectifiers to produce the high-voltage DC required.

Paper or absolute: This filter uses a special form of paper made usually from woven glass fibre.

Adsorption: Makes use of activated carbon (charcoal), activated alumina or other chemicals to adsorb odours, gases and the like.

Air washers will also act as air cleaning devices, but are unlikely to be suitable for use in modern systems.

Materials used should be inherently non-flammable or so treated that they retain such qualities through their life. Viscous filter liquids should have a flash point of not less than 177°C. All such material should generate a minimum quantity of smoke and toxic gas.[5]

Relative efficiencies

Figure 22.12 indicates trends of efficiencies which may be expected from the first four types of particulate filter referred to above.

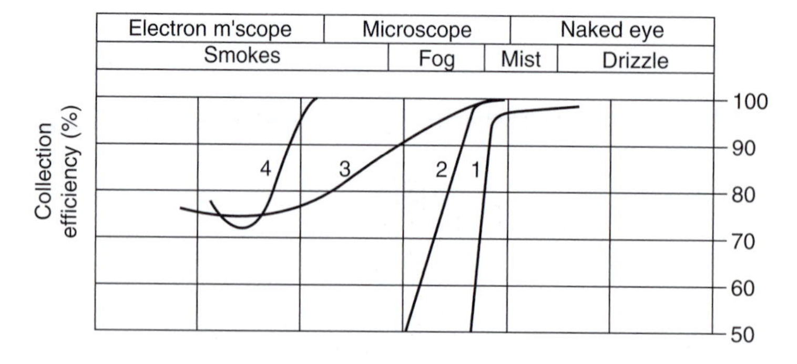

Figure 22.12 Efficiencies of air filters: (1) viscous coated metal, (2) fabric, (3) electrostatic and (4) paper absolute

Table 22.4 Typical characteristics of particulate filter media

Filter type	*Face velocity* (m/s)	*Resistance* (Pa) *Clean*	*Resistance* (Pa) *Dirty*	*Approximate efficiency/ arrestance* (%)
Viscous impingement				
Cleanable panel	1.5–2.5	20–60	100–160	65–80 arrestance
Automatic curtain	2–2.5	30–60	100–190	80 arrestance
Dry fabric or fibrous				
Cleanable panel	1.5–2	30–50	75–100	70–80 arrestance
Disposable panel	1.5–3.5	45–90	150–250	70–90 arrestance
Low efficiency bag	1.5–3.5	25–50	200–250	30–50 dust spot
Medium efficiency bag	1.5–3.5	55–140	200–350	50–90 dust spot
High efficiency bag	1.5–2.5	55–140	200–350	Up to 95 dust spot
Automatic roll	2.5–3.5	30–80	160–200	30–45 dust spot
Electrostatic, plus bag filter	1.5–2.5	120–200	250–400	Up to 95 dust spot
HEPA				
Low efficiency	Up to 1.5	up to 150	Up to 400	95 sodium flame
Medium efficiency	Up to 1.5	up to 280	Up to 625	99.7 sodium flame
High efficiency	Up to 1.5	up to 280	Up to 625	99.997 sodium flame

It will be seen that filter 1, depending on impact, has no arrestance on small particle sizes. Filter 2 has a dust spot test efficiency of about 50 per cent for particles of about 1 μm, diminishing rapidly for the smaller particles. The electrostatic filter, curve 3, has a dust spot test efficiency of about 90 per cent while the paper filter 4 achieves virtually 100 per cent for particles of 0.1 nm and above.

All filters vary in efficiency according to the velocity of air through them. Typical velocities and resistances of various filters are given in Table 22.4. Following the recently introduced requirement for SFP the face velocities should be considered with a view to reducing the pressure losses.

In the case of fabric and paper filters, owing to the large surface areas involved, designs are usually based on forming the material into a *zigzag* formation.

Filter cleaning

When heavily charged with dirt, the resistance of most filters rises sharply, thus reducing air flow, and hence ventilation rate. The cleaning of filters is achieved in a variety of ways as follows:

Washable filters: These, consisting of a foam plastic element, are contained in metal frames with wire retaining grids as shown in Fig. 22.13. The material is flame retardant and, when fouled, is washed in detergent and reused. The media have a long life.

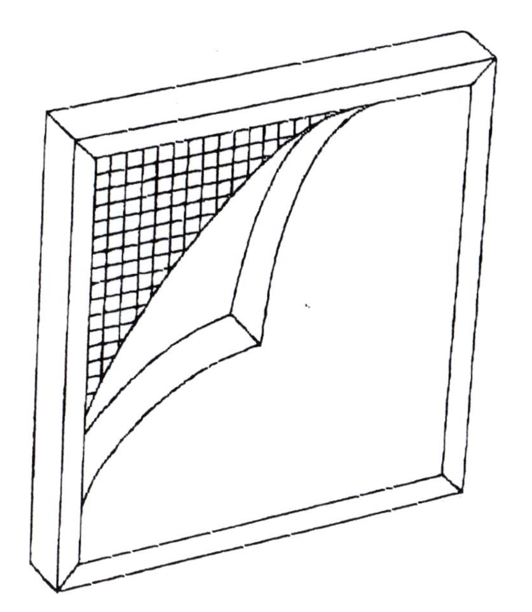

Figure 22.13 Washable filter (ACE)

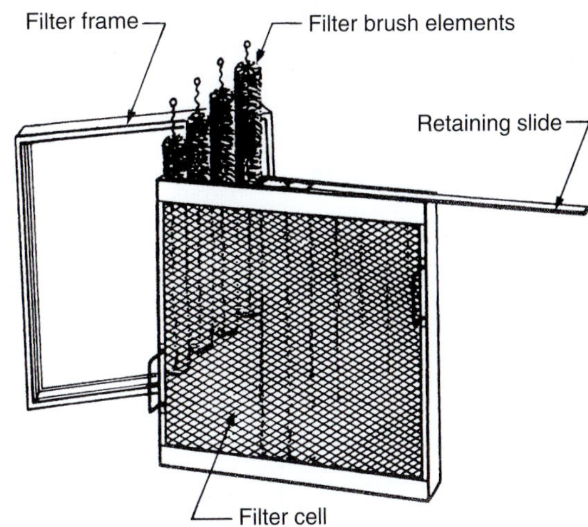

Figure 22.14 Brush filter (Universal filters)

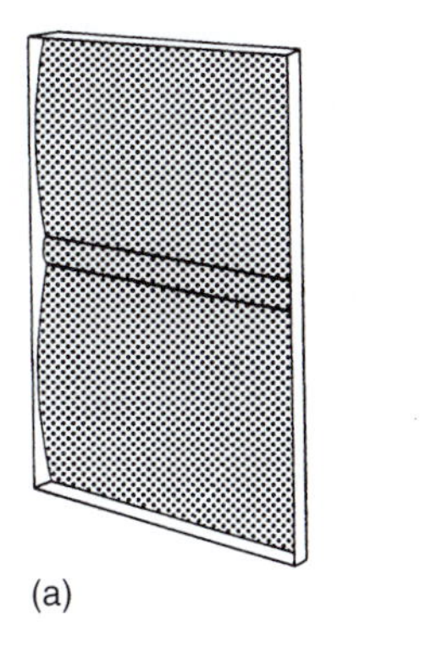

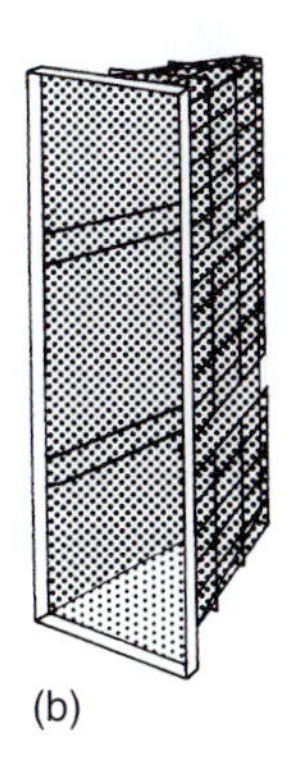

Figure 22.15 Fabric filter, single cell (Trox): (a) flat panel and (b) wedge or 'V'

Brush filters: Eliminating the necessity for any special cleaning, these consist of 'flue-brush' elements contained in a segmented frame as shown in Fig. 22.14. The element of hair, nylon, steel or nickel silver wire is removed for either vacuum or other simple cleaning and washed before replacement. The media have an indefinite life.

Fabric filters: In panel or wedge form, as in Fig. 22.15 these are mounted in frames and when dirty are thrown away. In another type, the filter material is of a glass fibre or other suitable base, and is supplied in rolls. The roll is horizontal and is gradually wound from one spool onto another on the principle of a camera film, see Fig. 22.16; this movement is achieved by motor drive controlled from a pressure differential switch across the filter, or from a timing device. Fresh filtering medium is unrolled only as it is required. The medium may pass over open mesh drums, as shown in the diagram, or be flat and held in place by retaining bars on both upstream and downstream faces. Such equipment may, alternatively, operate with the rolls vertical but the edge sealing arrangements are then less effective and the drive may be less positive due to mechanical problems.

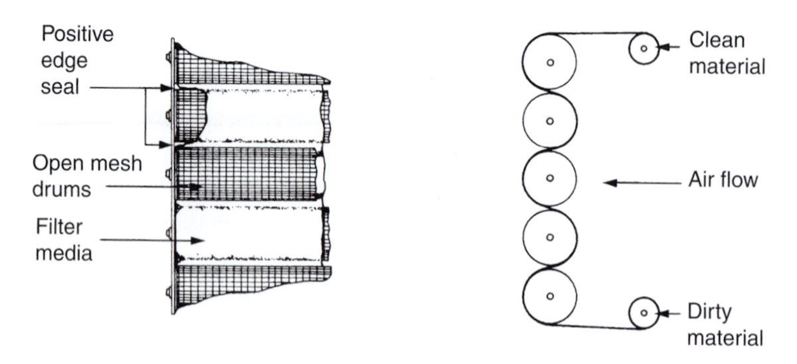

Figure 22.16 Automatic roll-type fabric filter (Ozonair)

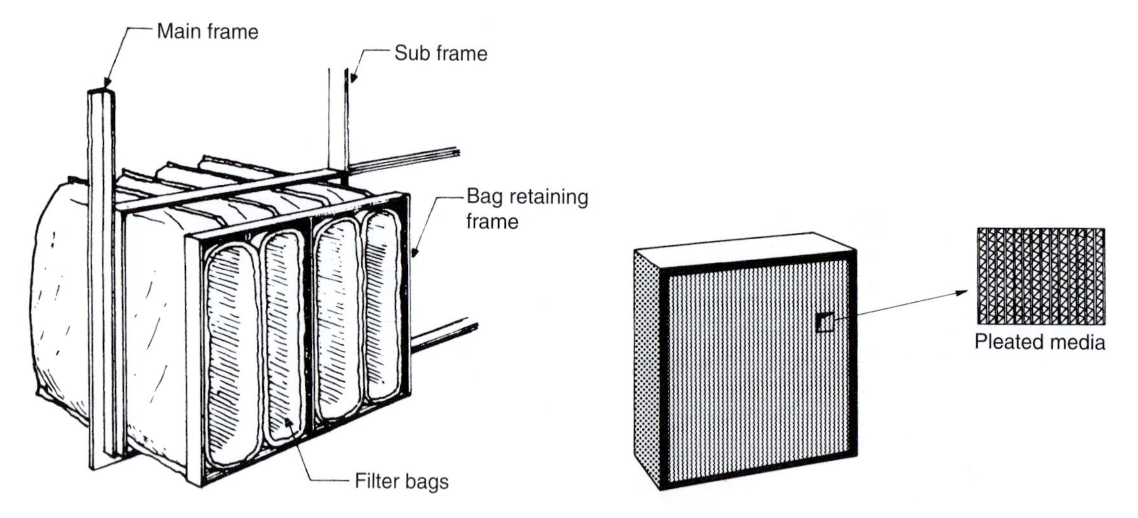

Figure 22.17 Bag-type fabric filter

Figure 22.18 HEPA filter, single cell (AAF)

This type of filter will, in theory, require no labour for maintenance, excepting at the long intervals of possibly 3–6 months for the changing of a complete roll.

Bag filters: The bag filter has a replaceable medium and this has largely taken the place of the 'roll' fabric filter since it has better performance characteristics and requires less skill in maintenance. A common arrangement is shown in Fig. 22.17.

HEPA filters: High efficiency particulate air filters are supplied in panel form and are discarded when dirty and replaced. To achieve high efficiency the air velocity through the medium is kept relatively low. It is essential that efficient pre-filters are installed upstream of HEPA filters to extend the life of the media. A typical cell is shown in Fig. 22.18. Special attention must be given to sealing the cells into the frames.

Viscous impingement filters: These may be in the form of cells comprising viscous coated corrugated plates removed for cleaning by hand, which are washed, re-oiled and replaced. They have also been developed on a crude self-cleaning principle and, in one form see Fig. 22.19, the zigzag plates are vertical, oil flowing over them from a pump drawing from the base tank at intervals. In another form the cells are on an endless chain dipping into the oil in the base tank and returning for reuse. This type of filter would normally be used where an extremely dirty atmosphere exists, such as found in a heavy industrial area.

Electrostatic filter: In this type of filter the dirt collects on the earthed plates and its removal is accomplished by washing with hot water jets; adequate drainage is therefore necessary. This may be done by hand or

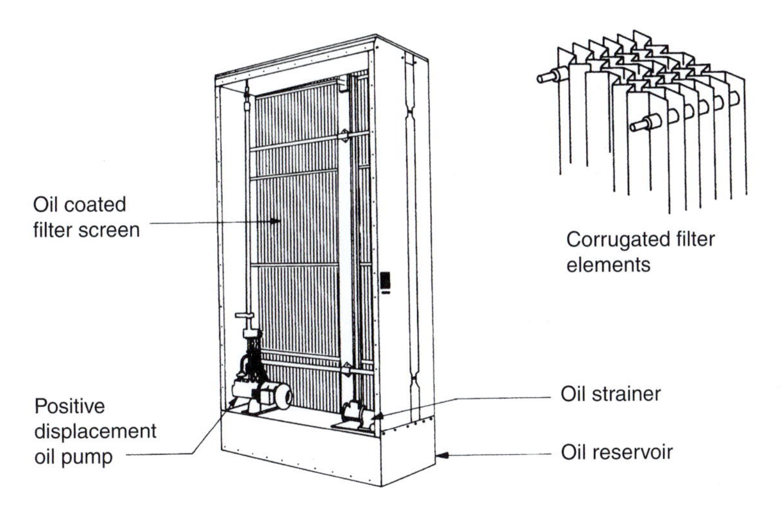

Figure 22.19 Automatic viscous roller-type filter (Ozonair)

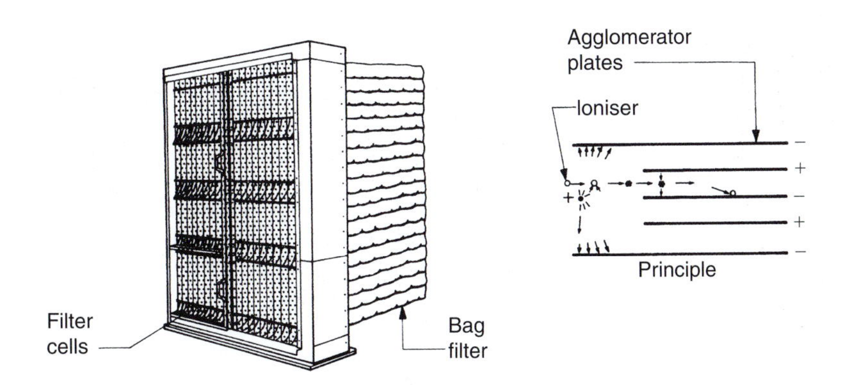

Figure 22.20 Electrostatic filter (AAF)

automatically, the latter being preferred. Another type uses oiled plates to increase the dust retention properties. After cleaning, a period of drying is necessary before the filter can be put back into use. Pre-filtering is recommended and consideration should be given to an after-filter for use when the electrostatic section is being cleaned or during failure.

Another pattern, which is more compact and requires less maintenance, allows the dust particles to collect on the plates in a thickening layer to be swept off by the air stream as agglomerated flakes on to a roll or bag filter which is an integral part of the filter assembly (see Fig. 22.20).

For clearance of fog only, the electrostatic and absolute filters are effective, or a combination of electrostatic plus fabric. Excessive collection of moisture from fog may cause an electrostatic filter to cut out due to short circuit. A special heater may be used ahead of the filter to ensure that air entering is dry, thus avoiding this trouble.

Adsorption filter: The service life of an activated carbon filter is between 6 months and 2 years, depending on the type of installation and the concentration of contaminants being adsorbed. An efficient particulate filter is recommended for use upstream. Normally, the cells are re-chargeable. Typical cell arrangements are shown in Fig. 22.21. The pressure drop across the unit is about 70 Pa and the face velocity of the order 1.5 m/s.

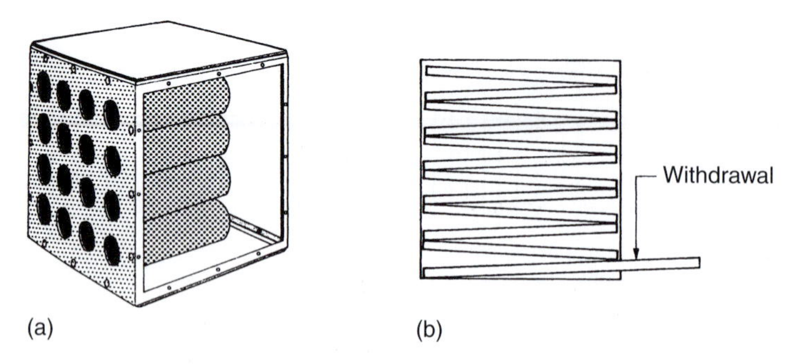

Figure 22.21 Adsorption filter, single cell: (a) cylindrical and (b) panel (section)

Table 22.5 Maximum final pressure loss (dirty filter)

Filter class	*Typical initial pressure drop* (Pa)	*Final pressure drop* (Pa)
G1–G4	70	150
F5–F7	80	200
F8–F9	120	300

It should be noted that there is potential for a significant increase in pressure loss as the filters become progressively loaded in use and the fan would normally be selected at the required volume flow against the average of the initial and final pressure losses. The recommended maximum final pressure loss values are indicated in Table 22.5.

British Standard BSEN 13779: 2004 defines SFP with clean filters. This does not represent the normal operation of the plant over time with the filters operating between the clean and dirty condition. However this situation cannot be demonstrated in a new system and as a consequence the clean (or near clean) condition is favoured in Part L of the Building Regulations.

Air filters most commonly in use today are dry fabric filters in the form of flat panel, cells and bag arrangement. The classification and efficiency of the grades of filter most commonly used are included in Table 22.6.

Air humidification

During winter months the outside air has a low moisture content and when introduced to a conditioned space tends to reduce the relative humidity. To counter this effect, it is often necessary to introduce moisture into the space to maintain the required conditions for comfort or for a process.

Air humidifiers may be classed as either direct or indirect.

Direct types: Introduce moisture into the space to be treated and their use is normally limited to industrial or horticultural applications: these types are outside the scope of this book.

Indirect types: Introduce moisture into the supply air either at the plant or within the ductwork, dependent upon the type of system.

A further sub-classification of the indirect type is *storage* or *non-storage*. The storage types incorporate a small tank under the humidifying apparatus from which water is introduced into the air stream, in one of a variety of ways. The water level in the tank is maintained by a water supply via a ball valve. With the non-storage type moisture, in the form of steam or atomised water particles, is injected directly into the air stream.

Concern as to humidifier-related illness including Legionellae has changed the approach to selection and use of the associated equipment quite radically. Storage-type units are no longer acceptable for application

Table 22.6 Filter classifications

Class	*Typical application*	*Performance*	*Standard*	*Title, category*
Coarse	Pre-filtration against insects, textile fibres, course particulates	50–80% arrestance	BSEN 779	G1, G2
	Medium level pre-filtration, protection against pollens, fan coil units, single ventilation units	80–90% arrestance	BSEN 779	G3
	High level pre-filtration	>90% arrestance	BSEN 779	G4
Fine	Supply ventilation for schools, restaurants	40–60% dust spot	BSEN 779	F5
	General laboratories, offices, computer rooms, hospital general areas	60–90% dust spot	BSEN 779	F6, F7
	Hospital general theatres, intensive care, pharmacy	90–95% dust spot	BSEN 779	F8, F9
HEPA	Filters for specific high efficiency air quality control	85% MPPS	BSEN 1822	H10
		95% MPPS	BSEN 1822	H11
		99.5% MPPS	BSEN 1822	H12
		99.95% MPPS	BSEN 1822	H13
	Ultra clean operating theatres, pharmacy aseptic suite	99.995% MPPS	BSEN 1822	H14
ULPA	Filters to provide the highest levels of air quality beyond the scope of conventional building services ventilation	99.9995% MPPS	BSEN 1822	U15
		99.99995% MPPS	BSEN 1822	U16
		99.999995% MPPS	BSEN 1822	U17

to systems in hospitals, where steam injection is recommended, and now commonly used in commercial buildings. Where storage types exist or are envisaged, particular care in cleaning, maintenance and inspection is required at frequent intervals. Further recommendations relate to the need for biocide treatment of the tank water and drainage thereof to waste, at regular intervals, preferably daily. The tank must be drained and cleaned when the humidifier is to be out of use for any extended period.[6]

The types of indirect humidifier may be summarised as follows:

- *Non-storage*
  –Steam injection
  –Mechanical separators
- *Storage*
  –Spray washers
  –Capillary washers
  –Pan humidifiers
  –Sprayed coils
  –Ultrasonic atomisers

To mitigate the risk of humidifier-related illness storage-type humidifiers are no longer used in current designs.

Steam injection

Controlled steam injection may be supplied either from a central boiler plant via a sparge pipe, Fig. 22.22, or from a local packaged unit as in Fig. 22.23. The former, unless served from an independent steam generator for the purpose, has limited use due to the odour and oil traces characteristic of central boiler installations. The latter type of equipment is mounted on or close to the air-conditioning plant and consists, in essence, of a small electrode steam generator, sized to suit the moisture requirement of the plant, controls and a water supply: steam heated units are also available. The standard equipment available is not suitable where the static

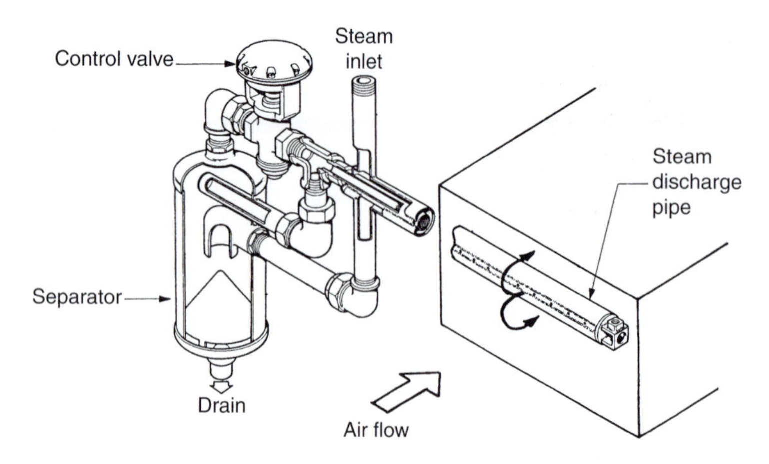

Figure 22.22 Mains steam humidifier, injection type (Spirax Sarco)

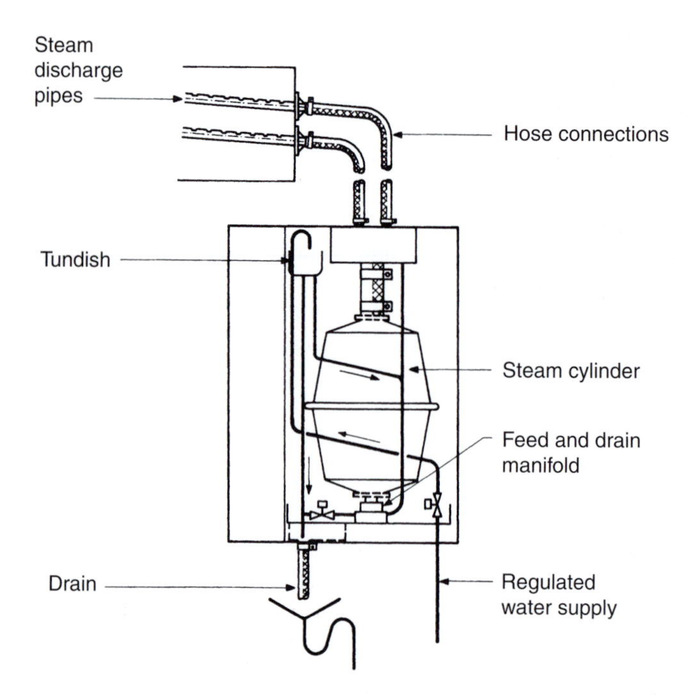

Figure 22.23 Packaged steam humidifier, injection type (Eaton Williams)

pressure in the duct is over 1250 Pa. Good access is required for all humidifiers and sparge distribution pipe-work should be arranged to be self-draining. Saturation efficiency is about 80 per cent.

Mechanical separators

These commonly operate via use of spinning discs, as illustrated in Fig. 22.24. It must be remembered that, in the process of atomisation and evaporation, any mineral salts dissolved or suspended in the water supply will be released and deposited: some water treatment may be required. Saturation efficiency is claimed to be as high as 90 per cent.

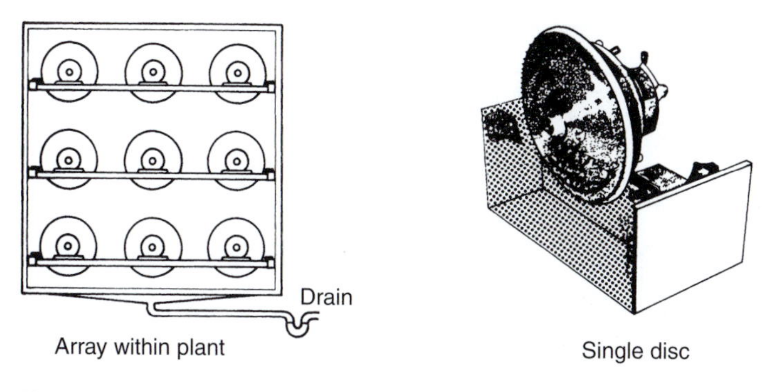

Figure 22.24 Spinning disc-type humidifier

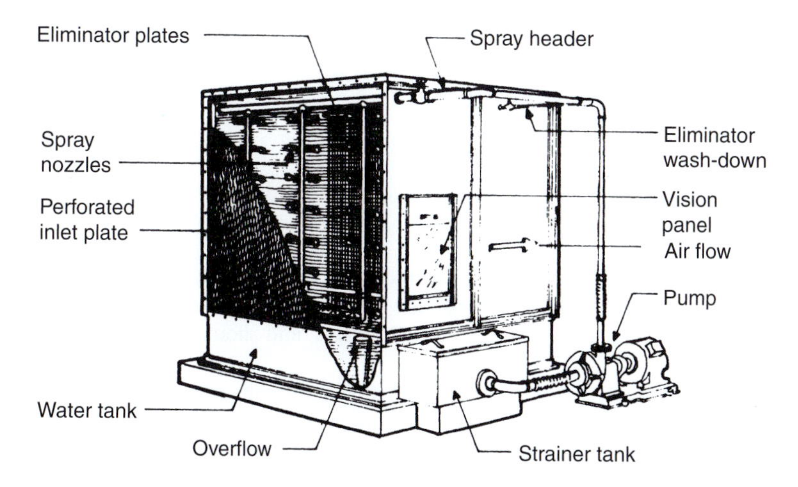

Figure 22.25 Air washer, single bank spray type

Spray washers

This type of equipment is seldom used in current designs, but since many may be still in use a full description is given, together with a figure of appropriate vintage.

An air washer, see Fig. 22.25, consists of a casing with a tank formed in the base to contain water. Spray nozzles mounted on vertical pipes connected to a header deliver water in the form of a fine mist. The spray is projected either with or against the air current, or, where two banks of sprays are provided, in both directions, one with and one against the air stream. The water for the sprays is delivered by a pump, under a gauge pressure of between 0.2 and 0.3 kPa, the water being drawn from the tank through a filter.

The casing has an access door and internal illumination, and may be of galvanised steel, or constructed of brick or concrete, asphalted inside.

As the mist-laden air is drawn through the chamber it requires to have the free moisture removed, and for this purpose eliminator plates of zigzag formation are provided to arrest the water droplets. Some makes precede these with scrubber plates, down which water is caused to run by a spray pipe at the top in order to flush down dirt which has collected from the air. Galvanised eliminator plates are liable to rapid deterioration; better materials are copper and stainless steel. Glass plates with serrated ribs have been used successfully, and they are permanent.

A single bank air washer will not saturate the air more than that corresponding to about 70 per cent of the *wet bulb depression*. A double bank washer may reach 90 per cent. The recirculated water may be cooled or

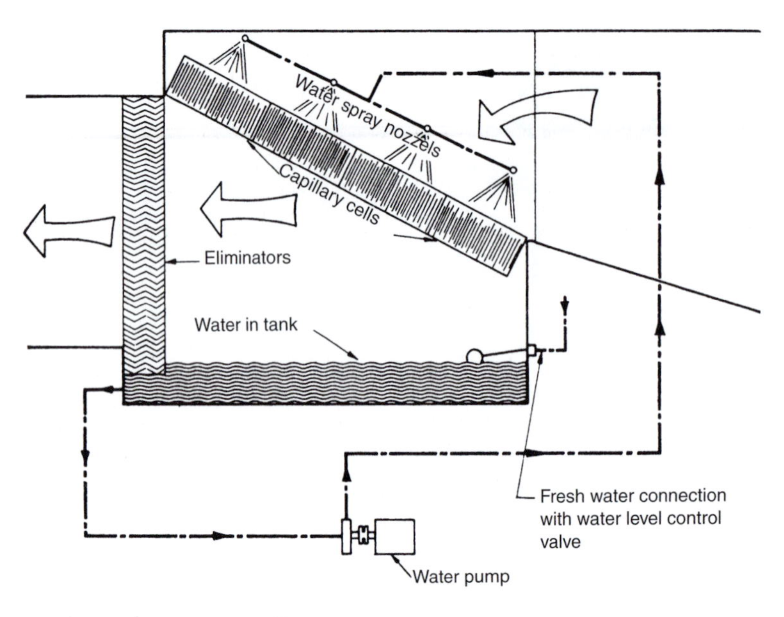

Figure 22.26 Air washer, capillary type

heated for full air-conditioning and dehumidification. Where it is desired to saturate at the dew point, a two bank washer is generally necessary with large spray nozzles to pass the necessary quantity of water for the temperature rise allowed. Spray nozzles vary in capacity from 0.05 to 0.25 litre/s, and should be easily cleanable and of non-corrodible metal.

The air velocity through an air washer is usually 2.5–3.0 m/s. The length is normally about 2.5 m, but is increased with the two or more banks of sprays needed for cooling – sometimes to as much as 4.0 m. On the inlet side straightening vanes, or a perforated grille, are required in order to distribute the air evenly over the whole area. The tank is kept filled by a ball valve, with a hand valve for quick filling, and there is in addition a drain and overflow pipe. The pump is of normal type, preferably with flexible connections where noise may cause trouble, and arranged to be flooded by the water in the tank, in order to avoid the need for priming.

Apart from removing heavier and gritty material a spray washer has a low air filtration efficiency, but is effective in adsorbing certain gases, such as sulphur dioxide, from the air.

Capillary washers

One arrangement of this type is shown in Fig. 22.26. Typically such units comprise one or a series of cells inclined at an angle, or vertical, and containing corrugated aluminium or filaments of glass or similar material which provide an extended surface for water and air contact. Water at low pressure is caused to flow over the cells by flooding nozzles from a pump of low power consumption. The water and air have to negotiate the *striations* of the fill together, and are thus intimately mixed so that saturation up to 90 per cent may be achieved. At the same time it may be taken that the filtering efficiency is of a reasonably high order down to 3 μm. It may equally be used with refrigerated water for a full air-conditioning system. A maximum velocity through the cells of 2 m/s is recommended. Although this form of humidifier is still available from manufacturers, it is not in common use today.

Pan humidifiers

A simple pan humidifier consists of a shallow water tray, replenished by a ball valve and heated to improve its effectiveness by an electric immersion element, or a piped heating supply. Efficiencies are low, and since the heated water would be at a temperature that would present a high risk of bacteriological contamination, this type of humidifier cannot be recommended. This type of equipment is not in common use.

Sprayed coils

Cooling coils fitted in air handling plant, sprayed with water from low pressure nozzles, can provide a convenient and effective means of humidification. This arrangement is described later under the heading 'Sprayed cooling coils'.

Ultrasonic atomisers

A recent development generates water vapour from a cold water source using high frequency vibrations to atomise water droplets into the air stream which are then evaporated. Electronic oscillating circuits power transducers, matched at their resonant frequency, to produce high frequency mechanical vibrations just below the water surface which cause water particles to be released from the surface. It is claimed that these humidifiers use less than 10 per cent of the power that would be absorbed by a steam raising system.

General

Indirect humidifiers using a water spray or equivalent should be fitted with eliminator plates downstream to prevent moisture carry-over into the ductwork. Treatment of water may be necessary where supplies contain a high degree of temporary hardness or calcium salts. It is worthy of re-emphasis that humidifiers having water storage are subject to risk of bacteria growth and therefore require special attention. Components should be completely drained when not in use for more than a few days. Adequate provision for access, inspection and maintenance is of paramount importance.

Currently humidification is not so widely applied as in the past. The need for the provision of humidification in most normal ventilation and air-conditioning systems needs to be justified in terms of initial and recurring energy and maintenance costs against potential benefits which may generally be restricted to artefact storage or pharmaceutical applications. Where a case can be made for humidification and to mitigate risks to health it is recommended that only steam injection be used either from local electric- or gas-fired units or from sparge pipes inserted into the plant served from a centralised steam supply.

Air heating and cooling coils

Air heating coils

When originally introduced these were commonly of plain tubing, as shown in Fig. 22.27(a). In order to economise in space and achieve greater output from a given amount of metal, finned heaters are now generally used as in Fig. 22.27(b). Construction in the latter case is commonly with copper tubes and aluminium fins but better practice requires that fins also are of copper in order to avoid corrosion. Plain tubes are less likely to become choked with dirt than finned – a fate all too common. Plain tube heaters are suitable for use in fresh air intakes to prevent wet fog collection on fabric or electrostatic filters. They should in this case be of galvanised steel or other non-corrodible metal, but not copper. Heaters are arranged in stacks or batteries with automatic controls preferably of modulating type to give a steady temperature of output.

Where the heater is warmed by hot water, a constant temperature supply is required from the boiler or calorifier with a pumped circulation. The flow required being relatively large, a control valve of diverter type will preserve the circulation irrespective of the demands of the heater. If the heater is fed by steam, it will require the usual stop valve and steam trap and a means of balancing the pressures so that the condensate may not be held up by a vacuum caused by control valve shut-off. Heaters may be enclosed in sheet-steel casings, 'packaged' with other elements into a prefabricated unit, or built into builders' work enclosures, although this final method cannot be recommended unless the builders' work is lined with sheet metal.

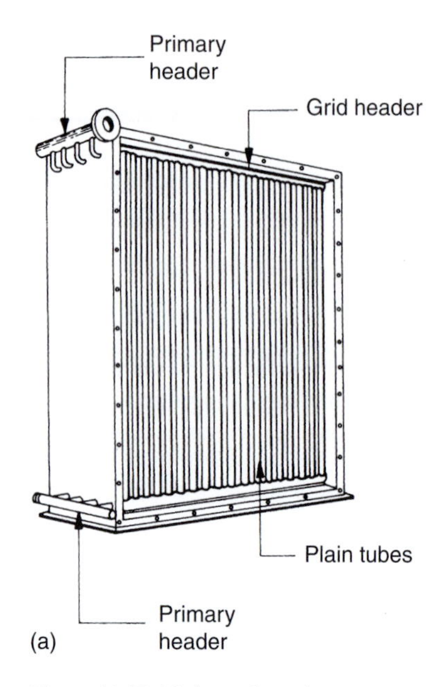

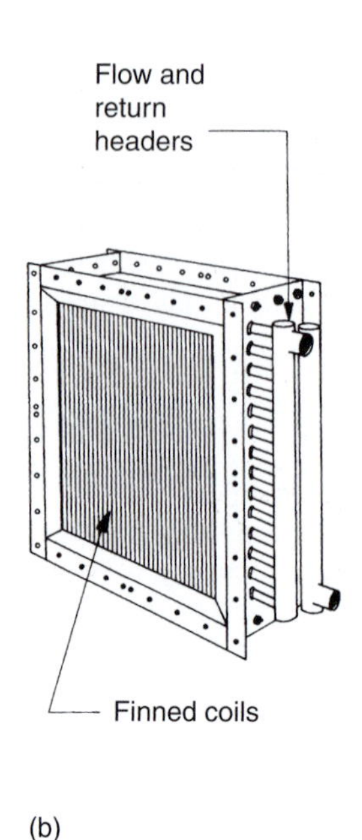

Figure 22.27 Air heater batteries

The number of rows of tubes depends on temperature rise, temperature and nature of heating medium, i.e. whether steam or hot water, and air speed. The face area depends on air volume and free area between tubes. This is again determined by the velocity, and it is usual to fix this arbitrarily beforehand, generally between 4 and 6 m/s through free area, or 2.5–3.5 m/s face velocity. For sizing of heaters reference is necessary to makers' data.

Direct electric air heaters are available and suitable for use as *trim heaters* on supplies to individual rooms where a water or steam heating media cannot be provided economically. Electric heaters are also provided on some packaged room air-conditioning units for computer rooms and the like. The case for direct electric heating normally can be made only where the annual demand is relatively small. Heat output is by any number of controlled steps.

Indirect gas-fired heaters may also be considered where close control of temperature is not important and where long distribution routes for heating water or steam would otherwise exist, leading to high initial cost and considerable distribution heat loss. A good example of their use is for serving a shopping centre. A flue to atmosphere is, of course, required.

Air cooling coils

Cooling coil surfaces as used in air-conditioning generally perform two functions – to remove sensible heat, and to remove moisture or latent heat: paradoxically, however, they may be sprayed with recirculated water and then used to add moisture or latent heat. The sizing and temperature of operation depend on the sensible: latent ratio. If a low relative humidity is required, the dew point will be low, and hence the water temperature must be low. If sensible cooling only is required, the coil surface temperature must be kept above dew point.

In any coil a certain proportion of air fails to come in contact with the cold surfaces, and thus a part may be chilled and dehumidified and a part remain unchanged. By correct selection of coil form the desired ratio may be obtained.

In direct expansion systems, the actual refrigerant gas from the compressor is passed direct into the coils, which form the evaporator of the refrigerating plant.

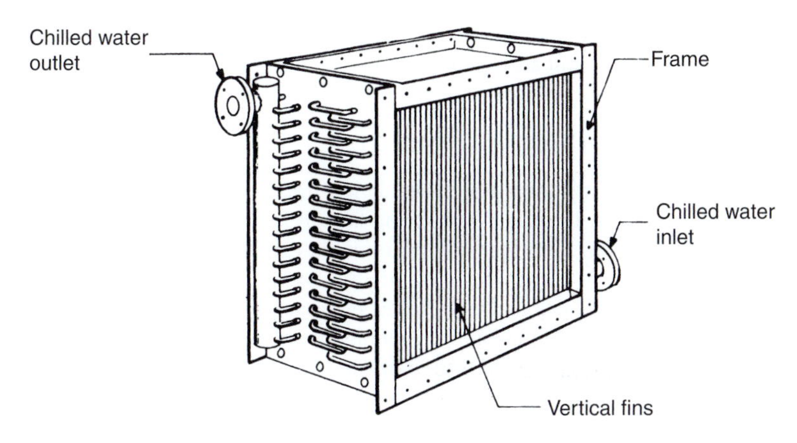

Figure 22.28 Cooling coil (six-row arrangement)

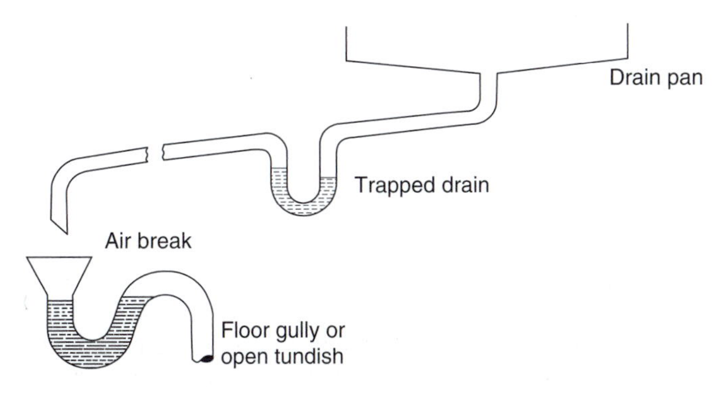

Figure 22.29 Drain arrangements from cooling coil

In a chilled water system, water is circulated through the coil, and this may be used down to about 3.3°C. Below this an anti-freezing mixture such as calcium chloride brine becomes necessary.

In cooling by refrigeration, the higher the evaporator temperature the less power is consumed for a given duty, hence it is desirable to keep the cooling coil temperature as high as possible consistent with the final air temperature required to meet design conditions. This applies whether the coil is used for direct expansion or with chilled water or brine.

A cooling coil surface is therefore usually designed for small temperature differences between water and air; for instance, for air cooled from 24°C to 13°C, water may be at 10°C inlet and 14.5°C outlet (i.e. leaving above the air-outlet temperature). An air washer cannot give a performance comparable to this. Coil face velocities are limited to 2.5 m/s; above 2.25 m/s eliminator plates are fitted to reduce moisture carry-over on coils performing a dehumidification function.

Coolers are generally of finned or block type (as in Fig. 22.28) comprising banks of small bore tubes threaded through plates between which the air passes, the whole being of copper which may be tinned after fabrication or, more normally in present day commercial practice, made up with copper tubes and aluminium fins, although the life of this form will be less than the all copper alternative. For applications where coils are to be water sprayed, as discussed later, they must of course be all copper and preferably be tinned. Coils are frequently arranged horizontally, with fins vertical, so as to facilitate drainage of condensation when dehumidifying. A drain pan should be provided and be graded towards a bottom outlet to prevent stagnant water. The connecting drain pipe should have a 'U' trap to maintain a water seal to prevent loss or ingress of air (depending whether the coil is on the suction or discharge side of the fan), see Fig. 22.29. The pipe should run to drain via an air gap. It is also a requirement that in certain applications, for example operating theatres, the drain pan is removable for cleaning purposes. Air handling unit access for drain pan removal must also be provided.

Sprayed cooling coils

In this application a cooling coil, arranged as normal for a water circulation through the tube internals, is mounted over a water tank and has headers and spray nozzles fitted, facing the upstream face. The nozzles are supplied at low pressure with water drawn by a pump from the tank and returned thereto by gravity after wetting the coil surfaces. A spray pump capacity of 0.75–1.0 litre/s per m^2 of coil face area is usual and, to make good any loss due to evaporation, the tank is provided with make-up from a mains supply via a ball valve.

In winter, during which season the water chilling plant may not be running, use of the spray will enable the coil to act as an adiabatic humidifier and in many respects this is the principal application. As a result of the small air to water temperature differences for which a cooling coil is designed, the surface area available is large and this, when wetted by the sprays, provides for a high saturation efficiency of 80–90 per cent.

With some system configurations, it may further be useful, in winter, to keep the water circulation within the coil active to cater for those circumstances when the air-on conditions are below say 5°C and air to water heat exchange takes place. As a result, a modicum of free cooling will become available in the chilled water circuit for use elsewhere in the system. For certain mid-season operations, the spray may be used to enhance the cooling and dehumidifying performance of the coil when active.

It must be emphasised that care is necessary on the part of the plant owner and operator to ensure that maintenance standards are adhered to and that the necessary precautions against biological and other contamination of the spray water, and thence the conditioned air, are taken. For this reason, sprayed coils are less popular than they were.

Air coolers dehumidifiers using desiccants

As an alternative to using refrigeration for both cooling and dehumidification, air may be dried by an *absorption* process using a liquid hygroscopic chemical, such as lithium chloride or, in the case of most air-conditioning applications, by an *adsorption* process using a solid material such as silica gel.

A process, sometimes known as desiccant cooling, involves the adsorption of water vapour into a desiccant material and for continuous operation a heat source, usually a thermal wheel, is required to remove the moisture from the desiccant material.

In the cooling application at design conditions, warm moist outside air at 28°C DB and 20°C WB is passed through the desiccant wheel leaving at approximately 60°C but at a much lower moisture content. This air is then cooled sensibly by the thermal wheel to a dry bulb temperature of approximately 23°C with the final stage of cooling of the supply air by conventional means. These systems particularly suited to displacement ventilation systems where the supply air temperature is approximately 19°C.

For the regeneration of the desiccant wheel to take place extract air returns to the plant at approximately 25°C DB having gained moisture from the occupied space. It is first passed through an evaporative humidifier to reach a condition of 17°C DB and 94 per cent RH. This air then picks up sensible heat only from the thermal wheel from which it exits at 55°C DB and 12 per cent RH. Finally this air is then reheated to approximately 65°C and 7 per cent RH before it enters the moisture laden side of the desiccant wheel. Having regenerated the desiccant material the air is discharged to atmosphere. *Guide Section B2* provides further information on desiccant cooling systems.

In winter operation the thermal wheel simply operates in reclaim mode using the extract air to heat the supply air to the desired temperature.

Specialised packaged plant is manufactured for the process and whilst use has been limited historically to industrial process applications, the system described above has been used increasingly for comfort air-conditioning applications. Examples of office buildings exist where the combination of the technologies of desiccant treatment with other low energy cooling or heating techniques has produced some interesting highly energy efficient solutions.

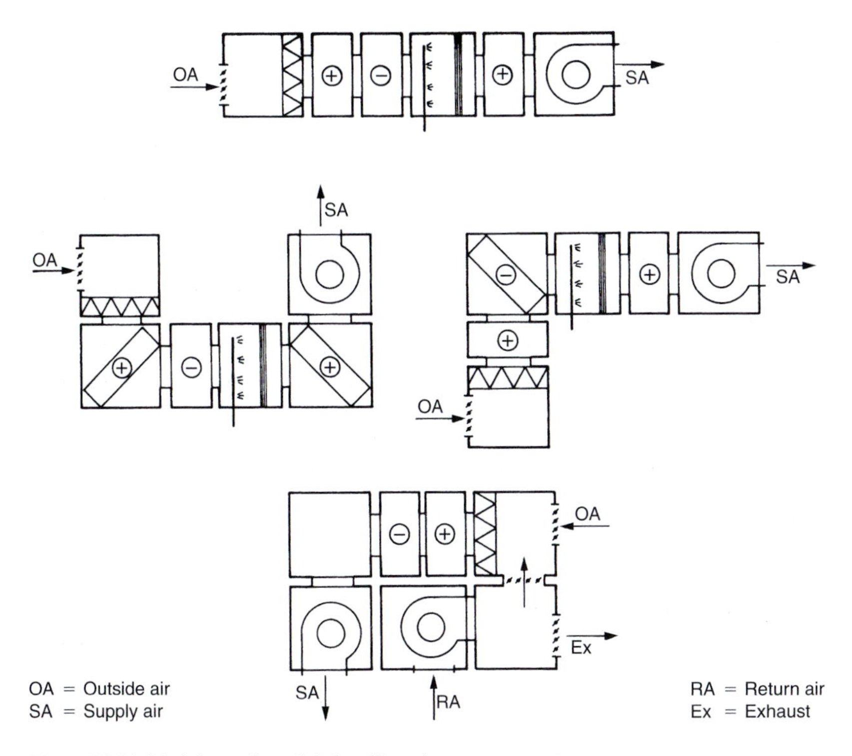

Figure 22.30 Modular packaged air handling plant arrangements

Packaged air handling plant

The various components previously described may be built up to provide a complete plant in a number of ways. They may be connected using sheet metal ducts or by incorporation within a masonry chamber: the former method tends to be untidy and space consuming and the latter is somewhat cumbersome and suspect as to permanent airtightness unless lined with sheet metal.

Current practice favours the use of factory packaged plant where individual components are housed in modular casings for site assembly. The better designs for such modules are so arranged, as shown in Fig. 22.30, that they may be assembled in a number of different ways and thus fitted to the building space available.

All the arrangements in Fig. 22.30 indicate the supply fan at the discharge of the unit creating 'draw through' solutions. Whilst this arrangement is the most common in applications where high levels of filtration are required there can be benefit in placing the fan closer to the air intake, creating a 'blow through' solution with air handling unit under positive pressure.

A recent and logical extension of this principle has been the production of these modules in weather protected form such that they, themselves, are plant rooms for, say, roof mounting without the need for architectural enclosure.

In addition to the components shown, packaged equipment is available with air-to-air heat recovery devices built in, and with packaged air-cooled refrigeration plant to serve direct expansion cooling coils (these are sometimes operated as heat pumps).

Adequate access for maintenance must be provided between the component elements of the plant; this is particularly important where the process uses water, such as in humidifiers, or where dehumidification may occur on cooling coils. Space must be allowed along the length of the plant to enable control sensing devices to be fitted and sufficient space alongside the enclosure for components to be withdrawn; the width for withdrawal being equivalent to the width of the plant itself unless provision is made for components to be removed in sections.

Substantial heat loss from the building can be prevented by the use of automatic dampers behind the intake and exhaust air louvres which close when the ventilation plant shuts down.

Heat recovery systems

It is now a requirement to meet carbon emission targets required by Part L and to achieve these it may be necessary that heat recovery systems are employed on ventilation systems to minimise the energy required to raise or lower the temperature of the incoming outside air. Care needs to be taken since manufacturers' efficiency figures for the various types of equipment can be misleading or ambitious on occasion. The potential for energy recovery may be reduced where frost coils are used upstream of the pre-filters to prevent them freezing and the consequences of omitting them need to be considered.

Heat recovery systems should be designed to achieve increasing levels of efficiency according to the capacity and annual hours run of the system. BSEN 13053: 2006, Ventilation for Buildings, air handling units, rating and performance etc., provides further information indicating that where high levels of recovery are required this may only be possible by air-to-air recovery devices and recognises that additional allowance for pressure drop are associated with this method of heat transfer.

Air-to-air heat exchangers

Any ventilation or air-conditioning system which takes in outside air, heats and/or cools it and then discharges an equivalent or lesser quantity to waste, offers potential for energy saving. The simplest of plenum ventilation plants, arranged to recirculate 60 or 70 per cent of the air handled, will nevertheless require a heat source to raise the temperature of the remaining outside air supply to whatever level the application may demand. At the other extreme, a sophisticated air-conditioning plant will consume energy in preheaters, reheaters and zone heaters plus that which will be required as a result of an adiabatic or similar humidification process. In either case, some proportion of the treated air delivered to the building will, by design, quite properly be discarded.

It has been emphasised in earlier chapters that the admission of certain minimum quantities of outside air is necessary for human occupancy and prevention of condensation. For commercial and industrial premises where noxious fumes are generated, in however low a concentration, additional outside air quantities must be supplied beyond the minimum, equivalent to the volume collected for discharge to atmosphere.

To bring the quantities of heat so wasted into perspective, consider a small office block with a floor area of 100 m^2 which has a mechanical ventilation plant to provide a low average quantity of outside air at 2.5 litre/s per m^2, the plant running for 60 hours per week. Over an average winter season, a heat supply of about 280 GJ (equivalent to that provided by burning about 8 tonnes of oil) would be necessary to do no more than raise the ventilation air supply to a degree or so *less* than room temperature. The associated extract ventilation plant would then reject a similar air quantity, at room temperature or warmer, back to outside.

In order to overcome the nuisance of heat gain to rooms from lighting, modern practice allows extracted air to pass over and through luminaires, taking with it up to 70 per cent or so of the electrical input thereto. In our hypothetical example, therefore, the temperature of the air discharged might well be several degrees *above* that held in the office space proper – say 22°C or 23°C.

For applications such as hospitals and similar buildings, which ventilate or air-condition without recirculation, in order to avoid contamination, and swimming pools which may be similarly served in order to reduce risk of condensation, the air quantity rejected is far greater per m^2 of floor area. Further, in kitchens and industrial premises, where process heat gain may be high, air will be exhausted at temperatures much greater than those quoted above.

It will be obvious that great scope for energy conservation exists if the heat in exhaust air can be reclaimed and applied, in part at least, as a source of energy to raise the temperature of the outside air used in the parallel supply plants. These same comments apply of course to economies to be achieved in cooling capacity during the summer months since the temperature of air discharged from an air-conditioned building may then be *less* than that of the outside ambient: it may, furthermore, carry less moisture. This aspect assumes more importance in climates which produce extremes of summer temperature.

Available equipment

A variety of types of equipment is available for air-to-air exchange in ventilation or air-conditioning plants. These fall under one of the following headings:

(1) Plate heat exchangers
(2) Glass tube heat exchangers
(3) Thermal wheels
(4) 'Heat pipe' heat exchangers
(5) Run-around coils with water circulation
(6) Run-around coils using refrigeration.

Of these six types, the first three effect heat exchange directly from air-to-air whereas the remainder employ an intermediate circulating medium.

Efficiency of heat reclaim

Before describing the various types of equipment in more detail, the matter of their efficiency in operation requires definition. The data normally quoted derive *temperature* efficiency from the expression:

$$\eta = \left(\frac{t_3 - t_1}{t_2 - t_1}\right)100$$

where the supply and exhaust flow rates are equal, and t_1 is the temperature of the outside air, t_2 is the temperature of the exhaust air from which heat is to be reclaimed, t_3 is the temperature of the supply air after it has been passed through the heat exchanger and η is the efficiency (%). As will be noted, the expression makes use of dry bulb temperatures and thus, strictly speaking, relates only to sensible heat recovery.

Consider the case of a heat exchanger unit, having a quoted efficiency of 76 per cent, which is applied to a winter reclaim situation handling equal quantities of outside air at −1°C, saturated, and exhaust air at 22°C DB, 15.5°C WB.

Thus,

$$t_3 = [0.76(22 + 1)] - 1 = 16.5°\text{C}$$

For use in summer, when the stated efficiency might have fallen to, say, 66 per cent[7] taking the same exhaust condition and the outside air supply at 28°C DB, 20°C WB, then:

$$t_3 = 28 - [0.66(28 - 22)] = 24°\text{C}$$

If these data are plotted on a psychrometric chart as in Fig. 22..31(a), the various significant properties may be read as summarised in Table 22.7.

From the expression for efficiency given above and replacing temperatures by the respective enthalpy values the *enthalpy efficiency* may be established. Using the same example the expression may be rewritten in terms of enthalpy thus:

$$\text{Winter} \quad 100\left(\frac{S_1 - O_1}{R - O_1}\right) = 100\left(\frac{25.3 - 7.7}{43.4 - 7.7}\right) = 49.2 \text{ per cent}$$

$$\text{Summer} \quad 100\left(\frac{O_2 - S_2}{O_2 - R}\right) = 100\left(\frac{57.0 - 52.9}{57.0 - 43.4}\right) = 30.0 \text{ per cent}$$

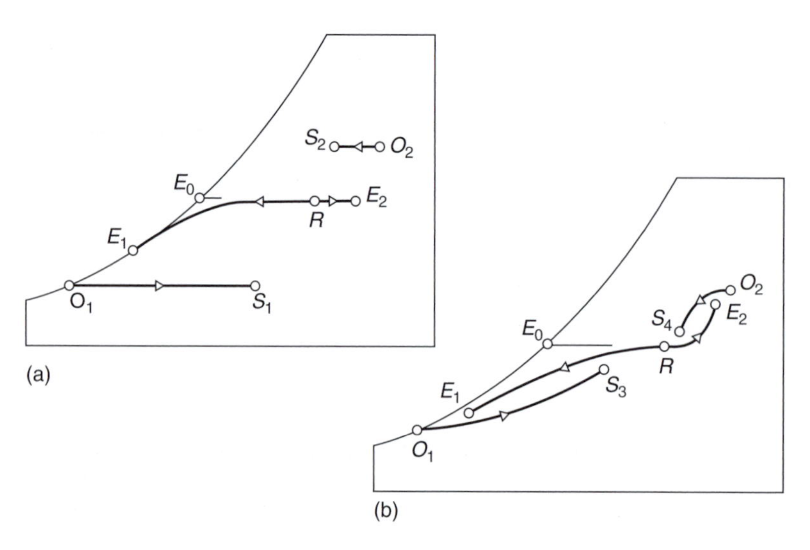

Figure 22.31 Performance of an air-to-air heat exchanger

Table 22.7 Physical quantities related to state points indicated in Fig. 22.31

Position from Fig. 22.31			*Per kg dry air*	
Point	*Statement*	*DB temperature* (°C)	*Enthalpy* (kJ/kg)	*Moisture* (kg/kg)
O_1	Winter, outside	−1	7.7	0.0035
O_2	Summer, outside	28	57.0	0.0113
R	Room exhaust	22	43.4	0.0084
S_1	Unit output, winter	16.5	25.3	0.0035
S_2	Unit output, summer	24	52.9	0.0113
S_3	Unit output, winter	16.5	34.8	0.0072
S_4	Unit output, summer	23.5	46.7	0.0090

Since temperature and enthalpy-based calculations may produce results which are significantly different in terms of efficiency, it is important to examine critically any data produced by manufacturers in this respect. Where a heat exchanger has equal ability to recover both sensible and latent heat, as illustrated in Fig. 22.31(b), the efficiencies in enthalpy terms become:

$$\text{Winter} \quad 100\left(\frac{S_3 - O_1}{R - O_1}\right) = 100\left(\frac{34.8 - 7.7}{43.4 - 7.7}\right) = 76 \text{ per cent}$$

$$\text{Summer} \quad 100\left(\frac{O_2 - S_4}{O_2 - R}\right) = 100\left(\frac{57.0 - 46.7}{57.0 - 43.4}\right) = 76 \text{ per cent}$$

Some types of equipment have different characteristics with respect to sensible and latent heat exchange. Taking the previous example and allowing for latent transfer to be at 50 per cent efficiency, the winter and summer results in term of enthalpy would be 67 and 58 per cent, respectively.

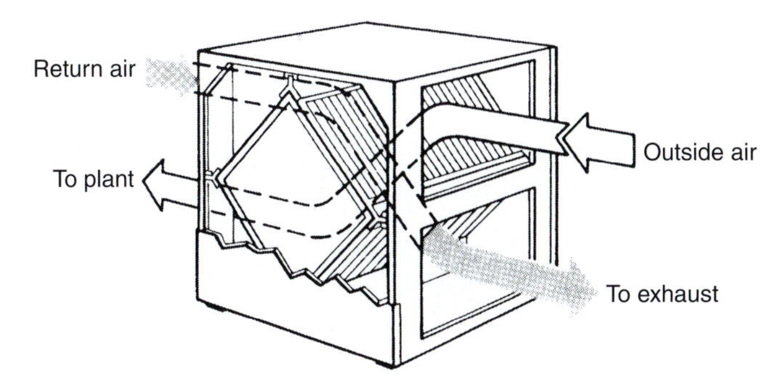

Figure 22.32 Plate-type heat exchanger

Where supply and exhaust air quantities are not equal, the notional efficiency will change *pro rata* to the mass ratio. That is, if the supply air quantity were twice that of the exhaust air, the efficiency would be about halved; in the converse case, the efficiency would increase by about 25 per cent; precise figures should be obtained from the manufacturer.

Plate-type heat exchangers

Having no moving parts, this is probably the most simple type of equipment. The two air streams are directed in cross- or counterflow through a casing which is compartmented to form narrow passages carrying, alternately, exhaust and supply air. Energy is transferred by conduction through the separating plates and contamination of one air stream by the other thus avoided. The form of the casing is arranged to suit the configuration of the transfer surface and to provide for convenience of air duct connections; one example is shown in Fig. 22.32. Condensation may occur in the return air passageways and drains are therefore required.

Since the separating plates are normally of metal (aluminium or stainless steel) moisture transfer is not possible and sensible heat only is exchanged. An epoxy or vinyl coating may be applied to aluminium plates for use in mildly corrosive atmospheres such as swimming pools.

Units are available to handle air quantities in the range 60–24 000 litre/s and may be built up in modular fashion to suit individual requirements. Due to the relatively low rate of heat transfer per unit area, the plate surface necessary is large but unit sizes are reasonably compact since air flow passages are kept to minimum width. Temperature efficiencies in the range of 50–80 per cent are claimed, and resistance to air flow is 140–300 Pa at a face velocity of 3 m/s.

This type of heat exchanger offers no method of control and therefore a bypass section may be needed to avoid, for example, heating the outside air above the required supply temperature only to have to cool it down again.

Glass tube heat exchangers

The operation is similar to that of the plate heat exchanger; normally the 'clean' supply air would be passed through the tubes and the 'contaminated' exhaust around them to allow for easier cleaning. These units are particularly suited to handling corrosive fumes from laboratories, metal treatment shops, fume cupboards, and for swimming pools.

Units may be obtained to handle up to 16 000 litre/s and with a temperature efficiency of up to 80 per cent. Pressure drop would be of the order of 250 Pa.

Thermal wheels

Constructed on the lines illustrated in Fig. 22.33 the *thermal wheel* or *regenerative* heat exchanger consists of a shallow drum containing appropriate packing which is arranged to rotate slowly between two axial air

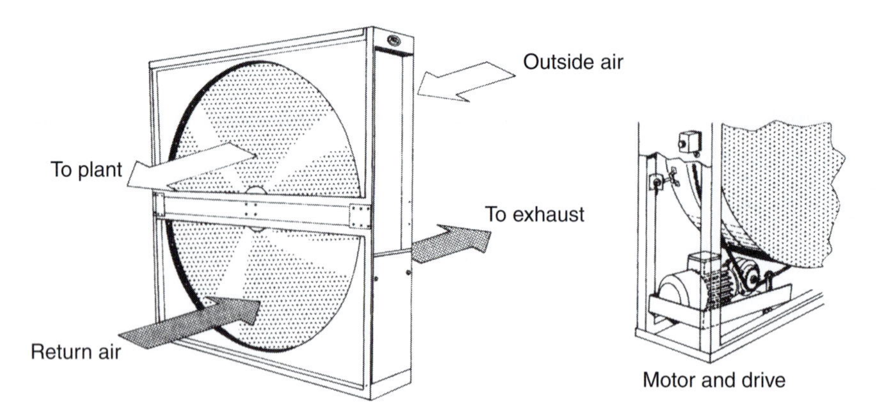

Figure 22.33 Thermal wheel-type heat exchanger

streams, transferring energy between the two. The wheel is mounted in a supporting structure and motor driven at approximately 20 rev/min: the speed may be varied as a means of controlling output.

The media and the form of the heat transfer surface vary as between manufacturers and determine the characteristics of the energy transfer. Sensible heat transfer is obtained from media formed by alternate flat and corrugated metal sheets of aluminium or stainless steel. As for plate exchangers, a protective coating may be applied for use in swimming pool applications and the like. To achieve both latent and sensible heat transfer a corrugated inorganic hygroscopic material may be used; typically these would be produced by an etching process or a lightweight coating of a hygroscopic salt.

Cross-contamination between the two air streams is minimised by so arranging the respective fan positions that the supply air pressure at the recuperator is greater than that of the exhaust stream. By using suitable labyrinth seals and incorporating a *purge sector* which allows for the matrix to be scavanged before supply air passes to the building, it is claimed that contamination is kept to less than 0.1 per cent. *Lithium bromide* as used for treatment of the hygroscopic-type matrix is stated to be *bacteriostatic*, i.e. it inhibits the propagation of bacteria. It should be noted, however, that the hygroscopic material may absorb toxic gases, or similar vapours, from the exhaust air which would not be completely removed by purging and hence present a risk of contaminating the incoming air.

Wheels are available in sizes up to about 5.5 m diameter to handle air quantities in the range 300–30 000 litre/s but multiple units in the middle of the size range are often more convenient to handle large air quantities. Efficiency may be as high as 85 per cent in sensible heat reclaim and up to 88 per cent is claimed by some manufacturers for transfer of total heat in hygroscopic types; however, efficiencies higher than 85 per cent should be viewed with caution. Resistance to air flow at a face velocity of 3 m/s will be about 150 Pa. The power required to rotate the wheel is quite small being between 60 and 1100 W.

'Heat pipe' heat exchanger

As in the case of the plate type, heat pipe units have no moving parts and are simple in concept. A working fluid is however employed to effect heat transfer. Construction consists of a box enclosure having a dividing partition to separate the supply and exhaust air streams, through which an array of finned heat pipes is assembled.

The 'heat pipe' itself is a by-product of nuclear research developed further in connection with the space programme: in essence it is no more than a super-conductor of sensible heat. Each individual conductor is a sealed tube, pressure and vacuum tight, provided with an internal wick of woven glass fibre normally as a concentric lining to the tube. During manufacture, a working fluid is introduced in sufficient quantity to saturate the wick. The actual fluid used is selected to suit the temperature range required and would typically be one of the common refrigerants.

In operation, heat applied to one end of the pipe will cause the liquid to evaporate and the resultant vapour will travel to the 'cool' end where it will condense, surrendering energy, and the liquid will return through

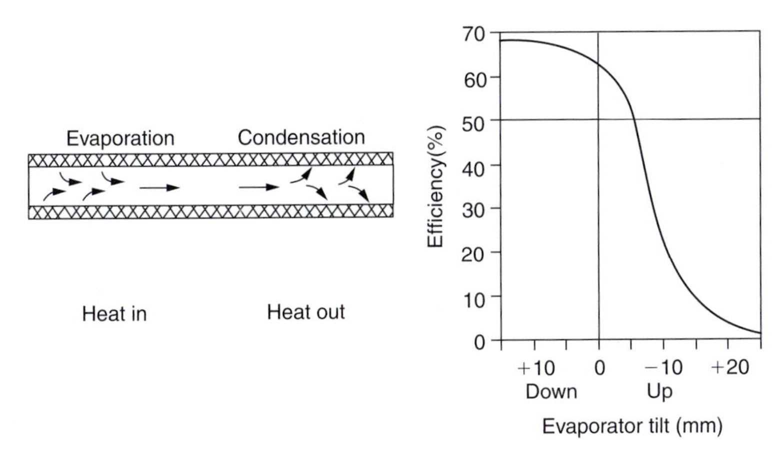

Figure 22.34 Heat pipe orientation and performance (in counterflow)

the wick by capillary action to the 'hot' end. Figure 22.34 illustrates this process and shows also how the heat transfer capacity of a pipe may be adjusted by varying the angle to the horizontal at which it lies. This characteristic may be used to match capacity to a given application or, by automation, to provide a means of control. Where the angle of tilt is used in this way, a facility must be provided to reverse the action when the season changes, winter to summer.

An alternative arrangement is where the pipes are installed vertically to transfer heat from a warm lower duct to a cool duct above. With this configuration, movement of the heat transfer fluid is by phase change; liquid in the lower section absorbs heat and changes to a gas, which condenses, releasing heat in the upper section, causing the liquid to drop to the lower end. Vertical units will not function where the cool duct is below the warmer one.

The capacity of a built-up heat exchanger of given overall dimensions will vary according to the number of rows of heat pipes, the fin spacing and the air velocity. Typically, a six-row unit having 55 fins per 100 mm would, for equal supply and exhaust air quantities, have an efficiency for sensible heat exchange of up to 80 per cent at a face velocity of 3 m/s and with a resistance to air flow of 200 Pa. Module sizes range from 150 to 36 000 litre/s. Efficiency of heat exchange is dependent upon the relative direction of air flow in the two ducts. Counterflow gives the higher performance, which is the basis for most published data; parallel air flow will reduce efficiency by about one-fifth of the quoted percentage. Subject to the effectiveness of the division plate and seals between the two air streams there should be no cross-contamination between supply and extract air. This method of heat exchange is seldom used due to the relatively high cost.

Run-around coils (water circulation)

This approach to the problem has the merit of extreme flexibility and is, moreover, founded upon a well understood technology. As shown in Fig. 22.35, the basis of the system is a pair of conventional finned tube heating/cooling coils, one fitted in each air stream, connected by a pipework system for pumped circulation of the working fluid, often a 25 per cent solution *of glycol anti-freeze* in water. Table 22.8 gives the freezing point and specific heat capacity of water and ethylene glycol solutions in concentrations 0–40 per cent glycol by mass. The specific heat capacity of the solution affects the efficiency of heat transfer; a reduction in specific heat capacity giving lower efficiency.

The flexibility of the system derives from the obvious ease by which the coils may be connected together; there is no need to disturb the routes of what may be large air ducts to bring them, inlet and outlet for both supply and exhaust, to the heat exchanger. Furthermore, coils may be fitted to any number of exhaust ducts and the heat therefrom collected and distributed to any number of similar coils fitted to supply air ducts. Diversity of energy availability and energy demand between air handling plants may thus be used to best advantage.

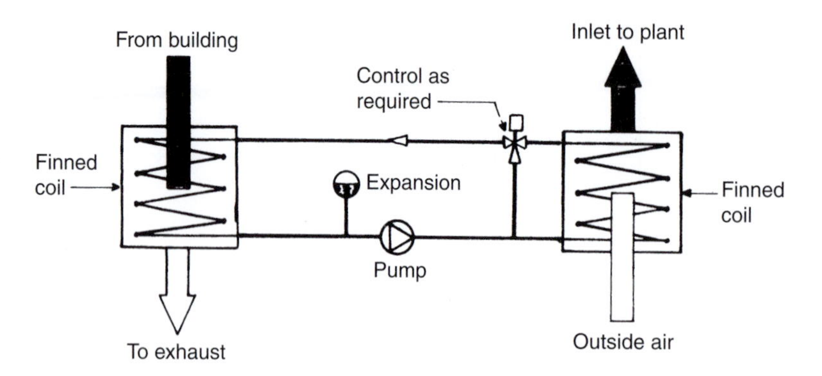

Figure 22.35 Run-around coils with water circulation

Table 22.8 Properties of water/ethylene glycol solutions

Glycol concentration (%)	*Freezing point* (°C)	*Specific heat capacity* (at 10°C) (kJ/kg K)
0	0	4.2
10	−5	4.1
20	−10	3.9
25	−13	3.8
30	−16	3.7
40	−25	3.5

There are, of course, compensating disadvantages, those of most consequence being the need for double heat exchange (exhaust air to fluid and fluid to supply air), the relatively small temperature differentials available for such energy transfer, the need for water pumping power and the matter of heat loss and gain from and to the pipework system.

Direct transfer of latent heat is not possible with this system, but in winter the coil in the exhaust air stream would run wet as would that in the supply air stream during some summer conditions: energy transfer would thus be assisted and efficiency improved as in the case of plate-type heat exchangers. In the context of what has been said before, however, heat transfer would be sensible only.

Little more needs to be added with regard to this type other than to emphasise that the small temperature differentials between either air stream and the working fluid will result in deep coils (typically 6–8 rows) and high resistance to air flow with consequent penalties in fan power requirement. The overall efficiency, ignoring fan and pump power, some of which will be recovered in winter but will be a penalty in summer, is not likely to be more than 40–65 per cent at best.

Run-around coils (using refrigeration)

If one considers an exhaust and a supply air duct, separate but not too distant, it is obvious that the evaporation and condensing elements of a refrigeration plant could be fitted within the respective air streams, Fig. 22.36. By such means, one of the disadvantages of a water circulating system, i.e. small temperature differentials, could be overcome: in fact, using this 'heat pump' principle (see Chapter 24), the supply air temperature may be raised above that of the exhaust air. The energy required to drive the compressor imposes a penalty but much of this would be recovered as heat to the supply air stream.

Many types of packaged air-conditioning plant designed for roof mounting incorporate not only supply and exhaust fans but also air-cooled refrigeration equipment for summer use. In some cases facilities are provided whereby air paths may be redirected during the winter and some part of the refrigeration capacity used as a heat pump drawing energy from exhaust air.

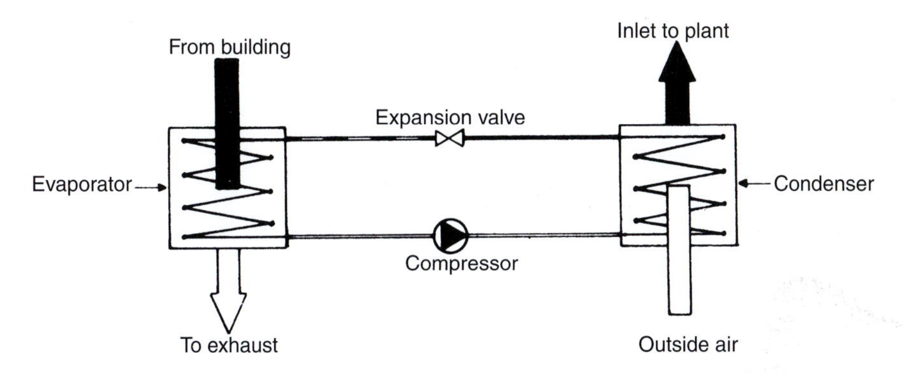

Figure 22.36 Run-around coils using a refrigeration cycle

Systems compared

The principal disadvantages of any air-to-air heat recovery system which does not make use of an intermediate fluid for heat transfer is, as previously explained, that two air ducts which may be quite large must be brought together at the heat exchanger and then the intake and exhaust air connections separated at the building perimeter to avoid the risk of recirculation. Other problems occur also in analysis but these are amenable to technical rather than spatial solution:

- Any heat exchange element which may at some periods of the year operate with wetted surfaces must be provided with a condensation collection tray and drain facilities. It is important that the configuration of the exchange surfaces permits flow and collection of moisture.
- In extreme weather conditions, a small amount of preheat may be required to prevent freezing of condensed moisture since this could lead to damage to the equipment and an unacceptable resistance to air flow. Preheating may also be desirable for thermal wheel installations to avoid excessive moisture exchange in winter.
- Heat exchangers which have small passages presented to air flow will soon become clogged with dirt unless pre-filters are provided: wash down may be required at intervals. Some manufacturers of thermal wheels claim that the reversal of air flow which occurs in normal operation acts to maintain cleanliness.

In considering the relative economics of alternative methods, account must be taken of annual mean rather than peak load efficiencies and of available means for control of heat exchange, since under certain outside conditions which normally occur during mid-season, the maximum rate of heat transfer available may increase, not decrease, energy consumption. As has been mentioned, the performance of a thermal wheel may be varied by speed change and that of a heat pipe unit by automation of the angle of tilt. The water circuit of run-around coils may be fitted with motorised valves as required, but for plate heat exchangers an arrangement of face and bypass dampers will be necessary.

Resistance to air flow and consequent increases in fan power must be considered as must the energy consumption by auxiliaries, drive motors, pumps, etc. These, of course, are likely to remain running even when the enthalpy of the supply and exhaust air streams is so close as to lead to minimal interchange. Table 22.9 presents a summary comparing the performance of the various types of equipment based upon the design conditions listed in the first three lines of Table 22.7 with supply and exhaust air quantities equal at a rate of 5000 litre/s.

Unitary equipment

The previously described equipment is generally associated with central plant serving an entire building or zone within a building.

Table 22.9 Comparative performance of various types of air-to-air heat exchanger

		Energy reclaim efficiency (%)			
		Temperature		*Enthalpy*	
Type of equipment	*Make*	*Winter*	*Summer*	*Winter*	*Summer*
Parallel plate metal	A	62	62	40	26.5
	B	65	59	41	26
	C	57	53	37	20
Glass tube	C	61	57	40	23
Thermal wheel	A	74	74	48.5	30
Non-hygroscopic	B	75	75	72	44
	C	74	74	52	37
Hygroscopic	A+C	74	74	74	74
	B	70	70	70	70
Heat pipe	C	64	64	43	29
Run-around coil water	B	67	61	–	–
Water/glycol (25%)	B	64	60	–	–
Water/glycol (25%)	C	55	50	35	30

Note
For the winter and summer design conditions listed in Table 22.7. Supply and exhaust air quantity = 5000 litre/s.

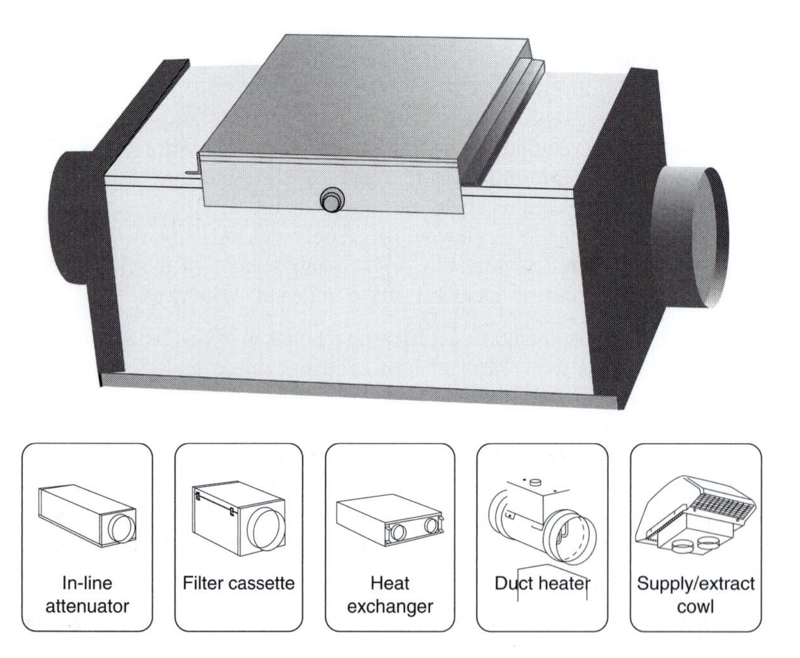

Figure 22.37 Cased fan unit and accessories (Nuaire)

There are applications where the provision of main plant is not possible or unnecessary requiring the ventilation equipment to be within or local to the area to be served. Unitary equipment is available which is designed for mounting above false ceilings.

- Figure 22.37 shows a typical cased unit with supply or extract fan with a range of optional accessories including air filter, heater battery and sound attenuator.

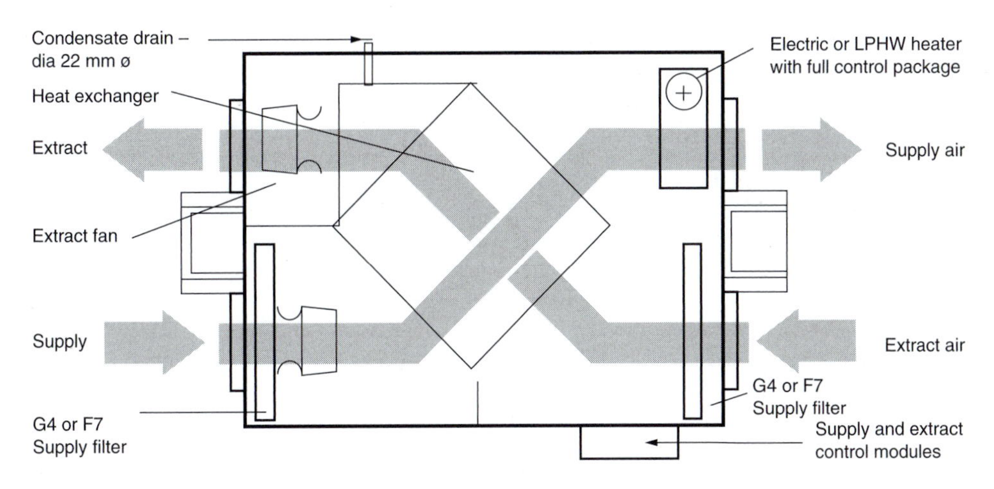

Figure 22.38 Typical ventilation unit with heat recovery (Nuaire)

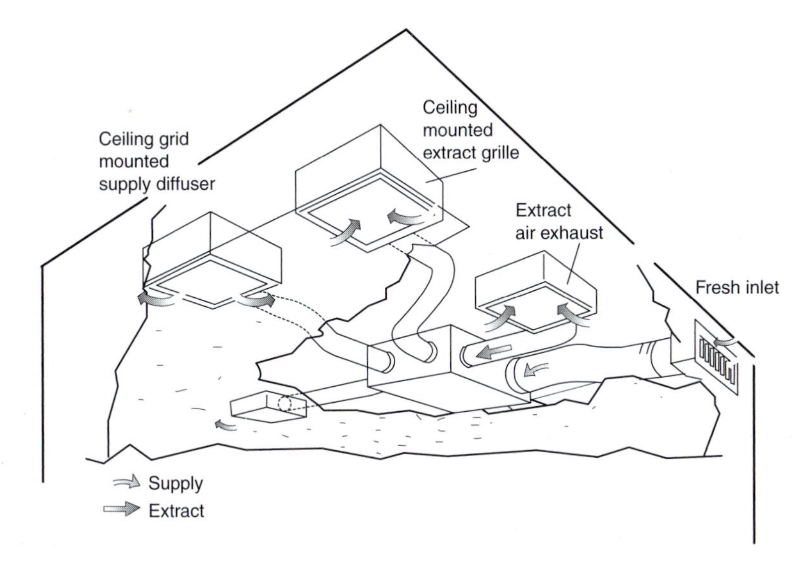

Figure 22.39 Typical arrangement of duct connections to a ventilation unit with heat recovery (Nuaire)

Packaged ventilation and heat recovery units typically up to 400 litre/s are available which can also be mounted above a false ceiling. These comprise supply and extract fans, recuperator-type heat exchanger with air filters and reheater coil.

- Figure 22.38 shows a typical ventilation heat recovery unit with integral fans, filters and supplementary heater.
- Figure 22.39 shows a typical arrangement for a heat recovery unit with room side and atmosphere side ductwork.

Notes

1. For details of standard fan testing see BS 848: Part 1: 1992.
2. AC, alternating current (mains supply); and DC, direct current (supply to motor).

3. Warren Spring Laboratory, on behalf of the Department of Trade and Industry.
4. BS 6540: Part 1: 1985.
5. Test Methods for Ignitability, Smoke and Toxicity of Air Filters. London Scientific Services, 1990.
6. CIBSE, *Minimising the Risk of Legionnaires' Disease*, CIBSE Technical Memorandum TM13, 1991 (revised 2000). CIBSE, *Legionellosis (Interpretation of Approved Code of Practice: The Prevention or Control of Legionellosis*, CIBSE Guidance Note GN3, 1993.
7. In winter, as may be seen from Fig. 22.31(a), the exhaust side of the heat exchange surface will be wetted for part of the process. In the case of the hygroscopic thermal wheel, however, under the same winter conditions as illustrated, there will be an increase in the moisture content of the incoming air stream, as indicated by S_3 in Fig. 22.31(b).

Further reading

CIBSE Guide B section 2.5 – Ventilation and Air Conditioning Equipment.

Chapter 23

Calculations for air-conditioning design

It is now proposed to consider the fundamental principles underlying the design of an air-conditioning system. These principles are the same no matter what particular form the system may take, but the degree of accuracy necessary to be achieved will depend on the application and the sophistication of the controls to be provided. For instance, a microchip manufacturing plant may require minimum tolerance in conditions, whereas less strict limits would be acceptable for comfort conditioning in the case of a department store.

First to be considered here is the general case, as applied to a central air-conditioning system for a single large space, and this is followed by some notes on how these general principles may be applied to certain of the specific types of apparatus already discussed. Design data have been built up around each of the particular forms of equipment mentioned and it would be beyond the scope of this book to explore each one in detail.

Heat gains

The various factors which contribute to the heat gains and losses which occur in a conditioned space have been outlined in Chapters 4 and 5. When designing an air-conditioning system the principal concern is directed towards heat gains, especially during the summer months, although the same system will most probably provide a heating service in mid-seasons and winter also. The reason for this approach is that heat gains present more searching demands than do heat losses.

Sensible heat gains

The *quantity* of conditioned air which must be provided to combat sensible heat gains is directly proportional to the difference between the supply air temperature and that to be maintained in the space. The temperature rise which may be permitted will probably be limited to 6 or 8 K owing to the difficulty of mixing cool entering air with warmer room air without producing draughts. The mass of air flow required to maintain a desired room temperature is thus arrived at very simply by the use of an expression similar to that noted in Chapter 18, where:

$$M = H/(cAt)$$

where

M = mass flow of entering air (kg/s)
H = sensible heat gains (kW)
C = specific heat capacity of air (kJ/kg K)
At = design temperature rise (K)

Latent heat gains

These do not affect the *quantity* of conditioned air required since they do not cause a rise in temperature. The latent gains are treated quite separately from the sensible gains. The mass flow of air required to deal with the latter will usually be found to produce no more than a small increment in humidity, but in an extreme case, limitation of that increment to an acceptable figure may require that the mass be increased and the design temperature rise be reduced in consequence.

Psychrometry

Psychrometry is a subject concerned with the behaviour of mixtures of air and water vapour and knowledge of it is necessary in order to perform any air-conditioning calculations. Some of the general principles were referred to in Chapter 1 but a complete study is outside the scope of this book and is dealt with in many text-books on thermodynamics and several excellent specialist works.[1]

Most of the terms which relate to mixtures of air and water vapour were defined in Chapter 1 of this book but a brief list recapitulating those items which are particularly relevant here would include:

Dry bulb temperature, *DB* (°C)
Wet bulb temperature, *WB* (°C)
Dew-point temperature, *DP* (°C)

Vapour pressure (kPa)

Relative humidity, *RH*, and percentage saturation (%); absolute humidity or moisture content (kg/kg of dry air)

Total heat, 777, or specific enthalpy (kg/kg of dry air)

Specific volume (m^3/kg of dry air)

Barometric pressure

The standard level of atmospheric pressure is 101.325 kPa exactly, corresponding to 760 mm of mercury at 0°C and standard gravity (9.80665 m/s^2). This level, equating to 1.01325 bar, was the barometric pressure used in the calculation of the psychrometric properties of air and water vapour presented in the *Guide Section Cl* and referred to later in this chapter. For practical purposes in air-conditioning design, the data so presented are accurate for situations having barometric pressures between 95 and 105 kPa but, in circumstances which are outside these limits, use must be made of other published data.[2]

Psychrometric chart[3]

The relationship between the various properties of a mixture of air and water vapour may be presented in the form of tables, such as those published in the *Guide Section Cl*, or as a chart based thereon as in Fig. 23.1. The principal advantage provided by tables is the level of accuracy offered, but this is achieved sometimes only by tedious interpolation. Since it is necessary in design to visualise stages in a process, a chart is often to be preferred as a working tool although a numerical check via tables will serve to 'polish' the conclusions. The arrangement of the co-ordinates on the chart and the method of use are indicated in Fig. 23.2, any single *state point* representing values of a number of properties:

Dry bulb temperature: The base of the chart has an evenly spaced scale with divisions at 0.5°C. A vertical line drawn through the state point downwards will meet that scale.

Wet bulb temperature: The saturation line has an evenly spaced scale with divisions at 1°C. A line drawn through the state point, sloping upwards to the left, will meet that scale.

Dew-point temperature: This is read using the wet bulb scale via a horizontal line drawn through the state point to the left, to intersect the saturation line.

Moisture content: The right-hand side of the chart has an evenly spaced scale with divisions at 0.001 kg/kg of dry air. A horizontal line drawn through the state point to the right will meet that scale.

Percentage saturation: From a scale at the head of the chart, the curves downwards to the left are at 10 per cent intervals. Interpolation is necessary where a state point falls between the curved lines.

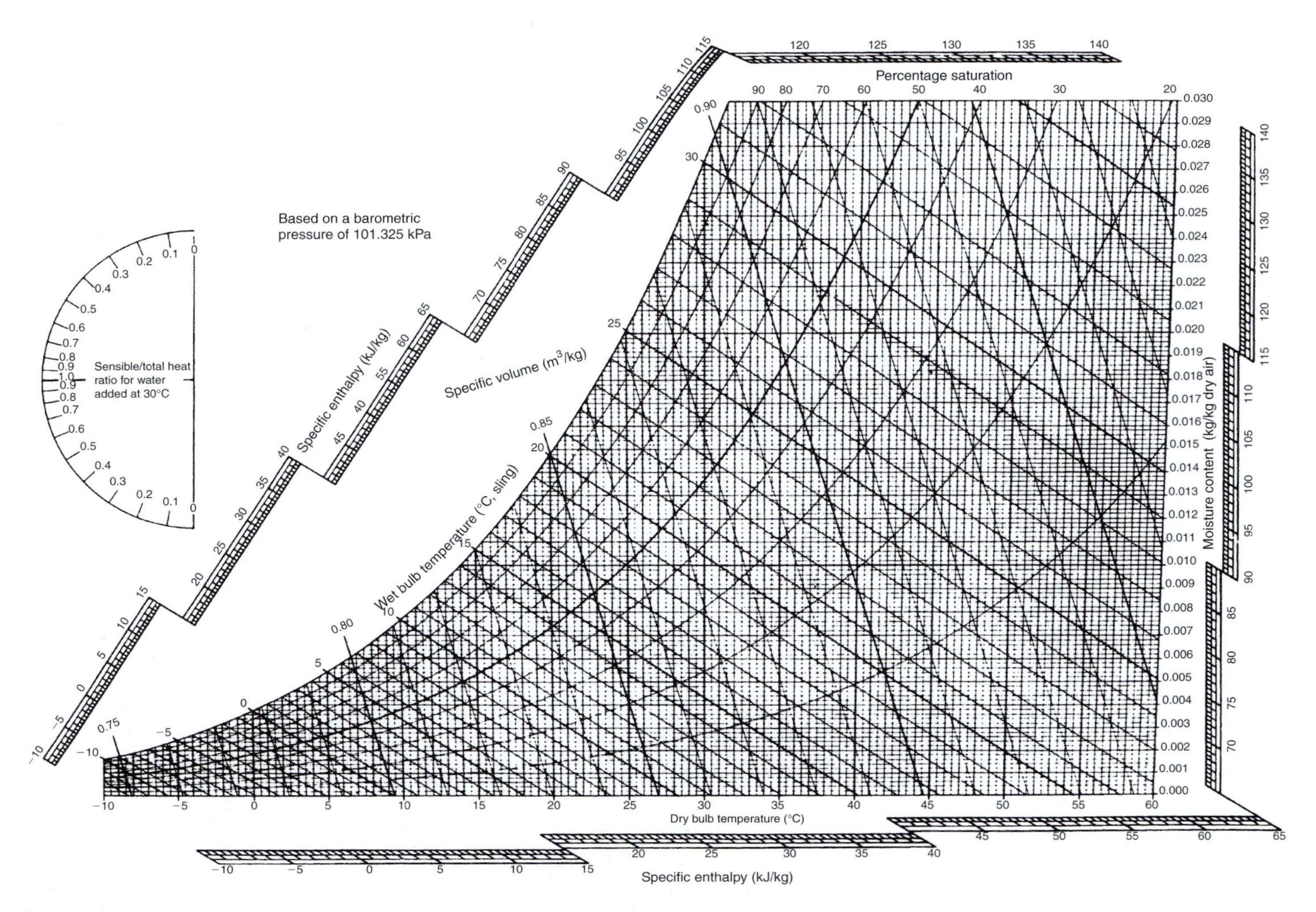

Figure 23.1 Psychrometric chart (as developed for *CIBSE Guide*)

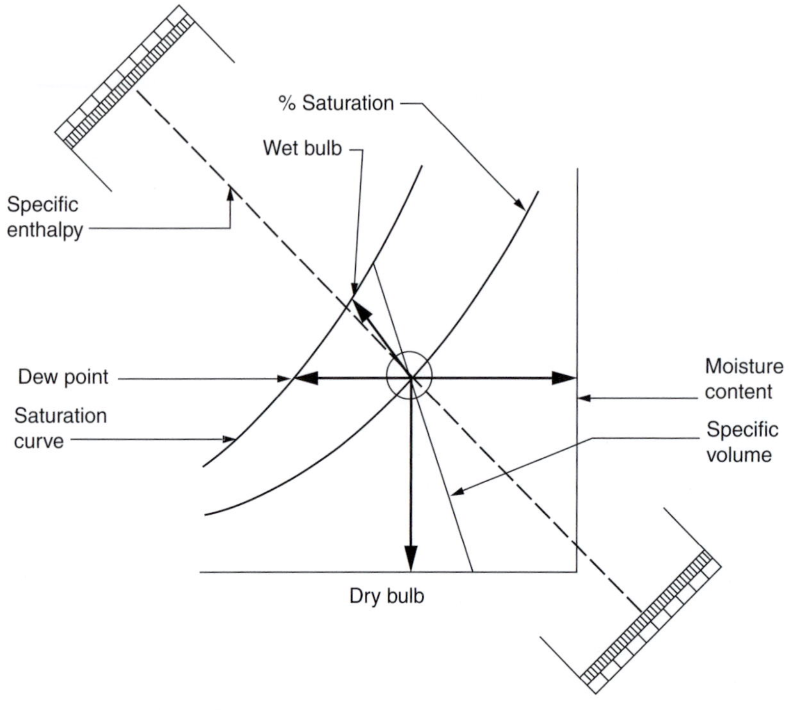

Point marked is 20°C DB. 50% Sat. other data read from the chart are:

W.B.	=	13.9°C
Dew point	=	9.4°C
Moisture content	=	0.0074 kg/kg dry air
Specific enthalpy	=	38.8 kJ/kg
Specific volume	=	0.84 m^3/kg dry air

Figure 23.2 Format and method of use for psychrometric chart

Specific enthalpy (total heat): Detached 'saw-tooth' scales to the left and top and to the right and base are evenly spaced with divisions at 0.5 kJ/kg of dry air. It is necessary to align a rule across both scales and through the state point to obtain a value.

Specific volume: Above the saturation line is a widely spaced scale with divisions at 0.05 m^3/kg of dry air. A line drawn through the state point, parallel to the scale lines will allow interpolation.

Application

Having arrived at the maximum hourly heat gain for the space or spaces to be served, probably the most searching of the seasonal conditions to be met, it is necessary to calculate the conditions to be maintained in the plant and the capacity of the various components (fans, cooling coils, heater batteries, water chillers and so on).

The routine calculations necessary are best illustrated by application to an example such as the all-purpose hall, in use for concerts and other activities, as shown in Fig. 5.12.

This is repeated here for convenience of reference as Fig. 23.3. The building characteristics and the bases for the design of the system are:

Building

Seating capacity	= 500 persons
Lighting load	= 20 kW
Volume of hall	= 6000 m^3

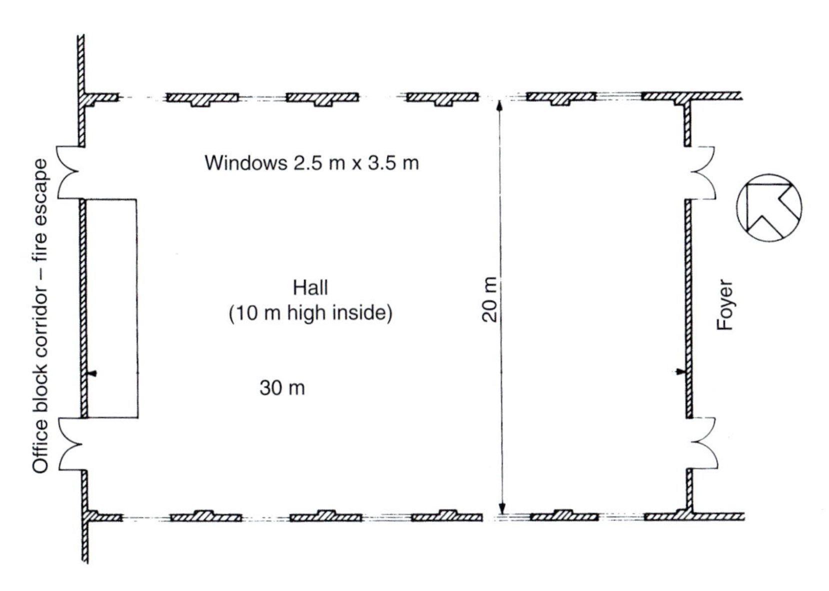

Figure 23.3 Plan of building used in the example

External conditions	*Summer*	*Winter*
Temperature	= 32°C DB	−2°C saturated
	= 17.4°C WB	
Enthalpy	= 62.90 kJ/kg	5.99 kJ/kg
Internal conditions[4]	*Summer*	*Winter*
Temperature	= 22°C DB	21°C DB
50% saturation	= 11.3°C WB	10.4°C WB
Enthalpy	= 43.39 kJ/kg	41.08 kJ/kg
	Outside air per occupant, year round	
Table 1.9	= 10 litre/s	

Summer cooling and dehumidification

The orientation of the example building presents two faces to solar heat gain and it was made clear in Chapter 5 that the peak gain occurred at 16.00 hours BST in September.

Building fabric heat gain

The scope of the calculations set out in the earlier chapter was restricted to an analysis of the heat gains arising from conduction through the structural elements and from solar glazing. The criteria were the outside and inside air temperatures noted above and the results were as follows:

Total heat gain from solar and conduction (from Chapter 5) = 18.2 kW

Infiltration heat gain

To continue with the example, the matter of infiltration arises which may best be dealt with from experience as an assumed rate of air change.

Allow half an hour change per hour

Room volume = (30 × 20 × 10) = 6000 m^3

From Table 23.1, specific volume of dry air at room condition = 0.83 m^3/kg

Thus mass of infiltrated air:
(6000 × 0.5)/(0.83 × 3600) = 1.0 kg/s

From Table 23.1, enthalpy of dry air:
at external condition (32°C) = 32.20 kJ/kg
at internal condition (22°C) = 22.13 kJ/kg

Thus, by difference, sensible heat gain:
1.0(32.20 − 22.13) = 10.07 kW

From *Basis of Design*, enthalpy:
at external condition = 62.90 kJ/kg
at internal condition = 43.39 kJ/kg

Thus, by difference, latent heat gain:
[1.0(62.90 − 43.39)]−10.07 = 9.44 kW

Internal heat gain[5]

This arises due to matters which occur within the room. In this case they are confined to heat gains from occupants and to lighting but might, in other circumstances, include computers, motors, or other sources of heat.

From Fig. 5.8, for persons at rest:
sensible gain per occupant = 70 W
latent gain per occupant = 45 W

Table 23.1 Properties of dry and saturated air at various temperatures

Temperature (°C)	*Specific volume* (m^3/kg *dry air*)	*Saturation moisture content* (kg/kg *dry air*)	*Specific enthalpy* (*total heat*) (kJ/kg *dry air*)	
			Dry air	*Saturated air*
0	0.77	0.0038	0.00	9.47
2	0.78	0.0046	2.01	12.98
4	0.79	0.0050	4.02	16.70
6	0.79	0.0058	6.04	20.65
8	0.80	0.0067	8.05	24.86
10	0.80	0.0076	10.06	29.35
12	0.80	0.0087	12.07	34.18
14	0.81	0.0100	14.08	39.37
16	0.82	0.0114	16.10	44.96
18	0.82	0.0129	18.11	51.01
20	0.83	0.0147	20.11	57.55
22	0.83	0.0167	22.13	64.65
24	0.84	0.0189	24.14	72.37
26	0.84	0.0214	26.16	80.78
28	0.85	0.0242	28.17	89.96
30	0.86	0.0273	30.18	99.98
32	0.86	0.0307	32.20	111.0
34	0.87	0.0346	34.21	123.0
36	0.87	0.0389	36.22	136.2
38	0.88	0.0437	38.24	150.7
40	0.89	0.0491	40.25	166.6

Source
The *Guide Section* at atmospheric pressure of 101.325 kPa.

Thus, sensible gain = (500 × 70) = 35.0 kW
and latent gain = (500 × 45) = 22.5 kW

From *Basis of Design,* sensible gain from lighting = 20.0 kW

Supply air quantity to room

This is calculated from the total of the individual sensible heat gains set out above. It will be noted that no inclusion is made for the gain arising from the supply of outside ventilation air to the room for the benefit of the occupants. This latter is a load on the plant and thus does not affect that air quantity.

Sensible heat gain in room:
(18.2 + 10.1 + 35.0 + 20.0) = 83.3 kW

Design criteria now assumed:
air temperature rise from supply to room = 8 K
hence, supply air temperature = (22 − 8) = 14°C

From Table 1.1, for dry air at 15°C:
specific heat capacity = 1.01 kJ/kg K

Thus, total air supply mass required:
83.3/(8 × 1.01) = 10.31 kg/s

Ventilation air

Criteria for the supply of outside air were set out in Table 1.8 and it is assumed that smoking is not allowed in this hall.

From *Basis of Design*:
number of occupants = 500
outside air quantity per person = 10 litre/s

From Table 23.1, specific volume of dry air at room condition = 0.83 m^3/kg

Thus, mass flow of ventilation air:
(500 × 10)/(1000 × 0.83) = 4.15 kg/s

Air mixture entering plant

This quantity is the air supply mass required to deal with the sensible heat gain and which includes both the ventilation air and the exhaust air from the hall which is recirculated. The mixture therefore has components having different properties of temperature and enthalpy, etc.

From the results above, by difference:
mass of air recirculated = (10.31 − 4.15) = 6.16 kg/s

Thus, temperature of air mixture, dry bulb:
[(6.16 × 22) + (4.15 × 32)]/10.31 = 26.0°C

and enthalpy of air mixture:
[(6.16 × 43.39) + (4.15 × 62.90)]/10.31 = 51.24 kJ/kg

Air mixture leaving plant

It is now necessary to take account of latent heat gain in the room inasmuch as this will affect the ability of the air mixture leaving the plant to absorb unwanted moisture in the room.

Latent heat gain in the room as previously:
calculated = (9.44 + 22.5) = 31.94 kW

From Chapter 1, approximate latent heat of vaporisation of water at 22°C = 2450 kJ/kg

Thus, moisture increment in room:
31.94/(10.31 × 2450) = 0.00126 kg/kg

From Fig. 23.1, moisture content of air at room condition = 0.0084 kg/kg

Thus, moisture of air leaving plant:
(0.0084 − 0.00126) = 0.00714 kg/kg

The energy used to drive the supply fan will be converted into heat and a proportion of this will be transferred to the air stream. Also, the walls of the supply duct between the plant and the room inlet, although thermally insulated, will allow some heat ingress. As a result, the dry bulb temperature of the air leaving the plant must be lower than that required at the room inlet. Such heat gains may be assessed in detail but, for this example, may be taken as 10 per cent of the sensible gain, thus accounting for a 1.0 K temperature rise.

Hence, at the inlet to the supply fan, from Figs 23.5(a) and (b):
dry bulb temperature = (14 − 1) = 13°C

and, from Fig. 23.1, air at 13°C dry bulb moisture content = 0.00714 kg/kg
Then
wet bulb temperature = 9.8°C
and enthalpy = 28.50 kJ/kg

Cooling capacity

From the various values now available, the amount of cooling capacity required to meet the required conditions for peak gains in summer may be calculated.

Difference in enthalpy, entering to leaving:
(51.24 − 28.5) = 22.74 kJ/kg

Thus, cooling capacity:
(22.74 × 10.31) = 235 kW

Plant duties

From the results calculated, it is now possible to arrive at the capacities of the various items of equipment which, together, will go to make up the plant for summer use. These are, excluding any margin which might be added in practice:

Supply fan, handling air at 13°C, and specific volume at 0.81 m^3/kg:
(10.31 × 0.81) = 8.35 m^3/s = 8350 litre/s
Exhaust fan, handling air at 22°C, and specific volume 0.83 m^3/kg
(4.15 kg/s exhaust to outside and 6.16 kg/s recirculation):
(10.31 × 0.83) = 8.56 m^3/s = 8560 litre/s

Cooling coil, 8350 litre/s with:
air on = 26.0°C DB and 14.0°C WB
air off = 13.0°C DB and 9.8°C WB
water chiller, excluding any calculated allowance for heat gains
in pumps and water circulating pipes, etc. = 235 kW

Figures 23.4 (a) and (b) show, graphically on a small section of the chart, the psychrometric changes taking place during the processes outlined in the example: (a) for summer and (b) for winter conditions. Figures 23.5(a) and (b) illustrate the corresponding conditions on a diagram of the plant and system.

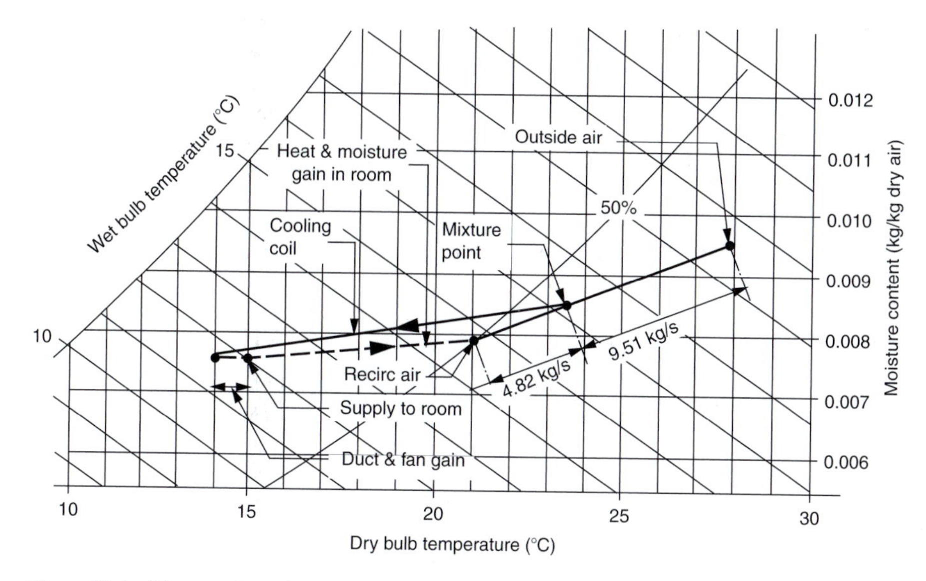

Figure 23.4a Diagram of psychrometric changes for summer with the hall occupied (see example calculation)

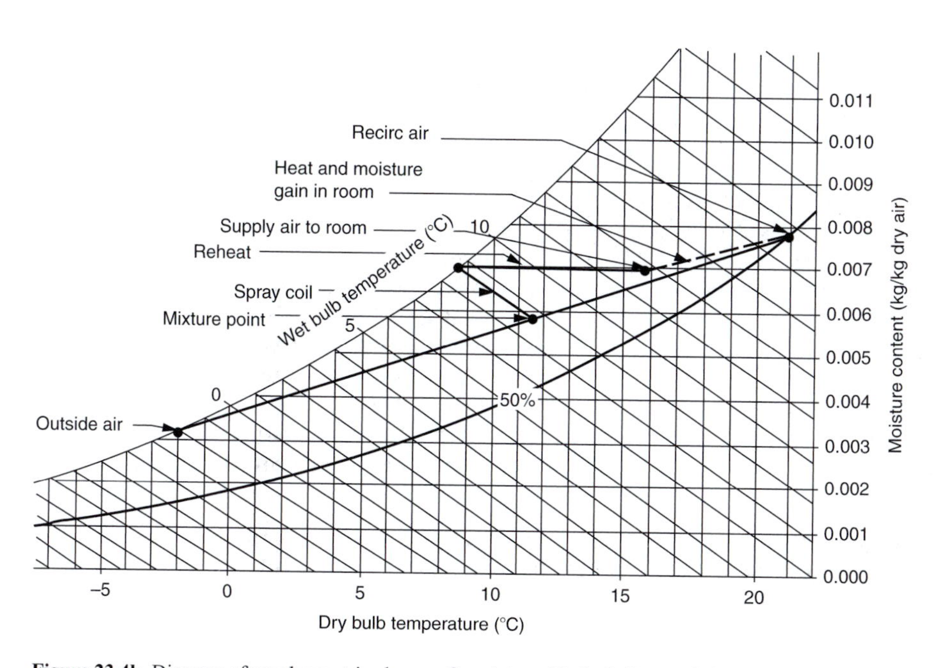

Figure 23.4b Diagram of psychrometric changes for winter with the hall occupied (see example calculation)

Cooling coils

The example above has assumed that the medium which is circulated through the cooling coil is *chilled water* and that a suitable coil could be found to meet the required duty. Use of chilled water avoids cooling all the air down to a low dew point for dehumidification purposes and then, subsequently, reheating.

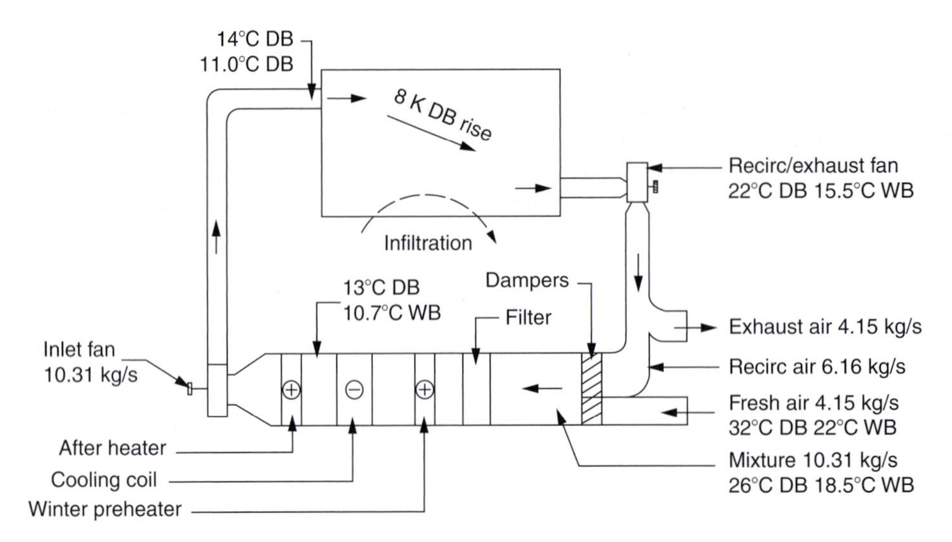

Figure 23.5a Plant conditions when cooling with the hall occupied (see example)

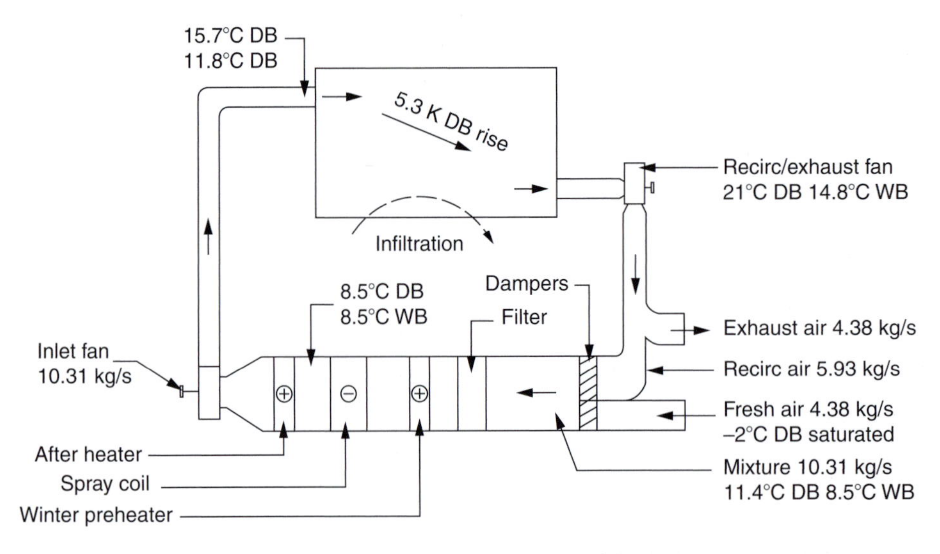

Figure 23.5b Plant conditions when heating with the hall occupied (see example)

Further, advantage is taken of the characteristic of such a coil such that moisture may be deposited on the chilled surface without the bulk of the air being reduced to the same temperature. This is achieved by keeping the surface at the lowest possible temperature consistent with absence of freezing. Had the coil been cooled by direct expansion of a refrigerant within it, a *DX coil,* the calculation would have been similar but the surface temperature of the coil would have been lower.

In some circumstances, a coil might be arranged to have unchilled and recirculated water sprayed on to the surface in order to improve performance and to provide for adiabatic saturation of the air flow and means for humidification in winter. Concern as to humidifier-related illness has led to suspicion of sprayed coils, since doubt exists whether plant owners and operators will ensure that the means provided to prevent bacteriological contamination are maintained. Humidification by steam injection is thus recommended, with saturation efficiencies of up to 80 per cent.

Winter heating and humidifying

The operation of the plant in winter might be with partial recirculation, as in summer, or with 100 per cent outside air. It would be best arranged so as to be suitable for the latter method of operation in average spring and autumn weather when neither heating nor mechanical cooling is required. In colder weather, recirculation might be used to promote economy in running cost. The latter arrangement is often made by providing an outside sensor to control motorised dampers and vary the proportion of outside to recirculated air: this is often referred to as an *economy cycle.* In order to avoid the complication of introducing another variable, the two calculations which follow assume that heat losses through the building fabric and those arising from infiltration are offset by a separate system such as hot water radiators. In consequence, the supply air leaving the plant would be at the room temperature, i.e. 21°C.

Hall empty, preconditioning for occupation

The following calculation assumes that the air entering the plant is a mixture of outside and recirculated air as for summer operation with the ambient temperature being −2°C. In this example, it is assumed that humidification will be by steam (*isothermal process*) using a packaged injection-type steam humidifier.

Dry bulb temperature and other conditions:

in room	= 21°C
outside air	= −2°C
mass flow of air	= 10.31 kg/s
mass flow of recirculated air	= 6.16 kg/s
mass flow of fresh air	= 4.15 kg/s
specific heat of air	= 1.01 kJ/kg

Temperature of mixed air entering plant:
$[(6.16 \times 21) + (4.15 \times (-2))]/10.31 \quad = 11.74°C$

and hence, after-heater duty:
$(10.31 \times 1.01)(11.74-(-2)) \quad = 143$ kW

From Fig. 23.1, moisture content of:

air at room condition	= 0.0079 kg/kg
outside air (no additional moisture in recirculated air)	= 0.0032 kg/kg

Thus, moisture added by humidifier:
$(0.0079 - 0.0032) \times 10.31 = 0.049$ kg/s = 178 litre/h
and energy absorbed in humidification:
(0.049×2450) = 120 kW

Hence, total energy absorbed excluding all losses:
$(143 + 120)$ = 263 kW

As an alternative, if permissible, the cooling coil would be pump sprayed and thus would provide adiabatic saturation to the air flow. For the purpose of the example, the figuring assumes 100 per cent coil efficiency, i.e. saturating the air to room dew point but a level of, say, 80–90 per cent would probably arise in practice.

From Fig. 23.1, at room condition of 21°C 50% saturation,
Dew point = 10.4°C
Enthalpy at dew point = 30.40 kJ/kg
mixed condition of 11.74°C and moisture content of 0.0032 kg/kg

The purpose of preheater is to raise the dry bulb temperature of the supply air from the mixed condition to a temperature such that after the air has passed through the spray coil (*adiabatic process*) it is at the dew point of the room condition, i.e. 10.4°C. As the hall is unoccupied and no moisture is added to the recirculated air the off coil condition for the preheater from Fig. 23.1 is 22.4°C with an unchanged moisture content of 0.0032 kg/kg and an enthalpy equal to room dew point of 30.40 kJ/kg.

The duty of preheater is therefore:
(30.40 − 19.40) × 10.31 = 113 kW

and that of after heater:
(41.08 − 30.40) × 10.31 = 110 kW

Total energy capacity absorbed excluding all losses:
(113 + 110) = 223 kW

Moisture added at sprayed coil (as before):
0.049 kg/s = 178 litre/h

Hall occupied

The following calculation assumes that the air entering the plant is a mixture of outside and recirculated air as for summer operation, outside air conditions being as above. In this example, again it is assumed that humidification will be by steam using a packaged injection-type steam humidifier. The internal heat gains, (35 + 20) = 55 kW sensible and 22.5 kW latent, are those due to occupancy and lighting only.

Dry bulb temperature and other conditions:
in room = 21.0°C
mass flow of air = 10.31 m^3/s
specific heat of air = 1.01 kJ/kg

Temperature of mixed air entering plant:
[(6.16 × 21) + (4.15 × (−2))]/10.31 = 11.74°C

Air temperature rise in room from supply due to 55 kW sensible gain:
55/(10.31 × 1.01) = 5.3°C

Thus, air temperature leaving plant:
(21 − 5.3) = 15.7°C

and main heater duty:
(10.31 × 1.01)(15.7 − 11.74) = 41.2 kW
From Fig. 23.1, moisture content of:
air at room condition = 0.0079 kg/kg
air entering plant
[(6.16 × 0.0079) + (4.15 × 0.0032)]/10.31 = 0.00601 kg/kg

and rise in moisture content in room due to 22.5 kW latent gain:
22.5/(10.31 × 2450) = 0.00089 kg/kg

Thus, moisture content of supply air:
(0.0079 − 0.00089) = 0.00701 kg/kg

and moisture added by humidifier:
(0.00701 − 0.00601) × 10.31 = 0.0103 kg/s = 37 litre/h

Energy absorbed in humidification:
(0.0103 × 2450) = 25.24 kW

Hence, total energy absorbed:
(41.2 + 25.24) = 66.44 kW

As an alternative, once again if permissible, the cooling coil would be pump sprayed and thus would provide adiabatic saturation to the air flow. From Fig. 23.1, at minimum fresh air the enthalpy of the mixed air is higher than the enthalpy of the dew point of the supply air entering the room. The percentage of recirculated air will therefore be adjusted automatically to provide air entering the spray coil at the correct condition as follows:

Temperature of air entering spray coil = 11.4°C
Moisture content of air mixture = 0.00585 kg/kg
Proportion of recirculated air is 57.5% with 42.5% fresh air.

Enthalpy of air mixture:
$[(5.93 \times 41.08) + (4.38 \times 5.99)]/10.31$ = 26.17 kJ/kg
Air temperature rise from supply to room due to 55 kW sensible gain (as before) = 5.3°C
Thus, air temperature leaving plant (as before) = 15.7°C
Rise in moisture content from supply to room due to 22.5 kW latent gain (as before) = 0.00089 kg/kg
Hence, moisture content of supply air (as before) = 0.00701 kg/kg

From Fig. 23.1 as the enthalpy of the mixed air is equal to the enthalpy of the dew point of the supply air entering the room, preheat is not required.

The duty of preheater is therefore zero.

From Fig. 23.1 enthalpy of supply air entering the room at 15.7°C and moisture at 0.00701 kg/kg = 34.00 kJ/kg

Thus, difference in enthalpy through plant:
$(34.00 - 26.17)$ = 7.83 kJ/kg

and after-heater capacity:
(7.83×10.31) = 81 kW

Moisture added at spray coil is difference between that of mixture and supply air:
$(0.00701 - 0.00585) \times 10.31 = 0.0120$ kg/s = 53 litre/h

It will be noted that, despite the different methods used for humidification and the differences in calculation routine, there are no significant differences between the predictions of energy absorbed or in the amount of water used. Indeed, the differences existing there may have arisen as a result of the difficulty in reading accurate results from Fig. 23.1.

Design calculations for other systems

The design calculations for the various other types of systems follow a routine, not dissimilar from that which has been outlined earlier, but are adapted to suit the particular characteristics in each case. The following paragraphs provide brief summaries of the similarities and differences in approach.

Single-duct terminal reheat systems

The central plant is designed to provide full conditioned air quantities to all areas served, to meet peak demands. No allowance for diversity of load can be made. The terminal reheat equipment is designed to cater for local differences between maximum and minimum load conditions.

Whilst this system has the advantage of offering good temperature control, it is inherently wasteful of energy. Low humidity conditions may arise.

Fan coil systems, ducted outside air

These are designed on similar lines to those applied to induction systems in cases where the fresh air is delivered to the room space via the fan coil unit.

Where fresh air is ducted independently, it may be supplied year round at constant temperature at, or a degree or two less than, room condition. In this case the fan coil unit, which should be of the four-pipe type (Fig. 19.17), will require to be designed to deal with the whole of the heat gains or losses occurring within the space served.

Fan coil system, local outside air

Such systems which are, in effect, no more than individual room heaters/coolers must be arranged to deal with the whole conditioning load, outside air being provided either directly to the units or via opening windows, etc. No direct control of room humidity can be achieved although dehumidification will occur.

Induction systems

The primary air supply, conditioned at the central plant, is designed to provide an adequate volume for ventilation purposes and humidity control. Temperature is varied to suit external weather conditions and may be further adjusted to take account of solar gain.

The coil or coils in the room units are designed to deal with local sensible heat gains or losses. Chilled water temperature should be selected to reduce the risk of condensation on the coil.

Double duct systems

The central plant is designed to provide a supply of both cool and warm air which are distributed in parallel to individual terminal units. The cool air duct provides an air quantity adequate in volume and temperature to meet the maximum anticipated cooling load of heat gain to the building fabric and from the outside air supply.

The warm air duct provides a supply adequate in temperature to meet the design heat loss, the volume usually being allowed as 75 per cent of that in the cool duct.

Variable volume systems

In the case of the true variable volume system, the air quantity provided by the central plant will be limited to that required to meet the maximum coincident load. It will be reduced from that design volume for all part load conditions. Supply temperature may be constant and determined by the peak cooling load, varied to suit external conditions, or varied to suit the loads actually occurring in the space. This last option requires control feedback from the terminal devices to indicate the actual operating mode.

Normal practice suggests that the air supply to individual rooms should not be reduced by more than 40 per cent and, if minimum load is less than this then some form of temperature adjustment will be necessary, possibly at the central plant but more probably by reheat on a zonal or local basis. The all-air VAV system may be supplemented by fan-assisted units, normally installed as an integral part of the control terminal and served from a chilled water supply. The chilled water temperature must be selected such that any risk of condensation on the coil is reduced.

System diagrams and automatic controls

For diagrams of a variety of air-conditioning systems and notes on the application of automatic controls to them, the reader is referred to Chapter 28.

Notes

1. Goodman, W., *Air Conditioning Analysis.* Macmillan, New York, 1947. Jones, W.P., *Air Conditioning Engineering.* Butterworth Heinemann, 2001.
2. *M-C Psychometric Charts for a Range of Barometric Pressures.* Northwood Publications, 1972.
3. Copies of the chart in pad form are obtainable from CIBSE.
4. Dry bulb temperature is used here in preference to operative temperature. See Chapter 5.
5. Sensible and latent heat gains arising from occupants assume here that they are seated at rest. In other circumstances, the gain per occupant might be more but the number, equally, might be less.

Chapter 24

Refrigeration and heat rejection

For full air-conditioning, some means of cooling and dehumidification is necessary and this, in the majority of cases, is provided by use of a mechanical refrigeration machine or machines. This equipment may be similar to the plant used for ice-making and cold storage work, etc., except that the temperature to be produced is likely to be higher than that required for such applications.

The energy balance of a refrigeration cycle is such that it may be thought of as a 'thermal transformer', taking in energy at a (relatively) low temperature and discarding it at some (relatively) higher temperature. Where cooling is required, it is the energy at low temperature which is used and that at a higher temperature discarded: in the converse sense, where heating is required, the energy at the higher temperature is used and that at the lower temperature discarded. The latter application is that of a *heat pump* which is often applied in modern practice to make use of a low grade energy source which would otherwise go to waste.

Mechanical refrigeration

This depends upon the principle that a liquid may be made to boil at a chosen low temperature if it is held at a pressure which is reduced to an appropriate level.[1] To produce boiling, the liquid must be supplied with heat from an external source and this source will thus lose energy and be cooled. Given a suitable liquid, the temperature of boiling may be chosen to suit the required conditions, without resort to unduly low pressures, and although many different substances have been used, complex hydrocarbons have been found to offer the most suitable characteristics as refrigerants. The vapour given off in boiling is compressed, which process adds heat, and the hot vapour is then liquified by removal of that heat, the pressure still being maintained. A sudden release of the pressure is then arranged and, in consequence, the fluid returns to the state in which it began, ready once again to boil at a low temperature. This sequence of events is known as the *vapour compression cycle.*

A refrigeration plant, working on the vapour compression principle, as shown in Fig. 24.1 thus comprises these principal components:

- A *compressor* to apply pressure to the refrigeration medium.
- A *condenser* to receive the compressed gas and liquefy it, the latent heat being taken away from the circuit by some external means. One method is to cool the condenser by an air current. Alternatively, the condenser may be cooled with a water circulation which may pass to a cooling tower.
- An *expansion device* by which the pressure of the liquid is reduced.
- An *evaporator* in which the medium re-evaporates, extracting heat from whatever surrounds it, e.g. from cooling water or air in an air-conditioning plant or from a water/ethylene glycol mix where temperatures below the freezing point of water are needed.

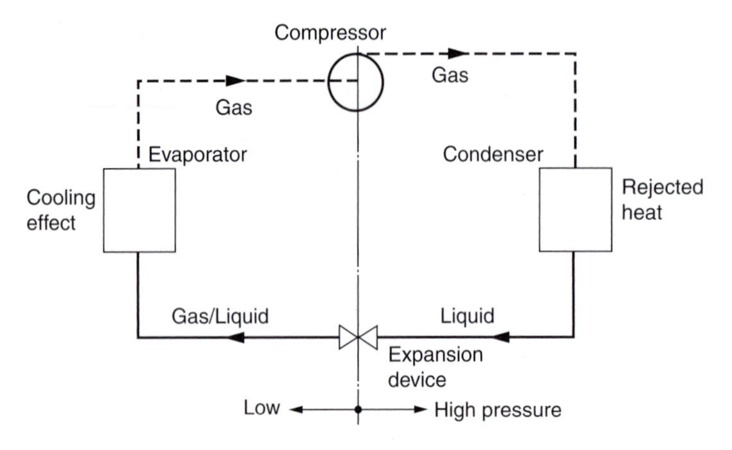

Figure 24.1 Principle of vapour compression cycle

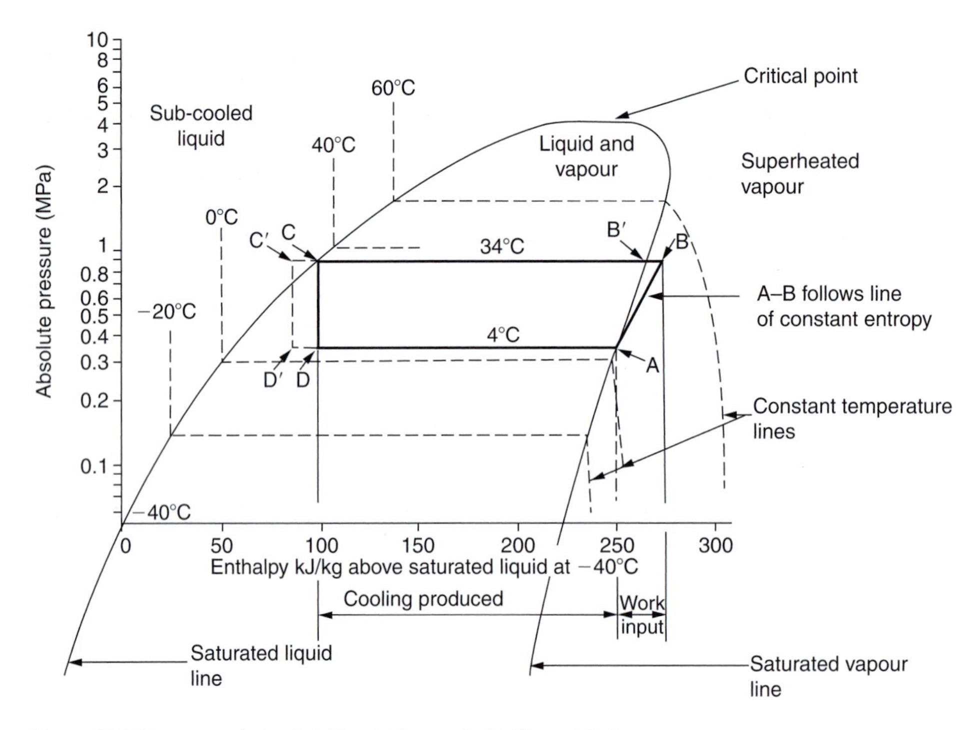

Figure 24.2 Pressure-enthalpy (total heat) diagram for Refrigerant 134a

Refrigeration cycle

The vapour compression cycle may best be considered on a pressure-enthalpy (total heat) diagram, as Fig. 24.2 which is drawn for the fluid *Refrigerant 134a* (see later text).

Inside the curved envelope, the medium exists as a mixture of vapour and liquid and the increase in enthalpy from left to right on any pressure line within the envelope represents an increase in latent heat. Further, within the envelope, lines of equal temperature (isotherms) are horizontal. Outside the envelope to the left, the medium exists as a liquid below its saturation temperature and there the isotherms are almost vertical. To the right of the saturated vapour curve, the medium exists in the form of a superheated vapour and the isotherms curve downwards. The *critical point* is that at which latent heat ceases to exist: it is not possible to liquify a gas by pressure alone if it is above the critical temperature.

On the diagram, the refrigeration cycle is represented by the outline *A–B–C–D*, the components being:

- *A–B* here the gas is compressed causing a rise in pressure and enthalpy which equals the energy put into the gas by the compressor, all in the superheat region. This takes place at constant entropy.
- *B–B′* represents cooling of the superheated gas in the condenser down to the saturated vapour temperature.
- *B′–C* here latent heat is removed, also in the condenser, and the gas is condensed to liquid.
- *C–D* this is the pressure drop through the expansion device without any change in enthalpy (adiabatic).
- *D–A* represents vaporisation to a dry saturated state in the evaporator, latent heat – represented by increasing enthalpy – being drawn from the water, the air or other medium being cooled. This is the *cooling effect.*

If the condenser were arranged to 'sub-cool' the liquid, say to point C′, each unit mass of refrigerant in circulation would produce a greater cooling effect (*D′–A*) and the cycle would be more efficient in consequence.

The ratio of the cooling effect (as *D–A*) to the energy input (as *A–B*), in terms of enthalpy, is known as the *coefficient of performance (COP).* The smaller the range of pressure (and hence temperature) over which the cycle operates, the less will be the energy expended for a given cooling effect. Hence, for economy in running, it is desirable to design for:

- Evaporator temperatures as *high* as is consistent with other considerations (such as dew point temperature in an air-conditioning application).
- Condenser temperatures as *low* as possible. When cooling is to be atmospheric, weather records will decide the safe minimum level to be assumed. Maximum cooling is generally required in the hottest summer weather when the condensing arrangements are least efficient and caution is thus necessary in selecting an appropriate temperature.
- Should the reader wish to pursue this matter further, a selection of charts depicting the properties of refrigerants in common use is provided in the *Guide Section B4.*

Application of refrigeration

For application to air-conditioning, using a cooling coil, water is pump-circulated through a closed system returning to the evaporator of the refrigeration plant at a temperature which is generally between 7°C and 12°C, depending upon the dew point to be maintained: in passing through the evaporator, this water temperature will be lowered by about 4–6 K. In order that the necessary heat transfer may take place, the refrigerant must be at some temperature below that of the leaving water but, at the same time, it must generally be slightly above freezing point. Thus, in a typical case, the following conditions might obtain:

Apparatus dew point	10°C
Cooling coil outlet	12°C
Cooling coil inlet	6°C
Water at evaporator outlet	5.5°C

The refrigerant in the evaporator would in this case be maintained at about 1°C, giving a differential for 4.5 K for heat transfer. As will be appreciated, this small temperature potential means that the cooling surface of a simple tubular type would need to be very extensive: a variety of devices has been developed to augment the transfer rate.

An ethylene glycol solution may be used in cooling coils in order to allow lower air temperatures to be obtained (e.g. to achieve a low dew point condition): the temperatures of the fluid circulating may be −7°C from the evaporator and −3°C returning to it, or lower as required. To achieve such conditions, it is necessary to consider the strength of the ethylene glycol solution, data for which are available from standard tables.

In instances where cooling for an air-conditioning system is provided from a refrigeration machine by *direct expansion*, the refrigerant is piped directly to cooling coils in the air stream which thus become the evaporator. The surface temperature of the coils is a function of the leaving air temperature required, the form of the coil surface and the velocity of the air flow. Refrigerant temperatures much below freezing point are inadmissible owing to the risk of build-up of ice on the coil surface when dehumidification is taking place. An apparatus dew point of 3°C is normally considered as the practical minimum for such coils if frosting is to be avoided.

Refrigerating media

The factors affecting the choice of a refrigerant from a thermal point of view will now be clear. A substance is required which can be liquefied at a moderate pressure and which has a high latent heat of evaporation. By these means the size of the compressor will be kept to a minimum and the mass of the refrigerant circulated kept relatively small for a given amount of cooling. In addition, compatibility with the type of compressor and the refrigerant system, cost, environmental issues and safety have to be taken into account. Environmentally, the requirements of the following regulations must be met: EC Regulation no 2037/2000 on ozone depleting substances: as well as phasing-out and controlling use of chlorofluorocarbons (CFCs) and hydrochlorofluorocarbon (HCFC) refrigerants this regulation also includes legal requirements for the minimisation and avoidance of refrigerant emissions and leakage. The Environmental Protection (Controls on Substances that Deplete the Ozone Layer) Regulations 1996, places a legal duty on owners and operators to comply. The containment and recovery of refrigerants is covered by the F-Gas Regulation EC842/2006. The guidance given in CIBSE GN1 should also be followed.

> The UK is party to a number of international agreements including the Montreal Protocol and the Kyoto Protocol. The Montreal Protocol is implemented through the EC Regulation 2037/2000. The Kyoto Protocol addresses the issue of the emission of man-made greenhouse gases including many refrigerants.
>
> The risks associated with the escape of refrigerant and the risks of systems bursting or exploding due to over-pressure of refrigerant or equipment failure should be minimised by complying with relevant regulations, codes and standards. In addition, it is CIBSE policy that the requirements of BS EN 378: *Specification for refrigerating systems and heat pumps. Safety and environmental requirements* should also be complied with.
>
> BS EN 378 limits the hazards from refrigerants by stipulating the maximum charge of refrigerant for given occupancy categories and refrigerant safety groups. The standard summarises the maximum refrigerant charge and other restrictions for chillers (indirect closed systems) and direct (DX) systems. The most recent version of BS EN 378 should always be consulted for full details. Maximum refrigerant charge is related to the 'practical limit' or maximum allowable short term refrigerant concentration should the entire charge be released into the space or the room occupied by the system; this does not apply to systems located outside. The Institute of Refrigeration's Safety Codes provide specific guidance on the requirements of BS EN 378.
>
> The designer should take account of the requirements of the Health and Safety at Work etc. Act 1974 and all related regulations, UK health and safety regulations with specific requirements for refrigeration and heat rejection. Specific guidance on meeting the requirements of these regulations for vapour compression refrigeration systems is given in the Institute or Refrigeration's Safety Codes. The codes also give guidance on health and safety risk assessments for refrigeration systems. Guidance on compliance with the regulations with respect to the risk of exposure to *Legionella* bacteria is given in HSC Approved Code of Practice and Guidance L8: *Legionnaires' disease – the control of legionella bacteria in water systems.*

The refrigerating media available include ammonia, carbon dioxide and numerous manufactured gases. Ammonia (NH_3), while high in efficiency and low in cost, has not been thought suitable for many air-conditioning applications due to its toxic nature and the serious results which might attend a burst or leak in the system. However, ammonia is widely used in the food refrigeration and cold storage industries on account of its excellent refrigeration properties and is being considered increasingly for air-conditioning applications. To date ammonia has proven to be a very effective and safe refrigerant where plant has been designed, installed and maintained in accordance to relevant safety standards and codes of practice.

Carbon dioxide (CO_2) is an excellent refrigerant from an environmental and safety point of view. The main barriers to its use is that it results in low energy efficiency and that it operates at higher pressures (around 100 bar) and has a substantially higher volumetric capacity than most other refrigerants, this means that equipment designs are unsuitable. Considerable development is taking place on CO_2 for small refrigeration systems including car air-conditioning.

A range of synthetic refrigerants, halogenated hydrocarbons sometimes referred to as *freons*, which are colourless, non-inflammable, non-corrodent to most metals and generally non-toxic, are those now in common use. They may be categorised as falling within one of three chemical forms:

CFCs (chlorofluorocarbons): These have a high ozone-depleting potential (ODP) contributing to the breakdown of the ozone layer, are *banned* by the Montreal Protocol and ceased manufacture in the European Community by January 1995. Examples are Rll, R12 and R114.

HCFCs (hydrochlorofluorocarbons): These have limited ODP, are classified under the Montreal Protocol as *transitional substances* and are currently being phased out. Examples are R22, R123 and R124. R123 is available as a 'retrofit' refrigerant for Rll. In this context, 'retrofit' means a fluid which may be substituted into an existing system but will require material changes to equipment.

HFCs (hydrofluorocarbons): These contain no chlorine and therefore have zero ODP and in consequence are not controlled by the Montreal Protocol. However, there has been uncertainty about HFCs ever since the Kyoto Protocol. The UK Government's position is that HFCs are not a sustainable technology in the long term and should only be used where other safe, technically feasible, cost effective and more environmentally acceptable alternatives do not exist. In the meantime a range of hydrofluorocarbons has been developed and is available commercially to replace the long established CFC and HCFC refrigerants. Some are blends of two or more HFCs and are designed to approximately mimic the properties of an existing CFC or HCFC, or to counteract undesirable properties of a single HFC. Examples are R407C and R410A R407C is a blend of HFC-32, HFC-125 and HFC-134a and is designed to be a close match for R22. R410A is another HFC blend intended for use in applications that formerly used R22 and is a better choice for smaller systems. R134a is a pure fluid and is a 'drop-in' refrigerant for R12. Here, 'drop-in' means a fluid which can be substituted directly, requiring replacement of some serviceable components only. When changing refrigerants, the type of oil used by the machine needs to be checked for compatibility with the new refrigerant. R134a is emerging as the preferred refrigerant for larger air-conditioning applications.

Hydrocarbons: Hydrocarbons such as propane (R290) and isobutane (R600a) have good refrigeration properties and are compatible with standard materials and components. However, because hydrocarbons are highly flammable, specific safety precautions are necessary. Generally systems with small refrigerant charges, or indirect systems with refrigerant containing parts outside the building, are most suitable for hydrocarbon refrigerants.

In addition to an ODP classification, refrigerants are also given ratings for global warming potential (GWP), an index providing a simple comparison with carbon dioxide which has an index rating of unity. The properties and values for the ODP and GWP indices of common refrigerants are given in Table 24.1 together with those for ammonia as a comparison.

In addition, the table gives values for the occupational exposure limit (OEL), in ppm, which reflect the toxicity level of the various refrigerants. In the knowledge of these allowable concentration levels in occupied areas, it is necessary that adequate rates of ventilation be provided in plant rooms and recommended that refrigerant leak detection be installed.

It is mandatory to limit the discharge of any refrigerant to atmosphere and, in consequence, provision should be made for *pump-down* (removal) from machines during maintenance activities. In addition, consideration should be given to limiting the volume of refrigerant gas in a system together with improved standards of design and installation for refrigerant pipework in order to reduce the risk of leakage.[2]

The future availability of replacement refrigerant for topping-up and servicing should be considered because many refrigerants are subject to ozone and global warming related regulations. CFCs have already been phased out and HCFCs are currently in the process of being phased out. The long-term availability of HFC refrigerants depends on whether safe, practical and economic replacements with lower GWP will become available. The designer and building operator should keep up to date with current and developing regulations and standards.

Air also may be used as a refrigerating medium. One method is to compress it to an absolute pressure of about 1.4 MPa (14 bar) and then, after removing the heat of compression, allow it to expand through a valve. In aircraft, the air cooling cycle is used in quite small turbo equipment running at very high speeds taking advantage of the extremely low temperature of the surrounding ambient for use in the condensing side. For normal land use, bearing in mind the relative cost of plant required, air is not a practical choice as a refrigerant for air-conditioning.

Water is also used as a refrigerant in a system termed *steam jet* described later.

The refrigerants most suitable for direct expansion into coils in the airway are R407C, R134a and certain refrigerant blends, the others all being objectionable owing to their smell, toxicity, inflammability, or inefficiency.

Table 24.1 Properties of refrigerants

	Refrigerant				
Properties	*Ammonia*	*R22*	*R123*	*R134a*	*R407C*
Gauge pressure (kPa)					
condenser (30°C)	+1060	+1100	+8.2	+670	1160
evaporator (−15°C)	+155	+195	−85.4	+62.6	189
evaporator (−5°C)	+404	+427	−60.5	+249	321
Boiling point (°C) (standard pressure)	−33.3	−40.8	+27.8	−26.1	−43.8
Critical temperature (°C)	1333	96	184	101	87.3
Volume of vapour at −15°C (m^3/kg)	0.509	0.078	0.873	0.121	0.09
Latent heat of evaporation at −15°C(kJ/kg)	1320	218	175	187	212
Theoretical energy input per unit/energy output (kW/kW)	0.211	0.216	0.203	0.217	–
Coefficient of performance (27°C to −15°C)	4.75	4.65	4.93	4.61	–
Ozone depleting potential (ODP)	0	0.05	0.014	0	0
Global warming potential (GWP)	0	510	29	1600	1980
Occupational exposure limit (OEL)[b] (ppm)	25	1000	10[a]	1000[a]	1000

Notes
a Provisional recommendation by refrigerant manufacturers.
b Health and Safety Executive. Occupational Exposure Limits. EH40/93.

Types of refrigeration plant

Vapour compression

Vapour compression plant is normally classified by the compressor type and thus includes reciprocating, rotary, scroll, screw or centrifugal, usually directly coupled to a prime mover, normally an electric motor. An *open* compressor has the motor connected through a coupling requiring an external shaft and seal to contain the refrigerant. *Hermetic* compressors have the motor and compressor as a self-contained assembly with the motor in contact with the refrigerant within the casing, no shaft seal being required. *Semi-hermetic* compressors are similar to the hermetic type, but with access to the compressor for repairs: motor failure would require the assembly to be repaired off-site. *Sealed units* are of the hermetic type with the whole assembly contained within a welded steel shell, used extensively in refrigerators and freezers, and in packaged chiller units for air-conditioning.

Reciprocating

A positive displacement piston machine which can operate over a wide range of conditions, the compressor may have up to 16 cylinders arranged in V or W formation, see Fig. 24.3. A typical sealed compressor is illustrated in Fig. 24.4

Capacity control is normally provided by cylinder 'unloading' in steps or by switching multiple compressor and refrigerant circuits; hot gas bypass and speed regulation are also available.

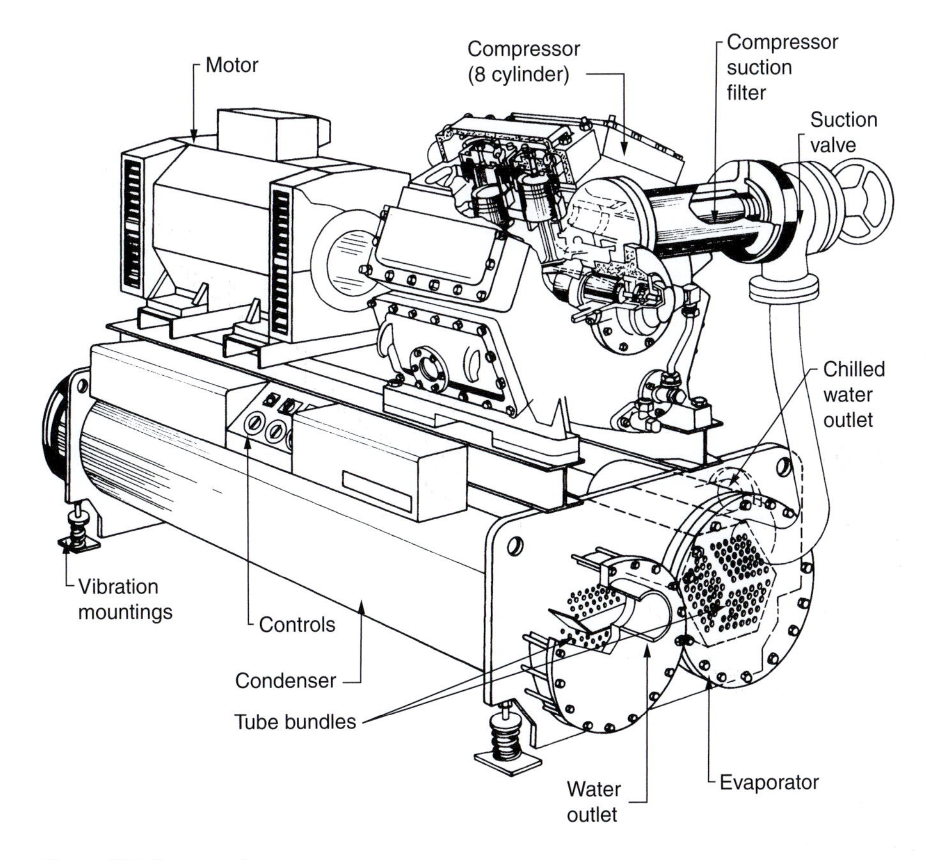

Figure 24.3 Reciprocating compressor

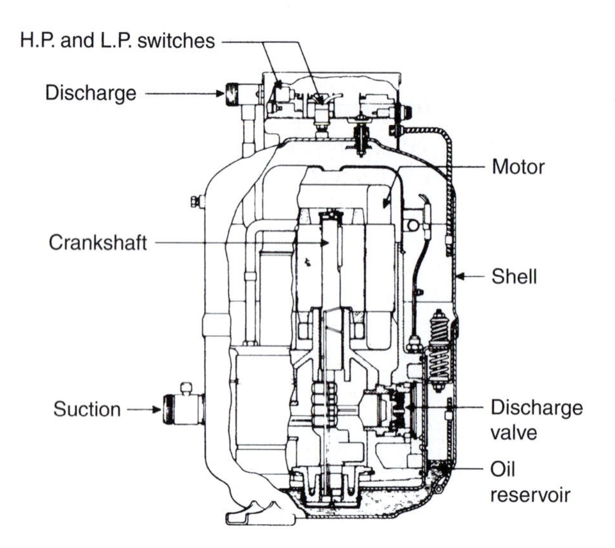

Figure 24.4 Sealed (hermetic) compressor

In installations where duty and standby machines are required for security in operation, benefits in efficiency may be obtained by the use of two-speed plant; typically the compressor total efficiency at low speed will be 70 per cent compared with 60 per cent when running at high speed.

Reciprocating machines are available in the range 55–1400 kW cooling duty using R134a and R407C. However, for most current applications, these compressors have now been superseded by screw machines. They may be connected to direct expansion air coils or water chillers on the suction side and to air cooled, water cooled or evaporative condensers on the compressor discharge side.

Hermetic piston-type compressors are widely available up to 120 kW using R407C and are used in combination to make larger capacities. Such machines are obtainable built as a weatherproof unit complete with an air-cooled condenser. A low silhouette arrangement is available for installation on flat roofs. In split systems, where the evaporator and condenser are located apart from the compressor or each other, the length of the connecting refrigerant lines must be kept to within practical limits.

Rotary

In effect another form of piston compressor, but without valves, this type was developed for small-scale applications. The assembly is contained in a cylindrical casing and depends for its operation on a shaft eccentric to the cylinder carrying a rotor which when rotated produces the compression effect.

Scroll

A robust and efficient type of machine, based upon the compression effect obtained when an involute spiral is rotated within a second fixed volute and contained within fixed plates at either face, the gas being compressed as the volume is reduced closer to the centre of the scroll, Fig. 24.5. Units are available in the range 50–170 kW, operating on R134a and R407C, for either air or water cooling.

Screw

Compression is produced by rotating helical screws, the seal being achieved by oil. A retractable vane enables load variation in a simple manner over a wide range. This form of compressor, Fig. 24.6, may be used with a wide range of refrigerants from quite small machines to a cooling capacity in excess of 4.5 MW and offers

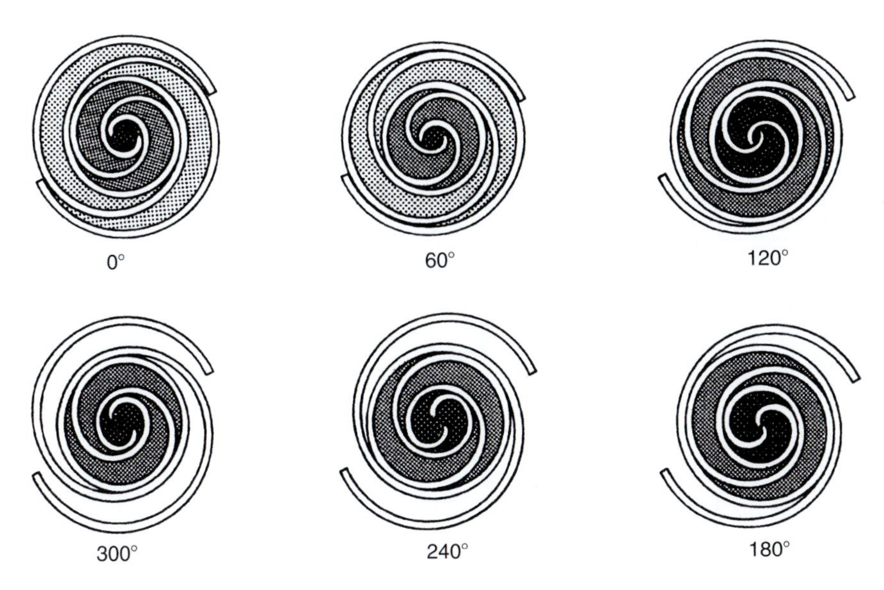

Figure 24.5 Scroll compressor, operating principle

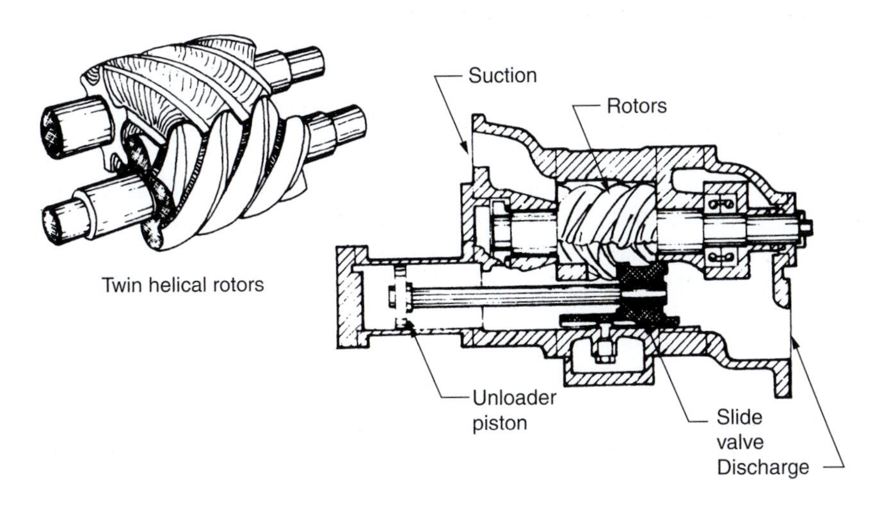

Figure 24.6 Screw compressor

the advantages of minimum vibration and low noise level. Open, hermetic or semi-hermetic machines are available, depending on the size and the manufacturer. In the smaller sizes, screw compressors are manufactured as sealed units with capacity control by a simple slide valve mechanism down to 25 per cent of full load.

Centrifugal

The centrifugal compressor is used chiefly where large duties are required. The advantages are:

- Saving of space as compared with screw and reciprocating machines.
- Reduced vibration (thus suitable for a roof-top plant chamber).

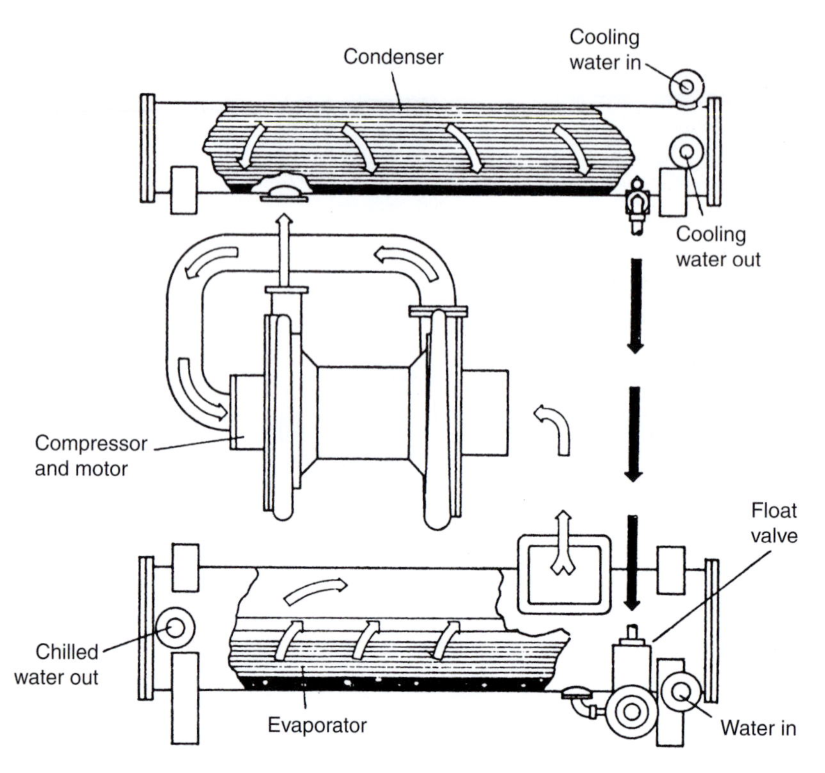

Figure 24.7 Centrifugal system

- Reduced maintenance due to there being no wearing or reciprocating parts.
- Efficient part load operation with a control range down to 10 per cent.

The most common refrigerant used in centrifugal compressors is R134a, replacing those using CFCs Rll and R12 in earlier models, which for air-conditioning temperatures operate over a small pressure range, thus reducing slip losses between blades and casing.

An important consideration is the turn-down range, i.e. the ability to follow load variations. The two-stage machine shown in Fig. 24.7 has a turn-down to 10 per cent of full load, thus making it highly flexible in operation. Single- or multi-stage machines are available, some driven through gearing to achieve the required speed of rotation. These machines are available in the capacity range of 700 kW to 4.5 MW and are used invariably for water or ethylene glycol solution chilling and may be water or air cooled. Capacity control is by throttling at the suction inlet or speed control. These machines are normally of the semi-hermetic or open type with the condenser and evaporator close-coupled to the compressor as one complete unit.

Absorption plant

A single stage absorption plant is shown in Fig. 24.8. This type uses either ammonia as the refrigerant and water as the absorption medium, or water as the refrigerant and lithium bromide as the absorption medium. The equipment has no moving parts except pumps. Two-stage plant is also available, incorporating a second concentrator which utilises heat recovery from the first stage of the cycle to provide improved efficiency. The source of energy being either steam, medium temperature hot water or gas, this type of plant is suited to

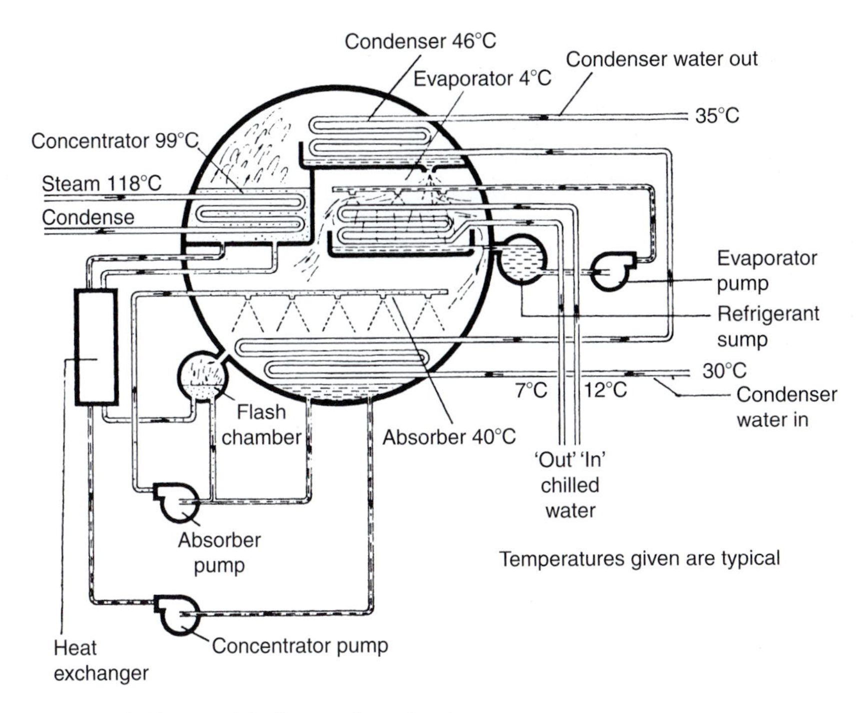

Figure 24.8 Diagram of single stage absorption system

installation where there is a heat source of a suitable grade available, such as with combined heat and power (CHP) systems. Such plant works under a high vacuum and is available in a range from 10 kW to over 5 MW cooling capacity. The heat to be removed by the condenser water with this system is about double that of a vapour compression plant of equivalent capacity. For example, an absorption chiller of 350 kW capacity will use a nominal steam quantity of 220 g/s, with heat rejection of some 900 kW. The equipment is usually very heavy and does need a level surface for installation.

Absorption chillers do not use refrigerants or other substances that can cause ozone depletion or contribute to global warming. However, many of the substances are toxic and safe handling and disposal procedures should be complied with in accordance with the manufacturer's instructions.

Steam-jet plant

An interesting type of refrigerator, using water as the medium, is that shown in Fig. 24.9. Its operation depends on the possibility of causing water to boil at low temperatures under high vacua. As noted previously, water at 7°C boils at an absolute pressure of 1 kPa, i.e. about one-hundredth of atmospheric pressure. The absence of any special refrigerant is an advantage both economically and environmentally.

The energy input with this equipment is much greater than with the positive compression types owing to the inefficiency of jet compression and the requirement for condensing water flow is about 5 times that of a vapour compression plant.

A variation of the same system, but using a centrifugal compressor in place of the steam-jet compressor, has also been developed and this avoids the above-mentioned disadvantage of high energy input. The great difficulty with both types is the maintenance of the extraordinarily high vacuum for long periods. This method is not widely used, but has been found economical in industrial plants where exhaust steam is available.

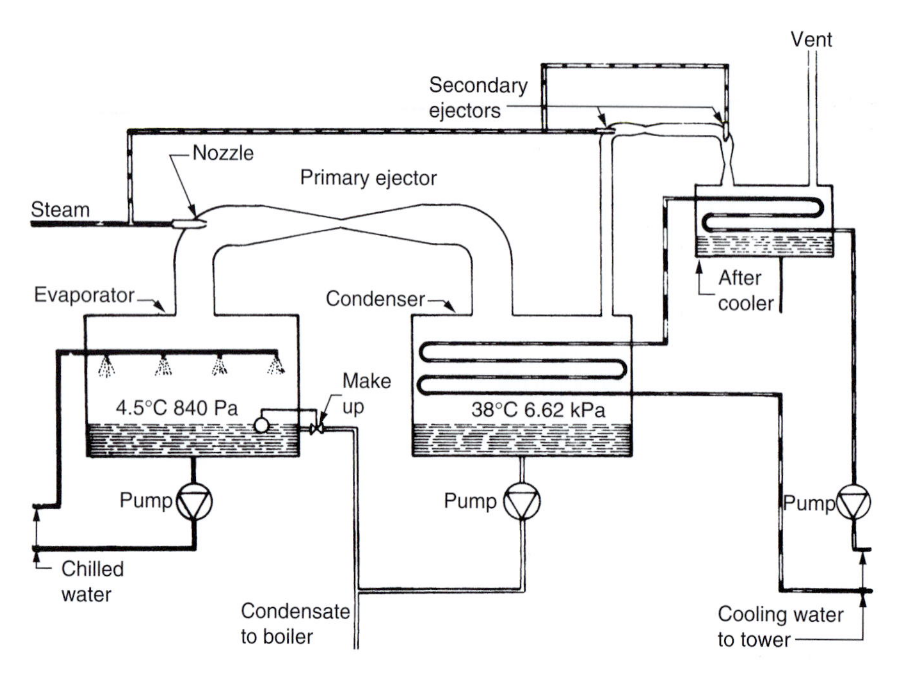

Figure 24.9 Diagram of steam-jet system

Choice of refrigeration plant

The selection of refrigeration plant will depend upon a number of factors, among which are: efficiency, the location of suitable space; the availability of waste heat; the importance laid upon plant noise and whether condenser heat must be rejected at a distance from the compressor.

Where a refrigeration plant serves a single air-conditioning system of the *all-air* variety, and the two can be located close together, a direct expansion coil or coils may be used subject to the practicality of matching the control characteristics of this type of heat exchange surface to the system load. In other circumstances, for single systems of any *air/water* variety and for all systems which consist of a number of distributed air-handling units, a chilled water system is used to transport cooling energy from the evaporator. In this latter case the refrigeration plant as a whole is often referred to as a *water chiller*. A typical arrangement of the principal components of a chilled water system is shown in Fig. 24.10.

Current practice on very large installations is to use centrifugal or screw compressors. Large air-conditioned buildings with central chiller plant are often designed with multiple chillers. Multiple chillers offer operational flexibility, some standby capacity and less disruptive maintenance. The chillers can be sized to handle a base load and increments of a variable load and, with a suitable sequencing control strategy, may achieve better energy efficiency than a single chiller installation. Where multiple refrigeration machines are installed, machine sizing should be related to the cooling demand profiles in preference to installing a number of equal-sized machines. Good control provisions in such cases are essential.

On medium to large plant either scroll, reciprocating or screw semi-hermetic multi-compressors are used. These can be either equally or unequally sized. For example, four compressors are often employed in a packaged unit, each compressor representing one step of capacity. This arrangement whilst energy-saving compared to other methods of capacity control, can result in frequent compressor cycling when small variations in load occur. A combination of this arrangement using one compressor with cylinder unloading or inverter control may provide a more reliable, cost effective and energy efficient alternative. Another alternative is the application of four unequal sized compressors which could also have the addition of individual capacity control

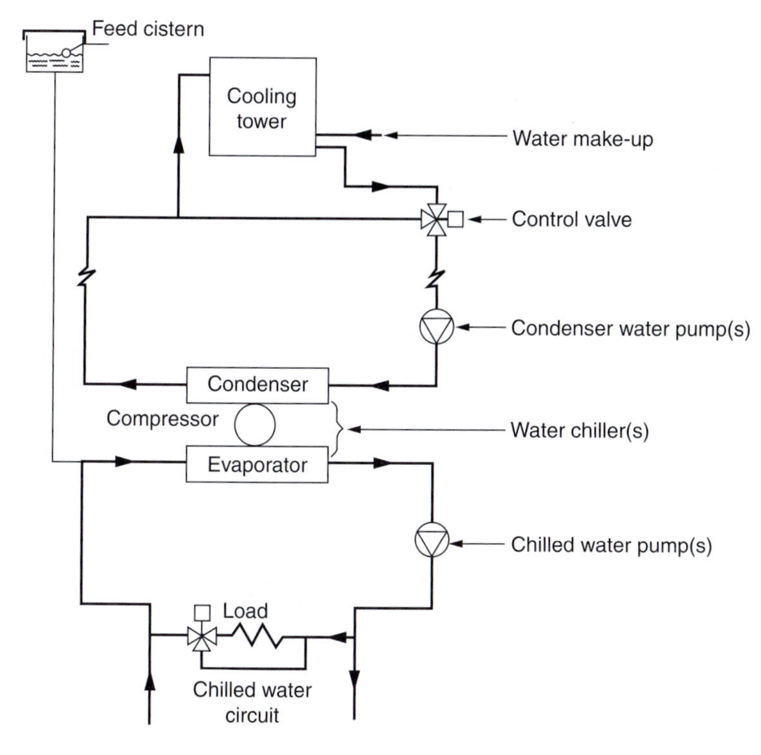

Figure 24.10 Chilled water system

or variable speed motors with inverter control. On small- to medium-sized installations, scroll or screw machines are now used extensively. The use of reciprocating equipment is generally becoming less frequent with exception to areas where maintenance capability of screw equipment is not readily available.

Where a vapour compression system is being considered, the designer should minimise the direct global warming effect from possible refrigerant emissions by selecting a refrigerant with a low GWP and by selecting a machine with a low specific refrigerant charge (kg of refrigerant per kW cooling capacity). *Total Equivalent Warming Impact* (TEWI) is a way of assessing the overall impact of refrigeration systems from the direct refrigerant-related and indirect fuel-related emissions. Designers should seek to minimise TEWI through the selection of an appropriate refrigeration machine and refrigerant and by optimising equipment selection and system design for the best energy efficiency.

When selecting refrigeration plant, ensure good refrigeration efficiency is achieved through the selection of an efficient machine and by minimising the refrigeration 'lift' (i.e., the difference between the temperature of the cooling medium (usually air or chilled water) and the heat sink (usually ambient air).

The Building Regulations Part L 2006 requires the *Seasonal Energy Efficiency Ratio* (SEER) to be calculated. The SEER is the ratio of the total amount of cooling energy provided, divided by the total energy input to the cooling plant summed over the year. For chillers, the energy efficiency ratio (EER), is the ratio of cooling energy delivered by the cooling system divided by the energy input. The EER is provided by the manufacturers of the equipment based upon performance against parameters detailed in BS EN 14511.

The reliability, security of supply, maintenance, and backup of the refrigeration or heat rejection system is a major design consideration the importance of which will depend upon the nature of the end user's business operation. A distinction must be made between critical and highly resilient operations such as machine equipment rooms and telecommunications centres, and standard office comfort cooling applications. The financial consequences of the loss of cooling to a dealer floor may be considerable, whereas loss of comfort cooling may be tolerated for short periods.

The risk of an outbreak of *Legionnaires' disease* have led designers to favour the use of air-cooled condensers, in preference to any water cooling methods, wherever it is possible to so site the refrigeration plant that the necessary large quantities of outside air are freely available for circulation. However, there is an energy and space penalty for using air cooling, due to the resulting higher condensing temperature, which may be of the order of an additional 20 per cent power absorbed by the compressor motor. Since a cooling tower may be located at a considerable distance from the refrigeration plant, horizontally or vertically, such a combination has obvious attractions as far as flexibility is concerned and it is common practice for the machinery to be sited in a basement and the cooling tower at roof level. It is of course necessary that an open circuit cooling tower should be positioned sufficiently above the refrigeration plant to prevent any drainage or priming problems. Where this cannot be achieved, an intermediate heat exchanger may be installed.

Multiple chillers

In large air-conditioning systems, it is a common practice to split the refrigeration capacity between multiple machines in parallel with chilled water control. Unless, careful attention is given to low load condition, frequent compressor on/off cycling can occur, exceeding the manufacturers limits. It is also essential to coordinate the design of the control of the air-handling equipment with that of the refrigerating machines; the choice being between a constant flow and a variable flow chilled water system.

Large systems may require the use of several chillers, either to meet the required capacity and/or to provide plant redundancy. In these cases the following circuit arrangements should be considered (see Fig. 24.11).

(a) *Parallel evaporators*: Parallel circuits allow multi-pass heat exchangers at a relatively low water pressure drop, consequently a lower pump power is required than for a series circuit. However, a slightly higher compressor power is required than for a series circuit, due to both machines having the same evaporating temperature. The designer should design the controls carefully to avoid frequent on/off cycling under partial load conditions. There is also a danger of freezing one evaporator when the other is switched off and control is by a common thermostat downstream of the evaporators.

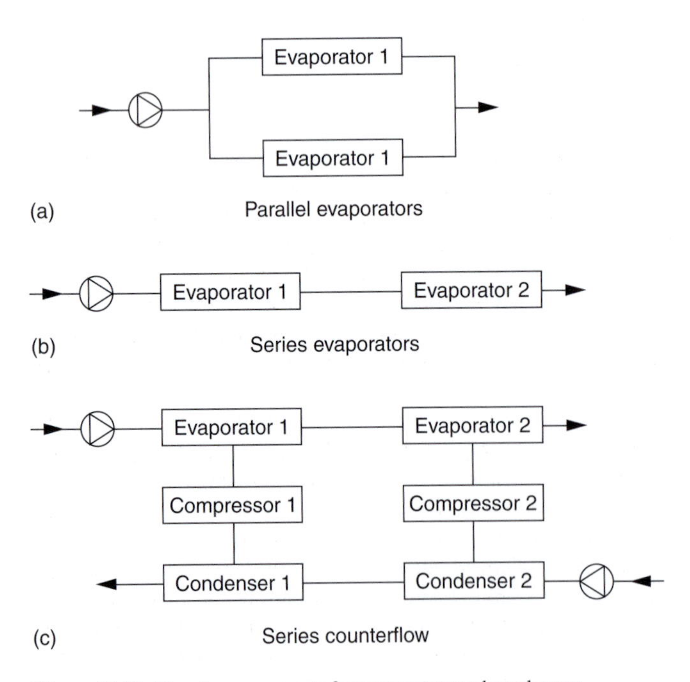

Figure 24.11 Circuit arrangements for evaporators and condensers

(b) *Series evaporators*: Compared to the parallel arrangements, systems of this type impose a higher chilled water pressure drop and, therefore, a higher pump power is required. Consequently a single pass evaporator may well be necessary. The compressor power is slightly lower than for a parallel arrangement, as the upstream machine will have a higher evaporating temperature. Chilled water temperature is generally easier to control than with the parallel arrangement.
(c) *Series counterflow*: Although having a higher pressure drop, this arrangement of evaporators and condensers can result in a lower compressor power than either parallel or series arrangements for multiple machine installations, particularly for heat reclaim schemes.

Of the above options (a) and (b) are the most popular.

Storage systems

An annual cooling load profile, plotted for almost any air-conditioned building, would show that the maximum load only occurs on a few occasions. This is due to the fact that many of the components of the total are not only seasonal but transitory hour-by-hour. Thus, where the central plant is sized to match the peak load, it will operate at considerably less than its full capacity for the majority of the time. An alternative to sizing a refrigeration plant to meet the peak cooling load would, therefore, be to provide some means to store a cooled medium (water, ethylene glycol solution, ice, etc.) and thus allow a somewhat smaller plant to run for longer periods at full output. The stored energy would then be available, as required, to make up the deficit between the plant capacity and the peak requirement. The advantages of such an arrangement are:

- Reduced size of refrigeration plant reduced.
- Maximum demand on power supply.
- Reduced unit energy charge, if run off-peak inherent.
- Standby capacity in the energy store stability.
- Available for control.

The parallel disadvantages are:

- higher capital cost,
- increased space requirement,
- possibility of lower COP with low temperature or ice storage systems.

The storage system concept may be applied in varying degrees. It is not uncommon in the UK for a limited quantity of chilled water to be provided for *peak lopping* and, in this case, capacity would be provided to supplement a water chilling plant for 1 or 2 hours in the day, at the time of maximum load. At the other extreme, it would not be impossible to provide sufficient storage to meet a total peak-day cooling load, the plant running during night hours only during the period when an off-peak tariff applies. A further option would be to run the plant continuously over 24 hours and to provide a limited level of storage such that during the whole period of demand for cooling, both the plant output and the stored energy would be used in parallel. Figure 24.12 illustrates these three basic operating modes and indicates the relationship between plant size and stored water quantity.

Whereas the design of a hot water storage system (Chapter 8,) is able to limit the required volume of the vessels by applying pressure and storing water at an elevated temperature, parallel action (i.e. storage at a temperature much lower than that of usage) is not really practical in the case of chilled water. A temperature of about 4°C is that most usually chosen as the practical minimum but, even at this level, design problems arise since the specific mass of water at that temperature is critical and stratification is unstable.

The use of alternative media such as ice and phase-change chemicals must be considered if storage volumes are to be of practical size. As a result, there has been an increasing use of ice storage systems over the past few years, particularly large city centre office developments where roof space is at a premium and quite often there is a surplus of lower value basement space. Another advantage to the office occupier is that because there is a significant, sometimes by 50 per cent, reduction in the refrigeration capacity required the

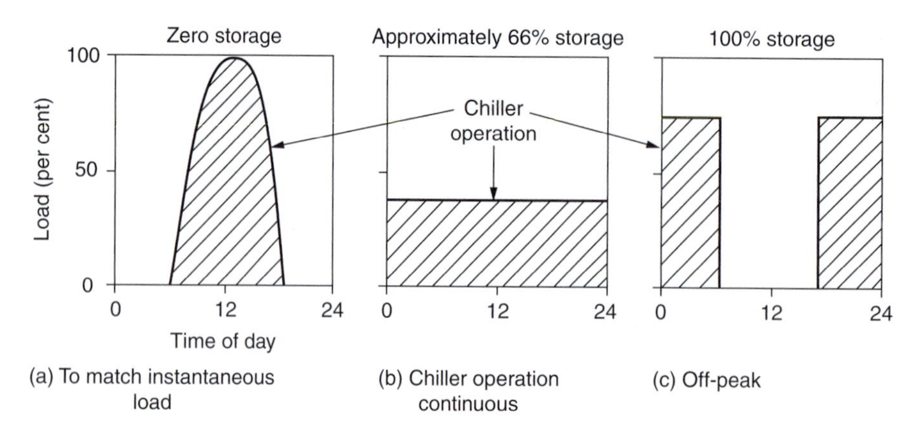

Figure 24.12 Chilled water storage options

use of dry air coolers in lieu of wet cooling towers becomes more practical. The systems also have inherent stored cooling capacity but care must be taken where there is a 24-hour cooling requirement, e.g. computer equipment rooms.

CIBSE Technical Memorandum TM18:1994 Ice storage provides comprehensive information on the use of ice for cool thermal storage. Whilst there are various systems available in commercial office schemes it is more usual to incorporate indirect ice-on-coil systems with a control strategy providing chiller priority. In this case, the cooling load for the building is met initially by the chillers with ice store topping up any shortfall in capacity, then, if advantageous from an energy tariff point of view, the chillers will be turned off at a time that allows the ice store to meet the cooling load and be depleted by the end of the day. The advantages of this arrangement are that the chillers will be operating more often at their maximum capacity and hence more efficiently and the size of the ice store is usually less.

Free cooling

Use may be made of the principle of evaporative cooling in order to produce water at a temperature low enough to be suitable for cooling in an air-handling plant. A cooling tower, provided to reject condenser heat in summer, may be usefully employed to provide a source of cooling at times when the outside wet bulb temperature is low enough. A basic arrangement is shown in Fig. 24.13 where a closed circuit water cooler is piped into the chilled water circuit: an alternative arrangement would be to use an open circuit tower with a heat exchanger interposed between the open circuit and the closed chilled water circuit, in order to prevent fouling in the closed circuit.

A system of *dry coil free cooling* may be used to advantage, in particular with systems which operate continuously, such as those which serve computer suites. A water and glycol mixture is circulated by a pump between a fan-assisted dry cooler, similar in form to an air-cooled condenser, and a cooling coil in the air stream. Typically, such a system would be designed such that, when the outside air temperature fell to 5°C or below, full cooling would be achieved by this method. As the outside air temperature rose above 5°C, cooling would be introduced progressively from a conventional refrigeration plant to supplement the free cooling effect. A development of this principle is to use the glycol/water circuit and fan-assisted cooler in a dual mode, as a source of free cooling in winter and as a means of heat rejection from the refrigeration machine in summer. The principles of these approaches are shown in Fig. 24.14, the equipment being available in packaged form from specialist manufacturers. It is claimed that a pay-back period of 2 to 3 years may be achieved on the additional capital cost of the equipment.

The most efficient form of free cooling in this context is by use of the *thermosyphon*[3] principle. This may be applied using either dry coolers or evaporative condensers, the latter being more effective since they cool the

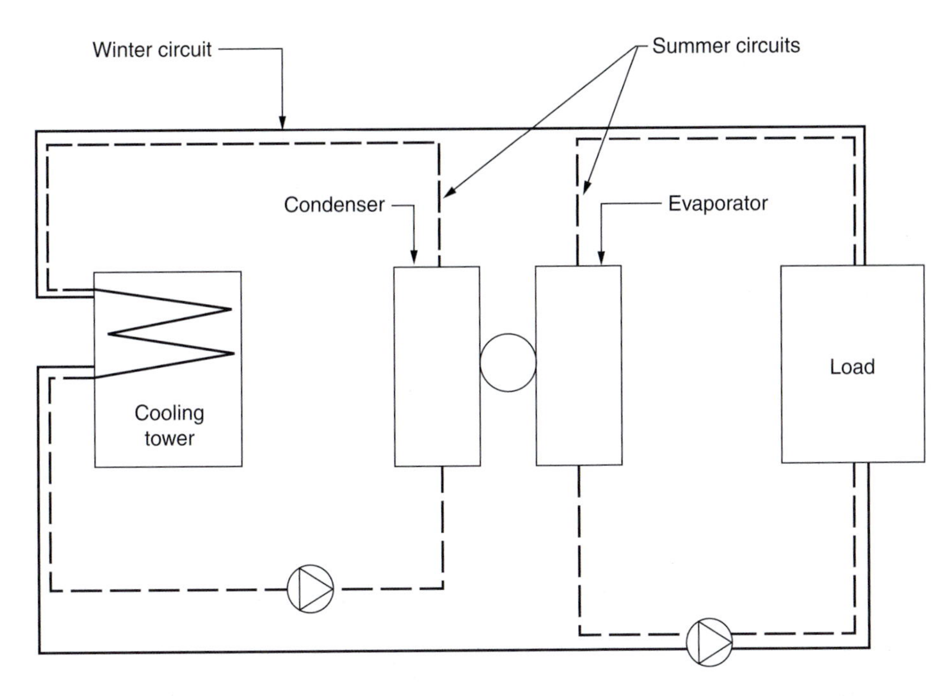

Figure 24.13 Free cooling using an evaporative cooling tower

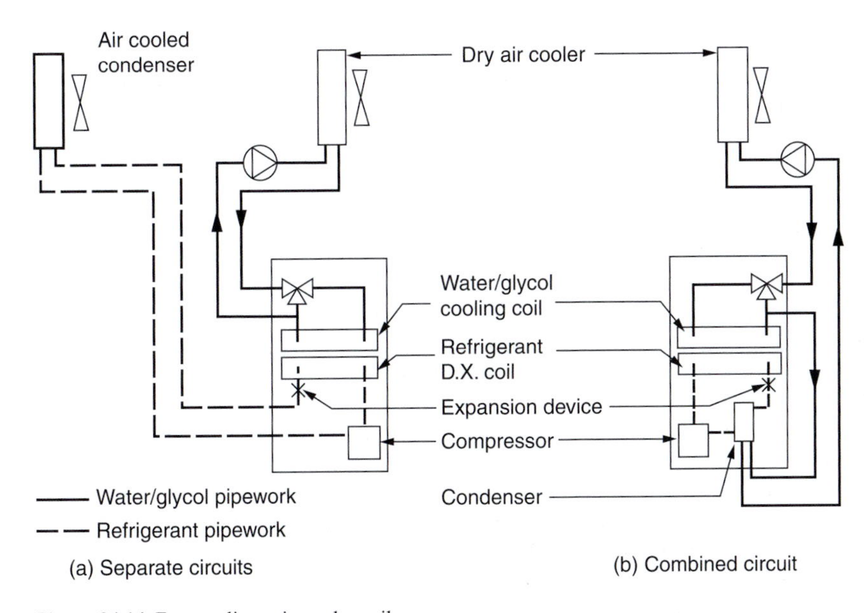

Figure 24.14 Free cooling using a dry coil

refrigerant towards the ambient wet bulb temperature rather than the dry bulb, as is the case with dry coolers. The refrigerant is piped in the normal manner, via the compressor, between a purpose-designed evaporator and the condenser, which must be sited at a higher level than the remainder of the system. A bypass facility is provided around the compressor.

When outside ambient conditions are suitable, the compressor may be switched off and the bypass put into use, allowing refrigerant to flow between the evaporator and condenser. In this mode, the evaporated refrigerant from the chilled water heat exchanger is drawn up to the condenser, by pressure difference, where it condenses and returns by gravity to the evaporator. In a multi-chiller installation, using this method of operation when outside ambient conditions permit and with a constant cooling demand throughout the year, energy savings of about 50 per cent may be achieved. The additional capital cost of the installation would be of the order of 20 per cent.

Capacity controls for mechanical refrigeration

Reciprocating compressors

Capacity control of reciprocating equipment is achieved as follows:

Multi-modular: Several compressors are incorporated in a chiller package, each compressor representing a step of capacity. It is important that the chiller control system is compatible with the compressor safety controls, which should be set to avoid frequent cycling.

Cylinder unloading: Several methods are available but is most common for the suction valve on one or more cylinders to be maintained in a raised position by hydraulic pressure, so allowing the refrigerant gas to pass back and forth without check and thereby reducing the mass flow through the compressor. A minimum gas flow must be maintained to minimise overheating and ensure adequate oil return. Insufficient oil return will adversely affect the operation of compressors or contribute to nuisance tripping of the oil safety switch. It is recommended that long hours of operation with unloaded cylinders is avoided.

Speed variation: The output of a reciprocating compressor is directly proportional to the speed of shaft rotation, which may be changed by varying the speed of the prime mover. A certain minimum speed must be maintained for lubrication to be effective. Two-speed compressors have been used in the past but variable speed utilising inverter control is becoming more common. Whilst accepting that the majority of compressors used will be semi-hermetic or hermetic, it is important to note that with multi-speed and inverter control applications, there may be problems with damage to windings during operation. This may be due to fluctuations ('spikes') in the electricity supply or to the compressor power requirement during speed changes. It is important that the designer gives careful consideration to this issue.

Hot gas bypass: The load on the compressor is maintained whilst the evaporator capacity is varied. The most effective arrangement is for the hot refrigerant gas to bypass the condenser and inject the refrigerant into the system down stream of the expansion valve and upstream of the evaporator. It should be noted that this method of capacity control offers no energy economies at part load, and depending on method chosen, can result in high discharge temperatures and therefore should be avoided. Extensive operation can cause compressor damage.

Evaporator pressure regulator: This is a means of maintaining the evaporator pressure by throttling the flow of gas to the suction of the compressor. Energy efficiency is impaired and therefore this method should be avoided.

Cylinder unloading and speed variation are more economical due to the great reduction in power consumption arising at part load compared with the small or zero reduction arising from using hot gas or evaporated pressure regulation.

Centrifugal compressors

In most centrifugal applications, the machine must respond to two basic variables:

(1) refrigeration load
(2) entering condensing water temperature.

A centrifugal compressor is, for a given speed, a relatively constant volume device compared with a multi-cylinder reciprocating compressor where cylinders can be deactivated progressively to accommodate load changes.

The control system must be able to alter both the head and flow output of the compressor in response to load changes. This is possible by using one of the following methods:

(1) refrigerant flow control by variable inlet guide vanes
(2) variable speed control.

Speed control is generally the most efficient method. However, its use is limited to drives whose speed can be economically and efficiently varied and to applications where the discharge pressure (head requirement) falls with a decrease in load (this restriction only applies to centrifugal compressors).

The most generally accepted method of flow control, particularly for hermetic centrifugal compressors, is that of variable inlet guide vanes. The vanes are usually located just before the inlet to the impeller wheel (or first impeller wheel in multi-stage compressors) of the compressor and are controlled by the temperature of the water leaving the evaporator. This method offers good efficiency over a wide range of capacity.

Hot gas bypass is useful to extend the control range of a machine to very low loads, particularly where the system head requirements (condenser pressure) remain high, thereby avoiding compressor surge. Instead of discharging gas into the compressor inlet, which can cause high-temperature problems, the hot condenser gas is passed through a pipe and valve to the bottom of the cooler, thereby providing a 'false' load on the cooler. In this manner the compressor experiences a constant load. However, this technique should not be used continuously but only for occasional part load conditions. Designers should be aware that centrifugal compressor manufacturers often quote a time limit for continuous part load operations.

Surge is caused by flow breakdown in the impeller passageways, the impeller can no longer maintain the required system pressure and a periodic partial or complete flow reversal through the impeller occurs. Surge is characterised by a marked increased in the operating noise level and by the wide fluctuations in discharge pressure eventually leading to shut-down. For this reason it is often not practical to run a centrifugal compressor at part load under high summer condenser temperature conditions. Designers should be aware that for the same cooling capacity, different sizes of compressor have different surge lines.

Screw compressors

Capacity control is normally obtained by varying the compressor displacement using a sliding valve to retard the point at which compression begins and, at the same time, reducing the sizes of the discharge port to obtain the desired volume ratio. This typically allows 10 to 100 per cent capacity control although below 60 per cent of full load the compressor efficiency is very low. Variable motor speed control using an inverter is also increasingly used and at low loads offers higher efficiency than the slide valve method. Another form of capacity control is the use of multiple compressors.

Scroll compressors

Capacity control can be obtained using two speed motors for multiple compressors. Although variable motor speed has been used it is not ideally suited for scroll compressors because it is not compatible with the method of radial compliance usually employed in scroll compressors which prevents damage by small quantities of liquid refrigerant or solid particles passing through the compressor. Speed control also creates difficulties with the compressor lubrication system. A relatively new form of capacity control uses an electronic modulating system that momentarily separates the scrolls axially and can provide between 10 and 100 per cent capacity variation. Because the shaft continues to rotate at full speed the compressor lubrication system is not affected.

Refrigeration plant components

Refrigerant pipework

Some plant configurations will involve the design of pipework carrying the refrigerant between components in the cycle. Special attention must be given to this important aspect of the installation because the fluid in circulation, being volatile, may be in gas or liquid form or a mixture of the two. When dealing with such design, reference should be made to specialist works on the subject, to ensure that the essential basic criteria are met in order to:

- Minimise oil loss from the compressor.
- Ensure satisfactory refrigerant flow to the evaporator.
- Prevent liquid refrigerant or oil in slugs from entering the compressor (during operation or when inoperative).
- Prevent oil from collecting in any part of the system.
- Avoid excessive pressure loss.
- Maintain the system clean and free from water.

All pipework or plant carrying refrigerant at low temperatures (low pressure/evaporator part of the circuit) or chilled water must be insulated to prevent the formation of condensation and to reduce heat gains. To this end an effective vapour barrier must be applied, or be integral with the insulation, on the outer surface. BS 5422: 1990 gives the recommended thicknesses for given system operating temperatures. Particular care should be exercised in the selection of pipework and system components when using ammonia to avoid copper and copper alloys.

Refrigerant detection

Refrigerant detectors and alarms are required in all refrigeration equipment plant rooms to prevent the exposure of workers to refrigerant concentrations higher than the Health and Safety Executive (HSE) OELs and to warn of higher toxic concentrations. Detectors are also required in plant rooms that contain hydrocarbon or ammonia systems to start emergency ventilation and shut down any electrical equipment that is not suitable for operation in explosive atmospheres. Refrigerant detectors can also be used as a means of detecting refrigerant leaks although their effectiveness is affected by how well the plant room is ventilated. Refrigerant detectors are unlikely to be effective for detecting leaks from equipment installations outside.

The location of refrigerant detectors should take account of the density of the refrigerant. HCFCs, HFCs and hydrocarbons are heavier than air and refrigerant detectors should therefore be located at low level. Ammonia vapour is lighter than air and detectors are therefore usually located above the refrigeration equipment.

Common types of detector include semiconductor sensors and infrared analysers. Electrochemical sensors are also used for ammonia detection. Semiconductor and electrochemical sensors are sensitive to other gases, including some cleaning chemicals, whilst infrared systems are not selective but are also more expensive. Detectors may be discrete single-point devices or aspirated systems that may have several sensing points connected to the sensors by air sampling tubes and a small air pump. All types of detector require periodic recalibration and electrochemical sensors have a short lifetime.

Evaporators

In its most simple form, the evaporator is tubular, the tubes containing the refrigerant and the whole array immersed in the liquid to be cooled. In the *dry system* the evaporator coils are filled with vapour having little liquid present but when operated on the *flooded system*, the liquid refrigerant discharges into a cylinder feeding the coils by gravity. As evaporation takes place, the gas returns to the top of the cylinder and from there returns through the suction pipe to the compressor.

A common form of evaporator for application to air-conditioning practice is the *shell and tube* type, used with a closed circuit chilled water system serving cooling coils. The water is contained in the tubes and the refrigerant in the shell. A development of the shell and tube type is the direct-expansion shell type evaporator in which the refrigerant is in the tubes and the water in the shell. The tubes may be arranged in two or more circuits, each with its own expansion device and magnetic valve on the liquid inlet to allow step control. Different arrangements of baffles in the shell control the water velocity over the tubes in order to improve heat transfer. Plate heat exchangers are also used extensively and these comprise of either a gasketed or brazed type. The gasketed plate heat exchanger consists of a pack of corrugated metal plates with portholes for the passage of the two fluids between which heat transfer will take place. The plate pack is compressed between a frame plate and a pressure plate by means of tightening bolts. These plates are fitted with a gasket which seals the channel and directs the fluids into alternative channels. This arrangement allows additional plates to be easily added to increase the duty of the heat exchanger. The channels formed between the plates are arranged so that the refrigerant flows in one channel and the coolant in the other. Very low coolant and refrigerant temperature differences are possible (less than 2°C) making plate heat exchangers very suitable for systems that employ 'free cooling' techniques. The advantages of this type of heat exchanger include very low refrigerant charge (the internal volume is only about 10 per cent of that for flooded evaporators) and the heat transfer coefficients can be 3 to 4 times greater than that of a shell and tube heat exchanger. The disadvantages include oil fouling occurring which can affect the heat transfer and freezing due to a low mass of coolant inside the heat exchanger.

The brazed plate heat exchanger is a variant of the gasketed plate heat exchanger except that it cannot be dismantled for cleaning. It is composed of a number of 'herringbone' corrugated plates brazed together. The plates are normally stainless steel coated with copper on one side. The assembly is clamped together with end plates and heated in a vacuum oven until the copper melts and forms a brazed joint. The channels between the two plates can be varied in their cross-sectional dimensions to achieve the optimum heat transfer coefficient for the required application. The advantages and disadvantages of brazed plate heat exchangers are the same as for gasketed plate heat exchangers.

With the exception of refrigerant blends, it is generally not necessary to resort to ethylene glycol solutions for air-conditioning purposes, since chilled water at about 4°C satisfies all normal requirements. Precautions against accidental freezing of the water in the evaporators include low suction pressure cut-outs and low water temperature cut-outs as well as water-flow switches. Care should be taken if using refrigerant blends such as R407C to avoid risk of icing due to low-temperature glide. The extent of the temperature glide is dependent upon the boiling points and proportions of the individual constituents.

In instances where the cooling load may be less than the minimum output of the refrigeration machine and where some severe limitation exists as to the number of starts per hour, the use of an inertia vessel, sometimes referred as a buffer vessel, may be necessary.

In order to restrict the number of times that an electric motor starts in an hour due to the heating that takes place on start-up. This heat, caused by the initial low reactance and consequent high inrush current, must have time to dissipate, otherwise the motor may burn out. Although important for all motors, it is especially so for those used in hermetic chillers when cooling is restricted and consequently the safe number of starts per hour are less than those for an open motor. Typical figures would be:

Motor type	*Starts per hour*
Open motors	10
Motors for hermetic Reciprocating compressors	4–6
Motors for hermetic Centrifugal chillers	2

The parameters which influence the frequency of starting are primarily the minimum duty of the chiller, the mass of water in the system and the temperature range over which the return water is allowed to vary. Of secondary significant is the flow rate of the water passing through the chiller and hence the temperature drop across the chiller.

Considering a system comprising a chiller operating at minimum capacity, a cooling coil and pump. If the load on the coil is equal to the minimum capacity of the chiller, the water temperature at the inlet to the evaporator will be constant and the chiller will run without stopping. If, however, the load on the coil is less than the chiller capacity, heat will be removed from the coil and also from the water, gradually lowering the temperature of the water returning to the evaporator until the thermostat opens and shuts down the compressor. If:

Q = the minimum compressor duty (kW)
$q(1)$ = average load during pulldown (kW)

$q(2)$ = average load during heat-up (kW)
t = variation in water temperature at the chiller inlet: °C (design LChWT – minimum possible LChWT)
$T(1)$ = cool down time (seconds)
$T(2)$ = heat-up time (seconds)
W = the mass of water in the primary system (kg)
F = the frequency of starts per hour
c = the specific heat capacity of water (4.187 kJ/kg°C)

Then during pulldown the heat removed from the water equals $Q-q(1)$, and the time taken for the water to drop in temperature,

$$T(1) = \frac{W\Delta tc}{Q - q(1)}$$

Once the compressor has switched off, the temperature of the water will rise due to the heat being removed from the air by the coil: $q(2)$, and the time taken for this to happen,

$$T(2) = \frac{W\Delta tc}{q(2)}$$

$T(1)$ and $T(2)$ are a minimum, and hence F is a maximum, when $T(1)$ equals $T(2)$ and the load applied is 50 per cent of the minimum chiller capacity.

$$\text{The cycle time} = T(1) + T(2) = \frac{W\Delta tc}{Q - q(1)} + \frac{W\Delta tc}{q(2)} = W\Delta tc\left\{\frac{1}{Q - q(1)} + \frac{1}{q(2)}\right\}$$

The frequency

$$F = \frac{3600}{T(1) + T(2)}$$

When $T(1) = T(2)$ then $Q - q(1) = q(2)$

$$\text{And } F = \frac{3600}{W\Delta tc\dfrac{2}{q^{(2)}}} = \frac{1800q(2)}{W\Delta tc}$$

Although it is possible that $q(2)$ will be less than $q(1)$ the difference on a large project is not likely to be significant and thus for most instances $q(1)$ may be assumed to be equal to $q(2)$ and in which case Q equals 2 $q(2)$ thus F may be expressed as

$$F = \frac{900Q}{W\Delta tc}$$

And as c may be assumed equal to 4.187 kJ/kg °C

$$F = \frac{900Q}{4.18W\Delta t} = \frac{214.95Q}{W\Delta t} = \text{say}\frac{215Q}{W\Delta t}$$

Δt has been defined as the allowable variation in water temperature at the inlet to the evaporator. The maximum temperature will be the maximum allowed to pass to the coil and minimum that at which it is still safe to operate the chiller. Although, there will be some overshoot of temperature at the inlet to the evaporator the requirements of both the coil and the chiller will still be protected by the thermostat. This overshoot will tend to decrease the frequency of starting, but as it is determined by the coil and control characteristics for each particular project it is not possible to put an actual figure to the eventual frequency of starting, safe to say that it tends to increase the safety margin of the unit operation. The determination of the maximum temperature of water flowing to the coil must be made by the designer taking into account the load pattern to those areas supplied by the chiller. The minimum temperature on to the chiller at switch off depends on the compressor/evaporator performance at minimum duty.

Evaporators in the form of *direct expansion (DX)* cooling coils in the air stream are described in Chapter 22. The refrigerant flow rate is controlled by a thermostatic expansion valve, or a capillary tube on smaller plant; the sensor to control the valve being located in the suction line to the compressor to maintain the correct degree of superheat at the compressor intake and to avoid the damage which would occur if liquid entered. Coils handling volatile refrigerants present fluid distribution problems and to ensure equal flow of refrigerant through each circuit in the cooling coil assembly, the fluid is passed through a distributor at the coil inlet to divide the flow equally.

Condensers

The evaporative condenser, consisting of coils of piping over a tank from which water is drawn and then circulated to drip over the pipes through which the refrigerant is passed, is the simplest form of condenser. Its use is restricted to cases where the compressor can be sited near to the condenser, to avoid long lines of piping containing refrigerant under pressure. Evaporative condensers are available in sizes up to 1 MW. They give an increase in efficiency because any intermediate exchange of heat is eliminated as compared with a water-cooled condenser connected to a cooling tower. The design must be carefully considered where capacity control is required and where winter operation is envisaged.

Often for air-conditioning applications, the condenser takes the form of a refrigerant-to-water heat exchanger of the shell and tube multi-pass type. Circulating water piping from the condenser to the water cooler with these types traverses the building, and all equipment containing refrigerant is confined to the plant room. Water cooling systems, however, are less popular today because the temperatures at which these operate, around 30°C, are such as to increase the risk of active development of the bacteria of *Legionnaires' disease.* With regular and thorough maintenance, combined with a suitable water treatment, the risk of harmful bacteria development is small.[4]

This risk has led to an increase in the use of air-cooled condensers and dry air coolers for all sizes of plant, where the building occupants are particularly susceptible to the disease, as may occur in hospitals. Air-cooled condensers are in any case used in most small self-contained or split-package systems and in duties of up to 350 kW in capacity, Fig. 24.15. They are often found to be economical in first cost up to about 100 kW cooling capacity. For larger plant, the running cost consequence of dry air coolers compared with that of evaporative water-cooled condensers must be considered. The higher condensing temperature that results from air cooling leads to the compressor doing more work to produce the same cooling effect, giving a lower coefficient of performance and in consequence a higher running cost.

Packaged units comprising evaporator, compressor and multiple-fan air-cooled condenser are available in large sizes, up to 1.6 MW cooling, limited only by the largest size that can be transported on a low-load lorry. A typical example is shown in Fig. 24.16.

Care must also be exercised when arriving at the external design conditions when sizing heat rejection equipment. Actual air temperatures can exceed the selected heat rejection design temperature because of climate change, city-wide heat island effect, local (roof-top) heat island effect and because of localised air re-circulation. For example, when located on a roof with side-screens, this can increase the local external temperature by as much as 5°C. This is as a result of solar gain into the area and a percentage, however small, of inevitable shortcircuiting of the cooling air. The effect on the capacity of the heat rejection equipment can be dramatic because the temperature difference between the condensing temperature or the condenser water temperature and the ambient condition is usually small. If the actual outside air dry bulb temperature exceeds the selected heat rejection design temperature, there is a strong chance that a refrigeration machine will trip out on high condensing pressure unless some kind of unloading device is fitted to the evaporator/compressor combination. This will either cause the failure of the whole cooling system in the case of tripping or result in

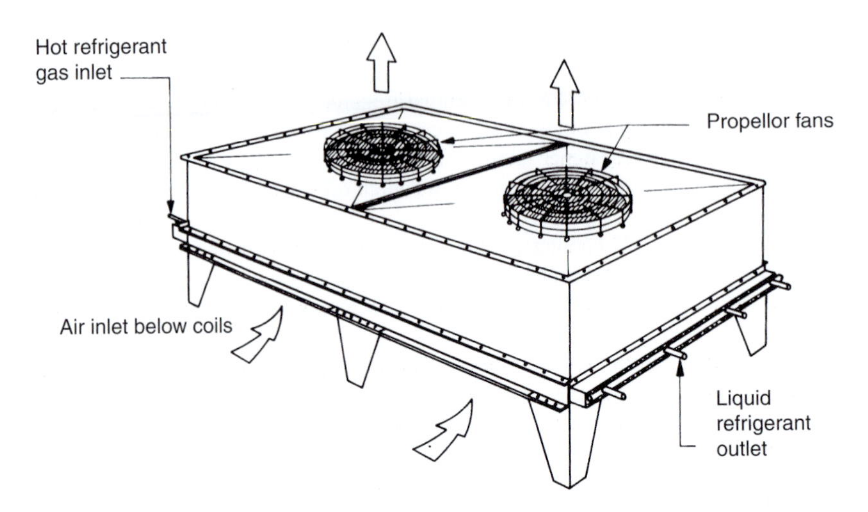

Figure 24.15 Air-cooled condenser

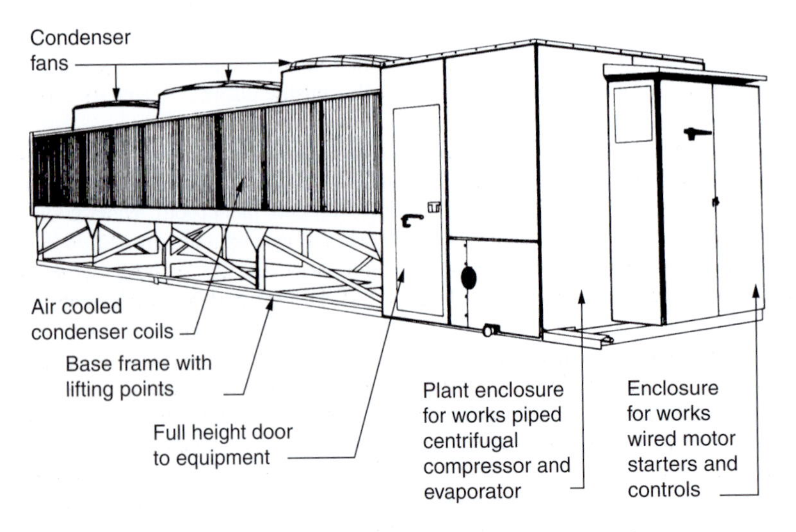

Figure 24.16 Packaged air-cooled centrifugal plant (Trane)

increased internal conditions in the case of unloading. Detailed below are suggested external design temperatures for the selection of heat rejection equipment.

Location	*Recorded air dry bulb temperature*	*Suggested baseline for heat rejection equipment selection*
Belfast	26	30 or 35
Birmingham	35	40
Cardiff	31	35
Edinburgh	34	35 or 40
Glasgow	29	35
London	38	40
Manchester	34	40
Plymouth	29	35

Increasing the design selection temperature will probably make the condenser coils bigger, heavier, more expensive and may make units noisier because of increased air flow. For most of the year, the condenser will be oversized and will provide lower condensing temperatures for refrigeration equipment (unless fan speed control is provided) – this will produce improved COP/EER and will improve energy consumption.

A development of the dry air cooler, which requires a smaller heat exchange surface and a lower air quantity, is realised by precooling of the upstream air using an *adiabatic effect*. The approach air stream is sprayed with a fine mist of mains water, or a pump pressurised supply, the quantity injected being controlled carefully to ensure that it is totally absorbed with no residue. The cooled and humidified air is then drawn over the air cooler by the fan(s) in the usual way. As a further precaution against *L. pneumophilia*, the spray water may be passed through an ultraviolet irradiation unit to provide disinfection. In hard water areas, ionic scale prevention should be used.

Air-cooled condensers are often used in the tropics, despite the fact that they are bulky, in circumstances where the outside air temperature is high. This apparent paradox arises from the simple fact that evaporative condensers and cooling towers require some measure of skilled maintenance, whereas the air-cooled condenser requires little attention other than simple basic cleaning and, of course, some lubrication. A further advantage is that it requires no water supply in those places where such may happen to be scarce.

Evaporative coolers

The heat extracted by the refrigerating machine (cooling effect), together with the heat equivalent of the power input to the compressor, raises the temperature of the condenser water by an amount which is dependent upon the quantity of the water which is circulated through the condenser.

The lower the temperature of the condenser water the less power will be required to produce a given cooling effect; and it also follows, conversely, that with a given size of plant the greater will be the amount of cooling possible.

Water from a well or borehole or from the main supply will always be the coldest, the former at 12°C and the latter at about 18°C in summer. In the case of mains supply, the quantity to be wasted, however, generally rules out this method. For instance, a 700 kW plant with power input of about 120 kW with a 10 K rise through the condenser would require a flow of $(700 + 120)/(10 \times 4.2) = 20$ litre/s and it is likely that water charges would exceed the cost of current for running the compressor many times over. An alternative is river water, which, subject to water authority approval, may be used directly for condenser cooling purposes. Before such a solution is adopted, however, consideration must be given to the need for filtration, the quality of the water and whether the composition is aggressive to the system materials; maintenance will need to be of the highest quality.

Applications exist outside the British Isles, however, where well water is more freely and cheaply available, which produce economical solutions. In one known instance in Europe, such well water is first passed through a precooling coil integral to the air-handling plant prior to use in the condenser. Similarly, in some parts of the Caribbean, clear sea water is available via fissures in the coral which may be pumped in quantities of up to 1000 litre/s through specially designed condensers.

However, cooling the water by evaporation is more generally adopted. Evaporative coolers depend on the ability of water to evaporate freely when in a finely divided state, extracting the latent heat necessary for the process from the main body of water, which is then returned, cooled, to the condenser. In the case stated above, the consumption of water with an evaporative cooler (assuming no loss of spray by windage) would be only $820/2258 = 0.36$ litre/s.

Evaporative coolers divide themselves into two categories, i.e. *natural draught* and *fan draught*. The former is represented by:

The spray pond: In this type the water to be cooled is discharged through sprays over a shallow pond in which the water is collected and returned to the plant. To prevent undue loss by windage the pond is usually surrounded by a louvred screen. Owing to the large area needed for the spray pond system and, moreover, the consequent probability of pollution, this type of equipment is seldom possible for air-conditioning applications.

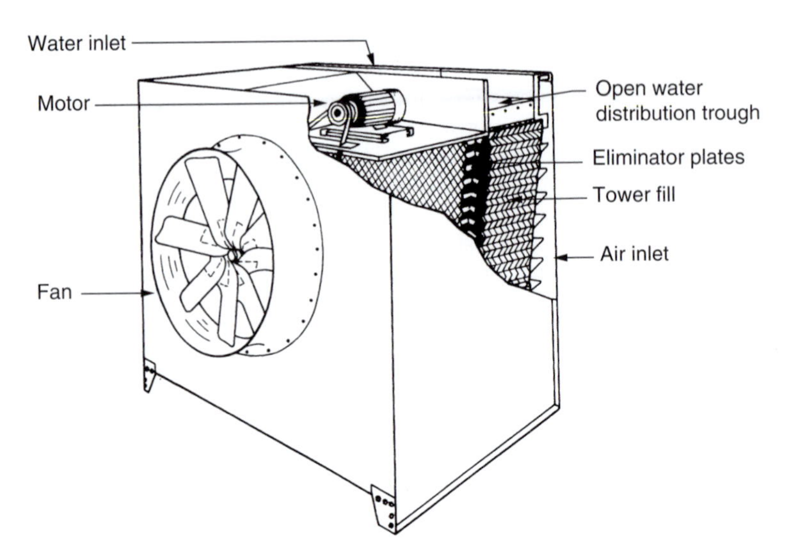

Figure 24.17 Low silhouette cross-flow cooling tower (SPX Marley)

Natural draught cooling tower: In this, water is pumped to the top of a tower which contains a specially designed 'packing' of plastic or other material which will not support microbiological growth, arranged so as to split up the water stream and present as large an area as possible to the air, which is drawn upwards due to the temperature difference, and by wind. The base of the tower is formed into a shallow tank to collect the water for return to the plant. Again, owing to its size and height, this type of cooler is not frequently used for air-conditioning.

Condenser coil type (evaporative condenser): Where the refrigerating machine is near the point where an outdoor cooler may be used, the condenser heat exchanger may be dispensed with and the refrigerant delivered to coils outside, over which water is dripped by a pump. The water collects in a tank at the base and is recirculated. A louvred screen is usually necessary surrounding the coils.

Fan draught systems are more commonly used owing to the compact space into which they may be fitted. Where possible they are placed on the roof, but if this is impracticable they may be used indoors, or in a basement, with ducted connections for suction and discharge to outside. All fan-assisted coolers must be fitted with effective moisture eliminators at the air discharge to minimise the carry-over of moisture which may be contaminated. The following methods are typical:

Cross-flow type, Fig. 24.17: Low silhouette with the water pumped to the top of the tower and discharged over a fill material. Air is drawn in horizontally across the fill to be discharged vertically or horizontally by a fan located at the outlet.

Induced draught cooling tower, Fig. 24.18: In this case the water is delivered to the top by the condenser pump, and cascades over a packing as with the natural draught type. An axial fan is arranged to draw air upwards over the slats at high velocity so that a much reduced area of contact is required.

Forced draught cooling tower, Fig. 24.19: This type is similar in function to the induced draught type except that the fan, normally centrifugal, discharges air into the tower at low-level pressurising the shell slightly and forcing the air upwards through the packing.

Film cooling tower: Similar to the induced draught type; in this case the water is not sprayed or broken up in any way, but is allowed to fall from top header troughs down slats arranged in egg-crate form. The water remains as a film on the surface of the slats. Higher air velocities than usual are permissible without risk of carrying over of water, and yet air resistance is low due to the open nature of the surfaces.

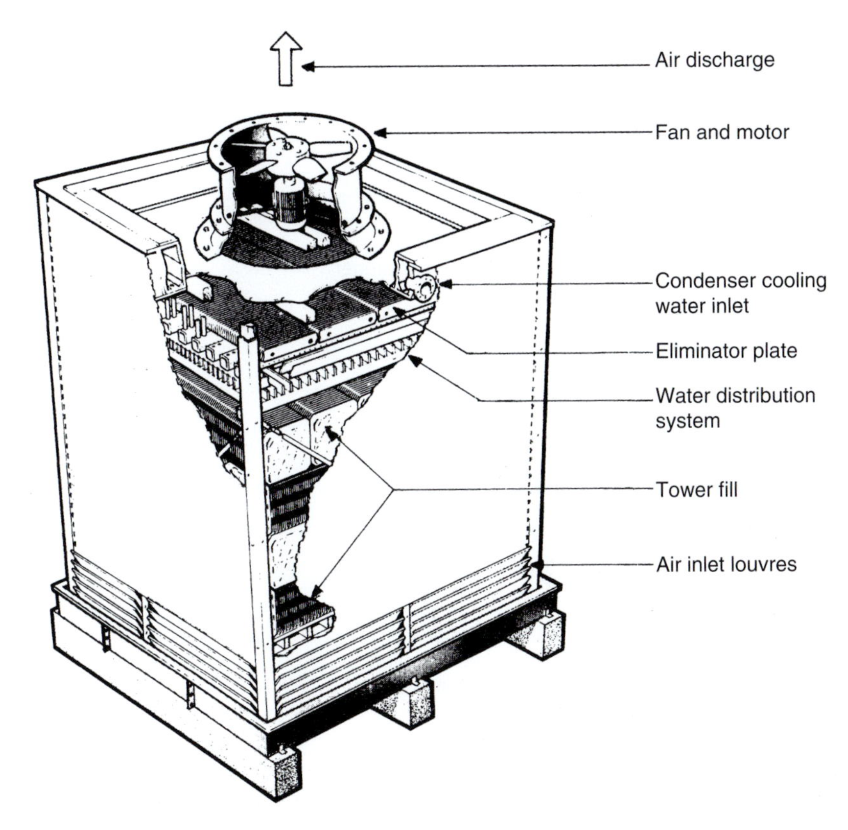

Figure 24.18 Vertical induced draught cooling tower (SPX Marley)

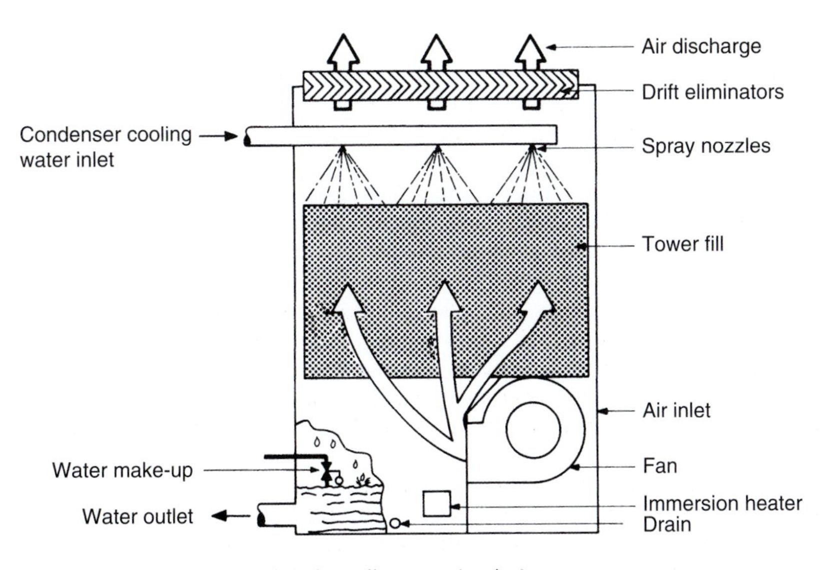

Figure 24.19 Vertical forced draught cooling tower (section)

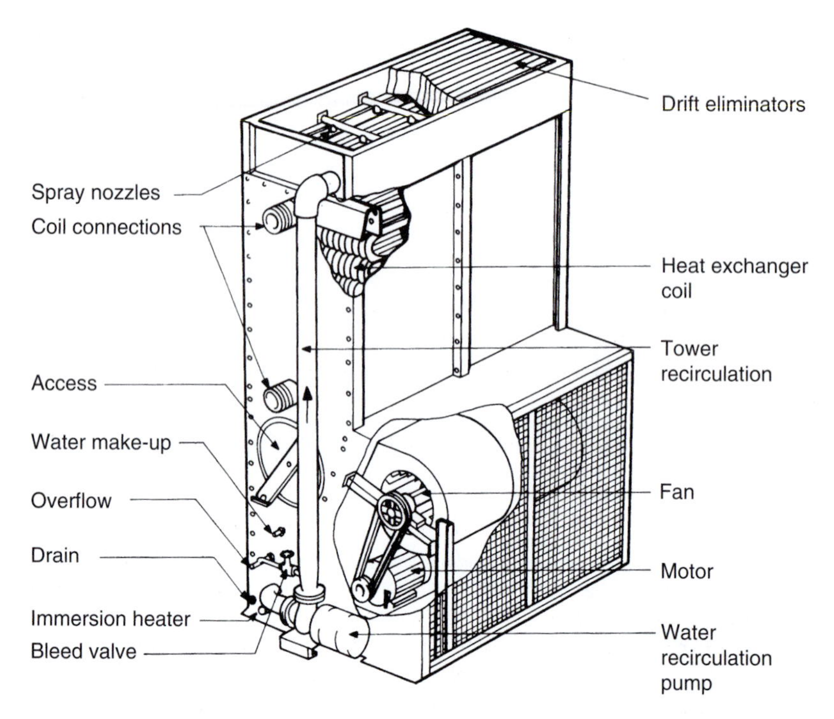

Figure 24.20 Closed circuit water cooler

Closed circuit water coolers, Fig. 24.20: In this case the condenser water is not evaporated but circulated through tube bundles which act, as it were, as the packing within the tower. In most other respects such coolers resemble induced draught or forced draught units. An independent water supply is required for the local evaporation circuit which is pump re-circulated over the tube bundles. Such coolers are useful for application to heat recovery systems where fouling of condenser water must be avoided.

where the cooler is below the level of the refrigerating apparatus; or where the cooler is used to provide 'free cooling'. Due to the additional heat exchange inherent in the circuit and consequent loss of efficiency, this type of tower tends to be relatively large in size.

Regardless of the type of evaporative cooler, its location must be carefully selected such that the air discharge, which could be contaminated, is not carried into air intakes, or into openable windows, or across public access routes. A high quality of maintenance is essential with such cooling devices and provision must be incorporated for inspection, maintenance, cleaning, water treatment and taking water samples, registration with local authorities may also be required. To reduce the quantity of water droplets carried from evaporative coolers, eliminator plates should be fitted at the air discharge position.

Rating of cooling towers

The heat to be removed from the condenser cooling water by the cooling tower is equal to the sum of the cooling load plus the heat equivalent of power absorbed by the compressor. An approximate figure of 1.2 kW per kW of refrigeration may be used.

The quantity of water to be passed through the tower is dependent on the temperature drop allowable between inlet and outlet. A usual figure is 5 K and the inlet water flow will then be $1.2/(5 \times 4.2) = 0.057$ litre/s per kW of refrigeration.

The temperature to which the cooling tower may be expected to reduce the condenser water depends on the maximum wet bulb of the atmosphere and the design of the tower: the higher the efficiency the closer will be the water outlet to the wet bulb temperature. A good efficiency will give a difference of about 3 K, so that, if the external design wet bulb is taken at 20°C (refer to Chapter 5), the cooling water will be brought down to 23°C, and, with a 5 K temperature rise through the condenser, the outlet will be at 28°C. This temperature then, in turn, forms the basis for design of the compressor and condenser (e.g. condensing at about 37°C).

Condenser water treatment

As evaporation takes place, there is a continuous concentration of scale-forming solids which may build up to such a degree as to foul the condenser tubes. Similarly, the evaporative surfaces of the cooler suffer a build-up of deposit. In order to overcome this problem:

- A constant bleed-off of the water is required, the rate of which may be calculated from the known analysis of the water, the evaporation rate and the maximum concentration admissible.
- As is common practice, the water may be treated by a regular chemical dosage which also generally contains biocides to prevent microbiological growth.

In the case of the closed circuit water cooler, the condenser tubes and the tower tube bundles do not suffer internal fouling but the external surfaces of the latter will, in time, become coated with solids. The need for water treatment of the spray water thus remains.

The total water loss from an evaporative cooler is the sum of evaporation loss, bleed rate and windage loss (water droplets carried from the tower by natural or by induced air movement) which would normally be between 3 and 5 per cent of the condenser water flow rate. It is normal for water storage equivalent to 24-hour usage to be held on site as a reserve against mains failure.

Water treatment to prevent scaling, microbiological contamination and corrosion is a legal requirement for any evaporative heat rejection system. The capital and running costs of water treatment may be significant and must be taken into account in calculating life cycle costs to assess the economics of alternative refrigeration and heat rejection systems.

Packaged cooling towers

Whereas past practice was to use structural containment to form the tower proper and to use purpose-built internals, water distribution arrangements, etc., cooling towers are now commonly factory fabricated for delivery to site in a minimum number of parts. Whilst this development reduces site works, there is a tendency to design to minimum size also with the result that the depth and capacity of the base tank is inadequate. This condition may lead to problems of overflow when the condenser pumps are stopped or, more seriously, to the formation of a vortex at tank outlet due to an insufficient water depth. To reduce the risk of this problem, only the minimum length of condenser cooling water pipework should be installed above the water level in the cooling tower tank, thereby reducing the quantity of water that will drain into the tank when the pump circulation is stopped; excessive drain-back will cause water loss from the system via the tank overflow. In all evaporative coolers, water loss by whatever cause is made up by 'fresh' water through a float valve.

Heat recovery

Air-conditioned buildings offer an inherent opportunity for some form of heat recovery system, utilising the heat to be rejected from the condenser of a refrigeration plant. The inner and core areas of such buildings are heat producing on a significant scale, due to lighting and other electrical loads, whilst the perimeter areas, in winter, require the addition of heat. Thus, if the air-conditioning system is such that the refrigeration plant operates in winter to produce chilled water, as in the case of a four-pipe fan coil system (see Chapter 19, low grade heat is available for immediate use or for storage and use at some later time. The progression of design development is shown in Fig. 24.21. Part (a) represents the conventional separatist approach where surplus heat extracted at the condenser of a refrigeration plant is rejected by a cooling tower, a simultaneous demand for heating being provided by a boiler plant. In circumstances where heat rejection plus compressor power match heat demand, the cooling tower and boiler plant could be done away with and the circuit of part (b) substituted: except for the addition here of water circuits.

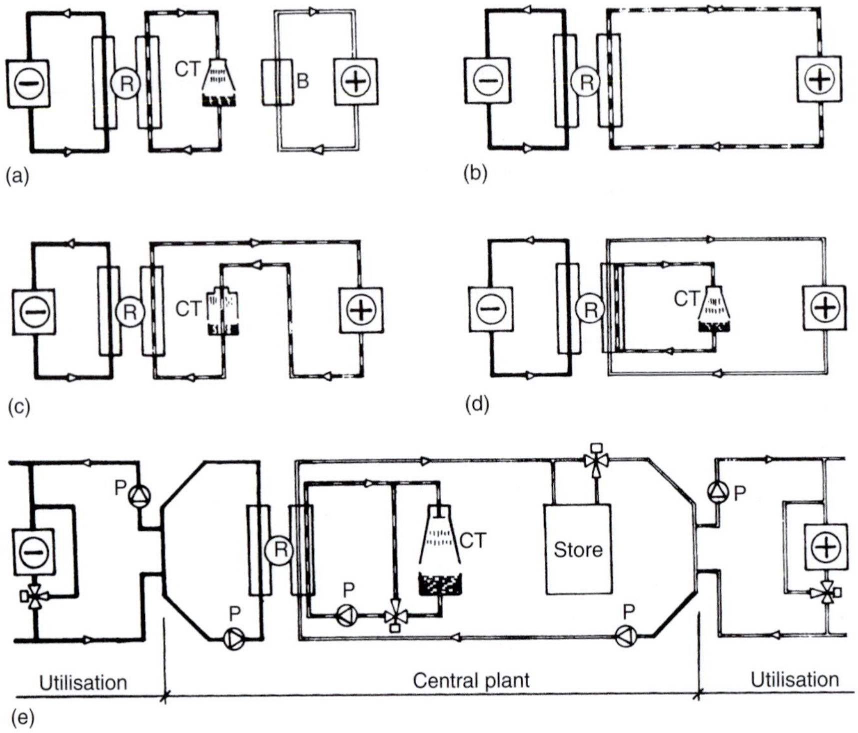

Figure 24.21 Centralised heat recovery systems – development of the concept

In practice, of course, heat rejection would not match heat demand and, in consequence, the components deleted would have to have alternatives substituted for them to cater for out-of-balance conditions, as shown in part (c) of Fig. 24.21. Since water from an open cooling tower carries atmospheric pollutants, it could not be circulated at large throughout the building and thus a closed circuit evaporative cooler is illustrated. The disadvantages of this type of equipment, in terms of efficiency and physical size, are such, however, that what is called a 'double bundle' condenser is introduced as illustrated in part (d). This consists of an oversized shell which contains two quite separate sets of tubes. Water to an open cooling tower circuit passes through one set in the conventional manner and to the heat recovery circuit through the other.

For the type of commercial building to which central heat recovery systems are applied, there is commonly a heat surplus over the 24 hours of operation and some means of storing the excess, when available, for those hours when a deficit occurs is thus needed. Heavily insulated water vessels may meet this need and could, in turn, be 'topped up' by immersion heaters or pipe coils from a boiler plant if necessary. Figure 24.21(e) shows one form of the finally developed system, including automatic control valves, circulating pumps, etc.

Use of multiple machines

In the case of larger-scale projects, not all of the refrigeration plant provided for water chilling purposes would necessarily be equipped with double bundle condensers and used for energy recovery purposes; selection would depend upon circumstances. Taking, e.g. a building with which the authors were concerned, the total capacity of the cooling plant was 11.4 MW but only one of the three centrigual machines was used for heat recovery purposes. The equipment chosen produced chilled water at 5.6°C from return at 12.8°C and condenser water was made available at 41.5°C from return at 32.2°C. The power input was approximately 850 kW.

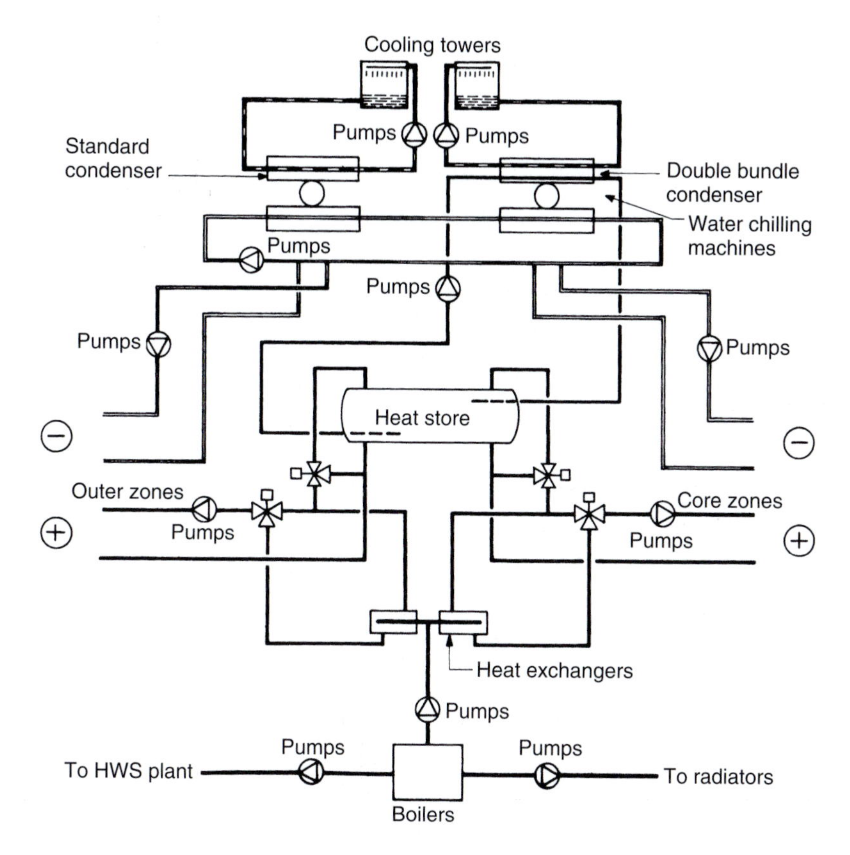

Figure 24.22 Arrangement of central plant heat recovery

Figure 24.22 shows a typical arrangement of plant that may be developed for a large-scale installation.

The principle of heat recovery from a refrigerating machine is used in the reversed cycle heat pump system described in Chapter 19 but in that case the refrigeration equipment is distributed through the building interconnected by a water circuit.

Heat pumps

The operating principles of refrigeration equipment have been described where the evaporator is utilised as a source of cooling whilst heat produced at the condenser is either rejected or recovered for use. When similar equipment is utilised in the reverse sense, drawing energy from a low-temperature source at the evaporator with the specific purpose of making use of the higher grade output at the condenser, the apparatus is described as a *heat pump*, see Fig. 24.23. As may be seen from this diagram the cycle is the same as for the refrigeration machine (Fig. 24.1) with the component parts described rather differently, but essentially performing in exactly the same manner. Reference should be made to the description of the refrigeration cycle to follow the principle of operation of a heat pump; in which case the compressed gas is passed to the condenser where heat is removed for use, and in the evaporator the refrigerant absorbs heat at a relatively low temperature from the *heat source*.

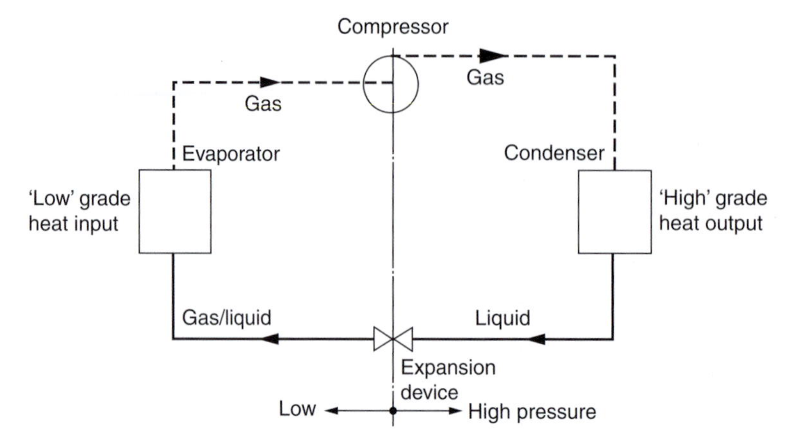

Figure 24.23 The heat pump cycle

Coefficient of performance (COP)

This is the term used to describe the *advantage* offered by operation of a heat pump: use of the parallel term *efficiency* is inappropriate when the ideal of 100 per cent is exceeded, such a condition being impossible of achievement. In strict theoretical terms, the COP is defined as:

$$COP = T_1/(T_1 - T_2)$$

where T_1 is the condensing, and T_2 the evaporating temperature of the thermodynamic cycle, in degrees kelvin. It is more usual however, in practical terms, to express this coefficient as the simple ratio of energy output to energy input; but it is necessary in this respect to be aware whether the consumption of all appropriate auxiliaries is included in the calculation.

It must be remembered of course that, in comparison with a cooling application, the energy in driving the compressor is an asset as far as a heat pump is concerned; the equation being transposed algebraically:

- Cooling evaporator capacity = condenser capacity − compressor power
- Heating condenser capacity = evaporator capacity + compressor power

It will be appreciated that it is practicable for the same equipment to have a dual function, providing cooling in summer and heating in winter. Such a unit is described in Chapter 19. One technical reservation which must be appreciated is that, in designing the components for optimum operation for this dual function, these may not be those best suited to either the cooling or heating mode alone and in consequence if the period in one mode far exceeds the other the unit should be selected with a bias towards the more extended period of usage. Typically, a heat pump operating from a source temperature of 5°C and designed for heating only would have a COP of about 3, whereas a reversible machine for the same conditions would have a COP of about 2.6.

As may be seen from Fig. 24.24, which represents the output of a reciprocating semi-hermetic machine, performance varies with evaporating and condensing temperature, the smaller the temperature difference, the better the performance characteristic. In application to practical problems, it must be remembered that the theoretical performance shown in Fig. 24.2 is distorted in practice by superheat and sub-cooling effects: these affect predicted performance which may well be only two-thirds of the theoretical.

In consequence of the requirement that the temperature difference between the condensing and evaporating temperatures (and hence between heat utilisation medium and heat sink) be minimised, applications to space heating are restricted. In winter, where heat output is required to be at maximum, heat sources such as outside air, river water or the earth's crust are likely to be at their annual minimum temperature. This is illustrated in

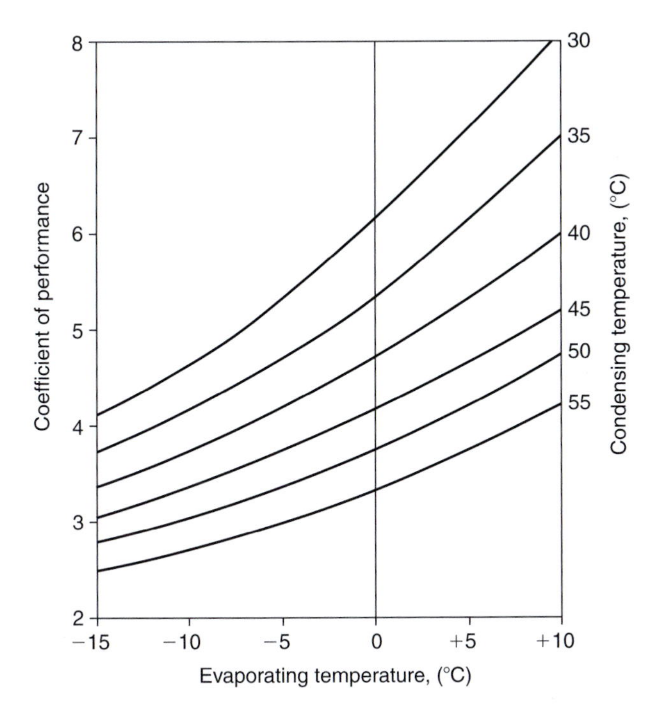

Figure 24.24 Typical heat pump performance

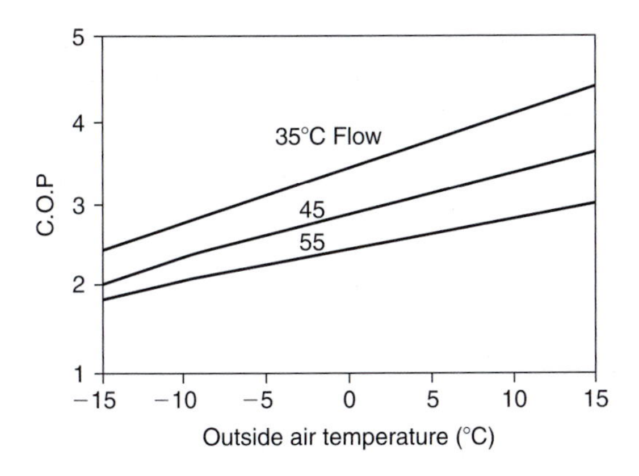

Figure 24.25 Variation of COP with source temperature

Fig. 24.25 which gives the likely order of variation in COP for an air-source heat pump. Industrial effluents and power station waste are, however, relatively constant as to temperature throughout the year and it is to these that attention should be directed as well as to any advantage that may arise from collection of solar energy.

The reduction in performance at low source temperature may lead to the need for plant of high capacity, and therefore cost, to meet the heating demand under extreme conditions, in consequence of which, a heat pump may be selected to match only a part of the full design load, say 70 per cent, the remainder being met by a secondary heating source. A further consideration with air-source plant is the requirement to defrost the evaporator coil during cold weather, which may account for an overall loss of heat output of up to 10 per cent.

Table 24.2 Typical temperature ranges of ambient energy sources

Source	*Temperature* (°C)
Outside air	−10 to 20
Ground water	8 to 12
Surface water	2 to 6
Ground coils	6 to 12

In selecting the alternative energy sources, it is necessary to consider the flexibility of a combination of a direct supply (electricity or gas) with a storage fuel (oil or a renewable solid fuel source) with respect to the future reliability of supply of either category of energy source.

There are four basic types of heat pump suitable for use in heating and air-conditioning which may be defined according to the heat sources and method of transporting heat to point of use, namely:

(1) air-to-air (air source to air system)
(2) air-to-water
(3) water-to-water
(4) water-to-air.

If the heat source originates from a natural ambient supply, the temperature will vary: typical ranges are shown in Table 24.2.

Compressor drive

Whilst it is common practice to consider the heat pump as being electrically driven, this is not necessarily the case except where small production units are concerned. If a broad approximation of 25 per cent be considered as representative of the conversion of primary energy to electrical power, it is obvious that a heat pump so driven, having a practical COP of, say, 2.5, will equate as follows to, say, a boiler fired by natural gas:

Electrical heat pump
COP (primary energy) $= 0.25 \times 2.5 = 0.625$
Gas-fired boiler COP approximately (primary energy) $= 0.62$

In consequence, it is proper to consider the case of a heat pump driven by a prime mover such as a gas or a diesel engine in circumstances where the waste heat from the power source may be recovered. Commonly quoted data for prime movers are, at full load:

Shaft power	33 per cent
Waste heat to oil and jacket coolers	30 per cent
Waste heat to exhaust	32 per cent

If a recovery potential of 50 per cent were to be applied to the sources of waste heat and an average 5 per cent energy overhead considered for oil or gas supply, then the following equation results:

Motive power	0.33×2.5	$= 0.825$
Waste heat	$0.62 \times 0.5 \times 0.95$	$= 0.295$
Combined COP		$= 1.12$

In consequence, a clear case exists for further examination of the use of heat pumps driven by other than electricity, the ratio of advantage being of the order of 1.8:1 in favour of this approach. A further advantage of an engine driven heat pump is that both low and high grade heat are available; the former from the heat pump

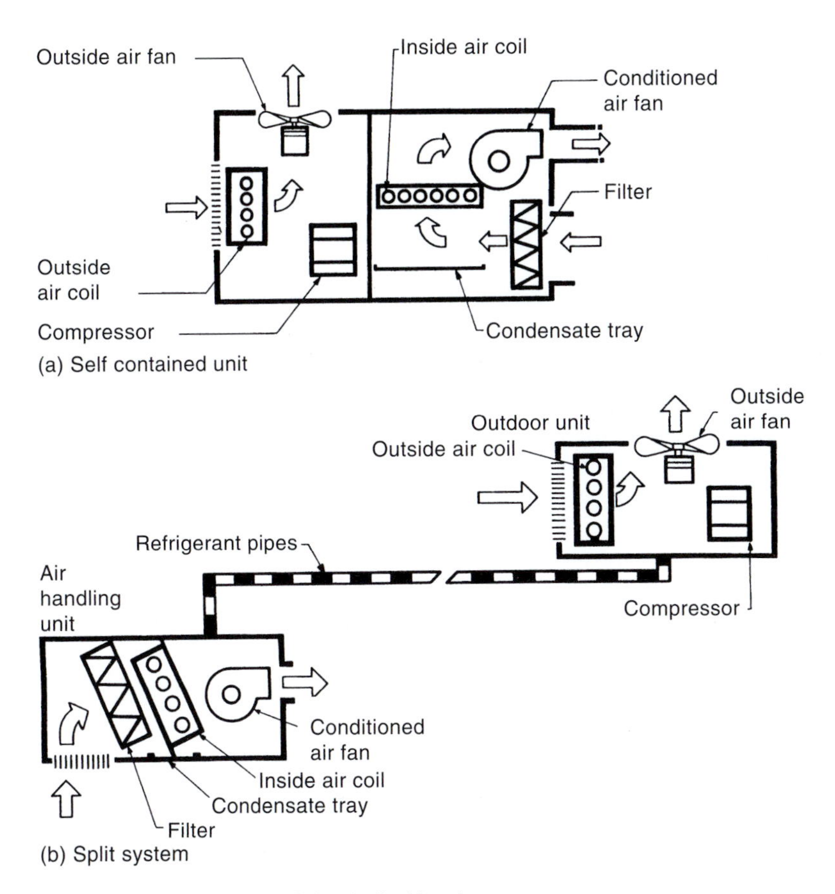

Figure 24.26 Configuration of electrically driven heat pumps

condenser and the latter from cooling of the engine jacket. Notwithstanding these apparent advantages, electrically driven machines have become more popular in commercial applications due mainly to their simplicity, proven record in use and requirement for no more than the very low level of maintenance skills currently available.

Packaged heat pumps

These are available in a range from small domestic scale units (3 kW) through to equipment having an output in excess of 4 MW heating and 3 MW cooling. At the domestic level, the seasonal COP of an air-source device of around 2.6, coupled with relatively high installation costs and poor reliability, is not an attractive alternative to conventional domestic heating methods for the householder. Equipment is available in a number of configurations, some of the more common types of electrically driven machines being illustrated in Fig. 24.26.

On a commercial scale, there is a good case for heat pumps in certain applications, such as in swimming pools retail stores and industrial drying, where their installation is now commonplace. In swimming pools the heat pump is used as a heat recovery device, dehumidifying and cooling the warm and moist extract air as the energy source to heat incoming air (during cold weather), pool water and hot water supplies.

The application in retail stores normally takes one of two forms:

(1) An air-to-air unit, often a reverse cycle device, using outside air as the source to heat or cool supply air as required.
(2) A water-to-air (or -water) device, extracting heat from condenser cooling water connected to freezer cabinets or cold stores.

In office systems, where air-conditioning is required, the use of distributed reverse cycle equipment, referred to earlier, is more popular than that of the less adaptable equivalent central plant; however, both have been used with satisfactory results. For further reference, the now defunct Electricity Council has published numerous valuable case studies on heat pump applications.

Notes

1. Water at atmospheric pressure boils at 100°C: if that pressure were to be reduced to 1 kPa then the boiling point would be 7°C.
2. CFCs, HCFCs and Halons. Professional and Practical Guidance on Substances which Deplete the Ozone Layer. CIBSE Guidance Note GN 1, 2000.
3. Pearson, S.F., *Thermosyphon Cooling.* Institute of Refrigeration, Session 189–190.
4. *Minimising the Risk of Legionnaires' Disease*. CIBSE Technical Memorandum TM13, 2002. *Legionellosis (Interpretation of Approved Code of Practice: The Prevention or Control of Legionellosis).*

Chapter 25

Hot water supply systems

The provision of hot water for baths, showers, basins, sinks and other points of draw-off may be dealt with either locally or from a central system. In this respect it echoes the 'direct' and 'indirect' methods of providing heating service.

Local systems use either gas or electricity as a heat source, the heater being placed near to the point of consumption. In the case of a central system, the water is heated at some convenient position, which may be relatively remote from the point of consumption, and distributed as may be appropriate by pipework. The heat supply to a central system may be either solid, liquid or gaseous fuel or electrical energy, and can also incorporate solar or combined heat and power (CHP) as a primary or preheat source. For further reference to solar and CHP as primary heat medium to hot water systems refer to Chapters 14 and 15.

Pipework losses

In order to provide a satisfactory level of service to the user and ensure that the hot water flow temperature discharges from the draw-off point within a maximum of 30 seconds, a central system of any size must be arranged to circulate hot water through a two-pipe system of flow distribution and secondary return pipework. This circulation is achieved by means of a circulator pump which will continue winter and summer, although means will normally be provided to stop it when the building is unoccupied subject to legionella considerations. Heat will be lost from this circulating pipework, although insulated, and this loss will persist whether or not any water is drawn off. The ratio between the energy expended in heating up the water supply from cold and that lost through the circulating system represents the efficiency of the system. It is of prime importance, therefore, that the distribution system be kept as compact as is practicable.

An alternative to utilising a two-pipe pumped circulation system is the adoption of single-pipe flow distribution pipework incorporating onto the pipework, self-regulating electrically powered trace heating tape that is designed to replace the system heat losses. This trace heating tape is attached directly to the flow pipework along its whole length and the two are encased in pipework insulation of an appropriate thickness. The tape is designed to be self-regulating with the heat input modulating to provide sufficient heat per metre run of pipe to compensate for the pipework heat losses to ensure the set flow temperature is maintained up to the draw-off point.

Because the trace heating system adjusts its output to the requirements of the hot water system it is considered to be economical in operation and provides space saving in installation.

Water conservation

Whichever pipework system is adopted, one of the initial issues which the designer should consider is the environmental sustainability concern of water conservation. This issue can be addressed in the design of the hot water system by either limiting the flow rate of water from the outlet or the time period the outlet is open or a combination of the two.

A typical device to control the water flow rate is a flow regulator. This device can be either installed within the pipework connection to a draw-off point or within the outlet of the draw-off tap itself. These devices are preset to a selected maximum flow rate that is not exceeded irrespective of the system pressure and/or the tap being fully open.

The selection of PIR (passive infrared) operated taps which opens the outlet only when the user's hands are in close proximity to the outlet will restrict the time period the draw-off outlet is open and thereby the volume of water discharged. This type of tap or the spring loaded outlet controlled taps also overcomes the problem of user misuse in leaving the outlet running following use.

Taking the secondary return pipe or the trace heating tape to a point as close as practicable to the hot water outlet can also reduce the volume of water used, by reducing the time period the outlet has to run cooled water to waste before the hot water discharges.

By adopting such or similar water controlling methods over a period of time significant water consumption savings and consequently energy savings can be achieved.

Choice of system

For a single isolated draw-off point, some type of local heater would be the obvious choice and, equally, a central system would be preferred for a hotel having bathrooms closely planned both back-to-back and vertically on several floors. As a generality, most commercial and institutional buildings, such as hotels, hospitals and educational establishments (whether planned for efficiency or not), are likely to be best served by a central system of some sort because this will make available a bulk reserve supply of hot water to deal with peak demands and will at the same time avoid dispersal of those scarce maintenance activities which may be available.

In the case of many other applications, however, the choice is less easy to make and a single building could quite well be served by a mixture of systems. As a simple example, consider a centrally heated eight-storey building consisting of a dozen lock-up shops at ground floor level with six floors of offices above, all occupied by a single tenant, and a large penthouse flat leased separately.

Obviously, all points throughout the building could be supplied from a central plant in the boiler house but this would mean providing a source of heat for 365 days per annum to serve the flat. It would also mean that service to the shops would have to be charged individually to the tenants which might be uneconomic if one or more of them used hot water for cleaning cars (a not unknown habit!).

A sensible solution might be to provide a central system from the boiler plant for the offices, a separate mini-central system for the various hot water needs of the flat and a local heater for each shop. The energy supply to the flat and shops would be via an individual meter per tenant.

Local systems

There are two basic types of local hot water system, fundamentally different in concept. The first makes use of some form of instantaneous heater and has no hot water storage capacity. Except for fuel consumption by a pilot flame in the case of gas firing, this type has no associated heat loss when not in use. The second type incorporates hot water storage, adequate in capacity to meet the local demand, and however well insulated the storage vessel may be, a standing heat loss will occur.

Apart from any question of heat losses, the two types of system will for a given hot water production consume equal amounts of energy. In terms of rate of energy supply, however, an instantaneous heater will impose a greater load since the water outflow must be brought up to the temperature of use, from cold, during the short time of actual demand. With a storage unit, no such limitation exists and the rate of energy supply required will be a function of vessel size and demand pattern.

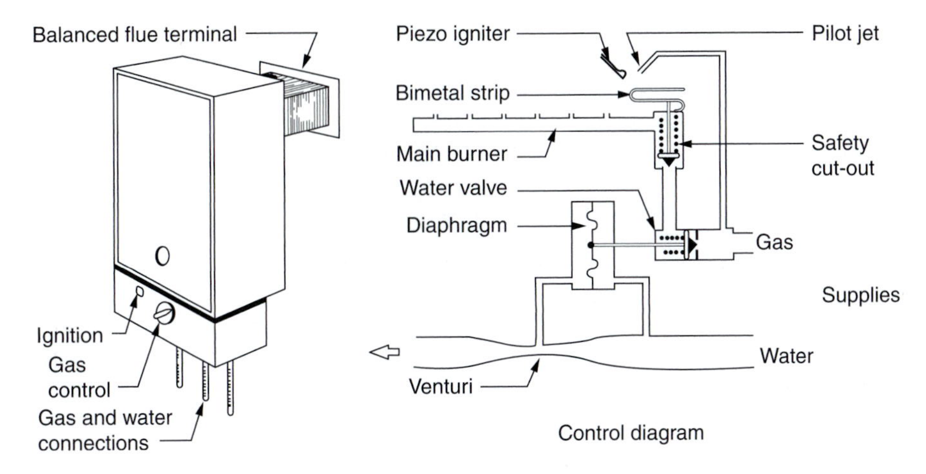

Figure 25.1 Instantaneous multi-point gas water heater (Ascot)

Instantaneous heaters

The earliest type of heater in this category was the old fashioned bathroom geyser, gas fired. The much refined modern equivalent is made in single- and multi-point form and Fig. 25.1 shows an example of the latter type. A number of draw-off points may be served provided that these are within the limit of pipe length and hot water discharge time period (see Table 25.7). A gas pilot burns continuously and when a hot water tap is opened water begins to flow and, by the pressure difference across the venturi, the main gas valve is opened, whereby the gas burners are ignited. A safety device cuts off the gas should the pilot light be extinguished. The control diagram in the figure shows the principle of operation.

For all such heaters, whether used continuously or intermittently and wherever fitted, a flue with an outside terminal is essential. Existing heaters which discharge the products of combustion into the room where fitted are lethal and should, when found, be replaced. Models having balanced flues, with inlet and outlet ducts communicating with the outside are typically used in most installations.

A now familiar component is the single-point electric instantaneous heater, designed to serve a wash basin or a shower as shown in Fig. 25.2. Various models are available which will provide varying levels of outflow. Such units are provided with thermostatic controls and a diaphragm pressure switch which permits current to be available only when water is flowing. Preset cut-out and other safety devices are incorporated. Physically, the casing of a unit of this type is small being only 200–300 mm square with a depth of perhaps only 60–80 mm. Electrical loadings are high, as from 3 to 9 kW, and since usage of water is immediately adjacent it follows that particular care must be taken in providing electrical protection and good earthing facilities.

In most applications instantaneous heaters, gas or electric, are designed to be connected directly to the cold water main supply. Where supply via a cold water storage cistern is envisaged, a static head of at least 10 m, equivalent to a pressure of about 98 kPa, is necessary. The rate of water flow from all such heaters is restricted and, in the case of multi-point units, it should not be assumed that a good supply may be obtained from more than one tap at a time. In the case of single-point units, water control is normally on the inlet connection and outlet is via an inconvenient swivel spout, permanently open to atmosphere.

Storage heaters (on-peak)

For single-point draw-off, local storage heaters are available in the capacity range of about 7–70 litre and are normally provided with 3 kW electrical heating elements. In this application, the simplest form of heater is the free outlet or non-pressure type illustrated in Fig. 25.3. It is the cold water inlet which is the point of control,

Figure 25.2 Instantaneous single-point electric water heater (Heatrae Sadia)

Figure 25.3 Non-pressure-type electric water heater (Heatrae Sadia)

discharge of hot water being by displacement as the cold supply is admitted. The storage is always open to atmosphere, normally via a swivel spout, and, as the contents expand on heating, a drip may occur at the outlet. To overcome the inconvenience of a swivel arm, the outlet may be piped to a special combined cold water inlet hot outlet tap as shown in Fig. 25.4(a) and (b).

Over recent years following modification of the water regulations, the use of mains fed unvented water heaters has become popular. For local multi-point use, up to three outlets, an electric storage heater of 10–15 litre capacity can be chosen.

These units adopt the option of two methods to accommodate hot water expansion as shown in Fig. 25.5. Option 1 is to allow the expanded heated water to reverse flow up the cold feed supply pipe, providing that the nearest cold water draw-off on the cold supply pipe is a minimum of 2.8 m, for 10 litre storage capacity, and 4.2 m for 15 litre storage capacity distance away from the heater.

Option 2 is for the storage heater's cold feed supply adjacent to the heater to be fitted with safety fittings incorporating a suitable sized expansion vessel, check valve, pressure relief and a combined pressure and temperature relief valve on the heater.

For multi-point local use, as might arise in each of a number of dispersed toilets in an office building, both gas-fired and electrical storage heaters are available. Packaged gas-fired units are, in effect, cylinders incorporating a stainless steel heat exchanger, direct fired as shown in Fig. 25.6, and are insulated and cased in sheet metal. Such heaters come with various storage capacity and heat input, and have a typical water capacity of about 80 litre and a heat input of 8 kW: the products of combustion must be discharged to outside via a conventional or balanced flue arrangement.

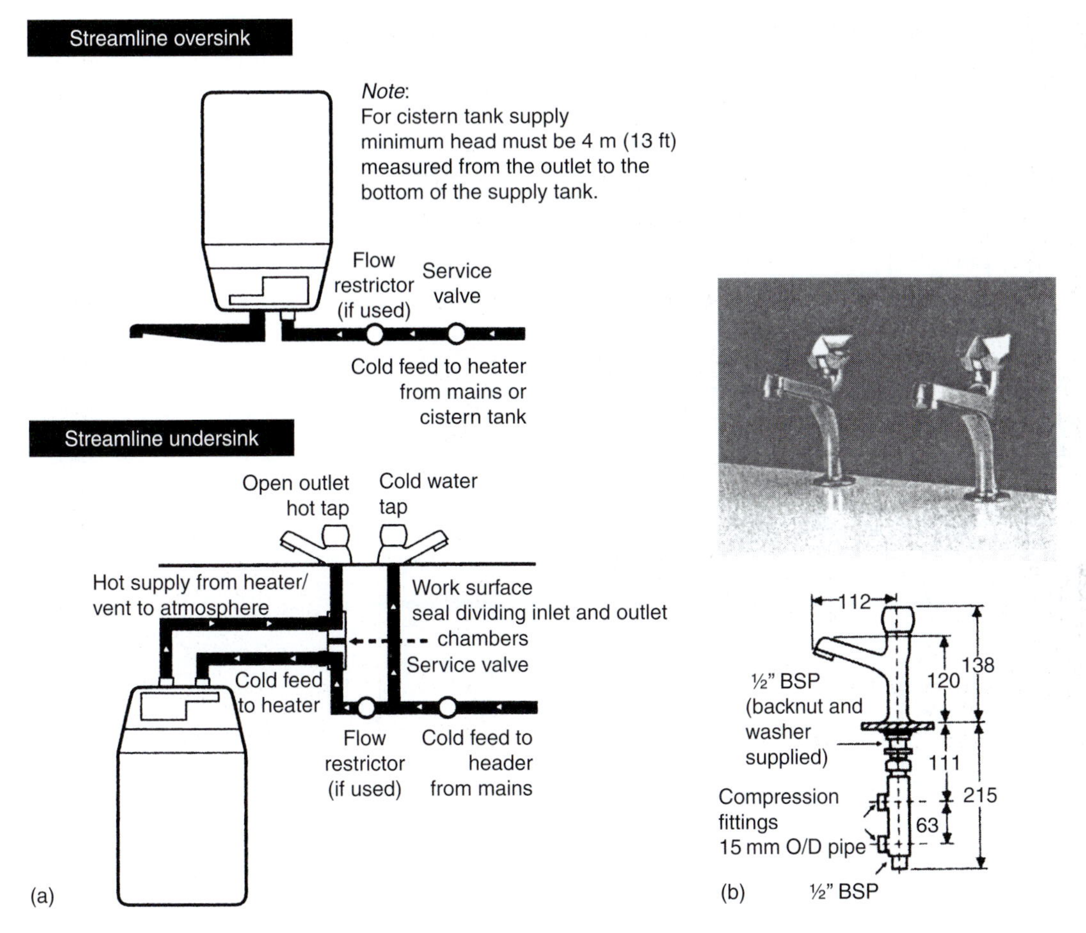

Figure 25.4 (a) Swivel spout and combined cold inlet/hot outlet detail (Heatrae Sadia) and (b) combined cold inlet/hot outlet tap (Heatrae Sadia)

With electrical local multi-point supply this is conveniently arranged using a so-called combination-type unit where heater and cold water cistern are fitted in one casing as shown in Fig. 25.7. The heater may be cylindrical or rectangular for easy wall fixing, and capacities in the range 20–140 litre are available, normally with a 3 kW heating element. The pressure available at the draw-off points will be meagre if the combined unit cannot be fixed well above the level of the taps.

For applications where larger storage or heating capacities are necessary, or where a combination unit cannot be fitted high enough above the draw-off point, vented pressure-type equipment is used, as shown in Fig. 25.8. All vented pressure-type units require a cold water supply from a remote storage cistern and an open vent. Storage volumes cover the range 70–300 litre with heater element capacities to suit.

Unvented models are also available and can be considered where the cold water supply pressure and flow rate are suitable for the system. As with all unvented water heaters, hot water expansion is to be catered for. Unlike the small 10–15 litre multi-points previously mentioned, only one method of expansion is generally used. This being the incorporation of an expansion vessel on the cold feed with the associated safety fitting as identified in Option 2 for unvented multi-points.

However, a variation to the expansion vessel is the incorporation of an internal air gap within the vessel itself utilising a float baffle to maintain the integrity of the air gap as shown in Figs 25.9(a) and 25.9(b).

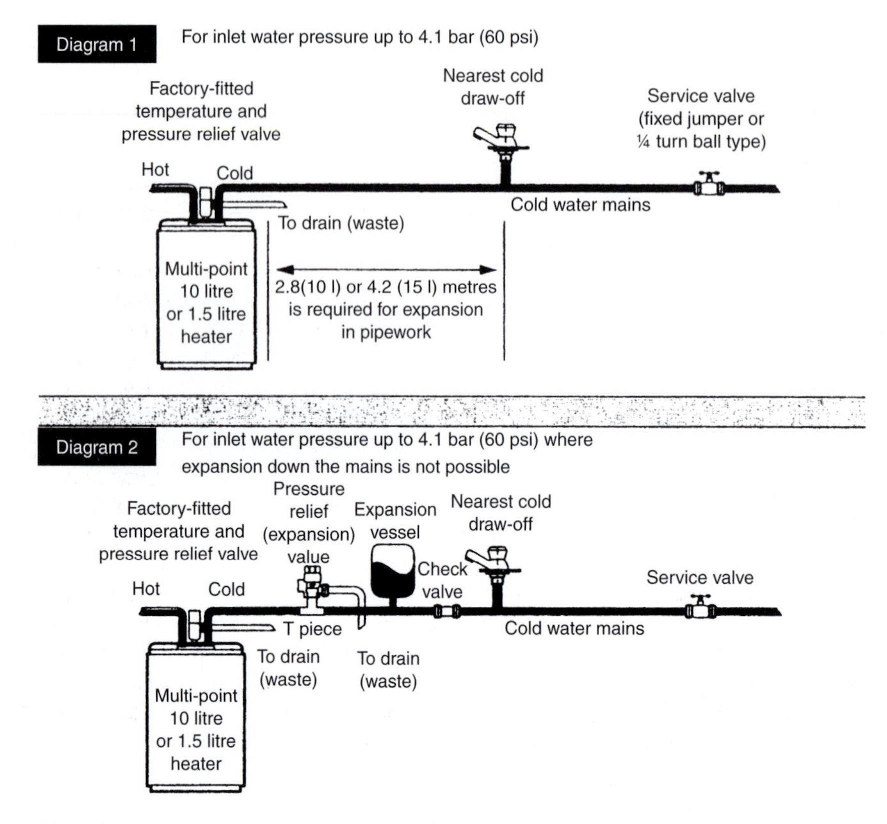

Figure 25.5 Hot water expansion detail 10–15 litre heaters (Heatrae Sadia)

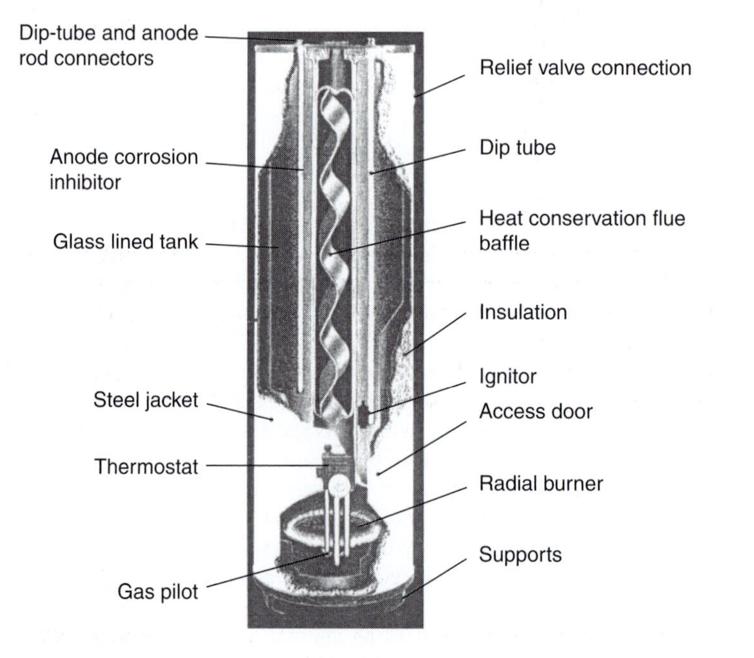

Figure 25.6 Storage-type heater, gas fired (Andrews)

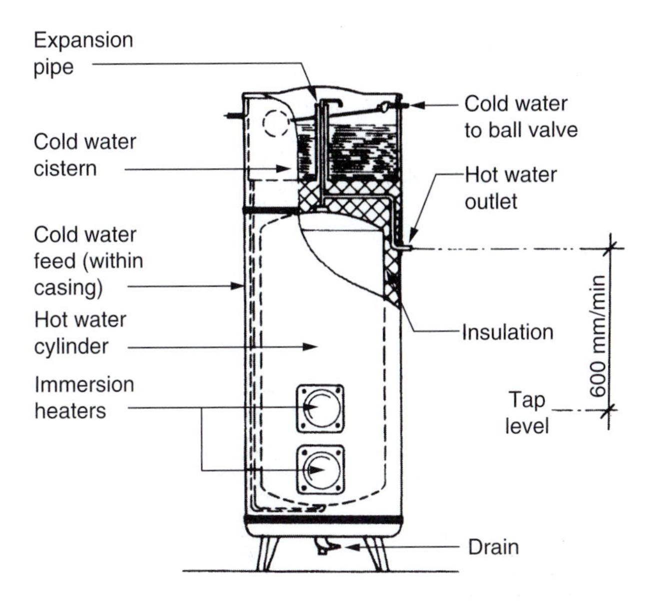

Figure 25.7 Combination electric storage heater with cistern (Sadia)

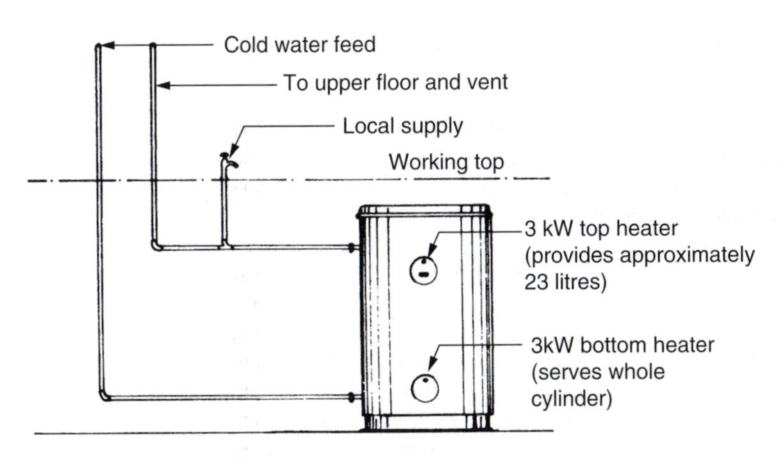

Figure 25.8 Pressure-type electric storage heater (Heatrae Sadia)

Local storage heaters are well insulated and are normally fitted with an insulation jacket with an enamelled sheet steel protective cover, ready for mounting in an exposed position. In comparison with an instantaneous heater, at least 10 times the space is needed for the same level of service but if sited in a cupboard the difference may not be relevant. A suitably sized storage heater does however offer the advantage of using off-peak current which is not possible with the alternative type.

To make use of the cheap off-peak rate, a storage unit should be fitted with two sets of immersion heaters. One set, fitted near the bottom of the vessel, will charge the store overnight and the other, fitted near the top, will be arranged to heat a smaller volume of water at the end of the day when the overnight charge has been exhausted. On a domestic scale, an existing 120 litre cylinder (900 mm × 450 mm diameter) might be converted by using a dual heater as shown in Fig. 25.10(a) but the arrangement as shown in Fig. 25.10(b) is to be preferred. Better still, for a new installation, would be a 210 litre cylinder (1450 mm × 450 mm diameter) as in Fig. 25.10(c).

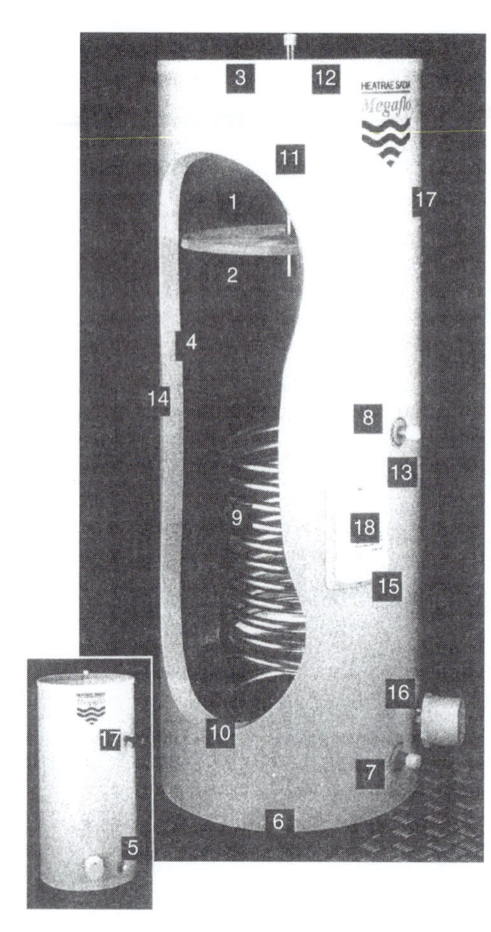

1 Air gap for water expansion
2 Float baffle for maintaining air gap
3 Hot water outlet
4 Stainless steel cylinder
5 Cold feed inlet
6 Support plinth
7 Primary flow
8 Primary return
9 Indirect primary heating coil
10 Cold feed inlet baffle
11 Plastic coated outer casing
12 Moulded top
13 Sealing grommets to pipework
14 Thermal insulation
15 Capacity label
16 Immersion heater
17 Temperature and pressure relief valve
18 Thermal controls

Figure 25.9(a) Unvented storage heater with internal air gap (Heatrae Sadia)

The megaflo internal air gap system

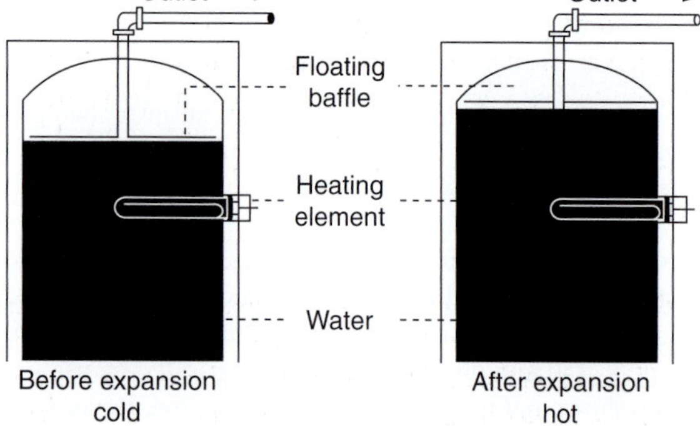

Figure 25.9(b) Integral air gap system detail (Heatrae Sadia)

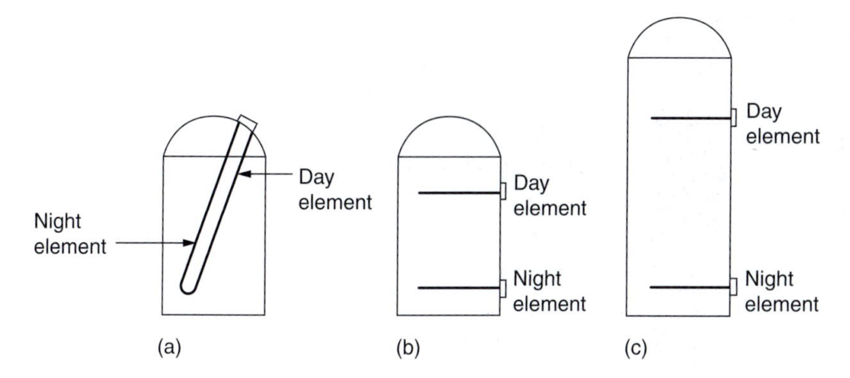

Figure 25.10 Arrangement for electric immersion heaters

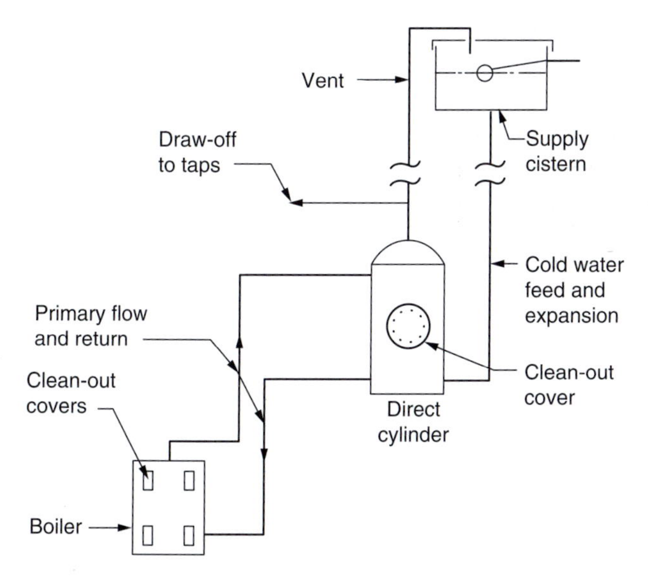

Figure 25.11 Direct hot water system

Central systems

A central system will usually consist of a boiler or water heater in some form, coupled by circulating piping to a storage vessel or vessels. The combination of the two will be so proportioned as to provide adequate service to the draw-off points to match the predetermined pattern of usage. For instance, in a hospital there may be a continuous demand for hot water all day and in this case a small storage capacity with a rapid recovery period (large boiler power) is probably appropriate. Conversely, for a sports pavilion where there may be a single sudden demand following a game, a large storage capacity and a long recovery period (small boiler power) may be adequate.

Direct systems

In the recent past, but historic now, many systems were direct following a traditional pattern in which the water drawn off at the various taps, etc. was the same water as that which circulated through the boiler, as shown in Fig. 25.11. If the water were hard then a scale deposit would occur on the heating surfaces and thus a special type of boiler was necessary, of a simple type with large waterways and bolt-on access to ample cleaning 'mud-holes'.

On the other hand, if the water were soft, then an ordinary boiler of cast iron or mild steel would cause discolouration as a result of rusting of internal surfaces. Special boilers were thus made of cast iron subjected

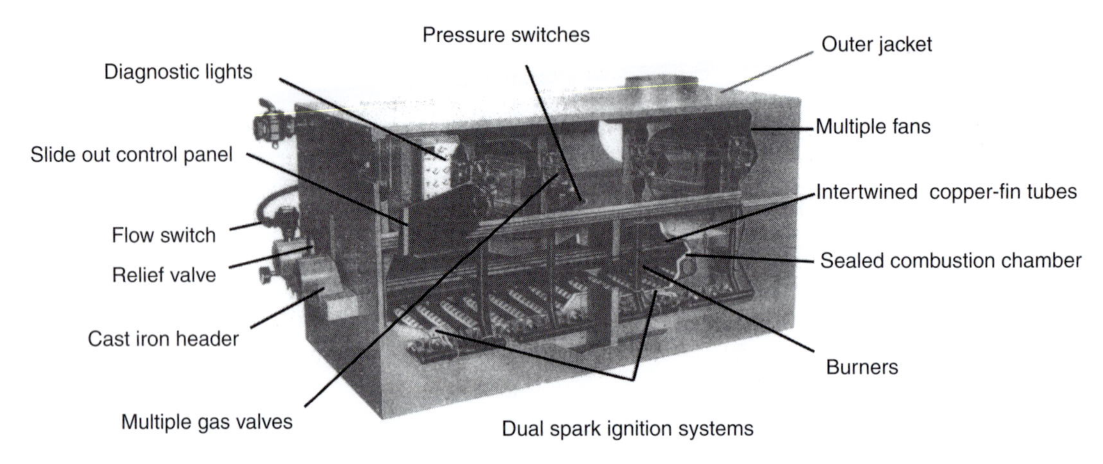

Figure 25.12 High-output direct fired non-storage water heater (Lochinvar)

to an anti-corrosion treatment such as *bowerbarifing* or, where made in mild steel, were treated internally with some form of vitreous enamel coating.

Modern systems utilise either direct fired storage water heaters similar to the unit shown in Fig. 25.6 or where larger demands are required direct fired non-storage multi-tube hot water boilers as shown in Fig. 25.12. This type of boiler is constructed of a horizontal bank of water carrying finned tubes mounted above the gas burners producing a highly efficient instantaneous water boiler. In hard water areas scale build-up in the tubes is inhibited by the scouring action of the water through the tubes. This type of unit can be married to a hot water storage vessel to increase the hot water volume where infrequent peak demands exceed the instantaneous capability of the water boiler.

More efficient boilers/water heaters have now been developed. These are termed 'condensing' or high-efficiency boilers, and work on the principle of recovering as much waste heat as possible which is normally rejected to atmosphere through the exhaust flue.

This higher efficiency is achieved by utilising larger heat exchangers or sometimes two heat exchangers which maximises heat transfer from the burners as well as recovering waste heat which is normally lost within the flue gas discharge. As a result when in condensing mode the flue gases leaving a condensing boiler are typically in the range of 50–60°C compared with conventional boiler discharges of 120–130°C. More detailed information on this type of boiler/water heater can be found in Chapter 13.

Indirect systems

The total separation of boiler water from that drawn off at taps, etc., and a solution to the associated problem of cleaning the internal surfaces of the boiler, are both found when an indirect system is adopted. Figure 25.13 illustrates such an arrangement in which any type of heating boiler may be used. The storage cylinder becomes indirect as a result of the provision of heating surface within it to contain the (primary) water circulated from the boiler. The (secondary) water to be heated is outside the heating surface and is thus isolated from the boiler. It will be noted that this separation introduces the need for an additional feed and expansion cistern.

The primary water within the heating surface is recirculated and thus will remain unchanged with the result that scale deposition internally is likely to be negligible. Further, when this primary water is kept at a temperature below about 90°C then, except in areas where the supply has an extremely high *temporary* hardness, there will be little deposit on the external secondary surfaces.

Combined systems

The water content of a boiler serving a direct hot water supply arrangement must, for obvious reasons, be quite separate from that within any plant supplying an adjacent heating system. The result is that, unless a

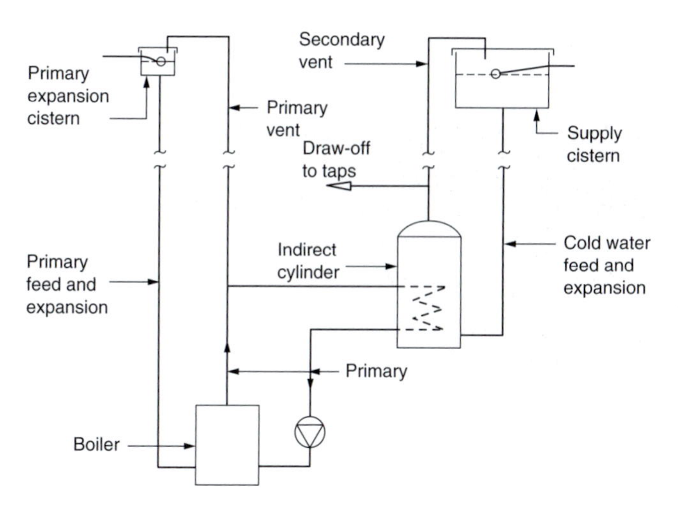

Figure 25.13 Indirect hot water supply system

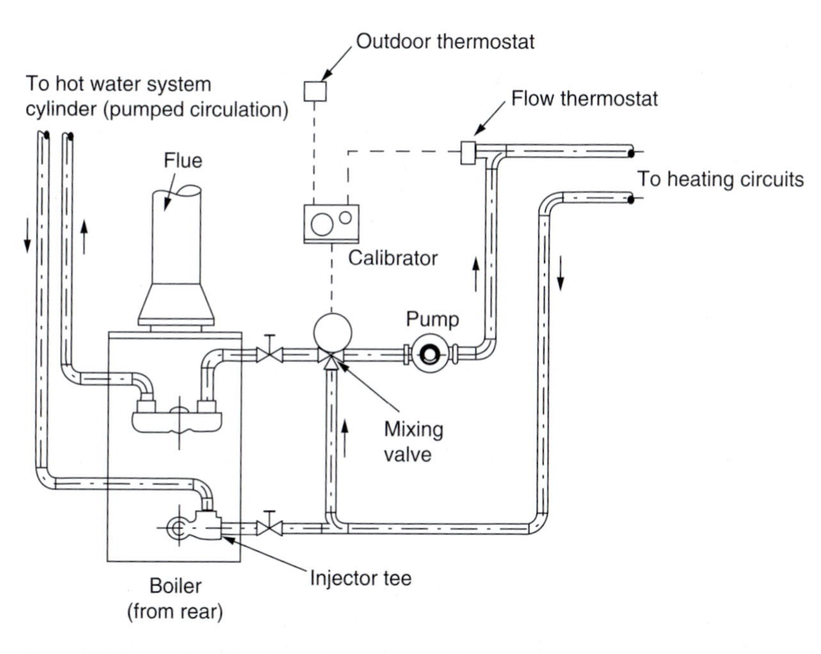

Figure 25.14 Combined heating and hot water supply system

duplicate boiler is provided, service will be lost in the event of breakdown and during the necessary annual cleaning. With an *indirect* system however it is common practice to arrange for an indirect cylinder to be served from the same boiler, or group of boilers, which supply the heating system.

Where only one boiler exists, a problem arises in that the primary water supply to the indirect cylinder will be required at a constant temperature year-round whereas the parallel supply to the heating system will need to vary in temperature with the seasons. This difficulty is overcome, as shown in Fig. 25.14, by running the boiler at a temperature to suit the cylinder and supplying the heating system through a mixing valve as discussed in Chapter 28.

During the summer, when heating is not required, the boiler will be oversized and problems due to cycling and low combustion efficiencies may arise. For very small combined systems, it may be sensible to use

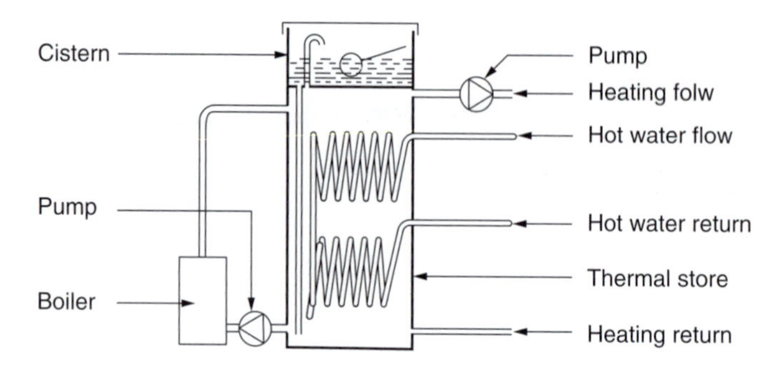

Figure 25.15 Arrangement of domestic combined system

another source of energy such as electricity during periods of low overall load: for larger systems, the provision of more than one boiler is to be recommended.

In the case of a group of boilers, one of them might be chosen to be of a size matching the hot water load and be dedicated to serve the indirect cylinder in normal circumstances, although connected in parallel with the other boilers. In the event of breakdown of the dedicated boiler, supply to the cylinder could come from one of the heating units. Conversely, should a heating boiler break down in exceptional winter weather, the hot water boiler could be used to provide additional capacity. As will have been understood from the discussion regarding boiler margins in Chapter 13, no hard and fast rules can be laid down for plant selection since patterns of use, availability of space and other extraneous circumstances may intervene.

Combination systems

Reference to boilers so described has been made in Chapter 13 but it is the provision of hot water output to a combined *system* which is considered here. Where both heating and hot water are supplied from a single boiler, it is unlikely that both systems will require their full share of output simultaneously, even in mid-winter. For the remainder of the year, as stated above, the boiler will be too large, with attendant penalties.

Despite this situation, however, interaction between the two systems may be such that response by the boiler to demand from either will appear sluggish. The introduction of some form of thermal store between the boiler and the two varying demands should thus improve not only operating efficiency but also response rate.

The design concept used for one make of domestic unit is illustrated in Fig. 25.15, the heat source being a conventional boiler. The output of the heat source is pump circulated to the store which has a capacity, varied to suit the size of house served, of between 120 and 200 litre. From the store, a quite separate circulation is directed to the heating system and this feature is turned to advantage when, following a nighttime shut-down, restarting the heating pump brings the radiators to store temperature in a matter of minutes.

The two coils immersed within the store have external fins to aid heat transfer and are connected in series. Since their water content is small, they are permitted to be supplied with a cold water *feed from the incoming main supply* with the result that outflow pressure is more than adequate to serve a shower head. Provision is made for the hot water discharge to be blended with cold water so that the final supply is at a safe temperature for children. In effect, the store acts as a batch-production instantaneous heater able to provide a hot water outflow, at a rate of about 0.2–0.3 litre/s, for a period of 5 minutes followed by a recovery period of 20–40 minutes in preparation for a further similar outflow.

Cylinders, indirect cylinders and calorifiers

Whilst, in the present context, it is unnecessary to define a cylinder, other than to say that typical sizes for such vessels are as set out in Table 25.1 the difference between an *indirect cylinder* and a *calorifier* is less clearly established. Early textbooks and catalogues seem to use the first of the two terms to distinguish heat exchangers for domestic application from those fitted elsewhere. This usage is confused by the fact that the

Table 25.1 Typical capacities and dimensions of direct and indirect cylinders

	Dimensions (mm)			*Heating surface* (m)	
Pattern	*Height*	*Diameter*	*Capacity* (litre)	*Coil*	*Annulus*
Direct	300	1600	98	–	–
	350	900	74	–	–
	400	900	98	–	
	400	1050	116	–	
	450	675	86	–	–
	450	750	98	–	–
	450	825	109	–	–
	450	900	120	–	
	450	1050	144	–	–
	450	1200	166	–	–
	450	1500	210	–	
	500	1200	200	–	
	500	1500	255	–	–
	600	1200	290	–	–
	600	1500	370	–	–
	600	1800	450	–	
Indirect	300	1600	96	0.35	
	350	900	72	0.27	
	400	900	96	0.35	
	400	1050	114	0.42	
	450	675	84	0.31	–
	450	750	95	0.35	–
	450	825	106	0.40	–
	450	900	117	0.44	–
	450	1050	140	0.52	–
	450	1200	162	0.61	–
	450	1500	206	0.79	–
	500	1200	190	0.75	–
	500	1500	245	0.87	–
	600	1200	280	1.10	–
	600	1500	360	1.40	–
	600	1800	440	1.70	–
Single feed	400	1050	104	0.42	0.63
	450	750	56	0.35	0.52
	450	900	108	0.44	0.66
	450	1050	130	0.52	0.78
	450	1200	152	0.61	0.91
	450	1500	196	0.79	1.18
	500	1200	180	0.75	1.13

Note
Based upon a transmission coefficient of 30 W/m2 K, the heating surface listed is stated to be adequate to raise the temperature of the cylinder contents by 55 K in 1 hour.

former had, almost always, an annular heat exchanger whereas the latter were in one form or another of shell and tube construction. Although the annular-type unit has now all but disappeared, subsequent references here will draw the traditional distinction.

Indirect cylinders

With an annular heat exchanger, an indirect cylinder could be installed either horizontally or vertically. Those cylinders fitted with a helical coil, as in Fig. 25.16, may be preferred from the point of view of differential pressure, primary to secondary, but there remains some doubt whether or not they suffer from a deficiency in heating surface.

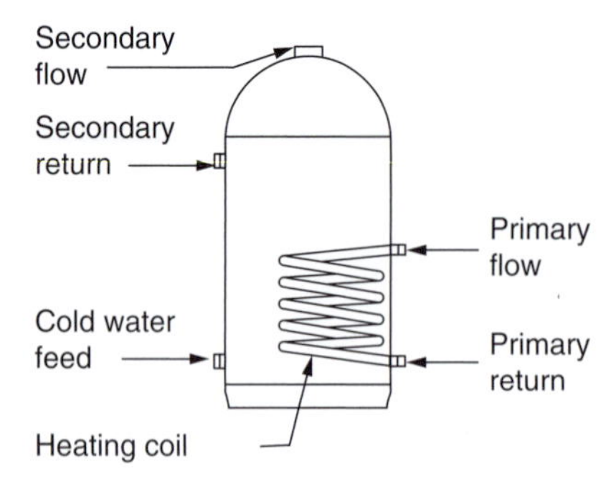

Figure 25.16 Indirect cylinder with helical coil

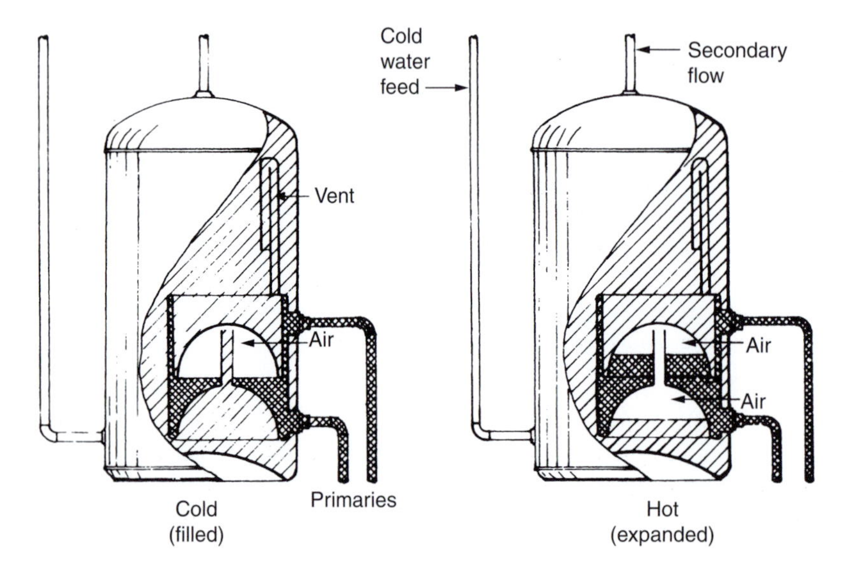

Figure 25.17 Indirect cylinder, single-feed type (Primatic)

Single-feed indirect cylinders

The requirement that a separate feed and expansion cistern be provided to serve the primary circuit of a conventional indirect cylinder led to the development in the 1970s of a special type as illustrated in Fig. 25.17. This pattern of unit had no moving parts and relied upon air cushions to separate the primary water content from the secondary store, using naturally balancing pressures. Once filled, the air cushions expand or contract in tandem as the primary circuit heats or cools: any excess air is vented via the secondary water content.

However in use failure of the air bubble caused the discoloured central heating water to mix with the domestic hot water resulting in contamination. Therefore the use of these units has been phased out.

Water-to-water calorifiers

For larger units, as covered within this heading, a tubular 'hairpin' pattern heat exchanger or heater battery of the type shown in Fig. 25.18 is used. The calorifier shell is provided with a flanged neck which supports a tube plate and a domed cover, the latter having a partition within it to route the primary water flow in and out of the tubes. Means for withdrawal of the heater battery on a runway are incorporated so that the tubes may be cleaned. The primary pipework connections should be arranged so that withdrawal may be achieved with minimum disturbance to insulation, etc.

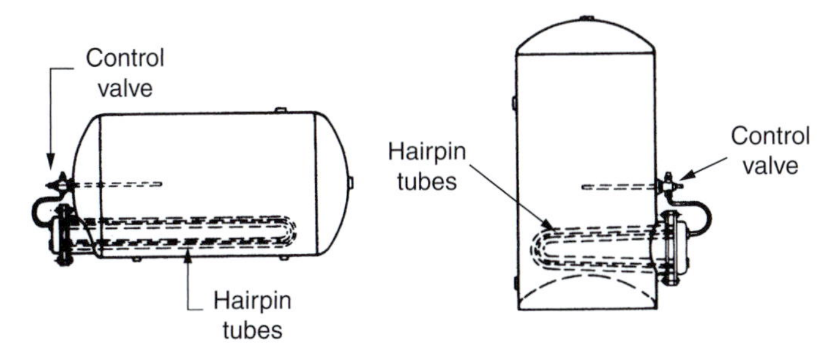

Figure 25.18 Tubular-type calorifier for low-temperature hot water and steam primaries

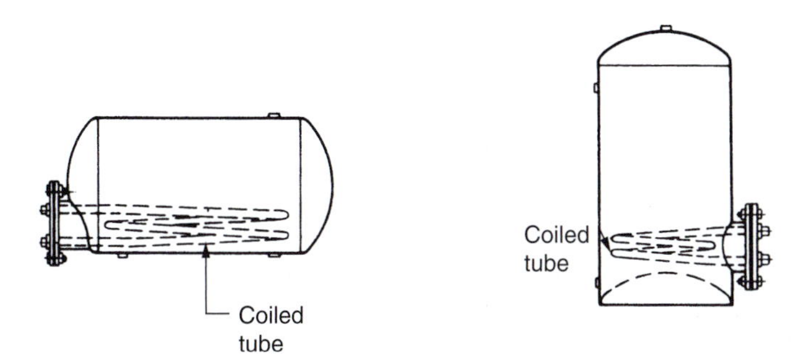

Figure 25.19 Tubular-type calorifiers for medium- and high-temperature hot water primaries

For application to medium- and high-temperature hot water systems, a calorifier will frequently take the form shown in Fig. 25.19 in which the primary water follows a single pass in order to maintain a high velocity with the comparatively small volume of water which is in circulation with this type of system. Provision for withdrawal of the heater battery is made as before, the neck being enlarged to suit the alternative tube configuration.

Steam-to-water calorifiers

The configuration generally adopted when the heating medium is steam is similar in most respects to that shown in Fig. 25.18, with hairpin tubes. The domed cover, here known as a *steam chest*, is provided with an upper connection for the steam supply and, usually, a pressure gauge. Condensate, when formed, leaves by a lower connection and passes to a steam trap set.

Bubble-top calorifiers

A calorifier of this type consists of a vertical cylinder, not unlike the pattern shown in Fig. 25.20, but having the shell extended above the hot water outlet to form a sealed air space at the top, equivalent to about 15 per cent of the water volume. The cold water feed derives directly from a pressure main and, when the cylinder is first filled, air is trapped and compressed in the shell extension or 'bubble-top'.

When the water content is heated, it expands, and the volume of expansion is accommodated by further compression of the air within the 'bubble-top'. Numerous safety features are incorporated, as shown in Fig. 25.20, those on the primary side being varied to suit the heating medium which may be either hot water or steam.

This principle of using an air pocket at the top of the cylinder is adopted for domestic systems by a proprietary unit as shown in Fig. 25.9(a), the expansion pocket being complimented with safety feature valves and thermal cut-out to meet the requirements of Building Regulation G3 document.

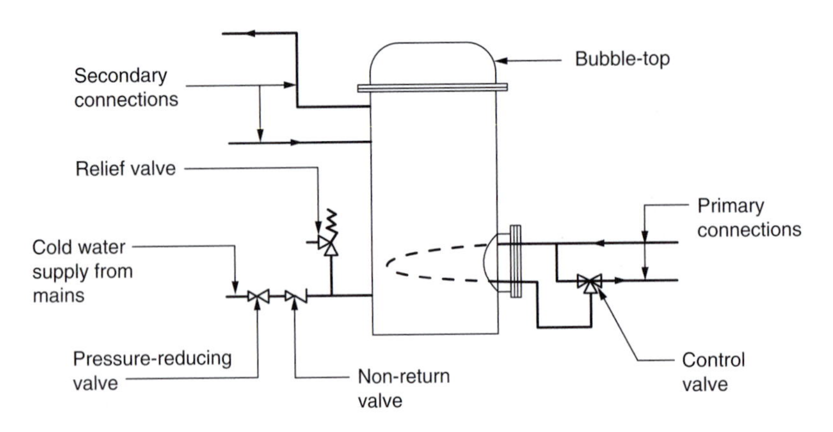

Figure 25.20 'Bubble-top' calorifier

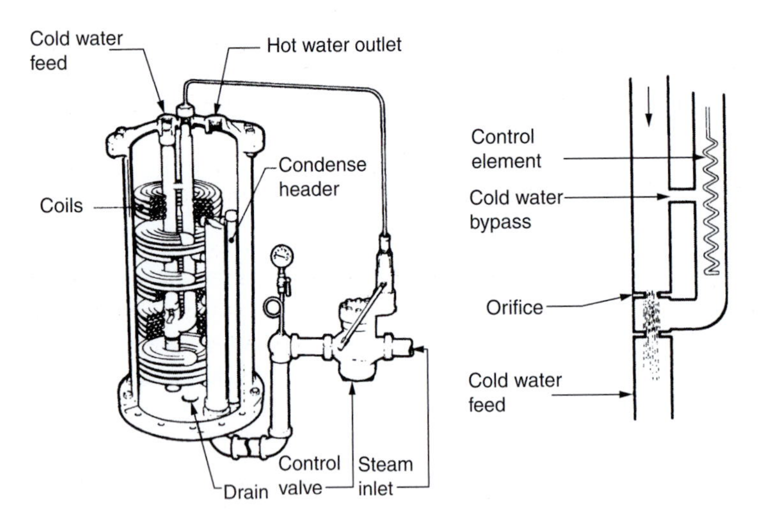

Figure 25.21 Angelery hot water generator

Angelery hot water generators

This equipment is suitable for use with either steam or hot water as the primary medium and deals with the process of supplying hot water in a manner which differs from the conventional. As may be seen from Fig. 25.21, the heat exchange surface consists of a series of interlacing coils which provide a high capacity in a small space and an ability to shed scale as they flex in expansion and contraction. While the generator is not strictly an instantaneous heater in the normally accepted sense, it is possible in many applications to dispense with bulk storage completely, the small volume of hot water retained in the shell being adequate to meet an initial demand for a minute or two only. In instances where batch or surge loads occur, storage in an 'accumulator' may be required but, for variations in a normal load, the generator draws on surplus boiler capacity rather than storage. Success in operation depends largely upon the performance of the temperature control valve, the sensing element of which is so mounted as to take account of both temperature and rate of demand.

Plate-type heaters

For process applications and for use in hotels and similar institutional buildings such as hospitals, where large quantities of hot water at a constant temperature are required, an application may exist for a true instantaneous

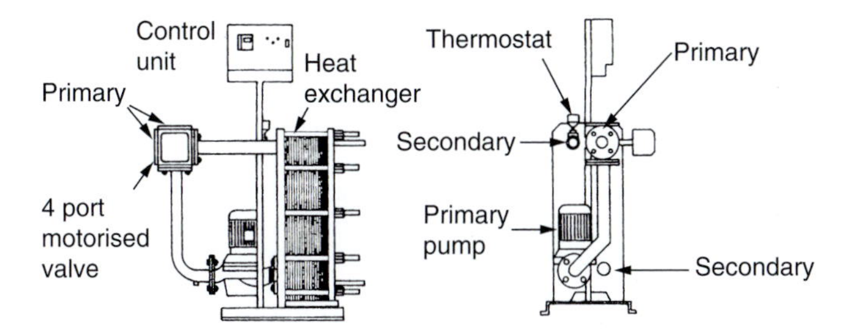

Figure 25.22 Plate-type water heater

heater such as the plate-type equipment shown in Fig. 25.22. The components of this package include the heat exchanger, a primary circulating pump, a motorised valve and the pipework interconnections. The valve is controlled thermostatically from the secondary side to maintain a constant temperature output at 60°C with a primary water input at 82°C.

The heat exchanger, built up as a 'sandwich' from a number of corrugated plates which provide passages through which the separated primary and secondary water circuits flow, is mounted to a frame and has cover plates which are pulled together by long end-to-end compression bolts: flanges on one cover plate provide for pipework connections. The output of the unit is determined by the size and number of plates making up the heat exchanger and a range of ratings up to 1115 kW is available.

The addition of an accumulator (buffer vessel) interconnected to the plate heat exchanger by a continuously pumped pipework circulation circuit will enhance the output to meet peak surge demand above the normal demand requirements.

Since the efficiency of such a unit depends largely upon the high flow velocity of both primary and secondary water, the passages between the plates are necessarily restricted and any deposition on the drowned surfaces will affect performance. The manufacturers claim that the heat exchanger is easy to dismantle for maintenance, as a result of the bolted construction, but use of such equipment in a hard water area with an untreated supply requires some deliberation.

Rating of calorifiers

The heating surface (tube area) required is a function of the temperature difference between the primary water or steam and the secondary water and of the heat transfer coefficient, W/m^2 K. The question now arises as to what is the true temperature difference bearing in mind that the only known conditions are probably those of the entering primary flow and the entering cold water supply.

The notional design conditions, taking Fig. 25.13 for reference, might be, say, 80°C flow and 70°C return for the primary circuit, 70°C for the secondary outflow and 10°C for the entering cold water feed. The mean temperatures would thus seem to be 75°C for the primary and 40°C for the secondary, giving a difference of 35 K. However, this level of difference is very unlikely to remain stable for any length of time in practice.

Following a period of heavy draw-off, the whole of the hot water content of the cylinder may have been used and the difference then might be 60 K or more. At the other extreme, when use of hot water is negligible, and storage temperature has reached 70°C, the difference then might be as little as 5 K. As to the heat transfer coefficient, this will be a function of tube size, spacing and of the level of external scaling: it will vary also according to the vigour of the convection velocities within the shell which will be temperature dependent.

In consequence of all the variables, it is best to consult the calorifier manufacturers and not to rely upon a theoretical approach alone.

Requirements for storage capacity and boiler power

A decision as to the volume of hot water storage and the capacity of the associated boiler plant depends upon considerations which are quite different from those which apply to a heating system. There, as a rule, the quantity of heat which is required under design conditions may be calculated with a fair degree of accuracy

Table 25.2 Capacity of various standard fittings

Fitting	*Capacity* (litre)	*Required temperature* (°C)
Bath, average	80–120	40–45
Sink	12–18	50–60
Basin, normal fill	5	40–50
Shower	0.15–0.2 litre/s	40

and, after adjustments have been made to suit particular circumstances, the use of meteorological data may be applied to determine the 24-hour pattern of demand.

A hot water supply system, except in the case of an industrial process having a finite time cycle, has to fill a demand which depends upon a user pattern which can be projected only empirically and is, at best, intermittent. For example, the daily usage of hot water in a school, a hotel and an office block with a canteen might be identical for given sizes but the peak demands would differ as to both magnitude and timing. Furthermore, usage in each would not be the same per head, per unit of floor area or per any other criterion.

A natural relationship exists for any particular type of demand between the volume of hot water stored and the boiler power provided to heat it. In general terms, the more generous the storage the smaller the boiler capacity needed since it will have a long time to restore the temperature following draw-off. Nevertheless, such an arrangement carried to excess would lead to an unreasonable delay in recovery of storage temperature if this were to have fallen below normal, due to nighttime shut-down for example. Conversely, the combination of a small storage volume and a comparatively large boiler would provide quick recovery of temperature after draw-off but probably would be inadequate to meet a sustained heavy demand. A compromise between the extremes must be chosen.

Approximate methods

In many cases a storage capacity equal to the maximum draw-off in any 1 hour at peak load conditions will be an adequate provision. The associated boiler power may then be sized on the basis that this volume of water will be heated from cold over some longer period such as 2 or 3 hours. A rule of thumb basis of this sort is, however, acceptable only in circumstances where no parallel experience with a similar load exists or where no general statistical data are available. A digest of collected data is given in Table 25.2 for the capacities of certain fittings and in Table 25.3 of daily and hourly consumptions of hot water in various types of building. These latter values are necessarily averages but may be used to settle a first approximation for capacity.

The approximate method probably overstates the requirement and an approach based on data established from field studies is included in the CIBSE Guide G. There, a series of figures represents the relationship between storage capacity and boiler power for a number of building types. This information may be presented in a rather simpler manner, without any loss of accuracy, by the listings of Table 25.4. The values given do not include any allowance for the loss of effective storage capacity which results as incoming cold water mixes with the hot water held in the vessel. To provide for this situation, an addition of 25 per cent should be made to the total of the volume calculated but not, of course, to the equivalent total representing the output required of the associated boiler or electrical immersion heater.

Use of electrical energy

Further values are also provided in CIBSE Guide G providing storage capacities associated with off-peak electric heating.

Allowance for heat loss

None of the values listed in Tables 25.4 include any allowance for boiler or heater capacity to make good the heat lost from storage vessels and circulating pipework, primary or secondary. In all cases these losses should

Table 25.3 Daily hot water demands (adapted from Institute of Plumbing Guide)

Type of building	*Daily demand* (litre/ person)	*Storage per 24 hour demand* (litre)	*Recovery period* (hour)
Dwellings	115	115	Bedroom
1 bedroom	15	115	Bedroom
2 bedroom	15	115	Bedroom
3+ bedrooms	55	115	Bedroom
Student en suite	70	20	Bedspace
Student, comm.	70	20	Bedspace
Nurses home	70	20	Bedroom
Children's home	70	25	Bedspace
Elderly sheltered	70	25	Bedroom
Elderly care home	90	25	Bedspace
Prison			Inmate
Hostels			
Budget	115	35	Bedroom
Travel Inn/Lodge	115	35	Bedroom
4/5 Star Luxury	135	45	Bedroom
Offices and general work places			
with canteen	15	5	Person
without canteen	10	5	Person
Shops			
with canteen	15	5	Person
without canteen	10	5	Person
Factory			
with canteen	15	5	Person
without canteen	10	5	Person
Schools			
Nursery	15	5	Pupil
Primary	15	5	Pupil
Secondary	15	5	Pupil
6th form college	15	5	Pupil
Boarding	114	25	Pupil
Hospitals			
District General	200	50	Bed
Surgical ward	110	50	Bed
Medical ward	110	50	Bed
Paediatric ward	125	70	Bed
Geriatric ward	70	40	Bed
Sports changing			
Sports hall	20	20	Person
Swimming pool	20	20	Person
Field sports	35	35	Person
All weather pitch	35	35	Person
Places of assembly (excl. staff)			
Art gallery	2	1	Person
Library	2	1	Person
Museum	1	1	Person
Theatre	1	1	Person
Cinema	1	1	Person
Bars	2	2	Person
Nightclub	1	1	Person
Restaurant	6	6	Cover

be calculated and, where storage vessels are served by a conventional boiler plant, they should be added to the output requirement determined from Table 25.4. For a vessel which is to be heated with an off-peak supply, however, it has been suggested that the calculated total may be dealt with more conveniently if it be converted to an equivalent stored volume of hot water. This may be calculated from:

Table 25.4 Hot water storage at 65°C and boiler power per person and per meal

Building and use	*Validity of data* (persons or meals/day)	*Required storage when the capacity (litre per recovery period is person or as follows litre per meal)* (hours) ½	*1*	*2*	*3*	*4*
Hostel						
Service	80–320	3.0	5.0	7.2	9.7	12.2
Hotel						
Service	80–320	7.5	11.0	15.0	16.8	18.5
Catering	140–840	2.4	3.4	4.7	5.5	5.9
Office						
Service	110–660	0.7	0.9	1.2	1.6	2.0
Catering	40–370	2.0	3.5	5.0	6.1	7.1
Restaurant						
Service	100–1010	0.3	0.4	0.5	0.6	0.7
Catering	100–1010	0.4	0.6	0.9	1.1	1.3
School						
Service	360–1600	0.7	0.9	1.1	1.2	1.5
Catering	240–1200	1.1	1.6	2.3	2.8	3.2
Shop						
Service	50–220	1.5	1.7	1.9	2.4	2.8
Catering	60–540	1.0	1.3	1.5	1.8	2.2
Boiler output, kW, required to achieve the recovery rates stated is storage capacity multiplied by:		0.128	0.064	0.032	0.021	0.016

Notes
1. When storage is to be heated electrically, the rating of the immersion heaters should correspond to the boiler output listed.
2. When occupancy levels or meals served are less than the minimum of the validity band, both storage capacity and boiler output must be increased. The following multiplying factors are proposed: 75% of minimum × 1.2; 50% of minimum × 1.4; 25% of minimum × 1.6.

$$V = 3600(Lhp)/c(t_s - t_c)$$

where

V = notional equivalent stored volume (litre)
L = daytime heat loss from vessels, etc. (kW)
H = daytime running hours (hour)
P = proportion of annual load taken off-peak (per cent)
c = specific heat capacity of water = 4.2 kJ/kg K
t_s = temperature of stored hot water (°C)
t_c = temperature of cold water feed (°C) = 10°C

Examples

- A school has 500 pupils and serves 400 meals each day. The water for cloakrooms, etc. is required to be at 55°C and for kitchens, etc. at 65°C. Heat loss is ignored for this example. Recovery periods to be 2 hours for service and 1 hour for catering. The demands are to be met by two separate plants.

  Cloakrooms
  From Table 25.4
  Storage required = 1.1 litre/person
  Boiler rating = 1.1 × 0.032 = 0.035 kW/person

Thus

Storage = 1.1×500 = 550 litre
With margin for mixing = 1.25×550 = 687 litre
Boiler rating = $0.035 \times 500[(55–10)/(65–10)]$ = 14.3 kW

Catering
From Table 25.4
Storage required
Boiler rating = 1.6 litre/meal = 0.10 kW/meal

Thus

Storage = $1.6 \times 400 = 1.25 \times 640$ = 640 litre
With margin for mixing = 0.10×400 = 800 litre
Boiler rating = 40 kW

Storage temperature

It should be noted that the temperatures listed in Table 25.2 are for hot water use and do not represent those which are required at the storage vessel. Taking no account of other relevant factors, it is obviously reasonable to store hot water at a temperature higher than that required at the point of use since a smaller volume in storage is then necessary.

It was for many years considered to be good practice to design for, and control, hot water storage temperatures at 65°C (150°F) for all but special applications. Dishwashers, for instance, required an elevated temperature of about 85°C, whereas service to primary schools, old people's homes and prisons was provided at some 15–20 K lower than the normal level. Propositions were made during the early 1970s, with energy conservation in mind, that the traditional temperature levels were not actually needed and figures of the order of 45°C were proposed, with seemingly little thought given to the increased storage volume – and increased heat loss from larger vessels – which would have been required.

These unfortunate suggestions have however been overtaken by the results of investigations into the incidence of Legionnaires' disease and the conclusion that it is related to aerosol contamination, some of which has originated in hot water systems. The legionella bacterium can be present in most mains water supplies but is dormant at temperatures below about 20°C. It multiplies rapidly between 25°C and 45°C and is killed instantly at a temperature of 70°C. Recommendations are, in consequence, that hot water should not be stored at a temperature less than 55°C and, where it has been held at a lower and more critical temperature for any length of time, i.e. overnight, it should be reheated to 55–60°C for an hour prior to exposure and use. Thus, if complex and fallible control cycles are to be avoided, a practical solution is to maintain a storage temperature of 60–65°C whenever the plant is in operation, particularly where night shut-down or any other intermittent heating routine is adopted.

In instances where higher temperatures are necessary, as in the case of dishwashers as mentioned previously, practice in recent years has been to fit such equipment with local electrical or other booster heaters supplied with make-up from the normal hot water system. Any requirements for water at temperatures lower than 55°C may be met by mixing of hot and cold water, which process should be arranged as near to the point of use as is practicable.

The control of the legionella bacterium is an important feature in the design of hot water services and guidance in addition to that mentioned above is provided in the HSE, Approved Code of Practice and Guidance L8, CIBSE TM13 and Hospital Technical Memorandum 2040.

Temperature control

Since the volume of hot water stored acts as a cushion, close control of secondary water temperature is neither possible nor necessary. In common with the familiar rod-type thermostat switching arrangements associated with electrical immersion heaters, the simplest forms of control for water temperature are direct acting, which is to say that expansion of the sensing element provides the motive power for the controller.

In Fig. 25.18, a simple and robust type of direct-acting valve is shown, the primary outlet pipework commonly being works mounted: this pattern may be used whether the primary supply is water or steam. The next stage in complexity is a separated direct-acting valve which may be connected by a capillary to the sensing element mounted in the storage vessel. Apart from these simple types, thermostats and control valves may be electrical, electronic or pneumatic and be either two- or three-way pattern.

A disadvantage inherent to both types of direct-acting valve, and of other types when misapplied, is that they provide *proportional* control. Thus, with the sensing element immersed in the water store, the valve will be wide open to the primary medium when the store is cold and will close progressively as the temperature rises. Since the temperature difference between the primary medium and the stored water will decrease as the valve closes, sluggish recovery of the last few degrees of storage temperature is inevitable. Any form of on/off control is to be preferred for this particular application.

Feed cisterns

For a conventional vented hot water supply system, a cold water storage cistern is required to supply the hot water storage vessel and replace the water drawn off at taps, etc. A cistern is necessary also, as in the case of a conventional vented heating system, to accommodate the increase in volume as the water content of the storage vessel and pipework is heated. This expansion takes place via the feed pipe connecting the cistern to the storage vessel and not, as some publications assert through the vent pipe!

To prevent the expanded hot water entering the cold water storage cistern with the possible consequence of contamination by the legionella bacterium, hospital technical notes advise that consideration should be given to installing a non-return valve on the cold feed adjacent to the hot water storage vessel, together with an increase in the open vent pipe size.

The cold water feed pipe from the cistern (or pressurising equipment as noted later) to the storage vessel should, ideally, be so connected to the latter that mixing of the cold supply with the hot water stored is prevented and disturbance to stratification of the latter is minimised. A number of devices have been produced from time to time to achieve this end but with only limited success.

In order that an adequate pressure be provided at the points of use, the cistern should be sited as far above the highest such point as is practicable: in a building of any significant size, a purpose-built tank room having easily cleaned surfaces, and not used as an overflow store, is to be preferred.

Following the modification of the water regulation in 1989 with the acceptance of the use of unvented hot water systems (refer to the proceeding section), it has become a common feature to provide the cold water storage cistern at low level within a building and boost the water, by a multi-pump cold water booster set, to the cold water outlets and cold feed supply to the hot water storage vessel(s) and thereby onto the hot water outlets (refer to Fig. 25.23). The provision of adequate pressure at the points of use being easily achieved by the selection of the pressure developed by the cold water booster set.

It is common practice to provide within the cold water cistern water storage for one day's interruption of mains supply. However, when it is known that the mains supply is reliable and the water supply company's reaction time for re-instatement of supply is short, consideration should be given to reducing the water storage quantity.

It is essential, to prevent contamination of the stored water, that all cisterns be provided with rigid, close fitting covers which exclude light and that they should have properly made insect screens at all openings. Where pipe connections penetrate the cover, the holes should be drilled and tightly fitting to the pipe. The water inflow supply and the outflow connection should be so positioned, generally opposite to each other, such that the stored water flows through the cistern ensuring that there are no areas where the water may stagnate.

The cistern and pipework should be suitably insulated to prevent heat gain thereby reducing the possibility of the temperature of the stored water to rise.

Unvented hot water systems

Past editions of Water Byelaws have included requirements which have inhibited design development of unvented storage-type hot water systems in the British Isles. Direct connection to a mains pressure service pipe has not been permitted and provision of an open vent to atmosphere has been mandatory. The use of the bubble-top calorifiers which have been described earlier was generally confined to sites protected by Crown immunity.

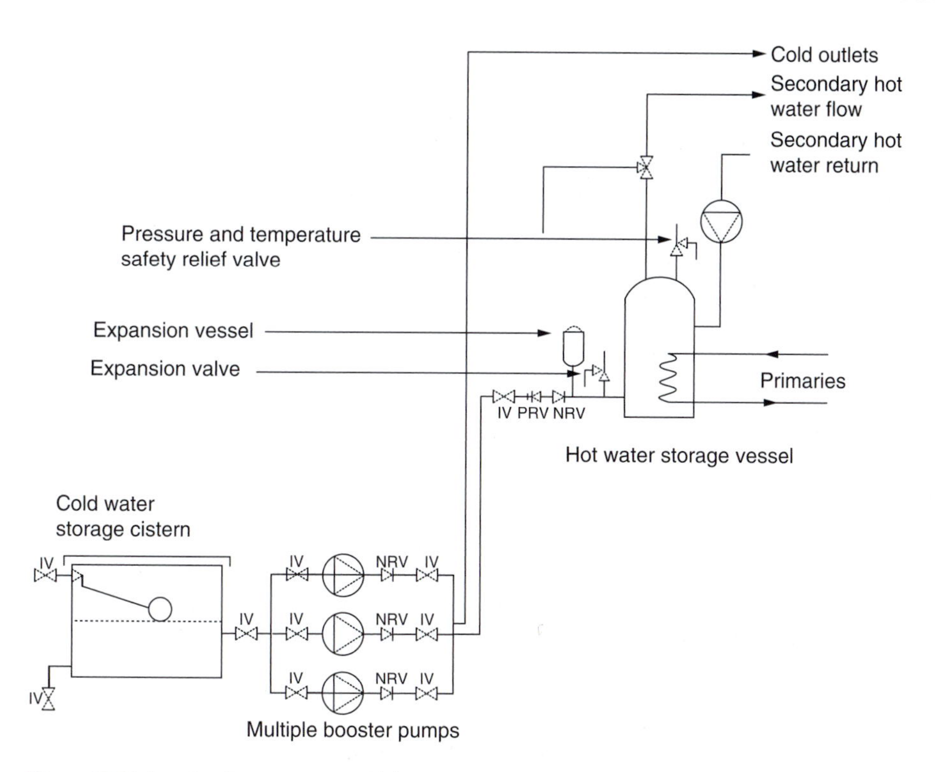

Figure 25.23 Low-level water storage with booster set to hot and cold water systems

The 1986–1989 edition of the Byelaws and the current Water Supply (Water Fittings) Regulations no longer proposes that there should be a prohibition of connection to mains pressure service pipes and, as a result, water companies have relaxed their previous attitudes and the use of unvented hot water storage systems became acceptable. The requirement and guidance on the unvented units and their installation are defined in the Building Regulations G3.

These regulations cover unvented systems with a capacity of:

- A system up to 500 litre and a power input not exceeding 45 kW.
- A system over 500 litre or having a heat input of over 45 kW.

It defines that the unvented storage units of 500 litre capacity and not exceeding 45 kW are to be in the form of proprietary units approved by a suitable certifying body.

A system of over 500 litre and 45 kW is to be individually designed for specific applications and incorporate all the necessary design and safety features cited in the Building Regulations G3. The design of the units is to be undertaken by an appropriately qualified engineer. It is also a requirement of these Regulations and also of the Water Regulations that these unvented systems can only be fitted by a suitably trained and certified installer.

For unvented systems with a capacity of 15 litre or less the Building Regulations are not applicable.

Control packages

The safety devices defined in the Regulations are shown in Fig. 25.24 and are listed below:

Protective

- A thermostat to control the energy source. This should not be set at above 75°C and preferably at 5 K lower.
- A temperature operated cut-out acting on the energy source. This must be factory set at 85°C and be of the manual reset type.

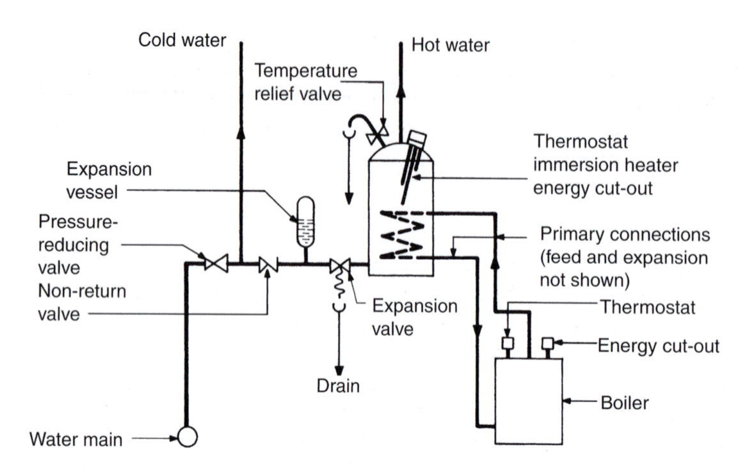

Figure 25.24 Unvented hot water secondary supply system

- A temperature and pressure operated relief valve, factory set to open at 90–95°C. The outlet from this valve must be to a tun-dish and piped thence to a point where discharge of very hot water will not be dangerous. The discharge pipework sizes, related to length of pipe run and number of bends, are indicated in Section G3 of the Building Regulations.

Functional

- A pressure controller at the connection to the service pipe is required to deal with fluctuations in the supply and to determine the system operating pressure. Good control characteristics at low flow rates and the ability to accept a wide range of inlet pressures are necessary. Either a pressure reducing valve or a pressure limiting valve may be used, factory set.
- A non-return valve in the feed to the storage vessel is necessary in order to avoid back-flow of hot water into the cold water supply.
- A diaphragm-type expansion vessel to cater for the expected increase in volume (nominally 4 per cent).
- An expansion relief valve which should not open during normal system operation. The outlet from the valve must be visible and piped to a convenient point.

Water treatment

A short discussion was included in Chapter 11 to cover the incidence of corrosion in heating systems which have the advantage of reusing and recirculating the same water content. In the present context of hot water supply, the secondary water content in indirect systems and the whole content of direct systems are subject to continual change. In general terms, the problem in this case relates more usually to the formation of scale rather than to the effects of corrosion. In this respect it is of interest to note that public water supplies to the mainland of the British Isles, south and east of a line drawn between Hull and Bristol (with local exceptions), all fall into either the 'hard' or 'very hard' categories.

A complete discussion of this subject would be beyond the scope of this book and the brief notes which follow are intended only as an introduction to the methods commonly used to treat cold water supplies to hot water systems. Further information may be found in the CIBSE Guide G but specialist advice should be sought where particular problems are known to exist.

Hardness of water

The calcium and magnesium salts present in raw water exist in two forms, as bicarbonates which form what is known as temporary hardness and as sulphates, chlorides and nitrates which form what is known as permanent hardness. The salts within the temporary category fall out of solution when the water temperature is raised to about 70°C which situation occurs adjacent to heat exchange surfaces even when the storage temperature is below that level. The salts of permanent hardness remain in solution at the temperatures generally encountered in indirect systems but calcium sulphate may cause problems in direct systems at boiler surfaces or where the primary medium of an indirect system is at high temperature.

Commercial installations

Here, dependent upon circumstances, it is probable that a full-scale water treatment plant will be installed. This is most likely to be a base- or ion-exchange process which makes use of beds of either natural or synthetic *zeolites* (sodium aluminium silicates). These minerals contain sodium in combination and have the property of exchanging this with the calcium and magnesium salts in a raw water to form sodium bicarbonate which is not hard-scale forming. Natural zeolites (greensands) are impermeable and have a lower capacity for exchange, per unit volume, than the synthetic type. The latter, as a result of their porous structure, must not be used with an unfiltered raw water supply and have a shorter life than the natural material.

The exchange process does not continue indefinitely since the capacity of the zeolite is said to be *exhausted* when all the sodium content has been exchanged: it is then necessary to regenerate the mineral by slow flushing with a strong solution of common salt (sodium chloride) followed by brisk backwash to remove the brine residual and dispose of the calcium and magnesium. It must be emphasised that the process of base exchange, whilst reducing scale deposition, does not reduce the total content of dissolved solids in the water.

Dosing

Internal water treatment by dosing boiler make-up water has been a familiar process in the case of industrial plant where the facilities exist to allow the effect to be monitored and the quantities of additive adjusted accordingly. For use at domestic level, it has been necessary to develop simple methods for automatic metering of dosage which require no skill to operate and will not permit overfeed. A number of devices which allow suitable quantities of phosphates and polyphosphates to be injected into an incoming mains water supply are now available for use at domestic level. This injection modifies the hardness salts such that, following treatment, they remain suspended in the water, instead of bonding one to another and forming scale when temperature is increased. The process requires only that the bulk supply of chemicals introduced be renewed from time to time, in the form of powder or spheroid concentrates.

Magnetic/electromagnetic conditioning, etc.

As an alternative to dosing, use may be made of magnetic/electromagnetic conditioning which have no chemical effect upon the various hardness salts and do not change the composition of the water. The process involves introduction of a pipeline unit which incorporates a magnetic circuit through which the water passes. As a result of this exposure, the particles of hardness salts do not agglomerate but remain freely suspended in the water and thus do not form scale on heated surfaces.

An alternative but not dissimilar process results from exposure of the incoming water to an electrochemical device within which water flow past a zinc–copper combination acts to produce an electrolytic cell. This, in sequence, provides an electrical potential which discourages the particles of scaling salts from attracting one another and thus precludes formation of a layer of scale.

Materials, etc.

Vessels and pipework

The two materials most commonly used in construction of hot water systems have been galvanised steel and copper. Selection of one or the other has often been made following study of data such as that set out in Table 25.5, the results representing the limits within which steel could be used. As a result of the ease with which it

Table 25.5 Limits for use of galvanised steel

	Temporary hardness greater than:			
pH value	mg/litre	*Parts per 100 000*	*Grains per Imperial gallon*[a]	*Grains per US gallon*
7.3	210	21	15.0	12.3
7.4	150	15	10.5	8.8
7.5	140	14	9.8	8.2
7.6	110	10	7.7	6.4
7.7	90	9	6.3	5.3
7.8	80	8	5.6	4.7
7.9 and over	70	7	4.9	4.1

Note
a 1 grain per Imperial gallon = 1° Clark.

is manipulated and the simpler fabrication techniques which are associated with it, copper piping is now used almost universally. Hence, since a mixture of metals is particularly undesirable where the associated water is subject to continuous change, with a consequent continuous release of dissolved oxygen, the use of galvanised steel for storage vessels is no longer usual. Copper vessels having thin shells are now very often provided with sacrificial aluminium anodes to protect them against corrosion.

Over recent years the use of other materials has become available to the designer such as CPVC. This pipework material removes the problems associated with corrosion. However, the greater expansion rates of this material must be taken into consideration by the system designer.

For hot water storage vessels the use of stainless steel or glass lined vessels has been adopted by manufacturers.

Insulation

Heat losses from all the components of a hot water system continue throughout the year, often day and night. To suggest that, for many systems, as much energy is dissipated in this way as is used to heat the water drawn off at taps is no exaggeration. In consequence, a very good standard of thermal insulation to pipework is recommended. Storage vessels for domestic use may be factory insulated with sprayed urethane foam or fitted with a standard quilted jacket. Larger size vessels for the commercial sector should be insulated to at least as good a standard, and be finished with a protective sheet metal casing. The insulation in all cases should as a minimum comply with the requirements of the Building Regulations.

Piping design for central hot water supply systems

Piping arrangements for hot water supply systems fall naturally into four categories and are best considered within these, as follows:

(1) The primary pipework which provides for circulation of water between the energy source and the storage vessel or vessels. This circulation is related to the maintenance of the required temperature in the store and, except in that sense, is independent of the rate of outflow at draw-off points.
(2) The secondary outflow pipework which is required to pass hot water in the quantity demanded by the draw-off points when taps, etc. are opened.
(3) The cold water feed pipe to the hot water store which is required to pass make-up water equivalent to the hot water quantity drawn off. This pipe, originating from either an elevated cistern, boosted water supply from low-level storage or a mains pressure service pipe, conveys the motive force which causes hot water to flow at draw-off points.
(4) The secondary circulating pipework which, in conjunction with the outflow pipework, provides means whereby hot water is constantly available near to the draw-off points.

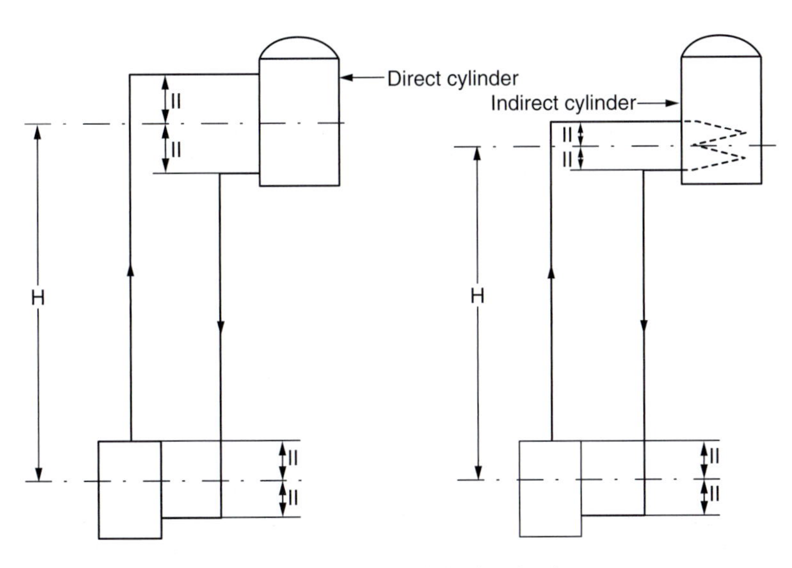

Figure 25.25 Effective heights for gravity circulation in primaries

Table 25.6 Pipe sizes and energy flow (kW) for primary circuits in copper (BS EN 1057 R250) and galvanised steel (BS 1387: heavy) pipes

Copper pipes			*Steel pipes*		
	System type			*System type*	
Pipe size	*Direct*	*Indirect*	*Pipe size*	*Direct*	*Indirect*
28	5	7	25	4	7
35	8	10	32	8	13
42	12	18	40	11	19
54	22	31	50	19	31
67	35	51	65	35	59
76	47	66	80	53	82

Primary circulation pipework – direct gravity system

As indicated in the preceding text direct gravity circulation systems and in fact gravity circulation in general are no longer adopted and do not meet the requirements of the Building Regulations Part L. However for the reader's interest the following paragraphs describe the design method for gravity circulation systems.

To ensure that circulation between energy source and storage is brisk, taking account of the probability that the available pressure difference will be small, the primary flow and return pipework should be sized generously. In addition, it is sensible in hard water areas to make allowance for the inevitable reduction in diameter due to scale deposition and thus, even for the smallest system, piping should not be less than 25 mm galvanised steel or 28 mm copper.

The pipe size required may be determined using exactly the same methods of calculation as those described in Chapter 12 for a heating system operating with gravity circulation, taking the circulating head as the difference in pressure between the hot rising and the cold falling columns. The effective height is from the centre of the boiler to the centre of the cylinder, as shown in Fig. 25.25, and a temperature difference, flow to return, of 20 K may be assumed for purposes of calculation. As a first approximation, the rates of energy flow through given sizes of pipe, as listed in Table 25.6, may be used. These are based upon a boiler flow temperature of 60°C and a height of 3 m, the travel being 15 m and incorporating eight bends. The tabulated figures for energy flow are 20 per cent less than the theoretical in order to allow for scaling in the pipes.

It will be appreciated that, although the flow temperature from the boiler may be controlled at a relatively constant level, the temperature of water returning to it from the store will vary according to the rate of draw-off and, in consequence, the circulating head will change also. The most active circulation will occur during and after a period of heavy draw-off, i.e. when the lower part of the store is certain to be cold, and the least active after a period of no draw-off when the contents of the store are nearing boiler flow temperature.

This situation suggests that supply and demand are in equilibrium at all times, which is generally the case, and that the variation in circulating head will offer a form of temperature control. The exception arises when, following a period of 'stagnation' at near to the required storage temperature, a quick response to sudden heavy draw-off is required and the necessary rapid circulation is delayed by system inertia.

Where a direct fired type of water heater is used no primary pipework is needed and the response rate to demand is always extremely rapid.

Primary circulation pipework – indirect gravity system

While the advisability of sizing pipework generously, where circulation is to be by gravity, applies to indirect systems also, in this case no allowance need be made for scale deposition when selecting pipe sizes since the water content of the primary circuit is constantly recirculated and does not change. The need for a brisk circulation remains, nevertheless, since the temperature of the stored secondary water is raised through some form of heat exchanger whereas, in a direct system, it is heated within a boiler.

Since the primary circuit will, in all probability, originate from a boiler plant which also serves a heating system, it is also probable that the flow temperature therefrom is fixed at 80°C. In any event, however, it will be advantageous to provide a primary flow temperature at a reasonably high level in order to economise in heat exchanger surface. The temperature difference, flow to return, may again be about 20 K.

The pipe sizes required may again be determined using the same methods of calculation as those described in Chapter 12. As a first approximation, the data in Table 25.6 may be used, these being based upon a flow temperature of 80°C and other particulars as before, plus an allowance equivalent to 3 m of pipe for the coil within the storage cylinder. The pipework arrangement and effective height are shown in Fig. 25.25.

Primary circulation pipework – pumped systems

Pumped primary circuits are now required for compliance with Building Regulations Part L. Pumped primary circuits may either be direct or indirect, although direct systems are very seldom installed these days. It is necessary for pumped primary circuits to be arranged to be quite separate from any space heating circuit served from the same energy source. The temperature difference, flow to return, may again be about 10–15 K. For a low-rise institutional building or group of buildings, it may be more economic to provide a number of dispersed hot water stores fed from a constant temperature primary distribution system, as shown in Fig. 25.26, rather than a widespread secondary hot water circuit. Such an arrangement may offer the facility to serve other equipment requiring

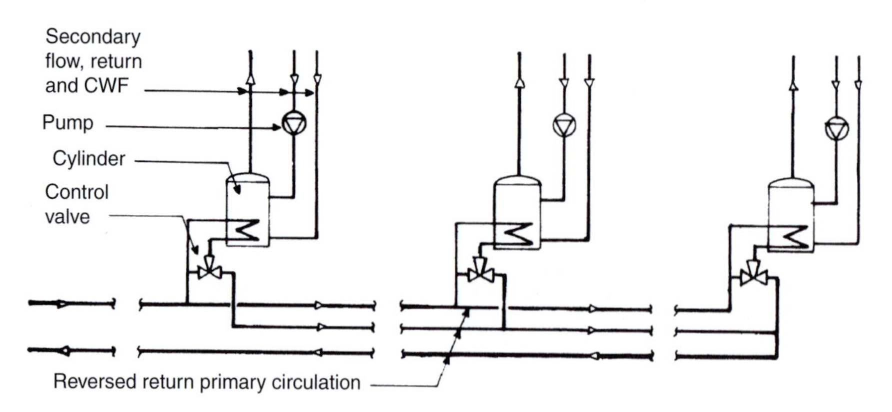

Figure 25.26 Dispersed indirect cylinders with central-source primaries

a constant temperature supply such as ventilation heater batteries, etc., and would be quite separate from any parallel circuit which, at a temperature varied to suit weather conditions, might serve heating apparatus. A separate cold water supply and hot water expansion arrangement will, of course, be required for each of the dispersed stores.

Sizing of the pump and pipework is similar to that described for heating circuits as covered elsewhere in Chapter 12.

Secondary outflow and return pipework

The simplest form of outflow pipework from a secondary system is a series of dead legs as shown in Fig. 25.27. The length of these must however be strictly limited for three very good reasons:

(1) If the length were too great, an undue time would elapse before hot water reached the tap and, furthermore, cold water would have to be run to waste.
(2) Following draw-off, a long dead leg would be full of hot water which, being unused, would be an energy wastage.
(3) If draw-off were intermittent, an excessive quantity of water would stagnate in the pipework and might be held on occasions within the critical temperature band for rapid development of legionella.

The current Water Regulations require that a hot water system is designed to ensure that a minimum temperature of not less than 50°C is achieved within 30 seconds after a tap is fully open. For energy conservation the Water Regulations also recommend the maximum lengths of uninsulated hot water pipework indicated in Table 25.7.

Fundamental to the design of secondary outflow pipework is the quantity of hot water required at the various points of draw-off. Table 25.8 gives details of the capacity of a variety of standard fittings and the rates of water flow which are necessary to produce this capacity in a reasonable time.

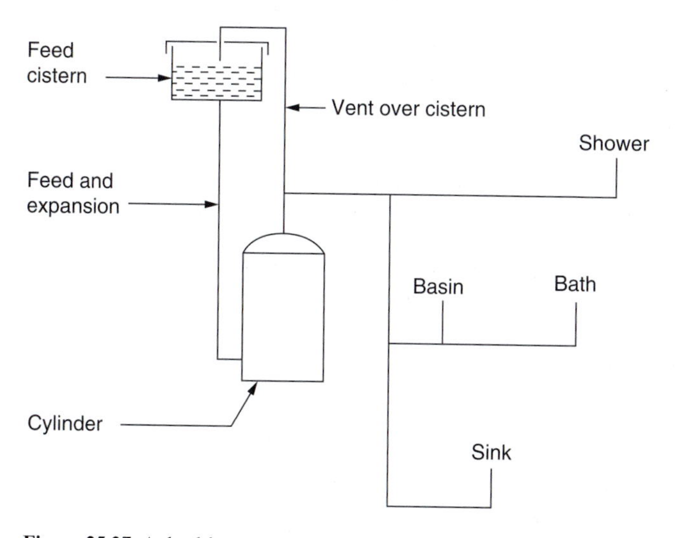

Figure 25.27 A dead-leg system

Table 25.7 Maximum insulated lengths of uninsulated hot water pipes

Outside diameter of pipe (mm)	*Maximum length* (m)
12	20
Over 12 and up to and including 22	12
Over 22 and up to and including 28	8
Over 28	3

Table 25.8 Discharge rates and pipe sizes for draw-off points

Fitting	*Rate of flow* (litre/s)	*Size of connection* (mm)
Bath		
Private	0.3	15
Institutional	0.5	20
Basin		
Separate taps	0.15	15
mixer tap	0.08	8 mm × 2
Shower		
Spray	0.15	15
Sink	0.3	15

Table 25.9 Comparative demand units

	Category of application		
Fitting	*Private*	*Public*	*Congested*
Basin	1	2	4
Bath	4	8	16
Sink	2	5	10
Shower	2	3	6

Where there are only a few draw-off points to be served, as in a private house, it is reasonable to assume that at some time all the taps, etc. will be in use simultaneously, although this is unlikely. With any sizeable system it would be extravagant to make such an assumption since, as the number of taps served increases, it becomes less probable that they will all be open at once.

For instance, a hot water tap at a wash basin may be open for 30 seconds but at least a minute will elapse before the tap is opened again. In a row of 10 basins, this would mean that a maximum of about three taps is likely to be open simultaneously even if people are queuing to use them. The type of building, the type of fitting and the pattern of use are all important matters for consideration. Thus, a group of showers in a sports club house, at the end of a match, is likely to be in full use at the same time but, in a hotel, it is probable that only a small proportion of the baths provided will be filling at the same time. While special cases must always be considered on their merits, a generalised approach for application to a wide spectrum of buildings is necessary.

Various attempts have been made to establish a basis which may be used with confidence and a method based upon the theory of probability is shown in Table 15 of the Institute of Plumbing Design Guide 2003 which, adopts the concept of a scale of *demand units*. The type of application is taken into account in this scale by weighting the units according to assumed intervals of use varying from 5 to 20 minutes for a basin and from 20 to 80 minutes for a bath. From this approach, a much simplified list of 'demand units' may be produced as Table 25.9, with three categories of use:

'*Congested*' (where times of draw-off are regulated).
'*Public*' (normal random usage).
'*Private*' (infrequent or spasmodic).

Using a simple diagram of the outflow system, the individual and the cumulative 'demand unit' totals may be marked up against each section of the pipework and then, using Fig. 25.28, which is based upon probability of simultaneous use, these may be converted to flow rates in litre/s.

Similarly, flow to fittings such as wash fountains and washing machines which may require a specific pattern of supply must be dealt with separately according to type and the maker's flow requirements.

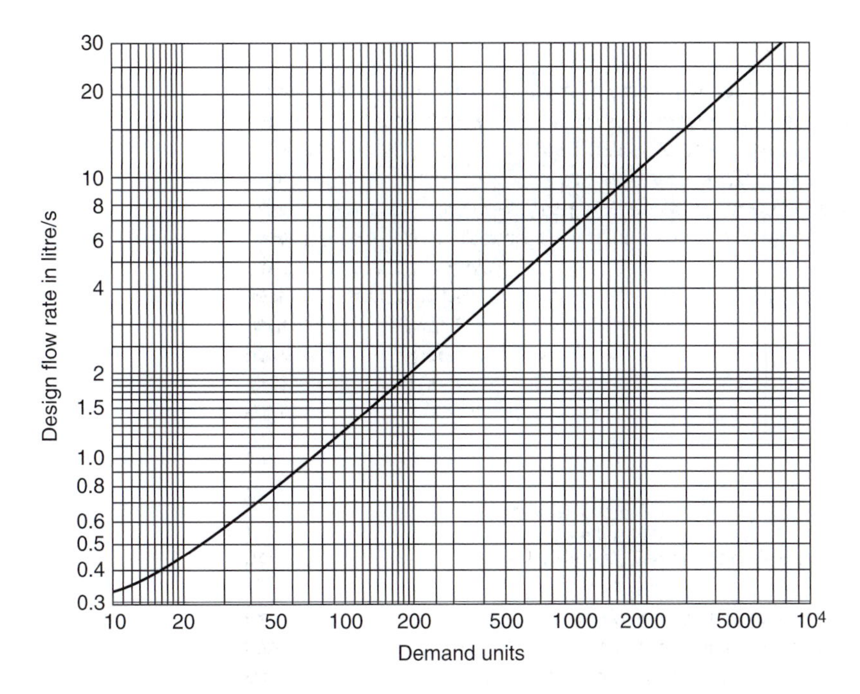

Figure 25.28 Demand units and flow probability

Table 25.10 Pressure in kPa and head of water in m at 10°C

Head (m)	*0*	*0.1*	*0.2*	*0.3*	*0.4*	*0.5*	*0.6*	*0.7*	*0.8*	*0.9*
0	0	0.98	1.96	2.94	3.92	4.90	5.88	6.86	7.84	8.82
1	9.8	10.8	11.8	12.7	13.7	14.7	15.7	16.7	17.6	18.6
2	19.6	20.6	21.6	22.5	23.5	24.5	25.5	26.5	27.4	28.4
3	29.4	30.4	31.4	32.4	33.3	34.3	35.3	36.3	37.3	38.2
4	39.2	40.2	41.2	42.2	43.1	44.1	45.1	46.1	47.1	48.0
5	49.0	50.0	51.0	52.0	52.9	53.9	54.9	55.9	56.9	57.8
6	58.8	59.8	60.8	61.8	62.7	63.7	64.7	65.7	66.7	67.7
7	68.6	69.6	70.6	71.6	72.5	73.5	74.5	75.5	76.5	77.4
8	78.4	79.4	80.4	81.4	82.4	83.3	84.3	85.3	86.3	87.3
9	88.2	89.2	90.2	91.2	92.2	93.1	94.1	95.1	96.1	97.1

It cannot be over-emphasised that a method such as that noted above is a statistical approach to a human problem and is therefore no more than a useful indication of a route to follow in design. There is no substitute for either a broad experience or a specific knowledge of the pattern of use in other similar applications.

Cold water feed pipework

The cold water feed to a hot water store is an integral part of the secondary outflow pipework. However, it is convenient to deal with it separately here, since it may originate either from an elevated storage cistern or as permitted by the Water Regulations from a mains pressure service pipe or boosted water supply from low-level storage. In either case, the head of water (or the mains pressure) provides the motive force which produces outflow at the points of draw-off. The relationship between these alternatives is given in Table 25.10.

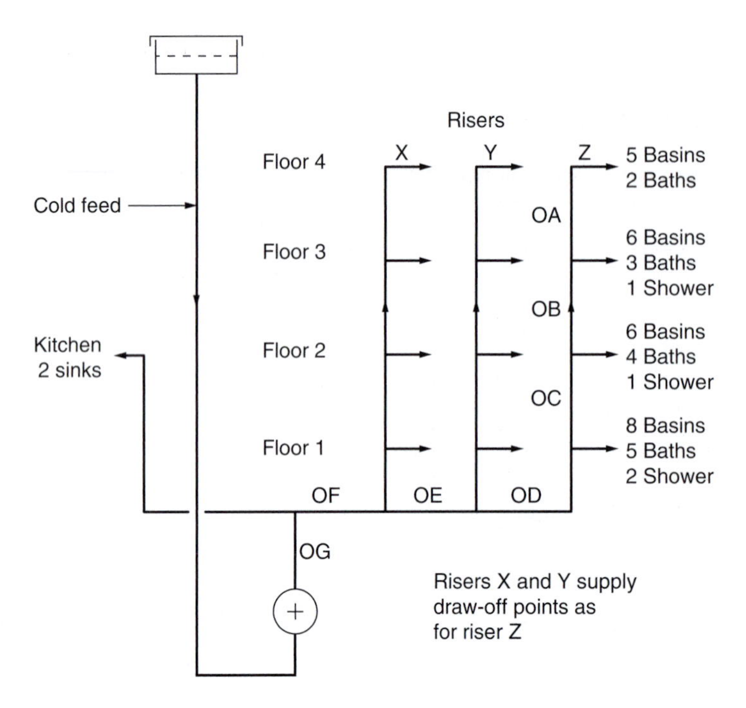

Figure 25.29 Pipe sizing example: outflow fittings served

Table 25.11 Application of demand unit calculation: worked example

Floor	*Fittings*	*Demand units (public)*				*Flow* (litre/s) *From Fig. 25.28*	*Flow* (litre/s) *Totals*
4	5 basins	× 2	=	10	26	0.5	0.5
	2 baths	× 8	=	16			
3	6 basins	× 2	=	12	36	0.6	
	3 baths	× 8	=	24			
	1 shower	× 0.15 litre/s				0.15	0.75
2	6 basins	× 2	=	12	44	0.75	
	4 baths	× 8	=	32			
	1 shower	× 0.15 litre/s				0.15	0.9
1	8 basins	× 2	=	16	56	0.9	
	5 baths	× 8	=	40			
	2 showers	× 0.15 litre/s				0.30	1.2
Kitchen	2 sinks	× 0.3 litre/s				0.6	

Item	*Section*	*Demand units (cumulative)*	*Flow* (litre/s) *From Fig. 25.28*		*Showers and sinks*		*Totals*
Riser	OA	26	0.5		–		0.5
	OB	62	0.9	+	1 × 0.15	=	1.05
	OC	106	1.3	+	2 × 0.15	=	1.6
Mains	OD	162	1.7	+	4 × 0.15	=	2.3
	OE	324	3.1	+	8 × 0.15	=	4.3
	OF	486	4.0	+	12 × 0.15	=	5.8
Total	OG		5.8	+	2 × 0.3	=	6.4

Table 25.12 Flow of water at 75°C in copper pipes (BS EN 1057 R250)

Head loss (m of water per m run)	*Water flow* (litre/s) *in pipes of stated outside diameter* (mm)									*Pressure loss* (kPa per m run)
	15	*22*	*28*	*35*	*42*	*54*	*67*	*76*	*108*	
0.01	–	0.13	0.25	0.46	0.77	1.56	2.80	3.97	10.4	0.1
0.02	–	0.19	0.37	0.67	1.13	2.29	4.11	5.81	15.2	0.2
0.03	–	0.23	0.47	0.84	1.42	2.87	5.14	7.26	18.9	0.3
0.04	–	0.27	0.55	0.99	1.66	3.36	6.02	8.50	22.1	0.4
0.05	0.11	0.31	0.62	1.12	1.88	3.80	6.80	9.60	25.0	0.5
0.06	0.12	0.34	0.69	1.24	2.08	4.20	7.51	10.6	27.6	0.6
0.07	0.13	0.38	0.75	1.35	2.26	4.57	8.17	11.6	30.0	0.7
0.08	0.14	0.40	0.81	1.45	2.44	4.92	8.78	12.4	32.2	0.8
0.09	0.15	0.43	0.86	1.55	2.60	5.24	9.37	13.2	34.3	0.9
0.10	0.16	0.46	0.92	1.64	2.75	5.56	9.92	14.0	36.3	1.0
0.12	0.18	0.51	1.01	1.82	3.04	6.14	11.0	15.5	40.1	1.2
0.14	0.19	0.55	1.10	1.98	3.31	6.67	11.9	16.8	43.6	1.4
0.16	0.21	0.59	1.19	2.13	3.56	7.18	12.8	18.1	46.8	1.6
0.18	0.22	0.63	1.27	2.27	3.80	7.65	13.6	19.2	49.8	1.8
0.20	0.23	0.67	1.34	2.40	4.02	8.10	14.4	20.4	–	2.0
Equivalent lengths for $\zeta = 1.0$	0.6	1.0	1.5	2.0	2.5	3.6	4.7	5.6	8.9	Equivalent lengths for $\zeta = 1.0$

Notes
Values of ζ for fittings: bends, tees, reducers, enlargements = 1.0.
Screw down valve or tap = 10.0; connections to cistern and cylinder = 1.5.

Piping design for outflow

Taking the system illustrated in Fig. 25.29 as an example of feed from an elevated cistern, the hot water flow rate through the cold water feed and the secondary outflow pipework would be determined as shown in Table 25.11. Many designers mark up the 'demand units' and flow rates on a diagram of the system but the sequence of working is more easily illustrated by the tabular approach used here.

If the cold water feed were from a mains pressure service pipe, the procedure followed would be just the same. In all probability, the cold water supply arrangements would be served from the same intake into the building and the duty of the pressure-reducing valve would thus be based upon the total combined demand from all draw-off points, hot and cold.

Once the water quantities to be carried by each section of outflow pipework have been established, the next step is to consider either the head of water available from the cistern or the residual of the working pressure selected where feed is from a mains service pipe.

In either event, the worst case is that presented by the highest draw-off point in the building. Progressing down a multi-storey building, a greater head or pressure will be available on lower floors which may lead to splashing problems there unless steps are taken to throttle the supply.

Tables 25.12 and 25.13 have been compiled from data included in a now superceded version of the CIBSE *Guide Section C4* for water flowing at 75°C in copper pipes to BS EN 1057 R250 and in galvanised steel pipes to BS 1387: heavy respectively (although the data is still relevant today). Bearing in mind the other potential variables in this application, errors arising from applying these tables to designs where water at other temperatures or other grades of copper or galvanised pipe are to be used are negligible. It will be noted that flow rates may be read from these tables against either head loss (m water per m run) or pressure loss (kPa per m run), whichever is more convenient for the application. Table 25.14 gives the equivalent data for water flow at 65°C in PVC pipes.

Table 25.13 Flow of water at 75°C in galvanised steel pipes (BS 1387: heavy)

Head loss (m of water per m run)	*Water flow* (litre/s) *in pipes of stated internal diameter* (mm)									*Pressure loss* (kPa per m run)
	15	*20*	*25*	*32*	*40*	*50*	*65*	*80*	*100*	
0.01	–	0.10	0.19	0.41	0.63	1.21	2.47	3.85	7.84	0.1
0.02	–	0.14	0.27	0.59	0.91	1.72	3.52	5.48	11.1	0.2
0.03	–	0.18	0.33	0.72	1.11	2.12	4.33	6.74	13.7	0.3
0.04	–	0.20	0.38	0.84	1.28	2.46	5.01	7.79	15.8	0.4
0.05	0.10	0.23	0.43	0.94	1.44	2.75	5.61	8.72	17.7	0.5
0.06	0.11	0.25	0.47	1.03	1.58	3.02	6.15	9.56	19.4	0.6
0.07	0.11	0.27	0.51	1.11	1.70	3.26	6.65	10.3	21.0	0.7
0.08	0.12	0.29	0.54	1.19	1.82	3.49	7.11	11.1	22.4	0.8
0.09	0.13	0.31	0.57	1.26	1.94	3.70	7.55	11.7	23.8	0.9
0.10	0.14	0.32	0.61	1.33	2.04	3.90	7.96	12.4	25.1	1.0
0.12	0.15	0.36	0.66	1.46	2.24	4.28	8.71	13.5	27.5	1.2
0.14	0.16	0.39	0.73	1.61	2.47	4.71	9.59	14.9	30.3	1.4
0.16	0.17	0.40	0.76	1.67	2.55	4.87	9.93	15.4	31.3	1.6
0.18	0.18	0.44	0.81	1.80	2.75	5.25	10.7	16.5	33.8	1.8
0.20	0.19	0.46	0.86	1.89	2.90	5.54	11.3	17.5	35.6	2.0
Equivalent lengths for $\zeta = 1.0$	0.4	0.6	0.8	1.1	1.4	1.9	2.7	3.3	4.7	Equivalent lengths for $\zeta = 1.0$

Notes
Values of ζ for fittings: bends, tees, reducers, enlargements = 1.0.
Screw down valve or tap = 10.0; connections to cistern and cylinder = 1.5.

Table 25.14 Flow of water at 65°C in PVC pipes

Head loss (m of water per m run)	*Water flow* (litre/s) *in pipes of stated outside diameter* (mm)										*Pressure loss* (kPa per m run)
	16	*20*	*25*	*32*	*40*	*50*	*63*	*75*	*90*	*110*	
0.008	–	0.05	0.10	0.19	0.35	0.64	1.20	2.44	3.97	6.76	0.08
0.009	–	0.06	0.11	0.21	0.38	0.69	1.28	2.60	4.23	7.21	0.09
0.010	0.03	0.06	0.11	0.22	0.40	0.73	1.36	2.75	4.48	7.63	0.10
0.020	0.05	0.09	0.17	0.32	0.59	1.07	1.99	4.02	6.54	11.11	0.20
0.030	0.06	0.11	0.21	0.40	0.73	1.34	2.49	5.01	8.14	13.82	0.30
0.040	0.07	0.13	0.25	0.47	0.86	1.56	2.91	5.86	9.50	16.13	0.40
0.050	0.08	0.15	0.28	0.54	0.97	1.76	3.28	6.60	10.71	18.17	0.50
0.060	0.08	0.17	0.31	0.59	1.07	1.95	3.62	7.28	11.81	20.03	0.60
0.070	0.09	0.18	0.33	0.64	1.17	2.12	3.93	7.91	12.82	21.74	0.70
0.080	0.10	0.19	0.36	0.69	1.25	2.28	4.23	8.50	13.76	23.33	0.80
0.090	0.11	0.21	0.38	0.74	1.34	2.43	4.50	9.05	14.65	24.84	0.90
0.100	0.11	0.22	0.41	0.78	1.42	2.57	4.77	9.57	15.50	26.26	1.00
0.120	0.12	0.24	0.45	0.86	1.56	2.83	5.25	10.55	17.07	28.92	1.20
0.140	0.14	0.26	0.49	0.94	1.70	3.08	5.70	11.44	18.52	31.36	1.40
0.160	0.15	0.28	0.52	1.01	1.82	3.30	6.12	12.28	19.87	33.64	1.60
Equivalent lengths for $\zeta = 1.0$	0.5	0.7	0.9	1.3	1.8	2.3	3.2	4.4	5.6	7.2	Equivalent lengths for $\zeta = 1.0$

Notes
Values of ζ for fittings: bends, tees, reducers, enlargements = 1.0.
Screw down valve or trap = 10.0; connections to cisterns and cylinders = 1.5.

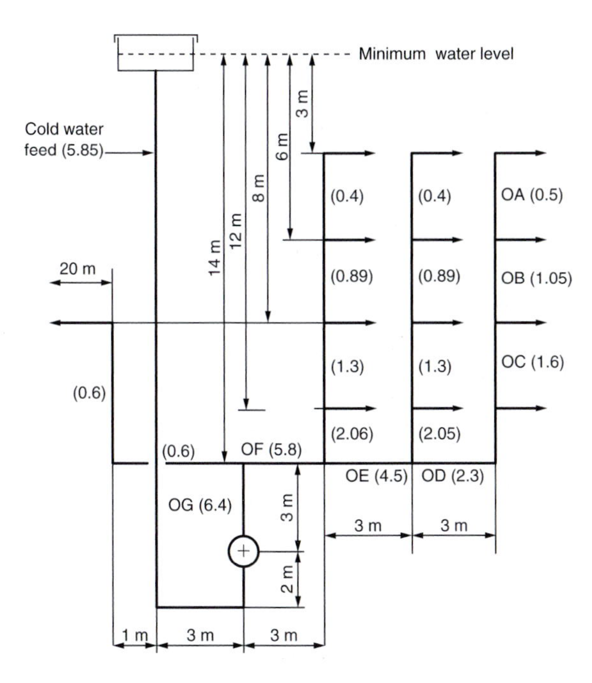

Figure 25.30 Pipe sizing example: dimensions

Note: Figures in brackets are flow rates in litre/s from Table 25.11

Example

Taking the system shown in Fig. 25.29 as an example, physical data and the flow rates previously calculated may be added as shown in Fig. 25.30.

Clearly, the worst case is that of the fourth-floor taps on riser Z which are both the most distant from the cistern and only 3 m below minimum water level, i.e. they have only 3 m head of water available to cover frictional resistance and provide outflow. The travel to this point is:

Cold water feed	=	24 m	
allow for fittings	=	3 m	27 m
Outflow main	=	12 m	
allow for fittings	=	2 m	14 m
Riser Z	=	11 m	
allow for fittings	=	2 m	13 m
Total	=		54 m

Hence, the available unit head = 3/54 = 0.055 m/m run and, interpolating from Tables 25.12 and 25.13, the following listing may be made:

	Pipe size (mm)	
Riser Z	*Copper*	*Steel*
Branch	28	25
OA	28	25
OB	35	32
OC	42	40
OD	54	50
OE	67	65
OF	67	65
OG	67	65
Cold water feed	67	65

Table 25.15 Theoretical heat emission from horizontal steel (BS 1387) and copper (BS EN 1057 R250) pipes

Pipe size (mm)		*Emission* (W/m run)			
		Bare			
Steel	*Copper*	*Steel*	*Copper*	*Insulated*	*Thickness of* (mm)
15	15	42	31	8	25
20	22	51	43	9	25
25	28	62	53	9	32
32	35	75	64	10	32
40	42	84	75	12	32
50	54	102	93	13	32
65	67	125	112	16	32
80	76	143	125	17	32
100	108	179	171	18	38

Notes
Mean temperature difference = 40 K.
Insulation conductivity = 0.04 W/m K.

The example could be continued by considering the lower branches of riser Z since a greater head of water will be available progressively. The head lost due to friction on this longest run, as far as the junction to each branch, would thus be deducted from the total available and the surplus used to size the pipes of the branch. The same process would then be applied to the other risers X and Y and to the branches from them, the run to each being shorter than that to the furthest branch on riser Z.

Re-examination is particularly appropriate for the long branch to sinks in the kitchen since the head available there is 9 m and the loss through the cold water feed pipe and main section OG is only about 2 m ($33 \times 0.055 = 1.82$ m). Thus a head of 7 m is available at the branch which, allowing for 5 m equivalent length there for fittings, represents a unit availability of $7/(29 + 5) = 0.2$ m/m run. The pipe size required, in consequence, for a flow of 0.6 litre/s, would be 22 mm in copper or 25 mm in galvanised steel, one size less in each case than if the original unit figure of 0.055 m/m run had been used.

For a system of modest size, such refinement in calculation may not be worthwhile bearing in mind the many variables. An experienced designer may be content to use a single value for unit head loss except where this would lead to obviously uneconomic design. However, in the case of large systems or those in buildings having either inadequate cistern height or a variety of floor levels, exhaustive examination is necessary.

Piping design for secondary returns

Note has already been made of the limitations which should be placed upon the length of dead legs and thus, where hot water draw-off points are dispersed widely, it is necessary to consider whether to provide:

- A local instantaneous heater to some or all draw-off points.
- A number of dispersed storage vessels with a dead-leg distribution from each. (These might be served either by local energy sources or from a primary distribution system as shown in Fig. 25.26)
- A secondary circulation system.
- Single pipe trace heated.

It is not possible to generalise as to either the most practical or the most economic solution since each application must be considered on its merits. A secondary circulation, with which we are concerned here, offers facilities for heating towel rails and linen cupboard coils, etc. at times when a heating system may be shut down but, since there are continuous heat losses from the circulating pipes even when well insulated, is therefore wasteful of energy.

The water quantity to be circulated through secondary pipework is a function of the heat emission from the system and of a chosen temperature drop, normally 10–12 K for a gravity circulation and about 5 K where a pump is to be used. The heat emission derives from both outflow and return pipework plus the useful output of any towel rails or linen cupboard coils connected to them. Piping emission may be calculated using data from Table 25.15

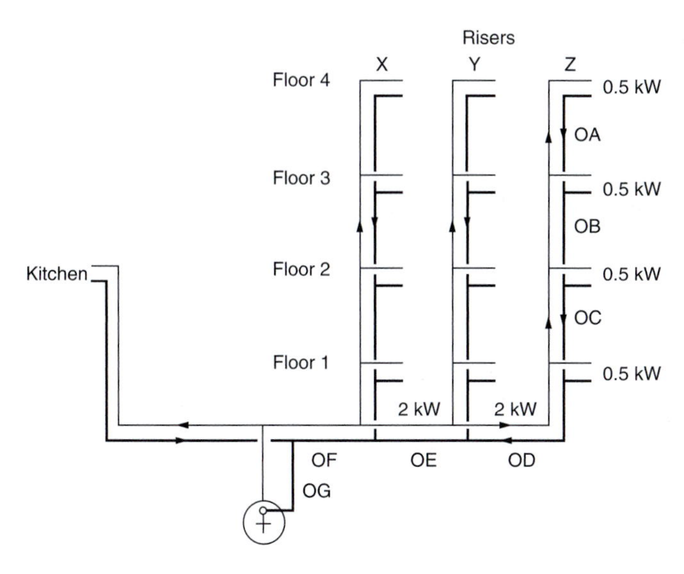

Figure 25.31 Pipe sizing example: secondary returns

Table 25.16 Heat emissions from piping: worked example

Section	*Pipe size*	*Emission* (W/m)	*Length* (m)	*Emission* (W) *Sectional*	*Total*	
OA	28	9	3	27		
OB	35	10	3	30		
OC	42	12	3	36		
OD	54	13	2	26	119	
Riser X					119	
Riser Y					119	
OD	54	13	3	39		
OE	67	16	3	48		
OF	67	16	3	48		
OG	67	16	3	48	183	
						540
				Returns as $^2/_3$ of flows		360
					Total (W)	900

and the sectional emissions are collected and totalled to produce sectional loadings in exactly the same way as for a heating system.

The sizes of the outflow piping will be known as a result of an earlier calculation but the water quantities representing heat emission will be small in comparison with those required for draw-off. Thus, the pressure loss arising from circulation in these pipes is likely to be low and, in conventional design, the return pipes will be a size or two smaller than the equivalent flow pipes. Thus, emission from the returns may be taken either from assumed smaller sizes or, by way of an approximation, as being two-thirds of that of the associated flows.

If a gravity circulation appears practicable, reference would be made to data such as that listed in Chapter 12 and, for flow at storage temperature (65°C) and return at say 55°C, the available circulating head would be 50.25 Pa/m height. Alternatively, if a pumped circulation were necessary as a result of either system size or a preponderance of mains pipework below cylinder level, then a unit pressure drop of about 60 Pa/m of total travel may be taken as a first approximation. This figure includes a 25 per cent allowance for bends and other resistances and takes account of the 'over-size' flow pipework.

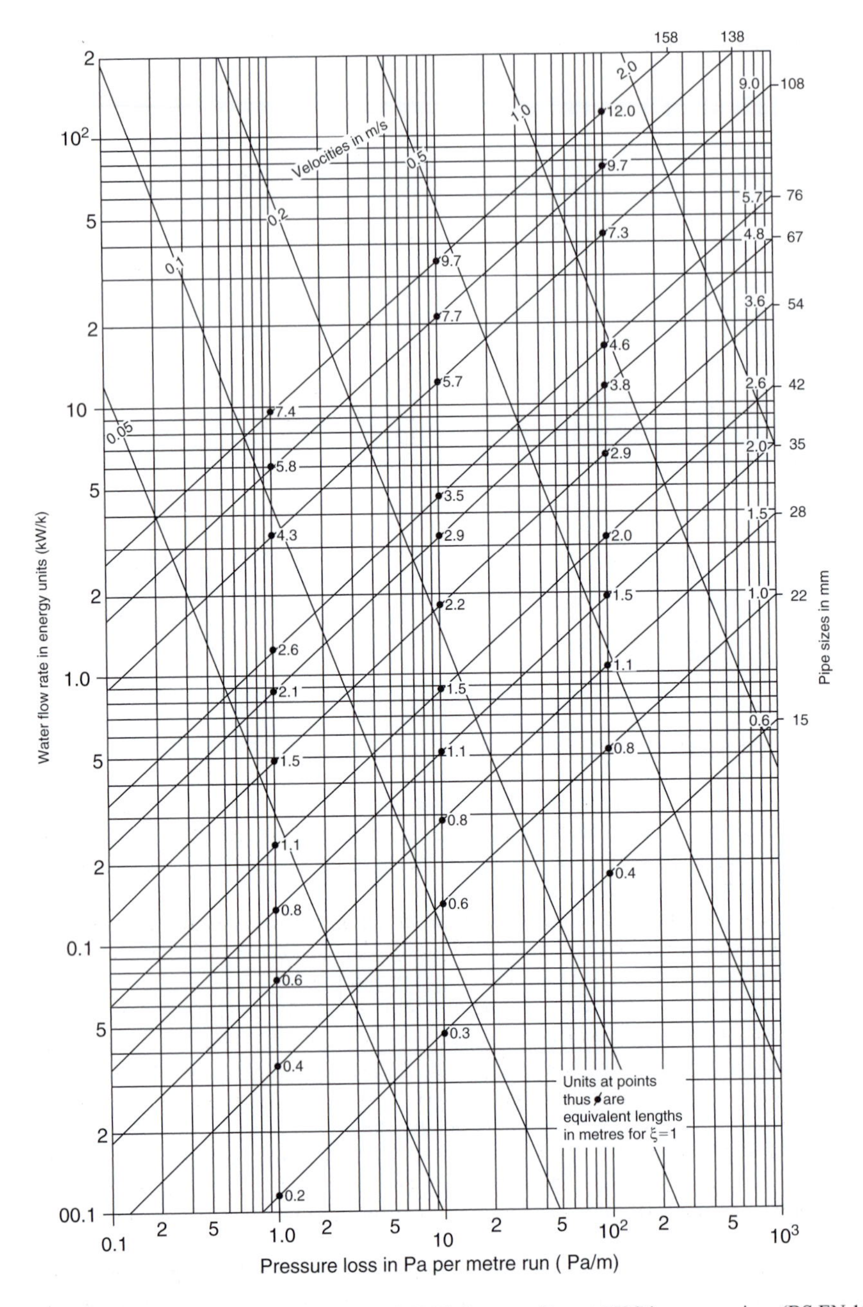

Figure 25.32 Sizing chart in energy units (kW/K) for water flow at 75°C in copper pipes (BS EN 1057 R250)

Example

Using Fig. 25.31 as an example, this being the same system as was shown in Fig. 25.30 but with return pipework added. Each branch circulation serves a towel rail and a linen cupboard coil per floor, rated together at 0.5 kW, and a drying coil at 200 W is fitted in the kitchen (6.2 kW in all). The sizes of the outflow pipework have already been calculated and heat emission may thus be set out as shown in Table 25.16.

Table 25.17 Pressure loss in secondary circulation: worked example

1	*2*	*3*	*4*	*5*	*6*	*7*	*8*	*9*	*10*	*11*
			Flow piping				*Return piping*			
				Pressure loss				*Pressure loss*		
Section	*Loading* (kW/K)	*Length* (m)	*Pipe size* (mm)	*Unit* (Pa/m)	*Section* (Pa)	*Total* (Pa)	*Pipe size* (mm)	*Unit* (Pa/m)	*Section* (Pa)	*Total* (Pa)
OG	0.69	5	67	0.75	3.75	3.75	42	6.5	32.5	32.5
OF	0.67	5	67	0.7	3.5	7.25	42	6.2	31.0	63.5
OE	0.45	5	67	0.25	1.25	8.5	35	8	40	104
OD	0.22	7	54	0.3	2.1	10.6	28	7	49	153
OC	0.17	4	42	0.7	2.8	13.4	28	4	16	169
OB	0.11	4	35	0.3	1.2	14.6	15	42.5	170.0	339
OA	0.06	4	28	0.7	2.8	17.4	15	15.0	60.0	399
Floor 1	0.06	5	28	0.7	3.5		15	15.0	75.0	
OG–OD	–	22	–	–	10.6	14.1	–	–	153	228
Floor 2	0.06	5	28	0.7	3.5		15	15.0	75.0	
OG–OC	–	26	–	–	13.4	16.9		–	169	244
Floor 3	0.06	5	28	0.7	3.5		15	15.0	75.0	
OG–OB	–	30	–	–	14.6	18.1	–	–	339	414
Floor 4	0.06	5	28	0.7	3.5		15	15.0	75.0	
OG–OA	–	34	–	–	17.4	20.9	–	–	399	474

Notes
1. Pressure loss as read from Fig. 25.32.
2. Equivalent lengths for pipe fittings are assessed on a per cent basis.

For an extensive circulation system, it would be necessary to make a full calculation in order to apportion the mains emission, using the method described in Chapter 12. In the present case, however, the towel rails are uniformly disposed and the emission represents only 0.9/7.1 = 13 per cent of the useful heat output. It would, in consequence, be reasonable to apportion the pipe losses *pro rata*.

By reference to Fig. 25.32, an energy-based chart for water flow in copper pipes, prepared on the same basis as Fig. 12.2, the listings as columns 5, 6 and 7 of Table 25.17 are produced, those in column 7 being the proportion of the available pressure which is absorbed in the flow mains to each floor.

Assuming circulation to be by gravity, the available pressure for floor 1, which is the worst case in this context, would be $5 \times 50.25 = 251$ Pa and, by way of comparison, that for floors 2, 3 and 4 would be 402, 553 and 704 Pa, respectively. From these pressures, the small proportion (Table 25.17) absorbed in the flow pipework must be deducted, e.g. the balance available at floor 1 is $251 - 14 = 237$ Pa. Hence, the unit pressure available for the return pipework from floor 1 is $237/27 = 8.8$ Pa/m run.

This unit availability may be used as for the flow piping and the pipe sizes noted in column 8 of Table 25.17 selected. As a result, again by reference to Fig. 25.32, the listings in columns 9–11 in Table 25.17 are produced, those in column 11 being the proportion of the available pressure which is absorbed in the return mains to each floor.

It will be noted that an exact balance is not possible between available pressure (237 Pa) and pressure loss (228 Pa) for either floor 1 or any other floor. It is rarely practicable to achieve a precise match using commercial sizes of pipe but the consequences, in this application, are not very significant when of the magnitude illustrated. The result would be that the system came into hydraulic balance at a temperature difference varying by a degree or so from the arbitrary 10 K chosen.

For a pumped circulation system similar to that considered here, a temperature drop of 5 K might have been chosen, thus doubling the water quantities previously calculated. The index circuit in this case would then be the most distant, that at floor 4 of riser Z, and the pressure loss through the flow pipes to that extremity would then be about 130 Pa. Assuming that the return pipes were one size smaller than those previously selected, with a minimum of 15 mm, the pressure loss through them would be 1960 Pa giving a total of 2.1 kPa.

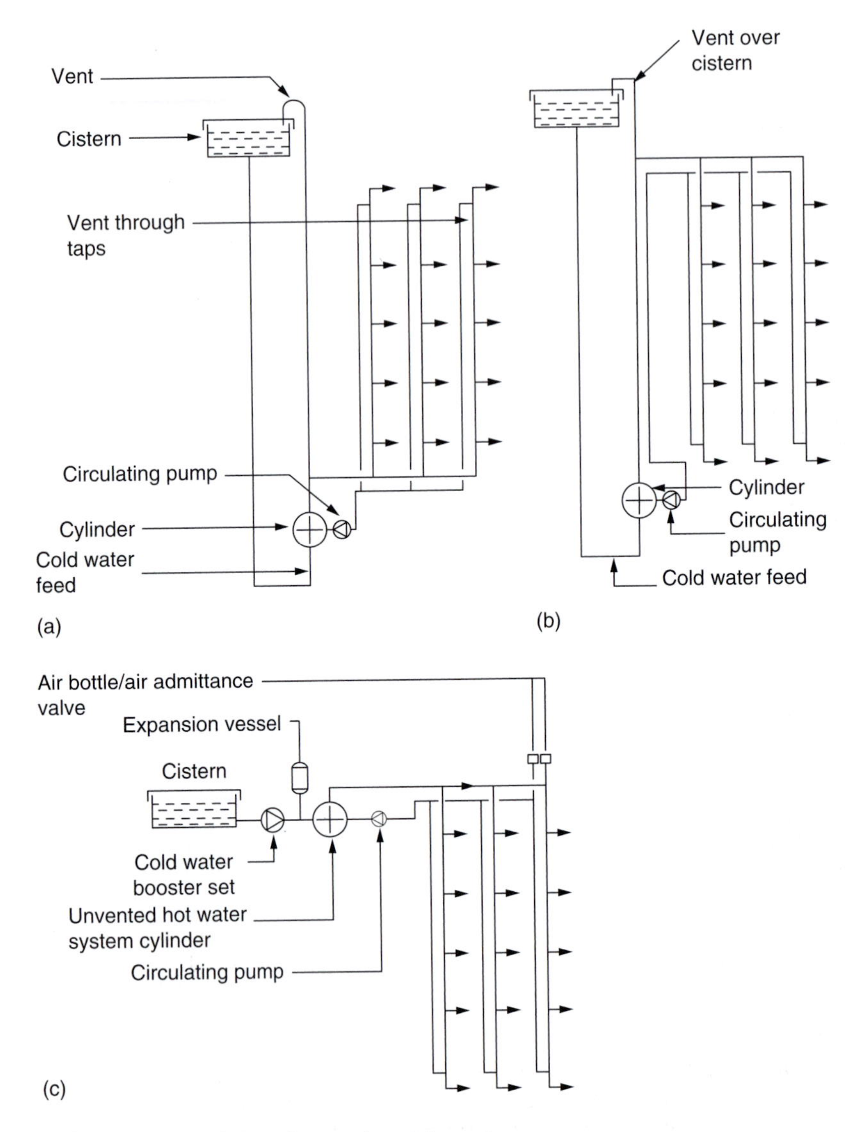

Figure 25.33 Arrangements for secondary piping systems

It might be thought that the disparity between the pressure loss totals for flow and return is too great and that larger pipes should be chosen for the latter. However, a circulating pump rated to produce the required delivery of 0.33 litre/s against a pressure head of (say) 2.5 kPa represents the lowest end of the scale of performance offered by many manufacturers of small domestic heating pumps. This situation serves only to emphasise that the temperature differential across a pumped secondary circulation is in many cases an unsatisfactory design criterion.

System arrangements

In very many instances, the arrangement of a piping system is dictated by the configuration of the building rather than technical considerations. Furthermore, it is very often the case that hybrid arrangements suit the

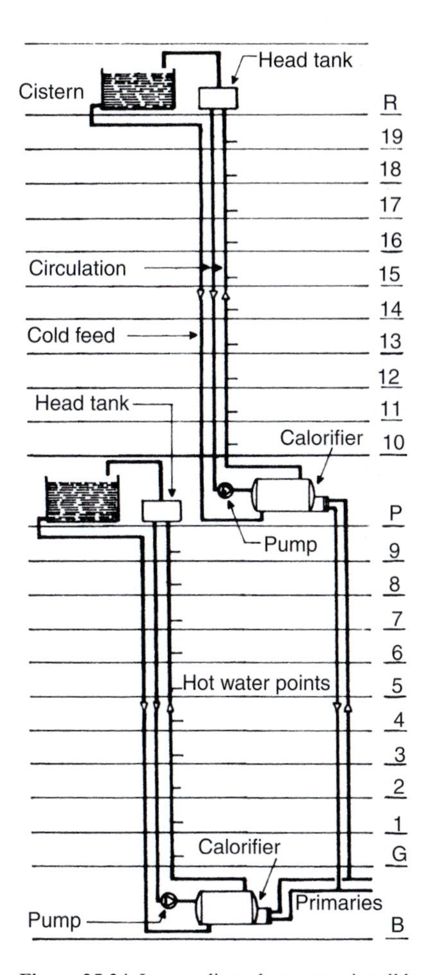

Figure 25.34 Intermediate plant rooms in tall building

needs of particular areas in a single building. There is, in principle, nothing wrong with a mixture of arrangements provided that the design follows an easily discerned logic, that no hydraulic interactions take place, that problems related to excessive pressure at draw-off points are avoided and that facilities for air venting and drainage are considered as the design is developed. It is worth emphasis that the presence of ai, and the necessity to dispose of it are perennial problems with any hot water system. More practical difficulties arise in use from air venting, or the lack of it, than from almost any other cause.

Various specific system arrangements can be identified and these are illustrated in Fig. 25.33, the two fundamental sub-categories being up-feed and down-feed, vented and un-vented.

Up-feed systems

For the sake of easy comparison, item (a) in the figure shows a simple up-feed system as instanced also by the example considered earlier in this chapter. This is probably the most common arrangement but has a disadvantage in that the longest runs of pipe very often serve the draw-off points subject to the least pressure head. Air venting at the top of each riser is sometimes possible but otherwise reliance must be placed upon release of air through the highest tap of each riser which, in that case, must be above the end of the circulation, as shown.

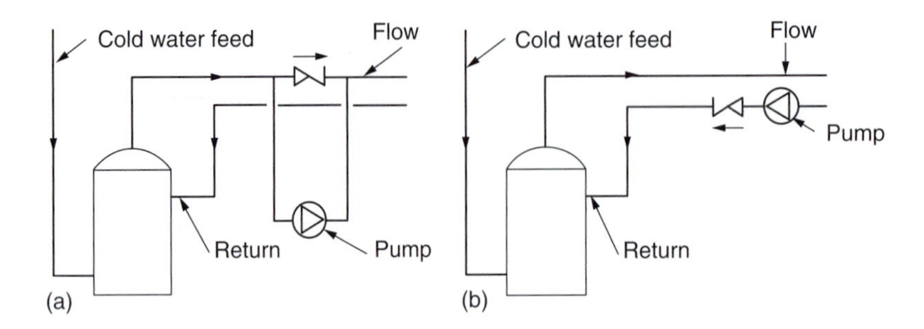

Figure 25.35 Circulating pumps for secondary systems

Down-feed systems

The least satisfactory of such arrangements is as item (b) of the diagram, and arises from building arrangements which require that both the outflow and the return main are run not far below the cold water cistern. At times of heavy demand, the outflow main will be emptied of water, air will be admitted through the vent and a spluttering spasmodic supply will result. This configuration should be avoided except in circumstances where it is possible to fit the cistern very much higher than the circulating mains.

Item (c) of the diagram shows an un-vented down-feed system. In this instance the high-level flow and return pipework cannot be vented via the draw-off points. To overcome this lack of venting, automatic air valves and/or air bottles should be provided.

Tall buildings

Bearing in mind the comments made earlier regarding problems of excess pressure at draw-off points, the case of very tall buildings is worth consideration. The preferred design solution is, as for other building services, the introduction of intermediate plant rooms at intervals over the height of the building. Plant for the hot water system might thus be disposed as shown in Fig. 25.34.

Circulating pumps

Secondary circulation pumps may be fitted, as was the case for a heating circuit pump, in either the flow or return pipework, Fig. 25.35(a) and (b), to achieve the required hydraulic performance, however, the latter is recommended.

With the circulation pump positioned on the flow, the quantity of water flow to serve the draw-off at peak simultaneous demand is generally many times greater than that required in circulation to counter heat loss. It is necessary, therefore, to introduce a bypass, incorporating a non-return valve, around the pump to avoid outflow being inhibited. During times other than of high demand, water will pass through the pump creating stagnant conditions in the bypass line, thereby allowing the possible development of legionella. Positioning the circulation pump on the return without a bypass obviates this problem.

As in the case of heating systems, centrifugal-type pumps are best suited to the circulation of the comparatively small quantities of water involved. The pump body should be of a pattern which may be opened for cleaning and removal of any scale and should preferably be made of a copper alloy which is compatible with the pipe materials used in the system. Current practice is to provide a single duty pump with no standby pump fitted in line. This is to prevent a dead leg through the section of piping and the standby pump where legionella could possibly develop. A standby pump can be stored adjacent to the working pump, for rapid replacement in the event of pump failure.

Chapter 26

Noise control

Introduction

Building services plant, be it located inside or outside buildings, generates noise which invariably must be controlled if it is not to cause disturbance to building occupiers or neighbours. This chapter sets down some background information on noise and acoustics to provide the reader with the basic understanding required in order to consider the noise generated by building services equipment and how it may be controlled. Further guidance can be found in documents such as the CIBSE Guide B and other reference books.[10,1,3]

Sound pressure versus sound power

Both sound power and sound pressure are used to describe the noise emission of building services equipment. It is therefore essential to appreciate which is being referred to and the fundamental difference between the two.

The sound power level of a source, say a fan, is described using the terminology L_w and represents the inherent acoustic energy produced by the fan. The sound power only changes as a result of the operating characteristics of the fan itself, e.g. if it is rotated more quickly or if more power is applied to the fan. In this respect it is analogous to an electric fire. A 1 kW electric fire always has an inherent power of 1 kW.

The sound pressure level, L_p, of the fan describes the sound level measured at a particular position away from the fan. The room conditions in which the fan sits therefore influence this sound pressure level, but not the sound power level which is an inherent property of the fan. The sound pressure level therefore describes how the sound power of the fan is influenced by the environment around it. When considering the electric fire analogy, temperature is analogous to the sound pressure level. The temperature will vary depending upon how far away it is measured from the heater and the size of the room that it is placed in.

The sound power level of a device cannot be directly measured, it is the variation in sound pressure that can be measured and with a full understanding of the conditions which surround the source, the sound power can be determined from the sound pressure measurements. Similarly, knowing the sound power level of a source, so the sound pressure level at a particular listening position can be established given an understanding of how the environment will modify the noise propagation.

Terminology

There is a wide variety of terminology used to describe the acoustic properties of materials and constructions, the acoustic environment within a room, and a noise level and its characteristics. Precise use of terminology is essential to avoid misunderstanding and reference pressures or powers should always be used where they will provide clarity.

dB: The decibel is a descriptor of ratios. When considering the fluctuation of sound pressure, which is the method by which our ears detect sound, it is the difference in pressure, or ratio, between the smallest variation we can usually hear and that which creates the sound of interest, which is considered. The logarithm to base 10 of this ratio is multiplied by 20 to give a value in decibels. Generally the lowest sound which is considered to be audible is 0 dB whilst at about 120 dB the level of noise usually gives rise to pain.

The sound pressure level is therefore defined by the equation:

$$L_p = 20\log\left(\frac{P}{P_0}\right)$$

where

P = the sound pressure level in Pa
P_0 = the reference pressure level 2×10^{-5} Pa

In a similar way decibels are also used to describe sound power:

$$L_w = 10\log\left(\frac{W}{W_0}\right)$$

where

W = the sound power level in Watts
W_0 = the reference power level of 10^{-12} W

Sound levels from any source can be measured in frequency bands[4] in order to provide detailed information about the spectral content of the noise, i.e. whether is it high pitched, low pitched or with no distinct tonal character. These measurements are usually undertaken in octave or 1/3 octave frequency bands. Within building services the most commonly used octave band frequency range is 31.5–8 kHz.

dBA: If the energy in each frequency band is logarithmically summed, a single dB figure is obtained often called the linear sound level. This is usually not very helpful as it simply describes the total amount of acoustic energy measured and does not take any account of the ear's ability to hear certain frequencies more readily than others.

Instead, the dBA figure is used, as this is found to relate better to the loudness of the sound heard. The dBA figure is obtained by subtracting an appropriate correction, as shown in Fig. 26.1, which represents the variation in the ear's ability to hear different frequencies, from the individual octave or 1/3 octave band values, before logarithmically summing them. As a result the single dBA value provides a good representation of how loud a sound is. An 'A' weighted sound pressure level would have the nomenclature L_{pA} whilst an 'A' weighted sound power level would have the nomenclature L_{wA}.

L_{eq}: As almost all sounds vary or fluctuate with time it is helpful instead of having an instantaneous value to describe the noise event, to have an average of the total acoustic energy experienced over its duration. The $L_{eq,(07:00-19:00)}$, for example, describes the equivalent continuous noise level over the 12-hour period between 7 a.m. and 7 p.m. During this time period the L_p at any particular time is likely to have been either greater or lower than the $L_{eq,(07:00-19:00)}$. The L_{eq} may be used to describe both internal and external noise levels.

$L_{max,FAST}$: The $L_{max,FAST}$ is the loudest instantaneous noise level. This is usually the loudest 125 ms measured during any given period of time.

L_{An}: Another method of describing, with a single value, a noise level which varies over a given time period is, instead of considering the average amount of acoustic energy, to consider the length of time for which a particular noise level is exceeded. If a level of x dB is exceed for say 6 minutes within 1 hour, that level can be described as being exceeded for 10 per cent of the measurement period. This is denoted as the $L_{10,1\ hour} = x$ dB.

The L_{A10} index is often used to describe road traffic noise whilst the L_{A90}, the noise level exceeded for 90 per cent of the time, is the usual descriptor of the underlying background noise. L_{A1} in addition to L_{Amax} are common descriptors of construction noise.

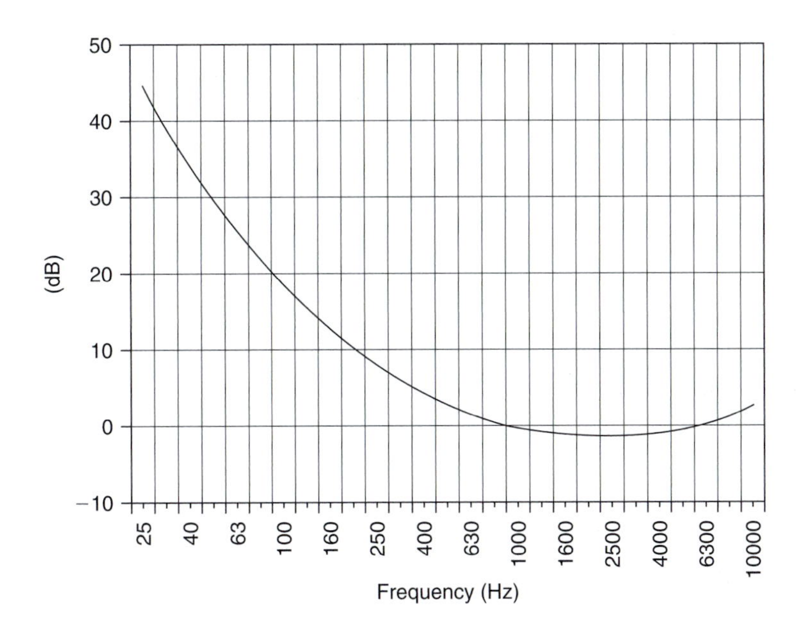

Figure 26.1 A weighting attenuation filter (subtracted from linear values to give an A weighted value)

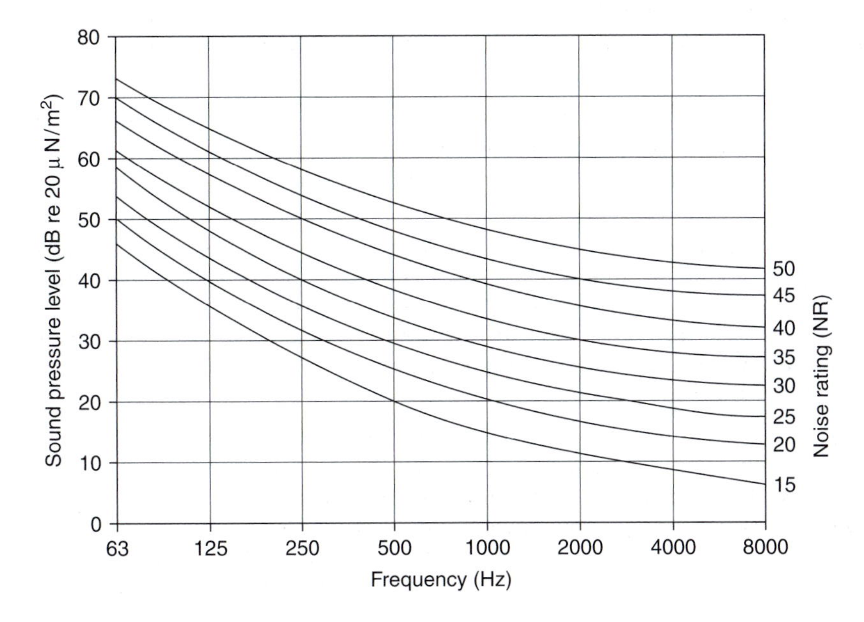

Figure 26.2 NR criterion curves

NR/NC: The 'A' weighted sound pressure level cannot be used to define a spectrum or to compare sounds of different frequencies. For most acoustic design purposes it is more useful to make use of a curve that defines the noise level in each of a number of frequency bands. The curve is described using a single number with one of many prefixes including noise rating (NR) and noise criteria (NC). The NR, see Fig. 26.2, and NC curves are commonly used to describe noise from building services systems in the UK but a number of other descriptors are more commonly used overseas including room noise criterion (RC) and preferred noise criterion (PNC).

Table 26.1 Constants of the NR curve calculation

	Octave band centre frequency (Hz)								
	31.5	*63*	*125*	*250*	*500*	*1000*	*2000*	*4000*	*8000*
a	55.4	35.4	22.0	12.0	4.8	0.00	−3.5	−6.1	−8.0
b	0.681	0.790	0.870	0.930	0.974	1.000	1.015	1.025	1.030

In practice the difference between the NR and NC is quite small. NR curves are slightly more relaxed at low frequencies and slightly more onerous at higher frequencies when compared with NC. Typical NR levels are given below:

NR 25 very quiet
NR 30 normal living space
NR 35 spaces with some activity
NR 40 busy spaces
NR 45 light industry
NR 50 heavy industry

The CIBSE *Guide Section A*[2] gives recommended NR values for the design of ventilation systems for various specific applications, the following are selected examples:

Auditoria	*NR*	*20–30*
Conference/board rooms		25–30
Department stores/shops		35–40
Dwellings		
Bedrooms		25
Living rooms		30
Hotel bedrooms		20–30
Lecture halls		25–35
Offices		
Executive		30
General/open plan		35
TV studios		25
Restaurant/dining rooms		35–40

The NR curves are generated from the following formula and constants:

$$L = a + bn$$

where

L = the octave band sound level for NR level n and the constants a and b are frequency band dependent and given by Table 26.1.

Measurement of sound

The speed of development of electronic equipment means that the measurement of sound has advanced rapidly over the years. That said, the method of recording the variation in pressure level, the microphone capsule still remains broadly similar to that used 20 years ago, it is the opportunity to sample, store and manipulate data far more effectively and in much greater quantities that has brought about the greatest advances in sound level meter design. Today laptop computers are regularly used to acquire, store and manipulate data although the traditional microphone capsule is still used as the transducer to capture the sound signal. Dedicated sound level meters however still remain the mainstay of most site-based building services measurement work and can readily be expected to have 1/3 octave filters, the ability to measure L_{eq}, L_{max} and all of the other statistical parameters and possibly measure the reverberation time of the room. Figure 26.3 shows a modern sound level meter.

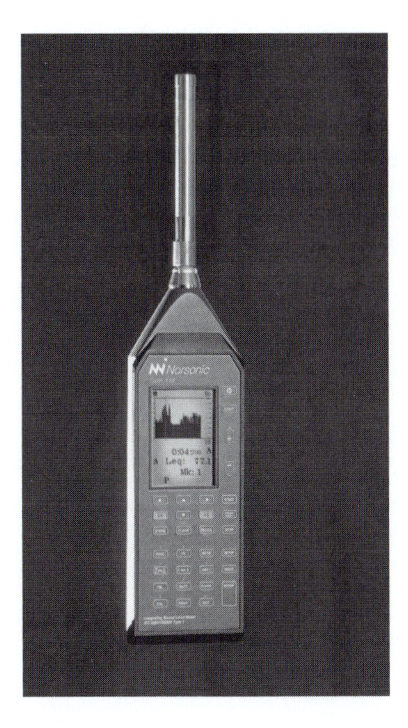

Figure 26.3 Modern sound level meter (picture courtesy of Norsonic Instruments)

Sound power is not measured directly, it is determined from the results of sound pressure level measurements under known conditions and therefore is usually the subject of precise methodologies undertaken in controlled or laboratory surroundings.

Another form of measurement which although not commonly used has been available for many years is sound intensity measurement. This method measures sound pressure level using two precisely located microphones so that both phase and amplitude of the sound level can be determined, this then enables the vector properties of the sound level to be established and hence not only can the level of sound be determined, but also the direction from which it radiates. This has great benefits when undertaking diagnostic tests to establish which item or part of an item of plant may be generating most noise.

Laboratory testing – methodology/standards

The majority of items of building services plant generate noise and therefore manufacturers are expected to provide information about the noise generating characteristics of their equipment. This includes all ventilation equipment, heat rejection equipment and some hydraulic systems.

Laboratory testing is usually undertaken within reverberant and sometimes anechoic laboratory chambers in accordance with recognised measurement standards, which prescribe the precise measurement conditions, number of measurements, etc.

Where particularly large items of plant are to be tested such as chillers or cooling towers and there is a need to simulate a cooling load or similar so that compressors and fans operate realistically, such tests are often undertaken outside and less accurate engineering test methods are used, although these are still described in measurement standards.

The measurement of building services noise on site, either during the construction period, during commissioning or as a problem-solving exercise is more *ad hoc* in nature. The only relevant measurement methodology for measuring noise from building services within buildings is set down in the Association of Noise Consultants Guidelines,[5] however these methodologies may not suit all measurement circumstances.

Room acoustics

Sound in a room

The level of noise and its character that we hear within a space are a function of both the noise source and the influence that the room has on the sound generated. The most commonly used descriptor of the acoustics of a space and hence how this will modify the sound from a particular source is the reverberation time.

The reverberation time (T_{60}) is the length of time in seconds it would take for a sound to decay by 60 dB and it is therefore a measure of the echo within the room. This reverberation time is often referred to as the T_{60}, however it is often impractical to measure a 60 dB noise level decay and so the reverberation time is often based upon the T_{20} or T_{30} which relate to the decay over 20 or 30 dB normalised to a decay of 60 dB. The reverberation time can be calculated as follows:

$$T = \frac{0.16V}{A}$$

where

T = reverberation time in seconds
V = the room volume in m^3
A = the total absorption in the room in m^2

and where

$$A = S\alpha$$

where

S = the average surface area of the room in m^2
α = the average absorption coefficient of the room finishes

Preferred reverberation times in different types of rooms are set down in various texts.[6,7] For a living room the reverberation time might be 0.5 seconds whilst in a large church it may be significantly longer at 3 seconds.

All materials possess the property of acoustic absorption whereby some of the acoustic energy which impinges upon the surface is not reflected but absorbed. This property is described by the coefficient α. The greater α, the greater the acoustic absorption.

Consider a noise source such as a fan within a simple room and a listener position some metres away but in the same room, the acoustic energy from the source will reach the listener by a number of different pathways. The sound which radiates directly from the source to the listener is the direct component, whilst the sound which is reflected off the walls, ceiling and floor before it reaches the listener is called the reverberant component. To assess the total amount of energy reaching the listener both contributions must be added together.

$$L_{p\ total} = 10\log\left(10^{\frac{L_{p\ direct}}{10}} + 10^{\frac{L_{p\ reverberant}}{10}}\right)$$

where (for spherical source propagation):

$$L_{p\ direct} = L_w - 10\log d - 11$$

$$L_{p\ reverberant} = L_w - 10\log V + 10\log T + 14$$

and

d = distance between source and listener position, m

The direct component is a function of the distance between the source and listener and any specific directivity characteristics of the source. The reverberant component is a function of the influence the room will have on the sound reaching the listener, it is therefore a function of the volume of the room and its reverberation time.

At some position away from the source the direct component of the noise level and the reverberant component will be numerically the same. This distance is called the room radius r and is given by the equation:

$$r = \sqrt{\left(\frac{A}{16\pi}\right)}$$

where

$$A = 0.163\frac{V}{T}$$

The room radius can be helpful when assessing whether the noise reaching a listening position from a particular source is principally composed of the direct sound component or the reverberant component. If for example it was principally the direct component, control of the reverberant component would do little to minimise the overall noise level at the listening position.

Plant noise

Introduction

Noise is generated by almost all building services systems including ventilation, air-conditioning, hydraulic systems, heating, lifts, standby generators, transformers, switchgear, even lighting on occasion. Noise is not only generated by the primary items of plant such as fans, but also by the services distribution be it air movement in ductwork or water movement in pipework.

The noise generating aspects of various building services items are discussed further below.

Chillers/heat rejection plant

Heat rejection plant generally contains some form of rotating or reciprocating machinery and therefore the noise generated usually contains a tonal component related to the frequency of rotation of the compressor. In addition where the condensing coils and compressor are located together in the same piece of equipment, noise is generated by the fans which force air across the condensing coils. This noise is usually broadband. It is possible to attenuate the noise from heat rejection plant and this is usually done by means of lagging, enclosing compressors and reducing air movement noise by using passive silencers in the air path.

Boilers

The principal source of noise from large commercial boilers is the combustion noise from the burners. This is almost always controlled by the use of shrouds which are usually readily removable to enable access for maintenance. An associated noise from boilers is that of any associated flue dilution system which serves to force air into the boiler flue and so dilute the exhaust gasses allowing more localised discharge. Flue dilution systems usually utilise a bifurcated in line fan within the plant room, close to the boilers. These are universally noisy fans as the motor is outside of the air stream and they will often generate noise levels greater than those of the boiler itself.

Fans/air handling units

The noise characteristics of fans are dependent upon the fan and motor configuration. Backward curved and forward curved centrifugal fans exhibit different noise levels as do axial, bifurcated and variable pitch fans. A simple formula for providing an estimate of the fan sound power level based upon the volumetric duty and total fan pressure is:

$$L_w = 10\log Q + 20\log P + 40$$

Table 26.2 Fan sound power spectrum corrections to be applied to single figure sound power level

	Spectrum corrections							
Fan type	*63*	*125*	*250*	*500*	*1 K*	*2 K*	*4 K*	*Hz*
Forward curved centrifugal	−2	−7	−12	−17	−22	−27	−32	dB
Backward curved centrifugal	−7	−8	−7	−12	−17	−22	−27	dB
Axial flow	−5	−5	−6	−7	−8	−10	−13	dB
Mixed flow	−12	−11	−10	−10	−13	−17	−22	dB

where

Q = volume flow (m^3/s)
P = static pressure (Pa)
L_W = overall sound power level (dB re 10^{-12} W)

Fan sound power spectrum correction factors to be applied to the single figure sound power level are given in Table 26.2.

Air handling units usually incorporate one or more centrifugal fans, the noise emission from the inlet and discharge to the air handling unit is influenced by the effects of the coils, filters and other elements within the air stream. Proprietary silencers are often incorporated within the air handling unit.

Terminal units/fan coil units

Noise from fan coil units is principally related to the noise from the fan itself although there may be air movement noise and noise radiated by the unit casing unless the fan is effectively isolated. When undertaking assessments of noise intrusion into rooms, the casing radiated noise and discharge noise are usually considered separately. Noise from fan coil units is usually dominant in the 250 Hz octave band.

Other terminal devices without fans such as induction units, grilles or diffusers may generate noise as a result of their introduction into the air stream. In this respect they must be carefully sized so that air velocities are within a specific range.

Reverse cycle heat pump units

These differ from fan coil units in that they contain a compressor in order that they can provide either heating or cooling as required. In most other respects they do not differ significantly from fan coil units, the fan being the principal noise generator, however the additional compressor can create increased noise often with intermittent operation making these units unsuitable, for particularly noise sensitive applications.

Computer room units

Computer room units or close control air-conditioning units are air handling units reconfigured for internal use within an IT room. The units may be arranged so that they supply air into the floor and return air via a grille on the front, on the top or from the ceiling void. Alternatively they may supply air from the top whilst return air enters through a grille on the bottom of the unit. The units usually move significant air quantities and hence result in high noise levels (room noise levels of NR60–NR75 may be typical). Attenuation can often be provided within the air path using silencers within the floor or ceiling voids, however casing radiated noise must also be considered and this often limits the noise reduction achievable.

Sources of information on noise levels

The best source of noise level data for building services systems is invariably the equipment manufacturers. The market generally demands accurate information about all aspects of the performance of building services plant

Table 26.3 Approximate attenuation of unlined sheet metal ducts at octave frequencies

		Attenuation (dB/m) *for stated octave band* (Hz)			
Duct section	*Mean dimension or diameter* (mm)	*63*	*125*	*250*	*500 and above*
Rectangular	<300	1.0	0.7	0.3	0.3
	300–450	1.0	0.7	0.3	0.2
	450–900	0.6	0.4	0.3	0.1
	>900	0.5	0.3	0.2	0.1
Circular	<900	0.1	0.1	0.1	0.1
	>900	0.03	0.03	0.03	0.06

Table 26.4 Approximate attenuation of unlined and lined square elbows without turning valves

	Attenuation (dB)	
Frequency × *width* (kHz · mm)	*Unlined*	*Lined*
<50	0	0
50–100	1	1
100–200	5	6
200–400	8	11
400–800	4	10
>800	3	10

Table 26.5 Approximate attenuation of unlined and lined square elbows with turning valves

	Attenuation (dB)	
Frequency × *width* (kHz · mm)	*Unlined*	*Lined*
<50	0	0
50–100	1	1
100–200	4	4
200–400	6	7
>400	4	7

and the acoustic performance is no exception although it should be recognised that owing to its specialised nature the quality of acoustic information on products is not always what it could be. Generic information, for example the noise level of boilers based upon their kW rating or the noise level of fans based upon their operating pressure and volume, can be of assistance at the early assessment stages of a project, but it is likely to be inaccurate and is no substitute for specific data relevant to the precise item of equipment to be installed.

Often manufacturers will only test the acoustic performance of one model of a range, for instance their smallest and largest fan coil units, and interpolate results for the mid-size units from this data. On some occasions this level of accuracy may be adequate however on others, particularly when the noise output is critical, it will not be and additional acoustic testing prior to installation will be required.

Manufacturers use various descriptors for the noise emission from their products, in some cases the inherent sound power of the device, sometimes a value for the sound pressure level at a particular distance from the equipment and sometimes an overall room noise level for a 'typical' installation. In all cases data must be considered with caution to make sure it describes the noise level at the operating conditions of interest, e.g. when a chiller is operating at full, not part, load.

Noise in ventilation systems

An estimate of the sound power of the fan in a ventilation system can be obtained from Table 26.2 and its associated equations.

Duct fittings including bends, tees, plena, grilles, dampers and diffusers all influence the level of noise reaching the room served by a ventilation system. The variations in attenuation provided by some of these fittings across the frequency spectrum are given in Tables 26.3–26.5. It is important to appreciate that the attenuation suggested is for the fitting in isolation. Four bends in close proximity are unlikely to give 4 times as much attenuation as one bend because in practice the fitting will also influence the noise level upstream as well as downstream hence the overall effect is seldom a true summation of the attenuation of each fitting.

Table 26.6 Maximum air velocities in ductwork for particular room criteria

Room criterion NR/NC	*20*	*25*	*30*	*35*	*40*	*45*	*50*
Main duct (m/s)	4.5	5	6	7.5	9	10	10
Branch run (m/s)	2.5	3	3.5	4.5	6	7	7
Final run outs, grilles and diffusers (m/s)	1.5	1.5	2	2.5	3	3.5	4

Ductwork provides attenuation of sound simply because the duct wall is excited and radiates the acoustic energy hence the same amount of acoustic energy is attenuated within the duct as breaksout of the sides of the duct. Circular ductwork is stiffer than rectangular ductwork and hence the attenuation to ductborne noise is less but breakout noise from the sides of the ductwork is also less, as can be seen in Table 26.3.

Openings at the ends of ducts, with or without grilles, provide attenuation by reflecting the sound back down the duct as a result in the sudden change of duct to room volume. This effect is pronounced at low frequencies whilst there is little worthwhile attenuation at higher frequencies.

Branches and tees divide the acoustic energy passing down the duct in proportion to the volume of air passing down each branch.

It is essential to remember that although each of these fittings has the ability to attenuate some of the fan noise within the duct, each fitting also has the ability to regenerate noise as a result of air movement across the fitting.

The noise generated by air movement in ductwork or across fittings increases at a rate of 18 dB for every doubling of the air velocity. Consequently, it is important that the air velocity in ductwork is not only correct from a pressure drop and energy point of view, but it is also essential if noise problems are to be avoided as it influences the overall level of noise which will be achieved at the end of the system.

Typically in the design of low-pressure air systems, the maximum velocities shown in Table 26.6 are adopted for different desired room noise levels, in order that the regenerated noise does not add to the fan noise and increase the overall room noise level.

Silencers

Proprietary silencers are used in ductwork systems to control both fan noise and regenerated noise upstream of the silencer. Such silencers can be rectangular or circular in section although the rectangular silencers usually provide significantly greater attenuation for a given cross-section and length. Rectangular silencers usually comprise a section of ductwork, linings to two sides and internal splitters. The ductwork is fitted with 50–100 mm lining of mineral wool to two parallel internal faces. This lining is usually retained with perforated or expanded galvanised sheet steel to retain the mineral wool. The lining is returned back to the ductwork at the front and back of the silencer. Additional splitters, 200 mm mineral wool retained to both sides by the acoustically transparent perforated lining, may be restrained with channels in the middle of the ductwork between the two side linings to increase the attenuation. Typical examples of silencers are shown in Fig. 26.4.

Introducing these silencers into a ventilation system will result in a significant pressure drop across the silencer as a result of the reduction in free area created by the linings and splitters. Consequently, silencers must have an increased cross-section relative to the adjacent ductwork if this pressure drop (and the potential for increased velocity noise) is to be avoided. In low-pressure ventilation systems a silencer may be typically twice the cross-section of the adjacent ductwork and will have a pressure drop of about 50 Pa.

Where proprietary silencers may be inappropriate, too large, or uneconomic, flexible attenuating ductwork may be appropriate. This is very similar to insulated flexible ductwork only the inner flexible lining is acoustically transparent. Flexible acoustic ductwork is often found in the ductwork between fan coil units and diffusers where it would be impractical to install conventional rectangular silencers.

The performance of a silencer is described by its static or dynamic insertion loss which is the reduction in noise level as a result of its introduction into a ductwork system. The static insertion loss is measured using a loudspeaker whilst the dynamic insertion loss uses a fan source and therefore includes air movement noise which reduces the apparent performance of the silencer under specific conditions.

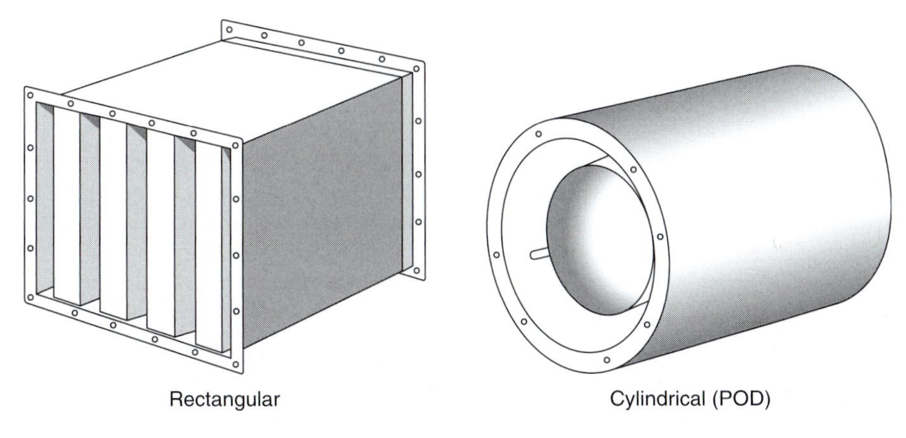

Figure 26.4 Examples of silencers

Figure 26.5 Acoustic louvres (courtesy of Allaway Acoustics)

Acoustic louvres

Acoustic louvres, as illustrated by Fig. 26.5, are also sometimes used on the atmospheric side of a system instead of a proprietary silencer to reduce fan noise. The acoustic louvre replaces a conventional architectural louvre, however they do present relatively high-pressure drops and do not provide as much attenuation as an in-duct silencer.

Dampers

The introduction of dampers or any other devices within ventilation systems generates noise as a result of turbulence caused by the fitting in the air stream. Generally the more the damper is closed the greater the noise generated and if the system serves a noise sensitive room then the position of the dampers can be critical. It is therefore helpful where possible to design ductwork distribution systems such that they are balanced and require minimal adjustment of dampers in order to balance the various branches off against each other. Lack of thought in the layout of ductwork systems can produce significant imbalances between various branches resulting in excessive noise being generated where dampers must be throttled to achieve the necessary air volumes.

The noise generated by air movement in ductwork or across fittings increases at a rate of 18 dB for every doubling of the air velocity. Consequently, it is important that the air velocity in ductwork is not only correct from a pressure drop and energy point of view, it is also essential if noise problems are to be avoided.

Typically in the design of low-pressure air systems, the following maximum velocities are adopted for different desired room noise levels, in order that the regenerated noise does not add to the fan noise and increase the overall room noise level.

Air in pipework

When air is entrained in water distribution systems it is circulated with the fluid medium and results in a rattling noise.

To avoid the entrainment of air, pipework distribution systems should be designed following good practice to avoid high points or other pipework configurations which trap air preventing it from being effectively vented.

Removing air, particularly from closed systems such as chilled water distribution, can be very difficult and experience suggests that complete removal of air from chilled water systems takes an extended period and often is not undertaken effectively. Vacuum step de-aerators can be effective at removing air, otherwise regular manual bleeding of the system or effective automatic air vents are required.

Good pipework design and effective venting of a system will always be far more effective at controlling noise than any noise control measures such as lagging applied to the pipework.

Water hammer

Water hammer occurs when the fluid passing down a pipe is abruptly stopped owing to closure of a valve or other device. The confined liquid is forced to stop and a pressure wave passes down the fluid medium. This energy is dissipated by exciting the pipe walls which, if not adequately restrained, radiate noise.

Efficient pipework design calls for greater water volumes and increased pressure within pipework. Automatic valves operated by solenoids are often quick acting. These all militate against reducing pipework noise and avoiding water hammer. Measures to control water hammer are all essentially the same, they introduce a compressible elastic material in the pipework to compensate for the incompressibility of the liquid itself and may include surge tanks and relief valves.

Plant room noise

Where plant is contained within a plant room there is usually a need to establish the noise level likely to be present in the plant room as a result of all the normally operating plant. This is usually done by assessing the reverberant noise level within the room considering the contribution from each separate item of plant.

Airborne noise transmission to adjacent rooms can then be assessed taking account of the sound reduction properties of the plant room walls, floor and soffit, and taking account of any doors or penetrations.

Where services distribution in the form of ducts, pipes, cable trays or trunking passes from the plant room to adjacent areas it is essential that these pathways do not allow unwanted noise transmission from the plant room. Consequently it is essential that the penetrations in the walls or slabs are made good and the penetrations are suitably sleeved and sealed to prevent both airborne noise transmission around the pipe or duct and prevent structure borne noise transmission as a result of the duct coming into direct contact with the wall or slab.

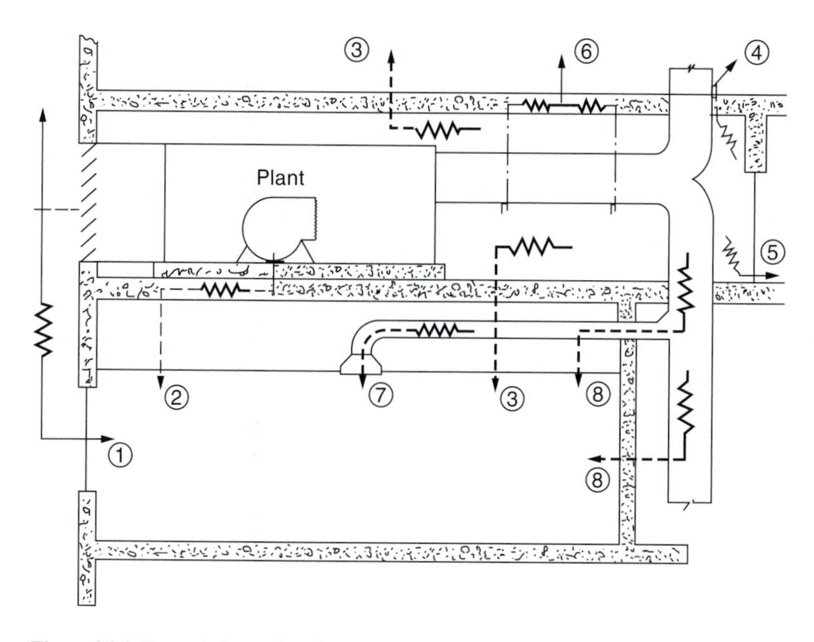

Figure 26.6 Transmission paths of noise and vibration: 1, airborne noise via intake louvres; 2, plant vibration via structure; 3, airborne noise via openings; 4, airborne noise via poor seals; 5, airborne noise via openings; 6, plant vibration via solid duct supports; 7, ductborne noise to terminals; 8, ductborne noise through duct walls

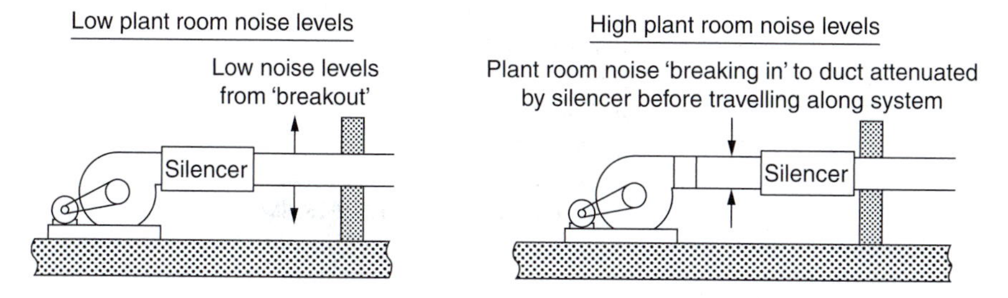

Figure 26.7 Diagram of the positioning of silencers to avoid plant room noise breakout. From Woods Practical Guide to Noise Control

To control structure borne noise transmission all services within the plant room must be effectively isolated to prevent acoustic energy being transmitted into the building structure (see vibration isolation) consideration often needs to be given not only to the plant itself, but also to the distribution systems and control gear.

All paths of airborne sound propagation must be considered as seen in Fig. 26.6.

Sound transmission via ductwork systems is controlled by the positioning of proprietary silencers within the ductwork systems. The precise position of the silencer should be selected depending upon whether plant room noise break-in back into the ductwork is likely to occur. Where the plant room is very noisy and break-in may occur the silencer should be positioned close to the plant room wall. Where this may be difficult owing to the position of fire dampers or access doors the intervening ductwork between the silencer and the plant room wall should be lagged with a suitable proprietary acoustic lagging product. Figure 26.7 illustrates the principles.

Noise emission to atmosphere

As well as considering noise from building services systems affecting occupants of the building it is also important to consider noise emission from the building services systems to neighbouring buildings. There is also the potential for noise from externally mounted plant to break back into the building itself via open windows and the like.

The setting of limits to atmospheric noise emission for a building is usually addressed during the planning application process with the local authority. The local planning authority will usually refer to guidance within a number of documents which provide guidance on levels of acceptable noise from buildings. These guidance documents include BS 4142,[8] PPG 24[9] and BS 8233.[10] The planning authority would also seek guidance from the local Environmental Health Department who usually have responsibility for policing noise disturbance created by building services noise from buildings.

Conditions to the planning approval process are usually imposed by the local authority which limit noise emission to a specific level or such that it does not increase the prevailing background noise level outside any neighbouring buildings. These conditions usually relate to all noise generating items of plant including externally located heat rejection plant as well as ventilation intakes, discharges and noise from plant rooms.

The need to control the noise from building services systems to the atmosphere clearly depends upon the noise emission targets selected. In some cases onerous limits will require significant noise control measures to be introduced, in other instances simply making careful selections when choosing the make and model of the plant can be sufficient to enable the noise emission targets to be achieved.

Typically if the listening position of interest is a comparatively long way away from the noise generating item of plant (say a neighbouring residential window is 30 m away from a roof-mounted packaged air-cooled chiller) then the following equation can be used to predict the noise level at the listening position.

Sound pressure level

$$L_{\mathrm{p}} \,@\, \text{listener} = L_p \,@\, 1\,\text{m from equipment} - 20 \log r + 10 \log N - S + R$$

where

R = distance to boundary (m)
N = number of sources
S = screening effect (dB)
R = reflection effect in (dB)

Effect of screening

Screening, as illustrated in Fig. 26.8, can be an effective way of controlling noise from externally located plant to neighbouring buildings, although care should always be taken to ensure the screen will be effective at protecting all of the potential listening positions. Visual screens positioned around external plant are often used, however if close to the plant they may incorporate openings or louvres to allow free air movement around the equipment. Such openings and louvres often negate any acoustic screening effects.

The commissioning of building services systems

Acoustic commissioning usually takes place following installation and is usually undertaken to demonstrate that the installations comply with the original building specification. Usually noise levels will be measured in the unoccupied building with all systems operating at their normal duty. A methodology for undertaking these noise level measurements is set down in the Association of Noise Consultants Guidelines – Measurement of Noise in Buildings.[5] The extent of such measurements is usually either set down in the building contract or is at the discretion of the commissioning engineer.

Usually it is not until the commissioning of air and water systems that it becomes clear that there is a potential noise problem. The usual reasons for the high noise levels commonly experienced immediately after installation of the building services systems can be many fold. Typical reasons include:

- Air entrained within chilled water and hot water systems
- Incorrect operating speed of pumps
- Incorrect or inappropriate balancing of water systems
- Dirty/clogged strainers
- Dirty filters

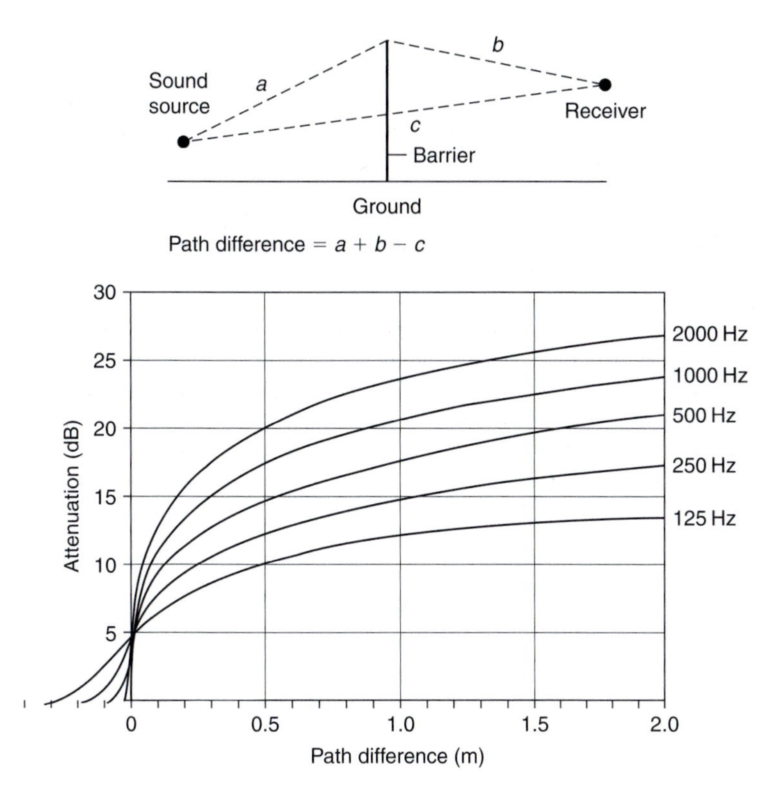

Figure 26.8 Attenuation by a noise barrier as a function of path difference. Taken from DfES BB93

- Incorrectly adjusted anti-vibration mountings
- Fans operating above the scheduled duty
- Air systems unbalanced/incorrectly balanced
- Supply and extract air volumes mismatched
- Services distribution rigidly fixed to partitions and slabs when they should be isolated
- Control systems working incorrectly.

In the majority of cases excessive noise from recently installed building services systems is as a result of incorrectly commissioned systems. Once the above potential causes have been eliminated it is then necessary to consider whether the cause may be incorrectly sized equipment such as too small pipework or incorrectly positioned components such as dampers too close to terminals. In all cases a methodical approach to establish the source of the noise should be adopted to consider and eliminate each component of the system from the fan through to the terminal or from the boiler and pump through to the emitter. This in conjunction with reviewing whether plant or equipment has been correctly installed should always be done prior to reviewing whether plant and equipment are incorrectly selected.

The precise specification of the noise level will determine the detail with which measurements are undertaken. If the noise level specification is described using NR or NC curves then frequency-based measurements will be required, which may not be the case if the specification is in terms of a simple dBA.

Vibration control

In the majority of building applications the criterion for vibration is usually whether it is perceptible to occupants of the building. In some cases such as specialist laboratories or hospitals more onerous standards may be required where microscopes or other vibration sensitive apparatus are used.

Building services systems must be designed so as to prevent the build-up and transmission of unwanted vibration and this is usually done by mounting vibration generating equipment such as motors, pumps and other plant on anti-vibration mountings.

Types of anti-vibration mount

Three types of vibration isolator are usually used either independently or combined in a specific mounting. These are steel helical springs, neoprene in compression or neoprene engineered so as to act in shear. Most anti-vibration mounts not only provide isolation of the vibration, but also damping in order that the movement is minimised. Usually damping is provided at the expense of optimum isolation. The mountings need to be selected so that they provide the desired level of isolation at the frequencies of interest and hence can only be selected with full knowledge of the vibrating frequency.

Inertia bases are often used in conjunction with plant to optimise the vibration isolation. This is achieved by increasing the mass of the vibrating equipment, equalising the centre of gravity of the equipment over the mountings and moving the mounting positions away from the centre of gravity. These measures all enable a more stable and effective isolation system to be utilised. Consequently inertia bases tend to be used on large pumps and fans. Figure 26.9 shows the various forms of anti-vibration mountings.

In many cases effective isolation of the fan or pump alone is not sufficient to prevent energy being transmitted into the building structure and pipework, and ductwork often also needs to be supported resiliently. In these instances anti-vibration mountings are introduced into the pipe or ductwork supports.

Natural ventilation

Natural ventilation solutions for buildings which may include open windows or other air paths to outside, and hybrid systems which draw air through openings in the façade and eject the vitiated air mechanically also present acoustic problems although these are usually different to those associated with full ventilation or air-conditioning solutions.

The use of open windows or other forms of ventilation opening within the building façade clearly increase the potential for external noise intrusion into the building from road, rail, aircraft and industrial noise sources. This is an important consideration, particularly in building types such as schools and residential accommodation where there is a particular emphasis towards natural ventilation solutions and where the indoor noise level can be critical. Methods for mitigating external noise intrusion include using attenuated air paths, and there are a large number of acoustic trickle ventilators on the market which provide this function for residential developments. Alternatively, labyrinth-type air pathways between the panes of double-glazed windows can also prove effective. The inside to outside sound reduction provided by an open window is usually taken to be no more than about 10 dB.

Thermal mass

There is an increasing move towards omitting suspended ceilings within concrete framed buildings in order to expose the soffit to the space and thus gain benefit from the moderating effects on internal temperatures brought about by the thermal mass of the exposed slabs.

This clearly has an impact on the acoustic environment of the space as omitting the suspended ceiling removes a substantial area of usually highly effective acoustic absorption. This is exacerbated by the fact that the soffit (and floor) usually presents the greatest surface area in a space and suspended ceilings, because of the void behind, tend to be extremely efficient absorbers. Depending upon the function of the space this absence of an acoustic ceiling can be readily accommodated where alternative absorbing finishes can be introduced or it can be prohibitive for instance because there are no available wall surfaces for acoustic treatment.

In the open plan office-type environment guidance suggests that a slight increase in reverberation times owing to a reduction in ceiling absorption is acceptable up to a point. This is usually a factor of the communication distances over which people will speak. A normal office without a suspended ceiling and no substantial alternative

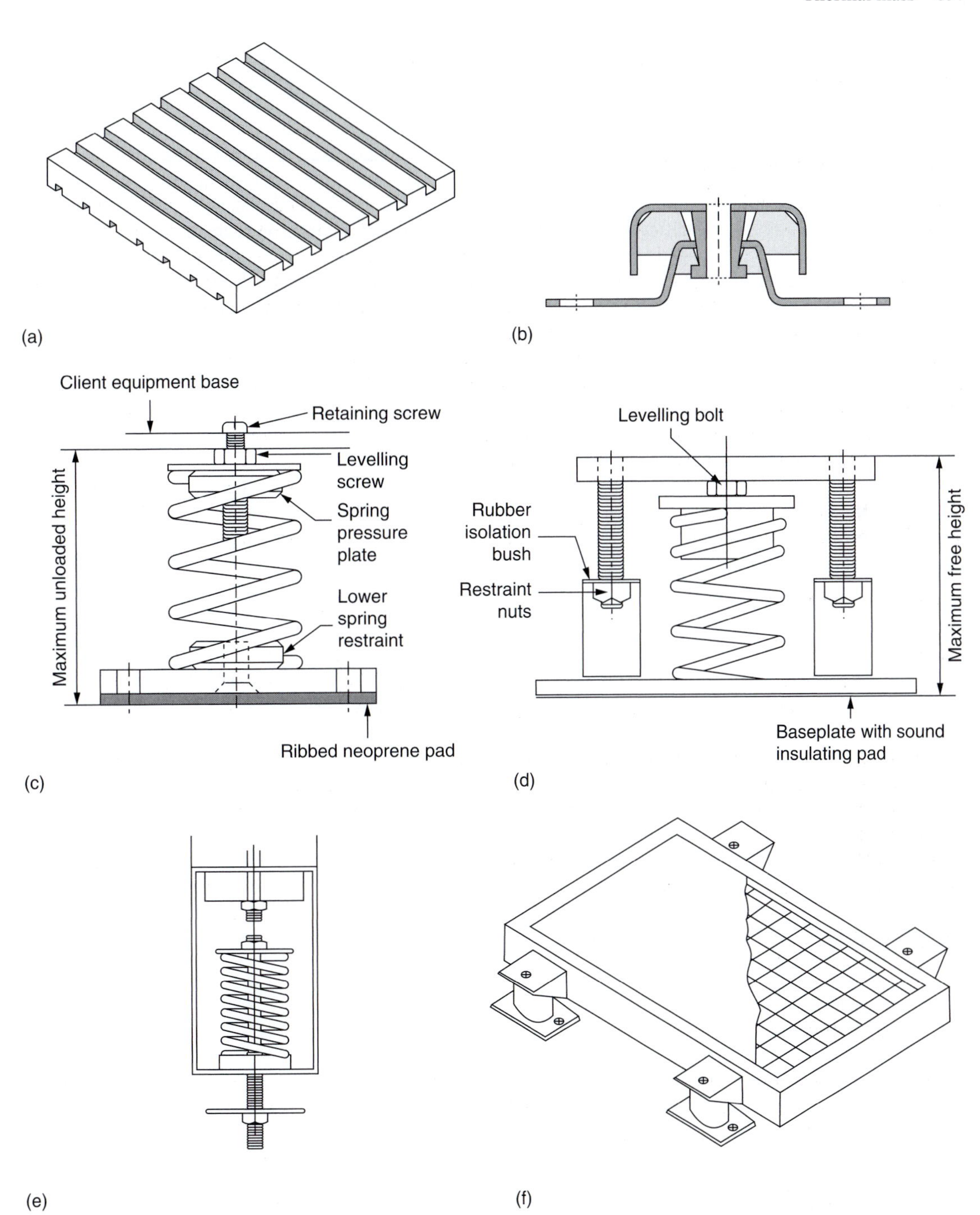

Figure 26.9 Typical vibration isolators: (a) ribbed mat, (b) elastomer in shear, (c) open spring isolator, (d) restrained spring isolator, (e) pipe/duct hanger, (f) formwork for inertia base (courtesy of Eurovib (Acoustic Products Ltd). Taken from CIBSE Guide B

acoustically absorbing finishes will result in normal conversation being possible only over a metre or so before it becomes unintelligible. Furthermore the absence of absorption results in deterioration of sound quality for telephone users and a build-up of background noise levels resulting in an even less pleasant acoustic environment.

In smaller spaces such as cellular offices or study rooms the absence of a suspended ceiling is less noticeable, indeed traditionally many such spaces had plaster ceilings as opposed to acoustic ceilings. However telephone usage and particularly teleconferencing can be impaired if there is not a certain quantity of absorption in the room.

In lecture theatres or large meeting rooms with more than about 20 people the room acoustics are critical for providing acceptable listening and speaking conditions. This is even more important in spaces used for teaching as reinforced by the extensive guidance on the room acoustics of schools.[7] Although acoustic absorption on the soffit may not always be the preferred location, from an acoustic point of view, it is inevitably the only available surface when walls are covered with storage, whiteboards or fenestration.

Overall, exposed soffits need not preclude the achievement of acceptable acoustics but alternative sources of absorption need to be introduced. Wall panels are seldom as acoustically efficient as suspended ceilings and furniture systems are not usually designed with acoustic absorption in mind. Depending upon the intended usage, suspended baffles, wall panels or floated acoustic elements are all likely to be necessary considerations if acceptable acoustic conditions are to be achieved.

Notes

1. Sharland, Ian (1972), *Woods Practical Guide to Noise Control*, First Edition, Woods of Colchester Limited.
2. CIBSE (2006), *Guide A: Environmental Design*. CIBSE.
3. Sound Research Laboratories Ltd (1988), *Noise Control in Building Services*, Pergamon Press.
4. Beranek, Leo L. (1988), *Noise and Vibration Control Revised Edition*, Institute of Noise Control Engineering.
5. Association of Noise Consultants (1997). *ANC Guidelines – Noise Measurements in Buildings, Part 1: Noise from Building Services.* Association of Noise Consultants.
6. BS 8233: 1999, *Sound Insulation and Noise Reduction for Buildings – Code of Practise*. British Standards Institution.
7. DfES (2003), *Building Bulletin 93: Acoustic Design of Schools 'A Design Guide'*. Department for Skills and Education.
8. BS 4142: 1997, *Method for: Rating Industrial Noise Affecting Mixed Residential and Industrial Areas.* British Standards Institution.
9. PPG 24: September 1994, *Planning Policy Guidance: Planning and Noise*. Department of the Environment.
10. CIBSE (2002), *Guide B5: Noise and Vibration Control for HVAC*. CIBSE.

Chapter 27

Motor drives, starting methods and control

Introduction

The electric motor is the most common source of rotational energy used in heating and air-conditioning systems, this chapter discusses the types of motors used, methods of starting together with the methods of control.

Basic electrical equations

The following basic electric equations are applicable to the application of motors, under steady state operating conditions:

a.c. Single phase currents and powers:

$$I_1 = \frac{P_m}{\eta V_p \cos\phi} \times 100 \text{ A} \tag{27.1}$$

$$P_e = V_p I_1 \cos\phi \text{ W} \tag{27.2}$$

a.c. Three phase currents and powers:

$$I_1 = \frac{P_m}{\eta \sqrt{3} V_1 \cos\phi} \times 100 \text{ A} \tag{27.3}$$

$$P_e = \sqrt{3} V_1 I_1 \cos\phi \text{ W} \tag{27.4}$$

Motor speed

$$n_s = \frac{120 f}{p} \tag{27.5}$$

$$S = \frac{n_s - n_r}{n_s} \tag{27.6}$$

Motor powers and torque:

$$\eta = \frac{P_m}{P_e} \times 100\% \tag{27.7}$$

$$P_e = \frac{P_m}{\eta} \times 100 \text{ W} \tag{27.8}$$

$$T_m = \frac{P_m}{\omega_r} \text{ N.m} \tag{27.9}$$

$$\omega_r = \frac{2\pi n_r}{60} \text{ rad / s} \tag{27.10}$$

Conversion factors:

1 h.p. (horsepower) = 745.7 W

1 ch (cheval or metric horsepower) = 736 W

where:

Cos ϕ	*Power factor*
f	Frequency of the a.c. supply voltage (Hz)
I_l	Supply phase current (A)
η	Motor efficiency (%)
n_r	Motor rotor speed (rpm)
n_s	Motor synchronous speed (rpm)
p	The number of pairs of poles
P_e	Motor input electrical power (W)
P_m	Motor mechanical output power/rated power (W)
S	Motor per unit slip (pu)
T_m	Motor output rotational torque (N.m)
V_l	Line voltage, between two phases (V)
V_p	Phase voltage, between phase and earth (V)
ω_r	Motor rotor angular speed (rad/s)

Types of motors

Electric motors fall within two main groups characterised by the nature of their supply voltage either direct current (d.c.) or alternating current (a.c.) where the magnitude of voltage varies sinusoidally with time. A.C. voltages are generally either single phase or three phase comprising of three sinusoidal voltages separated by 120 electrical degrees. Two phase motors exist but the use of these is uncommon. Figure 27.1 illustrates the types of motors available within each of these groups.

This chapter concentrates on a.c. induction motors and their methods of control and in particular the squirrel-cage motor which is the most common form of electric motor used in heating and air-conditioning applications in buildings (Fig. 27.2). Historically d.c. motors have not been commonly used for heating and air-conditioning applications. However, d.c. motors are beginning to be introduced for applications such as variable volume fan coil units, due to their ease of speed control and improved efficiency compared with fractional-horsepower single phase a.c. motors. The methods of starting and speed control for d.c. motors are quite different to those applied to a.c. motors.

Concepts of motors

Motors are constructed with a stationary member (stator) and rotating member (rotor) and have field windings which produce the magnetic flux and an armature winding in which the electromotive force (emf) is induced.

The field can be physically situated on either the stator or the rotor and whilst permanent magnets may be used in small motors as the primary source of flux, in the majority of motors the field is electromagnetic. In *salient pole* motors (refer to Fig. 27.3) the field circuits are concentrated and wound around protruding poles (these are only used where the field is supplied by d.c.). Salient pole motors are characterised by a non-uniform air gap. *Non-salient pole* field coils are distributed in slots cut into a cylindrical magnetic structure providing a uniform air gap between the stator and rotor (refer to Fig. 27.4).

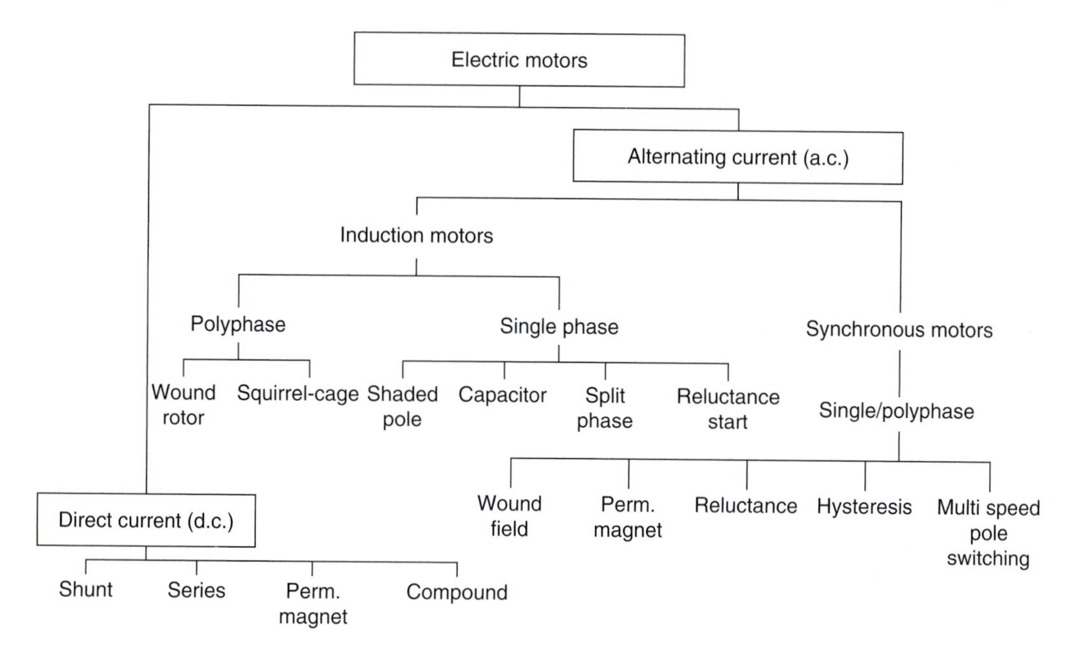

Figure 27.1 Electric motor family

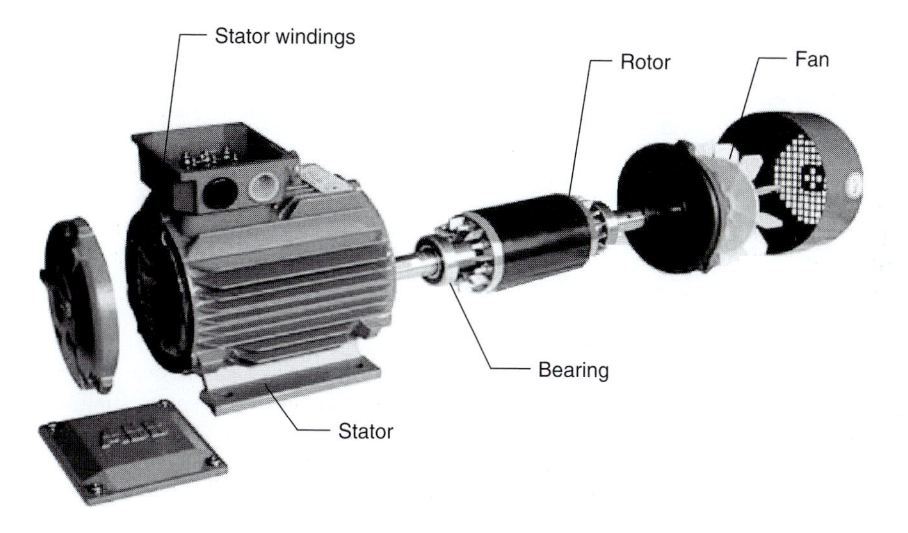

Figure 27.2 Exploded view of an ABB squirrel-cage motor (see colour plate section at the end of the book)

For d.c. motors the armature is located on the rotor and the stator provides the field structure. For most a.c. motors the armature windings are located on the stator and the field winding is on the rotor which may be salient pole or non-salient pole.

Where windings are located on the rotor special arrangements must be provided to make electrical contact to the rotating member, such connections are made through carbon brushes bearing on either slip rings or a commutator mounted on but insulated from the rotor shaft.

North and south poles are provided in pairs e.g. for a three phase induction motor a group of L1, L2 and L3 windings would represent one pair.

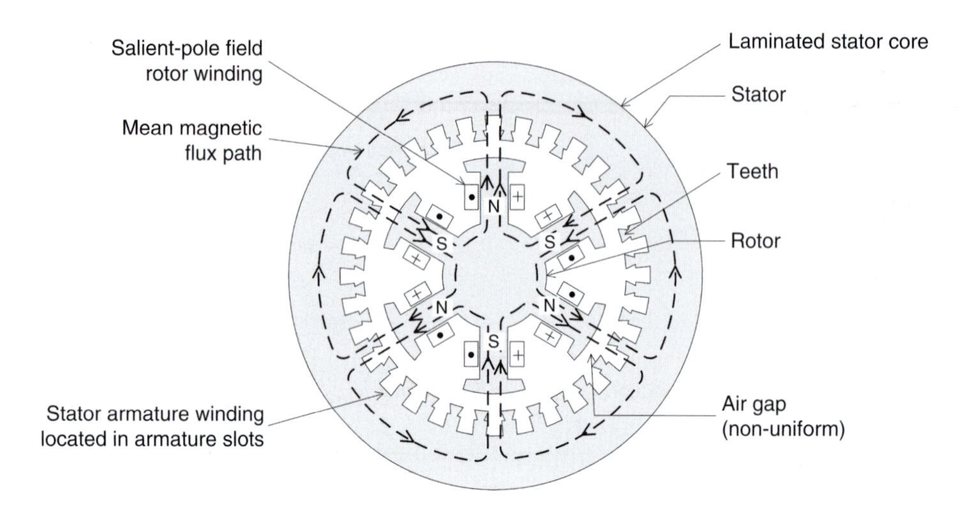

Figure 27.3 Diagram of a salient pole 6 pole synchronous motor

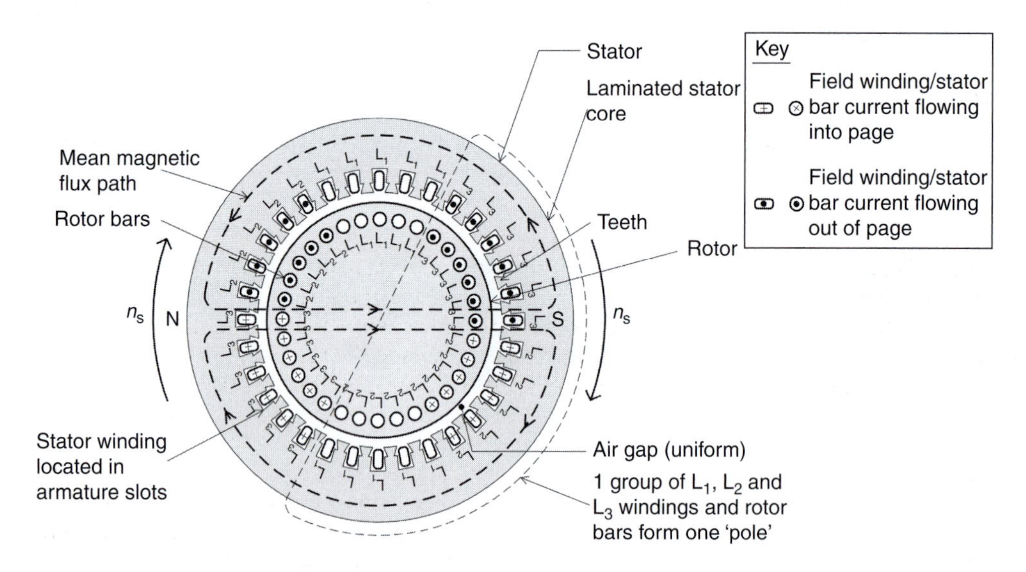

Figure 27.4 Diagram of a 2 pole squirrel-cage induction motor

Alternating current motors

There are two principle types of a.c. motor, the synchronous motor and the induction motor and these are differentiated by the nature of their field winding and characterised by the rotor speed which for the synchronous motor rotates at *synchronous speed.*

Synchronous motors

Synchronous motors have either a field winding on their rotor which is supplied with d.c. or a permanent magnet and a three phase armature winding on its stator (refer to Fig. 27.3). The interaction between the rotor magnetic field and that produced by the stator winding results in torque on the rotor which tries to align the

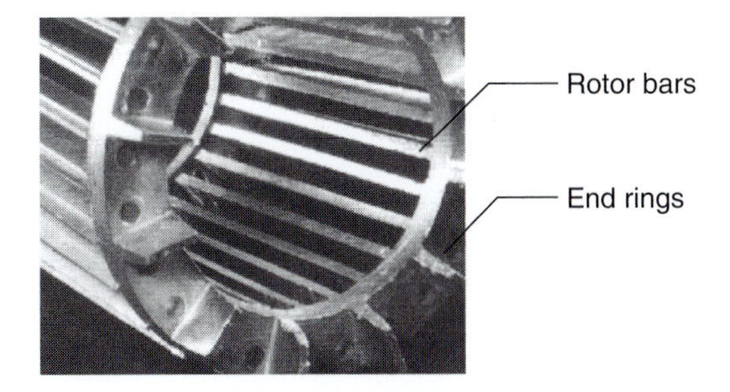

Figure 27.5 Photograph of a 'squirrel cage'

rotor field with the stator field. The speed of rotation of the rotor is the same as that of the rotating field produced by the stator winding i.e. at synchronous speed (n_s).

With no mechanical load applied to the rotor then, ignoring losses, the magnetic fields produced by the stator and rotor are aligned. When a mechanical load is applied to the rotor the resulting torque causes the rotor to decelerate until the resulting misalignment of the stator and rotor fields produces the torque necessary to balance the applied load.

Most synchronous motors require a d.c. field voltage and as such they are not commonly used in the heating and air-conditioning of building applications.

Induction motors

The name 'induction motor' comes from the a.c. 'induced' into the rotor via the magnetic flux produced in the stator.

Polyphase squirrel-cage motors

The stator structure is manufactured from steel laminations shaped to form poles and a three phase field distributed winding is wound around these poles, each displaced by 120 electrical degrees. The field winding is connected in star or delta configuration to the voltage source to produce a rotating magnetic field.

The rotor is manufactured from steel laminations over a steel shaft core and radial slots around the periphery house rotor bars made from either cast aluminum or copper conductors shorted by end rings at each end of the rotor. This arrangement of the rotor bars looks like a squirrel cage; hence, the term, *squirrel-cage induction motor* (Fig. 27.5).

Motor torque is developed from the interaction of currents flowing in the rotor bars and the stators' rotating magnetic field (which rotates at synchronous speed). In actual operation, rotor speed always lags the magnetic field's speed, allowing the rotor bars to cut magnetic lines of force and produce useful torque. This speed difference is called slip speed. Slip also increases with load and it is necessary for producing torque.

Polyphase wound rotor motors

The construction of the wound rotor motor is similar to the squirrel-cage motor except the rotor bars are replaced by a three phase distributed winding with the same number of poles as the stator. The ends of the windings are normally brought out to slip rings to allow external resistances to be added to the rotor for speed control purposes, particularly during starting. With the development of variable voltage and frequency drives, these motors are now rarely used in heating and air-conditioning applications in buildings.

Single phase induction motors

The construction of the single phase induction motor is similar to that of the polyphase squirrel-cage motor except a single phase field winding replaces the three phase winding (Fig. 27.6).

Unlike three phase field windings which produce a unidirectional rotating field, a single phase winding produces an oscillating field but the axis of the stator field remains fixed in space. If a squirrel-cage rotor is turning

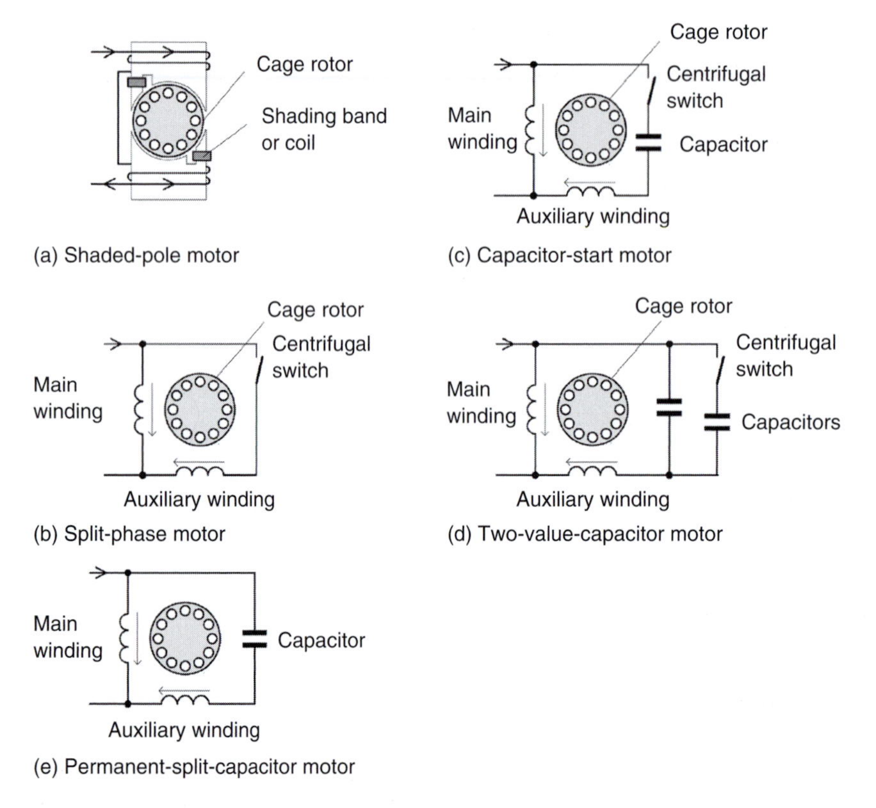

Figure 27.6 Schematic diagrams for single phase a.c. motor types

in such a field a pulsating torque is developed but the problem in single phase induction motors is to start the rotor rotating. The following common types of single phase motor employ different methods of achieving this.

Shaded pole single phase motors

These are the least expensive of the fractional-horsepower motors, they have salient stator poles, with one-coil-per-pole main windings. The auxiliary winding consists of a shorted heavy copper coil wound around half of one (or rarely both) salient stator poles. Induced currents in the shorted turn delay the build up of magnetic flux and cause the flux in the shaded portion of the pole to lag the flux in the other portion in time. The result is like a rotating field moving in direction from the unshaded to the shaded portion of the pole. These motors produce a low starting torque.

Split-phase or resistance-split-phase motors

Split-phase motors employ two separate windings (a main winding and an auxiliary winding) having different reactance–resistance ratios resulting in their axes being displaced by 90 electrical degrees in space. The current reaches its maximum in the high reactance coil winding at a later time and the rotor experiences a shift in magnetic field that provides the necessary starting torque. When the motor is almost up to speed the high-resistance winding is disconnected by a centrifugal switch.

Capacitor motors

Capacitor motors employ a capacitor in series with an auxiliary winding to provide the necessary phase shift, these are available in three types: capacitor-start, two-value-capacitor and permanent-split-capacitor. As the names of the first two types imply, these use a centrifugal switch to reduce the size of the capacitor when the

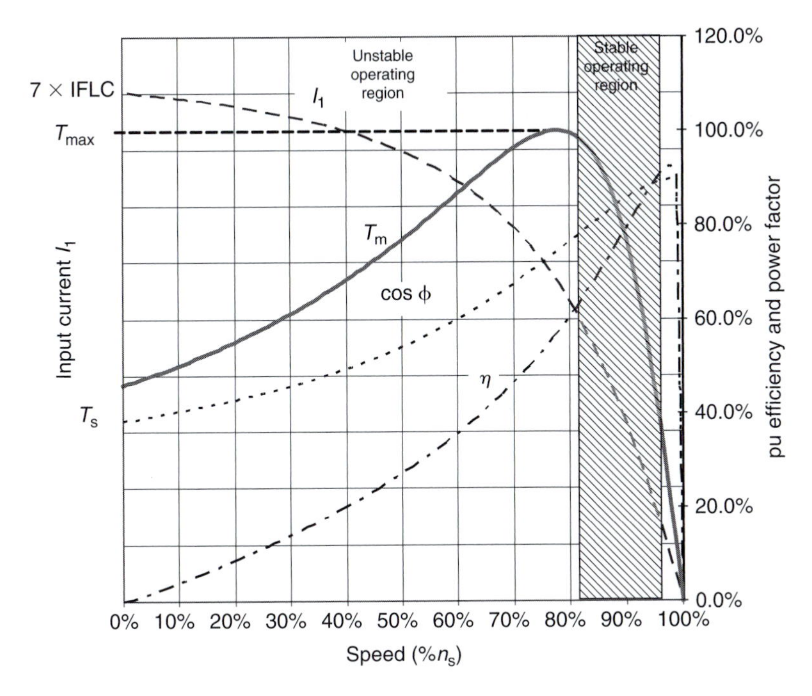

Figure 27.7 Typical steady state induction motor characteristics

motor is almost up to speed. The two-value-capacitor uses two capacitors to improved performance, the larger provides good starting torque and is then switched out by a centrifugal switch, the smaller capacitor remains in circuit to improve the power factor (refer to motor characteristics).

Induction motor characteristics and specification

Typical steady state characteristics for a three phase induction motor are illustrated in Fig. 27.7 this shows the relationship between the motor input current (I_1), motor torque (T_m), power factor ($\cos \phi$) and efficiency (η) against motor speed (n_r expressed as a % n_s).

The starting torque (T_s) should be many times greater than the starting torque required by the load, the greater the difference in torques, the faster the motor accelerates.

The maximum torque (T_{max}) also called breakdown torque or pull-out torque, is the maximum torque a motor can deliver without stalling.

The maximum power factor ($\cos \phi$) of squirrel-cage induction motors differs with motor rating between approximately 0.60 and 0.93 lagging and varies with speed, peaking at 100 per cent rated load. Power factor is the ratio between real power (P) and apparent power (S). Low power factor loads increase the losses in a power distribution system resulting in increased energy costs and electricity. Utility suppliers apply a penalty to consumers who operate with a low-power factor.

The motor should be selected to suit the nature of the load so that it is operated within the stable operating region.

The characteristics indicate that the efficiency (η) of the motor is approximately proportional to $(1-S)$, where S is the per unit slip (refer to Eq. 27.6).

Rated output

The rated output of a motor is the mechanical power available at the shaft expressed in watts, in some countries this is expressed in horsepower. Table 27.1 provides a summary of the 'standard' ratings for polyphase squirrel-cage induction motors.

Table 27.1 Standard squirrel cage induction motor ratings

	Aluminum frame							*Steel frame*				*Cast iron frame*			
	1 Ph (230 V)							*3 Ph* (400 V)							
Motor rated output (kW)	2 *pole*	4 *pole*	6 *pole*	2 *pole*	4 *pole*	6 *pole*	8 *pole*	2 *pole*	4 *pole*	6 *pole*	8 *pole*	2 *pole*	4 *pole*	6 *pole*	8 *pole*
0.055							•								
0.06					•										
0.09				•	•	•	•								
0.12		•		•	•	•	•								
0.15						•									
0.18	•	•	•	•	•	•	•							•	•
0.25	•	•	•	•	•	•	•						•	•	•
0.32			•			•									
0.37	•	•	•	•	•	•	•					•	•	•	•
0.55	•	•	•	•	•	•	•					•	•	•	•
0.65			•												
0.75	•	•		•	•	•	•					•	•	•	•
0.85			•												
1.1	•	•		•	•	•	•					•	•	•	•
1.4	•														
1.5	•	•		•	•	•	•					•	•	•	•
1.7		•													
1.85		•													
2.2		•		•	•	•						•	•	•	•
3				•	•	•	•					•	•	•	•
4				•	•	•	•					•	•	•	•
5.5				•	•	•	•					•	•	•	•
7.5				•	•	•	•					•	•	•	•
11				•	•	•	•					•	•	•	•
15				•	•	•	•					•	•	•	•
18.5				•	•	•	•					•	•	•	•
22				•	•	•	•					•	•	•	•
30				•	•	•	•					•	•	•	•
37				•	•	•	•				•	•	•	•	•
45				•	•	•				•	•	•	•	•	•
55				•	•					•	•	•	•	•	•
75				•	•			•	•	•	•	•	•	•	•
90				•	•			•	•	•	•	•	•	•	•
95				•	•									•	
110								•	•	•	•	•	•	•	•
132								•	•	•	•	•	•	•	•
160								•	•	•	•	•	•	•	
200								•	•	•		•	•		
250								•	•	•		•	•		
280								•							
315								•	•	•					
355								•	•	•					
400								•	•	•					
450									•	•					
500									•	•					
560									•						
630									•						

Ratings based on 50 Hz general purpose (single speed) squirrel-cage motors taken from ABB catalogue for low-voltage general purpose motors.

Motor speeds

In Europe where the frequency of the supply voltage is 50 Hz this results in the following synchronous and typical full-load speeds for motors. Full-load speeds vary with motor rating.

Induction motor speeds

No. of poles	*Synchronous speed* (rpm)	*Typical full-load speed* (rpm)
2 pole	3000	2900
4 pole	1500	1440
6 pole	1000	960
8 pole	750	700

For heating and air-conditioning applications it is common to use either 2 or 4 pole motors.

Motor losses and efficiency

Motor efficiency is a measure of how well a motor converts electrical energy to useful work, energy lost in the process is emitted as heat.

Motor losses can be divided into five major areas, all of which are influenced by design and construction decisions e.g. the size of the air gap between the stator and the rotor. Large air gaps minimise manufacturing costs. In general smaller air gaps improve efficiency and power factor. Figure 27.8 illustrates typical losses for a squirrel-cage induction motor.

A European Scheme established through co-operation between CEMEP (European Committee of Manufacturers of Electrical Machines and Power Electronics) and the European Commission designates energy efficiency classes for low-voltage a.c. motors to improve energy efficiency and thus reduce CO_2 emissions (Fig. 27.9).

The scheme defines three classes of efficiency EFF1, EFF2 and EFF3 which apply to 2 and 4 pole, three phase squirrel-cage induction motors rated for 400 V, 50 Hz with S1 duty type (continuous running duty) with the output 1.1 to 90 kW.

Rated voltage

In Europe and for heating and air-conditioning applications, the normal rated voltages for a.c. low-voltage (lv) motors are 230 V for single phase motors and 230 V/400 V for three phase motors.

It becomes economic to use medium-voltage (mv) motors on large centrifugal or screw chillers with single compressor ratings greater than 3.8 MW_e at 3.3 kV and 8 MW_e at 11 kV.

Environmental ratings

Unless otherwise specified motors are suitable for the following environmental conditions:

Maximum altitude	1000 m above sea level
Maximum ambient air temperature	40°C
Minimum ambient air temperature	−15°C (600 W $< P_m >$ 3.3 MW

Insulation system

The thermal classes applied to insulation systems used for motor windings (in accordance with IEC 62114) are as follows:

Insulation class	*Maximum ambient temperature*	*Maximum permissible temperature rise*	*Hotspot temperature margin*
Class B insulation system	40°C	80°K	+10°K
Class F insulation system	40°C	105°K	+10°K
Class H insulation system	40°C	125°K	+15°K

It is normal practice to manufacture motors with a maximum temperature rise at a class below that of the insulation system to provide a safety margin. A common arrangement is a Class F insulation system (155°C)

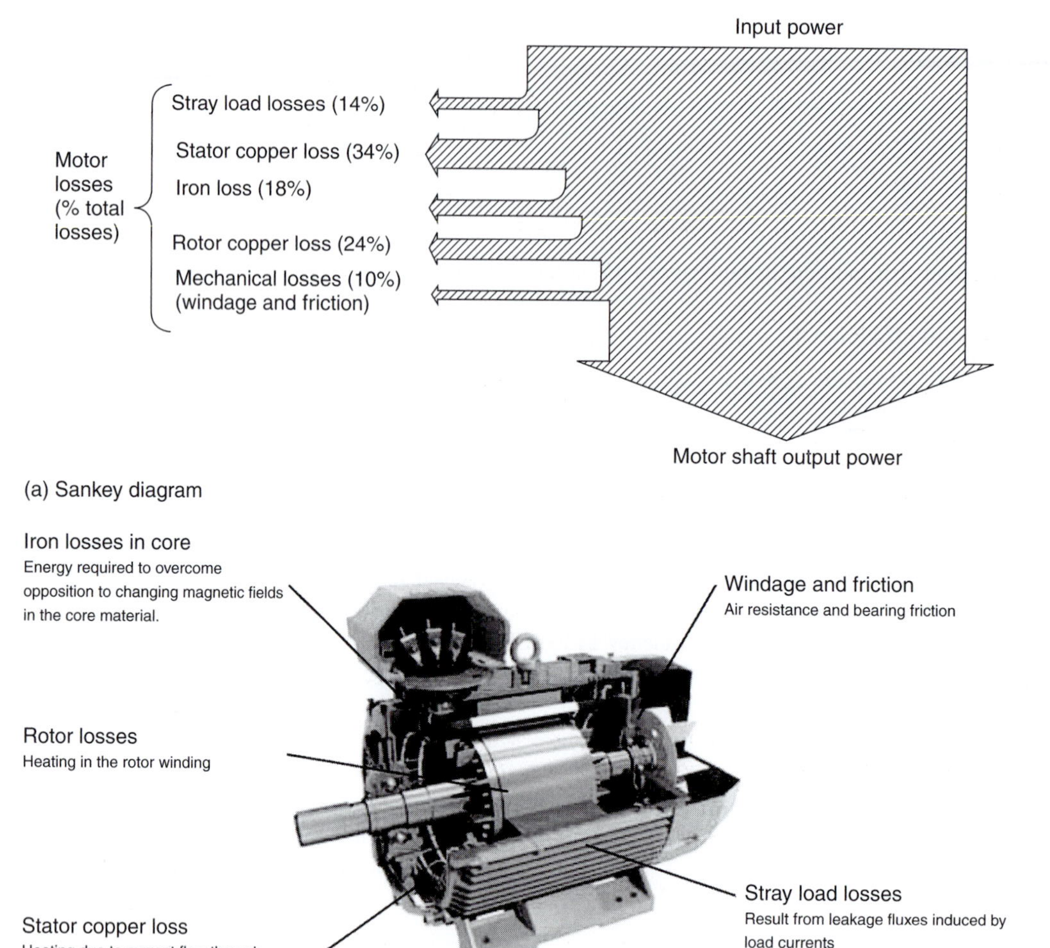

Figure 27.8 Squirrel-cage induction motor losses (see colour plate section at the end of the book)

with a Class B temperature rise (40°C) which provides a 15°C safety margin by limiting the operating temperature to 120°C.

This safety margin can be used to increase the loading by up to 12 per cent for limited periods, to operate at higher ambient temperatures or altitudes, or with greater voltage and frequency tolerances. It can also be used to extend insulation life.

Motors for powered smoke and heat exhaust ventilation systems should comply with BS EN 12101-3 for the correct classification for their application. These are often mounted in the air stream inside the ventilation duct where the air stream can be utilised to cool the motor. Due to the high air flow and speed from the fan the motor rating can be increased, while still maintaining the normal temperature rise limitations (Class F insulation system with Class B temperature rise), although for larger motor ratings Class H insulation system may be necessary.

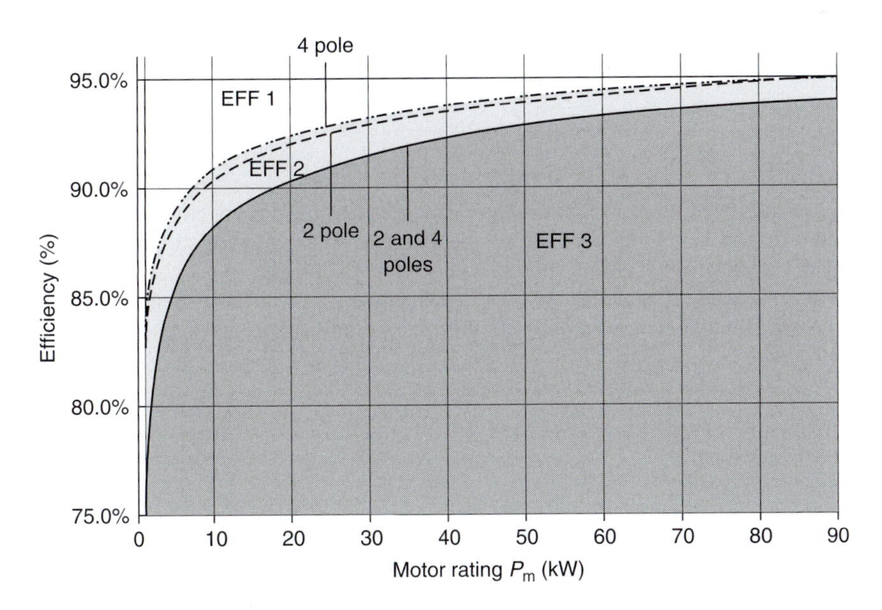

Figure 27.9 CEMEP motor efficiency bands

Degrees of protection

Motor enclosures should be suitable for the environment that the motor is installed, these are classified by degree of Ingress Protection (IP protection) to BS EN 60529 and resistance to external impact (IK code) to BS EN 50102.

In relation to motors the common degrees of protection/IK codes are as follows:

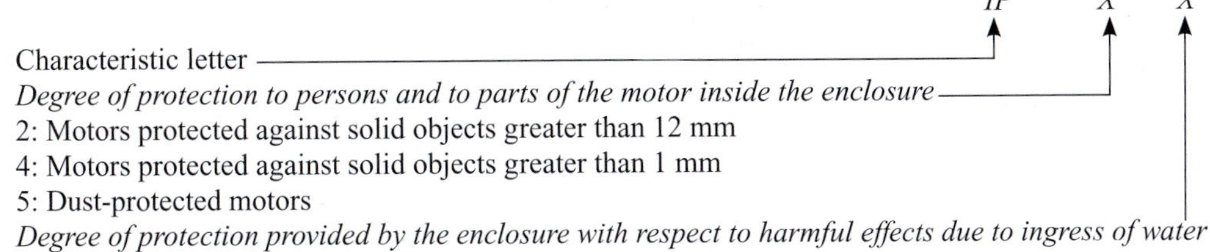

3: Motors protected against spraying water
4: Motors protected against splashing water
5: Motors protected against water jets
6: Motors protected against heavy seas

For general purpose motors the 'standard' degrees of ingress protection are Open 'Drip proof' IP23, Totally enclosed IP55.with an Impact Code of IK08 providing protection to external impacts of up to 5 J.

Speed control of induction motors

The induction motor is basically a constant speed machine. However, there are applications where variations in speed are desirable as part of an overall control system e.g. static pressure control on variable air volume systems and variable speed pumping. Traditional volume control methods for ventilation and hydraulic systems involve the introduction of resistances into the system to control flow rates, this introduces in efficiencies into the systems and a more efficient approach is to control the speed of the motor.

The speed of an induction motor (n_r) is determined by the synchronous speed (n_s) and the per unit slip (S) of the rotor Eqs 27.5 and 27.6 can be rearranged to provide the following equation:

$$n_r = \frac{120f}{p}(1 - S)\text{rpm}$$

This relationship suggests that the speed of an induction motor can be varied by varying either the slip the supply frequency or the number of poles. Any method of control that depends on variation of slip is inherently inefficient since the efficiency of the induction motor is approximately equal to $(1-S)$.

Dual speed motors which are provided with two stator windings arranged with different number of pair poles are available, these are used in ventilation applications but they only offer discrete and stepped variation in motor speed.

For a given motor the following methods of variable speed control are available:

- Variable voltage, constant frequency control.
- Variable voltage, variable frequency control.
- Regulation of slip power (rotor I^2R control).

Regulation of slip power is achieved by the connection of external resistors in series with the rotor field windings, this type of control can only be applied to wound rotor motors and is not commonly used in heating and air-conditioning applications.

Variable voltage constant frequency speed control

The torque speed curves in Fig. 27.10 indicate the motor output torque produced for a motor with 100, 80, 60 and 40 per cent of full input voltage and constant frequency applied, these indicate that the motor output torque (T_{max}) is approximately proportional to the square of the motor voltage (e.g. at 80 per cent of the voltage T_{max} is 64 per cent of that with 100 per cent of the voltage).

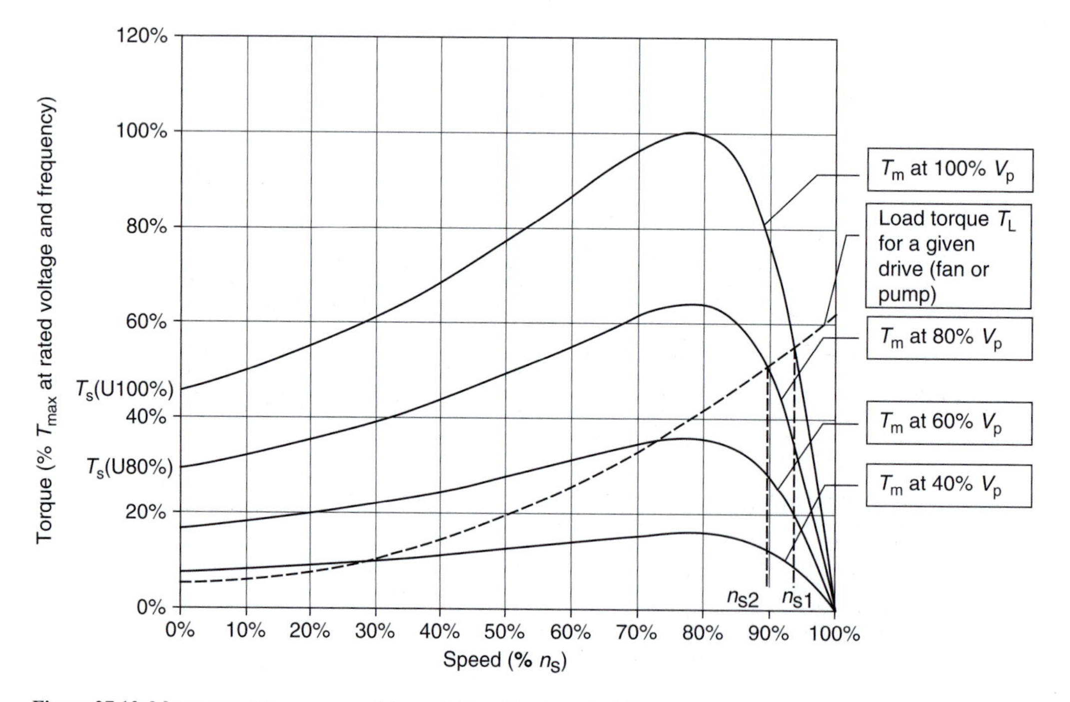

Figure 27.10 Motor output torque v speed for variable voltage constant frequency

The starting torque (T_s) is greater than the load torque (T_L) and the motor accelerates due to ($T_s - T_L$) and continues to do so until $T_m = T_L$ where the equilibrium speed (n_{s1}) is attained. At the lower voltage, the intersection between the motor torque and load torque occurs at a different speed (n_{s2}) and the motor continues to run at this speed.

This method of speed control is simple and economical, but the range of control is limited (Fig. 27.10 indicates that at lower voltages, operation is outside of the motor's stable operating range and the motor torque may be insufficient during starting), this makes it suitable for fan type loads where $T_L \propto n_r^2$.

For small single phase motors a method of 'fixed' voltage speed control is achieved by using a transformer with tapped secondary voltages, this method of control is often used in fan coil units. A simple method of variable voltage speed control can be provided by the use of phase-controlled thyristors or triacs, but these produce significant harmonics and should be avoided in large quantities.

Variable voltage constant frequency control is employed by *soft-starters* (refer to later section for details of soft-starters), to reduce the current drawn by a motor during starting.

Variable voltage, variable frequency control

By varying the frequency in proportion to the voltage a constant air-gap flux density is maintained and a family of torque v speed curves are obtained with almost constant starting torque but different synchronous speeds (Fig. 27.11). This provides a method of matching the required load torque at speeds close to the synchronous speed thus maintaining high efficiencies.

Variable voltage and frequency control is employed by *variable speed drives* (VSDs), detailed later in this chapter.

Motors starters and drives

This section describes the common methods of starting a.c. squirrel-cage induction motors, these fall into two categories: full-voltage (or across-the-line starters) and reduced voltage starters.

Full-voltage starters apply full line voltage (400 V for polyphase motors in Europe) directly to the motor terminals, generally only the *direct-on-line* (DOL) starter falls into this category.

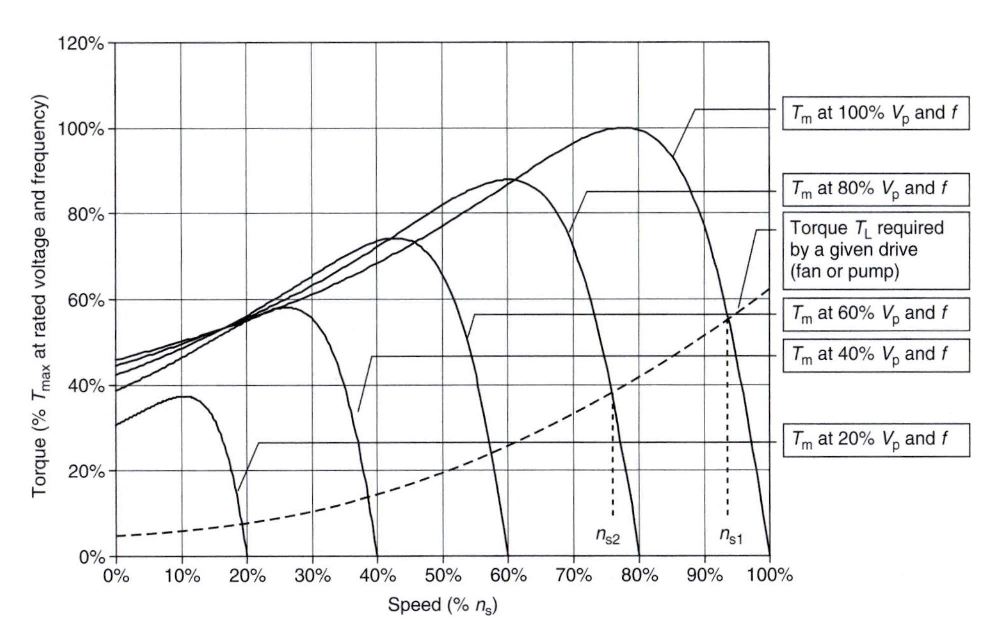

Figure 27.11 Motor output torque v speed for variable voltage variable frequency

Reduced voltage, single speed motor starters are appropriate where the application requires a gentle start and smooth acceleration up to full speed or limitation of the starting current. Many starters apply reduced voltage to the windings, such as *star–delta, part-wind, autotransformer* and *electronic starters* such as soft-starters and VSDs.

Depending upon the method of starting, motors draw currents significantly greater than their full-load running current during starting and this has the effect of reducing the voltage to other equipment connected to the same source of supply resulting from the voltage drop in the common sections of the network (comprising the supply transformer and supply cables). The magnitude of the reduction in voltage is related to the impedance of the common sections of the network and is proportional to the starting current and this can result in flicker of lights or malfunction of other electrical equipment not only within the building but also in use by other consumers.

The object of the motor starter is to provide the following functions:

- To connect and disconnect the motor safely from the electricity supply.
- To prevent the motor restarting after a stoppage due to a drop in the voltage or an interruption in the electricity supply.
- To protect the motor against abnormal overload conditions. BS 7671 'Requirements for Electrical Installations', requires motors with a rating exceeding 0.35 kW to be provided with control equipment incorporating means of protection against overload of the motor.

Optional functions:

- To limit the current during starting.
- To control the torque of the motor during the starting period.
- To reverse the rotation of the motor.
- To apply braking torque to the motor.

Direct-on-line starter

This is the simplest form of starting. In this method (refer to Fig. 27.12) the starter contactor contacts connect the three (or single) phase supply lines to the motor terminals. For three phase motors the motor windings are delta connected. The initial (starting) current drawn by the motor can be between 5 and 7 times the full-load current. Due to the high initial current, the use of DOL starting is normally limited to motors up to 7.5 kW. Where there

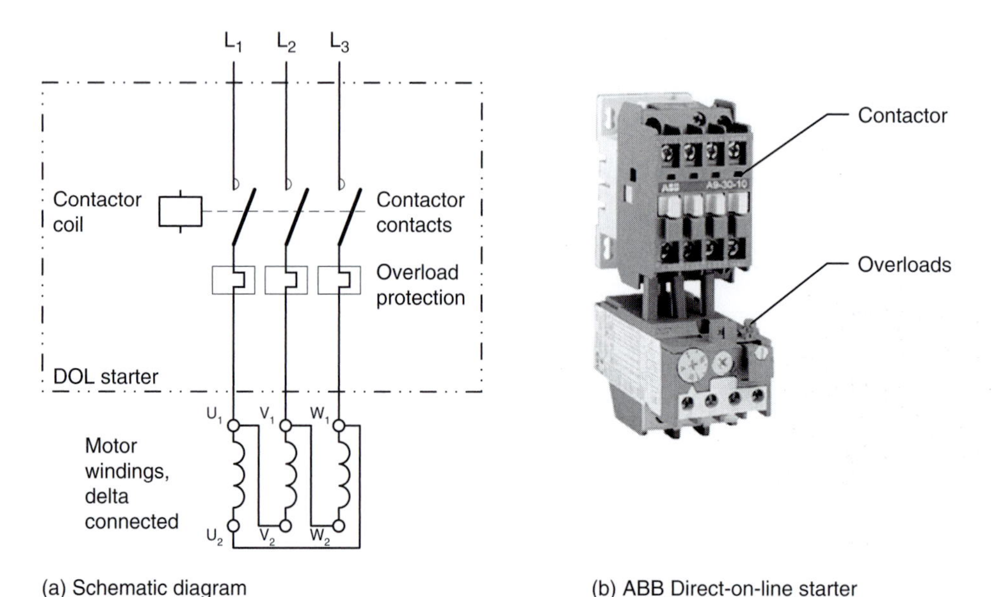

(a) Schematic diagram

(b) ABB Direct-on-line starter

Figure 27.12 DOL starter (see colour plate section at the end of the book)

is no limitation on starting currents, motors up to 120 kW may be started DOL. However, the torque may rise to 200 per cent of the full-load torque. The combined effect of high torque and acceleration may cause damage to belt or shaft drives.

Star–delta (wye-delta) starter

Where DOL starting is unsuitable the starting current and torque may be reduced by applying a lower voltage to the motor windings. With a star–delta starter (refer to Fig. 27.13) the three phase windings of the motor are initially connected in star by closing the main contactor (M) and the star contactor. In this configuration the phase voltage (230 V in Europe) is applied across each motor winding. Consequently the starting current is reduced to about 3 times the full-load current and a corresponding reduction in torque to about 33 per cent of the full-load torque. Once the motor has achieved approximately 85 per cent of its rated speed, the starter connects the windings in a delta configuration, applying full-line voltage (400 V in Europe) across each winding and allowing the motor to accelerate to full-load speed as in DOL configuration. The transfer between star and delta configuration is initiated by a timer and it is performed as an open transition (the star contactor is opened prior to the delta contactor closing).

Star delta starting is not recommended for large squirrel-cage motors due to the magnitude of current transients produced by the switching process when changing from the star to delta configuration. During the

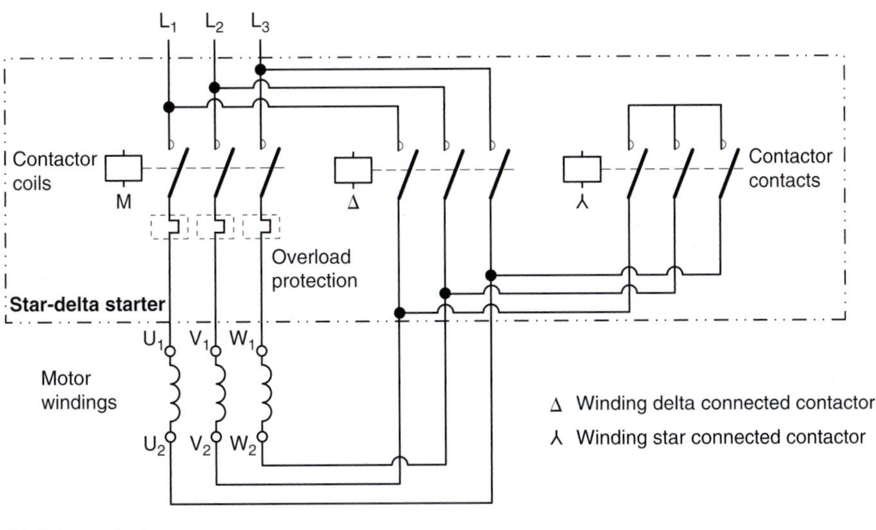

(a) Schematic diagram

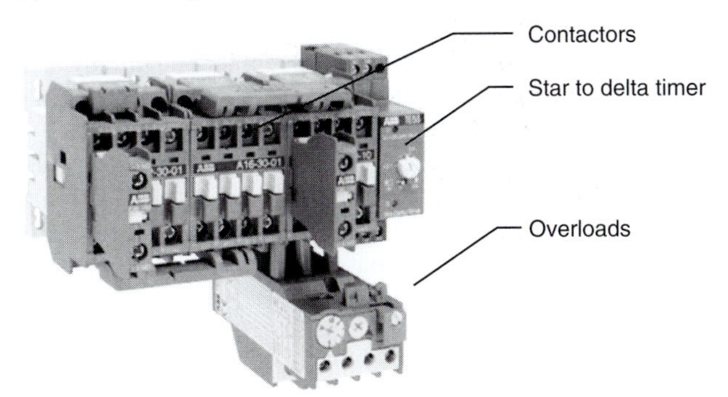

(b) ABB Star-delta starter

Figure 27.13 Star–delta starter (see colour plate section at the end of the book)

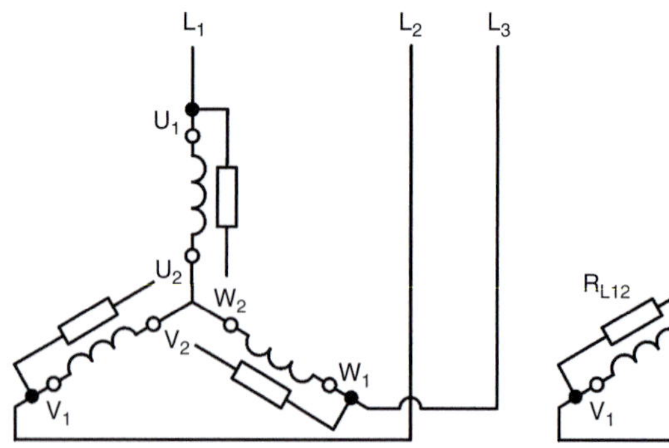

Step 1 – Motor windings star connected and resistors out of circuit.

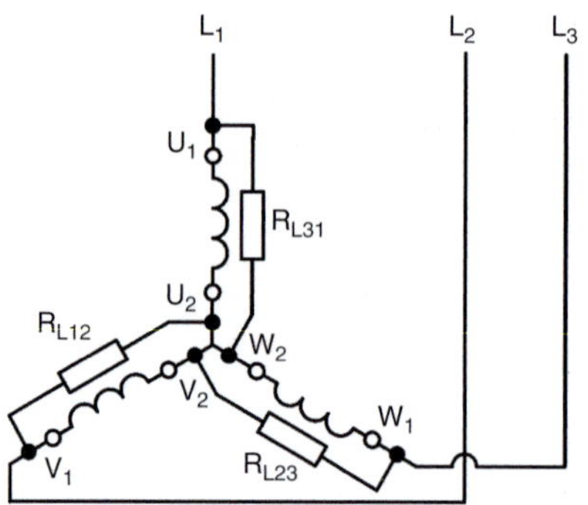

Step 2 – Motor windings star connected, resistors connected in parallel with each winding.

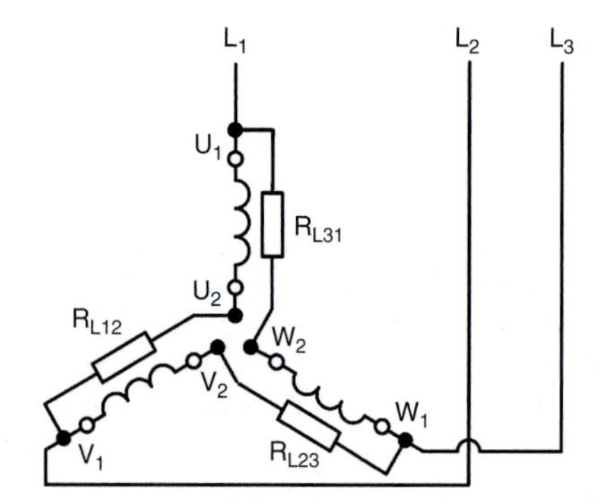

Step 3 – Star contactor open, motor windings delta connected in series with resistors, % phase voltage.

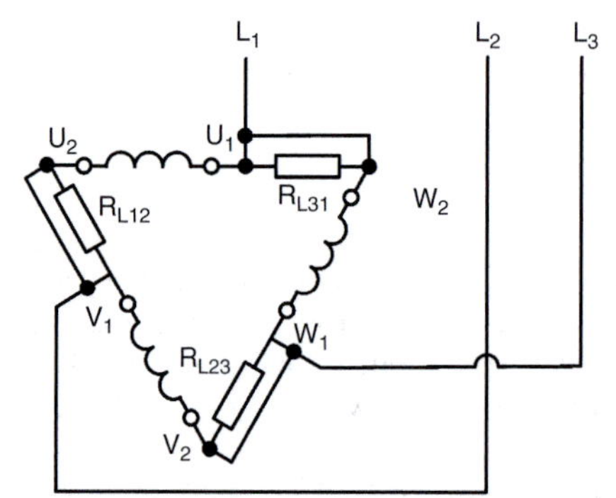

Step 4 – Resistors short-circuited, motor windings delta connected across the full phase voltage.

Figure 27.14 Winding/resistor configurations for closed transition star–delta starting

changeover period the squirrel-cage motor is disconnected from the supply and as the secondary winding is short-circuited, the magnetic field does not collapse immediately and it may be up to 3 seconds before the residual voltage of the stator falls to zero. When the motor is reconnected to the supply the phase position of the residual voltage may be out of phase with respect to the supply voltage, thus causing a large transient current to flow in the stator winding. Depending upon the phase displacement of the two voltages, the current transient may exceed 20 times the full-load current of the motor. This transient current, although only a few cycles in duration cause mechanical and electrical stresses within the motor and high-frequency electromagnetic radiation can be generated along the supply cable, which may cause interference with other systems.

The control circuit must include interlocks to ensure that the motor starter is interrupted so that it is reset to star configuration whenever the motor is stopped to avoid starting the motor in delta which would result in the starting currents the magnitude of those obtained using DOL.

For star–delta starting both ends of each motor winding must be brought out to the terminal box.

Closed transition star–delta (wauchope) starter

The closed transition star–delta starter (also referred to as the Wauchope starter) which is a modified form of a star–delta starter, introduces series resistances which dampen the step voltage increase changing from star to delta configuration. The net effect is that the peak current is no greater than that at full voltage thus avoiding the current transients (Fig. 27.14).

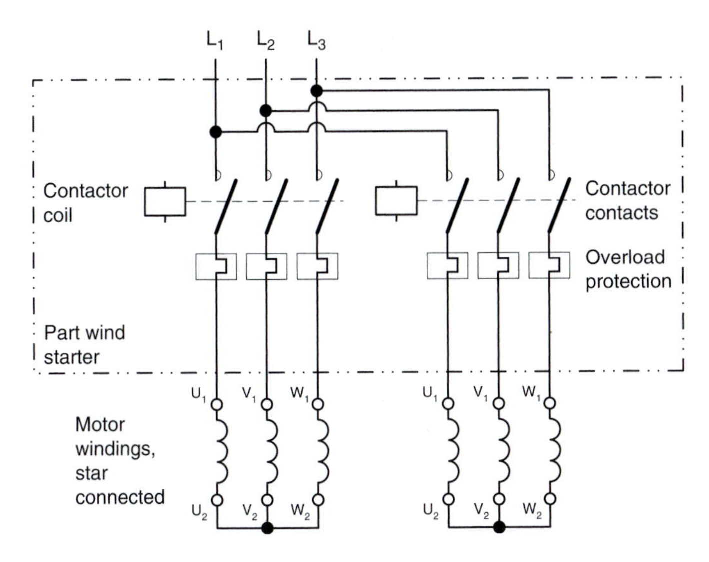

Figure 27.15 Schematic of a part-wind starter

Initially the motor is connected in star to the supply. During the next step a resistor is connected in parallel with each phase winding of the motor. The line current is increased by the amount taken by the resistor, but the current in the winding remains the same. This is a preparatory step and is only in operation for a fraction of a second.

The star point of the winding is now opened which has the effect of connecting the motor windings in delta with the resistor in series with each phase winding. Finally the resistors are then short-circuited and the motor is connected to the supply in delta.

This method not only prevents transient peak currents but also the motor develops a continuous torque during the starting period without any fall in speed during the interval of changeover.

Part-wind starter

Part-wind starters can only be used with motors that have two separate sets of windings, these are wound to provide either 50:50 or 60:40 speed ratios. The starter operates similar to a DOL starter except the windings are generally star-connected, the high-speed winding contactor is closed first and then within 0.5 to 0.75 seconds the low-speed contactor is closed for the run condition (Fig. 27.15).

Typical starting current is 4.5 times the motor's full-load current and starting torque is approximately 40 per cent of what would be developed at full voltage.

Part-wind starters are only suitable for loads which have low starting torque requirement such as low-inertia fans and some compressors since the load must be able to be accelerated from zero to full rated speed with only a proportion of the motor capacity. If the load cannot reach full speed before the second winding energises, the motor's torque and inrush current will jump to the DOL values and defeat the purpose of the starter.

Autotransformer starter

Autotransformer starters can be connected as either open transition or closed transition (also known as the Korndorffer starter), the later being preferred and is indicated in Fig. 27.16, refer to discussion under star–delta starters regarding current transients during the switching process.

During starting line voltage is applied to the transformer via the 'start' contactor, the 'run' contactor is open and the motor is connected to a transformer tap (65 per cent indicated), the star contactor is closed. When the motor has accelerated to full speed the 'star' contactor is opened effectively converting the

(a) Schematic of an autotransformer starter

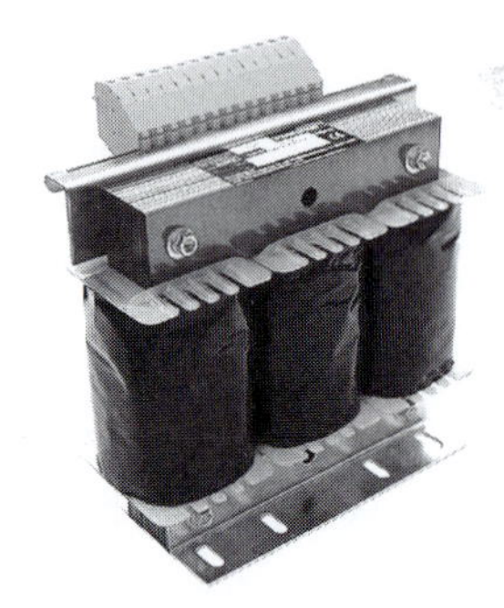

(b) Motor starting autotransformer

Figure 27.16 Autotransformer starter (see colour plate section at the end of the book)

transformer starter into a primary reactance starter, next the 'run' contactor closes bridging the primary reactor and applying full-line voltage to the motor.

The autotransformer provides a reduced voltage to the motor terminals during starting using taps on a three phase autotransformer which allows adjustment for a range of current and torque requirements (to suit the nature of the load). The standard taps are 50, 65 and 80 per cent of the full line voltage resulting in starting torque of 25, 42 and 64 per cent and line currents of 25, 42 and 64 per cent of the full voltage value of line current, respectively.

Soft start starter

In this method of starting a variable (increasing) voltage is applied to the motor.

The variable voltage is produced by a pair of thyristors (connected in each phase connection to the motor), controlled by a microprocessor. The thyristors are triggered (thyristors conduct) each half cycle by a pulse generated by the microprocessor and commutation (thyristors switched off) occurs when the current waveform passes through zero. The rms voltage is varied by adjusting the angle that the thyristors are triggered, the higher the trigger angle the lower the output voltage.

During starting the firing angle is advanced so that more and more of the full sine wave of each phase voltage is conducted. The motor accelerates up to full speed as the starter applies the full voltage sine wave to the motor.

The bypass contactor can be provided integral to or external to the soft-starter, it is closed once the motor reaches full speed, bypassing the soft-starter to improve the operating efficiency of the starter.

As an alternative to the star connection illustrated in Fig. 27.17, the soft-starter can be connected inside the delta connection (e.g. connected between U_1 and V_2, V_1 and W_2 and W_1 and U_2), this allows a lower rating starter to be used for a motor since the starter only now carries $\sqrt{3}$ of the full-load current.

When applying this method of starting to centrifugal fans, care must be taken to ensure that the control of the applied voltage on starting does not limit the accelerating torque to the extent that the fan fails to run up to speed.

Variable speed drives

VSDs provide a variable voltage and frequency output to the motor (Fig. 27.18).

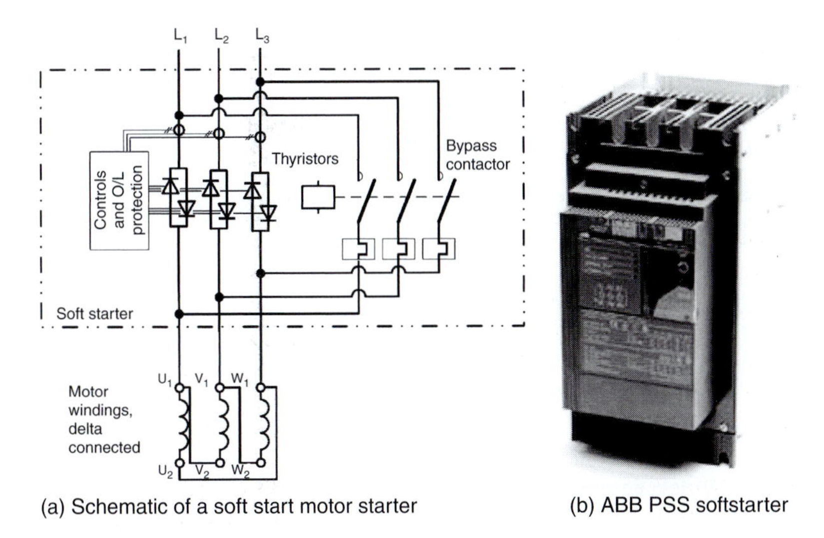

(a) Schematic of a soft start motor starter (b) ABB PSS softstarter

Figure 27.17 Soft-start starter (see colour plate section at the end of the book)

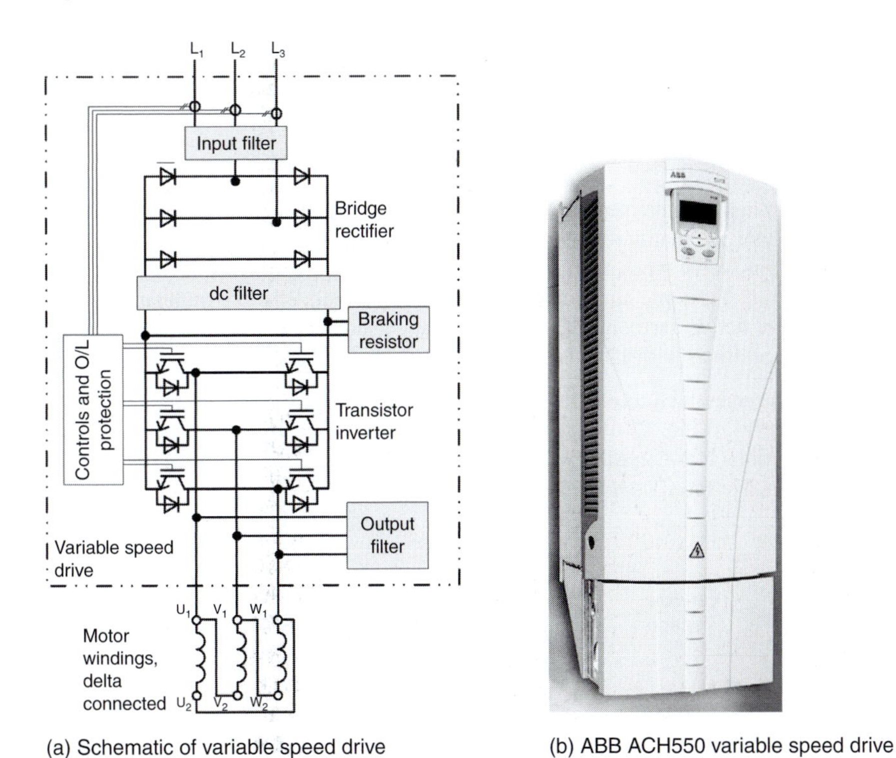

(a) Schematic of variable speed drive (b) ABB ACH550 variable speed drive

Figure 27.18 Variable speed drives (see colour plate section at the end of the book)

The 50 Hz a.c. supply voltage is converted into d.c. by the rectifier. The d.c. is supplied to an inverter, via a smoothing circuit and is converted into an a.c. voltage with variable magnitude and frequency (the voltage magnitude is varied in proportion to the frequency) which is supplied to the motor.

The inverters generally use pulse width modulation (PWM) which combine both voltage control and frequency control. The PWM inverter synthesises the motor current by switching the transistors (older inverters utilised thyristors) on and off at high frequencies typically greater than 5 kHz (500–2500 Hz for thyristors). The output voltage waveform is not a true sinusoid and as such includes harmonics which are attenuated by the output filter and motor inductance.

In addition to the variable speed control, savings in energy occur as the energy consumption of the motor is related to its load (torque) and speed. Other benefits include:

- Smooth starting – an inverter may be used in place of a 'soft starter'
- No inrush current on starting
- Smooth motor acceleration
- May be remotely controlled
- One inverter may control a number of fans.

Not all general purpose motors are suitable for VSDs and when selecting general purpose motors for use with VSDs the following shall be taken into consideration:

- The voltage supplied by the frequency converter is not a pure sinusoid, this may increase the losses, vibration and noise of the motor. This change in distribution of losses may affect the temperature rise of the motor.
- The operating speed of the motor may deviate considerably from its nominal speed, when operating at higher speeds, it should be ensured that the highest permissible rotational speed is not exceeded. Guide values for the maximum speeds are available from motor manufactures.
- At low-speed operation the cooling capacity of the fan decreases, which may cause higher temperature risers in the motor and affect the performance of the bearing grease.
- Lubrication, life of sealed bearing can be reduced compared with operation on direct line starters.
- As a general rule, for 400 V, motors ⩾100 kW should be fitted with insulated bearings and for motors ⩾350 kW the VSD output shall additionally be fitted with a common mode filter.

VSDs produce the following effects which need to be taken account of in the installation:

- Conducted and radiated electromagnetic interference can be generated by the cable interconnections between the drive and the motor and the manufacturers guidance should be followed strictly in respect of the provision of dV/dt limiting filters, cable types and installation methods.
- The input rectifier of VSDs generate harmonic currents onto the supply current. Where a number of drives are being installed the provision of active harmonic filters should be considered or where large drives are installed (>100 kW) then the use of drives with 12 pulse input rectifiers should be considered.

Current developments include the manufacture of integrated motors and controllers where the inverter is mounted directly onto the motor (refer to Fig. 27.19), this arrangement simplifies wiring, reduces radiated EMC (generated by the interconnections between the inverter output terminals and the motor). Figure 27.20 shows a pump with an integrated variable speed controller.

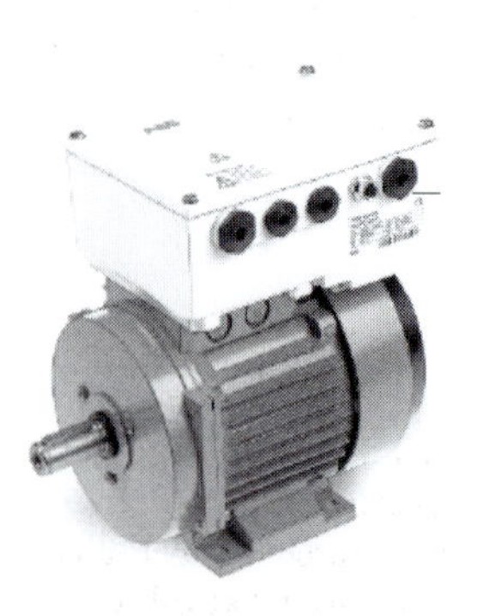

Figure 27.19 Grundfos E pump

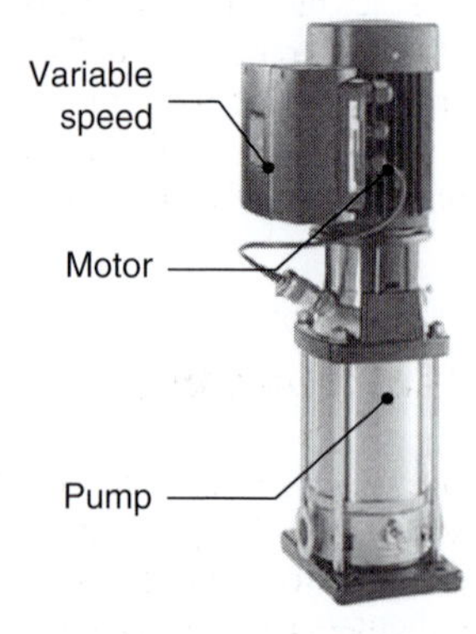

Figure 27.20 ABB Drive[IT] motor with integrated VSD

Summary of motor starters

Method of starting	*Starting current (% FLC)*	*Starting torque (% T_{max})*	*Typical heating and air-conditioning applications*	*Comments*
DOL	550% to 700%	100%	Pumps, small fans up to approximately 7.5 kW, if no limits due to voltage drops maximum of 120 kW. Sprinkler pumps.	Lowest cost. Highest starting torque. Can be used with standard motor. Most reliable. High starting torque may damage driven load.
Star–delta	200% to 300%	33%	Pumps fans and chiller compressors requiring a low starting torque.	Requires delta wound motors. Acceleration time approximately 7 s.
Open transition			Maximum of 110 kW due to stresses caused by switching transients.	
Closed transition			Large pumps and chiller compressors above approximately 70 kW	Reduces motor stress and transients during star to delta transition.
Autotransformer	150% to 450% (depending upon tapping)	25% to 64% (depending upon tapping)	Pumps and fans.	Not suitable for frequently started loads (approximately >5 per hour). High cost. Adjustable in the field. Least strain on the motor.
Part wind	450%	42%	Only suitable for unloaded or lightly loaded applications e.g. some types of chiller compressor.	Low cost. Dual wound motor required. Low pull out torque.
Soft-start	150% to 400%	10% to 85%	Pumps, fans and chiller compressors.	Preferred replacement for mechanical reduced voltage starters.
VSD	<100%	<100%	Pumps and fans serving variable volume systems.	Special attention to installation arrangements required. Generates harmonics on the input current. Provides energy savings when operating below full load.

Figure 27.21 indicates the range of ratings available for each type of motor starter together with the 'rule of thumb' recommended use range. The 'rule of thumb' recommended range should be used where the limitation of starting current is the primary selection factor, outside of this range the effect of currents on the electrical system should be considered. For all motor installations details of the maximum starting currents should be advised to the electricity distribution network operator (DNO) for agreement.

Maintenance of motors and plant incorporating motors

To prevent injury to personnel during maintenance, every fixed electric motor shall be provided with an efficient means of switching off, readily accessible, easily operated and so placed as to prevent danger.

The requirements for switching off falls into two categories: 'isolation' to remove the source of electricity to prevent electric shock during mechanical maintenance and 'emergency switching' to remove a hazard (generally to prevent injury to personnel).

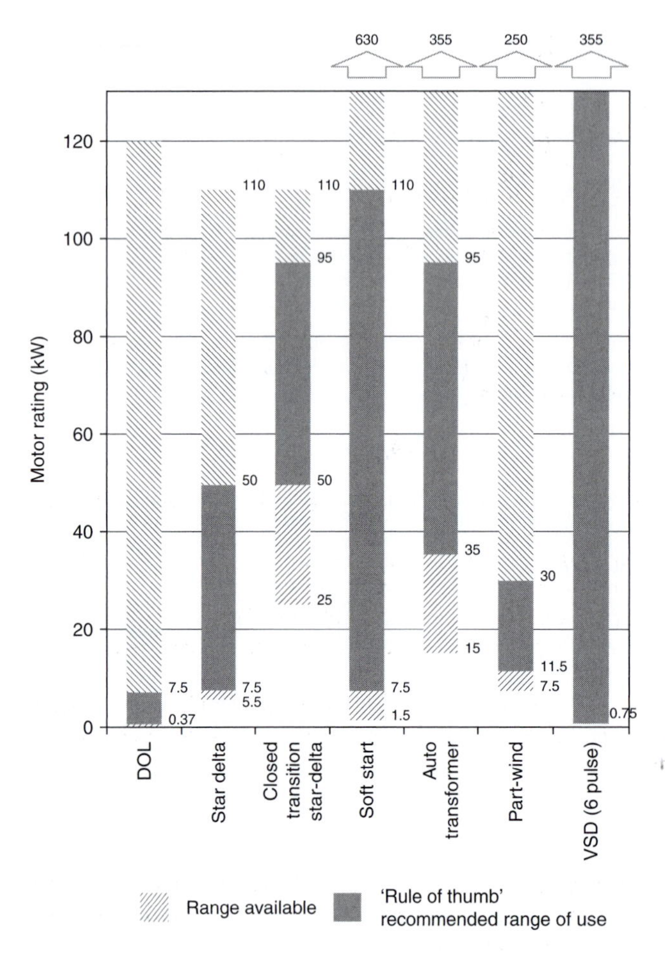

Figure 27.21 Diagram indicating available and 'Rule of thumb' recommended use of starters

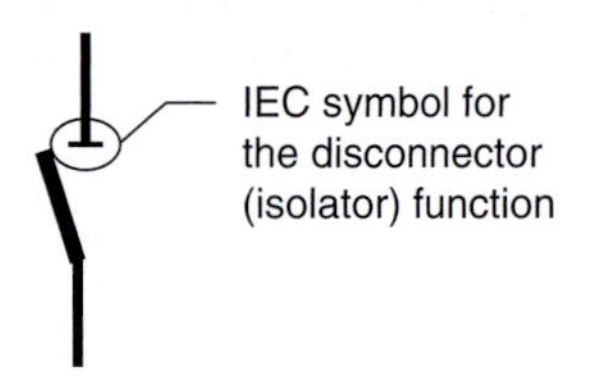

Figure 27.22 IEC symbol for disconnector

The function of isolation must be performed by a suitable device, the requirements of which are covered by BS EN 60947-1 (contact gap, reliable indication of contact position, means of securing the operator in the open position, etc.) (Fig. 27.22). Such devices must be marked with the IEC symbol for a disconnector and would be satisfied by either of the following functional devices (manufactured in accordance with BS EN 60947):

- Disconnector
- Switch-disconnector

- Fuse-switch-disconnector
- Some circuit breakers.

A means of interrupting the supply for the purpose of emergency switching shall be performed by a device requiring a single action resulting in the removal of the hazard by cutting off the appropriate supply. The device shall be manually operated directly interrupting the main circuit to the motor, the operating device should be clearly identifiable and preferably colored red. This may be performed by the device provided for isolation subject to it being installed in a readily accessible position, where the hazard might occur. Alternatively it is normal practice to install an e-stop (emergency power off button) adjacent to the drive connected in series with the motor starter, switching off the motor starter when operated.

Motor starter field connections/controls interfaces

Controls interfaces

Motor starters are required to interface with local controls and or central control systems, to provide a means of manual/automatic control and monitoring of motors.

Central control systems may include the building management systems (BMS) (refer to Chapter 28), energy management systems, fire detection systems or generator load control systems.

Tables 27.2 and 27.3 provide an overview of typical motor starter controls interfaces (input control and output signals) together with their application to each type of starter, including typical interfaces with a BMS system.

Motor run and trip conditions derived from the motor starter status are normally combined with another form of verification such as a flow switch or differential pressure switch to provide plant status indication and initiate auto changeover operations of duty/standby plant.

Sub-metering

Whilst sub-metering of energy is a statutory requirement for some buildings in the UK under Parts L2A and L2B of the Building Regulations, sub-metering is good practice and provides significant benefits which can be realised outside of the scope of the Regulations and outside the UK.

Sub-metering end use categories such as pumps and fans provide a means of identifying where and when energy is being wasted. The strategy for provision of sub-metering on a project should be developed as part of the metering strategy in accordance with CIBSE TM39 'Building Energy Metering'. The metering strategy is likely to entail being able to separately identify the energy consumed by the following end use categories:

- Ventilation systems
- Heating systems
- Cooling systems.

Electricity consumption (measured in kiloWatt hours kWh) may be arrived at using direct power consumption meters or estimating the power consumption using the run-hours of a constant known load. Reasonable provision sub-metering would be to provide additional meters such that the following consumptions can be directly measured or reliably estimated:

- Heating plant with a rated power input greater than 50 kW.
- Chiller installations comprising one or more chillers feeding a common distribution circuit with a power input greater than 20 kW.
- Electric humidification with a power input greater than 10 kW.
- Motor Control Centers providing power to fans and pumps with an input power greater than 10 kW.
- Individual fans or pumps with an input power greater than 10 kW.

Table 27.2 Typical motor starter controls interfaces

Control type	*Control/signal*	*Function(s)*
Manual controls	HAND-OFF-AUTO control switch	Selects whether the motor runs under manual control (subject to any safety interlocks being satisfied) is in a 'control' OFF mode (this does not satisfy the requirements for 'isolation') or automatically from a control signal input (e.g. time control or standby operation in the event of failure of the duty motor).
	Duty selection switch	Applicable to duty/standby pairs of motors and provides manual selection of the 'duty' motor/sharing run-hours between multiple drives, not applicable where a BMS is provided and duty sharing is scheduled by the BMS.
	E-stop (emergency power off) button	Switches off the motor as a means of emergency switching.
	Starter interlock	De-energises star–delta starters or bypass contactors of soft-starters in the event of motor local isolator being opened.
Automatic controls	Drive enable signal	Controls/safety interlock to ensure the status of any related control or motor is correct prior to energising the motor.
	Winding thermistor	Protects motor windings against over temperature.
Soft starters	Top of ramp (TOR)	Indicates when drive has reached full output voltage and used to close an external bypass contactors.
	Event relay	The quantity of output signals and nature of conditions available vary depending upon drive manufacturer and product range, the following are typical conditions available; overload, fault, high current, thyristor overload, locked rotor, underload, phase imbalance, high-/low-current warning, shunt fault SCR (rectifier).
	Analogue voltage or current outputs	Provides an input to a BMS/PLC or drives an external meter. The outputs available vary depending upon drive manufacturer and product range, the following are typical; motor current, main voltage, active power, reactive power, apparent power, motor temperature, SCR (rectifier) temperature, power factor.
Variable speed drives	Analogue voltage or current inputs	Motor speed control
	Analogue voltage output	Output frequency Output current
	Digital inputs	Run enable (deactivation stops drive) Constant speed (operates drive at a constant speed e.g. used for smoke extract control).
	Event relays	The quantity of output signals and nature of conditions available vary depending upon drive manufacturer and product range, the following are typical conditions available; start, run and fault.
Local time or optimum start controller	Run signal	Operates plant to provide environmental conditions to suit hours of occupancy.
Building Management System (BMS)	Run signal	Operates plant to provide environmental conditions to suit hours of occupancy.
Fire detection system	Fire shut-down signal	Stops selected plant in the event of a fire alarm condition, normally associated with ventilation systems. Can also be interfaced with Fire Service Ventilation Controls.
	Fire run signal (zoned to suit areas served by plant)	Runs selected plant in the event of a fire alarm condition irrespective of status of time control inputs, normally associated with zonal hot smoke extract systems and atria. Where variable speed drives are used fan speed ramps up to required volume. Can also be interfaced with Fire Service Ventilation Controls.
Fire Service Ventilation Controls	AUTO-OFF-EXTRACT ONLY control switch	Generally only used in conjunction with smoke extract and pressurisation systems. In AUTO fan runs normally or in response to signals from the fire detection system. OFF fan in 'control' OFF mode. EXTRACT ONLY fan runs irrespective of status of time control inputs. Where variable speed drives are used fan speed ramps up to required volume.

Table 27.3 Application of typical motor starter controls interfaces

	Inputs				*Outputs*			
		BMS				*BMS*		
Starter type(s)	*Control signal*	*Signal*	*AO*	*DO*	*Signal*	*Signal*	*AI*	*DI*
Direct-on-line autotransformer part wind	HAND 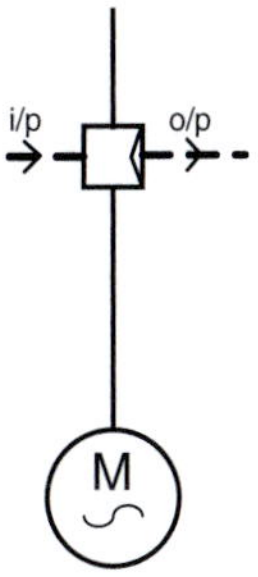							•
	OFF							•
	AUTO	Run		•				
	e-stop							
	Enable							
	Fire shut-down							•
	Fire run							•
					Motor run			•
					Motor tripped			•
Star–delta	HAND 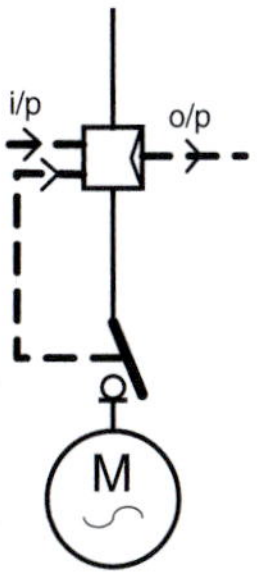							•
	OFF							•
	AUTO	Run		•				
	e-stop							
	Enable							
	Fire shut-down							•
	Fire run							•
	Starter interlock							
					Motor run			•
					Motor tripped			•
Soft-start	HAND 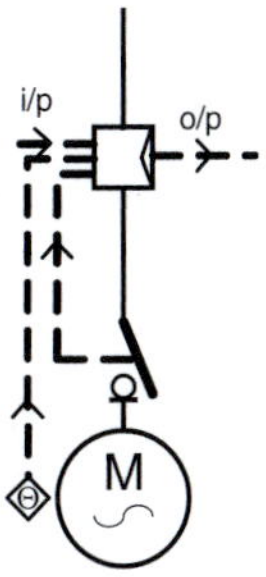							•
	OFF							•
	AUTO	Run		•				
	e-stop							
	Enable							
	Fire shut-down							•
	Fire run							•
	Starter interlock							
	PTC winding thermistor							•
					Motor run			•
					Fault			•
					Event	Refer to Table		•
					Output	27.2 for types	•	
Variable speed drive	HAND 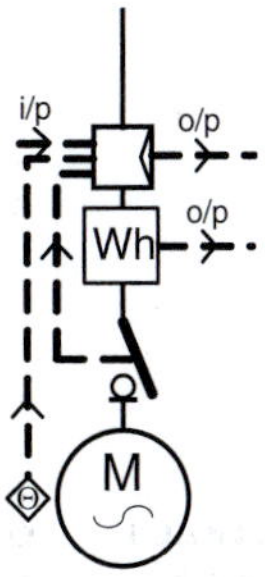							•
	OFF							•
	AUTO	Run		•				
	e-stop							
	Enable							
	Fire shut-down							•
	Fire run (constant speed)							•
	PTC winding thermistor							•
		Motor speed	•					
					Motor run			•
					Fault			•
					Output frequency		•	
					kWh			•

Table 27.4 Summary of motor control center and control panel characteristics

Type	*Examples of typical applications*	*Advantages*
Form 2 'Wardrobe' MCCs	General office ventialtion, cooling and heating systems	Least cost option. Most space efficient.
Form 4 'Cubicle' MCCs	Ventilation or cooling systems serving critical areas e.g. trading floors, apparatus or equipment rooms Environmental systems associated with process applications Hot smoke extract systems Pressurisation systems	Permits maintenance of a single drive without isolation of all motors supplied from MCC.
Local motor starter control panels (MCPs)	Form 4 applications plus the following: Systems utilising distributed intelligence with distributed outstations	Failure or maintenance of MCP generally only affects motor supplied from MCP.

On large buildings, automatic meter reading and data collection facilities should be provided, Part L2A and Part L2B of the Building Regulations makes this a requirement for buildings other than dwellings with a total floor area >1000 m^2. This requires power consumption meters to be provided with either a kWh pulsed output or analogue output connected to the BMS or an energy management system.

Motor control centres and control panels

For heating and air-conditioning applications motor starters and associated controls are combined within a single assembly referred to as a Motor Control Centre (MCC) or a Motor Control Panel (MCP). In MCCs the motor starters and controls for a number of motors are grouped together into a common assembly, MCPs generally relate to a single motor or a packaged set e.g. a duty changeover pair of pumps/fans or a booster set.

MCCs differ in construction to provide different degrees of separation between starters, busbars, and controls and BS EN 60439-1 'Low-voltage switchgear and control gear assemblies – Type-tested and partially type-tested assemblies' describes a system for classifying Forms of separation. The common Forms of separation for MCCs are Form 2 (referred to as wardrobe type) and Form 4 (cubicle type), MCPs are a type of Form 2 construction. Further details on each of these types of MCC is provided in the following sections and Table 27.4 provides a summary of typical application and advantages for each.

Form 2 'wardrobe' type MCCs

Form 2 'wardrobe' type MCCs comprise of a power section which houses the distribution equipment and the motor starters and a controls section which houses all of the controls supplies, controllers, control relays, etc., equipment within this section is generally supplied at 24 V a.c. Although the live parts of electrical equipment within the power section would be insulated, any maintenance on equipment within this section requires the entire MCC and all motors supplied from the MCC to be isolated first. Separation of the controls into a separate section or compartment allows maintenance or user adjustment of controls set points without affecting the operation of all motors supplied from the MCC (Figs 27.23 and 27.24).

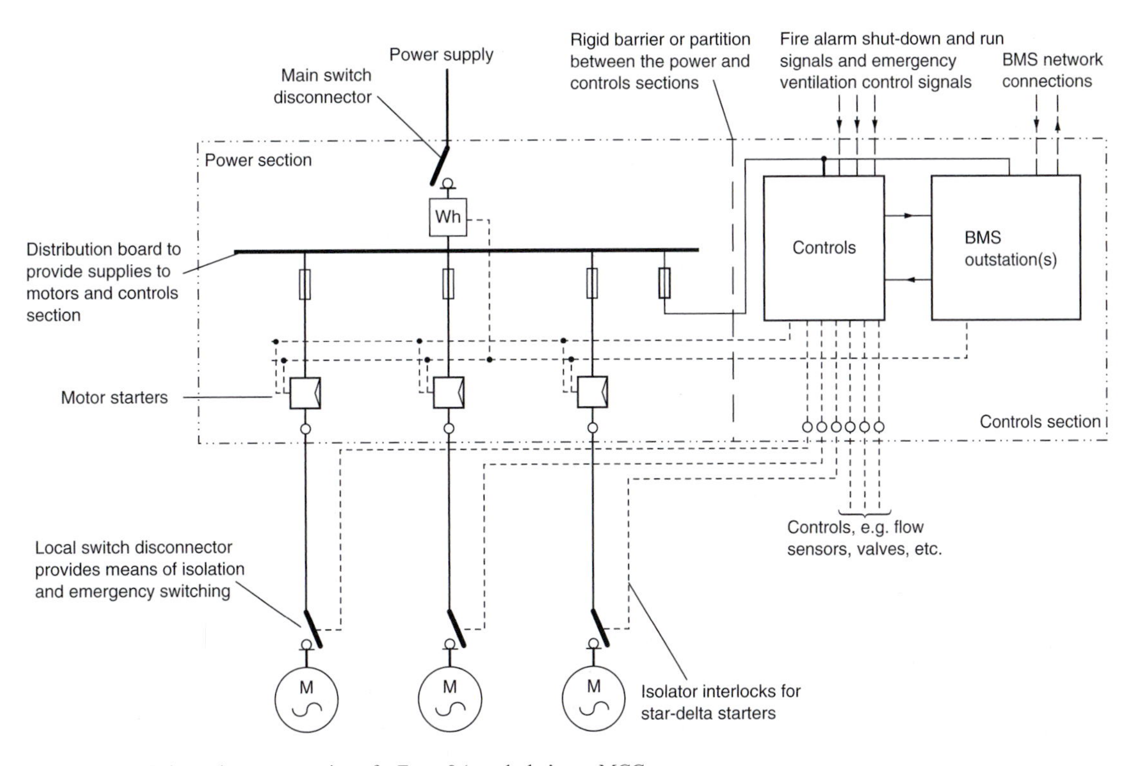

Figure 27.23 Schematic representation of a Form 2 'wardrobe' type MCC

Figure 27.24 Views of a Form 2 'Wardrobe' MCC with the doors closed and open (photograph courtesy of Integrated Control Systems Limited) (see colour plate section at the end of the book)

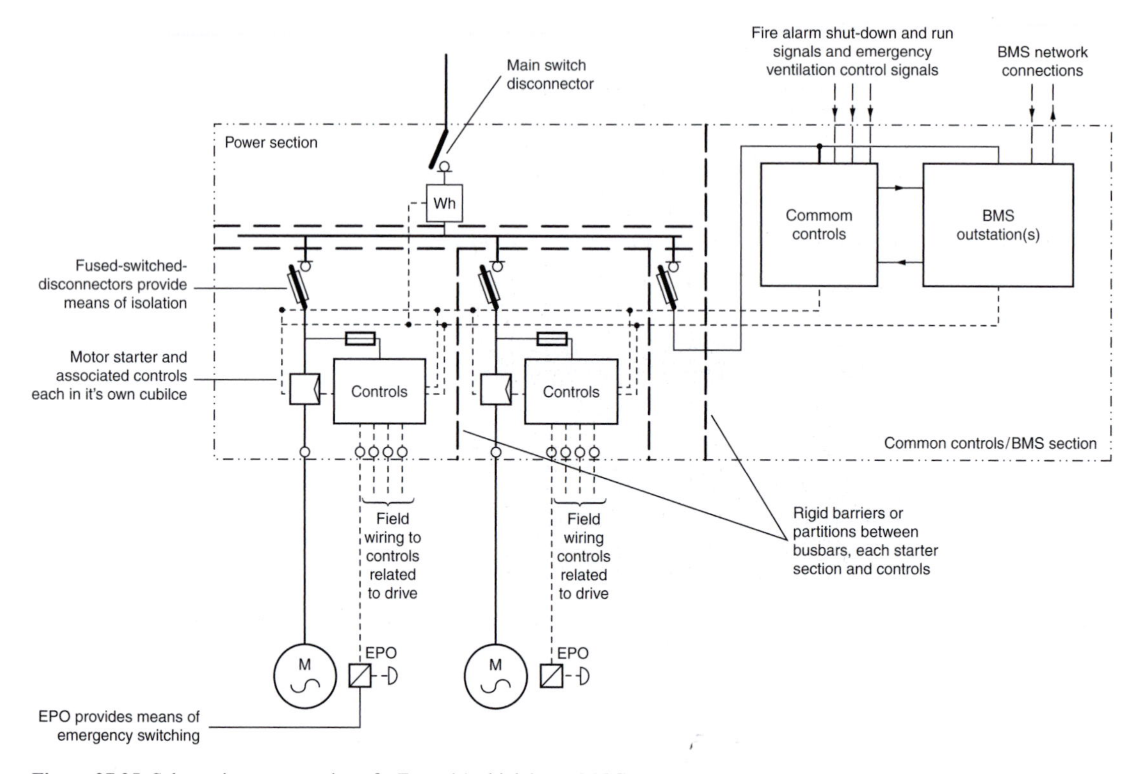

Figure 27.25 Schematic representation of a Form 4 'cubicle' type MCC

Form 4 'cubicle' type MCCs

Form 4 'cubicle' type MCCs are separated to provide separate sections (or cubicles) to house each motor starter together with its associated controls relays, etc. Again controls are generally supplied at 24 V a.c. and a common controls section is still provided to house common controls equipment such as power supplies and BMS outstations. Each motor starter is normally supplied from a fused-switch-disconnector (or circuit breaker) which satisfies the requirement for means of disconnection for the motor, this allows the use of e-stop buttons (EPOs) for each drive to provide the local means of emergency switching (Figs 27.25 and 27.26).

This arrangement allows maintenance to be performed on a single starter without requiring other motors supplied from the MCC to be isolated first.

For applications where the mean time to repair a motor starter is critical, each motor starter and its associated controls can be mounted to a withdrawable tray which is connected to the internal power and control wiring using a plug and socket arrangement. A typical Form 4 MCC motor starter withdrawable tray is illustrated in Fig. 27.27.

Different types of Form 4 MCCs are available (these are detailed in BS EN 60439-1) which define the termination and shrouding of the field power cables between the MCC and the motors. These types are relevant to applications where it is a requirement to be able to connect new motors to an MCC without having to isolate the MCC.

Local motor control panels

Local MCPs generally comprise of a single cubicle which houses a single or number of motor starters (the later would apply to an item of packaged plant) together with associated controls/controllers and a BMS outstation (Fig. 27.28).

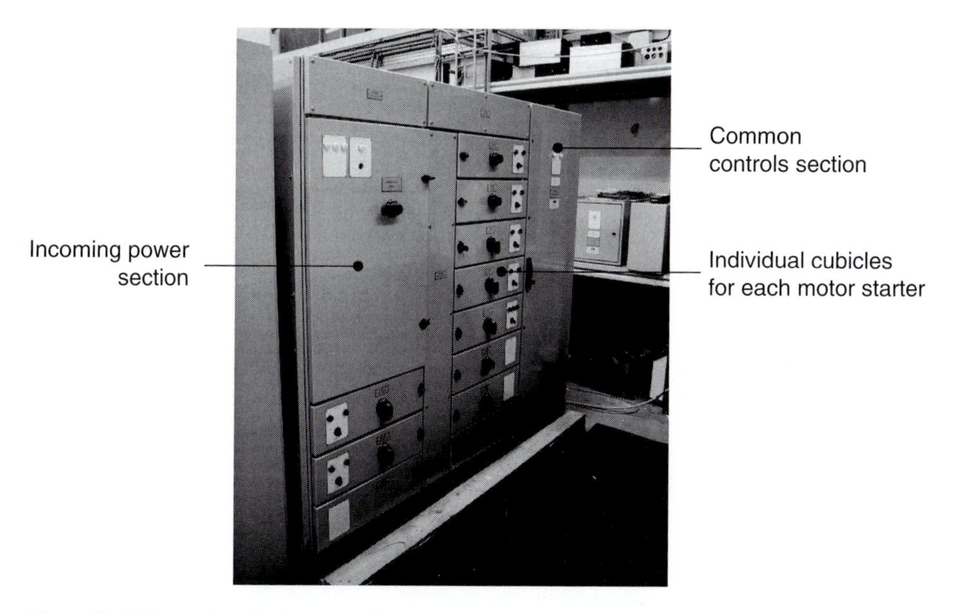

Figure 27.26 Front view of a Form 4 'cubicle' MCC with doors closed (photograph courtesy of Integrated Control Systems Limited) (see colour plate section at the end of the book)

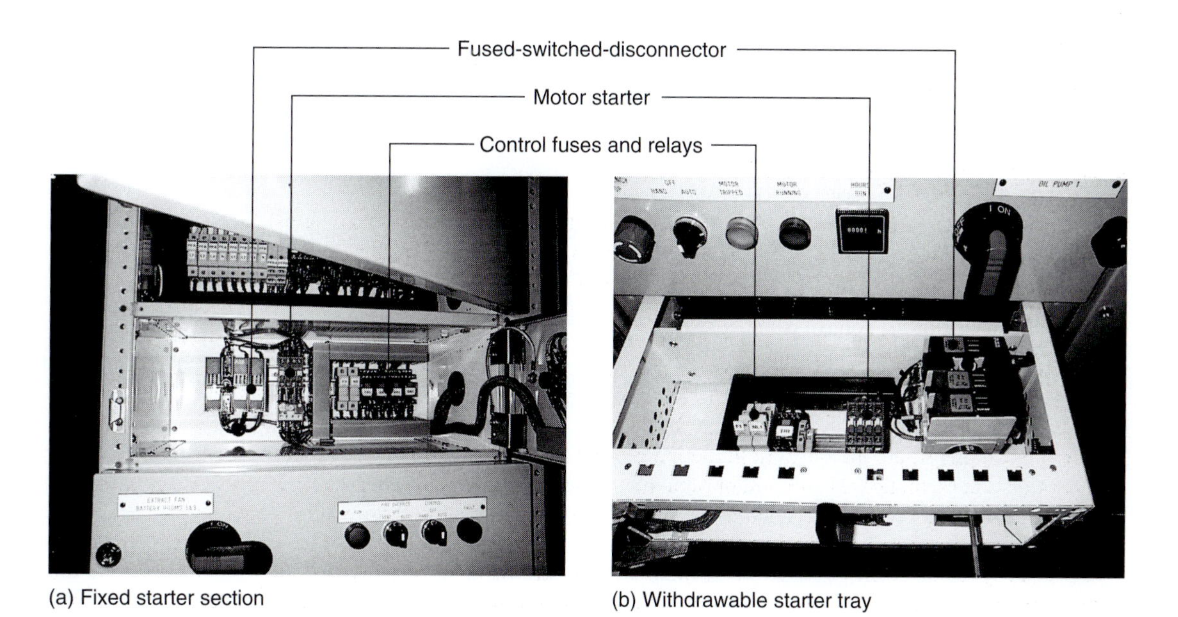

Figure 27.27 Form 4 MCC, fixed and withdrawable motor starter cubicles (photographs courtesy of Sapphire Controls Limited) (see colour plate section at the end of the book)

The control panel would be located local to the item of plant controlled and the power supply would be derived from a switchboard or distribution board (not indicated in the schematic representation), possibly dedicated to serve a number of local MCPs.

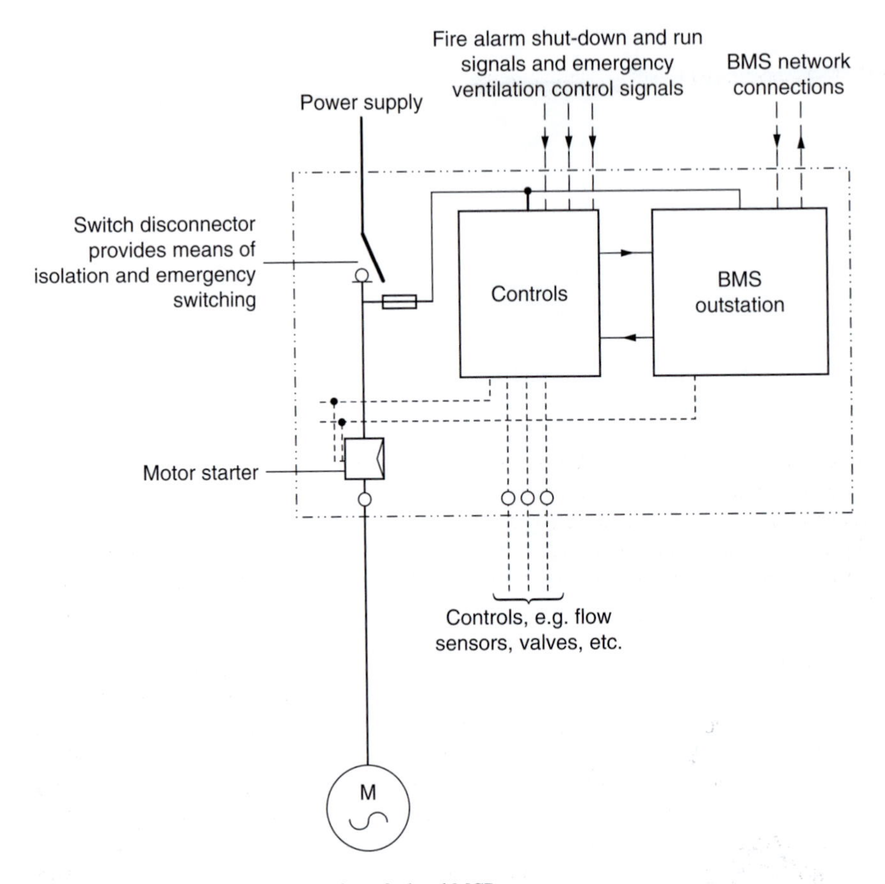

Figure 27.28 Schematic representation of a local MCP

This arrangement is particularly applicable where a BMS system utilises distributed intelligence with the use of intelligent outstations (refer to Chapter 28). When applied to a single motor local MCPs provide the same advantages as the Form 4 MCCs but offer greater resilience since only a single motor is affected by failure or maintenance of the control panel.

Chapter 28

Controls and building management systems

Controls theory and practice is an extremely broad and complex subject. It is possible here to give only a general overview of the more important principles and to outline some of the more common practices. For further details, the reader is recommended to consult either a specialist text or a more comprehensive digest.

The controls philosophy must be developed together with that of the systems design since, in the final installation, the two have to operate in harmony. However accurate the selection of equipment for a given application, if the controls are not correctly designed, installed, commissioned or maintained then the performance of the overall system is unlikely to meet the design requirements.

The popular adage *keep it simple* applies to controls perhaps more particularly than to any other aspect of heating and air-conditioning. There is no benefit gained if the designer produces a system of controls which is beyond the comprehension of those personnel responsible for the operation and maintenance of the completed installations. By way of example, a full building management system might be operated as no more than an expensive time switch if that is the level of understanding of the operators: a control system must be designed for the user and not to satisfy the ambitions of the designer. However, increasingly difficult targets for energy consumption often require sophistication in the control philosophy and the optimisation abilities of a well-designed control system can reap considerable savings for the operators. The designer must generally avoid the systems trying to maintain too precise a control of the environment, as this is almost always unnecessary and has large energy penalties.

There has been a very considerable development in the application of controls since the introduction of microprocessor-based control systems.

This has been particularly marked in the case of building and energy management systems, where significant benefits in the conduct of plant operation, maintenance and economy in running are available. The topic is dealt with towards the end of this present chapter.

The main objectives of a controls system may be summarised as:

- Safe plant operation
- Protection to the building and system components
- Maintenance of desired conditions
- Economy in operation.

It is essentially the desire to achieve energy savings that may lead sometimes to a proliferation of controls; nevertheless, such an objective is fundamental to good practice and may include:

- Limiting plant operating periods
- Economical control of space conditions
- Efficient plant operation to match the load/optimisation of part load efficiencies
- Reduction of standing losses.

The main objectives of the building energy management part of the system may be summarised as:

- Monitoring system performance
- Optimisation of system interrelationships with the building and its occupants
- Centralisation of fault reporting.

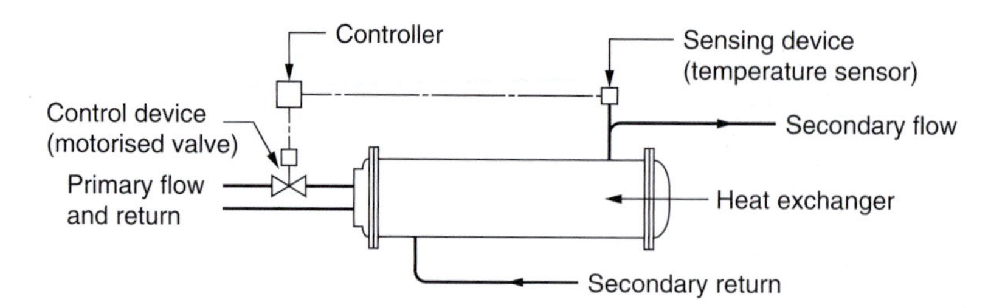

Figure 28.1 Simple heat exchanger controls

The importance of economical use of energy is emphasised by some of the basic provisions incorporated into recent legislation.

Elementary components

The nature of heating and air-conditioning systems is such that, for the majority of the period of operation, plant and system capacity will exceed demand and the order of this excess varies with time: steady state conditions may be assumed never to occur. It follows therefore that, if the plant were to be uncontrolled, the conditions in the occupied space would be outside the desired range in consequence of which some means of control is a fundamental requirement.

A simple control system, Fig. 28.1, comprises a *sensing device*, to measure the variable, a *controller*, to compare the measured variable with the desired set-point and to send a signal to the *control device*, which in turn regulates the input. In such a system, a temperature sensor (sensing device) in the flow pipe from a heat exchanger measures the temperature of the water (controlled variable) and signals the information to the controller. The controller compares the flow temperature with the desired temperature (set-point) and passes a signal to the control valve (control device) to open or close, thereby regulating the amount of heat introduced to the heat exchanger. This is an example of *closed-loop* control, where feedback from the controlled variable is used to provide a control action to limit deviation from the set-point. An *open-loop* system has no feedback from the controlled variable; an example of this is given later in connection with heating system controls.

System types

There are many classes of control system, but they may be grouped conveniently under headings:

- Direct acting
- Electric/electronic
- Pneumatic.

The simplest form of controller is *direct acting*, comprising a sensing element, say in a room or in the water flow and which, by liquid expansion or vapour pressure through a capillary, transmits power to a bellows or diaphragm operating a valve spindle. The most common example is the thermostatic radiator valve (TRV) which, when installed in the supply pipe to a radiator, convector or other heat emitter (as in Fig. 28.2), serves the purpose of an individual room controller.

Direct acting thermostats have little power and their control band can be somewhat wide, although they have been considerably improved in recent years. Direct acting thermostatic equipment gives gradual movement of the controlling device and thus may be said to *modulate*. In order to reduce the extent to which the TRV has to close to control the output of the emitter under part load conditions, it is normal to reduce the temperature of the heating water supplied to it under part load conditions. Direct acting controls are often used on critical functions where the processing time of a building energy management system (BEMS) or its complexity are required to be avoided (e.g. high temperature close off protection valves).

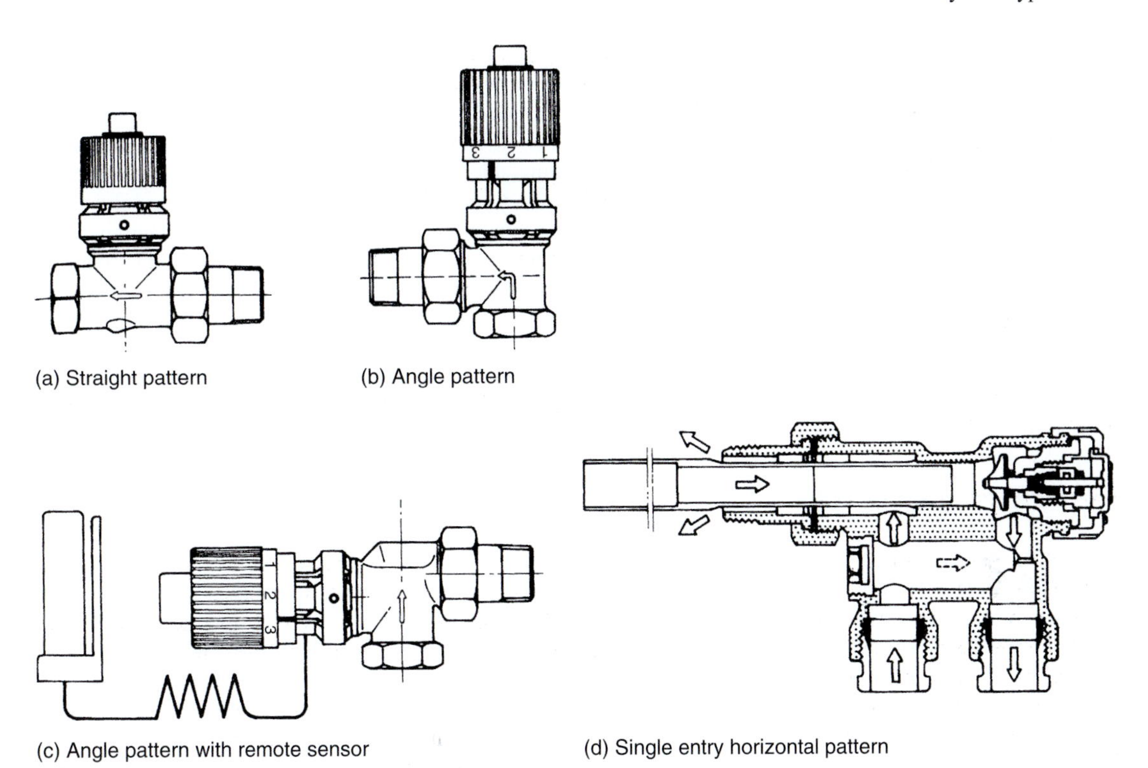

Figure 28.2 Thermostatic control valves for radiators, etc.

The most common control system is the *electric/electronic*, which may be found in domestic applications (thermostat and motorised valve) through to large commercial and industrial installations. The basic functions in an electric system are switching and resistance variation. Switching may be achieved by closing metallic contacts, using a tilting mercury switch, or a relay (electromagnetic switch) which is an electrical holding device that uses the magnetic effect produced by an energised/de-energised coil.

To produce a variable signal, a resistance in a sensing device such as a temperature sensor, changes with the temperature sensed. This signal is processed by the controller which then provides the signal and power, for example, to provide movement of a modulating control valve or damper. The simplest control device is the solenoid valve consisting of a coil within which an iron 'plunger' slides to give linear movement.

Electronic controls operate at 24 V or less and use smaller strength signals from the sensing elements; typically thermocouples or thermistors which have no moving parts. The circuitry incorporates amplifiers to magnify the signal or software to adapt non-linear characteristics. Electronic systems provide accurate control and, being free from mechanical parts, are very reliable. With the reduction in the cost of microprocessors, software logic, or 'intelligence' is often built into some field-mounted sensors themselves. Where software logic is built into the controller the system may be described as *direct digital control* (*DDC*).

For large installations, *electric/electronic* systems of control are by far the most common, and may be applied to perform any number of desired functions in a great variety of ways. The simplest electrical system comprises an on/off thermostat connected to an electrically operated valve or damper such that, when the thermostat calls for heat, the valve or damper is opened and, likewise when satisfied, the valve or damper is closed.

More sophisticatedly, modulating controls enable, for instance, a mixing valve to be so adjusted automatically that it finds some mid-position and so supplies the desired flow temperature to the system. For example, for a quick heat up in the morning, the control system may be so arranged as to call for full heat for the first hour or so according to external weather, after which the water temperature will be reduced proportionately to the external temperature.

The wiring to electrical controls and to the circuit boards for electronic controls, together with the associated switches, indicator lights and displays can become complex and depending on the manufacturers range it is presently usual to centralise these components in a pre-wired control panel. However with the advent of distributed totally intelligent controllers the industry is likely to move to a large number of individual control panels incorporating 'peer to peer' communication (i.e. being able to communicate directly with each other).

This panel may also accommodate remote temperature and humidity indicators and recorders and other instruments if required; motor starters may also be installed into such panels. It is essential that all components within and on the face of control panels are permanently and accurately labelled, and that all items required to be read for normal operation, can be read without needing to open the panel. With the decreasing cost of electronic processing power, many controls manufacturers are moving to *intelligent controllers* which can be connected by a data bus, whilst being located locally to the item being controlled, in order to reduce wiring (power and/or controls) costs. This allows the information for the operators to be centralised by the data bus.

Less popular these days *are pneumatic controls*. Pneumatic systems would in modern practice be considered only for industrial plant or when either tight closure or large actuator forces (say for rapid closing or opening) are required. This is due to the inherently safe 'stall characteristics' of pneumatic actuators which, unlike electrical or electronic ones can safely operate at full power against a load for long periods of time.

Sensing devices

Siting of sensing elements is critical to the achievement of good control. In pipework or ductwork, sensors must be so arranged that the active part of the device is immersed fully in the fluid and that the position senses the average conditions. Where necessary, averaging devices serpentined across the full cross-section of a duct should be used. Sensors should be protected from the radiant effects of local heat exchangers, such as heater batteries in an air handling unit.

In sensing space conditions, the device must not be in direct solar radiation or be located on a surface not representative of the space conditions such as on a poorly insulated outside wall. Local effects from heat sources, radiators or office equipment for example, will also give unsatisfactory results. Where necessary aspirated sensors should be used to improve sensitivity to *air* temperature.

Temperature

Thermal expansion of metal or gas or a change in electrical characteristics due to temperature variation are the common methods of detection. Figure 28.3(a) shows a simple bi-metallic type thermostat having closing point contacts; a two-wire type would, for example, open or close a circuit to stop or start a motor; a three-wire type would, typically, open one circuit and close another to start a motor or operate a valve in reverse direction.

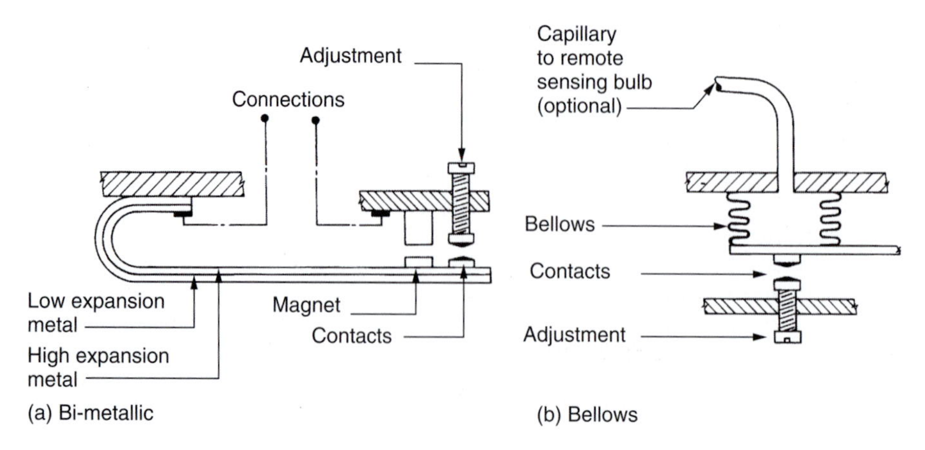

Figure 28.3 Electrical bi-metallic and sealed bellows thermostats

The sealed bellows type, Fig. 28.3(b), is filled with a gas, vapour or liquid, which responds to change in temperature by variation in volume and pressure causing expansion or contraction. Remote sensing elements operate on this principle, the remote sensing bulb being connected to the bellows by capillary tube.

Electronic sensing elements have no moving parts. The resistance bulb type, normally a coil of nickel, copper or platinum wire around a core, produces a variation in electrical resistance with change in temperature. *Thermistors*, which are semiconductor devices, also produce a change in resistance, but inversely with respect to temperature change such that resistance decreases with increase in temperature; the non-linear output may be corrected using linearising resistors in the circuit, or by using a 'look-up table' in the controller or sensor software. *Thermocouples* comprise two dissimilar metal wires joined at one end; a voltage proportional to the temperature difference between the junction and the free ends results.

Development of the simple *Wheatstone bridge* circuit, Fig. 28.4(a), is used in electronic controls. When the resistances are in balance there is zero output but as one or more of the resistances changes, the circuit becomes unbalanced which results in an output signal proportional to the change. A bridge circuit showing calibration and set-point adjustment is shown in Fig. 28.4(b).

Dead-band controllers have been developed to reduce energy use, on the principle that for comfort applications a variation in temperature, of perhaps 3–4°C, is acceptable. This type of controller has a wide dead-band through which a change in temperature produces no change in the output. Figure 28.5 shows an example of such an application.

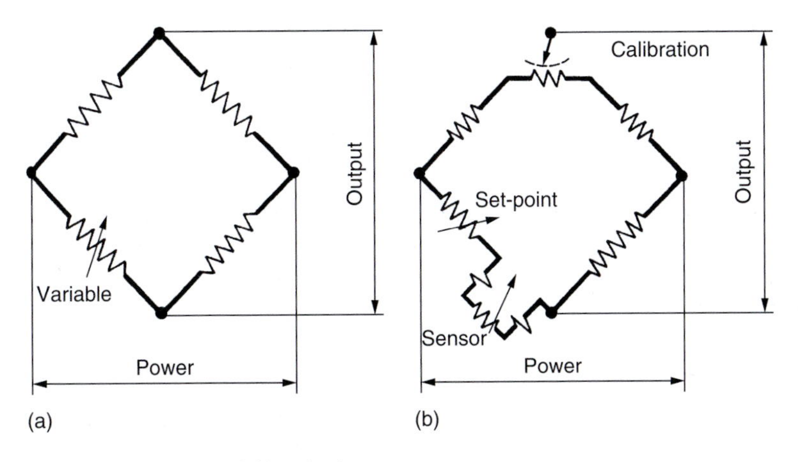

Figure 28.4 Wheatstone bridge circuits

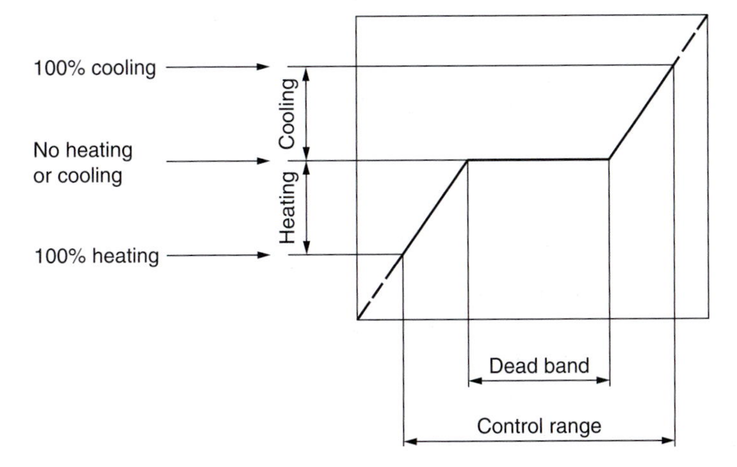

Figure 28.5 Dead-band operation

Humidity

Fabrics which changed dimension with humidity variation, such as hair, nylon or wood, were initially used. These days, coated polymer tape is now used. The change in resistance of the sensor is translated to a varying output signal by electronics integral to the sensor.

For electronic applications, solid-state sensors use polymer film elements ('bulk polymer') to produce variations in resistance or capacitance.

Wet and dry bulb thermostats working on a differential principle were once popular but are no longer used.

Humidity remains the most difficult parameter to measure cheaply and accurately and the design and control set-points and tolerances should reflect this uncertainty, especially where condensation is possible, as with chilled ceilings.

Pressure

Bellows, diaphragms and Bourdon tubes are typical of the sensing devices used and, of these, bellows and diaphragms acting against a spring are the most common. Such equipment can be sensitive to small changes in pressure, typically 10 Pa. The pressure sensing motion may then be transmitted directly to an electric or pneumatic control device.

In electronic systems, the diaphragm or the sensing element is connected to either a solid-state device having the characteristic that, when distorted, its resistivity changes; this is known as the *piezo-electric* effect, or to a capacitance or inductance transducer.

Flow

There are many methods used to detect fluid flow rate. In water systems the most common is to detect the pressure difference across a restriction to flow, such as an orifice plate or a venturi and extensive use is made of calibrated valves.

Various devices are available to sense air velocity in ductwork. In larger cross-sections, where the velocity may vary across the duct area, an array of sensing devices is required to establish an average value.

Paddle blade switches are used in water circuits, normally as a safety feature. For example, a flow switch of this type would normally be incorporated in the chilled water circuit connected to the evaporator of a refrigeration plant and be arranged in such a manner that the plant would not start until the switch sensed that water flow was established. By this means, damage to the equipment would be avoided consequent upon freezing of the static contents of the evaporator. However it should be noted that unless there is laminar flow throughout the pipe or duct, flow switches can be difficult to set up. Care needs to taken where systems are variable volume.

Enthalpy

Normally only used in air-conditioning plant, temperature and humidity sensors feed signals to a controller, the output from which is a signal proportional to the enthalpy of the air. Control devices are available which accept signals corresponding to the enthalpies of two air streams, typically outside air and exhaust air and, depending upon the relationship between these two values, a control action on dampers or heat exchangers is initiated. These are especially useful in the optimisation of recirculated air and the optimisation of 'free cooling'.

Control devices

The most common components used in the field are control valves, for water (or steam), and control dampers for air systems. The selection and sizing requires an understanding of both the devices and of the system characteristics. The system to be controlled and the associated flow rates would be sized at the peak design load, but would operate for most of the time at some partial load. The control device, therefore, has to provide stable control over the full range of operating conditions.

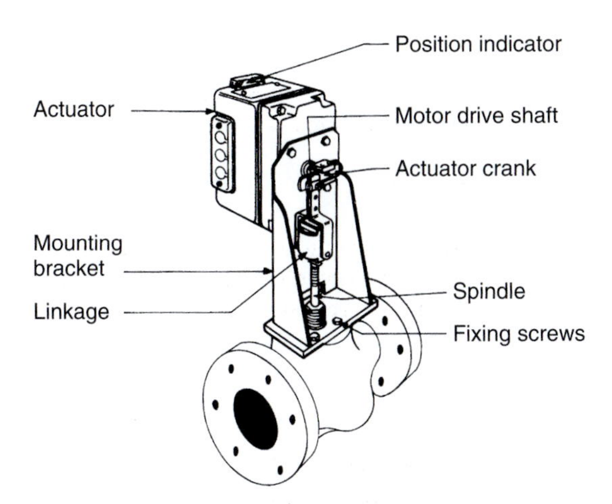

Figure 28.6 Electrical control valve

The movement of a valve or damper is determined by an actuator, which is the component that responds to the signal from the controller. The actuator characteristics which are of importance are *torque* (the ability to cause movement of the control device) and *stroke period* which is the period of movement between the limiting positions (open to closed and vice versa). Selection of the actuator type will depend upon the choice of control system.

Electric motors

For control mechanisms, these range between 5 and 50 VA and operate via reduction gearing to give a high torque/low speed characteristic: they normally operate using a single phase supply at 240 V or lower (normally 24 V) to suit the system. Two-position motors are of the unidirectional spring return, or unidirectional three-wire signal type, and are used when the speed of movement produced by a solenoid is too fast.

Modulating control requires a reversible motor that can be held at any position through the movement: either reversible induction or shaded-pole motors may be used. A typical arrangement driving a control valve is illustrated in Fig. 28.6. The motor may be provided with cam operated auxiliary switches to open or close at any set position of the movement. An auxiliary potentiometer may also be driven to provide a signal to another control device.

Some lower cost thermic actuators work on the basis of the expansion of an electrically heated wax capsule which then moves the valve spindle. Their slow response is appropriate for most building controls.

For the operation of dampers, the power required may be considerable (as discussed earlier under 'system types') and the motor must be chosen to suit. Sometimes the damper will require to be sectionalised, each section being worked by one motor with linkage. Again, such damper motors may be on/off, i.e. open/closed, or they may modulate to give settings in any intermediate position.

Solenoids

These devices operate on the electromagnetic principle with the armature directly connected to the control device, normally two positions and suitable for small sizes (Fig. 28.7). Modulating control may be achieved by applying a variable voltage to the coil, the spindle movement being proportional to the supply voltage.

Pneumatic actuators

Pneumatic actuators (Fig. 28.8) operate by means of a diaphragm or copper bellows connected to the valve spindle. The air supply connects to the top of the chamber (in the direct acting type) and has a small orifice plate, or needle valve, allowing only a small quantity of air to pass. The diaphragm top also connects to the thermostat or other sensing device and when the pilot valve in the latter shuts, the air pressure builds up on top of the diaphragm

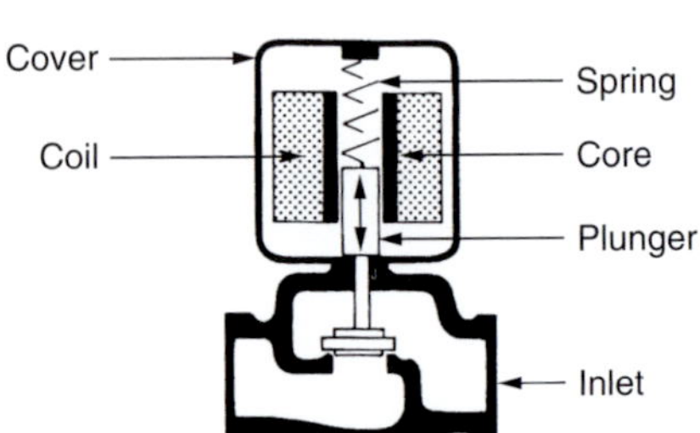

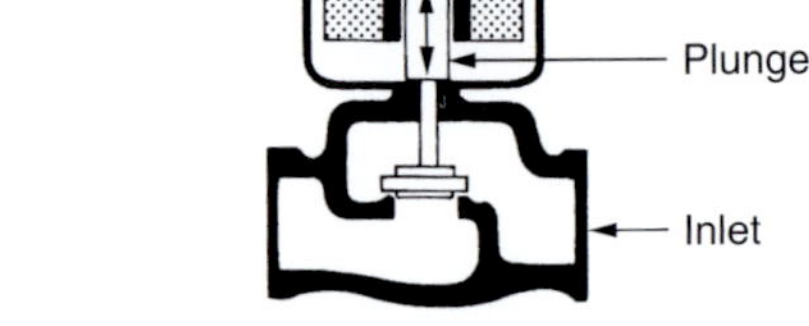

Figure 28.7 Solenoid valve

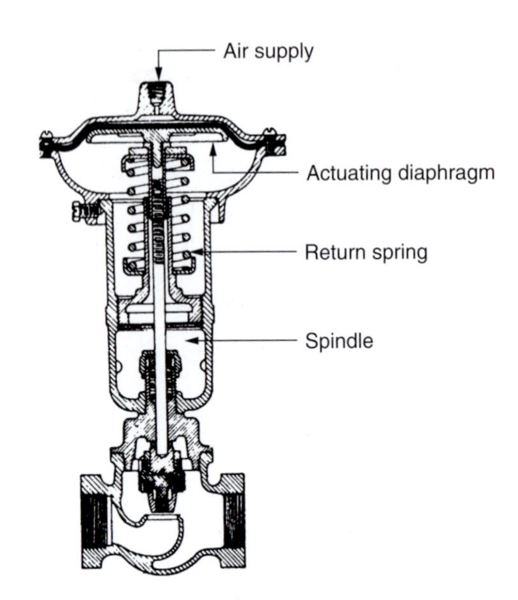

Figure 28.8 Pneumatic diaphragm valve

and depresses the valve spindle to close or (in the reverse acting type) to open the valve. When the pilot valve opens, the pressure on the diaphragm is released and the spring around the valve spindle forces the valve up again.

This form of actuator is relatively rarely used in modern building services, but can be useful where high torque or very tight shut-off is required. Where tight shut-off is required, the pneumatic actuator's ability to operate consistently on full power close off (as opposed to an electronic or electrical actuator), can be a major benefit.

The air from the pilot valve discharges to atmosphere. Any movement of the pilot valve causes the main valve to find a similar intermediate position, so giving a 'floating' control over the complete range. A pneumatic three-port valve operates in the same way.

Each valve should have a control cock and pressure gauge. These are usually centralised on a board along with the thermometers and other instruments, so that the operator may see the condition at a glance. The pressure gauges are connected to the diaphragm top in each case, so giving an indication as to whether the valve is open, shut, or in a mid-position. A main pressure gauge on the air supply will show whether or not the compressor is functioning.

Valves

Selection of the correct type of valve to provide for stable control requires detailed knowledge of the application and is specified in terms of the *valve characteristic* and *authority*. Valve characteristic, determined by the design of the plug, (Fig. 28.9[1]), is a function of valve lift and flow rate; the four main characteristics are shown in Fig. 28.10.

The choice of valve characteristic depends upon the output characteristic of the controlled equipment. Heat exchangers having water as the primary medium have a characteristic as shown by Fig. 28.11 which shows the heat output to primary heating water flow rate for a typical air heater battery. Steam applications, however have a very different characteristic, since heat output is effectively directly proportional to flow rate, a *linear* characteristic valve would normally be suitable.

An *equal percentage* valve would be most suitable for this type of application, producing a closer relationship between valve lift and heat output.

Valve authority is a function of the pressure drop across the valve and of the pressure drop across the remainder of the circuit as given by

$$N = P_1/(P_1 + P_2)$$

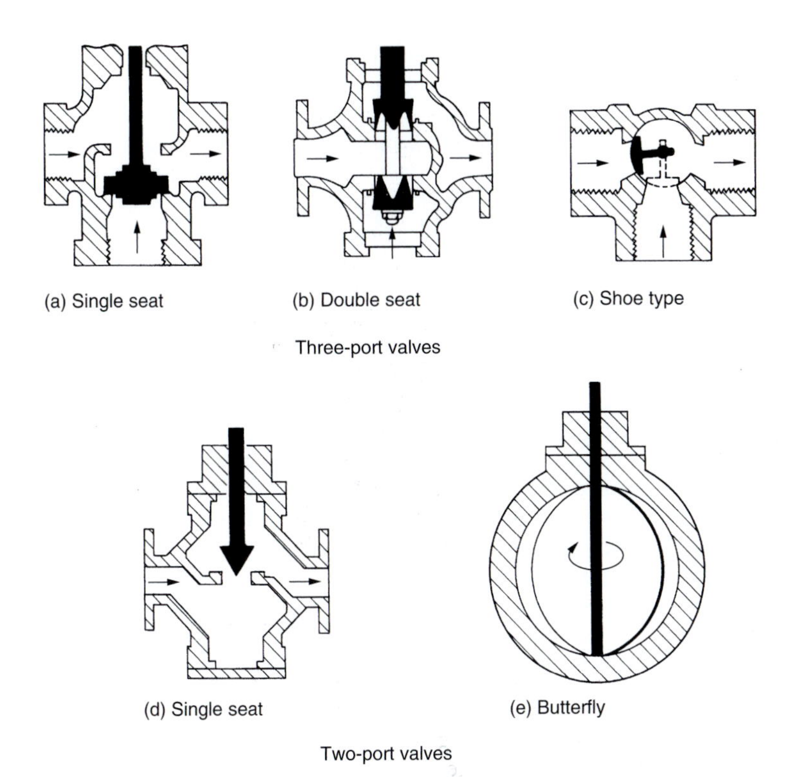

Figure 28.9 Typical valve types

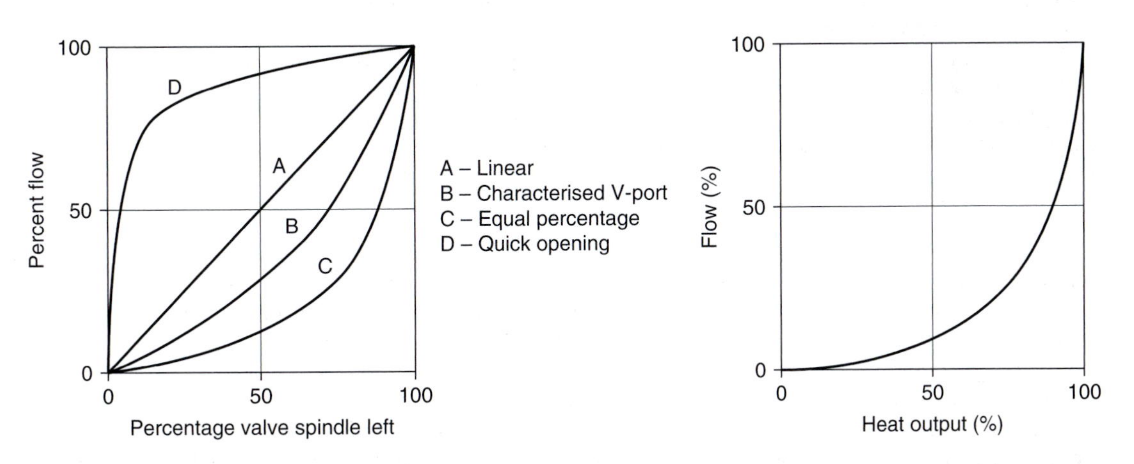

Figure 28.10 Control valve characteristics

Figure 28.11 Typical hot water air heater battery characteristic

where

N = valve authority
P_1 = pressure drop across the open valve
P_2 = pressure drop across the remainder of the circuit.

The most common applications are represented in Fig. 28.12. For stable control, the valve authority should be not less than the following values:

Mixing	= 0.3
Diverting	= 0.5
Throttling	= 0.5

Application of the above criteria would normally lead to control valves being selected at one, or maybe two, pipe sizes smaller than the pipeline diameter (depending upon the fluid temperature difference being used). Three-port valves should always be fitted in the *mixing* mode, i.e. with two inlets and one outlet. As illustrated in Fig. 28.12, such valves may be applied to either a mixing or a diverting duty but many of the patterns available are not suitable for use in the *diverting* mode, i.e. with one inlet and two outlets.

Selection of two-port valves follows similar criteria as for the three-port variety. However, more detailed consideration is necessary in connection with the remainder of the circuit, because as the valves close the flow rate and pressure distribution around the circuit will vary: this is particularly important since the installation is likely to have many two-port valves in the same pumped circuit. As the control valves throttle the flow through the controlled circuits the pump, if operated at constant speed, would tend to develop an increased pressure which in turn would affect the authority of the valves. Added to which, this increase in pressure might affect the ability of the valves to open or close fully and it may be necessary, therefore, to use either variable flow pumping or flow bypass valves with a system controlled by two-port valves, see Chapters 11 and 29.

Steam control is a special application of the two-port valve. As the valve closes, the area under the seat decreases causing an increase in steam velocity, which tends to counter the decrease in volume flow. The limiting condition, for practical purposes, is when the downstream absolute pressure is 60 per cent of the upstream pressure. When sizing a steam valve, therefore, the pressure drop at full load is taken as 40 per cent of the upstream absolute pressure, except when this pressure is less than 100 kPa, when impractical conditions arise

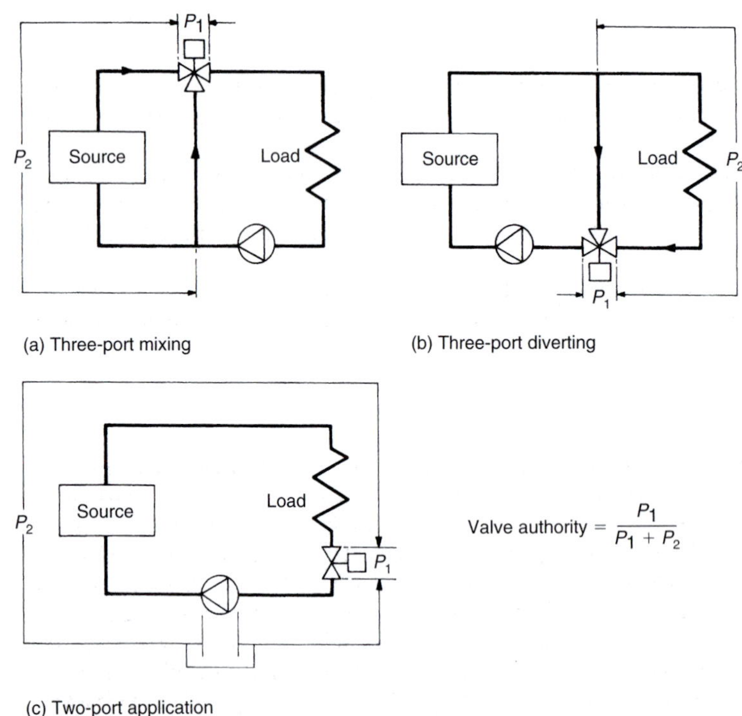

Figure 28.12 Basic control valve applications

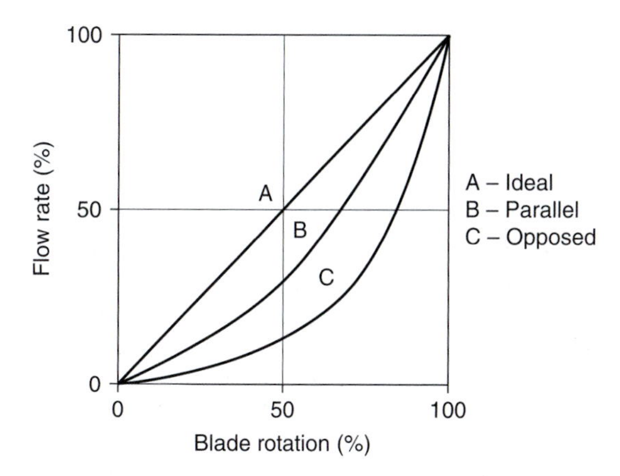

Figure 28.13 Control damper characteristics

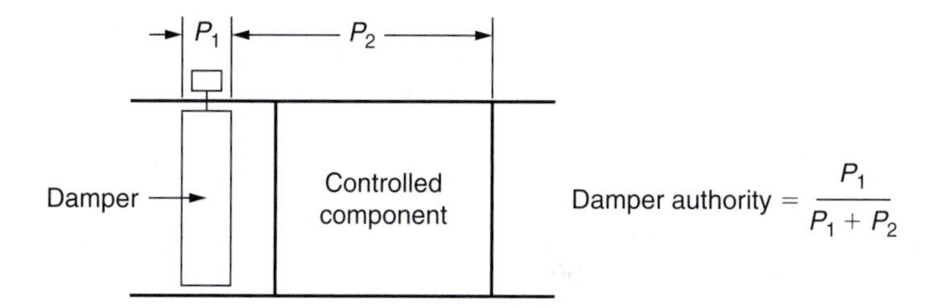

Figure 28. 14 Basic control damper application

from this rule of thumb and a smaller pressure drop is taken, with resulting loss in effective control; reference to manufacturers' data is necessary in these circumstances. The heat exchanger surface area must, in consequence, be suitable for a steam supply lower than the mains pressure.

Butterfly valves are, in effect, two-port control devices, but have poor control characteristics equivalent to that illustrated as *quick opening* in Fig. 28.10. Application is normally limited to two positions (open/closed) but standard equipment cannot be guaranteed to give total shut-off: special valves with liners are available for this purpose.

Noise emanating from control valves should not be overlooked in the selection of components and in the design of the configuration of connecting pipework. The cause may be either unique or a combination of mechanical vibration, turbulence or cavitation.

Dampers

Control of air flow follows the same principles as for water systems, and the function may be either modulating or two (or more) positions. There are two forms of control damper, parallel and opposed blade type, and their inherent characteristics are shown in Fig. 28.13. Dependent upon the damper authority (defined above for valves) the damper characteristic will vary when installed in a system (Fig. 28.14). The closest to linearity is obtained where, for opposed blade dampers, the authority is 0.05 and, for parallel blade type, 0.2. To limit the pressure drop in a system, therefore, opposed blade dampers, sized for an authority of about 0.05, would normally be selected.

When used for controlling outside/exhaust/recirculating air in an air handling plant (Fig. 28.15) the outside air damper (A), and the exhaust damper (B), would normally be opposed blade type, dimensioned at the duct size. However, for abnormal applications where, for example, high resistance plant components are to be installed, a check should be made from first principles. The function of the recirculation damper (C) is to impose a resistance in the bypass duct equivalent to the static pressure difference between the supply and

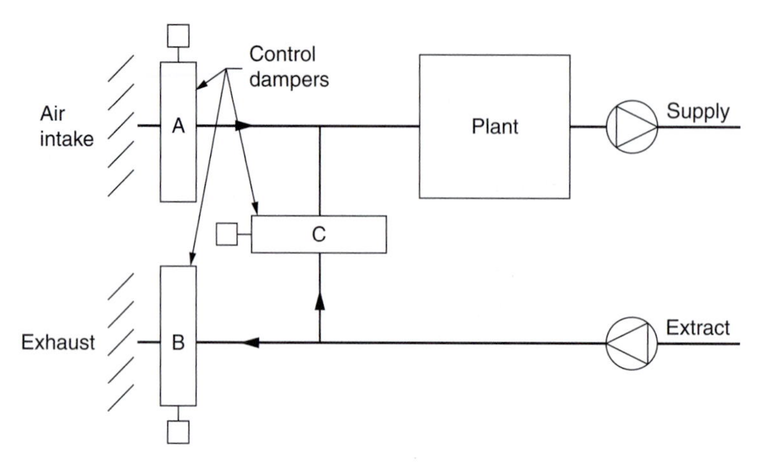

Figure 28.15 Outside/exhaust/recirculating air control

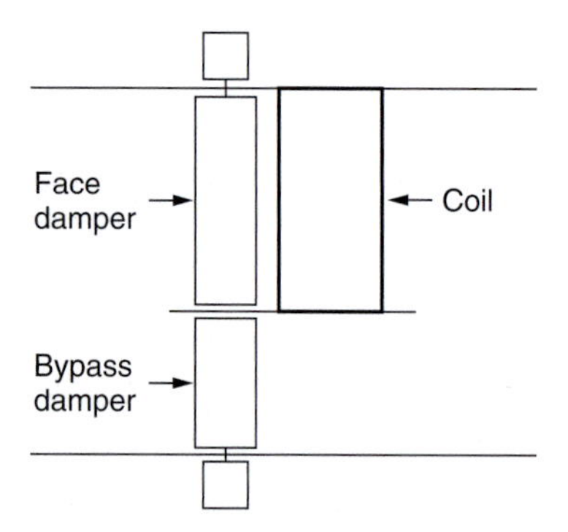

Figure 28.16 Face and bypass control

extract ducts and to have a characteristic to complement that of the outside air damper in maintaining constant the mixed flow to the fan. This damper will of necessity have a high authority, up to 100 per cent, and since the characteristic of the parallel blade type is better at higher authority (pressure drop in this case having no influence on running costs) the parallel blade type would be preferred. Where the pressure to be absorbed by the damper is high, due to the operating conditions of the plant, an additional resistance, such as a perforated plate, may be incorporated in the bypass to increase the authority of the damper.

Another common application of control dampers is to modulate the rate of flow through a cooling coil to provide what is termed *face and bypass* control. This is most commonly used where high amounts of dehumidification are required of a cooling coil relative to the sensible temperature decrease required across it. The principle is illustrated in Fig. 28.16. In sizing the dampers, the same logic as above may be applied, to provide constant pressure drop across the combination throughout the range of operation and hence not affect the flow rate in the system. As for outside/recirculating air dampers, the characteristics of the face and bypass dampers must complement one another. It would normally be the case that the face damper would be the same size as the coil, but the bypass damper will usually be smaller than the associated recirculation duct.

Controller modes of operation

There are various ways in which a controller can cause a control device to operate in response to a signal from a sensing device. The most common modes that may apply are described here.

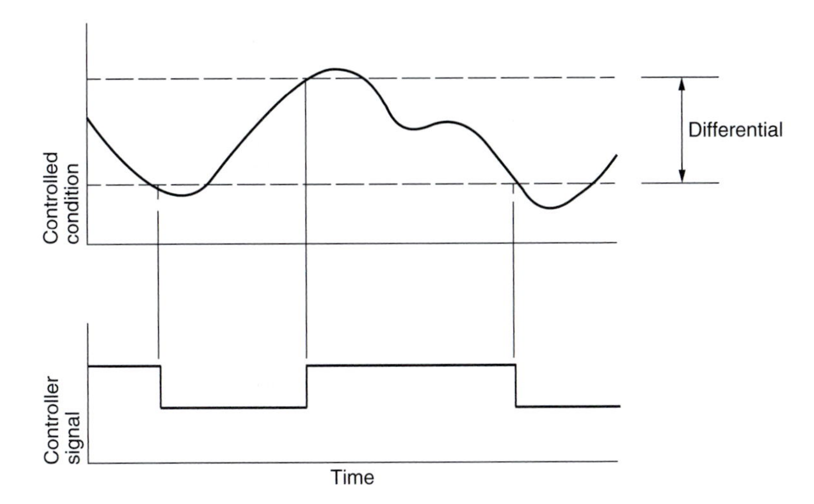

Figure 28.17 Two-position control

Two-position control

A typical application is on/off switching, in which the sensing device, perhaps a simple switching thermostat, provides two signals, e.g. opening contacts below a set-point and closing them above a set-point. The device may be arranged to operate in the opposite way and this is then called reverse acting.

The interval between the switching actions, an inherent characteristic of the device, is normally referred to as the differential gap. Where a differential gap is too wide, an accelerator (in the case of a thermostat, a heating element energised when the thermostat calls for heat thereby anticipating the response) may be used to reduce the range, or swing, of the controlled condition. The swing between the controlled space temperature may be greater than the differential gap in the sensing device due to slow response of the controls and the thermal inertia inherent in the building. Figure 28.17 illustrates the performance of such a method of control.

On/off control would give quite acceptable results where the controlled variable has large thermal inertia, such as a hot water service storage calorifier or a space heated by a mainly radiant source, where close control is not critical.

Step control

It is sometimes necessary to operate a series of switching operations in sequence from one sensing device. For example, when multiple refrigeration compressors have to be started in turn, with increasing cooling load as sensed by a change in chilled water return temperature, or when air heating is accomplished by an electrical heater with multiple elements switched in steps, see Fig. 28.18.

Proportional control

With this form of control the output signal from the controller is proportional to the input signal from the sensor. Figure 28.19 illustrates proportional control action from a start condition and shows that, initially, the controlled variable will be driven towards the set-point at such a rate as to cause overshoot. On the return cycle, the overshoot will be less and this oscillation will continue until stable conditions exist; but if the system is unstable the hunting will continue indefinitely. With proportional control there will nearly always be an offset from the desired set-point. The proportional band, the deviation in the controlled variable necessary to produce the full range of control action, may normally be varied on the controller.

Proportional control is to be preferred to an on/off approach where the thermal capacity of the controlled variable is low and where the primary rate of heat exchange is fast, as in the case of air heating and cooling batteries and non-storage water heat exchangers.

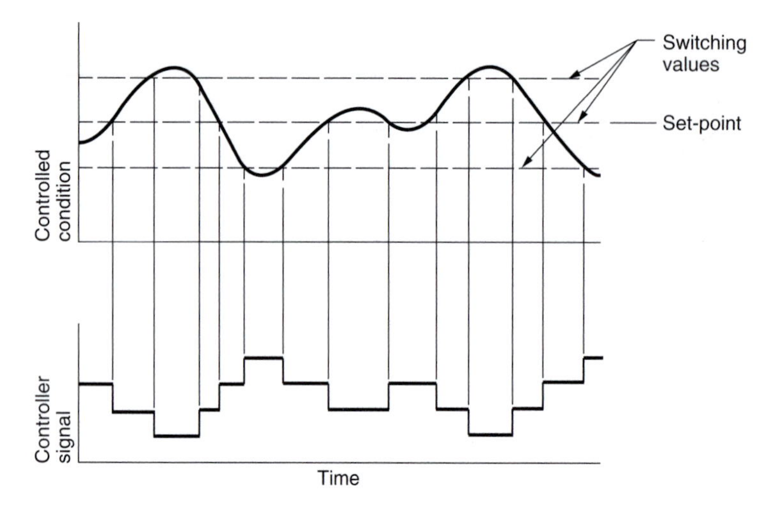

Figure 28.18 Step control

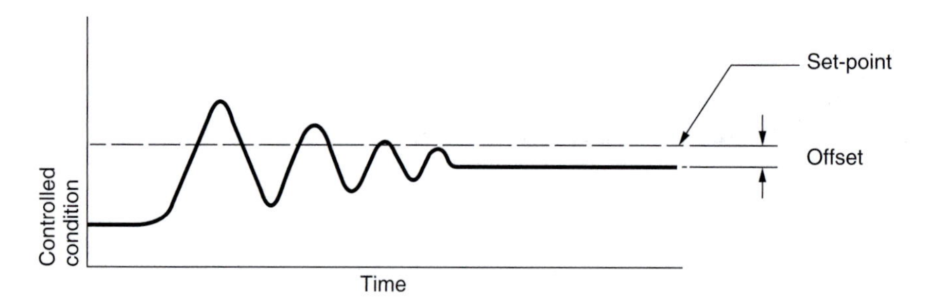

Figure 28.19 Proportional control

Floating control

With floating control, there is normally a neutral zone around the set-point, within which no control action occurs: the control device remains in the last controlled position. When the variable moves outside the neutral zone, a signal causes the actuator to move at a constant rate in a direction corresponding to whether the variable is above or below the set-point, see Fig. 28.20. An example would be a control device motor running at a constant but slow speed; the full stroke being at least 2 minutes. A multi-step controller might also be used to initiate a multi-speed actuator action. Other forms of modulating control avoid offset and may be preferred to proportional or floating control in some applications. The use of floating control can give benefits in terms of lower energy use.

Integral control

Seldom used alone, this is an important addition to other forms of control, particularly to the proportional mode. With integral action there is continuous movement whilst deviation from the set-point persists such that the *rate* of movement is a function of the amount of deviation from the set-point.

Derivative control

This mode involves a further development of integral action such that the controller output is a function of the rate of change of the controlled variable. This form of control, like the integral mode, would not normally be used alone, but in combination with others.

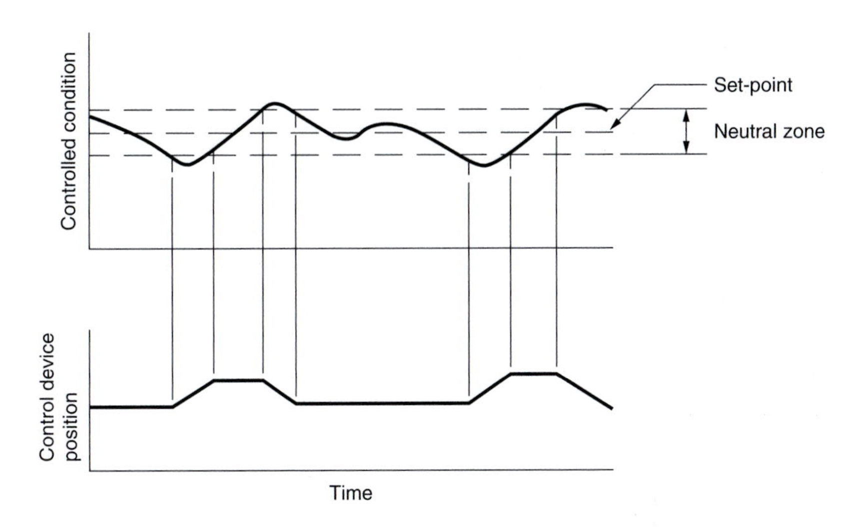

Figure 28.20 Floating control

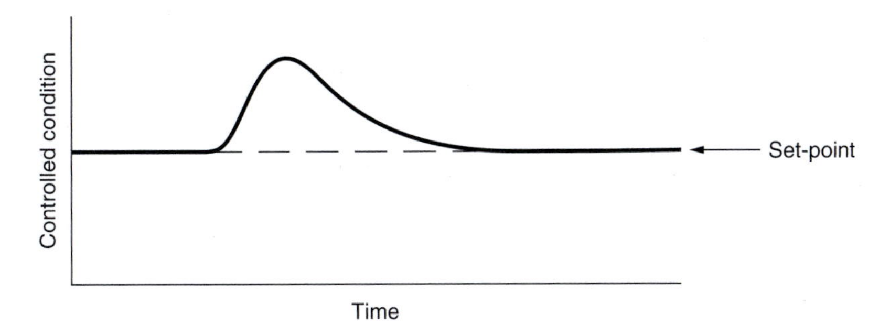

Figure 28.21 Proportional plus integral control

Proportional plus integral

Sometimes referred to as proportional with reset (or abbreviated to PI) this combination gives stable control with zero offset, as shown in Fig. 28.21. So long as there is deviation from the set-point, the controller will continue to signal a change until zero error exists. This approach would be applied, for example, to space temperature control in circumstances where the load fluctuated widely over relatively short periods of time and where, in consequence, a simpler type of controller could not produce a stable condition without the proportional band being wide beyond acceptable limits. In addition, this mode is used more generally for applications where close control is required.

Proportional plus integral plus derivative

Abbreviated to PID, this mode of control would be used where there are sudden and significant load changes and where zero offset from the desired set-point is required. There is seldom a case for such control in heating and air-conditioning work.

Systems controls

Common to all systems is the requirement, often by legislation, that they be started and stopped by some form of time switch. There are two basic types; the simple on/off switch and optimum start (and stop) control.

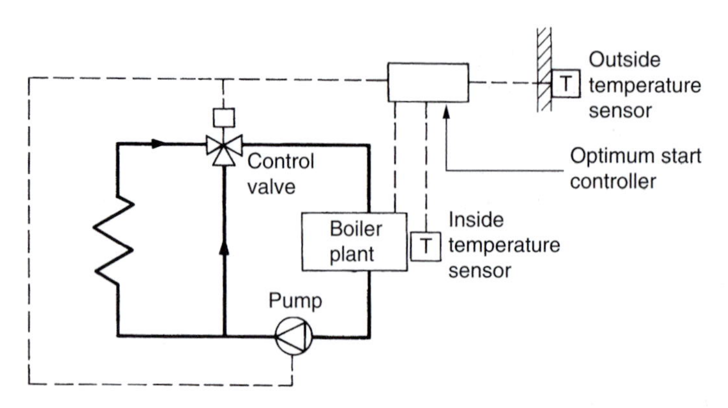

Figure 28.22 Optimum start control system

On/off time switch

Simple devices of this type are suitable for small systems, those where the number of days of use is limited, such as church heating, or those where the thermal inertia of the building is so high that little practical purpose would be served by varying the system start-up time with respect to external temperature. Time switches for other than the simplest domestic scale buildings would normally be of the 7-day type, enabling each day of the week to be programmed separately. Microprocessor-based systems can provide for a complete year, including bank holidays, daylight saving change over, etc., to be entered to memory at one time. To reduce the heat-up period, an additional feature to time switch control, known *as fixed time boosted start*, is the ability to run the system at maximum output for a period until either the desired internal space conditions are achieved or the end of a fixed preheat period is reached, at which time the controls change to normal mode under thermostatic control.

Optimum start control

This system serves to delay the start time of a heating or air-conditioning system until the latest possible time to give the shortest preheat period, normally at full output, which will achieve the desired conditions at the start of the occupancy period. The system of control monitors the inside and outside temperatures, see Fig. 28.22, and takes account of the thermal response of the building and the system. Modern equipment is self-adapting, in that it can monitor its own performance (in achieving correct conditions at the start of the working period) and take corrective action to improve these from one day to the next. Optimum stop control is also available to switch plant off as soon as possible, consistent with maintaining acceptable temperatures at the end of the occupied period. This is quite satisfactory for heating systems, but there is concern over its application to all air-conditioning systems where stopping the plant in advance of occupants leaving the building also stops the supply of outside air to the space.

The characteristics of the optimum start principle are shown in Fig. 28.23 from which it can be seen that the switch 'on' time is related to the fall in internal temperature. It also shows that the outside temperature affects the rate of fall of inside temperature, and also the rate at which the heating system can heat up the space. Generally, wet heating systems have a slow response compared with air-based systems, and in consequence need to be switched 'on' earlier. In hotter climates optimum start may be applied to space precooling, but this facility is generally less applicable in the climate of the British Isles.

Regardless of the method employed for system start-up, separate frost protection is an essential feature for all systems in order to protect the conditioned spaces from condensation risk (normally 10°C is the lowest acceptable temperature) and to protect plant components from freezing. As moving water is less prone to freezing than when static, it is normal where parts of the system are subject to outside temperatures, that the first stage of a frost protection strategy, is to switch on the circulating pumps.

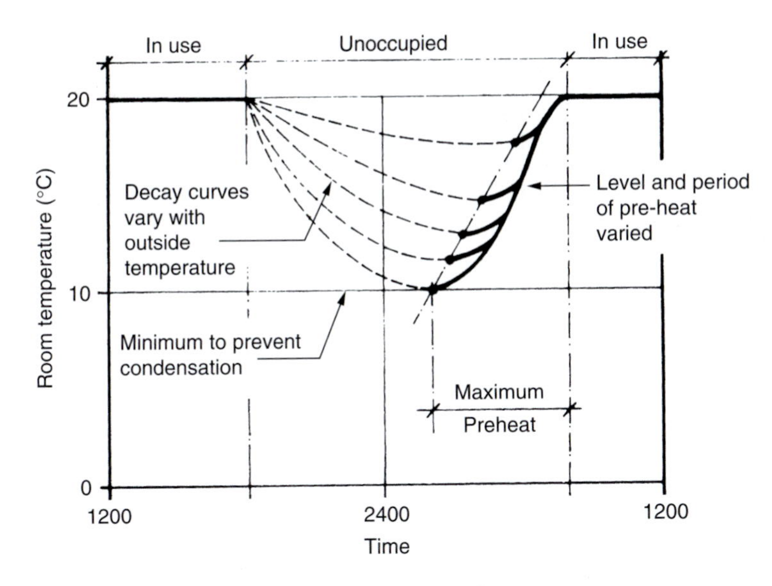

Figure 28.23 Characteristics of optimum start control

Heating system controls

Figure 28.24 illustrates a simple heating system with optimum start, plus conventional heating *compensator* controls. In addition to optimum start provisions, as shown in Fig. 28.22 this includes temperature sensors to sense external temperature and that of the main water flow to the system. By means of calibrating mechanisms, the temperature of water leaving the control valve is adjusted to suit the outside air temperature. In cold weather, the balance between the two temperature sensors will require the water to be at a high temperature and, conversely, in warm weather balance will be achieved with water at a lower temperature. At any point between the two extremes the water temperature is adjusted proportionately to the outdoor temperature. Refinements have been introduced into this system by way of a heating element in the external unit, so that the effect of wind can be taken into account. The compensator controller effectively resets the desired temperature set-point for the system flow water appropriate to the outside conditions obtaining. If there are a number of main circuits to various parts of the building, it is usual for each to have its own compensator, circulating pump and mixing valve. This is an example of *open-loop control*, where there is no feedback to the water temperature from the controlled variable, namely the space temperature. It would then be usual to provide individual *closed-loop control* of output at the area served. With modern digital control, a closed-loop option which can utilised (with suitable boilers) is for the boiler controller to regularly (say every 15 or 30 minutes) interrogate the controllers on the individual loads served (i.e. heater batteries or emitters) and to dynamically schedule the boiler water temperatures to the lowest which can serve the loads at that time. This not only reduces standing losses to their absolute minimum, but also forces the boilers into the deepest condensing possible at any one time.

Zone control

Control of the heating system for a building or series of buildings by *zone control is*, based, in effect, on the assumption that certain areas have common characteristics and hence similar heat requirements. For instance, in a rectangular building with north and south aspects on the long face, it may be that all rooms on the south aspect will have similar heating requirements and may be served from one zone, likewise north facing rooms may be served from a second zone. The southerly aspects will require less heat than the northerly at times and therefore will need independent mixing valve control to provide different flow water temperatures. Similarly, in a tall building, due to the greater exposure of the upper floors, it might well be desirable to zone the upper

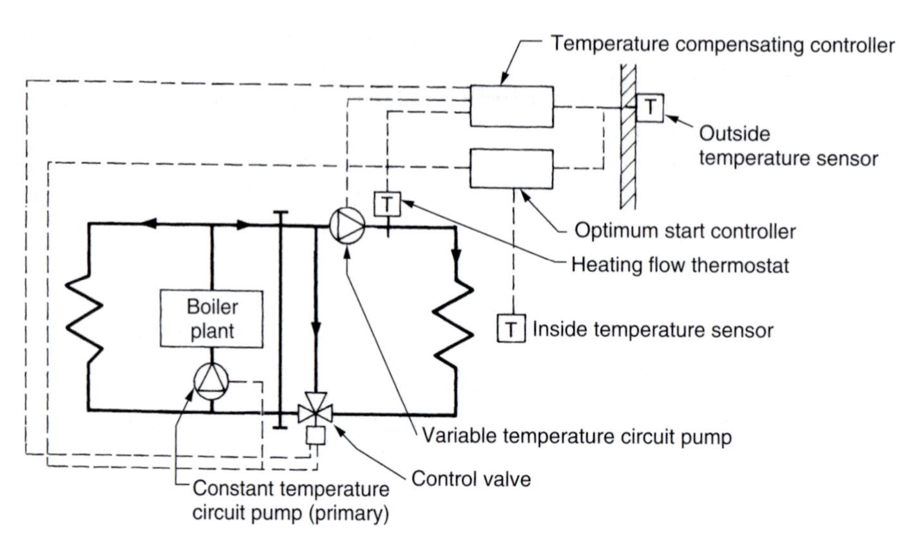

Figure 28.24 Basic heating

floors separately from the lower ones. Further zoning may be required depending on the elevational treatment, leading to four or more zones for the building.

With the increasing cost of energy/environmental awareness and the reducing cost of terminal controls, such coarse control is rarely in practice used alone.

Individual room control

Present good practice is to be achieved by control of the heat emitter in each room. In buildings where there is significant variation in room loads, due perhaps to machinery or occupancy levels, individual room control will be required to enable uniform temperatures to be maintained. The alternative is, of course, to assume that the occupants can control the output from heat emitters manually, dependent upon conditions at any one time: in practice, however, this is seldom satisfactory because openable windows are most often used as the control device with the consequent waste of energy.

A combination of compensator control plus individual room control has considerable merit, since the zone compensation provides a limit to energy use (by reducing standing losses from the distribution system) and the room controls give good control of comfort due to a capability to offset local heat gains. Clearly, with such a combination system zoning may not be critical and larger areas may be served from each zone, resulting in initial cost savings.

Whereas compensated temperature circuits may be required for systems which serve radiators, radiant panels and natural convectors, it may be undesirable to serve heater batteries from these circuits as their load characteristics may differ from other loads. In the case of fan convectors, care has to be taken to avoid only mildly heated air feeling cool to occupants due to the effects of 'wind chill'. A constant temperature circuit is required for systems serving the primary heat source to hot water service storage calorifiers. There is therefore a conflict between the calorifiers need for an almost constant temperature of heating water and the wish for the heating water to be as cool as possible in order to minimise standing losses. It is for this reason, that it is increasingly common to separate these two functions (i.e. separate heating boiler systems and directly heated hot water supply).

Boilers

Control of the output of the boiler is achieved by control of the burner unit (for gas- or oil-fired boilers). Burners may have, on/off, high/low or fully modulating burner control.

Where boilers are designed to work satisfactorily should condensing occur with the resulting greater overall thermal efficiencies, then systems should be designed for the water temperatures through the boiler to drop to as low as possible under changing load conditions. Where, unusually on new installations, the boilers are not designed to work satisfactorily in a condensing mode, with the subsequent formation of acids within the boiler, the output of the boiler should be controlled to ensure that it runs at a constant high temperature. Heating water, to loads requiring lower temperature water can then be mixed as described earlier.

The control of the boiler will usually be under the influence of a time switch and frost and optimum start control, as described previously. In other than very small systems, multiple boilers would be installed, which provide standby in case of failure of one unit and allow each unit, when in use, to operate closer to full output and peak efficiency, rather than on low load with consequent inefficiency.

Boilers are commonly provided by the manufacturer complete with integral safety and thermostatic controls. Multiple boiler installations are piped in parallel and controlled in sequence; more boilers operating as the load increases. Provision should be made for the sequence to be varied to give equal use of all units through the life of the plant. Although straightforward in concept, this form of control can present problems during partial load and it is important to consider the operating temperatures through the full range of output. Some of the points to consider with multiple boilers are those related to ensuring that:

- Return water temperature is high enough to prevent condensation occurring in the flue-ways of non-condensing boilers.
- Where loads tolerant of low heating water temperatures occur with those which are not (e.g. hot water service heating), that the systems are arranged to ensure suitable water temperatures are available at the different parts of the systems, or separate systems are provided.
- Resulting flow temperatures from sequence operation are within acceptable limits (excessively high temperatures off individual boilers can occur).
- Flow rates through individual units are at or above the manufacturers' recommended minimum.
- Flow rates are balanced through each unit, under full and part load conditions.
- Heat losses are minimised from 'off-load' boilers.

It may be necessary to introduce individual pumps to each boiler; or to use a primary circulation through the boilers, from which secondary circuits are connected to the various heat emitters or heat exchangers. The plant controls must always be considered in conjunction with the remainder of the heating system.

Domestic heating

Such evidence as is available suggests that the most effective method for control of domestic heating is the TRV coupled, of course, with full use of a time programmer to provide for intermittent boiler shut-down at night or during other periods when the dwelling is unoccupied. TRVs should not be provided on all emitters unless a facility is introduced to maintain water flow through the system or to turn the pump off when there is no/low load.

Heat exchangers

Storage calorifiers and non-storage heat exchangers would normally be provided with a modulating valve, fitted to the primary supply, and controlled via a sensor located in the shell in the case of a storage calorifier, or in the secondary flow pipe from a non-storage heat exchanger. In the case of a heat exchanger serving a heating system, it is possible to link these controls with a compensated controller to vary the flow temperature according to outside conditions. A direct acting control valve may suffice where simple constant temperature is required.

Air-conditioning system controls

Automatic controls are an essential part of any air-conditioning system. The wide variety of purposes for which such plants are required, the number of different systems which it is possible to select, and the multitude of types of control equipment available require a complete book in themselves to cover adequately.

It is possible here to indicate only the main principles involved, some of the chief component items of apparatus commonly used being described elsewhere (Chapter 22).

From this it may be possible to understand some of the problems concerned and how they may be tackled. Other permutations and combinations of system and equipment may be built up to suit any variety of circumstances.

The main types of plant will be considered and reference to the diagrams will be necessary along with the following descriptions. For further information the reader is directed to the CIBSE H.

Central system with cooling coil

The control arrangement in this case is illustrated in Fig. 28.25 which shows a typical array of sensing devices, control valves and motorised dampers, including some of the desired safety features, to provide control over temperature and humidity in the conditioned space under all outside air conditions. This figure relates in principle to the example air-conditioning calculation in Chapter 23 with humidification by steam injection.

In an all-air system, the quantities of outside and recirculated air, respectively, may be varied in any combination between full outside air, with no recirculation, and a mixture in any ratio subject to the provision of the minimum quantity of outside air which will satisfy the ventilation requirements of the occupancy. The mixture ratio is controlled to provide the most economic level of plant operation as determined by the condition of the outside air relative to the required condition of the supply air.

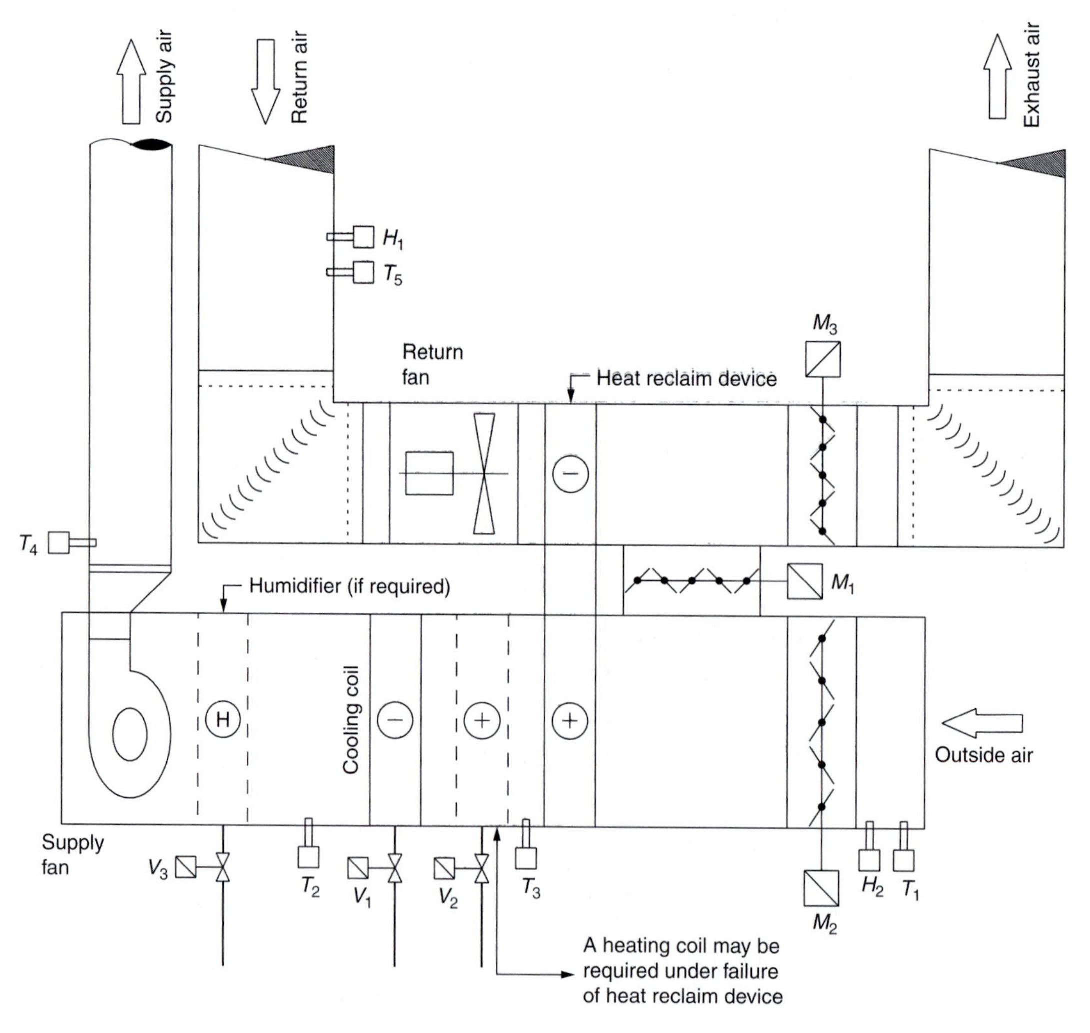

Figure 28.25 Control principles of a central plant system with cooling coil

A dry bulb thermostat or an enthalpy sensor is used to sense the outside air condition, the latter being used more generally nowadays in preference to a wet bulb thermostat. Modulating dampers in the recirculated, exhaust and outside air ducts are controlled to provide the desired mixing ratio and, for reasons of energy economy, the plant may operate on full recirculation during building heat up in winter.

In summer, the dampers will normally be set to introduce a minimum outside air quantity and the cooling coil will be controlled to maintain the required dew point for the supply air and, during this season, moisture will be removed from the air for the majority of outside conditions. Rarely where very close control is required, and the energy penalty accepted, the final dry bulb temperature will be controlled by the after-heater. During mid-seasons, full outside air may be used to take advantage of the available 'free cooling' potential which is available using air at outside temperature.

In winter, the dampers may modulate to mix outside and recirculated air to provide the most economical ratio such that, whenever outside conditions permit, the mixture will provide air at the desired supply temperature and thus eliminate the need for preheating. The humidifier will be operated as required to provide the correct moisture content in the supply as required by the space, by sensing the humidity content of the return/exhaust air.

Central system with cooling coil bypass ('face and bypass')

This is a development of the previous system where, in order to achieve the dehumidification required, whilst avoiding wasteful use of cooling and reheating in the event of prolonged periods of less than peak load, a fixed bypass is introduced around the cooling coil as shown in Fig. 28.26.

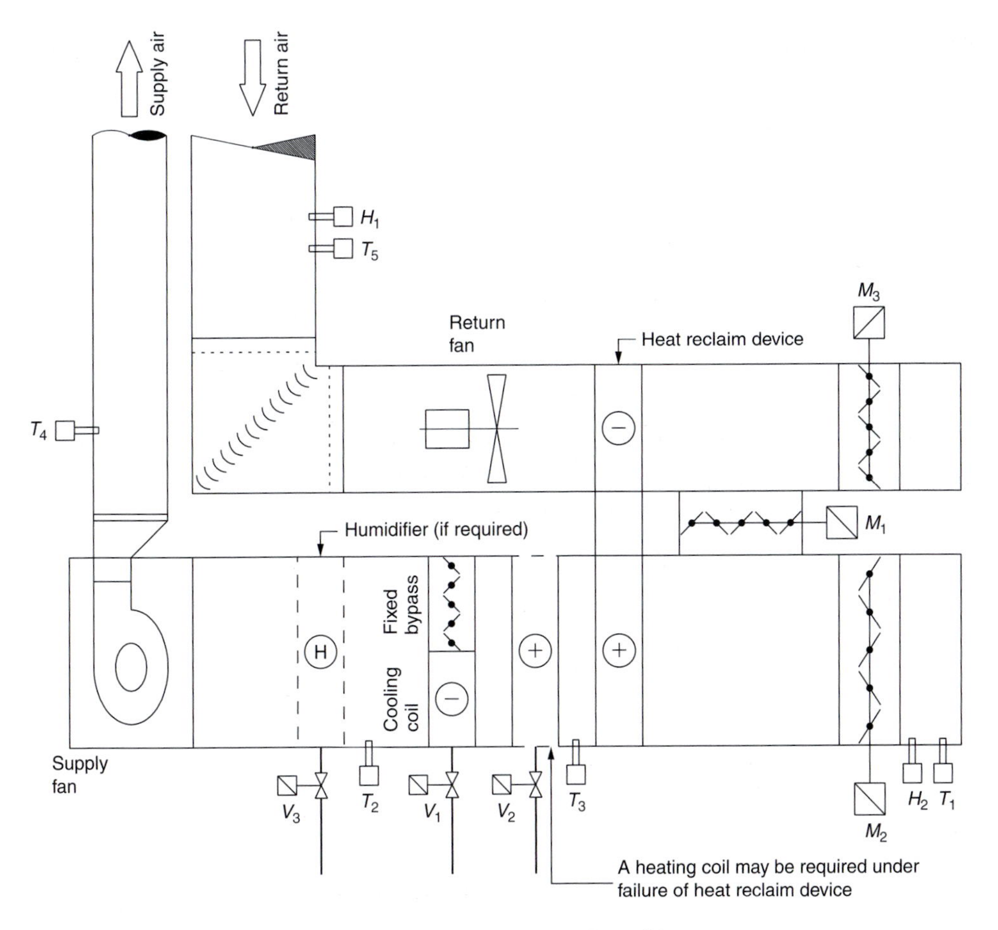

Figure 28.26 Control principles of a central plant system with cooling coil bypass

Zoned system

In this case the central plant delivers air at a constant outlet condition and the zones are controlled individually.

Terminal reheat system

In this case, all or the bulk of the required reheating to achieve control of dry bulb temperature is transferred from the central plant to individual rooms within the conditioned space. Thus, the output of the central plant is controlled to meet the peak cooling load likely to occur at any one time in any individual room.

Within each room, a thermostat will be arranged to control admission of heat to the reheater incorporated in the terminal unit. This reheater would usually take the form of a hot water coil provided with a control valve, but electric resistance heaters are sometimes used.

This form of system is inherently wasteful of energy and is generally to be avoided. However, it is sometimes possible that the amount of reheating can be acceptably minimised by the use of a control system which controls the central plant under the dictates of constantly interrogating the terminal load controllers to always achieve a supply condition which minimises the wasteful reheat required at the terminals.

Fan coil systems

The principal difference between those systems previously described and a fan coil, arrangement is that where a ducted air supply is provided, this handles outside air only and not the full conditioned air quantity.

Depending on the requirement for humidification in winter and dehumidification in summer the temperature of the outside air supplied to the space may be varied to minimise reheating at the fan coil unit (FCU). Controls will thus be as described for a central plant system. Each FCU will be provided with cooler and reheater coils fitted with control valves and these will be operated separately or, more usually, in sequence by a thermostat in the fan coil inlet.

The loads can be controlled by varying the heating and cooling water through the coils as shown in Fig. 28.28a (waterside control) or by arranging the coils in parallel and varying the air flow over them (airside control) as shown in Fig. 28.28b, valves in the water connections which may be two, three or four pipe. Manual speed adjustment and on/off control to the fan is often provided. Automatically controlled fan switching, however, may cause annoyance due to the change in noise level.

Airside control units utilise one air control damper rather than water control valves. As small control valves are susceptible to blockages they may be chosen where there are maintenance concerns, although they do incur greater running costs as the dampers generally cannot achieve the tight close-off of the valves, with the consequent wastage with heat transfer between the heating and cooling coil.

Where individual units are provided it would be possible to introduce ventilation air from the central plant into the space via terminals independent of the units. However this would minimise the ability of the system to optimise outside air for free cooling, due to the discomfort that introducing cold air directly into the space would cause.

Passive chilled beam systems

The control arrangements for a system which uses passive chilled beams to cool the space is usually the same as for that serving a FCU system, i.e. the central plant supplies a minimum amount of outside air to the space. The primary difference on the requirements of the air from the central plant, is, as passive chilled beam cannot heat then the air supplied will usually be supplied at the room design condition (i.e. load neutral to the room), and there is no ability to optimise any free cooling capacity in the outside air (unless the designer can be sure that all areas served will have similar load patterns).

Active chilled beam systems

Using active chilled beams, the amount of air supplied from the central plant usually has to be greater than the minimum fresh air requirement. If the difference between the air volume required and the minimum air

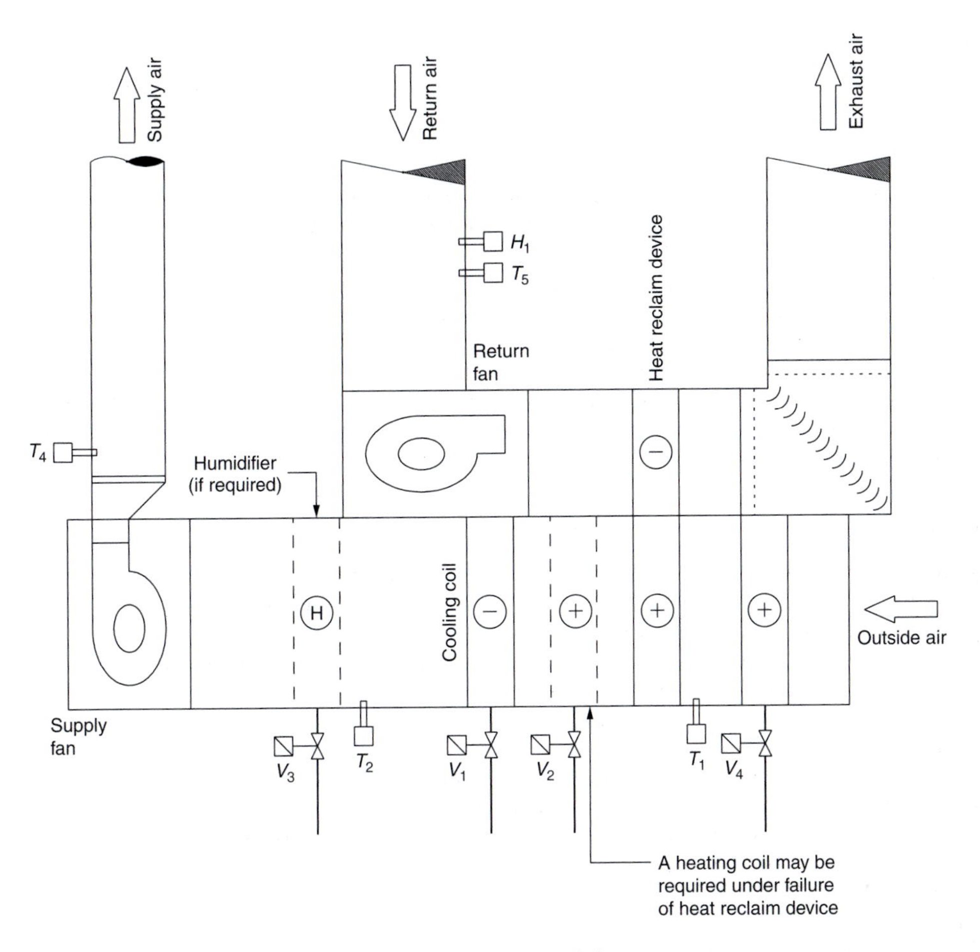

Figure 28.27 Control principles of a central plant without recirculation

volume is small, then it would be usual that the system is arranged as shown in Fig. 28.27. Where this volume difference is such that recirculation of a portion of the return air is worthwhile, then the central plant would be arranged and controlled as Fig. 28.25, or 28.26.

Variable volume system

A true variable volume system, in its simplest form, will have a central plant controlled in most respects in a manner similar to that described earlier. Particular attention will be paid however, in the interests of energy saving, to the control of the fan operation such that volume variation at terminal units is reflected in volume reduction at the central plant. Fan volume control is achieved by sensing the static pressure in the supply duct and controlling the fan to maintain that pressure constant. The best position for locating the sensor can only be found by testing during the commissioning process by simulating varying load conditions, but a practical rule-of-thumb for most systems is two-thirds along the index run from the supply fan. The methods of fan control are dealt with in Chapter 22.

In the simple form of the true system, control of space conditions and in particular those in internal zones of deep plan buildings is achieved by variation in the quantity of air supplied rather than by any changes in

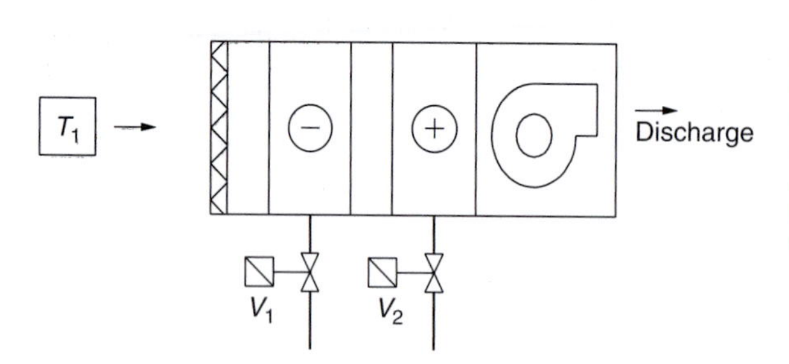

In heating mode V_1 is closed, and V_2 is open and modulates to maintain a room or return air temperature.

In cooling mode V_2 is closed, and V_1 modulates to maintain a room or return air temperature.

For stability of operation and energy efficiency there is usually a clear band between the heating and cooling set points.

(A) Principles of operation for a water side control fan coil unit

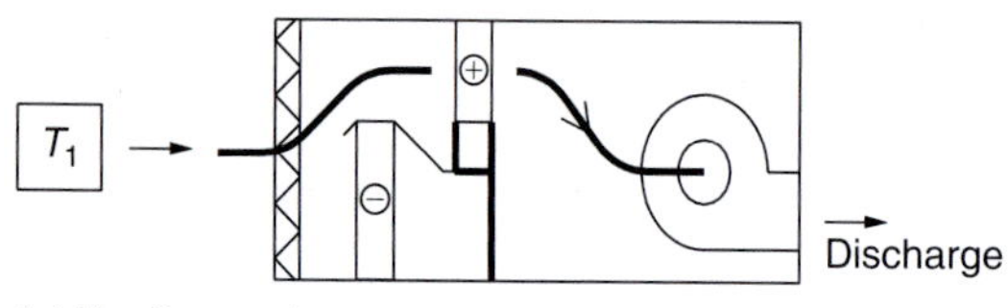

(a) Heating mode

In heating mode the cooling damper is closed and air passes through the heating coil.

Cooling pick-up is minimised by the cooling duct and damper design.

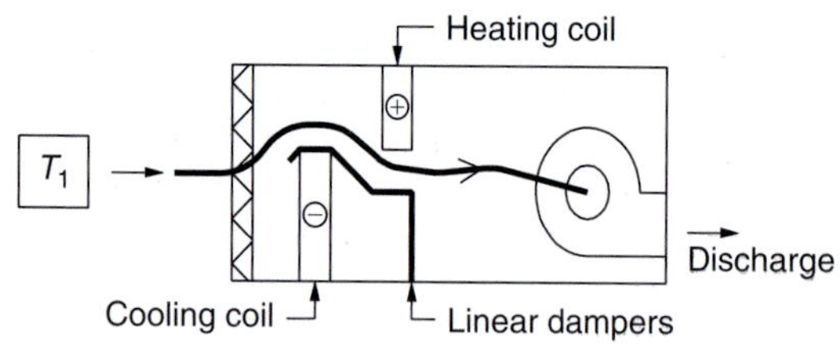

(b) Bypass mode (no load)

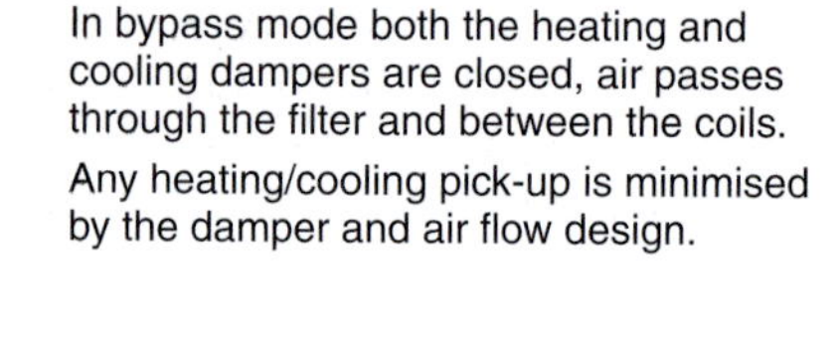

In bypass mode both the heating and cooling dampers are closed, air passes through the filter and between the coils.

Any heating/cooling pick-up is minimised by the damper and air flow design.

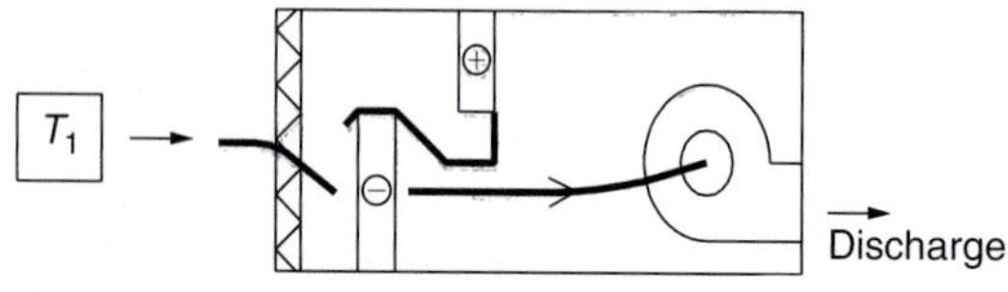

(c) Cooling mode

In cooling mode the heating damper is closed, directing air through the cooling coil.

Heat pick-up is minimised by the damper design.

(B) Principles of operation for an air side control fan coil unit (Lennox)

Figure 28.28 Principles of operation of FCUs

temperature. This control, by room thermostat operation upon electrical or pneumatic devices, acts to reduce the volume of conditioned air delivered either to a zone or an individual room where the load is at less than the design peak. Such devices may be integral to the actual terminal diffusers or incorporated in regulators serving a number of such diffusers.

Good practice suggests that the volume of air supplied to meet the designed full load cannot be reduced under control by more than about 40 per cent if room distribution difficulties are to be avoided, unless variable geometry supply diffusers are used to maintain air velocities. In practice the minimum air volume acceptable may be greater than this is order to maintain the minimum outside air requirements to the space. In consequence, control by terminal reheat may be necessary should local demand for cooling fall below that level: hence, room or zone terminals may be supplied from a dual-duct system or be themselves provided with reheat batteries supplied by heat reclaimed hot water or, rarely, where use is very intermittent, with electrical resistance heaters. The degree of wasteful cooling and then reheating can be minimised by scheduling the supply air to the variable air volume (VAV) boxes. The optimum supply temperature can be determined if the individual VAV box controllers are interrogated by the central plant controller.

For perimeter zones, the necessary reheat may be controlled via any form of convective heating system that is installed if this can be arranged to provide adequate response. Modern control systems allow the perimeter heating and variable volume box controls to be sequenced, room by room. Of particular importance is the supply of adequate quantities of outside air to each room. Unless additional controls are incorporated, the outside air quantity will vary proportionally with the supply quantity, which may lead to inadequate outside air provision and high running costs. To overcome this, an air velocity sensor is installed in the intake duct to control motorised dampers and ensure that the ventilation air quantity is maintained at or above the minimum requirement.

In Chapter 19, the concept of fan assisted control terminals was introduced as a means to provide additional cooling to supplement that inherent in the air supply. In this case, a valve in the chilled water supply to the cooling coil, and in some parallel arrangements, the fan, is controlled by a thermostat sensing room temperature.

It is also necessary to control the extract fan as a 'slave' to the supply fan in order to maintain the correct balance between the supply and extract air streams over the full operating range and to achieve maximum energy savings.

Systems which are controlled to vary the volume of air delivered to rooms or zones while the central plant volume remains constant do not offer the same facilities for energy conservation.

Induction system

There were two principal methods of control applied to induction systems, the 'changeover' and the 'non-changeover'. The former was more appropriate to climates having distinct seasons and need not concern us here. Two-, three- and four-pipe units were available to provide progressively better availability for control of space conditions: the two-pipe system was the more common in Great Britain. However, induction unit systems are seldom used these days; the preference being for either fan coil or active chilled beam systems.

Dual-duct system

Figure 28.29 illustrates the principle of the control system applying to a dual-duct system where two fans are included, one for the warm duct and one for the cool.

The temperature of the cool duct is varied according to the external temperature from say 7°C in summer to say 16°C in winter, or to a more limited schedule where the internal gains dominate the room loads. In summer, it is cooled by the chilled water coil: in winter the heater comes into use as necessary. The temperature of the warm duct is also varied according to the outside temperature from, say, 21°C in summer to, say, 38°C in winter.

This form of system has historically been used where constant air flows and pressures are required for reasons other than temperature control (e.g. hospitals and laboratories). In their constant volume role they are inherently wasteful of energy and should be avoided. However where the terminal device modulates the volume of both warm, cold and total volume, then the wastage is lessened. Both forms however require considerable distribution space (see Chapter 19).

Reverse cycle heat pump system

The control of individual heat pump units is described in Chapter 19. To supplement these local units, it is normal to install a central air handling plant to provide ventilation air to the conditioned spaces; this may be taken to be of the same configuration as for fan coil systems.

Displacement ventilation systems

In this form of system air is supplied to the floor void of the areas served at a constant temperature all year around. When the outside air is below the design supply temperature, this allows 'free cooling' and mechanical

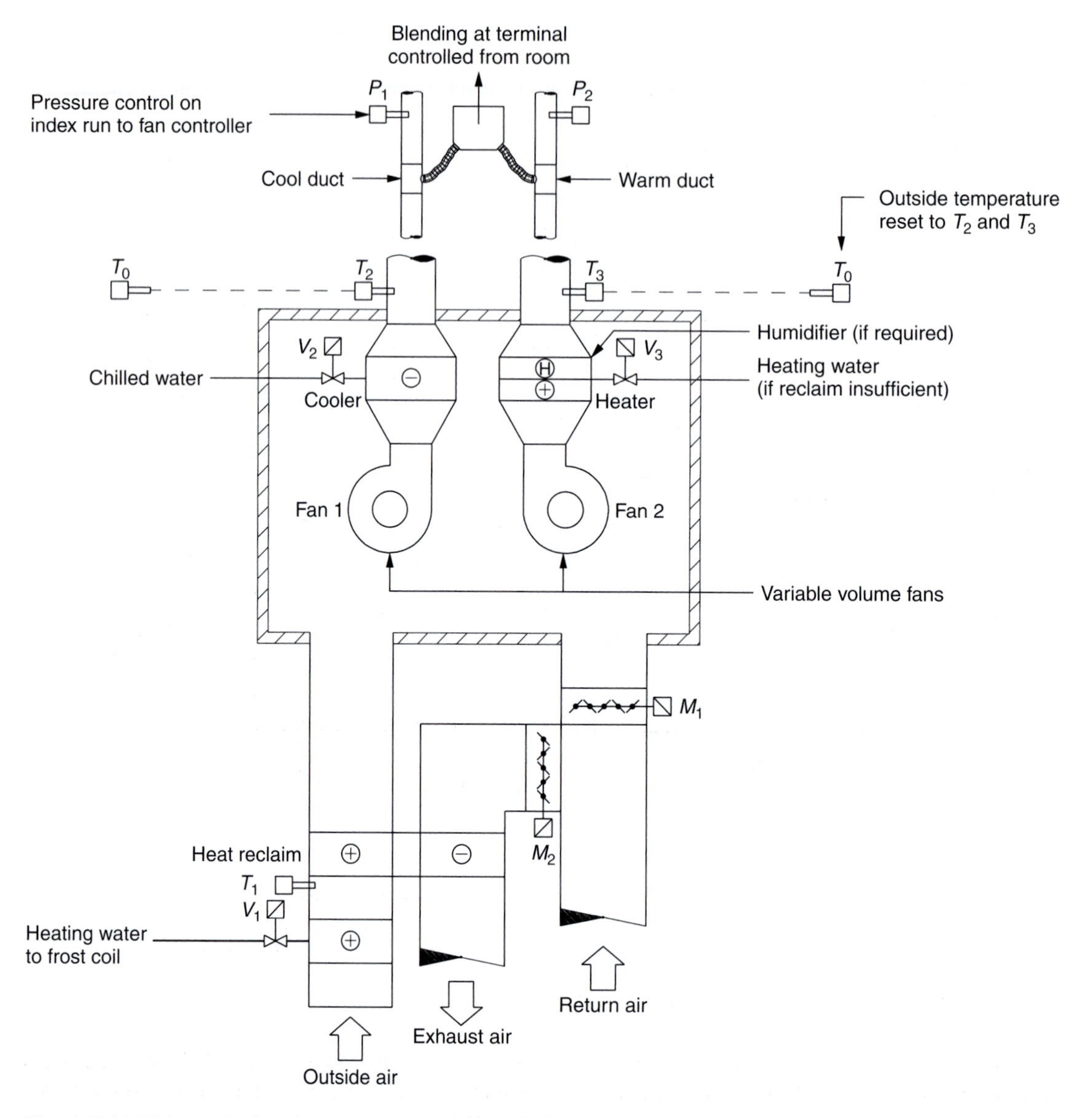

Figure 28.29 Diagram of a dual-duct system showing principles of control

cooling is only required as the outside ambient rises above the design supply temperature (minus fan and system temperature gain). Systems are generally arranged as shown on Fig. 28.30.

Natural ventilation systems

Where areas of a building are to be tempered in summer by the use of natural ventilation, the control of the opening (usually a window) can be either manual or automatic. Manual operation is obviously the easiest and cheapest 'system'. However increased degrees of ventilation are often optimally required before the temperature in the room has risen above design, and experience shows that humans are often not good at foreseeing this condition. As nighttime temperatures are almost always lower than those during the day, if security allows, then utilising nighttime ventilation is often advantageous. However care must be taken to avoid overcooling

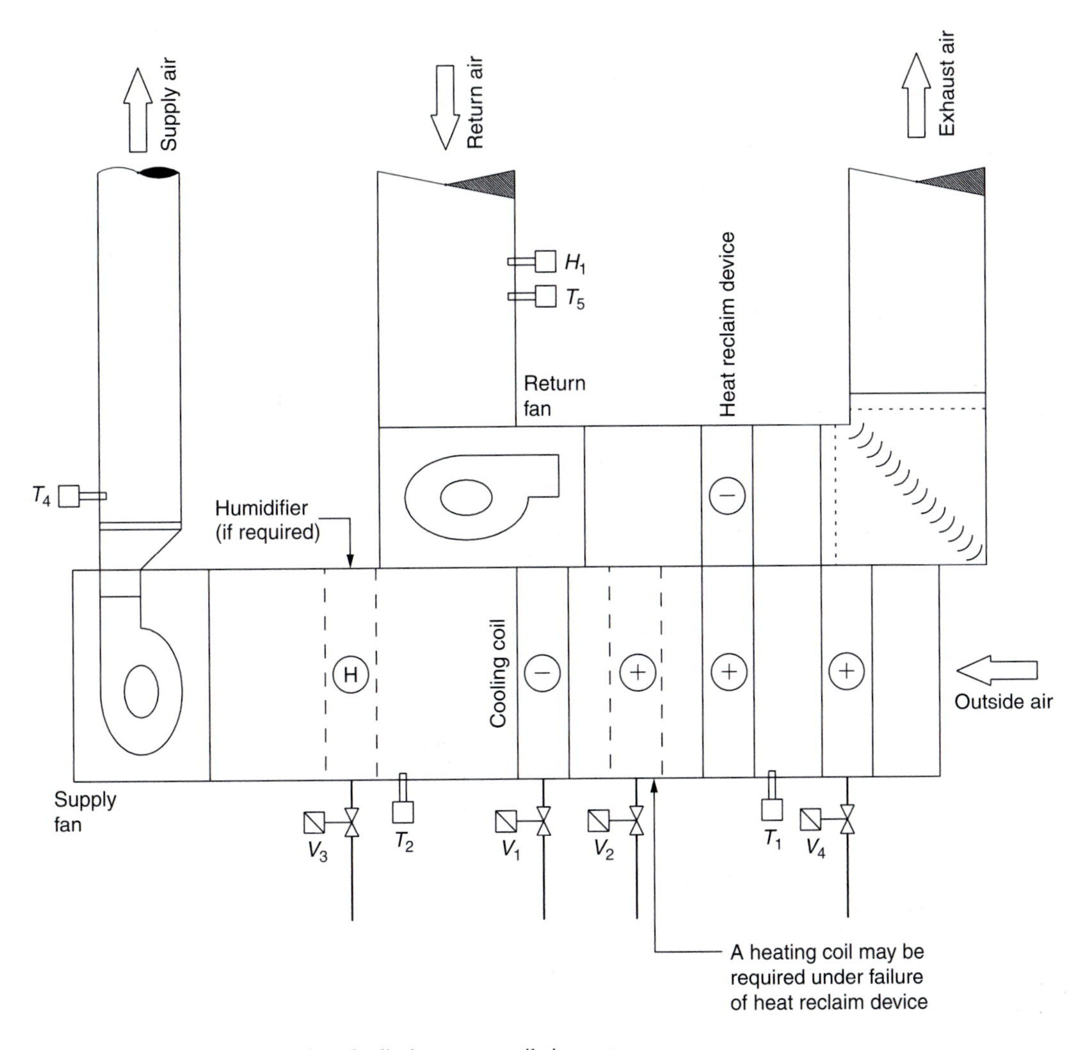

Figure 28.30 Principle of operation of a displacement ventilation system

the space, which would result in occupants either requiring heating or being uncomfortable in the early hours of a summer morning. For these reasons, automatic control of the openings (or as a minimum, those required to achieve nighttime ventilation), may be advantageous. Care must be taken to ensure that actuators can achieve tight close off, in order to avoid draughts and unnecessary heat loss in winter.

Utilisation of the thermal mass of the building

As described under 'natural ventilation', the thermal mass of the building fabric, can be used to utilise the diurnal range in outside temperature to moderate the internal environment. Where simply opening the windows does not provide sufficient moderating effect or is unacceptable for security reasons, or where the windows cannot be opened for other reasons such as acoustics then outside air can be centrally supplied to the building, intentionally using parts of the fabric as the distribution system. The most popular application of this in UK is to utilise the airways in precast floor planks as the final ductwork run to the room served. In this system cool nighttime air is supplied to the rooms during the night, (described in Chapter 19 as a 'hollow floor system'). The air cools the concrete (itself being warmed by the concrete). The control system has to optimise precooling all of the rooms (as some will have been hotter than others at the end of occupancy), with avoiding overcooling any of the rooms. During the day, the control system has to optimise the supply air volume to the space in order to control the release of 'coolth' from the concrete whilst supplying a minimum air volume to the occupants. This requires the control system to have sophisticated self-learning routines which compare, on say 15 minute increments, the outside and inside air temperatures and air supply rate, today with yesterday, and to vary the supply volume by an increment based upon what was achieved previously under similar conditions. The central air supply units usually need to have a full recirculation feature which is required on start up (daily, weekly) to allow temperatures within the system to stabilise. The principle of this type of system and its control is shown in Fig. 28.31.

This form of system is limited in its capacity when compared with a conventional heating and cooling system, and so requires a high performance envelope which limits winter heat loss and summer heat gain.

Mixed-mode systems

The term mixed mode is used, where the Buildings systems are designed to operate differently at different times of the year. Examples of this may be:

- Natural ventilation in summer, with the windows intended to be closed in winter when mechanical supply and extract are provided in order to utilise heat recovery.
- Heating by displacement ventilation systems utilised in winter, then in mid-season natural ventilation, and then in summer, cooling by displacement ventilation.

Whenever the systems are designed to operate in more than mode throughout the year:

- The different modes need to be clearly thought through
- The control system needs clear 'rules' when to utilise each mode
- The system must have the facility for tuning and adjustment by the operations staff
- The occupants must be aware of the variation in how the building works and what is expected of them.

Smoke control

Although plant start and stop functions have been dealt with previously in this chapter, an additional feature to be incorporated, common to ventilation and air-conditioning systems where air is distributed to and from occupied spaces, is provision for smoke control during a fire situation. The requirements vary dependent upon the fire authority concerned and the particular characteristics of the building, but common to the majority of installations is the need for the plant:

- To stop under a fire condition (other than special situations such as say, operating theatres or fume cupboards).
- To provide the facility for the fire brigade to be able to switch the plant 'override on' and 'override off' from a convenient position close to a main entrance to the building.

This is called the 'Fireman's Panel'. Separate control over the switching of supply and extract systems is usually required. In a fire situation, as noted above the air handling plant would normally be arranged to stop automatically, but could be programmed to switch into predetermined 'fire' modes. As this is life safety switching, it is usually required to be 'hard wired', rather than be a function of the control system/BEMS. With its fire-rated cabling and much more comprehensive error checking it may be acceptable that this fire switching is carried out by the fire alarm system.

Chilled water system control

Contrary to the general practice with heating circuits, it is not usual to vary the temperature of the chilled water flow from the evaporator to cooling coils. However, there can be very large variation in cooling loads at zone or central plants dependent upon the time of day as the sun moves around the building or the operation of lighting and IT loads. As a result, depending on the magnitude of this variation, will often be more economical to use a variable water flow system in preference to constant flow. Where constant flow is used, control of

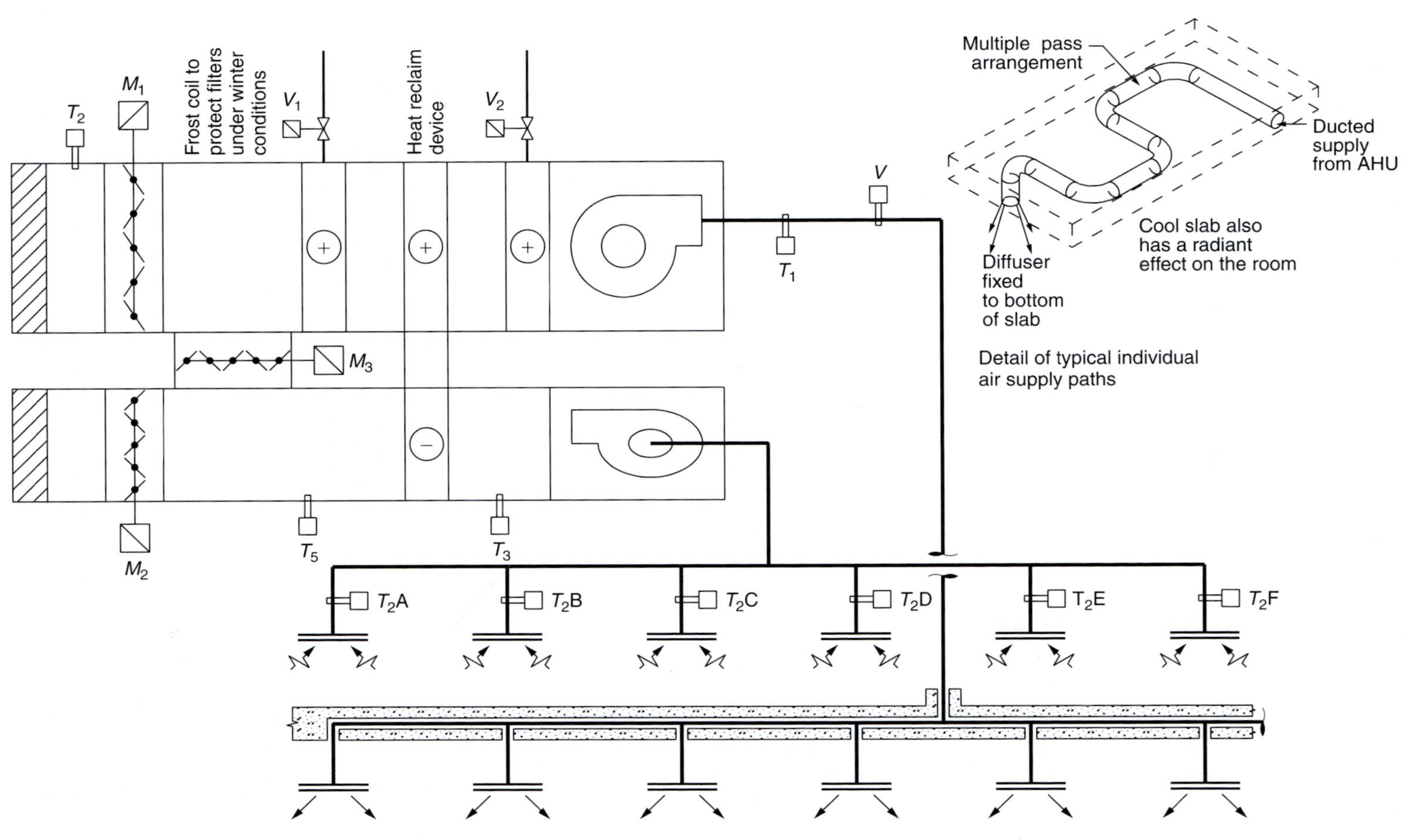

Figure 28.31 Principles of an air supply system utilising the thermal mass of the structure

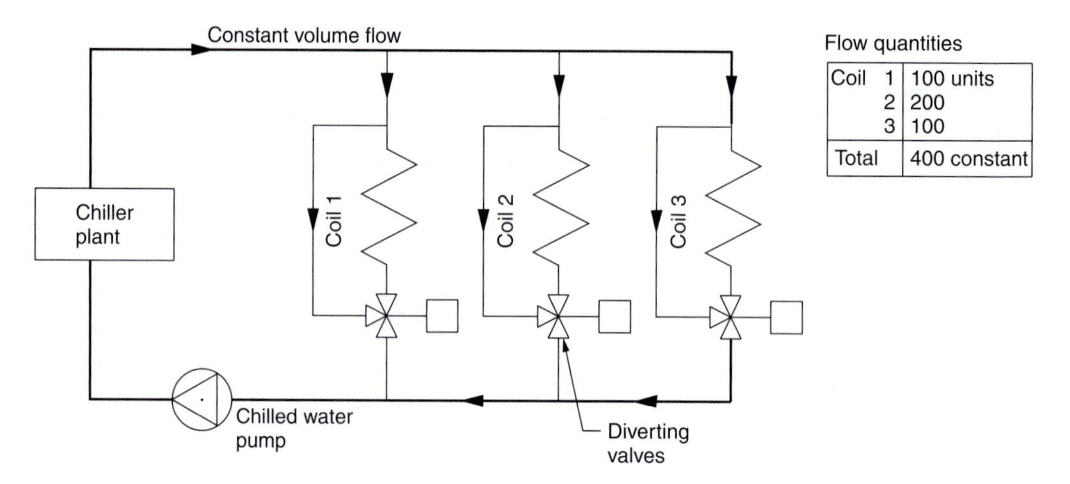

Coil		
Coil	1	100 units
	2	200
	3	100
Total		400 constant

Figure 28.32 Three-way valve control to a cooling coil

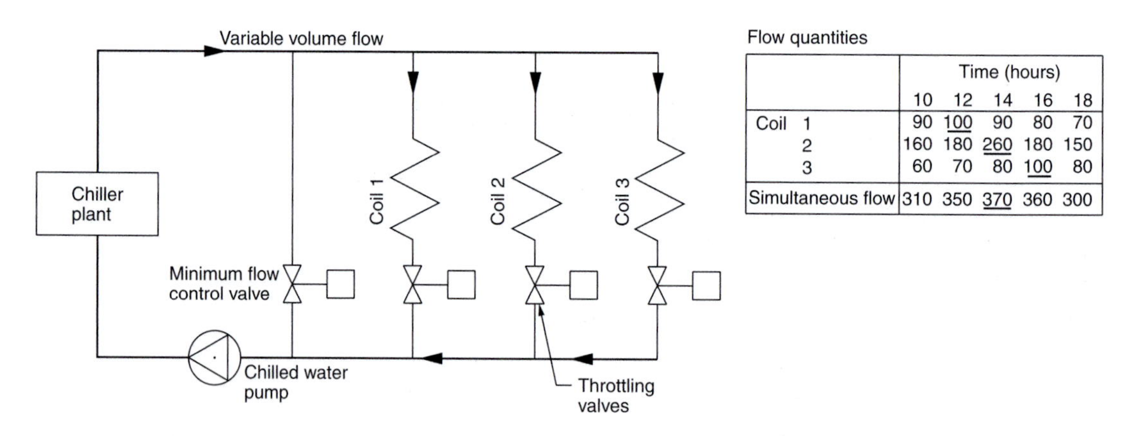

	Time (hours)				
	10	12	14	16	18
Coil 1	90	100	90	80	70
2	160	180	260	180	150
3	60	70	80	100	80
Simultaneous flow	310	350	370	360	300

Figure 28.33 System with two-way control valves

chilled water to the coil will be by three-way modulating control valves as Fig. 28.32, but where variable flow is preferred, two-way modulating valves would be used, but care has to be taken to achieve minimum flows required at the chiller(s) in order to maintain acceptable flow through the evaporator and to avoid freezing.

A typical system incorporating two-way control valves is shown in Fig. 28.33, from which it can be seen that the maximum pumped water quantity is proportional to the maximum simultaneous cooling load, and not to the sum of the maximum coil loads: in consequence, smaller pumps may be used thus reducing both installation and running.

Chiller control

Single chillers will normally be supplied by the manufacturer with integral safety and capacity controls. Multiple chillers are normally arranged in parallel with respect to the chilled water circuit, see Fig. 28.34. Where, rarely very tight control of supply chilled water temperature is required, a series arrangement could be used. The points to be aware of when selecting the appropriate arrangement are that:

- Since the frequency of starting must be limited, the minimum output (turn down) of one machine must be less than the lowest operating load of the system, or inertia vessels incorporated into the chilled water system to reduce the frequency of starting.

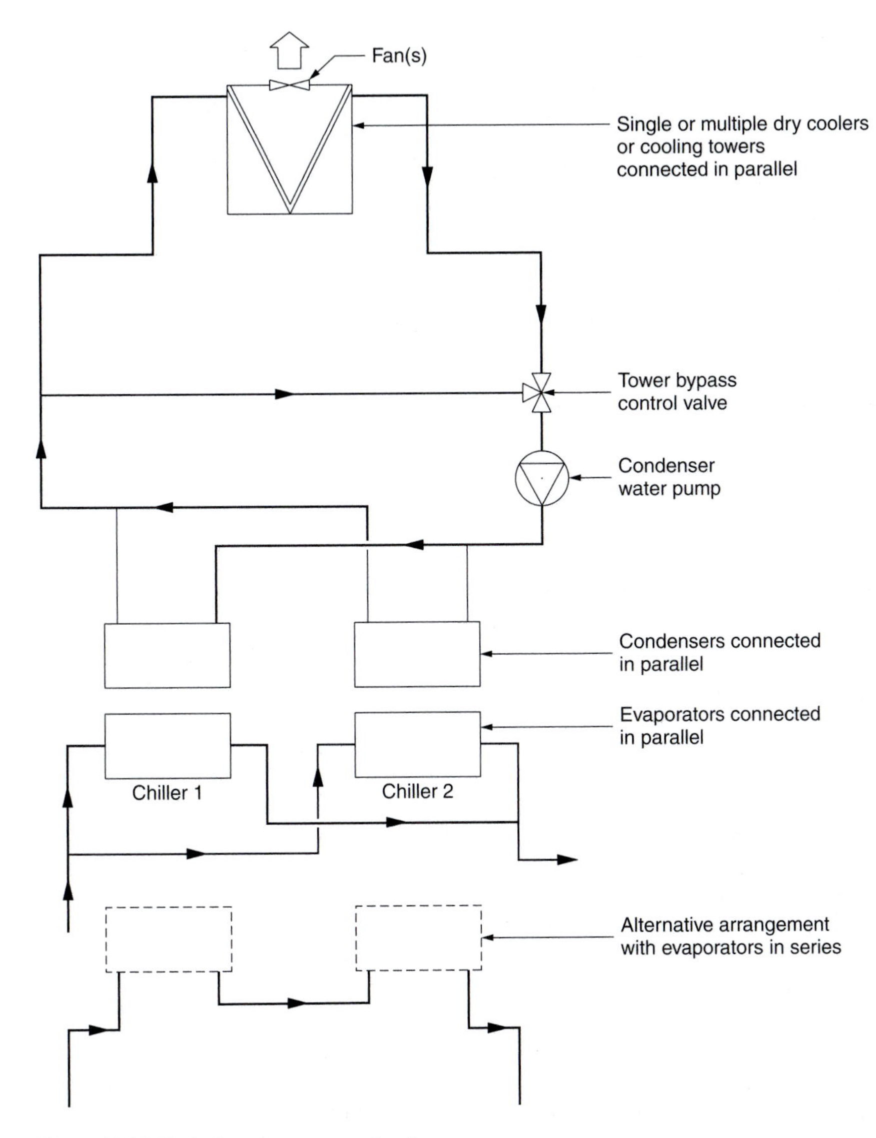

Figure 28.34 Typical condenser water circuit

- In a parallel arrangement, it must be established that the sequence control will not lead to sub-cooling and that the water flow rates through each evaporator are regulated to the correct quantity. Difficulties with part load operation of multiple chiller can normally be best overcome by the use of the chillers manufacturers own multiple chiller control system or 'chiller manager'. With modern digital control of chillers, this will often allow varying combinations of compressor (rather than whole chillers) to be sequenced.
- Although a series arrangement may permit control of water flow temperature to be more stable, the hydraulic resistance of the system will increase with consequent higher pumping costs. (It is not usual to provide arrangements where more than two machines operate in series.)

Normally, with reciprocating machines, capacity would be adjusted by proportional control, sensing the temperature of the return water. With screw or centrifugal machines, PI control would normally be used.

Cooling towers

A typical arrangement for a condenser water circuit, connecting chiller condensers to either cooling towers or an equivalent, is shown in Fig. 28.34. The cooling effect of the towers is controlled either by sequence switching or speed control of the fans. As well as resulting in less bearing and belt wear, the use of speed control will result in less annoyance to neighbours, as it will remove the obvious noise of repeated fan starts.

A bypass valve is normally provided in the heat-rejection circuit, which operates in conjunction with the fan controls on start-up and sometimes during colder weather, in order to maintain a constant return water temperature to the condensers (to avoid a low temperature lock out at the chillers). To ensure economy in energy, the condenser water temperature may be varied with respect to the imposed load and thus retain a practical minimum between the condensing and evaporating temperatures of the chillers. Attention must be given to the hydraulics of the condenser water circuit to avoid unintended cavitation and all water-based cooling plant must be provided with frost protection.

Where water is used for condenser cooling, a number of chilling machines may be connected together, in parallel, into a single circuit to be pumped over multiple cooling towers or their equivalent. Where air-cooled condensers are used, however, each chilling machine will have a dedicated cooler. Fan speed control, fan switching or modulating dampers may be used to control either the condensing pressure or temperature. Normally, PI action would be installed for control purposes.

Where water-cooled chillers are used, the option exists to use the heat rejection device alone during periods of sufficiently low outside ambient, in order to provide 'free cooling'. This is expanded upon in Chapter 24.

Building management systems

Continuing increases in processing power and reduction in unit cost of processing has led to flexibility in how systems are configured and also to different approaches to application and this, in turn, has given rise to descriptions such as *energy management system*, *building energy management system*, *building automation system*, *supervisory and control system*; the authors prefer BEMS as the generic title. The BEMS can be thought of as sitting above the 'controls'. With almost all modern BEMS, were the BEMS itself to fail, the fundamental 'control' elements of the system would continue to function.

The basic functions of BEMS may include:

- Continuous monitoring of systems
- Warning of *out of limit* conditions (alarm)
- Initiation of emergency sequences
- Logging of significant parameters
- Monitoring and recording energy use
- Condition monitoring and fault analysis
- Planned maintenance and other housekeeping functions
- Tenant billing.

Systems may be further enhanced by the use of modern database software, graphics and word processing techniques to provide opportunities for applying BEMS to total building management functions. The BEMS, therefore, can provide many benefits but these are dependent upon the capability of the systems operators and their motivation to make use of the data available. The BEMS should of course be designed in the knowledge of the operator's objectives and of the level of skill available: unfortunately this is not always possible. Some of the benefits which can result, however, are:

- Lower energy consumption
- Improved system reliability
- Savings from programmed maintenance
- Reduced number of watch keeping operatives
- Improved building management.

To balance these advantages, the initial cost of a BEMS may be high but this is very much linked to the extent of the systems and the degree of sophistication. If utilised correctly, BEMS features can reduce the cost of operational staff. The largest single benefit is likely to be the ability to analyse the performance of the building and its system and to optimise its energy use. These savings are very hard to quantify when the cost of the system is initially being considered. Comprehensive training of the operators is vital if the potential of the BEMS is to be realised and this is likely to require basic software skills where a facility for creating new routines for fine tuning of the control functions is incorporated. The more the interactions between the building itself and the building services are part of the design (low energy/mixed-mode systems) the greater the need for these tools to give the buildings operational staff the information required to optimise the systems.

Hardware

The principal components of a BEMS may be described briefly as follows.

Outstations

The electronic device actually processing the control algorithms is generically known as an 'outstation'. Historically these were 'stations' out in the field, i.e. not at the central console. These days the processing may be at the level of an embedded controller (such as with a chiller) as illustrated in Fig. 28.35 and 28.36. Modern systems use 'peer' to 'peer' communication so that generally each outstation on the system can 'see' and 'talk to' every other point on the system so as to ease diagnostic tasks for maintenance personnel in plant rooms, and ensure that failure of a 'head end' has minimal effect upon the overall system.

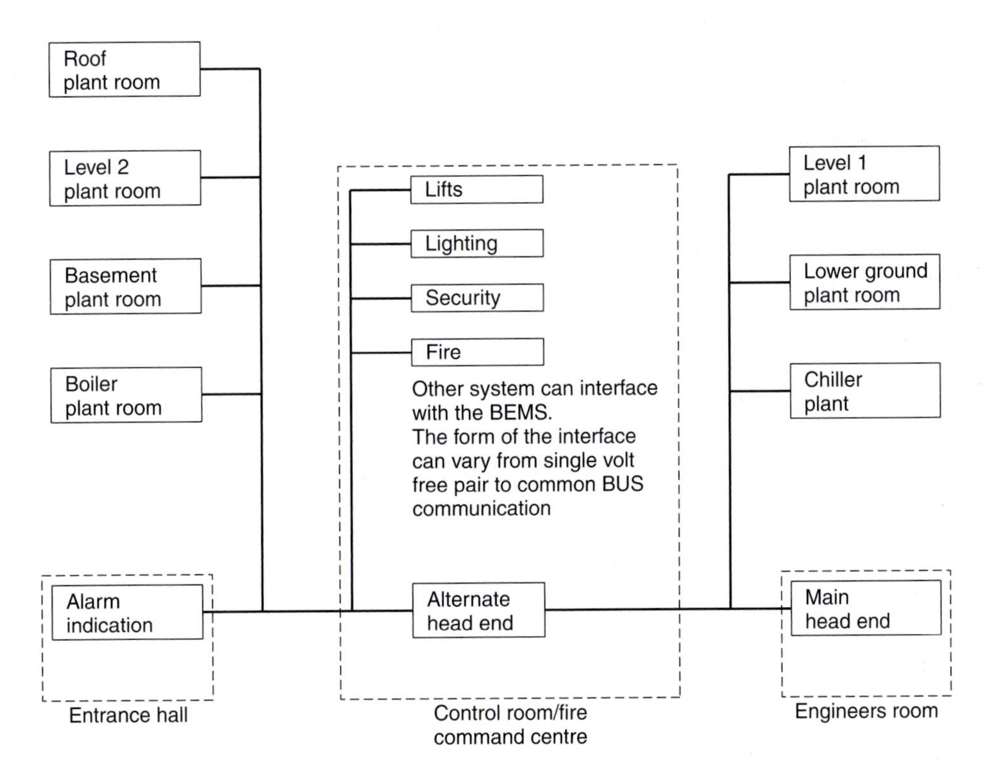

Figure 28.35 Typical BEMS network

Sensing devices

For every point to be monitored, a sensing device, switch, auxiliary contact or the like will be required.

Data inputs

Digital: two-position switching signals for status and alarm.
Analogue: values measured by varying voltage or current in the form of a pulsed signal.
Control: signals to component driving mechanisms.

Then in order to achieve centralisation of information, the outstations or controllers need to be networked together.

Transmission network

The network between outstations or controllers can be a separate wiring system for the controls/BEMS or be amalgamated onto the buildings main data 'highway' links sometimes known as local area networks (LAN). When considering the relative advantages of the use of such an integrated approach, factors such as confidentiality and programming/commissioning constraints need to be considered. Critical links are normally duplicated along alternative routes for security reasons, or if life safety, hard wired.

On very large sites, or where client wide monitoring (e.g. county or state) is required it is increasingly easier to use the greater internet to network buildings together. There are a number of 'Standard' protocols (transmission language) available, and at present the industry is slowly rationalising/standardising on the market dominant options. The use of such 'open' protocols (such as BACnet), allows the client greater commercial freedom in the future for both maintenance an alterations to the system, which is an important factor to consider at the initial procurement stage.

All BEMS cabling should be protected from extraneous electrical interference and fibre optic technology may prove beneficial for data transmission.

Head end

The early systems had central intelligence, in which all signals were transmitted and received through a central processor. Almost all of systems today have 'distributed intelligence' with the head end providing a management facility.

There can be multiple head ends on a system. Whilst a 'head end' could merely comprise a keypad on one outstation, it would be normal, and preferable that it comprises:

- Computer, screen(s) keyboard, mouse
- Alarm printer
- Report printer.

All systems run under a graphical interface (although the ability for experienced users to use text commands may be an advantage on large systems). An example of a typical installation network is given in Fig. 28.35.

Benefits in use

One of the major difficulties experienced on many projects, particularly the larger and more complex, is the achievement of satisfactory completion including correct commissioning and performance testing of the building services systems. At best, it is likely that the systems will be commissioned to satisfy the bases of design. Controls set-points, for example, would normally be based upon theoretical steady state calculations, modified perhaps by the judgement of the commissioning engineer, to reflect the dynamic response of the plant under the particular operating conditions obtaining at the time of the commissioning. Where dynamic thermal modelling has been undertaken in the design process, then more accurate commissioning set-points could be provided.

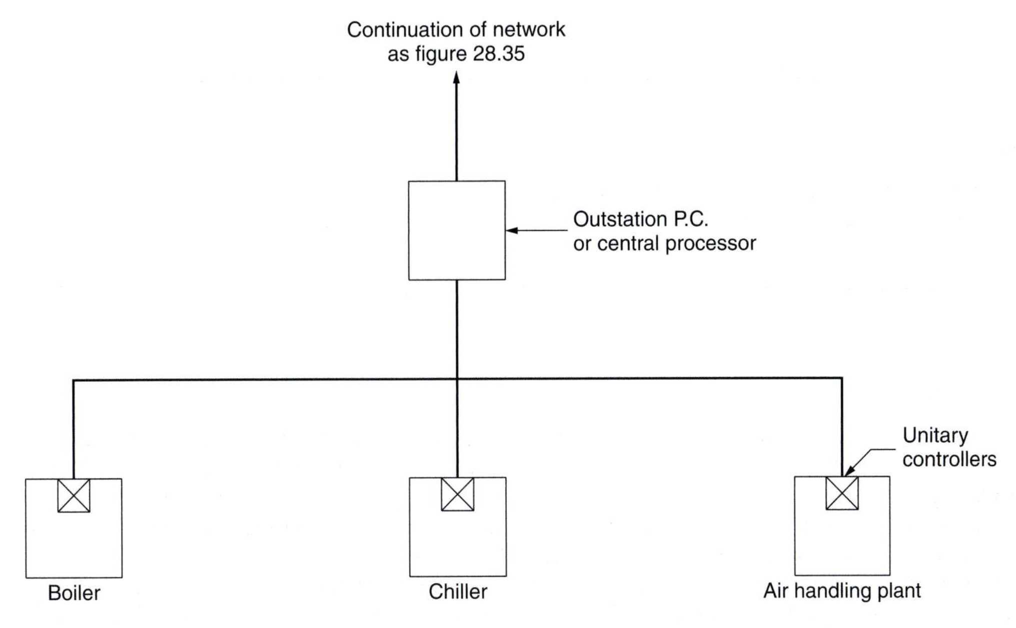

Figure 28.36 Unitary controllers

Using a BEMS, it is possible for the systems and controls to be finely tuned with the actual usage patterns of the building and to the actual thermal response of the building elements. This may be undertaken on site or at a remote location, utilising telephone/data links, which may provide an opportunity for the designer to monitor performance. This work may be assisted by the use of computer simulation techniques,[2] to check the performance against design parameters and to test alternative controls strategies. The system performance over a range of different control philosophies may be simulated through the weather conditions and operating periods for minimum energy use. It is only after such work is complete that the client will gain maximum benefit from the significant investment in a building management system.

The use of computer design methods for system sizing is standard and many of these design programs have been extended into the more complex area of performance simulation. One aim of these models is to simulate the performance of the building envelope and also the way in which the building, with its engineering systems, interacts with climate and occupancy patterns in a real situation, as described in Chapter 6. The increasing availability of relatively low-cost but very powerful microprocessors will enable BEMS eventually to have on-board simulation facilities running in real time. This will have two major benefits: firstly, it will enhance significantly the monitoring capability of the BEMS in being able to raise alarms immediately if there should be significant departure from the design performance concept and, secondly, in self-learning adaptive control.

The magnitude of potential energy savings arising from the use of a BEMS is dependent, of course, upon the type and condition of the installations before the addition. Energy savings of up to 40 per cent may be achieved where the BEMS is introduced into an existing installation which was poorly controlled and maintained. Even so, compared with an efficiently operated system without BEMS, the addition may offer potential savings of up to 10 per cent. Table 28.1 provides an indication of the order of savings related to a range of operating functions. Whilst the use of a BEMS is now usually taken to be a requirement in order to operate a building efficiently, many studies have been undertaken to examine their impact on otherwise well-run installations.[3]

The Implementation of the Energy Performance of Buildings Directive within the EU requires owner/operators to monitor the performance of their buildings (see Chapter 31). This alone will probably make BEMS financially viable on all new buildings within the EU.

Table 28.1 Energy saving using a building management system

	Energy saving (%) relating to	
Item	*Inefficient system*	*Efficient system*
Part load efficiency	9.5	2.0
Optimum start	7.5	2.0
Temperature control	7.0	2.0
Optimum stop	5.0	1.0
Holiday scheduling	4.0	–
Pumping/distribution	2.0	0.5
Miscellaneous	2.0	0.5
Staff awareness	5.0	1.0

Source
DOE/PSA M and E Engineering Guide, Volume 2, May 1987.

Future developments may be expected in all areas of BEMS systems. Transmission protocols are standardising across the industry and in respect of hardware, improvements are anticipated to operator terminals in graphic displays, touch responsive screens and the use of active video images which will allow users the facility to see on screen the results of keyboard commands.

Notes

1. Reproduced from CIBSE Guide H, section 3.3.4.1.
2. BRE, IEA (21C), *Empirical Validation*, Ref. IEA 21RN327/93, January 1993.
3. Energy Efficiency Office Best Practice Programme Case Study 202, Energy savings in NHS hospitals suggested year on year savings of 3 per cent being achieved due to BEMS-based monitoring and targeting.

Further reading

Building Control Systems, CIBSE Guide H 2000
Understanding Controls, CIBSE Knowledge Series, KS04

Chapter 29

Commissioning and handover

General

It is essential to fully commission and correctly set up the building services installation prior to handing the building over to the occupier. This includes all operational aspects of the systems, briefing the building operator and the provision all handover documentation. This chapter sets out the basic activities involved with commissioning and handing over building services installations.

Commissioning is normally carried out by the M&E installer as a sub-contract to the building main contractor. The M&E installer could have their own in-house commissioning specialists or they often employ separate commissioning firms. In some cases commissioning managers may be employed to oversee, plan and programme the commissioning operation.

Commissioning

Commissioning is the all encompassing term for the advancement of the installation from the stage of initial static completion, to full and satisfactory working order to meet the specified requirements. The Building Regulations Part L2A requires that 'the person carrying out the work shall give the local authority a notice confirming that the fixed building services have been commissioned'. The purpose of commissioning building services is to demonstrate that the installed equipment has been selected, installed and commissioned in compliance with the design intent and specifications.

Commissioning therefore:

- Proves the installation
- Proves the design
- Ensures plant and equipment are working at optimum performance to meet the client's requirements and achieve the predicted energy targets
- Provides documented records at handover.

Once the building services systems have been individually commissioned, it is necessary to set the building controls to automatic and performance test the systems as a whole to demonstrate maintenance of the required environmental control throughout the year.

The main elements requiring commissioning are:

- Water systems regulation and pump flow rates (CIBSE Code W)
- Air systems regulation and fan flow rates (CIBSE Code A)
- Room air distribution and acoustics
- Boilers and fuel supplies (CIBSE Code B)
- Chillers and associated heat rejection (CIBSE Code R)

- Automatic controls and building management system (BMS) (CIBSE Code C)
- Water treatment
- Lighting (CIBSE Code L).

The commissioning process includes:
- Design for commissioning
- Planning for commissioning
- Pre-commissioning and setting to work
- Balancing and regulation of distribution systems
- Equipment/plant commissioning
- Witnessing
- Proving system performance
- Documentation of the installation and performance.

Design for commissioning

It is important that a review of the project specification and design drawings is carried out by a competent person who understands the requirements for commissioning. It is also important that as the detailed designs are developed, systems are designed to be commissioned and that this continues to be monitored during the various stages of construction, i.e. production of installation details, installation and inspection.

This review should include assessment of the commissioning requirements with respect to:
- Regulation
- Measurement
- Monitoring
- Controlling
- Access
- Maintenance
- Health and safety

In addition to fan and pump sizing, the design should consider the arrangement of branches, positioning of measuring and regulating stations and selection of terminal units and their pressure drops. For both water and air systems it is a bad practice to have units directly served from mains where the terminal unit valve or damper will be required to remove excessive pressure to achieve flow rate.

In the case of water systems, regulating valves should be positioned and selected to give good *authority* at the design flow without the requirement for excessive regulation. This will often result in valves less than the line pipe size. The manufacturers' installation recommendations must be checked and regulating valves and measuring stations positioned to give the correct straight pipe lengths to avoid turbulence which would distort the readings.

Similarly the test points on air systems must be positioned to avoid turbulence and the type of balancing damper should be specified to suit the accuracy required by the system. Iris dampers would normally be required on air systems where greater accuracy is required.

Planning for commissioning

The planning for commissioning should include a detailed commissioning programme setting out the timing and sequencing of:

- Testing
- Cleaning
- Flushing

- Pre-commissioning
- Balancing
- Witnessing
- System proving
- Documentation
- Instruction/training

Commissioning Method Statements should be prepared for each system and an overall method statement prepared co-ordinating the interfaces between systems and plant.

Pre-commissioning

A number of activities should be carried out to ensure that the system is ready for commissioning to take place. For pipework systems, these include:

- A general check of the installation and components, e.g. adequate venting, positioning of flow measurement devices, regulating valves installed in correct direction, provision and positioning of strainers, etc.
- The system should be cleaned, flushed and chemically treated to remove debris and to minimise the risk of corrosion and biofilm development. This is an extensive exercise usually carried out by specialists; guidance is given in BSRIA Application Guide AG 1/2001.1. After flushing and cleaning, the water quality within the system will be tested to confirm acceptable levels of suspended solids and bacteria (pseudomonas) are achieved. When acceptable levels are achieved the system will be filled and treated with inhibitor to provide on-going protection but the level of inhibitor must be regularly checked and topped up as necessary by maintenance personnel.
- Prior to setting the system to work, the commissioning specialist should check that the system is fully vented, pumps are set correctly, automatic control and manual valves are operational but are set to their normal open position.
- After setting to work a general check should be made on the operation of the pumps, condition of strainers and after stopping the system, venting should be further checked.

Similar preparation is required for ductwork systems, including internal cleaning which is described in Chapter 21.

Balancing and regulation of water systems

The measurement and regulation of flow rates in systems is to ensure that design figures are achieved. The system designer should specify the tolerances required for the system and select the measuring and balancing equipment to suit the accuracy required. Figure 29.1 shows a number of measuring and balancing valves used in water systems (abbreviations are explained in the following text).

DRV/OV: The combination of *double regulating valve* and fixed *orifice valve* is commonly known as a measuring station. The water flow can be regulated down whilst the manometer attached to the tapping's provided either side of the orifice plate, measures the pressure drop across the fixed orifice. The orifice plate is simply a circular metal plate with an accurately machined, centrally positioned hole (orifice). The square law principle can be used to predict flow once a flow and pressure drop relationship has been established.

Single set venturi valve: The valve is similar to the *DRV/OV* above in regulation but instead of measuring across an accurately machined orifice, the manometer measures across a venturi orifice. The venturi orifice provides more accurate flow measurement than a traditional orifice plate because it recovers more pressure (70–80 per cent compared to 40–60 per cent). The square law principle can be used to predict flow once a flow and pressure drop relationship has been established.

DROV: The *double regulating orifice valve* has pressure tapping's across the valve for measuring the valve pressure drop. This valve is less accurate than the two previous valves as the valve orifice becomes variable

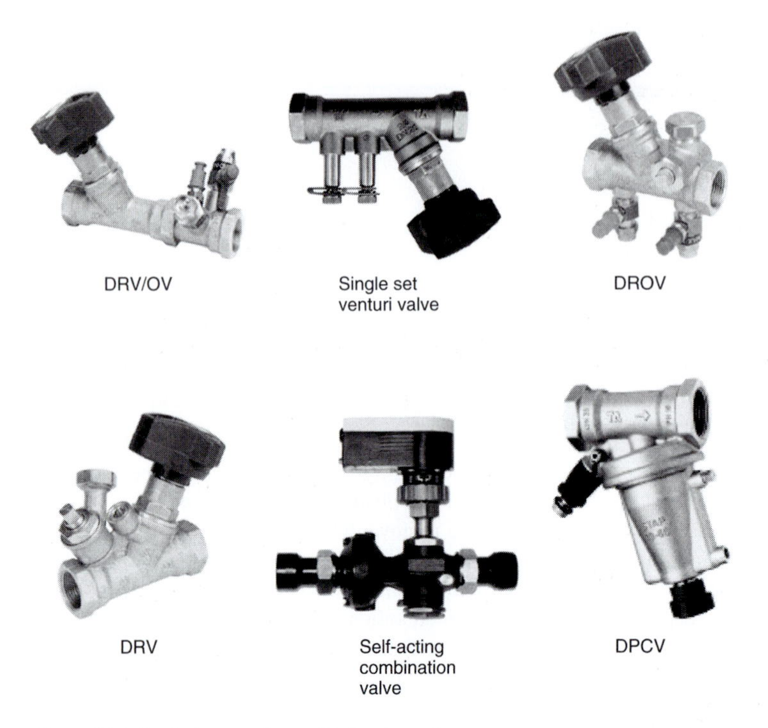

Figure 29.1 Typical measuring and balancing valves

as the valve is adjusted from full open. Reference to manufacture's performance curves is necessary to establish flow rates. The square law principle cannot be used with this valve.

DRV: This type of regulating valve used on its own will give good control on regulation, but cannot offer measurement. A typical application would be in the return of a three-way control valve, where measurement would be made on the common measuring station in the return pipe.

Self acting combination valve: This type of valve can measure integrated flow, temperature and differential pressure, providing stable and accurate control of variable volume systems, offering on/off or modulating control.

DPCV: *Differential pressure control valves* are used in variable volume systems to maintain a constant pressure within a system or branch of outlet to achieve good valve authority during the changing dynamics of a variable volume system.

Traditionally the majority of HVAC systems have been designed with fixed speed primary and secondary pumps to provide constant volume flow with the application of three port valves for controlling the loads in sub-circuits. A typical system is shown in Fig. 29.2.

For these circuits, operation of the circulating pump is always at constant speed irrespective of whether the system is operating at full load or part load conditions. Therefore irrespective of the control valve position, the amount of energy being circulated in a heating or cooling system with constant flow temperature control also remains the same, i.e. when control valves are in bypass the system is still pumping in full flow.

The main commissioning procedure therefore is generally one of water regulation where the flow rates are *proportionally balanced* in the respective branches and sub-branches and terminals, to satisfy the design requirements.

Whereas control and balancing is maintained at part load, there are no savings from reduced pumping unless the heating is switched off in the summer and conversely the cooling in the winter or the designer has provided smaller pumps for the minimum seasonal duties.

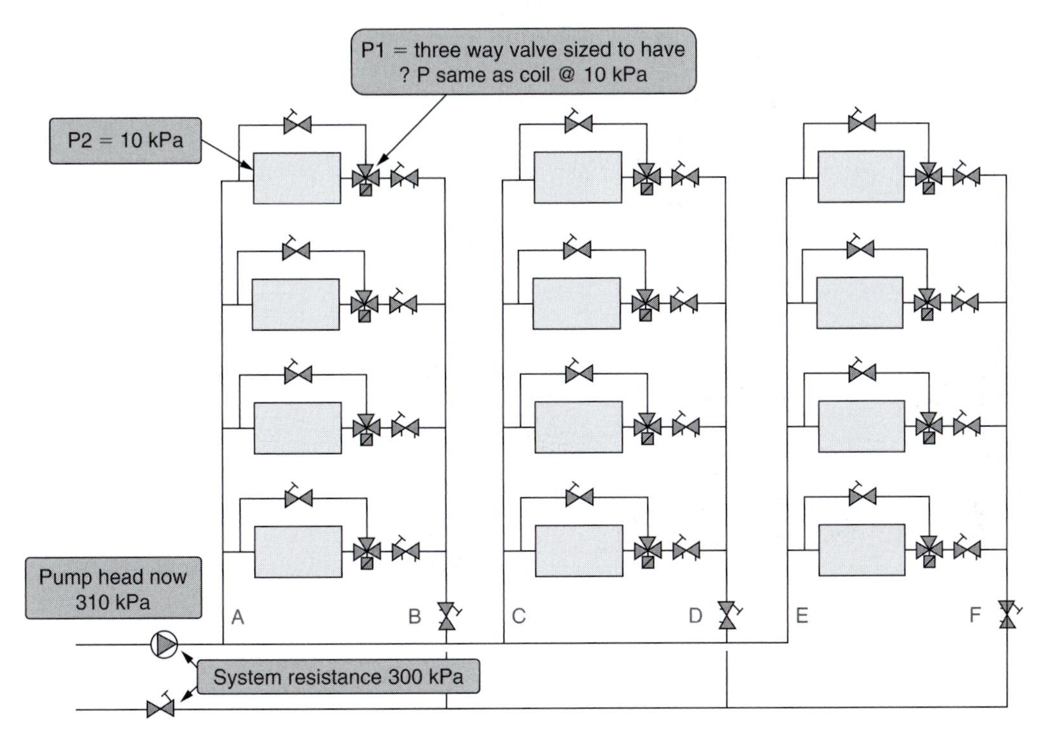

Figure 29.2 Constant volume system

The regulation and commissioning of constant volume systems is a long established procedure and widely documented in the various CIBSE and BSRIA codes and guides. The carrying out of initial scans of the systems, prior to regulation, to establish the system distribution, the most favoured and least favoured terminals/branches with respect to pressure drops is equally important when approaching regulation of systems whether they be constant or variable flow.

Control valve *authority* is defined as the ratio of the pressure drop across the valve when fully open to that across the circuit being controlled, including the valve. In the case shown in Fig. 29.2 the *authority* is:

$$\text{Valve authority} \quad \frac{\text{PI}}{\text{P1} + \text{P2}} = N$$

$$\text{Therefore} \quad N = \frac{10}{10 + 10} = 0.5$$

An *authority* of 0.3–0.5 is required to give good control of the load.

With the need to address and reduce energy usage within buildings, designers are developing more innovative ways of controlling the water distribution and energy usage. This has resulted in variable volume systems with two port control valves instead of three port.

Figure 29.3 shows a typical variable volume system with inverter controlled pump and DPCVs on the branches to terminal units.

The closer the DPCV is to the load the better valve authority can be maintained and the more energy savings are gained. Refer to *CIBSE KS/7 Knowledge series for design guidance* and *KS/9 for commissioning of variable volume systems.* In Fig. 29.3, each of the branches is controlled by a DPCV and can therefore be balanced independently of the other branches.

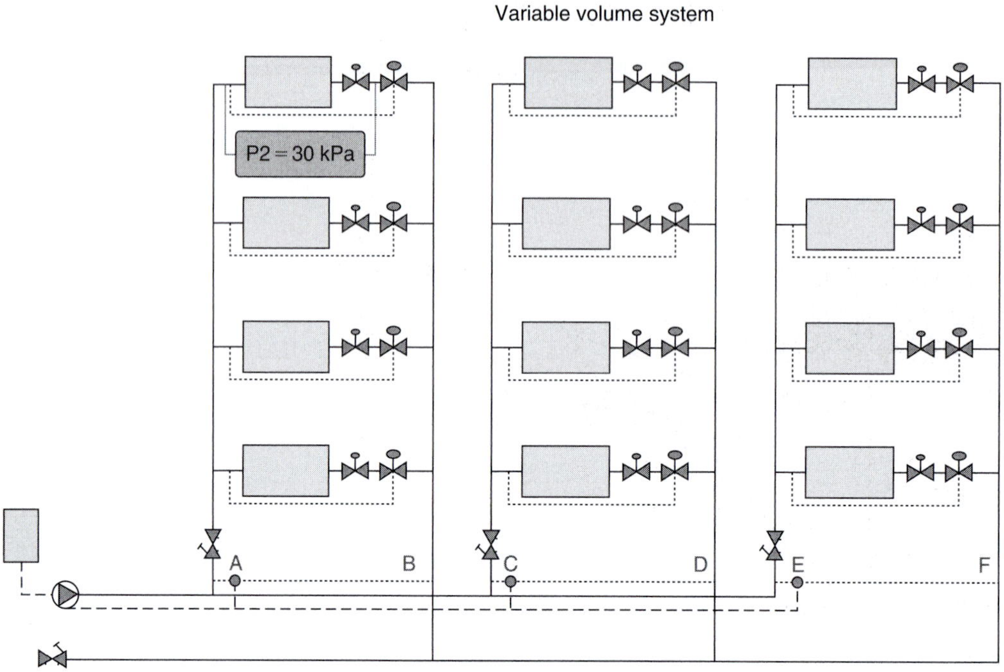

Variable volume system

P2 = 30 kPa

A B C D E F

Figure 29.3 Variable volume system – pump inverter and DPCV

Measuring instruments and balancing dampers for air systems

Commissioning involves setting the plant items to work and regulating the air flow rates within acceptable tolerances. The procedures to be followed in commissioning are given in CIBSE Code A. The following describes some of the air flow measuring instruments in use today.

Pitot tube

In its standard form, this comprises two co-axial tubes; the centre tube faces the air stream and receives velocity plus static pressure (total pressure), and the outer tube, which has a series of small holes around the wall, measures static pressure alone. Used with a manometer, as shown in Fig. 29.4, velocity pressure may be measured.

When measuring air speeds in a duct, however, it must be remembered that the velocity varies over the duct cross-section. In air flow unaffected by fittings and other components the velocity will be greatest at the centre and least at the periphery; such conditions being suitable for flow measurement. Where the flow has a non-uniform velocity, profile measurement will be subject to error. In a round duct it is necessary to take readings at a number of points in concentric rings of approximately equal area. The average speed multiplied by the area of the duct will give the volume of air passing. In the case of a rectangular duct the method is similar, except that the duct is divided into equal rectangles. Further details are provided in *BSRIA Application Guide AG 3/89.2.*

The rotating vane anemometer

This instrument (Fig. 29.5(a)) measures air speed by vanes which revolve as the air impinges on them. The instrument, which is calibrated in metres, serves only to count the revolutions over a given time, such as 1 minute, taken by a stop-watch. After each measurement the reading is returned to zero. The instrument requires to be calibrated periodically. The standard instrument is too insensitive for use below about 0.5 m/s and some instruments are unsuitable for use above about 15 m/s.

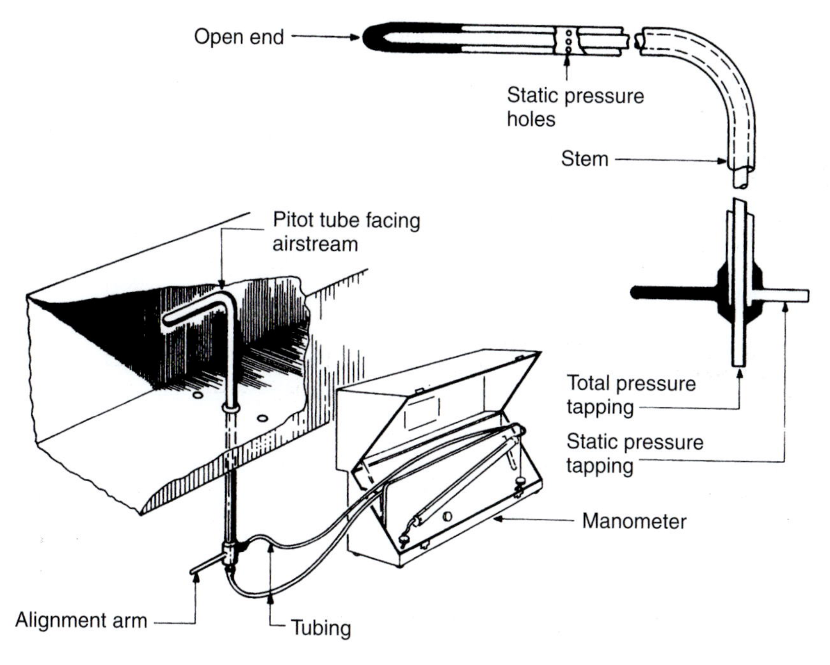

Figure 29.4 Pitot tube in use

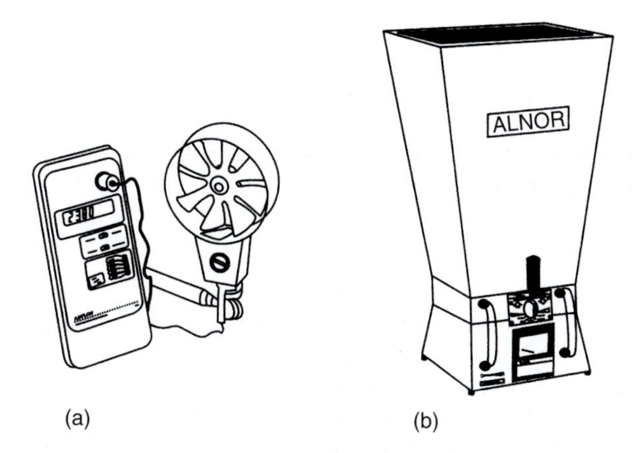

Figure 29.5 (a) Rotating vane and (b) anemometer hood

For general work it is a useful instrument, chiefly for measuring the air speed from or to openings, etc. In such cases the instrument is placed about 30 mm from the grille and the speed is multiplied by the whole face area (regardless of free area) to obtain the volume in m^3/s. Readings are taken at various points and averaged. It is not so useful for measurements in ducts, as the anemometer has to be introduced through a hole in the duct wall and is then difficult to manipulate.

It must be admitted that the anemometer requires very careful handling, and in the hands of an unskilled operator entirely erroneous results can be obtained. Electronic types are available and may be used for remote reading, where the indicating instrument is connected by cable to the rotating vane. Various measuring head diameters are available, suitable for a range of measurement applications. Anemometer readings increase in accuracy when used in conjunction with a calibrated catchment hood (Fig. 29.5(b)).

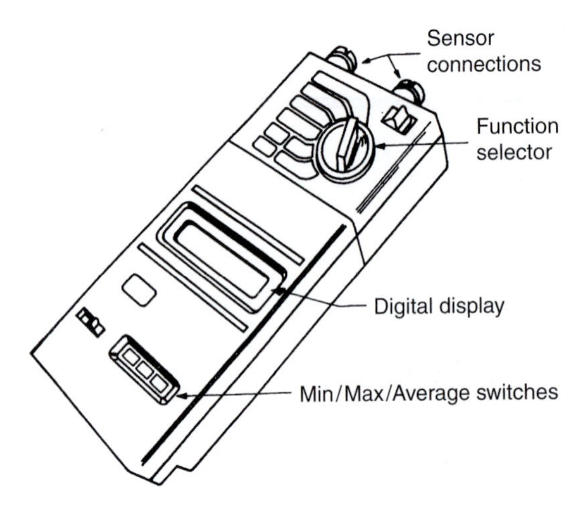

Figure 29.6 Digital multi-purpose instrument

The hot wire anemometer

Originally developed for laboratory use, this type of instrument is particularly valuable for measurement of low velocities such as those that occur in occupied spaces. The measuring head consists of a fine platinum wire which is heated electrically and inserted in the air stream. Air movement cools the wire and, by calibration, current flow may be metered to indicate the velocity sensed. As the instrument is not sensitive to the direction of flow, it is not used for measurement at grilles.

Diaphragm air pressure gauge

This instrument has a diaphragm and linkage to provide indication of static pressure on a dial. Typically, the scale range is up to 0.25 kPa. It is often helpful to measure static pressure in commissioning induction units, terminal control boxes and across components to check on the pressure loss.

Digital instruments

Modern instruments are battery operated (optional mains power connections for long periods of use are also available) and provide direct digital displays of the measured variable. Some makes are capable of giving dual output for two measured variables, together with minimum, maximum and average of readings taken over a period of time. Typically, humidity, temperature, air speed, pressure and rotational speed measurement are available by fitting different heads to a single instrument, as in Fig. 29.6. With memory facilities to provide functions such as data logging, input codes for location referencing and volume flow rates calculated from velocity measurements, this form of instrument is very convenient for site use.

Air balancing dampers

Figure 29.7 shows a number of measuring and balancing dampers used in air systems.

Plate: The plate damper constructed as a rotating plate on a central spindle is commonly used in small diameter ductwork. Unfortunately, due to its simple crude construction, the plate damper can give downstream distortion of the air stream when partially closed.

Opposed blade: This type of damper is made up of a series of plate dampers arranged so that they will swivel in alternate directions. Blades can either be flat or aerofoil section, the latter giving less turbulence of the airflow and is, therefore, quieter. It should be noted that where aerofoil blades are used, the regulating mechanism usually has an arrow, which must point in the direction of airflow.

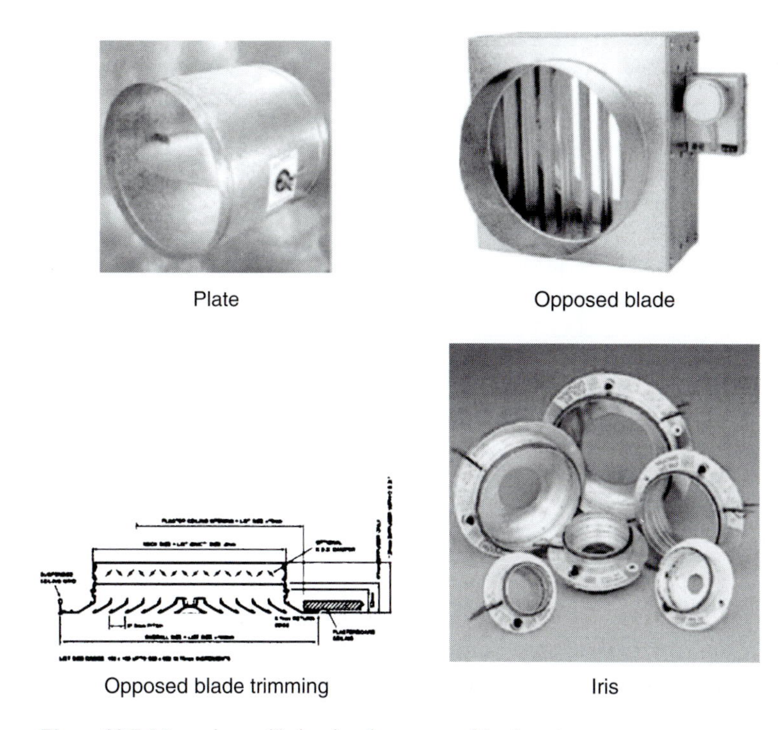

Figure 29.7 Measuring and balancing dampers used in air systems

Opposed blade trimming: Another application for opposed blade dampers is at terminal diffusers or grilles for trimming the air flow rate. It should be noted that excessive trimming at the terminal may give rise to noise problems.

Iris: Iris dampers for use in circular ducts are made up of a diaphragm of overlapping leaves connected to an exterior ring. Turning the ring in one direction opens the diaphragm, whilst turning the ring in the other direction closes the diaphragm. This type of damper can be calibrated with performance graphs produced by the manufacturers.

A simple air distribution system is shown in Fig. 29.8. Proportional balance is achieved by balancing each terminal unit in turn to the index terminal unit in each branch, i.e. the least favoured terminal. This is repeated for each branch and then the measurements at the terminal units are used to progressively balance the branch dampers to index branch, see Fig. 29.9.

Figure 29.9 shows a damper at the fan but as systems are now usually specified with inverter drives, this can be used to regulate the fan speed and therefore the air flow rather than the main dampers. This is more energy efficient than reducing the airflow by adjusting the main damper. This damper can, therefore, be omitted.

Equipment, plant and controls commissioning

The commissioning of plant and controls is generally carried out by the suppliers of the plant and, in the case of controls and BMS, by the system specialist supplier/installer.

CIBSE provide guidance on procedures for specialist equipment, e.g. for boilers, see Commissioning Code B; refrigeration systems, see Code R, lighting, see Code L. The commissioning of control systems can be a lengthy process and is often underestimated as this will involve checking the functioning of all equipment, control panels and outstations, checking correct operation of all control devices, calibration of sensors, checking of software and training of operatives, refer to Code C.

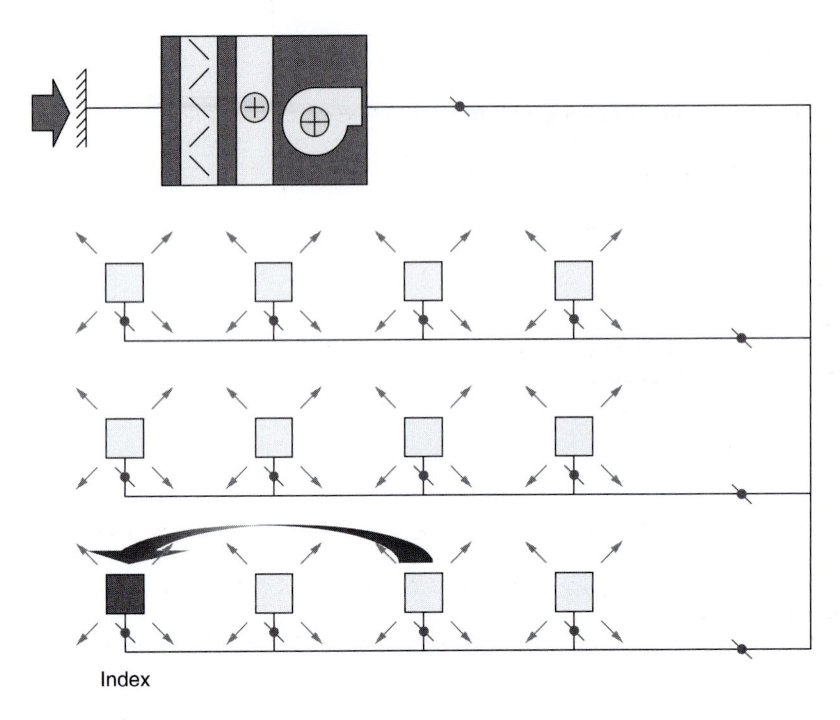

Figure 29.8 Simple air distribution system

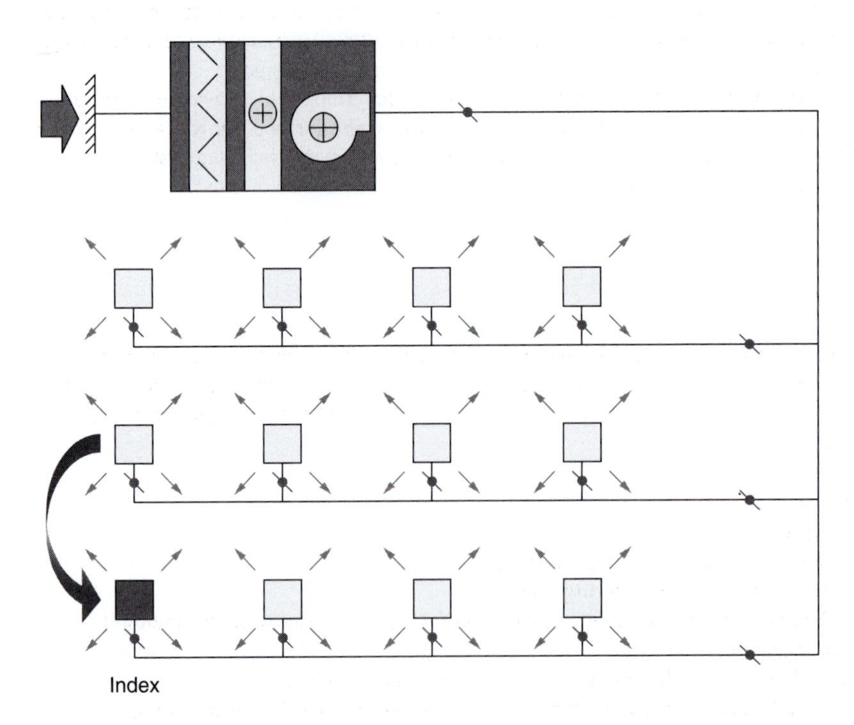

Figure 29.9 Proportional balancing of branches

Witnessing and proving system performance

Witnessing of sample commissioning results is carried out on behalf of the client to confirm the accuracy of the values recorded in the documentation prepared by the commissioning specialist. The witnessing would normally be carried out by the designer or the commissioning management organisation, if appointed.

The designer should specify the accepted tolerance on flow rates; these would normally be as follows:

- Heating mains −0/+10 %
- Heating branches and terminal units −10/+10 %
- Chilled water mains, branches and terminal units −0/+10 %
- Air flow in mechanical ventilation, comfort cooling systems −5/+10 %
- Air flow in air-conditioning, pressurisation systems −0/+10 %
- Air flow in close control air-conditioning systems −0/+5 %

The witnessing process would include:

- Checking calibration dates for instruments used for measurements.
- Checking the flow rate at selected points within the system to compare with the values recorded in the commissioning documentation issued.
- Checking sample control functions, sequence of plant operation, plant start-up and close down.
- Checking performance and documentation for all plant and equipment.
- Signing to confirm acceptance of the commissioning documentation based upon the sample results witnessed.

In critical applications it may be necessary to carry out performance testing of the system. This would involve the simulation of the peak design loads but should also have the flexibility to model the mid-season loads where operational performance of control systems may be the overriding factor. This is, however, a costly and time-consuming activity and performance testing by return visits by the commissioning engineer and designer at the appropriate time of the year is often the accepted procedure. This has the added benefit of possible fine-tuning adjustments to suit feedback from the occupants or facilities manager.

A further procedure that is often carried out is known as a 'black building test'. This is when a mains electric supply failure is simulated and a test is carried out to check generator operation and performance of other standby systems such as UPS (uninterrupted power supply), etc. The test would also be used to show that the building services plant restarts in the correct sequence when the mains supplies are re-instated.

Documentation of performance

The complete commissioning documentation must be included within the Operating and Maintenance Manuals which are handed to the client as part of the Health and Safety File which is provided at handover. The documentation would normally be provided on standard forms which would set out the appropriate data for the system or plant commissioned.

Data provided would normally include:

- Details of the measuring instrument used.
- Measuring device type, position and unique reference number.
- Details of design flow rate required at each measuring point.
- Record of measurements taken at first scan and then subsequent measurements following balancing.
- Record of settings on balancing devices.
- Comparison of final values achieved with design, normally expressed as a percentage.
- Equipment details such as manufacturer, model number, pump or fan settings, performance charts, etc.
- Record of control set points.

Handover

The importance of the handover stage of a project is not always fully appreciated. This is the stage when the client will have occupation of the building for the first time and in taking occupation he will take on the responsibility of understanding, operating and maintaining the building services systems. Responsibility for handover of a building services installation rests with the contractor; however, the designer should have input to aid the process and ensure that documentation is comprehensive and of a suitable quality. It is also beneficial if the building manager who will be responsible for management and operation of the services has a pro-active role in the handover process. At the very least, the management team should be present on site to observe the systems testing and commissioning activities.

To achieve successful handover of the building, the following is required:

- The building and services installation must be substantially complete with only minor defects listed with agreed actions to rectify.
- Commissioning must be complete and witnessed.
- Client training carried out in the operation of systems, emergency procedures and maintenance requirements.
- All handover documentation complete.

The handover documentation would include the Health and Safety File and the Building Log Book. The Health and Safety File is a requirement of the CDM Regulations and the Log Book is a requirement of the Building Regulations Approved Document Part L2A.

Health and safety file

The Construction (Design and Management) Regulations 2007, known as CDM Regulations, set down procedures to improve the management and co-ordination of health and safety throughout all stages of a construction project which is more fully described in Chapter 3. These regulations require a Health and Safety File to be produced during the construction period and handed over to the building operator at practical completion of the contract works.

The CDM Regulations give a list of information that should be included in the Health and Safety File, including:

- Record (as built) drawings
- Design criteria
- Operating and Maintenance Manuals for all systems, plant and equipment (incorporating all commissioning documentation)
- Details of the construction methods and materials used
- Fire Safety Manual
- Asset register
- Equipment and product warranties (running from the time of practical completion)

Record drawings for building services should include to scale co-ordination drawings (plans, elevations and sections) of all of the installed systems, schematic drawings and schedules of key components identifying their location within the building.

Operating and Maintenance manuals should contain full details, produced by specialist contractors and suppliers, of all installed plant and equipment to provide details for the successful operation and maintenance. In addition, complete records of the systems testing and commissioning activities should be included, together with the controls and BMS function details.

Building log book

Provision of a Building Log Book is now a requirement of Building Regulations Part L2A. A guide to assist in the preparation of the Log Book is given in CIBSE TM31 which also includes examples for a number of typical buildings. The Log Book is separate from the health and Safety File in that its purpose is that of a 'Users Manual' rather than that of an 'Operators Manual', which is more the function of the Health and Safety File.

The Log Book provides the following:

- Provision for annual overview and updates.
- Description of the building.
- Basis of design.
- System descriptions and schematics.
- Description of controls and operation.
- Metering, monitoring and targeting philosophy.
- Building and energy performance objectives.
- Summary of equipment and plant performance.
- Details of systems commissioning.
- Building and equipment certification records.

A Log Book should be used as an on-going record of key changes, monitoring performance of the systems and energy consumption, user feedback, training, etc. and as a consequence its format should enable regular updates. It should be readily available for quick and easy reference by the building management team and ideally, ownership of the Log Book should reside with a named individual who takes responsibility for its content and regular review.

Further reading

ASHRAE Handbook 2003: HVAC applications. Chapter 42 – *New building commissioning*.

BRE 2003: Digest 474 – HOBO protocol: Handover of office building operations (incorporating March 2003 amendment).

BRE 2003: Digest 478 – Building Performance Feedback (POE).

BRE 2004: Whole building commissioning – clients: Information Paper 8/04 Part 1.

BRE 2004: Whole building commissioning – designers: Information Paper 8/04 Part 2.

BRE 2004: Whole building commissioning – specifiers: Information Paper 8/04 Part 3.

BRE 2004: Whole building commissioning – facilities managers: Information Paper 8/04 Part 4.

BSRIA 1995: Technical Note TN 15/95-Handover information for building services.

BSRIA 1997: Application Guide 9/97 – Standard specification for the CDM Regulations Health and Safety File.

BSRIA 1998: Commissioning of building service installations – a guide for designers, contractors and facilities managers. *Construction Quality Forum special report*: December 1998.

BSRIA 2001: Application Guide 3/89.3 – Commissioning air systems: Application procedures for buildings. 3rd edition.

BSRIA 2002: Application Guide 2/89.3 – Commissioning water systems application principles.

BSRIA 2002: Application Guide 5/2002 – Commissioning management: How to achieve a fully functioning building.

BSRIA 2002: Technical Memoranda 1/88.1 0 Commissioning HVAC systems division of responsibilities.

BSRIA 2002: Application Guide 16/2002 – Variable flow water systems: Design, installation and commissioning guidance.

BSRIA: Application Guide AG1/87.1 – Operating and Maintenance Manuals for building services installations.

BSRIA 2004: Building Application Guide BG 2/2004 – Computer based operating and maintenance manuals.

CIBSE 1996: Commissioning Codes A – Air distribution systems.

CIBSE 2001: Commissioning Codes C – Automatic Controls.

CIBSE 2002: Commissioning Codes R – Refrigeration systems.
CIBSE 2002: Commissioning Codes B – Boilers.
CIBSE 2003: Commissioning Codes L-Lighting.
CIBSE 2003: Commissioning Codes M – Commissioning Management.
CIBSE 2003: Commissioning Codes W – Water distribution systems.
CIBSE 2006: TM31 – Building Log Book toolkit.
CIBSE 2007: KS9 – Commissioning variable flow pipework systems.

Chapter 30

Safety in design

Introduction

Construction covers a wide range of activities, processes, contracts and employs people with wide range of abilities and intellect. This means that the management of health and safety is a fairly complex issue made more difficult by the fact that each construction site is unique, with its own set of problems that need to be solved as they arise. The statistics for construction show that each year many people die in accidents and even more contract chronic ill health. Consequently, it may be concluded that to date the management of health and safety on sites has not been very successful.

The Construction (Design and Management) Regulations 2007 (CDM Regulations) are intended to improve these management failings by requiring a systematic treatment of risks to deliver, as far it is possible to do so, construction sites that are safe. Designers have a significant role to play in delivering this objective.

'Designers are in a unique position to reduce the risks that arise during construction work and have a key role to play in CDM', opens the section dealing with designers in the Approved Code of Practice (ACoP)[1] that supports the Construction (Design and Management) Regulations 2007. Thus designers become the first intervention: if they eliminate as many of the hazards as they can, there will be less for a contractor to manage on the site.

As explained in Chapter 3, designers have specific duties under the Construction (Design and Management) Regulations 2007. These are statutory duties, which must be delivered. These Regulations extend designers duties beyond the design phase, by requiring them to consider risks to health and safety of workers who will have to work on buildings to construct, inspect, maintain, repair and eventually demolish the buildings they have designed. The enforcement authorities believe that a failure to address these issues at the design stage has, in the past, led to difficulties with devising safe systems of work for operations of the kind listed above. In discharging their duties designers are required to:

(1) eliminate hazards (when they could give rise to risks);
  – only when (1) is not feasible,
(2) reduce risks from the remaining hazards; and then
(3) provide information to help a contractor to manage the remaining hazards.

The message implicit in the requirements set out above is quite simple: Designers must not produce designs that cannot be constructed safely. This is a fairly onerous duty.

Although the number of designers that have been prosecuted for a breach of the CDM Regulations since 1994 are few, there have been some and the fines have been significant. But the actual fine is not the end of the story. There are costs associated with preparation for and appearing in court and, in many cases, these costs are far in excess of the fines that have been handed down by the courts. Added to this is the effect that a high-profile prosecution could have on a designer's reputation and it rapidly becomes apparent that a breach of health and safety law could constitute a significant risk to a business.

Therefore, conforming to relevant health and safety legislation should be a very important activity that must underpin the design, construction, operation and maintenance of buildings. Understanding how the applicable legislation applies to what designers do is, therefore, an essential part of the process of design that building services engineers carry out. This chapter summarises the main legislative requirements that must be considered, and provides pointers to information that the building professional will find useful in meeting the legal requirements.

Some preliminaries

In order to discharge their duties fully, designers must, as the first step in a hazard reduction process, know the hazards covered by health and safety law and when workers would be exposed to them. Without this fundamental knowledge, they cannot begin the process required by the CDM Regulations.

Hazards covered by health and safety law and the type of work in which they may be encountered in the context of building services are listed in Table 30.1.

Designers create the hazardous work situation by requiring construction workers to operate in the work situations in column (1) in Table 30.1.

Fortunately, designers do not need to consider every hazard, because this can deflect them from their real duty, which is to identify and then attempt to deal with the *significant hazards*. Significant hazards have been defined by the HSE as those hazards that are:

- not likely to be obvious to a competent contractor or other designers;
- unusual; or
- difficult to manage.

This is the test that designers should apply to hazards that they identify. Therefore, a good understanding of what these terms mean is the next step in the successful delivery of a designer's duties.

Hazards not likely to be obvious

A designer is allowed to assume that a competent contractor will recognise most hazards on a construction site. For example, when working in a trench, it is obvious that there is a hazard to be managed: collapse of the trench, and the contractor duly puts structures in place to protect the workers in the trench from this hazard. However, if the same trench was in contaminated ground, the contaminants would not be obvious unless the contractor was told about it. If the contaminants were hazardous, this would become a significant hazard. It is this sort of hidden danger associated with excavations that needs to be brought into the open so that a contractor can arrange to deal with them.

In the context of mechanical services a hazard that is not obvious could be the contents of pipes, which can be from deposits in the pipe that may be hazardous to health. For example, in one reported incident a worker on the decommissioning of a derelict factory was removing some pipes. He set about the job using a grinder to cut through the pipes. Unknown to him, the pipes had contained flammable fluids and the action of grinding generated enough heat to ignite the residue in the pipes, which cascaded out of the pipes and onto the worker who suffered severe burns. It could be argued that the flammable residue in the pipes was a hazard that was not obvious. Consequently, the contractor did not put processes in place to protect the worker from this hazard.

Similarly, to a worker carrying out maintenance on pipes in amongst a cluster of other service runs, the fact that a pipe that he will be working close to is hot may not be obvious. Therefore, he may not take adequate precautions against preventing contact with such a pipe. This has caused serious accidents in the past.

Hazards that are unusual

These are rare, because they are often linked to new processes and the effect of something going wrong are not known, simply because it has not happened before or that it was not predictable. Buried apparatus that is very susceptible to even minor vibrations falls under this category.

Table 30.1 Hazards in construction

(1) Hazardous work situation	*(2) Hazard associated with work situation*	*(3) Operation in which hazard could be encountered (common construction operations)*
Working at height	Falling	Installing services at high level Maintaining services at high level Any work off ladders
Working on or with incomplete structures	Collapse of structure	Installing services in trenches Removing sections of duct during maintenance Installing ducts and pipes
Working close to high-energy sources	Electric shock, impact, explosion	Installing services Commissioning buildings Working close to: gas, electric-buried and overhead, water mains, etc. Connecting into existing supplies Maintaining live services – to be avoided if at all possible
Working close to or over water	Drowning	Installing services in trenches below or close to water table Installing services close to or over water, especially fast-flowing water
Working close to or with materials hazardous to health	Inhaling toxic materials	Using chemicals: adhesives, sealants Using processes which could release airborne particles and fumes, e.g. heat, abrasives, cutting, hammering, etc. Maintaining services: deposits in ducts, pipes, asbestos, etc. Dismantling services: deposits in ducts, pipes, asbestos, etc.
Working close to or with flammable materials	Fire and explosion	Using solvent-based sealants Using materials with ignition or flash point below 32°C Ignition sources (work that could create sparks) in explosive atmospheres, e.g. welding in trenches, inappropriate lighting
Working with or close to hazardous materials	Ill health	Working with asbestos, cement, solvents, irritants: acids and alkalis Creating dust: disc-cutting-bricks and block, chases in masonry, concrete Dismantling old services – toxic deposits in ducts, pipes, etc.
Handling heavy or awkward components	Musculo-skeletal injury	Lifting/manoeuvring heavy loads: installing service ducting and pipes Working with hard to grasp components: large diameter pipes, large valves
Working close to live traffic	Impact with vehicle	Installing services in trenches on highways Working in circulation areas, e.g. fork lift trucks
Working with cranes	Collapse of crane	Lifting items of (heavy) plant on to roof or into basements Lifting ducts and pipes into location
Working with or close to noisy equipment	Noise	Working near construction plant and machinery Using power tools, especially in enclosed spaces, e.g. ceiling voids
Using hand-held rotary or percussive tools	Hand–arm vibration	Cutting ducting and pipes to create deviations and bends
Working in an awkward position	Musculo-skeletal injury	Installing service ducts Undoing valves located in hard to access Installing and maintaining services close to structural members
Working in confined spaces	Asphyxiation	Working inside: box-girders, sewers, ceiling spaces, sumps Operations requiring maintenance using fuel driven plant or machinery
Working close to ionising radiation	Biological effects of radiation absorption	Very rare, but a possibility on certain sites

Dealing with hazardous liquids (other than utilities) being conveyed in buried conduits, e.g. liquid oxygen under pressure, can also be unusual, simply because it is very rarely encountered. This means that the particular contractor may not have developed a method of dealing with this hazard. Consequently, it becomes a significant hazard.

Hazards that are difficult to manage

These are hazards that although protection against them exists, such protection is hard to provide because one of or a combination of position or site constraints makes the provision of protection very difficult. An example of this type of hazard is erecting temporary supports in fast flowing water. It is difficult to manage for many reasons, not least of which is the prediction of what the water might bring down with it to impact the temporary supports.

Consider the installation of plant rooms onto the roof of a building. If the intention is to lift these items using a crane, a regulation site operation can become very difficult to manage if the plant rooms are heavy and the crane has to lift over a large radius, because this will require a very large crane. Finding room for large cranes can present problems, especially on a congested site.

Maintaining plant on a fragile roof also comes under this heading. The provision of protective measures for this kind of operation is fairly a complex operation, not least of which are the hazards associated with working on a fragile roof to install the protective measures. A significant number of maintenance workers have fallen through fragile roofs.

Some significant hazards to consider

HSG 224: Managing Health and Safety in Construction, the ACoP that supported the 1994 Regulations, sets out some hazards for designers to consider. The lists covered four areas:

(1) Areas over which the designer has direct influence.
(2) Special risks associated with occupied buildings and refurbishment work.
(3) Some hazards that should always be considered significant.
(4) Understanding the construction processes.

These lists were not retained in the ACoP supporting the 2007 Regulations. Nevertheless, this does not invalidate their significance as hazards worthy of consideration by designers and they are reproduced, with appropriate modifications, below.

Areas over which the designer has direct influence

The hazards or hazardous work situations that the Health and Safety Executive considers designers should try to eliminate (or reduce) include:

- Risks from site hazards by selecting the position of plant to minimise the chances of contact with:
  - buried services, including gas pipelines;
  - overhead cables;
  - traffic moving to, from and around the site.
- Health hazards, by, for example:
  - specifying less hazardous materials, e.g. solvent free or low solvent adhesives and water-based paints;

  - avoiding processes that create hazardous fumes, vapours, dust, noise or vibration, including disturbance of existing asbestos, cutting chases in brickwork and concrete, flame cutting or sanding areas coated with lead paint or cadmium;
  - specifying materials that are easy to handle, e.g. lighter valves and components.
- Safety hazards, by, for example:
  - eliminating the need to work at height, particularly where it would involve work from ladders, or where safe means of access and a safe place of work is not provided;
  - not locating plant on fragile roofing assemblies;
  - not requiring deep and long excavations in public areas or on highways;
  - not specifying materials that could create a significant fire risk during construction.
- Hazardous work by designing plant to allow pre-installation work to be carried out in more controlled conditions off site including, for example:
  - design elements such as process plant, so that sub-assemblies can be erected at ground level and then safely lifted into place;
  - arrange for cutting to size to be done off site, under controlled conditions, to reduce the amount of dust released.
- The risk of falling/injury where it is not possible to avoid work at height, by, for example, designing to allow early installation of:
  - anchorages, such as proprietary cast-in anchors, to reduce the use of ladders;
  - edge protection or other features that increase the safety of access and construction.
- Unsafe construction, by, for example:
  - providing lifting points and marking the weight, centre of gravity of heavy or awkward items requiring slinging both on drawings and on the items themselves;
  - making appropriate allowance for temporary works required during construction;
  - designing joints in vertical components so that bolting up can easily be done by someone standing on a permanent floor;
  - designing connections to minimise the risk of unsafe assembly.
- Unsafe maintenance and cleaning work, by, for example:
  - making provision for safe permanent access;
  - designing plant rooms to allow safe access to plant and for its removal and replacement;
  - designing safe access for roof-mounted plant, and roof maintenance;
  - making provision for safe temporary access to allow for painting and maintenance, etc., which might involve allowing for access by mobile elevating work platforms or for the erection of scaffolding.
- Unsafe demolition by identifying hazards for inclusion in the health and safety file, including:
  - sources of substantial stored energy;
  - unusual stability concepts;
  - alterations that have changed the structure.

Special risks associated with occupied buildings and refurbishment work

The special risks associated with working in occupied buildings or sites and on refurbishment can often be avoided or reduced, but only if they are identified and addressed at the design stage. Work such as creating openings in load-bearing elements can threaten the stability of structures by substantially weakening them or because of faults in the original construction, or subsequent work. Deciding the design strategy, timing and sequence of the work requires good communication and co-operation between all parties.

Where the roof of the existing building is to be retained, craning in items of plant may not be a feasible option. In such circumstances, transporting items of plant from the point of delivery to the point of installation will have to be given careful consideration. In addition, the possibility of and extent to which components can be stored on existing floors will need to be confirmed.

Designers must include adequate health and safety information with the design. This includes information about hazards that remain in the design, and the resulting risks. They need to make clear to CDM Coordinators, or whoever is preparing the *Safety Plans*, any assumptions about working methods or precautions, so that people carrying out the construction work can take them into account.

Some hazards that should always be considered significant

Some construction operations that, if they were to go wrong, would have such serious consequences that HSE considers that where the design requires them, designers should always provide information that will help a contractor to manage them. These include:

- Hazards that could cause multiple fatalities to the public, such the use of a crane close to busy public place, major road or railway.
- Temporary works, required to ensure stability during the construction alteration or demolition of the whole or any part of plant, e.g. bracing during construction of steel or concrete frame buildings.
- Hazardous or flammable substances specified in the design, e.g. epoxy grouts, fungicidal paints or those containing isocyanates.
- Features of the design and sequences of assembly or disassembly that are crucial to safe working.
- Specific problems and possible solutions, e.g. arrangements to enable the removal of large items of plant from the basement of the building.
- Plant that creates particular access problems.
- Heavy or awkward prefabricated elements likely to create risks in handling.

Understanding the construction processes

In order to deliver their duties in respect of the hazards outlined above and in Table 30.1, designers will need to understand the processes required to install services and plant in a building and then to clean, maintain and, if necessary, remove them safely. This will require designers to:

- Take full account of the risks that can arise during the proposed construction processes, giving particular attention to new or unfamiliar processes, and to those that may place large numbers of people at risk.
- Consider the stability of partially erected plant and, where necessary, providing information to show how temporary stability could be achieved during construction.
- Consider the effect of proposed work on the integrity of existing plant, particularly during refurbishment.
- Ensure that the overall design takes full account of any temporary works, e.g. temporary propping, which may be needed, no matter who is to develop those works.
- Ensuring that there are suitable arrangements (e.g. access and hard standing) for cranes, and other heavy equipment, if required.

Recording your decisions

Although there is no legal requirement to record any CDM-related decisions, a brief record of what the decision was and why it was made will help to:

- Stop a decision being reversed for the wrong reasons.
- Allow responsibility for a decision to be placed with the appropriate person.
- Demonstrate that you have exercised reasonable professional judgement.

This record of discussions should form the basis of the design stage documentation that designers are required to supply, in the form of information about residual hazards.

Design stage documentation

Designers are required to supply other designers with information about residual hazards: hazards that it has been unable to eliminate at the design stage (remember, only significant hazards need to be included in this information – see above). The purpose of this information is to allow the other designers to review each other's assessments of risks to health and safety, and to address them at an early stage in the design process, when it should be possible to alter a design without too many implications for cost.

At these reviews, it is important that critical health and safety-related design decisions are not reversed, without very good reasons for doing so.

Construction stage information

The purpose of the construction information about residual hazards is to stop contractors discovering hazards on site. Giving a contractor early warning about the existence of significant residual hazards allows it to plan for them.

Post-construction information

This is contained in the Health and Safety File, which is a very important legal document. It is needed for future construction, maintenance or demolition work, because it is the only record of residual hazards that workers will have to consider in the future.

The File for a project must contain relevant information to allow workers in the future to plan the work. Designers are expected to make a major contribution to the File.

Note

1. An ACoP has special legal status. It gives practical advice on how to comply with the law. If you follow the advice you will be doing enough to comply with the law in respect of those specific matters on which the Code gives advice.

Further reading

Managing Health and Safety in Construction: Construction (Design and Management) Regulations 2007. CDM Approved Code of Practice; HSE Series Code L Series No. 144; HSE Books, 2007.

Chapter 31

The building in operation

Introduction

The operation of buildings and their services installations requires a significant breadth of knowledge to provide reliable and cost-effective operation to achieve the desired conditions and a safe environment for the building occupants. In summary, this includes:

- An understanding of the services installations and their mode of operation
- Means to effect energy conservation
- Energy supply tariffs
- System running costs
- Maintenance requirements
- Health and safety
- Legislation.

Previous chapters in this book address the first two points above, particularly in respect of the heating, ventilation and air-conditioning of buildings, and the associated controls and building management systems.

The operating and maintenance manuals, made available at the completion of installation works, together with record drawings of the installations and the Building Log Book, provide the basic information necessary to enable an understanding of the building services systems and their operation and maintenance requirements.

Legislation and good practice

Legislation, standards and codes of practice impose a discipline on building operators to ensure compliance with relevant legal requirements and other recommendations for good and safe working practices. Coupled with the demands for compliance is the need for accurate records of the work and inspections carried out. These are summarised in the CIBSE publication in its Knowledge Series 'Managing your building services' and in the BSRIA key fact sheets available from their website www.bsria.co.uk/legislation.

The principal legislation relating to the operation of building services are:

- Health and Safety at Work etc. Act 1974
- The Building Act 1984
- Building Regulations (refer to Chapter 3)
- Confined Spaces Regulations 1997
- Construction Regulations, including Construction (Design and Management) Regulations 1994 (CDM); revised 2007
- Control of Asbestos at Work Regulations 2002
- Control of Substances Hazardous to Health Regulations 2002 (COSHH)

- Electricity at Work Regulations 1989
- Emissions into the atmosphere
- Environmental Protection Act 1990
- Fire Precautions Act 1971 and Fire Precautions (Workplace) Regulations 1997 (amended 1999)
- F-Gas Regulations 2006, effective from July 2007
- Health and Safety (Display Screen Equipment) Regulations 1992
- Lifts, lifting equipment and escalators
- Lightning protection systems
- Maintaining portable and transportable electrical equipment
- Management of Health and Safety at Work Regulations 1999
- Manual Handling Operations Regulations 1992
- Pressure Systems Safety Regulations 2000
- Reporting of Injuries, Diseases and Dangerous Occurrences Regulations 1995 (RIDDOR).

Associated with this legislation there are a number of routine inspections and testing which need to be done with records maintained as evidence that work has been carried out satisfactorily, for example:

- Fire Precautions (Workplace) Regulations 1997
- Gas Safety (Installation and Use) Regulations 1998
- Water quality inspections
- Lifts and lifting equipment
- The provision and Use of Work Equipment Regulations 1998
- Ventilation duct hygiene.

The maintenance and operation of building services is highly regulated and there are many areas of this work where the operatives and service providers are required to be registered; the most common being, for example, the Council of Registered Gas Installers (CORGI) and the National Inspection Council for Electrical Installation Contracting (NICEIC).

On a more general level it is fundamental that maintenance staff are properly trained, instructed and competent to carry out the work. Those responsible for maintenance functions should be conversant with the requirements and legislation of competency of both staff and companies contracted to undertake maintenance and other work on installed plant and systems.

Running costs

The cost of operation of any system providing space heating, ventilation, air-conditioning or hot water supply will depend upon a number of variables including:

- Fossil fuel consumption
- Power consumption and maximum demand
- Water consumption
- Materials and consumables
- Labour
- Insurance and similar on-costs
- Interest on capital and depreciation.

When selecting systems for a building it is necessary that both the initial cost of the installation and the operating costs be calculated for all the options to establish the most appropriate balance to suit the client's particular circumstances. In some cases initial costs may be the principal factor, whereas for others low operating costs may be the priority; in most cases, however, there will be an optimum choice somewhere between these two extremes. In addition, there are other factors to be taken into account such as reliability, health and safety and environmental issues. The CIBSE Guide to Ownership, Operation and Maintenance of Building Services provides a comprehensive reference to the factors to be accounted for in calculating and monitoring the running costs of building services.

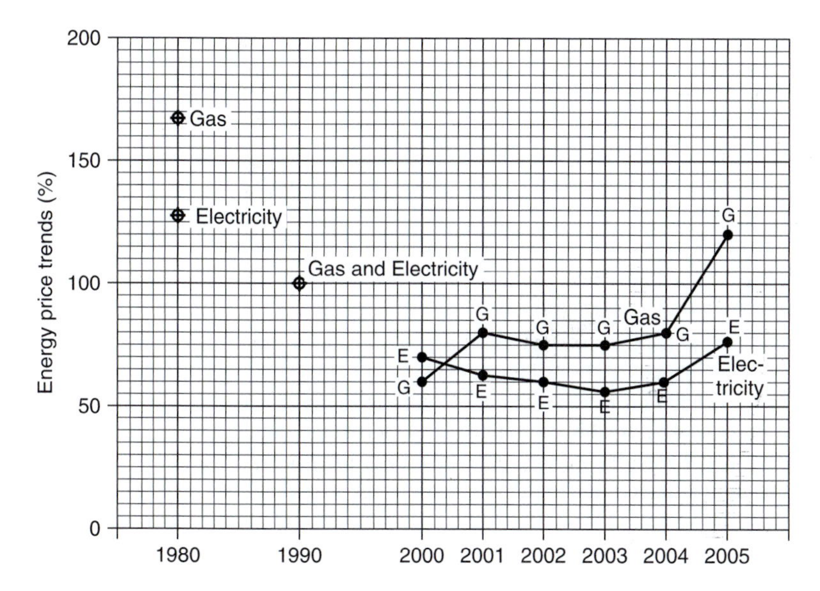

Figure 31.1 Industrial consumer energy price trends in real terms since 1980

Energy costs over the last few years have risen after a short period of relative stability. Notwithstanding the taxes that may be applied to energy use in the interests of reducing carbon emissions, since fossil fuel is a finite resource, it is likely that energy prices will continue to rise in real terms in the foreseeable future, but there is no way of predicting when and by how much prices will rise. Figure 31.1 shows trends in energy prices since 1980, based on information produced by the Department of Trade and Industry, UK Energy statistic published in July 2006.

It has been estimated that approximately 55 per cent of national annual consumption of primary energy in the British Isles is used in building services and that about 50 per cent of this relates to the domestic sector.

The UK Government has acknowledged that climate change (resulting in the main from fossil fuel burning causing the greenhouse effect and in consequence global warming) has to be addressed.

The Climate Change Levy, a tax on energy use, was announced in March 1999 and became effective from April 2001. The levy applies to commercial and industrial users, not domestic, at rates reflecting the carbon intensity of the various fuels, which until 2008 will be at the following levels and thereafter are expected to rise in line with inflation:

Electricity	0.43 p/kWh
Gas	0.15 p/kWh
Coal	0.15 p/kWh
LPG	0.07 p/kWh

In 2006, for large commercial users the price of electricity would be of the order 6.0–7.5 p/kWh and for gas 1.75–2.5 p/kWh; clearly these can only be indicative and will vary depending upon individual consumers' negotiations with the supply companies.

Various capital allowances and exemptions from the levy are available as part of the government's package to encourage investment in energy efficient installations and use of renewable energy sources. For example, for the first period the following are eligible for *Enhanced Capital Allowances* enabling 100 per cent offset in the first year, subject to meeting the eligibility criteria set by the DEFRA; a more comprehensive summary is given in Chapter 2:

- Combined heat and power
- Boiler plant, equipment and controls

- Efficient motors
- Variable speed drives
- Refrigeration plant, equipment and controls
- Thermal screens
- Pipe insulation
- Lighting products and controls.

Accepting a new installation

A description of the process and deliverables that are normally a prerequisite to practical completion of a services installation contract is given in Chapter 29. Prior to handover, there is merit in those to be responsible for managing and operating the new installation to witness the commissioning and functional testing of the systems to aid the familiarisation process. The Health and Safety File, Operating and Maintenance Manuals, and Building Services Log Book are key documents to the successful on-going operation of the services, the content of which is described in the earlier chapter. It should be noted that the Log Book provides the occupier a facility for the on-going monitoring of the performance of a building, including CO_2 emissions.

The Energy Performance of Buildings Directive

The Energy Performance of Buildings Directive (EPBD) is an EC directive intended to substantially increase the awareness and investment in energy efficiency in buildings. The requirements cover energy performance standards for new and existing buildings, energy certification and plant inspections, and are described more fully in Chapters 2 and 3. The procedures to be dealt with by building operators include:

- Compliance with Part L for an extension, installation of fixed services (e.g. air-conditioning) or an increase in the installed capacity of building services plant.
- Preparation of an Energy Performance Certificate (EPC), at the point of construction or when they are sold or rented out.
- The requirement for Display Energy Certificates (DEC) for public buildings over 1000 m^2.
- Regular inspection of boilers and heating/cooling installations.

The EPBD requires reports to be produced which are intended to provide the owner and occupier of a building guidance on measures that could be taken to improve the energy efficiency of the building. The basis of those reports will in many cases (but not all) be based on a 'walk round' inspection of the building; in some cases, particularly in simple buildings, recommendation reports might be based on an automated filtering of generic lists of measures. The regulations do not require a detailed energy audit but one could be requested by the owner/occupier if required. The specific requirements are for reports on cost-effective improvements to accompany an EPC and a DEC. The output of the requirement to inspect air-conditioning systems is also a written report based on an inspection regime outlined in CIBSE TM44, Inspection of Air Conditioning Systems.

Energy metering

To facilitate efficiency of operation, low running costs and to reduce the carbon emissions for a building it is necessary to understand where all forms of energy are being used, and wasted. Most buildings have energy supply meters for billing purposes but these are unlikely to give sufficiently detailed information for the purposes of effective energy monitoring and in identifying opportunities for energy savings.

Sub-metering end use, such as heating, hot water, cooling, fan and pump power, lighting and small power, will enable building managers to log actual consumption and to compare this with benchmarks.

New buildings will incorporate as a minimum energy metering in compliance with Part L requirements and designers should consider specifying additional metering at all points of major energy usage. CIBSE TM39, Building Energy Metering, gives details of a methodology to be followed in respect of a structured approach to metering and energy conservation, and recommends that at least 90 per cent of energy usage in a building should be sub-metered.

Automated monitoring and targeting systems are available which will considerably reduce man-power in reading meters distributed throughout a building.

Typically 5–10 per cent savings can be achieved using a targeted metering and conservation programme. As a consequence, installation of meters and associated time in reading and analysing the results are likely to be cost effective. In addition, *Enhanced Capital Allowances* are available for capital spend on monitoring and targeting equipment.

A detailed record of energy consumption should be maintained in the form of an easily read schedule, updated on a regular basis at the very least on an annual cycle, and retained in the Building Log Book.

Energy benchmarks

Benchmarks and guidance on good practice for energy consumption in buildings are given in publications by the CIBSE and The Carbon Trust, ECG series. Examples of benchmarks are given in Chapter 2. The building manager may use data obtained from meters installed on various systems to plot building actual performance against good and typical practice as shown in Fig. 31.2.

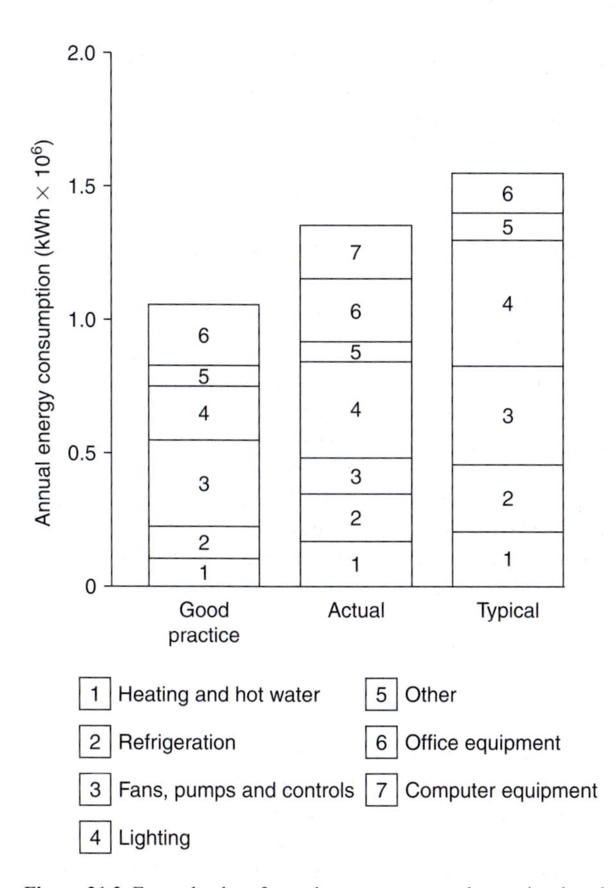

Figure 31.2 Example plot of actual energy consumption against benchmark values

Estimating energy use

Various methods are available for making energy consumption calculations, ranging from manual estimates through to detailed thermal modelling using computer simulation techniques, see Chapter 6. It is generally accepted that heating, hot water service and ventilation systems may be assessed with reasonable accuracy using manual methods, but that air-conditioning systems require calculation by some more refined form of analysis.

The various methods available may be used with reasonable confidence to compare design solutions. Use of the results to forecast actual energy use must be approached, however, with caution. There are numerous factors which may affect such a forecast adversely, such as the relevance of the weather data assumed, the method of operation or control of the systems, how the occupants use the space – such as the hours of use and level of internal heat gain and the quality of the maintenance (or lack of it). All these aspects will have a marked effect on actual energy use. For residential buildings, the *BRE Domestic Energy Model* (BREDEM) provides the standard method of estimating energy use in dwellings and, to assess the effectiveness of design solutions, increasing use is made of *Energy Ratings*. Two national rating schemes, supported by the Energy Efficiency Office of the Department of the Environment, are the *National Home Energy Rating* (NHER) scheme and the MVM *Starpoint* scheme. Both are computer models which provide a simple numerical energy rating, from 1 to 10 for NHER and from 1 to 5 for Starpoint, based upon the energy cost. The higher the rating, the more energy efficient the design.

For all other types of building the degree-day method which has been in use for many years is still the standard manual calculation for heated and naturally ventilated buildings and provides results of acceptable accuracy. Particular care must be taken, however, where the systems vary significantly from the accepted traditional designs, in which case recourse to computer modelling methods would be the preferred alternative. The degree-day method is a function of heat loss, the ratio of normal-to-peak load which will apply over the period considered, hours of use, internal heat gains, thermal characteristics of the building fabric and system efficiencies.

Heat losses

The totals calculated for design purposes will be in excess of those used as a basis for an estimate of energy consumption. This is due to the fact that if one considers any building, heating design must be such that on each external aspect, sufficient warmth may be provided to maintain a satisfactory internal temperature. In practice, air infiltration resulting from wind will occur only on the windward side; other aspects, i.e. the leeward side, will exfiltrate.

The improvement in the thermal transmission properties of the building fabric in recent years has reduced heat losses by conduction and, in consequence, infiltration losses are a higher proportion of the total; between 40 and 50 per cent of the total in the example in Chapter 4. For such application, the heat loss used for energy calculations may thus be taken at 15–20 per cent lower than the total calculated for design purposes.

Any losses from piping in a central system which do not contribute to the building heat requirements must be calculated as a separate exercise and added to the net heat loss figure. It should be remembered that, with certain types of system, mains losses may be constant throughout the heating season and thus disproportionately high. An example would be a fan-convector system, controlled by room thermostat switching of the fan motor, fed from constant temperature heating mains.

Proportion of full-load operation

This will be a variable factor depending on the weather. Obviously no system will be called upon to operate at 100 per cent output (based on winter external design temperature) during the whole season. The routine of establishing the proportion of this full load which may be assumed for the purpose of calculation, and how it varies for different parts of the country, is the basis of the method to be described.

Degree-days

The concept of the degree-day provides a measure of the duration and severity of weather above or below a given base temperature. The heating degree-day method is well established and may be used to estimate

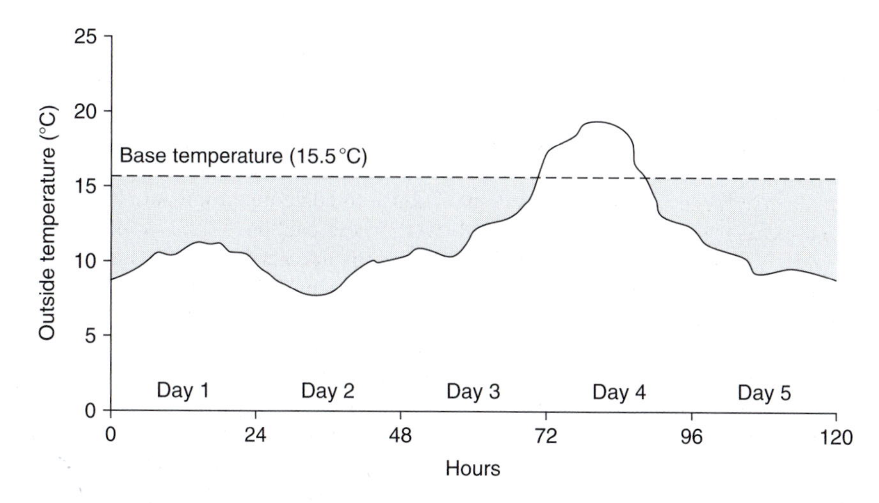

Figure 31.3 Degree-days illustrated by shaded area

the likely energy used to heat a building and to monitor trends in energy performance and the effectiveness of energy conservation measures. Although cooling degree-days are published the methodology is not well developed and is not recommended for use in connection with cooling or air-conditioning of buildings.

Recent publications by the CIBSE, Technical Memorandum, TM41, 2006 Degree-Days theory and application and The Carbon Trust, Good Practice Guide, GPG 310, Degree-Days for energy management – a practical introduction, together provide a current view of the theory and application of heating degree-days.

The British degree-day assumes that in a building maintained at 18.3°C, no heating is required when the external temperature is 15.5°C or over. The difference between the external daily mean temperature and 15.5°C is then taken as the number of degree-days for the day in question. Figure 31.3 illustrates how the degree-day value is established over a period, from which it is noted that when the outside temperature is above the base temperature, degree-days are taken as zero. Thus, for an external daily mean temperature of 5.5°C, there are 10 degree-days and if this mean were constant for 7 days then 70 degree-days would be accumulated. Monthly totals are published for various locations which may be used to compare trends from month-to-month or be summed to give the total for the season or for the year. Degree-days for use with hospital installations are used to a base temperature of 18.5°C. Improved thermal insulation standards should be considered when establishing the base temperature to be adopted; *Guide A* provides degree-days for a range of base temperatures between 10°C and 20°C for selected sites.

The degree-day should not be regarded as an absolute unit, nor should it be pressed beyond its usefulness for comparative purposes. For instance, the effect of low night temperatures may give exaggerated degree-day readings even though the buildings may not be heated at night.

If it were assumed that full load on a heating system takes place when the outside temperature is at −2°C, the maximum degree-day may be taken as 17.5 in any one day. For a heating season of 30 weeks (October/April), there is a possible total of 3710 degree-days, and for 39 weeks (September/May) a possible of 5778 degree-days. Traditionally the 30-week season has been used but, with rising standards, the 39 (or nominally 40) week season is often considered to be more appropriate, particularly for application to residential heating.

Degree-day data are compiled monthly by the Meteorological Office and are published by various organisations. The values listed in Table 31.1 are for a 20-year average to 2006 for stations in the UK.

Period of use

Here it is necessary to make assessments of the period of occupancy of a building and the length of time during which the heating system will be at work. It is clear that a finite distinction must be made between the two

Table 31.1 Degree-day totals for the UK, 20-year averages (1987–2006), for a base temperature of 15.5°C

Area	*Annual total*	*Seasonal totals*	
		September/May	*October/April*
Thames Valley (Heathrow)	1816	1742	1605
South Eastern (Gatwick)	2086	1970	1788
Southern (Hurn)	2032	1902	1718
South Western (Plymouth)	1735	1635	1478
Severn Valley (Filton)	1792	1717	1577
Midlands (Elmdon)	2204	2072	1871
West Pennines (Ringway)	2142	2009	1814
North Western (Carlisle)	2346	2173	1937
Borders (Boulmer)	2376	2168	1899
North Eastern (Leeming)	2311	2151	1921
East Pennines (Finningley)	2190	2055	1854
East Anglia (Honington)	2193	2062	1869
West Scotland (Abbotsinch)	2412	2219	1968
East Scotland (Leuchars)	2498	2291	2071
NE Scotland (Dyce)	2570	2342	2053
NW Scotland (Stornaway)	2497	2222	1922
Wales (Aberporth)	2074	1919	1710
N. Ireland (Aldergrove)	2279	2106	1870

Source
www.vesma.com.

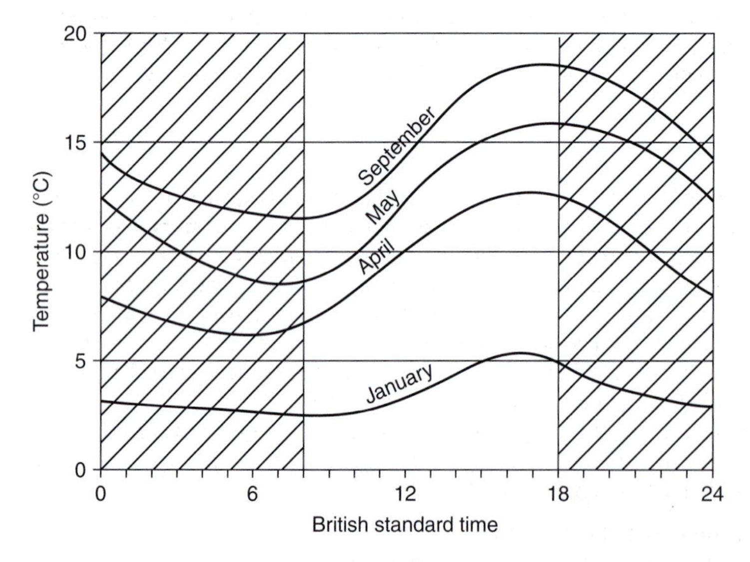

Figure 31.4 Typical variations in outside temperature

and although the former may be defined with ease, the latter will depend upon the particular characteristics of each individual building structure and the system associated with it. It is the response of the structure to heat input and the varying abilities of any one of a variety of systems to match that response which is important.

For the heating system, the hours of use may be considered in terms of periods per year, per week or per day. For the yearly use, as has been explained previously, a figure of 30 weeks is the traditional period assumed for commercial and industrial premises. Since, as illustrated in Fig. 31.4, the diurnal external temperature range in May and September is such that early morning and evening heating may well be required in houses, flats, hotels and hostels, etc., a 40-week season is likely to be more appropriate in these instances.

Table 31.2 Approximate temperature rise due to internal heat gains

Building characteristics	*Temperature increment* (K)
Large areas of external glazing; much heat-producing equipment; dense occupancy	5–6
One or two of the above characteristics	4–5
Traditional with normal levels of glazing, equipment and occupancy	3–4
Small glazed areas; little or no heat producing equipment; sparse occupancy	2–3
Residential and dwellings	5–8

Note
These data should not be used for purposes outside this present context.

The weekly use again depends upon the kind of building and type of occupancy. Domestic premises, hotels and the like require service over 7 days; commercial and industrial buildings need not be heated for more than 5 days and no more than 1- or 2-day usage is normally considered for churches, village halls and the like. For daily use, similar conditions apply and it will be appreciated that continuous heating over 24 hours each day is very seldom required. The shorter the period when service is provided, however, the greater will be the no-load losses or, where plant operation is to be intermittent, the greater the requirement for preheating.

Hence, if energy supply is to be by oil or gas firing or by some form of direct electrical heating, on-peak, no nighttime operation of the system will be necessary except perhaps in the most severe weather when heat might be needed to prevent condensation. The plant will be required nevertheless to provide preheating prior to occupation. On the other hand, with solid fuel firing by an automatic means, some level of intermittent operation will usually be necessary in order to keep the fire alive overnight and a no-load loss will thus arise with little heat input to the building. This loss will not be absolute of course and some minor reduction in pre-heat time may be possible simply because the water content of the boiler will not be completely cold.

Equivalent full-load hours

The period during which the heating system is in use, as distinct from the period of occupancy of the building, may be expressed in terms of *equivalent hours of full-load operation*. A number of different methods leading to evaluation of the quantity by manual calculation have been devised and an outline of one of these is presented in the *Guide Section B18*.[1] The underlying theory takes account of variations in the level of adventitious heat gains within a building, the period of occupation, the thermal capacity of the structure and the ability of the heating system to respond to demand. Allowance is made also for the fact that the degree-day data compiled by the Meteorological Office relate to the maintenance of an internal temperature of 18.3°C, as has already been explained: for other internal temperatures, an adjustment is necessary.

As far as the internal heat gains are concerned, it is suggested that buildings may be classified as shown in Table 31.2 and that the thermal capacity of the structure be categorised as light, medium or heavy, although it is appreciated that judgement will be required to interpret these distinctions:

- *Light*: Single-storey factory-type buildings with little partitioning.
- *Medium*: Single-storey buildings of masonry or concrete with solid partitions.
- *Heavy*: Multi-storey buildings with a lightweight facade and solid partitions and floors.

With regard to hours of system use, the method allows for continuous running over 24 hours or, alternatively, for operation as is appropriate to periods of occupation from 4 to 16 hours per day.

The routine set out in the *Guide Section B18* may be simplified, with no loss of accuracy, by selection of the appropriate degree-day figure from Table 31.1 and an estimated value of internal heat gain from Table 31.2. The former is then corrected by multiplication using values chosen in turn from each of Tables 31.3–31.5 in order to represent the various aspects of the installation.

For example, consider a building in East Anglia having a traditional heavyweight structure and provided with a slow response heating system such as hot water radiators. The system is operated intermittently over

Table 31.3 Factors relating building characteristics to inside and outside design temperatures

Temperature increment from Table 31.2 (K)	*Inside design temperature* (°C)											
	19				*20*				*21*			
	Outside design temperature (°C)											
	−1	*−2*	*−3*	*−4*	*−1*	*−2*	*−3*	*−4*	*−1*	*−2*	*−3*	*−4*
2	1.40	1.34	1.28	1.22	1.46	1.40	1.34	1.28	1.52	1.45	1.39	1.33
3	1.27	1.21	1.16	1.11	1.34	1.28	1.22	1.17	1.40	1.34	1.28	1.23
4	1.14	1.09	1.04	0.99	1.21	1.16	1.11	1.06	1.28	1.22	1.17	1.12
5	1.01	0.96	0.92	0.88	1.09	1.04	0.99	0.95	1.16	1.11	1.06	1.02
6	0.88	0.83	0.80	0.76	0.96	0.92	0.87	0.84	1.04	0.99	0.95	0.91
7	0.76	0.72	0.69	0.66	0.83	0.80	0.76	0.73	0.92	0.88	0.84	0.81

Table 31.4 Factors relating structural mass to mode of plant operation and system type

	Continuous plant operation (24 hours) system		*Intermittent plant operation*[a]			
			Slow response system		*Fast response system*	
Building structure	*7-day week*	*5-day*[a] *week*	*7-day week*	*5-day*[b] *week*	*7-day week*	*5-day*[b] *week*
Heavy	1.0	0.85	0.95	0.81	0.85	0.71
Medium	1.0	0.80	0.85	0.68	0.70	0.56
Light	1.0	0.75	0.70	0.53	0.55	0.41

Notes
a With night-time shutdown.
b With weekend shutdown.

Table 31.5 Factors relating structural mass to period of occupation

	Period of actual occupation			
Building structure	*4 hours*	*8 hours*	*12 hours*	*16 hours*
Heavy	0.96	1.0	1.03	1.05
Medium	0.82	1.0	1.13	1.23
Light	0.68	1.0	1.23	1.40

30 weeks per annum for a 5-day week and occupation during a 12-hour day. The inside and outside design temperatures are 20°C and −3°C, respectively, and thus:

Degree-days (Table 31.1)	=	1869
Temperature increment (Table 31.2)	=	3°C
Factor for building type, use and temperatures (Table 31.3)	=	1.22
Factor for intermittent use over 5 days of a slow response system in a heavy building (Table 31.4)	=	0.81
Factor for occupation over 12 hours in a heavy building (Table 31.5)	=	1.03
Hence, Equivalent full-load operation	=	1869 × 1.22 × 0.81 × 1.03
	=	1902 hours per annum

Table 31.6(a) Heating systems (%)

	Intermittent			*Continuous*		
System description	*Heat conversion efficiency*	*Utilisation efficiency*	*Seasonal efficiency*	*Heat conversion efficiency*	*Utilisation efficiency*	*Seasonal efficiency*
Automatic central radiator or convector systems	65	97	63	70	100	70
Automatic central warm air systems	65	93	60	70	100	70
Fan-assisted electric off-peak heaters	100	90	90			
Direct electric floor and ceiling systems, non-storage	100	95	95	100	95	95
District heating warm air radiators/ convectors	75	90	67.5	75	100	75
Electric storage radiators				100	75	75
Electric floor storage systems				100	70	70

Notes
Allow for rekindling on intermittent solid fuel plant.
Allow for fuel oil preheating where required.

Table 31.6(b) Hot water systems (%)

System description	*Heat conversion efficiency*	*Utilisation efficiency*	*Seasonal efficiency*
Gas circulator/storage cylinder[a]	65	80	52
Gas- and oil-fired boiler/storage cylinder[a]	70	80	56
Off-peak electric storage with cylinder and immersion heater	100	80	80
Instantaneous gas multi-point heater	65	95	62
District heating with local calorifier[a,b]	75	80	60
District heating with central calorifiers and distribution[a,b]	75	75	56

Notes
a Make separate allowance for mains losses.
b Heat conversion efficiency in summer may reduce depending on sizing of heat generators.

It is a simple step from this point to calculate the net annual heat requirement, i.e. for a heat loss of 500 kW:

Annual load $= (1902 \times 500)/1000 = 951$ MWh

or

$= 951 \times 3.6 = 3424$ GJ

Annual fuel or energy consumption

In order to determine annual fuel consumption, and hence cost, it is necessary in each case to take account of both the properties of the fuel and the *seasonal heat conversion efficiency*. The latter, which includes allowance for plant operation at less than full load, may be read from that part of Tables 31.6(a) and (b) appropriate to the plant under consideration. In addition, consideration must be given to the *utilisation efficiency* of the heating or other system (as distinct from that of the plant) which will depend upon the facilities offered for control, the disposition of equipment and other kindred aspects. Suggested values, based upon data included in the *Guide Section B18*, are given also in the two parts of Tables 31.6(a) and (b). It will be appreciated that these efficiency values are necessarily figures for guidance: they should be varied in circumstances where it is known that either plant or system performance is significantly better or worse than the values in the table. For example, modern boilers complying with Part L of the Building Regulation, 2006, have a minimum heat conversion efficiency much higher than those given in Tables 31.6(a) and (b); 84 per cent for boilers installed in new non-domestic buildings, as described in Chapter 7.

Taking, as a single example, a boiler plant fired by light grade fuel oil (class E), which operates to serve the system considered above, the relevant facts may be marshalled as follows:

Calorific value of fuel (Table 16.4)	=	40.1 MJ/kg
Specific gravity of fuel (Table 16.4)	=	0.94
Annual load (previously calculated)	=	3424 GJ
Routine of system operation	=	intermittent
System type	=	automatic central radiator
Heat conversion efficiency (Table 31.6(a))	=	0.65
Utilisation efficiency (Table 31.6(a))	=	0.97

Thus

$$\text{Consumption} = (3424 \times 1000)/(40.1 \times 0.94 \times 0.65 \times 0.97) = \text{say } 144\,000 \text{ litre per annum}$$

Running of auxiliaries

In addition to the fuel consumption, allowance must be made for usage of electrical power by a variety of auxiliary equipment such as circulating pumps and pressurising equipment, etc. In particular, boiler ancillaries must be considered; oil preheaters, transfer pumps, burners and fans; gas boosters and burners; solid fuel stokers, transporters and ash handling equipment. In making estimates of the running times of such equipment in instances where the heat generating plant is operated intermittently, an inclusion must be made for the preheating period which will average about 2 hours where optimum start controls are used and 2–3 hours or more in the case of simple time switching.

As a rule, for the lower range of boiler sizes up to say 1 MW, the cost of the electric current for burner operation is not very significant and is often ignored. For a boiler of this size, when burning oil requiring preheating:

Hourly fuel consumption	=	about	0.03 kg/s
Temperature rise (say 10–80°C)	=		70 K
Specific heat capacity	=	about	2 kJ/kg

Thus, approximately

Heat required = 0.03 × 70 × 2	=		4.2 kW
Burner fan and pump, etc.	=	say	2.0 kW
Induced draught fan (if any)	=	say	3.0 kW
Transfer pumps and sundries	=	say	1.0 kW
Total			10.2 kW

Running hours for the auxiliaries will be greater than those of boiler output but even if the total were to be half as much again, the consumption noted above would represent only about 1 per cent of the overall energy input. In the case of solid fuel firing, the same order of power consumption for auxiliaries would apply but for gas firing, using a packaged burner with an associated pressure booster, the power consumed would be only about a third of that amount.

Similarly, the current used for driving circulating pumps, etc. is of relatively small magnitude and indeed most of the energy paid for emerges as heat somewhere in the system and so is not altogether lost. A boiler plant of 1 MW capacity would probably be associated with a heating system requiring pumping power of the order of 2 or 3 kW, running continuously throughout the season and thus perhaps equivalent to another 1 per cent of the overall energy input.

As a very broad approximation, pending a proper calculation from manufacturers' ratings of the various drives, etc., an allowance for a power consumption equivalent to 5 per cent of the overall energy input as fuel should cover all auxiliary equipment.

Direct heating systems

For direct systems of convective type, methods similar to those used for calculation of energy consumption in central plants may be used, using appropriate values from Tables 31.6(a) and (b), provided that equivalent arrangements for thermostatic and time switch control are provided. If the system were to be of radiant type, a true comparison must be based upon equation of demand to maintenance of operative temperature. An estimate may then be made of running cost taking the time when the heaters are likely to be in use and multiplying this by their rated capacity. There is often no true temperature control, in the accepted sense, provided for such systems, and data from comparable installations are probably the best guide.

Energy performance using degree-days

Since energy used in space heating is directly related to degree-days, this can be used to assess the energy performance of a building and its heating system. Various aspects of performance may be monitored using this simple relationship, including:

- Changes in efficiency of operation
- Identification of system faults
- Assessment of energy conservation measures
- Setting of running cost budgets.

As an alternative to relying upon published degree-days, buildings with building management systems can be used to establish the degree-days for the particular building using data collected from the site. Where this is done, care must be taken to check that the measuring devices are regularly calibrated.

The first step is to establish a database of energy consumption against degree-days, always ensuring that measurements are accurate and are for the time period of the degree-day information. Energy consumption can be established in a number of ways:

- Using automated meter reading, via a building management system or a monitoring and targeting system
- Manual meter reading
- Monthly bills.

From the monthly energy consumption and degree-days a graph can be plotted, typically as given in Fig. 31.5. Using the statistical line of best fit for the individual points a *performance line* for the building is established. The spread of the points is an indication of measurement accuracy, which should be examined critically to be satisfied that the data is a reasonable basis against which to monitor on-going system performance. The slope of the performance line is a measure of fuel consumed for an increase in degree-days, whilst the intercept at the vertical axis, which is normal for most buildings, represents the non-weather-related energy usage, for example in heating for hot water, humidification and kitchen appliances. Where the performance line intercepts the horizontal axis, this indicates that internal heat gains, perhaps from equipment, are higher than normal, or that the building is particularly well insulated and sealed against air infiltration. Performance lines having a distinct curvature indicates that the heating system is not following normal trends and should be investigated to ensue that the controls are functioning correctly.

Clearly, energy usage may be monitored against the performance line. Significant deviations or trends at variance from the norm would indicate a change in system performance which would warrant investigation of changes in usage patterns within the building or of component malfunction. The effects of energy conservation measures may also be established in a similar manner.

Degree-days can also be used to assist the setting of budgets for energy consumption. Recognising that weather patterns vary from one year to the next suggests that simply using the previous year's figures to predict next year's consumption is liable to considerable error. Also, using the 20-year average degree-day figure

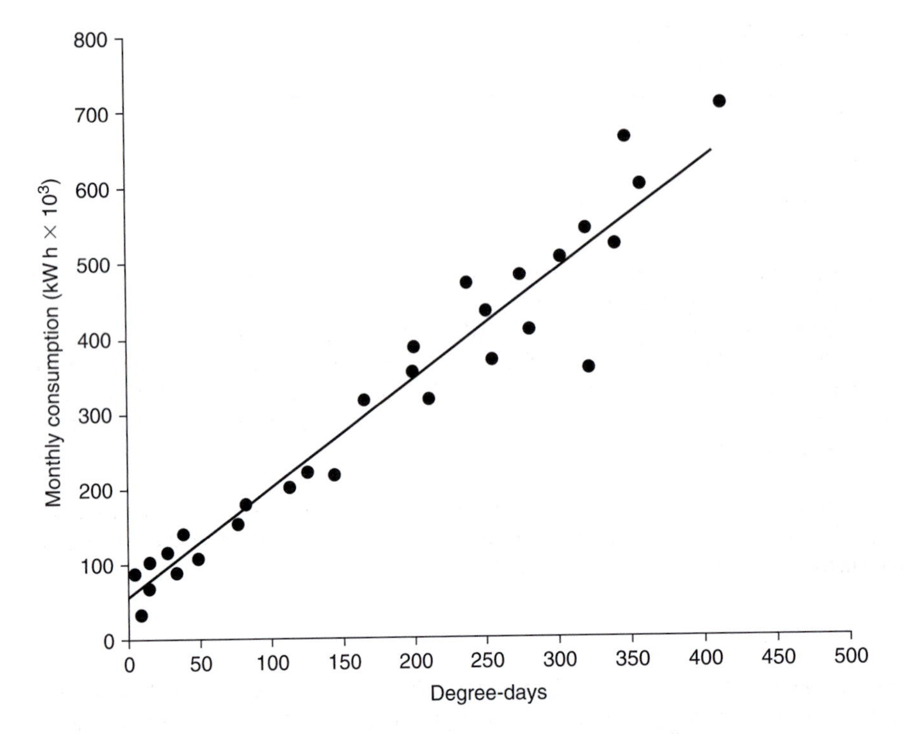

Figure 31.5 Plot of monthly energy consumption; showing the performance line

will carry considerable risk (50 per cent) of an underestimate. To reduce this risk consideration should be given to using either the upper quartile (75th percentile likely to be exceeded 1 year in four), or the 90th percentile (exceeded 1 year in ten). A view may also be taken on trends arising from global warming, but for budgeting purposes perhaps this introduces unnecessary further risk.

Mechanically ventilated buildings

Essentially, the thermal energy requirements for mechanically ventilated buildings may be calculated following similar procedures to those adopted for the equivalent heated and naturally ventilated spaces, with limited correction and with some additions. It is assumed that no cooling or humidification is to be provided.

Heat requirements

The following items should be considered in turn:

- Reduction in air infiltration rate, possibly due to sealed windows.
- Additional load due to ventilation air, i.e. the quantity of outside air introduced by the ventilation system.
- Reduction in load due to incidental gain to the air from the fan (fan gain).
- Allowance for the benefit arising from the use of any heat recovery device.
- Correction to the effective length of the heating season.

The first two items above will be self-explanatory.

A mechanical ventilation system operating in a heating mode will benefit from heat gain to the air as the electrical power driving the fan is down-graded to heat. The quantity of heat dissipated to the air stream will depend upon whether the fan motor is within or outside of the air stream. Table 31.7 gives approximations

Table 31.7 Approximation of fan heat gain to an air stream

Location of motor	*Motor size* (kW)	*Heat gain to air stream* (%)
In air stream	All	100
Out of air stream	Up to 4	75
Out of air stream	Above 4	85

adequate for the purpose. These values should be adjusted, as required, to reflect the limitation to specific fan power given by Part L of the 2006 Building Regulations, as described in Chapter 22.

The application of heat recovery devices to air systems, together with the thermal efficiencies of available equipment, is described in Chapter 22. The benefit in thermal energy terms is in effect to reduce the additional load imposed upon the supply plant due to the introduction of outside air. Any increase in system pressures arising from the imposition of heat exchange equipment in the supply and extract systems must be included in the calculations together with any power requirements for additional pumps or drive motors associated with the heat recovery devices.

Since the ventilation plant may run year-round, the effective period over which heating is introduced may be extended to offset the effect of cold days outside of the normal heating season. In consequence, the degree-day totals may have to be increased, perhaps to the annual totals (see Table 31.1).

Power requirements

The system components driven by electric motors would normally include the supply and extract fans and any drives associated with the heat recovery equipment. Taking the supply fan as an example the power requirement is given by:

$$W = KP_t/(1000\eta)$$

where

W = absorbed power (kW)
V = volume flow rate (litre/s)
P_t = fan total pressure (kPa)
η = fan, motor and drive efficiency

The annual power consumption for the supply fan is given by $W \times$ number of hours in operation which will, of course, include for the preheat periods where appropriate.

Application

It must be appreciated that manual calculations of the type outlined earlier in this chapter for heating and ventilating plants, relying upon application of degree-day data, etc. cannot be expected to provide an accuracy in absolute terms of better than *plus or minus 20 per cent*. The variables encountered and the assumptions made preclude a greater precision. The results however are valuable in a comparative sense and may be used with a much improved level of confidence when considering the relative merits of a variety of energy sources: this is the application for which they are intended.

Air-conditioned buildings

Energy predictions for air-conditioned buildings take on a different dimension from those for heated only buildings. As well as heat energy, the designer is concerned with cooling, dehumidification and humidification and in consequence, simple averaged temperature relationships between inside and outside are no longer an adequate basis for calculation. It is necessary to take account of the coincident values for dry-bulb and wet-bulb temperatures and of solar radiation; wind speed may be important but its variation with time is not normally taken into account since it is less significant than the other parameters.

It would be possible to make manual calculations but, for practical purposes, these would be limited to 12 average monthly conditions. However, averages over such long periods may give misleading results since, for example, a month's average conditions may indicate no heating or cooling; a similar and equally important situation may exist in respect of humidification and dehumidification.

Such energy calculations can only be carried out with any accuracy and consistency by computer simulation methods, which are described in some detail in Chapter 6.

Hot water supply

The consumption of energy for a central system may be considered in two parts:

(1) Heat losses from storage vessel(s) (Table 31.8), circulating pipework (Table 25.15), towel airers, linen cupboard coils, dry coils, etc.
(2) Actual hot water drawn off by the users.

It is worthy of note that these two components of the total are often of the same order of magnitude.

A knowledge of the type of building will determine whether heat losses are continuous for 24 hours per day, as in a hospital, or for some lesser period such as 8 hours per day in a school. Similarly the days per annum will vary. For intermittent operation, the heat loss from storage vessels and pipework will continue after the use of the system has ceased, perhaps until the water temperature is in equilibrium with the surrounding air depending upon the period of close-down.

The heat consumption of the water actually drawn off, taken from cold at say 10°C to hot at say 60°C, will be derived from data of water demand of comparable buildings. Table 25.3 provides data in this respect. The total of the two components will then form the basis of the sum in which the calorific value of fuel and the seasonal efficiency are taken into account as for heating (see Table 31.6(b)).

Energy audits

An energy audit is a management tool to control energy use and costs. It is essential, in the first place, to analyse energy use; where, how much and in what form it is being expended. Then, to monitor consumption at regular intervals, monthly being often the most convenient period. The results should then be compared with a suitable indicator for the same periods, such as degree-days or energy benchmarks for the type of building,

Table 31.8 Heat loss from storage cylinders (insulated with 75 mm glass fibre sited in an ambient temperature of 25°C

	Heat loss from cylinder	
Capacity of cylinder (litre)	Watt	MJ/annum
150	40	1300
250	55	1800
300	60	2000
450	80	2500
650	100	3300
1000	135	4300
1500	180	5600
2500	250	7600
3000	280	8800
4500	360	11 300

as described earlier in this chapter and in Chapter 2. Once the operator has a full understanding of how energy is being used, targets may be set with the aim of reducing usage, followed by detailed studies with intent to identify further savings. At this point it may be beneficial to consult an expert in the field.

In carrying out such an audit, it is necessary to establish that, commensurate with the cost of meeting the objective, the following criteria are met in the most energy efficient manner:

- The most economical source of energy is used and at the best commercial tariff.
- All energy is converted efficiently.
- Distribution losses are minimised.
- Patterns of demand are optimised and plant controls are compatible.
- Energy recovery equipment is provided.

Examples of reducing energy demands are given in Chapter 2.

A graphical representation of readings taken will identify significant variations from an established trend. Figure 31.6 shows a typical example of fuel oil usage against degree-days for two heating seasons, and as can be seen a straight line graph should be obtained. An alternative approach is to calculate for each month a *litre/degree-day* which should give a constant figure, subject to allowable practical tolerances. An annual litre/degree-day ratio, or an equivalent for different fuels, will provide also a means to compare the performance of one building with another. This will quickly identify poor performance, for whatever reason, and establish where improvements should be investigated as a first priority.

It is generally accepted that the energy consumption in the majority of buildings can be reduced by a minimum of 20 per cent by the application of good management and cost-effective improvements. The process for implementing a programme of energy-saving measures may be summarised as:

- Establish use against accepted benchmarks
- Identify and prioritise opportunities for savings
- Produce an action plan
- Obtain buy-in from management and staff
- Implement the plan and monitor results
- Evaluate against benchmarks
- Set new targets.

A *monitoring and targeting* system may usefully be set up to assist this process, either by implementing a simple manual method or for larger energy users, say with annual energy bills in excess of £10 000, a computer-based software system should be used. Reference should be made to the CIBSE Guide F, Energy Efficiency in Buildings for further information on achieving energy savings in building services systems.

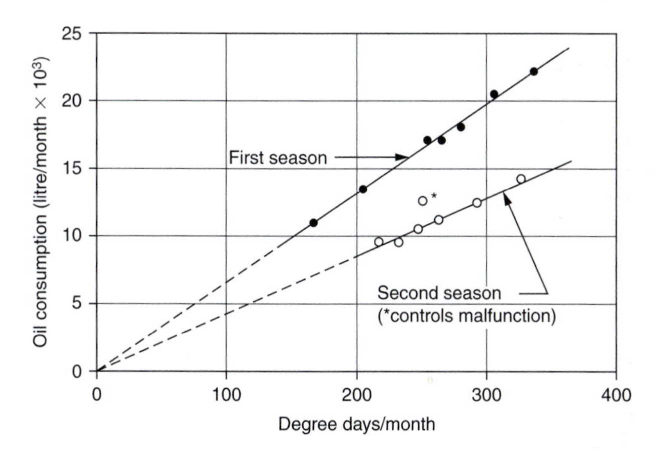

Figure 31.6 Fuel oil consumption relative to degree-days

Maintenance

There are various levels of maintenance which may be applied to building services, the most common being:

- *Corrective*: The majority of operations are carried out on breakdown or fall off in performance, backed up sometimes with specific tasks undertaken on a regular basis. This approach carries a high risk of plant failure and inefficient operation.
- *Preventative*: Planned procedures are undertaken at regular intervals related to statistical failure rates of equipment with intent to preserve the life of the plant overall to a maximum and to minimise the risk of breakdown. Work is carried out to a predetermined schedule enabling resources and material purchase to be planned in advance.
- *Conditional*: Electrical power absorbed, vibration, temperature profiles or even condition of lubricating oils may be monitored on discrete items of equipment and compared against models for optimal running conditions. Deviation from predetermined parameters indicates the need for maintenance action. This form is best suited to critical equipment where failure will result in consequential losses. Although there are increased costs associated with sensor installation and analysis, overall cost reduction may be significant in eliminating unnecessary maintenance.

Planned maintenance is programmed in relation to the task to be undertaken and its frequency, typically weekly, monthly, quarterly or annually, including for example inspections, cleaning, safety checks, testing, replacement of consumables and parts, and refurbishment or replacement of plant. In addition is the work specifically associated with efficiency and reliability, such as function testing, calibration, meter reading, condition monitoring, energy monitoring, controls fine tuning. Provision must also be made for the inevitable unplanned activities.

There are considerable benefits in establishing an early stage in the design process a maintenance policy taking into account resilience, hours of operation, level of engineering skill available and other matters particular to a given installation. To assist the process reference to Fig. 31.7 (from CIBSE publications) will indicate the options available and factors to be considered.

It should also be recognised that the role of the maintenance team has a significant impact upon many important aspects of running a building, including:

- Health and safety, e.g. electrical safety, ductwork cleaning and avoiding risks from legionella.
- Reliability of plant operation, to support the business objectives.
- Reducing energy use, water consumption, running costs and carbon emissions.
- Ensuring longevity of the installation, by protection, cleaning and water treatment.
- Facilitating churn, i.e. changes of use, with minimum disruption to the building function.

Post-occupancy evaluation

It is important that a building satisfies the needs of the people who use it. A post-occupancy evaluation (POE) is a systematic appraisal of the users' opinions to assess levels of satisfaction in a wide range of interests and identify improvements to the building, accommodation and its performance. Users would typically include occupiers, visitors, operation and management teams, the disabled and others with special interests.

The outcome of a POE will likely have a range of benefits for new and existing premises relating to planning and management, and including operation of the heating and air-conditioning to improve occupant comfort, system reliability and reduce energy use, running costs and carbon emissions.

Contract maintenance

Average maintenance of small installations is often covered by a maintenance contract involving monthly visits costing a nominal sum per annum.

In the medium range of installations it is now common practice to employ one of the maintenance contractor firms to include for example safety checks, cleaning of boilers, lubrication, cleaning of calorifiers, changing filters on air handling systems, cleaning and checking operation of air treatment equipment, adjustment

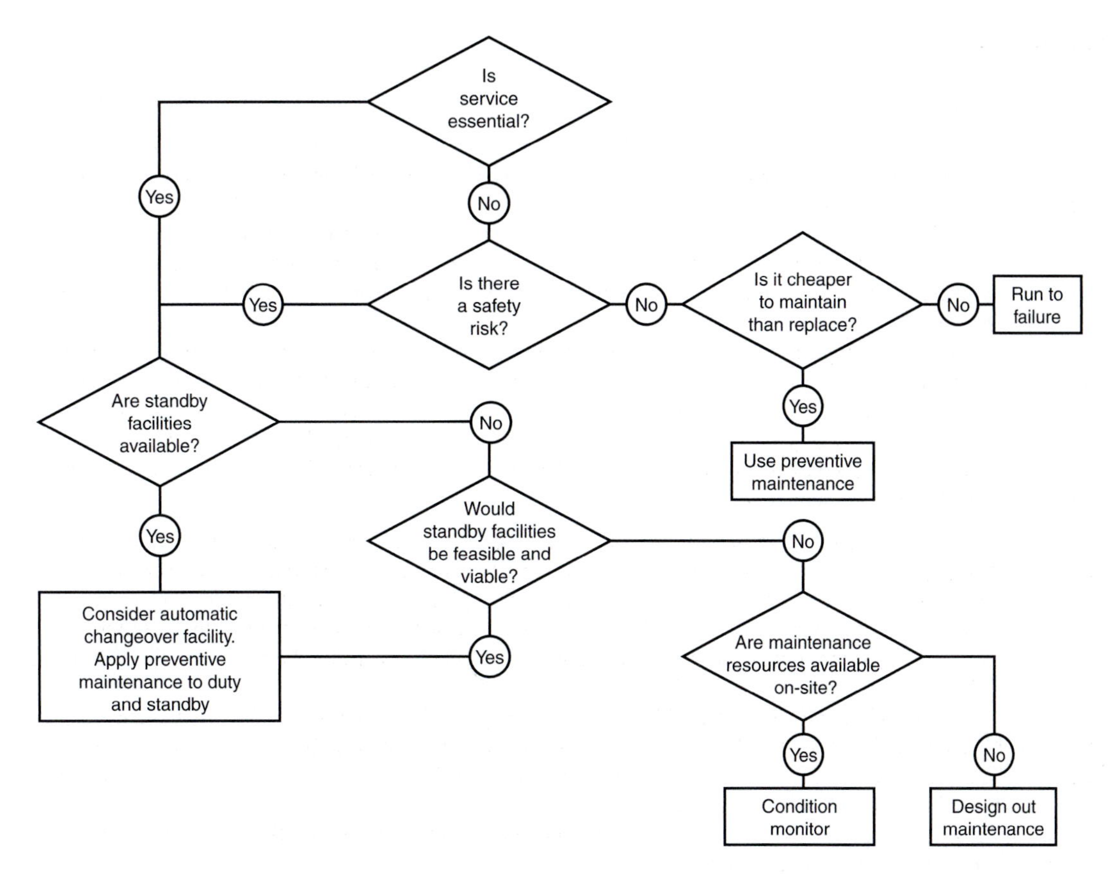

Figure 31.7 Example of maintenance decision making

and attention to controls and building management systems, testing electrical components, inspection and routine activities associated with refrigeration plant and general overhaul. It is often the case that a specialist contractor will be employed to look after items such as water treatment, controls and refrigeration equipment. Annual charges will vary according to the size and complexity of the installation. In the larger-scale plants such as are dealt with by hospital trusts and industrial concerns, the maintenance staff may be part of the organisation working to a programme of planned maintenance. Building management systems may be used for plant asset records and in planning, scheduling and recording maintenance activities.

Taken by and large, maintenance of building services plant based on a *preventative* approach may be expected to be covered by an allowance of the order of £7–15/m^2 (gross building area) per annum at 2006 labour costs. Clearly the cost will depend to a large extent on the level of sophistication of the systems and the relative ease of access to plant. Reference may be made to industry maintenance specification guides for further details.

Insurance

Currently, insurance and annual inspection is required for plant including steam boilers and receivers, pressure vessels, certain types of ventilation equipment and associated electrical power distribution. (Other buildings' life safety installations must also be included.) The same requirements do not apply to low-pressure heating systems nor to general ventilation and air-conditioning plant, but this matter is under review by the Health and Safety Commission. Nevertheless, it is customary for the owner of any sizeable system to take out insurance as a matter of self-protection, and to cover against fracture, burn-out, flooding from water pipework and storage and accidents of all kinds.

Investment appraisals

All major decisions concerning proposals for schemes relating capital expenditure to revenue return, ranging from new installations, to modernisation and to energy conservation measures, should be subject to some form of economic appraisal. The capital cost may relate to all sizes and conditions of building, from the dwelling to the industrial complex and from present stock, to concepts as yet undeveloped from a basic client brief.

Similarly, revenue return may arise from static measures such as orientation, building form and materials of construction or from the dynamic characteristics, life cycle and maintenance needs of equipment. For a full understanding of the impact of engineering systems upon the building, space for plant, fuel storage, service shafts and the like should be a part of the analysis. Obviously, the effect of taxation cannot be ignored, nor indeed can the impact of investment grants, capital allowances, accelerated depreciation allowances and so on. Each of these facets of the problem acts and interacts with the remainder.

Whole life costs

Whole life costing methods, sometimes referred to as life cycle costing, may be used to compare alternative solutions and would include the total initial cost of the installation, operating costs through the life of the plant, including energy and maintenance, and any disposal cost at the end of its useful life. Analysis may involve component replacement at intervals, if the total period used for comparison exceeds the expected life of any component part. Many economic appraisal methods may be used to produce long-term cost forecasts to assist in the decision-making process. Recourse to this method of analysis would be made where there is a high initial cost difference, or high-energy cost difference, or where there is significant disparity between the expected life of the options. The same financial appraisal may be applied to major replacement or repair studies.

The financial return from *static* energy conservation inclusions, orientation, building form and shape, solar exclusion, thermal insulation and the like may be assumed to relate directly to the remaining whole life of the building. In the case of *dynamic* action, however, different criteria obtain, but there is limited information available of a historical nature with respect to this subject area; in particular on operating costs. One important aspect is the expected life of plant and equipment. Table 31.9 brings together information from various sources, mainly CIBSE Guide to Ownership, Operation and Maintenance of Building Services, for a range of systems and components. The life of any component will depend to a large extent upon the working environment, the quality of manufacture, level of maintenance, hours of use and its suitability to the mode of operation. As a consequence, a spread of years is offered for guidance in the table. The periods offered here are based upon 'economic life', which may be defined as the estimated number of years until the item no longer represents the least expensive method of performing its function.

Table 31.9 Economic life of equipment

Type	*Item*	*Life* (years)
Boiler plant	*Steam and HTHW*	
	Shell and tube	20–25
	Water tube	25–30
	Electrode	15–20
	Medium and LPHW	
	Steel	15–20
	Cast iron sectional	15–25
	Electrode	15–25
	Condensing	15–20
	Incinerators	15–20
Boiler auxiliaries	Combustion controls	15–20
	Boiler electrodes	5–10
	Feed pumps/feed water treatment	15–20
	Oil burner	15–20
	Atmospheric gas burner	20–25
	Forced air gas burner	15–20
	Solid fuel handling	10–15

Table 31.9 (Continued)

Type	*Item*	*Life* (years)
	Oil storage tanks, external	15–20
	Oil storage tanks, underground	10–20
	High-temperature fans	15–25
	Instrumentation	10–20
Flues	Stainless steel	15–25
	Mild steel	8–15
Heating equipment	Calorifiers	20–25
	Radiators, cast iron	20–25
	Radiators, steel	10–15
	Radiant heating panels	20–30
	Unit heaters, gas or electric	10–15
	Unit heaters, hot water or steam	15–20
	Fan convectors	12–18
	Natural convectors	15–25
	Underfloor heating, plastic pipe in concrete	25–35
	Underfloor heating, steel pipe in concrete	25–40
	Underfloor electric heating	20–30
Liquid distribution	Pipework, closed system	25–30
	Pipework, open system	20–25
	Pumps, base mounted	20–25
	Pumps, pipeline	15–20
	Pumps, sump	8–12
	Pumps, condensate	10–15
	Water tanks, cast iron, plastic	30–35
	Water tanks, galvanised steel cistern	10–15
	Valves (glandless)	20–25
	Flexible hoses, connections	8–15
	Water treatment	15–20
Refrigeration plant	Reciprocating, large	15–20
	Reciprocating, medium or small	10–15
	Centrifugal	15–20
	Screw	15–25
	Absorption	15–20
	DX split cooling units	10–15
	Cooling towers, galvanised	10–15
	Cooling towers, plastic coated metal	20–25
	Air-cooled condensers	15–20
	Evaporative condenser	15–20
Heat pumps		10–15
Air-handling plant	Package units, external	10–15
	Package units, internal	15–20
	Cooling coils/heating coils, aluminium fins	15–20
	Cooling coils, copper fins	25–30
	Humidifiers, steam direct	10–15
	Humidifiers, electric generated	5–10
	Fans, centrifugal (heavy duty)	20–25
	Fans, axial, centrifugal	15–20
Air distribution	Ductwork, galvanised	25–30
	Dampers	15–20
	Fans, propellor	10–15
	Fans, roof mounted	15–20
	Grilles, diffusers, etc.	25–30
	VAV or DD control terminals	15–20
	Fan coil units	12–15
	Variable refrigerant units	10–15
	Chilled ceiling, beams	20–25
Controls	Pneumatic	15–20
	Electric/electronic	12–20
	Self-contained	8–10
	Sensors	3–10
	BMS outstations	5–15
	Actuators	10–15
Electric Motors	Motors and starters	15–20

BMS, building management systems; DD, dual duct DX, direct expansion; HTHW, high-temperature hot water; LPHW, low-pressure hot water; VAV, variable air volume.

Payback period

Probably the most common term used when assessing the viability of a proposal to replace an existing scheme with one to be justified on grounds of running cost savings, such as energy-saving methods, is *simple payback period*, i.e. the number of years required for a capital expenditure to be recovered through annual income or, in the context of energy conservation, annual savings. Nevertheless, using raw cost data, payback is a rather crude concept for use in decision making since it takes no account of the fact that capital, if invested elsewhere, would earn interest. For example, if an energy conservation measure to reduce annual running cost by £2000 required capital expenditure totalling £8000 this would give a simple payback period of £8000/2000, i.e. 4 years. Also, the value of future revenue costs or savings is affected by inflation; this interest on capital may be considered together in respect of the 'time value of money'.

Present value

The concept *of present value* (PV) has much to commend it since it is easy to understand and can handle staged expenditure or known changes in the pattern of annual saving. In brief, PV analysis converts all outgoings – including capital expenditure, running costs, repairs and where appropriate all income, to their equivalent values as measured at a single point in time, usually the present. The analysis relies upon the fact that a pound today is worth more than a pound tomorrow – inflation aside – since if today's pound were invested it would have earned interest by tomorrow. Thus, stating the converse, a pound at some future date is worth less than a pound today: the value will have been reduced – or discounted – in proportion to the rate of interest assumed.

Most relevant works of reference contain tables listing PV over a wide range of time periods for selection of interest or discount rates. Organisations will set a level which is most appropriate to their business objectives. Whilst the required rate of return on an investment may be set at a level higher than the discount rate, it is common practice for the same figure to be adopted for an assessment of viability and for a comparison of options. There are two important aspects which require emphasis, the first being that the method assumes constant money value in real terms over the period considered, i.e. that there is either zero inflation or that inflation is at a common level across the board. The second is that the analysis is sensitive to both the discount rate and the life cycle of the component parts. Examples of the application of PV techniques are given in the *Guide Section B18*. It is normal for the private sector to expect a payback on investment of 5 years or less; perhaps related to their business plan cycle.

The formulae for calculating PV, discounted cash flow rates, etc., can be found in many publications, including CIBSE Guide to Ownership, Operation and Maintenance of Building Services, and there is little benefit in reproducing these here. Calculations these days are most commonly carried out by the use of simple spreadsheets, allowing parameters to be varied to test the sensitivity to input data. More detailed models may be developed to accommodate different inflation rates for future costs.

Energy-savings viability charts

It is possible to use the tabulated figures for PV to produce what, for want of a better description, may be called viability charts, plotting the ratio between capital cost and annual savings due to conservation measures against a time base. Figure 31.8(a) shows how such a chart would appear for a number of different discount rates from 0 to 20 per cent. Figure 31.8(b) takes a discount rate of 10 per cent as a base and shows how the relationship will vary if energy costs inflate disproportionately to the general pattern.[2]

Three curves are shown, representing zero, plus 3.5 and plus 7 per cent per annum. An excess rise of 3.5 per cent per annum represents a doubling in 20 years and an excess rise of 7 per cent per annum represents a doubling in 10 years.

To conclude this brief attempt to explain the impact of economic factors upon energy conservation measures, the authors feel bound to add that they cannot accept that cost equations are the end of the story. These take no account of amenity values, thermal and visual comfort, aesthetics, contentment, productivity and quality of life. Some at least of these aspects should be quantified and introduced as weighting to the results produced by soulless mathematics. There remain, furthermore, the fundamental issues of national and international

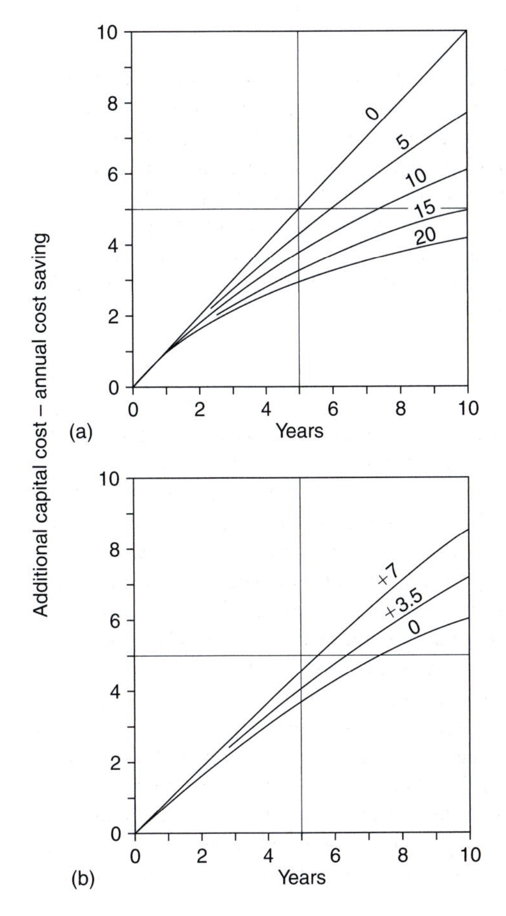

Figure 31.8 Viability charts showing effect of changes in (a) discount rates and (b) energy cost premium

importance that the remaining reserves of energy be husbanded and that CO_2 emissions be limited to minimum practical levels. Economic considerations should perhaps be regarded as of secondary importance.

Notes

1. CIBSE *Guide Section B*18 has not been revised since 1986, but remains probably the most relevant collection of data available in respect of degree-day calculations.
2. At the time of compiling this revision, energy prices have remained effectively in line with the level of general inflation for a number of years. Relative high-energy cost inflation rates could return in the future, and in consequence it is considered valid to include the methodology for carrying out cost appraisals.

Further reading

Building Services Component Life Manual (from work by the Building Performance Group), published by Blackwell Science Ltd.

CIBSE Application Manual No. 5, Energy Audits and Surveys.

CIBSE Energy Efficiency in Buildings, 1998.
CIBSE Guide to Ownership, Operation and Maintenance of Building Services, 2000.
CIBSE TM31 Building Log Book Toolkit.
HAPM Components Life Manual (Commissioned by Housing Association Property Mutual Ltd.), published by E&FN Spon.
Standard Maintenance Specification for Mechanical and Electrical Services in Buildings, SMG 90, published by HVCA.
BSRIA Condition Based Maintenance – an evaluation guide for building services, AG5/2001.
BSRIA Condition Based Maintenance using non-destructive testing, AG1/2003.
BSRIA Toolkit for building operation audits, AG13/2000.
BSRIA Commissioning Application Guides.
BSRIA Specification FMS 8.
HVCA SFG 20: Standard Maintenance Specification for Mechanical Services in Buildings, Volumes 1 to 5; SMG 90 series.
Fire Protection Association: Electricity at Work Regulations 1989 – compliance for firms without electrical staff.
Health and Safety Executive HS(G) 38: Lighting at work.
Institute of Refrigeration Code of Practice.
NHS Estates Concode: Guide to building, engineering and grounds maintenance contracts for the NHS estate.
ECA website – www.eca.gov.uk.
The Carbon Trust website – www.thecarbontrust.co.uk.
BRE Digest 474 HOBO Protocol, Handover of Office Building Operations.
Carbon Trust Energy Consumption Guides:
ECG019: Energy use in offices
ECG036: Energy efficiency in hotels – a guide for owners and managers
ECG054: Energy efficiency in further and higher education – cost-effective low-energy buildings
ECG072: Energy consumption in hospitals
ECG073: Saving energy in schools. A guide for head teachers, governors, premises managers and school energy managers
ECG075: Energy use in Ministry of Defence establishments
ECG078: Energy in sports and recreation buildings
ECG081: Energy efficiency in industrial buildings and sites
Carbon Trust Good Practice Guides
GPG 348: Building Log Books – a user's guide
GPG 310: Degree-days for energy management
'Online toolkit for Managing and Occupying Buildings Sustainability' – www.mobs.org.uk; DTI Partners in Innovation Programme.
Energy Saving Guide: A blueprint for steps to energy efficiency, British Property Federation.

Appendix I

SI unit symbols

Quantity	*Unit*	*Symbol*	*Common multiples or sub-multiples*	*Multiplier*	*Symbol*
Length	metre	m	kilometre	m × 10^3	km
			millimetre	m × 10^{-3}	mm
			micron	m × 10^{-6}	μm
Area	square metre	m^2	hectare	m^2 × 10^5	ha
			sq. millimetre	m^2 × 10^{-9}	mm^2
Volume	cubic metre	m^3	litre	m^3 × 10^{-3}	(litre)[a]
			cu. millimetre	m^3 × 10^{-9}	mm^3
Time	second	s	hour	s × 3600	h
Velocity	metre per second	m/s			
Acceleration	metre/sec^2	m/s^2			
Frequency	hertz (cycle per sec)	Hz			
Rotational frequency	revolutions per sec	s^{-1}			
Mass	kilogram	kg	tonne	kg × 10^3	t
			gram	kg × 10^{-3}	g
			milligram	kg × 10^{-6}	mg
Density (specific mass)		kg/m^3			
Specific volume		m^3/kg			
Mass flow rate		kg/s			
Volume flow rate		m^3/s	litre per second	(m^3/s) × 10^{-3}	litre/s
Momentum		kg m/s			
Force	newton	N	meganewton	N × 10^6	MN
			kilonewton	N × 10^3	kN
Torque		Nm			
Pressure (and stress)	pascal	Pa	megapascal	Pa × 10^6	MPa
			kilopascal	Pa × 10^3	kPa
			bar	Pa × 10^5	b
			millibar	Pa × 10^2	mb
Viscosity					
Dynamic	pascal second	Pa s	centipoise	Pa s × 10^{-3}	cP
Kinematic	centimetre^2/sec	cm^2/s	centistoke	(cm^2/s) × 10^{-2}	cSt
Temperature	Kelvin	K			
	degree Celsius	°C			
Heat					
Energy			gigajoule	J × 10^9	GJ
Work	joule	J	megajoule	J × 10^6	MJ
Quantity of heat			kilojoule	J × 10^3	kJ
Heat flow rate (power)	watt	W	gigawatt	W × 10^9	GW
			megawatt	W × 10^6	MW
			kilowatt	W × 10^3	kW
Thermal conductivity		W/mK			
Thermal resistivity		mK/W			
Specific heat capacity		kJ/kg K			
Latent heat		kJ/kg			

Note
a This book does not use l as a symbol for litre.

Appendix II

Conversion factors

Imperial units to SI

Unit	Imperial	SI Exact	SI Approximate
Length	1 inch	25.4 mm	25 mm
	1 foot	0.3048 m	0.3 m
	3.28 feet	1 m	
	1 yard	0.9144 m	0.9 m
	1 mile	1.609 km	1.6 km
Area	1 sq. in	645.2 mm^2	
	1 sq. ft	0.092 m^2	
	10.77 sq. ft	1 m^2	
	1 sq. yd	0.836 m^2	
	1 acre	4046.9 m^2	
Volume	1 cu. in	16.39 mm^3	16 mm^3
	1 cu. ft	28.32 litre	28 litre
	35.32 cu. ft	1 m^3	
	1 pint	0.568 litre	0.6 litre
	1 gallon	4.546 litre	4.5 litre
Mass	1 pound	0.4536 kg	
	2.205 pounds	1 kg	
	1 tonne	1.016 tonne	1 tonne
Density	1 lb/cu. ft	16.02 kg/m^3	
Volume flow rate	1 gall/minute (g.p.m.)	0.076 litre/s	0.075 litre/s
	1 cu.ft/minute (c.f.m.)	0.472 litre/s	0.5 litre/s
Velocity	1 foot/minute	0.0051 m/s	
	197 feet/minute	1.0 m/s	
	1 mile/hour	0.447 m/s	0.5 m/s
Temperature	1 degree Fahrenheit	0.556°C	
	$t = 32$°F	$t = 0$°C	
Heat	1 British thermal unit (Btu)	1.055 kJ	1 kJ
	1 'Old' therm (100 000 Btu)	105.5 MJ	100 MJ
	1 Unit of electricity (kWh)	3600 kJ	
Heat flow rate	1 Btu/hour	0.2931 W	0.3 W
	1 horsepower	745.7 W	750 W
	1 tonne refrigeration (12 000 Btu/hour)	3.516 kW	3.5 kW

(*Continued*)

Imperial units to SI (continued)

Unit	*Imperial*	*SI* *Exact*	*SI* *Approximate*
Intensity of heat flow rate	1 Btu/sq. ft hour	3.155 W/m^2	3 W/m^2
Transmittance (U value)	$\frac{1 \text{ Btu}}{\text{sq. ft hour°F}}$	5.678 W/m^2K	6 W/m^2K
Conductivity (k value)	$\frac{1 \text{ Btu inch}}{\text{sq. ft hour°F}}$	0.1442 W/mK	
Resistivity ($1/k$)	$\frac{1 \text{ sq. ft hour°F}}{\text{Btu inch}}$	6.934 m K/W	
Calorific value	1 Btu/lb 1 Btu/cu. ft	2.326 kJ/kg 37.26 J/litre or 37.26 kJ/m^3	2.5 kJ/kg
Pressure	1 pound force per sq. in (lb f/sq. in)	6895 Pa or 68.95 mbar	7000 Pa or 70 mbar
	1 inch w.g. (at 4°C)	249.1 Pa or 2.491 mbar	250 Pa or 2.5 mbar
	1 inch mercury (at 0°C)	33.86 mbar	34 mbar
	1 mm mercury	1.333 mbar	
	1 atmosphere (standard)	101 325 Pa	1 bar
Pressure drop	1 inch w.g./100 ft	8.176 Pa/m	
Latent heat of steam (atmospheric pressure)	970 Btu/lb	2258 kJ/kg	2300 kJ/kg
Latent heat of fusion of ice	144 Btu/lb	330 kJ/kg	
Steam flow rate	1 lb/hour 8 lb/hour	0.126 g/s –	 1 g/s
Heat content	1 Btu/lb 1 Btu/gall 1 Btu/cu. ft	2.326 kJ/kg 0.2326 kJ/litre 0.0372 kJ/litre	
Thermal diffusivity	1 ft^2/hour	$2.581 \times 10^{-5} m^2/s$	
Moisture content	1 lb/lb 100 grains/lb	1 kg/kg 0.014 kg/kg	

Index

Colour Plate Section

Energy Performance Certificate
Non-Domestic Building

HM Government

Jubilee House
High Street
Anytown
A1 2CD

Certificate Reference Number:
1234-1234-1234-1234

This certificate shows the energy rating of this building. It indicates the energy efficiency of the building fabric and the heating, ventilation, cooling and lighting systems. The rating is compared to two benchmarks for this type of building: one appropriate for new buildings and one appropriate for existing buildings. There is more advice on how to interpret this information on the Government's website www.communities.gov.uk/epbd.

Energy Performance Asset Rating

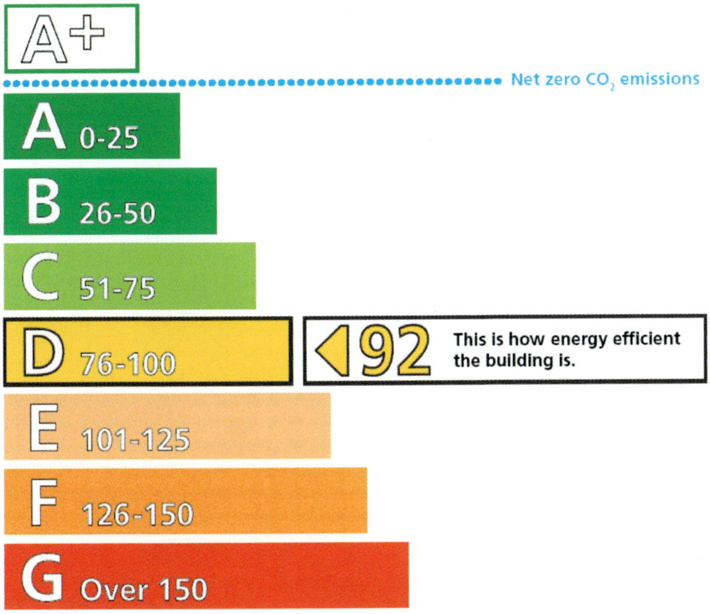

More energy efficient

Net zero CO_2 emissions

A 0-25

B 26-50

C 51-75

D 76-100 — 92 This is how energy efficient the building is.

E 101-125

F 126-150

G Over 150

Less energy efficient

Technical information

Main heating fuel: Gas
Building environment: Air Conditioned
Total useful floor area (m^2): 2927
Building complexity (NOS level): 4

Benchmarks

Buildings similar to this one could have ratings as follows:

58 If newly built

94 If typical of the existing stock

Plate 3.3 An illustrative energy performance certificate

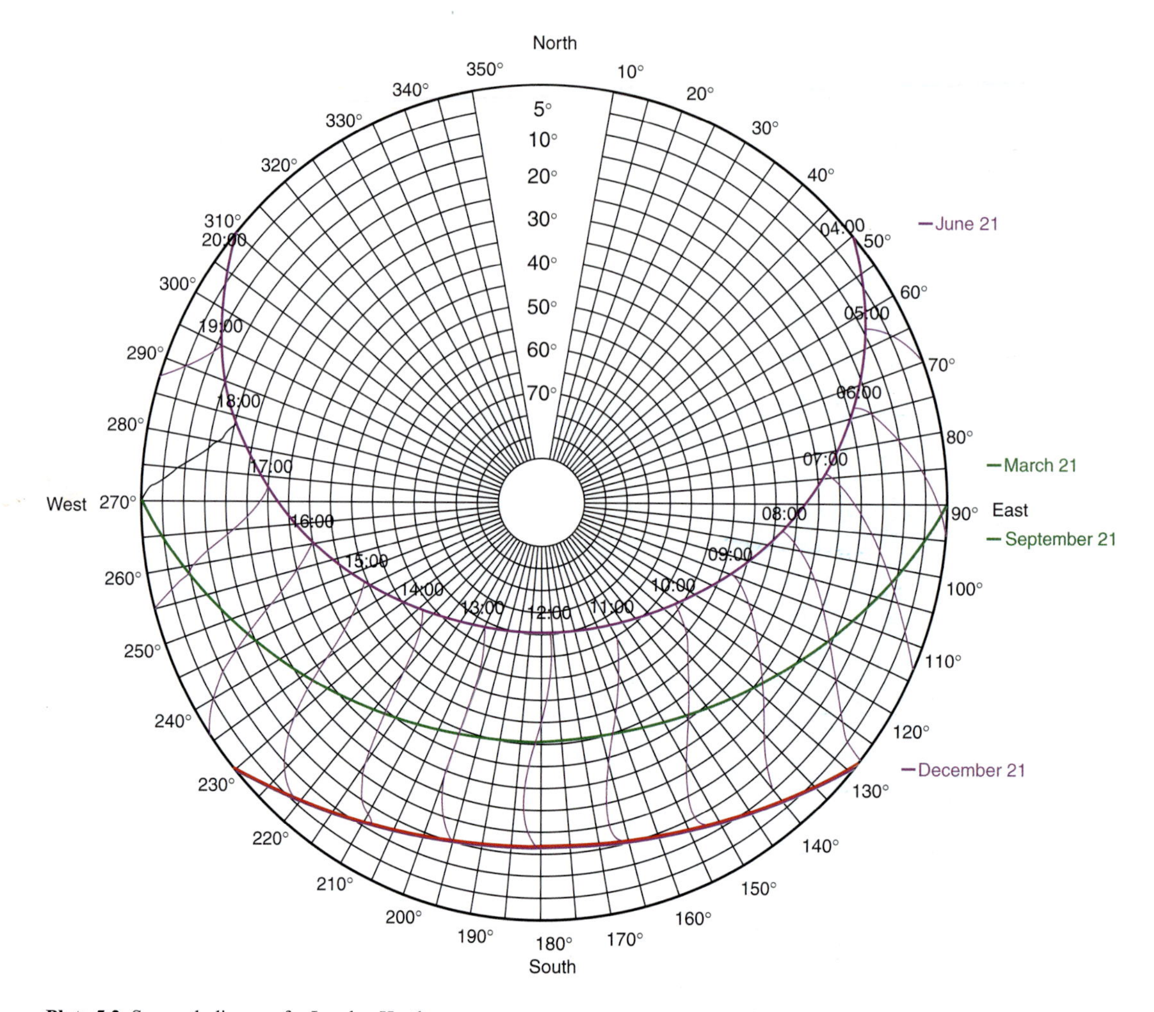

Plate 5.2 Sun-path diagram for London Heathrow

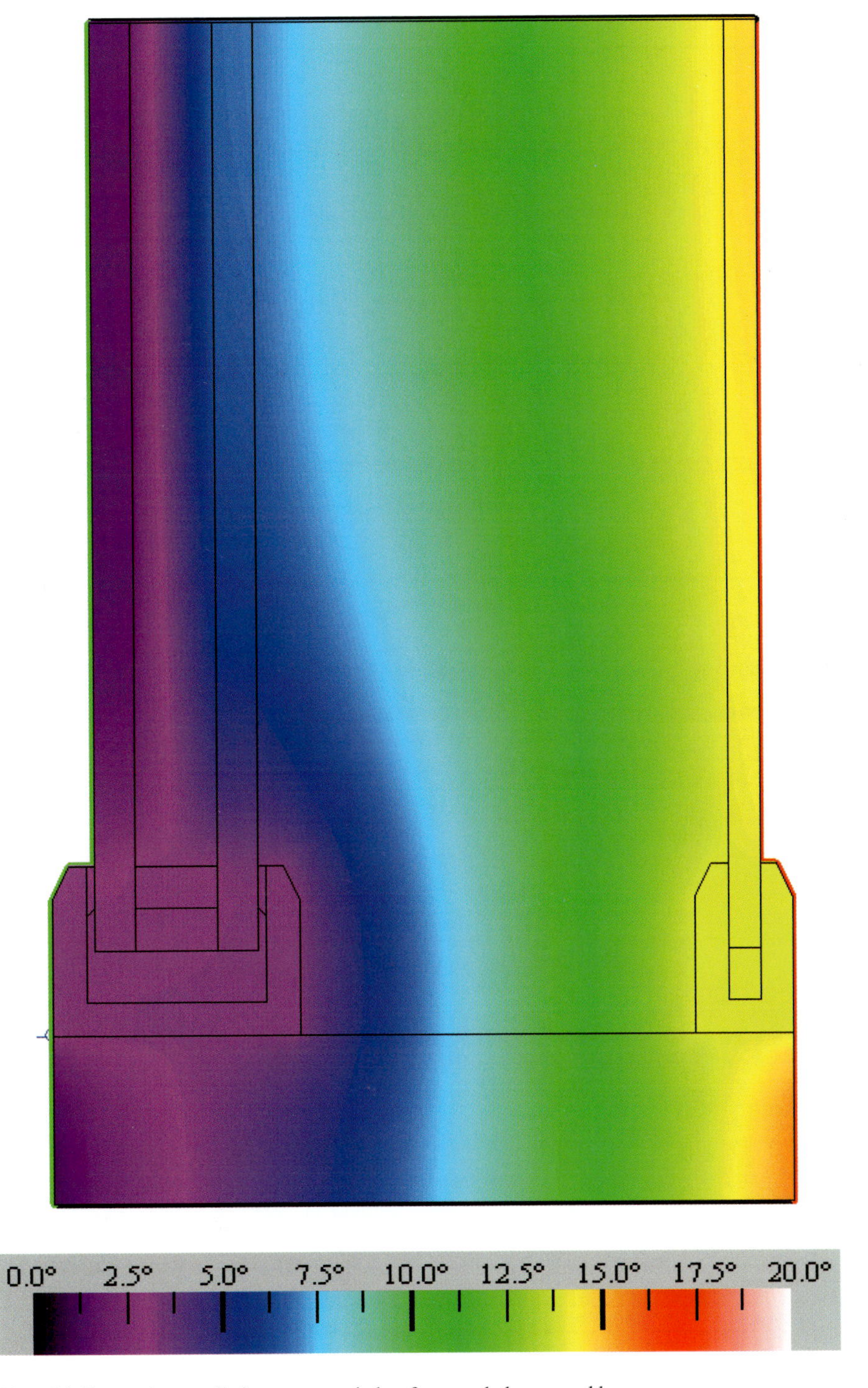

Plate 6.1 Temperature prediction across a window frame and glass assembly

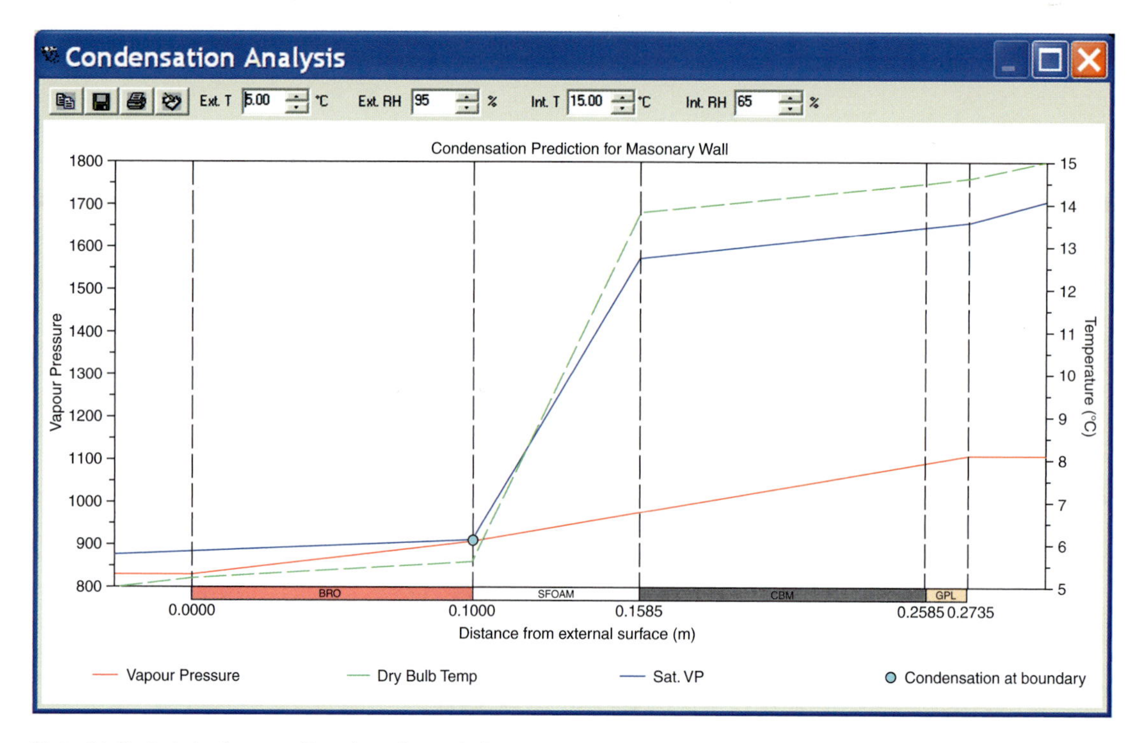

Plate 6.2 Typical plot from a wall condensation analysis

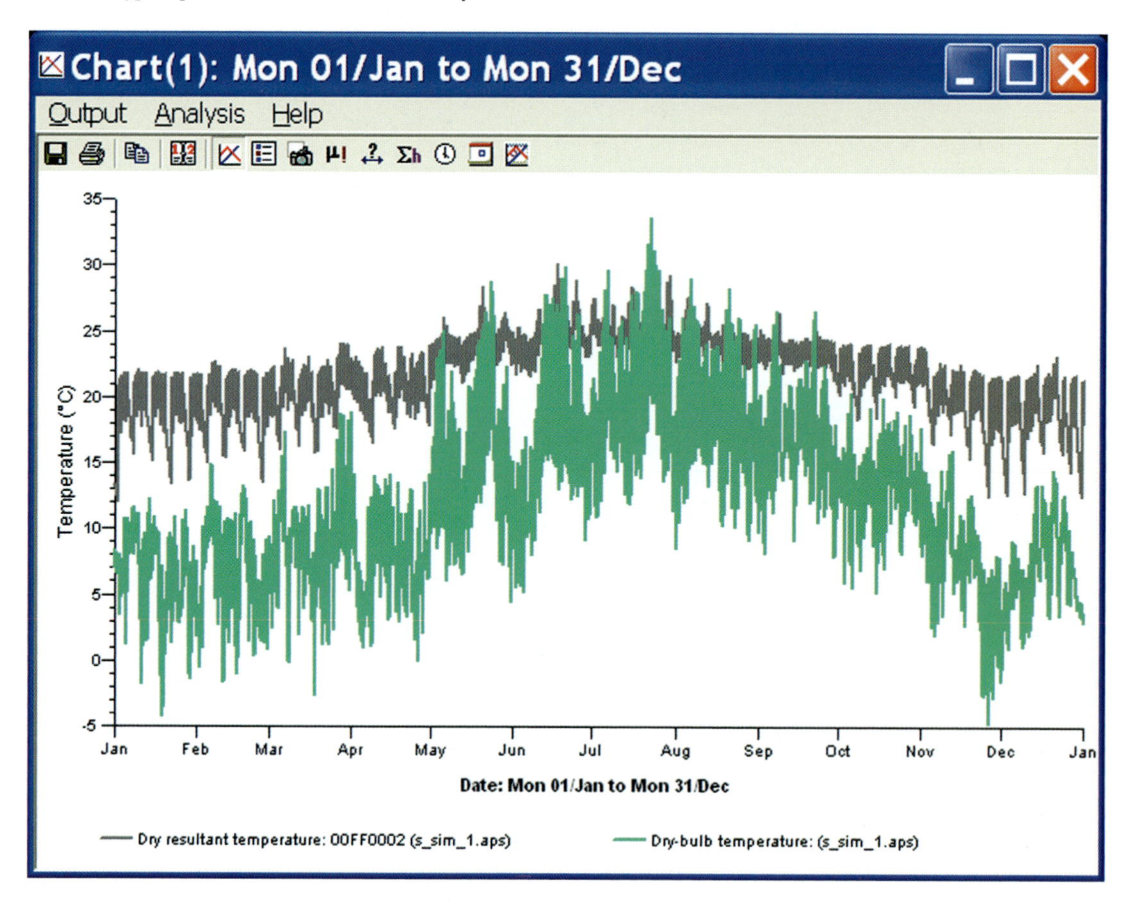

Plate 6.8 Temperature predictions using the DSY weather data

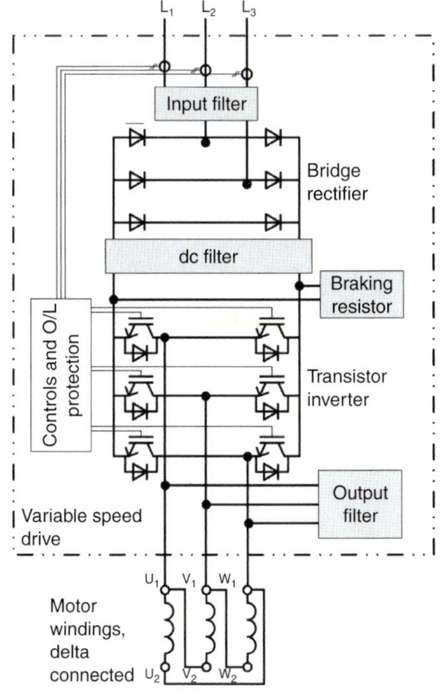

(a) Schematic of variable speed drive

(b) ABB ACH550 variable speed drive

Plate 27.18 Variable speed drives

Plate 27.24 Views of a Form 2 'Wardrobe' MCC with the doors closed and open (photograph courtesy of Integrated Control Systems Limited)

Plate 27.26 Front view of a Form 4 'cubicle' MCC with doors closed (photograph courtesy of Integrated Control Systems Limited)

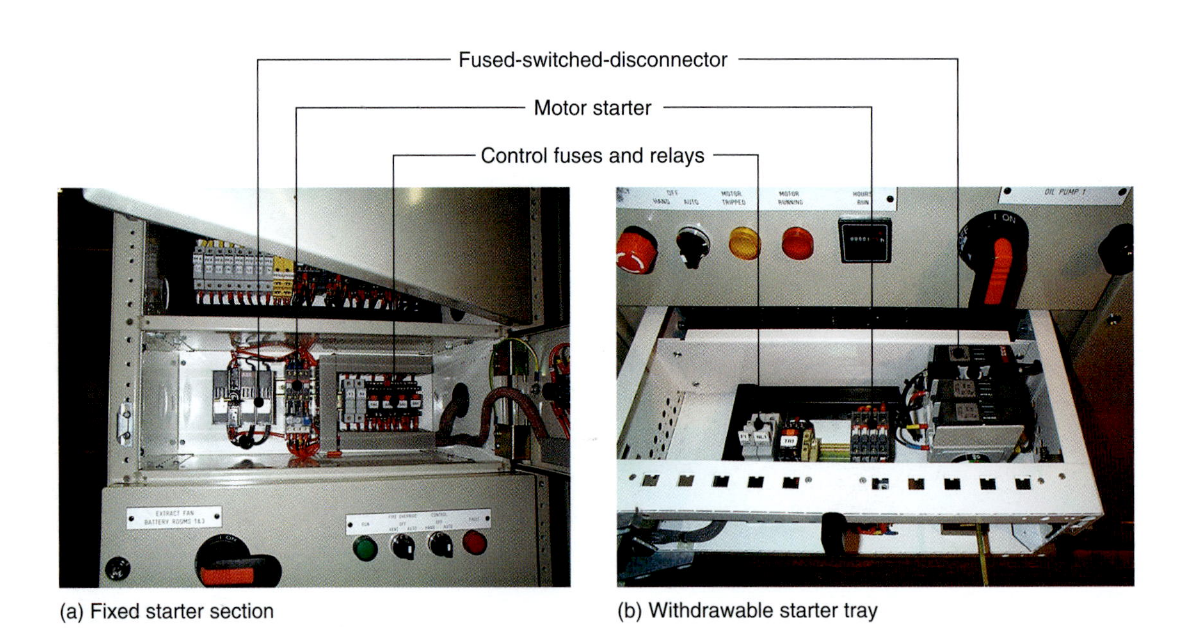

(a) Fixed starter section

(b) Withdrawable starter tray

Plate 27.27 Form 4 MCC, fixed and withdrawable motor starter cubicles (photographs courtesy of Sapphire Controls Limited)